Op Amp Applications Handbook

Op Amp Applications Handbook

ANALOG DEVICES

Walt Jung, Editor Emeritus

with the technical staff of Analog Devices

A Volume in the Analog Devices Series

ELSEVIER

AMSTERDAM • BOSTON • HEIDELBERG • LONDON
NEW YORK • OXFORD • PARIS • SAN DIEGO
SAN FRANCISCO • SINGAPORE • SYDNEY • TOKYO

Newnes is an imprint of Elsevier

Newnes

Newnes is an imprint of Elsevier
30 Corporate Drive, Suite 400, Burlington, MA 01803, USA
Linacre House, Jordan Hill, Oxford OX2 8DP, UK

First Elsevier printing November, 2004.

Library of Congress Cataloging-in-Publication Data

Jung, Walter G.
Op Amp applications handbook / by Walt Jung.
p. cm – (Analog Devices series)
ISBN-13: 978-0-7506-7844-5 ISBN-10: 0-7506-7844-5
1. Operational amplifiers—Handbooks, manuals, etc. I. Title. II. Series.

TK7871.58.O618515 2004
621.39'5--dc22 2004053842

ISBN 13: 978-0-7506-7844-5
ISBN 10: 0-7506-7844-5

British Library Cataloguing-in-Publication Data
A catalogue record for this book is available from the British Library.

For information on all Newnes publications
visit our website at www.books.elsevier.com

Printed and bound by CPI Group (UK) Ltd, Croydon, CR0 4YY
Transferred to Digital Print 2011

Contents

Foreword

The signal-processing products of Analog Devices, Inc. (ADI), along with those of its worthy competitors, have always had broad applications, but in a special way: they tend to be used in critical roles making possible—and at the same time limiting—the excellence in performance of the device, instrument, apparatus, or system using them.

Think about the *op amp*—how it can play a salient role in amplifying an ultrasound wave from deep within a human body, or measure and help reduce the error of a feedback system; the *data converter*—and its critical position in translating rapidly and accurately between the world of tangible physics and the world of abstract digits; the *digital signal processor*—manipulating the transformed digital data to extract information, provide answers, and make crucial instant-by-instant decisions in control systems; *transducers*, such as the life-saving MEMS accelerometers and gyroscopes; and even *control chips*, such as the one that empowers the humble thermometric junction placed deep in the heart of a high-performance—but very vulnerable—microcomputer chip.

From its founding two human generations ago, in 1965, ADI has been committed to a leadership role in designing and manufacturing products that meet the needs of the existing market, anticipate the near-term needs of present and future users, and envision the needs of users yet unknown—and perhaps *unborn*—who will create the markets of the future. These existing, anticipated and envisioned "needs" must perforce include far more than just the design, manufacture and timely delivery of a physical device that performs a function reliably to a set of specifications at a competitive price.

We've always called a product that satisfies these needs "the augmented product," but what does this mean?

The *physical* product is a highly technological product that, above all, requires *knowledge* of its possibilities, limitations and subtleties. But when the earliest generations—and to some extent later generations—of such a product appear in the marketplace, there exist few (if any) school courses that have produced graduates proficient in its use. There are few knowledgeable designers who can foresee its possibilities. So we have the huge task of creating awareness; teaching about principles, performance measures, and existing applications; and providing ideas to stimulate the imagination of those creative users who will provide our next round of challenges.

This problem is met by deploying people and publications. The *people* are Applications Engineers, who can deal with user questions arriving via phone, fax, and e-mail—as well as working with users in the field to solve particular problems. These experts also spread the word by giving seminars to small and large groups for purposes from inspiring the creative user to imbuing the system, design, and components engineer with the nuts-and-bolts of practice. The *publications*—both in hard copy and on-line—range from authoritative handbooks, such as the present volume, comprehensive data sheets, application notes, hardware and software manuals, to periodic publications, such as "Solutions Bulletins" and our unique *Analog Dialogue*—the sole survivor among its early peers—currently in its 38th year of continuous publication in print and its 6th year of regular publication on the Internet.

This book is the ultimate expression of product "augmentation" as it relates to operational amplifiers. In some senses, it can be considered a descendant of two early publications. The first is a 1965 set of *Op Amp*

Notes (Parts 1, 2, 3, and 4), written by Analog Devices co-founder Ray Stata, with the current text directly reflecting these roots. Much less directly would be the 1974 first edition of the *IC Op Amp Cookbook*, by Walter Jung. Although useful earlier books had been published by Burr-Brown, and by Dan Sheingold at Philbrick, these two timely publications were seminal in the early days of the silicon era, advocating the understanding and use of IC op amps to a market in the process of growing explosively. Finally, and perhaps more important to current students of the op amp art, would be the countless contributions of ADI design and applications engineers, amassed over the years and so highly evident within this new book.

Operational amplifiers have been marketed since 1953, and practical IC op amps have been available since the late 1960s. Yet, half a century later, there is still a need for a book that embraces the many aspects of op amp technology—one that is thorough in its technical content, that looks forward to tomorrow's uses and back to the principles and applications that still make op amps a practical necessity today. We believe that this is such a book, and we commend Walter Jung for "augmenting" the op amp in such an interesting and accessible form.

Ray Stata
Daniel Sheingold
Norwood, Massachusetts, April 28, 2004

Preface

Op Amp Applications Handbook is another book on the operational amplifier, or *op amp*. As the name implies, it covers the application of op amps, but does so on a broader scope. Thus it would be incorrect to assume that this book is simply a large collection of app notes on various devices, as it is far more than that. Any IC manufacturer in existence since the 1960s has ample application data on which to draw. In this case, however, Analog Devices, Inc. has had the benefit of applications material with a history that goes back beyond early IC developments to the preceding period of solid-state amplifiers in modular form, with links to the even earlier era of vacuum tube op amps and analog computers, where the operational amplifier began.

This book brings some new perspectives to op amp applications. It adds insight into op amp origins and historical developments not available elsewhere. Within its major chapters it also offers fundamental discussions of basic op amp operation; the roles of various device types (including both op amps and other specialty amplifiers, such as instrumentation amplifiers); the procedures for optimal interfacing to other system components such as ADCs and DACs, signal conditioning and filtering in data processing systems, and a wide variety of signal amplifiers. The book concludes with practical discussions of various hardware issues, such as passive component selection, printed circuit design, modeling and breadboarding, etc. In short, while this book does indeed cover op amp applications, it also covers a host of closely related design topics, making it a formidable toolkit for the analog designer.

The book is divided into 8 major chapters, and occupies nearly 900 pages, including index. The chapters are outlined as follows:

Chapter 1, *Op Amp Basics*, has five sections authored by James Bryant, Walt Jung, and Walt Kester. This chapter provides fundamental op amp operating information. An introductory section addresses their ideal and non-ideal characteristics along with basic feedback theory. It then spans op amp device topologies, including voltage and current feedback models, op amp internal structures such as input and/or output architectures, the use of bipolar and/or FET devices, single supply/dual supply considerations, and op amp device specifications that apply to all types. The two final sections of this chapter deal with the operating characteristics of precision and high-speed op amp types. This chapter, itself a book-within-a-book, occupies about 115 pages.

Chapter 2, *Specialty Amplifiers*, has three sections authored by Walt Kester, Walt Jung, and James Bryant. This chapter provides information on those commonly used amplifier types that use op amp-like principles, but aren't op amps themselves—instead they are specialty amplifiers. The first section covers the design and application of differential input, single-ended output amplifiers, known as instrumentation amplifiers. The second section is on programmable gain amplifiers, which are op amp or instrumentation amplifier stages, designed to be dynamically addressable for gain. The final section of the chapter is on isolation amplifiers, which provide galvanic isolation between sections of a system. This chapter occupies about 48 pages.

Chapter 3, *Using Op Amps with Data Converters*, has five sections authored by Walt Kester, James Bryant, and Paul Hendriks. The first section is an introductory one, introducing converter terms and the concept of minimizing conversion degradation within the design of an op amp interface. The second section covers ADC and DAC specifications, including such critically important concepts as linearity, monotonicity, missing codes. The third section covers driving ADC inputs in both single-ended and differential signal modes, op amp stability and settling time issues, level shifting, etc. This section also includes a discussion of dedicated differential driver amplifier ICs, as well as op amp-based ADC drivers. The fourth section is concerned with driving converter reference inputs, and optimal use of sources. The fifth and final section covers DAC output buffer amplifiers, using both standard op amp circuits as well as differential driver ICs. This chapter occupies about 50 pages.

Chapter 4, *Sensor Signal Conditioning*, has five sections authored by Walt Kester, James Bryant, Walt Jung, Scott Wurcer, and Chuck Kitchin. After an introductory section on sensor types and their processing requirements, the remaining four sections deal with the different sensor types. The second section is on bridge circuits, covering the considerations in optimizing performance with respect to bridge drive mode, output mode, and impedance. The third section covers strain, force, pressure, and flow measurements, along with examples of high performance circuits with representative transducers. The fourth section, on high impedance sensors, covers a multitude of measurement types. Among these are photodiode amplifiers, charge amplifiers, and pH amplifiers. The fifth section of the chapter covers temperature sensors of various types, such as thermocouples, RTDs, thermistor and semiconductor-based transducers. This chapter occupies about 78 pages.

Chapter 5, *Analog Filters*, has eight sections authored by Hank Zumbahlen. This chapter could be considered a stand-alone treatise on how to implement modern analog filters. The eight sections, starting with an introduction, include transfer functions, time domain response, standard responses, frequency transformations, filter realizations, practical problems, and design examples. This chapter is more mathematical than any other within the book, with many response tables as design aids. One key highlight is the design example section, where an online filter-builder design tool is described in active filter implementation examples using Sallen-Key, multiple feedback, state variable, and frequency dependent negative resistance filter types. This chapter, another book-within-a-book, occupies about 110 pages.

Chapter 6, *Signal Amplifiers*, has six sections authored by Walt Jung and Walt Kester. These sections are audio amplifiers, buffer amplifiers/driving capacitive loads, video amplifiers, communication amplifiers, amplifier ideas, and composite amplifiers. In the audio, video, and communications amplifier sections, various op amp circuit examples are shown, with emphasis in these sections on performance to high specifications— audio, video, or communications, as the case may be. The "amplifier ideas" section is a broad-range collection of various amplifier applications, selected for emphasis on creativity and innovation. The final section, on composite amplifiers, shows how additional discrete devices can be added to either the input or output of an op amp to enhance net performance. This book-within-a-book chapter occupies about 180 pages.

Chapter 7, *Hardware and Housekeeping Techniques*, has seven sections authored by Walt Kester, James Bryant, Walt Jung, Joe Buxton, and Wes Freeman. These sections are passive components, PCB design issues, op amp power supply systems, op amp input and output protection, thermal considerations, EMI/RFI considerations, and the final section, simulation, breadboarding and prototyping. All of these practical topics have a commonality that they are not completely covered (if at all) by the op amp data sheet. But, most importantly, they can be just as critical as the device specifications towards achieving the final results. This book-within-a-book chapter occupies about 154 pages.

Chapter 8, the *History* chapter has four sections authored by Walt Jung. It provides a detailed account of not only the beginnings of operational amplifiers, but also their progress and the ultimate evolution into the IC form known today. This began with the underlying development of feedback amplifier principles, by Harold Black and others at Bell Telephone Laboratories. From the first practical analog computer feedback amplifier building blocks used during World War II, vacuum tube op amps later grew in sophistication, popularity and diversity of use. The first solid-state op amps were "black-brick" plug in modules, which in turn were followed by hybrid IC forms, using chip semiconductors on ceramic substrates. The first monolithic IC op amp appeared in the early 1960s, and there have been continuous developments in circuitry, processes and packaging since then. This chapter occupies about 64 pages, and includes several hundred literature references.

The book is concluded with a thorough index with three pointer types: subject, ADI part number, and standard part numbers.

Acknowledgments

A book on a scale such as **Op Amp Applications Handbook** isn't possible without the work of many individuals. In the preparation phase many key contributions were made, and these are here acknowledged with sincere thanks. Of course, the first "Thank you" goes to ADI management, for project encouragement and support.

Hearty thanks goes next to Walt Kester of the ADI Central Applications Department, who freely offered his wisdom and counsel from many years of past ADI seminar publications. He also commented helpfully on the manuscript throughout. Special thanks go to Walt, as well as the many other named section authors who contributed material.

Thanks go also to the ADI Field Applications and Central Applications Engineers, who helped with comments and criticism. Ed Grokulsky, Bruce Hohman, Bob Marwin and Arnold Williams offered many helpful comments, and former ADI Applications Engineer Wes Freeman critiqued most of the manuscript, providing valuable feedback.

Special thanks goes to Dan Sheingold of ADI, who provided innumerable comments and critiques, and special insights from his many years of op amp experience dating from the vacuum tube era at George A. Philbrick Researches.

Thanks to Carolyn Hobson, who was instrumental in obtaining many of the historical references.

Thanks to Judith Douville for preparation of the index and helpful manuscript comments.

Walt Jung and Walt Kester together prepared slides for the book, and coordinated the stylistic design. Walt Jung did the original book page layout and typesetting.

Specific-to-Section Acknowledgments

Acknowledged here are focused comments from many individuals, specific to the section cited. All were very much appreciated.

Op Amp History; Introduction:

Particularly useful to this section of the project was reference information received from vacuum tube historian Gary Longrie. He provided information on early vacuum tube amplifiers, the feedback experiments of B. D. H. Tellegen at N. V. Philips, and made numerous improvement comments on the manuscript.

Mike Hummel provided the reference to Alan Blumlein's patent of a negative feedback amplifier.

Dan Sheingold provided constructive comments on the manuscript.

Bob Milne offered many comments towards improvement of the manuscript.

Op Amp History; Vacuum Tube Op Amps:

Particularly useful were numerous references on differential amplifiers, received from vacuum tube historian Gary Longrie. Gary also reviewed the manuscript and made numerous improvement comments. Without his enthusiastic inputs, the vacuum tube related sections of this narrative would be less complete.

Dan Sheingold supplied reference material, reviewed the manuscript, and made numerous constructive comments. Without his inputs, the vacuum tube op amp story would have less meaning.

Bel Losmandy provided many helpful manuscript inputs, including his example 1956 vacuum tube op amp design. He also reviewed the manuscript and made many helpful comments.

Paul de R. Leclercq and Morgan Jones supplied the reference to Blumlein's patent describing his use of a differential pair amplifier.

Bob Milne reviewed the manuscript and offered various improvement comments.

Steve Bench provided helpful comment on several points related to the manuscript.

Op Amp History; Solid-State Modular and Hybrid Op Amps:

Particularly useful was information from two GAP/R alumni, Dan Sheingold and Bob Pease. Both offered many details on the early days of working with George Philbrick, and Bob Pease furnished a previously unpublished circuit of the P65 amplifier.

Dick Burwen offered detailed information on some of his early ADI designs, and made helpful comments on the development of the narrative.

Steve Guinta and Charlie Scouten provided many of the early modular op amp schematics from the ADI Central Applications Department archival collection.

Lew Counts assisted with comments on the background of the high speed modular FET amplifier developments.

Walt Kester provided details on the HOS-050 amplifier and its development as a hybrid IC product at Computer Labs.

Op Amp History; IC Op Amps:

Many helpful comments on this section were received, and all are very much appreciated. In this regard, thanks go to Derek Bowers, JoAnn Close, Lew Counts, George Erdi, Bruce Hohman, Dave Kress, Bob Marwin, Bob Milne, Reza Moghimi, Steve Parks, Dan Sheingold, Scott Wurcer, and Jerry Zis

Op Amp Basics; Introduction:

Portions of this section were adapted from Ray Stata's "Operational Amplifiers - Part I," **Electromechanical Design**, September, 1965.

Bob Marwin, Dan Sheingold, Ray Stata, and Scott Wayne contributed helpful comments.

Op Amp Basics; Structures:

Helpful comments on various op amp schematics were received from ADI op amp designers Derek Bowers, Jim Butler, JoAnn Close, and Scott Wurcer.

Signal Amplifiers; Audio Amplifiers:

Portions of this section were adapted from Walt Jung, "Audio Preamplifiers, Line Drivers, and Line Receivers," Chapter 8 of Walt Kester, **System Application Guide**, Analog Devices, Inc., 1993, ISBN 0-916550-13-3, pp. 8-1 to 8-100.

During the preparation of this material the author received helpful comments and other inputs from Per Lundahl of Lundahl Transformers, and from Arne Offenberg of Norway.

Signal Amplifiers; Amplifier Ideas:

Helpful comments were received from Victor Koren and Moshe Gerstenhaber.

Signal Amplifiers, Composite Amplifiers:

Helpful comments were received from Erno Borbely, Steve Bench, and Gary Longrie.

Hardware and Housekeeping Techniques; Passive Components and PCB Design Issues:

Portions of these sections were adapted from Doug Grant and Scott Wurcer, "Avoiding Passive Component Pitfalls," originally published in **Analog Dialogue 17-2**, 1983.

Hardware and Housekeeping Techniques; EMI/RFI Considerations:

Eric Bogatin made helpful comments on this section.

The above acknowledgments document helpful inputs received for the Analog Devices 2002 Amplifier Seminar edition of ***Op Amp Applications****.*

While reasonable efforts have been made to make this work error-free, some inaccuracies may have escaped detection. The editors accept responsibility for error correction within future editions, and will appreciate errata notification(s).

Walt Jung, Editor Emeritus
Op Amp Applications Handbook

Op Amp History Highlights

1928

Harold S. Black applies for patent on his feedback amplifier invention.

1930

Harry Nyquist applies for patent on his regenerative amplifier (patent issued in 1933).

1937

U.S. Patent No. 2,102,671 issued to H.S. Black for "Wave Translation System."

B.D.H. Tellegen publishes a paper on feedback amplifiers, with attributions to H.S. Black and K. Posthumus.

Hendrick Bode files for an amplifier patent, issued in 1938.

1941

Stewart Miller publishes an article with techniques for high and stable gain with response to dc, introducing "cathode compensation."

Testing of prototype gun director system called the T10 using feedback amplifiers. This later leads to the M9, a weapon system instrumental in winning WWII.

Patent filed by Karl D. Swartzel Jr. of Bell Labs for a "Summing Amplifier," with a design that could well be the genesis of op amps. Patent not issued until 1946.

1946

George Philbrick founds company, George A. Philbrick Researches, Inc. (GAP/R). His work was instrumental in op amp development.

1947

Medal for Merit award given to Bell Labs's M9 designers Lovell, Parkinson, and Kuhn. Other contributors to this effort include Bode and Shannon.

Operational amplifiers first referred to by name in Ragazzini's key paper "Analysis of Problems in Dynamics by Electronic Circuits." It references the Bell Labs work on what became the M9 gun director, specifically referencing the op amp circuits used.

Bardeen, Brattain, and Shockley of Bell Labs discover the transistor effect.

1948

George A. Philbrick publishes article describing a single-tube circuit that performs some op amp functions.

1949

Edwin A. Goldberg invents chopper-stabilized vacuum tube op amp.

1952

Granino and Theresa Korn publish textbook *Electronic Analog Computers*, which becomes a classic work on the uses and methodology of analog computing, with vacuum tube op amp circuits.

1953

First commercially available vacuum tube op amp introduced by GAP/R.

1954

Gordon Teal of Texas Instruments develops a silicon transistor.

1956

GAP/R publishes manual for K2-W and related amplifiers, that becomes a seminal reference.

Nobel Prize in Physics awarded to Bardeen, Brattain, and Shockley of Bell Labs for the transistor.

Burr-Brown Research Corporation formed. It becomes an early modular solid-state op amp supplier.

1958

Jack Kilby of Texas Instruments invents the integrated circuit (IC).

1959

Jean A. Hoerni files for a patent on the planar process, a means of stabilizing and protecting semiconductors.

1962

George Philbrick introduces the PP65, a square outline, 7-pin modular op amp which becomes a standard and allows the op amp to be treated as a *component*.

1963

Bob Widlar of Fairchild designs the μA702, the first generally recognized monolithic IC op amp.

1965

Fairchild introduces the milestone μA709 IC op amp, also designed by Bob Widlar.

Analog Devices, Inc. (ADI) is founded by Matt Lorber and Ray Stata. Op amps were their first product.

1967

National Semiconductor Corp. (NSC) introduces the LM101 IC op amp, also designed by Bob Widlar, who moved to NSC from Fairchild. This device begins a second generation of IC op amps.

Analog Dialogue magazine is first published by ADI.

1968

The μA741 op amp, designed by Dave Fullagar, is introduced by Fairchild and becomes the standard op amp.

1969

Dan Sheingold takes over as editor of *Analog Dialogue* (and remains so today).

1970

Model 45 high speed FET op amp introduced by ADI.

1972

Russell and Frederiksen of National Semiconductor Corp. introduce an amplifier technique that leads to the LM324, the low cost, industry-standard general-purpose quad op amp.

1973

Analog Devices introduces AD741, a high-precision 741-type op amp.

1974

Ion implantation, a new fabrication technique for making FET devices, is described in a paper by Rod Russell and David Culmer of National Semiconductor.

1988

ADI introduces a high speed 36V CB process and a number of fast IC op amps. High performance op amps and op amps designed for various different categories continue to be announced throughout the 1980s and 1990s, and into the twenty-first century.

Chapter 8 provides a detailed narrative of op amp history.

CHAPTER 1

Op Amp Basics

- Section 1-1: Introduction
- Section 1-2: Op Amp Topologies
- Section 1-3: Op Amp Structures
- Section 1-4: Op Amp Specifications
- Section 1-5: Precision Op Amps
- Section 1-6: High Speed Op Amps

CHAPTER 1

Op Amp Basics

James Bryant, Walt Jung, Walt Kester

Within Chapter 1, discussions are focused on the basic aspects of op amps. After a brief introductory section, this begins with the fundamental *topology* differences between the two broadest classes of op amps, those using *voltage feedback* and *current feedback*. These two amplifier types are distinguished more by the nature of their internal circuit topologies than anything else. The voltage feedback op amp topology is the classic structure, having been used since the earliest vacuum tube based op amps of the 1940s and 1950s, through the first IC versions of the 1960s, and includes most op amp models produced today. The more recent IC variation of the current feedback amplifier has come into popularity in the mid-to-late 1980s, when higher speed IC op amps were developed. Factors distinguishing these two op amp types are discussed at some length.

Details of op amp input and output *structures* are also covered in this chapter, with emphasis on how such factors potentially impact application performance. In some senses, it is logical to categorize op amp types into performance and/or application classes, a process that works to some degree, but not altogether.

In practice, once past those obvious application distinctions such as "high speed" versus "precision," or "single" versus "dual supply," neat categorization breaks down. This is simply the way the analog world works. There is much crossover between various classes, i.e., a high speed op amp can be either single or dual-supply, or it may even fit as a precision type. A low power op amp may be precision, but it need not necessarily be single-supply, and so on. Other distinction categories could include the input stage type, such as FET input (further divided into JFET or MOS, which, in turn, are further divided into NFET or PFET and PMOS and NMOS, respectively), or bipolar (further divided into NPN or PNP). Then, all of these categories could be further described in terms of the type of input (or output) stage used.

So, it should be obvious that categories of op amps are like an infinite set of analog gray scales; *they don't always fit neatly into pigeonholes, and we shouldn't expect them to*. Nevertheless, it is still very useful to appreciate many of the aspects of op amp design that go into the various structures, as these differences directly influence the optimum op amp choice for an application. Thus structure differences are application drivers, since we choose an op amp to suit the nature of the application—for example, single-supply.

In this chapter various op amp performance *specifications* are also discussed, along with those specification differences that occur between the broad distinctions of voltage or current feedback topologies, as well as the more detailed context of individual structures. Obviously, op amp specifications are also application drivers; in fact, they are the most important since they will determine system performance. We choose the best op amp to fit the application, based on the required bias current, bandwidth, distortion, and so forth.

SECTION 1-1

Introduction

Walt Jung

As a precursor to more detailed sections following, this introductory chapter portion considers the most basic points of op amp operation. These initial discussions are oriented around the more fundamental levels of op amp applications. They include: *Ideal Op Amp Attributes, Standard Op Amp Feedback Hookups, The Non-Ideal Op Amp, Op Amp Common-Mode Dynamic Range(s),* the various *Functionality Differences of Single and Dual-Supply Operation*, and the *Device Selection* process.

Before op amp applications can be developed, some requirements are in order. These include an understanding of how the fundamental op amp operating modes differ, and whether dual-supply or single-supply device functionality better suits the system under consideration. Given this, then device selection can begin and an application developed.

First, an *operational amplifier* (hereafter simply op amp) is a differential input, single-ended output amplifier, as shown symbolically in Figure 1-1. This device is an amplifier intended for use with *external feedback elements*, where these elements determine the resultant function, or *operation*. This gives rise to the name "operational amplifier," denoting an amplifier that, by virtue of different feedback hookups, can perform a variety of operations.[1] At this point, note that there is no need for concern with any actual technology to implement the amplifier. Attention is focused more on the behavioral nature of this building block device.

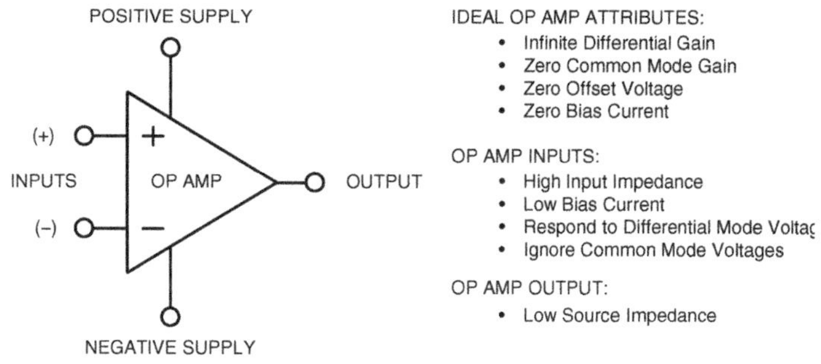

Figure 1-1: The ideal op amp and its attributes

An op amp processes small, differential mode signals appearing between its two inputs, developing a single-ended output signal referred to a power supply common terminal. Summaries of the various ideal op amp attributes are given in Figure 1-1. While real op amps will depart from these ideal attributes, it is very helpful for first-level understanding of op amp behavior to consider these features. Further, although these initial discussions talk in idealistic terms, they are also flavored by pointed mention of typical "real world" specifications—for a beginning perspective.

[1] The actual *naming* of the operational amplifier occurred in the classic Ragazinni, et al paper of 1947 (see Reference 1). However, analog computations using op amps as we know them today began with the work of the Clarence Lovell-led group at Bell Labs, around 1940 (acknowledged generally in the Ragazinni paper).

It is also worth noting that this op amp is shown with five terminals, a number that happens to be a minimum for real devices. While some single op amps may have more than five terminals (to support such functions as frequency compensation, for example), none will ever have fewer. By contrast, those elusive ideal op amps don't require power, and symbolically function with just four pins.[2]

Ideal Op Amp Attributes

An ideal op amp has infinite gain for *differential* input signals. In practice, real devices will have quite high gain (also called *open-loop gain*) but this gain won't necessarily be precisely known. In terms of specifications, gain is measured in terms of V_{OUT}/V_{IN}, and is given in V/V, the dimensionless numeric gain. More often, however, gain is expressed in decibel terms (dB), which is mathematically dB = 20 • log (numeric gain). For example, a numeric gain of 1 million (10^6 V/V) is equivalent to a 120 dB gain. Gains of 100 dB – 130 dB are common for precision op amps, while high speed devices may have gains in the 60 dB – 70 dB range.

Also, an ideal op amp has zero gain for signals *common* to both inputs, that is, *common-mode* (CM) signals. Or, stated in terms of the rejection for these common-mode signals, an ideal op amp has infinite *CM rejection* (CMR). In practice, real op amps can have CMR specifications of up to 130 dB for precision devices, or as low as 60 dB–70 dB for some high speed devices.

The ideal op amp also has zero *offset voltage* ($V_{OS} = 0$), and draws zero *bias current* ($I_B = 0$) at both inputs. Within real devices, actual offset voltages can be as low as 1 µV or less, or as high as several mV. Bias currents can be as low as a few fA, or as high as several µA. This extremely wide range of specifications reflects the different input structures used within various devices, and is covered in more detail later in this chapter.

The attribute headings within Figure 1-1 for INPUTS and OUTPUT summarize the above concepts in more succinct terms. In practical terms, another important attribute is the concept of *low source impedance*, at the output. As will be seen later, low source impedance enables higher useful gain levels within circuits.

To summarize these idealized attributes for a signal processing amplifier, some of the traits might at first seem strange. However, it is critically important to reiterate that op amps simply are never intended for use without overall feedback. In fact, as noted, the connection of a suitable *external* feedback loop defines the *closed-loop* amplifier's gain and frequency response characteristics.

Note also that all real op amps have a positive and negative power supply terminal, but rarely (if ever) will they have a separate ground connection. In practice, the op amp output voltage becomes referred to a power supply common point. *Note: This key point is further clarified with the consideration of typically used op amp feedback networks.*

The basic op amp hookup of Figure 1-2 applies a signal to the (+) input, and a (generalized) network delivers a fraction of the output voltage to the (–) input terminal. This constitutes *feedback*, with the op amp operating in *closed-loop* fashion. The feedback network (shown here in general form) can be resistive or reactive, linear or nonlinear, or any combination of these. More detailed analysis will show that the circuit gain characteristic as a whole follows the inverse of the feedback network transfer function.

The concept of feedback is both an essential and salient point concerning op amp use. With feedback, the net closed-loop gain characteristics of a stage such as Figure 1-2 become primarily dependent upon a set of *external components* (usually passive). Thus behavior is less dependent upon the relatively unstable amplifier open-loop characteristics.

---

[2] Such an op amp generates its own power, has two input pins, an output pin, and an output common pin.

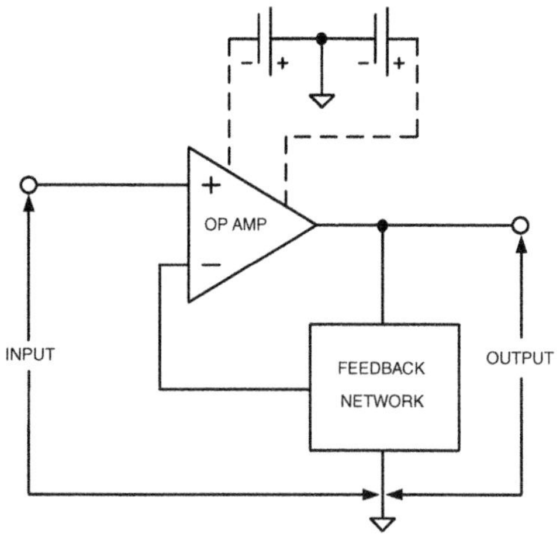

Figure 1-2: A generalized op amp circuit with feedback applied

Note that within Figure 1-2, the input signal is applied between the op amp (+) input and a *common* or *reference point*, as denoted by the ground symbol. It is important to note that this reference point is also common to the output and feedback network. By definition, the op amp stage's output signal appears between the output terminal/feedback network input, and this common ground. This single relevant fact answers the "Where is the op amp grounded?" question so often asked by those new to the craft. The answer is simply that it is grounded *indirectly*, by virtue of the commonality of its input, the feedback network, and the power supply, as is shown in Figure 1-2.

To emphasize how the input/output signals are referenced to the power supply, dual supply connections are shown dotted, with the ± power supply midpoint common to the input/output signal ground. But do note, while all op amp application circuits may not show full details of the power supply connections, every *real* circuit will always use power supplies.

Standard Op Amp Feedback Hookups

Virtually all op amp feedback connections can be categorized into just a few basic types. These include the two most often used, *noninverting* and *inverting* voltage gain stages, plus a related *differential* gain stage. Having discussed above just the attributes of the ideal op amp, at this point it is possible to conceptually build basic gain stages. Using the concepts of infinite gain, zero input offset voltage, zero bias current, and so forth, standard op amp feedback hookups can be devised. For brevity, a full mathematical development of these concepts isn't included here (but this follows in a subsequent section). The end-of-section references also include such developments.

The Noninverting Op Amp Stage

The op amp noninverting gain stage, also known as a *voltage follower with gain*, or simply *voltage follower*, is shown in Figure 1-3.

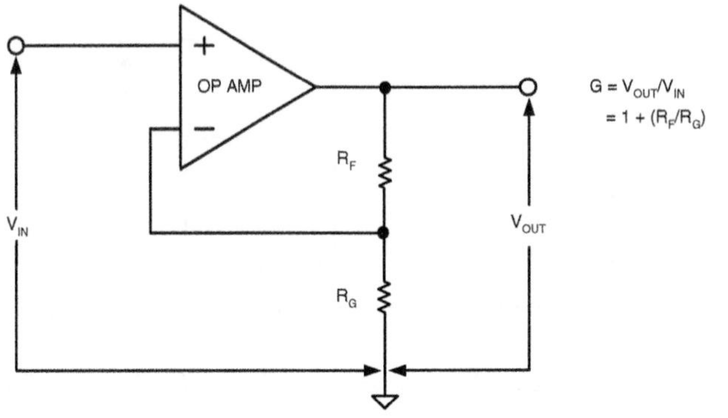

Figure 1-3: The noninverting op amp stage (voltage follower)

This op amp stage processes the input V_{IN} by a gain of G, so a generalized expression for gain is:

$$G = \frac{V_{OUT}}{V_{IN}} \qquad \text{Eq. 1-1}$$

Feedback network resistances R_F and R_G set the stage gain of the follower. For an ideal op amp, the gain of this stage is:

$$G = \frac{R_F + R_G}{R_G} \qquad \text{Eq. 1-2}$$

For clarity, these expressions are also included in the figure. Comparison of this figure and the more general Figure 1-2 shows R_F and R_G here as a simple feedback network, returning a fraction of V_{OUT} to the op amp (–) input. *(Note that some texts may show the more general symbols Z_F and Z_G for these feedback components—both are correct, depending upon the specific circumstances.)*

In fact, we can make some useful general points about the network $R_F - R_G$. We will define the transfer expression of the network as seen from the top of R_F to the output across R_G as β. Note that this usage is a general feedback network transfer term, *not* to be confused with bipolar transistor forward gain. β can be expressed mathematically as:

$$\beta = \frac{R_G}{R_F + R_G} \qquad \text{Eq. 1-3}$$

So, the feedback network returns a fraction of V_{OUT} to the op amp (–) input. Considering the ideal principles of zero offset and infinite gain, this allows some deductions on gain to be made. The voltage at the (–) input is forced by the op amp's feedback action to be equal to that seen at the (+) input, V_{IN}. Given this relationship, it is relatively easy to work out the ideal gain of this stage, which in fact turns out to be simply the inverse of β. This is apparent from a comparison of Eqs. 1-2 and 1-3.

Thus an ideal noninverting op amp stage gain is simply equal to $1/\beta$, or:

$$G = \frac{1}{\beta} \qquad \text{Eq. 1-4}$$

This noninverting gain configuration is one of the most useful of all op amp stages, for several reasons. Because V_{IN} sees the op amp's high impedance (+) input, it provides an ideal interface to the driving source. Gain can easily be adjusted over a wide range via R_F and R_G, with virtually no source interaction.

A key point is the interesting relationship concerning R_F and R_G. Note that to satisfy the conditions of Eq. 1-2, only their *ratio* is of concern. In practice this means that stable gain conditions can exist over a range of actual $R_F - R_G$ values, so long as they provide the same ratio.

If R_F is taken to zero and R_G open, the stage gain becomes unity, and V_{OUT} is then exactly equal to V_{IN}. This special noninverting gain case is also called a *unity gain follower*, a stage commonly used for buffering a source.

Note that this op amp example shows only a simple resistive case of feedback. As mentioned, the feedback can also be reactive, i.e., Z_F, to include capacitors and/or inductors. In all cases, however, it must include a dc path, if we are to assume the op amp is being biased by the feedback (which is usually the case).

To summarize some key points on op amp feedback stages, we paraphrase from Reference 2 the following statements, which will always be found useful:

> *The* summing point *idiom is probably the most used phrase of the aspiring analog artificer, yet the least appreciated. In general, the inverting (−) input is called the* summing point, *while the noninverting (+) input is represented as the* reference terminal. *However, a vital concept is the fact that, within linear op amp applications,* the inverting input (or summing point) assumes the same absolute potential as the noninverting input or reference (within the gain error of the amplifier). *In short, the amplifier tries to servo its own summing point to the reference.*

The Inverting Op Amp Stage

The op amp inverting gain stage, also known simply as the *inverter*, is shown in Figure 1-4. As can be noted by comparison of Figures 1-3 and 1-4, the inverter can be viewed as similar to a follower, but with a transposition of the input voltage V_{IN}. In the inverter, the signal is applied to R_G of the feedback network and the op amp (+) input is grounded.

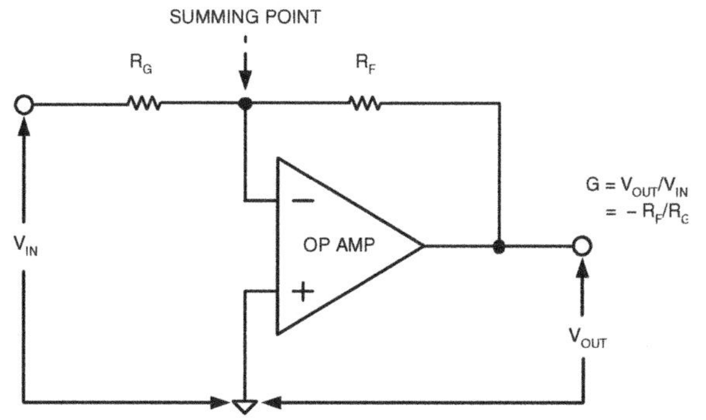

Figure 1-4: The inverting op amp stage (inverter)

The feedback network resistances R_F and R_G set the stage gain of the inverter. For an ideal op amp, the gain of this stage is:

$$G = -\frac{R_F}{R_G} \qquad \text{Eq. 1-5}$$

For clarity, these expressions are again included in the figure. Note that a major difference between this stage and the noninverting counterpart is the input-to-output sign reversal, denoted by the minus sign in Eq. 1-5. Like the follower stage, applying ideal op amp principles and some basic algebra can derive the gain expression of Eq. 1-5.

The inverting configuration is also one of the more useful op amp stages. Unlike a noninverting stage, however, the inverter presents a relatively low impedance input for V_{IN}, i.e., the value of R_G. This factor provides a finite load to the source. While the stage gain can in theory be adjusted over a wide range via R_F and R_G, there is a practical limitation imposed at high gain, when R_G becomes relatively low. If R_F is zero, the gain becomes zero. R_F can also be made variable, in which case the gain is linearly variable over the dynamic range of the element used for R_F. As with the follower gain stage, the gain is ratio dependent, and is relatively insensitive to the exact R_F and R_G values.

The inverter's gain behavior, due to the principles of infinite op amp gain, zero input offset, and zero bias current, gives rise to an effective node of zero voltage at the (–) input. The input and feedback currents sum at this point, which logically results in the term *summing point*. It is also called a *virtual ground*, because of the fact it will be at the same potential as the grounded reference input.

Note that, technically speaking, *all* op amp feedback circuits have a summing point, whether they are inverters, followers, or a hybrid combination. The summing point is always the feedback junction at the (–) input node, as shown in Figure 1-4. However in follower type circuits this point isn't a virtual ground, since it follows the (+) input.

A special gain case for the inverter occurs when $R_F = R_G$, which is also called a *unity gain inverter*. This form of inverter is commonly used for generating complementary V_{OUT} signals, i.e., $V_{OUT} = -V_{IN}$. In such cases it is usually desirable to match R_F to R_G accurately, which can readily be done by using a well-specified matched resistor pair.

A variation of the inverter is the *inverting summer*, a case similar to Figure 1-4, but with input resistors R_{G2}, R_{G3}, etc (not shown). For a summer individual input resistors are connected to additional sources V_{IN2}, V_{IN3}, and so forth, with their common node connected to the summing point. This configuration, called a *summing amplifier*, allows linear input current summation in R_F.[3] V_{OUT} is proportional to an inverse sum of input currents.

The Differential Op Amp Stage

The op amp differential gain stage (also known as a *differential amplifier*, or *subtractor*) is shown in Figure 1-5.

Paired input and feedback network resistances set the gain of this stage. These resistors, R_F–R_G and R_F'–R_G', *must be matched as noted,* for proper operation. Calculation of individual gains for inputs V_1 and V_2 and their linear combination derives the stage gain.

[3] The very first general-purpose op amp circuit is described by Karl Swartzel in Reference 3, and is titled "Summing Amplifier." This amplifier became a basic building block of the M9 gun director computer and fire control system used by Allied Forces in World War II. It also influenced many vacuum tube op amp designs that followed over the next two decades.

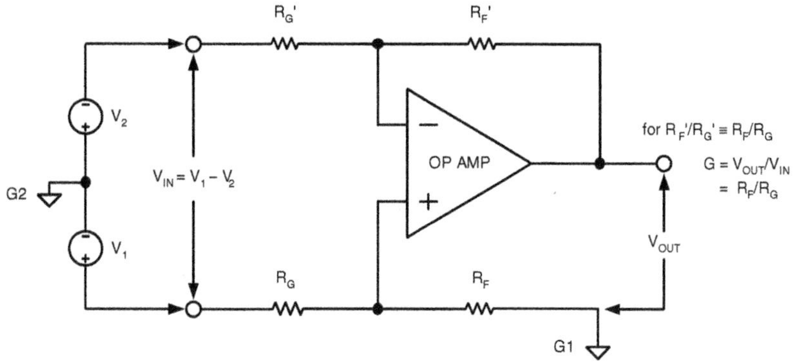

Figure 1-5: The differential amplifier stage (subtractor)

Note that the stage is intended to amplify the *difference* of voltages V_1 and V_2, so the net input is $V_{IN} = V_1 – V_2$. The general gain expression is then:

$$G = \frac{V_{OUT}}{V_1 - V_2} \qquad \text{Eq. 1-6}$$

For an ideal op amp and the resistor ratios matched as noted, the gain of this differential stage from V_{IN} to V_{OUT} is:

$$G = \frac{R_F}{R_G} \qquad \text{Eq. 1-7}$$

The great fundamental utility that an op amp stage such as this allows is the property of rejecting voltages *common* to $V_1 – V_2$, i.e., common-mode (CM) voltages. For example, if noise voltages appear between grounds G1 and G2, the noise will be suppressed by the common-mode rejection (CMR) of the differential amp. The CMR however is only as good as the matching of the resistor ratios allows, so in practical terms it implies precisely trimmed resistor ratios are necessary. Another disadvantage of this stage is that the resistor networks load the $V_1 – V_2$ sources, potentially leading to additional errors.

The Nonideal Op Amp—Static Errors Due to Finite Amplifier Gain

One of the most distinguishing features of op amps is their staggering magnitude of dc voltage gain. Even the least expensive devices have typical voltage gains of 100,000 (100 dB), while the highest performance precision bipolar and chopper stabilized units can have gains as high as 10,000,000 (140 dB), or more. Negative feedback applied around this much voltage gain readily accomplishes the virtues of closed-loop performance, making the circuit dependent only on the feedback components.

As noted above in the discussion of ideal op amp attributes, the behavioral assumptions follow from the fact that negative feedback, coupled with high open-loop gain, constrains the amplifier input error voltage (and consequently the error current) to infinitesimal values. The higher this gain, the more valid these assumptions become.

In reality, however, op amps *do* have finite gain and errors exist in practical circuits. The op amp gain stage of Figure 1-6 will be used to illustrate how these errors impact performance. In this circuit the op amp is ideal except for the finite open-loop dc voltage gain, A, which is usually stated as A_{VOL}.

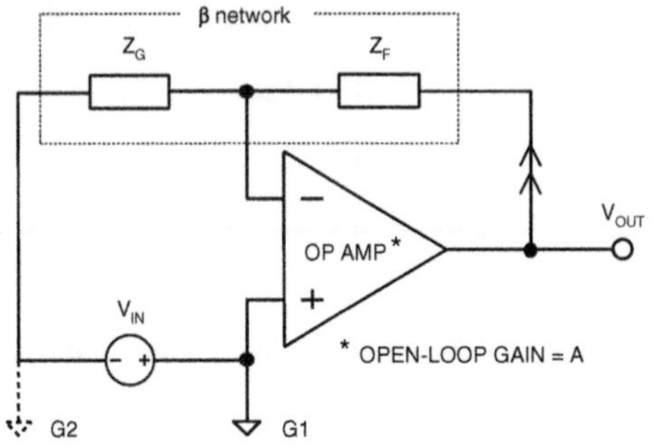

Figure 1-6: Nonideal op amp stage for gain error analysis

Noise Gain (NG)

The first aid to analyzing op amps circuits is to differentiate between *noise gain* and *signal gain*. We have already discussed the differences between noninverting and inverting stages as to their signal gains, which are summarized in Eqs. 1-2 and 1-4, respectively. But, as can be noticed from Figure 1-6, the difference between an inverting and noninverting stage can be as simple as where the reference ground is placed. For a ground at point G1, the stage is an inverter; conversely, if the ground is placed at point G2 (with no G1) the stage is noninverting.

Note, however, that in terms of the feedback path, *there are no real differences*. To make things more general, the resistive feedback components previously shown are replaced here with the more general symbols Z_F and Z_G, otherwise they function as before. The feedback attenuation, β, is the same for both the inverting and noninverting stages:

$$\beta = \frac{Z_G}{Z_G + Z_F} \qquad \text{Eq. 1-8}$$

Noise gain can now be simply defined as: *The inverse of the net feedback attenuation from the amplifier output to the feedback input.* In other words, the inverse of the β network transfer function. This can ultimately be extended to include frequency dependence (covered later in this chapter). Noise gain can be abbreviated as NG.

As noted, the inverse of β is the ideal noninverting op amp stage gain. Including the effects of finite op amp gain, a modified gain expression for the noninverting stage is:

$$G_{CL} = \frac{1}{\beta} \times \left[\frac{1}{1 + \frac{1}{A_{VOL}\beta}} \right] \qquad \text{Eq. 1-9}$$

where G_{CL} is the finite-gain stage's closed-loop gain, and A_{VOL} is the op amp open-loop voltage gain for loaded conditions.

It is important to note that this expression is identical to the ideal gain expression of Eq. 1-4, with the addition of the bracketed multiplier on the right side. Note also that this right-most term becomes closer and closer to unity, as A_{VOL} approaches infinity. Accordingly, it is known in some textbooks as the *error multiplier* term, when the expression is shown in this form.[4]

It may seem logical here to develop another finite gain error expression for an inverting amplifier, but in actuality there is no need. Both inverting and noninverting gain stages have a common feedback basis, which is the noise gain. So Eq. 1-9 will suffice for gain error analysis for both stages. Simply use the β factor as it applies to the specific case.

It is useful to note some assumptions associated with the rightmost error multiplier term of Eq. 1-9. For $A_{VOL}\beta >> 1$, one assumption is:

$$\frac{1}{1+\frac{1}{A_{VOL}\beta}} \approx 1-\frac{1}{A_{VOL}\beta} \qquad \text{Eq. 1-10}$$

This in turn leads to an estimation of the percentage error, ε, due to finite gain A_{VOL}:

$$\varepsilon(\%) \approx \frac{100}{A_{VOL}\beta} \qquad \text{Eq. 1-11}$$

Gain Stability

The closed-loop gain error predicted by these equations isn't in itself tremendously important, since the ratio Z_F/Z_G could always be adjusted to compensate for this error.

But note however that closed-loop gain *stability* is a very important consideration in most applications. Closed-loop gain instability is produced primarily by variations in open-loop gain due to changes in temperature, loading, and so forth.

$$\frac{\Delta G_{CL}}{G_{CL}} \approx \frac{\Delta A_{VOL}}{A_{VOL}} \times \frac{1}{A_{VOL}\beta} \qquad \text{Eq. 1-12}$$

From Eq. 1-12, any variation in open-loop gain (ΔA_{VOL}) is reduced by the factor $A_{VOL}\beta$, insofar as the effect on closed-loop gain. This improvement in closed-loop gain stability is one of the important benefits of negative feedback.

Loop Gain

The product $A_{VOL}\beta$, which occurs in the above equations, is called *loop gain*, a well-known term in feedback theory. The improvement in closed-loop performance due to negative feedback is, in nearly every case, proportional to loop gain.

The term "loop gain" comes from the method of measurement. This is done by breaking the closed feedback loop at the op amp output, and measuring the total gain around the loop. In Figure 1-6 for example, this could be done between the amplifier output and the feedback path (see arrows). To a first

---

[4] Some early discussions of this finite gain error appear in References 4 and 5. Terman uses the open-loop gain symbol of A, as we do today. West uses Harold Black's original notation of μ for open-loop gain. The form of Eq. 1-9 is identical to Terman's (or to West's, substituting μ for A).

approximation, closed-loop output impedance, linearity error, and gain instability are all reduced by $A_{VOL}\beta$ with the use of negative feedback.

Another useful approximation is developed as follows. A rearrangement of Eq. 1-9 is:

$$\frac{A_{VOL}}{G_{CL}} = 1 + A_{VOL}\beta \qquad \text{Eq. 1-13}$$

So, for high values of $A_{VOL}\beta$,

$$\frac{A_{VOL}}{G_{CL}} \approx A_{VOL}\beta \qquad \text{Eq. 1-14}$$

Consequently, in a given feedback circuit the loop gain, $A_{VOL}\beta$, is approximately the numeric ratio (or difference, in dB) of the amplifier open-loop gain to the circuit closed-loop gain.

This loop gain discussion emphasizes that, indeed, loop gain is a very significant factor in predicting the performance of closed-loop operational amplifier circuits. The open-loop gain required to obtain an adequate amount of loop gain will, of course, depend on the desired closed-loop gain.

For example, using Eq. 1-14, an amplifier with A_{VOL} = 20,000 will have an $A_{VOL}\beta \approx 2000$ for a closed-loop gain of 10, but the loop gain will be only 20 for a closed-loop gain of 1000. The first situation implies an amplifier-related gain error on the order of ≈ 0.05%, while the second would result in about 5% error. Obviously, the higher the required gain, the greater will be the required open-loop gain to support an $A_{VOL}\beta$ for a given accuracy.

Frequency Dependence of Loop Gain

Thus far, it has been assumed that amplifier open-loop gain is independent of frequency. Unfortunately, this isn't the case. Leaving the discussion of the effect of open-loop response on bandwidth and dynamic errors until later, let us now investigate the general effect of frequency response on loop gain and static errors.

The open-loop frequency response for a typical operational amplifier with superimposed closed-loop amplifier response for a gain of 100 (40 dB), illustrates graphically these results in Figure 1-7. In these Bode plots, subtraction on a logarithmic scale is equivalent to normal division of numeric data.[5] Today, op amp open-loop gain and loop gain parameters are typically given in dB terms, thus this display method is convenient.

A few key points evolve from this graphic figure, which is a simulation involving two hypothetical op amps, both with a dc/low frequency gain of 100 dB (100 kV/V). The first has a gain-bandwidth of 1 MHz, while the gain-bandwidth of the second is 10 MHz.

- The open-loop gain A_{VOL} for the two op amps is noted by the two curves marked 1 MHz and 10 MHz, respectively. Note that each has a –3 dB corner frequency associated with it, above which the open-loop gain falls at 6 dB/octave. These corner frequencies are marked at 10 Hz and 100 Hz, respectively, for the two op amps.
- At any frequency on the open-loop gain curve, the numeric product of gain A_{VOL} and frequency, f, is a constant (10,000 V/V at 100 Hz equates to 1 MHz). This, by definition, is characteristic of a constant gain-bandwidth product amplifier. All *voltage feedback* op amps behave in this manner.

[5] The log-log displays of amplifier gain (and phase) versus frequency are called *Bode* plots. This graphic technique for display of feedback amplifier characteristics, plus definitions for feedback amplifier stability were pioneered by Hendrick W. Bode of Bell Labs (see Reference 6).

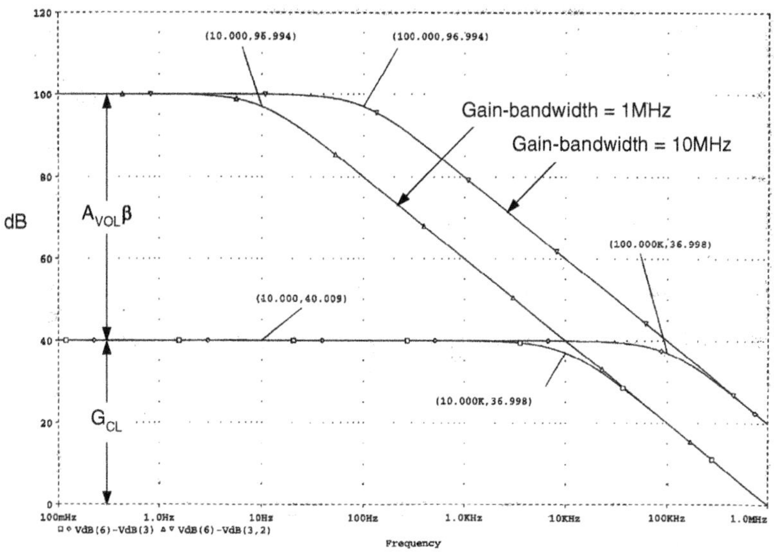

Figure 1-7: Op amp closed-loop gain and loop gain interactions with typical open-loop responses

- $A_{VOL}\beta$ in dB is the difference between open-loop gain and closed-loop gain, as plotted on log-log scales. At the lower frequency point marked, $A_{VOL}\beta$ is thus 60 dB.
- $A_{VOL}\beta$ decreases with increasing frequency, due to the decrease of A_{VOL} above the open-loop corner frequency. At 100 Hz for example, the 1 MHz gain-bandwidth amplifier shows an $A_{VOL}\beta$ of only 80 db – 40 db = 40 dB.
- $A_{VOL}\beta$ also decreases for higher values of closed-loop gain. Other, higher closed-loop gain examples (not shown) would decrease $A_{VOL}\beta$ to less than 60 dB at low frequencies.
- G_{CL} depends primarily on the ratio of the feedback components, Z_F and Z_G, and is relatively independent of A_{VOL} (apart from the errors discussed above, which are inversely proportional to $A_{VOL}\beta$). In this example $1/\beta$ is 100, or 40 dB, and is so marked at 10 Hz. Note that G_{CL} is flat with increasing frequency, up until that frequency where G_{CL} intersects the open-loop gain curve, and $A_{VOL}\beta$ drops to zero.
- At this point where the closed-loop and open-loop curves intersect, the loop gain is by definition zero, which implies that beyond this point there is no negative feedback. Consequently, closed-loop gain is equal to open-loop gain for further increases in frequency.
- Note that the 10 MHz gain-bandwidth op amp allows a 10× increase in closed-loop bandwidth, as can be noted from the –3 dB frequencies; that is 100 kHz versus 10 kHz for the 10 MHz versus the 1 MHz gain-bandwidth op amp.

Figure 1-7 illustrates that the high open-loop gain figures typically quoted for op amps can be somewhat misleading. As noted, beyond a few Hz, the open-loop gain falls at 6 dB/octave. Consequently, closed-loop gain stability, output impedance, linearity and other parameters dependent upon loop gain are degraded at higher frequencies. One of the reasons for having dc gain as high as 100 dB and bandwidth as wide as several MHz, is to obtain adequate loop gain at frequencies even as low as 100 Hz.

A direct approach to improving loop gain at high frequencies, other than by increasing open-loop gain, is to increase the amplifier open-loop bandwidth. Figure 1-7 shows this in terms of two simple examples. It should be borne in mind however that op amp gain-bandwidths available today extend to the hundreds of MHz, allowing video and high-speed communications circuits to fully exploit the virtues of feedback.

Op Amp Common-Mode Dynamic Range(s)

As a point of departure from the idealized circuits above, some practical basic points are now considered. Among the most evident of these is the allowable input and output dynamic ranges afforded in a real op amp. This obviously varies with not only the specific device, but also the supply voltage. While we can always optimize this performance point with device selection, more fundamental considerations come first.

Any real op amp will have a finite voltage range of operation, at both input and output. In modern system designs, supply voltages are dropping rapidly, and 3 V – 5 V total supply voltages are now common. This is a far cry from supply systems of the past, which were typically ±15 V (30 V total). Obviously, if designs are to accommodate a 3 V – 5 V supply, careful consideration must be given to maximizing dynamic range, by choosing a correct device. Choosing a device will be in terms of exact specifications, but first and foremost it should be in terms of the basic topologies used within it.

Output Dynamic Range

Figure 1-8 is a general illustration of the limitations imposed by input and output dynamic ranges of an op amp, related to both supply rails. Any op amp will always be powered by two supply potentials, indicated by the positive rail, $+V_S$, and the negative rail, $-V_S$. We will define the op amp's input and output CM range in terms of how closely it can approach these two rail voltage limits.

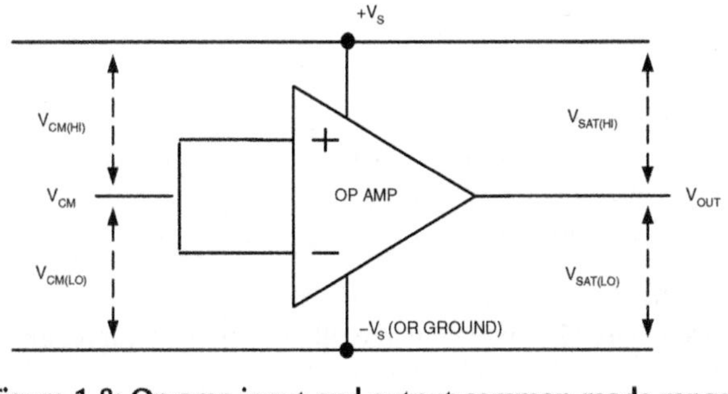

Figure 1-8: Op amp input and output common-mode ranges

At the output, V_{OUT} has two rail-imposed limits, one high or close to $+V_S$, and one low, or close to $-V_S$. Going high, it can range from an upper saturation limit of $+V_S - V_{SAT(HI)}$ as a positive maximum. For example if $+V_S$ is 5 V, and $V_{SAT(HI)}$ is 100 mV, the upper V_{OUT} limit or positive maximum is 4.9 V. Similarly, going low it can range from a lower saturation limit of $-V_S + V_{SAT(LO)}$. So, if $-V_S$ is ground (0 V) and $V_{SAT(HI)}$ is 50 mV, the lower limit of V_{OUT} is simply 50 mV.

Obviously, the internal design of a given op amp will impact this output CM dynamic range, since, when so necessary, the device itself must be designed to minimize both $V_{SAT(HI)}$ and $V_{SAT(LO)}$, to maximize the output dynamic range. Certain types of op amp structures are so designed, and these are generally associated with designs expressly for *single-supply* systems. This is covered in detail later within the chapter.

Input Dynamic Range

At the input, the CM range useful for V_{IN} also has two rail-imposed limits, one high or close to $+V_S$, and one low, or close to $-V_S$. Going high, it can range from an upper CM limit of $+V_S - V_{CM(HI)}$ as a positive maximum. For example, again using the $+V_S = 5$ V example case, if $V_{CM(HI)}$ is 1 V, the upper V_{IN} limit or positive CM maximum is $+V_S - V_{CM(HI)}$, or 4 V.

Figure 1-9 illustrates by way of a hypothetical op amp's data how $V_{CM(HI)}$ could be specified, as shown in the upper curve. This particular op amp would operate for V_{CM} inputs *lower* than the curve shown.

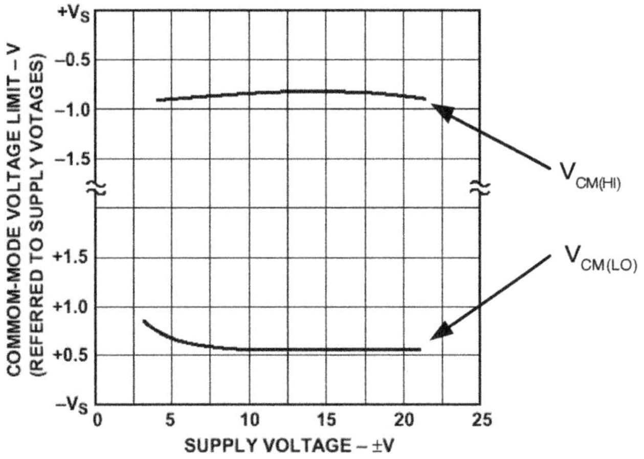

Figure 1-9: A graphical display of op amp input common mode range

In practice the input CM range of real op amps is typically specified as a *range of voltages*, not necessarily referenced to $+V_S$ or $-V_S$. For example, a typical ±15 V operated dual supply op amp would be specified for an operating CM range of ±13 V. Going low, there will also be a lower CM limit. This can be generally expressed as $-V_S + V_{CM(LO)}$, which would appear in a graph such as Figure 1-9 as the lower curve, for $V_{CM(LO)}$. If this were again a ±15 V part, this could represent typical performance.

To use a single-supply example, for the $-V_S = 0$ V case, if $V_{CM(LO)}$ is 100 mV, the lower CM limit will be 0 V + 0.1 V, or simply 0.1 V. Although this example illustrates a lower CM range within 100 mV of $-V_S$, it is actually much more typical to see single-supply devices with lower or upper CM ranges, which *include* the supply rail.

In other words, $V_{CM(LO)}$ or $V_{CM(HI)}$ is 0 V. There are also single-supply devices with CM ranges that include *both* rails. More often than not, however, single-supply devices will not offer graphical data such as Figure 1-9 for CM limits, but will simply cover performance with a tabular range of specified voltage.

Functionality Differences of Dual-Supply and Single-Supply Devices

There are two major classes of op amps, the choice of which determines how well the selected part will function in a given system. Traditionally, many op amps have been designed to operate on a dual power supply system, which has typically been ±15 V. This custom has been prevalent since the earliest IC op amps days, dating back to the mid-sixties. Such devices can accommodate input/output ranges of ±10 V (or slightly more), but when operated on supplies of appreciably lower voltage, for example ±5 V or less, they suffer either loss of performance, or simply don't operate at all. This type of device is referenced here as a *dual-supply* op amp design. This moniker indicates that it performs optimally on dual voltage systems only, typically ±15 V. It may or may not also work at appreciably lower voltages.

Figure 1-10 illustrates in a broad overview the relative functional performance differences that distinguish the dual-supply versus single-supply op amp classes. This table is arranged to illustrate various general performance parameters, with an emphasis on the contrast between single-and dual-supply devices. Which particular performance area is more critical will determine which type of device will be the better system choice.

PERFORMANCE PARAMETER	DUAL SUPPLY	SINGLE SUPPLY
SUPPLY LIMITATIONS	Best >10V, Limited <10V	Best <10V, Limited >10V
OUTPUT V RANGE	– Limited	+ Greatest
INPUT V RANGE	– Limited	+ Greatest
TOTAL DYNAMIC RANGE	+ Greatest	– Least
V & I OUTPUT	+ Greater	– Less
PRECISION	+ Greatest	– Less (growing)
LOAD IMMUNITY	+ Greatest	– Least
VARIETY AVAILABLE	+ Greater	– Less (growing)

Figure 1-10: Comparison of relative functional performance differences between single and dual-supply op amps

More recently, with increasing design attention to lower overall system power and the use of single rail power, the single-supply op amp has come into vogue. This has not been without good reason, as the virtues of using single supply rails can be quite compelling. A review of Figure 1-10 illustrates key points of the dual versus single supply op amp question.

In terms of *supply voltage limitations*, there is a crossover region in terms of overall utility, which occurs around 10 V of total supply voltage.

For example, single-supply devices tend to excel in terms of their *input and output voltage dynamic ranges.* Note that in Figure 1-10 a maximum range is stated as a percentage of available supply. Single-supply parts operate better in this regard, because they are internally designed to maximize these respective ranges. For example, it is not unusual for a device operating from 5 V to swing 4.8 V at the output, and so on.

But, rather interestingly, such devices are also usually restricted to lower supply ranges (only), so their upper dynamic range in absolute terms is actually more limited. For example, a traditional ±15 V dual-supply device can typically swing 20 V p-p, or more than four times that of a 5 V single-supply part. If the *total dynamic range* is considered (assuming an identical input noise), the dual-supply operated part will have four times (or 12 dB) greater dynamic range than that of the 5 V operated part. Or, stated in another way, the input errors of a real part such as noise, drift, and so forth, become four times more critical (relatively speaking), when the output dynamic range is reduced by a factor of 4. Note that these comparisons do not involve any actual device specifications, *they are simply system-based observations*. Device specifications are covered later in this chapter.

In terms of total *voltage and current output*, dual-supply parts tend to offer more in absolute terms, since single-supply parts are usually designed not just for low operating voltage ranges, but also for more modest current outputs.

In terms of *precision*, the dual-supply op amp has long been favored by designers for highest overall precision. However, this status quo is now beginning to be challenged, by such single-supply parts as the truly excellent chopper-stabilized op amps. With more and more new op amps being designed for single-supply use, high precision is likely to become an ever-increasing strength of this category.

Load immunity is often an application problem with single-supply parts, as many of them use common-emitter or common-source output stages, to maximize signal swing. Such stages are typically much more load sensitive than the classic common-collector stages generally used in dual-supply op amps.

There is now a greater *variety* of dual-supply op amps available. However, this is at least in part due to the ~30-year head start they have been enjoying. Currently, new op amp designs are increasingly oriented around one or more aspects of single-supply compatibility, with strong trends toward lower supply voltages, smaller packages, and so forth.

Device Selection Drivers

As the op amp design process is begun, it is useful to keep in mind the fact that there are several selection *drivers*, which can dictate priorities. This is illustrated by Figure 1-11.

FUNCTION	PERFORM	PACKAGE	MARKET
Single, Dual, Quad	Precision	Type	Cost
Single Or Dual Supply	Speed	Size	Availability
Supply Voltage	Distortion, Noise	Footprint	
	Low Bias Current		
	Power		

Figure 1-11: Some op amp selection drivers

Actually, any *single* heading along the top of this chart can, in fact, be the dominant selection driver and take precedence over all of the others. In the early days of op amp design, when such things as supply range, package type, and so forth, were fairly narrow in spread, performance was usually the major driver. Of course, it is still very much so and will always be. But, today's systems are much more compact and lower in power, so things like package type, size, supply range, and multiple devices can often be major drivers of selection. As one example, if the only available supply voltage is 3 V, look at 3 V compatible devices first, and then fill other performance parameters as you can.

As another example, one coming from another perspective, sometimes all-out performance can drive everything else. An ultralow, non-negotiable input current requirement can drive not only the type of amplifier, but also its package (a FET input device in a glass-sealed hermetic package may be optimum). Then, everything else follows from there. Similarly, high power output may demand a package capable of several watts dissipation; in which case, find the power handling device and package first, and then proceed accordingly.

At this point, the concept of these "selection drivers" is still quite general. The following sections of the chapter introduce device types, which supplement this with further details of a realistic selection process.

Classic Cameo

Ray Stata Publications Establish ADI Applications Work

In January of 1965 Analog Devices Inc. (ADI) was founded by Matt Lorber and Ray Stata. Operating initially from Cambridge, MA, modular op amps were the young ADI's primary product. In those days, Ray Stata did more than administrative tasks. He served in sales and marketing roles, and wrote many op amp applications articles. Even today, some of these are still available to ADI customers.

One very early article set was a two part series done for **Electromechanical Design**, which focused on clear, down-to-earth explanation of op amp principles.[1]

A second article for the new ADI publication **Analog Dialogue** was titled "Operational Integrators," and outlined various errors that plague integrators (including capacitor errors).[2]

A third impact article was also done for **Analog Dialogue**, titled "User's Guide to Applying and Measuring Operational Amplifier Specifications."[3] As the title denotes, this was a comprehensive guide to aid the understanding of op amp specifications, and also showed how to test them.

Ray authored an Applications Manual for the 201, 202, 203 and 210 series of chopper op amps.[4]

Ray was also part of the **EEE** "Speaks Out" series of article-interviews, where he outlined some of the subtle ways that op amp specs and behavior can trap unwary users (above photo from that article).[5]

Although ADI today makes many other products, those early op amps were the company's roots.

---

[1] Ray Stata, "Operational Amplifiers – Parts I and II," **Electromechanical Design**, Sept., Nov., 1965.

[2] Ray Stata, "Operational Integrators," **Analog Dialogue,** Vol. 1, No. 1, April, 1967. See also ADI AN357.

[3] Ray Stata, "User's Guide to Applying and Measuring Operational Amplifier Specifications," **Analog Dialogue,** Vol. 1, No. 3, September 1967. See also ADI AN356.

[4] Ray Stata, **Applications Manual for 201, 202, 203 and 210 Chopper Op Amps**, ADI, 1967.

[5] "Ray Stata Speaks Out on 'What's Wrong with Op Amp Specs'," **EEE**, July 1968.

References: Introduction

1. John R. Ragazzini, Robert H. Randall and Frederick A. Russell, "Analysis of Problems in Dynamics by Electronic Circuits," **Proceedings of the IRE**, Vol. 35, May 1947, pp. 444–452.
2. Walter Borlase, **An Introduction to Operational Amplifiers (Parts 1–3)**, September 1971, Analog Devices Seminar Notes, Analog Devices, Inc.
3. Karl D. Swartzel, Jr. "Summing Amplifier," **US Patent 2,401,779,** filed May 1, 1941, issued June 11, 1946.
4. Frederick E. Terman, "Feedback Amplifier Design," **Electronics**, Vol. 10, No. 1, January 1937, pp. 12–15, 50.
5. Julian M. West, "Wave Amplifying System," **US Patent 2,196,844,** filed April 26, 1939, issued April 9, 1940.
6. Hendrick W. Bode, "Relations Between Attenuation and Phase In Feedback Amplifier Design," **Bell System Technical Journal**, Vol. 19, No. 3, July, 1940. See also: "Amplifier," **US Patent 2,173,178,** filed June 22, 1937, issued July 12, 1938.
7. Ray Stata, "Operational Amplifiers-Parts I and II," **Electromechanical Design**, September, November 1965.
8. Dan Sheingold, Ed., **Applications Manual for Operational Amplifiers for Modeling, Measuring, Manipulating, and Much Else**, George A. Philbrick Researches, Inc., Boston, MA, 1965. See also **Applications Manual for Operational Amplifiers for Modeling, Measuring, Manipulating, and Much Else, 2nd Ed.,** Philbrick/Nexus Research, Dedham, MA, 1966, 1984.
9. Walter G. Jung, **IC Op Amp Cookbook, 3rd Ed.**, Prentice-Hall PTR, 1986, 1997, ISBN: 0-13-889601-1.
10. Walt Kester, Editor, **Linear Design Seminar**, Analog Devices, Inc., 1995, ISBN: 0-916550-15-X.
11. Sergio Franco, **Design With Operational Amplifiers and Analog Integrated Circuits, 2nd Ed.** (Sections 1.2 – 1.4), McGraw-Hill, 1998, ISBN: 0-07-021857-9.

Op Amp Topologies

Walt Kester, Walt Jung, James Bryant

The previous section examined op amps without regard to their internal circuitry. In this section the two basic op amp topologies—voltage feedback (VFB) and current feedback (CFB)—are discussed in more detail, leading up to a detailed discussion of the actual circuit structures in Section 1-3.

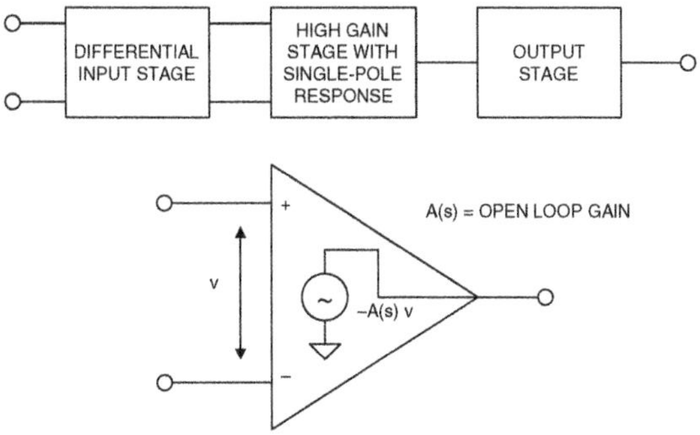

Figure 1-12: Voltage feedback (VFB) op amp

Although not explicitly stated, the previous section focused on the voltage feedback op amp and the related equations. In order to reiterate, the basic voltage feedback op amp is repeated here in Figure 1-12 (without the feedback network) and in Figure 1-13 (with the feedback network).

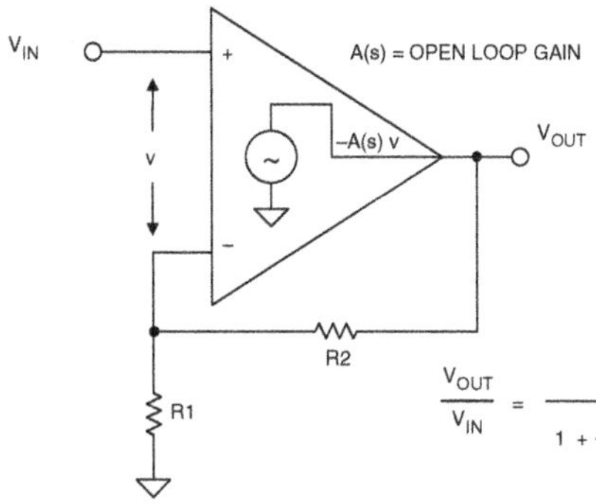

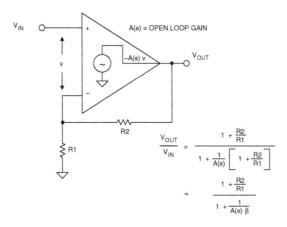

$$\frac{V_{OUT}}{V_{IN}} = \frac{1+\frac{R2}{R1}}{1+\frac{1}{A(s)}\left[1+\frac{R2}{R1}\right]} = \frac{1+\frac{R2}{R1}}{1+\frac{1}{A(s)\,\beta}}$$

Figure 1-13: Voltage feedback op amp with feedback network connected

It is important to note that the error signal developed because of the feedback network and the finite open-loop gain A(s) is in fact a small voltage, v.

Current Feedback Amplifier Basics

The basic *current feedback* amplifier topology is shown in Figure 1-14. Notice that within the model, a unity gain buffer connects the noninverting input to the inverting input. In the ideal case, the output impedance of this buffer is zero ($R_O = 0$), and the error signal is a small current, i, which flows into the inverting input. The error current, i, is mirrored into a high impedance, T(s), and the voltage developed across T(s) is equal to $T(s) \cdot i$. (The quantity T(s) is generally referred to as the *open-loop transimpedance gain.*)

This voltage is then buffered and connected to the op amp output. If R_O is assumed to be zero, it is easy to derive the expression for the closed-loop gain, V_{OUT}/V_{IN}, in terms of the R1-R2 feedback network and the *open-loop* transimpedance gain, T(s). The equation can also be derived quite easily for a finite R_O, and Figure 1-14 gives both expressions.

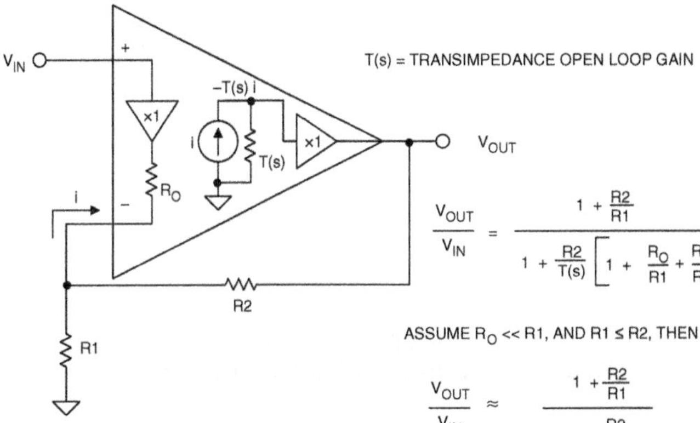

$$\frac{V_{OUT}}{V_{IN}} = \frac{1 + \frac{R2}{R1}}{1 + \frac{R2}{T(s)}\left[1 + \frac{R_O}{R1} + \frac{R_O}{R2}\right]}$$

$$\frac{V_{OUT}}{V_{IN}} \approx \frac{1 + \frac{R2}{R1}}{1 + \frac{R2}{T(s)}}$$

Figure 1-14: Current feedback (CFB) op amp topology

At this point it should be noted that current feedback op amps are often called *transimpedance* op amps, because the *open-loop* transfer function is in fact an impedance as described above. However, the term *transimpedance amplifier* is often applied to more general circuits such as current-to-voltage (I/V) converters, where either CFB or VFB op amps can be used. Therefore, some caution is warranted when the term *transimpedance* is encountered in a given application. On the other hand, the term *current feedback op amp* is rarely confused and is the preferred nomenclature when referring to op amp topology.

From this simple model, several important CFB op amp characteristics can be deduced.

- Unlike VFB op amps, CFB op amps *do not have balanced inputs.* Instead, the noninverting input is high impedance, and the inverting input is low impedance.
- The open-loop gain of CFB op amps is measured in units of Ω (transimpedance gain) rather than V/V as for VFB op amps.
- For a fixed value feedback resistor R2, the closed-loop gain of a CFB can be varied by changing R1, without significantly affecting the closed-loop bandwidth. This can be seen by examining the simplified equation in Figure 1-14. The denominator determines the overall frequency response; and if R2 is constant, then R1 of the numerator can be changed (thereby changing the gain) without affecting the denominator—hence the bandwidth remains relatively constant.

The CFB topology is primarily used where the ultimate in high speed and low distortion is required. The fundamental concept is based on the fact that in bipolar transistor circuits currents can be switched faster than voltages, all other things being equal. A more detailed discussion of CFB op amp ac characteristics can be found in Section 1-5.

Figure 1-15 shows a simplified schematic of an early IC CFB op amp, the AD846—introduced by Analog Devices in 1988 (see Reference 1). Notice that full advantage is taken of the complementary bipolar (CB) process which provides well matched high f_t PNP and NPN transistors.

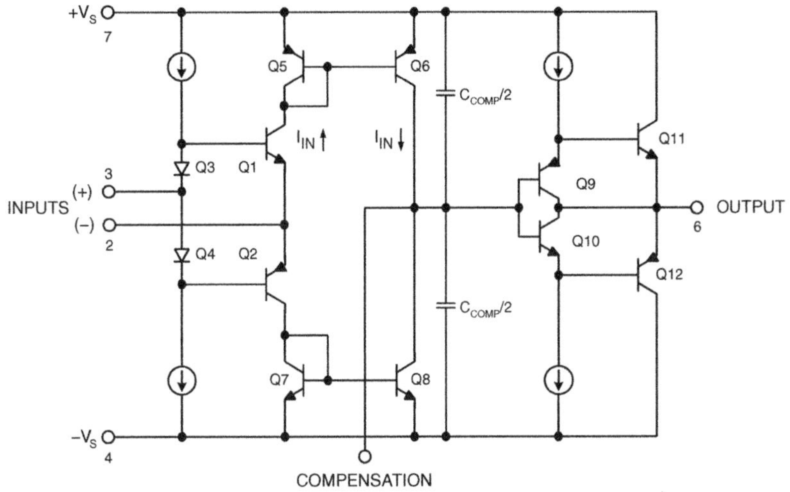

Figure 1-15: AD846 current feedback op amp (1988)

Transistors Q1–Q2 buffer the noninverting input (Pin 3) and drive the inverting input (Pin 2). Q5–Q6 and Q7–Q8 act as current mirrors that drive the high impedance node. The C_{COMP} capacitor provides the dominant pole compensation; and Q9, Q10, Q11, and Q12 comprise the output buffer. In order to take full advantage of the CFB architecture, a high speed complementary bipolar (CB) IC process is required. With modern IC processes, this is readily achievable, allowing direct coupling in the signal path of the amplifier.

However, the basic concept of current feedback can be traced all the way back to early vacuum tube feedback circuitry, which used negative feedback to the input tube cathode. This use of the cathode for feedback would be analogous to the CFB op amp's low impedance (–) input, in Figure 1-15.

Current Feedback Using Vacuum Tubes

Figure 1-16 is an adaptation from a 1937 article on feedback amplifiers by Frederick E. Terman (see Reference 2). Notice that the ac-coupled R2 feedback resistor for this two-stage amplifier is connected to the low impedance cathode of T1, the pentode vacuum tube input stage. Similar examples of early tube circuits using cathode feedback can be found in Reference 3.

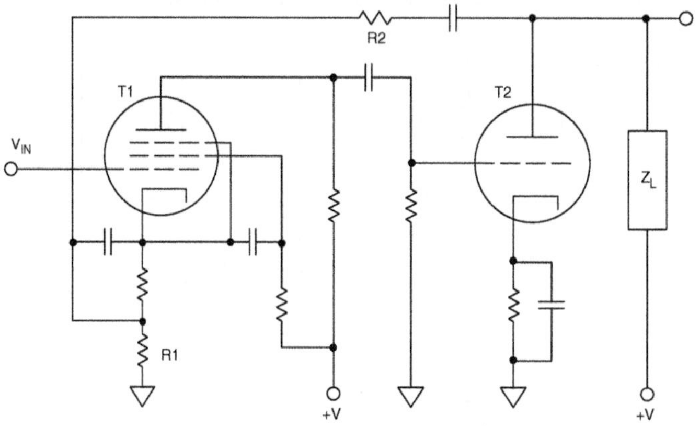

Figure 1-16: A 1937 vacuum tube feedback circuit designed by Frederick E. Terman, using current feedback to the low impedance input cathode (adapted from Reference 2)

Dc-coupled op amp design using vacuum tubes was difficult for numerous reasons. One reason was a lack of suitable level shifters. Multistage op amps either required extremely high supply voltages or suffered gain loss because of resistive level shifters. In a 1941 article, Stewart E. Miller describes how to use gas discharge tubes as level shifters in several vacuum tube amplifier circuits (see Reference 4). A circuit of particular interest is shown in Figure 1-17.

In the Figure 1-17 reproduction of Miller's circuit, the R2 feedback resistor and the R1 gain setting resistor are labeled for clarity, and it can be seen that feedback is to the low impedance cathode of the input tube. The author suggests that the closed-loop gain of the amplifier can be adjusted from 72 dB–102 dB, by varying the R1 gain-setting resistor from 37.4 Ω to 1.04 Ω.

What is really interesting about the Miller circuit is its frequency response, which is reproduced in Figure 1-18. Notice that the closed-loop bandwidth is nearly independent of the gain setting, and the circuit certainly does not exhibit a constant gain-bandwidth product as would be expected for a traditional VFB op amp.

For a gain of 72 dB, the bandwidth is about 30 kHz, and for a gain of 102 dB (30 dB increase), the bandwidth only drops to ~15 kHz. With a 72 dB gain at 30 kHz VFB op amp, bandwidth would be expected to drop 5 octaves to ~0.9 kHz for 102 dB of gain.

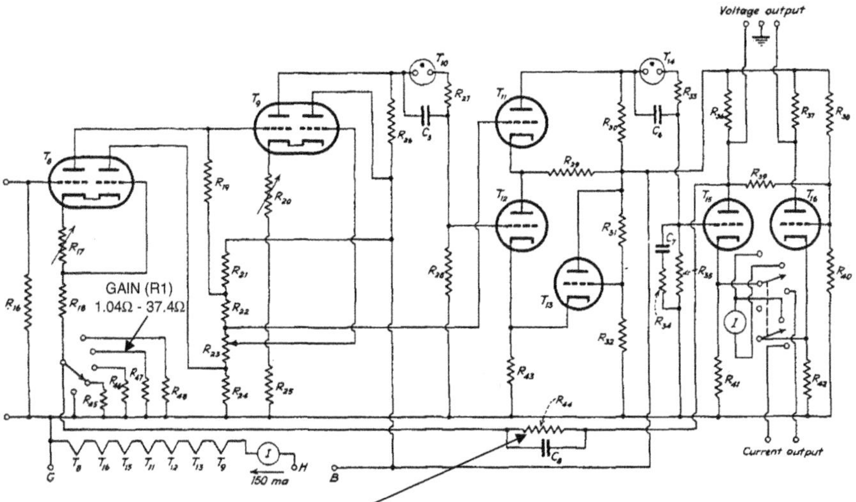

Adapted from: Stewart E. Miller, "Sensitive DC Amplifier with AC Operation," Electronics, November 1941, pp. 27–31, 105–109

Figure 1-17: A 1941 vacuum tube feedback circuit using current feedback

To clarify this point on bandwidth, a standard VFB op amp 6 dB/octave (20 dB/decade) slope has been added to Figure 1-18 for reference.

Although there is no mention of the significance of this within the text of the actual article, it nevertheless illustrates a popular application of CFB behavior, in the design of high speed programmable gain amplifiers with relatively constant bandwidth.

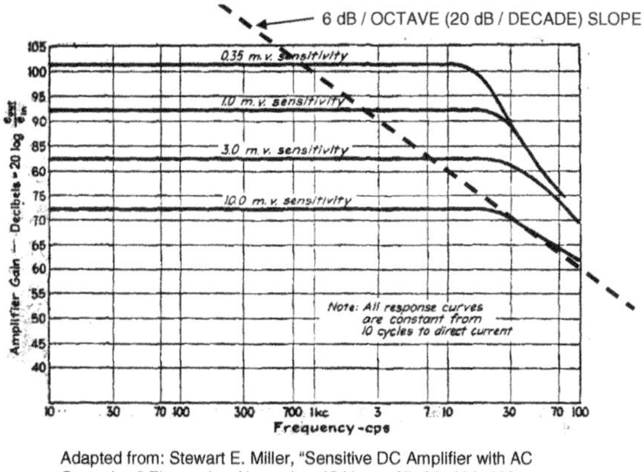

Adapted from: Stewart E. Miller, "Sensitive DC Amplifier with AC Operation," Electronics, November 1941, pp. 27–31, 105–109

Figure 1-18: A 1941 feedback circuit shows characteristic CFB gain-bandwidth relationship

When transistor circuits ultimately replaced vacuum tube circuits between the late 1950s and the mid-1960s, the current feedback architecture became popular for certain high speed op amps. Figure 1-19 shows a fast-settling op amp designed at Bell Labs in 1965, for use as a building block in high speed A/D converters (see Reference 5).

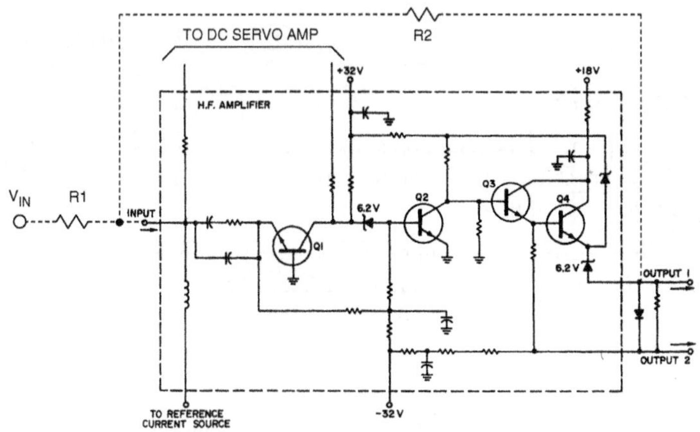

Adapted from: J. O. Edson and H. H. Henning, "Broadband Codecs for an Experimental 224Mb/s PCM Terminal," BSTJ, Vol. 44, No. 9, November 1965, pp. 1887–1950

Figure 1-19: A 1965 solid- state current feedback op amp design from Bell Labs

The circuit shown is a composite amplifier containing a high speed ac amplifier (shown inside the dotted outline) and a separate dc servo amplifier loop (not shown). The feedback resistor R2 is ac coupled to the low impedance emitter of transistor Q1. The circuit design was somewhat awkward because of the lack of good high frequency PNP transistors, and it also required zener diode level shifters, and nonstandard supplies.

Hybrid circuit manufacturing technology, which was well established by the 1980s, allowed the use of fast, relatively well-matched NPN and PNP transistors, to realize CFB op amps. The Analog Devices' AD9610 and AD9611 hybrids were good examples of these devices introduced in the mid-1980s.

With the development of high speed complementary bipolar IC processes in the 1980s (see Reference 6) it became possible to realize completely dc-coupled current feedback op amps using PNP and NPN transistors such as the Analog Devices' AD846, introduced in 1988 (Figure 1-15). Device matching and clever circuit design techniques give these modern IC CFB op amps excellent ac and dc performance without a requirement for separate level shifters, awkward supply voltages, or separate dc servo loops.

Various patents have been issued for these types of designs (see References 7 and 8, for example), but it should be remembered that the fundamental concepts were established decades earlier.

References: Op Amp Topologies

1. Wyn Palmer, Barry Hilton, "A 500 V/µs 12-Bit Transimpedance Amplifier," **ISSCC Digest**, February 1987, pp. 176–177, 386.
2. Frederick E. Terman, "Feedback Amplifier Design," **Electronics**, January 1937, pp. 12–15, 50.
3. Edward L. Ginzton, "DC Amplifier Design Techniques," **Electronics**, March 1944, pp. 98–102.
4. Stewart E. Miller, "Sensitive DC Amplifier with AC Operation," **Electronics**, November 1941, pp. 27–31, 105–109.
5. J. O. Edson and H. H. Henning, "Broadband Codecs for an Experimental 224Mb/s PCM Terminal," **Bell System Technical Journal**, Vol. 44, No. 9, November 1965, pp. 1887–1950.
6. "Op Amps Combine Superb DC Precision and Fast Settling," **Analog Dialogue**, Vol. 22, No. 2, pp. 12–15.
7. David A. Nelson, "Settling Time Reduction in Wide-Band Direct-Coupled Transistor Amplifiers," **US Patent 4,502,020**, Filed October 26, 1983, Issued February 26, 1985.
8. Royal A. Gosser, "DC-Coupled Transimpedance Amplifier," **US Patent 4,970,470**, Filed October 10, 1989, Issued November 13, 1990.

SECTION 1-3

Op Amp Structures

Walt Kester, Walt Jung, James Bryant

This section describes op amps in terms of their *structures*, and Section 1-4 discusses op amp *specifications*. It is hard to decide which to discuss first, since discussion of specifications, to be useful, entails reference to structures, and discussion of structures likewise requires reference to the performance feature that they are intended to optimize.

Since the majority of readers will have at least some familiarity with operational amplifiers and their specifications, we shall discuss structures first, and assume that readers will have at least a first-order idea of the definitions of the various specifications. Where this assumption proves ill-founded, the reader should look ahead to the next section to verify any definitions required.

Because single-supply devices permeate practically all modern system designs, the related design issues are integrated into the following op amp structural discussions.

Single-Supply Op Amp Issues

Over the last several years, single-supply operation has become an increasingly important requirement because of market demands. Automotive, set-top box, camera/camcorder, PC, and laptop computer applications are demanding IC vendors to supply an array of linear devices that operate on a single-supply rail, with the same performance of dual supply parts. Power consumption is now a key parameter for line or battery-operated systems, and in some instances, more important than cost. This makes low voltage/low supply current operation critical; at the same time, however, accuracy and precision requirements have forced IC manufacturers to meet the challenge of "doing more with less" in their amplifier designs.

In a single-supply application, the most immediate effect on the performance of an amplifier is the reduced input and output signal range. As a result of these lower input and output signal excursions, amplifier circuits become more sensitive to internal and external error sources. Precision amplifier offset voltages on the order of 0.1 mV are less than a 0.04 LSB error source in a 12-bit, 10 V full-scale system. In a single-supply system, however, a "rail-to-rail" precision amplifier with an offset voltage of 1 mV represents a 0.8 LSB error in a 5 V full-scale system (or 1.6 LSB for 2.5 V full-scale).

To keep battery current drain low, larger resistors are usually used around the op amp. Since the bias current flows through these larger resistors, they can generate offset errors equal to or greater than the amplifier's own offset voltage.

Gain accuracy in some low voltage single-supply devices is also reduced, so device selection needs careful consideration. Many amplifiers with ~120 dB open-loop gains typically operate on dual supplies—for example OP07 types. However, many single-supply/rail-to-rail amplifiers for precision applications typically have open-loop gains between 25,000 and 30,000 under light loading (>10 kΩ). Selected devices, like the OP113/OP213/OP413 family, do have high open-loop gains (>120 dB), for use in demanding applications. Another example would be the AD855x chopper-stabilized op amp series.

Many trade-offs are possible in the design of a single-supply amplifier circuit—speed versus power, noise versus power, precision versus speed and power, and so forth. Even if the noise floor remains constant (highly unlikely), the signal-to-noise ratio will drop as the signal amplitude decreases.

Besides these limitations, many other design considerations that are otherwise minor issues in dual-supply amplifiers now become important. For example, signal-to-noise (SNR) performance degrades as a result of reduced signal swing. "Ground reference" is no longer a simple choice, as one reference voltage may work for some devices, but not others. Amplifier voltage noise increases as operating supply current drops, and bandwidth decreases. Achieving adequate bandwidth and required precision with a somewhat limited selection of amplifiers presents significant system design challenges in single-supply, low power applications.

Most circuit designers take "ground" reference for granted. Many analog circuits scale their input and output ranges about a ground reference. In dual-supply applications, a reference that splits the supplies (0 V) is very convenient, as there is equal supply headroom in each direction, and 0 V is generally the voltage on the low impedance ground plane.

In single-supply/rail-to-rail circuits, however, the ground reference can be chosen anywhere within the supply range of the circuit, since there is no standard to follow. The choice of ground reference depends on the type of signals processed and the amplifier characteristics. For example, choosing the negative rail as the ground reference may optimize the dynamic range of an op amp whose output is designed to swing to 0 V. On the other hand, the signal may require level shifting in order to be compatible with the input of other devices (such as ADCs) that are not designed to operate at 0 V input.

Very early single-supply "zero-in, zero-out" amplifiers were designed on bipolar processes, which optimized the performance of the NPN transistors. The PNP transistors were either lateral or substrate PNPs with much less bandwidth than the NPNs. Fully complementary processes are now required for the new breed of single-supply/rail-to-rail operational amplifiers. These new amplifier designs don't use lateral or substrate PNP transistors within the signal path, but incorporate parallel NPN and PNP input stages to accommodate input signal swings from ground to the positive supply rail. Furthermore, rail-to-rail output stages are designed with bipolar NPN and PNP common-emitter, or N-channel/P-channel common-source amplifiers whose collector-emitter saturation voltage or drain-source channel on resistance determine output signal swing as a function of the load current.

The characteristics of a single-supply amplifier input stage (common-mode rejection, input offset voltage and its temperature coefficient, and noise) are critical in precision, low voltage applications. Rail-rail input operational amplifiers must resolve small signals, whether their inputs are at ground, or in some cases near the amplifier's positive supply. Amplifiers having a minimum of 60 dB common-mode rejection over the entire input common-mode voltage range from 0 V to the positive supply are good candidates. It is not necessary that amplifiers maintain common-mode rejection for signals beyond the supply voltages. *But, what* is *required is that they do not self-destruct for momentary overvoltage conditions.* Furthermore, amplifiers that have offset voltages less than 1 mV and offset voltage drifts less than 2 μV/°C are also very good candidates for precision applications. Since *input* signal dynamic range and SNR are equally if not more important than *output* dynamic range and SNR, precision single-supply/rail-to-rail operational amplifiers should have noise levels referred-to-input (RTI) less than 5 μV p-p in the 0.1 Hz to 10 Hz band.

The need for rail-to-rail amplifier output stages is also driven by the need to maintain wide dynamic range in low supply voltage applications. A single-supply/rail-to-rail amplifier should have output voltage swings that are within at least 100 mV of either supply rail (under a nominal load). The output voltage swing is very dependent on output stage topology and load current.

Generally, the voltage swing of a good rail-to-rail output stage should maintain its rated swing for loads down to 10 kΩ. The smaller the V_{OL} and the larger the V_{OH}, the better. System parameters, such as "zero-scale" or "full-scale" output voltage, should be determined by an amplifier's V_{OL} (for zero-scale) and V_{OH} (for full-scale).

- Single Supply Offers:
  - Lower Power
  - Battery-Operated Portable Equipment
  - Requires Only One Voltage

- Design Trade-Offs:
  - Reduced Signal Swing Increases Sensitivity to Errors Caused by Offset Voltage, Bias Current, Finite Open-Loop Gain, Noise, etc.
  - Must Usually Share Noisy Digital Supply
  - Rail-to-Rail Input and Output Needed to Increase Signal Swing
  - Precision Less than the best Dual Supply Op Amps but not Required for All Applications
  - Many Op Amps Specified for Single Supply, but do not have Rail-to-Rail Inputs or Outputs

Figure 1-20: Single-supply op amp design issues

Since the majority of single-supply data acquisition systems require at least 12-to 14-bit performance, amplifiers which exhibit an open-loop gain greater than 30,000 for all loading conditions are good choices in precision applications. Single-supply op amp design issues are summarized in Figure 1-20.

Op Amp Input Stages

It is extremely important to understand input and output structures of op amps in order to properly design the required interfaces. For ease of discussion, the two can be examined separately, as there is no particular reason to relate them at this point.

Bipolar Input Stages

The very common and basic bipolar input stage of Figure 1-21 consists of a "long-tailed pair" built with bipolar transistors. It has a number of advantages: it is simple, has very low offset, the bias currents in the inverting and noninverting inputs are well-matched and do not vary greatly with temperature. In addition, minimizing the initial offset voltage of a bipolar op amp by laser trimming also minimizes its drift over temperature. This architecture was used in the very earliest monolithic op amps such as the μA709. It is also used with modern high speed types, like the AD829 and AD8021.

Although NPN bipolars are shown, the concept also applies with the use of PNP bipolars.

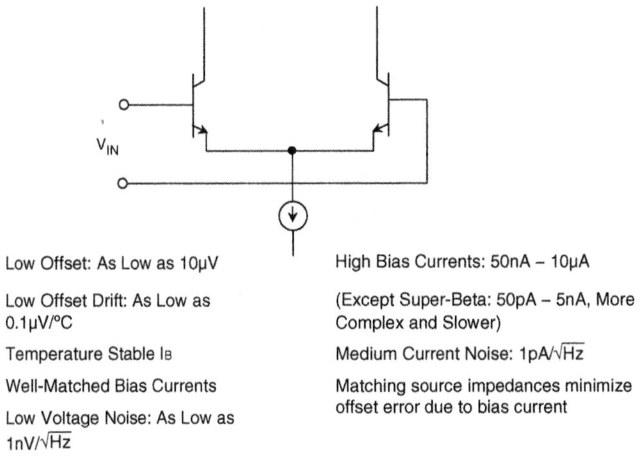

V_{IN}
Low Offset: As Low as 10μV
Low Offset Drift: As Low as 0.1μV/°C
Temperature Stable IB
Well-Matched Bias Currents
Low Voltage Noise: As Low as 1nV/√Hz
High Bias Currents: 50nA – 10μA
(Except Super-Beta: 50pA – 5nA, More Complex and Slower)
Medium Current Noise: 1pA/√Hz
Matching source impedances minimize offset error due to bias current

Figure 1-21: A bipolar transistor input stage

The AD829, introduced in 1990, is shown in Figure 1-22. This op amp uses a bipolar differential input stage, Q1–Q2, which drives a "folded cascode" gain stage which consists of a fast pair of PNP transistors, Q3–Q4 (see Reference 1). These PNPs drive a current mirror that provides the differential-to-single-ended conversion. The output stage is a two-stage complementary emitter follower.

The AD829 is a wideband video amplifier with a 750 MHz uncompensated gain-bandwidth product, and it operates on ±5 V to ±15 V supplies. For added flexibility, the AD829 provides access to the internal compensation node (C_{COMP}). This allows the user to customize frequency response characteristics for a particular application where the closed-loop gain is less than 20. The RC network connected between the output and the high impedance node helps maintain stability, when driving capacitive loads.

Input bias current is 7 μA maximum at 25°C, input voltage noise is $1.7\ nV/\sqrt{Hz}$, and input current noise is $1.5\ pA/\sqrt{Hz}$. Laser wafer trimming reduces the input offset voltage to 0.5 mV maximum for the "A" grade. Typical input offset voltage drift is 0.3 μV/°C.

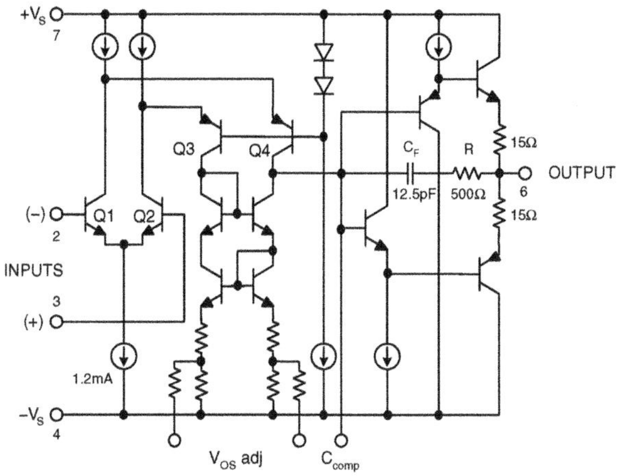

Figure 1-22: AD829 op amp simplified schematic

In an op amp input circuit such as Figure 1-22, the input bias current is the base current of the transistors comprising the long-tailed pair, Q1–Q2. It can be quite high, especially in high speed amplifiers, because the collector currents are high. It is typically ~3 µA, for the AD829. In amplifiers where the bias current is *uncompensated* (as true in this case), the bias current will be equal to one-half of the Q1–Q2 emitter current, divided by the H_{FE}.

The bias current of a simple bipolar input stage can be reduced by a couple of measures. One is by means of *bias current compensation*, to be described further below.

Another method of reducing bias current is by the use of *superbeta* transistors for Q1–Q2. Superbeta transistors are specially processed devices with a very narrow base region. They typically have a current gain of thousands or tens of thousands (rather than the more usual hundreds). Op amps with superbeta input stages have much lower bias currents, but they also have more limited frequency response.

Since the breakdown voltages of superbeta devices are quite low, they also require additional circuitry to protect the input stage from damage caused by overvoltage (for example, they wouldn't operate in the circuit of Figure 1-22).

Some examples of superbeta input bipolar op amps are the AD704/AD705/AD706 series, and the OP97/OP297/OP497 series (single, dual, quad). These devices have typical 25°C bias currents of 100 pA or less.

Bias Current Compensated Bipolar Input Stage

A simple bipolar input stage such as used in Figure 1-22 exhibits high bias current because the currents seen externally are in fact the base currents of the two input transistors.

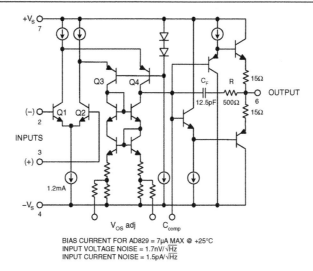

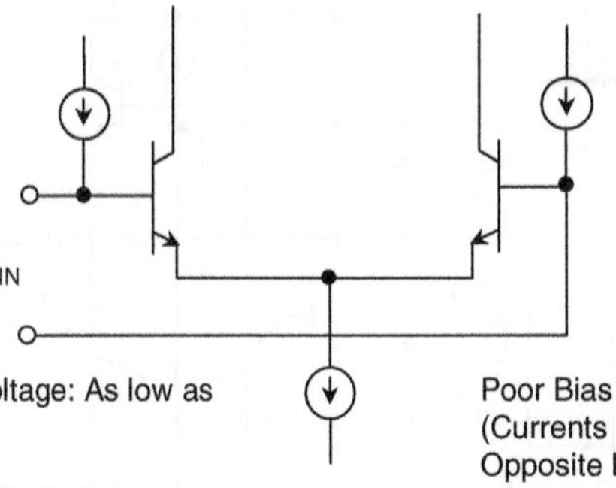

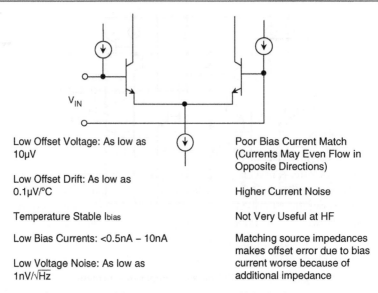

Low Offset Voltage: As low as 10µV

Low Offset Drift: As low as 0.1µV/°C

Temperature Stable Ibias

Low Bias Currents: <0.5nA – 10nA

Low Voltage Noise: As low as 1nV/√Hz

Poor Bias Current Match (Currents May Even Flow in Opposite Directions)

Higher Current Noise

Not Very Useful at HF

Matching source impedances makes offset error due to bias current worse because of additional impedance

Figure 1-23: A bias current compensated bipolar input stage

By providing necessary bias currents via an internal current source, as in Figure 1-23, the only *external* current then flowing in the input terminals is the difference current between the base current and the current source, which can be quite small.

Most modern precision op amps use some means of internal bias current compensation; examples would be the familiar OP07 and OP27 series.

The well-known OP27 op amp family is good example of bias-compensated op amps (see References 2 and 3). The simplified schematic of the OP27, shown in Figure 1-24, shows that the multiple-collector transistor Q6 provides the bias current compensation for the input transistors Q1 and Q2. The "G" grade of the OP27

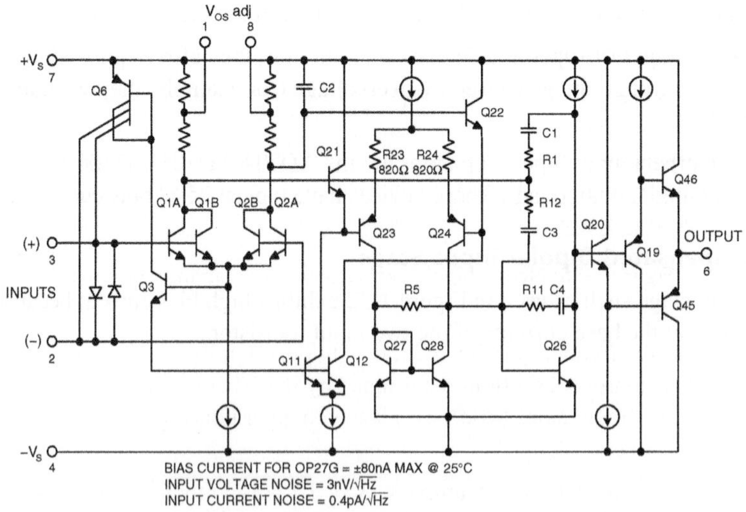

Figure 1-24: OP27 op amp uses bias current compensated input stage

has a maximum input bias current specification of ±80 nA at 25°C. Input voltage noise is $3\ nV/\sqrt{Hz}$, and input current noise $0.4\ pA/\sqrt{Hz}$. Offset voltage trimming by "Zener-zapping" reduces the input offset voltage of the OP27 to 50 µV maximum at 25°C for the "E" grade device (see Reference 4 for details of this trim method).

Bias-current-compensated input stages have many of the good features of the simple bipolar input stage, namely: low voltage noise, low offset, and low drift. Additionally, they have low bias current which is fairly stable with temperature. However, their current noise is not very good, and their bias current matching is poor.

These latter two undesirable side effects result from the external bias current being the *difference* between the compensating current source and the input transistor base current. Both of these currents inevitably have noise. Since they are uncorrelated, the two noises add in a root-sum-of-squares fashion (even though the dc currents subtract).

Since the resulting external bias current is the difference between two nearly equal currents, there is no reason why the net current should have a defined polarity. As a result, the bias currents of a bias-compensated op amp may not only be mismatched, they can actually flow in opposite directions. In most applications this isn't important, but in some it can have unexpected effects. (For example, the droop of a sample-and-hold [SHA] built with a bias-compensated op amp may have either polarity.)

In many cases, the bias current compensation feature is not mentioned on an op amp data sheet, and a simplified schematic isn't supplied. It is easy to determine if bias current compensation is used by examining the bias current specification. If the bias current is specified as a "±" value, the op amp is most likely compensated for bias current.

Note that this can easily be verified by examining the *offset current* specification (the difference in the bias currents). If internal bias current compensation exists, the offset current will be of the same magnitude as the bias current. Without bias current compensation, the offset current will generally be at least a factor of 10 smaller than the bias current. Note that these relationships generally hold, regardless of the exact magnitude of the bias currents.

It is also a well-known fact that, within an op amp application circuit, the effects of bias current on the output offset voltage of an op amp can often be cancelled by making the source resistances at the two inputs equal. However, there is an important caveat here. The validity of this practice holds true only for bipolar input op amps *without* bias current compensation; that is, where the input currents are well matched. In a case of an op amp using internal bias current compensation, adding an extra resistance to either input will usually make the output offset worse.

Bias Current Compensated Superbeta Bipolar Input Stage

As mentioned above, the OP97/297/OP497-series are high performance superbeta op amps, that also use input bias current compensation. As a result, their input bias currents are ±150 pA max at 25°C. Note that in this case the "±" prefix to the bias current magnitude indicates that the amplifier uses internal bias current compensation.

A simplified schematic of an OP97 (or one-quarter of the OP497) is shown in Figure 1-25. Note that the Q1–Q2 superbeta pair is protected against large destructive differential input voltages, by the use of both back-to-back diodes, and series current-limiting resistors. Note also that the Q1–Q2 superbeta pair is also protected against excessive collector voltage, by an elaborate bias and bootstrapping network.

As a result of these clamping and protection circuits, the input common-mode voltage of this op amp series can safely vary over the full range of the supply voltages used.

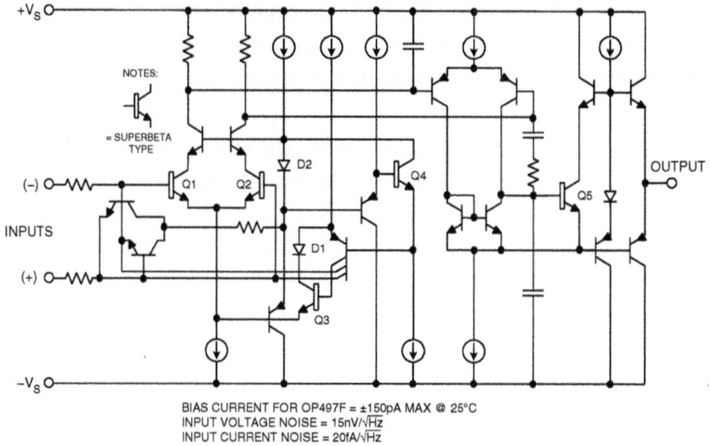

Figure 1-25: The OP97, OP297 and OP497 op amp series uses superbeta input stage transistors and bias current compensation

FET Input Stages

Field-Effect Transistors (FETs) have much higher input impedance than do bipolar junction transistors (BJTs) and would therefore seem to be ideal devices for op amp input stages. However, they cannot be manufactured on all bipolar IC processes, and when a process does allow their manufacture, they often have their own problems.

FETs have high input impedance, low bias current, and good high frequency performance. (In an op amp, the lower g_m of the FET devices allows higher tail currents, thereby increasing the maximum slew rate.) FETs also have much lower current noise.

On the other hand, the input offset voltage of FET long-tailed pairs is not as good as the offset of corresponding BJTs, and trimming for minimum offset does not simultaneously minimize drift. A separate trim is needed for drift, and as a result, offset and drift in a JFET op amp, while good, aren't as good as the best BJTs. A simplified trim procedure for an FET input op amp stage is shown in Figure 1-26.

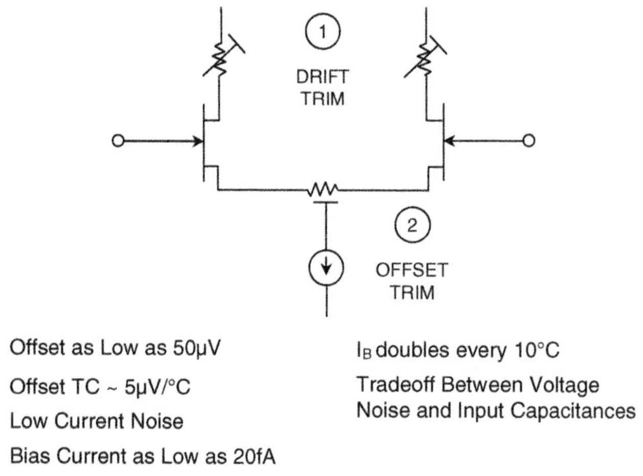

Figure 1-26: Junction field effect transistor (JFET) input op amp stage showing offset and drift trims

It is possible to make JFET op amps with very low voltage noise, but the devices involved are very large and have quite high input capacitance, which varies with input voltage, and so a trade-off is involved between voltage noise and input capacitance.

The bias current of an FET op amp is the leakage current of the gate diffusion (or the leakage of the gate protection diode, which has similar characteristics for a MOSFET). Such leakage currents double with every 10°C increase in chip temperature so that a FET op amp bias current is *one thousand times greater* at 125°C than at 25°C. Obviously this can be important when choosing between a bipolar or FET input op amp, especially in high temperature applications where bipolar op amp input bias current actually decreases.

Thus far, we have spoken generally of all kinds of FETs, that is, junction (JFETs) and MOS (MOSFETs). In practice, combined bipolar/JFET technology op amps (i.e., BiFET) achieve better performance than op amps using purely MOSFET or CMOS technology. While ADI and others make high performance op amps with MOS or CMOS input stages, in general these op amps have worse offset and drift, voltage noise, high-frequency performance than the bipolar counterparts. The power consumption is usually somewhat lower than that of bipolar op amps with comparable, or even better, performance.

JFET devices require more headroom than do BJTs, since their pinch-off voltage is typically greater than a BJTs base-emitter voltage. Consequently, they are more difficult to operate at very low power supply voltages (1–2 V). In this respect, CMOS has the advantage of requiring less headroom than JFETs.

Rail-Rail Input Stages

Today, there is common demand for op amps with input CM voltage that includes *both* supply rails, i.e., *rail-to-rail* CM operation. While such a feature is undoubtedly useful in some applications, engineers should recognize that there are still relatively few applications where it is absolutely essential. These applications should be distinguished from the many more applications where a CM range *close* to the supplies, or one that includes *one* supply, is necessary, but true input rail-to-rail operation is not.

In many single-supply applications, it is required that the input CM voltage range extend to one of the supply rails (usually ground). High side or low side current-sensing applications are examples of this. Many amplifiers can handle 0 V CM inputs, and they are easily designed using PNP differential pairs (or N-channel JFET pairs) as shown in Figure 1-27. The input CM range of such an op amp generally extends from about 200 mV below the negative rail ($-V_S$ or ground), to about 1 V–2 V of the positive rail, $+V_S$.

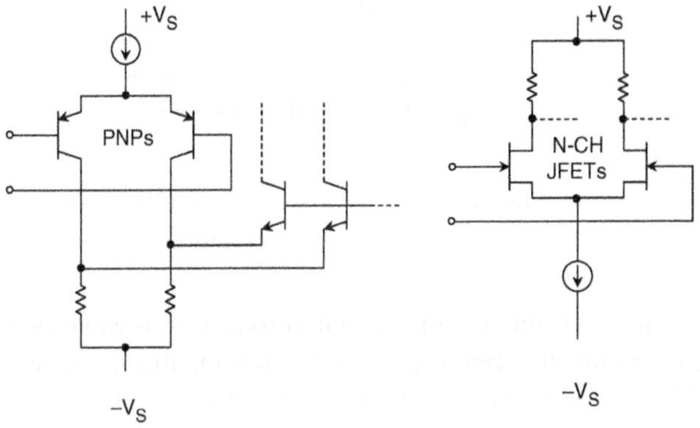

Figure 1-27: PNP or N-channel JFET stages allow CM inputs to the negative rail

An input stage could also be designed with NPN transistors (or P-channel JFETs), in which case the input CM range would include the positive rail, and go to within about 1 V–2 V of the negative rail. This requirement typically occurs in applications such as high-side current sensing. The OP282/OP482 input stage uses a P-channel JFET input pair whose input CM range includes the positive rail, making it suitable for high-side sensing.

The AD823 is a dual 16 MHz (G = +1) op amp with an N-channel JFET input stage (as in Figure 1-27, right). A simplified schematic of the AD823 is shown in Figure 1-28. This device can operate on single-supply voltages from +3 V to +36 V. This range also allows operation on traditional ±5 V, or ±15 V dual supplies if desired. Similar devices in a related (but lower power) family include the AD820, the AD822, and the AD824.

The AD823 JFET input stage allows the input common-mode voltage to range from 200 mV below the negative supply to within about 1.5 V of the positive supply. Input offset voltage is 0.8 mV maximum at 25°C, input bias current is 25 pA maximum at 25°C, offset voltage drift is 2 μV/°C, and input voltage noise is 16 $nV/\sqrt{Hz}$. Current noise is only 1 $fA/\sqrt{Hz}$. The AD823 is laser wafer trimmed for both offset voltage and offset voltage drift as described above.

A simplified diagram of a true rail-to-rail input stage is shown in Figure 1-29. Note that this requires use of *two* long-tailed pairs: one of PNP bipolar transistors Q1–Q2, the other of NPN transistors Q3–Q4. Similar input stages can also be made with CMOS pairs.

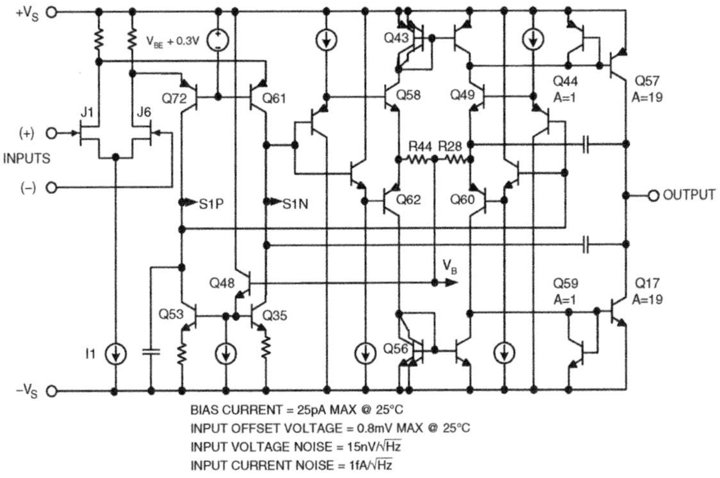

Figure 1-28: AD823 JFET input op amp simplified schematic

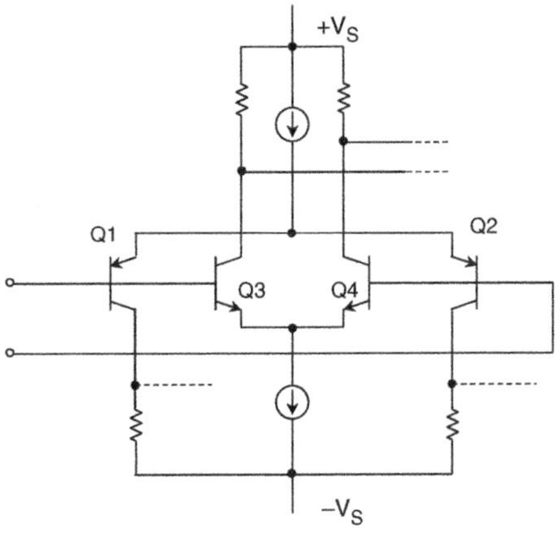

Figure 1-29: A true rail-to-rail bipolar transistor input stage

It should be noted that these two pairs will exhibit *different* offsets and bias currents, so when the applied CM voltage changes, the amplifier input offset voltage and input bias current does also. In fact, when both current sources remain active throughout most of the entire input common-mode range, amplifier input offset voltage is the *average* offset voltage of the two pairs. In those designs where the current sources are alternatively switched off at some point along the input common-mode voltage, amplifier input offset voltage is dominated by the PNP pair offset voltage for signals near the negative supply, and by the NPN pair offset voltage for signals near the positive supply. As noted, a true rail-to-rail input stage can also be constructed from CMOS transistors, for example as in the case of the CMOS AD8531/AD8532/AD8534 op amp family.

Amplifier input bias current, a function of transistor current gain, is also a function of the applied input common-mode voltage. The result is relatively poor common-mode rejection (CMR), and a changing common-mode input impedance over the CM input voltage range, compared to familiar dual-supply devices. These specifications should be considered carefully when choosing a rail-to-rail input op amp, especially for a noninverting configuration. Input offset voltage, input bias current, and even CMR may be quite good over *part* of the common-mode range, but much worse in the region where operation shifts between the NPN and PNP devices, and vice versa.

True rail-to-rail amplifier input stage designs must transition from one differential pair to the other differential pair, somewhere along the input CM voltage range. Some devices like the OP191/OP291/OP491 family and the OP279 have a common-mode crossover threshold at approximately 1 V below the positive supply (where signals do not often occur). The PNP differential input stage is active from about 200mV below the negative supply to within about 1 V of the positive supply. Over this common-mode range, amplifier input offset voltage, input bias current, CMR, input noise voltage/current are primarily determined by the characteristics of the PNP differential pair. At the crossover threshold, however, amplifier input offset voltage becomes the average offset voltage of the NPN/PNP pairs and can change rapidly.

Also, as noted previously, amplifier bias currents are dominated by the PNP differential pair over most of the input common-mode range, and change polarity and magnitude at the crossover threshold when the NPN differential pair becomes active.

Op amps like the OP184/OP284/OP484 family, shown in Figure 1-30, utilize a rail-to-rail input stage design where both NPN and PNP transistor pairs are active throughout most of the entire input CM voltage range. With this approach to biasing, there is no CM crossover threshold. Amplifier input offset voltage is the average offset voltage of the NPN and the PNP stages, and offset voltage exhibits a smooth transition throughout the entire input CM range, due to careful laser trimming of input stage resistors.

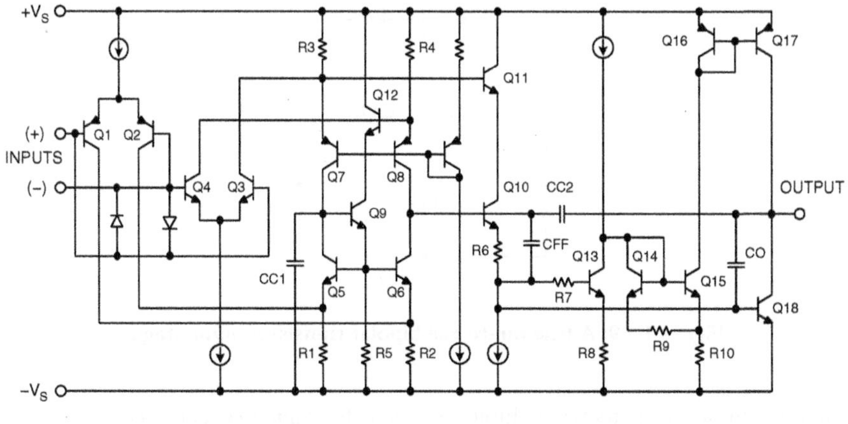

Figure 1-30: OP284 op amp simplified schematic shows true rail-to-rail input stage

In the same manner, through careful input stage current balancing and input transistor design, the OP184 family input bias currents also exhibit a smooth transition throughout the entire CM input voltage range. The exception occurs at the very extremes of the input range, where amplifier offset voltages and bias currents increase sharply, due to the slight forward-biasing of parasitic p-n junctions. This occurs for input voltages within approximately 1 V of either supply rail.

When *both* differential pairs are active throughout most of the entire input common-mode range, amplifier transient response is faster through the middle of the common-mode range by as much as a factor of 2 for bipolar input stages and by a factor of $\sqrt{2}$ for JFET input stages. This is due to the higher transconductance of two operating input stages.

Input stage g_m determines the slew rate and the unity-gain crossover frequency of the amplifier, hence response time degrades slightly at the extremes of the input common-mode range when either the PNP stage (signals approaching the positive supply rail) or the NPN stage (signals approaching the negative supply rail) are forced into cutoff. The thresholds at which the transconductance changes occur are approximately within 1 V of either supply rail, and the behavior is similar to that of the input bias currents.

In light of the many quirks of true rail-to-rail op amp input stages, applications that do require true rail-to-rail inputs should be carefully evaluated, and an amplifier chosen to ensure that its input offset voltage, input bias current, common-mode rejection, and noise (voltage and current) are suitable.

Don't Forget Input Overvoltage Considerations

In order to achieve the performance levels required, it is sometimes not possible to provide complete overdrive protection within IC op amps. Although most op amps have some type of input protection, care must still be taken to prevent possible damage against both CM and differential voltage stress.

This is most likely to occur, for example, when the input signal comes from an external sensor. Rather than present a cursory discussion of this topic here, the reader is instead referred to Chapter 7, Section 7-4 for a detailed examination of this important issue.

Output Stages

The earliest IC op amp output stages were NPN emitter followers with NPN current sources or resistive pull-downs, as shown in Figure 1-31A. Naturally, the slew rates were greater for positive-going than they were for negative-going signals.

While all modern op amps have push-pull output stages of some sort, many are still asymmetrical, and have a greater slew rate in one direction than the other. Asymmetry tends to introduce distortion on ac signals and generally results from the use of IC processes with faster NPN than PNP transistors. It may also result in an ability of the output to approach one supply more closely than the other in terms of saturation voltage.

In many applications, the output is required to swing only to one rail, usually the negative rail (i.e., ground in single-supply systems). A pull-down resistor to the negative rail will allow the output to approach that rail (provided the load impedance is high enough, or is also grounded to that rail), but only slowly. Using an FET current source instead of a resistor can speed things up, but this adds complexity, as shown in Figure 1-31B.

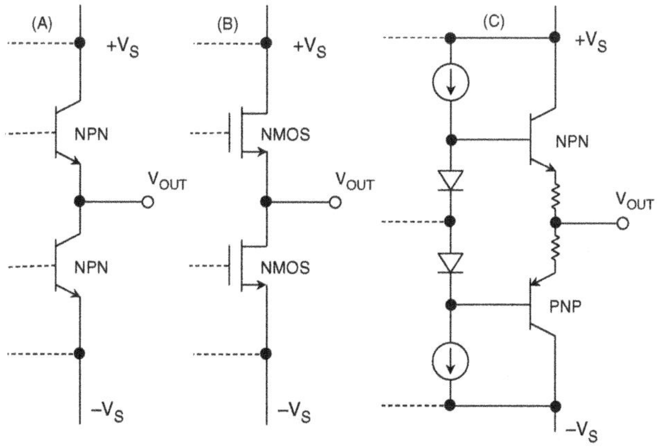

Figure 1-31: Some traditional op amp output stages

With modern complementary bipolar (CB) processes, well matched high speed PNP and NPN transistors are readily available. The complementary emitter follower output stage shown in Figure 1-31C has many advantages, but the most outstanding one is the low output impedance. However, the output voltage of this stage can only swing within about one V_{BE} drop of either rail. Therefore an output swing of 1 V to 4 V is typical of such a stage, when operated on a single 5 V supply.

The complementary common-emitter/common-source output stages shown in Figure 1-32A and B allow the op amp output voltage to swing much closer to the rails, but these stages have much higher open-loop output impedance than do the emitter follower-based stages of Figure 1-31C.

In practice, however, the amplifier's high open-loop gain and the applied feedback can still produce an application with low output impedance (particularly at frequencies below 10 Hz). What should be carefully evaluated with this type of output stage is the loop gain within the application, with the load in place. Typically, the op amp will be specified for a minimum gain with a load resistance of 10 kΩ (or more). Care should be taken that the application loading doesn't drop lower than the rated load, or gain accuracy may be lost.

It should also be noted these output stages can cause the op amp to be more sensitive to capacitive loading than the emitter-follower type. Again, this will be noted on the device data sheet, which will indicate a maximum of capacitive loading before overshoot or instability will be noted.

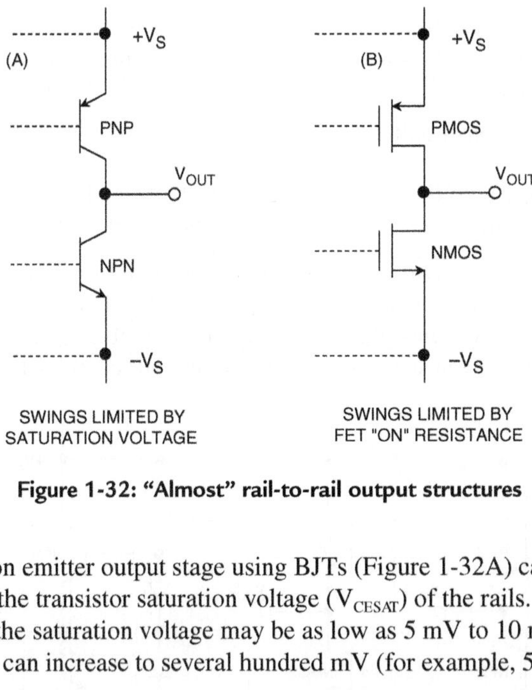

Figure 1-32: "Almost" rail-to-rail output structures

The complementary common emitter output stage using BJTs (Figure 1-32A) cannot swing completely to the rails, but only to within the transistor saturation voltage (V_{CESAT}) of the rails. For small amounts of load current (less than 100 µA), the saturation voltage may be as low as 5 mV to 10 mV, but for higher load currents, the saturation voltage can increase to several hundred mV (for example, 500 mV at 50 mA).

On the other hand, an output stage constructed of CMOS FETs (Figure 1-32B) can provide nearly true rail-to-rail performance, but only under no-load conditions. If the op amp output must source or sink substantial current, the output voltage swing will be reduced by the I × R drop across the FETs internal "on" resistance. Typically this resistance will be on the order of 100 Ω for precision amplifiers, but it can be less than 10 Ω for high current drive CMOS amplifiers.

For the above basic reasons, it should be apparent that there is no such thing as a *true* rail-to-rail output stage, hence the caption of Figure 1-32 ("Almost" Rail-to-Rail Output Structures). The best any op amp output stage can do is an almost rail-to-rail swing, when it is lightly loaded.

Op amps built on foundry CMOS processes have a primary advantage of low cost. Also, it is relatively straightforward to design rail-to-rail input and output stages with these CMOS devices, which will operate on low supply voltages.

Figure 1-33 shows a simplified schematic of the AD8531/AD8532/AD8534 (single/dual/quad) op amp, which is typical of these design types. The AD8531/AD8532/AD8534 operates on a single 2.7 V to 6.0 V supply and can drive 250 mA. Input offset voltage is 25 mV maximum at 25°C, and voltage noise 45 $nV/\sqrt{Hz}$.

This type of op amp is simple and cost effective, and the lack of high dc precision is often no disadvantage. To the contrary, the high output drive available can be an overriding plus, particularly in AC-coupled applications.

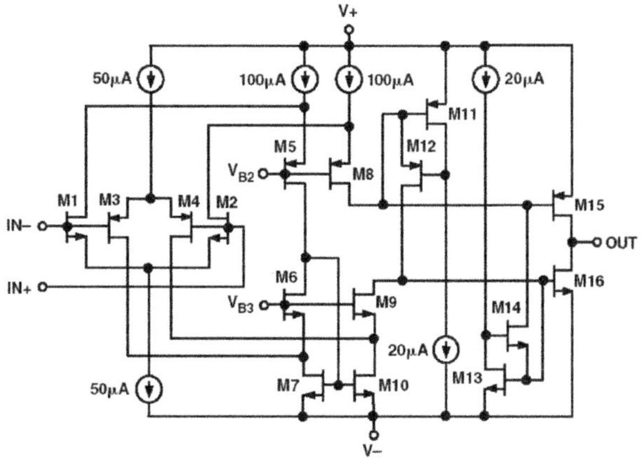

Figure 1-33: AD8531/AD8532/AD8534 CMOS rail-to-rail op amp simplified schematic

Output Stage Surge Protection

Most low speed, high precision op amps generally have output stages which are protected against short circuits to ground or to either supply. Their output current is limited to a little more than 10 mA. This has the additional advantage of minimizing self-heating of the chip (and thus minimizing dc errors due to chip temperature differentials).

If an op amp is required to deliver both high precision and a large output current, it is advisable to use a separate output stage (within the loop) to minimize self-heating of the precision op amp. A simple buffer amplifier such as the BUF04, or a section of a nonprecision op amp can be used.

Note that high speed op amps cannot have output currents limited to low values, as it would affect their slew rate and load drive ability. Thus most high speed op amps will source/sink between 50 mA–100 mA. Although many high speed op amps have internal protection for *momentary* shorts, their junction temperatures can be exceeded with sustained shorts. The user needs to be wary, and consult the specific device ratings.

Offset Voltage Trim Processes

The AD860x CMOS op amp family exploits the advantages of digital technology to minimize the offset voltage normally associated with CMOS amplifiers. Offset voltage trimming is done after the devices are packaged. A digital code is entered into the device to adjust the offset voltage to less than 1 mV, depending upon the grade. Wafer testing is not required, and the patented ADI technique called DigiTrim™ requires no extra pins to accomplish the function. These devices have rail-to-rail inputs and outputs (similar to Figure 1-33), and the NMOS and PMOS parallel input stages are trimmed separately using DigiTrim to minimize the offset voltage in both pairs. A functional diagram of the AD8602 DigiTrim op amp is shown in Figure 1-34.

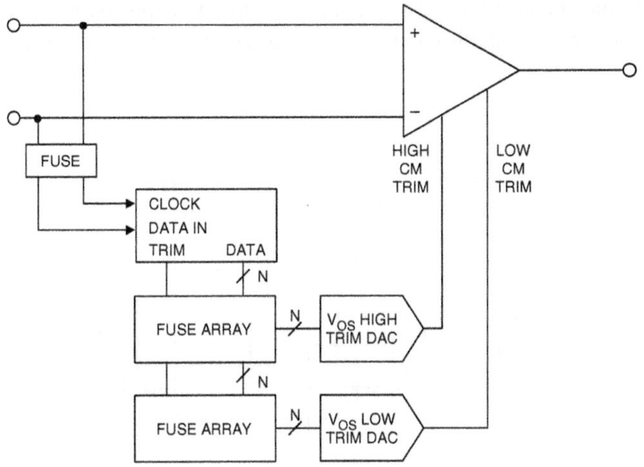

Figure 1-34: AD8602 (1/2) CMOS op amp showing DigiTrim™

DigiTrim adjusts the offset voltage by programming digitally weighted current sources. The trim information is entered through existing pins using a special digital sequence. The adjustment values can be temporarily programmed, evaluated, and readjusted for optimum accuracy before permanent adjustment is performed. After the trim is completed, the trim circuit is locked out to prevent the possibility of any accidental re-trimming by the end user.

The physical trimming, achieved by blowing polysilicon fuses, is very reliable. No extra pads or pins are required, and no special test equipment is needed to perform the trimming. The trims can be done after packaging so that assembly-related shifts can be eliminated. No testing is required at the wafer level because of high die yields

The first devices to use this new technique are the Analog Devices' AD8601/02/04 (single, dual, quad) rail-to-rail CMOS amplifiers. The offset is trimmed for both high and low common-mode conditions so that the offset voltage is under 500 μV over the full common-mode input voltage range. The bandwidth of the op amps is 8 MHz, slew rate is 5 V/μs, and supply current is only 640 μA per amplifier.

At this point it is useful to review the other popular trim methods. Analog Devices pioneered the use of thin film resistors and laser wafer trimming for precision amplifiers, references, data converters, and other linear ICs (see Reference 5). Up to 16-bit accuracy can be achieved with trimming, and the thin film resistors themselves are very stable with temperature and can add to the thermal stability and accuracy of a device, even without trimming. Thin film deposition and patterning are processes that must be tightly controlled. The laser trimming systems are also quite expensive. In-package trimming is not possible, so assembly-related shifts cannot be easily compensated. Nevertheless, thin film trimming at the wafer level provides continuous fine trim resolution in precision integrated circuits where high accuracy and stability are required.

Zener zapping uses a voltage to create a metallic short circuit across the base-emitter junction of a transistor to remove a circuit element (see References 4 and 6). The base-emitter junction is commonly referred to as a zener, although the mechanism is actually avalanche breakdown of the junction. During the avalanche breakdown across the base-emitter junction, the very high current densities and localized heating generate rapid metal migration between the base and emitter connections, leading to a metallic short across the junction. With proper biasing (current, voltage, and time), this short will have a very low resistance value. If a series of these base-emitter junctions are arranged in parallel with a string of resistors, zapping selected junctions will short out portions of the resistor string, thereby adjusting the total resistance value.

It is possible to perform zener zap trimming in the packaged IC to compensate for assembly-related shifts in the offset voltage. However, trimming in the package requires extra package pins. Alternately, trimming at the wafer level requires additional probe pads. Probe pads do not scale effectively as the process features shrink. Thus, the die area required for trimming is relatively constant, regardless of the process geometries. Some form of bipolar transistor is required for the trim structures, therefore a purely MOS-based process may not have zener zap capability. The nature of the trims is discrete since each zap removes a predefined resistance value. Increasing trim resolution requires additional transistors and pads or pins, which rapidly increase the total die area and/or package cost. This technique is most cost-effective for fairly large geometry processes where the trim structures and probe pads make up a relatively small percentage of the overall die area.

It was in the process of creating the industry standard OP07 in 1975 that Precision Monolithics Incorporated pioneered zener zap trimming (Reference 6). The OP07 and other similar parts must be able to operate from over ±15 V supplies. As a result, they utilize relatively large device geometries to support the high voltage requirements, and extra probe pads don't significantly increase die area.

Link trimming is the cutting of metal or poly-silicon links to remove a connection. In link trimming, either a laser or a high current is used to destroy a "shorted" connection across a parallel resistive element. Removing the connection increases the effective resistance of the combined element(s). Laser cutting is similar to laser trimming of thin films. The high local heat from the laser beam causes material changes that lead to a nonconductive area, effectively cutting a metal or conductive polysilicon connector.

The high-current link trim method works as an inverse to zener zapping—the conductive connection is destroyed, rather than created by a zener-zap.

Link trim structures tend to be somewhat more compact than laser trimmed resistor structures. No special processes are required in general, although the process may have to be tailored to the laser characteristics if laser cutting is used. With the high-current trimming method, testing at the wafer level may not be required if die yields are good. The laser cutting scheme doesn't require extra contact pads, but the trim structures don't scale with the process feature sizes. Laser cutting of links cannot be performed in the package, and requires additional probe pads on the die. In addition, it can require extra package pins for in-package high-current trims. Like zener zapping, link trimming is discrete. Resolution improvements require additional structures, increasing area and cost.

EEPROM trimming utilizes special, nonvolatile digital memory to store trim data. The stored data bits control adjustment currents through on-chip D/A converters. Memory cells and D/A converters scale with the process feature size. In-package trimming and even trimming in the customer's system is possible so that assembly-related shifts can be trimmed out. Testing at the wafer level is not required if yields are reasonable. No special hardware is required for the trimming beyond the normal mixed-signal tester system, although test software development may be more complicated.

Since the trims can be overwritten, it is possible to periodically reprogram the system to account for long-term drifts or to modify system characteristics for new requirements. The number of reprogram cycles possible depends on the process, and is finite. Most EEPROM processes provide enough rewrite cycles to handle routine recalibration.

This trim method does require special processing. Stored trim data can be lost under certain conditions, especially at high operating temperatures. At least one extra digital contact pad/package pin is required to input the trim data to the on-chip memory.

This technique is only available on MOS-based processes due to the very thin oxide requirements. The biggest drawback is that the on-chip D/A converters are large—often larger than the amplifier circuits they are adjusting. For this reason, EEPROM trimming is mostly used for data converter or system-level products where the trim D/A converters represent a much smaller percentage of the overall die area.

Figure 1-35 summarizes the key features of each ADI trim method. It can be seen from that all trim methods have their respective places in producing high performance linear integrated circuits.

PROCESS	TRIMMED AT:	SPECIAL PROCESSING	RESOLUTION
DigiTrim™	Wafer or Final Test	None	Discrete
Laser Trim	Wafer	Thin Film Resistor	Continuous
Zener Zap Trim	Wafer	None	Discrete
Link Trim	Wafer	Thin Film or Poly Resistor	Discrete
EEPROM Trim	Wafer or Final Test	EEPROM	Discrete

Figure 1-35: Summary of ADI trim processes

Op Amp Process Technologies

The wide variety of op amp processes is shown in Figure 1-36. The early 1960s op amps used standard NPN-based bipolar processes. The PNP transistors of these processes were extremely slow and were used primarily for current sources and level-shifting.

The ability to produce matching high speed PNP transistors on a bipolar process added great flexibility to op amp circuit designs. The first p-epi complementary bipolar (CB) process was introduced by ADI in the mid-1980s. The f_ts of the PNP and NPN transistors were approximately 700 MHz and 900 MHz, respectively, and had 30 V breakdowns. Since its original introduction in 1985, several additional CB processes have been developed at ADI designed for higher speeds and lower breakdowns. For example, a current 5 V CB process has 9 GHZ PNPs and 16 GHz NPNs. These CB processes are used in today's precision op amps, as well as those requiring wide bandwidths.

- BIPOLAR (NPN-BASED): This is Where it All Started
- COMPLEMENTARY BIPOLAR (CB): Rail-to-Rail, Precision, High Speed
- BIPOLAR + JFET (BiFET): High Input Impedance, High Speed
- COMPLEMENTARY BIPOLAR + JFET (CBFET): High Input Impedance, Rail-to-Rail Output, High Speed
- COMPLEMENTARY MOSFET (CMOS): Low Cost Op Amps (ADI DigiTrim Minimizes Offset Voltage and Drift in CMOS Op Amps)
- BIPOLAR (NPN) + CMOS (BiCMOS): Bipolar Input Stage adds Linearity, Low Power, Rail-to-Rail Output
- COMPLEMENTARY BIPOLAR + CMOS (CBCMOS): Rail-to-Rail Inputs, Rail-to-Rail Outputs, Good Linearity, Low Power, Higher Cost

Figure 1-36: Op amp process technology summary

The JFETs available on the Analog Devices complementary bipolar processes allow high input impedance op amps to be designed suitable for such applications as photodiode or electrometer preamplifiers. These processes are sometimes designated as *CBFET*.

CMOS op amps, generally have higher offset voltages and offset voltage drift than trimmed bipolar or BiFET op amps, however the Analog Devices DigiTrim process described above yields low offset voltage, while keeping costs low. Voltage noise for CMOS op amps tends to be larger, but the input bias current is very low. They offer low power and cost (foundry CMOS processes are typically used).

The addition of bipolar or complementary devices to a CMOS process (BiMOS or CBCMOS) adds greater flexibility, better linearity, and lower power as well as additional cost. The bipolar devices are typically used for the input stage to provide good gain and linearity, and CMOS devices for the rail-to-rail output stage.

In summary, there is no single IC process that is optimum for all op amps. Process selection and the resulting op amp design depends on the targeted applications and ultimately should be transparent to the customer.

References: Op Amp Structures

1. "Video Op Amp," **Analog Dialogue**, Vol. 24, No. 3, p. 19.
2. George Erdi, "Amplifier Techniques for Combining Low Noise, Precision, and High-Speed Perfomance," **IEEE Journal of Solid-State Circuits**, Vol. SC-16, December, 1981 pp. 653–661.
3. George Erdi, Tom Schwartz, Scott Bernardi, and Walt Jung, "Op Amps Tackle Noise and for Once, Noise Loses," **Electronic Design**, December 12, 1980.
4. George Erdi, "A Precision Trim Technique for Monolithic Analog Circuits," **IEEE Journal of Solid-State Circuits**, Vol. SC-10, December, 1975 pp. 412–416.
5. Richard Wagner, "Laser-Trimming on the Wafer," **Analog Dialogue**, Vol. 9, No. 3, pp. 3–5.
6. Donn Soderquist, George Erdi, "The OP07 UltraLow Offset Voltage Op Amp," **Precision Monolithics AN-13**, December, 1975.
7. Walt Kester, Editor, **Linear Design Seminar**, Analog Devices, 1995, ISBN: 0-916550-15-X.
8. Walt Kester, Editor, **High Speed Design Techniques**, Analog Devices, 1996, ISBN: 0-916550-17-6 (available for download at www.analog.com).
9. Walt Kester, Editor, **Practical Analog Design Techniques**, Analog Devices, 1995, ISBN: 0-916550-16-8. (available for download at www.analog.com).

SECTION 1-4

Op Amp Specifications

Walt Kester, Walt Jung, James Bryant

In this section, basic op amp specifications are discussed. The importance of any given specification depends of course upon the application. For instance offset voltage, offset voltage drift, and open-loop gain are very critical in precision sensor signal conditioning circuits, but not as important in high speed applications where bandwidth, slew rate, and distortion are the key specifications.

Most op amp specifications are largely topology independent. However, although voltage feedback and current feedback op amps have similar error terms and specifications, the application of each part warrants discussing some of the specifications separately. In the following discussions, this will be done where significant differences exist.

Input Offset Voltage, V_{OS}

Ideally, if both inputs of an op amp are at exactly the same voltage, the output should be at zero volts. In practice, a small differential voltage must be applied to the inputs to force the output to zero. This is known as the *input offset voltage*, V_{OS}.

Input offset voltage is modeled as a voltage source, V_{OS}, in series with the inverting input terminal of the op amp as shown in Figure 1-37. The corresponding output offset voltage (due to V_{OS}) is obtained by multiplying the input offset voltage by the dc noise gain of the circuit (see Figure 1-3 and Eq. 1-2).

Figure 1-37: Input offset voltage

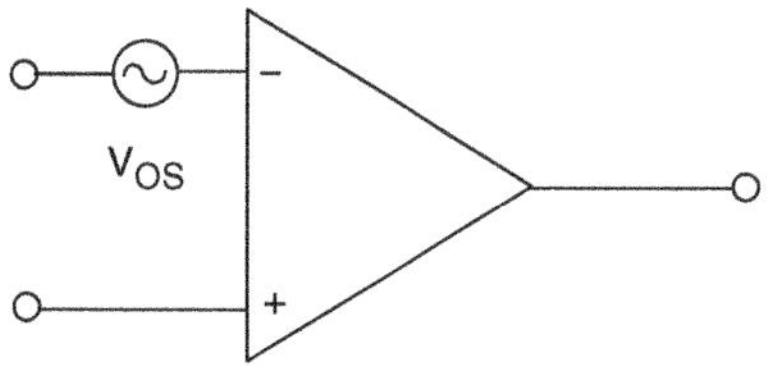

- Offset Voltage: The differential voltage that must be applied to the input of an op amp to produce zero output.
- Ranges:
  - Chopper-Stabilized Op Amps: <1µV
  - General-Purpose Precision Op Amps: 50-500µV
  - Best Bipolar Op Amps: 10–25µV
  - Best FET Op Amps: 100–1,000µV
  - High Speed Op Amps: 100–2,000µV
  - Untrimmed CMOS Op Amps: 5,000–50,000µV
  - DigiTrim CMOS Op Amps: <1,000µV

Chopper stabilized op amps have a V_{OS} that is less than 1 µV (AD8551 series), and the best precision bipolar op amps (super-beta or bias stabilized) can have offsets as low as 25 µV (OP177F). The very best trimmed FET types have about 100 µV of offset (AD8610B), and untrimmed CMOS op amps can range from 5 mV to 50 mV. However, the ADI DigiTrim CMOS op amps have offset voltages less than 1 mV (AD8605). Generally speaking, "precision" op amps will have V_{OS} < 0.5 mV, although some high speed amplifiers may be a little worse than this.

Measuring input offset voltages of a few microvolts requires that the test circuit does not introduce more error than the offset voltage itself. Figure 1-38 shows a standard circuit for measuring offset voltage. The circuit amplifies the input offset voltage by the noise gain of 1001. The measurement is made at the amplifier output using an accurate digital voltmeter. The offset referred to the input (RTI) is calculated by dividing the output voltage by the noise gain. The small source resistance seen by the inputs results in negligible bias current contribution to the measured offset voltage. For example, 2 nA bias current flowing through the 10 Ω resistor produces a 0.02 μV error referred to the input.

As simple as this circuit looks, it can give inaccurate results when testing precision op amps, unless care is taken in implementation. The largest potential error source comes from parasitic thermocouple junctions, formed where two different metals are joined. This thermocouple voltage can range from 2 μV/°C to more than 40 μV/°C. Note that in this circuit additional "dummy" resistors have been added to the noninverting input, in order to exactly match/balance the thermocouple junctions in the inverting input path.

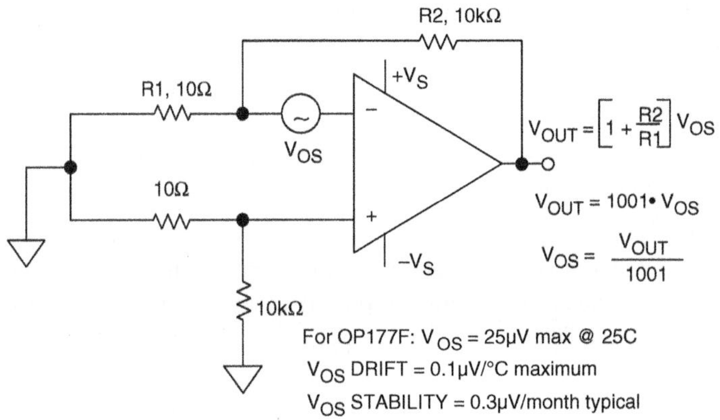

Figure 1-38: Measuring input offset voltage

The accuracy of the measurement also depends on the mechanical layout of the components and exactly how they are placed on the PC board. Keep in mind that the two connections of a component such as a resistor create two equal, but opposite polarity thermoelectric voltages (assuming they are connected to the same metal, such as the copper trace on a PC board). These will cancel each other, *assuming both are at exactly the same temperature*. Clean connections and short lead lengths help to minimize temperature gradients and increase the accuracy of the measurement. Note—see the Chapter 7 discussions on this general topic for more detail.

In the test circuit, airflow should be minimal so that all the thermocouple junctions stabilize at the same temperature. In some cases, the circuit should be placed in a small closed container to eliminate the effects of external air currents. The circuit should be placed flat on a surface so that convection currents flow up and off the top of the board, not across the components, as would be the case if the board were mounted vertically.

Measuring the offset voltage shift over temperature is an even more demanding challenge. Placing the printed circuit board containing the amplifier being tested in a small box or plastic bag with foam insulation prevents the temperature chamber air current from causing thermal gradients across the parasitic thermocouples. If cold testing is required, a dry nitrogen purge is recommended. Localized temperature cycling of the amplifier itself using a Thermostream-type heater/cooler may be an alternative, but these units tend to generate quite a bit of airflow that can be troublesome.

Generally, the test circuit of Figure 1-38 can be made to work for many amplifiers. Low absolute values for the small resistors (such as 10 Ω) will minimize bias current induced errors. An alternate V_{OS} measurement method is shown in Figure 1-39, and is suitable for cases of high and/or unequal bias currents (as in the case of current feedback op amps).

In this measurement method, an in amp is connected to the op amp input terminals through isolation resistors, and provides the gain for the measurement. The offset voltage of the in amp (measured with S closed) must then be subtracted from the final V_{OS} measurement. Also, the circuit shown in Figure 1-44, for measuring input bias currents, can also be used to measure input offset voltage independent of bias currents.

Figure 1-39: Alternate input offset voltage measurement using an in amp

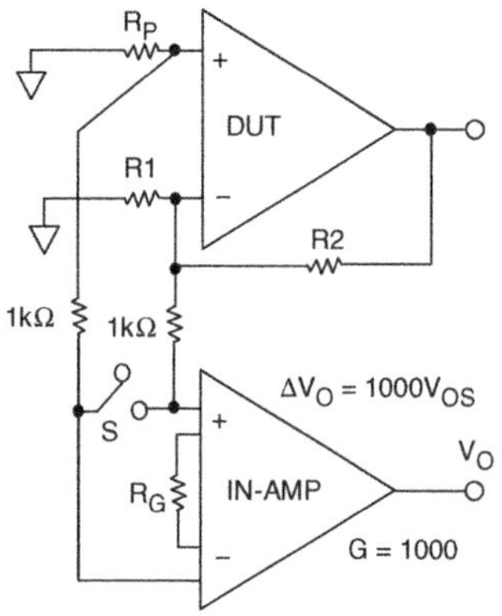

Offset Adjustment (Internal Method)

Many single op amps have pins available for optional offset null. To make use of this feature, two pins are joined by a potentiometer, and the wiper goes to one of the supplies through a resistor, as shown generally in Figure 1-40. Note that if the wiper is accidentally connected to the wrong supply, the op amp will probably be destroyed—this is a common problem, when one op amp type is replaced by another. The range of offset adjustment in a well-designed op amp is no more than two or three times the maximum V_{OS} of the

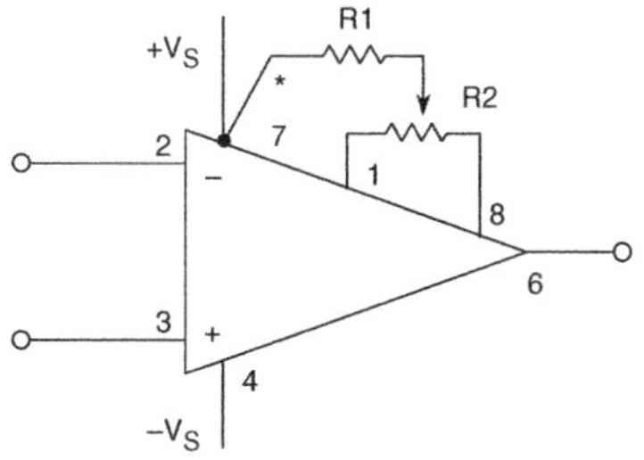

- * Wiper connection may be to either $+V_S$ or $-V_S$ depending on op amp
- R values depend on op amp. Consult data sheet
- Use to null out input offset voltage, not system offsets
- There may be high gain from offset pins to output — Keep them quiet
- Nulling offset causes increase in offset temperature coefficient, approximately 4μV/°C for 1mV offset null for FET inputs

Figure 1-40: Offset adjustment pins

lowest grade device, in order to minimize sensitivity. Nevertheless, the voltage gain of an op amp at its offset adjustment pins may actually be greater than the gain at its signal inputs. It is therefore very important to keep these pins noise free. Note that it is *never* advisable to use long leads from an op amp to a remote nulling potentiometer.

As mentioned above, the offset drift of an op amp with temperature will vary with the setting of its offset adjustment. The internal adjustment terminals should therefore be used only to adjust the op amp's own offset, *not to correct any system offset errors*, since doing so would be at the expense of increased temperature drift. The drift penalty for a FET input op amp is on the order of 4 μV/°C for each millivolt of nulled offset voltage. It is generally better to control offset voltage by proper device/grade selection.

Offset Adjustment (External Methods)

If an op amp doesn't have offset adjustment pins (popular duals and all quads do not), and it is still necessary to adjust the amplifier and system offsets, an external method can be used. This method is also most useful if the offset adjustment is to be done with a system programmable voltage, such as a DAC.

With an inverting op amp configuration, injecting current into the inverting input is the simplest method, as shown in Figure 1-41A. The disadvantage of this method is that some increase in noise gain is possible, due to the parallel path of R3 and the potentiometer resistance. The resulting increase in noise gain may be reduced by making $\pm V_R$ large enough so that the R3 value is much greater than R1||R2. Note that if the power supplies are stable and noise free, they can be used as $\pm V_R$.

Figure 1-41B shows how to implement offset trim by injecting a small offset voltage into the noninverting input. This circuit is preferred over Figure 1-41A, as it results in no noise gain increase (but it requires adding R_P). If the op amp has matched input bias currents, R_P should equal R1|| R2 (to minimize the added offset voltage). Otherwise, R_P should be less than 50 Ω. For higher values, it may be advisable to bypass R_P at high frequencies.

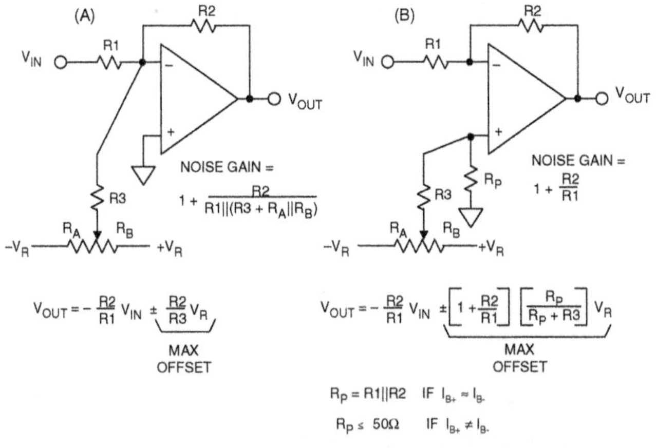

Figure 1-41: Inverting op amp external offset trim methods

The circuit shown in Figure 1-42 can be used to inject a small offset voltage when using an op amp in the noninverting mode. This circuit works well for small offsets, where R3 can be made much greater than R1. Note that otherwise, the signal gain might be affected as the offset potentiometer is adjusted. The gain may be stabilized, however, if R3 is connected to a fixed low impedance reference voltage source, $\pm V_R$.

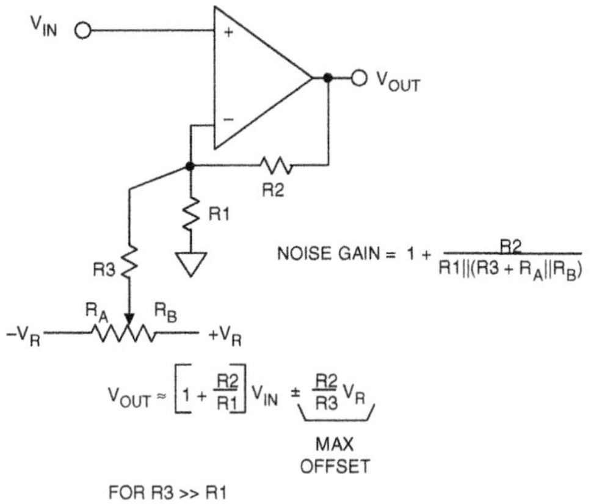

Figure 1-42: Noninverting op amp external offset trim methods

Input Offset Voltage Drift and Aging Effects

Input offset voltage varies with temperature, and its temperature coefficient is known as TCV_{OS}, or more commonly, *drift*. As we have mentioned, offset drift is affected by offset adjustments to the op amp, but when it has been minimized, it may be as low as 0.1 µV/°C (typical value for OP177F). More typical drift values for a range of general purpose precision op amps lie in the range 1–10 µV/°C. Most op amps have a specified value of TCV_{OS}, but instead, some have a second value of maximum V_{OS} that is guaranteed over the operating temperature range. Such a specification is less useful, because there is no guarantee that TCV_{OS} is constant or monotonic.

The offset voltage also changes as time passes, or *ages*. Aging is generally specified in µV/month or µV/1000 hours, but this can be misleading. Since aging is a "drunkard's walk" phenomenon it is proportional to the *square root* of the elapsed time. An aging rate of 1 µV/1000 hour therefore becomes about 3 µV/year (not 9 µV/year).

Long-term stability of the OP177F is approximately 0.3 µV/month. This refers to a time period *after* the first 30 days of operation. Excluding the initial hour of operation, changes in the offset voltage of these devices during the first 30 days of operation are typically less than 2 µV.

Input Bias Current, I_B

Ideally, no current flows into the input terminals of an op amp. In practice, there are always two *input bias currents*, I_{B+} and I_{B-} (see Figure 1-43).

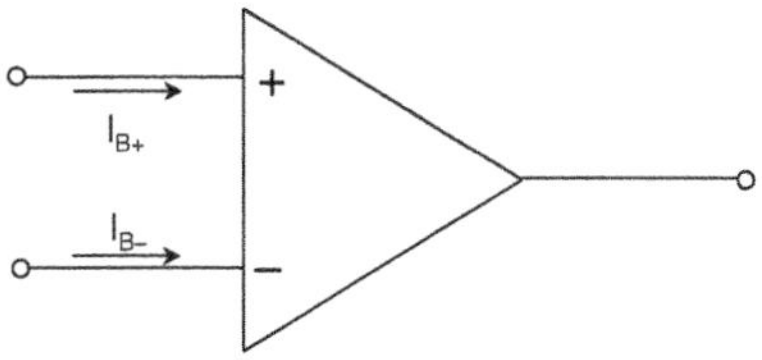

Figure 1-43: Op amp input bias current

- A very variable parameter
- I_B can vary from 60 fA (1 electron every 3 µs) to many µA, depending on the device.
- Some structures have well-matched I_B, others do not.
- Some structures' I_B varies little with temperature, but a FET op amp's I_B doubles with every 10°C rise in temperature.
- Some structures have I_B that may flow in either direction.

Values of I_B range from 60 fA (about one electron every three microseconds) in the AD549 electrometer, to tens of microamperes in some high speed op amps. Op amps with simple input structures using BJT or FET long-tailed pair have bias currents that flow in one direction. More complex input structures (bias-compensated and current feedback op amps) may have bias currents that are the difference between two or more internal current sources, and may flow in either direction.

Bias current is a problem to the op amp user because it flows in external impedances and produces voltages, which add to system errors. Consider a noninverting unity gain buffer driven from a source impedance of 1 MΩ. If I_B is 10 nA, it will introduce an additional 10 mV of error. This degree of error is not trivial in any system.

If the designer simply forgets about I_B and uses capacitive coupling, the circuit won't work—at all. Or, if I_B is low enough, it may work momentarily while the capacitor charges, giving even more misleading results. The moral here is not to neglect the effects of I_B, in any op amp circuit. The same admonition goes for in amp circuits.

Input bias current (or input offset voltage) may be measured using the test circuit of Figure 1-44. To measure I_B, a large resistance, R_S, is inserted in series with the input under test, creating an apparent additional offset voltage equal to $I_B \times R_S$. If the actual V_{OS} has previously been measured and recorded, the change in apparent V_{OS} due to the change in R_S can be determined, and I_B is then easily computed. This yields values for I_{B+} and I_{B-}. The rated value of I_B is the average of the two currents, or $I_B = (I_{B+} + I_{B-})/2$.

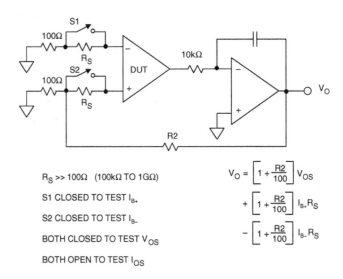

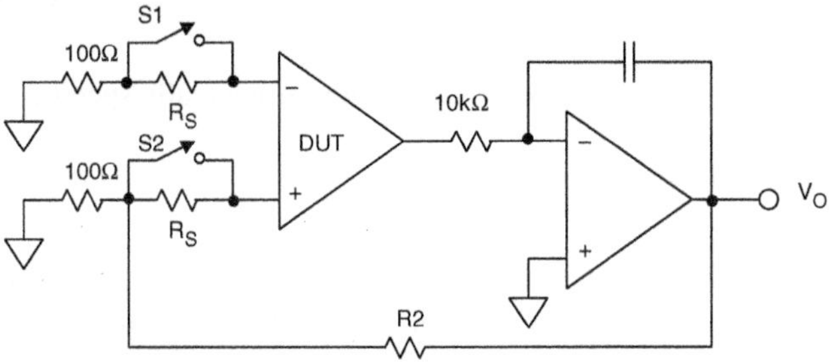

R_S >> 100Ω (100kΩ TO 1GΩ)

S1 CLOSED TO TEST I_{B+}

S2 CLOSED TO TEST I_{B-}

BOTH CLOSED TO TEST V_{OS}

BOTH OPEN TO TEST I_{OS}

$$V_O = \left[1 + \frac{R2}{100}\right] V_{OS} + \left[1 + \frac{R2}{100}\right] I_{B+} R_S - \left[1 + \frac{R2}{100}\right] I_{B-} R_S$$

Figure 1-44: Measuring input bias current

The *input offset current,* I_{OS}, may also be calculated, by taking the difference between I_{B-} and I_{B+}, or $I_{OS} = I_{B+} - I_{B-}$. Typical useful R_S values vary from 100 kΩ for bipolar op amps to 1000 MΩ for some FET input devices.

Note also that I_{OS} is only meaningful where the two individual bias currents are fundamentally reasonably well-matched, to begin with. This is true for most VFB op amps. However, it wouldn't, for example, be meaningful to speak of I_{OS} for a CFB op amp, as the currents are radically unmatched.

Extremely low input bias currents must be measured by integration techniques. The bias current in question is used to charge a capacitor, and the rate of voltage change is measured. If the capacitor and general circuit leakage is negligible (this is very difficult for currents under 10 fA), the current may be calculated directly from the rate of change of the output of the test circuit. Figure 1-45 illustrates the general concept. With one switch open and the opposite closed, either I_{B+} or I_{B-} is measured.

Figure 1-45: Measuring very low bias currents

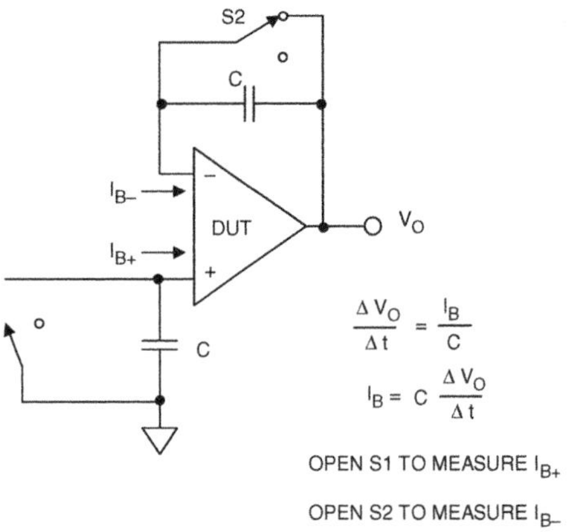

It should be obvious that only a premium capacitor dielectric can be used for C, for example Teflon or polypropylene types.

Canceling the Effects of Bias Current (External to the Op Amp)

When the bias currents of an op amp are well matched (the case with simple bipolar op amps, but *not* internally bias compensated ones, as noted previously), a bias compensation resistor, R3, (R3 = R1||R2) introduces a voltage drop in the noninverting input to match and thus compensate the drop in the parallel combination of R1 and R2 in the inverting input. This minimizes additional offset voltage error, as in Figure 1-46.

Note that if R3 is more than 1 kΩ or so, it should be bypassed with a capacitor to prevent noise pickup. Also note that this form of bias cancellation is useless where bias currents are not well-matched, and will, in fact, make matters worse.

Figure 1-46: Canceling the effects of input bias current within an application

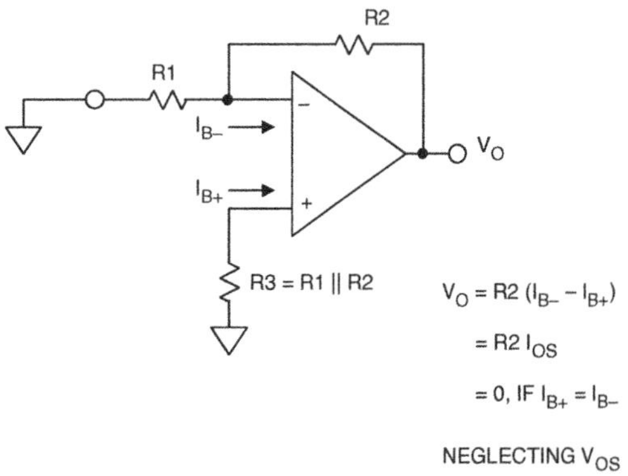

Calculating Total Output Offset Error Due to I_B and V_{OS}

The equations shown in Figure 1-47 are useful in referring all the offset voltage and induced offset voltage from bias current errors to either the input (RTI) or the output (RTO) of the op amp. The choice of RTI or RTO is a matter of preference.

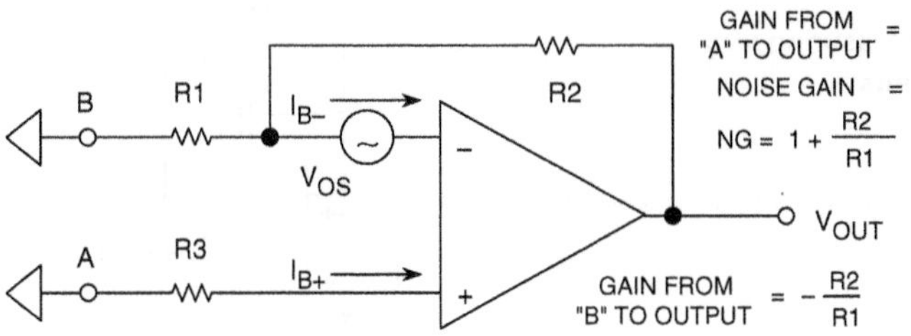

- $\text{OFFSET (RTO)} = V_{OS}\left[1+\frac{R2}{R1}\right] + I_{B+}\bullet R3\left[1+\frac{R2}{R1}\right] - I_{B-}\bullet R2$

- $\text{OFFSET (RTI)} = V_{OS} + I_{B+}\bullet R3 - I_{B-}\left[\frac{R1\bullet R2}{R1+R2}\right]$

FOR BIAS CURRENT CANCELLATION:

$\text{OFFSET (RTI)} = V_{OS}\quad \text{IF } I_{B+} = I_{B-}\ \text{ AND } R3 = \frac{R1\bullet R2}{R1+R2}$

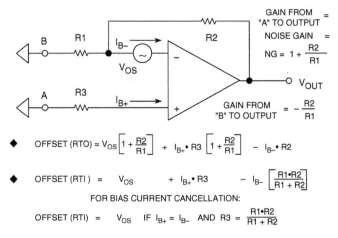

Figure 1-47: Op amp total offset voltage model

The RTI value is useful in comparing the cumulative op amp offset error to the input signal. The RTO value is more useful if the op amp drives additional circuitry, to compare the net errors with that of the next stage.

In any case, the RTO value is obtained by multiplying the RTI value by the stage noise gain, which is 1 + R2/R1.

Before departing the topic of offset errors, some simple rules towards minimization might bear repetition:

- Keep input/feedback resistance values low, to minimize offset voltage due to bias current effects.
- Use a bias compensation resistance with VFB op amps *not* using internal bias compensation. Bypass this resistance, for lowest noise pickup.
- If a VFB op amp *does* use internal bias current compensation, *don't* use the compensation resistance.
- When necessary, use *external* offset trim networks, for lowest induced drift.
- Select an appropriate precision op amp specified for low offset and drift, as opposed to trimming.
- For high performance low drift circuitry, watch out for thermocouple effects and used balanced, low thermal error layouts.

Input Impedance

VFB op amps normally have both differential and common-mode input impedances specified. Current feedback op amps normally specify the impedance to ground at each input. Different models may be used for different voltage feedback op amps, but in the absence of other information, it is usually safe to use the model in Figure 1-48. In this model, the bias currents flow into the inputs from infinite impedance current sources.

The common-mode input impedance data sheet specification (Z_{cm+} and Z_{cm-}) is the impedance from either input to ground (NOT from both to ground). The differential input impedance (Z_{diff}) is the impedance

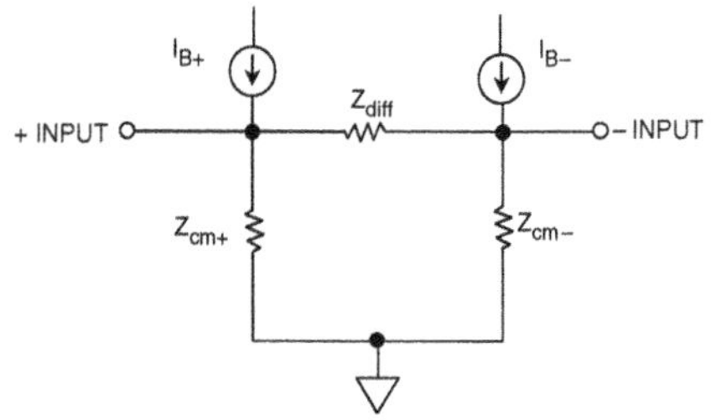

- Z_{cm+} and Z_{cm-} are the common-mode input impedance. The figure on the data sheet is for one, not both, but they are approximately equal. Z_{diff} is the differential input impedance.
- They are high resistance (10^5 – 10^{12} Ω) in parallel with a small
- shunt capacitance (sometimes as high as 25pF).
- In most practical circuits, Z_{cm-} is swamped by negative feedback.

Figure 1-48: Input impedance (voltage feedback op amp)

between the two inputs. These impedances are usually resistive and high (10^5 Ω–10^{12} Ω) with some shunt capacitance (generally a few pF, sometimes 20 pF–25 pF). In most op amp circuits, the inverting input impedance is reduced to a very low value by negative feedback, and only Z_{cm+} and Z_{diff} are of importance.

A current feedback op amp is even more simple, as shown in Figure 1-49. Z+ is resistive, generally with some shunt capacitance, and high (10^5 Ω–10^9 Ω) while Z– is reactive (L or C, depending on the device) but has a resistive component of 10 Ω–100 Ω, varying from type to type.

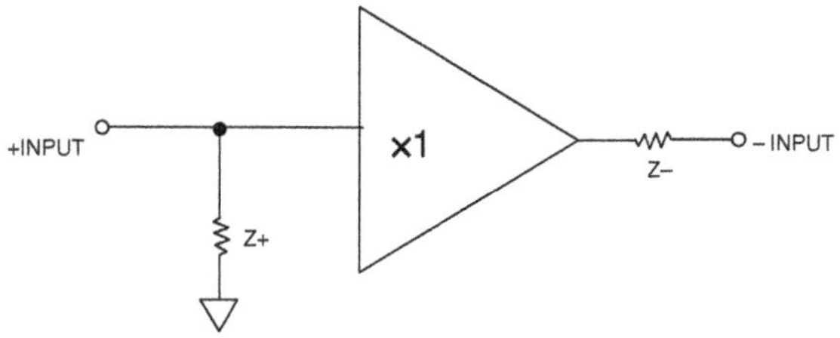

- Z+ is high resistance (10^5 – 10^9Ω) with little shunt capacitance.
- Z– is low and may be reactive (L or C). The resistive component is 10-100Ω.

Figure 1-49: Input impedance (current feedback op amp)

Manipulating Op Amp Noise Gain and Signal Gain

Consider an op amp and two resistors, R1 and R2, arranged as shown in the series of figures of Figure 1-50. Note that R1 and R2 need not be resistors; they could also be complex impedances, Z1 and Z2.

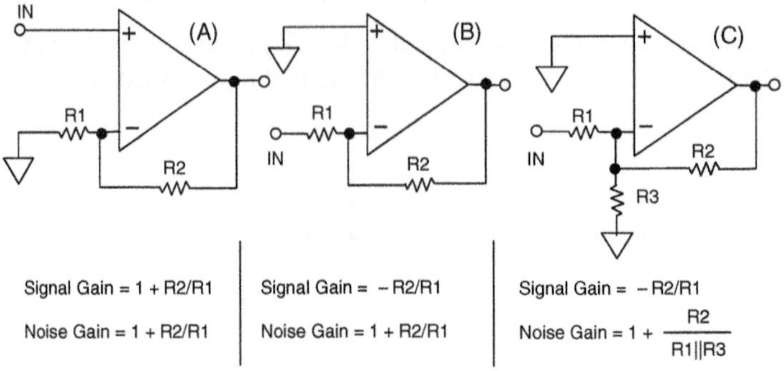

- Voltage Noise and Offset Voltage of the op amp are reflected to the output by the Noise Gain.
- Noise Gain, not Signal Gain, is relevant in assessing stability.
- Circuit C has unchanged Signal Gain, but higher Noise Gain, thus better stability, worse noise, and higher output offset voltage.

Figure 1-50: Manipulating op amp noise gain and signal gain

If we ground R1 and apply a signal to the noninverting input, we see a signal gain of 1 + R2/R1, as in Figure 1-50A. If we ground the noninverting input and apply the signal to R1, we see a signal gain of –R2/R1, as in Figure 1-50B. In both cases, the voltage noise of the op amp itself (as well as the input offset voltage) sees a gain of 1 + R2/R1, i.e., the *noise gain* of the op amp, as discussed earlier in this chapter.

This discussion is aimed at making the point that a stage's noise gain and signal gain *need not necessarily be equal*. Some times it can be to the user's advantage to manipulate them, so as to be somewhat independent of one another.

But, importantly, *it is the noise gain that is relevant in assessing stability*. It is sometimes possible to alter the noise gain, while leaving signal gain unaffected. When this is done, a marginally stable op amp stage can sometimes be made stable, with the same signal gain.

For example, consider the inverting amplifier of Figure 1-50B. If we add a third resistor to Figure 1-50B, it becomes Figure 1-50C. This dummy resistor R3, from the inverting input to ground, increases the noise gain to 1 + R2/(R1||R3). But, note the signal gain is unaffected; that is it is still –R2/R1.

This provides a means of stabilizing an unstable inverting amplifier—at a cost of worse signal-to-noise ratio, less loop gain, and increased sensitivity to input offset voltage. Nevertheless, it is still a sometimes useful trick.

Open-Loop Gain And Open-Loop Gain Nonlinearity

Open-loop voltage gain, usually called A_{VOL} (sometimes simply A_V), for most VFB op amps is quite high. Common values are 100,000 to 1,000,000, and 10 or 100 times these figures for high precision parts.

Some fast op amps have appreciably lower open-loop gain, but gains of less than a few thousand are unsatisfactory for high accuracy use. Note also that open-loop gain isn't highly stable with temperature. It can vary quite widely from device to device of the same type, so it is important that it be reasonably high.

Since a voltage feedback op amp operates as voltage in/voltage out, its open-loop gain is a dimensionless ratio, so no unit is necessary. However, data sheets sometimes express gain in V/mV or V/µV instead of V/V, for the convenience of using smaller numbers. Or, voltage gain can also be expressed in dB terms, as gain in dB = $20 \times \log A_{VOL}$. Thus an open-loop gain of 1 V/µV is equivalent to 120 dB, and so on.

CFB op amps have a current input and a voltage output, so their open-loop *transimpedance gain* is expressed in volts per ampere or ohms (or kΩ or MΩ). Values usually lie between hundreds of kΩ and tens of MΩ.

From basic feedback theory, it is understood that in order to maintain accuracy, a precision amplifier's dc open-loop gain, A_{VOL}, should be high. This can be seen by examining the closed-loop gain equation, including errors due to finite gain. The expression for closed loop gain with a finite gain error is:

$$G_{CL} = \frac{1}{\beta} \bullet \left[\frac{1}{1 + \frac{1}{A_{VOL}\beta}} \right] \qquad \text{Eq. 1-15A}$$

Since noise gain is equal to 1/β, there are alternate forms of this expression. Combining the two right side terms and using the NG expression, an alternate one is:

$$G_{CL} = \frac{NG}{1 + \frac{NG}{A_{VOL}}} \qquad \text{Eq. 1-15B}$$

Eqs. 1-15A and 1-15B are equivalent, and either can be used. As previously discussed, noise gain (NG) is simply the gain seen by a small voltage source in series with the op amp input, and is also the ideal amplifier signal gain in the noninverting mode. If A_{VOL} in Eqs. 1-15A and 1-15B is infinite, the closed-loop gain becomes exactly equal to the noise gain, 1/β.

However, for NG << A_{VOL} and finite A_{VOL}, there is a closed-loop gain error estimation:

$$\text{Closed loop error}\,(\%) \approx \frac{NG}{A_{VOL}} \bullet 100 \qquad \text{Eq. 1-16}$$

Note that the expression of Eq. 1-16 is equivalent to the earlier mentioned Eq. 1-11, when 1/β is substituted for NG. Again, either form can be used, at the user's discretion.

Notice from Eq. 1-16 that the percent gain error is directly proportional to the noise gain, therefore the effects of finite A_{VOL} are less for low gain. Some examples illustrate key points about these gain relationships.

In Figure 1-51, the first example for a NG of 1000 shows that for an open-loop gain of 2 million, the closed-loop gain error is about 0.05%. Note that if the open-loop gain stays constant over temperature and for various output loads and voltages, the 0.05% gain error can easily be calibrated out of the measurement, and then there is then no overall system gain error. If, however, the open-loop gain *changes*, the resulting closed-loop gain will also change. This introduces a *gain uncertainty*. In the second example, A_{VOL} drops to 300,000, which produces a gain error of 0.33%. This situation introduces a *gain uncertainty* of 0.28% in the closed-loop gain. In most applications, when using a good amplifier, the gain resistors of the circuit will be the largest source of absolute gain error, but it should be noted that gain uncertainty cannot be removed by calibration.

- "IDEAL" CLOSED LOOP GAIN = $1/\beta$ = NOISE GAIN (NG)
- ACTUAL CLOSED LOOP GAIN = $\frac{1}{\beta} \bullet \left[\frac{1}{1+\frac{1}{A_{VOL}\beta}} \right] = \frac{NG}{1+\frac{NG}{A_{VOL}}}$
- CLOSED LOOP GAIN ERROR (%) $\approx \frac{NG}{A_{VOL}} \times 100$ $(NG \ll A_{VOL})$

  – Ex. 1: Assume A_{VOL} = 2,000,000, NG = 1,000
  % Gain Error ≈ 0.05%

  – Ex. 2: Assume A_{VOL} Drops to 300,000
  % Gain Error ≈ 0.33%

  – Closed Loop Gain Uncertainty
  = 0.33% – 0.05% = 0.28%

Figure 1-51: Changes in open-loop gain cause closed-loop gain uncertainty

Changes in the output voltage level and output loading are the most common causes of changes in the open-loop gain of op amps. A change in open-loop gain with signal level produces a *nonlinearity* in the closed-loop gain transfer function, which also cannot be removed during system calibration. Most op amps have fixed loads, so A_{VOL} changes with load are not generally important. However, the sensitivity of A_{VOL} to output signal level may increase for higher load currents.

The severity of this nonlinearity varies widely from one device type to another, and generally isn't specified on the data sheet. The minimum A_{VOL} is always specified, and choosing an op amp with a high A_{VOL} will minimize the probability of gain nonlinearity errors. Gain nonlinearity can come from many sources, depending on the design of the op amp. One common source is thermal feedback (for example, from a hot output stage back to the input stage). If temperature shift is the sole cause of the nonlinearity error, it can be assumed that minimizing the output loading will help. To verify this, the nonlinearity is measured with no load, and then compared to the loaded condition.

An oscilloscope X-Y display test circuit for measuring dc open-loop gain nonlinearity is shown in Figure 1-52. The same precautions previously discussed relating to the offset voltage test circuit must also be observed in this circuit. The amplifier is configured for a signal gain of –1. The open-loop gain is defined as the change in output voltage divided by the change in the input offset voltage. However, for large values of A_{VOL}, the actual offset may change only a few microvolts over the entire output voltage swing. Therefore the divider consisting of the 10 Ω resistor and R_G (1 MΩ) forces the node voltage V_Y to be:

$$V_Y = \left[1+\frac{R_G}{10\Omega}\right] V_{OS} = 100{,}001 \bullet V_{OS} \cdot \qquad \text{Eq. 1-17}$$

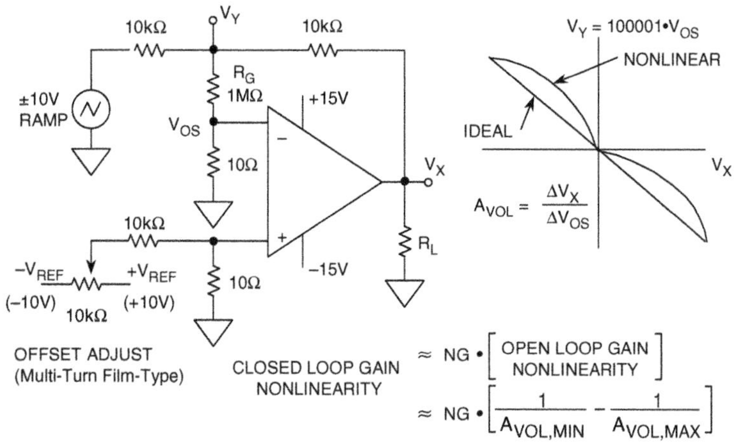

Figure 1-52: Circuit measures open-loop gain nonlinearity

The value of R_G is chosen to give measurable voltages at V_Y, depending on the expected values of V_{OS}.

The ±10 V ramp generator output is multiplied by the signal gain, −1, and forces the op amp output voltage V_X to swing from +10 V to −10 V. Because of the gain factor applied to the offset voltage, the offset adjust potentiometer is added to allow the initial output offset to be set to zero. The resistor values chosen will null an input offset voltage of up to ±10 mV. Stable 10 V voltage references such as the AD688 should be used at each end of the potentiometer to prevent output drift. *Note also that the ramp generator frequency must be quite low, probably no more than a fraction of 1 Hz because of the low corner frequency of the open-loop gain (0.1 Hz for the OP177).*

The plot on the right-hand side of Figure 1-52 shows V_Y plotted against V_X. If there is no gain nonlinearity, the graph will have a constant slope and A_{VOL} is calculated as follows:

$$A_{VOL} = \frac{\Delta V_X}{\Delta V_{OS}} = \left[1 + \frac{R_G}{10\,\Omega}\right]\left[\frac{\Delta V_X}{\Delta V_Y}\right] = 100{,}001 \bullet \left[\frac{\Delta V_X}{\Delta V_Y}\right]. \qquad \text{Eq. 1-18}$$

If there is nonlinearity, A_{VOL} will vary dynamically as the output signal changes. The approximate open-loop gain nonlinearity is calculated based on the maximum and minimum values of A_{VOL} over the output voltage range:

$$\text{Open-Loop Gain Nonlinearity} = \frac{1}{A_{VOL,MIN}} - \frac{1}{A_{VOL,MAX}}. \qquad \text{Eq. 1-19}$$

The closed-loop gain nonlinearity is obtained by multiplying the open-loop gain nonlinearity by the noise gain, NG:

$$\text{Closed-Loop Gain Nonlinearity} \approx NG \bullet \left[\frac{1}{A_{VOL,MIN}} - \frac{1}{A_{VOL,MAX}}\right]. \qquad \text{Eq. 1-20}$$

In an ideal case, the plot of V_{OS} versus V_X would have a constant slope, and the reciprocal of the slope is the open-loop gain, A_{VOL}. A horizontal line with zero slope would indicate infinite open-loop gain. In an actual op amp, the slope may change across the output range because of nonlinearity, thermal feedback, and so forth. In fact, the slope can even change sign.

Figure 1-53 shows the V_Y (and V_{OS}) versus V_X plot for an OP177 precision op amp. The plot is shown for two different loads, 2 kΩ and 10 kΩ. The reciprocal of the slope is calculated based on the end points, and the average A_{VOL} is about 8 million. The maximum and minimum values of A_{VOL} across the output voltage range are measured to be approximately 9.1 million, and 5.7 million, respectively. This corresponds to an open-loop gain nonlinearity of about 0.07 ppm. Thus, for a noise gain of 100, the corresponding closed-loop gain nonlinearity is about 7 ppm.

These nonlinearity measurements are, of course, most applicable to high precision dc circuits. But they are also applicable to wider bandwidth applications, such as audio. The X-Y display technique of Figure 1-52 will easily show for example, crossover distortion in a poorly designed op amp output stage.

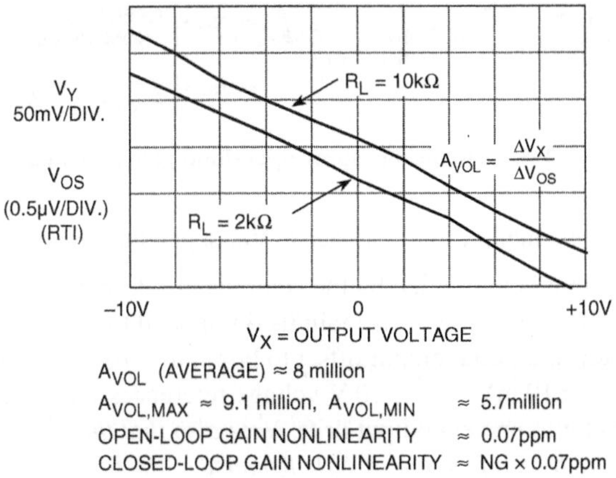

Figure 1-53: OP177 gain nonlinearity

Op Amp Frequency Response

There are a number of issues to consider when discussing the frequency response of op amps. Some are relevant to both voltage and current feedback op amp types, some apply to one or the other, but not to both. Issues that vary with type are usually related to small signal performance, while large-signal issues mostly apply to both.

A good working definition of "large-signal" is where the output voltage swing/frequency limit is set by the slew rate measured at the output stage, rather than the pole(s) of the small signal response. We shall therefore consider large signal parameters applying to both types of op amp before we consider those parameters where they differ.

Frequency Response—Slew Rate and Full-Power Bandwidth

The slew rate (SR) of an amplifier is the maximum rate of change of voltage at its output. It is expressed in V/s (or, more probably, V/μs). We have mentioned earlier why op amps might have different slew rates during positive-and negative-going transitions, but for this analysis we shall assume that good fast op amps have reasonably symmetrical slew rates.

If we consider a sine wave signal with a peak-to-peak amplitude of 2 V_p and of a frequency f, the expression for the output voltage is:

$$V(t) = V_p \sin 2\pi ft. \qquad \text{Eq. 1-21}$$

This sine wave signal has a maximum rate of change (slope) at the zero crossing. This maximum rate of change is:

$$\left.\frac{dV}{dt}\right|_{max} = 2\pi f V_p \qquad \text{Eq. 1-22}$$

To reproduce this signal without distortion, an amplifier must be able to respond in terms of its output voltage at this rate (or faster). When an amplifier reaches its maximum output rate of change, or *slew rate*, it is said to be *slew limiting* (sometimes also called rate limiting). Thus, we can see that the maximum signal frequency at which slew limiting *does not* occur is directly proportional to the signal slope, and inversely proportional to the amplitude of the signal. This allows us to define the *full power bandwidth* (FPBW) of an op amp, which is the maximum frequency at which slew limiting doesn't occur for rated voltage output. It is calculated by letting 2 V_p in Eq. 1-22 equal the maximum peak-to-peak swing of the amplifier, dV/dt equal the amplifier slew rate, and solving for f:

$$FPBW = Slew\,Rate / 2\pi V_p \qquad \text{Eq. 1-23}$$

It is important to realize that both slew rate and full-power bandwidth can also depend somewhat on the power supply voltage being used, and the load the amplifier is driving (particularly if it is capacitive).

The key issues regarding slew rate and full-power bandwidth are summarized in Figure 1-54. As a point of reference, an op amp with a 1 V peak output swing reproducing a 1MHz sine wave must have a minimum SR of 6.28 V/µs.

- Slew Rate = Maximum rate at which the output voltage of an op amp can change
- Ranges: A few volts/µs to several thousand volts/µs
- For a sinewave, $V_{OUT} = V_p \sin 2\pi ft$

  $dV/dt = 2\pi f V_p \cos 2\pi ft$

  $(dV/dt)_{max} = 2\pi f V_p$
- If $2V_p$ = full output span of op amp, then

  Slew Rate $= (dV/dt)_{MAX} = 2\pi \cdot FPBW \cdot V_p$

  FPBW = Slew Rate / 2 πV_p

Figure 1-54: Slew rate and full-power bandwidth

Realistically, for a practical circuit the designer would choose an op amp with an SR in excess of this figure, since real op amps show increasing distortion prior to reaching the slew limit point.

Frequency Response—Settling Time

The *settling time* of an amplifier is defined as the time it takes the output to respond to a step change of input and *come into, and remain within, a defined error band*, as measured relative to the 50% point of the input pulse, as shown in Figure 1-55.

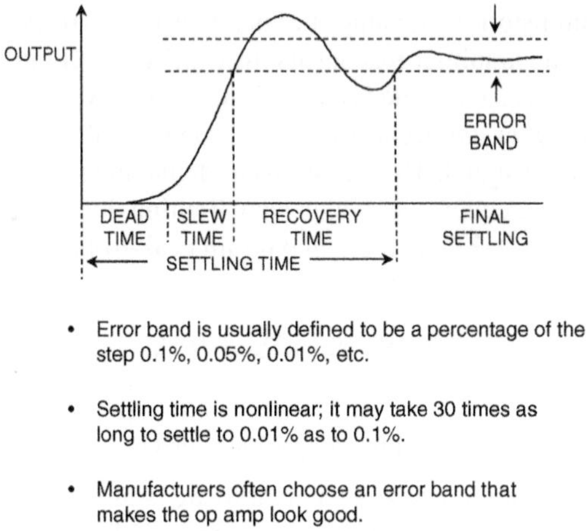
OUTPUT
ERROR
BAND
DEAD TIME
SLEW TIME
RECOVERY TIME
FINAL SETTLING
SETTLING TIME
• Error band is usually defined to be a percentage of the step 0.1%, 0.05%, 0.01%, etc.
• Settling time is nonlinear; it may take 30 times as long to settle to 0.01% as to 0.1%.
• Manufacturers often choose an error band that makes the op amp look good.

Figure 1-55: Settling time

Unlike a DAC device, there is no natural error band for an op amp (a DAC naturally has an error band of 1 LSB, or perhaps ±1 LSB). So, one must be chosen and defined, along with other definitions, such as the step size (1 V, 5 V, 10 V, and so forth). What is chosen will depend on the performance of the op amp, but since the value chosen will vary from device to device, comparisons are often difficult. This is true because settling is not linear, and many different time constants may be involved. Examples are early op amps using dielectrically isolated (DI) processes. These had very fast settling to 1% of full scale, but they took almost forever to settle to 10 bits (0.1 %). Similarly, some very high precision op amps have thermal effects that cause settling to 0.001% or better to take tens of ms, although they will settle to 0.025% in a few µs.

It should also be noted that thermal effects can cause significant differences between short-term settling time (generally measured in nanoseconds) and long-term settling time (generally measured in microseconds or milliseconds). In many ac applications, long-term settling time is not important; but if it is, it must be measured on a much different time scale than short-term settling time.

Measuring fast settling time to high accuracy is very difficult. Great care is required in order to generate fast, highly accurate, low noise, flat-top pulses. Large amplitude step voltages will overdrive many oscilloscope front ends, when the input scaling is set for high sensitivity.

The example test setup shown in Figure 1-56 is useful in making settling time measurements on op amps operating in the inverting mode. The signal at the "false summing node" represents the difference between the output and the input signal, multiplied by the constant k, i.e., the ERROR signal.

Many subtleties are involved in making this setup work reliably. The resistances should low in value, to minimize parasitic time constants. The back-back Schottky diode clamps help prevent scope overdrive, and allow high sensitivity. If R1 = R2, then k = 0.5. Thus the error band at the ERROR output will be 5 mV for 0.1% settling with a 10 V input step.

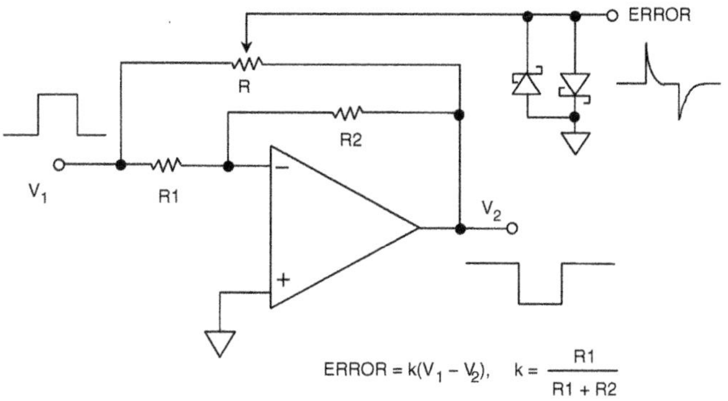

Figure 1-56: Measuring settling time using a "false summing node"

In some cases, a second (very fast) amplifier stage may be used after the false summing node, to increase the signal level. In any case, to ensure validity, testing of settling time must be done with a test setup identical to that used by the op amp manufacturer.

Many modern digitizing oscilloscopes are insensitive to input overdrive and can be used to measure the ERROR waveform directly—this must be verified for each oscilloscope by carefully examining the operating manual. Note that a direct measurement allows measurements of settling time in both the inverting and noninverting modes. An example of the output step response to a flat pulse input for the AD8039 op amp is shown in Figure 1-57. Notice that the settling time to 0.1% is approximately 18 ns.

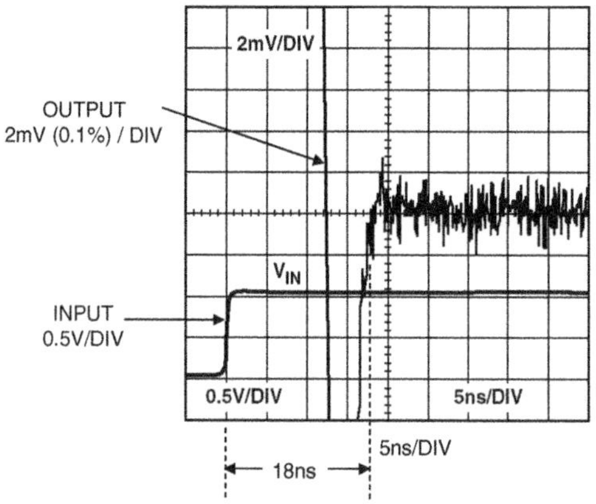

Figure 1-57: AD8039 G = +2 settling time measured directly

In making settling time measurements of this type, it is also imperative to use a pulse generator source capable of generating a pulse of sufficient flatness. In other words, if the op amp under test has a settling time of 20 ns to 0.1%, the applied pulse should settle to better than 0.05% in less than 5 ns.

This type of generator can be expensive, but a simple circuit, as shown in Figure 1-58, can be used with a reasonably flat generator to ensure a flat pulse output.

The circuit of Figure 1-58 works best if low capacitance Schottky diodes are used for D1–D2–D3, and the lead lengths on all the connections are minimized. A short length of 50 Ω coax can be used to connect the pulse generator to the circuit; however, best results are obtained if the test fixture is connected directly to the output of the generator. The pulse generator is adjusted to output a positive-going pulse at "A" which rises from approximately –1.8 V to +0.5 V in less than 5 ns (assuming the settling time of the DUT is in the order of 20 ns). Shorter rise times may generate ringing, and longer rise times can degrade the DUT settling time; therefore, some optimization is required in the actual circuit to get best performance. When the pulse generator output "A" goes above 0 V, D1 begins to conduct, and D2/D3 are reverse biased. The "0 V" region of the signal "B" at the input of the DUT is flat "by definition"—neglecting the leakage current and stray capacitance of the D2–D3 series combination. The D1 diode and its 100 Ω resistor help maintain an approximate 50 Ω termination during the time the pulse at "A" is positive.

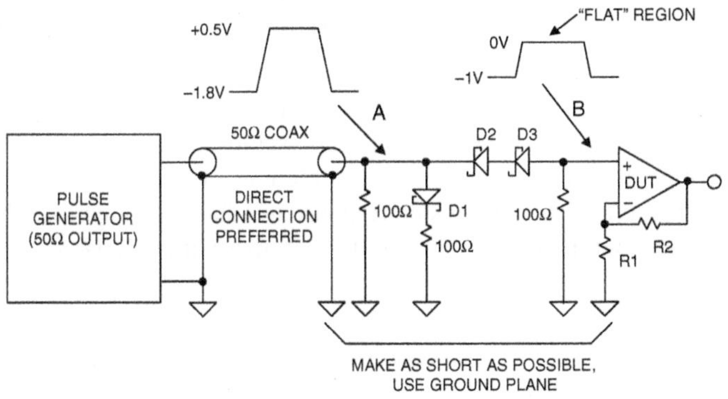

Figure 1-58: A simple flat pulse generator

Frequency Response—Voltage Feedback Op amps, Gain-Bandwidth Product

The open-loop frequency response of a voltage feedback op amp is shown in Figure 1-59. There are two possibilities: Figure 1-59A shows the most common, where a high dc gain drops at 6 dB/octave from quite a low frequency down to unity gain. This is a classic single pole response. By contrast, the amplifier in Figure 1-59B has two poles in its response—gain drops at 6 dB/octave for a while, and then drops at

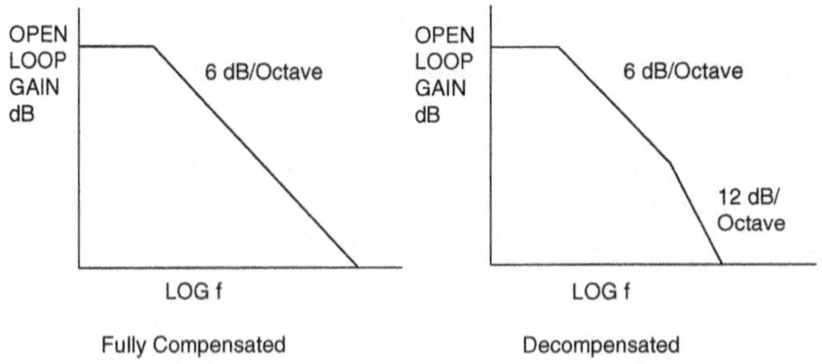

Figure 1-59: Frequency response of voltage feedback op amps

12 dB/octave. The amplifier in Figure 1-59A is known as an *unconditionally stable* or *fully compensated* type and may be used with a noise gain of unity. This type of amplifier is stable with 100% feedback (including capacitance) from output to inverting input.

Compare this to the amplifier in Figure 1-59B. If this op amp is used with a noise gain that is lower than the gain at which the slope of the response increases from 6 dB to 12 dB/octave, the phase shift in the feedback will be too great, and it will oscillate. Amplifiers of this type are characterized as "stable at gains ≥ X" where X is the gain at the frequency where the 6 dB/12 dB transition occurs. Note that here it is, of course, the *noise gain* that is being referenced. The gain level for stability might be between 2 and 25, typically quoted behavior might be "gain-of-five-stable," and so forth. These *decompensated* op amps do have higher gain bandwidth products than fully compensated amplifiers, all other things being equal. So they are useful, despite the slightly greater complication of designing with them. But, unlike their fully compensated op amp relatives, a decompensated op amp can never be used with direct capacitive feedback from output to inverting input.

The 6 dB/octave slope of the response of both types means that over the range of frequencies where this slope occurs, *the product of the closed-loop gain and the 3 dB closed-loop bandwidth at that gain is a constant* —this is known as the *gain bandwidth product* (GBW) and is a figure of merit for an amplifier.

For example, if an op amp has a GBW product of X MHz, its closed-loop bandwidth at a noise gain of 1 will be X MHz, at a noise gain of 2 it will be X/2 MHz, and at a noise gain of Y it will be X/Y MHz (see Figure 1-60). Notice that the *closed-loop* bandwidth is the frequency at which the noise gain plateau intersects the open-loop gain.

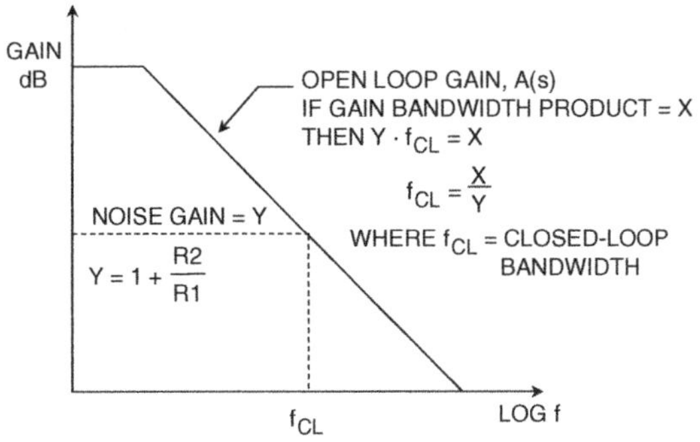

Figure 1-60: Gain-bandwidth product for voltage feedback op amps

In the above example, it was assumed that the feedback elements were resistive. This is not usually the case, especially when the op amp requires a feedback capacitor for stability.

Figure 1-61 shows a typical example where there is capacitance, C1, on the inverting input of the op amp. This capacitance is the sum of the op amp internal capacitance, plus any external capacitance that may exist. This always-present capacitance introduces a pole in the noise gain transfer function.

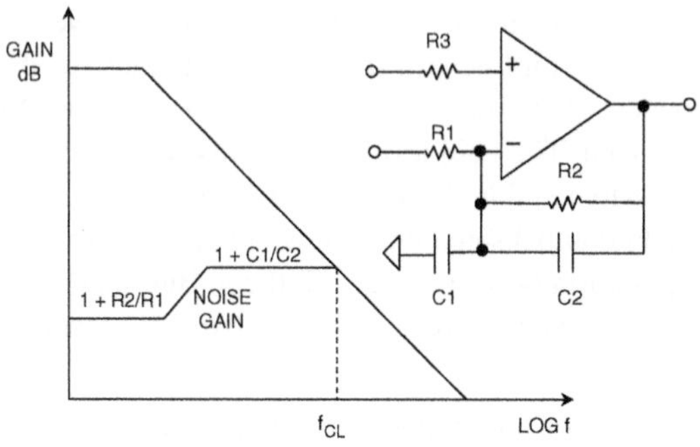

Figure 1-61: Bode plot showing noise gain for voltage feedback op amp with resistive and reactive feedback elements

The net slope of the noise gain curve and the open-loop gain curve, *at the point of intersection*, determines system stability. For unconditional stability, the noise gain must intersect the open-loop gain with a net slope of less than 12 dB/octave (20 dB per decade). Adding the feedback capacitor, C2, introduces a zero in the noise gain transfer function, which stabilizes the circuit. Notice that in Figure 1-61 the closed-loop bandwidth, f_{cl}, is the frequency at which the noise gain intersects the open-loop gain.

The Bode plot of the noise gain is a very useful tool in analyzing op amp stability. Constructing the Bode plot is a relatively simple matter. Although it is outside the scope of this section to carry the discussion of noise gain and stability further, the reader is referred to Reference 1 for an excellent treatment of constructing and analyzing Bode plots. Second-order systems related to noise analysis are discussed later in this section.

Frequency Response—Current Feedback Op Amps

Current feedback op amps do not behave in the same way as voltage feedback types. They are not stable with capacitive feedback, nor are they so with a short circuit from output to inverting input. With a CFB op amp, *there is generally an optimum feedback resistance for maximum bandwidth.* Note that the value of this resistance may vary with supply voltage—consult the device data sheet. If the feedback resistance is increased, the bandwidth is reduced. Conversely, if it is reduced, bandwidth increases, and the amplifier may become unstable.

In a CFB op amp, for a given value of feedback resistance (R2), *the closed-loop bandwidth is largely unaffected by the noise gain,* as shown in Figure 1-62. Thus it is not correct to refer to gain bandwidth product, for a CFB amplifier, because of the fact that it is not constant. Gain is manipulated in a CFB op amp application by choosing the correct feedback resistor for the device (R2), and then selecting the bottom resistor (R1) to yield the desired closed loop gain. The gain relationship of R2 and R1 is identical to the case of a VFB op amp (Figure 1-14).

Typically, CFB op amp data sheets will provide a table of recommended resistor values that provide maximum bandwidth for the device, over a range of both gain and supply voltage. It simplifies the design process considerably to use these tables.

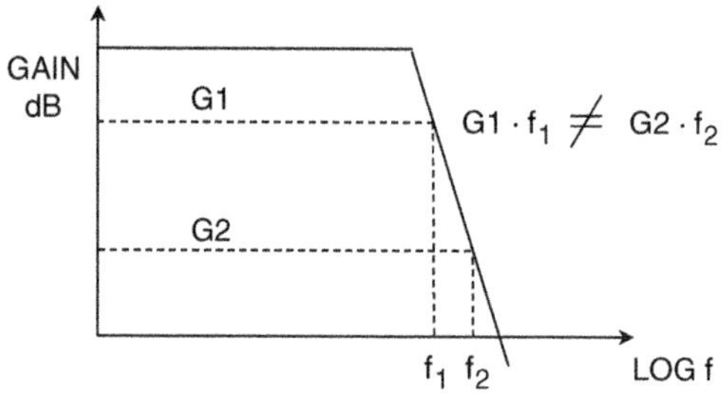

- Feedback resistor fixed for optimum performance. Larger values reduce bandwidth, smaller values may cause instability.
- For fixed feedback resistor, changing gain has little effect on bandwidth.
- Current feedback op amps do not have a fixed gain-bandwidth product.

Figure 1-62: Frequency response for current feedback op amps

Bandwidth Flatness

In demanding applications such as professional video, it is desirable to maintain a relatively flat bandwidth and linear phase up to some maximum specified frequency, and simply specifying the 3dB bandwidth isn't enough. In particular, it is customary to specify the *0.1 dB bandwidth*, or *0.1 dB bandwidth flatness*. This means there is no more than 0.1 dB ripple up to a specified 0.1 dB bandwidth frequency.

Video buffer amplifiers generally have both the 3 dB and the 0.1 dB bandwidth specified. Figure 1-63 shows the frequency response of the AD8075 triple video buffer.

Note that the 3 dB bandwidth is approximately 400 MHz. This can be determined from the response labeled "GAIN" in the graph, and the corresponding gain scale is shown on the left-hand vertical axis (at a scaling of 1 dB/division).

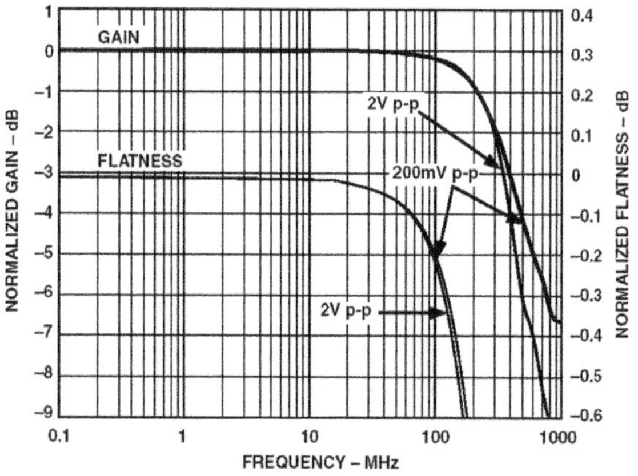

3dB BANDWIDTH ≈ 400MHz, 0.1dB BANDWIDTH ≈ 65MHz

Figure 1-63: 3 dB and 0.1 dB bandwidth for the AD8075, G = 2, triple video buffer, R_L = 150 Ω

The response scale for "FLATNESS" is on the right-hand vertical axis, at a scaling of 0.1 dB/division in this case. This allows the 0.1 dB bandwidth to be determined, which is about 65 MHz in this case. The general point to be noted here is the major difference in the applicable bandwidth between the 3 dB and 0.1 dB criteria. It requires a 400 MHz bandwidth amplifier (as conventionally measured) to provide the 65 MHz 0.1 dB flatness rating.

It should be noted that these specifications hold true when driving a 75 Ω source and load terminated cable, which represents a resistive load of 150 Ω. Any capacitive loading at the amplifier output will cause peaking in the frequency response, and must be avoided.

Operational Amplifier Noise

This section discusses the noise generated within op amps, not the external noise they may pick up. External noise is also important, and is discussed in detail in Chapter 7; in this section we are concerned solely with internal noise.

There are three noise sources in an op amp: a voltage noise, which appears differentially across the two inputs, and two current noise sources, one in each input. The simple voltage noise op amp model is shown in Figure 1-64. The three noise sources are effectively uncorrelated (independent of each other). There is a slight correlation between the two noise currents, but it is too small to need consideration in practical noise analyses. In addition to these three internal noise sources, it is necessary to consider the Johnson noise of the external gain-setting resistors that are used with the op amp.

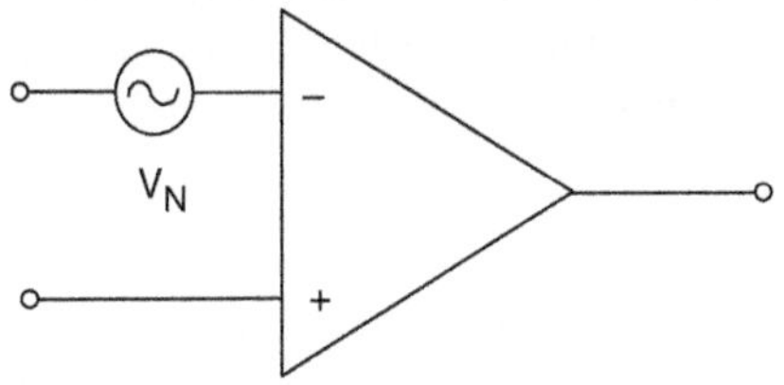

- Input Voltage Noise is bandwidth dependent and measured in nV/√Hz (noise spectral density)
- Normal Ranges are 1 nV/√Hz to 20 nV/√Hz

Figure 1-64: Input voltage noise

All resistors have a Johnson noise of $\sqrt{(4kTBR)}$ where k is Boltzmann's Constant (1.38 × 10^{-23}J/K), T is the absolute temperature, B is the bandwidth, and R is the resistance. Note that this is an intrinsic property—it is not possible to obtain resistors that do not have Johnson noise. The simple model is shown in Figure 1-65.

Uncorrelated noise voltages add in a "root-sum-of-squares" manner; i.e., noise voltages V_1, V_2, V_3 give a summed result of $\sqrt{(V_1^2 + V_2^2 + V_3^2)}$. Noise powers, of course, add normally. Thus, any noise voltage that is more than 3 to 5 times any of the others is dominant, and the others may generally be ignored. This simplifies noise assessment. The voltage noise of different op amps may vary from under 1 $nV/\sqrt{Hz}$ to 20 $nV/\sqrt{Hz}$, or even more. Bipolar op amps tend to have lower voltage noise than JFETs, although it is possible to make JFET op amps with low voltage noise (such as the AD743/AD745), at the cost of large input devices, and hence large input capacitance. Voltage noise is specified on the data sheet, and it isn't possible to predict it from other parameters.

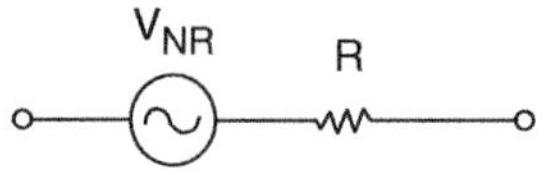

- ALL resistors have a voltage noise of $V_{NR} = \sqrt{(4kTBR)}$
- T = Absolute Temperature = T(°C) + 273.15
- B = Bandwidth (Hz)
- k = Boltzmann's Constant (1.38×10^{-23} J/K)
- A 1000 Ω resistor generates 4nV/√Hz @ 25°C

Figure 1-65: Johnson noise of resistors

Current noise can vary much more widely, dependent upon the input structure. It ranges from around $0.1\ \text{fA}/\sqrt{\text{Hz}}$ (in JFET electrometer op amps) to several $\text{pA}/\sqrt{\text{Hz}}$ (in high speed bipolar op amps). It isn't always specified on data sheets, but may be calculated in cases like simple BJT or JFETs, where all the bias current flows in the input junction, because in these cases it is simply the Schottky (or shot) noise of the bias current.

Shot noise spectral density is simply $\sqrt{(2I_Bq)}/\sqrt{\text{Hz}}$, where I_B is the bias current (in amps) and q is the charge on an electron (1.6×10^{-19} C). It can't be calculated for bias-compensated or current feedback op amps, where the external bias current is the *difference* of two internal currents. A simple current noise model is shown in Figure 1-66.

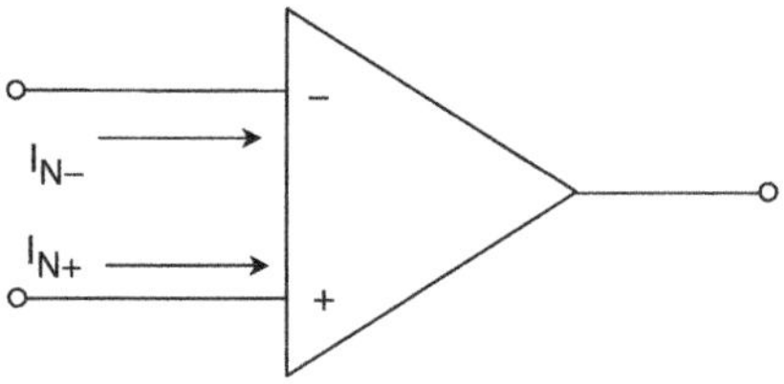

- Normal Ranges: 0.1fA/√Hz to 10pA/√Hz
- In Voltage Feedback op amps, the current noise in the inverting and noninverting inputs is uncorrelated (effectively) but roughly equal in magnitude.
- In simple BJT and JFET input stages, the current noise is the shot noise of the bias current and may be calculated from the bias current.
- In bias-compensated input stages and in current feedback op amps, the current noise cannot be calculated.
- The current noise in the two inputs of a current feedback op amp may be quite different. They may not even have the same 1/f corner.

Figure 1-66: Input current noise

Current noise is only important when it flows in an impedance, and thus generates a noise voltage. Maintaining relatively low impedances at the input of an op amp circuit contributes markedly to minimizing the effects of current noise (just as doing the same thing also aids in minimizing offset voltage).

It is logical therefore that the optimum choice of a low noise op amp depends on the impedances around it. This will be illustrated with the aid of some impedance examples, immediately below.

Consider for example an OP27, an op amp with low voltage noise (3 $nV/\sqrt{Hz}$), but quite high current noise (1 $pA/\sqrt{Hz}$). With zero source impedance, the voltage noise will dominate as shown in Figure 1-67 (left column). With a source resistance of 3 kΩ (center column), the current noise of (1 $pA/\sqrt{Hz}$) flowing in 3 kΩ will equal the voltage noise, but the Johnson noise of the 3 kΩ resistor is (7 $nV/\sqrt{Hz}$) and is dominant. With a source resistance of 300 kΩ (right column), the current noise portion increases $100\times$ to 300 $nV/\sqrt{Hz}$, voltage noise continues unchanged, and the Johnson noise (which is proportional to the resistance *square root*) increases tenfold. Current noise dominates.

The above example shows that the choice of a low noise op amp depends on the source impedance of the signal, and at high impedances, current noise always dominates.

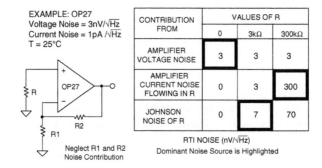

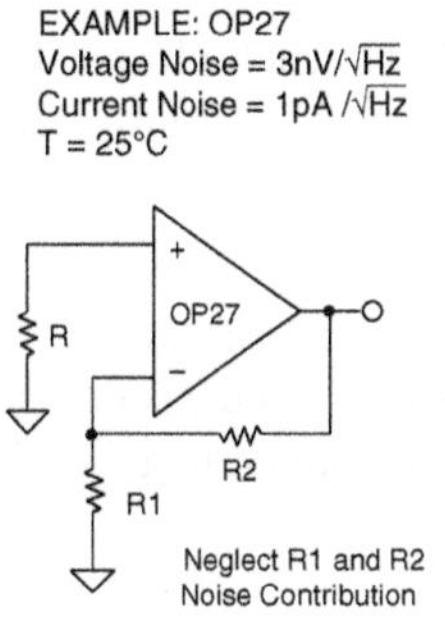

CONTRIBUTION FROM	VALUES OF R		
	0	3kΩ	300kΩ
AMPLIFIER VOLTAGE NOISE	**3**	3	3
AMPLIFIER CURRENT NOISE FLOWING IN R	0	3	**300**
JOHNSON NOISE OF R	0	**7**	70

RTI NOISE (nV/√Hz)
Dominant Noise Source is Highlighted

Figure 1-67: Different noise sources dominate at different source impedances

From Figure 1-68, it should be apparent that different amplifiers are best at different source impedances. For low impedance circuits, low voltage noise amplifiers such as the OP27 will be the obvious choice, since they are inexpensive, and their comparatively large current noise will not affect the application. At medium resistances, the Johnson noise of resistors is dominant, while at very high source resistance, we must choose an op amp with the smallest possible current noise, such as the AD549 or AD795.

Until recently, BiFET amplifiers tended to have comparatively high voltage noise (though very low current noise), and were thus more suitable for low noise applications in high rather than low impedance circuitry. The AD795, AD743, and AD745 have very low values of both voltage and current noise. The AD795 specifications at 10 kHz are 10 $nV/\sqrt{Hz}$ and 0.6 $fA/\sqrt{Hz}$, and the AD743/AD745 specifications at 10 kHz are 2.9 $nV/\sqrt{Hz}$ and 6.9 $fA/\sqrt{Hz}$. These make possible the design of low-noise amplifier circuits that have low noise over a wide range of source impedances.

The *noise figure* of an amplifier is the amount (in dB) by which the noise of the amplifier exceeds the noise of a perfect noise-free amplifier in the same environment. The concept is useful in RF and TV applications, where 50 Ω and 75 Ω transmission lines and terminations are ubiquitous, but is useless for an op amp that is used in a wide range of electronic environments. Noise figure related to communications applications is discussed in more detail in Chapter 6 (Section 6-4). Voltage noise spectral density and current noise spectral density are generally more useful specifications in most cases.

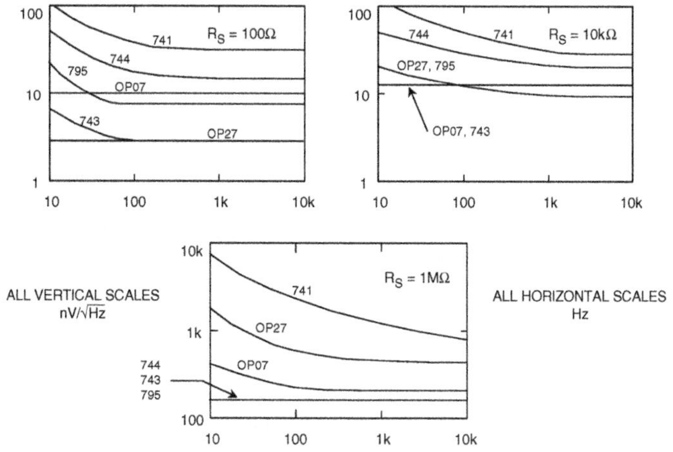

Figure 1-68: Different amplifiers are best at different source impedances

So far, we have assumed that noise is *white* (i.e., its spectral density does not vary with frequency). This is true over most of an op amp's frequency range, but at low frequencies the noise spectral density rises at 3 dB/octave, as shown in Figure 1-69. The power spectral density in this region is inversely proportional to frequency, and therefore the voltage noise spectral density is inversely proportional to the square root of the frequency. For this reason, this noise is commonly referred to as *1/f noise*. Note, however, that some textbooks still use the older term *flicker noise*.

The frequency at which this noise starts to rise is known as the *1/f corner frequency* (F_C) and is a figure of merit—the lower it is, the better. The 1/f corner frequencies are not necessarily the same for the voltage noise and the current noise of a particular amplifier, and a current feedback op amp may have three 1/f corners: for its voltage noise, its inverting input current noise, and its noninverting input current noise.

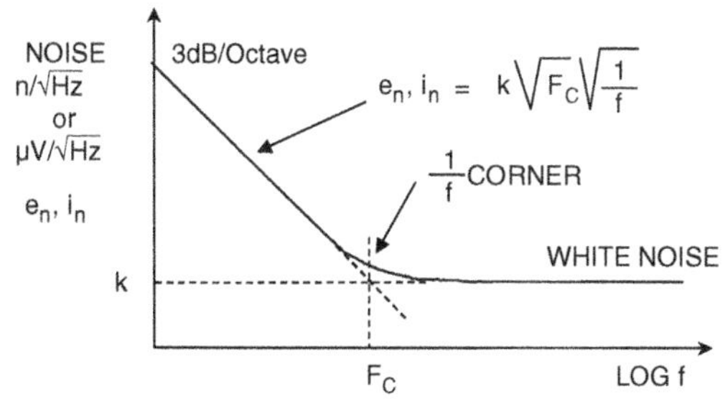

- 1/f Corner Frequency is a figure of merit for op amp noise performance (the lower the better)
- Typical Ranges: 2Hz to 2kHz
- Voltage Noise and Current Noise do not necessarily have the same 1/f corner frequency

Figure 1-69: Frequency characteristic of op amp noise

The general equation which describes the voltage or current noise spectral density in the 1/f region is

$$e_n, i_n, = k\sqrt{F_c}\sqrt{\frac{1}{f}}, \qquad \text{Eq. 1-24}$$

where k is the level of the "white" current or voltage noise level, and F_C is the 1/f corner frequency.

The best low frequency low noise amplifiers have corner frequencies in the range 1 Hz–10 Hz, while JFET devices and more general-purpose op amps have values in the range to 100 Hz. Very fast amplifiers however may make compromises in processing to achieve high speed which result in quite poor 1/f corners of several hundred Hz or even 1 kHz–2 kHz. This is generally unimportant in the wideband applications for which they were intended, but may affect their use at audio frequencies, particularly for equalized circuits.

Popcorn Noise

Popcorn noise is so-called because when played through an audio system, it sounds like cooking popcorn. It consists of random step changes of offset voltage that take place at random intervals in the 10+ millisecond timeframe. Such noise results from high levels of contamination and crystal lattice dislocation at the surface of the silicon chip, which in turn results from inappropriate processing techniques or poor quality raw materials.

When monolithic op amps were first introduced in the 1960s, popcorn noise was a dominant noise source. Today, however, the causes of popcorn noise are well understood, raw material purity is high, contamination is low, and production tests for it are reliable so that no op amp manufacturer should have any difficulty in shipping products that are substantially free of popcorn noise. For this reason, it is not even mentioned in most modern op amp textbooks.

RMS Noise Considerations

As was discussed above, noise spectral density is a function of frequency. In order to obtain the RMS noise, the noise spectral density curve must be integrated over the bandwidth of interest.

In the 1/f region, the RMS noise in the bandwidth F_L to F_C is given by

$$v_{n,rms}\left(F_L, F_C\right) = v_{nw}\sqrt{F_C}\sqrt{\int_{F_L}^{F_C}\frac{1}{f}df} = v_{nw}\sqrt{F_C \ln\left[\frac{F_C}{F_L}\right]} \qquad \text{Eq. 1-25}$$

where v_{nw} is the voltage noise spectral density in the "white" region, F_L is the lowest frequency of interest in the 1/f region, and F_C is the 1/f corner frequency.

The next region of interest is the "white" noise area which extends from F_C to F_H.

The RMS noise in this bandwidth is given by

$$v_{n,rms}\left(F_C, F_H\right) = v_{nw}\sqrt{F_H - F_C} \qquad \text{Eq. 1-26}$$

Eq. 1-25 and Eq. 1-26 can be combined to yield the total RMS noise from F_L to F_H:

$$v_{n,rms}\left(F_L, F_H\right) = v_{nw}\sqrt{F_C \ln\left[\frac{F_C}{F_L}\right] + \left(F_H - F_C\right)} \qquad \text{Eq. 1-27}$$

In many cases, the low frequency p-p noise is specified in a 0.1 Hz to 10 Hz bandwidth, measured with a 0.1 Hz to 10 Hz bandpass filter between op amp and measuring device.

The measurement is often presented as a scope photo with a time scale of 1s/div, as shown in Figure 1-70 for the OP213.

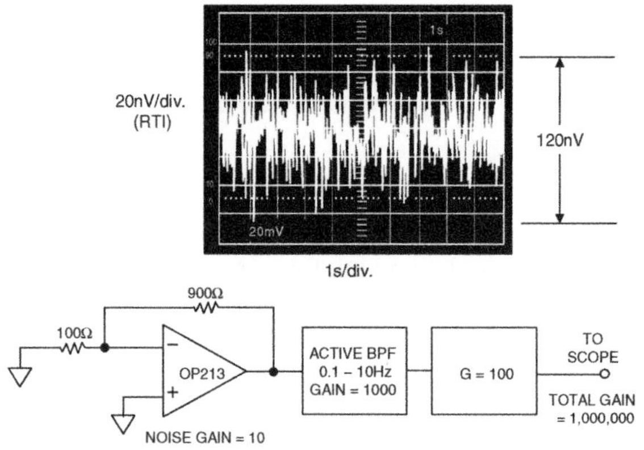

Figure 1-70: The peak-to-peak noise in the 0.1 Hz to 10 Hz bandwidth for the OP213 is less than 120 nV

It is possible to relate the 1/f noise measured in the 0.1 Hz to 10 Hz bandwidth to the voltage noise spectral density. Figure 1-71 shows the OP177 input voltage noise spectral density on the left-hand side of the diagram, and the 0.1 Hz to 10 Hz peak-to-peak noise scope photo on the right-hand side. Eq. 1-26 can be used to calculate the total RMS noise in the bandwidth 0.1 Hz to 10 Hz by letting $F_L = 0.1$ Hz, $F_H = 10$ Hz, $F_C = 0.7$ Hz, $v_{nw} = 10\ \text{nV}/\sqrt{\text{Hz}}$. The value works out to be about 33 nV RMS, or 218 nV peak-to-peak (obtained by multiplying the RMS value by 6.6—see the following discussion). This compares well to the value of 200 nV as measured from the scope photo.

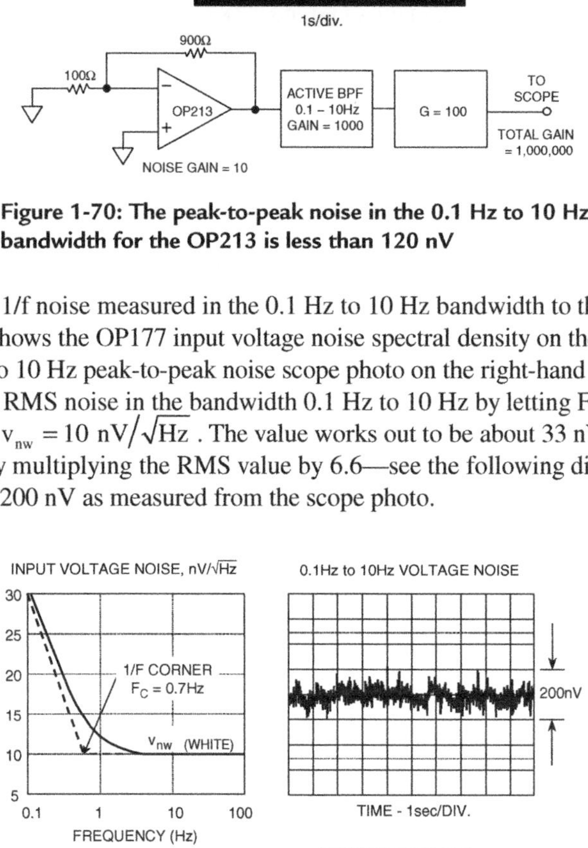

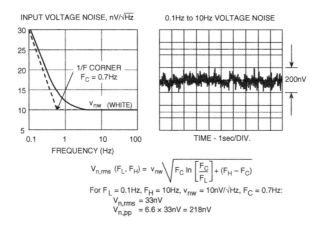

$$V_{n,rms}(F_L, F_H) = v_{nw}\sqrt{F_C \ln\left[\frac{F_C}{F_L}\right] + (F_H - F_C)}$$

For $F_L = 0.1\text{Hz}$, $F_H = 10\text{Hz}$, $v_{nw} = 10\text{nV}/\sqrt{\text{Hz}}$, $F_C = 0.7\text{Hz}$:

$V_{n,rms} = 33\text{nV}$

$V_{n,pp} = 6.6 \times 33\text{nV} = 218\text{nV}$

Figure 1-71: Input voltage noise for the OP177

It should be noted that at higher frequencies, the term in the equation containing the natural logarithm becomes insignificant, and the expression for the RMS noise becomes:

$$V_{n,rms}(F_H, F_L) \approx v_{nw}\sqrt{F_H - F_L} \qquad \text{Eq. 1-28}$$

And, if $F_H >> F_L$,

$$V_{n,rms}(F_H) \approx v_{nw}\sqrt{F_H} \qquad \text{Eq. 1-29}$$

However, some op amps (such as the OP07 and OP27) have voltage noise characteristics that increase slightly at high frequencies. The voltage noise versus frequency curve for op amps should therefore be examined carefully for flatness when calculating high frequency noise using this approximation.

At very low frequencies when operating exclusively in the 1/f region, $F_C >> (F_H - F_L)$, and the expression for the RMS noise reduces to:

$$V_{n,rms}(F_H, F_L) \approx V_{nw}\sqrt{F_C \ln\left[\frac{F_H}{F_L}\right]}. \qquad \text{Eq. 1-30}$$

Note that there is no way of reducing this 1/f noise by filtering if operation extends to dc. Making $F_H = 0.1$ Hz and $F_L = 0.001$ still yields an RMS 1/f noise of about 18 nV RMS, or 119 nV peak-to-peak.

The point is that averaging results of a large number of measurements over a long period of time has practically no effect on the RMS value of the 1/f noise. A method of reducing it further is to use a chopper-stabilized op amp, to remove the low frequency noise.

In practice, it is virtually impossible to measure noise within specific frequency limits with no contribution from outside those limits, since practical filters have finite roll-off characteristics. Fortunately, measurement error introduced by a single pole low-pass filter is readily computed. The noise in the spectrum above the single pole filter cutoff frequency, f_c, extends the corner frequency to 1.57 f_c. Similarly, a two pole filter has an apparent corner frequency of approximately 1.2 f_c. The error correction factor is usually negligible for filters having more than two poles. The net bandwidth after the correction is referred to as the filter *equivalent noise bandwidth* (see Figure 1-72).

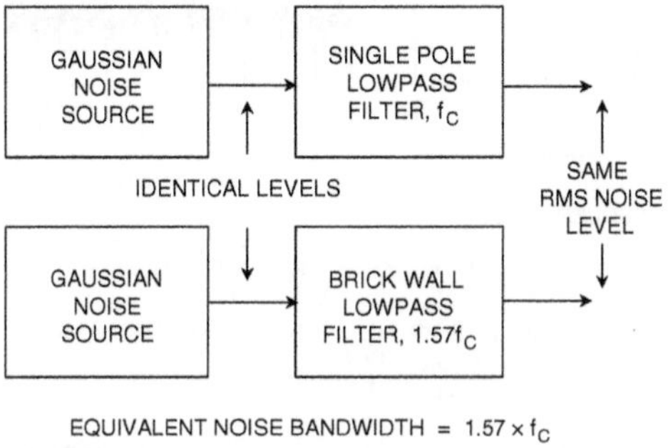

Figure 1-72: Equivalent noise bandwidth

It is often desirable to convert RMS noise measurements into peak-to-peak. In order to do this, one must have some understanding of the statistical nature of noise. For Gaussian noise and a given value of RMS noise, statistics tell us that the chance of a particular peak-to-peak value being exceeded decreases sharply as that value increases—but this probability never becomes zero.

Thus, for a given RMS noise, it is possible to predict the percentage of time that a given peak-to-peak value will be exceeded, but it is not possible to give a peak-to-peak value which will never be exceeded as shown in Figure 1-73.

Nominal Peak-to-Peak	% of the Time Noise will Exceed Nominal Peak-to-Peak Value
2 x rms	32%
3 x rms	13%
4 x rms	4.6%
5 x rms	1.2%
6 x rms	0.27%
6.6 x rms*	0.10%
7 x rms	0.046%
8 x rms	0.006%

*Most often used conversion factor is 6.6

Figure 1-73: RMS to peak-to-peak ratios

Peak-to-peak noise specifications, therefore, must always be written with a time limit. A suitable one is 6.6 times the RMS value, which is exceeded only 0.1% of the time.

Total Output Noise Calculations

We have already pointed out that any noise source that produces less than one-third to one-fifth of the noise of some greater source can be ignored, with little error. When so doing, both noise voltages must be measured at the same point in the circuit. To analyze the noise performance of an op amp circuit, we must assess the noise contributions of each part of the circuit, and determine which are significant. To simplify the following calculations, we shall work with noise spectral densities, rather than actual voltages, to leave bandwidth out of the expressions (the noise spectral density, which is generally expressed in $\mathrm{nV}/\sqrt{\mathrm{Hz}}$, is equivalent to the noise in a 1 Hz bandwidth).

If we consider the circuit in Figure 1-74, which is an amplifier consisting of an op amp and three resistors (R3 represents the source resistance at node A), we can find six separate noise sources: the Johnson noise of the three resistors, the op amp voltage noise, and the current noise in each input of the op amp. Each source has its own contribution to the noise at the amplifier output. Noise is generally specified RTI, or *referred to the input*, but it is often simpler to calculate the noise referred to the output (RTO) and then divide it by the *noise* gain (not the *signal* gain) of the amplifier to obtain the RTI noise.

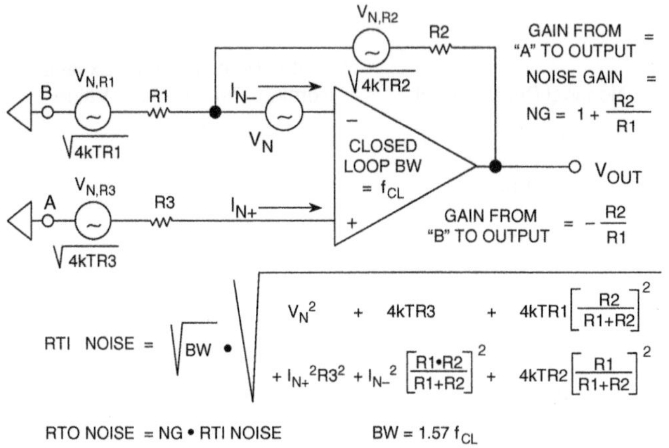

Figure 1-74: Op amp noise model for single-pole system

Figure 1-75 is a detailed analysis of how each of the noise sources in Figure 1-74 is reflected to the output of the op amp. Some further discussion regarding the effect of the current noise at the inverting input is warranted. This current, I_{N-}, does not flow in R1, as might be expected—the negative feedback around the amplifier works to keep the potential at the inverting input unchanged, so that a current flowing from that pin is forced, by negative feedback, to flow in R2 only, resulting in a voltage at the output of I_{N-} R2. We could equally well consider the voltage caused by I_{N-} flowing in the parallel combination of R1 and R2 and then amplified by the noise gain of the amplifier, but the results are identical—only the calculations are more involved.

NOISE SOURCE EXPRESSED AS A VOLTAGE	MULTIPLY BY THIS FACTOR TO REFER TO THE OP AMP OUTPUT
Johnson noise in R3: $\sqrt{(4kTR3)}$	Noise Gain = 1 + R2/R1
Noninverting input current noise flowing in R3: I_{N+} R3	Noise Gain = 1 + R2/R1
Input voltage noise: V_N	Noise Gain = 1 + R2/R1
Johnson noise in R1: $\sqrt{(4kTR1)}$	–R2/R1 (Gain from input of R1 to output)
Johnson noise in R2: $\sqrt{(4kTR2)}$	1
Inverting input current noise flowing in R2: I_{N-} R2	1

Figure 1-75: Noise sources referred to the output (RTO)

Notice that the Johnson noise voltage associated with the three resistors has been included in the expressions of Figure 1-75. All resistors have a Johnson noise of $\sqrt{(4kTBR)}$, where k is Boltzmann's Constant (1.38×10^{-23} J/K), T is the absolute temperature, B is the bandwidth in Hz, and R is the resistance in Ω. A simple relationship which is easy to remember is that a 1000 Ω resistor generates a Johnson noise of $4\,nV/\sqrt{Hz}$ at 25°C.

The analysis so far assumes that the feedback network is purely resistive and that the noise gain versus frequency is flat. This applies to most applications, but if the feedback network contains reactive elements (usually capacitors) the noise gain is not constant over the bandwidth of interest, and more complex techniques must be used to calculate the total noise (see in particular, Reference 2 and Chapter 4, Section 4-4 of this book).

The circuit shown in Figure 1-76 represents a second-order system, where capacitor C1 represents the source capacitance, stray capacitance on the inverting input, the input capacitance of the op amp, or any combination of these. C1 causes a breakpoint in the noise gain, and C2 is the capacitor that must be added to obtain stability.

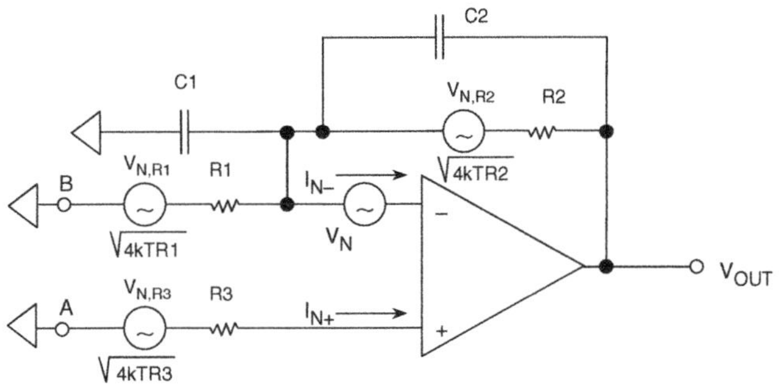

Figure 1-76: Op amp noise model with reactive elements (second-order system)

Because of C1 and C2, the noise gain is a function of frequency, and has peaking at the higher frequencies (assuming C2 is selected to make the second-order system critically damped). Textbooks state that a flat noise gain can be achieved if one simply makes R1C1 = R2C2.

But in the case of current-to-voltage converters, however, R1 is typically a high impedance, and the method doesn't work. Maximizing the signal bandwidth in these situations is somewhat complex and is treated in detail in Section 1-6 of this chapter and in Chapter 4, Section 4-4 of this book.

A dc signal applied to input A (B being grounded) sees a gain of 1 + R2/R1, the low frequency noise gain. At higher frequencies, the gain from input A to the output becomes 1 + C1/C2 (the high frequency noise gain).

The closed-loop bandwidth, f_{cl}, is the point at which the noise gain intersects the open-loop gain. A dc signal applied to B (A being grounded) sees a gain of –R2/R1, with a high frequency cutoff determined by R2-C2. Bandwidth from B to the output is $1/2\pi R2C2$.

The current noise of the noninverting input, I_{N+}, flows in R3 and gives rise to a noise voltage of I_{N+} R3, which is amplified by the frequency-dependent noise gain, as are the op amp noise voltage, V_N, and the Johnson noise of R3, which is $\sqrt{(4kTR3)}$. The Johnson noise of R1 is amplified by –R2/R1 over a bandwidth of

$1/2\pi R2C2$, and the Johnson noise of R2 is not amplified at all but is connected directly to the output over a bandwidth of $1/2\pi R2C2$. The current noise of the inverting input, I_{N-}, flows in R2 only, resulting in a voltage at the amplifier output of $I_{N-}R2$ over a bandwidth of $1/2\pi R2C2$.

If we consider these six noise contributions, we see that if R1, R2, and R3 are low, the effect of current noise and Johnson noise will be minimized, and the dominant noise will be the op amp's voltage noise. As we increase resistance, both Johnson noise and the voltage noise produced by noise currents will rise.

If noise currents are low, Johnson noise will take over from voltage noise as the dominant contributor. Johnson noise, however, rises with the square root of the resistance, while the current noise voltage rises linearly with resistance so, ultimately, as the resistance continues to rise, the voltage due to noise currents will become dominant.

These noise contributions we have analyzed are not affected by whether the input is connected to node A or node B (the other being grounded or connected to some other low impedance voltage source), which is why the noninverting gain (1 + Z2/Z1), which is seen by the voltage noise of the op amp, V_N, is known as the "noise gain."

Calculating the total output RMS noise of the second-order op amp system requires multiplying each of the six noise voltages by the appropriate gain and integrating over the appropriate frequency as shown in Figure 1-77.

NOISE SOURCE EXPRESSED AS A VOLTAGE	MULTIPLY BY THIS FACTOR TO REFER TO OUTPUT	INTEGRATION BANDWIDTH
Johnson noise in R3: $\sqrt{(4kTR3)}$	Noise Gain as a function of frequency	Closed-Loop BW
Noninverting input current noise flowing in R3: $I_{N+}R3$	Noise Gain as a function of frequency	Closed-Loop BW
Input voltage noise: V_N	Noise Gain as a function of frequency	Closed-Loop BW
Johnson noise in R1: $\sqrt{(4kTR1)}$	–R2/R1 (Gain from B to output)	$1/2\pi R2C2$
Johnson noise in R2: $\sqrt{(4kTR2)}$	1	$1/2\pi R2C2$
Inverting input current noise flowing in R2: $I_{N-}R2$	1	$1/2\pi R2C2$

Figure 1-77: Noise sources referred to the output for a second-order system

The root-sum-square of all the output contributions thus represents the total RMS output noise. Fortunately, this cumbersome exercise may be greatly simplified in most cases by making the appropriate assumptions and identifying the chief contributors.

Although shown, the noise gain for a typical second-order system is repeated in Figure 1-78. It is quite easy to perform the voltage noise integration in two steps, but notice that because of peaking, the majority of the output noise due to the input voltage noise will be determined by the high frequency portion where the noise gain is 1 + C1/C2. This type of response is typical of second-order systems.

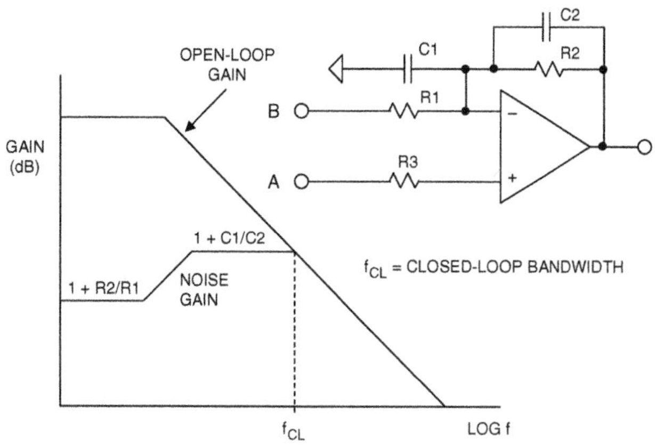

Figure 1-78: Noise gain of a typical second-order system

The noise due to the inverting input current noise, R1, and R2 is only integrated over the bandwidth $1/2\pi R2C2$.

Op Amp Distortion

Dynamic range of an op amp may be defined in several ways. The most common are to specify *harmonic distortion, total harmonic distortion* (THD), or *total harmonic distortion plus noise* (THD + N).

Other specifications related specifically to communications systems such as *intermodulation distortion* (IMD), *intercept points* (IP), *spurious free dynamic range* (SFDR), *multitone power ratio* (MTPR) and others are covered thoroughly in Chapter 6, Section 6-4. In this section, only harmonic distortion, THD, and THD + N will be covered.

The distortion components that make up total harmonic distortion are usually calculated by taking the root sum of the squares of the first five or six harmonics of the fundamental. In many practical situations, however, there is negligible error if only the second and third harmonics are included. The definition of THD and THD + N is shown in Figure 1-79.

- V_s = Signal Amplitude (RMS Volts)
- V_2 = Second Harmonic Amplitude (RMS Volts)
- V_n = nth Harmonic Amplitude (RMS Volts)
- V_{noise} = RMS value of noise over measurement bandwidth
- $$\text{THD + N} = \frac{\sqrt{V_2^2 + V_3^2 + V_4^2 + \ldots + V_n^2 + V_{noise}^2}}{V_s}$$
- $$\text{THD} = \frac{\sqrt{V_2^2 + V_3^2 + V_4^2 + \ldots + V_n^2}}{V_s}$$

Figure 1-79: Definitions of THD and THD + N

It is important to note that the THD measurement does not include noise terms, while THD + N does. The noise in the THD + N measurement must be integrated over the measurement bandwidth. In audio applications, the bandwidth is normally chosen to be around 100 kHz. In narrow-band applications, the level of the noise may be reduced by filtering.

On the other hand, harmonics and intermodulation products which fall within the measurement bandwidth cannot be filtered, and therefore may limit the system dynamic range.

Common-Mode Rejection Ratio (CMRR), Power Supply Rejection Ratio (PSRR)

If a signal is applied equally to both inputs of an op amp, so that the differential input voltage is unaffected, the output should not be affected. In practice, changes in common-mode voltage will produce changes in output. The op amp *common-mode rejection ratio* (CMRR) is the ratio of the common-mode gain to differential-mode gain. For example, if a differential input change of Y volts produces a change of 1 V at the output, and a common-mode change of X volts produces a similar change of 1 V, the CMRR is X/Y. When the common-mode rejection ratio is expressed in dB, it is generally referred to as common-mode rejection (CMR). Typical LF CMR values are between 70 dB and 120 dB, but at higher frequencies CMR deteriorates. Many op amp data sheets show a plot of CMR versus frequency, as shown in Figure 1-80 for an OP177 op amp.

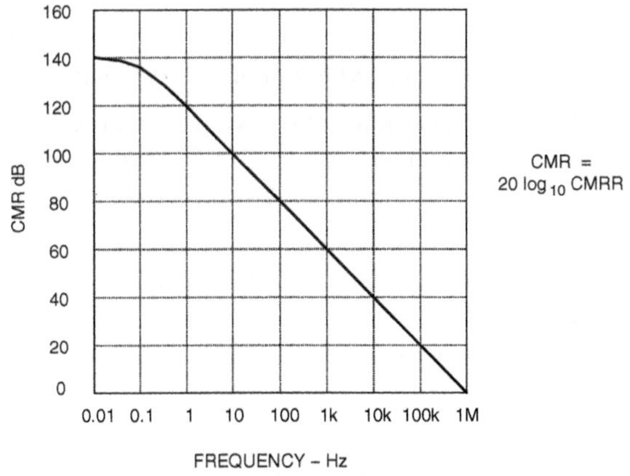

Figure 1-80: OP177 common-mode rejection (CMR)

CMRR produces a corresponding output offset voltage error in op amps configured in the noninverting mode as shown in Figure 1-81.

Figure 1-81: Calculating offset error due to common-mode rejection ratio (CMRR)

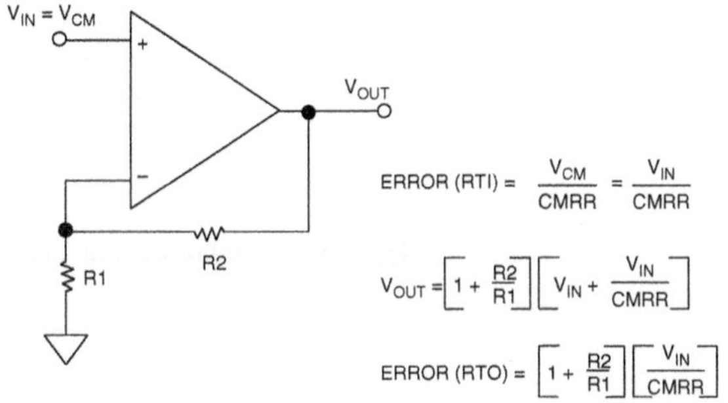

Note inverting mode operating op amps will have negligible CMRR error, as both inputs are held at a ground (or virtual ground), i.e., there is no CM dynamic voltage.

Common-mode rejection ratio can be measured in several ways. The method shown in Figure 1-82 uses four precision resistors to configure the op amp as a differential amplifier, a signal is applied to both inputs, and the change in output is measured—an amplifier with infinite CMRR would have no change in output. The disadvantage inherent in this circuit is that the ratio match of the resistors is as important as the CMRR of the op amp. A mismatch of 0.1% between resistor pairs will result in a CMR of only 66 dB—no matter how good the op amp. Since most op amps have a LF CMR of between 80 dB and 120 dB, it is clear that this circuit is only marginally useful for measuring CMRR (although it does an excellent job in measuring the matching of the resistors).

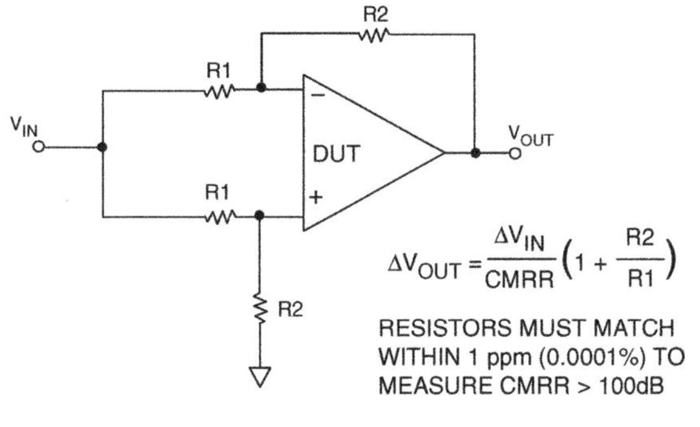

Figure 1-82: Simple common-mode rejection ratio (CMRR) test circuit

The slightly more complex circuit, shown in Figure 1-83, measures CMRR without requiring accurately matched resistors. In this circuit, the common-mode voltage is changed by switching the power supply voltages. (This is easy to implement in a test facility, and the same circuit with different supply voltage connections can be used to measure power supply rejection ratio.)

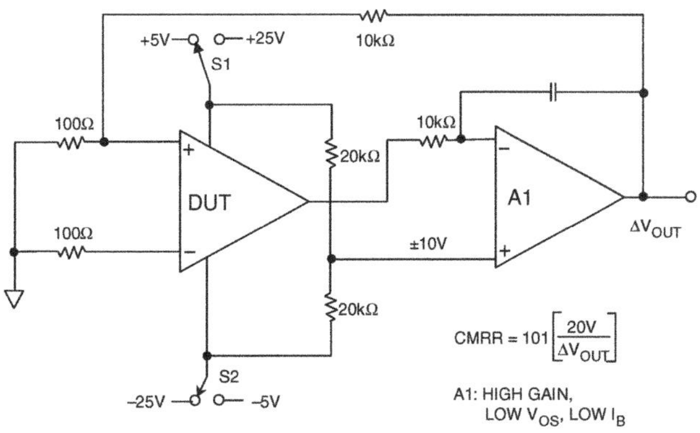

Figure 1-83: CMRR test circuit does not require precision resistors

The power supply values shown in the circuit are for a ±15 V DUT op amp, with a common-mode voltage range of ±10 V. Other supplies and common-mode ranges can also be accommodated by changing voltages, as appropriate. The integrating amplifier A1 should have high gain, low V_{OS}, and low I_B, such as an OP97 family device.

If the supply of an op amp changes, its output should not, but it does. The specification of *power supply rejection ratio* or PSRR is defined similarly to the definition of CMRR. If a change of X volts in the supply produces the same output change as a differential input change of Y volts, the PSRR on that supply is X/Y. The definition of PSRR assumes that both supplies are altered equally in opposite directions—otherwise the change will introduce a common-mode change as well as a supply change, and the analysis becomes considerably more complex. It is this effect that causes apparent differences in PSRR between the positive and negative supplies.

Typical PSR for the OP177 is shown in Figure 1-84.

The test setup used to measure CMRR may be modified to measure PSRR as shown in Figure 1-85.

The voltages are chosen for a symmetrical power supply change of 1 V. Other values may be used where appropriate.

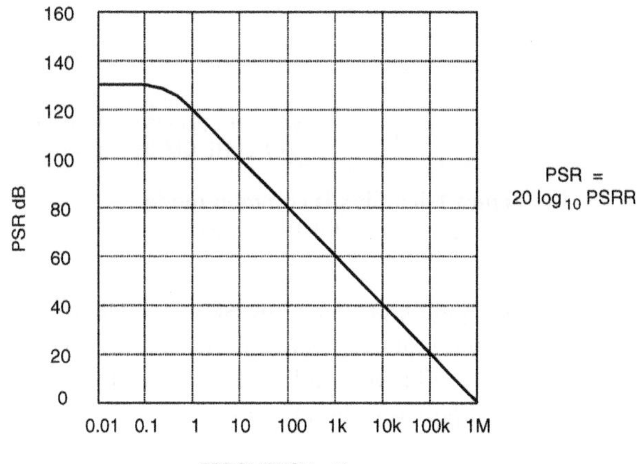

Figure 1-84: OP177 power supply rejection

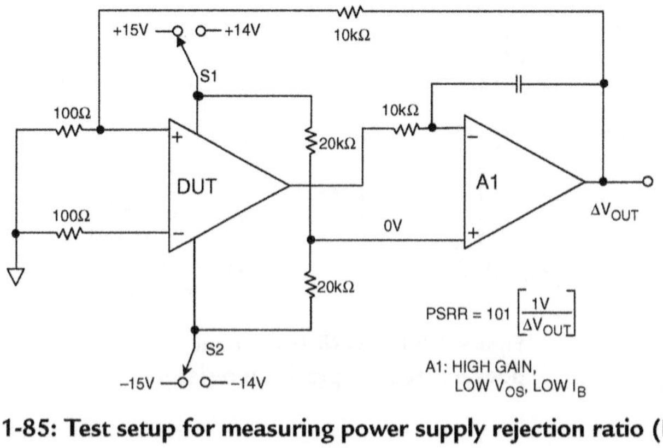

Figure 1-85: Test setup for measuring power supply rejection ratio (PSRR)

Power Supplies and Decoupling

Because op amp PSRR is frequency dependent, op amp power supplies must be well decoupled. At low frequencies, several devices may share a 10 µF–50 µF capacitor on each supply, provided it is no more than 10cm (PC track distance) from any of them.

At high frequencies, each IC should have the supply leads decoupled by a low inductance 0.1 µF (or so) capacitor with short leads/PC tracks. These capacitors must also provide a return path for HF currents in the op amp load. Typical decoupling circuits are shown in Figure 1-86. Further bypassing and decoupling information is found Chapter 7.

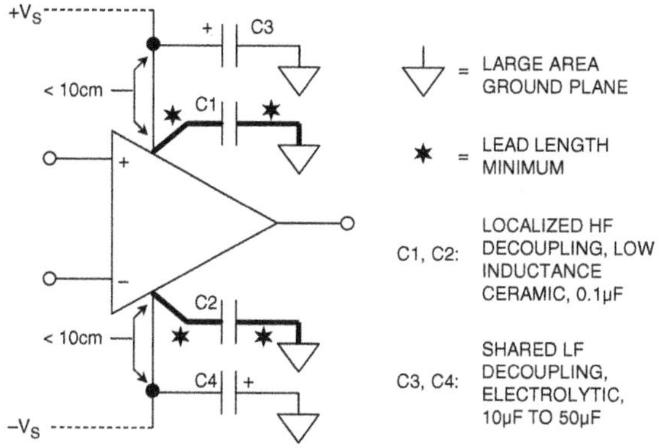

Figure 1-86: Proper low and high-frequency decoupling techniques for op amps

Power Supplies and Power Dissipation

Op amps have no ground terminal. Specifications of power supply are quite often in the form ±X Volts, but in fact it might equally be expressed as 2X Volts. What is important is where the CM and output ranges lie relative to the supplies. This information may be provided in tabular form or as a graph.

Data sheets will often advise that an op amp will work over a range of supplies (from +3 V to ±16.5 V for example), and will then give parameters at several values of supply, so that users may extrapolate. If the minimum supply is quite high, it is usually because the device uses a structure requiring a threshold voltage to function (zener diode).

Data sheets also give current consumption. Any current flowing into one supply pin will flow out of the other or out of the output terminal. When the output is open circuit, the dissipation is easily calculated from the supply voltage and current. When current flows in a load, it is easiest to calculate the total dissipation (remember that if the load is grounded to the center rail the load current flows from a supply to ground, not between supplies), and then subtract the load dissipation to obtain the device dissipation. Data sheets normally give details of thermal resistances and maximum junction temperature ratings, from which dissipation limits may be calculated knowing conditions. Details of further considerations relating to power dissipation, heatsinking, and so forth, can be found in Chapter 7, Section 7-5.

References: Op Amp Specifications

1. James L. Melsa and Donald G. Schultz, **Linear Control Systems**, McGraw-Hill, 1969, pp. 196–220, ISBN: 0-07-041481-5.
2. Lewis Smith and Dan Sheingold, "Noise and Operational Amplifier Circuits," **Analog Dialogue,** Vol. 3, No.1, pp. 1, 5–16. See also: **Analog Dialogue 25th Anniversary Issue**, pp. 19–31, 1991.
3. Thomas M. Frederiksen, **Intuitive Operational Amplifiers**, McGraw-Hill, 1988., ISBN: 0-07-021966-4.
4. Walter G. Jung, **IC Op Amp Cookbook, 3rd Ed.**, Prentice-Hall PTR, 1986, 1997, ISBN: 0-13-889601-1.
5. J. K. Roberge, **Operational Amplifiers-Theory and Practice**, John Wiley, 1975, ISBN: 0-471-72585-4.
6. D. Stout, M. Kaufman, **Handbook of Operational Amplifier Circuit Design**, New York, McGraw-Hill, 1976, ISBN: 0-07-061797-X.
7. J. Dostal, **Operational Amplifiers**, Elsevier Scientific Publishing, New York, 1981, ISBN: 0-444-99760-1.
8. Paul R. Gray and Robert G. Meyer, **Analysis and Design of Analog Integrated Circuits, 3rd Edition**, John Wiley, 1993, ISBN: 0-471-57495-3.
9. Sergio Franco, **Design With Operational Amplifiers and Analog Integrated Circuits, 2nd Ed.**, McGraw-Hill, 1998, ISBN: 0-07-021857-9.
10. Walt Kester, Editor, **Linear Design Seminar**, Analog Devices, Inc., 1995, ISBN: 0-916550-15-X.
11. Walt Kester, Editor, **Practical Analog Design Techniques**, Analog Devices, 1995, ISBN: 0-916550-16-8, (available for download at www.analog.com).
12. Walt Kester, Editor, **High Speed Design Techniques**, Analog Devices, 1996, ISBN: 0-916550-17-6, (available for download at www.analog.com).

SECTION 1-5

Precision Op Amps

Walt Kester, Walt Jung

This section examines in more detail some of the issues relating to amplifiers for use in precision signal conditioning applications. Although the OP177 op amp is used for the "gold standard" for precision in these discussions, more recent product introductions such as the rail-to-rail output OP777, OP727, and OP747, along with the OP1177, OP2177, and OP4177 offer nearly as good performance in smaller packages.

Precision op amp open-loop gains greater than 1 million are available, along with common-mode and power supply rejection ratios of the same magnitude. Offset voltages of less than 25 μV and offset drift less than 0.1 μV/°C are available in dual supply op amps such as the OP177, however, the performance in single-supply precision bipolar op amps may sometimes fall short of this performance. This is the trade-off that must sometimes be made in low power, low voltage applications. On the other hand, however, modern chopper stabilized op amps provide offsets and offset voltage drifts which cannot be distinguished from noise, and these devices operate on single supplies and provide rail-to-rail inputs and outputs. They, too, come with their own set of problems that are discussed later within this section.

It is important to understand that dc open-loop gain, offset voltage, power supply rejection (PSR), and common-mode rejection (CMR) alone shouldn't be the only considerations in selecting precision amplifiers. The ac performance of the amplifier is also important, even at "low" frequencies. Open-loop gain, PSR, and CMR all have relatively low corner frequencies; therefore, what may be considered "low" frequency may actually fall above these corner frequencies, increasing errors above the value predicted solely by the dc parameters. For example, an amplifier having a dc open-loop gain of 10 million and a unity-gain crossover frequency of 1 MHz has a corresponding corner frequency of 0.1 Hz. One must therefore consider the open-loop gain at the actual *signal* frequency. The relationship between the single-pole unity-gain crossover frequency, f_u, the signal frequency, f_{sig}, and the open-loop gain $A_{VOL(f_{sig})}$ (measured at the signal frequency) is given by:

$$A_{VOL(f_{sig})} = \frac{f_u}{f_{sig}} \qquad \text{Eq. 1-31}$$

In the example above, the open-loop gain is 10 at 100 kHz, and 100,000 at 10 Hz. Note that the constant gain-bandwidth product concept only holds true for VFB op amps. It doesn't apply to CFB op amps, but then they are rarely used in precision applications.

Loss of open-loop gain at the frequency of interest can introduce distortion, especially at audio frequencies. Loss of CMR or PSR at the line frequency or harmonics thereof can also introduce errors.

The challenge of selecting the right amplifier for a particular signal conditioning application has been complicated by the sheer proliferation of various types of amplifiers in various processes (Bipolar, Complementary Bipolar, BiFET, CMOS, BiCMOS, and so forth) and architectures (traditional op amps, instrumentation amplifiers, chopper amplifiers, isolation amplifiers, and so forth)

In addition, a wide selection of precision amplifiers which operate on single-supply voltages are now available which complicates the design process even further because of the reduced signal swings and voltage input and output restrictions. Offset voltage and noise are now a more significant portion of the input signal.

Selection guides and parametric search engines, which can simplify this process somewhat, are available on the World Wide Web (www.analog.com) as well as on CDROM. Some general attributes of precision op amps are summarized in Figure 1-87.

- Input Offset Voltage — <100μV
- Input Offset Voltage Drift — <1μV/°C
- Input Bias Current — <2nA
- Input Offset Current — <2nA
- DC Open-Loop Gain — >1,000,000
- Unity Gain Bandwidth Product, f_u — 500kHz – 5MHz
- Always Check Open Loop Gain at Signal Frequency
- 1/f (0.1Hz to 10Hz) Noise — <1μV p-p
- Wideband Noise — <10nV/√Hz
- CMR, PSR — >100dB
- Trade-offs:
  - Single supply operation
  - Low supply currents

Figure 1-87: Precision op amp characteristics

Precision Op Amp Amplifier DC Error Budget Analysis

In order to develop a concept for the magnitudes of the various errors in a high precision op amp circuit, a simple room temperature analysis for the OP177F is shown in Figure 1-88. The amplifier is connected in the inverting mode with a signal gain of 100. The key data sheet specifications are also shown in the diagram. We assume an input signal of 100 mV fullscale which corresponds to an output signal of 10 V. The various error sources are normalized to full-scale and expressed in parts per million (ppm). Note: parts per million (ppm) error = fractional error × 10^6 = % error × 10^4.

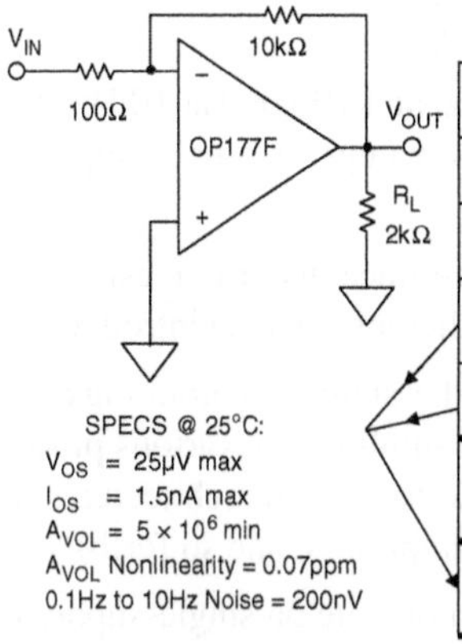

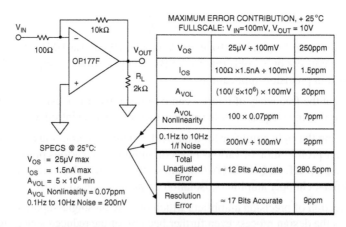

MAXIMUM ERROR CONTRIBUTION, + 25°C
FULLSCALE: V_{IN}=100mV, V_{OUT} = 10V

V_{OS}	25μV ÷ 100mV	250ppm
I_{OS}	100Ω ×1.5nA ÷ 100mV	1.5ppm
A_{VOL}	(100/ 5×10^6) × 100mV	20ppm
A_{VOL} Nonlinearity	100 × 0.07ppm	7ppm
0.1Hz to 10Hz 1/f Noise	200nV ÷ 100mV	2ppm
Total Unadjusted Error	≈ 12 Bits Accurate	280.5ppm
Resolution Error	≈ 17 Bits Accurate	9ppm

Figure 1-88: Precision op amp (OP177F) dc error budget analysis

Note that the offset errors due to V_{OS} and I_{OS} and the gain error due to finite A_{VOL} can be removed with a system calibration. However, the error due to open-loop gain nonlinearity cannot be removed with calibration and produces a relative accuracy error, often called *resolution error.*

A second contributor to resolution error is the 1/f noise. This noise is always present and adds to the uncertainty of the measurement. The overall relative accuracy of the circuit at room temperature is 9ppm, equivalent to ~17 bits of resolution.

It is also useful to compare the performance of a number of single-supply op amps to that of the "gold standard" OP177, and this is done in Figure 1-89 for some representative devices.

Note that the Figure 1-89 amplifier list does *not* include the category of chopper op amps, which excel in many of the categories. These are covered separately, next.

LISTED IN ORDER OF INCREASING SUPPLY CURRENT

PART NO.	V_{OS} max	V_{OS} TC	A_{VOL}min	NOISE (1kHz)	INPUT	OUTPUT	I_{SY}/AMP MAX
OP293	250µV	2µV/°C	200k	5nV/√Hz	0, 4V	5mV, 4V	20µA
OP196/296/496	300µV	2µV/°C	150k	26nV/√Hz	R/R	"R/R"	60µA
OP777	100µV	1.3µV/°C	300k	15nV√Hz	0, 4V	"R/R"	270µA
OP191/291/491	700µV	5µV/°C	25k	35nV/√Hz	R/R	"R/R"	350µA
*AD820/822/824	1000µV	20µV/°C	500k	16nV/√Hz	0, 4V	"R/R"	800µA
**AD8601/2/4	600µV	2µV/°C	20k	33nV/√Hz	R/R	"R/R"	1000µA
OP184/284/484	150µV	2µV/°C	50k	3.9nV/√Hz	R/R	"R/R"	1350µA
OP113/213/413	175µV	4µV/°C	2M	4.7nV/√Hz	0, 4V	5mV, 4V	3000µA
OP177F (±15V)	25µV	0.1µV/°C	5M	10nV/√Hz	N/A	N/A	2000µA

*JFET INPUT **CMOS

NOTE: Unless Otherwise Stated
Specifications are Typical @ 25°C
V_S = 5V

Figure 1-89: Precision single-supply op amp performance characteristics

Chopper Stabilized Amplifiers

For the lowest offset and drift performance, chopper-stabilized amplifiers may be the only solution. The best bipolar amplifiers offer offset voltages of 25 μV and 0.1 μV/°C drift. Offset voltages less than 5 μV with practically no measurable offset drift are obtainable with choppers, albeit with some penalties.

A basic chopper amplifier circuit is shown in Figure 1-90. When the switches are in the "Z" (auto-zero) position, capacitors C2 and C3 are charged to the amplifier input and output offset voltage, respectively. When the switches are in the "S" (sample) position, V_{IN} is connected to V_{OUT} through the path comprised of R1, R2, C2, the amplifier, C3, and R3. The chopping frequency is usually between a few hundred Hz and several kHz, and it should be noted that because this is a sampling system, the input frequency must be much less than one-half the chopping frequency in order to prevent errors due to aliasing. The R1-C1 combination serves as an antialiasing filter. It is also assumed that after a steady state condition is reached, there is only a minimal amount of charge transferred during the switching cycles. The output capacitor, C4, and the load, R_L, must be chosen such that there is minimal V_{OUT} droop during the auto-zero cycle.

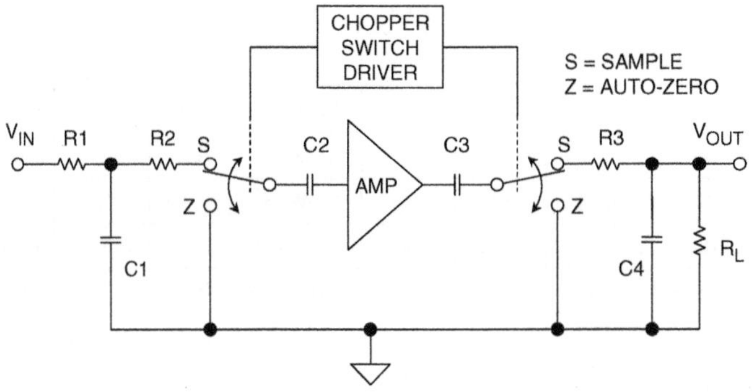

Figure 1-90: Classic chopper amplifier

The basic chopper amplifier of Figure 1-90 can pass only very low frequencies because of the input filtering required to prevent aliasing. In contrast to this, the *chopper-stabilized* architecture shown in Figure 1-91 is most often used in chopper amplifier implementations. In this circuit, A1 is the *main* amplifier, and A2 is

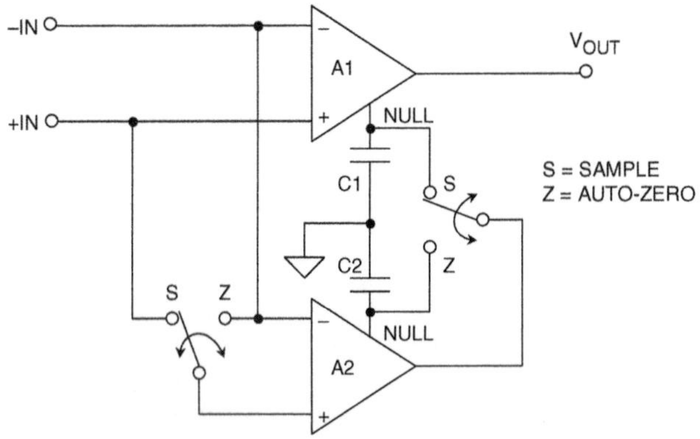

Figure 1-91: Modern chopper stabilized op amp

the *nulling* amplifier. In the sample mode (switches in "S" position), the nulling amplifier, A2, monitors the input offset voltage of A1 and drives its output to zero by applying a suitable correcting voltage at A1's null pin. Note, however, that A2 also has an input offset voltage, so it must correct its own error before attempting to null A1's offset. This is achieved in the auto-zero mode (switches in "Z" position) by momentarily disconnecting A2 from A1, shorting its inputs together, and coupling its output to its own null pin. During the auto-zero mode, the correction voltage for A1 is momentarily held by C1. Similarly, C2 holds the correction voltage for A2 during the sample mode. In modern IC chopper-stabilized op amps, the storage capacitors C1 and C2 are on-chip.

Note in this architecture that *the input signal is always connected to the output, through A1*. The bandwidth of A1 thus determines the overall signal bandwidth, and the input signal is not limited to less than one-half the chopping frequency as in the case of the traditional chopper amplifier architecture. However, the switching action does produce small transients at the chopping frequency, which can mix with the input signal frequency and produce intermodulation distortion.

A patented spread-spectrum technique is used in the AD8571/72/74 series of single-supply chopper-stabilized op amps, to virtually eliminate intermodulation effects.

These devices use a pseudorandom chopping frequency swept between 2 kHz and 4 kHz. Figure 1-92 compares the intermodulation distortion of a traditional chopper stabilized op amp (AD8551/52/54, left) that uses a fixed 4 kHz chopping frequency to that of the AD8571/72/74 (right) that uses the pseudorandom chopping frequency.

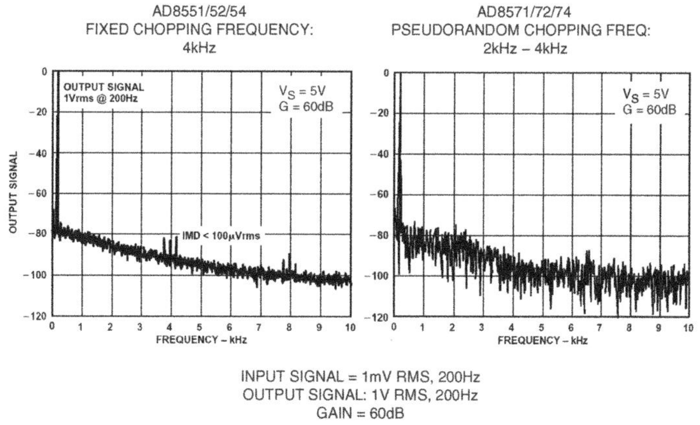

Figure 1-92: Intermodulation product: fixed versus pseudorandom chopping frequency

A comparison between fixed and pseudorandom chopping on the voltage noise is shown in Figure 1-93. Notice for the fixed chopping frequency, there are distinct peaks in the noise spectrum at the odd harmonics of 4 kHz, whereas with pseudorandom chopping, the spectrum is much more uniform, although the average noise level is higher.

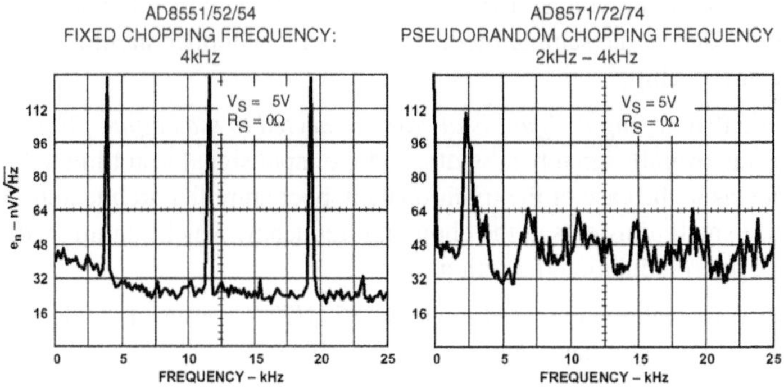

Figure 1-93: Voltage noise spectral density comparison: fixed versus pseudorandom chopping frequency

The AD8571/AD8572/AD8574 family of chopper-stabilized op amps offers rail-to-rail input and output single-supply operation, low offset voltage, and low offset drift. As discussed above, the pseudorandom chopping frequency minimizes intermodulation distortion with the input signal. The storage capacitors are internal to the IC, and no external capacitors other than standard decoupling capacitors are required. Key specifications for the devices are given in Figure 1-94.

- Single Supply: 2.7V to 5V
- 1µV Typical Input Offset Voltage
- 0.005µV/°C Typical Input Offset Voltage Drift
- 130dB CMR, PSR
- 750µA Supply Current/Op Amp
- 50µs Overload Recovery Time
- 50nV/√Hz Input Voltage Noise
- Pseudorandom Chopping Frequency
- 1.5MHz Gain-Bandwidth Product
- Single (AD8571), Dual (AD8572) and Quad (AD8574)

Figure 1-94: AD8571/72/74 chopper-stabilized rail-to-rail input/output amplifiers

It should be noted that extreme care must be taken when applying all of the chopper-stabilized devices. This is because in order to fully realize the full offset and drift performance inherent to the parts, parasitic thermocouple effects in external circuitry must be avoided. See Chapter 4, Section 4-5 for a general discussion of thermocouples, and Chapter 7, Section 7-1 related to passive components.

Noise Considerations for Chopper-Stabilized Op Amps

It is interesting to consider the effects of a chopper amplifier on low frequency 1/f noise. If the chopping frequency is considerably higher than the 1/f corner frequency of the input noise, the chopper-stabilized amplifier continuously nulls out the 1/f noise on a sample-by-sample basis. Theoretically, a chopper op amp therefore has no 1/f noise. However, the chopping action produces wideband noise which is generally much worse than that of a precision bipolar op amp.

Figure 1-95 shows the noise of a precision bipolar amplifier (OP177) versus that of the AD8571/72/74 chopper-stabilized op amp. The peak-to-peak noise in various bandwidths is calculated for each in the table below the graphs.

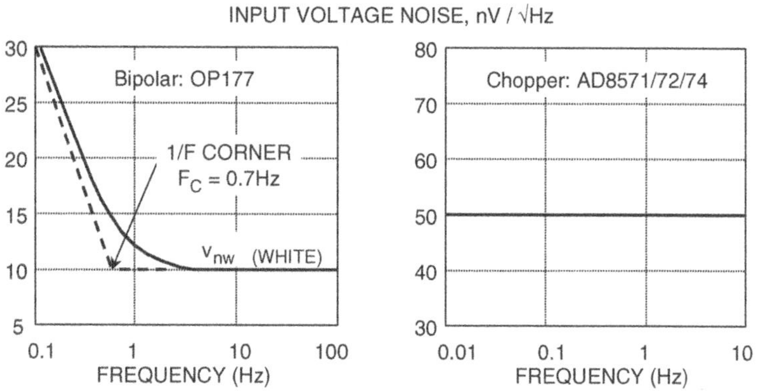

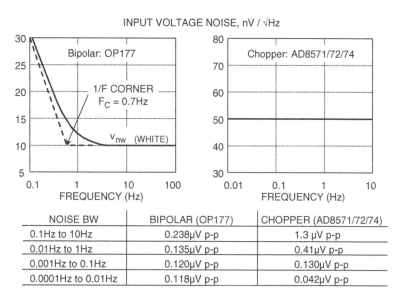

NOISE BW	BIPOLAR (OP177)	CHOPPER (AD8571/72/74)
0.1Hz to 10Hz	0.238μV p-p	1.3 μV p-p
0.01Hz to 1Hz	0.135μV p-p	0.41μV p-p
0.001Hz to 0.1Hz	0.120μV p-p	0.130μV p-p
0.0001Hz to 0.01Hz	0.118μV p-p	0.042μV p-p

Figure 1-95: Noise: bipolar versus chopper stabilized op amp

Note from the data that as the frequency is lowered, the chopper amplifier noise continues to drop, while the bipolar amplifier noise approaches a limit determined by the 1/f corner frequency and its white noise. Notice that only at very low frequencies (<0.01 Hz) is the chopper noise performance superior to that of the bipolar op amp.

In order to take advantage of the chopper op amp's lack of 1/f noise, much filtering is required—otherwise the total noise of a chopper will always be worse than a good bipolar op amp. Choppers should therefore be selected on the basis of their low offset and drift—not because of their lack of 1/f noise.

References: Precision Op Amps

1. Walt Kester, Editor, **1992 Amplifier Applications Guide**, Analog Devices, 1992, ISBN: 0-916550-10-9.
2. Walt Kester, Editor, **Practical Design Techniques for Sensor Signal Conditioning**, Analog Devices, 1999, ISBN: 0-916550-20-6.
3. Data Sheet for **AD8551/AD8552/AD8554 Zero-Drift, Single-Supply, Rail-to-Rail Input/Output Operational Amplifiers**, at www.analog.com.
4. Data Sheet for **AD8571/AD8572/AD8574 Zero-Drift, Single-Supply, Rail-to-Rail Input/Output Operational Amplifiers**, at www.analog.com.
5. Data Sheet for **OP777/OP727/OP747 Precision Micropower Single-Supply Operational Amplifiers**, at www.analog.com.
6. Data Sheet for **OP1177/OP2177/OP4177 Precision Low Noise, Low Input Bias Current Operational Amplifiers**, at www.analog.com.

High Speed Op Amps

Walt Kester

Introduction

High speed analog signal processing applications, such as video and communications, require op amps that have wide bandwidth, fast settling time, low distortion and noise, high output current, good dc performance, and operate at low supply voltages. These devices are widely used as gain blocks, cable drivers, ADC pre-amps, current-to-voltage converters, and so forth. Achieving higher bandwidths for less power is extremely critical in today's portable and battery-operated communications equipment. The rapid progress made over the last few years in high speed linear circuits has hinged not only on the development of IC processes but also on innovative circuit topologies.

The evolution of high speed processes using amplifier bandwidth as a function of supply current as a figure of merit is shown in Figure 1-96. (In the case of duals, triples, and quads, the current per amplifier is used.) Analog Devices BiFET process, which produced the AD712 (3 MHz bandwidth, 3 mA current) yields about 1 MHz per mA.

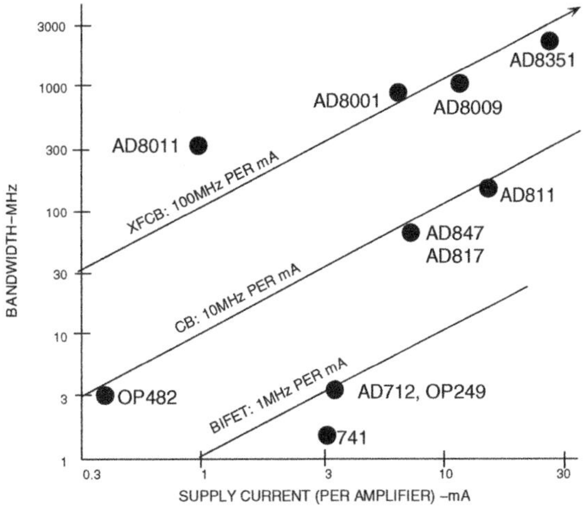

Figure 1-96: Amplifier bandwidth versus supply current for Analog Devices' processes

The CB (Complementary Bipolar) process (AD817, AD847, AD811, and so forth) yields about 10 MHz/mA of supply current. The f_ts of the CB process PNP transistors are about 700 MHz, and the NPNs about 900 MHz. The CB process at Analog Devices was introduced in 1985.

The next complementary bipolar process from Analog Devices was a high speed dielectrically isolated process called "XFCB" (eXtra Fast Complementary Bipolar) which was introduced in 1992. This process yields 3 GHz PNPs and 5 GHz matching NPNs, and coupled with innovative circuit topologies allows op amps to achieve new levels of cost-effective performance at astonishing low quiescent currents. The approximate figure of merit for this process is typically 100 MHz/mA, although the AD8011 op amp is capable of 300 MHz bandwidth on 1 mA of supply current due to its unique two-stage current-feedback architecture described later in this section.

Even faster CB processes have been developed at Analog Devices for low voltage supply products such as "XFCB 1.5" (5 GHz PNP, 9 GHz NPN), and "XFCB 2" (9 GHZ PNP, 16 GHz NPN). The AD8351 differential low distortion RF amplifier (shown on Figure 1-96) is fabricated on "XFCB 1.5" and has a bandwidth of 2 GHz for a gain of 12 dB. It is expected that newer complementary bipolar processes will be optimized for higher f_ts.

In order to select intelligently the correct high speed op amp for a given application, an understanding of the various op amp topologies as well as the trade-offs between them is required. The two most widely used topologies are voltage feedback (VFB) and current feedback (CFB). An overview of these topologies has been presented in a previous section, but the following discussion treats the frequency-related aspects of the two topologies in considerably more detail.

Voltage Feedback (VFB) Op Amps

A voltage feedback (VFB) op amp is distinguished from a current feedback (CFB) op amp by circuit topology. The VFB op amp is certainly the most popular in low frequency applications, but the CFB op amp has some advantages at high frequencies. We will discuss CFB in detail later, but first the more traditional VFB architecture.

Early IC voltage feedback op amps were made on "all NPN" processes. These processes were optimized for NPN transistors—the "lateral" PNP transistors had relatively poor performance. Some examples of these early VFB op amps using these poor quality PNPs include the 709, the LM101 and the 741 (see Chapter 8: "Op Amp History").

Lateral PNPs were generally used only as current sources, level shifters, or for other noncritical functions. A simplified diagram of a typical VFB op amp manufactur ed on such a process is shown in Figure 1-97.

The input stage is a differential pair (sometimes called a *long-tailed pair*) consisting of either a bipolar pair (Q1, Q2) or a FET pair. This "g_m" (transconductance) stage converts the small-signal differential input

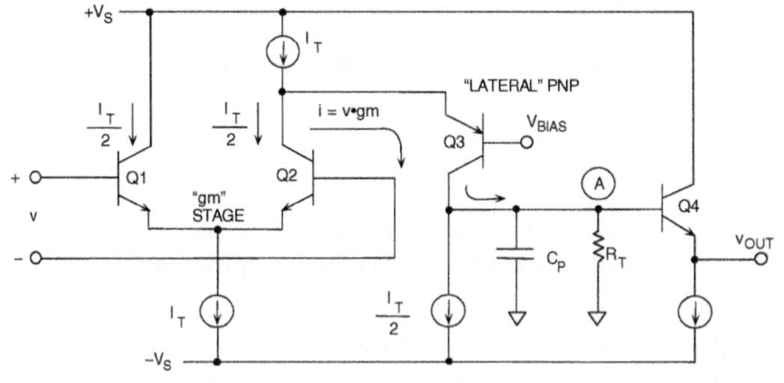

$$g_m = I_C \cdot \frac{q}{kT} \qquad V_{OUT} = \frac{i}{j\omega C_P} = v \cdot \frac{g_m}{j\omega C_P} \text{ @ HF}$$

Figure 1-97: Voltage feedback (VFB) op amp designed on an "all NPN" IC process

voltage, v, into a current, i, and its transfer function is measured in units of conductance, 1/Ω, (or mhos). The small-signal emitter resistance, r_e, is approximately equal to the reciprocal of the small-signal g_m.

The formula for the small-signal g_m of a single bipolar transistor is given by the following equation:

$$g_m = \frac{1}{r_e} = \frac{q}{kT}(I_C) = \frac{q}{kT}\left(\frac{I_T}{2}\right), \text{or} \qquad \text{Eq. 1-32}$$

$$g_m \approx \left(\frac{1}{26\text{mV}}\right)\left(\frac{I_T}{2}\right). \qquad \text{Eq. 1-33}$$

where I_T is the differential pair tail current, I_C is the collector quiescent bias current ($I_C = I_T/2$), q is the electron charge, k is Boltzmann's constant, and T is absolute temperature. At 25°C, $V_T = kT/q = 26$ mV (often called the *thermal voltage*, V_T).

As we will see shortly, the amplifier unity gain-bandwidth product, f_u, is equal to $g_m/2\pi C_P$, where the capacitance C_P is used to set the dominant pole frequency. For this reason, the tail current, I_T, is made proportional to absolute temperature (PTAT). This current tracks the variation in r_e with temperature thereby making g_m independent of temperature. It is relatively easy to make C_P reasonably constant over temperature.

The Q2 collector output of the g_m stage drives the emitter of a lateral PNP transistor (Q3). It is important to note that Q3 is not used to amplify the signal, only to level shift, i.e., the signal current variation in the collector of Q2 appears at the collector of Q3. The collector current of Q3 develops a voltage across high impedance node A, and C_P sets the dominant pole of the amplifier. Emitter follower Q4 provides a low impedance output.

The effective load at the high impedance node A can be represented by a resistance, R_T, in parallel with the dominant pole capacitance, C_P. The small-signal output voltage, v_{out}, is equal to the small-signal current, i, multiplied by the impedance of the parallel combination of R_T and C_P.

Figure 1-98 shows a simple model for the single-stage amplifier and the corresponding Bode plot. The Bode plot is conveniently constructed on a log-log scale.

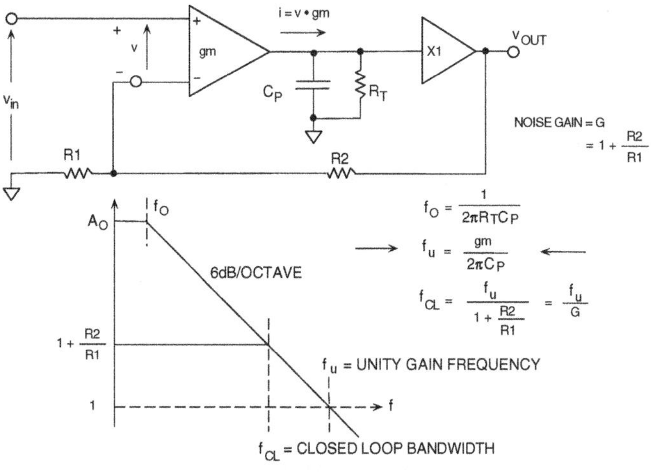

Figure 1-98: Model and Bode plot for a VFB op amp

The low frequency breakpoint, f_O, is given by:

$$f_o = \frac{1}{2\pi R_T C_P} \qquad \text{Eq. 1-34}$$

Note that the high frequency response is determined solely by g_m and C_P:

$$v_{out} = v \cdot \frac{g_m}{j\omega C_P} \qquad \text{Eq. 1-35}$$

The unity gain bandwidth frequency, f_u, occurs where $|v_{out}| = |v|$. Letting $\omega = 2\pi f_u$ and $|v_{out}| = |v|$, Eq. 1-35 can be solved for f_u,

$$f_u = \frac{g_m}{2\pi C_P} \qquad \text{Eq. 1-36}$$

We can use feedback theory to derive the closed-loop relationship between the circuit's signal input voltage, v_{in}, and its output voltage, v_{out}:

$$\frac{v_{out}}{v_{in}} = \frac{1+\dfrac{R2}{R1}}{1+\dfrac{j\omega C_P}{g_m}\left(1+\dfrac{R2}{R1}\right)} \qquad \text{Eq. 1-37}$$

At the op amp 3 dB closed-loop bandwidth frequency, f_{cl}, the following is true:

$$\frac{2\pi f_{cl} C_P}{g_m}\left(1+\frac{R2}{R1}\right) = 1, \text{ and hence} \qquad \text{Eq. 1-38}$$

$$f_{cl} = \frac{g_m}{2\pi C_P}\left(\frac{1}{1+\dfrac{R2}{R1}}\right), \text{ or} \qquad \text{Eq. 1-39}$$

$$f_{cl} = \frac{f_u}{1+\dfrac{R2}{R1}} \qquad \text{Eq. 1-40}$$

This demonstrates the fundamental property of VFB op amps: *The closed-loop bandwidth multiplied by the closed-loop gain is a constant*, i.e., the VFB op amp exhibits a constant gain-bandwidth product over most of the usable frequency range.

As noted previously, some VFB op amps (called *decompensated*) are not stable at unity gain, but designed to be operated at some minimum (higher) amount of closed-loop gain. However, even for these op amps, the gain bandwidth product is still relatively constant over the region of stability.

Now, consider the following typical example: $I_T = 100\ \mu A$, $C_P = 2$ pF. We find that:

$$g_m = \frac{I_T/2}{V_T} = \frac{50\,\mu A}{26\,mV} = \frac{1}{520\,\Omega} \qquad \text{Eq. 1-41}$$

$$f_u = \frac{g_m}{2\pi C_P} = \frac{1}{2\pi(520)(2\cdot 10^{-12})} = 153\,\text{MHz} \qquad \text{Eq. 1-42}$$

Now, we must consider the large-signal response of the circuit. The slew-rate, SR, is simply the total available charging current, $I_T/2$, divided by the dominant pole capacitance, C_P. For the example under consideration,

$$I = C\frac{dv}{dt}, \frac{dv}{dt} = SR,\ SR = \frac{I}{C} \qquad \text{Eq. 1-43}$$

$$SR = \frac{I_T/2}{C_P} = \frac{50\,\mu A}{2\,pF} = 25\,V/\mu s \qquad \text{Eq. 1-44}$$

The full-power bandwidth (FPBW) of the op amp can now be calculated from the formula:

$$FPBW = \frac{SR}{2\pi A} = \frac{25\,V/\mu s}{2\pi\cdot 1V} = 4\,\text{MHz} \qquad \text{Eq. 1-45}$$

where A is the peak amplitude of the output signal. If we assume a 2 V peak-to-peak output sinewave (certainly a reasonable assumption for high speed applications), then we obtain a FPBW of only 4 MHz, even though the small-signal unity gain bandwidth product is 153 MHz. For a 2 V p-p output sinewave, distortion will begin to occur much lower than the actual FPBW frequency. We must increase the SR by a factor of about 40 in order for the FPBW to equal 153 MHz. The only way to do this is to increase the tail current, I_T, of the input differential pair by the same factor. This implies a bias current of 4 mA in order to achieve a FPBW of 160 MHz. We are assuming that C_P is a fixed value of 2 pF and cannot be lowered by design. These calculations are summarized in Figure 1-99.

- Assume that $I_T = 100\mu A$, $Cp = 2pF$

  $$g_m = \frac{I_C}{V_T} = \frac{50\mu A}{26mV} = \frac{1}{520\Omega}$$

  $$f_u = \frac{g_m}{2\pi\ Cp} = 153MHz$$

- Slew Rate = SR =

  But for 2V peak-peak output (A = 1V)

  $$FPBW = \frac{SR}{2\pi\ A} = 4MHz$$

- Must increase I_T to 4mA to get FPBW = 160MHz
- Reduce g_m by adding emitter degeneration resistors

Figure 1-99: VFB op amp bandwidth and slew rate calculations

In practice, the FPBW of the op amp should be approximately 5 to 10 times the maximum output frequency in order to achieve acceptable distortion performance (typically 55 dBc–80 dBc @ 5 MHz to 20 MHz, but actual system requirements vary widely).

Notice, however, that increasing the tail current causes a proportional increase in g_m and hence f_u. In order to prevent possible instability due to the large increase in f_u, g_m can be reduced by inserting resistors in series with the emitters of Q1 and Q2 (this technique, called *emitter degeneration*, also serves to linearize the g_m transfer function and thus also lowers distortion).

This analysis points out that a major inefficiency of conventional bipolar voltage feedback op amps is their inability to achieve high slew rates without proportional increases in quiescent current (assuming that C_P is fixed, and has a reasonable minimum value of 2 pF or 3 pF).

This of course is not meant to say that high speed op amps designed using this architecture are deficient, just that circuit design techniques are available that allow equivalent performance at much lower quiescent currents. This is extremely important in portable battery-operated equipment where every milliwatt of power dissipation is critical.

VFB Op Amps Designed on Complementary Bipolar Processes

With the advent of complementary bipolar (CB) processes having high quality PNP transistors as well as NPNs, VFB op amp configurations such as the one shown in the simplified diagram in Figure 1-100 became popular.

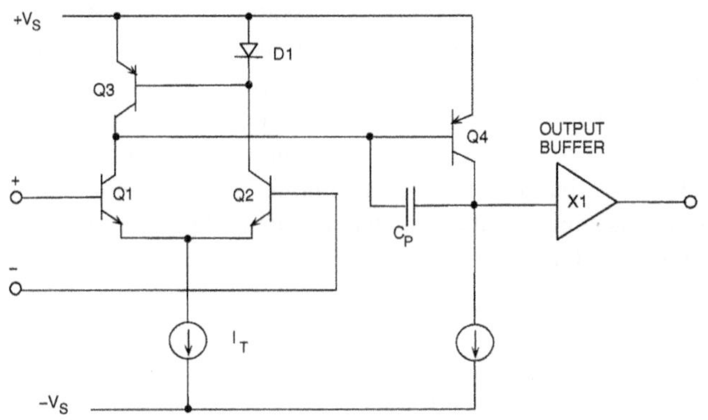

Figure 1-100: VFB op amp using two gain stages

Notice that the input differential pair (Q1, Q2) is loaded by a current mirror (Q3 and D1). We show D1 as a diode for simplicity, but it is actually a diode-connected PNP transistor (matched to Q3) with the base and collector connected to each other. This simplification will be used in many of the circuit diagrams to follow in this section. The common emitter transistor, Q4, provides a *second* voltage gain stage.

Since the PNP transistors are fabricated on a complementary bipolar process, they are high quality and matched to the NPNs, and therefore suitable for voltage gain. The dominant pole of the Figure 1-100 amplifier is set by C_P, and the combination of the gain stage, Q4, and local feedback capacitor C_p is often referred to as a *Miller Integrator*. The unity gain output buffer is usually a complementary emitter follower.

A model for this two-stage VFB op amp is shown in Figure 1-101. Notice that the unity gain bandwidth frequency, f_u, is still determined by the input stage g_m and the dominant pole capacitance, C_P. The second gain stage increases the dc open-loop gain, but maximum slew rate is still limited by the input stage tail current as: $SR = I_T/C_P$.

A two-stage amplifier topology such as this is widely used throughout the IC industry in VFB op amps, both precision and high speed. It can be recalled that a similar topology with a dual FET input stage was used in the early high speed, fast settling FET modular op amps (see Chapter 8: "Op Amp History").

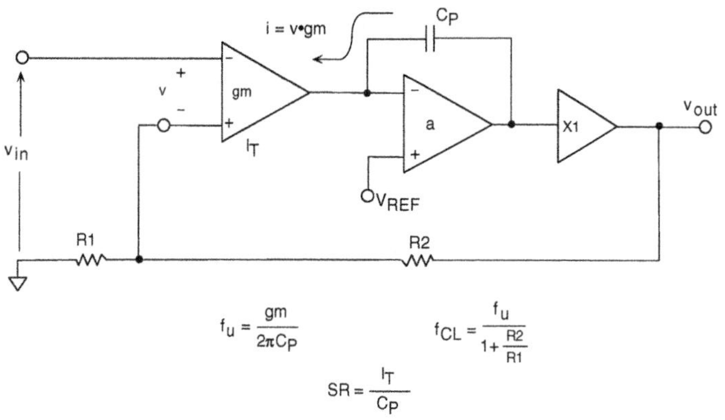

Figure 1-101: Model for two stage VFB op amp

Another popular VFB op amp architecture is the *folded cascode* as shown in Figure 1-102. An industry-standard video amplifier family (the AD847) is based on this architecture. This circuit also takes advantage of the fast PNPs available on a CB process. The differential signal currents in the collectors of Q1 and Q2 are fed to the emitters of a PNP cascode transistor pair (hence the term *folded cascode*). The collectors of Q3 and Q4 are loaded with the current mirror, D1 and Q5, and voltage gain is developed at the Q4–Q5 node. This single-stage architecture uses the junction capacitance at the high impedance node for compensation (C_{STRAY}). Some variations of the design bring this node to an external pin so that additional external capacitance can be added if desired.

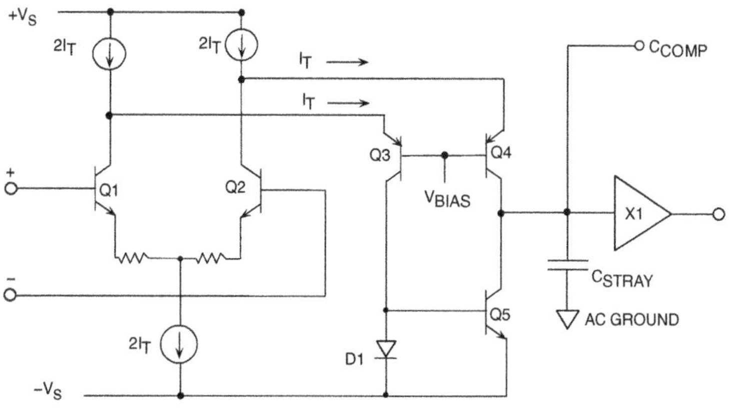

Figure 1-102: AD847 family folded cascode simplified circuit

With no emitter degeneration resistors in Q1 and Q2, and no additional external compensating capacitance, this circuit is only stable for high closed-loop gains. However, unity gain compensated versions of this family are available with the appropriate amount of emitter degeneration.

The availability of JFETs on a CB process allows not only low input bias current but also improvements in the slew rate trade-off, which must be made between g_m and I_T found in bipolar input stages. Figure 1-103 shows a simplified diagram of the AD845 16 MHz op amp. JFETs have a much lower g_m per mA of tail current than a bipolar transistor. This lower g_m of the FET allows the input tail current (hence the slew rate) to be increased, without having to increase C_P to maintain stability.

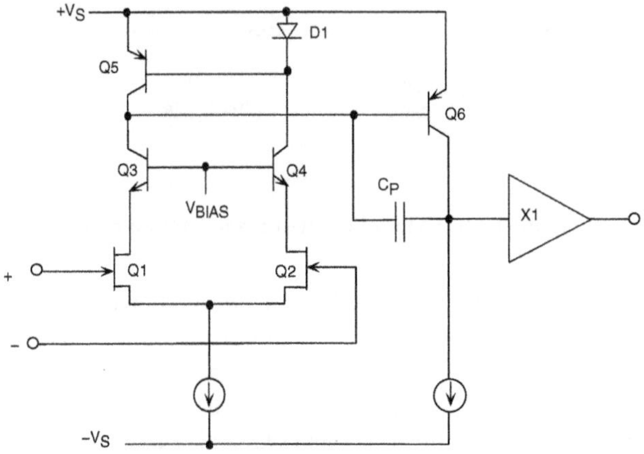

Figure 1-103: AD845 BiFET 16 MHz op amp simplified circuit

The unusual thing about this seemingly poor performance of the JFET is that it is exactly what is needed for a fast, high SR input stage. For a typical JFET, the value of gm is approximately $I_s/1$ V (I_s is the source current), rather than $I_c/26$ mV for a bipolar transistor, i.e., the FET gm is about 40 times lower. This allows much higher tail currents (and higher slew rates) for a given g_m when JFETs are used as the input stage.

A New VFB Op Amp Architecture for "Current-on-Demand" Performance, Lower Power, and Improved Slew Rate

Until recently, op amp designers had to make the above trade-offs between the input g_m stage quiescent current and the slew rate and distortion performance. ADI has patented a circuit core that supplies *current-on-demand*, to charge and discharge the dominant pole capacitor, C_P, while allowing the quiescent current to be small. The additional current is proportional to the fast slewing input signal and adds to the quiescent current.

A simplified diagram of the basic core cell is shown in Figure 1-104. The *quad-core* (g_m stage) consists of transistors Q1, Q2, Q3, and Q4 with their emitters connected together as shown. Consider a positive step voltage on the inverting input. This voltage produces a proportional current in Q1 that is mirrored into C_{P1} by Q5. The current through Q1 also flows through Q4 and C_{P2}.

At the dynamic range limit, Q2 and Q3 are correspondingly turned off. Notice that the charging and discharging current for C_{P1} and C_{P2} is not limited by the quad core bias current. In practice, however, small current-limiting resistors are required forming an "H" resistor network as shown. Q7 and Q8 form the second gain stage (driven differentially from the collectors of Q5 and Q6), and the output is buffered by a unity-gain complementary emitter follower (X1).

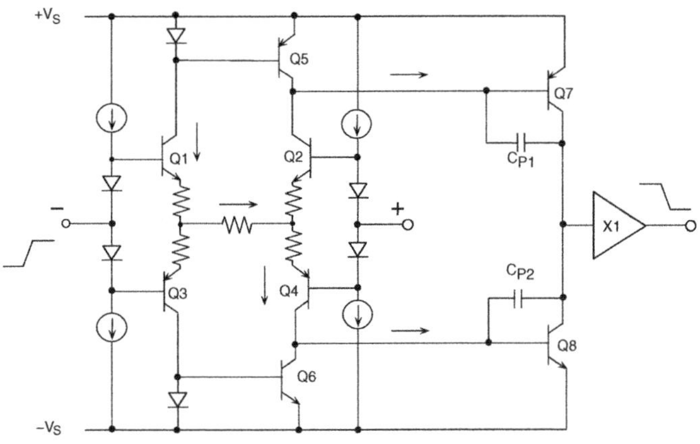

Figure 1-104: "Quad-Core" VFB g_m stage for current-on-demand

The quad core configuration is patented (see Reference 1), as well as the circuits that establish the quiescent bias currents (not shown in Figure 1-104). A number of new VFB op amps using this proprietary configuration have been released and have unsurpassed high frequency low distortion performance, bandwidth, and slew rate at the indicated quiescent current levels as shown in Figure 1-105.

LISTED IN ORDER OF DECREASING SUPPLY CURRENT

PART #	I_{SY} / AMP	BANDWIDTH	SLEWRATE	DISTORTION
AD9631/32 (1)	17mA	320MHz	1300V/µs	–72dBc@20MHz
AD8074/75 (3)	8mA	600MHz	1600V/µs	–62dBc@20MHz
AD8047/48 (1)	5.8mA	250MHz	750V/µs	–66dBc@5MHz
AD8041 (1)	5.2mA	160MHz	160V/µs	–69dBc@10MHz
AD8042 (2)	5.2mA	160MHz	200V/µs	–64dBc@10MHz
AD8044 (3)	2.8mA	150MHz	150V/µs	–75dBc@5MHz
AD8039 (2)	1.5mA	300MHz	425V/µs	–65dBc@5MHz
AD8031 (1)	0.75mA	80MHz	30V/µs	–62dBc@1MHz
AD8032 (2)	0.75mA	80MHz	30V/µs	–72dBc@1MHz

Number in ( ) indicates single, dual, triple, or quad

Figure 1-105: High speed VFB op amps

The AD9631, AD8074, and AD8047 are optimized for a gain of +1, while the AD9632, AD8075, and AD8048 for a gain of +2.

The same quad-core architecture is used as the *second* stage of the AD8041 rail-to-rail output, zero-volt input single-supply op amp. The input stage is a differential PNP pair which allows the input common-mode signal to go about 200 mV below the negative supply rail. The AD8042 and AD8044 are dual and quad versions of the AD8041.

Current Feedback (CFB) Op Amps

We will now examine in more detail the current feedback (CFB) op amp topology which is very popular in high speed op amps. As mentioned previously, the circuit concepts were introduced decades ago; however, modern high speed complementary bipolar processes are required to take full advantage of the architecture.

It has long been known that in bipolar transistor circuits, currents can be switched faster than voltages, other things being equal. This forms the basis of nonsaturating emitter-coupled logic (ECL) and devices such as current-output DACs. Maintaining low impedances at the current switching nodes helps to minimize the effects of stray capacitance, one of the largest detriments to high speed operation. The current mirror is a good example of how currents can be switched with a minimum amount of delay.

The current feedback op amp topology is simply an application of these fundamental principles of current steering. A simplified CFB op amp is shown in Figure 1-106. The noninverting input is high impedance and is buffered directly to the inverting input through the complementary emitter follower buffers Q1 and Q2. Note that the inverting input impedance is very low (typically 10 Ω to 100 Ω), because of the low emitter resistance (ideally, would be zero). This is a fundamental difference between a CFB and a VFB op amp, and also a feature that gives the CFB op amp some unique advantages.

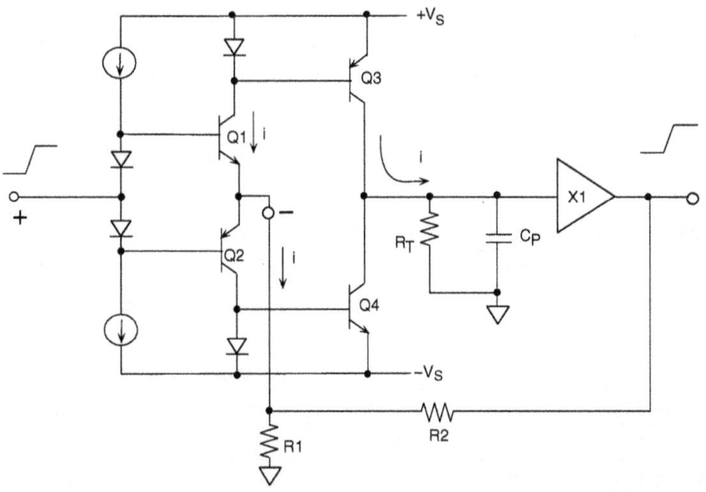

Figure 1-106: Simplified current feedback (CFB) op amp

The collector outputs of Q1 and Q2 drive current mirrors, which mirror the inverting input current to the high impedance node, modeled by R_T and C_P. The high impedance node is buffered by a complementary unity gain emitter follower. Feedback from the output to the inverting input acts to force the inverting input *current* to zero, hence the term *Current Feedback.* Note that in a ideal case, for zero inverting input impedance, no small-signal voltage can exist at this node, only small-signal current.

Now, consider a positive step voltage applied to the noninverting input of the CFB op amp. Q1 immediately sources a proportional current into the external feedback resistors creating an *error* current, which is mirrored to the high impedance node by Q3. The voltage developed at the high impedance node is equal to this current multiplied by the equivalent impedance. This is where the term *transimpedance op amp* originated, since the transfer function is an impedance, rather than a unitless voltage ratio as in a traditional VFB op amp.

Note also that the error current delivered to the high impedance node is not limited by the input stage tail current. In other words, unlike a conventional VFB op amp, *there is no slew rate limitation in an ideal CFB op amp*. The current mirrors supply *current-on-demand* from the power supplies. The negative feedback loop then forces the output voltage to a value that reduces the inverting input error current to zero.

The model for a CFB op amp is shown in Figure 1-107, along with the corresponding Bode plot. The Bode plot is plotted on a log-log scale, and the open-loop gain is expressed as a transimpedance, T(s), with units of ohms.

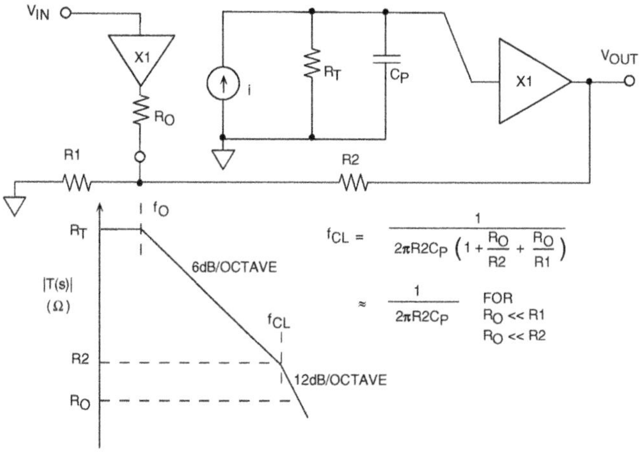

Figure 1-107: CFB op amp model and Bode plot

The finite output impedance of the input buffer is modeled by R_O. The input error current is *i*. By applying the principles of negative feedback, we can derive the expression for the op amp transfer function:

$$\frac{V_{out}}{V_{in}} = \frac{1+\frac{R2}{R1}}{1+j\omega C_P R2\left(1+\frac{Ro}{R2}+\frac{Ro}{R1}\right)} \qquad \text{Eq. 1-46}$$

At the op amp 3 dB closed-loop bandwidth frequency, f_{cl}, the following is true:

$$2\pi f_{cl} C_P R2\left(1+\frac{Ro}{R2}+\frac{Ro}{R1}\right)=1 \qquad \text{Eq. 1-47}$$

$$f_{cl} = \frac{1}{2\pi C_P R2\left(1+\frac{Ro}{R2}+\frac{Ro}{R1}\right)} \qquad \text{Eq. 1-48}$$

For the condition R_O << R2 and R1, the equation simply reduces to:

$$f_{cl} = \frac{1}{2\pi C_P R2} \qquad \text{Eq. 1-49}$$

Examination of this equation quickly reveals that *the closed-loop bandwidth of a CFB op amp is determined by the internal dominant pole capacitor, C_P, and the external feedback resistor, R2, and is independent of the gain-setting resistor, R1.* This ability to maintain constant bandwidth independent of gain makes CFB op amps ideally suited for wideband programmable gain amplifiers.

Because the closed-loop bandwidth is inversely proportional to the external feedback resistor, R2, a CFB op amp is usually optimized for a specific R2. Increasing R2 from its optimum value lowers the bandwidth, and decreasing it may lead to oscillation and instability because of high frequency parasitic poles.

The frequency response of the AD8011 CFB op amp is shown in Figure 1-108 for various closed-loop values of gain (+1, +2, and +10). Note that even at a gain of +10, the closed-loop bandwidth is still greater than 100 MHz. The peaking that occurs at a gain of +1 is typical of wideband CFB op amps used in the noninverting mode, and is due primarily to stray capacitance at the inverting input. This peaking can be reduced by sacrificing bandwidth, by using a slightly larger feedback resistor.

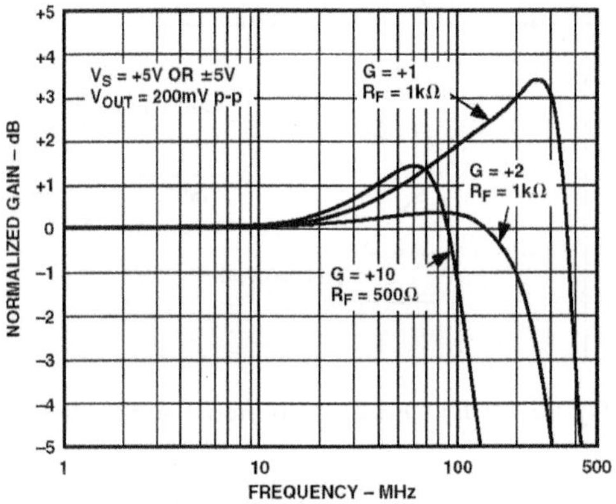

Figure 1-108: AD8011 frequency response, G = +1, +2, +10

The AD8011 CFB op amp (introduced in 1995) still represents state-of-the-art performance, and key specifications are shown in Figure 1-109.

- 1mA Power Supply Current (+5V or ±5V)
- 300MHz Bandwidth (G = +1)
- 2000 V/µs Slew Rate
- 29ns Settling Time to 0.1%
- Video Specifications (G = +2)
  Differential Gain Error 0.02%
  Differential Phase Error 0.06°
  25MHz 0.1dB Bandwidth
- Distortion
  –70dBc @ 5MHz
  –62dBc @ 20MHz
- Fully Specified for ±5V or +5V Operation

Figure 1-109: AD8011 key specifications

Traditional current feedback op amps have been limited to a single gain stage, using current-mirrors. The AD8011 (and also others in this family) unlike traditional CFB op amps, use a *two-stage gain configuration*, as shown in Figure 1-110.

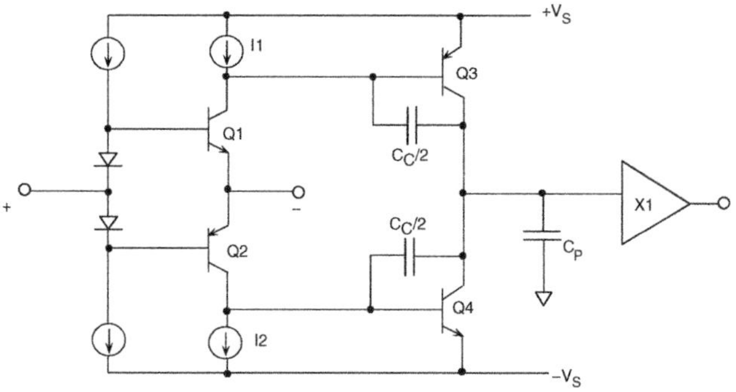

Figure 1-110: Simplified two-stage current feedback op amp

Until the advent of the AD8011, fully complementary two-gain stage CFB op amps had been impractical because of their high power dissipation. The AD8011 employs a patented (see Reference 2) second gain stage consisting of a pair of complementary amplifiers (Q3 and Q4). Note that they are not connected as current mirrors but as grounded-emitter gain stages. The detailed design of current sources (I1 and I2), and their respective bias circuits are the key to the success of the two-stage CFB circuit; they keep the amplifier's quiescent power low, yet are capable of supplying *current-on-demand* for wide current excursions required during fast slewing.

A further advantage of the two-stage amplifier is the higher overall bandwidth (for the same power), which means lower signal distortion and the ability to drive heavier external loads.

Thus far, we have learned several key features of CFB op amps. The most important is that for a given complementary bipolar IC process, *CFB generally yields higher FPBW (hence lower distortion) than VFB for the same amount of quiescent supply current.* This is because there is practically no slew-rate limiting in CFB. Because of this, the full power bandwidth and the small signal bandwidth are approximately the same.

The second important feature is that the *inverting input impedance of a CFB op amp is very low.* This is advantageous when using the op amp in the inverting mode as an I/V converter, because there is less sensitivity to inverting input capacitance than with VFB.

The third feature is that *the closed-loop bandwidth of a CFB op amp is determined by the value of the internal Cp capacitor and the external feedback resistor R2 and is relatively independent of the gain-setting resistor R1.*

The performance for a selected group of current feedback op amps is shown in Figure 1-111. Note that the op amps are listed in order of decreasing power supply current.

LISTED IN ORDER OF DECREASING SUPPLY CURRENT

PART #	I_{SY}/AMP	BANDWIDTH	SLEW RATE	DISTORTION
AD8009 (1)	14mA	1000MHz	5500V/μs	–80dBc@5MHz
AD8023 (3)	10mA	250MHz	1200V/μs	ΔG = 0.06%, Δφ = 0.02°
AD8001 (1)	5.0mA	600MHz	1200V/μs	–65dBc@5MHz
AD8002 (2)	5.0mA	600MHz	1200V/μs	–65dBc@5MHz
AD8004 (4)	3.5mA	250MHz	3000V/μs	–78dBc@5MHz
AD8013 (3)	4.0mA	140MHz	1000V/μs	ΔG = 0.02%, Δφ = 0.06°
AD8072 (2)	3.5mA	100MHz	500V/μs	ΔG = 0.05%, Δφ = 0.1°
AD8073 (3)	3.5mA	100MHz	500V/μs	ΔG = 0.05%, Δφ = 0.1°
AD8012 (2)	1.7mA	350MHz	2250V/μs	–66dBc@5MHz
AD8014 (1)	1.2mA	400MHz	4000V/μs	–70dBc@5MHz
AD8011 (1)	1.0mA	300MHz	5500V/μs	–70dBc@5MHz
AD8005 (1)	0.4mA	270MHz	1500/μs	–53dBc@5MHz

Number in ( ) Indicates Single, Dual, Triple, or Quad

Figure 1-111: Performance of selected CFB op amps

Figure 1-112 summarizes the general characteristics of CFB op amps.

- CFB yields higher FPBW and lower distortion than VFB for the same process and power dissipation
- Inverting input impedance of a CFB op amp is low, noninverting input impedance is high
- Closed-loop bandwidth of a CFB op amp is determined by the internal dominant-pole capacitance and the external feedback resistor, independent of the gain-setting resistor

Figure 1-112: Summary: CFB op amps

Effects of Feedback Capacitance in Op Amps

It is quite common to use a capacitor in the feedback loop of a VFB op amp, to shape the frequency response as in a simple single-pole lowpass filter shown in Figure 1-113A. The resulting noise gain is plotted on a Bode plot to analyze stability and phase margin. Stability of the system is determined by the net slope of the noise gain and the open-loop gain where they intersect.

For unconditional stability, the noise gain plot must intersect the open-loop response with a net slope of less than 12 dB/octave. In this case, the net slope where they intersect is 6 dB/octave, indicating a stable condition. Notice for the case drawn in Figure 1-113A, the second pole in the frequency response occurs at a considerably higher frequency than f_u.

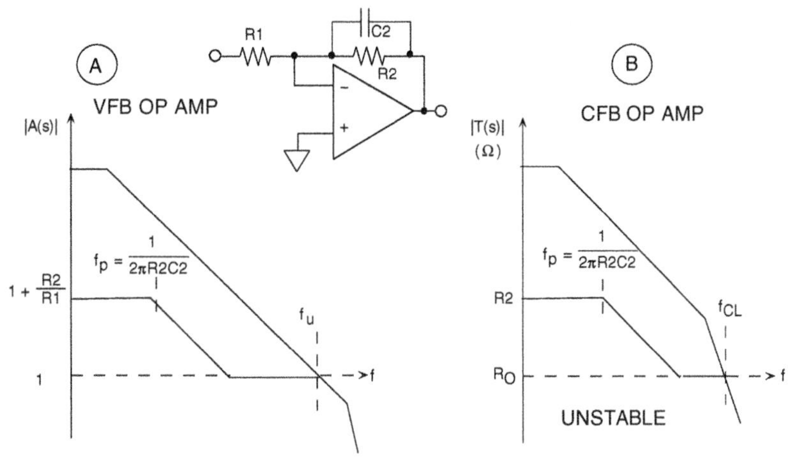

Figure 1-113: Noise gain stability analysis for VFB and CFB op amps with feedback capacitor

In the case of the CFB op amp (Figure 1-113B), the same analysis is used, except that the open-loop transimpedance gain, T(s), is used to construct the Bode plot.

The definition of *noise gain* (for the purposes of stability analysis) for a CFB op amp, however, must be redefined in terms of a *current* noise source attached to the inverting input as shown in Figure 1-114. This current is reflected to the output by an impedance, which we define to be the "current noise gain" of a CFB op amp:

$$\text{"CURRENT NOISE GAIN"} \equiv R_O + Z2\left(1+\frac{R_O}{Z1}\right) \qquad \text{Eq. 1-50}$$

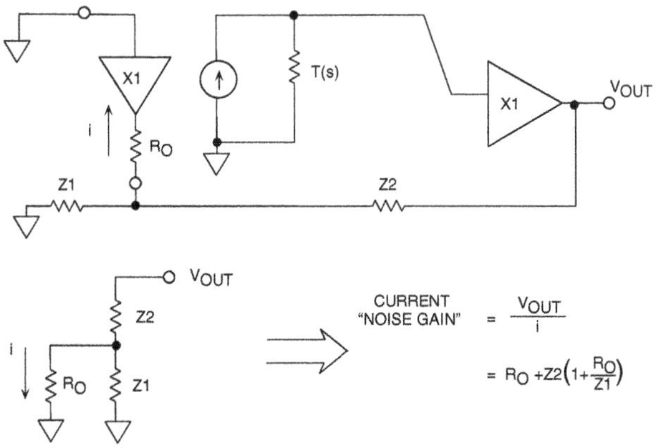

Figure 1-114: Current "noise gain" definition for CFB op amp for use in stability analysis

Now, return to Figure 1-113B, and observe the CFB *current noise gain* plot. At low frequencies, the CFB current noise gain is simply R2 (making the assumption that R_O is much less than Z1 or Z2. The first pole is determined by R2 and C2. As the frequency continues to increase, C2 becomes a short circuit, and all the inverting input current flows through R_O (again refer to Figure 1-114).

A CFB op amp is normally optimized for best performance for a fixed feedback resistor, R2. Additional poles in the transimpedance gain, T(s), occur at frequencies above the closed-loop bandwidth, f_{cl}, (set by R2). Note that the intersection of the CFB current noise gain with the open-loop T(s) occurs where the slope of the T(s) function is 12 dB/octave. This indicates instability and possible oscillation.

It is for this reason that *CFB op amps are not suitable in configurations that require capacitance in the feedback loop*, such as simple active integrators or low-pass filters.

They can, however, be used in certain active filters, such as the Sallen-Key configuration shown in Figure 1-115, which do not require capacitance in the feedback network.

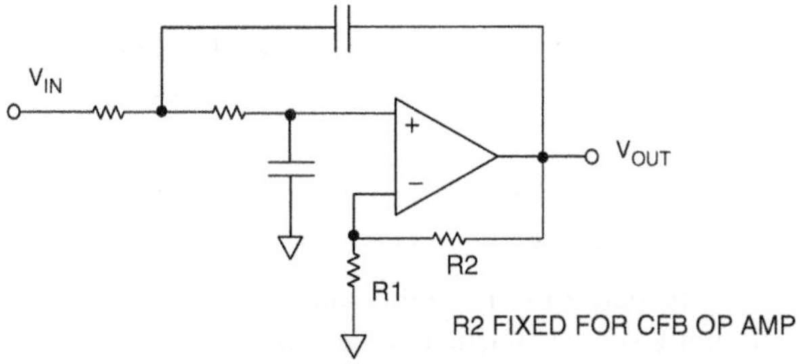

Figure 1-115: The Sallen-Key filter configuration

On the other hand, VFB op amps do make very flexible active filters. A multiple feedback 20 MHz low-pass filter example using an AD8048 op amp is shown in Figure 1-116.

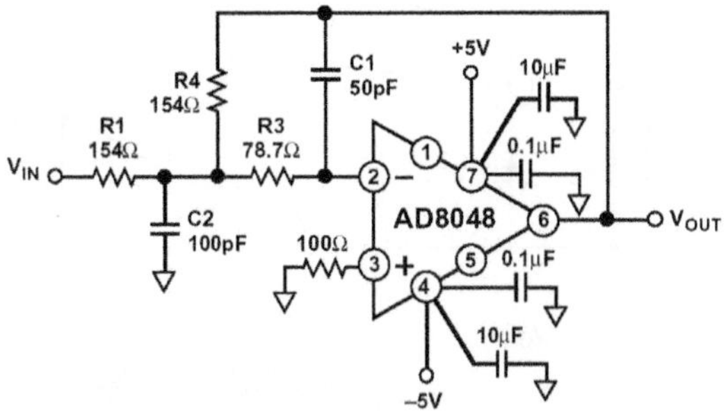

Figure 1-116: Multiple feedback 20 MHz low-pass filter using the AD8048 VFB op amp

In general, an active filter amplifier should have a bandwidth that is at least 10 times the bandwidth of the filter, if problems due to phase shift of the amplifier are to be avoided. (The AD8048 has a bandwidth of over 200 MHz in this configuration.)

Design details of the filter design can be found on the AD8048 data sheet. Further discussions on active filter design are included in Chapter 5 of this book.

High Speed Current-to-Voltage Converters, and the Effects of Inverting Input Capacitance

Fast op amps are useful as current-to-voltage converters in such applications as high speed photodiode preamplifiers and current-output DAC buffers. A typical application using a VFB op amp as an I/V converter is shown in Figure 1-117.

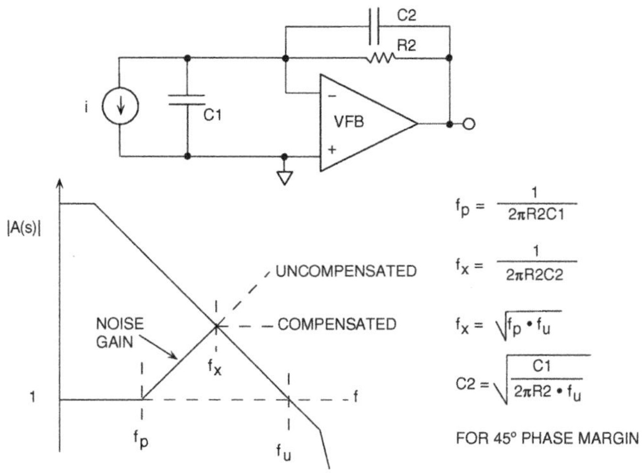

Figure 1-117: Compensating for input capacitance in a current-to-voltage converter using VFB op amp

The net input capacitance, C1, forms a pole at a frequency f_P in the noise gain transfer function as shown in the Bode plot, and is given by:

$$f_P = \frac{1}{2\pi R2C1} \qquad \text{Eq. 1-51}$$

If left uncompensated, the phase shift at the frequency of intersection, f_x, will cause instability and oscillation. Introducing a zero at f_x by adding feedback capacitor C2 stabilizes the circuit and yields a phase margin of about 45°.

The location of the zero is given by:

$$f_X = \frac{1}{2\pi R2C2} \qquad \text{Eq. 1-52}$$

Although the addition of C2 actually decreases the pole frequency slightly, this effect is negligible if C2 << C1. The frequency f_x is the geometric mean of f_p and the unity-gain bandwidth frequency of the op amp, f_u,

$$f_X = \sqrt{f_P \cdot f_u} \qquad \text{Eq. 1-53}$$

Combining Eq. 1-52 and Eq. 1-53 and solving for C2 yields:

$$C2 = \sqrt{\frac{C1}{2\pi R2 \cdot f_u}} \qquad \text{Eq. 1-54}$$

This value of C2 will yield a phase margin of about 45°. Increasing the capacitor by a factor of 2 increases the phase margin to about 65° (see Reference 3).

In practice, the optimum value of C2 may be optimized experimentally by varying it slightly, to optimize the output pulse response.

A similar analysis can be applied to a CFB op amp as shown in Figure 1-118. In this case, however, the low inverting input impedance, R_O, greatly reduces the sensitivity to input capacitance. In fact, an ideal CFB with zero input impedance would be totally insensitive to any amount of input capacitance.

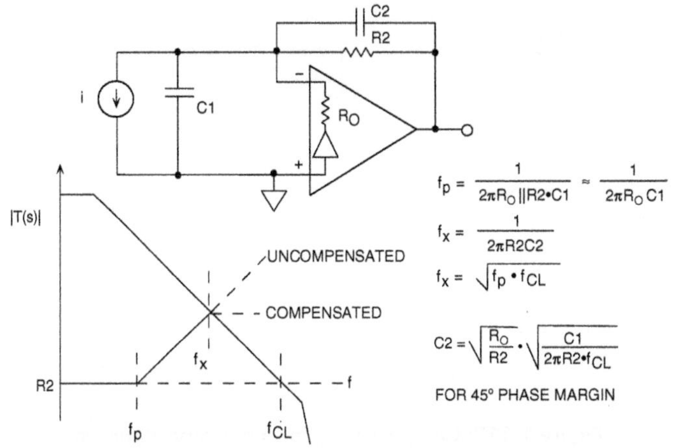

Figure 1-118: Current-to-voltage converter using a CFB op amp

The pole caused by C1 occurs at a frequency f_P:

$$f_P = \frac{1}{2\pi(R_O \parallel R2)C1} \approx \frac{1}{2\pi R_O C1} \qquad \text{Eq. 1-55}$$

This pole frequency will generally be much higher than the case for a VFB op amp, and the pole can be ignored completely if it occurs at a frequency greater than the closed-loop bandwidth of the op amp.

We next introduce a compensating zero at the frequency f_x by inserting the capacitor C2:

$$f_x = \frac{1}{2\pi R2C2} \qquad \text{Eq. 1-56}$$

As in the case for VFB, f_x is the geometric mean of f_p and f_{cl}:

$$f_x = \sqrt{f_P \cdot f_u} \qquad \text{Eq. 1-57}$$

Combining Eq. 1-56 and Eq. 1-57 and solving for C2 yields:

$$C2 = \sqrt{\frac{R_O}{R2}} \cdot \sqrt{\frac{C1}{2\pi R2 \cdot f_{cl}}} \qquad \text{Eq. 1-58}$$

There is a significant advantage in using a CFB op amp in this configuration as can be seen by comparing Eq. 1-58 with the similar equation for C2 required for a VFB op amp, Eq. 1-54. If the unity-gain bandwidth product of the VFB is equal to the closed-loop bandwidth of the CFB (at the optimum R2), then the size of the CFB compensation capacitor, C2, is reduced by a factor of $\sqrt{(R2/R_O)}$.

A comparison in an actual application is shown in Figure 1-119. The full scale output current of the DAC is 4 mA, the net capacitance at the inverting input of the op amp is 20 pF, and the feedback resistor is 500 Ω. In the case of the VFB op amp, the pole due to C1 occurs at 16 MHz. A compensating capacitor of 5.6 pF is required for 45° of phase margin, and the signal bandwidth is 57 MHz.

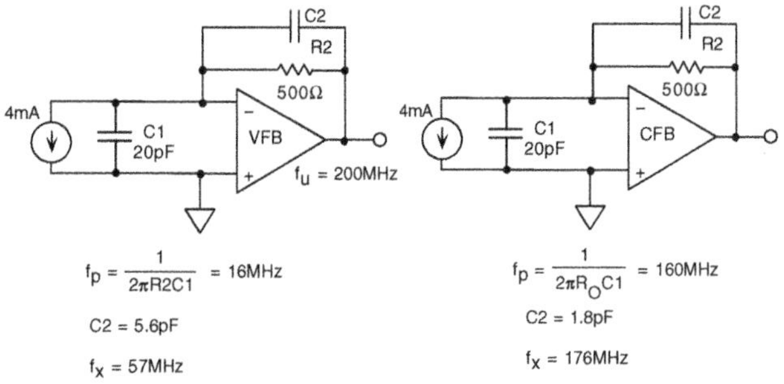

Figure 1-119: CFB op amp is relatively insensitive to input capacitance when used as an I/V converter

For the CFB op amp, however, because of the low inverting input impedance (R_O = 50 Ω), the pole occurs at 160 MHz, the required compensation capacitor is about 1.8 pF, and the corresponding signal bandwidth is 176 MHz. In practice, the pole frequency is so close to the closed-loop bandwidth of the op amp that it could probably be left uncompensated.

It should be noted that a CFB op amp's relative insensitivity to inverting input capacitance is when it is used in the inverting mode. In the noninverting mode, however, even a few picofarads of stray capacitance on the inverting input can cause significant gain peaking and potential instability.

Another advantage of the low inverting input impedance of the CFB op amp is when it is used as an I/V converter to buffer the output of a high speed current output DAC. When a step function current (or DAC switching glitch) is applied to the inverting input of a VFB op amp, it can produce a large voltage transient until the signal can propagate through the op amp to its output and negative feedback is regained. Back-to-back Schottky diodes are often used to limit this voltage swing as shown in Figure 1-120. These diodes must be low capacitance, small geometry devices because their capacitance adds to the total input capacitance.

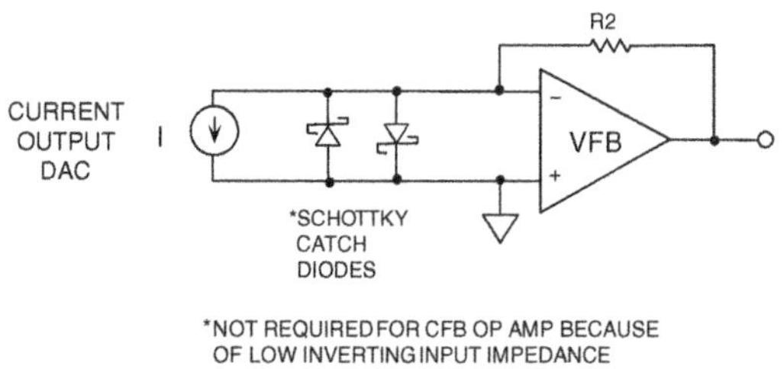

Figure 1-120: Low inverting input impedance of CFB op amp helps reduce effects of fast DAC transients

A CFB op amp, on the other hand, presents a low impedance (R_O) to fast switching currents even before the feedback loop is closed, thereby limiting the voltage excursion without the requirement of the external diodes. This greatly improves the settling time of the I/V converter.

Noise Comparisons between VFB and CFB Op Amps

In most applications of high speed op amps, it is generally the total output RMS noise that is of interest. Because of the high bandwidths involved, the chief contributor to the output RMS noise is therefore the white noise, and the 1/f noise is negligible.

Typical high speed op amps with bandwidths greater than 150 MHz or so, and bipolar VFB input stages, have input voltage noises ranging from about 2 nV to $20\ \text{nV}/\sqrt{\text{Hz}}$.

For a VFB op amp, the inverting and noninverting input current noise are typically equal, and almost always uncorrelated. Typical values for wideband VFB op amps range from $0.5\ \text{pA}/\sqrt{\text{Hz}}$ to $5\ \text{pA}/\sqrt{\text{Hz}}$. The input current noise of a bipolar input stage is increased when input bias-current compensation generators are added, because their current noise is not correlated, and therefore adds (in an RSS manner) to the intrinsic current noise of the bipolar stage. However, bias current compensation is rarely used in high speed op amps.

The input voltage noise in CFB op amps tends to be lower than for VFB op amps having the same approximate bandwidth. This is because the input stage in a CFB op amp is usually operated at a higher current, thereby reducing the emitter resistance and hence the voltage noise. Typical values for CFB op amps range from about 1 nV to $5\ \text{nV}/\sqrt{\text{Hz}}$.

The input current noise of CFB op amps tends to be larger than for VFB op amps because of the generally higher bias current levels. The inverting and noninverting current noise of a CFB op amp is usually different because of the unique input architecture, and are specified separately. In most cases, the inverting input current noise is the larger of the two. Typical input current noise for CFB op amps ranges from 5 pA to $40\ \text{pA}/\sqrt{\text{Hz}}$. This can often be dominant, except in cases of very high gain, when R1 is small.

The noise sources that dominate the output noise are highly dependent on the closed-loop gain of the op amp and the values of the feedback and feedforward resistors. For high values of closed-loop gain, the op amp voltage noise will tend be the chief contributor to the output noise. At low gains, the effects of the input current noise must also be considered, and may dominate, especially in the case of a CFB op amp.

Feedforward/feedback resistors in high speed op amp circuits may range from less than 100 Ω to more than 1 kΩ, so it is difficult to generalize about their contribution to the total output noise without knowing the specific values and the closed-loop gain.

The best way to make the noise calculations is to write a simple computer program that automatically performs the calculations, and include all the noise sources. The equation previously discussed can be used for this purpose (see Figure 1-74). In most high speed op amp applications, the source impedance noise can often be neglected for source impedances of 100 Ω or less.

Figure 1-121 summarizes the noise characteristics of high speed op amps.

- Voltage Feedback Op Amps:
  - Voltage noise: 2nV to 20nV/√Hz
  - Current noise: 0.5pA to 5pA/√Hz
- Current Feedback Op Amps:
  - Voltage noise: 1nV to 5nV/√Hz
  - Current noise: 5nV to 40pA/√Hz
- Noise Contribution from Source Negligible if < 100Ω
- Voltage Noise Usually Dominates at High Gains
- Reflect Noise Sources to Output and Combine (RSS)
- Errors Will Result if there is Significant High Frequency Peaking

Figure 1-121: High speed op amp noise summary

DC Characteristics of High Speed Op Amps

High speed op amps are optimized for bandwidth and settling time, not for precision dc characteristics as found in lower frequency precision op amps. In spite of this, however, high speed op amps do have reasonably good dc performance.

Input offset voltages of high speed bipolar input op amps are rarely trimmed, since offset voltage matching of the input stage is excellent, typically ranging from 1 mV to 3 mV, with offset temperature coefficients of 5 μV to 15 μV/°C.

Input bias currents on VFB op amps (with no input bias current compensation circuits) are approximately equal for (+) and (–) inputs, and can range from 1 μA to 5 μA. The output offset voltage due to the input bias currents can be nulled by making the effective source resistance, R3, equal to the parallel combination of R1 and R2.

As previously discussed, this scheme will not work with bias-current compensated VFB op amps that have additional current generators on their inputs. In this case, the net input bias currents are not necessarily equal or of the same polarity.

CFB op amps generally have unequal and uncorrelated input bias currents because the (+) and (–) inputs have completely different architectures. For this reason, external bias current cancellation schemes are also ineffective. CFB input bias currents range from 5 μA to 15 μA, being generally higher at the inverting input.

Figure 1-122 summarizes the offset considerations for high speed op amps.

- High Speed Bipolar Op Amp Input Offset Voltage:
  - Ranges from 1mV to 3mV for VFB and CFB
  - Offset TC ranges from 5µV to 15µV/°C
- High Speed Bipolar Op Amp Input Bias Current:
  - For VFB ranges from 1µA to 5µA
  - For CFB ranges from 5µA to 15µA
- Bias Current Cancellation Doesn't Work for:
  - Bias current compensated op amps
  - Current feedback op amps

Figure 1-122: High speed op amp offset voltage summary

References: High Speed Op Amps

1. Roy Gosser, "Wide-Band Transconductance Generator," **US Patent 5,150,074**, filed May 3, 1991, issued September 22, 1992.
2. Roy Gosser, DC-Coupled Transimpedance Amplifier, **US Patent 4,970,470**, filed October 10, 1989, issued November 13, 1990.
3. James L. Melsa and Donald G. Schultz, **Linear Control Systems**, McGraw-Hill, 1969, pp. 196–220, ISBN: 0-07-041481-5.
4. Thomas M. Frederiksen, **Intuitive Operational Amplifiers**, McGraw-Hill, 1988., ISBN: 0-07-021966-4.
5. Sergio Franco, **Design With Operational Amplifiers and Analog Integrated Circuits**, **2nd Ed.**, McGraw-Hill, 1998, ISBN: 0-07-021857-9.
6. Walt Kester, Editor, **High Speed Design Techniques**, Analog Devices, 1996, ISBN: 0-916550-17-6, (available for download at www.analog.com).
7. Data sheet for **AD8011 300 MHz, 1 mA Current Feedback Amplifier**, www.analog.com.

CHAPTER 2

Specialty Amplifiers

CHAPTER 2

Specialty Amplifiers

Walt Kester, Walt Jung, James Bryant

This chapter of the book discusses several popular types of *specialty* amplifiers, or amplifiers that are based in some way on op amp techniques. However, in an overall applications sense, they are not generally used as universally as op amps. Examples of specialty amplifiers include instrumentation amplifiers of various configurations, programmable gain amplifiers (PGAs), isolation amplifiers, and difference amplifiers.

Other types of amplifiers, for example such types as audio and video amplifiers, cable drivers, high-speed variable gain amplifiers (VGAs), and various communications-related amplifiers might also be viewed as specialty amplifiers. However, these applications are more suitably covered in Chapter 6, within the various signal amplification sections.

SECTION 2-1

Instrumentation Amplifiers

Walt Kester, Walt Jung

Probably the most popular among all of the specialty amplifiers is the *instrumentation amplifier* (hereafter called simply an *in amp*). The in amp is widely used in many industrial and measurement applications where dc precision and gain accuracy must be maintained within a noisy environment, and where large common-mode signals (usually at the ac power line frequency) are present.

Op Amp/In Amp Functionality Differences

An in amp is unlike an op amp in a number of very important ways. As already discussed, an op amp is a general-purpose gain block—user-configurable in myriad ways using external feedback components of R, C, and, (sometimes) L. The final configuration and circuit function using an op amp is truly whatever the user makes of it.

In contrast to this, an in amp is a more constrained device in terms of functioning, and also the allowable range(s) of operating gain. In many ways, it is better suited to its task than would be an op amp—even though, ironically, an in amp may actually comprise of a number of op amps within it. People also often confuse in amps as to their function, calling them "op amps." But the reverse is seldom (if ever) true. It should be understood that an in amp is *not* just a special type op amp; the function of the two devices is fundamentally different.

Perhaps a good way to differentiate the two devices is to remember that an op amp can be programmed to do almost anything, by virtue of its feedback flexibility. In contrast to this, an in amp *cannot* be programmed to do just anything. It can *only* be programmed for gain, and then over a specific range. An op amp is configured via a number of external components, while an in amp is configured either by one resistor, or by pin-selectable taps for its working gain.

In Amp Definitions

An in amp is a *precision* closed-loop gain block. It has a pair of differential input terminals, and a single-ended output that works with respect to a reference or common terminal, as shown in Figure 2-1. The input impedances are balanced and high in value, typically $\geq 10^9\ \Omega$. Again, unlike an op amp, an in amp uses an

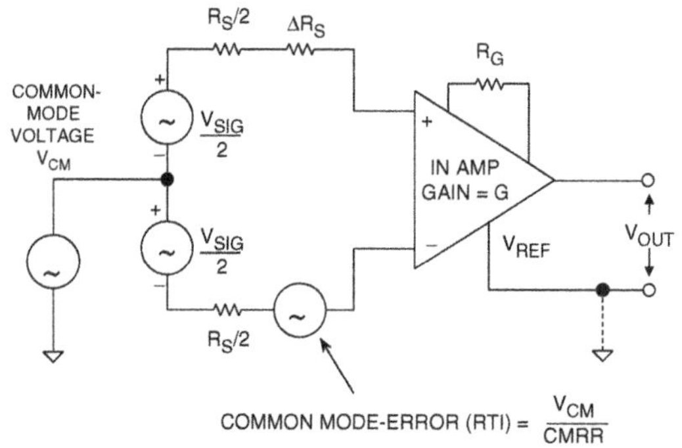

Figure 2-1: The generic instrumentation amplifier (in amp)

internal feedback resistor network, plus one (usually) gain set resistance, R_G. Also unlike an op amp is the fact that the internal resistance network and R_G are *isolated* from the signal input terminals. In amp gain can also be preset via an internal R_G by pin selection, (again isolated from the signal inputs). Typical in amp gains range from 1 to 1,000.

The in amp develops an output voltage that is referenced to a pin usually designated REFERENCE, or V_{REF}. In many applications, this pin is connected to circuit ground, but it can be connected to other voltages, as long as they lie within a rated compliance range. This feature is especially useful in single-supply applications, where the output voltage is usually referenced to mid-supply (i.e., 2.5 V in the case of a +5 V supply).

In order to be effective, an in amp needs to be able to amplify microvolt-level signals, while simultaneously rejecting volts of *common-mode* (CM) signal at its inputs. This requires that in amps have very high *common-mode rejection* (CMR). Typical values of in amp CMR are from 70 dB to over 100 dB, with CMR usually improving at higher gains.

It is important to note that a CMR specification for dc inputs alone isn't sufficient in most practical applications. In industrial applications, the most common cause of external interference is 50 Hz/60 Hz ac power-related noise (including harmonics). In differential measurements, this type of interference tends to be induced equally onto both in amp inputs, so the interference appears as a CM input signal. Therefore, specifying CMR over frequency is just as important as specifying its dc value. Note that imbalance in the two source impedances can degrade the CMR of some in amps. Analog Devices fully specifies in amp CMR at 50 Hz/60 Hz, with a source impedance imbalance of 1 kΩ.

Subtractor or Difference Amplifiers

A simple subtractor or difference amplifier can be constructed with four resistors and an op amp, as shown in Figure 2-2. It should be noted that this is *not* a true in amp (based on the previously discussed criteria), but it is often used in applications where a simple differential-to-single-ended conversion is required. Because of its popularity, this circuit will be examined in more detail, in order to understand its fundamental limitations before discussing true in amp architectures.

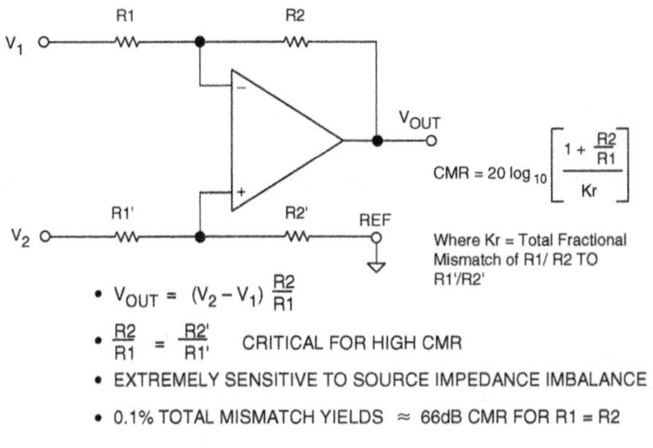

Figure 2-2: Op amp subtractor or difference amplifier

There are several fundamental problems with this simple circuit. First, the input impedance seen by V_1 and V_2 isn't balanced. The input impedance seen by V_1 is R1, but the input impedance seen by V_2 is R1' + R2'. The configuration can also be quite problematic in terms of CMR, since even a small source impedance imbalance will degrade the workable CMR. This problem can be solved with well-matched open-loop buffers in series with each input (for example, using a precision dual op amp). But, this adds complexity to a simple circuit, and may introduce offset drift and nonlinearity.

The second problem with this circuit is that the *CMR is primarily determined by the resistor ratio matching,* not *the op amp.* The resistor ratios R1/R2 and R1'/R2' must match extremely well to reject common mode noise—at least as well as a typical op amp CMR of ≥100 dB. Note also that the *absolute* resistor values are relatively unimportant.

Picking four 1% resistors from a single batch may yield a net ratio matching of 0.1%, which will achieve a CMR of 66 dB (assuming R1 = R2). But if one resistor differs from the rest by 1%, the CMR will drop to only 46 dB. Clearly, very limited performance is possible using ordinary discrete resistors in this circuit (without resorting to hand matching). This is because the best standard off-the-shelf RNC/RNR style resistor tolerances are on the order of 0.1% (see Reference 1).

In general, the worst-case CMR for a circuit of this type is given by the following equation (see References 2 and 3):

$$\text{CMR}(\text{dB}) = 20\log\left[\frac{1+\text{R}2/\text{R}1}{4\text{Kr}}\right], \qquad \text{Eq. 2-1}$$

where Kr is the *individual* resistor tolerance in fractional form, for the case where four discrete resistors are used. This equation shows that the worst-case CMR for a tolerance build-up for four unselected same-nominal-value 1% resistors to be no better than 34 dB.

A single resistor network with a net matching tolerance of Kr would probably be used for this circuit, in which case the expression would be as noted in the figure, or:

$$\text{CMR}(\text{dB}) = 20\log\left[\frac{1+\text{R}2/\text{R}1}{\text{Kr}}\right] \qquad \text{Eq. 2-2}$$

A net matching tolerance of 0.1% in the resistor ratios therefore yields a worst-case dc CMR of 66 dB using Eq. 2-2, and assuming R1 = R2. Note that either case assumes a significantly higher amplifier CMR (i.e., >100 dB). Clearly for high CMR, such circuits need four single-substrate resistors, with very high absolute and TC matching. Such networks using thick/thin-film technology are available from companies such as Caddock and Vishay, in ratio matches of 0.01% or better.

In implementing the simple difference amplifier, rather than incurring the higher costs and PCB real estate limitations of a precision op amp plus a separate resistor network, it is usually better to seek out a completely monolithic solution. The AMP03 is just such a precision difference amplifier, which includes an on-chip laser trimmed precision thin film resistor network. It is shown in Figure 2-3. The typical CMR of the AMP03F is 100 dB, and the small-signal bandwidth is 3 MHz.

Figure 2-3: AMP03 precision difference amplifier

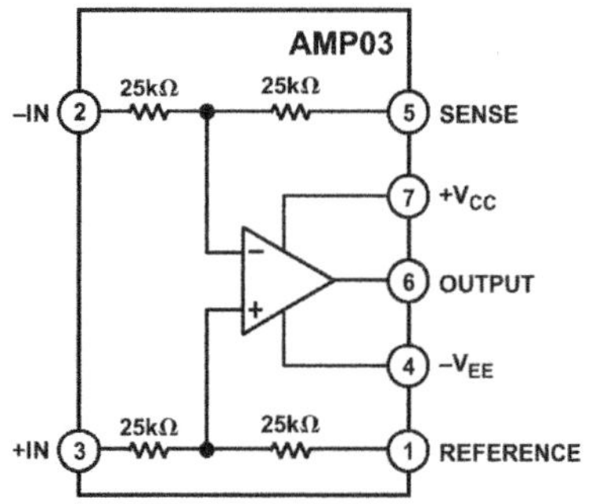

There are several devices related to the AMP03 in function. These are namely the SSM2141 and SSM2143 difference amplifiers. These sister parts are designed for audio line receivers (see Figure 2-4). They have low distortion, and high (pretrimmed) CMR. The net gains of the SSM2141 and SSM2143 are unity and 0.5, respectively. They are designed to be used with balanced 600 Ω audio sources (see the related discussions on these devices in the Audio Amplifiers section of Chapter 6).

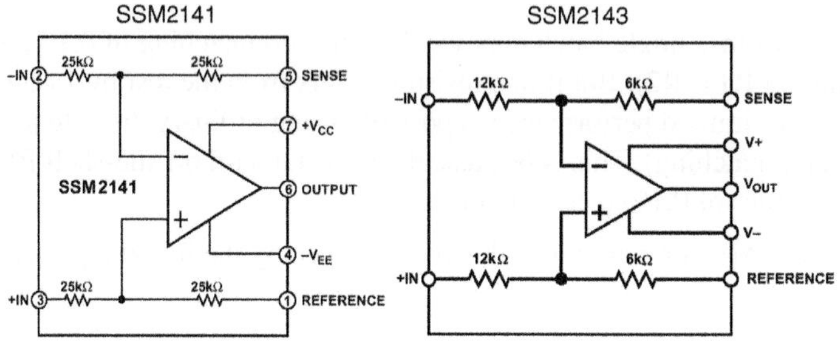

Figure 2-4: SSM2141 and SSM2143 difference amplifiers (audio line receivers)

Another interesting variation on the simple difference amplifier is found in the AD629 difference amplifier, optimized for high common-mode input voltages. A typical current-sensing application is shown in Figure 2-5. The AD629 is a differential-to-single-ended amplifier with a gain of unity. It can handle a common-mode voltage of ±270 V with supply voltages of ±15 V, with a small signal bandwidth of 500 kHz.

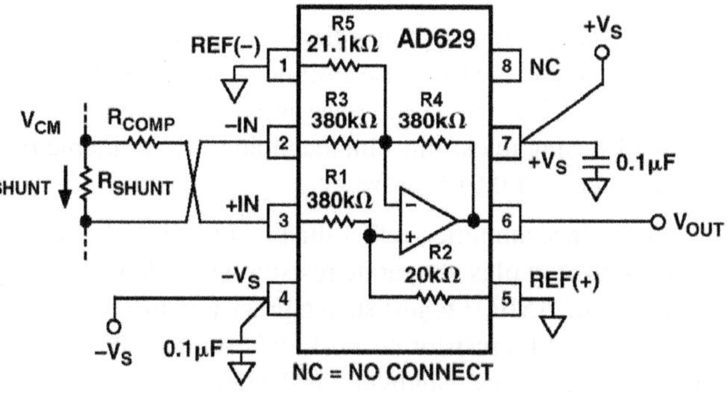

Figure 2-5: High common-mode current sensing using the AD629 difference amplifier

The high common-mode voltage range is obtained by attenuating the noninverting input (Pin 3) by a factor of 20 times, using the R1–R2 divider network. On the inverting input, resistor R5 is chosen such that R5||R3 equals resistor R2. The noise gain of the circuit is equal to 20 [1 + R4/(R3||R5)], thereby providing unity gain for differential input voltages. Laser wafer trimming of the R1–R5 thin film resistors yields a minimum CMR of 86 dB @ 500 Hz for the AD629B. Within an application, it is good practice to maintain balanced source impedances on both inputs, so dummy resistor R_{COMP} is chosen to equal to the value of the shunt sensing resistor R_{SHUNT}.

David Birt (see Reference 4) of the BBC has analyzed the simple line receiver topology in terms of loading presented to the source, and presented a modified and balanced form, shown as Figure 2-6. Here stage U1 uses a 4 resistor network identical to that of Figure 2-2, while feedback from the added unity gain inverter U2 drives the previously grounded R2' reference terminal. This has two overall effects; the input currents in the ± input legs become equal in magnitude, and the gain of the stage is halved.

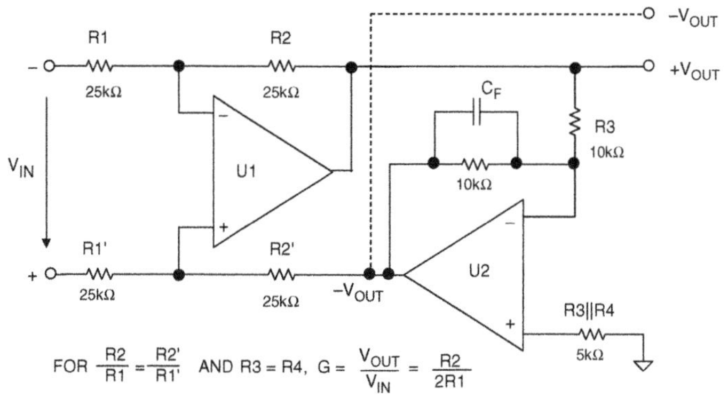

Figure 2-6: Balanced difference amplifier using push-pull feedback path

Compared to Figure 2-2, and for like resistor ratios, the Figure 2-6 gain from V_{IN} to V_{OUT} is one-half, or a gain of –6 dB (0.5) as shown. However the new circuit form also offers a complementary output from U2, $-V_{OUT}$.

The common-mode range of this circuit is the same as for Figure 2-2, but the CMR is about doubled with all resistors nominally equal (as measured to a single output). The inverter resistor ratio R3/R4 affects output balance, but not CMR. Like Figure 2-2, the gain of this circuit is not easily changed, as it involves precise resistor ratios.

Because of the two feedback paths, this circuit holds the inputs of U1 at a null for differential input signals. However CM signals are seen by U1, and the CM range of the circuit is $[1 + (R2'/R1')] \times V_{CM(U1)}$. Differential input resistance is R1 + R1'.

As can be noted from Figure 2-6, this circuit can be broken into a simple line receiver (left), plus an inverter (right). Thus existing line receivers like Figure 2-2 can be converted to the fully balanced topology, by simply adding an appropriate inverter, U2. This of course not only balances the *input* currents, but it also provides a balanced *output* signal.

For example, the SSM2141 line receiver and the OP275 are a good combination for implementing this approach. (See Reference 5, and the further discussions on these circuits in the Audio Amplifiers section of Chapter 6.)

In Amp Configurations

The simple difference amplifier circuits described above are quite useful (especially at higher frequencies) but lack the performance required for most precision applications. In many cases, true in amps are more suitable, because of their balanced and high input impedance, as well as their high common-mode rejection.

Two-Op-Amp In Amps

As noted initially, in amps are based on op amps, and there are two basic configurations that are extremely popular. The first is based on two op amps, and the second on three op amps. The circuit shown in Figure 2-7 is referred to as the *two-op-amp in amp*. Dual IC op amps are used in most cases for good matching, such as the OP297 or the OP284. The resistors are usually a thin film laser trimmed array on the same chip. The in amp gain can be easily set with an external resistor, R_G. Without R_G, the gain is simply 1 + R2/R1. In a practical application, the R2/R1 ratio is chosen for the desired minimum in amp gain.

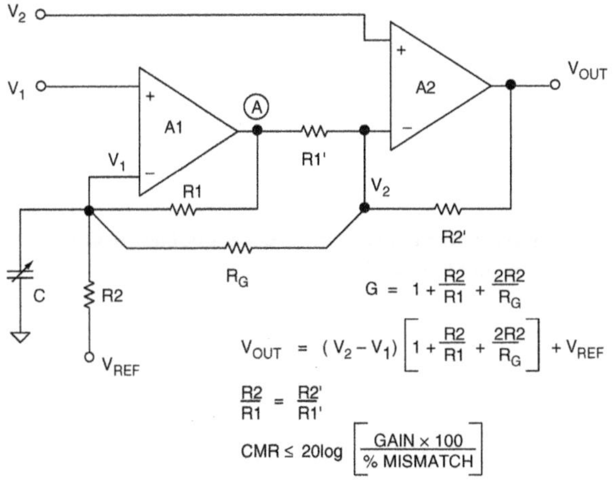

Figure 2-7: The two-op-amp instrumentation amplifier

The input impedance of the two-op-amp in amp is inherently high, permitting the impedance of the signal sources to be high and unbalanced. The dc common mode rejection is limited by the matching of R1/R2 to R1'/R2'. If there is a mismatch in any of the four resistors, the dc common mode rejection is limited to:

$$\text{CMR} \le 20\log\left[\frac{\text{GAIN}\times 100}{\%\text{MISMATCH}}\right]. \qquad \text{Eq. 2-3}$$

Notice that the net CMR of the circuit increases proportionally with the working gain of the in amp, an effective aid to high performance at higher gains.

IC in amps are particularly well suited to meeting the combined needs of ratio matching and temperature tracking of the gain-setting resistors. While thin film resistors fabricated on silicon have an initial tolerance of up to ±20%, laser trimming during production allows the ratio error between the resistors to be reduced to 0.01% (100 ppm). Furthermore, the tracking between the temperature coefficients of the thin film resistors is inherently low and is typically less than 3 ppm/°C (0.0003%/°C).

When dual supplies are used, V_{REF} is normally connected directly to ground. In single supply applications, V_{REF} is usually connected to a low impedance voltage source equal to one-half the supply voltage. The gain from V_{REF} to node "A" is R1/R2, and the gain from node "A" to the output is R2'/R1'. This makes the gain from V_{REF} to the output equal to unity, assuming perfect ratio matching. Note that it is critical that the source impedance seen by V_{REF} be low, otherwise CMR will be degraded.

One major disadvantage of the two-op-amp in amp design is that common mode voltage input range must be traded off against gain. The amplifier A1 must amplify the signal at V_1 by 1 + R1/R2. If R1 >> R2 (a low gain example in Figure 2-7), A1 will saturate if the V_1 common-mode signal is too high, leaving no A1 headroom to amplify the wanted differential signal. For high gains (R1<< R2), there is correspondingly more headroom at node "A," allowing larger common-mode input voltages.

The ac common-mode rejection of this configuration is generally poor because the signal path from V_1 to V_{OUT} has the additional phase shift of A1. In addition, the two amplifiers are operating at different closed-loop gains (and thus at different bandwidths). The use of a small trim capacitor "C" as shown in Figure 2-7 can improve the ac CMR somewhat.

A low gain (G = 2) single-supply two-op-amp in amp configuration results when R_G is not used, and is shown in Figure 2-8. The input common mode and differential signals must be limited to values that prevent saturation of either A1 or A2. In the example, the op amps remain linear to within 0.1 V of the supply rails, and their upper and lower output limits are designated V_{OH} and V_{OL}, respectively. These saturation voltage limits would be typical for a single-supply, rail-rail output op amp (such as the AD822, for example).

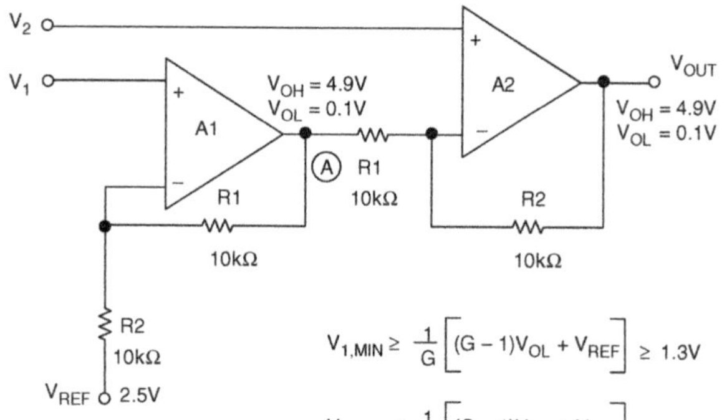

$$V_{REF} = \frac{V_{OH} + V_{OL}}{2} = 2.5V$$

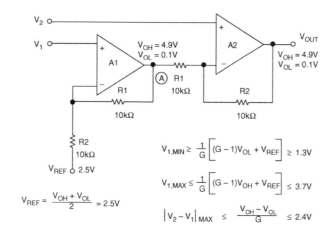

$$V_{1,MIN} \geq \frac{1}{G}\left[(G-1)V_{OL} + V_{REF}\right] \geq 1.3V$$

$$V_{1,MAX} \leq \frac{1}{G}\left[(G-1)V_{OH} + V_{REF}\right] \leq 3.7V$$

$$\left|V_2 - V_1\right|_{MAX} \leq \frac{V_{OH} - V_{OL}}{G} \leq 2.4V$$

Figure 2-8: Two-op-amp in amp single-supply restrictions for V_s = +5 V, G = 2

Using the Figure 2-8 equations, the voltage at V_1 must fall between 1.3 V and 2.4 V to prevent A1 from saturating. Notice that V_{REF} is connected to the average of V_{OH} and V_{OL} (2.5 V). This allows for bipolar differential input signals with V_{OUT} referenced to 2.5 V. A high gain (G = 100) single-supply two-op-amp in amp configuration is shown in Figure 2-9. Using the same equations, note that voltage at V_1 can now swing between 0.124 V and 4.876 V. V_{REF} is again 2.5 V, to allow for bipolar input and output signals.

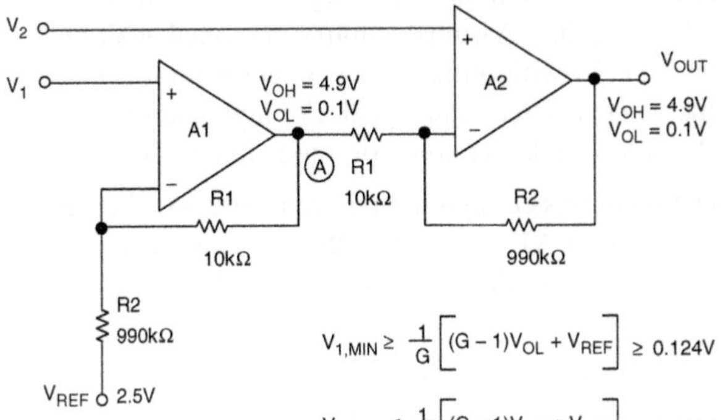

$$V_{REF} = \frac{V_{OH} + V_{OL}}{2} = 2.5V$$

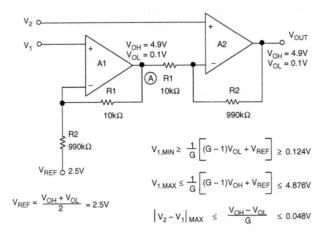

$$V_{1,MIN} \geq \frac{1}{G}\left[(G-1)V_{OL} + V_{REF}\right] \geq 0.124V$$

$$V_{1,MAX} \leq \frac{1}{G}\left[(G-1)V_{OH} + V_{REF}\right] \leq 4.876V$$

$$\left|V_2 - V_1\right|_{MAX} \leq \frac{V_{OH} - V_{OL}}{G} \leq 0.048V$$

Figure 2-9: Two-op-amp in amp single-supply restrictions for V_s = +5 V, G = 100

All of these discussions show that the conventional two-op-amp in amp architecture is fundamentally limited, when operating from a single power supply. These limitations can be viewed in one sense as a restraint on the allowable input CM range for a given gain. Or, alternately, it can be viewed as limitation on the allowable gain range, for a given CM input voltage.

Nevertheless, there are ample cases where a combination of gain and CM voltage cannot be supported by the basic two-op-amp structures of Figures 2-7 through 2-9, even with perfect amplifiers (i.e., zero output saturation voltage to both rails).

In summary, regardless of gain, the basic structure of the common two-op-amp in amp does not allow for CM input voltages of zero when operated on a single supply. The only route to removing these restrictions for single supply operation is to modify the in amp architecture.

The AD627 Single-Supply Two-Op-Amp In Amp

The above-mentioned CM limitations can be overcome with some key modifications to the basic two-op-amp in amp architecture. These modifications are implemented in the circuit shown in Figure 2-10, which represents the AD627 in amp architecture.

In this circuit, each of the two op amps is composed of a PNP common emitter input stage and a gain stage, designated Q1/A1, and Q2/A2, respectively. The PNP transistors not only provide gain but also level-shift the input signal positive by about 0.5 V, thereby allowing the common-mode input voltage to go to 0.1 V below the negative supply rail. The maximum positive input voltage allowed is 1 V less than the positive supply rail.

The AD627 in amp delivers rail-to-rail output swing, and operates over a wide supply voltage range (+2.7 V to ±18 V). Without the external gain setting resistor R_G, the in amp gain is a minimum of 5. Gains

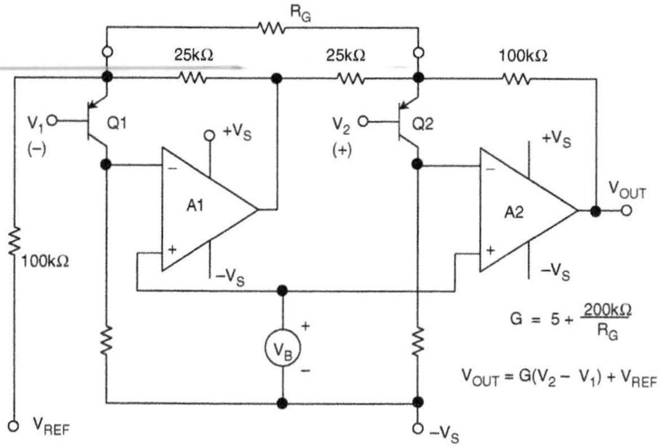

Figure 2-10: The AD627 in amp architecture

up to 1000 can be set with the addition of this external resistor. Common-mode rejection of the AD627B at 60 Hz with a 1 kΩ source imbalance is 85 dB when operating on a single 3 V supply and G = 5.

Even though the AD627 is a two-op-amp in amp, it is worthwhile noting that it is not subject to the same CM frequency response limitations as the basic circuit of Figure 2-7. A patented circuit keeps the AD627 CMR flat out to a much higher frequency than would otherwise be achievable with a conventional discrete two-op-amp in amp.

The AD627 data sheet has a detailed discussion of allowable input/output voltage ranges as a function of gain and power supply voltages (see Reference 7). In addition, interactive design tools that perform calculations relating these parameters for a number of in amps, including the AD627, are available on the ADI Web site.

Key specifications for the AD627 are summarized in Figure 2-11. Although it has been designed as a low power, single-supply device, the AD627 is capable of operating on traditional higher voltage supplies such as ±15 V, with excellent performance.

- Wide Supply Range: +2.7V to ±18V
- Input Voltage Range: $-V_S - 0.1V$ to $+V_S - 1V$
- 85μA Supply Current
- Gain Range: 5 to 1000
- 75μV Maximum Input Offset Volage (AD627B)
- 10ppm/°C Maximum Offset Voltage TC (AD627B)
- 10ppm Gain Nonlinearity
- 85dB CMR @ 60Hz, 1kΩ Source Imbalance (G = 5)
- 3μV p-p 0.1Hz to 10Hz Input Voltage Noise (G = 5)

Figure 2-11: AD627 in amp key specifications

Three-Op-Amp In Amps

A second popular in amp architecture is based on three op amps, and is shown in Figure 2-12. This circuit is typically referred to as the *three-op-amp in amp*.

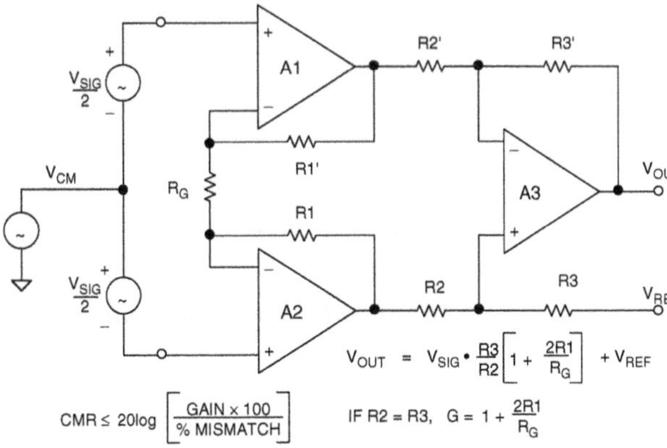

Figure 2-12: The three-op-amp in amp

Resistor R_G sets the overall gain of this amplifier. It may be internal, external, or (software or pin-strap) programmable, depending upon the in amp. In this configuration, CMR depends upon the ratio-matching of R3/R2 to R3'/R2'. Furthermore, common-mode signals are only amplified by a factor of 1, regardless of gain. (No common-mode voltage will appear across R_G, hence, no common-mode current will flow in it because the input terminals of an op amp will have no significant potential difference between them.)

As a result of the high ratio of differential-to-CM gain in A1–A2, CMR of this in amp theoretically increases in proportion to gain. Large common-mode signals (within the A1–A2 op amp headroom limits) may be handled at all gains. Finally, because of the symmetry of this configuration, common mode errors in the input amplifiers, if they track, tend to be canceled out by the subtractor output stage. These features explain the popularity of this three-op-amp in amp configuration—it is capable of delivering the highest performance.

The classic three-op-amp configuration has been used in a number of monolithic IC in amps (see References 8 and 9). Besides offering excellent matching between the three internal op amps, thin film laser trimmed resistors provide excellent ratio matching and gain accuracy at much lower cost than using discrete precision op amps and resistor networks. The AD620 (see Reference 10) is an excellent example of monolithic IC in amp technology. A simplified device schematic is shown in Figure 2-13.

The AD620 is a highly popular in amp and is specified for power supply voltages from ±2.3 V to ±18 V. Input voltage noise is only $9\,\text{nV}/\sqrt{\text{Hz}}$ @ 1 kHz. Maximum input bias current is only 1 nA, due to the use of superbeta transistors for Q1–Q2.

Overvoltage protection is provided by the internal 400 Ω thin-film current-limit resistors in conjunction with the diodes connected from the emitter-to-base of Q1 and Q2. The gain G is set with a single external R_G resistor, as noted by Eq. 2-4.

$$G = \left(49.4\,\text{k}\Omega/R_G\right) + 1 \qquad \text{Eq. 2-4}$$

As can be noted from this expression and Figure 2-13, the AD620 internal resistors are trimmed so that standard 1% or 0.1% resistors can be used to set gain to popular values.

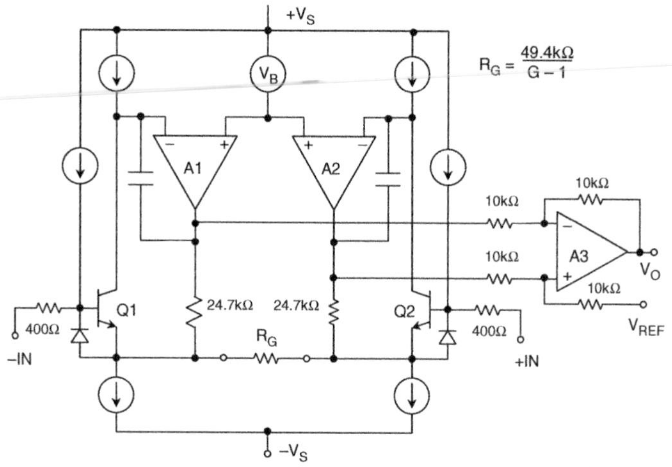

Figure 2-13: The AD620 in amp simplified schematic

As is true in the case of the two-op-amp in amp configuration, single supply operation of the three-op-amp in amp requires an understanding of the internal node voltages. Figure 2-14 shows a generalized diagram of the in amp operating on a single 5 V supply. The maximum and minimum allowable output voltages of the individual op amps are designated V_{OH} (maximum high output) and V_{OL} (minimum low output), respectively.

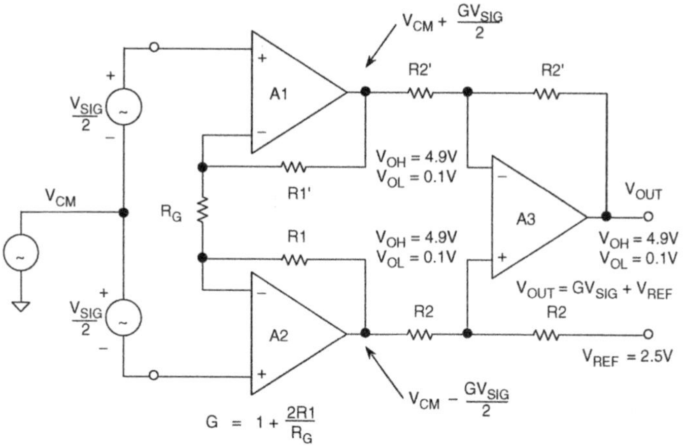

Figure 2-14: Three-op-amp in amp single 5 V supply restrictions

Note that the gain from the common-mode voltage to the outputs of A1 and A2 is unity. It can be stated that *the sum of the common-mode voltage and the signal voltage at these outputs must fall within the amplifier output voltage range*. Obviously this configuration cannot handle input common-mode voltages of either zero volts or 5 V, because of saturation of A1 and A2. As in the case of the two-op-amp in amp, the output reference is positioned halfway between V_{OH} and V_{OL} to allow for bipolar differential input signals.

While there are a number of good single-supply in amps, such as the AD627 discussed above, the highest performance devices are still among those specified for traditional dual-supply operation, i.e., the just-discussed AD620. For certain applications, even such devices as the AD620, which has been designed for dual supply operation, can be used with full precision on a single-supply power system.

Precision Single-Supply Composite In Amp

One way to achieve both high precision and single-supply operation takes advantage of the fact that many popular sensors (e.g. strain gauges) provide an output signal that is inherently centered around an approximate mid-point of the supply voltage (and/or the reference voltage). Taking advantage of this basic point allows the inputs of a signal conditioning in amp to be biased at "mid-supply." As a consequence of this step, the inputs needn't operate near ground or the positive supply voltage, and the in amp can still be used with all its precision.

Under these conditions, an AD620 dual-supply in amp referenced to the supply mid-point followed by an rail-to-rail op amp output gain stage provides very high dc precision. Figure 2-15 illustrates one such high performance in amp, which operates on a single 5 V supply.

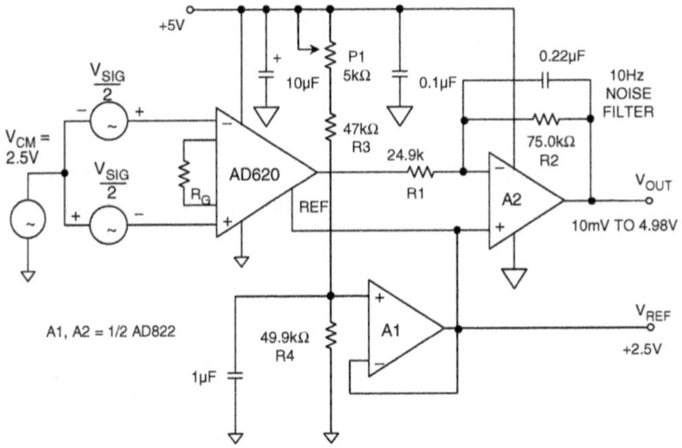

Figure 2-15: A precision single-supply composite in amp with rail-to-rail output

This circuit uses the AD620 as a low-cost precision in amp for the input stage, along with an AD822 JFET-input dual rail-to-rail output op amp for the output stage, comprised of A1 and A2. The output stage operates at a fixed gain of 3, with overall gain set by R_G.

In this circuit, R3 and R4 form a voltage divider which splits the supply voltage nominally in half to 2.5 V, with fine adjustment provided by a trimming potentiometer, P1. This voltage is applied to the input of A1, an AD822 voltage follower, which buffers it and provides a low impedance source needed to drive the AD620's reference pin as well as providing the output reference voltage V_{REF}. *Note that this feature allows a* bipolar *V_{OUT} to be measured with respect to this 2.5 V reference (not to GND).* This is despite the fact that the entire circuit operates from a single (unipolar) supply.

The other half of the AD822 is connected as a gain-of-3 inverter, so that it can output ±2.5 V, "rail-to-rail," with only ±0.83 V required of the AD620. This output voltage level of the AD620 is well within the AD620's capability, thus ensuring high linearity for the front end.

The general gain expression for this composite in amp is the product of the gain of the AD620 stage, and the gain of inverting amplifier:

$$\mathrm{GAIN} = \left(\frac{49.4\,\mathrm{k}\Omega}{R_G} + 1\right)\left(\frac{R2}{R1}\right). \qquad \text{Eq. 2-5}$$

For this example, an overall gain of 10 is realized with $R_G = 21.5\ \mathrm{k}\Omega$ (closest standard value). The table shown in Figure 2-16 summarizes various R_G gain values, and the resulting performance for gains ranging from 10 to 1000.

CIRCUIT GAIN	R_G (Ω)	V_{OS}, RTI (µV)	TC V_{OS}, RTI (µV/°C)	NONLINEARITY (ppm) *	BANDWIDTH (kHz)**
10	21.5k	1000	1000	< 50	600
30	5.49k	430	430	< 50	600
100	1.53k	215	215	< 50	300
300	499	150	150	< 50	120
1000	149	150	150	< 50	30

*Nonlinearity Measured Over Output Range: 0.1V < V_{OUT} < 4.90V
**Without 10Hz Noise Filter

Figure 2-16: Performance summary of the 5 V single-supply AD620/AD822 composite in amp

In this application, the allowable input voltage on either input to the AD620 must lie between 2 V and 3.5 V in order to maintain linearity. For example, at an overall circuit gain of 10, the common-mode input voltage range spans 2.25 V to 3.25 V, allowing room for the ±0.25 V full-scale differential input voltage required to drive the output ±2.5 V about V_{REF}.

The inverting configuration was chosen for the output buffer to facilitate system output offset voltage adjustment by summing currents into the A2 stage buffer's feedback summing node. These offset currents can be provided by an external DAC, or from a resistor connected to a reference voltage.

The AD822 rail-to-rail output stage exhibits a very clean transient response (not shown) and a small-signal bandwidth over 100 kHz for gain configurations up to 300. Note that excellent linearity is maintained over 0.1 V to 4.9 V V_{OUT}.

To reduce the effects of unwanted noise pickup, a filter capacitor is recommended across A2's feedback resistance to limit the circuit bandwidth to the frequencies of interest. This capacitor forms a first order low-pass filter with R2. The corner frequency is 10 Hz as shown, but this may be easily modified. The capacitor should be a high quality film type, such as polypropylene.

The AD623 In Amp

Like the two-op-amp in amp counterparts discussed previously, three-op-amp in amps require special design attention for wide CM range inputs on single power supplies. The AD623 single supply in amp configuration (see Reference 11), shown below in Figure 2-17 offers an attractive solution. In this device PNP emitter follower level shifters Q1 and Q2 allow the input signal to go 150 mV below the negative supply, and to within 1.5 V of the positive supply. The AD623 is fully specified for both single power supplies between 3 V and +12 V, and dual supplies between ±2.5 V and ±6 V.

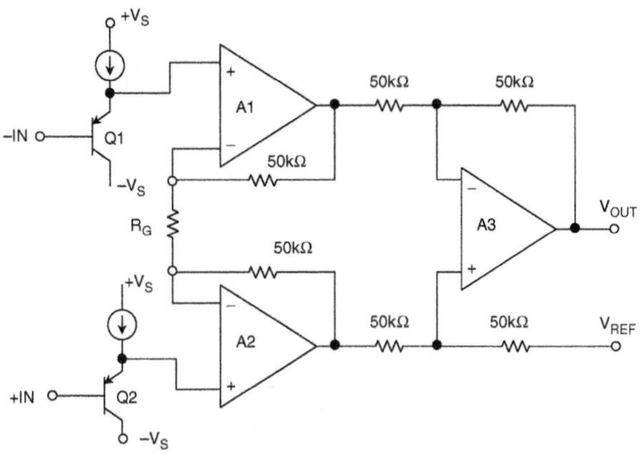

Figure 2-17: AD623 single-supply in amp architecture

The AD623 data sheet (Reference 11, again) contains excellent discussions and data on allowable input/output voltage ranges as a function of gain and power supply voltages. In addition, interactive design tools that perform calculations relating these parameters for a number of in amps, including the AD623, are available on the ADI Web site.

The key specifications of the AD623 are summarized in Figure 2-18.

- Wide Supply Range: +3V to ±6V
- Input Voltage Range: $-V_S - 0.15V$ to $+V_S - 1.5V$
- 575µA Maximum Supply Current
- Gain Range: 1 to 1000
- 100µV Maximum Input Offset Voltage (AD623B)
- 1µV/°C Maximum Offset Voltage TC (AD623B)
- 50ppm Gain Nonlinearity
- 105dB CMR @ 60Hz, 1kΩ Source Imbalance, G≥100
- 3µV p-p 0.1Hz to 10Hz Input Voltage Noise (G = 1)

Figure 2-18: AD623 in amp key specifications

In Amp DC Error Sources

The dc and noise specifications for in amps differ slightly from conventional op amps, so some discussion is required in order to fully understand the error sources.

The gain of an in amp is usually set by a single resistor. If the resistor is external to the in amp, its value is either calculated from a formula or chosen from a table on the data sheet, depending on the desired gain.

Absolute value laser wafer trimming allows the user to program gain accurately with this single resistor. The absolute accuracy and temperature coefficient of this resistor directly affects the in amp gain accuracy and drift. Since the external resistor will never exactly match the internal thin film resistor tempcos, a low TC (<25 ppm/°C) metal film resistor should be chosen, preferably with a 0.1% or better accuracy.

Often specified as having a gain range of 1 to 1000, or 1 to 10,000, many in amps will work at higher gains, but the manufacturer will not guarantee a specific level of performance at these high gains. In practice, as the gain-setting resistor becomes smaller, any errors due to the resistance of the metal runs and bond wires become significant. These errors, along with an increase in noise and drift, may make higher single-stage gains impractical. In addition, input offset voltages can become quite sizable when reflected to output at high gains. For instance, a 0.5 mV input offset voltage becomes 5 V at the output for a gain of 10,000. For high gains, the best practice is to use an in amp as a preamplifier, then use a post amplifier for further amplification.

In a *pin-programmable-gain* in amp such as the AD621, the gain-set resistors are internal, well matched, and the device gain accuracy and gain drift specifications include their effects. The AD621 is otherwise generally similar to the externally gain-programmed AD620.

The *gain error* specification is the maximum deviation from the gain equation. Monolithic in amps such as the AD624C have very low factory trimmed gain errors, with its maximum error of 0.02% at G = 1 and 0.25% at G = 500 being typical for this high quality in amp. Notice that the gain error increases with increasing gain. Although externally connected gain networks allow the user to set the gain exactly, the temperature coefficients of the external resistors and the temperature differences between individual resistors within the network all contribute to the overall gain error. If the data is eventually digitized and presented to a digital processor, it may be possible to correct for gain errors by measuring a known reference voltage and then multiplying by a constant.

Nonlinearity is defined as the maximum deviation from a straight line on the plot of output versus input. The straight line is drawn between the end points of the actual transfer function. Gain nonlinearity in a high quality in amp is usually 0.01% (100 ppm) or less, and is relatively insensitive to gain over the recommended gain range.

The total *input offset voltage* of an in amp consists of two components (see Figure 2-19). *Input* offset voltage, V_{OSI}, is the input offset component that is reflected to the output of the in amp by the gain G. *Output* offset voltage, V_{OSO}, is independent of gain.

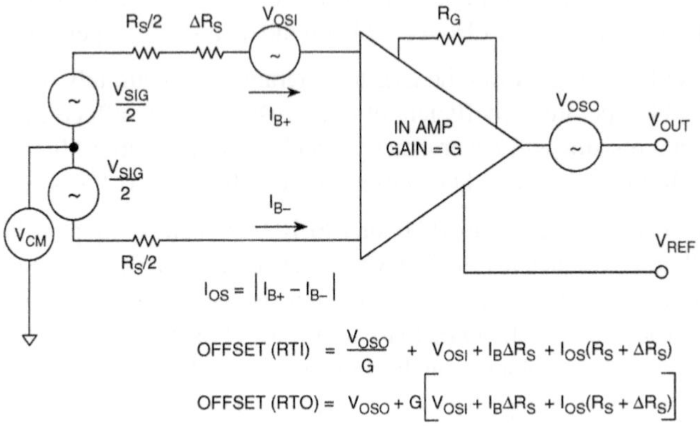

Figure 2-19: In amp offset voltage model

At low gains, output offset voltage is dominant, while at high gains input offset dominates. The output offset voltage drift is normally specified as drift at G = 1 (where input effects are insignificant), while input offset voltage drift is given by a drift specification at a high gain (where output offset effects are negligible).

The total output offset error, referred to the input (RTI), is equal to $V_{OSI} + V_{OSO}/G$. In amp data sheets may specify V_{OSI} and V_{OSO} separately, or give the total RTI input offset voltage for different values of gain.

Input bias currents may also produce offset errors in in amp circuits (Figure 2-19). If the source resistance, R_S, is unbalanced by an amount, ΔR_S, (often the case in bridge circuits), there is an additional input offset voltage error due to the bias current, equal to $I_B \Delta R_S$ (assuming that $I_{B+} \approx I_{B-} = I_B$). This error is reflected to the output, scaled by the gain G.

The input offset current, I_{OS}, creates an input offset voltage error across the source resistance, $R_S + \Delta R_S$, equal to $I_{OS}(R_S + \Delta R_S)$, which is also reflected to the output by the gain, G.

In amp *common-mode error* is a function of both gain and frequency. Analog Devices specifies in amp CMR for a 1 kΩ source impedance unbalance at a frequency of 60 Hz. The RTI common-mode error is obtained by dividing the common-mode voltage, V_{CM}, by the common-mode rejection ratio, CMRR.

Figure 2-20 shows the CMR for the AD620 in amp as a function of frequency, with a 1 kΩ source impedance imbalance.

Power supply rejection (PSR) is also a function of gain and frequency. For in amps, it is customary to specify the sensitivity to each power supply separately, as shown in Figure 2-21 for the AD620. The RTI power supply rejection error is obtained by dividing the power supply deviation from nominal by the power supply rejection ratio, PSRR.

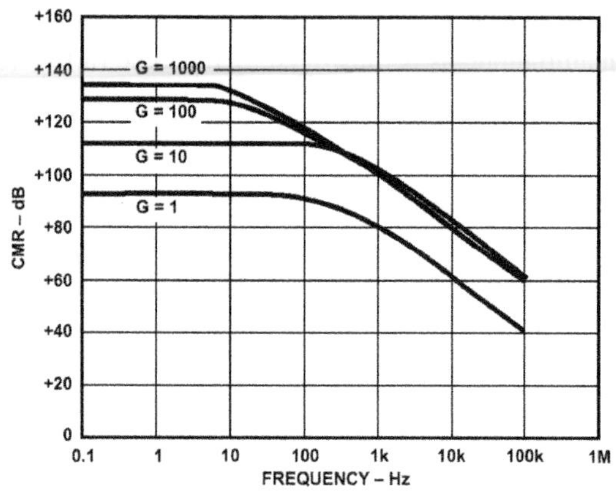

Figure 2-20: AD620 In amp common-mode rejection (CMR) versus frequency for 1 kΩ source imbalance

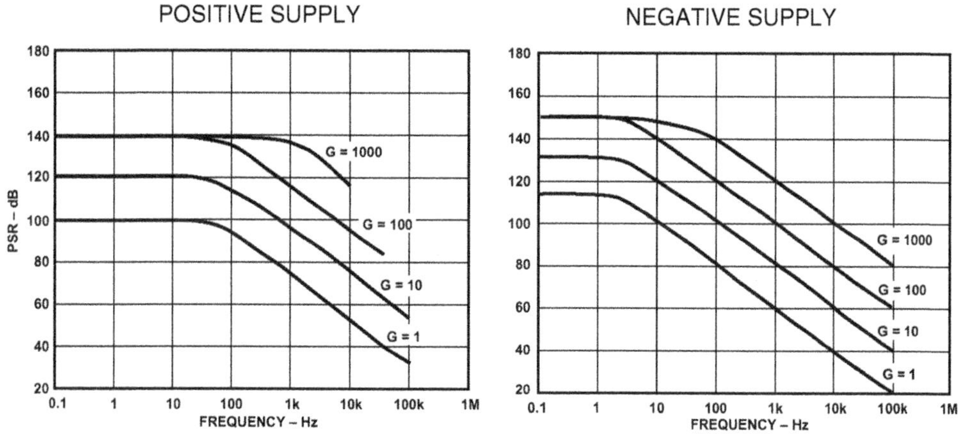

Figure 2-21: AD620 in amp power supply rejection (PSR) versus frequency

Because of the relatively poor PSR at high frequencies, decoupling capacitors are required on both power pins to an in amp. Low inductance ceramic capacitors (0.01 μF to 0.1 μF) are appropriate for high frequencies. Low ESR electrolytic capacitors should also be located at several points on the PC board for low frequency decoupling.

Note that these decoupling requirements apply to all linear devices, including op amps and data converters. Further details on power supply decoupling are found in Chapter 7.

Now that all dc error sources have been accounted for, a worst case dc error budget can be calculated by reflecting all the sources to the in amp input, as is illustrated by the table of Figure 2-22.

It should be noted that the dc errors can be referred to the in amp output (RTO), by simply multiplying the RTI error by the in amp gain.

ERROR SOURCE	RTI VALUE
Gain Accuracy (ppm)	Gain Accuracy × FS Input
Gain Nonlinearity (ppm)	Gain Nonlinearity × FS Input
Input Offset Voltage, V_{OSI}	V_{OSI}
Output Offset Voltage, V_{OSO}	$V_{OSO} \div G$
Input Bias Current, I_B, Flowing in ΔR_S	$I_B \Delta R_S$
Input Offset Current, I_{OS}, Flowing in R_S	$I_{OS}(R_S + \Delta R_S)$
Common Mode Input Voltage, V_{CM}	$V_{CM} \div CMRR$
Power Supply Variation, ΔV_S	$\Delta V_S \div PSRR$

Figure 2-22: In amp dc errors referred to the input (RTI)

In Amp Noise Sources

Since in amps are primarily used to amplify small precision signals, it is important to understand the effects of all the associated noise sources. The in amp noise model is shown in Figure 2-23.

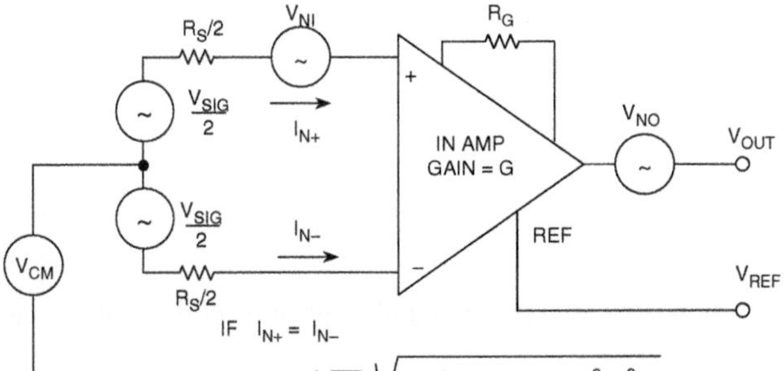

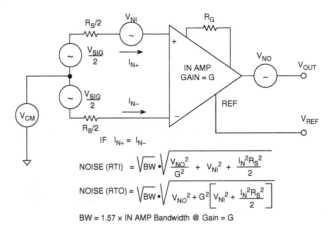

$$\text{NOISE (RTI)} = \sqrt{BW} \cdot \sqrt{\frac{V_{NO}^2}{G^2} + V_{NI}^2 + \frac{I_N^2 R_S^2}{2}}$$

$$\text{NOISE (RTO)} = \sqrt{BW} \cdot \sqrt{V_{NO}^2 + G^2 \left[V_{NI}^2 + \frac{I_N^2 R_S^2}{2} \right]}$$

BW = 1.57 × IN AMP Bandwidth @ Gain = G

Figure 2-23: In amp noise model

There are two sources of input voltage noise. The first is represented as a noise source, V_{NI}, in series with the input, as in a conventional op amp circuit. This noise is reflected to the output by the in amp gain, G. The second noise source is the output noise, V_{NO}, represented as a noise voltage in series with the in amp output. The output noise, shown here referred to V_{OUT}, can be referred to the input by dividing by the gain, G.

There are also two noise sources associated with the input noise currents I_{N+} and I_{N-}. Even though I_{N+} and I_{N-} are usually equal ($I_{N+} \approx I_{N-} = I_N$), they are *uncorrelated*, and therefore, the noise they each create must be summed in a root-sum-squares (RSS) fashion. I_{N+} flows through one-half of R_S, and I_{N-} the other half. This generates two noise voltages, each having an amplitude, $I_N R_S/2$. Each of these two noise sources is reflected to the output by the in amp gain, G.

The total output noise is calculated by combining all four noise sources in an RSS manner:

$$\text{NOISE}(\text{RTO}) = \sqrt{\text{BW}}\sqrt{V_{NO}{}^2 + G^2\left(V_{NI}{}^2 + \frac{I_{N+}{}^2 R_S{}^2}{4} + \frac{I_{N-}{}^2 R_S{}^2}{4}\right)} \qquad \text{Eq. 2-6}$$

If $I_{N+} = I_{N-} = I_N$,

$$\text{NOISE}(\text{RTO}) = \sqrt{\text{BW}}\sqrt{V_{NO}{}^2 + G^2\left(V_{NI}{}^2 + \frac{I_N{}^2 R_S{}^2}{2}\right)} \qquad \text{Eq. 2-7}$$

The total noise, referred to the input (RTI) is simply the above expression divided by the in amp gain, G:

$$\text{NOISE}(\text{RTI}) = \sqrt{\text{BW}}\sqrt{\frac{V_{NO}{}^2}{G^2} + \left(V_{NI}{}^2 + \frac{I_N{}^2 R_S{}^2}{2}\right)} \qquad \text{Eq. 2-8}$$

In amp data sheets often present the total voltage noise RTI as a function of gain. This noise spectral density includes both the input (V_{NI}) and output (V_{NO}) noise contributions. The input current noise spectral density is specified separately.

As in the case of op amps, the total in amp noise RTI must be integrated over the applicable in amp closed-loop bandwidth to compute an RMS value. The bandwidth may be determined from data sheet curves that show frequency response as a function of gain.

Regarding this bandwidth, some care must be taken in computing it, as it is often *not* constant bandwidth product relationship, as is true with VFB op amps. In the case of the AD620 in amp family for example, the gain-bandwidth pattern is more like that of a CFB op amp. In such cases, the safest way to predict the bandwidth at a given gain is to use the curves supplied within the data sheet.

In Amp Bridge Amplifier Error Budget Analysis

It is important to understand in amp error sources in a typical application. Figure 2-24 shows a 350 Ω load cell with a full-scale output of 100 mV when excited with a 10 V source. The AD620 is configured for a gain of 100 using the external 499 Ω gain-setting resistor. The table shows how each error source contributes to a total unadjusted error of 2145 ppm. Note, however, that the gain, offset, and CMR errors can all be removed with a system calibration. The remaining errors—gain nonlinearity and 0.1 Hz to 10 Hz noise—cannot be removed with calibration and ultimately limit the system resolution to 42.8 ppm (approximately 14-bit accuracy).

This example is of course just an illustration, but should be useful towards the importance of addressing performance-limiting errors such as gain nonlinearity and LF noise.

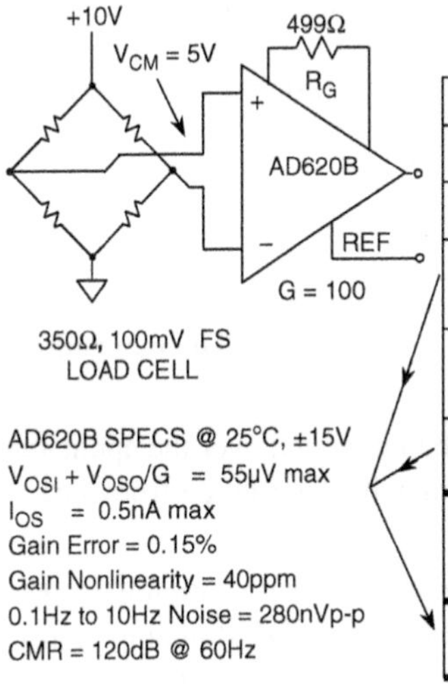

MAXIMUM ERROR CONTRIBUTION, +25°C
FULLSCALE: V_{IN} = 100mV, V_{OUT} = 10V

V_{OS}	55µV ÷ 100mV	550ppm
I_{OS}	350Ω × 0.5nA ÷ 100mV	1.8ppm
Gain Error	0.15%	1500ppm
Gain Nonlinearity	40ppm	40ppm
CMR Error	120dB 1ppm × 5V ÷ 100mV	50ppm
0.1Hz to 10Hz 1/f Noise	280nV ÷ 100mV	2.8ppm
Total Unadjusted Error	≈ 9 Bits Accurate	2145ppm
Resolution Error	≈ 14 Bits Accurate	42.8ppm

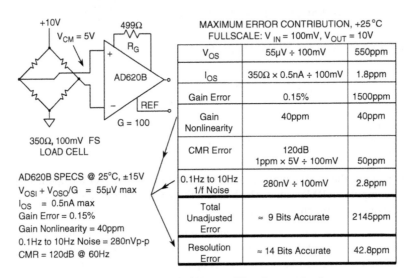

Figure 2-24: AD620B bridge amplifier dc error budget

In Amp Performance Tables

Figure 2-25 shows a selection of precision in amps designed primarily for operation on dual supplies. It should be noted that the AD620 is capable of single 5 V supply operation (see Figure 2-15), but neither its input nor its output are capable of rail-to-rail swings.

These tables allow at-a-glance inspection of key errors, which can be critical in getting the most performance from a system. From Figure 2-25 for example, it can be noted that the use of an AD621 in lieu of the AD620B in the gain-of-100 bridge circuit of Figure 2-24 allows reduction of the gain nonlinearity component of error by a factor of four times. It is also important to separate out errors that can be calibrated out as mentioned above, and those that can only be minimized by device specification improvements. Comparison of the AD620B and the AD622 specifications, for example, shows a higher V_{OS} for the latter. But, since V_{OS} can be calibrated out, the fact that it is higher for the AD622 isn't material to this particular application. The gain nonlinearity between the AD620 and AD622 is the same, so in an auto-cal system, they would likely perform comparably. On the other hand, the AD621B would be preferable for its lower gain nonlinearity, as noted.

	Gain Accuracy *	Gain Nonlinearity	V_{OS} Max	V_{OS} TC	CMR Min	0.1Hz to 10Hz p-p Noise
AD524C	0.5% / P	100ppm	50μV	0.5μV/°C	120dB	0.3μV
AD620B	0.5% / R	40ppm	50μV	0.6μV/°C	120dB	0.28μV
AD621B[1]	0.05% / P	10ppm	50μV	1.6μV/°C	100dB	0.28μV
AD622	0.5% / R	40ppm	125μV	1μV/°C	103dB	0.3μV
AD624C[2]	0.25% / R	50ppm	25μV	0.25μV/°C	130dB	0.2μV
AD625C	0.02% / R	50ppm	25μV	0.25μV/°C	125dB	0.2μV
AMP01A	0.6% / R	50ppm	50μV	0.3μV/°C	125dB	0.12μV
AMP02E	0.5% / R	60ppm	100μV	2μV/°C	115dB	0.4μV

*/P = Pin Programmable
*/R = Resistor Programmable

[1] G = 100
[2] G = 500

Figure 2-25: Precision in amps: data for V_s = ±15 V, G = 1000

In amps specifically designed for single supply operation are also shown, in Figure 2-26. It should be noted that although the specifications in this figure are given for a single 5 V supply, all of the amplifiers are also capable of dual supply operation and are specified for both dual and single supply operation on their data sheets. In addition, the AD623 and AD627 will operate on a single 3 V supply

	Gain Accuracy *	Gain Nonlinearity	V_{OS} Max	V_{OS} TC	CMR Min	0.1Hz to 10Hz p-p Noise	Supply Current
AD623B	0.5% / R	50ppm	100μV	1μV/°C	105dB	1.5μV	575μA
AD627B	0.35% / R	10ppm	75μV	1μV/°C	85dB	1.5μV	85μA
AMP04E	0.4% / R	250ppm	150μV	3μV/°C	90dB	0.7μV	290μA
AD626B[1]	0.6% / P	200ppm	2.5mV	6μV/°C	80dB	2μV	700μA

*/P = Pin Programmable
*/R = Resistor Programmable

[1] Differential Amplifier, G = 100

Figure 2-26: Single-supply in amps: data for V_s = 5 V, G = 1000

Note that the AD626 is not a true in amp, but is in fact a differential amplifier with a thin-film input attenuator that allows the common-mode voltage to exceed the supply voltages. This device is designed primarily for high- and low-side current-sensing applications. It will also operate on a single 3 V supply.

In Amp Input Overvoltage Protection

As interface amplifiers for data acquisition systems, in amps are often subjected to input overloads, i.e., voltage levels in excess of the full scale for the selected gain range. The manufacturer's "absolute maximum" input ratings for the device should be closely observed. As with op amps, many in amps have absolute maximum input voltage specifications equal to $\pm V_S$.

In some cases, external series resistors (for current limiting) and diode clamps may be used to prevent overload if necessary (see Figure 2-27). Some in amps have built-in overload protection circuits in the form of series resistors. For example, the AD620 series have thin film resistors, and the substrate isolation they provide allows input voltages that can exceed the supplies. Other devices use series-protection FETs; for example, the AMP02 and the AD524, because they act as a low impedance during normal operation, and a high impedance during overvoltage fault conditions. In any instance however, there are always finite safe limits to applied overvoltage (Figure 2-27).

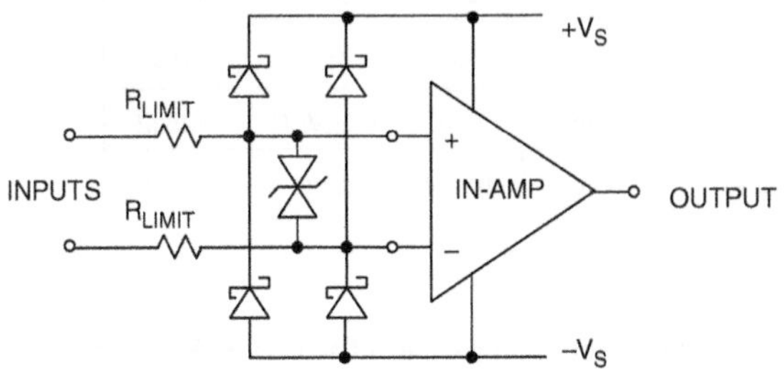

- Always Observe Absolute Maximum Data Sheet Specs
- Schottky Diode Clamps to the Supply Rails Will Limit Input to Approximately $\pm V_S$ ±0.3V, TVSs Limit Differential Voltage
- External Resistors (or Internal Thin-Film Resistors) Can Limit Input Current, but will Increase Noise
- Some In-Amps Have Series-Protection Input FETs for Lower Noise and Higher Input Overvoltages (up to ±60V, Depending on Device)

Figure 2-27: In amp input overvoltage considerations

In some instances, an additional Transient Voltage Suppressor (TVS) may be required across the input pins to limit the maximum differential input voltage. This is especially applicable to three-op-amp in amps operating at high gain with low values of R_G.

A more detailed discussion of input overvoltage and EMI/RFI protection can be found in Chapter 7 of this book.

In Amp Applications

Some representative in amp applications round out this section, illustrating how the characteristics lend utility and efficiency to a range of circuits.

In Amp Bridge Amplifier

In amps are widely used as precision signal conditioning elements. A popular application is a bridge amplifier, shown in Figure 2-28. The in amp is ideally suited for this application because the bridge output is fundamentally balanced, and the in amp presents it with a truly balanced high impedance load. The nominal resistor values in the bridge can range from 100 Ω to several kΩ, but 350 Ω is popular for most precision load cells.

Full-scale output voltages from a typical bridge circuit can range from approximately 10 mV to several hundred mV. Typical in amp gains in the order of 100 to 1000 are therefore ideally suited for amplifying these small voltages to levels compatible with popular analog-to-digital converter (ADC) input voltage ranges (usually 1 V to 10 V full scale).

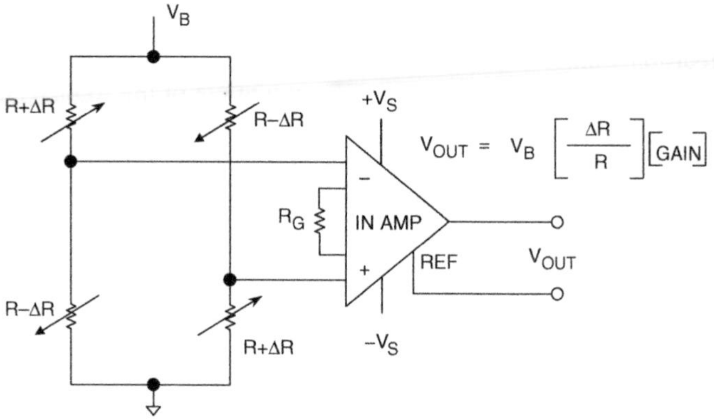

Figure 2-28: Generalized bridge amplifier using an in amp

In addition, the in amp's high CMR at power line frequencies allows common-mode noise to be rejected, when the bridge must be located remotely from the in amp.

Note that a much more thorough discussion of bridge applications can be found in Chapter 4 of this book.

In Amp A/D Interface

Interfacing bipolar signals to single-supply ADCs presents a challenge. The bipolar signal must be amplified and level-shifted into the input range of the ADC. Figure 2-29 shows how this translation can be achieved using the AD623 in amp, when interfacing a bridge circuit to the AD7776 10-bit, 2.5 μs ADC.

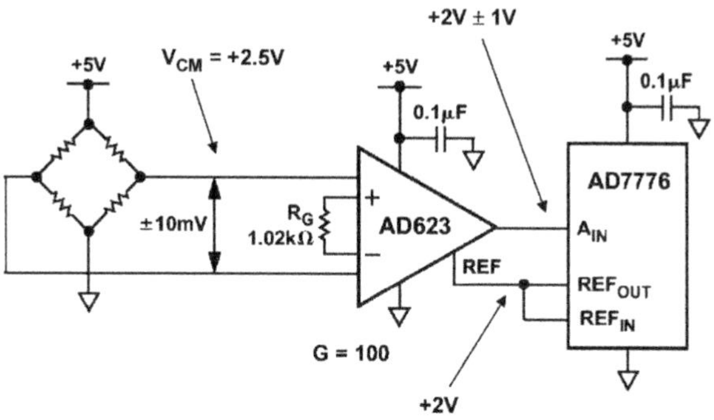

Figure 2-29: Single-supply data acquisition system

The bridge circuit is excited by a 5 V supply. The full-scale output from the bridge (±10 mV) therefore has a common-mode voltage of 2.5 V. The AD623 removes the common-mode component, and amplifies the bridge output by a factor of 100 (R_G = 1.02 kΩ).

This results in an output signal swing of ±1 V. This signal is level-shifted by connecting the REF pin of the AD623 to the 2 V REF_{OUT} of the AD7776 ADC. This sets the common-mode output voltage of the AD623 to 2 V, and the resulting signal into the ADC is +2 V ±1 V, corresponding to the input range of the AD7776.

In Amp Driven Current Source

Figure 2-30 shows a precision voltage-controlled current source using an in amp. The input voltage V_{IN} develops an output voltage, V_{OUT}, equal to GV_{IN} between the output pin of the AD620 and the REF pin. With the connections shown, V_{OUT} is also applied across sense resistor R_{SENSE}, thus developing a load current of V_{OUT}/R_{SENSE}. The OP97 acts as a unity gain buffer to isolate the load from the 20 kΩ impedance of the REF pin of the AD620. In this circuit the input voltage can be floating with respect to the load ground (as long as there exists a path for the in amp bias currents). The high CMR of the in amp allows high accuracy to be achieved for the load current, despite CM voltages.

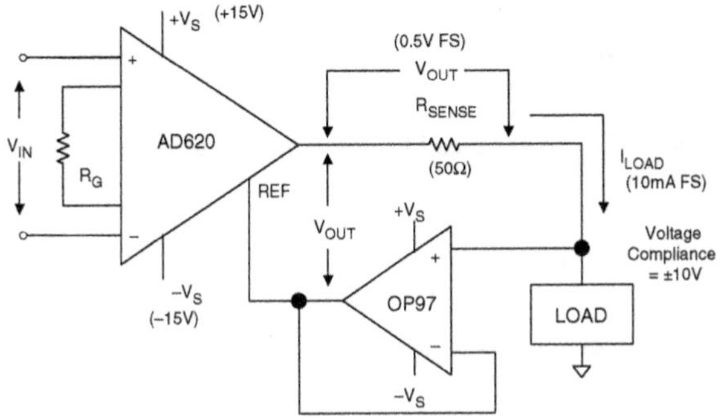

$$I_{LOAD} = \frac{V_{OUT}}{R_{SENSE}} = \frac{G\,V_{IN}}{R_{SENSE}}$$

Figure 2-30: Precision voltage-controlled current source using an in amp

The circuit will work for both large and small values of G in the AD620. The most simple form would be to let G = 1 with R_G open. In this case, $V_{OUT} = V_{IN}$, and I_{LOAD} is proportional to V_{IN}. But the gain factor of the in amp can readily be used to scale almost any input voltage to a desired current level.

The output load voltage compliance is typically ±10 V when operating on ±15 V power supplies, and load currents up to ±15 mA are allowable, limited by the AD620's drive. A typical operating condition might be a full scale load current of 10 mA, a full-scale $V_{OUT} = 0.5$ V, and $R_{SENSE} = 50\ \Omega$.

For small values of R_{SENSE}, the OP97 buffer could possibly be eliminated provided the resulting error incurred by the loading effect of the AD620 REF pin is acceptable. In this case the load and bottom R_{SENSE} node would be connected directly to the in amp REF pin.

Many other useful variations of the basic circuit exist, and can easily be added. For currents of up to 50 mA, a unity gain, low offset buffer can be added between the AD620 output and the top of R_{SENSE}. This will remove all load current from the AD620, allowing it to operate with greatest linearity.

The circuit is also very useful at very small currents. It will work well with the OP97 down to around 1 μA, before bias current of the op amp becomes a performance limitation. For even lower currents, a precision JFET op amp such as the AD8610 can easily be substituted. This step will allow precise low level currents, down to below 1 nA. Note that the AD8610 must be operated on supplies of ±13 V or less, but this isn't necessarily a problem (the AD620 will still operate well on supplies as low as ±2.5 V).

A factor that may not be obvious is that the output current capability of this current source is bilateral, as it is shown. This makes this form of current source a great advantage over a Howland type current source, which are always problematic with the numerous resistors required, which must be well-matched and stable for good performance. In contrast, the current source of Figure 2-30 is clean and efficient, requires no matched resistors, and is precise over very wide current ranges.

In Amp Remote Load Driver

Often remote loads present a problem in driving, when high accuracy must be maintained at the load end. For this type of requirement, an in amp (or simple differential amplifier) with separate SENSE/FORCE terminals can serve very well, providing a complete solution in one IC. Most of the more popular in amps available today have removed the separate SENSE/FORCE connections, due to the pin limitations of an 8-pin package (AD620, etc.). However, many classic in amps such as the AMP01 do have access to the SENSE/FORCE pins, and can perform remote sensing, as shown in Figure 2-31.

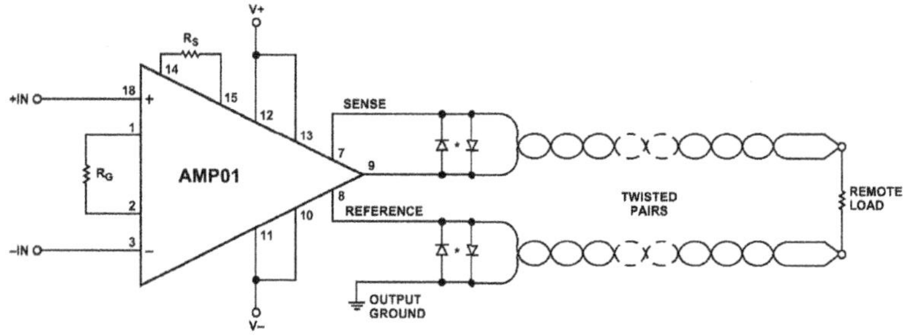

Figure 2-31: Precision in amp remote load driver using FORCE/SENSE connections

In this circuit a quad cable composed of two twisted pairs is used. One pair is dedicated to the load HIGH side, the other to the LO side. At the remote end, the load is connected as shown, with each twisted pair terminated at one end of the load.

Although the full load current still flows in the FORCE (AMP01 Pin 9) and OUTPUT GROUND connections, the resulting drop does not create an error, since the remote sensing of the second lead of each pair is returned back to the driver, and carries comparatively very little current. The reverse-parallel connected diodes are optional, and perform a "safety-valve" function, in case a sense line becomes open-circuit (100 Ω resistors might also be used).

The AMP01 is valuable to this function not simply because of the SENSE/FORCE capability, but because it also is capable of 50 mA output currents and is stable with the capacitive loading presented by a cable. Alternately, a precision differential amplifier like the AMP03 can also be used, at lower current levels.

For additional in amp background and reference material, see References 12 through 15.

References: Instrumentation Amplifiers

1. MIL-PRF-55182G, "Resistors, Fixed, Film, Nonestablished Reliability, Established Reliability, And Space Level, General Specification For," June 9, 1997.
2. Robert Demrow, "Narrowing the Margin of Error," **Electronics**, April 15, 1968, pp. 108–117.
3. Robert Demrow, "Evolution from Operational Amplifier to Data Amplifier," **Analog Devices Application Note**, September, 1968.
4. David Birt, "Electronically Balanced Analogue-Line Interfaces," **Proceedings of Institute of Acoustics Conference**, Windermere, U.K., Nov. 1990.
5. Walt Jung, "Op Amps in Line-Driver and Receiver Circuits, Part 1," **Analog Dialogue**, Vol. 26 No. 2, 1992.
6. W. Jung, A. Garcia, "Op Amps in Line-Driver and Receiver Circuits, Part 2," **Analog Dialogue**, Vol. 27, No. 1, 1993.
7. Data sheet for **AD627 Micropower, Single and Dual Supply Rail-to-Rail Instrumentation Amplifier**, www.analog.com.
8. S. Wurcer, L. Counts, "A Programmable Instrumentation Amplifier for 12-Bit Resolution Systems," **IEEE Journal of Solid-State Circuits**, Vol. SC-17 #6, Dec. 1982, pp. 1102–1111.
9. L. Counts, S. Wurcer, "Instrumentation Amplifier Nears Input Noise Floor," **Electronic Design**, June 10, 1982, p. 177.
10. Data sheet for **AD620 Low Cost, Low Power Instrumentation Amplifier**, www.analog.com
11. Data sheet for **AD623 Single Supply, Rail-to-Rail, Low Cost Instrumentation Amplifier**, www.analog.com.
12. Charles Kitchen and Lew Counts, **A Designer's Guide to Instrumentation Amplifiers**, Analog Devices, 2000.
13. Sections 2, 3, 4, of Walt Kester, Editor, **Practical Design Techniques for Sensor Signal Conditioning**, Analog Devices, Inc., 1999, ISBN: 0-916550-20-6.
14. Walter Borlase, "Application/Analysis of the AD520 Monolithic Data Amplifier," **Analog Devices Application Note**, 1972.
15. Jeff Riskin, **A User's Guide to IC Instrumentation Amplifiers, Analog Devices AN244**, January, 1978.

Classic Cameo

Robert Demrow's "Evolution from Operational Amplifier to Data Amplifier"

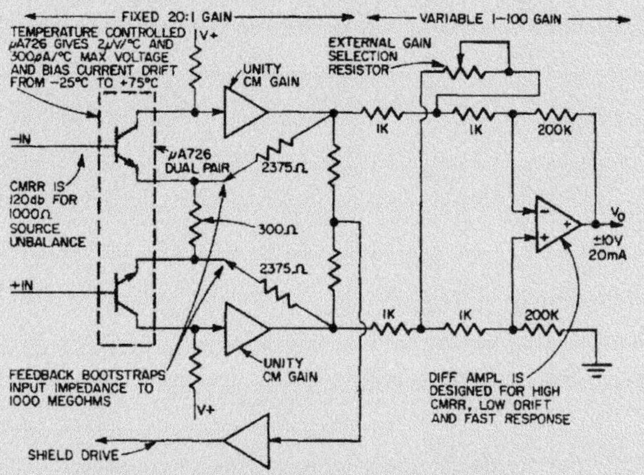

MODEL 601 WIDEBAND DIFFERENTIAL DC AMPLIFIER

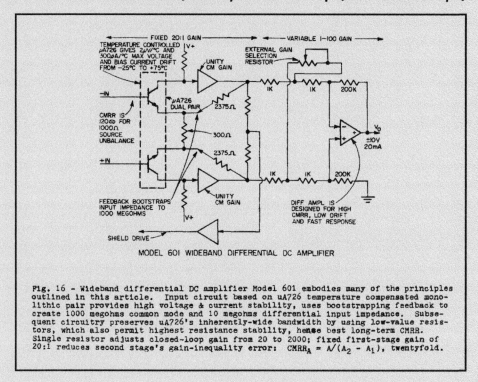

Fig. 16 - Wideband differential DC amplifier Model 601 embodies many of the principles outlined in this article. Input circuit based on uA726 temperature compensated monolithic pair provides high voltage & current stability, uses bootstrapping feedback to create 1000 megohms common mode and 10 megohms differential input impedance. Subsequent circuitry preserves uA726's inherently-wide bandwidth by using low-value resistors, which also permit highest resistance stability, hence best long-term CMRR. Single resistor adjusts closed-loop gain from 20 to 2000; fixed first-stage gain of 20:1 reduces second stage's gain-inequality error: $CMRR_A = A/(A_2 - A_1)$, twentyfold.

As applications engineering manager in the early years of ADI, Robert Demrow published numerous articles and application notes. It is a testament to the quality of these articles that most of them are still germane today—due in no small part to their lucid outlining of fundamental principles.

Demrow's 1968 application note, "Evolution from Operational Amplifier to Data Amplifier" outlined the relevant amplifier operating principles for retrieving analog signals from a noisy environment. It also introduced the ADI Model 601 *data amplifier* (above). Of course, a data amplifier is what we know today as an instrumentation amplifier. Within his Figure 16 can be seen several key operating principles: 1) dual high impedance inputs, as necessary for high CMR, 2) the use of a precision bipolar transistor differential pair front end, for low offset and drift (the μA726),[1] 3) a balanced, three amplifier stage topology.

It is interesting to note that some more popular IC in amps of 2004 utilize many of the same principles—for example, the AD620 family. Back in 1968, Robert Demrow outlined a host of sound design concepts, leading the way to later solid-state developments and the completely monolithic in amp ICs of today.

---

[1] James N. Giles, Editor, "The μA726 Temperature-Stabilized Transistor Pair," Chapter 8, **Fairchild Semiconductor Linear Integrated Circuits Handbook**, Fairchild Semiconductor, 1967.

Programmable Gain Amplifiers

Walt Kester, James Bryant

Most data acquisition systems with wide dynamic range need some method of adjusting the input signal level to the analog-to-digital-converter (ADC). Typical ADC full-scale input voltage ranges lie between 2 V and 10 V. To achieve the rated precision of the converter, the maximum input signal should be fairly near its full-scale voltage.

Transducers, however, have a very wide range of output voltages. High gain is needed for a small sensor voltage, but with a large output, a high gain will cause the amplifier or ADC to saturate. So, some type of predictably controllable gain device is needed. Amplifiers with programmable gain have a variety of applications, and Figure 2-32 lists some of them.

- Instrumentation
- Photodiode circuits
- Ultrasound preamplifiers
- Sonar
- Wide dynamic range sensors
- Driving ADCs (some ADCs have on-chip PGAs)
- Automatic gain control (AGC) loops

Figure 2-32: Programmable gain amplifier (PGA) applications

Such a device has a gain that is controlled by a dc voltage or, more commonly, a digital input. This device is known as a *programmable gain amplifier*, or PGA. Typical PGAs may be configured either for selectable *decade gains* such as 10, 100, 100, etc., or they might also be configured for *binary gains* such as 1, 2, 4, 8, etc. It is a function of the end system of course, which type might be the more desirable.

It should be noted that a factor common to the above application examples is that the different types of signals being handled is diverse. Some may require wide bandwidth, others very low noise, from either high or low impedance sources. The inputs may be single-ended, or they may be differential, crossing over into the realm of the just-discussed in amps.

The output from the PGA may be required to drive some defined input range of an ADC, or it may be part of a smaller subsystem, such as an AGC or gain-ranging loop. The circuits following fall into a range of categories addressing some of these requirements.

A PGA is usually located between a sensor and its ADC, as shown in Figure 2-33. Additional signal conditioning may take place before or after the PGA, depending on the application. For example, a photodiode needs a current-to-voltage converter between it and the PGA. In most other systems, it is better to place the gain first, and condition a larger signal. This reduces errors introduced by the signal conditioning circuitry.

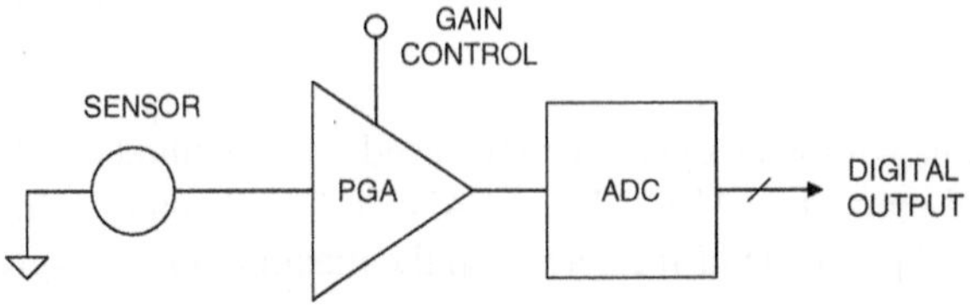

Used to increase the dynamic range of the system

A PGA with a gain of 1 to 2 theoretically increases the dynamic range by 6dB.

A gain of 1 to 4 gives a 12dB increase, etc.

Figure 2-33: PGAs in data acquisition systems

To understand the benefits of variable gain, assume an ideal PGA with two settings, gains of one and two. The dynamic range of the system is increased by 6 dB. Increasing the gain to a maximum four results in a 12 dB increase in dynamic range. If the LSB of an ADC is equivalent to 10 mV of input voltage, the ADC cannot resolve smaller signals, but when the gain of the PGA is increased to two, input signals of 5 mV may be resolved.

Thus, a central processor can combine PGA gain information with the digital output of the ADC to increase its resolution by one bit. Essentially, this is the same as adding additional resolution to the ADC. In fact, a number of ADCs now have on-chip PGAs for increased dynamic range (AD77xx-series, for example, covered later).

PGA Design Issues

In practice, PGAs aren't ideal, and their error sources must be studied and dealt with. A number of the various PGA design issues are summarized in Figure 2-34.

A fundamental PGA design problem is programming gain accurately. Electromechanical relays have minimal on resistance (R_{ON}), but are unsuitable for gain switching—slow, large, and expensive. CMOS switches are small, but they have voltage-/temperature-dependent R_{ON}, as well as stray capacitance, which may affect PGA ac parameters.

- How to switch the gain
- Effects of the switch on-resistance (R_{ON})
- Gain accuracy
- Gain linearity
- Bandwidth versus frequency versus gain
- Dc offset
- Gain and offset drift over temperature
- Settling time after switching gain

Figure 2-34: PGA design issues

To understand R_{ON}'s effect on performance, consider Figure 2-35, a poor PGA design. A noninverting op amp has four different gain-set resistors, each grounded by a switch, with an R_{ON} of 100 Ω–500 Ω. Even with R_{ON} as low as 25 Ω, the gain of 16 error would be 2.4%, worse than 8-bits. R_{ON} also changes over temperature and switch-switch.

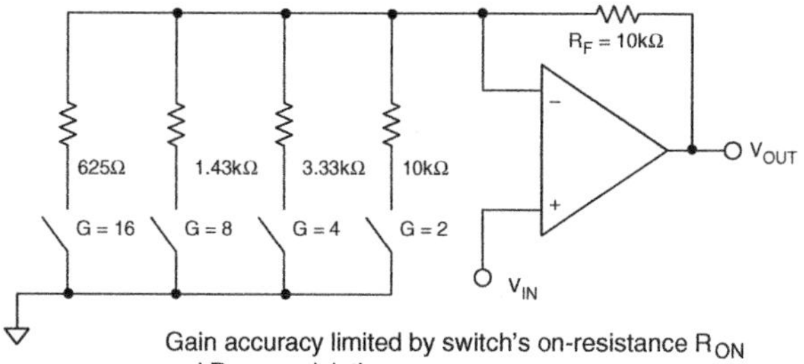

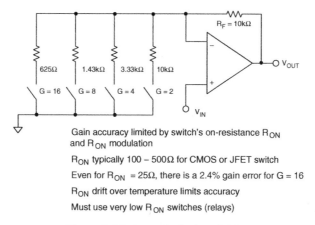

Gain accuracy limited by switch's on-resistance R_{ON} and R_{ON} modulation

R_{ON} typically 100 – 500Ω for CMOS or JFET switch

Even for R_{ON} = 25Ω, there is a 2.4% gain error for G = 16

R_{ON} drift over temperature limits accuracy

Must use very low R_{ON} switches (relays)

Figure 2-35: A poorly designed PGA

To attempt "fixing" this design, the resistors might be increased, but noise and offset could then be a problem. The only way to accuracy with this circuit is to use relays, with virtually no R_{ON}. Only then will the few mΩ of relay R_{ON} be a small error vis-à-vis 625 Ω.

It is much better to use a circuit *insensitive* to R_{ON}. In Figure 2-36, the switch is placed in series with the inverting input of an op amp. Since the op amp input impedance is very large, the switch R_{ON} is now irrelevant, and gain is now determined solely by the external resistors. Note: R_{ON} may add a small offset error if op amp bias current is high (if this is the case, it can readily be compensated with an equivalent resistance at V_{IN}).

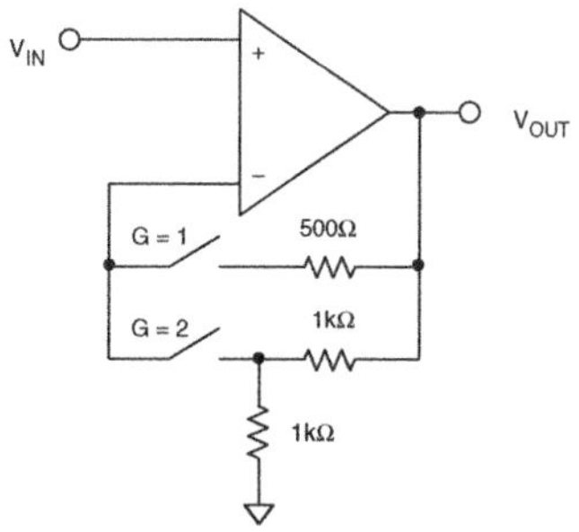

- R_{ON} is not in series with gain setting resistors
- R_{ON} is small compared to input impedance
- Only slight offset errors occur due to bias current flowing through the switches

Figure 2-36: Alternate PGA configuration minimizes the effects of R_{ON}

PGA Applications

The following section illustrates several PGA circuits using the above and other concepts.

AD526 Software Programmable PGA

The AD526 amplifier uses the just-described PGA architecture, integrating it onto a single chip, as diagrammed in Figure 2-37 (see References 1 and 2). The AD526 has five binary gain settings from 1 to 16, and its internal JFET switches are connected to the inverting input of the amplifier as in Figure 2-37. The gain resistors are laser trimmed, providing a maximum gain error of only 0.02%, and a linearity of 0.001%. The use of the FORCE/SENSE terminals connected at the load ensures highest accuracy (it also allows the use of an optional unity-gain buffer, for low impedance loads).

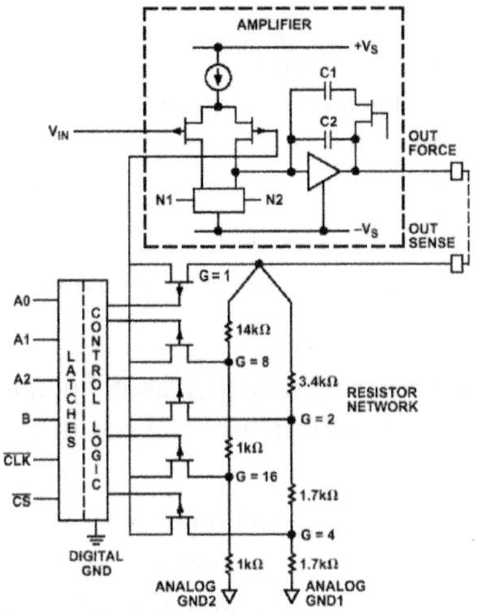

Figure 2-37: AD526 software programmable PGA simplified schematic

Functionally speaking, the AD526 is a programmable, precision, noninverting op amp gain stage, logic programmable over a range of 1 to 16 times V_{IN}. It typically operates from a ±15 V power supply, and has ±10 V output range (like a conventional op amp).

The key specifications for the AD526 are summarized in Figure 2-38.

- Software programmable binary gains from 1 to 16
- Low bias current JFET input stage
- Worst-case gain error: 0.02% (12-bit performance)
- Maximum gain nonlinearity: 0.001%
- Gain change settling time: 5.6µs (G = 16)
- Small signal bandwidth: 4MHz (G = 1), 0.35MHz (G = 16)
- Latched TTL-compatible control inputs

Figure 2-38: AD526 PGA key specifications

Low Noise PGA

This same design concepts can be used to build a low noise PGA as shown in Figure 2-39. It uses a single op amp, a quad switch, and precision resistors. The lower noise AD797 replaces the JFET input op amp of the AD526, but almost any voltage feedback op amp could be used in this circuit. The ADG412 was picked for its R_{ON} of 35 Ω.

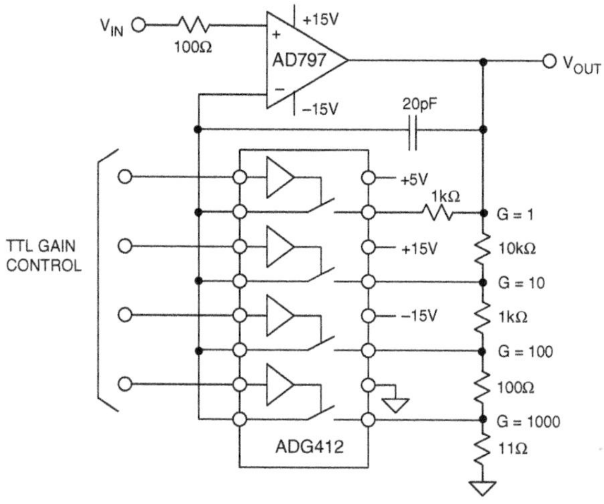

Figure 2-39: A very low noise PGA using the AD797 and the ADG412

The resistors were chosen to give decade gains of 1, 10, 100, and 1000, but if other gains are required, the resistor values may easily be altered. Ideally, a single trimmed resistor network should be used both for initial gain accuracy and for low drift over temperature. The 20 pF feedback capacitor ensures stability and holds the output voltage when the gain is switched. The control signal to the switches turns one switch off a few nanoseconds before the second switch turns on. During this break, the op amp is open-loop. Without the capacitor the output would start slewing. Instead, the capacitor holds the output voltage during switching. Since the time that both switches are open is very short, only 20 pF is needed. For slower switches, a larger capacitor may be necessary.

The PGA's input voltage noise spectral density at a gain of 1000 is only $1.65\,nV/\sqrt{Hz}$ at 1 kHz, only slightly higher than the noise performance of the AD797 alone. The increase is due to the ADG412 noise, and the current noise of the AD797 flowing through R_{ON}.

The accuracy of the PGA is important in determining the overall accuracy of a system. The AD797 has a bias current of 0.9 µA, which, flowing in a 35 Ω R_{ON}, results in an additional offset error of 31.5 µV. Combined with the AD797 offset, the total V_{OS} becomes 71.5 µV (max). Offset temperature drift is affected by the change in bias current and R_{ON}. Calculations show that the total temperature coefficient increases from 0.6 µV/°C to 1.6 µV/°C. Note that while these errors are small (and may not matter in the end) it is still important to be aware of them.

In practice, circuit accuracy and gain TC will be determined by the external resistors. Input characteristics such as common mode range and input bias current are determined solely by the AD797. A performance summary is shown below in Figure 2-40.

- R_{ON} adds additional input offset and drift:
  - $\Delta V_{OS} = I_b R_{ON} = (0.9\mu A)(35\Omega) = 31.5\mu V$ (max)
  - Total $V_{OS} = 40\mu V + 31.5\mu V = 71.5\mu V$ (max)
- Temperature drift due to R_{ON}:
  - At +85°C, $\Delta V_{OS} = (2\mu A)(45\Omega) = 90\mu V$ (max)
- Temperature coefficient total:
  - $\Delta V_{OS} / \Delta T = 0.6\mu V/°C + 1.0\mu V/°C = 1.6\mu V/°C$ (max)
  - Note: 0.6µV/°C is due to the AD797B
- RTI Noise : $1.65 nV/\sqrt{Hz}$ @ 1kHz, G = 1000
- Gain switching time < 1µs, G = 10

Figure 2-40: AD797/ADG412 PGA performance summary

DAC Programmed PGA

Another PGA configuration uses a DAC in the feedback loop of an op amp to adjust the gain under digital control, as shown in Figure 2-41. The digital code of the DAC controls its attenuation with respect to its reference input V_{REF}, acting functionally similar to a potentiometer. Attenuating the feedback signal increases the closed-loop gain.

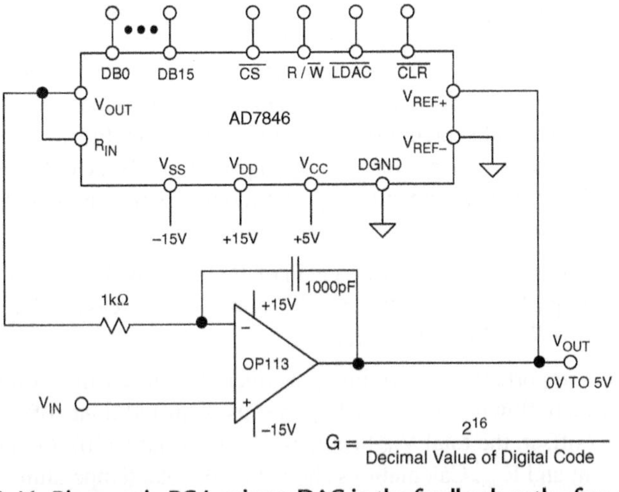

Figure 2-41: Binary gain PGA using a DAC in the feedback path of an op amp

A noninverting PGA of this type requires a multiplying DAC with a voltage mode output. Note that a multiplying DAC is a DAC with a wide reference voltage range, *which includes zero.* For most applications of the PGA, the reference input must be capable of handling bipolar signals. The AD7846 is a 16-bit converter that meets these requirements. In this application, it is used in the standard 2-quadrant multiplying mode.

The OP113 is a low drift, low noise amplifier, but the choice of the amplifier is flexible, and depends on the intended application. The input voltage range depends on the output swing of the AD7846, which is 3 V less than the positive supply, and 4 V above the negative supply. A 1000 pF capacitor is used in the feedback loop for stability.

The gain of the circuit is set by adjusting the digital inputs of the DAC, according to the equation given in Figure 2-41. $D_{0\text{-}15}$ represents the decimal value of the digital code. For example, if all the bits were set high, the gain would be 65,536/65,535 = 1.000015. If the eight least significant bits are set high and the rest low, the gain would be 65,536/255 = 257. The bandwidth of the circuit is a fairly high 4 MHz for a gain of +1. However, this does reduce with gain, and for a gain of 256, the bandwidth is only 600 Hz. If the gain-bandwidth product were constant, the bandwidth in a gain of 256 should be 15.6 kHz; but the internal capacitance of the DAC reduces the bandwidth to 600 Hz.

Performance characteristics of this binary PGA are summarized in Figure 2-42.

- Gain Accuracy:
  - 0.003% (G = +1)
  - 0.1% (G = +256)
- Nonlinearity: 0.001% (G = 1)
- Offset: 100µV
- Noise: 50nV/√Hz
- Bandwidth:
  - 4MHz (G = +1)
  - 600Hz (G = +256)

Figure 2-42: Binary gain PGA performance

The gain accuracy of the circuit is determined by the resolution of the DAC and the gain setting. At a gain of 1, all bits are on, and the accuracy is determined by the DNL specification of the DAC, which is ±1 LSB maximum. Thus, the gain accuracy is equivalent to 1 LSB in a 16-bit system, or 0.003%.

However, as the gain is increased, fewer of the bits are on. For a gain of 256, only bit 8 is turned on. The gain accuracy is still dependent on the ±1 LSB of DNL, but now that is compared to only the lowest eight bits. Thus, the gain accuracy is reduced to 1 LSB in an 8-bit system, or 0.4%. If the gain is increased above 256, the gain accuracy is reduced further. The designer must determine an acceptable level of accuracy. In this particular circuit, the gain was limited to 256.

Differential Input PGAs

There are often applications where a PGA with differential inputs is needed, instead of the single-ended types discussed so far. The AD625 combines an instrumentation amplifier topology similar to the AD620 with external gain switching capabilities to accomplish 12-bit gain accuracy (see Reference 3). An external switch is needed to switch between different gain settings, but its on resistance does not significantly affect the gain accuracy due to the unique design of the AD625.

The circuit in Figure 2-43 uses an ADG409 CMOS switch to switch the connections to an external gain-setting resistor network. In the example shown, resistors were chosen for gains of 1, 4, 16, and 64. Other features of the AD625 are 0.001% nonlinearity, wide bandwidth, and very low input noise.

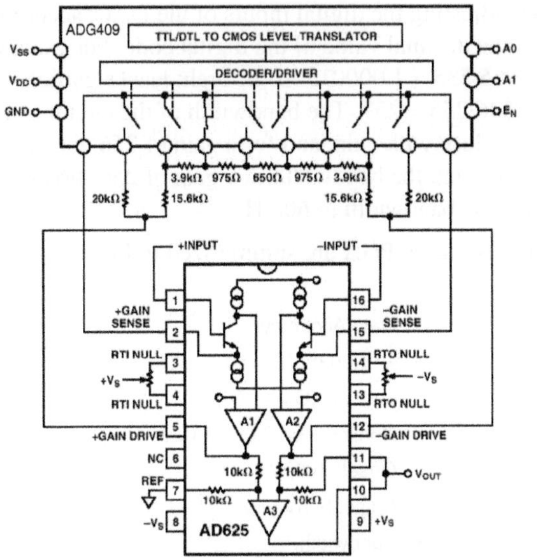

Figure 2-43: A software-programmable gain amplifier

The AD625 is uniquely designed so that the on resistance of the switches does not introduce significant error in the circuit. This can be understood by considering the simplified AD625 circuit shown Figure 2-44. The voltages shown are for an input of +1 mV on +IN and 0 V on –IN. The gain is set to 64 with $R_G = 635\ \Omega$ and the two resistors, $R_F = 20\ k\Omega$.

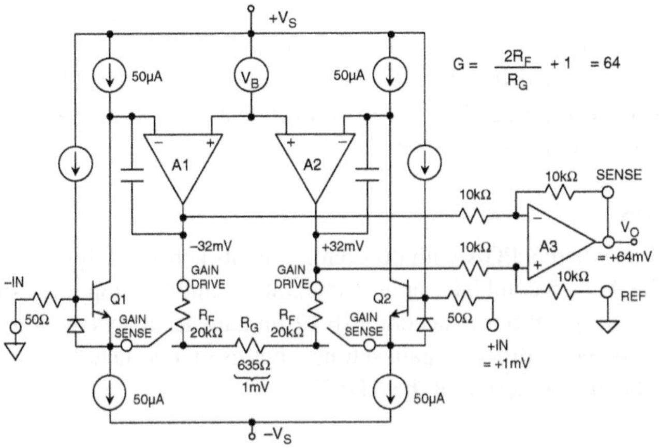

Figure 2-44: AD625 details showing external switches and gain-setting resistors for R_G = 635 Ω, +IN = 1 mV, –IN = 0 V

Since transistors Q1 and Q2 have 50 μA current sources in both their emitters and their collectors, negative feedback around A1 and A2, respectively, will ensure that no net current flows through either *gain sense* pin into either emitter. Since no current flows in the gain sense pins, no current flows in the gain setting switches, and their R_{ON} does not affect either gain or offset. In real life there will be minor mismatches, but the errors are well under the 12-bit level.

The differential gain between the inputs and the A1-A2 outputs is $2R_F/R_G + 1$. The unity-gain difference amplifier and matched resistors removes CM voltage, and drives the output.

Noninverting PGA circuits using an op amp are easily adaptable to single supply operation, but when differential inputs are desired, a single-supply in amp such as the AD623, AD627, or the AMP04 should be used. The AMP04 is used with an external CMOS switch in the single supply in amp PGA shown in Figure 2-45.

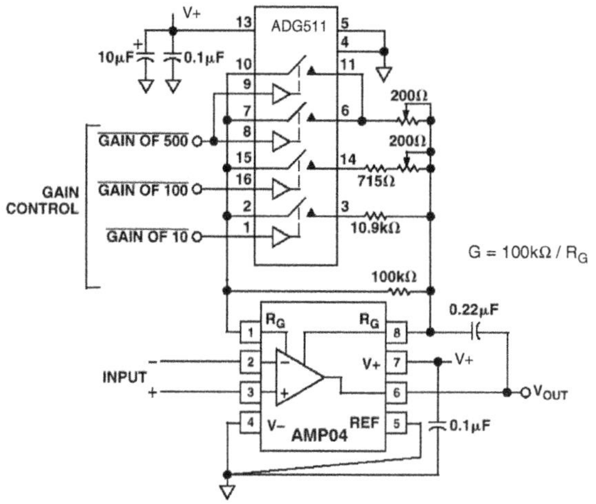

Figure 2-45: Single supply instrumentation PGA using the AMP04 in amp and the ADG511 switch

This circuit has selectable gains of 1, 10, 100, and 500, which are controlled by an ADG511. The ADG511 was chosen as a single supply switch with a low R_{ON} of 45 Ω. A disadvantage of this circuit is that the gain of this circuit is dependent on the R_{ON} of the switches. Trimming is required at the higher gains to achieve accuracy. At a gain of 500, two switches are used in parallel, but their resistance causes a 10% gain error in the absence of adjustment.

ADC with Onboard PGA

Certain ADCs (such as the AD77XX measurement series) have built-in PGAs and other conditioning circuitry. Circuit design with these devices is much easier, because an external PGA and its control logic are not needed. Furthermore, all the errors of the PGA are included in the specifications of the ADC, making error calculations simple.

The PGA gain is controlled over the common ADC serial interface, and the gain setting is factored into the conversion, saving additional calculations to determine input voltage.

This combination of ADC and PGA is very powerful and enables the realization of a highly accurate system, with a minimum of circuit design. As an example, Figure 2-46 shows a simplified diagram of the AD7730 sigma-delta measurement ADC which is optimized for digitizing low voltage bridge outputs directly (as low as 10 mV full scale) to greater than 16 bits noise free code resolution, without the need for external signal conditioning circuits.

Additional information and background reading on PGAs can be found in References 4 and 5.

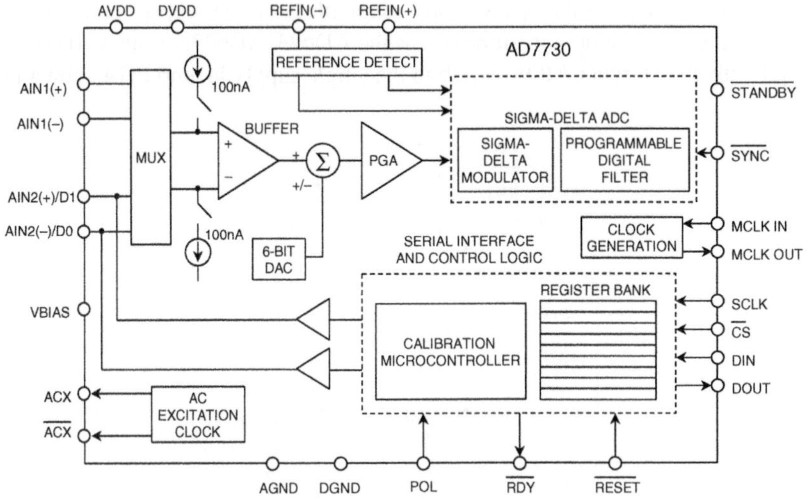

Figure 2-46: AD7730 sigma-delta measurement ADC with on-chip PGA

References: Programmable Gain Amplifiers

1. John Krehbiel, "Monolithic Software Programmable-Gain Amplifier," **Analog Dialogue**, Vol. 21, No. 2, 1987, pp. 12–13.
2. Data sheet for **AD526 Software Programmable Gain Amplifier**, www.analog.com.
3. Data sheet for **AD625 Programmable Gain Instrumentation Amplifier**, www.analog.com.
4. Chapters 2, 3, 8, Walt Kester, Editor, **Practical Design Techniques for Sensor Signal Conditioning**, Analog Devices, 1999, ISBN: 0-916550-20-6.
5. Chapter 2, Walt Kester, Editor, **Linear Design Seminar**, Analog Devices, 1995, ISBN: 0-916550-15-X.

Isolation Amplifiers

Walt Kester, James Bryant

Analog Isolation Techniques

There are many applications where it is desirable, or even essential, for a sensor to have no direct ("galvanic") electrical connection with the system to which it is supplying data. This might be in order to avoid the possibility of dangerous voltages or currents from one half of the system doing damage in the other, or to break an intractable ground loop. Such a system is said to be "isolated," and the arrangement that passes a signal without galvanic connections is known as an *isolation barrier.*

The protection of an isolation barrier works in both directions, and may be needed in either, or even in both. The obvious application is where a sensor may accidentally encounter high voltages, and the system it is driving must be protected. Or a sensor may need to be isolated from accidental high voltages arising downstream in order to protect its environment. Examples include the need to prevent the ignition of explosive gases by sparks at sensors and the protection from electric shock of patients whose ECG, EEG, or EMG is being monitored. The ECG case is interesting, as protection may be required in *both* directions: the patient must be protected from accidental electric shock, but if the patient's heart should stop, the ECG machine must be protected from the very high voltages (>7.5 kV) applied to the patient by the defibrillator that will be used to attempt to restart it. A summary of applications for isolation amplifiers (both analog and digital) is shown in Figure 2-47.

- Sensor is at a High Potential Relative to Other Circuitry (or may become so under Fault Conditions)
- Sensor May Not Carry Dangerous Voltages, Irrespective of Faults in Other Circuitry (e.g. Patient Monitoring and Intrinsically Safe Equipment for use with Explosive Gases)
- To Break Ground Loops

Figure 2-47: Applications for isolation amplifiers

Just as interference, or *unwanted* information, may be coupled by electric or magnetic fields, or by electromagnetic radiation, these phenomena may be used for the transmission of *wanted* information in the design of isolated systems.

The most common isolation amplifiers use *transformers*, which exploit magnetic fields, and another common type uses small high voltage capacitors, exploiting electric fields. *Optoisolators*, which consist of an LED and a photocell, provide isolation by using light, a form of electromagnetic radiation. Different isolators have differing performance: some are sufficiently linear to pass high accuracy analog signals across an isolation barrier. With others, the signal may need to be converted to digital form before transmission for accuracy is to be maintained (note this is a common V/F converter application).

Transformers are capable of analog accuracy of 12 to 16 bits and bandwidths up to several hundred kHz; their maximum voltage rating rarely exceeds 10 kV, and is often much lower. *Capacitively-coupled* isolation amplifiers have lower accuracy, perhaps 12 bits maximum, lower bandwidth, and lower voltage ratings—but they are low cost. Optical isolators are fast and cheap, and can be made with very high voltage ratings (4 kV – 7 kV is one of the more common ratings), but they have poor analog domain linearity and are not usually suitable for direct coupling of precision analog signals.

Linearity and isolation voltage are not the only issues to be considered in the choice of isolation systems. Operating power is of course, essential. Both the input and the output circuitry must be powered and, unless there is a battery on the isolated side of the isolation barrier (which is possible, but rarely convenient), some form of isolated power must be provided. Systems using transformer isolation can easily use a transformer (either the signal transformer or another one) to provide isolated power. It is, however, impractical to transmit useful amounts of power by capacitive or optical means. Systems using these forms of isolation must make other arrangements to obtain isolated power supplies—this is a powerful consideration in favor of choosing transformer-isolated isolation amplifiers: they almost invariably include an isolated power supply.

The isolation amplifier has an input circuit that is galvanically isolated from the power supply and the output circuit. In addition, there is minimal capacitance between the input and the rest of the device. Therefore, there is no possibility for dc current flow and minimum ac coupling. Isolation amplifiers are intended for applications requiring safe, accurate measurement of low frequency voltage or current (up to about 100 kHz) in the presence of high common-mode voltage (to thousands of volts) with high common-mode rejection. They are also useful for line-receiving of signals transmitted at high impedance in noisy environments, and for safety in general-purpose measurements, where dc and line-frequency leakage must be maintained at levels well below certain mandated minimums. Principal applications are in electrical environments of the kind associated with medical equipment, conventional and nuclear power plants, automatic test equipment, and industrial process control systems.

AD210 Three-Port Isolator

In a basic two-port form of isolator, the output and power circuits are not isolated from one another. A *three-port isolator* (input, power, output) is shown in Figure 2-48. Note that in this diagram, the input circuits, output circuits, and power source are all isolated from one another. This figure represents the circuit architecture of a self-contained isolator, the AD210 (see References 1 and 2).

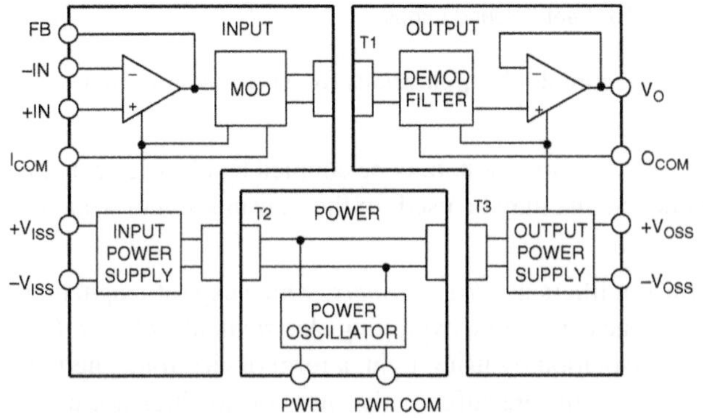

Figure 2-48: AD210 three-port isolation amplifier

An isolator of this type requires power from a two-terminal dc power supply (PWR, PWR COM). An internal oscillator (50 kHz) converts the dc power to ac, which is transformer-coupled to the shielded input section, then converted to dc for the input stage and the auxiliary power output.

The ac carrier is also modulated by the input stage amplifier output, transformer-coupled to the output stage, demodulated by a phase-sensitive demodulator (using the carrier as the reference), filtered, and buffered using isolated dc power derived from the carrier. The AD210 allows the user to select gains from 1 to 100, using external resistors with the input section op amp. Bandwidth is 20 kHz, and voltage isolation is 2500 V rms (continuous) and ±3500 V peak (continuous).

The AD210 is a three-port isolation amplifier, thus the power circuitry is isolated from both the input and the output stages and may therefore be connected to either (or to neither), without change in functionality. It uses transformer isolation to achieve 3500 V isolation with 12-bit accuracy.

Key specifications for the AD210 are summarized in Figure 2-49.

- Transformer Coupled
- High Common-Mode Voltage Isolation:
  - 2500V RMS Continuous
  - ±3500V Peak Continuous
- Wide Bandwidth: 20kHz (Full Power)
- 0.012% Maximum Linearity Error
- Input Amplifier: Gain 1 to 100
- Isolated Input and Output Power Supplies, ±15V, ±5mA

Figure 2-49: AD210 isolation amplifier key specifications

Motor Control Isolation Amplifier

A typical isolation amplifier application using the AD210 is shown in Figure 2-50. The AD210 is used with an AD620 instrumentation amplifier in a current-sensing system for motor control. The input of the AD210, being isolated, can be directly connected to a 110 V or 230 V power line without protection being necessary. The input section's isolated ±15 V powers the AD620, which senses the voltage drop in a small value current sensing resistor. The AD210 input stage op amp is simply connected as a unity-gain follower, which minimizes its error contribution. The 110 V or 230 V rms common-mode voltage is ignored by this isolated system.

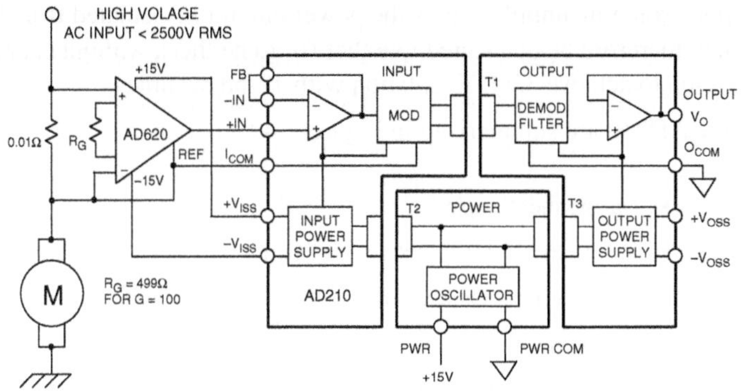

Figure 2-50: Motor control current sensing

Within this system the AD620 preamp is used as the system scaling control point, and will produce and output voltage proportional to motor current, as scaled by the sensing resistor value and gain as set by the AD620's R_G. The AD620 also improves overall system accuracy, as the AD210 V_{OS} is 15 mV, versus the AD620's 30 μV (with less drift also). Note that if higher dc offset and drift are acceptable, the AD620 may be omitted and the AD210 connected at a gain of 100.

Optional Noise Reduction Post Filter

Due to the nature of this type of carrier-operated isolation system, there will be certain operating situations where some residual ac carrier component will be superimposed upon the recovered output dc signal. When this occurs, a low impedance passive RC filter section following the output stage may be used (if the following stage has a high input impedance, i.e., nonloading to this filter). Note that will be the case for many high input impedance sampling ADCs, which appear essentially as a small capacitor. A 150 Ω resistance and 1 nF capacitor will provide a corner frequency of about 1 kHz. Note also that the capacitor should be a film type for low errors, such as polypropylene.

AD215 Two-Port Isolator

The AD215 is a high speed, two-port isolation amplifier, designed to isolate and amplify wide bandwidth analog signals (see Reference 3). The innovative circuit and transformer design of the AD215 ensures wide-band dynamic characteristics, while preserving dc performance specifications. An AD215 block diagram is shown in Figure 2-51.

The AD215 provides complete galvanic isolation between the input and output of the device, which also includes the user-available front-end isolated bipolar power supply. The functionally complete design,

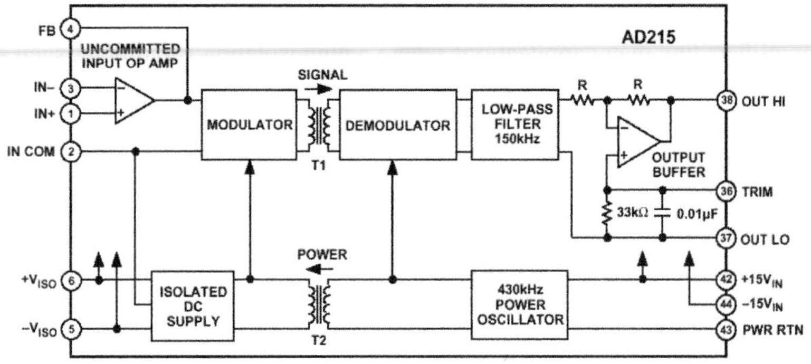

Figure 2-51: AD215 120kHz low distortion two-port isolation amplifier

powered by a ±15 V dc supply on the output side, eliminates the need for a user supplied isolated dc/dc converter. This permits the designer to minimize circuit overhead and reduce overall system design complexity and component costs.

The design of the AD215 emphasizes maximum flexibility and ease of use in a broad range of applications where fast analog signals must be measured under high common-mode voltage (CMV) conditions.

The AD215 has a ±10 V input/output range, a specified gain range of 1 V/V to 10 V/V, a buffered output with offset trim and a user-available isolated front end power supply which produces ±15 V dc at ±10 mA. The key specifications of the AD215 are summarized in Figure 2-52.

- Isolation voltage: 1500V rms
- Full power bandwidth: 120kHz
- Slew rate: 6V/µs
- Harmonic distortion: –80dB @ 1kHz
- 0.005% maximum linearity error
- Gain range: 1 to 10
- Isolated input power supply: ±15V @ ±10mA

Figure 2-52: AD215 isolation amplifier key specifications

Digital Isolation Techniques

Analog isolation amplifiers find many applications where a high isolation is required, such as in medical instrumentation. Digital isolation techniques provide similar galvanic isolation, and are a reliable method of transmitting digital signals without ground noise.

Optocouplers (also called optoisolators) are useful and available in a wide variety of styles and packages. A typical optocoupler based on an LED and a phototransistor is shown in Figure 2-53. A current of approximately 10 mA drives an LED transmitter, with light output is received by a phototransistor. The light produced by the LED saturates the phototransistor. Input/output isolation of 5000 V rms to 7000 V rms is common. Although fine for digital signals, optocouplers are too nonlinear for most analog applications. Also, since the phototransistor is being saturated, response times can range from 10 μs to 20 μs in slower devices, limiting high speed applications.

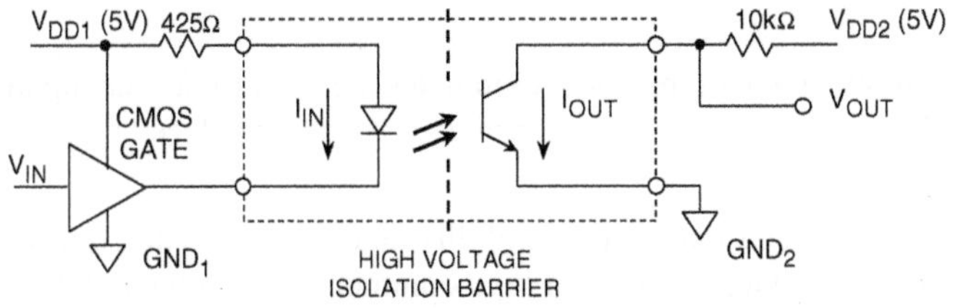

- Uses Light for Transmission over a High Voltage Barrier
- The LED is the Transmitter, and the Phototransistor is the Receiver
- High Voltage Isolation: 5000V to 7000V rms
- Nonlinear — Best for Digital or Frequency Information
- Rise and Fall times can be 10μs to 20μs in Slower Devices
- Example: Siemens ILQ-1 Quad (www.siemens.com)

Figure 2-53: Digital isolation using LED/phototransistor optocouplers

A much faster optocoupler architecture is shown in Figure 2-54, and is based on an LED and a photodiode. The LED is again driven with a current of approximately 10 mA.

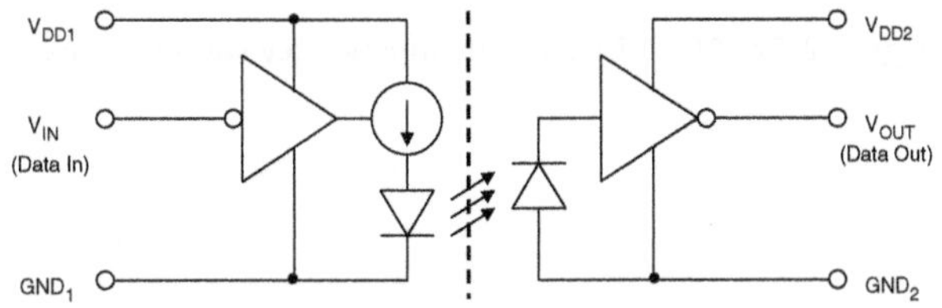

- 5V Supply Voltage
- 2500V rms I/O Withstand Voltage
- Logic Signal Frequency: 12.5MHz Maximum
- 25MBd Maximum Data Rate
- 40ns Maximum Propagation Delay
- 9ns Typical Rise/Fall Time
- Example: Agilent HCPL-7720
- (www.semiconductor.agilent.com)

Figure 2-54: Digital isolation using LED/photodiode optocouplers

This produces a light output sufficient to generate enough current in the receiving photodiode to develop a valid high logic level at the output of the transimpedance amplifier. Speed can vary widely between optocouplers, and the fastest ones have propagation delays of 20 ns typical, and 40 ns maximum, and can handle data rates up to 25 MBd for NRZ data. This corresponds to a maximum square wave operating frequency of 12.5 MHz, and a minimum allowable passable pulse width of 40 ns.

AD260/AD261 High Speed Logic Isolators

The AD260/AD261 family of digital isolators operates on a principle of transformer-coupled isolation (see Reference 4). They provide isolation for five digital control signals to/from high speed DSPs, microcontrollers, or microprocessors. The AD260 also has a 1.5 W transformer for a 3.5 kV rms isolated external dc/dc power supply circuit.

Each line of the AD260 can handle digital signals up to 20 MHz (40 MBd) with a propagation delay of only 14 ns which allows for extremely fast data transmission. Output waveform symmetry is maintained to within ±1ns of the input so the AD260 can be used to accurately isolate time-based pulsewidth modulator (PWM) signals.

A simplified schematic of one channel of the AD260/AD261 is shown in Figure 2-55. The data input is passed through a Schmitt trigger circuit, through a latch, and a special transmitter circuit which differentiates the edges of the digital input signal and drives the primary winding of a proprietary transformer with a "set-high/set-low" signal. The secondary of the isolation transformer drives a receiver with the same "set-hi/set-low" data, which regenerates the original logic waveform. An internal circuit operating in the background interrogates all inputs about every 5 μs, and in the absence of logic transitions, sends appropriate "set-hi/set-low" data across the interface. Recovery time from a fault condition or at power-up is thus between 5 μs and 10 μs.

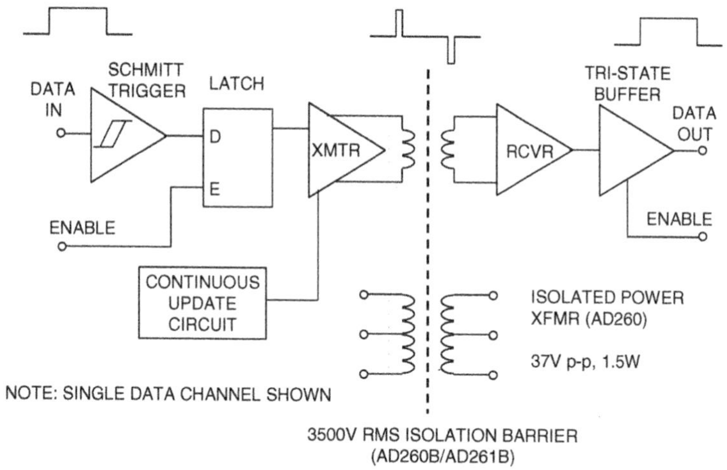

Figure 2-55: AD260/AD261 digital isolators

The power transformer (available on the AD260) is designed to operate between 150 kHz and 250 kHz and will easily deliver more than 1 W of isolated power when driven push-pull (5 V) on the transmitter side. Different transformer taps, rectifier and regulator schemes will provide combinations of ±5 V, 15 V, 24 V, or even 30 V or higher.

The transformer output voltage when driven with a low voltage-drop drive will be 37 V p-p across the entire secondary with a 5 V push-pull drive. Figure 2-56 summarizes the key specifications of the AD260/261 series.

- Isolation Test Voltage to 3500V rms (AD260B/AD261B)
- Five Isolated Digital Lines Available in six Input/Output Configurations
- Logic Signal Frequency: 20MHz Max.
- Data Rate: 40MBd Max.
- Isolated Power Transformer: 37V p-p, 1.5W (AD260)
- Waveform Edge Transmission Symmetry: ±1ns
- Propagation Delay: 14ns
- Rise and Fall Times < 5ns

Figure 2-56: AD260/AD261 digital isolator key specifications

The availability of low cost digital isolators such as those previously discussed solves most system isolation problems in data acquisition systems as shown in Figure 2-57. In the upper example, digitizing the signal first, then using digital isolation eliminates the problem of analog isolation amplifiers. While digital isolation can be used with parallel output ADCs provided the bandwidth of the isolator is sufficient, it is more practical with ADCs that have *serial* outputs. This minimizes cost and component count. A 3-wire interface (data, serial clock, framing clock) is all that is required in these cases.

An alternative (lower example) is to use a voltage-to-frequency converter (VFC) as a transmitter and a frequency-to-voltage converter (FVC) as a receiver. In this case, only one digital isolator is required.

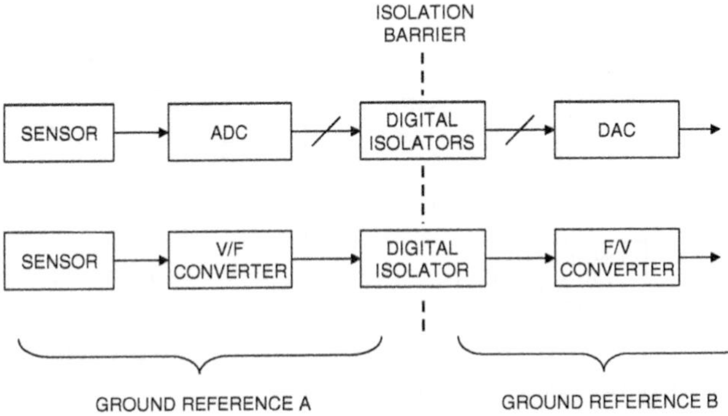

Figure 2-57: Practical application of digital isolation in data acquisition systems

References: Isolation Amplifiers

1. James Conant, "Precision Wideband Three-Port Isolation Amplifier," **Analog Dialogue**, Vol. 20, No. 2, 1986, p. 14.
2. Data sheet for **AD210 Precision, Wide Bandwidth 3-Port Isolation Amplifier**, www.analog.com.
3. Data sheet for **AD215 120 kHz Bandwidth, Low Distortion, Isolation Amplifier**, www.analog.com.
4. Data sheets for **AD260 and AD261 High Speed Logic Isolators**, www.analog.com.
5. Chapters 2, 3, 8, Walt Kester, Editor, **Practical Design Techniques for Sensor Signal Conditioning**, Analog Devices, 1999, ISBN: 0-916550-20-6.
6. Chapter 2, Walt Kester, Editor, **Linear Design Seminar**, Analog Devices, 1995, ISBN: 0-916550-15-X.
7. Chapter 10, Walt Kester, Editor, **Mixed-Signal and DSP Design Techniques**, Analog Devices, 2000, ISBN: 0-916550-23-0.

CHAPTER 3

Using Op Amps with Data Converters

Using Op Amps with Data Converters

Walt Kester, James Bryant, Paul Hendriks

SECTION 3-1

Introduction

Walt Kester

This chapter of the book deals with data conversion and associated signal conditioning circuitry involving the use of op amps. Data conversion is a very broad topic, and this chapter will provide only enough background material for the reader to make intelligent decisions regarding op amp selection. Much more material on the subject is available in the references (see References 1-5).

Figure 3-1 shows a generalized sampled data system and some possible applications of op amps. The analog input signal is first buffered and filtered before it is applied to the analog-to-digital converter (ADC). The buffer may or may not be required, depending upon the input structure of the ADC. For example, some ADCs (such as switched capacitor) generate transient currents at their inputs due to the internal conversion architecture, and these currents must be isolated from the signal source. A suitable buffer amplifier provides a low impedance drive and absorbs these currents. In some cases, an op amp is required to provide the appropriate gain and offset to match the signal to the input range of the ADC.

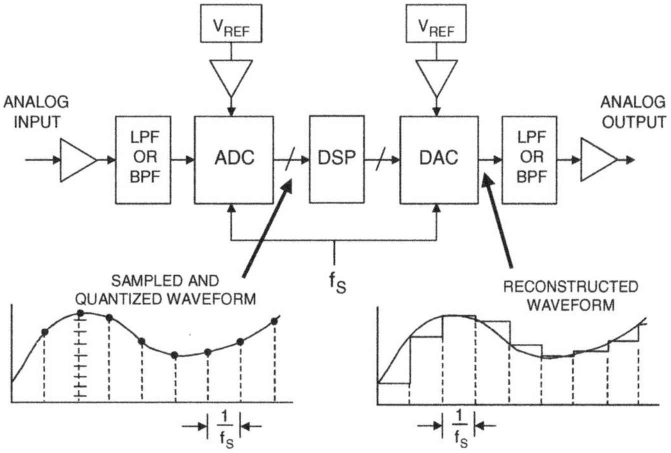

Figure 3-1: Typical sampled data system showing potential amplifier applications

Another key component in a sampled data system is the antialiasing filter which removes signals that fall outside the Nyquist bandwidth, $f_S/2$. Normally this filter is a low-pass filter, but it can be a band-pass filter in certain undersampling applications. If the op amp buffer is required, it may be located before or after the filter, depending on system considerations. In fact, the filter itself may be an active one, in which case the buffering function can be performed by the actual output amplifier of the filter. More discussions regarding active filters can be found in Chapter 5 of this book.

After the signal is buffered and filtered, it is applied to the ADC. The full-scale input voltage range of the ADC is generally determined by a voltage reference, V_{REF}. Some ADCs have this function on chip, while others require an external reference. If an external reference is required, its output may require buffering using an appropriate op amp. The reference input to the ADC may be connected to an internal switched capacitor network, causing transient currents to be generated at that node (similar to the analog input of such converters). Some references may therefore require a buffer to isolate these transient currents from the actual reference output. Other references may have internal buffers that are sufficient, and no additional buffering is needed in those cases.

The output of the ADC is then processed digitally by an appropriate processor, shown in the diagram as a digital signal processor (DSP). DSPs are processors that are optimized to perform fast repetitive arithmetic, as required in digital filters or fast Fourier transform (FFT) algorithms. The DSP output then drives a digital-to-analog converter (DAC) which converts the digital signal back into an analog signal.

- Gain setting
- Level-shifting
- Buffering ADC transients from signal source
- Buffering voltage reference outputs
- Buffering DAC outputs
- Active antialiasing filter before ADC
- Active anti-imaging filter after DAC

Figure 3-2: Data converter amplifier applications

The DAC analog output must be filtered to remove the unwanted image frequencies caused by the sampling process, and further buffering may be required to provide the proper signal amplitude and offset. The output filter is generally placed between the DAC and the buffer amplifier, but their positions can be reversed in certain applications. It is also possible to combine the filtering and buffering function if an active filter is used to condition the DAC output.

Trends in Data Converters

It is useful to examine a few general trends in data converters, to better understand any associated op amp requirements. Converter performance is first and foremost; maintaining that performance in a system application is extremely important. In low frequency measurement applications (10 Hz bandwidth signals or lower), sigma-delta ADCs with resolutions up to 24 bits are now quite common. These converters generally have automatic or factory calibration features to maintain required gain and offset accuracy. In higher frequency signal processing, ADCs must have wide dynamic range (low distortion and noise), high sampling frequencies, and generally excellent ac specifications.

In addition to sheer performance, other characteristics such as low power, single-supply operation, low cost, and small surface-mount packages also drive the data conversion market. These requirements result in application problems because of reduced signal swings, increased sensitivity to noise, and so forth. In addition, many data converters are now produced on low-cost foundry CMOS processes which generally make on-chip amplifier design more difficult and therefore less likely to be incorporated on-chip.

As has been mentioned previously, the analog input to a CMOS ADC is usually connected directly to a switched-capacitor sample-and-hold (SHA), which generates transient currents that must be buffered from the signal source. On the other hand, data converters fabricated on Bi-CMOS or bipolar processes are more likely to have internal buffering, but generally have higher cost and power than their CMOS counterparts.

It should be clear by now that selecting an appropriate op amp for a data converter application is highly dependent on the particular converter under consideration. Generalizations are difficult, but some meaningful guidelines can be followed.

- Higher sampling rates, higher resolution, higher ac performance
- Single supply operation (e.g., 5V, 3V)
- Lower power
- Smaller input/output signal swings
- Maximize usage of low cost foundry CMOS processes
- Smaller packages
- Surface-mount technology

Figure 3-3: Some general trends in data converters

The most obvious requirement for a data converter buffer amplifier is that it not degrade the dc or ac performance of the converter. One might assume that a careful reading of the op amp datasheets would assist in the selection process: simply lay the data converter and the op amp datasheets side by side, and compare each critical performance specification. It is true that this method will provide some degree of success; however, in order to perform an accurate comparison, the op amp must be specified under the exact operating conditions required by the data converter application. Such factors as gain, gain setting resistor values, source impedance, output load, input and output signal amplitude, input and output common-mode (CM) level, power supply voltage, and so forth, all affect op amp performance.

It is highly unlikely that even a well written op amp data sheet will provide an exact match to the operating conditions required in the data converter application. Extrapolation of specified performance to fit the exact operating conditions can give erroneous results. Also, the op amp may be subjected to transient currents from the data converter, and the corresponding effects on op amp performance are rarely found on datasheets.

Converter datasheets themselves can be a good source for recommended op amps and other application circuits. However this information can become obsolete as newer op amps are introduced after the converter's initial release.

Analog Devices and other op amp manufacturers today have on-line websites featuring parametric search engines, which facilitate part selection (see Reference 1). For instance, the first search might be for minimum power supply voltage, e.g., 3 V. The next search might be for bandwidth, and further searches on relevant specifications will narrow the selection of op amps even further. Figure 3-4 summarizes the selection process.

- The amplifier should not degrade the performance of the ADC/DAC
- Ac specifications are usually the most important
  - Noise
  - Bandwidth
  - Distortion
- Selection based on op amp data sheet specifications difficult due to varying conditions in actual application circuit with ADC/DAC:
  - Power supply voltage
  - Signal range (differential and common-mode)
  - Loading (static and dynamic)
  - Gain
- Parametric search engines may be useful
- ADC/DAC data sheets often recommend op amps (but may not include newly released products)

Figure 3-4: General amplifier selection criteria

While not necessarily suitable for the final selection, this process can narrow the search to a manageable number of op amps whose individual datasheets can be retrieved, then reviewed in detail before final selection.

From the discussion thus far, it should be obvious that in order to design a proper interface, an understanding of both op amps and data converters is required. References 2-6 provide background material on data converters.

The next section of this chapter addresses key data converter performance specifications without going into the detailed operation of converters themselves. The remainder of the chapter shows a number of specific applications of op amps with various data converters.

References: Introduction

1. ADI Web site, at www.analog.com.
2. Walt Kester, Editor, **Practical Analog Design Techniques**, Analog Devices, 1995, ISBN: 0-916550-16-8.
3. Walt Kester, Editor, **High Speed Design Techniques**, Analog Devices, 1996, ISBN: 0-916550-17-6.
4. Chapters 3, 8, Walt Kester, Editor, **Practical Design Techniques for Sensor Signal Conditioning**, Analog Devices, 1999, ISBN: 0-916550-20-6.
5. Chapters 2, 3, 4, Walt Kester, Editor, **Mixed-Signal and DSP Design Techniques**, Analog Devices, 2000, ISBN: 0-916550-23-0.
6. Chapters 4, 5, Walt Kester, Editor, **Linear Design Seminar**, Analog Devices, 1995, ISBN: 0-916550-15-X.

ADC/DAC Specifications

Walt Kester

ADC and DAC Static Transfer Functions and DC Errors

The most important thing to remember about both DACs and ADCs is that either the input or output is digital, and therefore the signal is *quantized.* That is, an N-bit word represents one of 2^N possible states, and therefore an N-bit DAC (with a fixed reference) can have only 2^N possible analog outputs, and an N-bit ADC can have only 2^N possible digital outputs. The analog signals will generally be voltages or currents.

The resolution of data converters may be expressed in several different ways: the weight of the Least Significant Bit (LSB), parts per million of full scale (ppm FS), millivolts (mV), and so forth. It is common that different devices (even from the same manufacturer) will be specified differently, so converter users must learn to translate between the different types of specifications if they are to successfully compare devices. The size of the least significant bit for various resolutions is shown in Figure 3-5.

RESOLUTION N	2^N	VOLTAGE (10V FS)	ppm FS	% FS	dB FS
2-bit	4	2.5 V	250,000	25	–12
4-bit	16	625 mV	62,500	6.25	–24
6-bit	64	156 mV	15,625	1.56	–36
8-bit	256	39.1 mV	3,906	0.39	–48
10-bit	1,024	9.77 mV (10 mV)	977	0.098	–60
12-bit	4,096	2.44 mV	244	0.024	–72
14-bit	16,384	610 μV	61	0.0061	–84
16-bit	65,536	153 μV	15	0.0015	–96
18-bit	262,144	38 μV	4	0.0004	–108
20-bit	1,048,576	9.54 μV (10 μV)	1	0.0001	–120
22-bit	4,194,304	2.38 μV	0.24	0.000024	–132
24-bit	16,777,216	596 nV*	0.06	0.000006	–144

*600nV is the Johnson Noise in a 10kHz BW of a 2.2kΩ Resistor @ 25°C

Note: 10 bits and 10V FS yields an LSB of 10mV, 1000ppm, or 0.1%.
All other values may be calculated by powers of 2.

Figure 3-5: Quantization: the size of a least significant bit (LSB)

As noted above (and obvious from this table), the LSB scaling for a given converter resolution can be expressed in various ways. While it is convenient to relate this to a full scale of 10 V, as in Figure 3-5, other full scale levels can be easily extrapolated.

Before we can consider op amp applications with data converters, it is necessary to consider the performance to be expected, and the specifications that are important when operating with data converters. The following sections will consider the definition of errors and specifications used for data converters.

The first applications of data converters were in measurement and control, where the exact timing of the conversion was usually unimportant, and the data rate was slow. In such applications, the dc specifications of converters are important, but timing and ac specifications are not. Today many, if not most, converters are used in *sampling* and *reconstruction* systems where ac specifications are critical (and dc ones may not be).

Figure 3-6 shows the transfer characteristics for a 3-bit unipolar ideal and nonideal DAC. In a DAC, both the input and output are quantized, and the graph consists of eight points—while it is reasonable to discuss a line through these points, it is critical to remember that the actual transfer characteristic is *not* a line, but a series of discrete points.

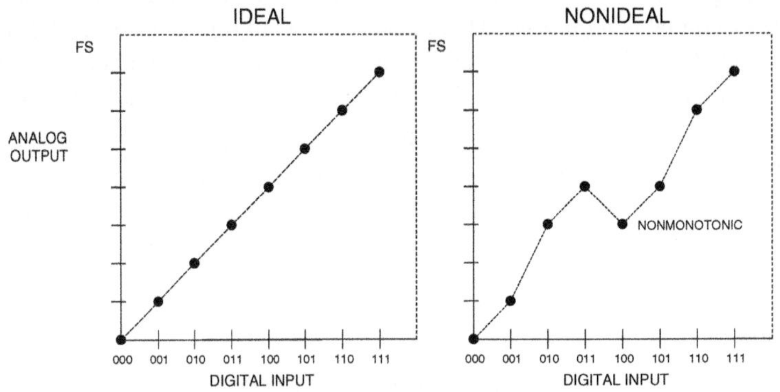

Figure 3-6: DAC transfer functions

Similarly, Figure 3-7 shows the transfer characteristics for a 3-bit unipolar ideal and nonideal ADC. Note that the input to an ADC is analog and is therefore *not* quantized, but its output *is* quantized.

The ADC transfer characteristic therefore consists of eight horizontal steps (when considering the offset, gain and linearity of an ADC we consider the line joining the midpoints of these steps).

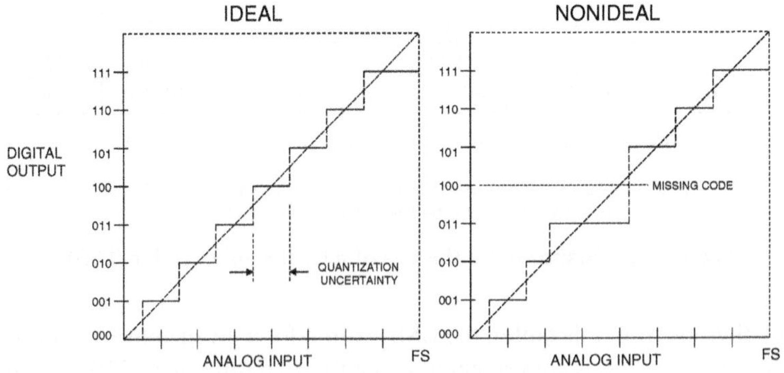

Figure 3-7: ADC transfer functions

The (ideal) ADC transitions take place at ½ LSB above zero, and thereafter every LSB, until 1½ LSB below analog full scale. Since the analog input to an ADC can take any value, but the digital output is quantized, there may be a difference of up to ½ LSB between the actual analog input and the exact value of the digital output. This is known as the quantization error or quantization uncertainty as shown in Figure 3-7. In ac (sampling) applications this quantization error gives rise to quantization noise which will be discussed shortly.

The *integral linearity* error of a converter is analogous to the linearity error of an amplifier, and is defined as the maximum deviation of the actual transfer characteristic of the converter from a straight line. It is generally expressed as a percentage of full scale (but may be given in LSBs). There are two common ways of choosing the straight line: *end point* and *best straight line.*

In the end point system, the deviation is measured from the straight line through the origin and the full scale point (after gain adjustment). This is the most useful integral linearity measurement for measurement and control applications of data converters (since error budgets depend on deviation from the ideal transfer characteristic, not from some arbitrary "best fit"), and is the one normally adopted by Analog Devices, Inc.

The best straight line, however, does give a better prediction of distortion in ac applications, and also gives a lower value of "linearity error" on a data sheet. The best fit straight line is drawn through the transfer characteristic of the device using standard curve fitting techniques, and the maximum deviation is measured from this line. In general, the integral linearity error measured in this way is only 50% of the value measured by end point methods. This makes the method good for producing impressive datasheets, but it is less useful for error budget analysis. For ac applications, it is even better to specify distortion than dc linearity, so it is rarely necessary to use the best straight line method to define converter linearity.

The other type of converter nonlinearity is *differential nonlinearity* (DNL). This relates to the linearity of the code transitions of the converter. In the ideal case, a change of 1 LSB in digital code corresponds to a change of exactly 1 LSB of analog signal. In a DAC, a change of 1 LSB in digital code produces exactly 1 LSB change of analog output, while in an ADC there should be exactly 1 LSB change of analog input to move from one digital transition to the next.

Where the change in analog signal corresponding to 1 LSB digital change is more or less than 1 LSB, there is said to be a DNL error. The DNL error of a converter is normally defined as the maximum value of DNL to be found at any transition.

If the DNL of a DAC is less than –1 LSB at any transition (Figure 3-6), the DAC is *nonmonotonic;* i.e., its transfer characteristic contains one or more localized maxima or minima. A DNL greater than +1 LSB does not cause nonmonotonicity, but is still undesirable. In many DAC applications (especially closed-loop systems where nonmonotonicity can change negative feedback to positive feedback), it is critically important that DACs are monotonic. DAC monotonicity is often explicitly specified on datasheets, but if the DNL is guaranteed to be less than 1 LSB (i.e., $|DNL| \leq 1LSB$) then the device must be monotonic, even without an explicit guarantee.

ADCs can be nonmonotonic, but a more common result of excess DNL in ADCs is *missing codes* (Figure 3-7). Missing codes (or nonmonotonicity) in an ADC are as objectionable as nonmonotonicity in a DAC. Again, they result from DNL > 1 LSB.

Quantization Noise in Data Converters

The only errors (dc or ac) associated with an ideal N-bit ADC are those related to the sampling and quantization processes. The maximum error an ideal ADC makes when digitizing a dc input signal is ±½ LSB. Any ac signal applied to an ideal N-bit ADC will produce quantization noise whose rms value (measured over the Nyquist bandwidth, dc to $f_s/2$) is approximately equal to the weight of the least significant bit (LSB), q, divided by $\sqrt{12}$. This assumes that the signal is at least a few LSBs in amplitude so that the ADC output always changes state. The quantization error signal from a linear ramp input is approximated as a sawtooth waveform with a peak-to-peak amplitude equal to q, and its rms value is therefore $q/\sqrt{12}$ (see Figure 3-8).

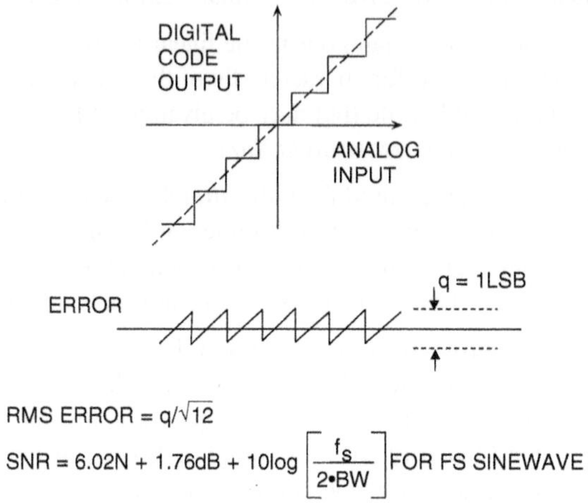

Figure 3-8: Ideal N-bit ADC quantization noise

It can be shown that the ratio of the rms value of a full scale sinewave to the rms value of the quantization noise (expressed in dB) is:

$$SNR = 6.02\,N + 1.76\,dB, \qquad \text{Eq. 3-1}$$

where *N* is the number of bits in the ideal ADC. *Note—this equation is only valid if the noise is measured over the entire Nyquist bandwidth from dc to $f_S/2$.* If the signal bandwidth, BW, is less than $f_S/2$, the SNR within the signal bandwidth BW is increased because the amount of quantization noise within the signal bandwidth is less.

The correct expression for this condition is given by:

$$SNR = 6.02\,N + 1.76\ dB + 10\log\left(\frac{f_S}{2 \cdot BW}\right). \qquad \text{Eq. 3-2}$$

The above equation reflects the condition called *oversampling*, where the sampling frequency is higher than twice the signal bandwidth. The correction term is often called *processing gain*. Notice that for a given signal bandwidth, doubling the sampling frequency increases the SNR by 3 dB.

ADC Input-Referred Noise

The internal circuits of an ADC produce a certain amount of wideband rms noise due to thermal and kT/C effects. This noise is present even for dc input signals, and accounts for the fact that the output of most wideband (or high resolution) ADCs is a distribution of codes, centered around the nominal dc input value, as is shown in Figure 3-9.

To measure its value, the input of the ADC is grounded, and a large number of output samples are collected and plotted as a histogram (sometimes referred to as a *grounded-input* histogram).

Since the noise is approximately Gaussian, the standard deviation (σ) of the histogram is easily calculated, corresponding to the effective input rms noise. It is common practice to express this rms noise in terms of LSBs, although it can be expressed as an rms voltage.

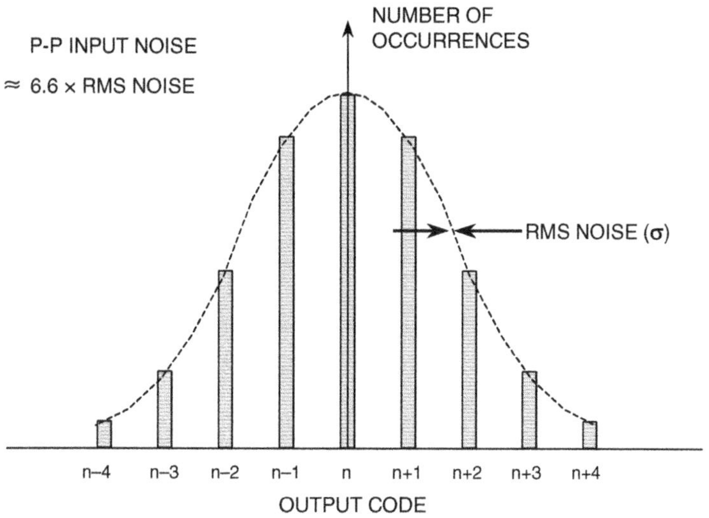

Figure 3-9: Effect of ADC input-referred noise on "grounded input" histogram

Calculating Op Amp Output Noise and Comparing it with ADC Input-Referred Noise

In precision measurement applications utilizing 16- to 24-bit sigma-delta ADCs operating on low frequency (<20 Hz, e.g.) signals, it is generally undesirable to use a drive amplifier in front of the ADC because of the increased noise due to the amplifier itself. If an op amp is required, however, the op amp output 1/f noise should be compared to the input-referred ADC noise. The 1/f noise is usually specified as a peak-to-peak value measured over the 0.1 Hz to 10 Hz bandwidth and referred to the op amp input (see Chapter 1 of this book). Op amps such as the OP177 and the AD707 (input voltage noise 350 nV p-p) or the AD797 (input voltage noise 50 nV p-p) are appropriate for high resolution measurement applications if required.

The general model for calculating the referred-to-input (RTI) or referred-to-output (RTO) noise of an op amp is shown in Figure 3-10. This model shows all possible noise sources. The results using this model are relatively accurate, provided there is less than 1 dB gain peaking in the closed loop frequency response. For higher frequency applications, 1/f noise can be neglected, because the dominant contributor is white noise.

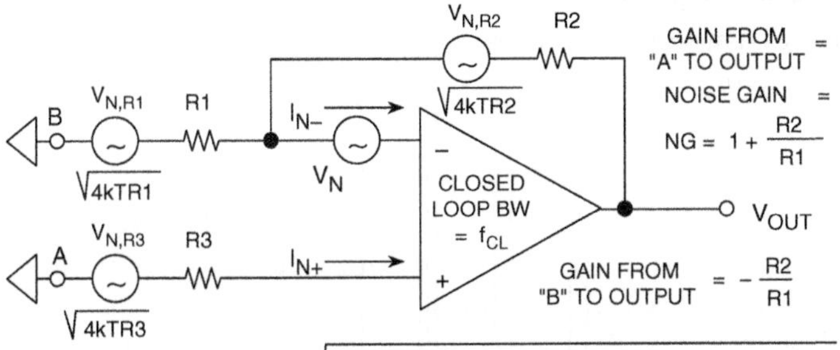

■ RTI NOISE = $\sqrt{BW} \bullet \sqrt{V_N^2 + 4kTR3 + 4kTR1\left[\frac{R2}{R1+R2}\right]^2 + I_{N+}^2 R3^2 + I_{N-}^2\left[\frac{R1 \bullet R2}{R1+R2}\right]^2 + 4kTR2\left[\frac{R1}{R1+R2}\right]^2}$

■ RTO NOISE = NG • RTI NOISE ■ BW = 1.57 f_{CL}

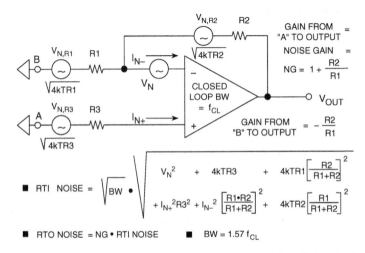

Figure 3-10: Op amp noise model for a first-order circuit with resistive feedback

An example of a practical noise calculation is shown in Figure 3-11. In this circuit, a wideband, low distortion amplifier (AD9632) drives a 12-bit, 25 MSPS ADC (AD9225). The input voltage noise spectral density of the AD9632 $\left(4.3\,\text{nV}/\sqrt{\text{Hz}}\right)$ dominates the op amp noise because of the low gain and the low values of the external feedback resistors. The noise at the output of the AD9632 is obtained by multiplying the input voltage noise spectral density by the noise gain of 2. To obtain the rms noise, the noise spectral density is multiplied by the equivalent noise bandwidth of 50 MHz which is set by the single-pole low-pass filter placed between the op amp and the ADC input.

Note that the closed-loop bandwidth of the AD9632 is 250 MHz, and the input bandwidth of the AD9225 is 105 MHz. With no filter, the output noise of the AD9632 would be integrated over the full 105 MHz ADC input bandwidth.

However, the sampling frequency of the ADC is 25 MSPS, thereby implying that signals above 12.5 MHz are not of interest, assuming Nyquist operation (as opposed to undersampling applications where the input signal can be greater than the Nyquist frequency, $f_s/2$). The addition of this simple filter significantly reduces noise effects.

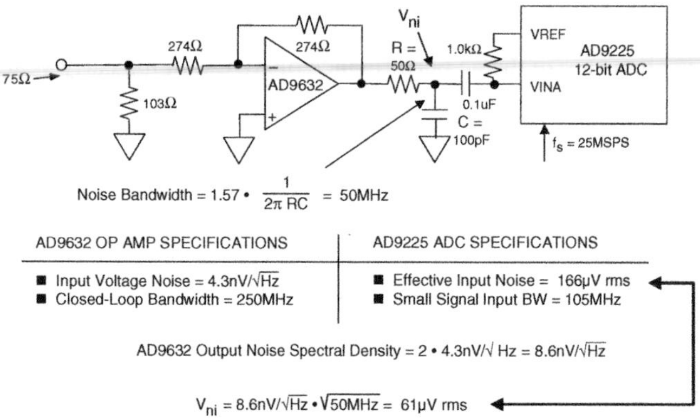

Figure 3-11: Noise calculations for the AD9632 op amp driving the AD9225 12-bit, 25 MSPS ADC

The noise at the output of the low-pass filter is calculated as approximately 61 μV rms which is less than half the effective input noise of the AD9225, 166 μV rms. Without the filter, the noise from the op amp would be about 110 μV rms (integrating over the full equivalent ADC input noise bandwidth of 1.57 × 105 MHz = 165 MHz).

This serves to illustrate the general concept shown in Figure 3-12. In most high speed system applications a passive antialiasing filter (either low-pass for baseband sampling, or band-pass for undersampling) is required, and placing this filter between the op amp and the ADC will serve to reduce the noise due to the op amp.

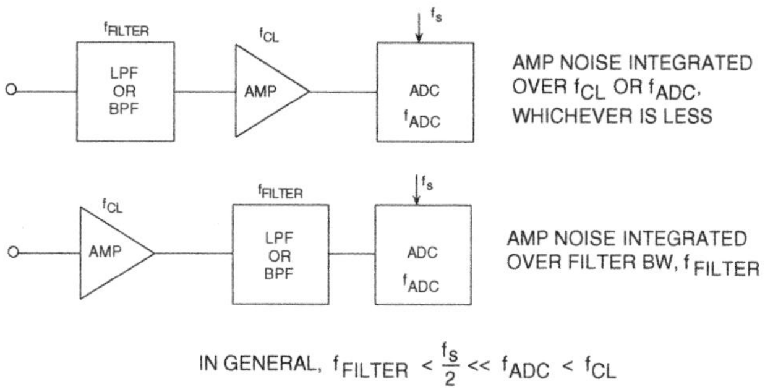

Figure 3-12: Proper positioning of the antialiasing filter will reduce the effects of op amp noise

Quantifying and Measuring Converter Dynamic Performance

There are various ways to characterize the ac performance of ADCs. In the early years of ADC technology (over 30 years ago) there was little standardization with respect to ac specifications, and measurement equipment and techniques were not well understood or available. Over nearly a 30-year period, manufacturers and customers have learned more about measuring the dynamic performance of converters, and the specifications shown in Figure 3-13 represent the most popular ones used today.

- Signal-to-Noise-and-Distortion Ratio (SINAD, or S/N +D)
- Effective Number of Bits (ENOB)
- Signal-to-Noise Ratio (SNR)
- Analog Bandwidth (Full-Power, Small-Signal)
- Harmonic Distortion
- Worst Harmonic
- Total Harmonic Distortion (THD)
- Total Harmonic Distortion Plus Noise (THD + N)
- Spurious Free Dynamic Range (SFDR)
- Two-Tone Intermodulation Distortion
- Multitone Intermodulation Distortion

Figure 3-13: Quantifying ADC dynamic performance

Practically all the specifications represent the converter's performance in the frequency domain, and all are related to noise and distortion in one manner or another.

ADC outputs are analyzed using fast Fourier transform (FFT) techniques, and DAC outputs are analyzed using conventional analog spectrum analyzers, as shown in Figure 3-14. In the case of an ADC, the input signal is an analog sinewave, and in the case of a DAC, the input is a digital sinewave generated by a direct digital synthesis (DDS) system.

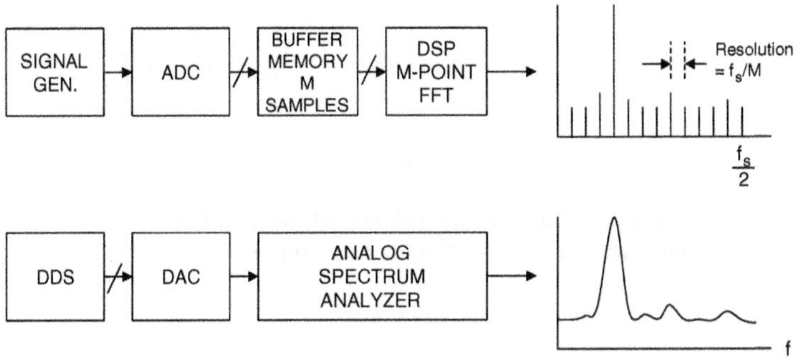

Figure 3-14: Measuring ADC/DAC dynamic performance

Signal-to-Noise-and-Distortion Ratio (SINAD), Signal-to-Noise Ratio (SNR), and Effective Number of Bits (ENOB)

SINAD and SNR deserve careful attention (see Figure 3-15), because there is still some variation between ADC manufacturers as to their precise meaning. *Signal-to-noise-and-Distortion* (SINAD, or S/N+D) is the ratio of the rms signal amplitude to the mean value of the root-sum-square (RSS) of all other spectral components, *including harmonics*, but excluding dc. SINAD is a good indication of the overall dynamic performance of an ADC as a function of input frequency, because it includes all components that make up noise (including thermal noise) and distortion. It is often plotted for various input amplitudes. SINAD is equal to THD+N if the bandwidth for the noise measurement is the same.

- SINAD (Signal-to-Noise-and-Distortion Ratio):
  - The ratio of the rms signal amplitude to the mean value of the root-sum-squares (RSS) of all other spectral components, including harmonics, but excluding dc.

- ENOB (Effective Number of Bits):

$$ENOB = \frac{SINAD - 1.76\ dB}{6.02}$$

- SNR (Signal-to-Noise Ratio, or Signal-to-Noise Ratio without Harmonics:
  - The ratio of the rms signal amplitude to the mean value of the root-sum-squares (RSS) of all other spectral components, excluding the first five harmonics and dc.

Figure 3-15: SINAD, ENOB, and SNR

The SINAD plot shows where the ac performance of the ADC degrades due to high-frequency distortion, and is usually plotted for frequencies well above the Nyquist frequency so that performance in undersampling applications can be evaluated.

SINAD is often converted to *effective-number-of-bits* (ENOB) using the relationship for the theoretical SNR of an ideal N-bit ADC: SNR = 6.02N + 1.76dB. The equation is solved for N, and the value of SINAD is substituted for SNR:

$$ENOB = \frac{SINAD - 1.76\ dB}{6.02} \qquad \text{Eq. 3-3}$$

Signal-to-noise ratio (SNR, or *SNR-without-harmonics*) is calculated the same as SINAD except that the signal harmonics are excluded from the calculation, leaving only the noise terms. In practice, it is only necessary to exclude the first five harmonics since they dominate. The SNR plot will degrade at high frequencies, but not as rapidly as SINAD because of the exclusion of the harmonic terms.

Many current ADC datasheets somewhat loosely refer to SINAD as SNR, so the design engineer must be careful when interpreting these specifications.

A SINAD/ENOB plot for the AD9220 12-bit, 10MSPS ADC is shown in Figure 3-16.

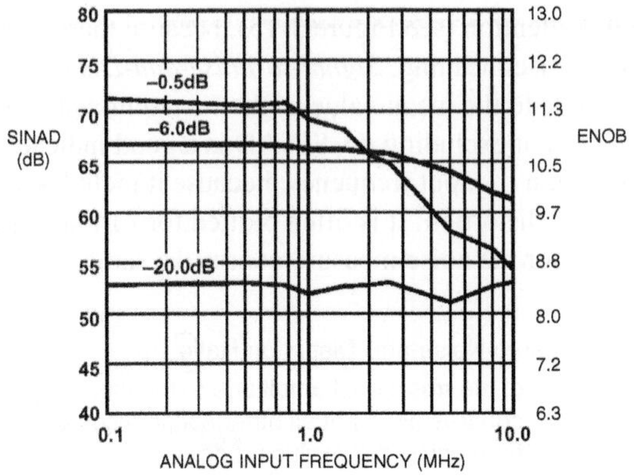

Figure 3-16: AD9220 12-bit, 10 MSPS ADC SINAD and ENOB for various input signal levels

Analog Bandwidth

The analog bandwidth of an ADC is that frequency at which the spectral output of the *fundamental* swept frequency (as determined by the FFT analysis) is reduced by 3 dB. It may be specified for either a small signal bandwidth (SSBW), or a full-scale signal (FPBW – full power bandwidth), so there can be a wide variation in specifications between manufacturers.

Like an amplifier, the analog bandwidth specification of a converter does not imply that the ADC maintains good distortion performance up to its bandwidth frequency. In fact, the SINAD (or ENOB) of most ADCs will begin to degrade considerably before the input frequency approaches the actual 3 dB bandwidth frequency. Figure 3-17 shows ENOB and full-scale frequency response of an ADC with a FPBW of 1 MHz; however, the ENOB begins to drop rapidly above 100 kHz.

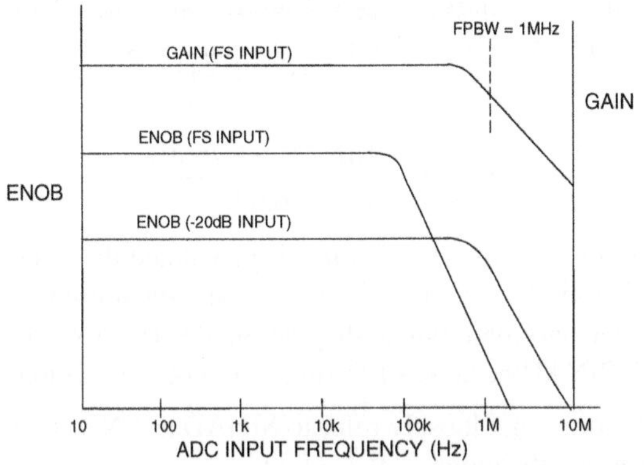

Figure 3-17: ADC Gain (bandwidth) and ENOB versus frequency shows importance of ENOB specification

Harmonic Distortion, Worst Harmonic, Total Harmonic Distortion (THD), Total Harmonic Distortion Plus Noise (THD + N)

There are a number of ways to quantify the distortion of an ADC. An FFT analysis can be used to measure the amplitude of the various harmonics of a signal. The harmonics of the input signal can be distinguished from other distortion products by their location in the frequency spectrum. Figure 3-18 shows a 7 MHz input signal sampled at 20 MSPS and the location of the first nine harmonics.

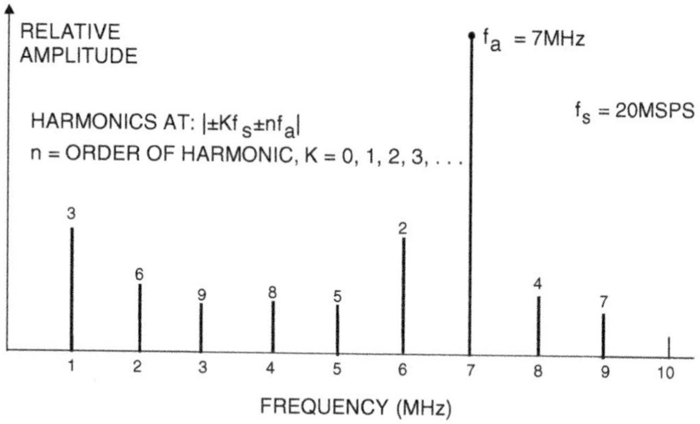

Figure 3-18: Location of harmonic distortion products: Input signal = 7 MHz, sampling rate = 20 MSPS

Aliased harmonics of f_a fall at frequencies equal to $|\pm Kf_s \pm nf_a|$, where n is the order of the harmonic, and K = 0, 1, 2, 3,... . The second and third harmonics are generally the only ones specified on a data sheet because they tend to be the largest, although some datasheets may specify the value of the *worst* harmonic.

Harmonic distortion is normally specified in dBc (decibels below *carrier*), although in audio applications it may be specified as a percentage. Harmonic distortion is generally specified with an input signal near full scale (generally 0.5 dB to 1 dB below full scale to avoid clipping), but it can be specified at any level. For signals much lower than full scale, other distortion products due to the DNL of the converter (not direct harmonics) may limit performance.

Total harmonic distortion (THD) is the ratio of the rms value of the fundamental signal to the mean value of the root-sum-square of its harmonics (generally, only the first five are significant). THD of an ADC is also generally specified with the input signal close to full scale, although it can be specified at any level.

Total harmonic distortion plus noise (THD + N) is the ratio of the rms value of the fundamental signal to the mean value of the root-sum-square of its harmonics plus all noise components (excluding dc). The bandwidth over which the noise is measured must be specified. In the case of an FFT, the bandwidth is dc to $f_s/2$. (If the bandwidth of the measurement is dc to $f_s/2$, THD + N is equal to SINAD.)

Spurious Free Dynamic Range (SFDR)

Probably the most significant specification for an ADC used in a communications application is its spurious free dynamic range (SFDR). The SFDR specification is to ADCs what the third order intercept specification is to mixers and LNAs.

SFDR of an ADC is defined as the ratio of the rms signal amplitude to the rms value of the *peak spurious spectral content* (measured over the entire first Nyquist zone, dc to $f_s/2$). SFDR is generally plotted as a function of signal amplitude and may be expressed relative to the signal amplitude (dBc) or the ADC full scale (dBFS) as shown in Figure 3-19.

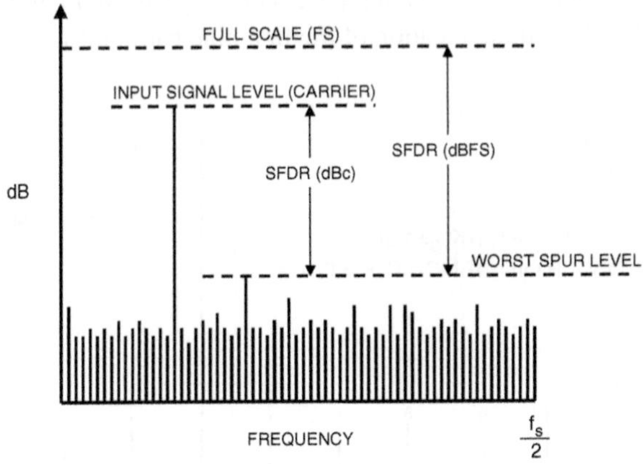

Figure 3-19: Spurious free dynamic range (SFDR)

For a signal near full scale, the peak spectral spur is generally determined by one of the first few harmonics of the fundamental. However, as the signal falls several dB below full scale, other spurs generally occur which are not direct harmonics of the input signal. This is because of the differential nonlinearity of the ADC transfer function as discussed earlier. Therefore, SFDR considers *all* sources of distortion, regardless of their origin.

Two-Tone Intermodulation Distortion (IMD)

Two-tone IMD is measured by applying two spectrally pure sinewaves to the ADC at frequencies f_1 and f_2, usually relatively close together. The amplitude of each tone is set slightly more than 6 dB below full scale so that the ADC does not clip when the two tones add in-phase. Second- and third-order product locations are shown in Figure 3-20.

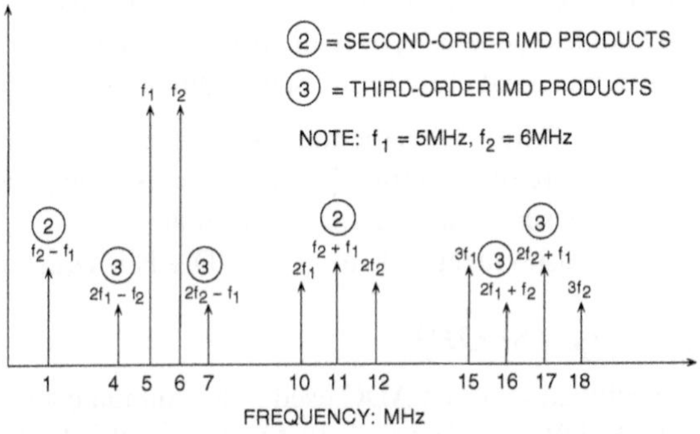

Figure 3-20: Second and third-order intermodulation products for f_1 = 5 MHz, f_2 = 6 MHz

Notice that the second-order products fall at frequencies which can be removed by digital filters. However, the third-order products $2f_2–f_1$ and $2f_1–f_2$ are close to the original signals, and are almost impossible to filter. Unless otherwise specified, two-tone IMD refers to these third-order products. The value of the IMD product is expressed in dBc relative to the value of *either* of the two original tones, and not to their sum.

Note, however, that if the two tones are close to $f_s/4$, the aliased third harmonics of the fundamentals can make the identification of the actual $2f_2–f_1$ and $2f_1–f_2$ products difficult. This is because the third harmonic of $f_s/4$ is $3f_s/4$, and the alias occurs at $f_s - 3f_s/4 = f_s/4$. Similarly, if the two tones are close to $f_s/3$, the aliased second harmonics may interfere with the measurement. The same reasoning applies here; the second harmonic of $f_s/3$ is $2\ f_s/3$, and its alias occurs at $f_s - 2\ f_s/3 = f_s/3$.

The concept of *second- and third-order intercept points* is not valid for an ADC, because the distortion products don't vary predictably (as a function of signal amplitude). The ADC doesn't gradually begin to compress signals approaching full scale (there is no 1 dB compression point); it acts as a *hard limiter* as soon as the signal exceeds the input range, producing extreme distortion due to clipping. Conversely, for signals much below full scale, the distortion floor remains relatively constant and is independent of signal level.

Multitone SFDR is often measured in communications applications. The larger number of tones more closely simulates the wideband frequency spectrum of cellular telephone systems such as AMPS or GSM. High SFDR increases the receiver's ability to capture small signals in the presence of large ones, and prevent the small signals from being masked by the intermodulation products of the larger ones.

References: ADC/DAC Specifications

1. Walt Kester, Editor, **Practical Analog Design Techniques**, Analog Devices, 1995, ISBN: 0-916550-16-8.
2. Walt Kester, Editor, **High Speed Design Techniques**, Analog Devices, 1996, ISBN: 0-916550-17-6.
3. Chapters 3, 8, Walt Kester, Editor, **Practical Design Techniques for Sensor Signal Conditioning**, Analog Devices, 1999, ISBN: 0-916550-20-6.
4. Chapters 2, 3, 4, Walt Kester, Editor, **Mixed-Signal and DSP Design Techniques**, Analog Devices, 2000, ISBN: 0-916550-23-0.
5. Chapters 4, 5, Walt Kester, Editor, **Linear Design Seminar**, Analog Devices, 1995, ISBN: 0-916550-15-X.

Driving ADC Inputs

Walt Kester, Paul Hendriks

Introduction

Op amps are often used as drivers for ADCs to provide the gain and level-shifting required for the input signal to match the input range of the ADC. An op amp may be required because of the antialiasing filter impedance matching requirements. In some cases, the antialiasing filter may be an active filter and include op amps as part of the filter itself. Some ADCs also generate transient currents on their inputs due to the conversion process, and these must be isolated from the signal source with an op amp. This section examines these and other issues involved in driving high performance ADCs.

To begin with, one shouldn't necessarily assume that a driver op amp is always required. Some converters have relatively benign inputs and are designed to interface directly to the signal source. There is practically no industry standardization regarding ADC input structures, and therefore each ADC must be carefully examined on its own merits before designing the input interface circuitry. In some applications, transformer drive may be preferable.

Assuming an op amp is required for one reason or another, the task of its selection is a critical one and not at all straightforward. Figure 3-21 lists a few of the constraints and variables. *The most important requirement is that the op amp should not significantly degrade the overall dc or ac performance of the ADC.* At first glance, it would appear that a careful comparison of an op amp data sheet with the ADC data sheet would allow an appropriate choice. However, this is rarely the case.

- Minimize degradation of ADC/DAC performance specifications
- Fast settling to ADC/DAC transient
- High bandwidth
- Low noise
- Low distortion
- Low power

- Note: Op amp performance must be measured under identical conditions as encountered in ADC/DAC application
  - Gain setting resistors
  - Input source impedance, output load impedance
  - Input / output signal voltage range
  - Input signal frequency
  - Input / output common-mode level
  - Power supply voltage (single or dual supply)
  - Transient loading

Figure 3-21: General op amp requirements in ADC driver applications

The problem is that the op amp performance specifications must be known for the *exact* circuit configuration used in the ADC driver circuit. Even a very complete data sheet is unlikely to provide all information required, due to the wide range of possible variables.

Although the op amp and ADC datasheets should definitely be used as a guide in the selection process, it is unlikely that the overall performance of the op amp/ADC combination can be predicted accurately without actually prototyping the circuit, especially in high performance applications.

Various tested application circuits are often recommended on either the op amp or the ADC data sheet, but these can become obsolete quickly as new op amps are released. In most cases, however, the ADC data sheet application section should be used as the primary source for tested interfaces.

Op Amp Specifications Key to ADC Applications

The two most popular applications for ADCs today are in either precision high-resolution measurements or in low distortion high speed systems. Precision measurement applications require ADCs of at least 16 bits of resolution, and sometimes up to 24 bits. Op amps used with these ADCs must be low noise and have excellent dc characteristics. In fact, high resolution measurement ADCs are often designed to interface directly with the transducer, eliminating the need for an op amp entirely.

If op amps are required, it is generally relatively straightforward to select one based on well understood dc specifications, as listed in Figure 3-22.

- Dc
  - Offset, offset drift
  - Input bias current
  - Open loop gain
  - Integral linearity
  - 1/f noise (voltage and current)

- Ac (Highly Application-Dependent)
  - Wideband noise (voltage and current)
  - Small and Large Signal Bandwidth
  - Harmonic Distortion
  - Total Harmonic Distortion (THD)
  - Total Harmonic Distortion + Noise (THD + N)
  - Spurious Free Dynamic Range (SFDR)
  - Third-Order Intermodulation Distortion
  - Third-Order Intercept Point

Figure 3-22: Key op amp specifications

It is much more difficult to provide a complete set of op amp ac specifications because they are highly dependent upon the application circuit. For example, Figure 3-23 shows some key specifications taken from the table of specifications on the data sheet for the AD8057/AD8058 high speed, low distortion op amp (see Reference 1). Note that the specifications depend on the supply voltage, the signal level, output loading, and so forth. It should also be emphasized that it is customary to provide only *typical* ac specifications (as opposed to *maximum* and *minimum* values) for most op amps. In addition, there are restrictions on the input and output CM signal ranges, which are especially important when operating on low voltage dual (or single) supplies.

Most op amp datasheets contain a section that provides supplemental performance data for various other conditions not explicitly specified in the primary specification tables. For instance, Figure 3-24 shows the

	V_S = ±5V	V_S = +5V
Input Common-Mode Voltage Range	–4.0V to +4.0V	0.9V to 3.4V
Output Common-Mode Voltage Range	–4.0V to +4.0V	0.9V to 4.1V
Input Voltage Noise	7nV/√Hz	7nV/√Hz
Small Signal Bandwidth	325MHz	300MHz
THD @ 5MHz, V_O = 2V p-p, R_L = 1kΩ	– 85dBc	– 75dBc
THD @ 20MHz, V_O = 2V p-p, R_L = 1kΩ	– 62dBc	– 54dBc

Figure 3-23: AD8057/AD8058 op amp key ac specifications, G = +1

AD8057/AD8058 distortion as a function of frequency for G = +1 and V_S = ±5 V. Unless it is otherwise specified, the data represented by these curves should be considered typical (it is usually marked as such).

Note however that the data in both Figure 3-24 (and Figure 3-25) is given for a dc load of 150 Ω. This is a load presented to the op amp in the popular application of driving a source and load-terminated 75 Ω cable. Distortion performance is generally better with lighter dc loads, such as 500 Ω – 1000 Ω (more typical of many ADC inputs), and this data may or may not be found on the data sheet.

Figure 3-24: AD8057/AD8058 op amp distortion versus frequency
G = +1, R_L = 150 Ω, V_S = ±5 V

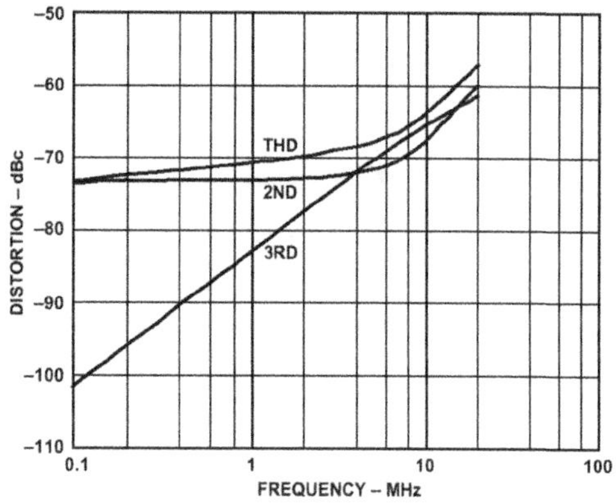

Figure 3-25: AD8057/AD8058 op amp distortion versus output voltage
G = +1, R_L = 150 Ω, V_S = ±5 V

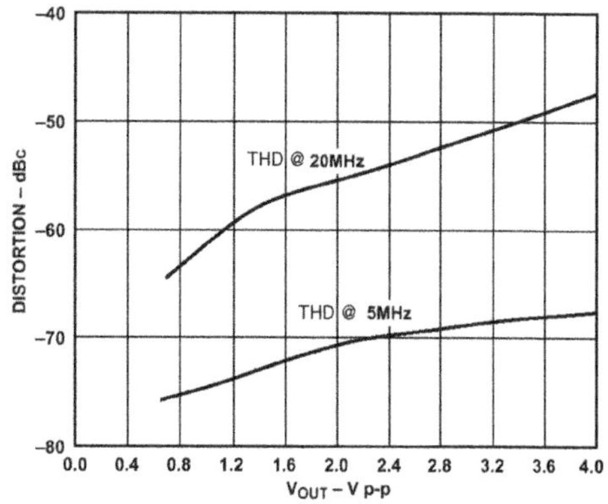

On the other hand, Figure 3-25 shows distortion as a function of output signal level for a frequencies of 5 MHz and 20 MHz.

Whether or not specifications such as those just described are complete enough to select an op amp for an ADC driver application depends upon the ability to match op amp specifications to the actually required ADC operating conditions. In many cases, these comparisons will at least narrow the op amp selection process. The following sections will examine a number of specific driver circuit examples using various types of ADCs, ranging from high resolution measurement to high speed, low distortion applications.

Driving High Resolution Sigma-Delta Measurement ADCs

The AD77XX family of ADCs is optimized for high resolution (16–24 bits) low frequency transducer measurement applications. Details of operation can be found in Reference 2, and general characteristics of the family are listed in Figure 3-26.

- Resolution: 16 – 24 bits
- Input signal bandwidth: <60Hz
- Effective sampling rate: <100Hz
- Generally Sigma-Delta architecture
- Designed to interface directly to sensors (< 1 kΩ) such as bridges with no external buffer amplifier (e.g., AD77xx series)
  - On-chip PGA and high resolution ADC eliminates the need for external amplifier
- If buffer is used, it should be precision low noise (especially 1/f noise)
  - OP177
  - AD707
  - AD797

Figure 3-26: High resolution low frequency measurement ADCs

Some members of this family, such as the AD7730, have a high impedance input buffer which isolates the analog inputs from switching transients generated in the front end programmable gain amplifier (PGA) and the sigma-delta modulator. Therefore, no special precautions are required in driving the analog inputs. Other members of the AD77XX family, however, either do not have the input buffer or, if one is included on-chip, it can be switched either in or out under program control. Bypassing the buffer offers a slight improvement in noise performance.

The equivalent input circuit of the AD77XX family without an input buffer is shown in Figure 3-27. The input switch alternates between the 10 pF sampling capacitor and ground. The 7 kΩ internal resistance, R_{INT}, is the on resistance of the input multiplexer. The switching frequency is dependent on the frequency of the input clock and also the internal PGA gain. If the converter is working to an accuracy of 20-bits, the 10 pF internal capacitor, CINT, must charge to 20-bit accuracy during the time the switch connects the capacitor to the input. This interval is one-half the period of the switching signal (it has a 50% duty cycle). The input RC time constant due to the 7 kΩ resistor and the 10 pF sampling capacitor is 70 ns. If the charge is to achieve 20-bit accuracy, the capacitor must charge for at least 14 time constants, or 980 ns. Any external resistance in series with the input will increase this time constant.

There are tables on the datasheets for the various AD77XX ADCs, which give the maximum allowable values of R_{EXT} in order to maintain a given level of accuracy. These tables should be consulted if the external source resistance is more than a few kΩ.

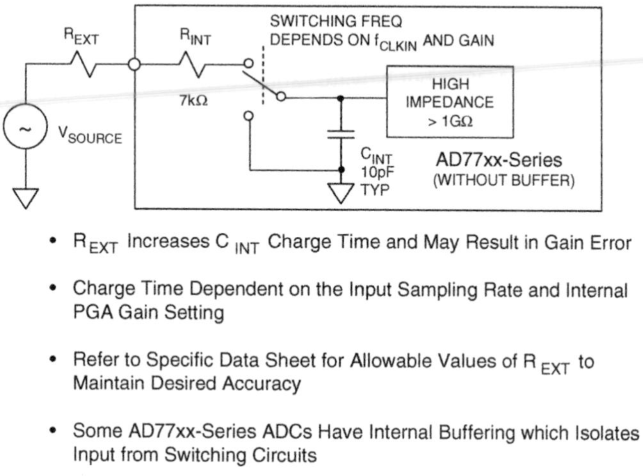

Figure 3-27: Driving unbuffered AD77XX-series Σ-Δ ADC inputs

Note that for instances where an external op amp buffer is found to be required with this type of converter, guidelines exist for best overall performance. This amplifier should be a precision low noise bipolar input type, such as the OP177, AD707, or the AD797.

Op Amp Considerations for Multiplexed Data Acquisition Applications

Multiplexing is a fundamental part of many data acquisition systems. Switches used in multiplexed data acquisition systems are generally CMOS-types shown in Figure 3-28. Utilizing P-Channel and N-Channel MOSFET switches in parallel minimizes the change of *on resistance* (R_{ON}) as a function of signal voltage. On resistance can vary from less than five to several hundred ohms, depending upon the device. Variation in on resistance as a function of signal level (often called *R_{ON}-modulation*) causes distortion if the multiplexer drives a load, therefore *R_{ON} flatness* is also an important specification.

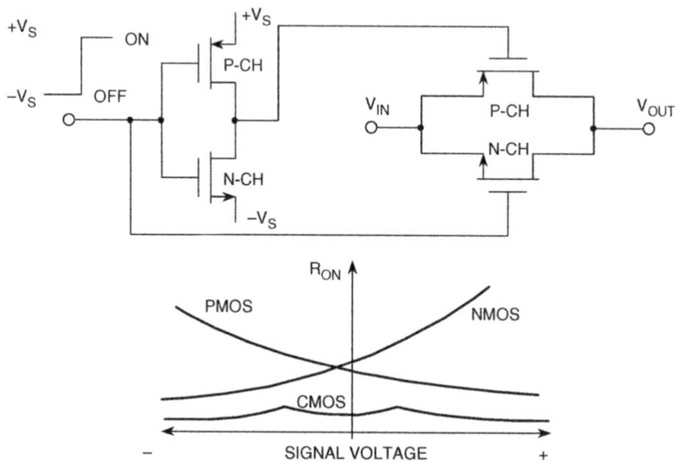

Figure 3-28: Basic CMOS analog switch

Because of the effects of nonzero R_{ON} and R_{ON}-modulation, multiplexer outputs should be isolated from the load with a suitable buffer op amp. A separate buffer is not required if the multiplexer drives a high input impedance, such as a PGA, SHA or ADC—but beware. Some SHAs and ADCs draw high frequency pulse current at their sampling rate and cannot tolerate being driven by an unbuffered multiplexer.

Key multiplexer specifications are *switching time, on resistance, on resistance flatness*, and *off-channel isolation, and crosstalk.* Multiplexer switching time ranges from less than 20 ns to over 1 µs, R_{ON} from less than 5 Ω to several hundred ohms, and off-channel isolation from 50 dB to 90 dB.

A number of CMOS switches can be connected to form an analog multiplexer, as shown in Figure 3-29. The number of input channels typically ranges from 4 to 16, and some multiplexers have internal channel-address decoding logic and registers, while with others, these functions must be performed externally. Unused multiplexer inputs *must* be grounded or severe loss of system accuracy may result. In applications requiring an op amp buffer, it should be noted that when the multiplexer changes channels it is possible to have a full-scale step function into the op amp and the ADC that follows it.

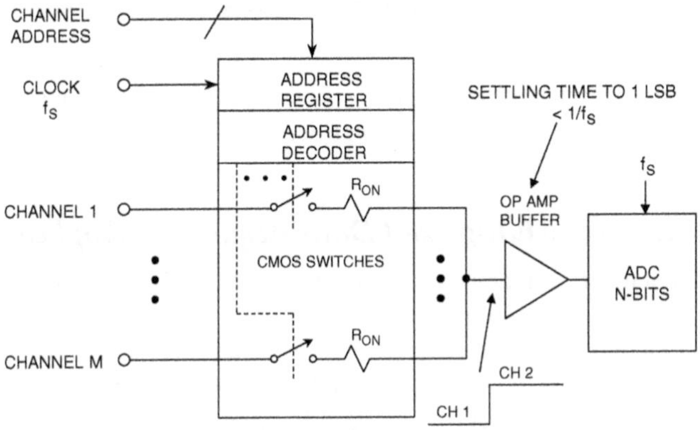

Figure 3-29: Typical multiplexed data acquisition system requires fast settling op amp buffer

Op amp settling time must be fast enough so that conversion errors do not result. It is customary to specify the op amp settling time to 1 LSB, and the allowed time for this settling is generally the reciprocal of the sampling frequency.

Driving Single-Supply Data Acquisition ADCs with Scaled Inputs

The AD789X and AD76XX family of single supply SAR ADCs (as well as the AD974, AD976, and AD977) includes a thin film resistive attenuator and level shifter on the analog input to allow a variety of input range options, both bipolar and unipolar.

A simplified diagram of the input circuit of the AD7890-10 12-bit, 8-channel ADC is shown in Figure 3-30. This arrangement allows the converter to digitize a ±10 V input while operating on a single +5 V supply.

Within the ADC, the R1/R2/R3 thin film network provides attenuation and level shifting to convert the ±10 V input to a 0 V to +2.5 V signal that is digitized. This type of input requires no special drive circuitry, because R1 isolates the input from the actual converter circuitry that may generate transient currents due to

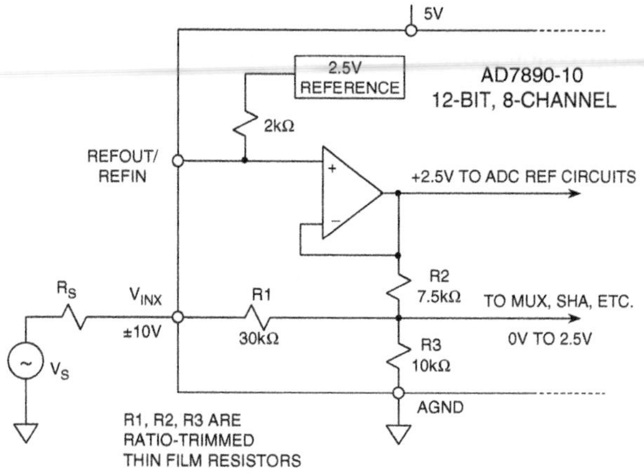

Figure 3-30: Driving single-supply data acquisition ADCs with scaled inputs

the conversion process. Nevertheless, the external source resistance R_S should be kept reasonably low, to prevent gain errors caused by the R_S/R1 divider.

Driving ADCs with Buffered Inputs

Some ADCs have on-chip buffer amplifiers on their analog input to simplify the interface. This feature is most often found on ADCs designed on either bipolar or BiCMOS processes. Conversely, input amplifiers are rarely found on CMOS ADCs because of the inherent difficulty associated with amplifier design in CMOS.

A typical input structure for an ADC with an input buffer is shown in Figure 3-31 for the AD9042 12-bit, 41 MSPS ADC. The effective input impedance is 250 Ω, and an external 61.9 Ω resistor in parallel with this internal 250 Ω provides an effective input termination of 50 Ω to the signal source. The circuit shows an ac-coupled input. An internal reference voltage of 2.4 V sets the input CM voltage of the AD9042.

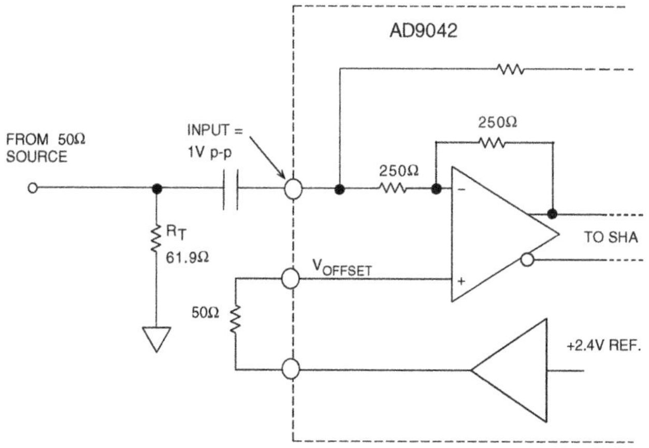

Figure 3-31: AD9042 ADC is designed to be driven directly from 50 Ω source for best SFDR

The input amplifier precedes the ADC sample-and-hold (SHA), and therefore isolates the input from any transients produced by the conversion process. The gain of the amplifier is set such that the input range of the ADC is 1 V p-p. In the case of a single-ended input structure, the input amplifier serves to convert the single-ended signal to a differential one, which allows fully differential circuit design techniques to be used throughout the remainder of the ADC.

Driving Buffered Differential Input ADCs

Figure 3-32 below shows two possible input structures for an ADC with buffered differential inputs. The input CM voltage is set with an internal resistor divider network in Figure 3-32A (left), and by a voltage reference in Figure 3-32B (right).

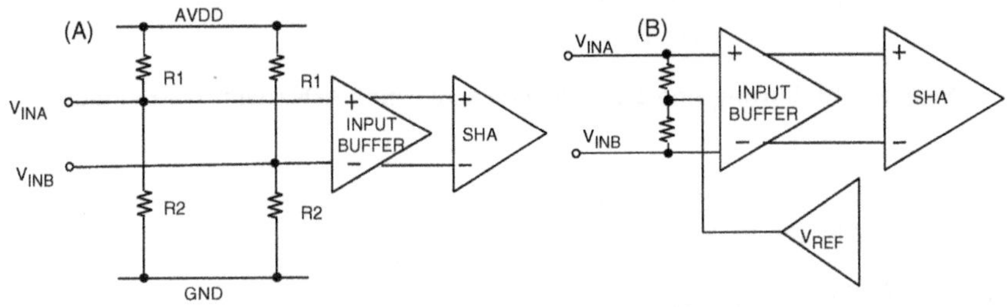

- Input buffers typical on BiMOS and bipolar processes
- Difficult on CMOS
- Simplified input interface – no transient currents
- Fixed common-mode level may limit flexibility

Figure 3-32: Simplified input circuit of typical buffered ADC with differential inputs

In single supply ADCs, the CM voltage is usually equal to one-half the power supply voltage, but some ADCs may use other values. Although the input buffers provide for a simplified interface, the fixed CM voltage may limit flexibility in some dc coupled applications.

It is worth noting that differential ADC inputs offer several advantages over single-ended ones. First, many signal sources in communications applications are differential (such as the output of a balanced mixer or an RF transformer). Thus, an ADC that accepts differential inputs interfaces easily in such systems. Second, maintaining balanced differential transmission in the signal path and within the ADC itself often minimizes even-order distortion products as well as improving CM noise rejection. Third, (and somewhat more subtly), a differential ADC input swing of say, 2 V p-p requires only 1 V p-p from twin driving sources. On low voltage and single-supply systems, this lower absolute level of drive can often make a real difference in the dual amplifier driver distortion, due to practical headroom limitations.

Given all of these points, it behooves the system engineer to operate a differential-capable ADC in the differential mode for best overall performance. This may be true even if a second amplifier must be added for the complementary drive signal, since dual op amps are only slightly more expensive than singles.

Driving CMOS ADCs with Switched Capacitor Inputs

CMOS ADCs are quite popular because of their low power and low cost. The equivalent input circuit of a typical CMOS ADC using a differential sample-and-hold is shown in Figure 3-33. While the switches are shown in the *track* mode, note that they open/close at the sampling frequency. The 16 pF capacitors represent the effective capacitance of switches S1 and S2, plus the stray input capacitance. The C_S capacitors (4pF) are the sampling capacitors, and the C_H capacitors are the hold capacitors. Although the input circuit is completely differential, this ADC structure can be driven either single-ended or differentially. Optimum performance, however, is generally obtained using a differential transformer or differential op amp drive.

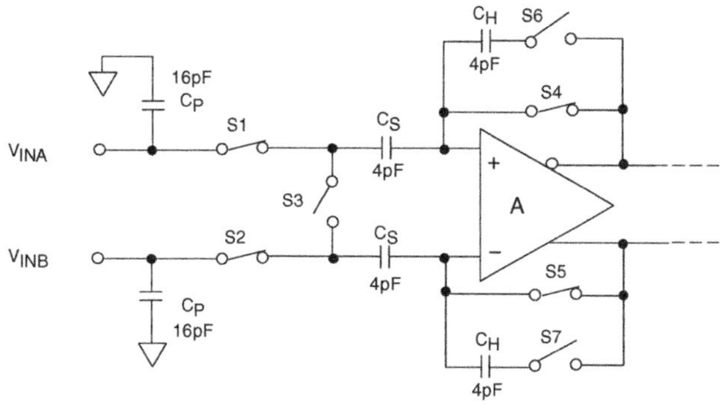

Figure 3-33: Simplified input circuit for a typical switched capacitor CMOS sample-and-hold

In the *track* mode, the differential input voltage is applied to the C_S capacitors. When the circuit enters the *hold* mode, the voltage across the sampling capacitors is transferred to the C_H hold capacitors and buffered by the amplifier A (the switches are controlled by the appropriate sampling clock phases). When the SHA returns to the *track* mode, the input source must charge or discharge the voltage stored on C_S to a new input voltage. This action of charging and discharging C_S, averaged over a period of time and for a given sampling frequency, f_s, makes the input impedance appear to have a benign resistive component. However, if this action is analyzed within a sampling period ($1/f_s$), the input impedance is dynamic, and certain input drive source precautions should be observed.

The resistive component to the input impedance can be computed by calculating the average charge that is drawn by C_H from the input drive source. It can be shown that if C_S is allowed to fully charge to the input voltage before switches S1 and S2 are opened, the average current into the input is the same as if there were a resistor equal to $1/(C_S f_S)$ connected between the inputs. Since C_S is only a few picofarads, this resistive component is typically greater than several kΩ for an $f_S = 10$ MSPS.

Over a sampling period, the SHA's input impedance appears as a dynamic load. When the SHA returns to the track mode, the input source should ideally provide the charging current through the R_{ON} of switches S1 and S2 in an exponential manner. The requirement of exponential charging means that the source impedance should be both low and resistive up to and beyond the sampling frequency.

The output impedance of an op amp can be modeled as a series inductor and resistor. When a capacitive load is switched onto the output of the op amp, the output will momentarily change due to its effective high frequency output impedance. As the output recovers, ringing may occur. To remedy this situation, a series resistor can be inserted between the op amp and the SHA input. The optimum value of this resistor is dependent on several factors including the sampling frequency and the op amp selected, but in most applications, a 25 Ω to 100 Ω resistor is optimum.

Single-Ended ADC Drive Circuits

Although most CMOS ADC inputs are differential, they can be driven single-ended with some ac performance degradation. An important consideration in CMOS ADC applications are the input switching transients previously discussed.

For instance, the input switching transient on one of the inputs of the AD9225 12-bit, 25 MSPS ADC is shown in Figure 3-34. This data was taken driving the ADC with an equivalent 50 Ω source impedance. During the sample-to-hold transition, the input signal is sampled when C_S is disconnected from the source. Notice that during the hold-to-sample transition, C_S is reconnected to the source for recharging. The transients consist of linear, nonlinear, and CM components at the sample rate. In addition to selecting an op amp with sufficient bandwidth and distortion performance, the output should settle to these transients during the sampling interval, $1/f_s$. The general circuit shown in Figure 3-35 is typical for this type of single-ended op amp ADC driver application.

- Hold-to-Sample Mode Transition – C_S Returned to Source for "recharging." Transient Consists of Linear, Nonlinear, and Common-Mode Components at Sample Rate.
- Sample-to-Hold Mode Transition – Input Signal Sampled when C_S is disconnected from Source.

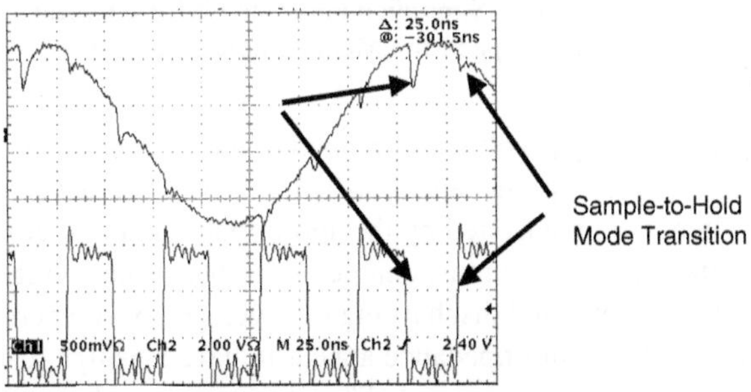

Figure 3-34: Single-ended input transient response of CMOS switched capacitor SHA (AD9225)

In this circuit, series resistor R_S has a dual purpose. Typically chosen in the range of 25 Ω–100 Ω, it limits the peak transient current from the driving op amp. Importantly, it also decouples the driver from the ADC input capacitance (and possible phase margin loss).

Another feature of the circuit are the dual networks of R_S and C_F. Matching both the dc and ac the source impedance for the ADC's V_{INA} and V_{INB} inputs ensures symmetrical settling of CM transients, for optimum noise and distortion performance. At both inputs, the C_F shunt capacitor acts as a charge reservoir and steers the CM transients to ground.

In addition to the buffering of transients, R_S and C_F also form a low-pass filter for V_{IN}, which limits the output noise of the drive amplifier into the ADC input V_{INA}. The exact values for R_S and C_F are generally

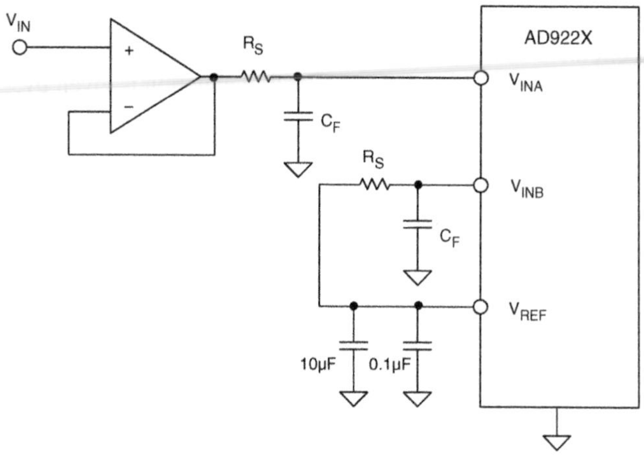

Figure 3-35: Optimizing single-ended switched capacitor ADC input drive circuit

optimized within the circuit, and the recommended values given on the ADC data sheet. The ADC data sheet information should also be consulted for the recommended drive op amp for best performance.

To enable best correlation of performance between environments, an ADC evaluation board should used (if available). This will ensure confidence when the ADC data sheet circuit performance is duplicated. Analog Devices makes evaluation boards available for many of their ADC and DAC devices (plus, of course, op amps), and general information on them is contained in Chapter 7 of this book.

Op Amp Gain Setting and Level Shifting in DC-Coupled Applications

In dc-coupled applications, the drive amplifier must provide the required gain and offset voltage, to match the signal to the input voltage range of the ADC. Figure 3-36 summarizes various op amp gain and level-shifting options. The circuit of Figure 3-36A operates in the noninverting mode, and uses a low impedance reference voltage, V_{REF}, to offset the output. Gain and offset interact according to the equation:

$$V_{OUT} = \left[1 + (R2/R1)\right] \bullet V_{IN} - \left[(R2/R1) \bullet V_{REF}\right] \qquad \text{Eq. 3-4}$$

Figure 3-36: Op amp gain and level shifting circuits

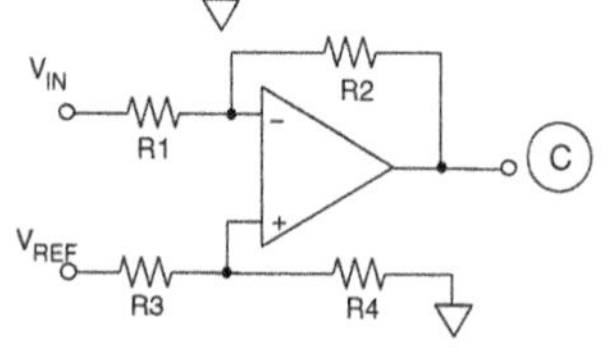

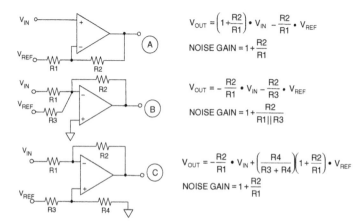

$$V_{OUT} = \left(1 + \frac{R2}{R1}\right) \bullet V_{IN} - \frac{R2}{R1} \bullet V_{REF}$$

$$\text{NOISE GAIN} = 1 + \frac{R2}{R1}$$

$$V_{OUT} = -\frac{R2}{R1} \bullet V_{IN} - \frac{R2}{R3} \bullet V_{REF}$$

$$\text{NOISE GAIN} = 1 + \frac{R2}{R1||R3}$$

$$V_{OUT} = -\frac{R2}{R1} \bullet V_{IN} + \left(\frac{R4}{R3 + R4}\right)\left(1 + \frac{R2}{R1}\right) \bullet V_{REF}$$

$$\text{NOISE GAIN} = 1 + \frac{R2}{R1}$$

The circuit in Figure 3-36B operates in the inverting mode, and the signal gain is independent of the offset. The disadvantage of this circuit is that the addition of R3 increases the noise gain, and hence the sensitivity to the op amp input offset voltage and noise. The input/output equation is given by:

$$V_{OUT} = -(R2/R1) \bullet V_{IN} - (R2/R3) \bullet V_{REF} \qquad \text{Eq. 3-5}$$

The circuit in Figure 3-36C also operates in the inverting mode, and the offset voltage V_{REF} is applied to the noninverting input without noise gain penalty. This circuit is also attractive for single-supply applications ($V_{REF} > 0$). The input/output equation is given by:

$$V_{OUT} = -(R2/R1) \bullet V_{IN} + \left[(R4/(R3+R4))(1+(R2/R1))\right] \bullet V_{REF} \qquad \text{Eq. 3-6}$$

Note that the circuit of Figure 3-36A is sensitive to the impedance of V_{REF}, unlike the counterparts in B and C. This is due to the fact that the signal current flows into/from V_{REF}, due to V_{IN} operating the op amp over its CM range. In the other two circuits the CM voltages are fixed, and no signal current flows in V_{REF}.

A dc-coupled single-ended op amp driver for the AD9225 12-bit, 25 MSPS ADC is shown in Figure 3-37. This circuit interfaces a ±2 V input signal to the single-supply ADC, and provides transient current isolation. The ADC input voltage range is 0 V to 4 V, and a dual supply op amp is required, since the ADC minimum input is 0 V.

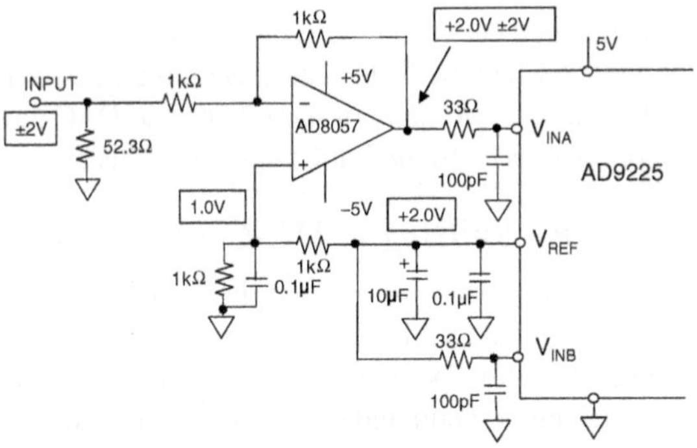

Figure 3-37: Dc-coupled single-ended level shifter and driver for the AD9225 12-bit, 25 MSPS CMOS ADC

The noninverting input of the AD8057 is biased at 1 V, which sets the output CM voltage at V_{INA} to 2 V for a bipolar input signal source. Note that the V_{INA} and V_{INB} source impedances are matched for better CM transient cancellation. The 100 pF capacitors act as small charge reservoirs for the input transient currents, and also form low-pass noise filters with the 33 Ω series resistors.

A similar level shifter and drive circuit is shown in Figure 3-38, operating on a single 5 V supply. In this circuit the bipolar ±1 V input signal is interfaced to the input of the ADC whose span is set for 2 V about a +2.5 V CM voltage. The AD8041 rail-to-rail output op amp is used. The 1.25 V input CM voltage for the AD8041 is developed by a voltage divider from the external AD780 2.5 V reference.

Note that single-supply circuits of this type must observe op amp input and output CM voltage restrictions, to prevent clipping and excess distortion.

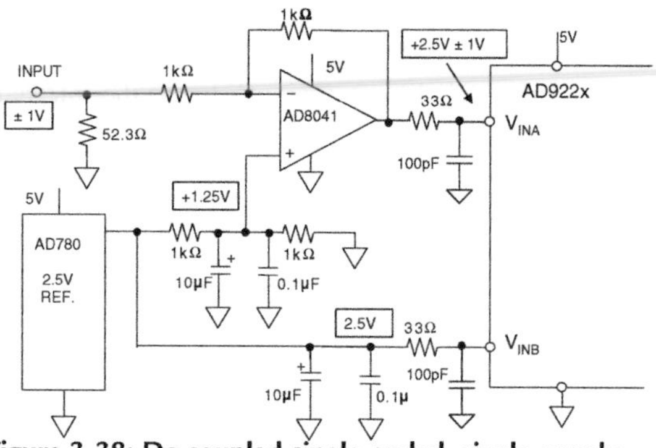

Figure 3-38: Dc-coupled single-ended, single-supply ADC driver/level shifter using external reference

Drivers for Differential Input ADCs

Most high performance ADCs are now being designed with differential inputs. A fully differential ADC design offers the advantages of good CM rejection, reduction in second-order distortion products, and simplified dc trim algorithms. Although they can be driven single-ended as previously described, a fully differential driver usually optimizes overall performance.

- High common-mode noise rejection
- Flexible input common-mode voltage levels
- Reduced input signal swings helps in low voltage, single-supply applications
- Reduced second-order distortion products
- Simplified dc trim algorithms because of internal matching
- Requires high performance differential driver

Figure 3-39: Differential input ADCs offer performance advantages

Waveforms at the two inputs of the AD9225 12-bit, 25 MSPS CMOS ADC are shown in Figure 3-40A, designated as V_{INA} and V_{INB}. The balanced source impedance is 50 Ω, and the sampling frequency is set for 25 MSPS. The diagram clearly shows the switching transients due to the internal ADC switched capacitor sample-and-hold. Figure 3-40B shows the difference between the two waveforms, $V_{INA}-V_{INB}$.

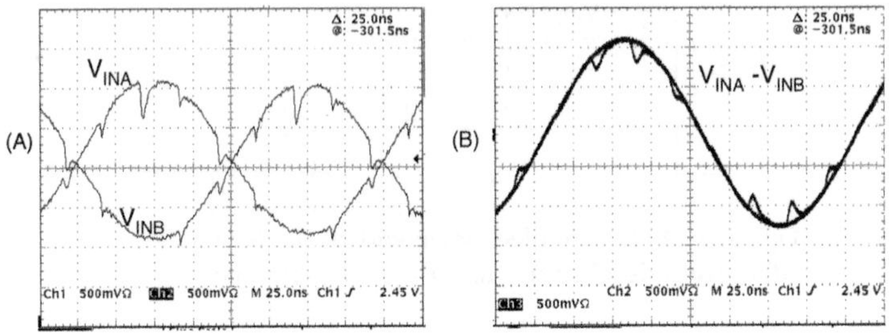

- Differential charge transient is symmetrical around midscale and dominated by linear component
- Common-mode transients cancel with equal source impedance

Figure 3-40: Differential input transient response of CMOS switched capacitor SHA (AD9225)

Note that the resulting differential charge transients are symmetrical about midscale, and that there is a distinct linear component to them. This shows the reduction in the CM transients, and also leads to better distortion performance than would be achievable with a single-ended input.

Transformer coupling into a differential input ADC provides excellent CM rejection and low distortion if performance to dc is not required. Figure 3-41 shows a typical circuit. The transformer is a Mini-Circuits RF transformer, model #T4-6T which has an impedance ratio of 4 (turns ratio of 2). The schematic assumes that the signal source has a 50 Ω source impedance. The 1:4 impedance ratio requires the 200 Ω secondary termination for optimum power transfer and low VSWR. The Mini-Circuits T4-6T has a 1 dB bandwidth

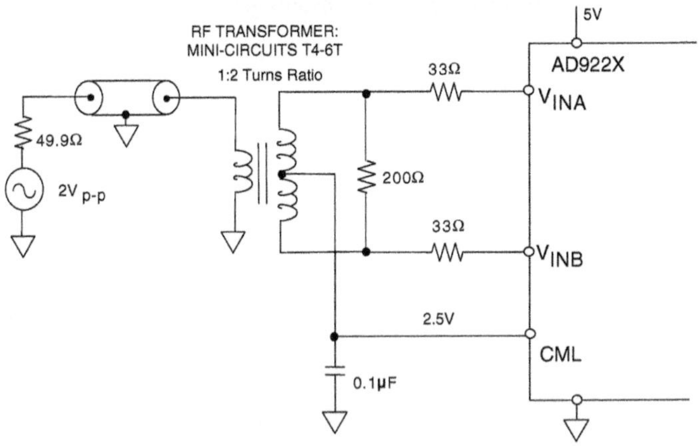

Figure 3-41: Transformer coupling into AD922X ADC

from 100 kHz to 100 MHz. The center tap of the transformer provides a convenient means of level shifting the input signal to the optimum CM voltage of the ADC. The AD922X CML (common-mode level) pin is used to provide the +2.5 CM voltage.

Transformers with other turns ratios may also be selected to optimize the performance for a given application. For example, a given input signal source or amplifier may realize an improvement in distortion performance at reduced output power levels and signal swings. Hence, selecting a transformer with a higher impedance ratio (i.e. Mini-Circuits #T16-6T with a 1:16 impedance ratio, turns ratio 1:4) effectively "steps up" the signal level thus reducing the driving requirements of the signal source.

Note the 33 Ω series resistors inserted between the transformer secondary and the ADC input. These values were specifically selected to optimize both the SFDR and the SNR performance of the ADC. They also provide isolation from transients at the ADC inputs. Transients currents are approximately equal on the V_{INA} and V_{INB} inputs, so they are isolated from the primary winding of the transformer by the transformer's CM rejection.

Transformer coupling using a CM voltage of +2.5 V provides the maximum SFDR when driving the AD922X series. By driving the ADC differentially, even-order harmonics are reduced compared with the single-ended circuit.

Driving ADCs with Differential Amplifiers

There are many applications where differential input ADCs cannot be driven with transformers because the frequency response must extend to dc. In these cases, op amps can be used to implement the differential drivers. Figure 3-42 shows how the dual AD8058 op amp can be connected to convert a single-ended bipolar signal to a differential one suitable for driving the AD922X family of ADCs. The input range of the ADC is set for a 2 V p-p signal on each input (4 V span), and a CM voltage of +2 V.

The A1 amplifier is configured as a noninverting op amp. The 1 kΩ divider resistors level shift the ±1 V input signal to +1 V ±0.5 V at the noninverting input of A1. The output of A1 is therefore +2 V ±1 V.

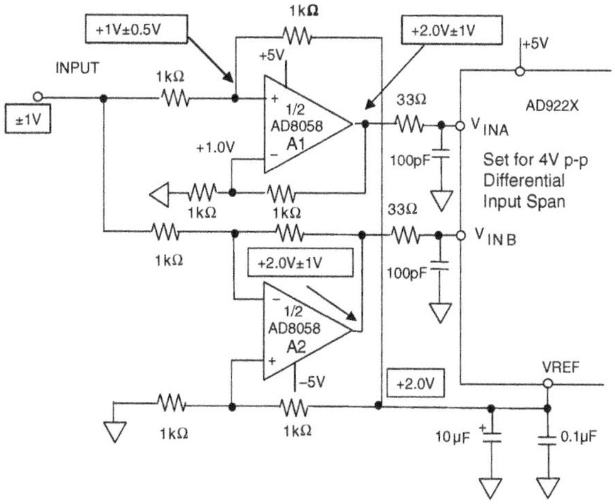

Figure 3-42: Op amp single-ended to differential dc-coupled driver with level shifting.

The A2 op amp inverts the input signal, and the 1 kΩ divider resistors establish a +1 V CM voltage on its noninverting input. The output of A2 is therefore +2 V ±1 V.

This circuit provides good matching between the two op amps because they are duals on the same die and are both operated at the same noise gain of 2. However, the input voltage noise of the AD8058 is $20\,\text{nV}/\sqrt{\text{Hz}}$, and this appears as $40\,\text{nV}/\sqrt{\text{Hz}}$ at the output of both A1 and A2 thereby introducing possible SNR degradation in some applications. In the circuit of Figure 3-42, this is mitigated somewhat by the 100 pF input capacitors which not only reduce the input noise but absorb some of the transient currents. It should be noted that because the input CM voltage of A1 can go as low as 0.5 V, dual supplies must be used for the op amps.

A block diagram of the AD813X family of fully differential amplifiers optimized for ADC driving is shown in Figure 3-43 (see References 3-5). Figure 3-43A shows the details of the internal circuit, and Figure 3-43B shows the equivalent circuit. The gain is set by the external R_F and R_G resistors, and the CM voltage is set by the voltage on the V_{OCM} pin. The internal CM feedback forces the V_{OUT+} and V_{OUT-} outputs to be balanced, i.e., the signals at the two outputs are always equal in amplitude but 180° out of phase per the equation,

$$V_{OCM} = \left(V_{OUT+} + V_{OUT-}\right)/2 \qquad \text{Eq. 3-7}$$

The circuit can be used with either a differential or a single-ended input, and the voltage gain is equal to the ratio of R_F to R_G.

The AD8138 has a 3 dB small-signal bandwidth of 320 MHz (G = +1) and is designed to give low harmonic distortion as an ADC driver. The circuit provides excellent output gain and phase matching, and the balanced structure suppresses even-order harmonics.

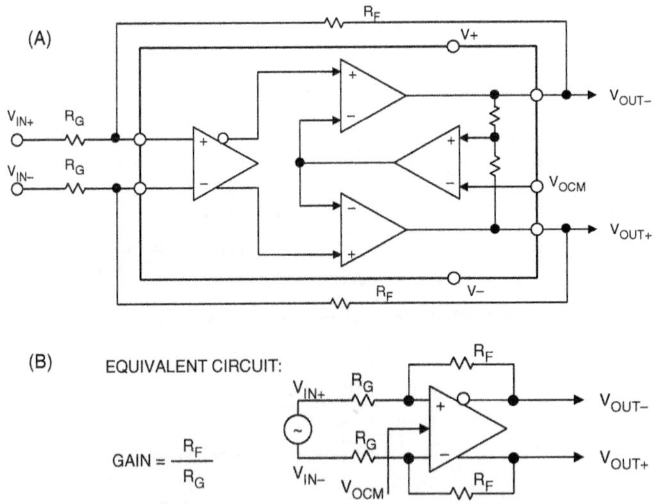

Figure 3-43: AD813X differential ADC driver functional diagram and equivalent circuit

Figure 3-44 shows the AD8138 driving the AD9203 10-bit, 40 MSPS ADC (see Reference 6). This entire circuit operates on a single 3 V supply. A 1 V p-p bipolar single-ended input signal produces a 1 V p-p differential signal at the output of the AD8138, centered around a CM voltage of 1.5 V (mid-supply).

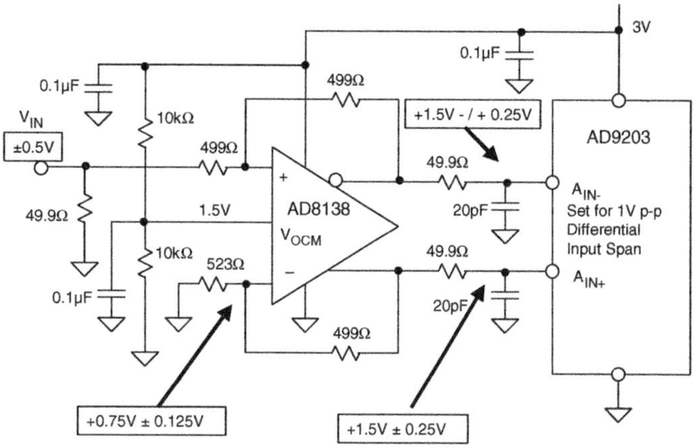

Figure 3-44: AD8138 driving AD9203 10-bit, 40 MSPS ADC

Each of the differential inputs of the AD8138 swing between +0.625 V and +0.875 V, and each output swings between 1.25 V and 1.75 V. These voltages fall within the allowable input and output CM voltage range of the AD8138 operating on a single 3 V supply.

The circuit as shown operates on a 1 V p-p single-ended bipolar input signal, and the input span of the AD9203 ADC is set for 1 V p-p differential. If the signal input amplitude is increased to 2 V p-p, the span of the AD9203 must be set for 2 V p-p differential. Under these conditions, each of the AD8138 inputs must swing between 0.5 V and 1 V, and each of the outputs between 1 V and 2 V.

As shown in Figure 3-45, increasing the amplitude in this manner offers a slight improvement in low frequency SINAD due to the improvement in low frequency SNR.

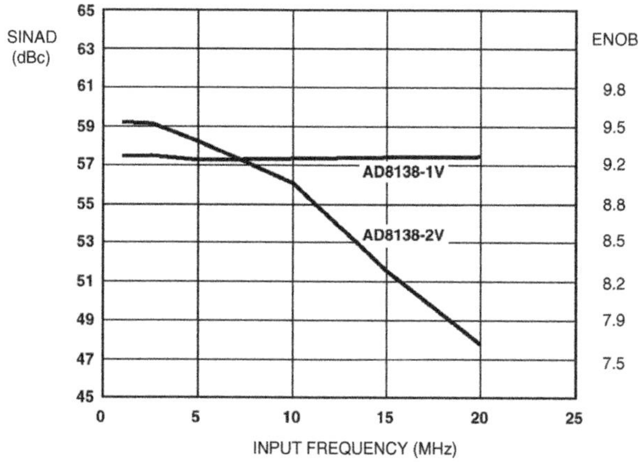

Figure 3-45: SINAD and ENOB for AD9203 12-bit, 40 MSPS ADC driven by AD8138 differential amplifier

At the same time however, a degradation occurs in the high frequency SINAD because of the larger distortion due to the larger signal swings.

Overvoltage Considerations

The input structures of most high performance ADCs are sensitive to overvoltage conditions because of the small geometry devices used in the designs. Although ADC inputs generally have ESD protection diodes connected from the analog input to each supply rail, these diodes are not designed to handle the large currents that can be generated from typical op amp drivers. Two good rules of thumb are to (1) limit the analog input voltage to no more than 0.3 V above or below the positive and negative supply voltages, respectively, and (2) limit the analog input current to 5 mA maximum in overvoltage conditions.

Several typical configurations for the drive amp/ADC interface are shown in Figure 3-46. In Figure 3-46A, the ADC requires no additional input protection because both the op amp and the ADC are driven directly from the same supply voltages. While the R_S resistor is not required for overvoltage protection, it does serve to isolate the capacitive input of the ADC from the output of the op amp.

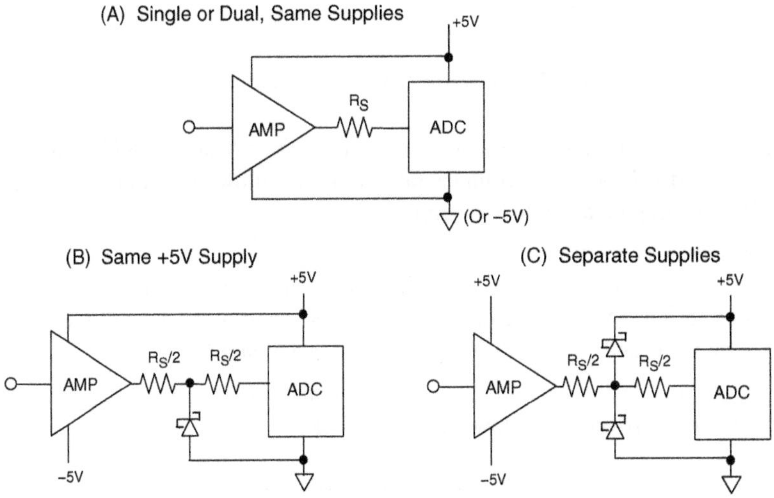

Figure 3-46: ADC input overvoltage protection circuits

Figure 3-46B shows a dual supply op amp driving a single supply ADC, with the 5 V supply is shared between the two devices. The diode protects the input of the ADC in case the output of the op amp is driven below ground. A Schottky diode is used because of its low forward voltage drop and its low capacitance. The R_S resistor is split into two equal resistors, and they are chosen to limit the ADC input fault current to 5 mA maximum. Note that the R_S resistor in conjunction with the ADC input capacitance forms a low-pass filter. If R_S is made too large, the input bandwidth may be restricted.

Figure 3-46C shows the condition where the op amp and the ADC are driven from separate supplies. Two Schottky diodes are required to protect the ADC under all power supply and signal conditions. As in 3-46A, the $R_S/2$ resistors limit ADC fault current.

References: Driving ADC Inputs

1. Data sheet for **AD8057/AD8058 Low Cost, High Performance Voltage Feedback, 325 MHz Amplifiers**, www.analog.com.
2. Chapter 8 of Walt Kester, Editor, **Practical Design Techniques for Sensor Signal Conditioning**, Analog Devices, 1999, ISBN: 0-916550-20-6.
3. Data sheet for **AD8131 Low-Cost, High-Speed Differential Driver**, www.analog.com.
4. Data sheet for **AD8132 Low-Cost, High-Speed Differential Amplifier**, www.analog.com.
5. Data sheet for **AD8138 Low Distortion Differential ADC Driver**, www.analog.com.
6. Data Sheet for **AD9203 10-Bit, 40 MSPS, 3 V, 74mW A/D Converter**, www.analog.com.
7. Chapters 3–6, Walt Kester, Editor, **Practical Analog Design Techniques**, Analog Devices, 1995, ISBN: 0-916550-16-8.
8. Chapters 4, 5, Walt Kester, Editor, **High Speed Design Techniques**, Analog Devices, 1996, ISBN: 0- 916550-17-6.
9. Chapters 3, 4, Walt Kester, Editor, **Mixed-Signal and DSP Design Techniques**, Analog Devices, 2000, ISBN: 0-916550-23-0.
10. Chapters 4, 5, Walt Kester, Editor, **Linear Design Seminar**, Analog Devices, 1995, ISBN: 0-916550-15-X.

Driving ADC/DAC Reference Inputs

Walt Jung, Walt Kester

It might seem odd to include a section on voltage references in a book devoted primarily to op amp applications, but the relevance will shortly become obvious. Unfortunately, there is little standardization with respect to ADC/DAC voltage references. Some ADCs and DACs have internal references, while others do not. In some cases, the dc accuracy of a converter with an internal reference can often be improved by overriding the internal reference with a more accurate and stable external one.

Although the reference element itself can be either a bandgap, buried zener, or XFET™ (see Reference 1), practically all references have some type of output buffer op amp. The op amp isolates the reference element from the output and also provides drive capability. However, this op amp must obey the general laws relating to op amp stability, and that is what makes the topic of references relevant to the discussion. Figure 3-47 summarizes voltage reference considerations.

- Data converter accuracy determined by the reference, whether internal or external
- Bandgap, buried zener, XFET generally have on-chip output buffer op amp
- Transient loading can cause instability and errors
- External decoupling capacitors may cause oscillation
- Output may require external buffer to source and sink current
- Reference voltage noise may limit system resolution

Figure 3-47: ADC/DAC voltage reference considerations

Note that a reference input to an ADC or DAC is similar to the analog input of an ADC, in that the internal conversion process can inject transient currents at that pin. This requires adequate decoupling to stabilize the reference voltage. Adding such decoupling might introduce instability in some reference types, depending on the output op amp design. Of course, a reference data sheet may not show any details of the output op amp, which leaves the designer in somewhat of a dilemma concerning whether or not it will be stable and free from transient errors. Fortunately, some simple lab tests can exercise a reference circuit for transient errors, and also determine stability for capacitive loading.

Figure 3-48 shows the transients associated with the reference input of a typical successive approximation ADC. The ADC reference voltage input must be stabilized with a sufficiently large decoupling capacitor, in order to prevent conversion errors. The value of the capacitor required as C_B may range from below 1 µF, to as high as 100 µF. This capacitor must of course have a voltage rating greater than the reference voltage. Physically, it will be of minimum size when purchased in a surface-mount style.

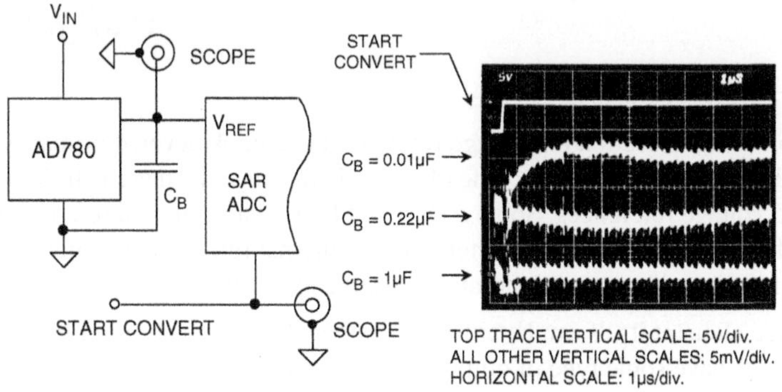

Figure 3-48: Successive approximation ADCs can present a transient load to the reference

Note that in this case, a 1 µF capacitor on the reference input is required to reduce the transients to an acceptable level. Note that the capacitor size can be electrically larger for further noise reduction—the tradeoff here is of course cost and PCB real estate. The AD780 will work with capacitors of up to 100 µF.

A well-designed voltage reference is stable with heavy capacitive decoupling. Unfortunately, some are not, as shown in Figure 3-49, where the addition of C_L to the reference output (a 0.01 µF capacitor) actually increases the amount of transient ringing. Such references are practically useless in data converter applications, because some amount of local decoupling is almost always required at the converter.

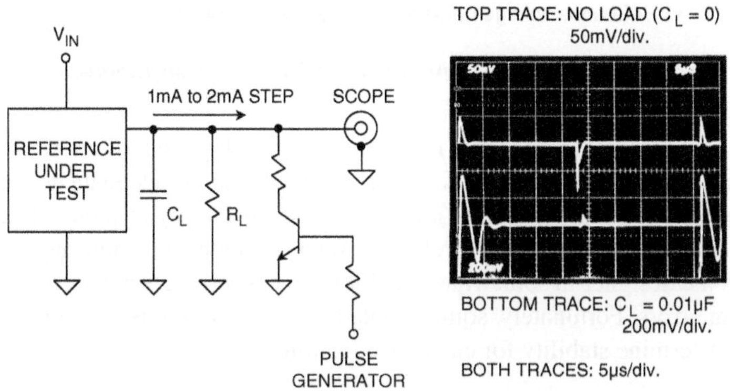

Figure 3-49: Make sure reference is stable with large capacitive loads

A suitable op amp buffer might be added between the reference and the data converter. Many good references are available (such as the AD780) which are stable with an output capacitor. This type of reference should be chosen for a data converter application, rather than incurring the further complication and expense of an op amp.

If very low noise levels are required from a reference, an additional low-pass filter followed by a low noise op amp can be used to achieve the desired performance. The reference circuit of Figure 3-50 is one such example (see References 2 and 3). This circuit uses external filtering and a precision low-noise op-amp to provide both very low noise and high dc accuracy. Reference U1 is a 2.5 V, 3.0 V, 5 V, or 10 V reference with a low noise buffered output. The output of U1 is applied to the R1–C1/C2 noise filter to produce a corner frequency of about 1.7 Hz.

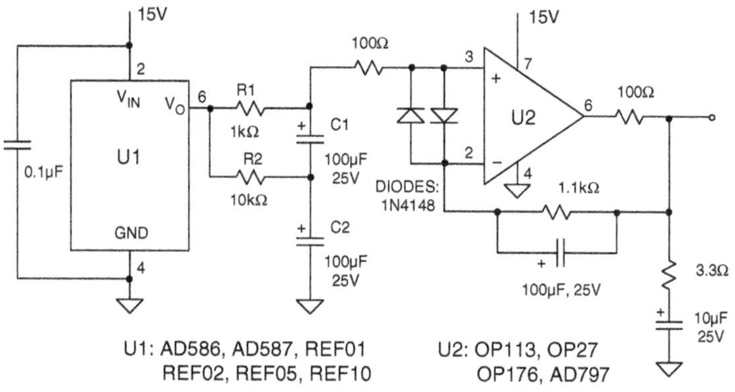

Figure 3-50: Low-noise op amp with filtering yields reference noise performance (1.5V to 5 nV/√HZ @ 1 kHz

Electrolytic capacitors usually imply dc leakage errors, but the bootstrap connection of C1 causes its applied bias voltage to be only the relatively small drop across R2. This lowers the leakage current through R1 to acceptable levels. Since the filter attenuation is modest below a few Hertz, the reference noise still affects overall performance at low frequencies (i.e., <10 Hz).

A precision low noise unity-gain follower, such as the OP113, then buffers the output of the filter. With less than ±150 µV of offset error and under 1 µV/°C drift, the buffer amplifier's dc performance will not seriously affect the accuracy/drift of most references. For example, an ADR292E for U1 will have a typical drift of 3 ppm/°C, equivalent to 7.5 µV/°C, higher than the buffer amplifier.

Almost any op amp will have a current limit higher than a typical IC reference, so this circuit allows greater current output. It also removes any load-related thermal errors that might occur when the reference IC is loaded directly.

Even lower noise op-amps are available, for 5–10 V use. The AD797 offers 1kHz noise performance less than $2\,nV/\sqrt{Hz}$ in this circuit, compared to about $5\,nV/\sqrt{Hz}$ for the OP113. With any buffer amplifier, Kelvin sensing can be used at the load point, a technique that eliminates I × R related output voltage errors.

References: Driving ADC/DAC Reference Inputs

1. Chapter 2, Walt Kester, Editor, **Practical Design Techniques for Power and Thermal Management**, Analog Devices, 1998, ISBN: 0-916550-19-2.
2. Walt Jung, "Build an Ultra-Low-Noise Voltage Reference," **Electronic Design Analog Applications Issue**, June 24, 1993.
3. Walt Jung, "Getting the Most from IC Voltage References," **Analog Dialogue**, Vol. 28, No. 1, 1994, pp. 13–21.

Buffering DAC Outputs

Walt Kester, Paul Hendriks

General Considerations

Another important op amp application is buffering DAC outputs. Modern IC DACs provide either voltage or current outputs. Figure 3-51 shows three fundamental configurations, all with the objective of using an op amp for a buffered output voltage.

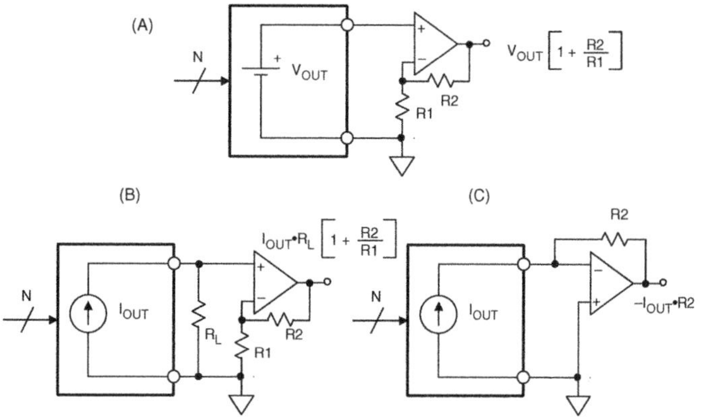

Figure 3-51: Buffering DAC outputs with op amps

Figure 3-51A shows a buffered voltage output DAC. In many cases, the DAC output can be used directly, without additional buffering. If an additional op amp buffer is needed, it is usually configured in a noninverting mode, with gain determined by R1 and R2.

There are two basic methods for dealing with a current output DAC. In Figure 3-51B, a voltage is simply developed across external load resistor, R_L. An external op amp can be used to buffer and/or amplify this voltage if required. Many DACs supply full-scale currents of 20 mA or more, thereby allowing reasonable voltages to be developed across fairly low value load resistors. For instance, fast settling video DACs typically supply nearly 30 mA full-scale current, allowing 1 V to be developed across a source and load terminated 75 Ω coaxial cable (representing a dc load of 37.5 Ω to the DAC output).

A direct method to convert the output current into a voltage is shown in Figure 3-51C, This circuit is usually called a current-to-voltage converter, or I/V. In this circuit, the DAC output drives the inverting input of an op amp, with the output voltage developed across the R2 feedback resistor. In this approach the DAC output always operates at virtual ground (which may give a linearity improvement vis-à-vis Figure 3-51B).

The general selection process for an op amp used as a DAC buffer is similar to that of an ADC buffer. The same basic specifications such as dc accuracy, noise, settling time, bandwidth, distortion, and so forth, apply to DACs as well as ADCs, and the discussion will not be repeated here. Rather, some specific application examples will be shown.

Differential to Single-Ended Conversion Techniques

A general model of a modern current output DAC is shown in Figure 3-52. This model is typical of the AD976X and AD977X TxDAC™ series (see Reference 1). Current output is more popular than voltage output, especially at audio frequencies and above. If the DAC is fabricated on a bipolar or BiCMOS process, it is likely that the output will sink current, and that the output impedance will be less than 500 Ω (due to the internal R/2R resistive ladder network). On the other hand, a CMOS DAC is more likely to source output current and have a high output impedance, typically greater than 100 kΩ.

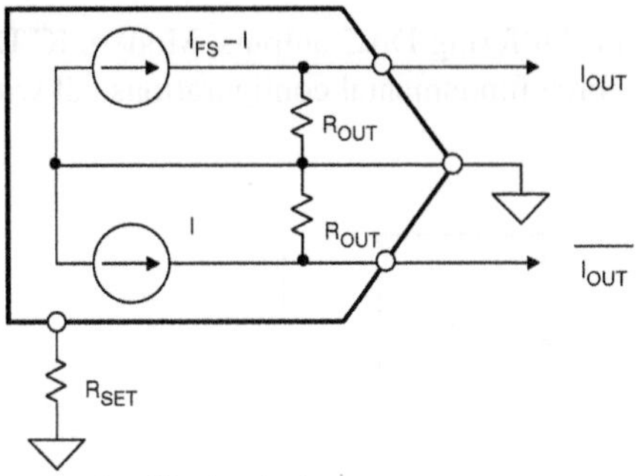

- I_{FS} 2 – 20mA typical
- Bipolar or BiCMOS DACs sink current, R_{OUT} < 500Ω
- CMOS DACs source current, R_{OUT} > 100kΩ
- Output compliance voltage < ±1V for best performance

Figure 3-52: Model of high speed DAC output

Another consideration is the output *compliance voltage*—the maximum voltage swing allowed at the output in order for the DAC to maintain its linearity. This voltage is typically 1 V to 1.5 V, but can vary depending upon the DAC. Best DAC linearity is generally achieved when driving a virtual ground, such as an op amp I/V converter.

Modern current output DACs usually have differential outputs, to achieve high CM rejection and reduce the even-order distortion products. Full-scale output currents in the range of 2 mA to 20 mA are common.

In most applications, it is desirable to convert the differential output of the DAC into a single-ended signal, suitable for driving a coax line. This can be readily achieved with an RF transformer, provided low frequency response is not required. Figure 3-53 shows a typical example of this approach. The high impedance current output of the DAC is terminated differentially with 50 Ω, which defines the source impedance to the transformer as 50 Ω.

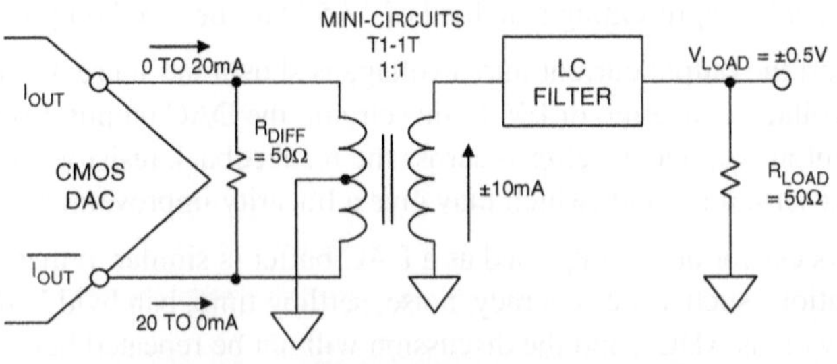

Figure 3-53: Differential transformer coupling

The resulting differential voltage drives the primary of a 1:1 RF transformer, to develop a single-ended voltage at the output of the secondary winding. The output of the 50 Ω LC filter is matched with the 50 Ω load resistor R_L, and a final output voltage of 1 V p-p is developed.

The transformer not only serves to convert the differential output into a single-ended signal, but it also isolates the output of the DAC from the reactive load presented by the LC filter, thereby improving overall distortion performance.

An op amp connected as a differential to single-ended converter can be used to obtain a single-ended output when frequency response to dc is required. In Figure 3-54 the AD8055 op amp is used to achieve high bandwidth and low distortion (see Reference 2). The current output DAC drives balanced 25 Ω resistive loads, thereby developing an out-of-phase voltage of 0 V to 0.5 V at each output. The AD8055 is configured for a gain of 2, to develop a final single-ended ground-referenced output voltage of 2 V p-p. Note that because the output signal swings above and below ground, a dual-supply op amp is required.

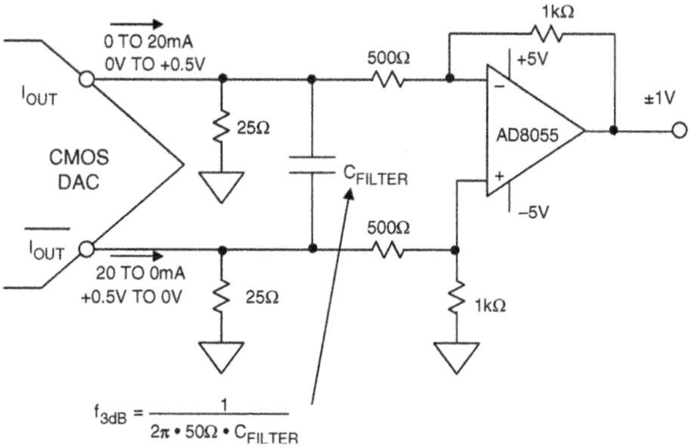

Figure 3-54: Differential dc-coupled output using a dual supply op amp

The C_{FILTER} capacitor forms a differential filter with the equivalent 50 Ω differential output impedance. This filter reduces any slew induced distortion of the op amp, and the optimum cutoff frequency of the filter is determined empirically to give the best overall distortion performance.

A modified form of the Figure 3-54 circuit can also be operated on a single supply, provided the CM voltage of the op amp is set to mid-supply (2.5 V). This is shown in Figure 3-55 below. The output voltage is 2 V p-p centered around a CM voltage of 2.5 V. This CM voltage can be either developed from the 5 V supply using a resistor divider, or directly from a 2.5 V voltage reference. If the 5 V supply is used as the CM voltage, it must be heavily decoupled to prevent supply noise from being amplified.

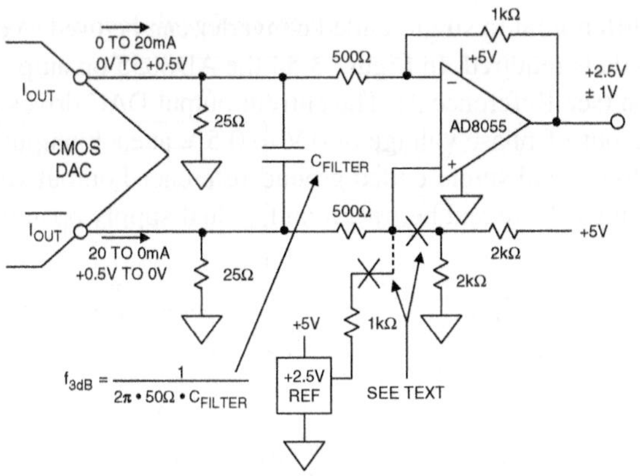

Figure 3-55: Differential dc-coupled output using a single-supply op amp

Single-Ended Current-to-Voltage Conversion

Single-ended current-to-voltage conversion is easily performed using a single op amp as an I/V converter, as shown in Figure 3-56. The 10 mA full-scale DAC current from the AD768 (see Reference 3) develops a 0 V to 2 V output voltage across the 200 Ω R_F.

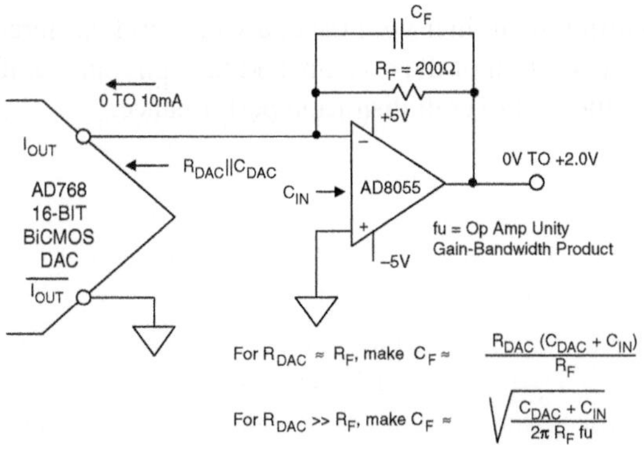

Figure 3-56: Single-ended I/V op amp interface for precision 16-bit AD768 DAC

Driving the virtual ground of the AD8055 op amp minimizes any distortion due to nonlinearity in the DAC output impedance. In fact, most high resolution DACs of this type are factory trimmed using an I/V converter.

It should be recalled, however, that compared to a differential operating mode using the single-ended output of the DAC in this manner will cause degradation in the CM rejection and increased second-order distortion products.

The C_F feedback capacitor should be optimized for best pulse response in the circuit. The equations given in the diagram should only be used as guidelines. A more detailed analysis of this circuit is given in Reference 6.

Differential Current-to-Differential Voltage Conversion

If a buffered differential voltage output is required from a current output DAC, the AD813X-series of differential amplifiers can be used as shown in Figure 3-57.

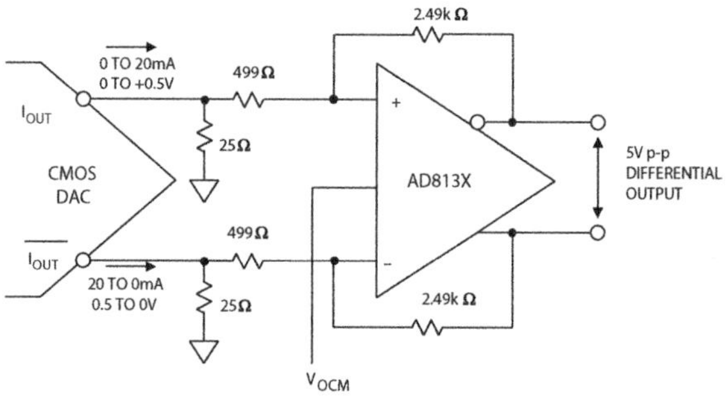

Figure 3-57: Buffering high speed DACs using AD813X differential amplifier

The DAC output current is first converted into a voltage that is developed across the 25 Ω resistors. The voltage is amplified by a factor of 5 using the AD813X. This technique is used in lieu of a direct I/V conversion to prevent fast slewing DAC currents from overloading the amplifier and introducing distortion. Care must be taken so that the DAC output voltage is within its compliance rating.

The V_{OCM} input on the AD813X can be used to set a final output CM voltage within the range of the AD813X. If transmission lines are to be driven at the output, adding a pair of 75Ω resistors will allow this.

An Active Low-Pass Filter for Audio DAC

Figure 3-58 shows an active low-pass filter which also serves as a current-to-voltage converter for the AD1853 sigma-delta audio DAC (see Reference 4). The filter is a 4-pole filter with a 3 dB cutoff frequency of approximately 75 kHz. Because of the high oversampling frequency (24.576 MSPS when operating the DAC at a 48 KSPS throughput rate), a simple filter is all that is required to remove aliased components above 12 MHz).

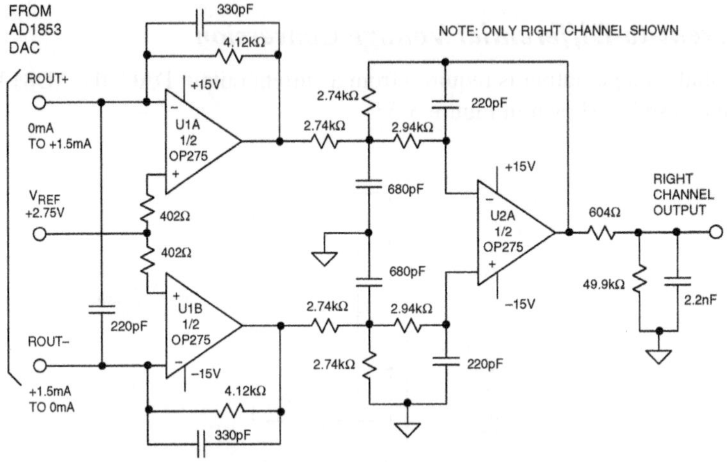

Figure 3-58: A 75 kHz 4-pole gaussian active filter for buffering the output of AD1853 stereo DAC

The diagram shows a single channel for the dual channel DAC output. U1A and U1B I/V stages form a 1-pole differential filter, while U2 forms a 2-pole multiple-feedback filter that also performs a differential-to-single-ended conversion.

A final fourth passive pole is formed by the 604 Ω resistor and the 2.2 nF capacitor across the output. The OP275 op amp was chosen for operation at U1 and U2, and for its quality audio characteristics (see Reference 5).

For further details of active filter designs, see Chapter 5 of this book.

References: Buffering DAC Outputs

1. Data sheet for **AD9772A 14-Bit, 160 MSPS TxDAC+® with** 2x **Interpolation Filter**, www.analog.com.
2. Data sheet for **AD8055/AD8056 Low Cost, 300 MHz Voltage Feedback Amplifiers**, www.analog.com.
3. Data sheet for **AD768 16-Bit, 30 MSPS D/A Converter**, www.analog.com.
4. Data sheet for **AD1853 Stereo, 24-Bit, 192 kHz, Multibit Σ-Δ DAC**, www.analog.com.
5. Data sheet for **OP275 Dual Bipolar/JFET, Audio Operational Amplifier**, www.analog.com.
6. Chapters 5, Walt Kester, Editor, **Practical Design Techniques for Sensor Signal Conditioning**, Analog Devices, 1999, ISBN: 0-916550-20-6.

CHAPTER 4

Sensor Signal Conditioning

CHAPTER 4

Sensor Signal Conditioning

Walt Kester, James Bryant, Walt Jung
Scott Wurcer, Chuck Kitchin

SECTION 4-1

Introduction

Walt Kester

This chapter of the book deals with various sensors and associated signal-conditioning circuitry involving the use of op amps and in amps. While the topic is generally very broad, the focus is to concentrate on circuit and signal processing applications of sensors rather than the details of the actual sensors themselves.

Strictly speaking, a *sensor* is a device that receives a signal or stimulus and responds with an electrical signal, while a *transducer* is a converter of one type of energy into another. In practice, however, the terms are often used interchangeably.

Sensors and their associated circuits are used to measure various physical properties such as temperature, force, pressure, flow, position, light intensity, and so forth. These properties act as the stimulus to the sensor, and the sensor output is conditioned and processed to provide the corresponding measurement of the physical property. We will not cover all possible types of sensors, only the most popular ones, and specifically, those that lend themselves to process control and data acquisition systems.

Sensors do not operate by themselves. They are generally part of a larger system consisting of signal conditioners and various analog or digital signal processing circuits. The *system* could be a measurement system, data acquisition system, or process control system, for example.

Sensors may be classified in a number of ways. From a signal-conditioning viewpoint it is useful to classify sensors as either *active* or *passive*. An *active* sensor requires an external source of excitation. Resistor-based sensors such as thermistors, resistance temperature detectors (RTDs), and strain gages are examples of active sensors, because a current must be passed through them and the corresponding voltage measured in order to determine the resistance value. An alternative would be to place the devices in a bridge circuit; however in either case, an external current or voltage is required.

On the other hand, *passive* (or *self-generating*) sensors generate their own electrical output signal without requiring external voltages or currents. Examples of passive sensors are thermocouples and photodiodes which generate thermoelectric voltages and photocurrents, respectively, which are independent of external circuits.

It should be noted that these definitions (*active* versus *passive*) refer to the need (or lack thereof) of external active circuitry to produce a sensor electrical output signal. It would seem equally logical to consider a thermocouple *active*, in the sense that it produces an output voltage without external circuitry; however the convention in the industry is to classify the sensor with respect to the external circuit requirement as defined above.

A logical way to classify sensors is with respect to the physical property the sensor is designed to measure. Thus we have temperature sensors, force sensors, pressure sensors, motion sensors, and so forth. However, sensors that measure different properties may have the same type of electrical output. For instance, Resistance Temperature Detector (RTD) is a variable resistance, as is a resistive strain gage. Both RTDs and strain gages are often placed in bridge circuits, and the conditioning circuits are therefore quite similar. In fact, bridges and their conditioning circuits deserve a detailed discussion. Figure 4-1 is an overview of basic sensor characteristics.

- Sensors:
  Convert a Signal or Stimulus (Representing a Physical Property) into an Electrical Output
- Transducers:
  Convert One Type of Energy into Another
- The Terms are often Interchanged
- Active Sensors Require an External Source of Excitation:
  RTDs, Strain-Gages
- Passive (Self-Generating) Sensors do not:
  Thermocouples, Photodiodes

Figure 4-1: An overview of sensor characteristics

The full-scale outputs of most sensors (passive or active) are relatively small voltages, currents, or resistance changes, and therefore their outputs must be properly conditioned before further analog or digital processing can occur. Because of this, an entire class of circuits has evolved, generally referred to as signal-conditioning circuits. Amplification, level translation, galvanic isolation, impedance transformation, linearization, and filtering are fundamental signal-conditioning functions that may be required. Figure 4-2 summarizes sensors and their outputs.

PROPERTY	SENSOR	ACTIVE/ PASSIVE	OUTPUT
Temperature	Thermocouple Silicon RTD Thermistor	Passive Active Active Active	Voltage Voltage/Current Resistance Resistance
Force / Pressure	Strain Gage Piezoelectric	Active Passive	Resistance Voltage
Acceleration	Accelerometer	Active	Capacitance
Position	LVDT	Active	AC Voltage
Light Intensity	Photodiode	Active	Current

Figure 4-2: Typical sensors and their output formats

Whatever form the conditioning takes, the circuitry and performance will be governed by the electrical character of the sensor and its output. Accurate characterization of the sensor in terms of parameters appropriate to the application, e.g., sensitivity, voltage and current levels, linearity, impedances, gain, offset, drift, time constants, maximum electrical ratings, stray impedances, and other important considerations can spell the difference between substandard and successful application of the device, especially where high resolution, precision, or low level measurements are necessary.

Higher levels of integration now allow ICs to play a significant role in both analog and digital signal conditioning. ADCs specifically designed for measurement applications often contain on-chip programmable-gain amplifiers (PGAs) and other useful circuits, such as current sources for driving RTDs, thereby minimizing the external conditioning circuit requirements.

To some degree or another, most sensor outputs are nonlinear with respect to the applied stimulus and as a result, their outputs must often be linearized in order to yield correct measurements. In terms of the design approach choice towards linearization, the designer can take a route along either of two major paths.

Analog is one viable route, and such techniques may be used to perform an "analog domain" linearization function.

However, the recent introduction of high performance ADCs now allows linearization to be done much more efficiently and accurately in software. This "digital domain" approach to linearization eliminates the need for tedious manual calibration using multiple and sometimes interactive analog trim adjustments.

A quite common application of sensors is within process control systems. One example would be control of a physical property, such as temperature. A sample block diagram of how this might be implemented is illustrated in Figure 4-3.

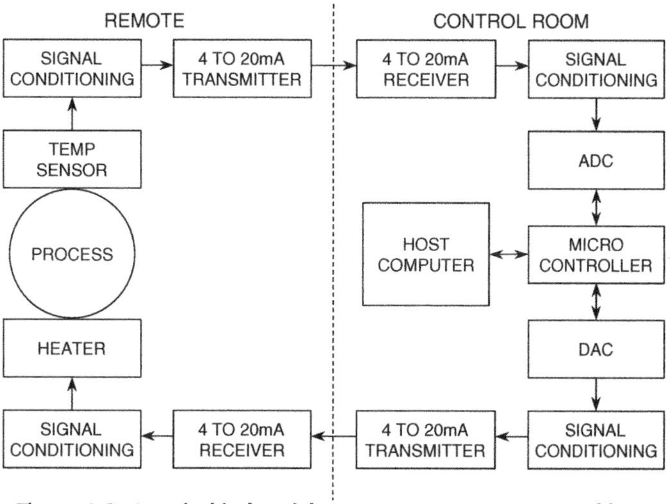

Figure 4-3: A typical industrial process temperature control loop

In this system, an output from a temperature sensor is conditioned, transmitted over some distance, received, and then digitized by an ADC. The microcontroller or host computer determines if the temperature is above or below the desired value, and outputs a digital word to the digital-to-analog converter (DAC).

The DAC output is conditioned and drives the remotely located *actuator*, in this case a heater. Notice that the interface between the control center and the remote process is via the industry-standard 4-20 mA loop.

Digital techniques are becoming more and more popular in processing sensor outputs in data acquisition, process control, and measurement. 8-bit microcontrollers (8051-based, for example) generally have sufficient speed and processing capability for most applications.

By including the A/D conversion and the microcontroller programmability on the sensor itself, a "smart sensor" can be implemented with self-contained calibration and linearization features among others. However, such digital techniques aren't a major focus of this text, so the section references may be consulted for further information.

The remaining sections of the chapter deal with analog signal conditioning methods for a variety of sensor types.

References: Introduction

1. Walt Kester, Bill Chestnut, and Grayson King, *Smart Sensors*, Chapter 9 of **Practical Design Techniques for Sensor Signal Conditioning**, Analog Devices, Inc., 1999, ISBN: 0-916550-20-6.
2. Compatibility of Analog Signals for Electronic Industrial Process Instruments, ANSI/ISA-S50.1-1982 (Rev. 1992), www.isa.org.
3. Editors, "Fieldbuses: Look Before You Leap," **EDN**, November 5, 1998, p. 197.
4. "MicroConverter Technology Backgrounder," Whitepaper, Analog Devices, Inc., www.analog.com.
5. Scott MacKenzie, **The 8051 Microcontroller, 3rd Ed.**, Prentice-Hall, 1999, ISBN: 0-13-780008-8.

SECTION 4-2

Bridge Circuits

Walt Kester

An Introduction to Bridges

This section of the chapter, 4-2, discusses more fundamental bridge circuit concepts. To gain greatest appreciation of these ideas, it should be studied along with those sections discussing precision op amps within Chapters 1 and 2. The next section (4-3) focuses on the detailed application circuits relating to strain-gage-based sensors. These sections can be read sequentially if the reader already understands the design issues related to precision op amp applications.

Resistive elements are some of the most common sensors. They are inexpensive, and relatively easy to interface with signal-conditioning circuits. Resistive elements can be made sensitive to temperature, strain (by pressure or by flex), and light. Using these basic elements, many complex physical phenomena can be measured, such as: fluid or mass flow (by sensing the temperature difference between two calibrated resistances), dew-point humidity (by measuring two different temperature points), and so forth.

Sensor element resistance can range from less than 100 Ω to several hundred kΩ, depending on the sensor design and the physical environment to be measured. Figure 4-4 indicates the wide range of sensor resistance encountered. For example, RTDs are typically 100 Ω or 1000 Ω. Thermistors are typically 3500 Ω or higher.

• Strain Gages	120Ω, 350Ω, 3500Ω
• Weigh-Scale Load Cells	350Ω – 3500Ω
• Pressure Sensors	350Ω – 3500Ω
• Relative Humidity	100kΩ – 10MΩ
• Resistance Temperature Devices (RTDs)	100Ω, 1000Ω
• Thermistors	100Ω – 10MΩ

Figure 4-4: Sensor resistances used in bridge circuits span a wide dynamic range

Resistive sensors such as RTDs and strain gages produce relatively small percentage changes in resistance, in response to a change in a physical variable such as temperature or force. For example, platinum RTDs have a temperature coefficient of about 0.385%/°C. Thus, in order to accurately resolve temperature to 1°C, the overall measurement accuracy must be much better than 0.385 Ω when using a 100 Ω RTD.

Strain gages present a significant measurement challenge because the typical change in resistance over the entire operating range of a strain gage may be less than 1% of the nominal resistance value. Accurately measuring small resistance changes is therefore critical when applying resistive sensors.

A simple method for measuring resistance is to force a constant current through the resistive sensor, and measure the voltage output. This requires both an accurate current source and an accurate means of measuring the voltage. Any change in the current will be interpreted as a resistance change. In addition, the power dissipation in the resistive sensor must be small and in accordance with the manufacturer's recommendations, so that self-heating does not produce errors. As a result, the drive current must be small, which tends to limit the resolution of this approach.

A *resistance bridge,* shown in Figure 4-5, offers an attractive alternative for accurately measuring small resistance changes. This is a basic Wheatstone bridge (actually developed by S. H. Christie in 1833), and is a prime example. It consists of four resistors connected to form a quadrilateral, a source of excitation voltage V_B (or, alternately, a current) connected across one of the diagonals, and a voltage detector connected across the other diagonal. The detector measures the difference between the outputs of the two voltage dividers connected across the excitation voltage, V_B. The general form of the bridge output V_O is noted in the figure.

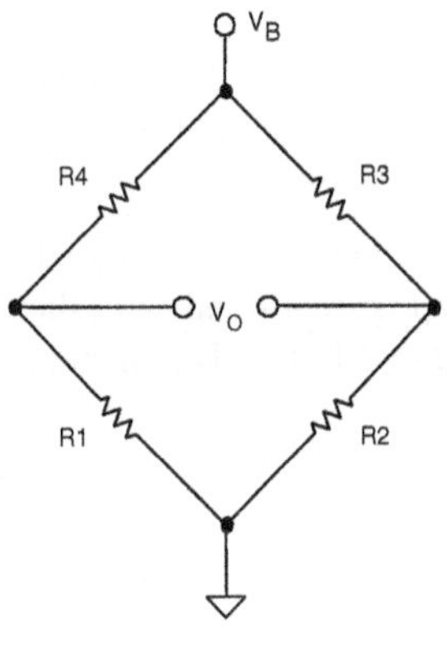

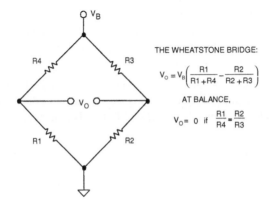

THE WHEATSTONE BRIDGE:

$$V_O = V_B\left(\frac{R1}{R1+R4} - \frac{R2}{R2+R3}\right)$$

AT BALANCE,

$$V_O = 0 \text{ if } \frac{R1}{R4} = \frac{R2}{R3}$$

Figure 4-5: The basic Wheatstone bridge produces an output null when the ratios of sidearm resistances match

There are two principal ways of operating a bridge such as this. One is by operating it as a null detector, where the bridge measures resistance indirectly by comparison with a similar standard resistance. On the other hand, it can be used as a device that reads a resistance difference directly, as a proportional voltage output.

When R1/R4 = R2/R3, the resistance bridge is said to be at a *null,* irrespective of the mode of excitation (current or voltage, ac or dc), the magnitude of excitation, the mode of readout (current or voltage), or the impedance of the detector. Therefore, if the ratio of R2/R3 is fixed at K, a null is achieved when R1 = K · R4. If R1 is unknown and R4 is an accurately determined variable resistance, the magnitude of R1 can be found by adjusting R4 until an output null is achieved. Conversely, in sensor-type measurements, R4 may be a fixed reference, and a null occurs when the magnitude of the external variable (strain, temperature, and so forth.) is such that R1 = K · R4.

Null measurements are principally used in feedback systems involving electromechanical and/or human elements. Such systems seek to force the active element (strain gage, RTD, thermistor, and so forth.) to balance the bridge by influencing the parameter being measured.

For the majority of sensor applications employing bridges, however, the deviation of one or more resistors in a bridge from an initial value is measured as an indication of the magnitude (or a change) in the measured variable. In these cases, the output voltage change is an indication of the resistance change. Because very small resistance changes are common, the output voltage change may be as small as tens of millivolts, even with the excitation voltage $V_B = 10$ V (typical for a load cell application).

In many bridge applications, there may not just be a single variable element, but two, or even four elements, all of which may vary. Figure 4-6 shows a family of four voltage-driven bridges, those most commonly suited for sensor applications. In the four cases the corresponding equations for V_O relate the bridge output voltage to the excitation voltage and the bridge resistance values. In all cases we assume a constant voltage drive, V_B. Note that since the bridge output is always directly proportional to V_B, the measurement accuracy can be no better than that of the accuracy of the excitation voltage.

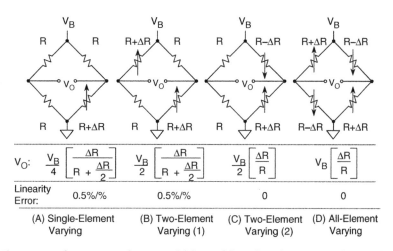

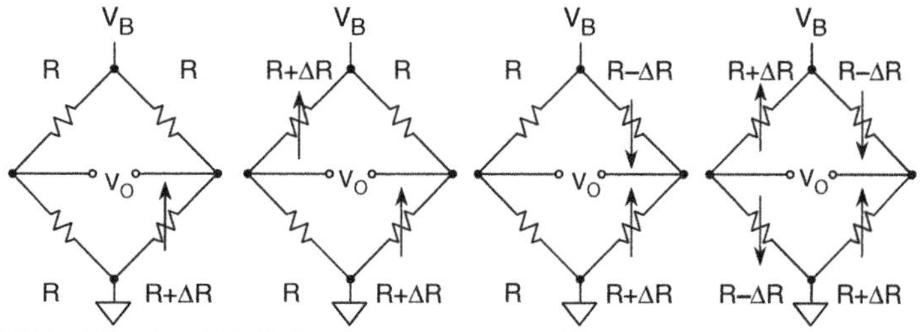

	(A) Single-Element Varying	(B) Two-Element Varying (1)	(C) Two-Element Varying (2)	(D) All-Element Varying
V_O:	$\frac{V_B}{4}\left[\frac{\Delta R}{R + \frac{\Delta R}{2}}\right]$	$\frac{V_B}{2}\left[\frac{\Delta R}{R + \frac{\Delta R}{2}}\right]$	$\frac{V_B}{2}\left[\frac{\Delta R}{R}\right]$	$V_B\left[\frac{\Delta R}{R}\right]$
Linearity Error:	0.5%/%	0.5%/%	0	0

Figure 4-6: The output voltage sensitivity and linearity of constant voltage drive bridge configurations differs according to the number of active elements

In each case, the value of the fixed bridge resistor "R" is chosen to be equal to the nominal value of the variable resistor(s). The deviation of the variable resistor(s) about the nominal value is assumed to be proportional to the quantity being measured, such as strain (in the case of a strain gage), or temperature (in the case of an RTD).

The *sensitivity* of a bridge is the ratio of the maximum expected change in the output voltage to the excitation voltage. For instance, if $V_B = 10$ V, and the full-scale bridge output is 10mV, then the sensitivity is 1mV/V. For the four cases of Figure 4-6, sensitivity can be said to increase going left-right, or as more elements are made variable.

The *single-element varying* bridge of Figure 4-6A is most suited for temperature sensing using RTDs or thermistors. This configuration is also used with a single resistive strain gage. All the resistances are nominally equal, but one of them (the sensor) is variable by an amount ΔR. As the equation indicates, the relationship between the bridge output and ΔR is not linear. For example, if R = 100 Ω and ΔR = 0.1 Ω (0.1% change in resistance), the output of the bridge is 2.49875 mV for $V_B = 10$ V. The error is 2.50000 mV – 2.49875 mV, or 0.00125 mV. Converting this to a % of full-scale by dividing by 2.5 mV yields an end-point linearity error in percent of approximately 0.05%. (Bridge end-point linearity error is calculated as the worst error in % FS from a straight line which connects the origin and the end point at FS, i.e., the FS gain error is not included). If $\Delta R = 1\ \Omega$, (1% change in resistance), the output of the bridge is 24.8756 mV, representing an end-point linearity error of approximately 0.5%. The end-point linearity error of the single-element bridge can be expressed in equation form:

Single-Element Varying

Bridge End-Point Linearity Error $\approx$ % Change in Resistance ÷ 2

It should be noted that the above nonlinearity refers to the nonlinearity of the bridge itself and not *the sensor.* In practice, most sensors themselves will exhibit a certain specified amount of nonlinearity, which must also be accounted for in the final measurement.

In some applications, the bridge nonlinearity noted above may be acceptable. If not, there are various methods available to linearize bridges. Since there is a fixed relationship between the bridge resistance change and its output (shown in the equations), software can be used to remove the linearity error in digital systems. Circuit techniques can also be used to linearize the bridge output directly, and these will be discussed shortly.

There are two cases to consider in the instance of a *two-element varying* bridge. In Case 1 (Figure 4-6B), both of the diagonally opposite elements change in the same direction. An example would be two identical strain gages mounted adjacent to each other, with their axes in parallel.

The nonlinearity for this case, 0.5%/%, the same as that of the single-element varying bridge of Figure 4-6A. However, it is interesting to note the sensitivity is now improved by a factor of 2, vis-à-vis the single-element varying setup. The two-element varying bridge is commonly found in pressure sensors and flow meter systems.

A second case of the two-element varying bridge, Case 2, is shown in Figure 4-6C. This bridge requires two identical elements that vary in *opposite* directions. This could correspond to two identical strain gages: one mounted on top of a flexing surface, and one on the bottom. Note that this configuration is now linear, and like two-element varying Case 1, it has twice the sensitivity of the Figure 4-6A configuration. Another way to view this configuration is to consider the terms $R + \Delta R$ and $R - \Delta R$ as comprising two sections of a linear potentiometer.

The *all-element varying* bridge of Figure 4-6D produces the most signal for a given resistance change, and is inherently linear. It is also an industry-standard configuration for load cells constructed from four identical strain gages. Understandably, it is also one of the most popular bridge configurations.

Bridges may also be driven from constant current sources, as shown in Figure 4-7, for the corresponding cases of single, dual, dual, and four active element(s). As with the voltage-driven bridges, the analogous output expressions are noted, along with the sensitivities.

Current drive, although not as popular as voltage drive, does have advantages when the bridge is located remotely from the source of excitation. One advantage is that the wiring resistance doesn't introduce errors in the measurement; another is simpler, less expensive cabling. Note also that with constant current excitation, all bridge configurations are linear except the single-element varying case of Figure 4-7A.

In summary, there are many design issues relating to bridge circuits, as denoted by Figure 4-8. After selecting the basic configuration, the excitation method must be determined. The value of the excitation voltage or current must first be determined, as this directly influences sensitivity. Recall that the full-scale bridge output is directly proportional to the excitation voltage (or current). Typical bridge sensitivities are 1mV/V to 10mV/V.

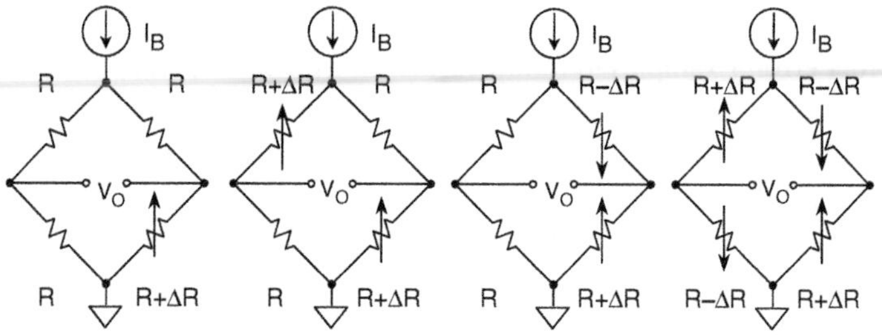

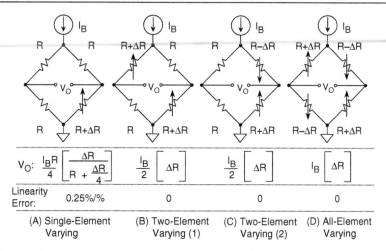

	(A) Single-Element Varying	(B) Two-Element Varying (1)	(C) Two-Element Varying (2)	(D) All-Element Varying
V_O:	$\frac{I_B R}{4}\left[\frac{\Delta R}{R+\frac{\Delta R}{4}}\right]$	$\frac{I_B}{2}\left[\Delta R\right]$	$\frac{I_B}{2}\left[\Delta R\right]$	$I_B\left[\Delta R\right]$
Linearity Error:	0.25%/%	0	0	0

Figure 4-7: The output voltage sensitivity and linearity of constant current drive bridge configurations also differs according to the number of active elements

- Selecting Configuration (1-, 2-, 4-Element Varying)
- Selection of Voltage or Current Excitation
- Ratiometric Operation
- Stability of Excitation Voltage or Current
- Bridge Sensitivity: FS Output / Excitation Voltage
  1mV/V to 10mV/V Typical
- Full-scale Bridge Outputs: 10mV – 100mV Typical
- Precision, Low Noise Amplification/Conditioning
  Techniques Required
- Linearization Techniques May Be Required
- Remote Sensors Present Challenges

Figure 4-8: A number of bridge considerations impact design choices

Although large excitation voltages yield proportionally larger full-scale output voltages, they also result in higher bridge power dissipation, and thus raise the possibility of sensor resistor self-heating errors. On the other hand, low values of excitation voltage require more gain in the conditioning circuits, and also increase sensitivity to low level errors such as noise and offset voltages.

Regardless of the absolute level, the stability of the excitation voltage or current directly affects the overall accuracy of the bridge output, as is evident from the V_B and I_B terms in the output expressions. Therefore stable references and/or *ratiometric* drive techniques are required, to maintain highest accuracy.

Here, ratiometric simply refers to the use of the bridge drive voltage of a voltage-driven bridge (or a current-proportional voltage, for a current-driven bridge) as the reference input to the ADC that digitizes the amplified bridge output voltage. In this manner the absolute accuracy and stability of the excitation voltage becomes a second-order error. Examples to follow further illustrate this point.

Amplifying and Linearizing Bridge Outputs

The output of a single-element varying bridge may be amplified by a single precision op amp connected as shown in Figure 4-9. Unfortunately this circuit, although attractive because of relative simplicity, has poor overall performance. Its gain predictability and accuracy are poor, and it unbalances the bridge due to loading from R_F and the op amp bias current. The R_F resistors must be carefully chosen and matched to maximize common mode rejection (CMR). Also, it is difficult to maximize the CMR while at the same time allowing different gain options. Gain is dependent upon the bridge resistances and R_F. In addition, the output is nonlinear, as the configuration does nothing to address the intrinsic bridge nonlinearity. In summary, the circuit isn't recommended for precision use.

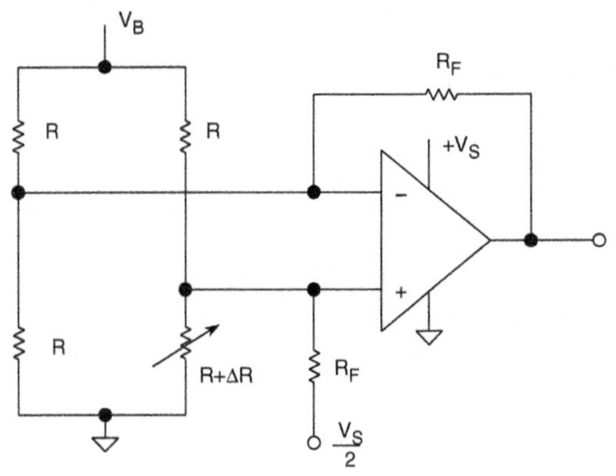

Figure 4-9: Using a single op amp as a bridge amplifier

However, a redeeming feature of this circuit is that it is capable of single supply operation, with a solitary op amp. Note that the R_F resistor connected to the noninverting input is returned to $V_S/2$ (rather than ground) so that both positive and negative ΔR values can be accommodated, with the bipolar op amp output swing referenced to $V_S/2$.

A much better approach is to use an *instrumentation amplifier* (in amp) for the required gain, as shown in Figure 4-10. This efficient circuit provides better gain accuracy, with the in amp gain usually set with a single resistor, R_G. Since the amplifier provides dual, high-impedance loading to the bridge nodes, it does not unbalance or load the bridge. Using modern in amp devices with gains ranging from 10–1000, excellent common mode rejection and gain accuracy can be achieved with this circuit.

However, due to the intrinsic characteristics of the bridge, the output is still nonlinear (see expression). As noted earlier, this can be corrected in software (assuming that the in amp output is digitized using an analog-to-digital converter and followed by a microcontroller or microprocessor).

The in amp can be operated on either dual supplies as shown, or alternately, on a single positive supply. In the figure, this corresponds to $-V_S = 0$. This is a key advantage, due to the fact that all such bridge circuits bias the in amp inputs at $V_B/2$, a voltage range typically compatible with amplifier bias requirements. In amps such as the AD620 family, the AD623, and AD627 can be used in single (or dual) supply bridge applications, provided their restrictions on the gain and input and output voltage swings are observed.

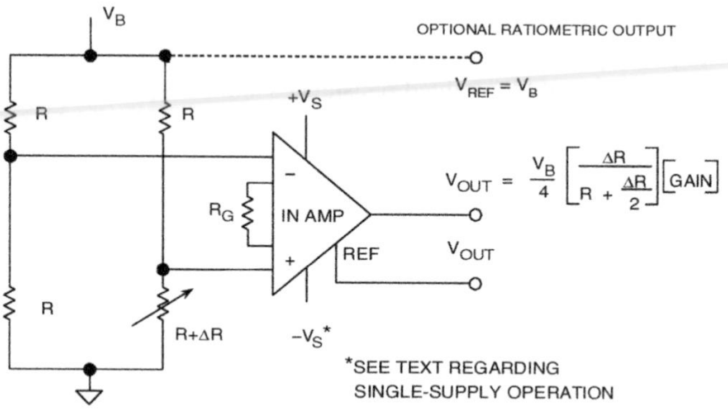

Figure 4-10: A generally preferred method of bridge amplification employs an instrumentation amplifier for stable gain and high CMR

The bridge in this example is voltage driven, by the voltage V_B. This voltage can optionally be used for an ADC reference voltage, in which case it also is an additional output, V_{REF}.

Various techniques are available to linearize bridge outputs, but it is important to distinguish between the linearity of the bridge equation (discussed earlier), and the sensor response linearity to the phenomenon being sensed. For example, if the active sensor element is an RTD, the bridge used to implement the measurement might have perfectly adequate linearity; *yet the output could still be nonlinear,* due to the RTD device's intrinsic nonlinearity. Manufacturers of sensors employing bridges address the nonlinearity issue in a variety of ways, including keeping the resistive swings in the bridge small, shaping complementary nonlinear response into the active elements of the bridge, using resistive trims for first-order corrections, and others. In the examples which follow, what is being addressed is the linearity error of the bridge configuration itself (as opposed to a sensor element within the bridge).

Figure 4-11 shows a single-element varying active bridge circuit, in which an op amp produces a forced bridge null condition. For this single-element varying case, only the op amp feedback resistance varies, with the remaining three resistances fixed.

As used here, the op amp output provides a buffered, ground referenced, low impedance output for the bridge measurement, effectively suppressing the $V_B/2$ CM bridge component at the op amp inputs.

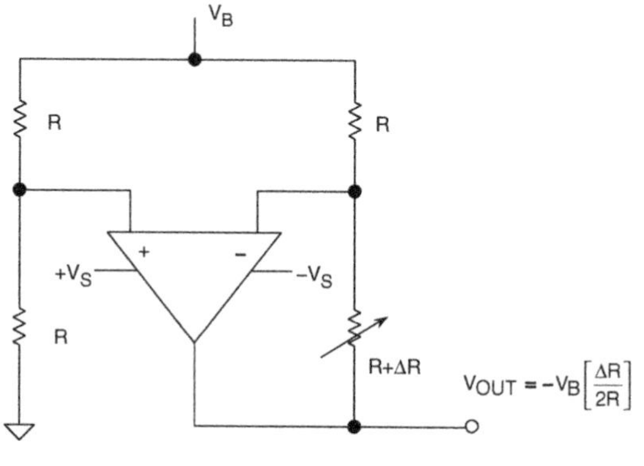

Figure 4-11: Linearizing a single-element varying bridge (Method 1)

The circuit works by adding a voltage in series with the variable resistance arm. This voltage is equal in magnitude and opposite in polarity to the incremental voltage across the varying element, and is linear with ΔR. As can be noted, the three constant "R" valued resistances and the op amp operate to drive a constant current in the variable resistance. This is the basic mechanism that produces the linearized output.

This active bridge has a sensitivity gain of two over the standard single-element varying bridge (Figure 4-6A). The key point is that the bridge's incremental resistance/voltage output becomes linear, even for large values of ΔR. However, because of a still relatively small output signal, a second amplifier must usually follow this bridge. Note also that the op amp used in this circuit requires dual supplies, because its output must go negative for conditions where ΔR is positive.

Another circuit for linearizing a single-element varying bridge is shown in Figure 4-12. The top node of the bridge is excited by the voltage, V_B. The bottom of the bridge is driven in complementary fashion by the left op amp, which maintains a constant current of V_B/R in the varying resistance element, R + ΔR.

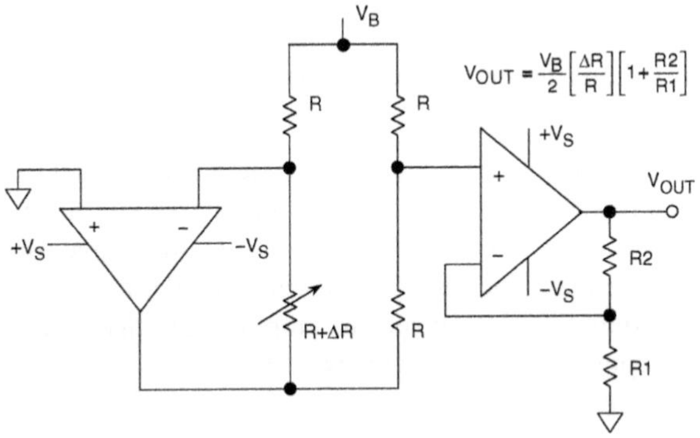

Figure 4-12: Linearizing a single-element varying bridge (Method 2)

Like the circuit of Figure 4-11, the constant current drive for the single-element variable resistance provides the mechanism for linearity improvement. Also, because of the fact that the bridge left-side center node is ground-referenced by the op amp, this configuration effectively suppresses CM voltages. This has the virtue of making the op amp selection somewhat less critical. Of course, performance parameters of high gain, low offset/noise, and high stability are all still needed.

The output signal is taken from the right-hand leg of the bridge, and is amplified by a second op amp, connected as a noninverting gain stage. With the scaling freedom provided by the second op amp, the configuration is very flexible. The net output is linear, and has a bridge-output referred sensitivity comparable to the single-element varying circuit of Figure 4-11.

The Figure 4-12 circuit requires two op amps operating on dual supplies. In addition, paired resistors R1-R2 must be ratio matched and stable types, for overall accurate and stable gain. The circuit can be a practical one using a dual precision op amp, such as an AD708, the OP2177 or the OP213.

A closely related circuit for linearizing a voltage-driven, *two-element* varying bridge can be adapted directly from the basic circuit of Figure 4-11. This form of the circuit, shown in Figure 4-13, is identical to the previous single-element varying case, with the exception that the resistance between V_B and the op amp (+) input is now also variable (i.e., both diagonal R + ΔR resistances vary, in a like manner).

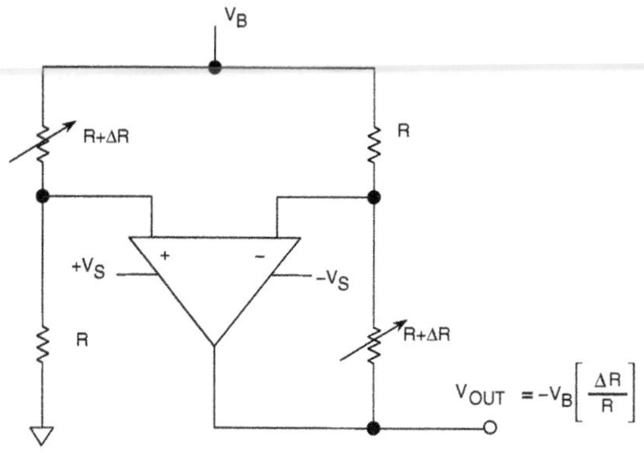

Figure 4-13: Linearizing a two-element varying voltage-driven bridge (Method 1)

For the same applied voltage V_B, this form of the circuit has twice the sensitivity, which is evident in the output expressions. A dual supply op amp is again required, and additional gain may also be necessary.

The two-element varying bridge circuit shown in Figure 4-14 uses an op amp, a sense resistor, and a voltage reference, set up in a feedback loop containing the sensing bridge. The net effect of the loop is to maintain a constant current through the bridge of $I_B = V_{REF}/R_{SENSE}$. The current through each leg of the bridge remains constant ($I_B/2$) as the resistances change, therefore the output is a linear function of ΔR. An in amp provides the additional gain.

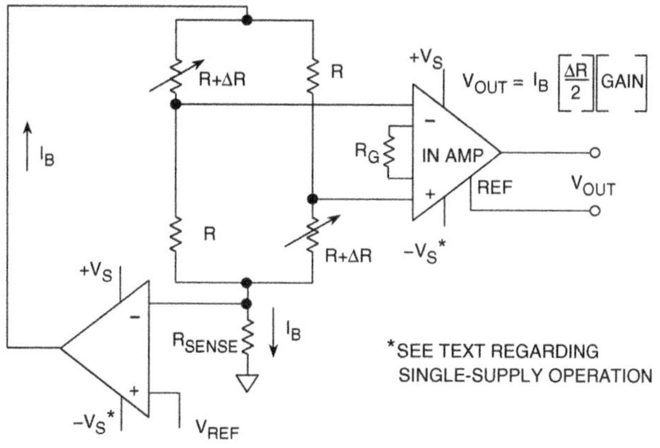

Figure 4-14: Linearizing a two-element varying current-driven bridge (Method 2)

This circuit can be operated on a single supply with the proper choice of amplifiers and signal levels. If ratiometric operation of an ADC is desired, the V_{REF} voltage can be used to drive the ADC.

Driving Remote Bridges

Wiring resistance and noise pickup are the biggest problems associated with remotely located bridges. Figure 4-15 shows a 350 Ω strain gage, which is connected to the rest of the bridge circuit by 100 feet of 30-gage twisted-pair copper wire. The resistance of the wire at 25°C is 0.105 Ω/ft, or 10.5 Ω for 100 ft. The total lead resistance in series with the 350 Ω strain gage is therefore 21 Ω. The temperature coefficient of the copper wire is 0.385%/°C. Now we will calculate the gain and offset error in the bridge output due to a 10°C temperature rise in the cable. These calculations are easy to make, because the bridge output voltage is simply the difference between the output of two voltage dividers, each driven from a 10 V source.

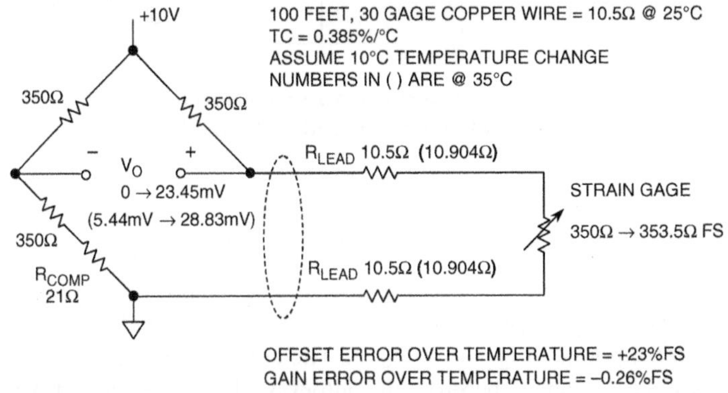

Figure 4-15: Wiring resistance related errors with remote bridge sensor

The full-scale variation of the strain gage resistance (with flex) above its nominal 350 Ω value is +1% (+3.5 Ω), corresponding to a full-scale strain gage resistance of 353.5 Ω which causes a bridge output voltage of +23.45 mV. Notice that the addition of the 21 Ω R_{COMP} resistor compensates for the wiring resistance and balances the bridge when the strain gage resistance is 350 Ω. Without R_{COMP}, the bridge would have an output offset voltage of 145.63 mV for a nominal strain gage resistance of 350 Ω. This offset could be compensated for in software just as easily but, for this example, we chose to do it with R_{COMP}.

Assume that the cable temperature increases 10°C above nominal room temperature. This results in a total lead resistance increase of +0.404 Ω (10.5 Ω × 0.00385/°C × 10°C) in each lead. *Note: The values in parentheses in the diagram indicate the values at 35°C.* The total additional lead resistance (of the two leads) is +0.808 Ω. With no strain, this additional lead resistance produces an offset of +5.44 mV in the bridge output. Full-scale strain produces a bridge output of +28.83 mV (a change of +23.39 mV from no strain). Thus the increase in temperature produces an offset voltage error of +5.44 mV (+23% full-scale) and a gain error of –0.06 mV (23.39 mV – 23.45 mV), or –0.26% full-scale. Note that these errors are produced solely by the 30-gage wire, and do not include any temperature coefficient errors in the strain gage itself.

The effects of wiring resistance on the bridge output can be minimized by the 3-wire connection shown in Figure 4-16. We assume that the bridge output voltage is measured by a high impedance device, therefore there is no current in the sense lead. Note that the sense lead measures the voltage output of a divider: the top half is the bridge resistor plus the lead resistance, and the bottom half is strain gage resistance plus the lead resistance. The nominal sense voltage is therefore independent of the lead resistance. When the strain gage resistance increases to full-scale (353.5 Ω), the bridge output increases to +24.15 mV.

Increasing the temperature to 35°C increases the lead resistance by +0.404 Ω in each half of the divider. The full-scale bridge output voltage decreases to +24.13 mV because of the small loss in sensitivity, but

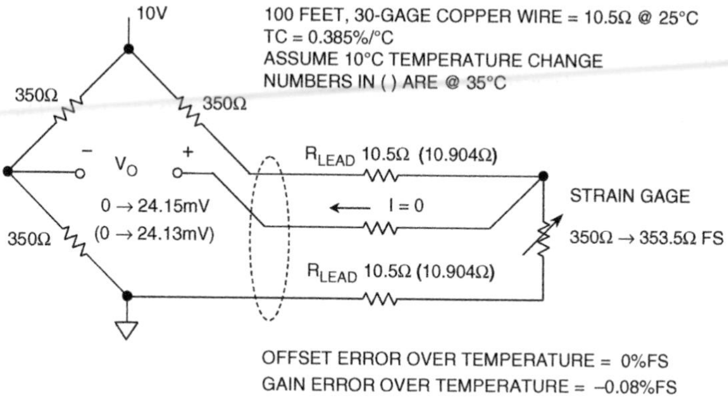

Figure 4-16: Remote bridge wiring resistance errors are reduced with 3-wire sensor connection

there is no offset error. The gain error due to the temperature increase of 10ºC is therefore only –0.02 mV, or –0.08% of full-scale. Compare this to the +23% full-scale offset error and the –0.26% gain error for the 2-wire connection shown in Figure 4-14.

The 3-wire method works well for remotely located resistive elements which make up one leg of a single-element varying bridge. However, all-element varying bridges are generally housed in a complete assembly, as in the case of a load cell. When these bridges are remotely located from the conditioning electronics, special techniques must be used to maintain accuracy.

Of particular concern is maintaining the accuracy and stability of the bridge excitation voltage. The bridge output is directly proportional to the excitation voltage, and any drift in the excitation voltage produces a corresponding drift in the output voltage.

For this reason, most all-element varying bridges (such as load cells) are six-lead assemblies: two leads for the bridge output, two leads for the bridge excitation, and two *sense* leads. To take full advantage of the additional accuracy that these extra leads allow, a method called Kelvin or 4-wire sensing is employed, as shown in Figure 4-17.

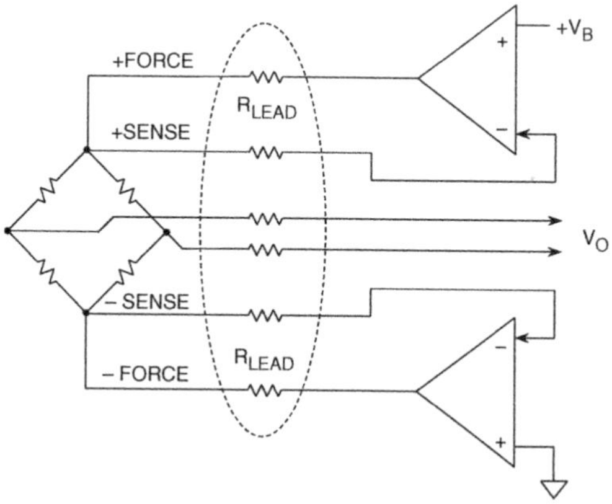

Figure 4-17: A Kelvin sensing system with a 6-wire voltage-driven bridge connection and precision op amps minimizes errors due to wire lead resistances

In this setup the drive voltage V_B is not applied directly to the bridge, but goes instead to the input of the upper precision op amp, which is connected in a feedback loop around the bridge (+) terminal. Although there may be a substantial voltage drop in the +FORCE lead resistance of the remote cable, the op amp will automatically correct for it, since it has a feedback path through the +SENSE lead. The net effect is that the upper node of the remote bridge is maintained at a precise level of V_B (within the capability of the op amp used, of course). A similar situation occurs with the bottom precision op amp, which drives the bridge (–) terminal to a ground level, as established by the op amp input ground reference. Again, the voltage drop in the –FORCE lead is relatively immaterial, because of the sensing at the –SENSE terminal.

In both cases, the sense lines go to high impedance op amp inputs, thus there is minimal error due to the bias current induced voltage drop across their lead resistance. The op amps maintain the required excitation voltage at the remote bridge, to make the voltage measured between the (+) and (–) sense leads always equal to V_B.

Note—a subtle point is that the lower op amp will need to operate on dual supplies, since the drive to the –FORCE lead will cause the op amp output to go negative. Because of relatively high current in the bridge (~30 mA), current buffering stages at the op amp outputs are likely advisable for this circuit.

Although Kelvin sensing eliminates errors due to voltage drops in the bridge wiring resistance, the basic drive voltage V_B must still be highly stable since it directly affects the bridge output voltage. In addition, the op amps must have low offset, low drift, and low noise. Ratiometric operation can be optionally added, simply by using V_B to drive the ADC reference input.

The constant current excitation method shown in Figure 4-18 is another method for minimizing the effects of wiring resistance on the measurement accuracy. This system drives a precise current I through the bridge, proportioned as per the expression in the figure. An advantage of the circuit in Figure 4-18 is that it only uses one amplifier.

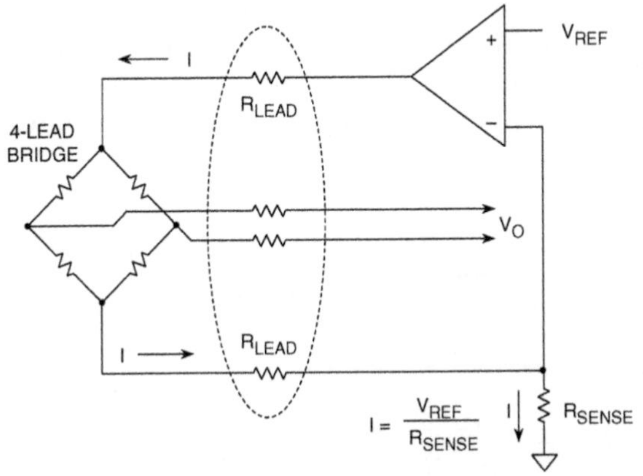

Figure 4-18: A 4-wire current-driven bridge scheme also minimizes errors due to wire lead resistances, plus allows simpler cabling

The accuracy of the reference, the sense resistor, and the op amp all influence the overall accuracy. While the precision required of the op amp should be obvious, one thing not necessarily obvious is that it may be required to deliver appreciable current, when I is more than a few mA (which it will be with standard 350 Ω bridges). In such cases, current buffering of the op amp is again in order.

Therefore for highest precision with this circuit, a buffer stage is recommended. This can be as simple as a small transistor, since the bridge drive is unidirectional.

System Offset Minimization

Maintaining an accuracy of 0.1% or better with a full-scale bridge output voltage of 20 mV requires that the sum of all offset errors be less than 20 μV. Parasitic thermocouples are cases in point and, if not given due attention, can cause serious temperature drift errors. All dissimilar metal-metal connections generate voltages between a few and tens of microvolts for a 1°C temperature differential, are basic thermocouple facts of life.

Fortunately, however, within a bridge measurement system the signal connections are differential; therefore this factor can be used to minimize the impact of parasitic thermocouples.

Figure 4-19 shows some typical sources of offset error that are inevitable in a system. Within a differential signal path, only those thermocouple pairs whose junctions are actually at different temperatures will degrade the signal. The diagram shows a typical parasitic junction formed between the copper printed circuit board traces and the kovar pins of an IC amplifier.

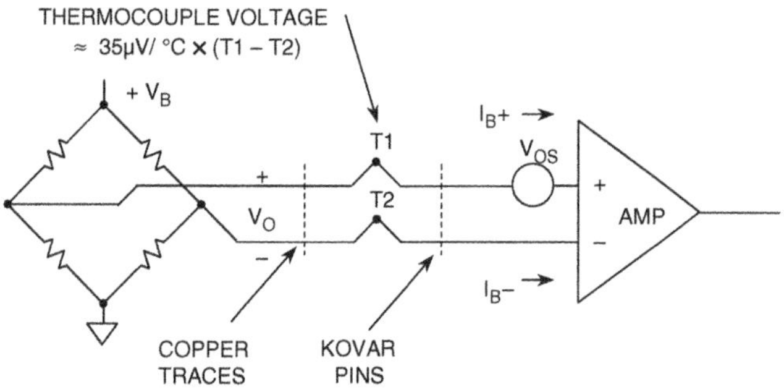

Figure 4-19: Typical sources of offset voltage within bridge measurement systems

This thermocouple voltage is about 35 μV/°C temperature differential. Note that this package-PC trace thermocouple voltage is significantly less when using a plastic package with a copper lead frame (recommended). Regardless of what package is used, all metal-metal connections along the signal path should be designed so that minimal temperature differences occur between the sides.

The amplifier offset voltage and bias currents are further sources of offset error. The amplifier bias current must flow through the source impedance. Any unbalance in either the source resistances or the bias currents produces offset errors. In addition, the offset voltage and bias currents are a function of temperature.

High performance low offset, low offset drift, low bias current, and low noise precision amplifiers such as the AD707, the OP177 or OP1177 are required. In some cases, chopper-stabilized amplifiers such as the AD8551/AD8552/AD8554 may be a solution. Ac bridge excitation such as that shown in Figure 4-20 can effectively remove offset voltage effects in series with a bridge output, V_O.

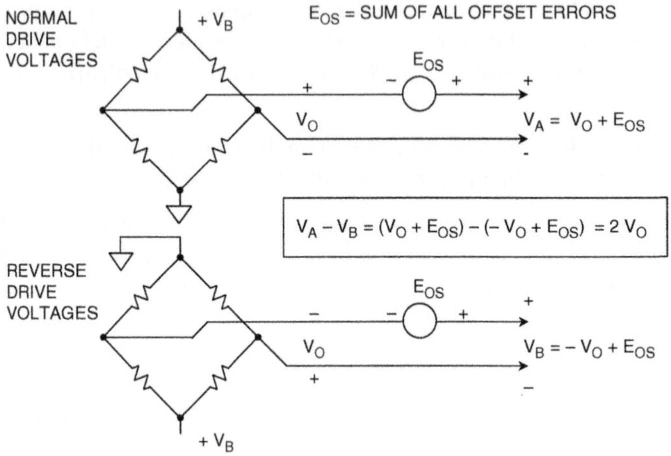

Figure 4-20: Ac bridge excitation minimizes system offset voltages

The concept is simple, and can be described as follows. The net bridge output voltage is measured under the two phased-sequence conditions, as shown. A first measurement (top) drives the bridge at the top node with excitation voltage V_B. This yields a first-phase measurement output V_A, where V_A is the sum of the desired bridge output voltage V_O and the net offset error voltage E_{OS}.

In the second measurement (bottom) the polarity of the bridge excitation is then reversed, and a second measurement, V_B, is made. Subtracting V_B from V_A yields 2 V_O, and the offset error term E_{OS} cancels as noted from the mathematical expression in the figure.

Obviously, a full implementation of this technique requires a highly accurate measurement ADC such as the AD7730 (see Reference 5) as well as a microcontroller to perform the subtraction.

Note that if a ratiometric reference is desired, the ADC must also accommodate the changing polarity of the reference voltage, as well as sense the magnitude. Again, the AD7730 includes this capability.

A very powerful combination of bridge circuit techniques is shown in Figure 4-21, an example of a high performance ADC. In Figure 4-21A is shown a basic dc operated ratiometric technique, combined with Kelvin sensing to minimize errors due to wiring resistance, which eliminates the need for an accurate excitation voltage.

The AD7730 measurement ADC can be driven from a single supply voltage of 5 V, which in this case is also used to excite the remote bridge. Both the analog input and the reference input to the ADC are high impedance and fully differential. By using the + and – SENSE outputs from the bridge as the differential reference voltage to the ADC, there is no loss in measurement accuracy if the actual bridge excitation voltage varies.

To implement ac bridge operation of the AD7730, an "H" bridge driver of P-Channel and N-Channel MOSFETs can be configured as shown in Figure 4-21B (note—dedicated bridge driver chips are available,

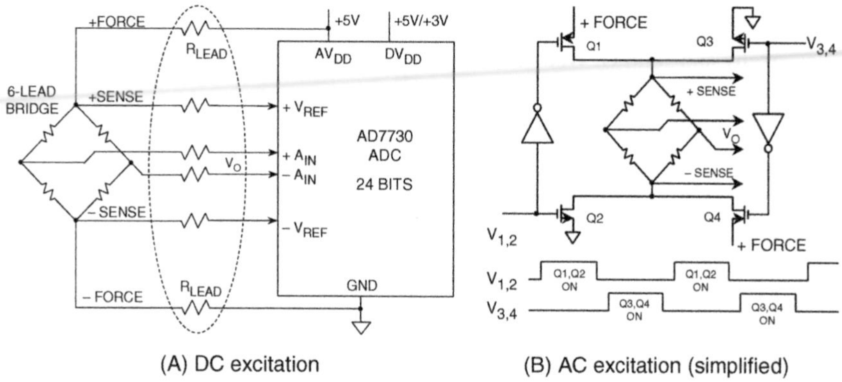

Figure 4-21: Ratiometric dc or ac operation with Kelvin sensing can be implemented using the AD7730 ADC

such as the Micrel MIC4427). This scheme, added to the basic functionality of the AD7730 configuration of 4-21A greatly increases the utility of the offset canceling circuit, as generally outlined in the preceding discussion of Figure 4-20.

Because of the on-resistance of the H-bridge MOSFETs, Kelvin sensing must also be used in these ac bridge applications. It is also important that the drive signals be nonoverlapping, as noted, to prevent excessive MOSFET switching currents. The AD7730 ADC has on-chip circuitry which generates the required nonoverlapping drive signals to implement this ac bridge excitation. All that needs adding is the switching bridge as noted in Figure 4-21B.

The AD7730 is one of a family of sigma-delta ADCs with high resolution (24 bits) and internal programmable gain amplifiers (PGAs) and is ideally suited for bridge applications. These ADCs have self- and system calibration features, which allow offset and gain errors due to the ADC to be minimized. For instance, the AD7730 has an offset drift of 5 nV/°C and a gain drift of 2 ppm/°C. Offset and gain errors can be reduced to a few microvolts using the system calibration feature.

References: Bridge Circuits

1. Ramon Pallas-Areny and John G. Webster, **Sensors and Signal Conditioning**, John Wiley, New York, 1991.
2. Dan Sheingold, Editor, **Transducer Interfacing Handbook**, Analog Devices, Inc., 1980, ISBN: 0-916550-05-2.
3. Sections 2, 3, Walt Kester, Editor, **1992 Amplifier Applications Guide**, Analog Devices, 1992, ISBN: 0-916550-10-9.
4. Sections 1, 6, Walt Kester, Editor, **System Applications Guide**, Analog Devices, 1993, ISBN: 0-916550-13-3.
5. Data sheet for **AD7730 Bridge Transducer ADC**, www.analog.com.

Strain, Force, Pressure and Flow Measurements

Walt Kester

Strain Gages

The most popular electrical elements used in force measurements include the resistance strain gage, the semiconductor strain gage, and piezoelectric transducers. The strain gage measures force indirectly by measuring the deflection it produces in a calibrated carrier. Pressure can be converted into a force using an appropriate transducer, and strain gage techniques can then be used to measure pressure. Flow rates can be measured using differential pressure measurements, which also make use of strain gage technology. These principles are summarized in Figure 4-22.

- Strain: Strain Gage, Piezoelectric Transducers
- Force: Load Cell
- Pressure: Diaphragm to Force to Strain Gage
- Flow: Differential Pressure Techniques

Figure 4.22 Strain gages are directly or indirectly the basis for a variety of physical measurements

The resistance-based strain gage uses a resistive element that changes in length, hence resistance, as the force applied to the base on which it is mounted causes stretching or compression. It is perhaps the most well-known transducer for converting force into an electrical variable.

An *unbonded* strain gage consists of a wire stretched between two points. Force acting upon the wire (area = A, length = L, resistivity = ρ) will cause the wire to elongate or shorten, which will cause the resistance to increase or decrease proportionally according to:

$$R = \rho L/A \qquad \text{Eq. 4-1}$$

and,

$$\Delta R/R = GF \cdot \Delta L/L \qquad \text{Eq. 4-2}$$

where GF = Gage factor (2.0 to 4.5 for metals, and more than 150 for semiconductors).

In this expression, the dimensionless quantity $\Delta L/L$ is a measure of the force applied to the wire and is expressed in *microstrains* ($1\ \mu\varepsilon = 10^{-6}$ cm/cm) which is the same as parts-per-million (ppm).

From Eq. 4-2, note that larger gage factors result in proportionally larger resistance changes, hence this implies greater strain gage sensitivity. These concepts are summarized in the drawing of Figure 4-23.

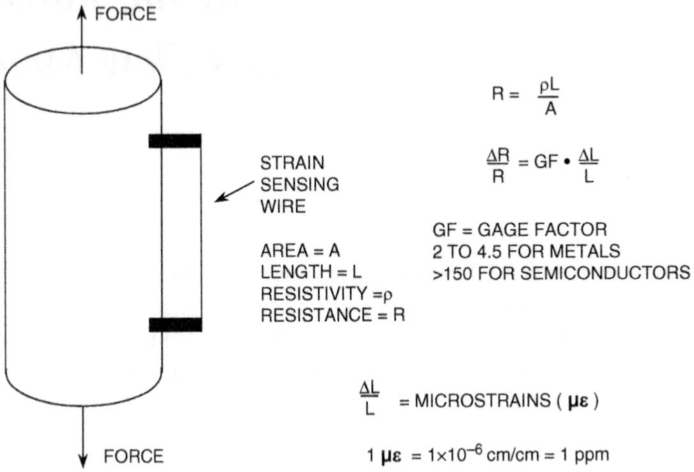

Figure 4-23: Operating principles of a basic unbonded strain gage

A *bonded* strain gage consists of a thin wire or conducting film arranged in a coplanar pattern and cemented to a base or carrier. The basic form of this type of gage is shown in Figure 4-24.

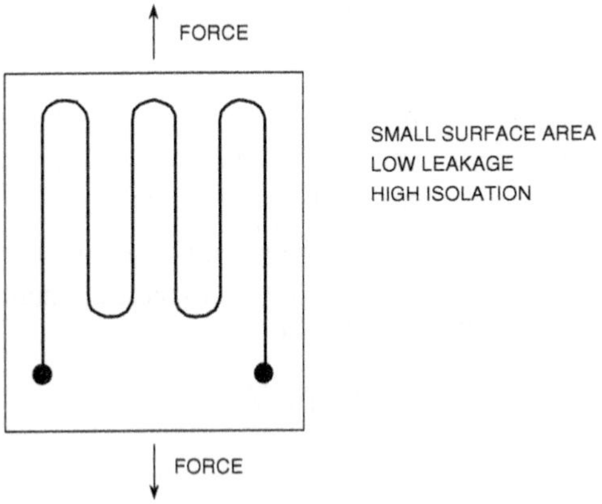

Figure 4-24: A bonded wire strain gage

This strain gage is normally mounted so that as much as possible of the length of the conductor is aligned in the direction of the stress that is being measured, i.e., longitudinally. Lead wires are attached to the base and brought out for interconnection. Bonded devices are considerably more practical and are in much wider use than are the aforementioned unbonded devices.

Perhaps the most popular version is the *foil-type gage*, produced by photo-etching techniques, and using similar metals to the wire types. Typical alloys are of copper-nickel (Constantan), nickel-chromium (Nichrome), nickel-iron, platinum-tungsten, and so forth. This strain gage type is shown in Figure 4-25.

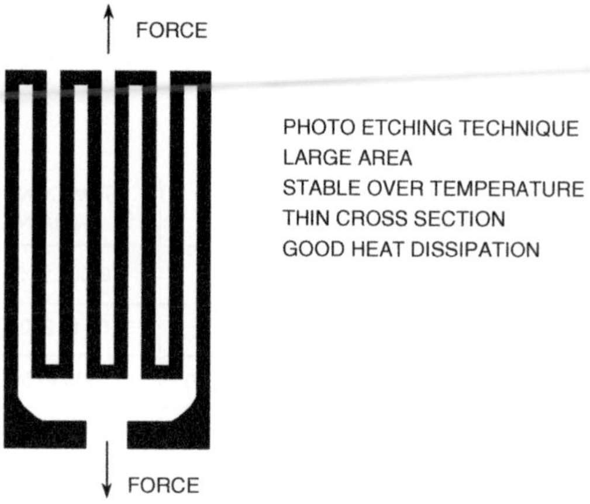

Figure 4-25: A metal foil strain gage

Gages having *wire sensing elements* present a small surface area to the specimen; this reduces leakage currents at high temperatures and permits higher isolation potentials between the sensing element and the specimen. Foil sensing elements, on the other hand, have a large ratio of surface area to cross-sectional area and are more stable under extremes of temperature and prolonged loading. The large surface area and thin cross section also permit the device to follow the specimen temperature and facilitate the dissipation of self-induced heat.

Semiconductor Strain Gages

Semiconductor strain gages make use of the piezoresistive effect in certain semiconductor materials such as silicon and germanium in order to obtain greater sensitivity and higher level output.

Semiconductor gages can be produced to have either positive or negative changes when strained. They can be made physically small while still maintaining a high nominal resistance.

Semiconductor strain gage bridges may have 30 times the sensitivity of bridges employing metal films, but are temperature-sensitive and difficult to compensate. Their change in resistance with strain is also nonlinear. They are not in as widespread use as the more stable metal-film devices for precision work; however, where sensitivity is important and temperature variations are small, they may have some advantage.

Instrumentation is similar to that for metal-film bridges but is less critical because of the higher signal levels and decreased transducer accuracy. Figure 4-26 summarizes the relative performance of metal and semiconductor strain gages.

PARAMETER	METAL STRAIN GAGE	SEMICONDUCTOR STRAIN GAGE
Measurement Range	0.1 to 40,000 με	0.001 to 3000 με
Gage Factor	2.0 to 4.5	50 to 200
Resistance, Ω	120, 350, 600, …, 5000	1000 to 5000
Resistance Tolerance	0.1% to 0.2%	1% to 2%
Size, mm	0.4 to 150 Standard: 3 to 6	1 to 5

Figure 4-26: A comparison of metal and semiconductor type strain gages

Piezoelectric force transducers are employed where the forces to be measured are dynamic (i.e., continually changing over the period of interest—usually of the order of milliseconds). These devices utilize the effect that changes in charge are produced in certain materials when they are subjected to physical stress. In fact, piezoelectric transducers are *displacement* transducers with quite large charge outputs for very small displacements, but they are invariably used as force transducers on the assumption that in an elastic material, displacement is proportional to force. Piezoelectric devices produce substantial output voltage in instruments such as accelerometers for vibration studies. Piezoelectric sensor output conditioning is discussed within Section 4-4 of this chapter.

Strain gages can be used to measure force, as shown in Figure 4-27, where a cantilever beam is slightly deflected by the applied force. Four strain gages are used to measure the flex of the beam, two on the top, and two on the bottom. The gages are connected in a four-element bridge configuration. Recall from Section 4-2 that this configuration gives maximum sensitivity and is inherently linear. This configuration also offers first-order correction for temperature drift in the individual strain gages.

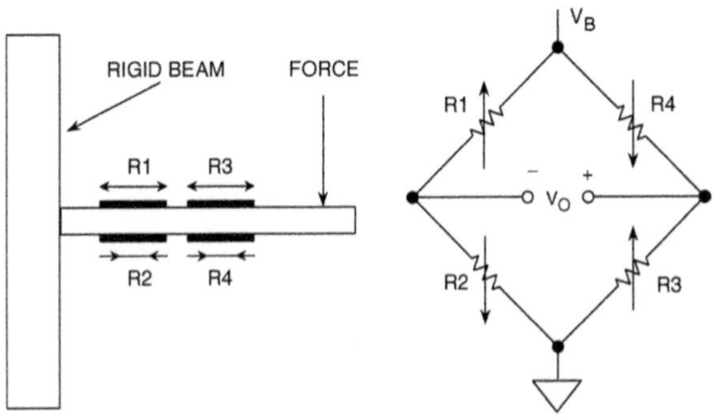

Figure 4-27: A beam force sensor using a strain gage bridge

Strain gages are low-impedance devices, consequently they require significant excitation power to obtain reasonable levels of output voltage. A typical strain-gage-based load cell bridge will have a 350 Ω impedance and is specified as having a sensitivity in a range 3 mV–10 mV full scale, per volt of excitation.

The load cell is composed of four individual strain gages arranged as a bridge, as shown in Figure 4-28. For a 10 V bridge excitation voltage with a rating of 3 mV/V, 30 mV of signal will be available at full-scale loading.

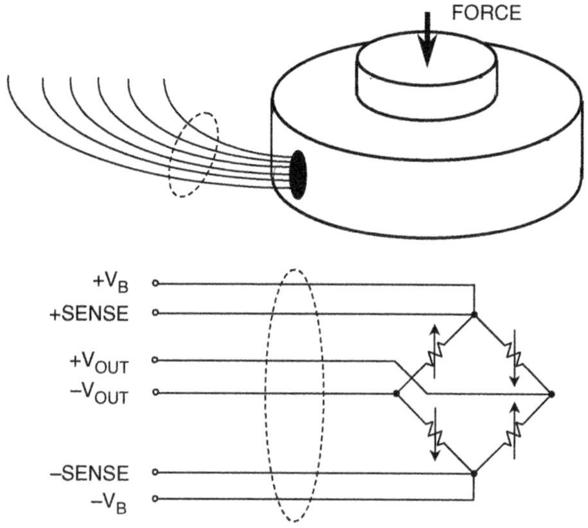

Figure 4-28: A load cell comprising four strain gages is shown in physical (top) and electrical (bottom) representations

While increasing the drive to the bridge can increase the output, self-heating effects are a significant limitation to this approach—they can cause erroneous readings, or even device destruction. One technique for evading this limitation is to use a low duty cycle pulsed drive signal for the excitation.

Many load cells have the ±"SENSE" connections as shown, to allow the signal-conditioning electronics to compensate for dc drops in the wires (Kelvin sensing as discussed in Section 4-2). This brings the wires to a total of six for the fully instrumented bridge. Some load cells may also have additional internal resistors for temperature compensation purposes.

Pressures in liquids and gases are measured electrically by a variety of pressure transducers. A number of mechanical converters (including diaphragms, capsules, bellows, manometer tubes, and Bourdon tubes) are used to measure pressure by measuring an associated length, distance, or displacement, and to measure pressure changes by the motion produced, as shown by Figure 4-29.

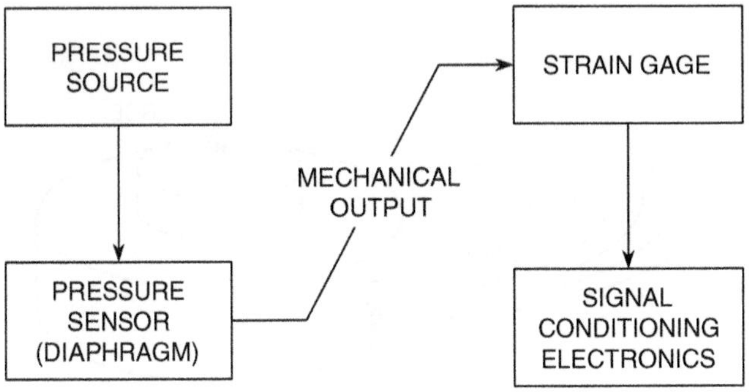

Figure 4-29: Pressure sensors use strain gages for indirect pressure measurement

The output of this mechanical interface is then applied to an electrical converter such as a strain gage, or piezoelectric transducer. Unlike strain gages, piezoelectric pressure transducers are typically used for high frequency pressure measurements (such as sonar applications, or crystal microphones).

There are many ways of defining flow (mass flow, volume flow, laminar flow, turbulent flow). Usually the *amount* of a substance flowing (mass flow) is the most important, and if the fluid's density is constant, a volume flow measurement is a useful substitute that is generally easier to perform. One commonly used class of transducers, which measures flow rate indirectly, involves the measurement of pressure.

Flow can be derived by taking the differential pressure across two points in a flowing medium—one at a static point and one in the flow stream. *Pitot tubes* are one form of device used to perform this function, where flow rate is obtained by measuring the differential pressure with standard pressure transducers.

Differential pressure can also be used to measure flow rate using the *venturi* effect by placing a restriction in the flow. Although there are a wide variety of physical parameters being sensed, the electronics interface is very often strain gage based.

Bridge Signal Conditioning Circuits

The remaining discussions of this section deal with applications that apply the bridge and strain gage concepts discussed thus far in general terms.

An example of an all-element varying bridge circuit is a fatigue monitoring strain sensing circuit, as shown in Figure 4-30. The full bridge is an integrated unit, which can be attached to the surface on which the strain or flex is to be measured. In order to facilitate remote sensing, current mode bridge drive is used. The remotely located bridge is connected to the conditioning electronics through a 4-wire shielded cable. The OP177 precision op amp servos the bridge current to 10 mA, being driven from an AD589 reference voltage of 1.235 V. Current buffering of the op amp is employed in the form of the PNP transistor, for lowest op amp self-heating, and highest gain linearity.

Figure 4-30: A precision strain gage sensor amplifier using a remote current-driven 1 kΩ bridge, a buffered precision op amp driver, and a precision in amp 100X gain stage

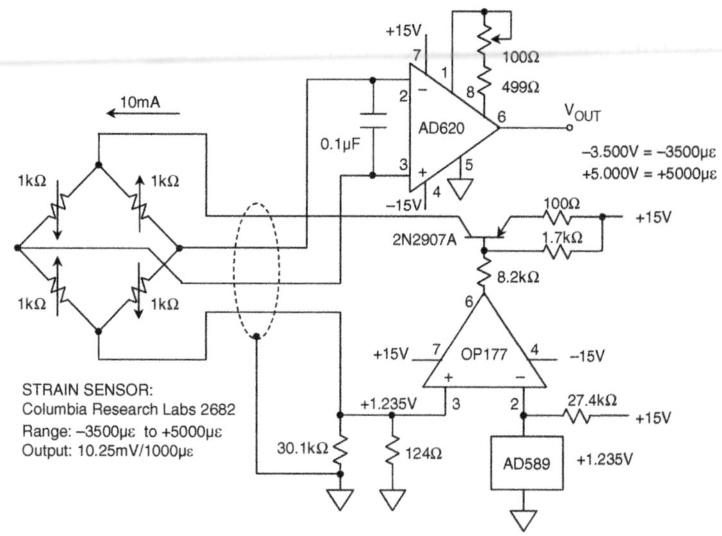

The strain gage produces an output of 10.25 mV/1000 µε. The signal is amplified by the AD620 in amp, which is configured for a gain of 100 times, via an effective R_G of 500 Ω. Full-scale voltage calibration is set by adjusting the 100Ω gain potentiometer such that, for a sensor strain of –3500 µε, the output reads –3.500 V; and for a strain of +5000 µε, the output registers +5.000 V. The measurement may then be digitized with an ADC which has a 10 V full-scale input range.

The 0.1 µF capacitor across the AD620 input pins serves as an EMI/RFI filter in conjunction with the bridge resistance of 1 kΩ. The corner frequency of this filter is approximately 1.6 kHz.

Another example is a load cell amplifier circuit, shown in Figure 4-31. This circuit is more typical of a bridge workhorse application. It interfaces with a typical 350Ω load cell, and can be configured to accommodate typical bridge sensitivities over a range of 3 mV–10 mV/V.

Figure 4-31: A precision 350 Ω load cell amplifier, using a buffered voltage-driven configuration with Kelvin sensing and a precision in amp

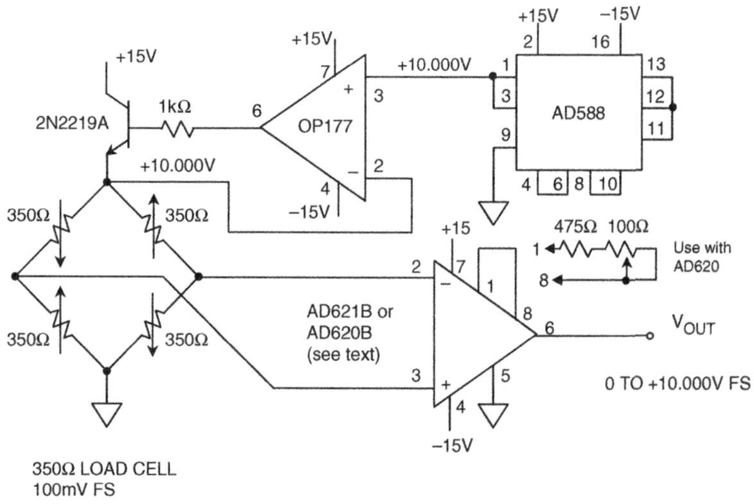

A 10.000 V bridge excitation excitation is derived from an AD588 10 V reference, with an OP177 and 2N2219A used as a buffer. The 2N2219A is within the OP177 feedback loop and supplies the necessary bridge drive current (28.57 mA). This ensures that the op amp performance will not be compromised. The Kelvin sensing scheme used at the bridge provides for low errors due to wiring resistances, and a precision zener diode reference, the AD588, provides lowest excitation drift and scaling with temperature changes.

To ensure highest linearity is preserved, a low drift instrumentation amplifier is used as the gain stage. This design has a minimum number of critical resistors and amplifiers, making the entire implementation accurate, stable, and cost effective. In addition to low excitation voltage TC, another stability requirement is minimum in amp gain TC. Both factors are critical towards insuring stable circuit scaling over temperature.

With the use of the AD621B in amp as shown, the scaling is for a precise gain of 100 (as set by the Pin 1–8 jumper), for lowest in amp gain TC. The AD621B is specified for a very low gain TC, only 5 ppm/°C. The gain of 100 translates a 100 mV full-scale bridge output to a nominal 10 V output. Alternately, an AD620B could also be used, with the optional gain network consisting of the fixed 475 Ω resistor, and 100 Ω potentiometer for gain adjustment. This will provide a 50 ppm/°C gain TC for the in amp, plus the TC of the external parts (which should have low temperature coefficients).

While the lowest TC is provided by the fixed gain AD621 setup, it doesn't allow direct control of overall scaling. To retain the very lowest TC, scaling could be accomplished via a software autocalibration routine. Alternately, the AD588 and OP177 reference/op amp stage could be configured for a variable excitation voltage (as opposed to a fixed 10.000 V as shown). Variable gain in the reference voltage driver will effectively alter the excitation voltage as seen by the bridge, and thus provide flexible overall system scaling. Of course, it is imperative that such a scheme be implemented with low TC resistances.

As shown previously, a precision load cell is usually configured as a 350 Ω bridge. Figure 4-32 shows a precision load cell amplifier, within a circuit possessing the advantage of being powered from just a single power supply.

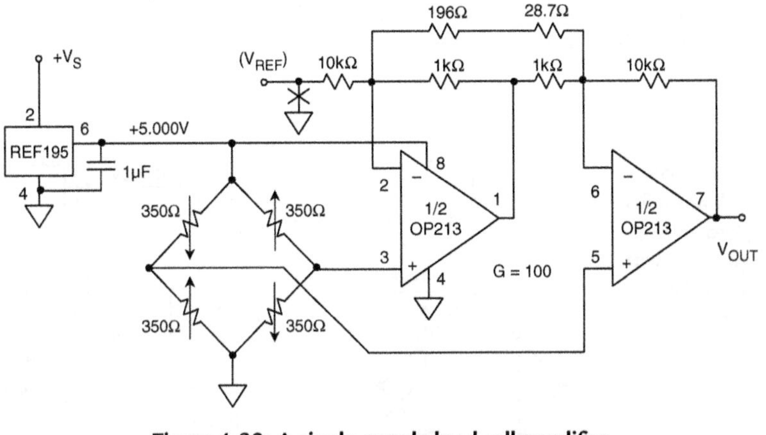

Figure 4-32: A single-supply load cell amplifier

As noted previously, the bridge excitation voltage must be both precise and stable, otherwise it can introduce measurement errors. In this circuit, a precision REF195 5 V reference is used as the bridge drive, allowing a TC as low as 5 ppm/°C. The REF195 reference can also supply more than 30 mA to a load, so it can drive a 350 Ω bridge (~14 mA) without need of a buffer. The dual OP213 is configured as a gain-of-100, two-op-amp in amp. The resistor network sets the gain according to the formula:

$$G = 1 + \frac{10\,k\Omega}{1\,k\Omega} + \frac{20\,k\Omega}{196\,\Omega + 28.7\,\Omega} = 100 \qquad \text{Eq. 4-3}$$

For optimum CMR, the 10 kΩ/1 kΩ resistor ratio matching should be precise. Close tolerance resistors (±0.5% or better) should be used, and all resistors should be of the same type.

For a zero volt bridge output signal, the amplifier will swing to within 2.5 mV of 0 V. This is the minimum output limit of the OP213. Therefore, if an offset adjustment is required, the adjustment should start from a positive voltage at V_{REF} and adjust V_{REF} downward until the output (V_{OUT}) stops changing. This is the point at which the amplifier limits the swing. Because of the single supply design, the amplifier cannot sense input signals that have negative polarity.

If linearity around or at zero volts input is required, or if negative polarity signals must be processed, the V_{REF} connection can be connected to a stable voltage that is mid-supply (i.e., 2.5 V) rather than ground. Note that when V_{REF} is not at ground, the output must be referenced to V_{REF}. An advantage of this type of referencing is that the output is now bipolar, with respect to V_{REF}.

The AD7730 24-bit sigma-delta ADC is ideal for direct conditioning of bridge outputs, and requires no interface circuitry (see Reference 10). A simplified connection diagram was shown in Figure 4.21A. The entire circuit operates on a single 5 V supply, which also serves as the bridge excitation voltage. Note that the measurement is ratiometric, because the sensed bridge excitation voltage is also used as the ADC reference. Variations in the 5 V supply do not affect the accuracy of the measurement.

The AD7730 has an internal programmable gain amplifier that allows a full-scale bridge output of ±10 mV to be digitized to 16-bit accuracy. The AD7730 has self- and system calibration features that allow offset and gain errors to be minimized with periodic recalibrations.

A "chop" or ac mode option minimizes the offset voltage and drift and operates similarly to a chopper-stabilized amplifier. The effective input voltage noise RTI is approximately 40nV rms, or 264 nV peak-to-peak. This corresponds to a resolution of 13 ppm, or approximately 16.5 bits. Gain linearity is also approximately 16 bits.

References: Strain, Force, Pressure And Flow Measurements

1. Ramon Pallas-Areny and John G. Webster, **Sensors and Signal Conditioning**, John Wiley, New York, 1991.
2. Dan Sheingold, Editor, **Transducer Interfacing Handbook**, Analog Devices, Inc., 1980, ISBN: 0-916550-05-2.
3. Sections 2, 3, Walt Kester, Editor, **1992 Amplifier Applications Guide**, Analog Devices, 1992, ISBN: 0-916550-10-9.
4. Sections 1, 6, Walt Kester, Editor, **System Applications Guide**, Analog Devices, 1993, ISBN: 0-916550-13-3.
5. Harry L. Trietley, **Transducers in Mechanical and Electronic Design**, Marcel Dekker, Inc., 1986.
6. Jacob Fraden, **Handbook of Modern Sensors, 2nd Ed.**, Springer-Verlag, New York, NY, 1996.
7. **The Pressure, Strain, and Force Handbook, Vol. 29**, Omega Engineering, One Omega Drive, P.O. Box 4047, Stamford CT, 06907-0047, 1995. www.omega.com.
8. **The Flow and Level Handbook, Vol. 29**, Omega Engineering, One Omega Drive, P.O. Box 4047, Stamford CT, 06907-0047, 1995, www.omega.com.
9. Ernest O. Doebelin, **Measurement Systems Applications and Design, 4th Ed.**, McGraw-Hill, 1990.
10. Data sheet for **AD7730 Bridge Transducer ADC**, www.analog.com.

High Impedance Sensors

Walt Kester, Scott Wurcer, Chuck Kitchin

Many popular sensors have output impedances greater than several megohms, and thus the associated signal-conditioning circuitry must be carefully designed to meet the challenges of low bias current, low noise, and high gain. Figure 4-33 lists a few examples of high impedance sensors.

A large portion of this section is devoted to a germane example, the analysis of a photodiode preamplifier. This application demonstrates many of the problems associated with high impedance sensor signal-conditioning circuits, and offers a host of practical solutions that can be applied to virtually all such sensors.

- Photodiode Preamplifiers
- Piezoelectric Sensors
- Humidity
- pH Monitors
- Chemical Sensors
- Smoke Detectors

Figure 4-33: High impedance sensors

Other examples of high impedance sensors to be discussed are piezoelectric sensors and charge-output sensors.

Photodiode Preamplifier Design

Photodiodes generate a small current that is proportional to the level of illumination. Their applications range from relatively low speed, wide dynamic range circuits to much higher speed circuits. Examples of the types of applications are precision light meters and high-speed fiber optic receivers.

One of the standard methods for specifying photodiode sensitivity is to state its short-circuit photocurrent (I_{sc}) for a given light level from a well-defined light source. The most commonly used source is an incandescent tungsten lamp running at a color temperature of 2850K.

At 100 fc (foot candles) of illumination (approximately the light level on an overcast day), the short-circuit current usually falls in a range of picoamps to hundreds of microamps for small area (less than 1 mm^2) diodes.

The equivalent circuit for a photodiode is shown in Figure 4-34. The short-circuit current is very linear over 6 to 9 decades of light intensity, and is therefore often used as a measure of absolute light levels. The open-circuit forward voltage drop across the photodiode varies logarithmically with light level but, because of its large temperature coefficient, the diode voltage is seldom used as an accurate measure of light intensity.

The shunt resistance R_{SH} is usually on the order of 1000 MΩ at room temperature, and decreases by a factor of 2 for every 10°C rise in temperature. Diode capacitance C_J is a function of junction area and the diode bias voltage. A value of 50 pF at zero bias is typical for small-area diodes.

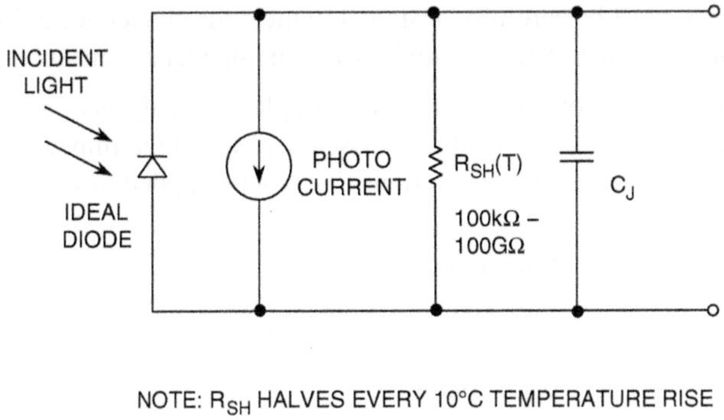

Figure 4-34: A photodiode equivalent circuit

Photodiodes may be operated in either of two basic modes, as shown in Figure 4-35. These modes are with zero bias voltage (*photovoltaic* mode, left) or with a reverse-bias voltage (*photoconductive* mode, right).

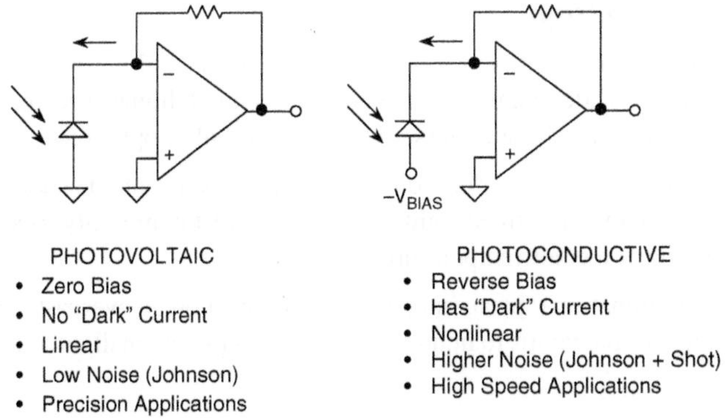

Figure 4-35: Photodiode operating modes

The most precise linear operation is obtained in the photovoltaic mode, while higher switching speeds can be realized when the diode is operated in the photoconductive mode at the expense of linearity. Under these reverse-bias conditions, a small amount of current called *dark current* will flow—even when there is no illumination.

There is no dark current in the photovoltaic mode. In the photovoltaic mode, the diode noise is basically the thermal noise generated by the shunt resistance. In the photoconductive mode, shot noise due to conduction is an additional source of noise. Photodiodes are usually optimized during the design process for use in either the photovoltaic mode or the photoconductive mode, but not both.

Figure 4-36 shows the photosensitivity for a small photodiode (Silicon Detector Part Number SD-020-12-001). This diode has a basic sensitivity of 0.03 μA/fc, and was chosen for the design example to follow. As this chart indicates, this photodiode's dynamic range covers six orders of magnitude.

ENVIRONMENT	ILLUMINATION (fc)	SHORT CIRCUIT CURRENT
Direct Sunlight	1000	30μA
Overcast Day	100	3μA
Twilight	1	0.03μA
Full Moonlit Night	0.1	3000pA
Clear Night / No Moon	0.001	30pA

Figure 4-36: Short circuit current versus light intensity for SD-020-12-001 photodiode (photovoltaic operating mode)

A convenient way to convert the photodiode current into a usable voltage is to use a low bias current op amp, configured as a current-to-voltage converter as shown in Figure 4-37. The diode bias is maintained at zero volts by the virtual ground of the op amp, and the short-circuit current is converted into a voltage. At maximum sensitivity the amplifier must be able to detect a diode current of 30 pA. This implies that the feedback resistor must be very large, and the amplifier bias current very small.

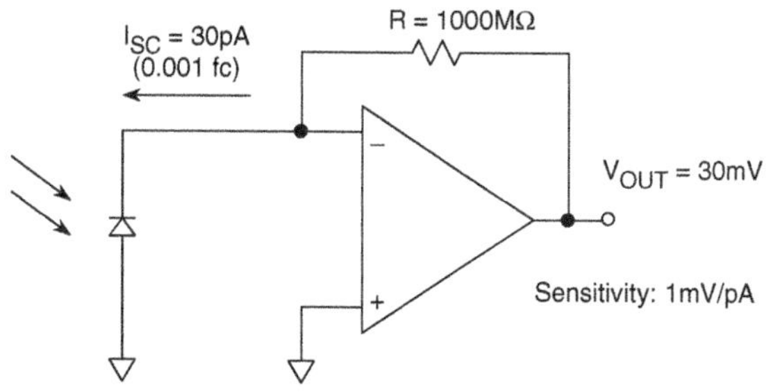

Figure 4-37: A simplified current-to-voltage converter uses a low bias current op amp and a high value feedback resistor

For example, 1000 MΩ will yield a corresponding voltage of 30 mV for this amount of current. Larger resistor values are impractical, so we will use 1000 MΩ for the most sensitive range. This will give an output voltage range of 10 mV for 10 pA of diode current and 10 V for 10 nA of diode current. This yields a range of 60 dB. For higher values of light intensity, the gain of the circuit must be reduced by using a smaller feedback resistor. For this range of maximum sensitivity, we should be able to easily distinguish between the light intensity on a clear, moonless night (0.001 fc), and that of a full moon (0.1 fc).

Notice that we have chosen to get as much gain as possible from one stage, rather than cascading two stages. This is in order to maximize the signal-to-noise ratio (SNR). If we halve the feedback resistor value, the signal level decreases by a factor of 2, while the noise due to the feedback resistor $\left(\sqrt{4kTR \cdot \text{Bandwidth}}\right)$ decreases by only $\sqrt{2}$. This reduces the SNR by 3 dB, assuming the closed loop bandwidth remains constant. Later in the analysis, we will find the resistors one of the largest overall output noise contributors.

To accurately measure photodiode currents in the tens of picoamps range, the bias current of the op amp should be no more than a few picoamps. This considerably narrows the choice. The industry-standard OP07 is an ultralow offset voltage (10 μV) bipolar op amp, but its bias current is 4 nA (4000 pA). Even superbeta bipolar op amps with bias current compensation (such as the OP97) have bias currents on the order of 100 pA at room temperature, but they might be suitable for very high temperature applications, as unlike FET amplifiers, the bias currents do not double for every 10ºC increase.

A JFET input electrometer-grade op amp is chosen for our photodiode preamp, since it must operate only over a limited temperature range. Figure 4-38 summarizes the performance of several popular "electrometer grade" FET input op amps.

PART #	V_{OS}, MAX*	TC V_{OS}, TYP	I_B, MAX*	0.1Hz TO 10Hz NOISE, TYP	PACKAGE
AD549K	250μV	5μV/°C	100fA	4μV p-p	TO-99
AD795JR	500μV	3μV/°C	3pA	1μV p-p	SOIC
AD820B	1000μV	2μV/°C	10pA	2μV p-p	SOIC, DIP

*25°C SPECIFICATION

Figure 4-38: Some JFET input electrometer grade op amps suitable for use in photodiode preamplifiers

As can be noted from this figure, the 25°C maximum bias current specification ranges from a few pA down to as low as 100 fA, and there are a number of packages types from which to choose. As will be seen shortly, the package finally chosen can and will affect the performance of the circuit in terms of the bias current realized within an application. This is due to relative ability to control the inevitable leakage currents in a design's production environment.

Of these devices, the AD549 and AD795 are fabricated on a BiFET process and use P-Channel JFETs as the input stage, as is shown in Figure 4-39. The rest of the op amp circuit is designed using bipolar devices. These BiFET op amps are laser trimmed at the wafer level, to minimize offset voltage and offset voltage drift. The offset voltage drift is minimized, by first trimming the input stage for equal currents in the JFET differential pair (drift trim resistors). A further trim of the JFET source resistors minimizes the input offset voltage (offset voltage trim resistors).

For these discussions, an AD795JR was selected for the photodiode preamplifier, with key specifications summarized in Figure 4-38. This allows high circuit performance in an SOIC packaged device.

Alternately, for even greater performance, the AD549 could be used. The AD549 uses the glass sealed TO-99 package, which allows the very highest performance in terms of low leakage. More on this follows.

Since the photodiode current is measured in terms of picoamperes, it should be understood that extremely close attention must be given to all potential leakage paths in the actual physical circuit. To put this in some

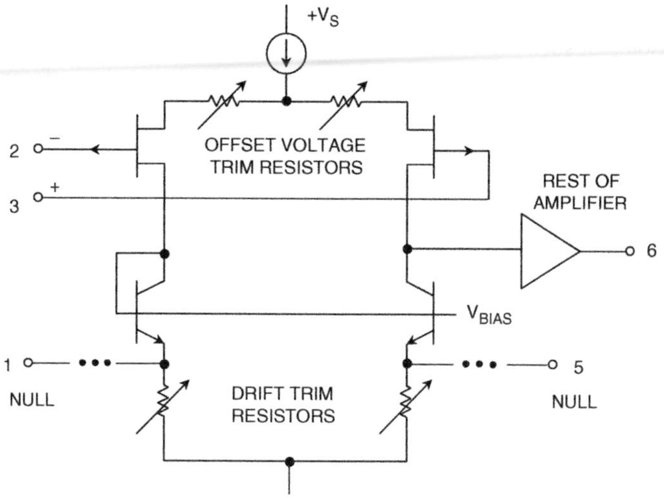

Figure 4-39: JFET input stage op amp with separate trims for offset voltage and drift

perspective, consider the a simple printed circuit card example with two parallel conductor traces on a high-quality, well-cleaned epoxy-glass PC board 0.05 inches apart and parallel for 1 inch. Such an insulator has a leakage resistance of approximately 10^{11} Ω at +125°C. By simple application of Ohm's law, 15 V of bias between these runs produces a 150 pA current—sufficient to mask all signal levels below this current. Obviously then, low-level photodiode circuitry needs to employ all possible means of minimizing such parasitic currents. Unfortunately, they can arise from numerous sources, some of which can be quite subtle in origin.

Figure 4-40 illustrates the circuit elements subject to leakage for the photodiode circuit, as enclosed within the dotted lines. The feedback resistor is highly critical, and should be a close tolerance (1%), low TC (50 ppm/°C) unit. Typical units suitable for R2 will be manufactured with thin film or metal oxide construction on ceramic or glass, with glass insulation. It should be readily apparent that any shunt conductive paths

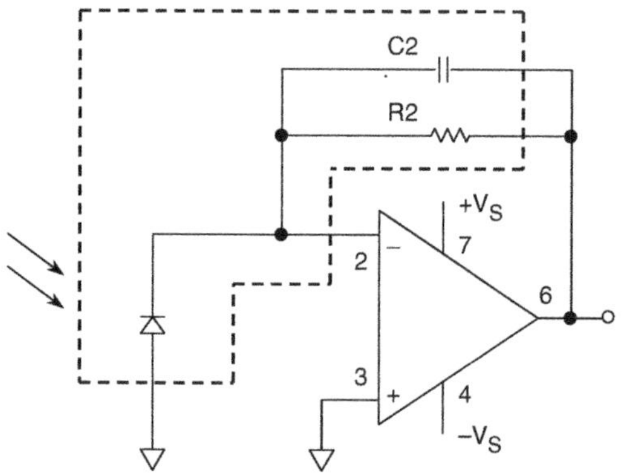

Figure 4-40: Critical leakage paths and components for a photodiode preamplifier circuit are those within the dotted line area

across this resistor's body can (and will) degrade or lower the net effective resistance, producing scaling errors. It is for this reason that such high value resistors are often glass enclosed, and can require special handling. Some sources of suitable high value resistors are listed in the section references. If used, compensation capacitor C2 should use the lowest loss dielectric possible. Typically this will mean a film type capacitor of Teflon, polypropylene, or polystyrene construction.

All connections to the op amp's summing junction should be kept short, clean, and free from manufacturing process chemicals and residues. In cases where an input cable is used to connect the photodiode to the preamp, it should be kept as short as possible, and should use Teflon or similar low loss dielectric insulation.

The above considerations deal mainly with the more obvious construction points towards optimizing accuracy and keeping leakage low. However, two of the more difficult leakage sources that can plague this circuit aren't quite as obvious. These are the op amp package-related parasitic leakages, which can occur from all op amp package pins adjacent to the input pin. Consider leakage as a high value resistance to Pin 2.

Although it doesn't show in this particular figure (since Pin 1 isn't actively used by the application), *any leakage from package Pin 1 can be very relevant.* Note also, that since Pin 3 is grounded, it prevents error from leakage between Pins 4 and 2. If however, Pin 1 has any significant voltage on it (which it does in the case of the offset null pin of the AD820BN DIP device) serious leakage will then occur between Pins 1 and 2. The AD795JR SOIC is immune to this leakage, as Pin 1 isn't connected internally. These comments serve to illustrate some of the subtleties of these leakage sources.

The situation just described for the AD820BN DIP packaged device is by no means unique, as Pin 1 is a standard offset trim pin on many op amps. This circumstance will always tend to leak current into any high impedance source seen at Pin 2. There are also cases for follower-connected stages where leakage is just as critical, if not more so. In such cases the leakage goes into Pin 3 as a high impedance, typically from Pin 4, which is $-V_S$. Fortunately however, there is a highly effective answer to controlling both of these leakage problems, and that is the use of circuit *guard* techniques.

Guarding is used to reduce parasitic leakage currents, by isolating a sensitive amplifier input from large voltage gradients across the PC board. It does this by interposing a conductive barrier or screen between a high voltage source and a sensitive input. The barrier intercepts the leakage which would otherwise flow into the sensitive node, and diverts it away. In physical terms, a guard is a low impedance conductor that completely surrounds an input line or node, and it is biased to a potential equal to the line's voltage.

Note that the low impedance nature of a guard conductor shunts leakage harmlessly away. The biasing of the guard to the same potential as the guarded pin reduces any possibility of leakage between the guard itself and the guarded node. The exact technique for guarding depends on the amplifier's mode of operation, i.e., whether the connection is inverting (like Figure 4-40), or a noninverting stage.

Figure 4-41 shows a PC board layout for guarding the inputs of the AD820 op amp, as operated within either an inverting (top) or a noninverting gain stage (bottom). This setup uses the DIP ("N") package, and would also be applicable to other devices where relatively high voltages occur at Pin 1 or 4. Using a standard 8-pin DIP outline, it can be noted that this package's 0.1" pin spacing allows a PC trace (the guard trace) to pass between adjacent pins. This is the key to implementing effective DIP package guarding—the complete surrounding of the guarded trace with a low impedance trace.

In the inverting mode (top), note that Pin 3 connected and grounded guard traces surround the op amp inverting input (Pin 2), and run parallel to the input trace. This guard would be continued out to and around the source device and feedback connection in the case of a photodiode (or around the input pad, in the case of a cable). In the follower mode (bottom), the guard voltage is the feedback divider tap voltage to Pin 2, i.e., the inverting input node of the amplifier. Although the feedback divider impedance isn't as low

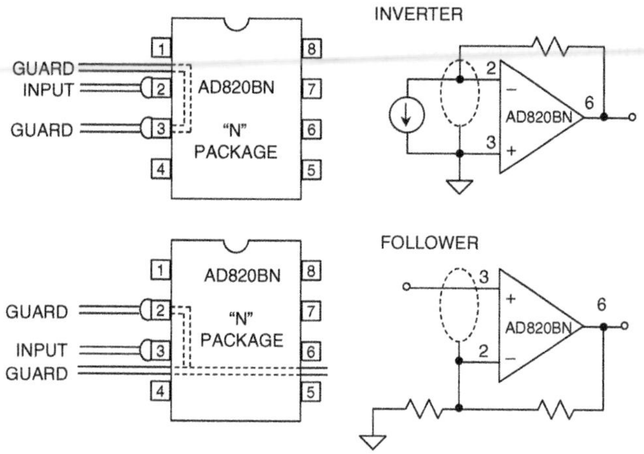

Figure 4-41: Guard techniques for inverting and noninverting op amp stages using DIP package devices

in absolute terms as a direct ground, it is still quite effective. Even a 1 kΩ or so impedance here will still be many orders of magnitude lower than the Pin 3 impedance. In both the inverting and the noninverting modes, the guard traces should be located on both sides of the PC board, with top and bottom side traces connected with several vias. Things become slightly more complicated when using guarding techniques with the SOIC surface mount ("R") package, because the 0.05" pin spacing doesn't allow routing of PC board traces between the pins. But there still is an effective guarding answer, at least for the inverting case. Figure 4-42 shows the preferred method.

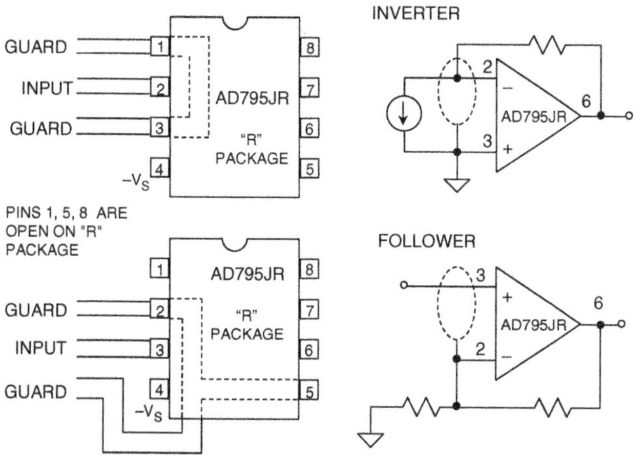

Figure 4-42: Guard techniques for inverting and noninverting op amp stages using SOIC package devices

In the AD795 SOIC "R" package, Pins 1, 5, and 8 are "no-connect" pins and can be used to route signal traces as shown. Thus in the case of the inverting stage (top), guarding is still completely effective, with dummy Pin 1 and Pin 3 acting as a grounded guard trace.

In the case of the follower stage (bottom), the guard trace must be routed around the $-V_S$ pin, and thus the Pin 4 to Pin 3 leakage is *not* fully guarded. For this reason, a very high impedance follower stage using an SOIC package op amp isn't generally recommended, as adequate guarding simply isn't possible. An exception to this caveat would apply the use of a *single-supply* op amp as a follower (for example, the AD820), in which case Pin 4 becomes grounded by default, and some degree of intrinsic guarding is established.

For extremely low bias current applications, such as for example with the AD549 with input bias current of 100 fA, the high impedance input signal connection of the op amp should be made to a virgin Teflon standoff insulator, as shown in Figure 4-43. Note—"virgin" Teflon is a solid piece of new Teflon material that has been machined to shape (as opposed to one welded together from powder or grains).

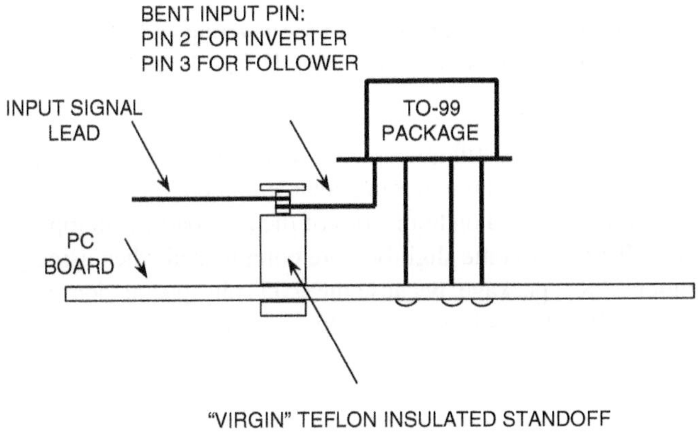

Figure 4-43: Input pin isolation technique using virgin Teflon standoff insulator

If mechanical and manufacturing considerations allow, the sensitive op amp input pin should be soldered directly to the Teflon standoff, rather than going through a PCB hole.

For TO-99 packaged devices, such as the AD549KH, two possible guarding choices present themselves. One method is to employ the device in a scheme like Figure 4-43, with the sensitive input pin going to the Teflon stand-off. Alternately, a round PCB layout scheme that is more amenable to the TO-99 package can be used, as shown in Figure 4-44.

This scheme uses a guard ring, which completely surrounds the input and feedback nodes, with the ring tied to the device's metal can through the Pin 8 connection. The guard ring is then also tied to either ground or the feedback divider, as suits the application. This setup can also be further modified, to use the more sensitive of the two inputs going to a *Teflon stand-off within the guard ring*, for the ultimate in performance.

Note that in all cases where control of leakage is critical, the PC board itself must be carefully cleaned and then sealed against humidity and dirt using a high quality conformal coating material. In addition to minimizing leakage currents, the entire circuit should be well shielded with a grounded metal shield to prevent stray signal pickup.

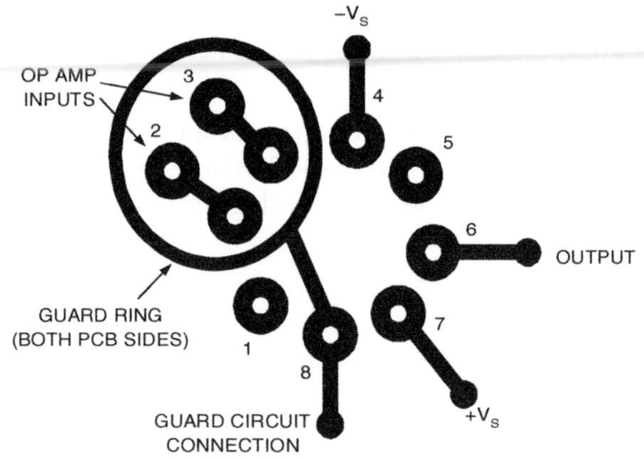

Figure 4-44: TO-99 package devices can use guard rings surrounding input pins 2 and 3 (PCB bottom view shown)

Preamplifier Offset Voltage and Drift Analysis

A photodiode preamp offset voltage and bias current model is shown in Figure 4-45. There are two important considerations in this circuit. First, the diode shunt resistance (R1) is a function of temperature—it halves every time the temperature increases by 10°C. At room temperature (25°C), R1 = 1000 MΩ, but at 70°C it decreases to 43 MΩ. This has a drastic impact on the circuit noise gain and hence the output offset voltage. In the example, at 25°C the dc noise gain is 2, but at 70°C it increases to 24.

The second circuit difficulty is that the input bias current doubles with every 10°C temperature rise. The bias current produces an output offset error equal to I_BR2. At 70°C bias current increases to 72 pA, compared to 3 pA at room temperature. Normally, the addition of a resistor (R3) between the noninverting input of the op amp and ground, with a value of R1||R2 would yield a first-order cancellation of this effect. However, because R1 changes with temperature, this method simply isn't effective. In addition, if R3 is

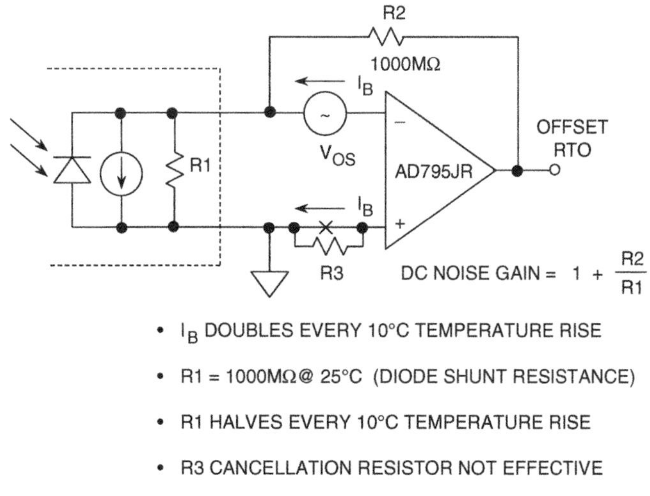

Figure 4-45: AD795JR photodiode preamplifier offset error model

used, the bias current then develops a voltage across it, which in turn would be applied to the photodiode as a parasitic bias. Such a bias would cause the diode response to become nonlinear, thus the use of R3 is also undesirable from a linearity point-of-view.

The total referred-to-output (RTO) offset voltage preamp errors are summarized in Figure 4-46. Notice that at 70°C the total error is 87.2 mV. This error is acceptable for the design under consideration. The primary contributor to the error at high temperature is of course the bias current.

	0°C	25°C	50°C	70°C
V_{OS}	0.575mV	0.500mV	0.575mV	0.635mV
Noise Gain	1.1	2	7	24
V_{OS} Error RTO	0.6mV	1.0mV	4.0mV	15.2mV
I_B	0.6pA	3.0pA	18.0pA	72.0pA
I_B Error RTO	0.6mV	3.0mV	18.0mV	72.0mV
Total Error RTO	1.2mV	4.0mV	22.0mV	87.2mV

Figure 4-46: AD795JR photodiode preamplifier offset error summary

Several steps can be taken to minimize amplifier temperature rise, and thus offset drift. Operating the amplifier at reduced supply voltages, minimizing the output drive requirements, and heat sinking are some ways to reduce this error. The addition of an external offset nulling circuit would minimize the initial input offset voltage error.

Thermoelectric Voltages as Sources of Input Offset Voltage

As discussed earlier in this chapter, thermoelectric potentials are generated by electrical connections that are made between different metals. For example, the copper PC board electrical contacts to the kovar input pins of a TO-99 IC package can create an offset voltage of up to 40 μV/°C, if the two bimetal junctions so formed are at different temperatures. Even ordinary solders, being composed of alloys different from PCB copper traces, can give rise to thermoelectric voltages. For example, common high tin content lead-tin solder alloys, when used with copper, create thermoelectric voltages on the order of 1 μV/°C to 3 μV/°C (see Reference 8). While some special cadmium-tin solders can reduce this voltage to 0.3 μV/°C, cadmium solders aren't in general use for health reasons. Another possible low thermal EMF solder is a low tin alloy such as Sn10Pb90.

The best general solution to minimizing this spurious thermocouple problem is to ensure that the connections to the inverting and noninverting input pins of the IC are made with the same material, and that the PC board thermal layout is such that these two pins remain at the same temperature. Everything should be balanced from a thermal standpoint. In the case where a Teflon standoff is used as an insulated connection point for the inverting input (as in the case of this preamp), prudence dictates that connections to the noninverting inputs also be made in a similar manner to minimize possible thermoelectric effects, and in keeping with the principle of thermal symmetry.

Preamplifier AC Design, Bandwidth, and Stability

The key to the preamplifier ac design is an understanding of the circuit noise gain as a function of frequency. Plotting gain versus frequency on a log-log scale makes the analysis relatively simple (see Figure 4-47). This type of plot is also referred to as a Bode plot. The noise gain is the gain seen by a small voltage source in series with one of the op amp input terminals. It is also the same as the noninverting signal gain (the gain from "A" to the output). In the photodiode preamplifier, the signal current from the photodiode passes through the C2/R2 network. It is important to distinguish between the signal gain and the noise gain, because it is the noise gain characteristic that determines the net circuit stability, regardless of where the signal is actually applied.

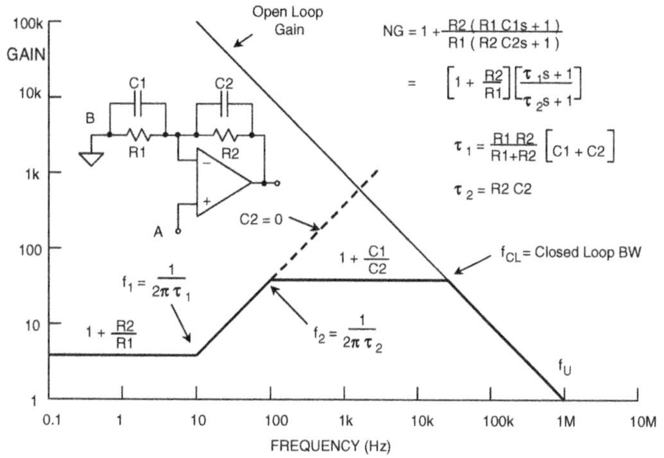

Figure 4-47: A generalized noise gain Bode plot

Note that the net slope between the noise gain and the open loop gain curves, *at the point where they intersect*, determines system stability. For unconditional stability, the noise gain curve must intersect the open loop response with a net slope of less than 12 dB/octave (or 20 dB per decade). In the figure, the dotted (C2 = 0) line shows a noise gain that intersects the open loop gain at a net slope of 12 dB/octave, indicating an unstable condition. This is what would occur in the circuit, without a feedback capacitor.

The general equations for determining the break points and gain values in the Bode plot are also given in Figure 4-47. It is useful to examine these gain characteristics with increasing frequency. At low frequencies, the circuit noise gain is 1 + R2/R1, as indicated by the lowest frequency shelf (below 10 Hz). There are two key time constants in this circuit, τ_1 and τ_2. The first comes into play as a zero in the noise gain transfer function, which occurs at a frequency of $f_1 = 1/2\pi\tau_1$, where $\tau_1 = R1||R2\ (C1 + C2)$. Stated simply, this frequency falls where the noise gain begins to increase to a new (higher) value from the low frequency gain of 1 + R2/R1 plateau. In the Figure 4-47 example f_1 occurs at 10 Hz.

Above f_1, gain increases towards a high frequency gain plateau where the gain is 1 + C1/C2, which is indicated as the highest frequency shelf (above 100 Hz). The second time constant, τ_2, comes into play as a pole of the transfer function, which occurs at a corner frequency, $f_2 = 1/2\pi\tau_2$, where $\tau_2 = R2C2$. It can also be noted that this is equal to the signal bandwidth, if the signal is applied at point "B."

Plotting the composite noise gain curve on the log-log graph is a simple matter of connecting the f_1 and f_2 breakpoints with a line having a 45° slope, after first sketching the flat low and high frequency gain plateaus. The point at which the high frequency noise gain intersects the op amp open loop gain is called the *closed loop bandwidth*. Notice that the *signal bandwidth* for a signal applied at point "B" is much less, and is 1/2 πR2C2.

Figure 4-48 shows the noise gain plot for the photodiode preamplifier using actual circuit values. The choice of C2 determines the actual signal bandwidth and also the phase margin. In the example, a signal bandwidth of 16 Hz was chosen. Notice that a smaller value of C2 would result in a higher signal bandwidth and a corresponding reduction in phase margin. It is also interesting to note that although the signal bandwidth is only 16 Hz, the closed loop bandwidth is 167 kHz. This will have important implications with respect to the output noise voltage analysis to follow.

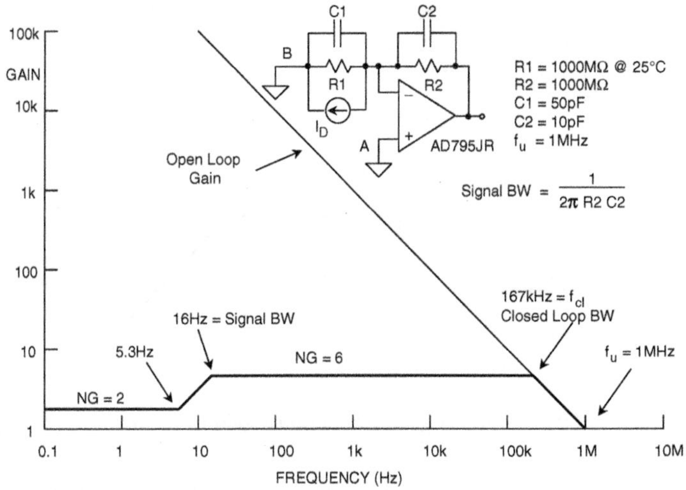

Figure 4-48: Noise gain of the AD795 photodiode preamplifier at 25°C

It is important to note that temperature changes do not significantly affect the stability of the circuit. Changes in R1 (the photodiode shunt resistance) affect only the low frequency noise gain and the frequency at which the zero in the noise gain response occurs. The high frequency noise gain is determined by the C1/C2 ratio.

Photodiode Preamplifier Noise Analysis

To begin a noise analysis, we first consider the AD795 input voltage and current noise spectral densities, as shown in Figure 4-49. The AD795 performance is truly impressive for a JFET input op amp: 1 μV p-p typical 0.1 Hz to 10 Hz noise, and a 1/f corner frequency of 12 Hz, comparing favorably with all but the best bipolar op amps. As shown in the (right) figure, the current noise is much lower than for bipolar op amps, a key factor making it an ideal choice for high impedance applications.

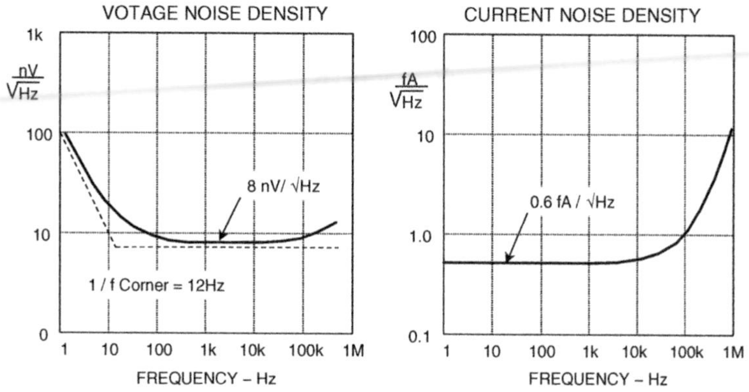

Figure 4-49: AD795 Voltage and current noise density performance

The complete noise model for an op amp is shown in Figure 4-50. This model includes the reactive elements C1 and C2. Each individual output noise contributor is calculated by integrating the square of its spectral density over the appropriate frequency bandwidth and then taking the square root, as:

$$\text{RMS OUTPUT NOISE DUE TO } V_1 = \sqrt{\int V_1(f)^2 df} \qquad \text{Eq. 4-4}$$

In most cases, this integration can be done by inspection of the graph of the individual spectral densities superimposed on a graph of the noise gain. The total output noise is then obtained by combining the individual components in a root-sum-squares manner. The table in the diagram shows how each individual source is reflected to the output, and the corresponding bandwidth for integration. The factor of 1.57 (π/2) is required to convert the single-pole bandwidth into its equivalent noise bandwidth.

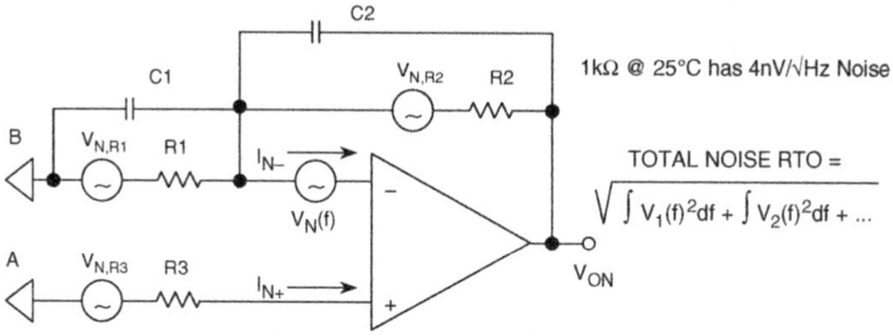

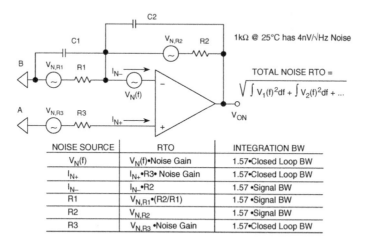

NOISE SOURCE	RTO	INTEGRATION BW
$V_N(f)$	$V_N(f)$•Noise Gain	1.57•Closed Loop BW
I_{N+}	I_{N+}•R3• Noise Gain	1.57•Closed Loop BW
I_{N-}	I_{N-}•R2	1.57 •Signal BW
R1	$V_{N,R1}$•(R2/R1)	1.57 •Signal BW
R2	$V_{N,R2}$	1.57 •Signal BW
R3	$V_{N,R3}$ •Noise Gain	1.57•Closed Loop BW

Figure 4-50: A noise model of preamp

The resistor Johnson noise spectral density V_R is given by:

$$V_R = \sqrt{4kTR} \quad \text{Eq. 4-5}$$

where *R* is the resistance in ohms, *k* is Boltzmann's constant (1.38×10^{-23} J/K), and *T* is the absolute temperature in Kelvins.

A simple way to compute this is to remember that the noise spectral density of a 1 kΩ resistor is $4\ \text{nV}/\sqrt{\text{Hz}}$ at +25°C. The Johnson noise of another resistor value can be found by multiplying by the square root of the ratio of the resistor value to 1000Ω. For example, a 4 kΩ resistor produces a noise density $\sqrt{4}$ times a 1 kΩ resistor, or $8\,\text{nV}/\sqrt{\text{Hz}}$ (at +25°C).

Finally, note that Johnson noise is broadband, and its spectral density is constant with frequency.

Input Voltage Noise

In order to obtain the output voltage noise spectral density plot due to the input voltage noise, the input voltage noise spectral density plot is multiplied by the noise gain plot. This is easily accomplished using the Bode plot on a log-log scale. The total RMS output voltage noise due to the input voltage noise is then obtained by integrating the square of the output voltage noise spectral density plot, and then taking the square root. In most cases, this integration may be approximated. A lower frequency limit of 0.01 Hz in the 1/f region is normally used. If the bandwidth of integration for the input voltage noise is greater than a few hundred Hz, the input voltage noise spectral density may be assumed to be constant. Usually, the value of the input voltage noise spectral density at 1 kHz will provide sufficient accuracy.

It is important to note that the input voltage noise contribution must be integrated over the entire closed-loop bandwidth of the circuit (the closed loop bandwidth, f_{cl}, is the frequency at which the noise gain intersects the op amp open loop response). This is also true of the other noise contributors that are reflected to the output by the noise gain (namely, the non inverting input current noise and the non inverting input resistor noise).

The inverting input noise current flows through the feedback network to produce a noise voltage contribution at the output. The input noise current is approximately constant with frequency, therefore, the integration is accomplished by multiplying the noise current spectral density (measured at 1 kHz) by the noise bandwidth which is 1.57 times the signal bandwidth ($1/2\ \pi R2C2$). The factor of 1.57 ($\pi/2$) arises when single-pole 3 dB bandwidth is converted to equivalent noise bandwidth.

Johnson Noise Due to Feedforward Resistor R1

The noise current produced by the feedforward resistor R1 also flows through the feedback network to produce a contribution at the output. The noise bandwidth for integration is also 1.57 times the signal bandwidth.

Noninverting Input Current Noise

The noninverting input current noise, I_{N+}, develops a voltage noise across R3 that is reflected to the output by the noise gain of the circuit. The bandwidth for integration is therefore the closed-loop bandwidth of the circuit. However, there is no contribution at the output if R3 = 0 (or, if R3 is used, but it is bypassed with a large capacitor). This will usually be desirable when operating the op amp in the inverting mode.

Johnson Noise Due to Resistor in Noninverting Input

The Johnson voltage noise due to R3 is also reflected to the output by the noise gain of the circuit. Again, if R3 is bypassed sufficiently, it makes no significant contribution to the output noise.

Summary of Photodiode Circuit Noise Performance

Figure 4-51 shows the output noise spectral densities for each of the contributors at 25°C. As can be noted, there is no contribution due to I_{N+} or R3, since the noninverting input of the op amp is grounded.

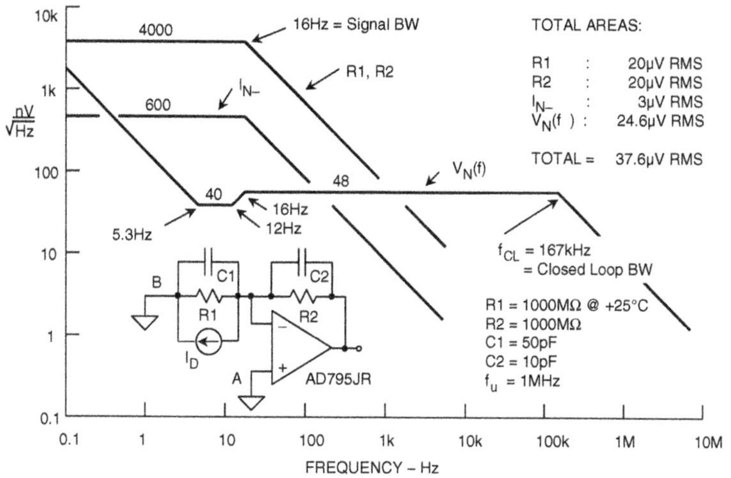

Figure 4-51: Preamp output spectral noise densities (nV/√Hz) @25°C

Noise Reduction Using Output Filtering

From the above analysis, the largest contributor to the output noise voltage at 25°C is the input voltage noise of the op amp reflected to the output by the noise gain. This contributor is large primarily because the noise gain over which the integration is performed extends to a bandwidth of 167 kHz (the intersection of the noise gain curve with the open-loop response of the op amp). If the op amp output is filtered by a single-pole low-pass filter with a 20 Hz cutoff frequency (τ = 7.95 ms), this contribution is reduced to less than 1 μVrms. The diagram for the final, filtered, optimized photodiode circuit design is shown in Figure 4-52.

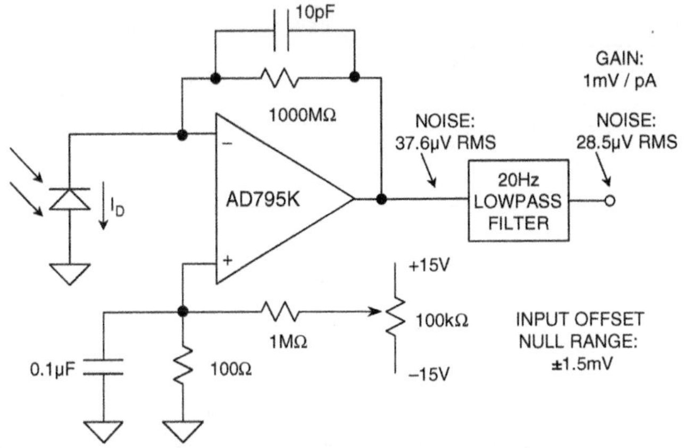

Figure 4-52: AD795K preamp with output filter and offset null option

Notice that the same results would not be achieved simply by increasing the feedback capacitor, C2. Increasing C2 lowers the high frequency noise gain, but the integration bandwidth becomes proportionally higher. Larger values of C2 may also decrease the signal bandwidth to unacceptable levels.

The addition of the post-filter stage reduces output noise to 28.5 μVrms; approximately 75% of its former value, and the resistor noise and current noise are now the largest contributors to output noise. Practically, this filter can be either active or passive. Care will need to be taken, of course, that the filter circuit does not add any significant noise of its own to the signal. Filter design is discussed in greater detail in Chapter 5 of this book. The final circuit also includes an offset trim arrangement that is capable of nulling op amp offsets of up to ±1.5 mV.

Summary of Circuit Performance

Performance characteristics are summarized in Figure 4-53. The total output voltage drift over 0 to 70ºC is 87.2 mV, corresponding to 87.2 pA of diode current. The offset nulling circuit shown on the noninverting input can be used to null out the room temperature offset. Note that this method is better than using the offset null pins because using the offset null pins will increase the offset voltage TC by about 3 μV/ºC for each millivolt nulled. In addition, the AD795 SOIC package does not have offset nulling pins.

The input sensitivity based on a total output voltage noise of 44 μV is obtained by dividing the output voltage noise by the value of the feedback resistor R2. This yields a minimum detectable diode current of 44 fA. If a 12-bit ADC is used to digitize the 10 V full-scale output, the weight of the least significant bit (LSB) is 2.5 mV. The output noise level is much less than this.

- Output Offset Error (0°C to 70°C) : 87.2mV
- Output Sensitivity: 1mV/pA
- Output Photosensitivity: 30V / foot-candle
- Total Output Noise @ 25°C : 28.5μV RMS
- Total Noise RTI @ 25°C : 44fA RMS, or 26.4pA p-p
- Range with R2 = 1000MΩ: 0.001 to 0.33 foot-candles
- Bandwidth: 16Hz

Figure 4-53: AD795JR photodiode preamp performance summary

Photodiode Circuit Trade-off

Many trade-offs could be made in the basic photodiode circuit design we have described. More signal bandwidth can be achieved in exchange for a larger output noise level. Reducing the feedback capacitor C2 to 1 pF increases the signal bandwidth to approximately 160 Hz. Further reductions in C2 are not practical because the parasitic capacitance is probably in the order of 1 pF to 2 pF. Some small amount of feedback capacitance is also required to maintain stability.

If the circuit is to be operated at higher levels of illumination (greater than approximately 0.3 fc), the value of the feedback resistor can be reduced, thereby resulting in further increases in circuit bandwidth and less resistor noise.

If gain-ranging is to be used to measure higher light levels, extreme care must be taken in the design and layout of the additional switching networks to minimize leakage paths and other parasitic effects.

Compensation of a High Speed Photodiode I/V Converter

A classical I/V converter is shown in Figure 4-54. Note that it is the same as the previous photodiode preamplifier, if we assume that R1 >> R2. The total input capacitance, C1, is the sum of the diode capacitance and the op amp input capacitance. Dynamically, this is a classical second-order system, and the following guidelines can be applied in order to determine the proper compensation.

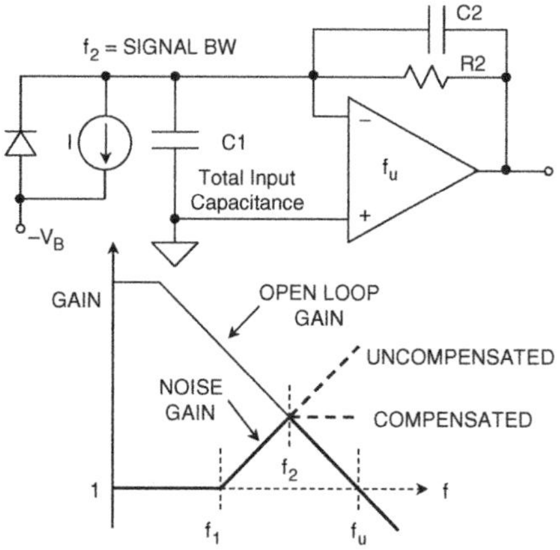

f_u = OP AMP UNITY GAIN BW PRODUCT

$$f_1 = \frac{1}{2\pi\, R2\, C1}$$

$$f_2 = \frac{1}{2\pi\, R2\, C2}$$

$$f_2 = \sqrt{f_1 \cdot f_u}$$

$$C2 = \sqrt{\frac{C1}{2\pi\, R2\, f_u}}$$

FOR 45° PHASE MARGIN

$$f_2 = \sqrt{\frac{f_u}{2\pi\, R2\, C1}}$$

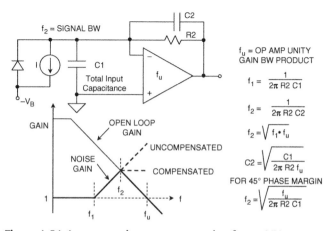

Figure 4-54: Input capacitance compensation for an I/V converter

The net input capacitance, C1, forms a zero at a frequency f_1 in the noise gain transfer function as shown in the Bode plot.

$$f_1 = \frac{1}{2\pi R2C1} \qquad \text{Eq. 4-6}$$

Note that we are neglecting the effects of the compensation capacitor C2 and are assuming that it is small relative to C1 and will not significantly affect the zero frequency f_1 when it is added to the circuit. In most cases, this approximation yields results that are close enough, considering the other variables in the circuit.

If left uncompensated, the phase shift at the frequency of intersection, f_2, will cause instability and oscillation. Introducing a pole at f_2 by adding the feedback capacitor C2 stabilizes the circuit and yields a phase margin of about 45 degrees.

$$f_2 = \frac{1}{2\pi R2C2} \qquad \text{Eq. 4-7}$$

Since f_2 is the geometric mean of f_1 and the unity-gain bandwidth frequency of the op amp, f_u,

$$f_2 = \sqrt{f_1 \cdot f_u} \qquad \text{Eq. 4-8}$$

These equations can be combined and solved for C2:

$$C2 = \sqrt{\frac{C1}{2\pi R2 \cdot f_u}} \qquad \text{Eq. 4-9}$$

This C2 value yields a phase margin of about 45 degrees; increasing it by a factor of 2 increases phase margin to about 65 degrees. In practice, an optimum C2 value should be determined experimentally, by varying it slightly to optimize the output pulse response.

Op Amp Selection for Wideband Photodiode I/V Converters

The op amp in the high speed photodiode I/V converter should be a wideband FET-input one in order to minimize the effects of input bias current and allow low values of photocurrents to be detected. In addition, if the equation for the 3 dB bandwidth, f_2, is rearranged in terms of f_u, R2, and C1, then

$$f_2 = \sqrt{\frac{f_u}{2\pi R2C1}} \qquad \text{Eq. 4-10}$$

where C1 is the sum of the diode capacitance, C_D, and the op amp input capacitance, C_{IN}. In a high speed application, the diode capacitance will be much smaller than that of the low frequency preamplifier design previously discussed—perhaps as low as a few pF.

By inspection of this equation, it is clear that in order to maximize f_2, the FET-input op amp should have both a high unity gain bandwidth product, f_u, and a low input capacitance, C_{IN}. In fact, the ratio of f_u to C_{IN} is a good figure-of-merit when evaluating different op amps for this application.

Figure 4-55 compares a number of FET-input op amps suitable for photodiode preamps. By inspection, the AD823 op amp has the highest ratio of unity gain bandwidth product to input capacitance, in addition to relatively low input bias current.

For these reasons, the AD823 op amp was chosen for the wideband design.

	Unity GBW Product f_u (MHz)	Input Capacitance C_{IN} (pF)	f_u/C_{IN} (MHz/pF)	Input Bias Current I_B (pA)	Voltage Noise @ 10kHz (nV/$\sqrt{Hz}$)
AD823	16	1.8	8.9	3	16
AD843	34	6	5.7	600	19
AD744	13	5.5	2.4	100	16
AD845	16	8	2	500	18
OP42	10	6	1.6	100	12
AD745*	20	20	1	250	2.9
AD795	1	1	1	3	8
AD820	1.9	2.8	0.7	10	13
AD743	4.5	20	0.2	250	2.9

*Stable for Noise Gains ≥5, Usually the Case, Since High Frequency Noise Gain = 1 + C1/C2, and C1 Usually ≥ 4C2

Figure 4-55: FET input op amps suitable for high speed photodiode preamps

High Speed Photodiode Preamp Design

The HP 5082-4204 PIN Photodiode will be used as an example for our discussion. Its characteristics are listed in Figure 4-56. It is typical of many PIN photodiodes.

- Sensitivity: 350μA @ 1mW, 900nm
- Maximum Linear Output Current: 100μA
- Area: 0.002cm^2 (0.2mm^2)
- Capacitance: 4pF @ 10V Reverse Bias
- Shunt Resistance: $10^{11}\Omega$
- Risetime: 10ns
- Dark Current: 600pA @ 10V Reverse Bias

Figure 4-56: HP 5082-4204 photodiode characteristics

As in most high speed photodiode applications, the diode will be operated in the reverse-biased or *photo-conductive* mode. This greatly lowers the diode junction capacitance, but causes a small amount of *dark current* to flow even when the diode is not illuminated (we will show a circuit that compensates for the dark current error later in the section). This photodiode is linear with illumination up to approximately 50 μA to 100 μA of output current.

The available dynamic range is limited by the total circuit noise, and the diode dark current (assuming no dark current compensation).

Using the circuit shown in Figure 4-57, assume that we wish to have a full scale output of 10 V for a diode current of 100 μA. This determines the value of the feedback resistor R2 to be 10 V/100 μA = 100 kΩ.

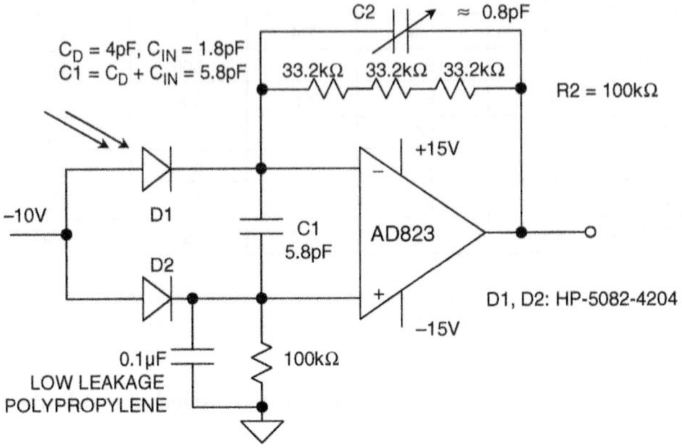

Figure 4-57: 2 MHz bandwidth photodiode preamp with dark current compensation

Using the diode capacitance, C_D = 4 pF, and the AD823 input capacitance, C_{IN} = 1.8 pF, the value of C1 = C_D + C_{IN} = 5.8 pF. Solving the above equations using C1=5.8pF, R2 = 100 kΩ, and f_u = 16 MHz, we find that:

$$f_1 = 274\text{ kHz}$$
$$C2 = 0.76\text{ pF}$$
$$f_2 = 2.1\text{ MHz.}$$

In the final circuit shown, notice at the 100 kΩ resistor is replaced with three 33.2 kΩ film resistors to minimize stray capacitance. The feedback capacitor, C2, is a variable 1.5 pF ceramic and is adjusted in the final circuit for best bandwidth/pulse response. The overall circuit bandwidth is approximately 2 MHz.

The full-scale output voltage of the preamp for 100 μA diode current is 10 V, and therefore the (uncompensated) error (RTO) due to the photodiode dark current of 600 pA is 60 μV.

This dark current error can be effectively canceled using a second photodiode, D2, of the same type. This diode is biased with a voltage identical to D1 and, with nominally matched characteristics, will tend to conduct a similar dark current. In the circuit, this "dummy" dark current drives the 100 kΩ resistance in the noninverting input of the op amp. This produces a dark current proportional bias voltage which, due to the CM rejection of the op amp, has an end result of suppressing the dark current effects.

High Speed Photodiode Preamp Noise Analysis

As in most noise analyses, only the key contributors need be identified. Because the noise sources combine in an RSS manner, any single noise source that is at least three or four times as large as any of the others will dominate.

In the case of the wideband photodiode preamp, the dominant sources of output noise are the input voltage noise of the op amp, V_N, and the resistor noise due to R2, $V_{N,R2}$ (see Figure 4-58). The input current noise of the FET-input op amp is negligible. The shot noise of the photodiode (caused by the reverse bias) is negligible because of the filtering effect of the shunt capacitance C1.

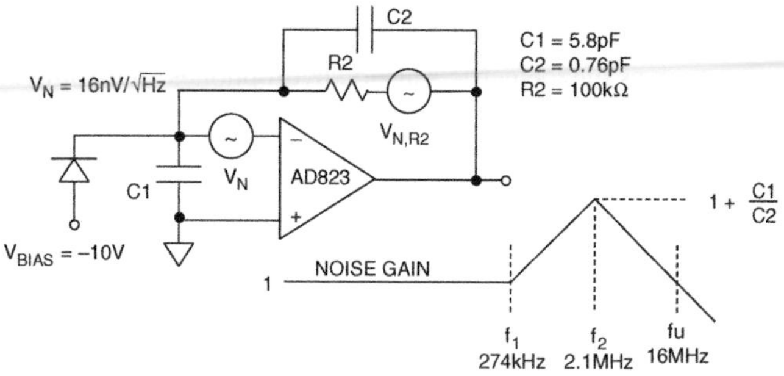

- V_N RTO NOISE $\approx V_N\left[1+\frac{C1}{C2}\right]\sqrt{1.57\,f_2}$ = 250µV RMS
- $V_{N,R2}$ RTO NOISE $\approx \sqrt{4kTR2 \bullet 1.57f_2}$ = 73µV RMS
- TOTAL RTO NOISE $=\sqrt{250^2+73^2}$ = 260µV RMS
- DYNAMIC RANGE $= 20\log\left[\frac{10V}{260\mu V}\right]$ = 92dB

Figure 4-58: Equivalent circuit of preamp for output noise analysis

The resistor noise is easily calculated by knowing that a 1 kΩ resistor generates about 4 $nV/\sqrt{z}$, therefore, a 100 kΩ resistor generates 40 $nV/\sqrt{z}$. The bandwidth for integration is the signal bandwidth, 2.1 MHz, yielding a total output rms noise of:

$$V_{N,R2}\text{RTO NOISE} = 40\sqrt{1.57\cdot 2.1\cdot 10^6} = 73\ \mu\text{V rms}$$

The factor of 1.57 converts the amplifier approximate single-pole bandwidth of 2.1 MHz into the *equivalent noise bandwidth*.

The output noise due to the input voltage noise is obtained by multiplying the noise gain by the voltage noise and integrating the entire function over frequency. This would be tedious if done rigorously, but a few reasonable approximations can be made which greatly simplify the math. Obviously, the low frequency 1/f noise can be neglected in the case of the wideband circuit. The primary source of output noise is due to the high frequency noise-gain peaking that occurs between f_1 and f_u. If we simply assume that the output noise is constant over the entire range of frequencies and use the maximum value for ac noise gain [1 + (C1/C2)], then

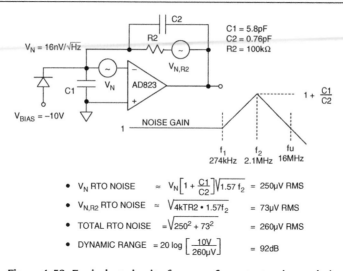

$$V_N\ \text{RTO NOISE} \approx V_N\left(1+\frac{C1}{C2}\right)\sqrt{1.57f_2} = 250\ \mu\text{V rms}$$

The total rms noise referred to the output is then the RSS value of the two components:

$$\text{TOTAL RTO NOISE} = \sqrt{(73)^2+(250)^2} = 260\ \mu\text{V rms}$$

The total output dynamic range can be calculated by dividing the 10 V full-scale output by the total 260 µ Vrms noise, and, converting to dB, yielding approximately 92 dB.

High Impedance Charge Output Sensors

High impedance transducers such as piezoelectric sensors, hydrophones, and some accelerometers require an amplifier that converts a transfer of charge into a voltage change. Due to the high dc output impedance of these devices, appropriate buffer amplifiers are required. The basic circuit for an inverting charge sensitive amplifier is shown in Figure 4-59.

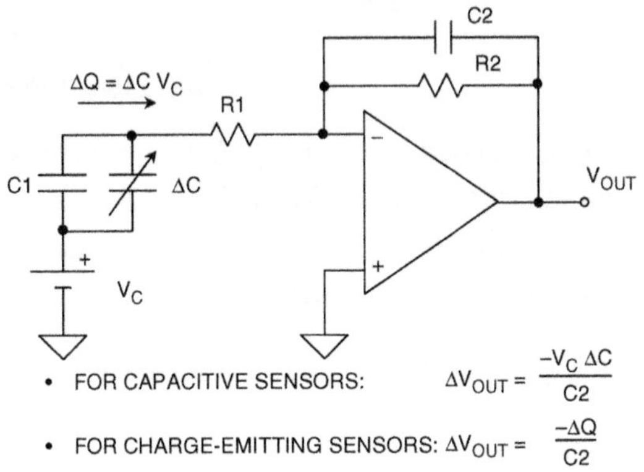

Figure 4-59: Charge amplifier basic principles

There are basically two types of charge transducers: capacitive and charge-emitting. In a capacitive transducer, the voltage across the capacitor (V_C) is held constant. The change in capacitance, ΔC, produces a change in charge, $\Delta Q = \Delta C V_C$. This charge is transferred to the op amp output as a voltage, $\Delta V_{OUT} = -\Delta Q/C2 = -\Delta C V_C/C2$.

Charge-emitting transducers produce an output charge, ΔQ, and their output capacitance remains constant. This charge would normally produce an open-circuit output voltage at the transducer output equal to ΔQ/C. However, since the voltage across the transducer is held constant by the virtual ground of the op amp (R1 is usually small), the charge is transferred to capacitor C_2 producing an output voltage $\Delta V_{OUT} = -\Delta Q/C2$. In an actual application, this charge amplifier only responds to ac inputs.

It should be noted that ac gain for this charge amplifier is determined by the *capacitance ratio*, not the resistances. This is unlike a conventional wideband amplifier, where the upper cutoff frequency is given by $f_2 = 1/2\ \pi R2C2$, and the lower by $f_1 = 1/2\ \pi R1C1$.

In the Figure 4-59 charge amplifier, with nominal transducer capacitance fixed, gain is set by C2. Typically, bias return resistor R2 will be a high value (≥1 megΩ), and R1 a much lower value. The resulting frequency response will be relatively narrow and bandpass shaped, with gain and frequency manipulated by the relative values, as suitable to SPICE analysis.

Low Noise Charge Amplifier Circuit Configurations

Figure 4-60 shows two ways to buffer and amplify the output of a charge output transducer. Both require using an amplifier which has a very high input impedance, such as the AD745. The AD745 provides both low voltage noise and low current noise. This combination makes this device particularly suitable in applications requiring very high charge sensitivity, such as capacitive accelerometers and hydrophones.

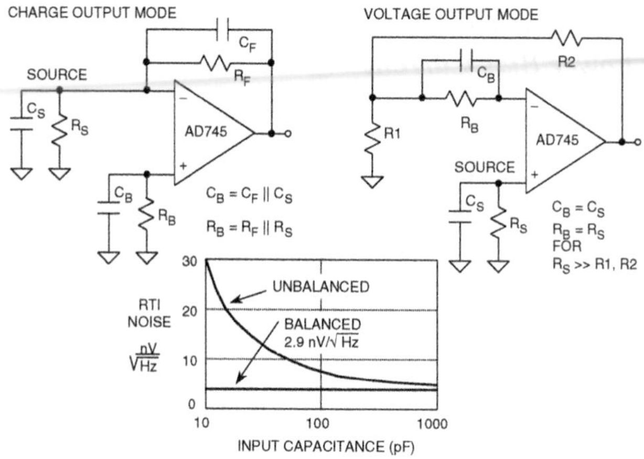

Figure 4-60: Two basic charge amplifier configurations using the AD745 low noise FET op amp

The first (left) circuit in Figure 4-60 uses the op amp in the inverting mode. Amplification depends on the principle of conservation of charge at the inverting input of the amplifier. The charge on capacitor C_S is transferred to capacitor C_F, thus yielding an output voltage of $\Delta Q/C_F$. The amplifier's input voltage noise will appear at the output amplified by the ac noise gain of the circuit, $1 + C_S/C_F$.

The second (right) circuit shown in Figure 4-60 is simply a high impedance follower with gain. Here the noise gain (1 + R2/R1) is the same as the gain from the transducer to the output. Resistor R_B, in both circuits, is required as a dc bias current return.

To maximize dc performance over temperature, source resistances should be balanced at the amplifier inputs, as represented by the resistor R_B shown in Figure 4-60. For best noise performance, the source capacitance should also be balanced with the capacitor C_B.

In general, it is good practice to balance the source impedances (both resistive and reactive) as seen by the inputs of precision low noise BiFET amplifiers such as the AD743/AD745. Balancing the resistive components will optimize dc performance over temperature, as balancing mitigates the effects of any bias current errors. Balancing the input capacitance will minimize ac response errors due to the amplifier's nonlinear common-mode input capacitance and, as shown in Figure 4-60, noise performance will be optimized. In any FET input amplifier, the current noise of the internal bias circuitry can be coupled to the inputs via the gate-to-source capacitances (20 pF for the AD743 and AD745) and appears as excess input voltage noise. This noise component is correlated at the inputs, so source impedance matching will tend to cancel out its effect. Figure 4-60 shows the required external components for both inverting and noninverting configurations. For values of C_B greater than 300 pF, there is a diminishing impact on noise, and C_B can then be simply a polyester bypass capacitor of 0.01 μF or greater.

40dB Gain Piezoelectric Transducer Amplifier Operates on Reduced Supply Voltages for Lower Bias Current

Figure 4-61 shows a gain-of-100 piezoelectric transducer amplifier connected in the voltage-output mode. Reducing the op amp power supplies to ±5 V reduces the effects of bias current in two ways: first, by lowering the total power dissipation and, second, by reducing the basic gate-to-junction leakage current. The addition of a clip-on heat sink such as the Aavid #5801 will further limit the internal junction temperature rise.

With ac coupling capacitor C1 shorted, the amplifier will operate over a range of 0°C to 85°C. If the optional ac coupling capacitor C1 is used, the circuit will operate over the entire –55°C to +125°C temperature range, but dc information is lost.

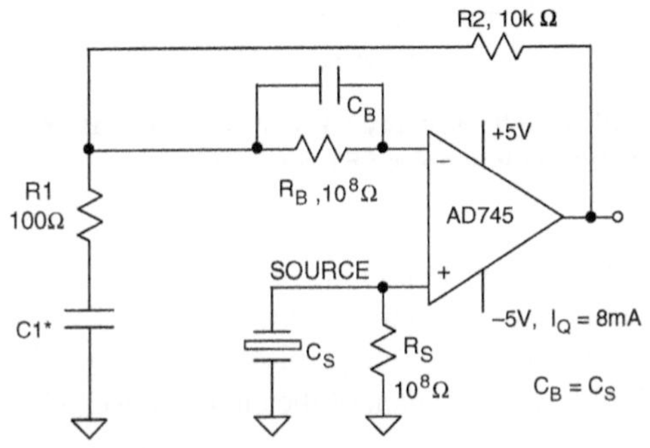

- ±5V Power Supplies Reduce I_B for 0°C to +85°C Operation, P_D = 80mW
- C1 Allows –55°C to +125°C Operation

Figure 4-61: A gain-of-100 piezoelectric transducer amplifier

Hydrophones

Interfacing the outputs of highly capacitive transducers such as hydrophones, some accelerometers, and condenser microphones to the outside world presents many design challenges. Previously designers had to use costly hybrid amplifiers consisting of discrete low-noise JFETs in front of conventional op amps to achieve the low levels of voltage and current noise required by these applications. Using AD743/AD745 monolithic ICs, designers can achieve almost the same level of performance of a hybrid approach.

In sonar applications, a piezo-ceramic cylinder is commonly used as the active element in the hydrophone. A typical cylinder has a nominal capacitance of around 6,000 pF with a series resistance of 10 Ω. The output impedance is typically 10^8 Ω or 100 MΩ.

Since the hydrophone signals of interest are inherently ac with wide dynamic range, noise is an overriding concern for sonar system designs. The noise floor of the hydrophone and its preamplifier together limit the system sensitivity, and thus the overall hydrophone usefulness. Typical hydrophone bandwidths are in the 1 kHz to 10 kHz range. The AD743 and AD745 op amps, with their low noise voltages of $2.9\text{nV}/\sqrt{\text{Hz}}$ and high input impedance of 10^{10} Ω (or 10 GΩ) are ideal for use as hydrophone amplifiers.

Op Amp Performance: JFET versus Bipolar

The AD743 and AD745 op amps are the first monolithic JFET devices to offer the low input voltage noise comparable to a bipolar op amp, *without* the high input bias currents typically associated with bipolar op amps. Figure 4-62 shows input voltage noise versus input source resistance of the OP27 (bipolar-input) and the JFET-input AD745 op amps. Note: the noise levels of the AD743 and the AD745 are identical.

From this figure (left), it is clear that at high source impedances, the low current noise of the AD745 provides lower overall noise than a high performance bipolar op amp such as the OP27. It is also important to note that, with the AD745, this noise reduction extends down to low source impedances. At high source impedances, the lower dc current errors of the AD745 also reduce errors due to offset and drift, as shown in Figure 4-62 (right).

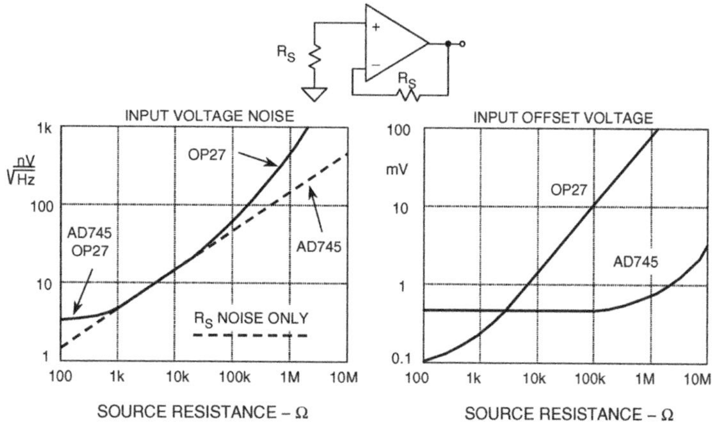

Figure 4-62: Total noise performance comparison, OP27 (bipolar) and AD745 (FET) op amps

The AD743 and AD745 are related, companion amplifiers, which differ in their levels of internal compensation. The AD743 is internally compensated for unity gain stability. The AD745, stable for noise gains of 5 or higher, has a much higher bandwidth and slew rate. This makes the latter device especially useful as a high-gain preamplifier where it provides both high gain and wide bandwidth. The AD743 and AD745 also operate with very low levels of distortion; less than 0.0003% and 0.0002% (at 1 kHz), respectively.

A pH Probe Buffer Amplifier

A typical pH probe requires a buffer amplifier to isolate its 10^6 to 10^9 Ω source resistance from external circuitry. Such an amplifier is shown in Figure 4-63. The low input current of the AD795JR minimizes the voltage error produced by the bias current and electrode resistance. The use of guarding, shielding, high insulation resistance standoffs, and other such standard picoamp methods previously discussed should be used to minimize leakage. Are all needed to maintain the accuracy of this circuit.

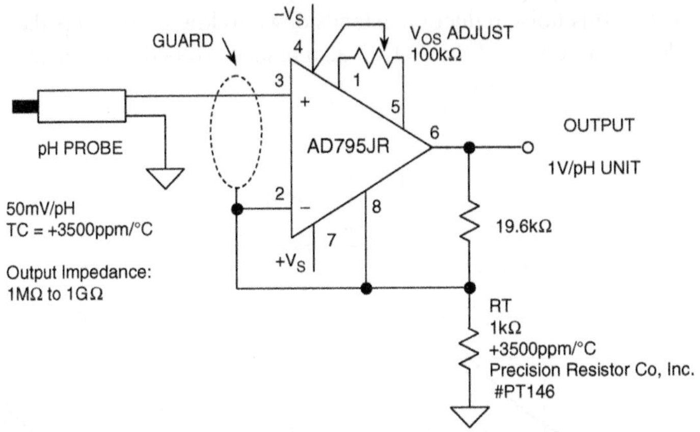

Figure 4-63: A gain of 20x pH probe amplifier allows 1 V/pH output scaling

The slope of the pH probe transfer function, 50 mV per pH unit at room temperature, has an approximate +3500 ppm/°C temperature coefficient. The buffer amplifier shown as Figure 4-63 provides a gain of 20, and yields a final output voltage equal to 1 V/pH unit. Temperature compensation is provided by resistor RT, which is a special temperature compensation resistor, 1 kΩ, 1%, +3500 ppm/°C, #PT146 available from Precision Resistor Co., Inc. (see Reference 15).

References: High Impedance Sensors

1. Ramon Pallas-Areny and John G. Webster, **Sensors and Signal Conditioning**, John Wiley, New York, 1991.
2. Dan Sheingold, Editor, **Transducer Interfacing Handbook**, Analog Devices, Inc., 1980, ISBN: 0-916550-05-2.
3. Section 3, Walt Kester, Editor, **1992 Amplifier Applications Guide**, Analog Devices, 1992, ISBN: 0-916550-10-9.
4. Walt Kester, Editor, **System Applications Guide**, Analog Devices, 1993, ISBN: 0-916550-13-3.
5. Walt Kester, Editor, **Linear Design Seminar**, Analog Devices, 1995, ISBN: 0-916550-15-X.
6. Walt Kester, Editor, **Practical Analog Design Techniques**, Analog Devices, 1995, ISBN: 0-916550-16-8.
7. Walt Kester, Editor, **High Speed Design Techniques**, Analog Devices, 1996, ISBN: 0-916550-17-6.
8. Jiri Dostal, Section 9.3.1, "Thermoelectric Voltages," **Operational Amplifiers 2nd Ed.**, Butterworth-Heinemann, 1993, pp. 265–268.
9. Optoelectronics Data Book, EG&G Vactec, St. Louis, MO, 1990.
10. Silicon Detector Corporation, Camarillo, CA, Part Number SD-020-12-001 Data Sheet.
11. Photodiode 1991 Catalog, Hamamatsu Photonics, Bridgewater, NJ.
12. Lewis Smith and Dan Sheingold, "Noise and Operational Amplifier Circuits," **Analog Dialogue 25th Anniversary Issue**, pp. 19–31, Analog Devices Inc., 1991.
13. James L. Melsa and Donald G. Schultz, **Linear Control Systems**, pp. 196–220, McGraw-Hill, 1969.
14. Jerald G. Graeme, **Photodiode Amplifiers: Op Amp Solutions**, McGraw-Hill, 1995.
15. Precision Resistor Co., Inc., 10601 75th St. N., Largo, FL, 33777-1427, 727 541-5771, www.precisionresistor.com.
16. **Ohmite Victoreen MAXI-MOX Resistors**, 3601 Howard Street, Skokie, IL 60076, 847 675-2600, www.ohmite.com/victoreen/.
17. **Vishay/Dale RNX Resistors**, 2300 Riverside Blvd., Norfolk, NE, 68701-2242, 402 371-0800, www.vishay.com.

Temperature Sensors

Walt Kester, James Bryant, Walt Jung

Temperature measurement is critical in many modern electronic devices, especially expensive laptop computers and other portable devices; their densely packed circuitry dissipates considerable power in the form of heat. Knowledge of system temperature can also be used to control battery charging, as well as to prevent damage to expensive microprocessors.

Compact high power portable equipment often has fan cooling to maintain junction temperatures at proper levels. In order to conserve battery life, the fan should only operate when necessary. Accurate control of the fan requires, in turn, knowledge of critical temperatures from an appropriate temperature sensor.

Accurate temperature measurements are required in many other measurement systems, for example within process control and instrumentation applications. Some popular types of temperature transducers and their characteristics are indicated in Figure 4-64. In most cases, because of low level and/or nonlinear outputs, the sensor output must be properly conditioned and amplified before further processing can occur.

THERMOCOUPLE	RTD	THERMISTOR	SEMICONDUCTOR
Widest Range: −184°C to +2300°C	Range: −200°C to +850°C	Range: 0°C to +100°C	Range: −55°C to +150°C
High Accuracy and Repeatability	Fair Linearity	Poor Linearity	Linearity: 1°C Accuracy: 1°C
Needs Cold Junction Compensation	Requires Excitation	Requires Excitation	Requires Excitation
Low-Voltage Output	Low Cost	High Sensitivity	10mV/K, 20mV/K, or 1μA/K Typical Output

Figure 4-64: Some common types of temperature transducers

Except for the semiconductor sensors of the last column, all of the temperature sensors shown have nonlinear transfer functions. In the past, complex analog conditioning circuits were designed to correct for the sensor nonlinearity. These circuits often required manual calibration and precision resistors to achieve the desired accuracy. Today, however, sensor outputs may be digitized directly by high resolution ADCs. Linearization and calibration can then be performed digitally, substantially reducing cost and complexity.

Resistance Temperature Devices (RTDs) are accurate, but require excitation current and are generally used within bridge circuits such as those described earlier. *Thermistors* have the most sensitivity, but are also the most nonlinear. Nevertheless, they are popular in portable applications for measurement of battery and other critical system temperatures.

Modern *semiconductor temperature sensors* offer both high accuracy and linearity over about a –55°C to +150°C operating range. Internal amplifiers can scale output to convenient values, such as 10 mV/°C. They are also useful in cold-junction compensation circuits for wide temperature range thermocouples. Semiconductor temperature sensors are also integrated into ICs that perform other hardware monitoring functions—an example is the imbedded transistor sensors used within modern µP chips.

Thermocouple Principles and Cold-Junction Compensation

Thermocouples comprise a more common form of temperature measurement. In operation, thermocouples rely on the fact that two dissimilar metals joined together produce a voltage output roughly proportional to temperature. They are small, rugged, relatively inexpensive, and operate over the widest range of contact temperature sensors.

Thermocouples are especially useful for making measurements at extremely high temperatures (up to +2300°C), and within hostile environments. Characteristics of some common types are shown in Figure 4-65.

JUNCTION MATERIALS	TYPICAL USEFUL RANGE (°C)	NOMINAL SENSITIVITY (µV/°C)	ANSI DESIGNATION
Platinum (6%)/ Rhodium-Platinum (30%)/Rhodium	38 to 1800	7.7	B
Tungsten (5%)/Rhenium-Tungsten (26%)/Rhenium	0 to 2300	16	C
Chromel-Constantan	0 to 982	76	E
Iron-Constantan	0 to 760	55	E
Chromel-Alumel	–184 to 1260	39	K
Platinum (13%)/Rhodium-Platinum	0 to 1593	11.7	R
Platinum (10%)/Rhodium-Platinum	0 to 1538	10.4	S
Copper-Constantan	–184 to 400	45	T

Figure 4-65: Some common thermocouples and their characteristics

However, thermocouples produce only millivolts of output, and thus they typically require precision amplification for further processing. They also require *cold-junction-compensation* (CJC) techniques, to be discussed shortly. They are more linear than many other sensors, and their nonlinearity has been well characterized.

The most common metals used for thermocouples are iron, platinum, rhodium, rhenium, tungsten, copper, alumel (composed of nickel and aluminum), chromel (composed of nickel and chromium) and constantan (composed of copper and nickel).

Figure 4-66 shows the voltage-temperature curves for three commonly used thermocouples, Types J, K, and S, referred to a 0°C fixed-temperature reference junction. Of these, Type J thermocouples are the most sensitive, producing the highest output voltage for a given temperature change, but over a relatively narrow temperature span. On the other hand, Type S thermocouples are the least sensitive, but can operate over a much wider range.

Figure 4-66: Type J, K, and S thermocouple output voltage versus temperature

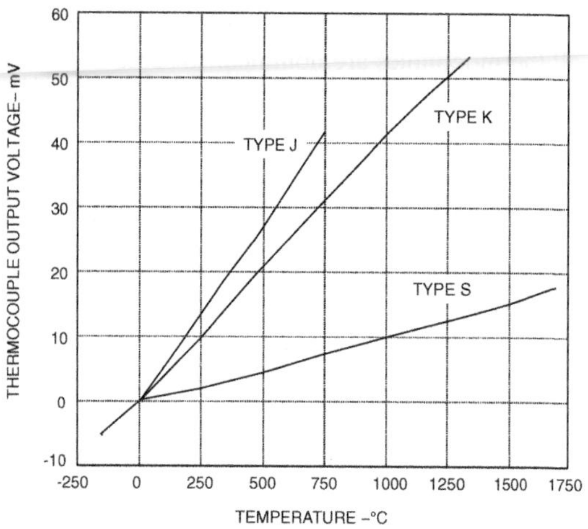

These characteristics are very important considerations when designing thermocouple signal conditioning circuitry. A main consideration is the fact that virtually any thermocouple employed will have a relatively low output signal, thus they generally require the careful application of stable, high gain amplifiers.

To understand thermocouple behavior more fully, it is also necessary to consider the nonlinearities in their response to temperature differences. As noted, Figure 4-66 shows the relationships between sensing junction temperature and voltage output for a number of thermocouple types (in all cases, the reference *cold* junction is maintained at 0°C).

While close scrutiny of these data may reveal the fact that none of the responses are quite linear, the exact nature of the nonlinearity isn't so obvious. What is needed is another perspective on the relationships displayed by these curves, to gain better insight into how the various devices can be best utilized.

Figure 4-67 shows how the thermocouple *Seebeck* coefficient varies with sensor junction temperature. The Seebeck coefficient is the *change* of output voltage with *change* of sensor junction temperature (i.e., the first derivative of output with respect to temperature). Note that we are still considering the case where the reference junction is maintained at 0°C.

Figure 4-67: Type J, K and S thermocouple Seebeck coefficient versus temperature

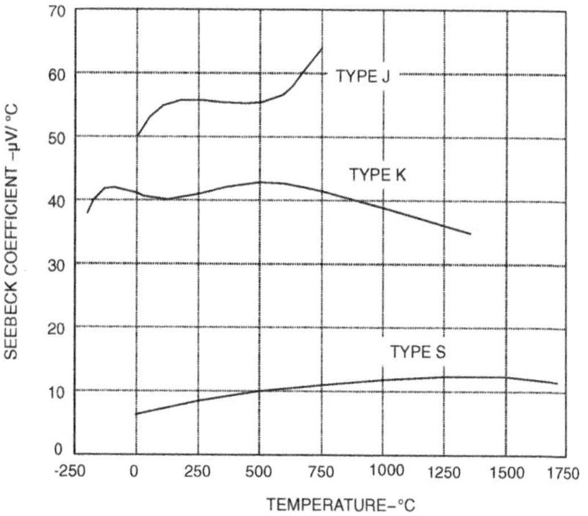

An ideal linear thermocouple would have a constant Seebeck coefficient with varying temperature, but in practice all thermocouples are nonlinear to some degree. In selecting a measurement thermocouple for a particular temperature range, we should therefore choose one whose Seebeck coefficient varies as little as possible over that range.

For example, a Type J thermocouple has a nominal Seebeck coefficient of 55 μV/°C, which varies by less than 1 μV/°C between 200°C and 500°C, making it ideal for measurements over this range (the flat region of the upper curve within Figure 4-67).

Presenting these data on thermocouples serves two purposes. First, Figure 4-66 illustrates the range and sensitivity of the three thermocouple types so that the system designer can, at a glance, determine that a Type S thermocouple has the widest useful temperature range, but a Type J thermocouple is the more sensitive of the three.

Second, the relative stability of the Seebeck coefficient over temperature provides a quick guide to a thermocouple's linearity. Using Figure 4-67, a system designer can choose a Type K thermocouple for its relatively linear Seebeck coefficient over the range of 400°C to 800°C, or a Type S over the range of 900°C to 1700°C. The behavior of a thermocouple's Seebeck coefficient is important in applications where variations of temperature rather than absolute magnitude are important. These data also indicate what performance is required of the associated signal-conditioning circuitry.

To successfully apply thermocouples we must also understand their basic operating principles. Consider the diagrams shown in Figure 4-68.

If we join two dissimilar metals, A and B, at any temperature above absolute zero, there will be a potential difference between them, i.e., their "thermoelectric EMF" or "contact potential," V1. This voltage is a function of the temperature of the measurement junction, T1, as is noted in Figure 4-68A. If we join two wires of metal A with metal B at two places, two measurement junctions are formed, T1 and T2 (Figure 4-68B). If the two junctions are at different temperatures, there will be a net EMF in the circuit, and a current I will flow, as determined by the EMF V1 – V2, and the total resistance R in the circuit (Figure 4-68B).

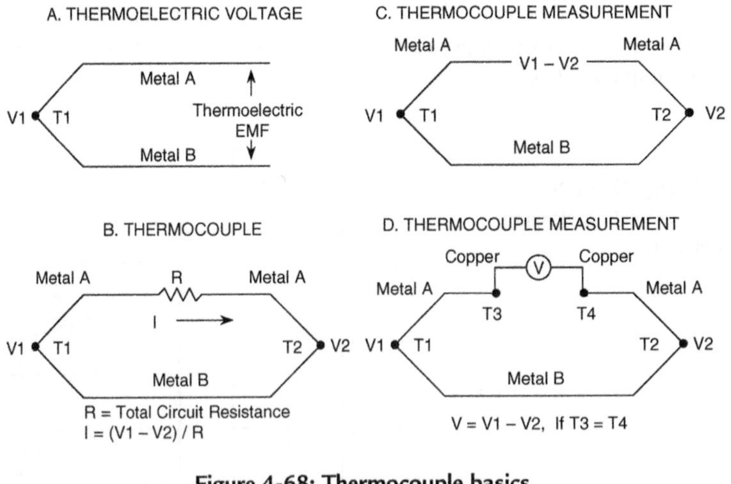

Figure 4-68: Thermocouple basics

If we open the circuit as in Figure 4-68C, the voltage across the break will be equal to the net thermoelectric EMF of the circuit; and if we measure this voltage, we can use it to calculate the temperature difference between the two junctions. We must remember that a thermocouple always measures the temperature *difference* between two junctions, not the absolute temperature at one junction. We can only measure the temperature at the measuring junction if we know the temperature of the other junction. This is the origin of the terms "reference" or "cold" junction.

But of course, it is not so easy to just measure the voltage generated by a thermocouple. *Any wire attached to a thermocouple is also a thermocouple itself*, and if care is not taken, errors can be introduced. Suppose that we attach a voltmeter to the circuit of Figure 4-68C, as shown in Figure 4-68D. The wires of the voltmeter will form further thermocouple junctions where they are attached (T3, T4). If both these additional junctions are at the same temperature (it does not matter exactly what temperature), then the "Law of Intermediate Metals" states that they will make no net contribution to the total EMF of the system. If they are at different temperatures, they will introduce errors.

Since *every pair of dissimilar metals in contact generates a thermoelectric EMF* (including copper/solder, kovar/copper [kovar is the alloy used for IC leadframes] and aluminum/kovar [at the bond inside the IC]), it is obvious that in practical circuits the problem is even more complex. In fact, it is necessary to take extreme care to ensure that both junctions of each junction pair in series with a thermocouple are at the same temperature, *except for the measurement and reference junctions themselves.*

Thermocouples generate a voltage, albeit a very small one, and don't require excitation for this most basic operation. As was shown in Figure 4-68D, however, two junctions are always involved (T1, the measurement junction temperature and T2, the reference junction temperature). If T2 = T1, then V2 = V1, and the output voltage V = 0. Thermocouple output voltages are often defined with respect to a reference junction temperature of 0ºC (hence the term *cold* or *ice point* junction). In such a system the thermocouple provides a convenient output voltage of 0 V at 0ºC. Obviously, to maintain system accuracy, the reference junction must remain at a well-defined temperature (but not necessarily 0ºC).

A conceptually simple approach to this need is shown in Figure 4-69, the ice-point reference, where junction T2 is kept at 0°C by virtue of being immersed in an ice water slurry. Although this ice water bath is relatively easy to define conceptually, it is quite inconvenient to maintain.

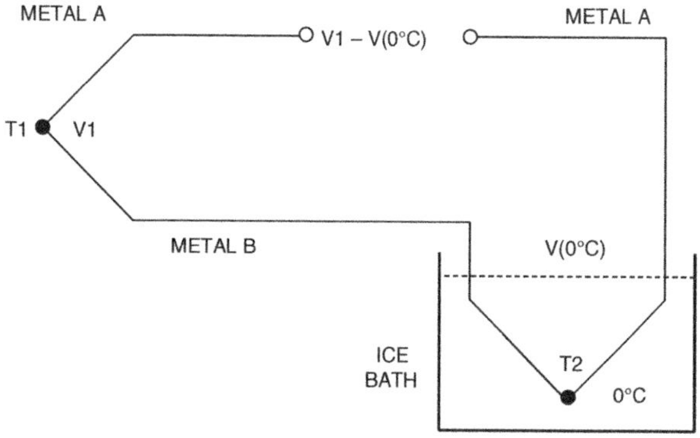

Figure 4-69: A thermocouple cold junction reference system using an ice-point (0°C) T2 reference

Today the inconvenient ice/water bath reference is replaced by an electronic equivalent. A temperature sensor of another sort (often a semiconductor sensor, sometimes a thermistor) measures the cold junction temperature and is used to inject a voltage into the thermocouple circuit which compensates for the difference between the actual cold junction temperature and its ideal value (usually 0°C) as shown in Figure 4-70.

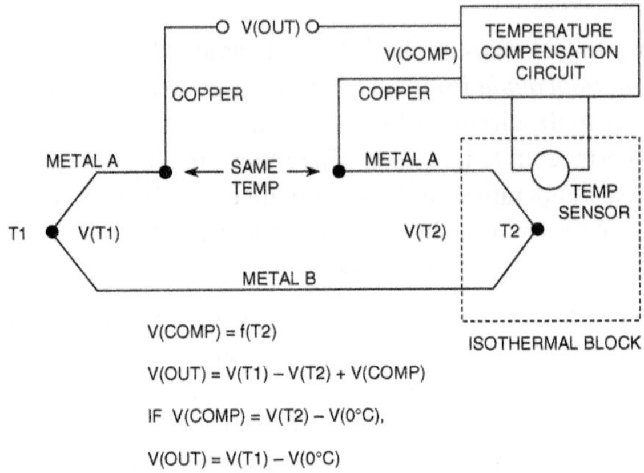

Figure 4-70: A semiconductor temperature sensor can be used to provide cold junction compensation

Ideally, the compensation voltage should be an exact match for the difference voltage required, which is why the diagram gives the voltage V(COMP) as f(T2) (a *function* of T2) rather than KT2, where K is a simple constant. In practice, since the cold junction is rarely more than a few tens of degrees from 0°C, and generally varies by little more than ±10°C, a linear approximation (V = KT2) to the more complex reality is usually sufficiently accurate and is often used. (Note—the expression for the output voltage of a thermocouple with its measuring junction at T°C and its reference at 0°C is a polynomial of the form $V = K_1T + K_2T^2 + K_3T^3 + ...$, but the values of the coefficients K_2, K_3, and so forth. are very small for most common types of thermocouple). References 7 and 8 give the values of these coefficients for a wide range of thermocouples.

When electronic cold-junction compensation is used, it is common practice to eliminate the additional thermocouple wire, and terminate the thermocouple leads in the isothermal block, as shown in the arrangement of Figure 4-71.

Figure 4-71: Termination of thermocouple leads directly to an isothermal block

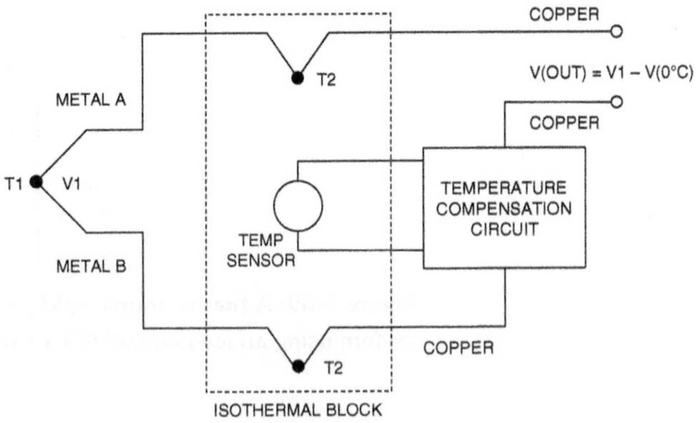

In Figure 4-71, the Metal A-Copper and the Metal B-Copper junctions, if at temperature T2, are equivalent to the Metal A-Metal B thermocouple junction in Figure 4-70.

Type K Thermocouple Amplifier and Cold Junction Compensator

The circuit in Figure 4-72 conditions the output of a Type K thermocouple, while providing cold-junction compensation, operating over temperatures between 0°C and 250°C. The circuit operates from single +3.3 V to +5.5 V supplies with the AD8551, and has been designed to produce a basic output voltage transfer characteristic of 10 mV/°C. A Type K thermocouple exhibits a Seebeck coefficient of approximately 40 μV/°C; therefore, at the cold junction, the TMP35 voltage output sensor with a temperature coefficient of 10 mV/°C is used with divider R1 and R2, to introduce an opposing cold-junction temperature coefficient of –40 μV/°C. This prevents the isothermal, cold-junction connection between the circuit's printed circuit board traces and the thermocouple's wires from introducing an error in the measured temperature. This compensation works extremely well for conditioning circuit ambient temperatures of 20°C to 50°C.

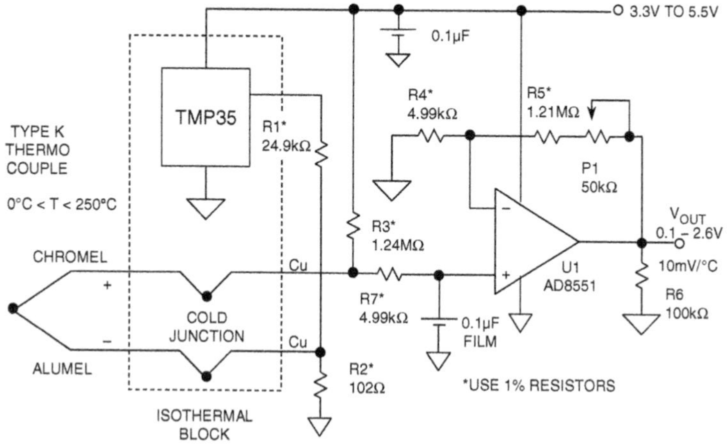

Figure 4-72: Using a TMP35 temperature sensor for cold junction compensation within a Type K thermocouple amplifier-conditioner

Over a 250°C measurement temperature range, the thermocouple produces an output voltage change of ~10 mV. Since the circuit's required full-scale output voltage change is 2.5 V, the required gain is ~250. Choosing R4 equal to 4.99 kΩ sets R5 ~1.24 MΩ. With a fixed 1% value for R5 of 1.21 MΩ, a 50 kΩ potentiometer is used with R5 for fine trim of the full-scale output voltage. The U1 amplifier should be a low drift, very high gain type. A chopper-stabilized AD8551 or an OP777 precision bipolar op amp is suitable for U1.

Both the AD8551 and the OP777 have rail-rail output stages. To extend low range linearity, bias resistor R3 is added to the circuit, supplying an output offset voltage of about 0.1 V (for a nominal supply voltage of 5 V). Note that this 10°C offset must be subtracted, when making final measurements referenced to the U1 output. Note also that R3 provides a useful open thermocouple detection function, forcing the U1 output to greater than 3 V should the thermocouple open. Resistor R7 balances the dc input impedance at the U1 (+) input, and the 0.1 μF film capacitor reduces noise coupling.

Single-Chip Thermocouple Signal Conditioners

While the construction of thermocouple signal conditioners with op amps and other discrete circuit elements offers great flexibility, this does come at the expense of component count. To achieve a greater level of integration in the thermocouple conditioning function, dedicated thermocouple signal conditioners can be used.

One such solution lies with the AD594 and AD595 (see Reference 9), which are complete, single-chip instrumentation amplifiers and thermocouple cold junction compensators, as shown in Figure 4-73. Suitable for use with either Type J (AD594) or Type K (AD595) thermocouples, these two devices combine an ice point reference with a precalibrated scaling amplifier. They provide a high level (10 mV/°C) output working directly from a thermocouple input, without additional precision parts. Pin-strapping options allow the devices to be used as a linear amplifier-compensator, or as a switched output set-point controller with fixed or remote setpoint control.

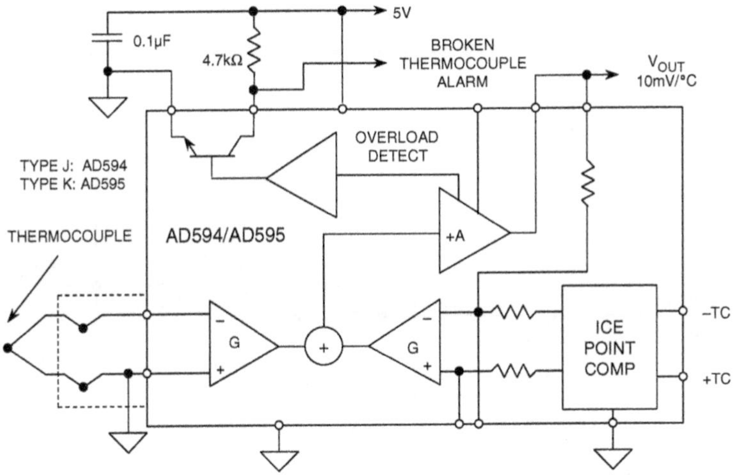

Figure 4-73: Functional diagram of AD594 and AD595 thermocouple signal-conditioning amplifiers

The AD594 and AD595 can be used to amplify the cold-junction compensation voltage directly, thereby becoming a standalone, 10 mV/°C output Celsius transducer. In such applications it is very important that the IC chip be at the same temperature as the cold junction of the thermocouple; this is usually achieved by keeping the two in close proximity and isolated from any heat sources.

The AD594/AD595 structure includes a flexible thermocouple failure alarm output, which provides broken thermocouple indication. The devices can be powered from either dual or single power supplies (as low as 5 V), but the use of a negative supply also allows temperatures below 0°C to be measured. To minimize self-heating, an unloaded AD594/AD595 operates with 160 μA of supply current, and can deliver ±5 mA to a load.

Although the AD594 is precalibrated by laser wafer trimming to match the characteristics of Type J thermocouples, and the AD595 for Type K, the temperature transducer voltages and gain control resistors are also made available at the package pins. So, if desired, the circuit can be recalibrated for other thermocouple types with the addition of external resistors. These terminals also allow more precise calibration for both thermocouple and thermometer applications. The AD594/AD595 are available in C and A performance grades, with calibration accuracies of ±1°C and ±3°C, respectively. Both are designed to be used with cold junctions between 0 to 50°C.

The 5 V powered, single-supply circuit shown of Figure 4-73 provides a scaled 10 mV/°C output capable of measuring a range of 0°C to 300°C. This can be from either a Type J thermocouple using the AD594, or a Type K with the AD595.

Resistance Temperature Detectors

The Resistance Temperature Detector (RTD), is a sensor whose resistance changes with temperature. Typically built of a platinum (Pt) wire wrapped around a ceramic bobbin, the RTD exhibits behavior which is more accurate and more linear than a thermocouple over wide temperature ranges.

Figure 4-74 illustrates the temperature coefficient of a 100 Ω RTD, and the Seebeck coefficient of a Type S thermocouple. Over the entire range (approximately –200°C to +850°C), the RTD is a more linear device. Hence, linearizing an RTD is less complex.

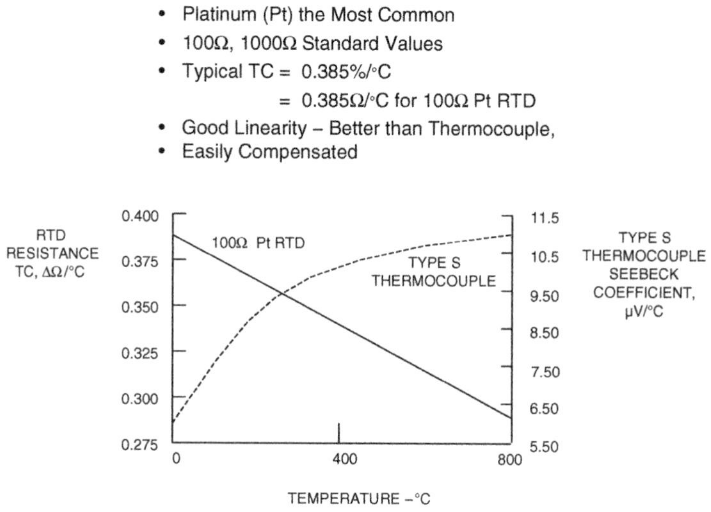

Figure 4-74: Resistance temperature detectors

Unlike a thermocouple, however, an RTD is a passive sensor, and a current excitation is required to produce an output voltage. The RTD's low temperature coefficient of 0.385%/°C requires high performance signal-conditioning circuitry similar to that used by a thermocouple. However, the typical voltage drop seen across an RTD is much larger than a thermocouple's output voltage. A system designer may opt for large value RTDs with higher output, but large-valued RTDs exhibit slow response times. Furthermore, although the cost of RTDs is higher than that of thermocouples, they use copper leads, and thermoelectric effects from terminating junctions do not affect their accuracy. And finally, because their resistance is a function of the absolute temperature, RTDs do not require cold-junction compensation.

Caution must be exercised with the level of current excitation applied to an RTD, because excessive current can cause self-heating. Any self-heating changes the RTD temperature, and therefore results in a measurement error. Hence, careful attention must be paid to the design of the signal-conditioning circuitry so that self-heating errors are kept below 0.5°C. Manufacturers specify self-heating errors for various RTD values and sizes, in both still and moving air. To reduce the error due to self-heating, the minimum current should be used to achieve the required system resolution, and the largest RTD value chosen that results in acceptable response time.

Another effect that can produce measurement error is voltage drop in RTD lead wires. This is especially critical with low value, 2-wire RTDs, because both the temperature coefficient and absolute value of the RTD resistance are small. If the RTD is located a long distance from the signal-conditioning circuitry, the connecting lead resistance can be ohms or tens of ohms. Even this small amount of lead resistance can contribute a significant error to the temperature measurement, as shown in Figure 4-75.

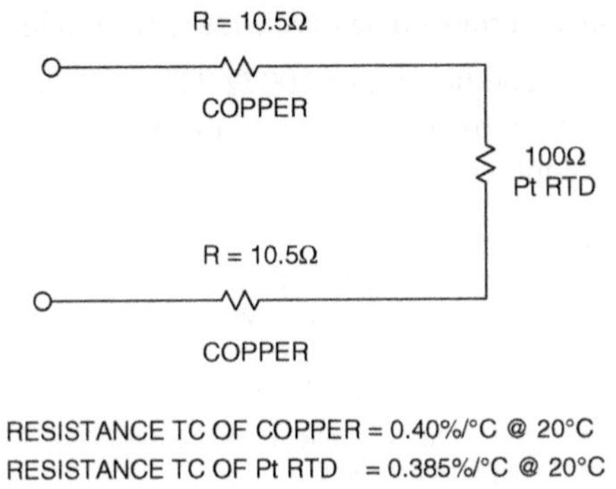

Figure 4-75: A 100 Ω Pt RTD with 100-foot #30 AWG lead wires

To illustrate this point, assume that a 100 Ω platinum RTD with 30-gage copper leads is located about 100 feet from a controller's display console. The resistance of 30-gage copper wire is 0.105 Ω/ft, and the two leads of the RTD will contribute a total 21 Ω to the network. Uncorrected, this additional resistance will produce a 55°C measurement error. Obviously, the temperature coefficient of the connecting leads can contribute an additional, and possibly significant, error to the measurement.

To eliminate the effect of the lead resistance, a 4-wire technique is used. In Figure 4-76, a 4-wire (Kelvin) connection is made to the RTD. A constant current, I, is applied though the FORCE leads of the RTD, and the voltage across the RTD itself is measured remotely, via the SENSE leads. The measuring device can be a DVM or an in amp, and high accuracy can be achieved provided that the measuring device exhibits high

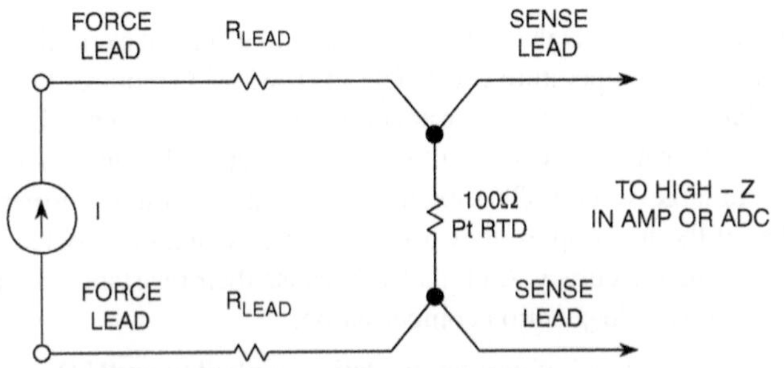

Figure 4-76: Use of Kelvin or 4-wire Pt RTD connections provides high accuracy

input impedance and/or low input bias current. Since the SENSE leads don't carry appreciable current, this technique is relatively insensitive to lead wire length. Some major sources of errors in this scheme are the stability of the constant current source, and the input impedance and/or bias currents in the amplifier or DVM, and the associated drift.

RTDs are generally configured in a four-resistor bridge circuit. The bridge output is amplified by an in amp for further processing. However, high resolution measurement ADCs allow the RTD output to be digitized directly. In this manner, linearization can be performed digitally, thereby easing the analog circuit requirements considerably.

For example, an RTD output can be digitized by one of the AD77XX series high resolution ADCs. Figure 4-77 shows a 100 Ω Pt RTD, driven with a 400 μA excitation current source. Note that the 400 μA RTD excitation current source also generates a 2.5 V reference voltage for the ADC, by virtue of flowing in a 6.25 kΩ resistor, R_{REF}, with the drop across this resistance being metered by the ADC's V_{REF} (+) and (–) input terminals.

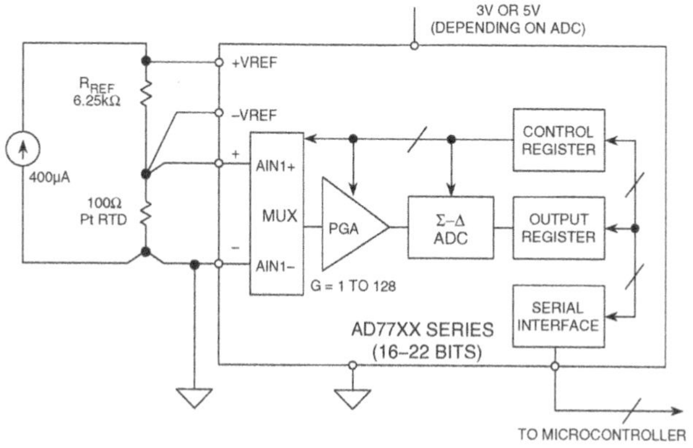

Figure 4-77: A Pt RTD interfaced to the AD77XX series of high resolution ADCs

It should be noted that this simple scheme has great benefits (beyond the obvious one of simplicity). Variations in the magnitude of the 400 μA excitation current do not affect circuit accuracy, since both the input voltage drop across the RTD as well as the reference voltage across R_{REF} vary ratiometrically with the excitation current. However, it should be noted that the 6.25 kΩ resistor must be a stable type with a low temperature coefficient, to avoid errors in the measurement. Either a wirewound resistor, or a very low TC metal film type is most suitable for R_{REF} within this application.

In this application, the ADC's high resolution and the gain of 1 to 128 input PGA eliminates the need for any additional conditioning. The high resolution ADC can in fact perform virtually all the conditioning necessary for an RTD, leaving any further processing such as linearization to be performed in the digital domain.

Thermistors

Similar in general function to RTDs, thermistors are low-cost temperature-sensitive resistors, constructed of solid semiconductor materials which exhibit a positive or negative temperature coefficient. Although positive temperature coefficient devices do exist, the most common thermistors are negative temperature coefficient (NTC) devices.

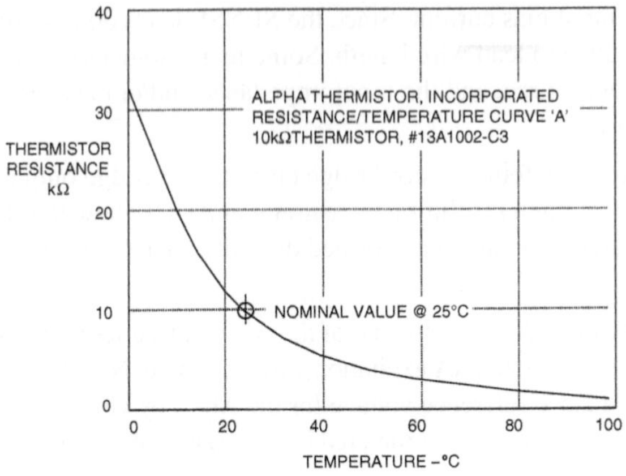

Figure 4-78: Resistance characteristics for a 10kΩ NTC thermistor

Figure 4-78 shows the resistance-temperature characteristic of a commonly used NTC thermistor. Although the thermistor is the most nonlinear of the three temperature sensors discussed, it is also the most sensitive.

The thermistor's very high sensitivity (typically – 44,000 ppm/°C at 25°C) allows it to detect minute temperature variations not readily observable with an RTD or thermocouple. This high sensitivity is a distinct advantage over the RTD, in that 4-wire Kelvin connections to the thermistor aren't needed for lead wire error compensation. To illustrate this point, suppose a 10 kΩ NTC thermistor with a typical 25°C temperature coefficient of –44,000 ppm/°C were substituted for the 100 Ω Pt RTD in the example given earlier. The total lead wire resistance of 21 Ω would generate less than 0.05°C error in the measurement, using the thermistor in lieu of the RTD. This is roughly a factor of 500 improvement in error sensitivity over an RTD.

However, the thermistor's high sensitivity to temperature does not come without a price. As previously shown in Figure 4-78, the temperature coefficient of thermistors does not decrease linearly with increasing temperature as with RTDs, and as a result linearization is required for all but the most narrow temperature ranges. Thermistor applications are limited to a few hundred degrees at best, because thermistors are also more susceptible to damage at high temperatures.

Compared to thermocouples and RTDs, thermistors are fragile in construction and require careful mounting procedures to prevent crushing or bond separation. Although a thermistor's response time is short due to its small size, its small thermal mass also makes it very sensitive to self-heating errors.

Thermistors are very inexpensive, highly sensitive temperature sensors. However, we have noted that a thermistor's temperature coefficient can vary, from –44,000 ppm/°C at 25°C, to –29,000 ppm/°C at 100°C. Not only is this nonlinearity the largest source of error in a temperature measurement, it also limits useful applications to very narrow temperature ranges without linearization.

As shown in Figure 4-79, a parallel resistor combination exhibits a more linear variation with temperature compared to the thermistor itself. This approach to linearizing a thermistor simply shunts it with a fixed, temperature-stable resistor. Paralleling the thermistor with a fixed resistor increases the linearity significantly. Also, the sensitivity of the combination still is high compared to a thermocouple or RTD. The primary disadvantage of the technique is that linearization is only effective within a narrow range. However, it is possible to use a thermistor over a wide temperature range, if the system designer can tolerate a lower net sensitivity, in order to achieve improved linearity.

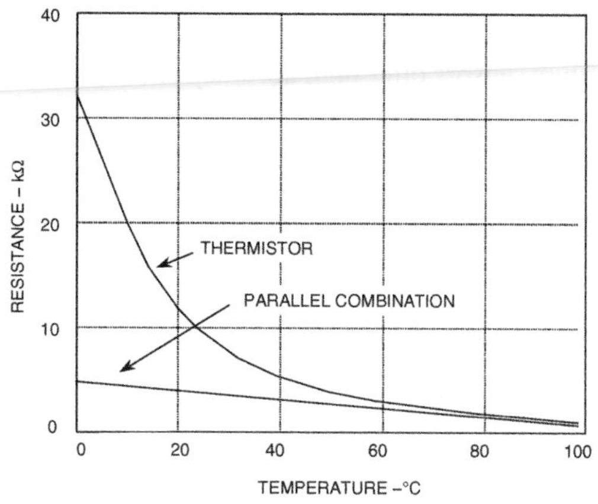

Figure 4-79: Linearization of NTC thermistor using a fixed shunt resistance

R, the value of the fixed shunt resistor, can be calculated from the following equation:

$$R = \frac{RT2 \cdot (RT1 + RT3) - 2 \cdot RT1 \cdot RT3}{RT1 + RT3 - 2 \cdot RT2} \qquad \text{Eq. 4-11}$$

where RT1 is the thermistor resistance at T1, the lowest temperature in the measurement range, RT3 is the thermistor resistance at T3, the highest temperature in the range, and RT2 is the thermistor resistance at T2, the midpoint, T2 = (T1 + T3)/2.

For a typical 10 kΩ NTC thermistor, RT1 = 32,650 Ω at 0°C, RT2 = 6,532 Ω at 35°C, and RT3 = 1,752 Ω at 70°C. This results in a value of 5.17 kΩ for R. The accuracy needed in the associated signal-conditioning circuitry depends on the linearity of the network. For the example given above, the network shows a nonlinearity of –2.3°C/+2.0°C.

The output of the network can be applied to an ADC for digital conversion (with optional linearization) as shown in Figure 4-80. Note that the output of the thermistor network has a slope of approximately –10 mV/°C, which implies that an 8- or 10-bit ADC easily has more than sufficient resolution with a full scale range of 1 V or less. The further linearization can be applied to the data in the digital domain, if desired.

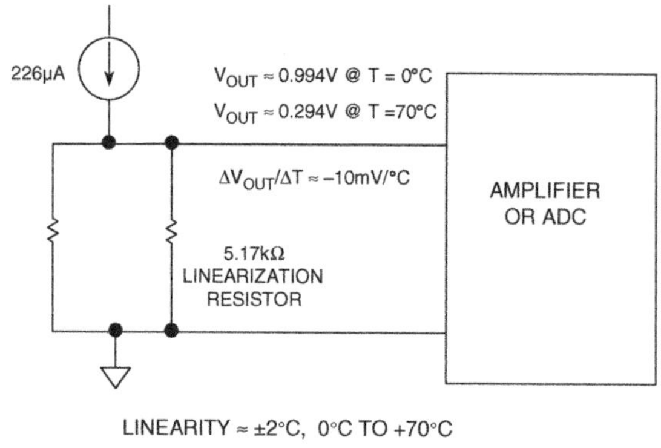

Figure 4-80: Linearized thermistor network with amplifier or ADC

Semiconductor Temperature Sensors

Modern semiconductor temperature sensors offer high accuracy and high linearity over an operating range of about –55°C to +150°C. Internal amplifiers can scale the output to convenient values, such as 10 mV/°C. They are also useful in cold-junction-compensation circuits for wide temperature range thermocouples.

All semiconductor temperature sensors make use of the relationship between a bipolar junction transistor's (BJT) base-emitter voltage to its collector current:

$$V_{BE} = \frac{kT}{q}\ln\left(\frac{I_C}{I_S}\right) \qquad \text{Eq. 4-12}$$

In this expression k is Boltzmann's constant, T is the absolute temperature, q is the charge of an electron, and I_S is a current related to the geometry and the temperature of the transistors. (The equation assumes a voltage of at least a few hundred mV on the collector, and ignores Early effects.)

If we take 'N' transistors identical to the first (see Figure 4-81) and allow the total current I_C to be shared equally among them, we find that the new base-emitter voltage applicable to this case is given by the equation

$$V_N = \frac{kT}{q}\ln\left(\frac{I_C}{N \cdot I_S}\right) \qquad \text{Eq. 4-13}$$

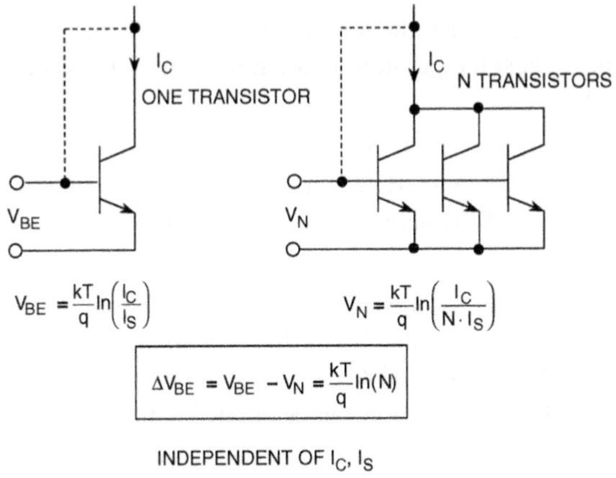

Figure 4-81: The basic relationships for BJT-based semiconductor temperature sensors

Neither of these circuits is of much use by itself, because of the strong temperature dependence of I_S. However, if we have equal currents flowing in one BJT, as well as the N similar BJTs, then the expression for the *difference* between the respective base-emitter voltages (or ΔV_{BE}) is proportional to absolute temperature, and it does not contain I_s.

This then leads to a far more useful relationship, developed as follows:

$$\Delta V_{BE} = V_{BE} - V_N = \frac{kT}{q}\ln\left(\frac{I_C}{I_S}\right) - \frac{kT}{q}\ln\left(\frac{I_C}{N \cdot I_S}\right)$$

$$\Delta V_{BE} = V_{BE} - V_N = \frac{kT}{q}\left[\ln\left(\frac{I_C}{I_S}\right) - \ln\left(\frac{I_C}{N \cdot I_S}\right)\right]$$

$$\Delta V_{BE} = V_{BE} - V_N = \frac{kT}{q}\ln\left[\left(\frac{I_C}{I_S}\right) \Big/ \left(\frac{I_C}{N \cdot I_S}\right)\right] = \frac{kT}{q}\ln(N)$$

This end result of this algebra is expressed in a single key equation, one worthy of restatement:

$$\Delta V_{BE} = \frac{kT}{q}\ln(N) \qquad \text{Eq. 4-14}$$

As one can note, Eq. 4-14 contains only the transistor emitter area ratio N and T as variables. Since N is fixed within a given design, it can be the basis of a transducer for T measurement. The circuit as shown in Figure 4-82 implements the above equation, and is popularly known as the "Brokaw Cell," after its inventor (see Reference 10).

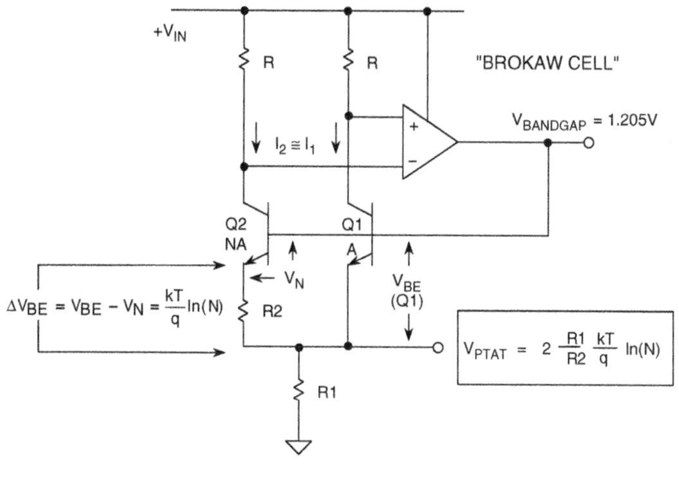

Figure 4-82: The "Brokaw Cell" is both a silicon bandgap voltage-based reference as well as a temperature sensor

The voltage $\Delta V_{BE} = V_{BE} - V_N$ appears across resistor R2, as noted. The emitter current in Q2 is therefore $\Delta V_{BE}/R2$. The op amp's servo loop and the two resistors 'R' force an identical current to flow through Q1. The Q1 and Q2 currents are equal, and they are summed, flowing in resistor R1.

The corresponding voltage developed across R1 is V_{PTAT}, a voltage proportional to absolute temperature (PTAT). This is given by:

$$V_{PTAT} = \frac{2R1(V_{BE} - V_N)}{R2} = 2\frac{R1}{R2}\frac{kT}{q}\ln(N)$$

Within the circuit, a voltage labeled as $V_{BANDGAP}$ appears at the base of Q1, and, as can be noted, is the sum of $V_{BE(Q1)}$ and V_{PTAT}. When this voltage is set by the design to be exactly equal to the bandgap voltage of silicon, it will then become independent of temperature. The voltage $V_{BE(Q1)}$ is complementary to absolute temperature (CTAT), and summing it with a properly proportioned V_{PTAT} from across R1 gives the desired end result; the bandgap voltage becomes constant with respect to temperature. Note that this assumes the proper choice of R1/R2 ratio and N, to make the summed voltage equal to $V_{BANDGAP}$, the silicon bandgap voltage (in this instance 1.205 V). This circuit has the virtues of dual application because of the above features. It is useful as a basic silicon bandgap temperature sensor, with either direct or scaled use of the voltage V_{PTAT}. It is also widely used a temperature stable reference voltage source, by suitable scaling of $V_{BANDGAP}$ to standard outputs of 2.500 V, 5.000 V, and so forth.

Current and Voltage Output Temperature Sensors

The concepts used in the bandgap voltage temperature sensor discussion above can also be used as the basis for a variety of IC temperature sensors, with linear, proportional-to-temperature outputs, of either current or voltage.

The AD592 device shown in Figure 4-83 is a two-terminal, current output sensor with a scale factor of 1 μA/K. This device does not require external calibration, and is available in several accuracy grades. The AD592 is a TO92 packaged version of the original AD590 TO52 metal packaged temperature transducer device (see Reference 11).

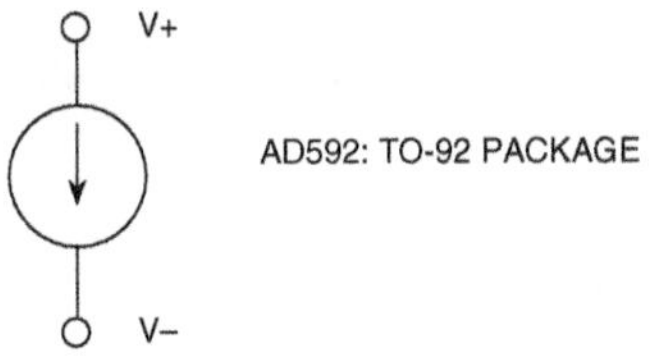

- 1μA/K Scale Factor
- Nominal Output Current @ 25°C: 298.2μA
- Operation from 4V to 30V
- ±0.5°C Max Error @ 25°C, ±1.0°C Error Over Temp, ±0.1°C Typical Nonlinearity (AD592CN)
- AD592 Specified from –25°C to +105°C

Figure 4-83: Current output absolute temperature sensor

The simplest operating mode for current mode temperature sensors is to load them with a precision resistor of 1% or better tolerance, and read the output voltage developed with either an ADC or a scaling amplifier/buffer. Figure 4-84 shows this technique with an ADC, as applicable to the AD592. The resistor load R1 converts the basic scaling of the sensor (1 μA/K) into a proportional voltage.

Choice of this resistor determines the overall sensitivity of the temperature sensor, in terms of V/K. For example, with a 1 kΩ precision resistor load as shown, the net circuit sensitivity becomes 1 mV/K. With a 5 V bias on the temperature sensor as shown, the AD592's full dynamic range is allowed with a 1 KΩ load. If a higher value R1 is used, higher bias voltage may be required, as the AD592 requires 4 V of operating headroom.

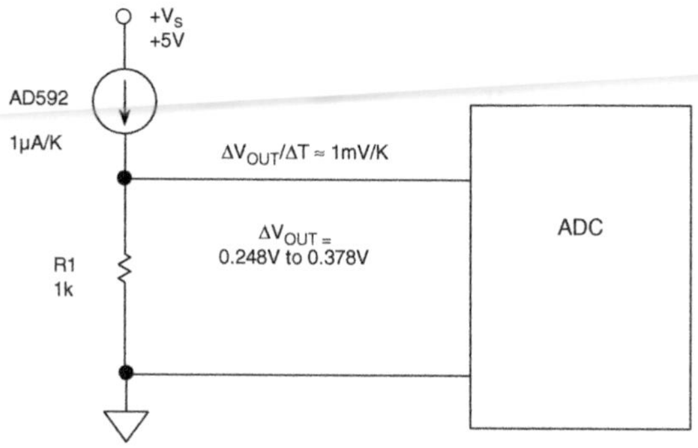

Figure 4-84: Current output temperature sensor driving a resistive load

The function just described is a Kelvin-scaled temperature sensor, so the ADC will be required to read the full dynamic range of voltage across R1. With an AD592, this span is from the range corresponding to –25°C (248 K) to 105°C (378 K), which is 0.248 V to 0.378 V. A 10-bit, 0.5 V scaled ADC can read this range directly with ≈0.5°C resolution.

If a centigrade-scaled reading is desired, two options present themselves. For a traditional analog approach, the common terminal of the ADC input can easily be biased with a reference voltage corresponding to 0°C, or 0.273 V. Alternately, the 0°C reference can be inserted in the digital domain, with the advantage of no additional hardware requirement.

The AD592 is available in three accuracy grades. The highest grade version (AD592CN) has a maximum error @ 25°C of ±0.5°C and ±1.0°C error from –25°C to +105°C, and a linearity error of ±0.35°C. The AD592 is available in a TO-92 package.

With regard to stand-alone digital output temperature sensors, it is worthy of note that such devices do exist, that is ADCs with built-in temperature sensing. The AD7816/AD7817/AD7818-series ADCs have on-board temperature sensors digitized by a 10-bit 9 µs conversion time switched capacitor SAR ADC. The device family offers a variety of input options for flexibility. The similar AD7416/AD7417/AD7418 have serial interfaces.

For a great many temperature sensing applications, a voltage mode output sensor is most appropriate. For this, there are a variety of standalone sensors that can be directly applied. In such devices the basic mode of operation is as a three-terminal device, using power input, common, and voltage output pins. In addition, some devices offer an optional shutdown pin.

The TMP35/TMP36 are low voltage (2.7 V to 5.5 V), SO-8 or TO-92 packaged voltage output temperature sensors with a 10 mV/°C scale factor, as shown in Figure 4-85. Supply current is below 50 μA, providing very low self-heating (less than 0.1°C in still air).

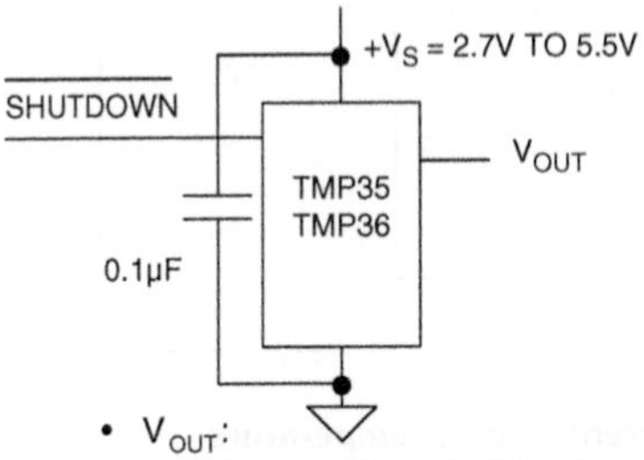

- V_{OUT}:
  - TMP35, 250mV @ 25°C, 10mV/°C (10°C to 125°C)
  - TMP36, 750mV @ 25°C, 10mV/°C (–40°C to +125°C)
- ±2°C Error Over Temp (Typical), ±0.5°C Nonlinearity (Typical)
- Specified –40°C to +125°C
- 50μA Quiescent Current, 0.5μA in Shutdown Mode

Figure 4-85: TMP35/36 absolute scaled voltage mode output temperature sensors with shutdown capability

Output scaling of this device family differs in range and 25°C offset. The TMP35 provides a 250 mV output at 25°C, and reads temperature from 10°C to 125°C. The TMP36 is specified from –40°C to +125°C. and provides a 750 mV output at 25°C. Both the TMP35 and TMP36 have an output scale factor of +10 mV/°C.

An optional shutdown feature is provided for the SO8 package devices, which reduces the standby current to 0.5 μA. This pin, when taken to a logic LOW, turns the device OFF, and the output becomes a high impedance state. If shutdown isn't used, the pin should be connected to $+V_S$.

The power supply pin of these voltage mode sensors should be bypassed to ground with a 0.1 μF ceramic capacitor having very short leads (preferably surface mount) and located as close to the power supply pin as possible. Since these temperature sensors operate on very little supply current and could be exposed to very hostile electrical environments, it is also important to minimize the effects of EMI/RFI on these devices. The effect of RFI on these temperature sensors is manifested as abnormal dc shifts in the output voltage due to rectification of the high frequency noise by the internal IC junctions. In those cases where the devices are operated in the presence of high frequency radiated or conducted noise, a large value tantalum electrolytic capacitor (>2.2 μF) placed across the 0.1 μF ceramic may offer additional noise immunity.

Ratiometric Voltage Output Temperature Sensors

In some cases, it is desirable for the output of a temperature sensor to be *ratiometric* with respect its supply voltage. A series of ADI temperature sensors have been designed to fulfill this need, in the form of the AD2210x series (see references 12–14). Of this series, the AD22103 illustrated in Figure 4-86 has an output that is ratiometric with regard its nominal 3.3 V supply voltage, according to the equation:

$$V_{OUT} = \frac{V_S}{3.3V} \times \left(0.25V + \frac{28mV}{°C} \times T_A\right). \qquad \text{Eq. 4-15}$$

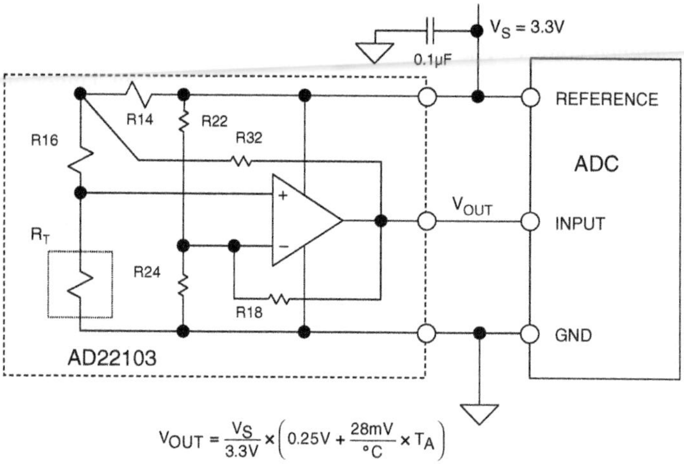

Figure 4-86: Ratiometric voltage output temperature sensors

Note also that the Figure 4-86 circuit uses the 3.3 V AD22103 power supply voltage as the reference input to the ADC. This step eliminates the need for a separate precision voltage reference source. Within a system, this key point potentially can have a positive impact.

Operation of the AD22103 is accomplished with an on-chip temperature sensing resistance R_T, which operates similarly to the RTD types discussed earlier in this section. This resistance is fed from a resistance network comprised of R14, R16, and R32. R14 and R16 provide a current drive component for R_T that is proportional to the supply voltage, a factor that gives the AD22103 a basic sensitivity that is proportional to the supply. R_T, like a classic platinum RTD, exhibits a nonlinear resistance versus temperature behavior. This nonlinear characteristic of R_T is corrected by a positive feedback loop, composed of R32 along with the Thevenin equivalent of resistances R14 and R16.

Gain scaling for the changing R_T output voltage is provided by the op amp negative feedback loop, R18, and the Thevenin equivalent of resistances R24 and R22. This references the gain network of the op amp to the supply voltage, instead of ground. The various resistance networks around the op amp are actively trimmed at temperature, to calibrate the sensor for its rated offset and scaling.

The net combination of these factors allows the device to behave in accordance with the relationship of Eq. 4-15. The AD22103 is specified over a range of 0°C to 100°C, and has an accuracy better than ±2.5°C, along with a linearity better than ±0.5% of full scale, i.e., 0.5°C over 100°C.

Since the AD22103 is a single-supply part, the sensing of low temperatures necessarily involves a positive output offset. For example, for 3.3 V operation of this example, the output offset is simply the 0.25 V term of Eq. 4-15. Accordingly, the 0 to 100°C temperature span translates to an output swing of 0.25 V to 3.05 V.

Should it be desired, operation of the AD22103 device is also possible at higher supply voltages. Because of the ratiometric operation feature, this will necessarily involve a change to both the basic sensitivity and the offset. For example, in operating the AD22103 at 5 V, the output expression changes to:

$$V_{OUT} = \frac{V_S}{5V} \times \left(0.378V + \frac{42.42mV}{°C} \times T_A\right) \qquad \text{Eq. 4-16}$$

However, it should be noted that the fact that the AD22103 is ratiometric does *not* preclude operating the part from a fixed reference voltage. The nominal current drain of the AD22103 is 500 μA, allowing a number of these sensors to be operated from a common reference IC without danger of overload (as well as other analog parts). For example, one such reference family is the REF19X series, which can supply output currents of up to 30 mA.

In addition to the above-described AD22103 3.3 V part, there is also a companion device, the AD22100. While basically quite similar to the AD22103, the AD22100 operates from a nominal 5 V power supply with reduced sensitivity, allowing operation over a range of –50 to +150°C. Over this range the rated accuracy of the AD22100 is 2% or better, and linearity error is 1% or less.

References: Temperature Sensors

1. Ramon Pallas-Areny and John G. Webster, **Sensors and Signal Conditioning**, John Wiley, New York, 1991.
2. Dan Sheingold, Editor, **Transducer Interfacing Handbook**, Analog Devices, Inc., 1980, ISBN: 0-916550-05-2.
3. Sections 2, 3, Walt Kester, Editor, **1992 Amplifier Applications Guide**, Analog Devices, 1992, ISBN: 0-916550-10-9.
4. Sections 1, 6, Walt Kester, Editor, **System Applications Guide**, Analog Devices, 1993, ISBN: 0-916550-13-3.
5. Dan Sheingold, Editor, **Nonlinear Circuits Handbook**, Analog Devices, Inc., 1974.
6. James Wong, "Temperature Measurements Gain from Advances in High-precision Op Amps," **Electronic Design**, May 15, 1986.
7. OMEGA Temperature Measurement Handbook, Omega Instruments, Inc.
8. **Handbook of Chemistry and Physics**, Chemical Rubber Co.
9. Joe Marcin, "Thermocouple Signal Conditioning Using the AD594/AD595," **Analog Devices AN369**.
10. Paul Brokaw, "A Simple Three-Terminal IC Bandgap Voltage Reference," **IEEE Journal of Solid-State Circuits**, Vol. SC-9, No. 6, December, 1974, pp. 388–393.
11. Mike Timko, "A Two-Terminal IC Temperature Transducer," **IEEE Journal of Solid-State Circuits**, Vol. SC-11, No. 6, December, 1976, pp. 784–788.
12. Data sheet for **AD22103 3.3 V Supply, Voltage Output Temperature Sensor with Signal Conditioning**, www.analog.com.
13. Data sheet for **AD22100 Voltage Output Temperature Sensor with Signal Conditioning**, www analog.com.
14. Adrian P. Brokaw, "Monolithic Ratiometric Temperature Measurement Circuit," **US Patent No. 5,030,849**, filed June 30, 1989, issued July 9, 1991.

Classic Cameo
AD590 Two-Terminal IC Temperature Transducer

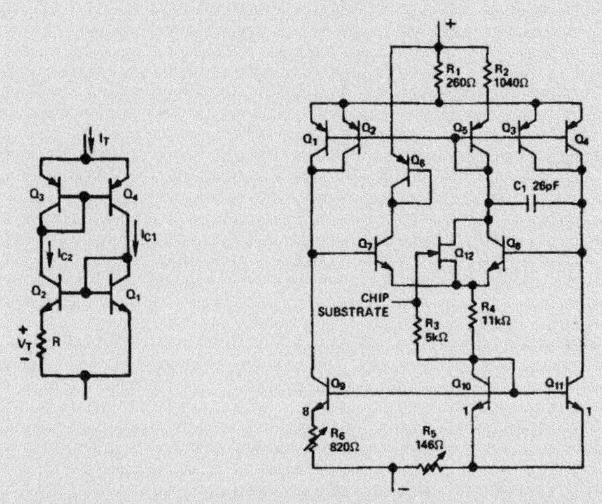

AD590 basic (left), and complete schematics (right)

Designed by Mike Timko, based on an original Paul Brokaw concept,[1] the AD590 [2,3,4] current mode IC temperature transducer was introduced in 1977. The AD590 established early an elegant method of accurate temperature measurement, based upon fundamental silicon transistor operating principles. It has been in ADI production since introduction, along with such related ICs as the AD592 discussed within this chapter.

The references below discuss operation in great detail, but suffice it to say that in the basic structure (left) a voltage proportional to absolute temperature, V_T, appears across resistor R. This makes the current I_T drawn from an external source proportional to absolute temperature. In the full circuit (right), trimmed thin film resistors implement a calibrated scaling for I_T of 1μA/K. Additional circuitry is added for startup and for increased accuracy, both with respect to applied voltage as well as against high temperature leakage.

A current-operated transducer such as this is quite convenient to operate, the output being impervious to long lead lengths, and also virtually noise-immune. The low scaling factor also makes the AD590 easy to operate on low voltage supplies without self-heating, yet high output impedance also holds calibration with higher applied voltages. Readout is simply accomplished with a single resistance, making a simple, two-component Kelvin-scaled thermometer possible.

---

1 Adrian P. Brokaw, "Digital-to-Analog Converter with Current Source Transistors Operated Accurately at Different Current Densities," **US Patent No. 3,940,760**, filed March 21, 1975, issued Feb. 24, 1976.

2 Mike Timko, "A Two-Terminal IC Temperature Transducer," **IEEE Journal of Solid-State Circuits**, Vol. SC-11, No. 6, December, 1976, pp. 784–788.

3 Mike Timko, Goodloe Suttler, "1μA/K IC Temperature-to-Current Transducer," **Analog Dialogue**, Vol. 12, No. 1, 1978, pp. 3–5.

4 Michael P. Timko, Adrian P. Brokaw, "Integrated Circuit Two Terminal Temperature Transducer," **US Patent No. 4,123,698**, filed July 6, 1976, issued October 31, 1978.

CHAPTER 5

Analog Filters

- Section 5-1: Introduction
- Section 5-2: The Transfer Function
- Section 5-3: Time Domain Response
- Section 5-4: Standard Responses
- Section 5-5: Frequency Transformations
- Section 5-6: Filter Realizations
- Section 5-7: Practical Problems in Filter Implementation
- Section 5-8: Design Examples

CHAPTER 5

Analog Filters

Hank Zumbahlen

SECTION 5-1

Introduction

Filters are networks that process signals in a frequency-dependent manner. The basic concept of a filter can be explained by examining the frequency-dependent nature of the impedance of capacitors and inductors. Consider a voltage divider where the shunt leg is a reactive impedance. As the frequency is changed, the value of the reactive impedance changes, and the voltage divider ratio changes. This mechanism yields the frequency dependent change in the input/output transfer function that is defined as the frequency-response.

Filters have many practical applications. A simple, single pole, low-pass filter (the integrator) is often used to stabilize amplifiers by rolling off the gain at higher frequencies where excessive phase shift may cause oscillations.

A simple, single pole, high pass filter can be used to block dc offset in high gain amplifiers or single supply circuits. Filters can be used to separate signals, passing those of interest, and attenuating the unwanted frequencies.

An example of this is a radio receiver, where the signal to be processed is passed through, typically with gain, while attenuating the rest of the signals. In data conversion, filters are also used to eliminate the effects of aliases in A/D systems. They are used in reconstruction of the signal at the output of a D/A as well, eliminating the higher frequency components, such as the sampling frequency and its harmonics, thus smoothing the waveform.

There are a large number of texts dedicated to filter theory. No attempt will be made to go heavily into much of the underlying math—Laplace transforms, complex conjugate poles and the like—although they will be mentioned.

While they are appropriate for describing the effects of filters and examining stability, in most cases examination of the function in the frequency domain is more illuminating.

An ideal filter will have an amplitude response that is unity (or at a fixed gain) for the frequencies of interest (called the *pass band*) and zero everywhere else (called the *stop band*). The frequency at which the response changes from pass band to stop band is referred to as the *cutoff frequency*.

Figure 5-1(A) shows an idealized low-pass filter. In this filter the low frequencies are in the pass band and the higher frequencies are in the stop band.

The functional complement to the low-pass filter is the high-pass filter. Here, the low frequencies are in the stop band, and the high frequencies are in the pass band. Figure 5-1(B) shows the idealized high-pass filter.

If a high-pass filter and a low-pass filter are cascaded, a *band-pass* filter is created. The band-pass filter passes a band of frequencies between a lower cutoff frequency, f_l, and an upper cutoff frequency, f_h. Frequencies below f_l and above f_h are in the stop band. An idealized band-pass filter is shown in Figure 5-1(C).

A complement to the band-pass filter is the *bandreject*, or *notch* filter. Here, the pass bands include frequencies below f_l and above f_h. The band from f_l to f_h is in the stop band. Figure 5-1(D) shows a notch response.

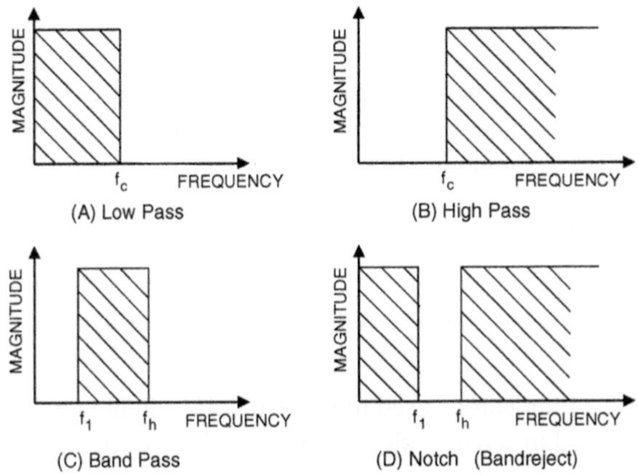

Figure 5-1: Idealized filter responses

Unfortunately, the idealized filters defined above cannot be easily built. The transition from pass band to stop band will not be instantaneous, but instead there will be a transition region. Stop band attenuation will not be infinite.

The five parameters of a practical filter are defined in Figure 5-2.

Figure 5-2: Key filter parameters

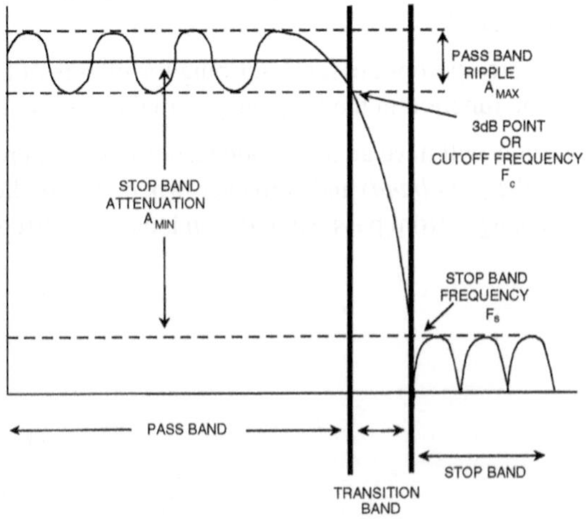

The *cutoff frequency* (F_c) is the frequency at which the filter response leaves the error band (or the –3 dB point for a Butterworth response filter). The *stop band frequency* (F_s) is the frequency at which the minimum attenuation in the stop band is reached. The *pass band ripple* (A_{max}) is the variation (error band) in the pass band response. The *minimum pass band attenuation* (A_{min}) defines the minimum signal attenuation within the stop band. The steepness of the filter is defined as the *order* (M) of the filter. M is also the number of poles in the transfer function. A pole is a root of the denominator of the transfer function. Conversely, a zero is a root of the numerator of the transfer function. Each pole gives a –6 dB/octave or –20 dB/decade response. Each zero gives a +6 dB/octave, or +20 dB/decade response.

Note that not all filters will have all these features. For instance, all-pole configurations (i.e., no zeros in the transfer function) will not have ripple in the stop band. Butterworth and Bessel filters are examples of all-pole filters with no ripple in the pass band.

Typically, one or more of the above parameters will be variable. For instance, if you were to design an antialiasing filter for an ADC, you will know the cutoff frequency (the maximum frequency to be passed), the stop band frequency, (which will generally be the Nyquist frequency (= ½ the sample rate)) and the minimum attenuation required (which will be set by the resolution or dynamic range of the system). Go to a chart or computer program to determine the other parameters, such as filter order, F_0, and Q, which determines the peaking of the section, for the various sections and/or component values.

It should also be pointed out that the filter will affect the phase of a signal, as well as the amplitude. For example, a single pole section will have a 90° phase shift at the crossover frequency. A pole pair will have a 180° phase shift at the crossover frequency. The Q of the filter will determine the rate of change of the phase. This will be covered more in depth in the next section.

The Transfer Function

The S-Plane

Filters have a frequency dependent response because the impedance of a capacitor or an inductor changes with frequency. Therefore the complex impedances:

$$Z_L = sL \qquad \text{Eq. 5-1}$$

and

$$Z_C = \frac{1}{sC} \qquad \text{Eq. 5-2}$$

$$s = \sigma + j\omega \qquad \text{Eq. 5-3}$$

are used to describe the impedance of an inductor and a capacitor, respectively, where σ is the Neper frequency in nepers per second (NP/s) and ω is the angular frequency in radians per sec (rad/s).

By using standard circuit analysis techniques, the transfer equation of the filter can be developed. These techniques include Ohm's law, Kirchoff's voltage and current laws, and superposition, remembering that the impedances are complex. The transfer equation is then:

$$H(s) = \frac{a_m s^m + a_{m-1} s^{m-1} + \ldots + a_1 s + a_0}{b_n s^n + b_{n-1} s^{n-1} + \ldots + b_1 s + b_0} \qquad \text{Eq. 5-4}$$

Therefore, $H(s)$ is a rational function of s with real coefficients with the degree of m for the numerator and n for the denominator. The degree of the denominator is the order of the filter. Solving for the roots of the equation determines the poles (denominator) and zeros (numerator) of the circuit. Each pole will provide a –6 dB/octave or –20 dB/decade response. Each zero will provide a +6 dB/octave or +20 dB/decade response. These roots can be real or complex. When they are complex, they occur in conjugate pairs. These roots are plotted on the s plane (complex plane) where the horizontal axis is σ (real axis) and the vertical axis is ω (imaginary axis). How these roots are distributed on the s plane can tell us many things about the circuit. In order to have stability, all poles must be in the left side of the plane. If there is a zero at the origin, that is a zero in the numerator, the filter will have no response at dc (high pass or band pass).

Assume an RLC circuit, as in Figure 5-3. Using the voltage divider concept it can be shown that the voltage across the resistor is:

$$H(s) = \frac{Vo}{Vin} = \frac{RCs}{LCs^2 + RCs + 1} \qquad \text{Eq. 5-5}$$

Substituting the component values into the equation yields:

$$H(s) = 10^3 \times \frac{s}{s^2 + 10^3 s + 10^7}$$

Factoring the equation and normalizing gives:

$$H(s) = 10^3 \times \frac{s}{\left[s - (-0.5 + j3.122) \times 10^3\right] \times \left[s - (-0.5 - j3.122) \times 10^3\right]}$$

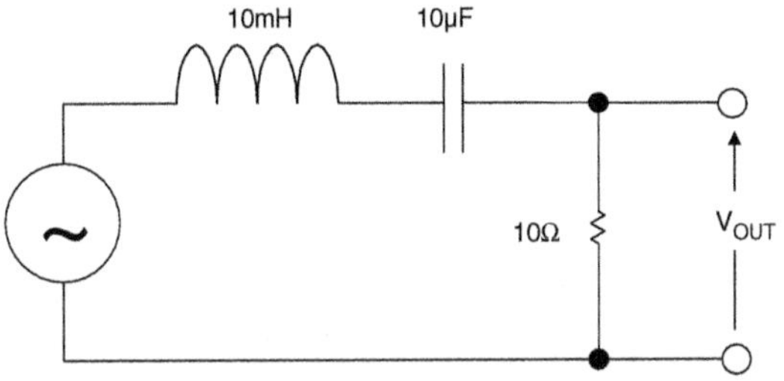

Figure 5-3: RLC circuit

This gives a zero at the origin and a pole pair at:

$$s = (-0.5 \pm j3.122) \times 10^3$$

Next, plot these points on the s plane as shown in Figure 5-4:

Figure 5-4: Pole and zero plotted on the s-plane

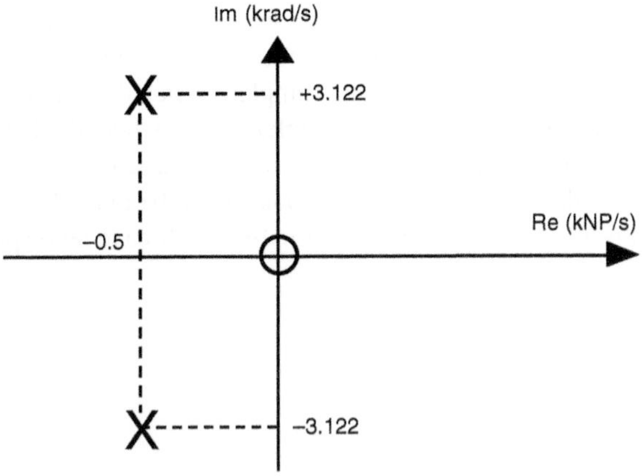

The above discussion has a definite mathematical flavor. In most cases there is more interest in the circuit's performance in real applications. While working in the s plane is completely valid, most of us don't think in terms of Nepers and imaginary frequencies.

F_o and Q

So if it is not convenient to work in the s plane, why go through the above discussion? The answer is that the groundwork has been set for two concepts that will be infinitely more useful in practice: F_0 and Q.

F_0 is the cutoff frequency of the filter. This is defined, in general, as the frequency where the response is down 3 dB from the passband. It can sometimes be defined as the frequency at which it will fall out of the pass band. For example, a 0.1 dB Chebyshev filter can have its F_0 at the frequency at which the response is down > 0.1 dB.

The shape of the attenuation curve (as well as the phase and delay curves, which define the time domain response of the filter) will be the same if the ratio of the actual frequency to the cutoff frequency is examined, rather than just the actual frequency itself. Normalizing the filter to 1 rad/s, a simple system for designing and comparing filters can be developed. The filter is then scaled by the cutoff frequency to determine the component values for the actual filter.

Q is the "quality factor" of the filter. It is also sometimes given as α where:

$$\alpha = \frac{1}{Q} \qquad \text{Eq. 5-6}$$

This is commonly known as the *damping ratio.* ξ is sometimes used where:

$$\xi = 2\alpha \qquad \text{Eq. 5-7}$$

If Q is > 0.707, there will be some peaking in the filter response. If the Q is < 0.707, roll-off at F_0 will be greater; it will have a more gentle slope and will begin sooner. The amount of peaking for a 2-pole low-pass filter versus Q is shown in Figure 5-5.

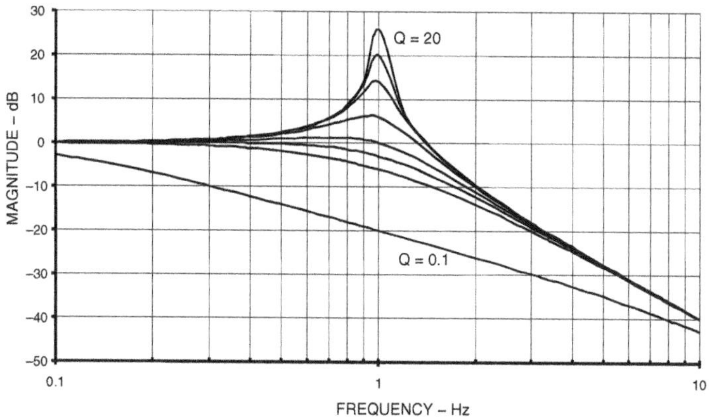

Figure 5-5: Low-pass filter peaking versus Q

Rewriting the transfer function H(s) in terms of ω_0 and Q:

$$H(s) = \frac{H_0}{s^2 + \frac{\omega_0}{Q}s + \omega_0^2} \qquad \text{Eq. 5-8}$$

where H_0 is the pass band gain and $\omega_0 = 2\pi F_0$.

This is now the *low-pass prototype* that will be used to design the filters.

High-Pass Filter

Changing the numerator of the transfer equation, H(s), of the low-pass prototype to H_0s^2 transforms the low-pass filter into a high-pass filter. The response of the high-pass filter is similar in shape to a low-pass, just inverted in frequency.

The transfer function of a high-pass filter is then:

$$H(s) = \frac{H_0 s^2}{s^2 + \frac{\omega_0}{Q}s + \omega_0^2} \qquad \text{Eq. 5-9}$$

The response of a 2-pole high-pass filter is illustrated in Figure 5-6.

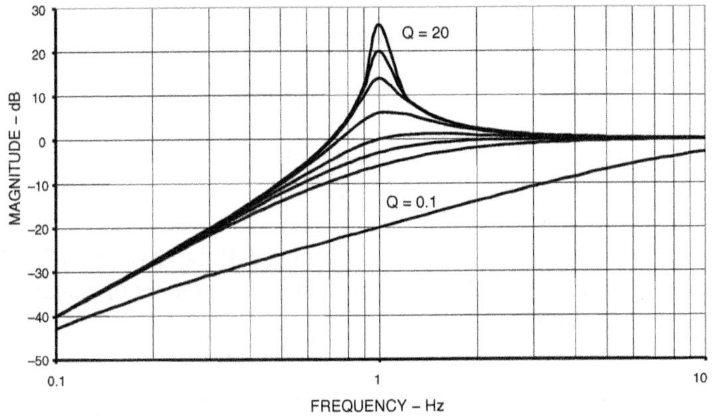

Figure 5-6: High-pass filter peaking versus Q

Band-Pass Filter

Changing the numerator of the low-pass prototype to $H_0\omega_0^2$ will convert the filter to a band-pass function.

The transfer function of a band-pass filter is then:

$$H(s) = \frac{H_0 \omega_0^2}{s^2 + \frac{\omega_0}{Q}s + \omega_0^2} \qquad \text{Eq. 5-10}$$

ω_0 here is the frequency ($F_0 = 2\,\pi\,\omega_0$) at which the gain of the filter peaks.

H_0 is the circuit gain and is defined:

$$H_0 = H/Q \qquad \text{Eq. 5-11}$$

Q has a particular meaning for the band-pass response. It is the selectivity of the filter. It is defined as:

$$Q = \frac{F_0}{F_H - F_L} \qquad \text{Eq. 5-12}$$

where F_L and F_H are the frequencies where the response is –3 dB from the maximum.

The bandwidth (BW) of the filter is described as:

$$BW = F_H - F_L \qquad \text{Eq. 5-13}$$

It can be shown that the resonant frequency (F_0) is the geometric mean of F_L and F_H, which means that F_0 will appear halfway between F_L and F_H on a logarithmic scale.

$$F_0 = \sqrt{F_H F_L} \qquad \text{Eq. 5-14}$$

Also, note that the skirts of the band-pass response will always be symmetrical around F_0 on a logarithmic scale.

The response of a band-pass filter to various values of Q are shown in Figure 5-7.

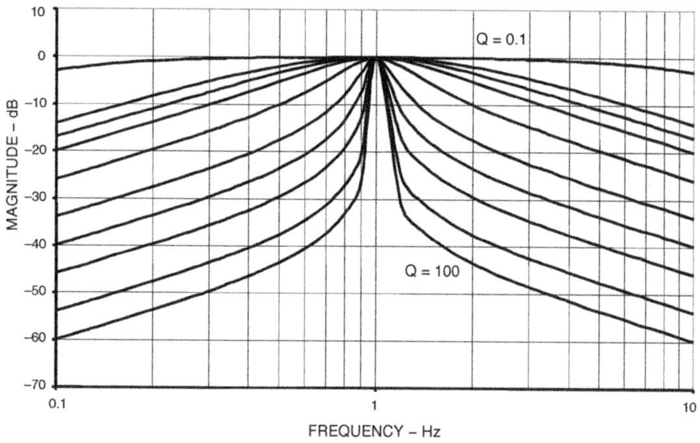

Figure 5-7: Band-pass filter peaking versus Q

A word of caution is appropriate here. Band-pass filters can be defined two different ways. The narrowband case is the classic definition that we have shown above.

In some cases, however, if the high and low cutoff frequencies are widely separated, the band-pass filter is constructed of separate high-pass and low-pass sections. Widely separated in this context means separated by at least two octaves (×4 in frequency). This is the wideband case.

Bandreject (Notch) Filter

By changing the numerator to $s^2 + \omega_z^2$, we convert the filter to a bandreject or notch filter. As in the band-pass case, if the corner frequencies of the bandreject filter are separated by more than an octave (the wideband case), it can be built out of separate low-pass and high-pass sections. The following convention will be adopted: A narrow band bandreject filter will be referred to as a *notch* filter and the wide band bandreject filter will be referred to as *bandreject* filter.

A notch (or bandreject) transfer function is:

$$H(s) = \frac{H_0\left(s^2 + \omega_Z^{\,2}\right)}{s^2 + \frac{\omega_0}{Q}s + \omega_0^{\,2}} \qquad \text{Eq. 5-15}$$

There are three cases of the notch filter characteristics. These are illustrated in Figure 5-8. The relationship of the pole frequency, ω_0, and the zero frequency, ω_z, determines if the filter is a standard notch, a low-pass notch or a high-pass notch.

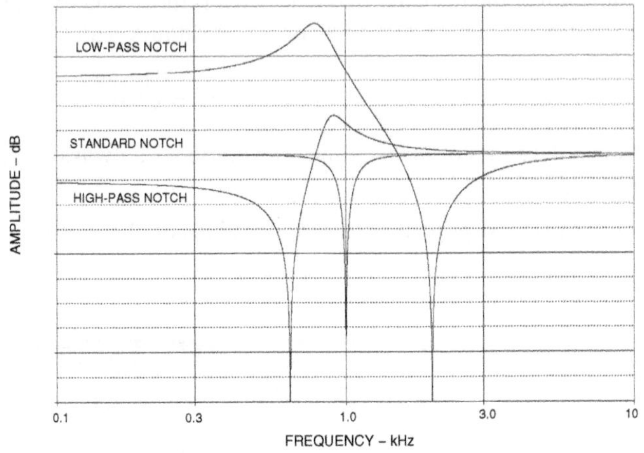

Figure 5-8: Standard, low-pass, and high-pass notches

If the zero frequency is equal to the pole frequency a standard notch exists. In this instance the zero lies on the $j\omega$ plane where the curve that defines the pole frequency intersects the axis.

A low-pass notch occurs when the zero frequency is greater than the pole frequency. In this case, ω_z lies outside the curve of the pole frequencies. What this means in a practical sense is that the filter's response below ω_z will be greater than the response above ω_z. This results in an elliptical low-pass filter.

A high-pass notch filter occurs when the zero frequency is less than the pole frequency. In this case, ω_z lies inside the curve of the pole frequencies. What this means in a practical sense is that the filter's response below ω_z will be less than the response above ω_z. This results in an elliptical high-pass filter.

The variation of the notch width with Q is shown in Figure 5-9.

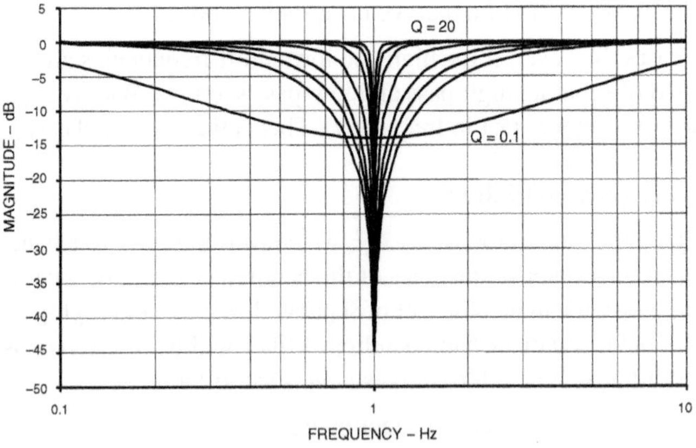

Figure 5-9: Notch filter width versus frequency for various Q values

All-Pass Filter

There is another type of filter that leaves the amplitude of the signal intact but introduces phase shift. This type of filter is called an *all-pass*. The purpose of this filter is to add phase shift (delay) to the response of the circuit. The amplitude of an all-pass is unity for all frequencies. The phase response, however, changes from 0° to 360° as the frequency is swept from 0 to infinity. The purpose of an all-pass filter is to provide phase equalization, typically in pulse circuits. It also has application in single sideband, suppressed carrier (SSB-SC) modulation circuits.

The transfer function of an all-pass filter is:

$$H(s) = \frac{s^2 - \frac{\omega_0}{Q}s + \omega_0^2}{s^2 + \frac{\omega_0}{Q}s + \omega_0^2} \qquad \text{Eq. 5-16}$$

Note that an allpass transfer function can be synthesized as:

$$H_{AP} = H_{LP} - H_{BP} + H_{HP} = 1 - 2H_{BP} \qquad \text{Eq. 5-17}$$

Figure 5-10 compares the various filter types.

Figure 5-10: Standard second-order filter responses

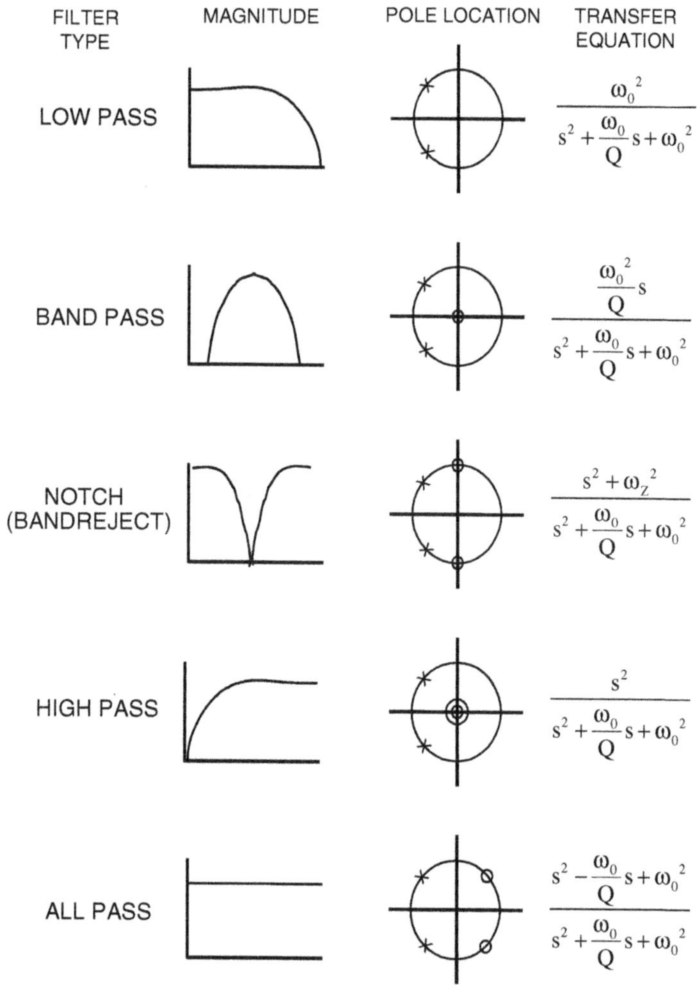
FILTER TYPE
MAGNITUDE
POLE LOCATION
TRANSFER EQUATION
LOW PASS
$\frac{\omega_0^2}{s^2 + \frac{\omega_0}{Q}s + \omega_0^2}$
BAND PASS
$\frac{\frac{\omega_0^2}{Q}s}{s^2 + \frac{\omega_0}{Q}s + \omega_0^2}$
NOTCH (BANDREJECT)
$\frac{s^2 + \omega_Z^2}{s^2 + \frac{\omega_0}{Q}s + \omega_0^2}$
HIGH PASS
$\frac{s^2}{s^2 + \frac{\omega_0}{Q}s + \omega_0^2}$
ALL PASS
$\frac{s^2 - \frac{\omega_0}{Q}s + \omega_0^2}{s^2 + \frac{\omega_0}{Q}s + \omega_0^2}$

Phase Response

As mentioned earlier, a filter will change the phase of the signal as well as the amplitude. The question is, does this make a difference? Fourier analysis indicates a square wave is made up of a fundamental frequency and odd order harmonics. The magnitude and phase responses of the various harmonics are precisely defined. If the magnitude or phase relationships are changed, the summation of the harmonics will not add back together properly to give a square wave. It will instead be distorted, typically showing overshoot and ringing or a slow rise time. This would also hold for any complex waveform.

Each pole of a filter will add 45° of phase shift at the corner frequency. The phase will vary from 0° (well below the corner frequency) to 90° (well beyond the corner frequency). The start of the change can be more than a decade away. In multipole filters, each of the poles will add phase shift, so that the total phase shift will be multiplied by the number of poles (180° total shift for a two-pole system, 270° for a three-pole system, and so forth).

The phase response of a single pole, low pass filter is:

$$\phi(\omega) = -\arctan\frac{\omega}{\omega_0} \qquad \text{Eq. 5-18}$$

The phase response of a low-pass pole pair is:

$$\phi(\omega) = -\arctan\left[\frac{1}{\alpha}\left(2\frac{\omega}{\omega_0} + \sqrt{4-\alpha^2}\right)\right] - \arctan\left[\frac{1}{\alpha}\left(2\frac{\omega}{\omega_0} - \sqrt{4-\alpha^2}\right)\right] \qquad \text{Eq. 5-19}$$

For a single pole high-pass filter the phase response is:

$$\phi(\omega) = \frac{\pi}{2} - \arctan\frac{\omega}{\omega_0} \qquad \text{Eq. 5-20}$$

The phase response of a high-pass pole pair is:

$$\phi(\omega) = \pi - \arctan\left[\frac{1}{\alpha}\left(2\frac{\omega}{\omega_0} + \sqrt{4-\alpha^2}\right)\right] - \arctan\left[\frac{1}{\alpha}\left(2\frac{\omega}{\omega_0} - \sqrt{4-\alpha^2}\right)\right] \qquad \text{Eq. 5-21}$$

The phase response of a band-pass filter is:

$$\phi(\omega) = \frac{\pi}{2} - \arctan\left(\frac{2Q\omega}{\omega_0} + \sqrt{4Q^2-1}\right) - \arctan\left(\frac{2Q\omega}{\omega_0} - \sqrt{4Q^2-1}\right) \qquad \text{Eq. 5-22}$$

The variation of the phase shift with frequency due to various values of Q is shown in Figure 5-11 (for low-pass, high-pass, band-pass, and all-pass) and in Figure 5-12 (for notch).

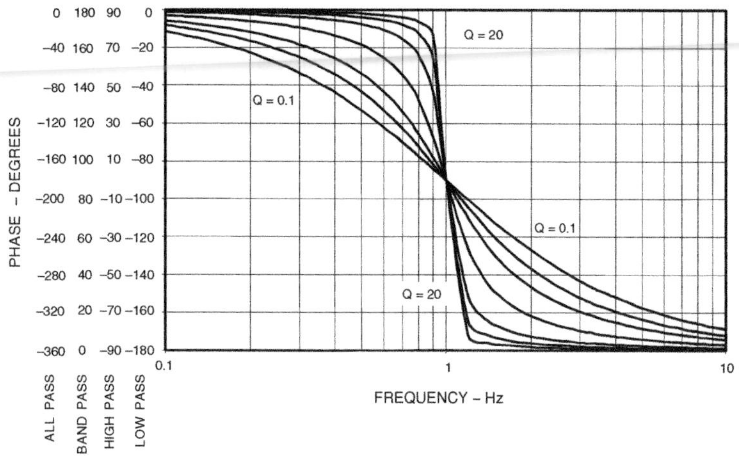

Figure 5-11: Phase response versus frequency

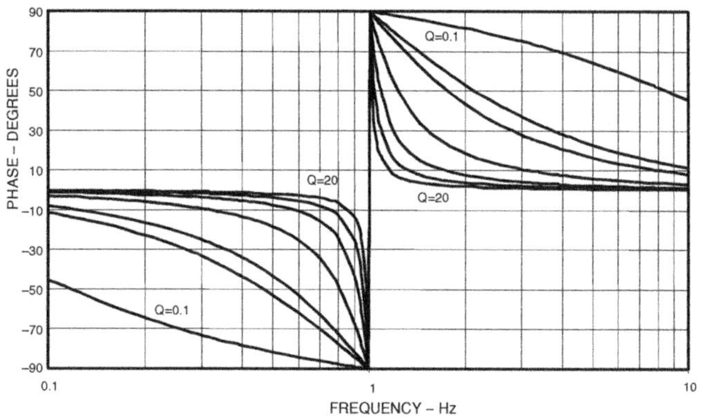

Figure 5-12: Notch filter phase response

It is also useful to look at the change of phase with frequency. This is the group delay of the filter. A flat (constant) group delay gives best phase response, but, unfortunately, it also gives the least amplitude discrimination. The group delay of a single low-pass pole is:

$$\tau(\omega) = -\frac{d\phi(\omega)}{d\omega} = \frac{\cos^2\phi}{\omega_0} \qquad \text{Eq. 5-23}$$

For the low-pass pole pair it is:

$$\tau(\omega) = \frac{2\sin^2\phi}{\alpha\omega_0} - \frac{\sin 2\phi}{2\omega} \qquad \text{Eq. 5-24}$$

For the single high-pass pole it is:

$$\tau(\omega) = -\frac{d\phi(\omega)}{d\omega} = \frac{\sin^2\phi}{\omega_0} \qquad \text{Eq. 5-25}$$

For the high-pass pole pair it is:

$$\tau(\omega)=\frac{2\sin^2\phi}{\alpha\omega_0}-\frac{\sin 2\phi}{2\omega} \qquad \text{Eq. 5-26}$$

And for the band-pass pole pair it is:

$$\tau(\omega)=\frac{2Q2\cos^2\phi}{\alpha\omega_0}+\frac{\sin 2\phi}{2\omega} \qquad \text{Eq. 5-27}$$

The Effect of Nonlinear Phase

A waveform can be represented by a series of frequencies of specific amplitude, frequency and phase relationships. For example, a square wave is:

$$F(t)=A\left(\frac{1}{2}+\frac{2}{\pi}\sin\omega t+\frac{2}{3\pi}\sin 3\omega t+\frac{2}{5\pi}\sin 5\omega t+\frac{2}{7\pi}\sin 7\omega t+\ldots\right) \qquad \text{Eq. 5-28}$$

If this waveform were passed through a filter, the amplitude and phase response of the filter to the various frequency components of the waveform could be different. If the phase delays were identical, the waveform would pass through the filter undistorted. If, however, the different components of the waveform were changed due to different amplitude and phase response of the filter to those frequencies, they would no longer add up in the same manner. This would change the shape of the waveform. These distortions would manifest themselves in what we typically call overshoot and ringing of the output.

Not all signals will be composed of harmonically related components. An amplitude modulated (AM) signal, for instance, will consist of a carrier and two sidebands at ± the modulation frequency. If the filter does not have the same delay for the various waveform components, then "envelope delay" will occur and the output wave will be distorted.

Linear phase shift results in constant group delay since the derivative of a linear function is a constant.

Time Domain Response

Until now the discussion has been primarily focused on the frequency domain response of filters. The time domain response can also be of concern, particularly under transient conditions. Moving between the time domain and the frequency domain is accomplished by the use of the Fourier and Laplace transforms. This yields a method of evaluating performance of the filter to a nonsinusoidal excitation.

The transfer function of a filter is the ratio of the output to input time functions. It can be shown that the impulse response of a filter defines its bandwidth. The time domain response is a practical consideration in many systems, particularly communications, where many modulation schemes use both amplitude and phase information.

Impulse Response

The impulse function is defined as an infinitely high, infinitely narrow pulse, with an area of unity. This is, of course, impossible to realize in a physical sense. If the impulse width is much less than the rise time of the filter, the resulting response of the filter will give a reasonable approximation actual impulse response of the filter response.

The impulse response of a filter, in the time domain, is proportional to the bandwidth of the filter in the frequency domain. The narrower the impulse, the wider the bandwidth of the filter. The pulse amplitude is equal to ω_c/π, which is also proportional to the filter bandwidth, the height being taller for wider bandwidths. The pulsewidth is equal to $2\pi/\omega_c$, which is inversely proportional to bandwidth. It turns out that the product of the amplitude and the bandwidth is a constant.

It would be a nontrivial task to calculate the response of a filter without the use of Laplace and Fourier transforms. The Laplace transform converts multiplication and division to addition and subtraction, respectively. This takes equations, which are typically loaded with integration and/or differentiation, and turns them into simple algebraic equations, which are much easier to deal with. The Fourier transform works in the opposite direction.

The details of these transforms will not be discussed here. However, some general observations on the relationship of the impulse response to the filter characteristics will be made.

It can be shown, as stated, that the impulse response is related to the bandwidth. Therefore, amplitude discrimination (the ability to distinguish between the desired signal from other, out-of-band signals and noise) and time response are inversely proportional. That is to say that the filters with the best amplitude response are the ones with the worst time response. For all-pole filters, the Chebyshev filter gives the best amplitude discrimination, followed by the Butterworth and then the Bessel.

If the time domain response were ranked, the Bessel would be best, followed by the Butterworth and then the Chebyshev. Details of the different filter responses will be discussed in the next section.

The impulse response also increases with increasing filter order. Higher filter order implies greater bandlimiting, therefore degraded time response. Each section of a multistage filter will have its own impulse response, and the total impulse response is the accumulation of the individual responses. The degradation in the time response can also be related to the fact that as frequency discrimination is increased, the Q of the individual sections tends to increase. The increase in Q increases the overshoot and ringing of the individual sections, which implies longer time response.

Step Response

The step response of a filter is the integral of the impulse response. Many of the generalities that apply to the impulse response also apply to the step response. The slope of the rise time of the step response is equal to the peak response of the impulse. The product of the bandwidth of the filter and the rise time is a constant. Just as the impulse has a function equal to unity, the step response has a function equal to 1/s. Both of these expressions can be normalized, since they are dimensionless.

The step response of a filter is useful in determining the envelope distortion of a modulated signal. The two most important parameters of a filter's step response are the overshoot and ringing. Overshoot should be minimal for good pulse response. Ringing should decay as fast as possible, so as not to interfere with subsequent pulses.

Real life signals typically aren't made up of impulse pulses or steps, so the transient response curves don't give a completely accurate estimation of the output. They are, however, a convenient figure of merit so that the transient responses of the various filter types can be compared on an equal footing.

Since the calculations of the step and impulse response are mathematically intensive, they are most easily performed by computer. Many CAD (Computer Aided Design) software packages have the ability to calculate these responses. Several of these responses are also collected in the next section.

SECTION 5-4

Standard Responses

There are many transfer functions to satisfy the attenuation and/or phase requirements of a particular filter. The one chosen will depend on the particular system. The importance of the frequency domain response versus the time domain response must be determined. Also, both of these considerations might be traded off against filter complexity, and thereby cost.

Butterworth

The Butterworth filter is the best compromise between attenuation and phase response. It has no ripple in the pass band or the stop band, and because of this is sometimes called a maximally flat filter. The Butterworth filter achieves its flatness at the expense of a relatively wide transition region from pass band to stop band, with average transient characteristics.

The normalized poles of the Butterworth filter fall on the unit circle (in the *s* plane). The pole positions are given by:

$$-\sin\frac{(2K-1)\pi}{2n}+j\cos\frac{(2K-1)\pi}{2n} \qquad K=1,2....n \qquad \text{Eq. 5-29}$$

where K is the pole pair number, and n is the number of poles.

The poles are spaced equidistant on the unit circle, which means the angles between the poles are equal.

Given the pole locations, ω_0, and α (or Q) can be determined. These values can then be used to determine the component values of the filter. The design tables for passive filters use frequency and impedance normalized filters. They are normalized to a frequency of 1 rad/sec and impedance of 1 Ω. These filters can be denormalized to determine actual component values. This allows the comparison of the frequency domain and/or time domain responses of the various filters on equal footing. The Butterworth filter is normalized for a –3 dB response at $\omega_o = 1$.

The values of the elements of the Butterworth filter are more practical and less critical than many other filter types. The frequency response, group delay, impulse response and step response are shown in Figure 5-15. The pole locations and corresponding ω_o and α terms are tabulated in Figure 5-26.

Chebyshev

The Chebyshev (or Chevyshev, Tschebyschev, Tschebyscheff or Tchevysheff, depending on how you translate from Russian) filter has a smaller transition region than the same-order Butterworth filter, at the expense of ripples in its pass band. This filter gets its name because the Chebyshev filter minimizes the height of the maximum ripple, which is the Chebyshev criterion.

Chebyshev filters have 0 dB relative attenuation at dc. Odd order filters have an attenuation band that extends from 0 dB to the ripple value. Even order filters have a gain equal to the pass band ripple. The number of cycles of ripple in the pass band is equal to the order of the filter.

The poles of the Chebyshev filter can be determined by moving the poles of the Butterworth filter to the right, forming an ellipse. This is accomplished by multiplying the real part of the pole by k_r and the imaginary part by k_I. The values k_r and k_I can be computed by:

$$K_r = \sinh A \qquad \text{Eq. 5-30}$$

$$K_I = \cosh A \qquad \text{Eq. 5-31}$$

where:

$$A = \frac{1}{n}\sinh^{-1}\frac{1}{\varepsilon} \qquad \text{Eq. 5-32}$$

where n is the filter order and:

$$\varepsilon = \sqrt{10^R - 1} \qquad \text{Eq. 5-33}$$

where:

$$R = \frac{R_{dB}}{10} \qquad \text{Eq. 5-34}$$

where:

$$R_{dB} = \text{pass band ripple in dB} \qquad \text{Eq. 5-35}$$

The Chebyshev filters are typically normalized so that the edge of the ripple band is at $\omega_0 = 1$. The 3 dB bandwidth is given by:

$$A_{3dB} = \frac{1}{n}\cosh^{-1}\left(\frac{1}{\varepsilon}\right) \qquad \text{Eq. 5-36}$$

This is shown in Table 1.

The frequency response, group delay, impulse response and step response are cataloged in Figures 5-16 to 5-20 on following pages, for various values of pass band ripple (0 .01 dB, 0.1 dB, 0.25 dB, 0.5 dB and 1 dB). The pole locations and corresponding ω_0 and α terms for these values of ripple are tabulated in Figures 5-27 to 5-31 on following pages.

ORDER	0.01dB	0.1dB	0.25dB	0.5dB	1dB
2	3.30362	1.93432	1.59814	1.38974	1.21763
3	1.87718	1.38899	1.25289	1.16749	1.09487
4	1.46690	1.21310	1.13977	1.09310	1.05300
5	1.29122	1.13472	1.08872	1.05926	1.03381
6	1.19941	1.09293	1.06134	1.04103	1.02344
7	1.14527	1.06800	1.04495	1.03009	1.01721
8	1.11061	1.05193	1.03435	1.02301	1.01316
9	1.08706	1.04095	1.02711	1.01817	1.01040
10	1.07033	1.03313	1.02194	1.01471	1.00842

Table 1: 3dB bandwidth to ripple bandwidth for Chebyshev filters

Bessel

Butterworth filters have fairly good amplitude and transient behavior. The Chebyshev filters improve on the amplitude response at the expense of transient behavior. The Bessel filter is optimized to obtain better transient response due to a linear phase (i.e., constant delay) in the pass band. This means that there will be relatively poorer frequency response (less amplitude discrimination).

The poles of the Bessel filter can be determined by locating all of the poles on a circle and separating their imaginary parts by:

$$\frac{2}{n} \qquad \text{Eq. 5-37}$$

where n is the number of poles. Note that the top and bottom poles are distanced by where the circle crosses the $j\omega$ axis by:

$$\frac{1}{n} \qquad \text{Eq. 5-38}$$

or half the distance between the other poles.

The frequency response, group delay, impulse response and step response for the Bessel filters are cataloged in Figure 5-21. The pole locations and corresponding ω_0 and α terms for the Bessel filter are tabulated in Figure 5-32.

Linear Phase with Equiripple Error

The linear phase filter offers linear phase response in the pass band, over a wider range than the Bessel, and superior attenuation far from cutoff. This is accomplished by letting the phase response have ripples, similar to the amplitude ripples of the Chebyshev. As the ripple is increased, the region of constant delay extends further into the stop band. This will also cause the group delay to develop ripples, since it is the derivative of the phase response. The step response will show slightly more overshoot than the Bessel and the impulse response will show a bit more ringing.

It is difficult to compute the pole locations of a linear phase filter. Pole locations are taken from the Williams book (see Reference 2), which, in turn, comes from the Zverev book (see Reference 1).

The frequency response, group delay, impulse response and step response for linear phase filters of 0.05° ripple and 0.5° ripple are given in Figures 5-22 and 5-23. The pole locations and corresponding ω_0 and α terms are tabulated in Figures 5-33 and 5-34.

Transitional Filters

A transitional filter is a compromise between a Gaussian filter, which is similar to a Bessel, and the Chebyshev. A transitional filter has nearly linear phase shift and smooth, monotonic rolloff in the pass band. Above the pass band there is a break point beyond which the attenuation increases dramatically compared to the Bessel, and especially at higher values of n.

Two transition filters have been tabulated. These are the Gaussian to 6 dB and Gaussian to 12 dB.

The Gaussian to 6 dB filter has better transient response than the Butterworth in the pass band. Beyond the breakpoint, which occurs at $\omega = 1.5$, the roll-off is similar to the Butterworth.

The Gaussian to 12 dB filter's transient response is much better than Butterworth in the pass band. Beyond the 12 dB breakpoint, which occurs at $\omega = 2$, the attenuation is less than the Butterworth.

As is the case with the linear phase filters, pole locations for transitional filters do not have a closed form method for computation. Again, pole locations are taken from Williams's book (see Reference 2). These were derived from iterative techniques.

The frequency response, group delay, impulse response and step response for Gaussian to 12 dB and 6 dB are shown in Figures 5-24 and 5-25. The pole locations and corresponding ω_0 and α terms are tabulated in Figures 5-35 and 5-36.

Comparison of All-Pole Responses

The responses of several all-pole filters, namely the Bessel, Butterworth and Chebyshev (in this case of 0.5 dB ripple) will now be compared. An 8-pole filter is used as the basis for the comparison. The responses have been normalized for a cutoff of 1 Hz. Comparing Figures 5-13 and 5-14, it is easy to see the trade-offs in the response types. Moving from Bessel through Butterworth to Chebyshev, notice that the amplitude discrimination improves as the transient behavior gets progressively poorer.

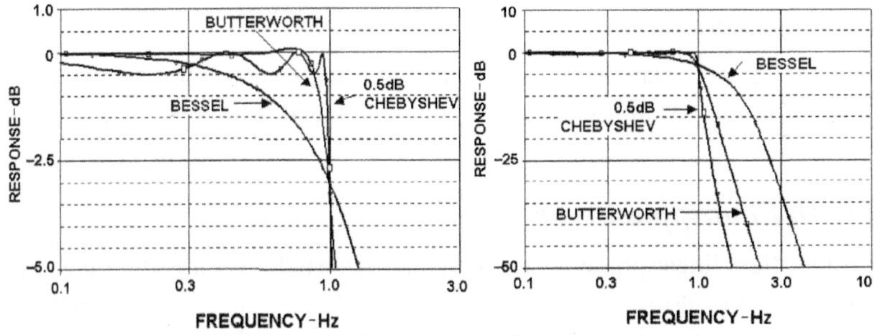

Figure 5-13: Comparison of amplitude response of Bessel, Butterworth and Chebyshev filters

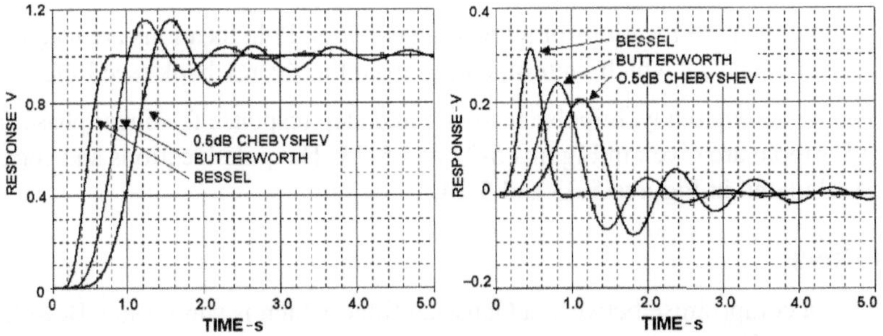

Figure 5-14: Comparison of Step and Impulse Responses of Bessel, Butterworth and Chebyshev

Elliptical

The previously mentioned filters are all-pole designs, which mean that the zeros of the transfer function (roots of the numerator) are at one of the two extremes of the frequency range (0 or ∞). For a low-pass filter, the zeros are at $f = \infty$. If finite frequency transfer function zeros are added to poles an Elliptical filter (sometimes referred to as Cauer filters) is created. This filter has a shorter transition region than the Chebyshev filter because it allows ripple in both the stop band and pass band. It is the addition of zeros in the stop band that causes ripple in the stop band but gives a much higher rate of attenuation, the most possible for a given number of poles. There will be some "bounceback" of the stop band response between the zeros. This is the stop band ripple. The Elliptical filter also has degraded time domain response.

Since the poles of an elliptical filter are on an ellipse, the time response of the filter resembles that of the Chebyshev.

An elliptical filter is defined by the parameters shown in Figure 5-2, those being A_{max}, the maximum ripple in the pass band, A_{min}, the minimum attenuation in the stop band, F_c, the cutoff frequency, which is where the frequency response leaves the pass band ripple and F_S, the stop band frequency, where the value of A_{max} is reached.

An alternate approach is to define a filter order n, the modulation angle, θ, which defines the rate of attenuation in the transition band, where:

$$\theta = \sin^{-1}\frac{1}{F_s} \qquad \text{Eq. 5-39}$$

and ρ which determines the pass band ripple, where:

$$\rho = \sqrt{\frac{\varepsilon^2}{1+\varepsilon^2}} \qquad \text{Eq. 5-40}$$

where ε is the ripple factor developed for the Chebyshev response, and the pass band ripple is:

$$R_{dB} = -10\log\left(1-\rho^2\right) \qquad \text{Eq. 5-41}$$

Some general observations can be made. For a given filter order n, and θ, A_{min} increases as the ripple is made larger. Also, as θ approaches 90°, F_S approaches F_C. This results in extremely short transition region, which means sharp roll-off. This comes at the expense of lower A_{min}.

As a side note, ρ determines the input resistance of a passive elliptical filter, which can then be related to the VSWR (Voltage Standing Wave Ratio).

Because of the number of variables in the design of an elliptic filter, it is difficult to provide the type of tables provided for the previous filter types. Several CAD (Computer Aided Design) packages can provide the design values. Alternatively several sources, such as Williams's (see Reference 2), provide tabulated filter values. These tables classify the filter by

$$C\ n\ \rho\ \theta$$

where the *C* denotes Cauer. Elliptical filters are sometime referred to as Cauer filters after the network theorist Wilhelm Cauer.

Maximally Flat Delay With Chebyshev Stop band

Bessel type (Bessel, linear phase with equiripple error and transitional) filters give excellent transient behavior, but less than ideal frequency discrimination. Elliptical filters give better frequency discrimination, but degraded transient response. A maximally flat delay with Chebyshev stop band filter takes a Bessel type function and adds transmission zeros. The constant delay properties of the Bessel type filter in the pass band are maintained, and the stop band attenuation is significantly improved. The step response exhibits no overshoot or ringing, and the impulse response is clean, with essentially no oscillatory behavior. Constant group delay properties extend well into the stop band for increasing n.

As with the elliptical filter, numeric evaluation is difficult. Williams's book (see Reference 2) tabulates passive prototypes normalized component values.

Inverse Chebyshev

The Chebyshev response has ripple in the pass band and a monotonic stop band. The inverse Chebyshev response can be defined that has a monotonic pass band and ripple in the stop band. The inverse Chebyshev has better pass band performance than even the Butterworth. It is also better than the Chebyshev, except very near the cutoff frequency. In the transition band, the inverse Chebyshev has the steepest roll-off. Therefore, the inverse Chebyshev will meet the A_{min} specification at the lowest frequency of the three. In the stop band there will, however, be response lobes which have a magnitude of:

$$\frac{\varepsilon^2}{(1-\varepsilon)} \qquad \text{Eq. 5-42}$$

where ε is the ripple factor defined for the Chebyshev case. This means that deep into the stop band, both the Butterworth and Chebyshev will have better attenuation, since they are monotonic in the stop band. In terms of transient performance, the inverse Chebyshev lies midway between the Butterworth and the Chebyshev.

The inverse Chebyshev response can be generated in three steps. First take a Chebyshev low-pass filter. Then subtract this response from 1. Finally, invert in frequency by replacing ω with $1/\omega$.

These are by no means all the possible transfer functions, but they do represent the most common.

Using the Prototype Response Curves

In the following pages, the response curves and the design tables for several of the low-pass prototypes of the all-pole responses will be cataloged. All the curves are normalized to a −3 dB cutoff frequency of 1 Hz. This allows direct comparison of the various responses. In all cases the amplitude response for the 2- through 10-pole cases for the frequency range of 0.1 Hz. to 10 Hz. will be shown. Then a detail of the amplitude response in the 0.1 Hz to 2 Hz. pass band will be shown. The group delay from 0.1 Hz to 10 Hz and the impulse response and step response from 0 seconds to 5 seconds will also be shown.

To use these curves to determine the response of real life filters, they must be denormalized. In the case of the amplitude responses, this is accomplished simply by multiplying the frequency axis by the desired cutoff frequency F_C. To denormalize the group delay curves, divide the delay axis by $2\pi F_C$, and multiply the frequency axis by F_C, as before. Denormalize the step response by dividing the time axis by $2\pi F_C$. Denormalize the impulse response by dividing the time axis by $2\pi F_C$ and multiplying the amplitude axis by the same amount.

For a high-pass filter, simply invert the frequency axis for the amplitude response. In transforming a low-pass filter into a high-pass (or bandreject) filter, the transient behavior is not preserved. Zverev (see Reference 1) provides a computational method for calculating these responses.

In transforming a low-pass into a narrowband band-pass, the 0 Hz axis is moved to the center frequency F_0. It stands to reason that the response of the band-pass case around the center frequency would then match the low-pass response around 0 Hz. The frequency response curve of a low-pass filter actually mirrors itself around 0 Hz, although negative frequency generally is not a concern.

To denormalize the group delay curve for a band-pass filter, divide the delay axis by πBW, where BW is the 3 dB bandwidth in Hz. Then multiply the frequency axis by BW/2. In general, the delay of the band-pass filter at F_0 will be twice the delay of the low-pass prototype with the same bandwidth at 0Hz. This is due to the fact that the low-pass to band-pass transformation results in a filter with order 2n, even though it is typically referred to it as having the same order as the low-pass filter from which it is derived. This approximation holds for narrowband filters. As the bandwidth of the filter is increased, some distortion of the curve occurs. The delay becomes less symmetrical, peaking below F_0.

The envelope of the response of a band-pass filter resembles the step response of the low-pass prototype. More exactly, it is almost identical to the step response of a low-pass filter having half the bandwidth. To determine the envelope response of the band-pass filter, divide the time axis of the step response of the low-pass prototype by πBW, where BW is the 3 dB bandwidth. The previous discussions of overshoot, ringing, and so forth can now be applied to the carrier envelope.

The envelope of the response of a narrowband band-pass filter to a short burst of carrier (that is where the burst width is much less than the rise time of the denormalized step response of the band-pass filter) can be determined by denormalizing the impulse response of the low pass prototype. To do this, multiply the amplitude axis and divide the time axis by πBW, where BW is the 3 dB bandwidth. It is assumed that the carrier frequency is high enough so that many cycles occur during the burst interval.

While the group delay, step, and impulse curves cannot be used directly to predict the distortion to the waveform caused by the filter, they are a useful figure of merit when used to compare filters.

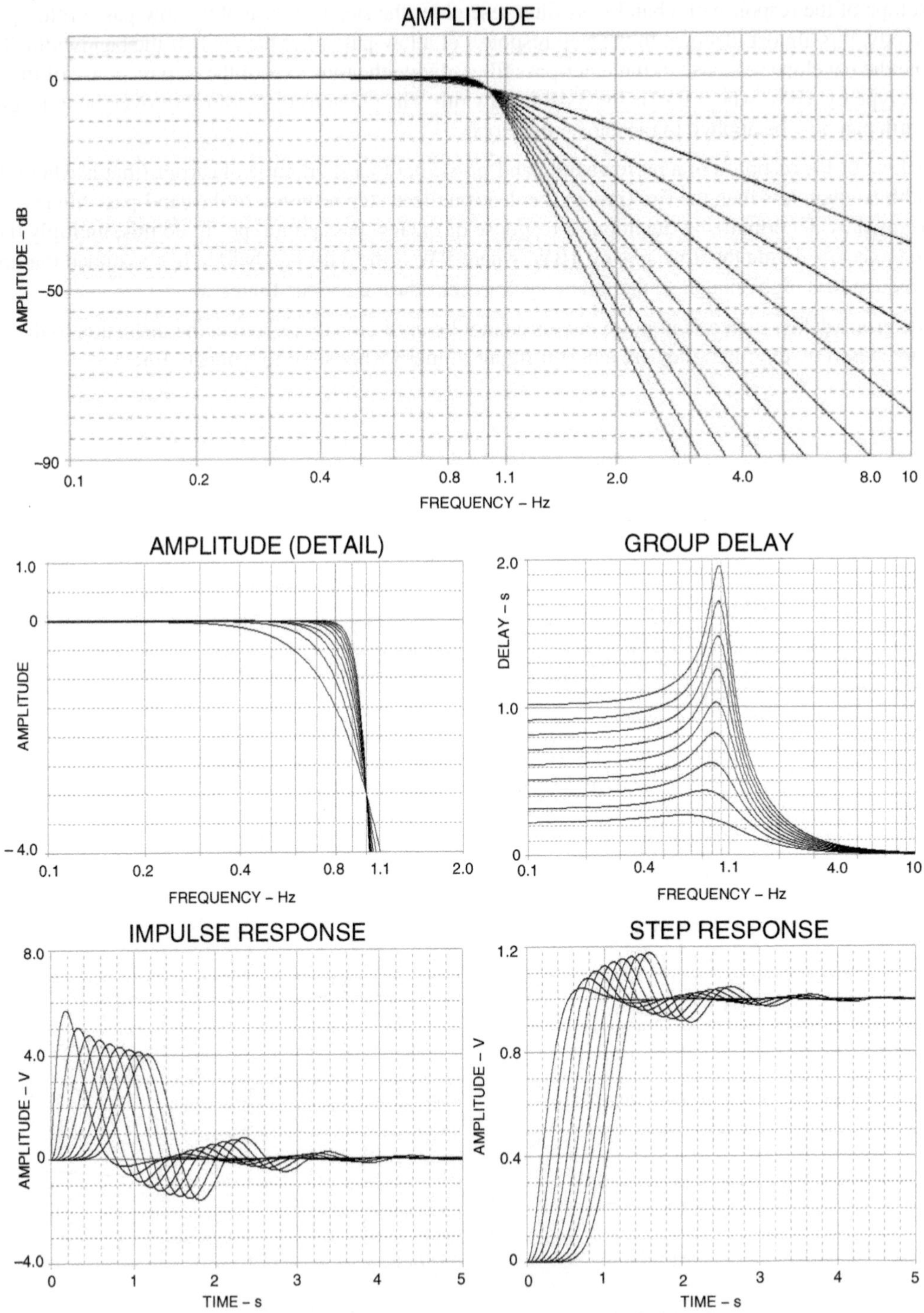

Figure 5-15: Butterworth response

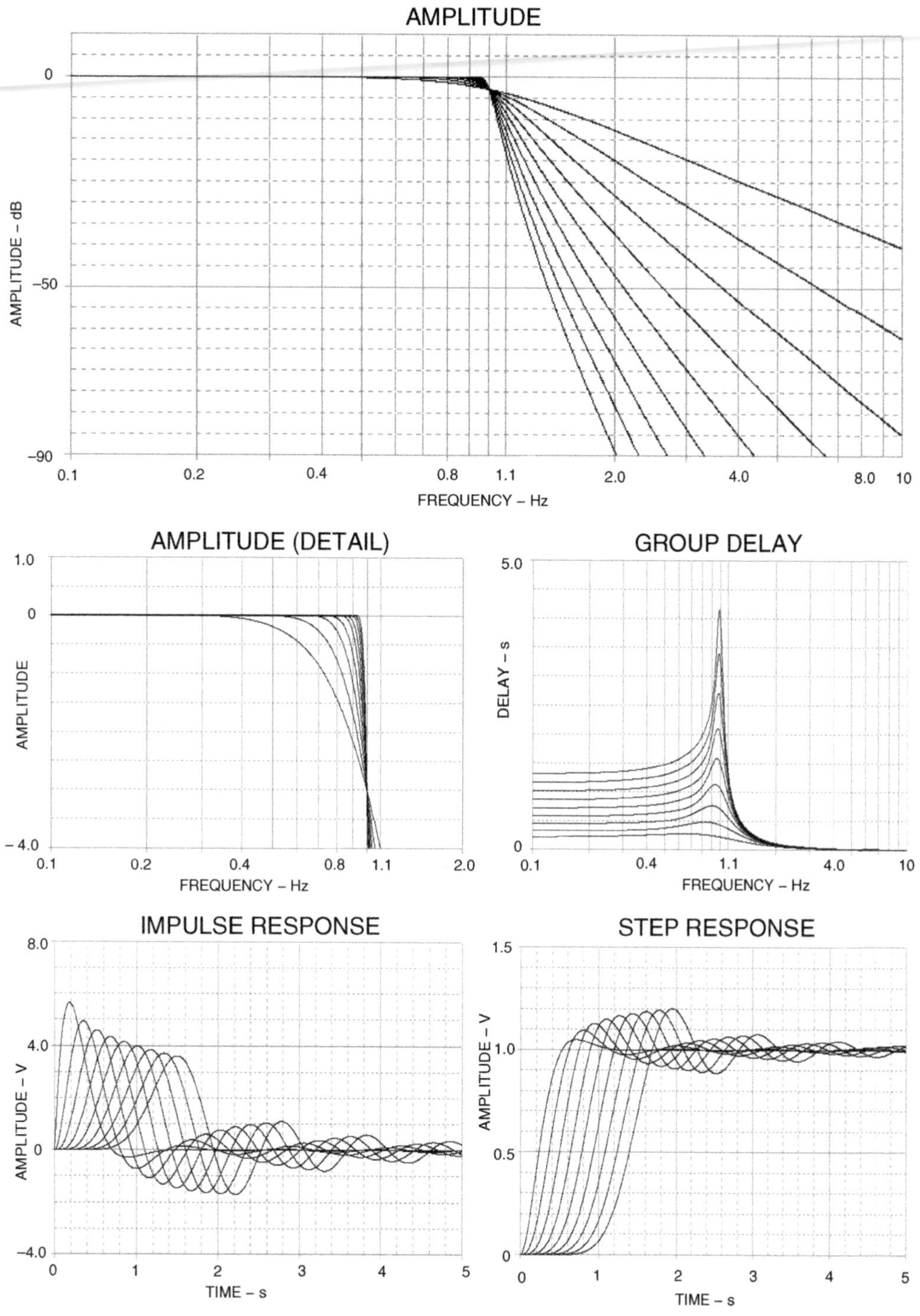

Figure 5-16: 0.01 dB Chebyshev response

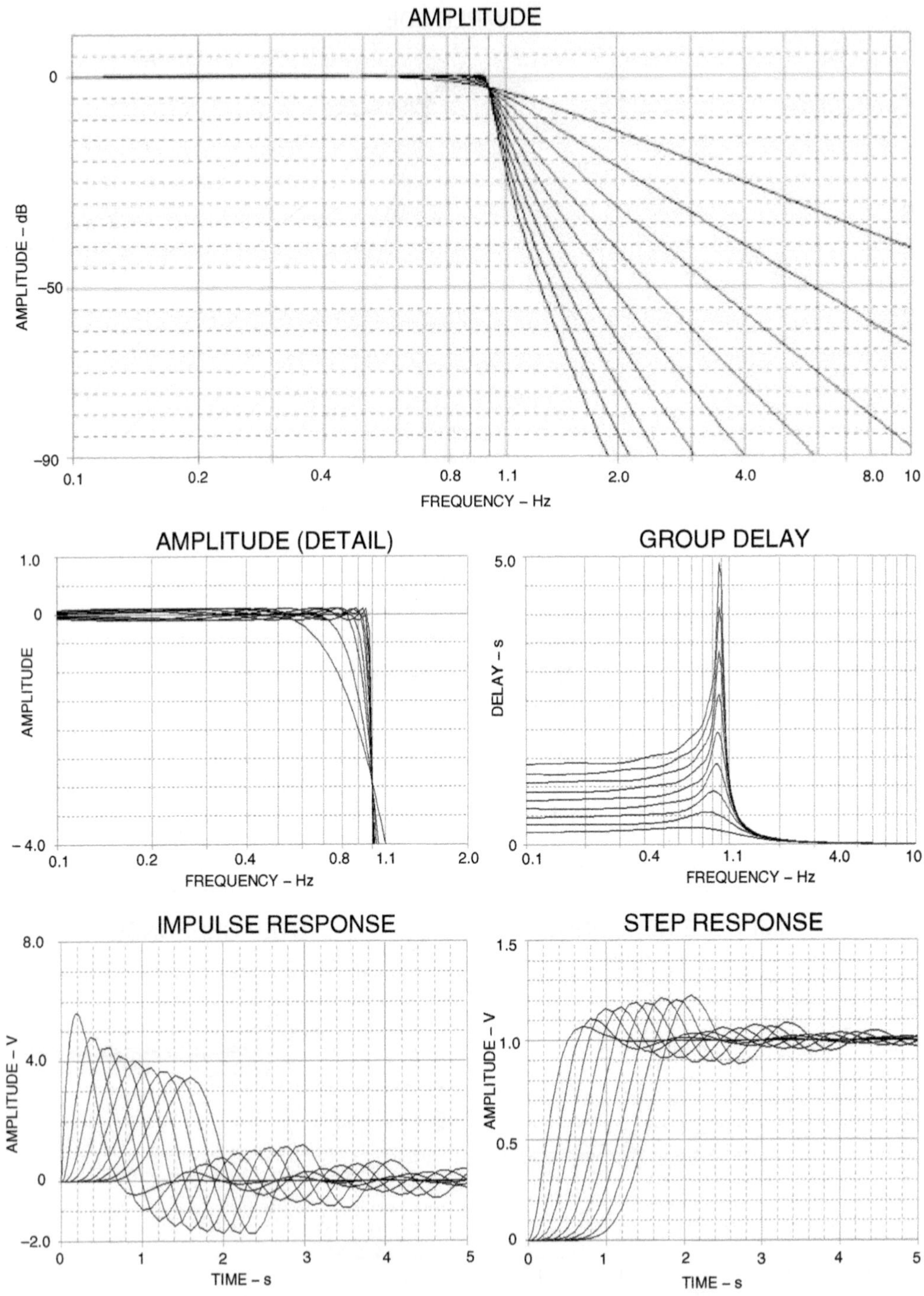

Figure 5-17: 0.1 dB Chebyshev response

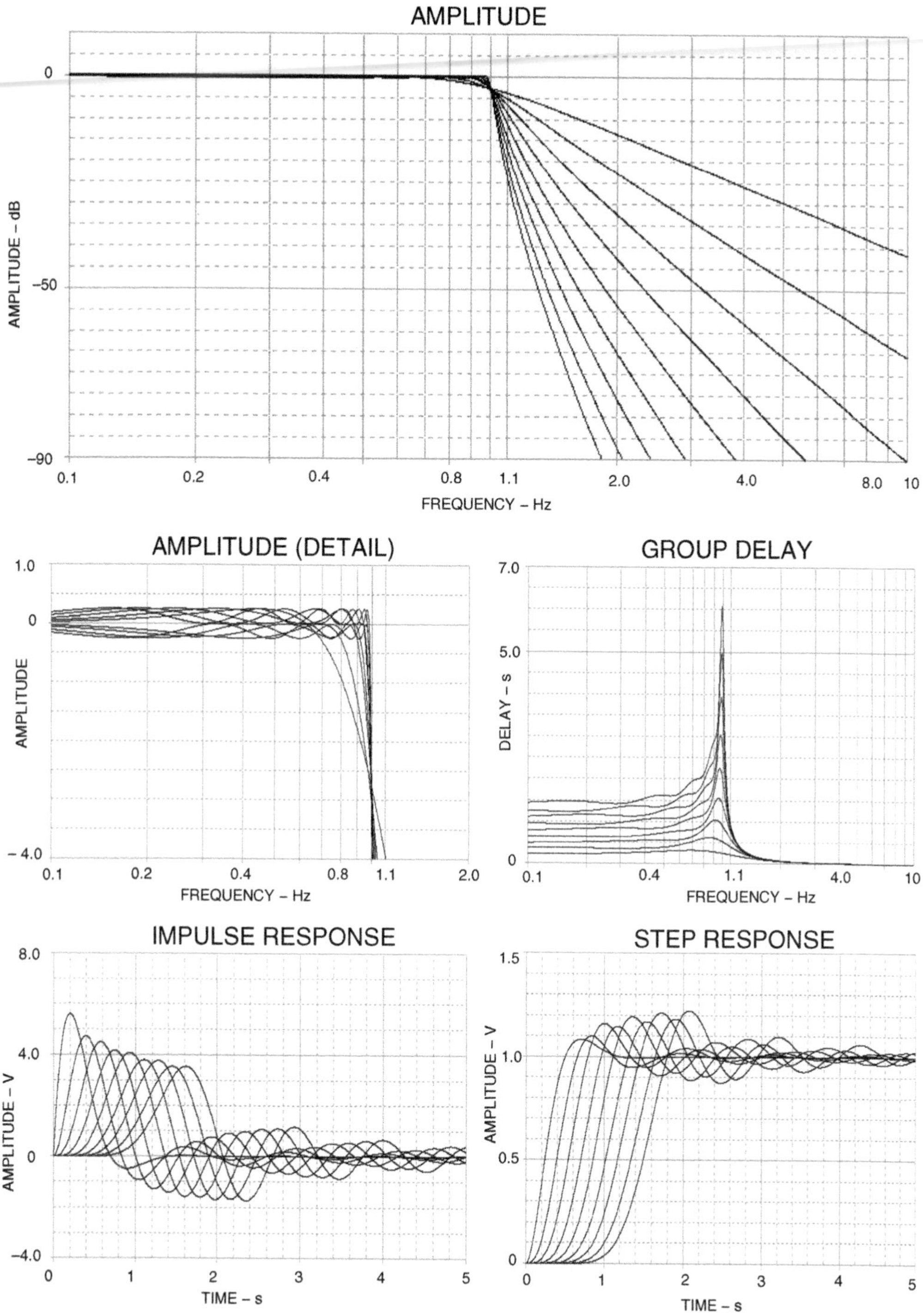

Figure 5-18: 0.25 dB Chebyshev response

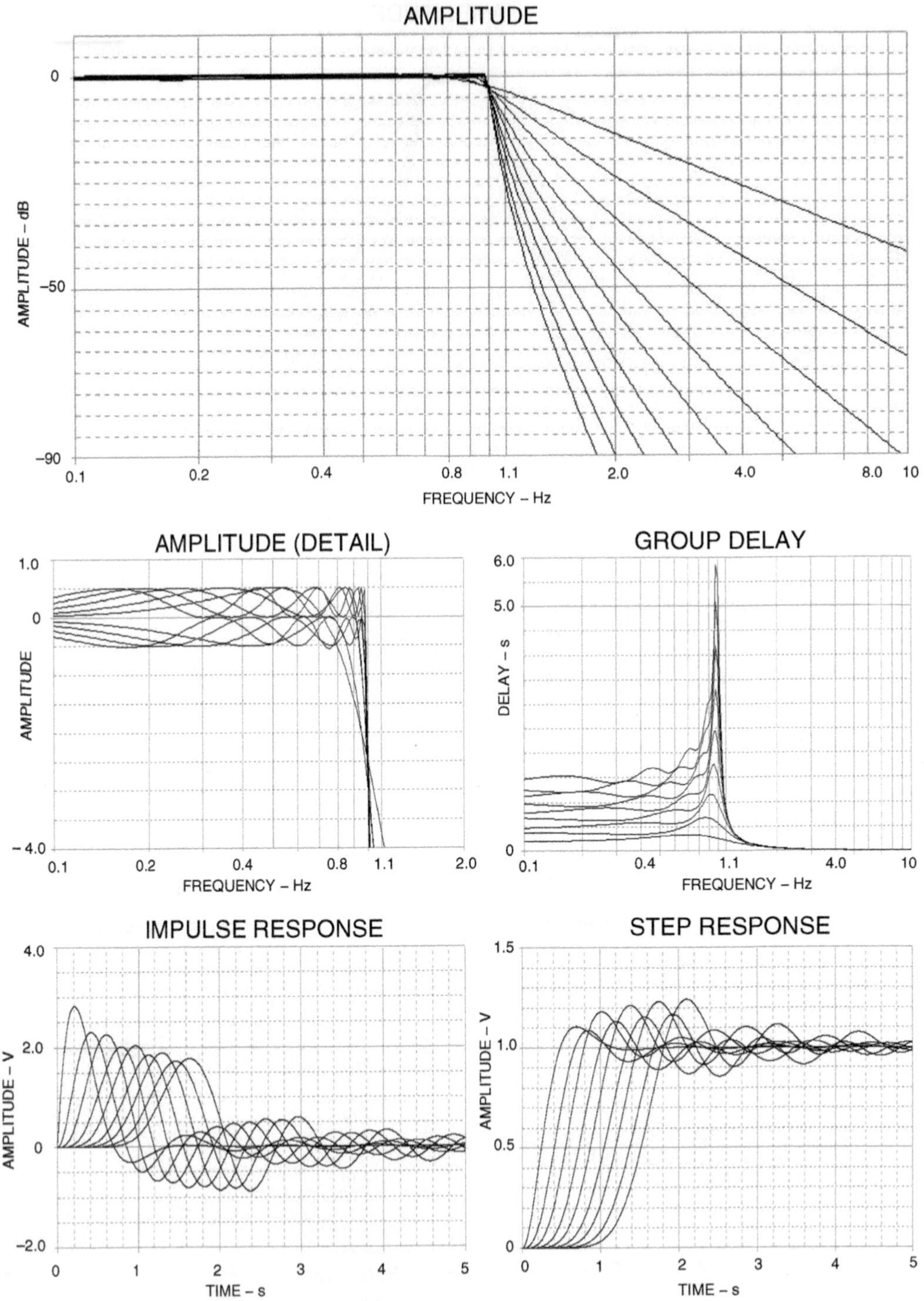

Figure 5-19: 0.5 dB Chebyshev response

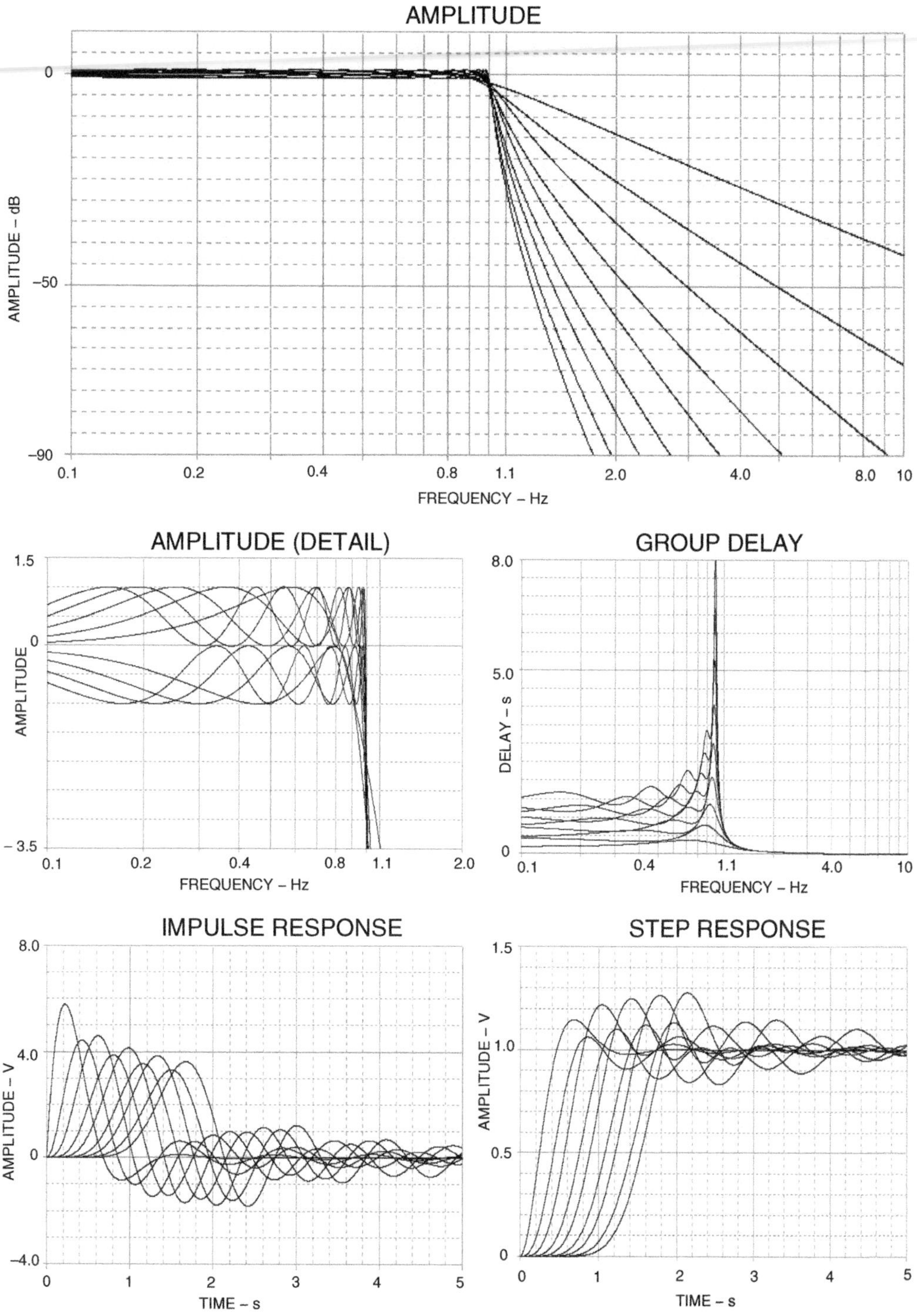

Figure 5-20: 1 dB Chebyshev response

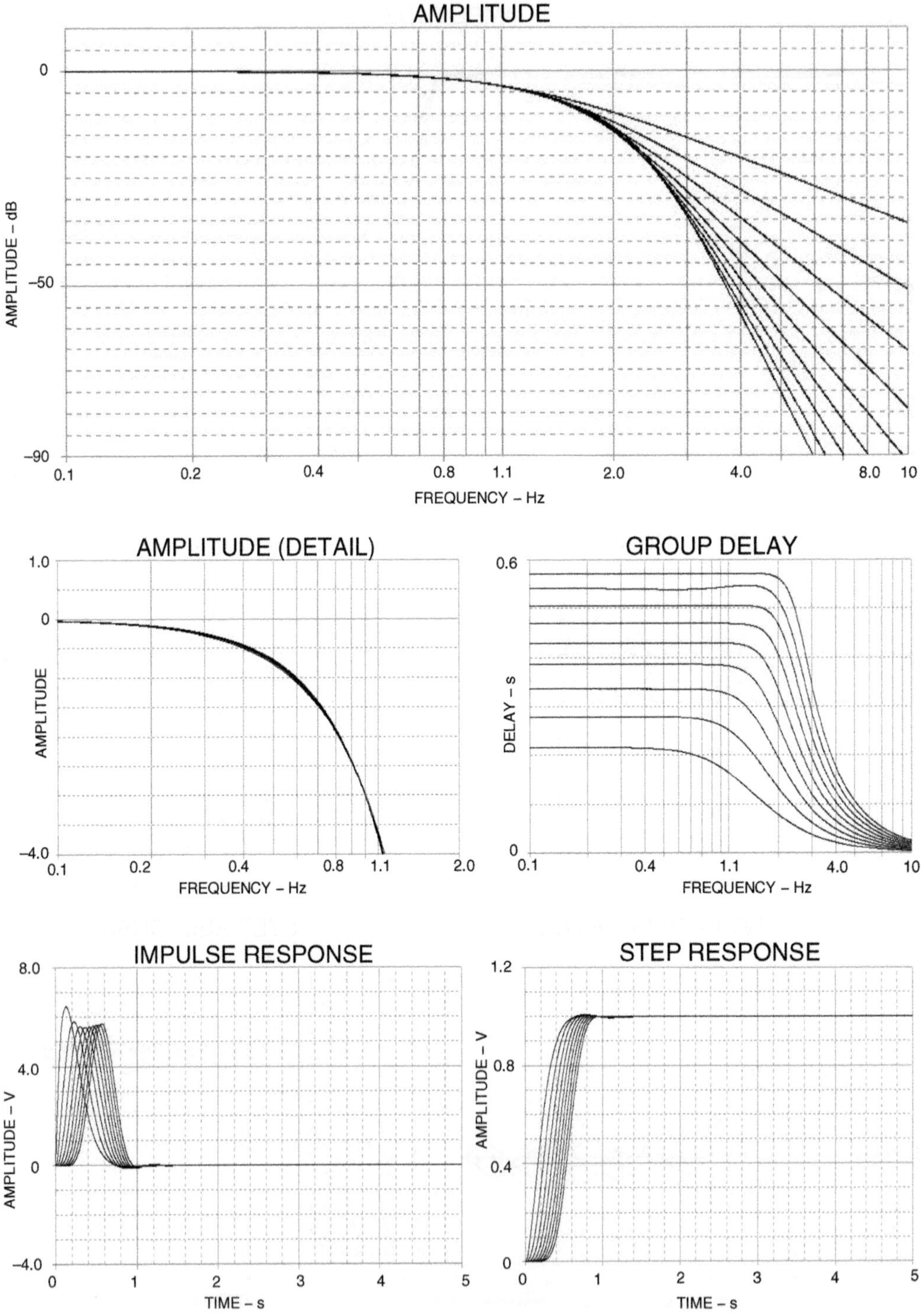

Figure 5-21: Bessel response

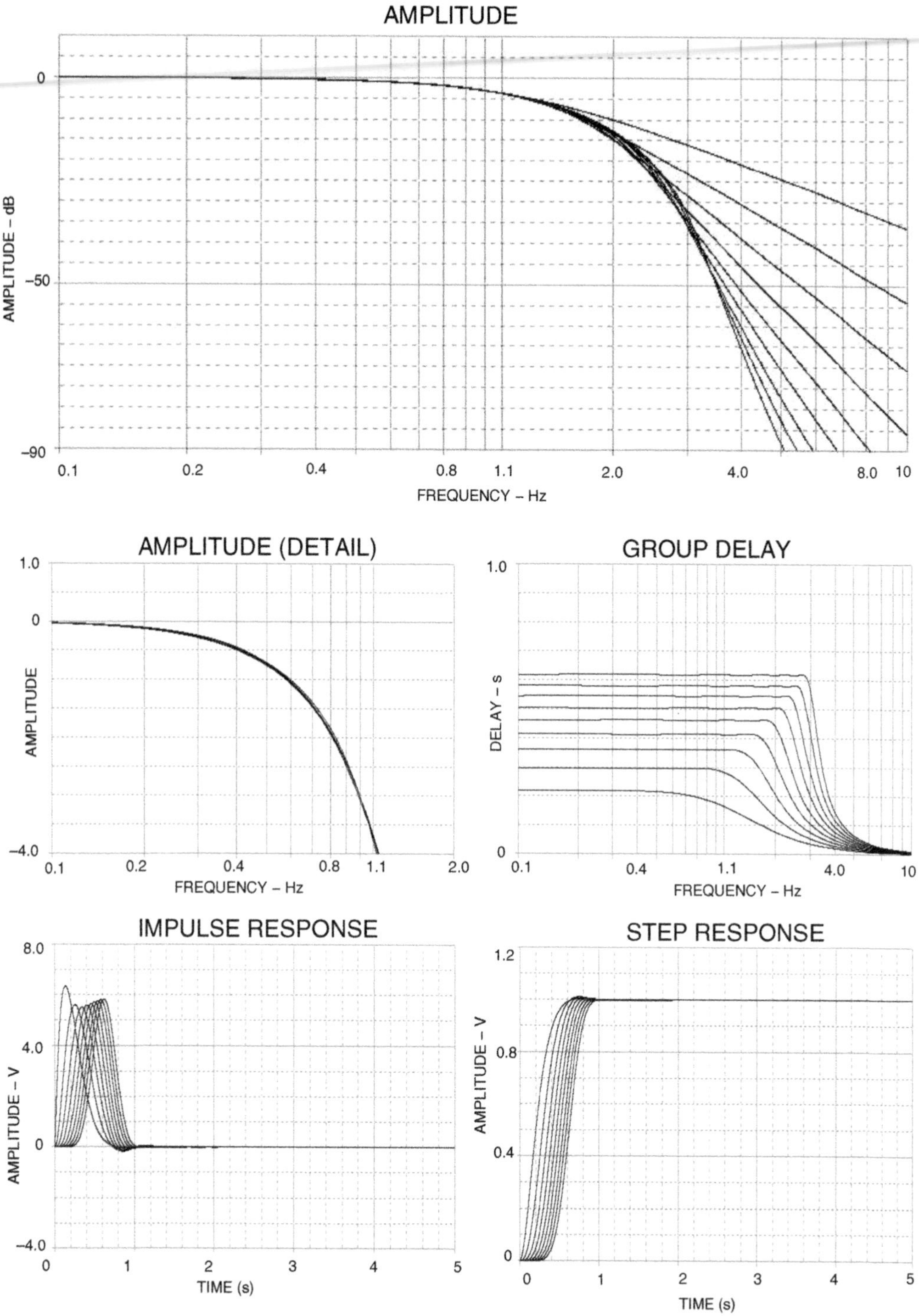

Figure 5-22: Linear phase response with equiripple error of 0.05°

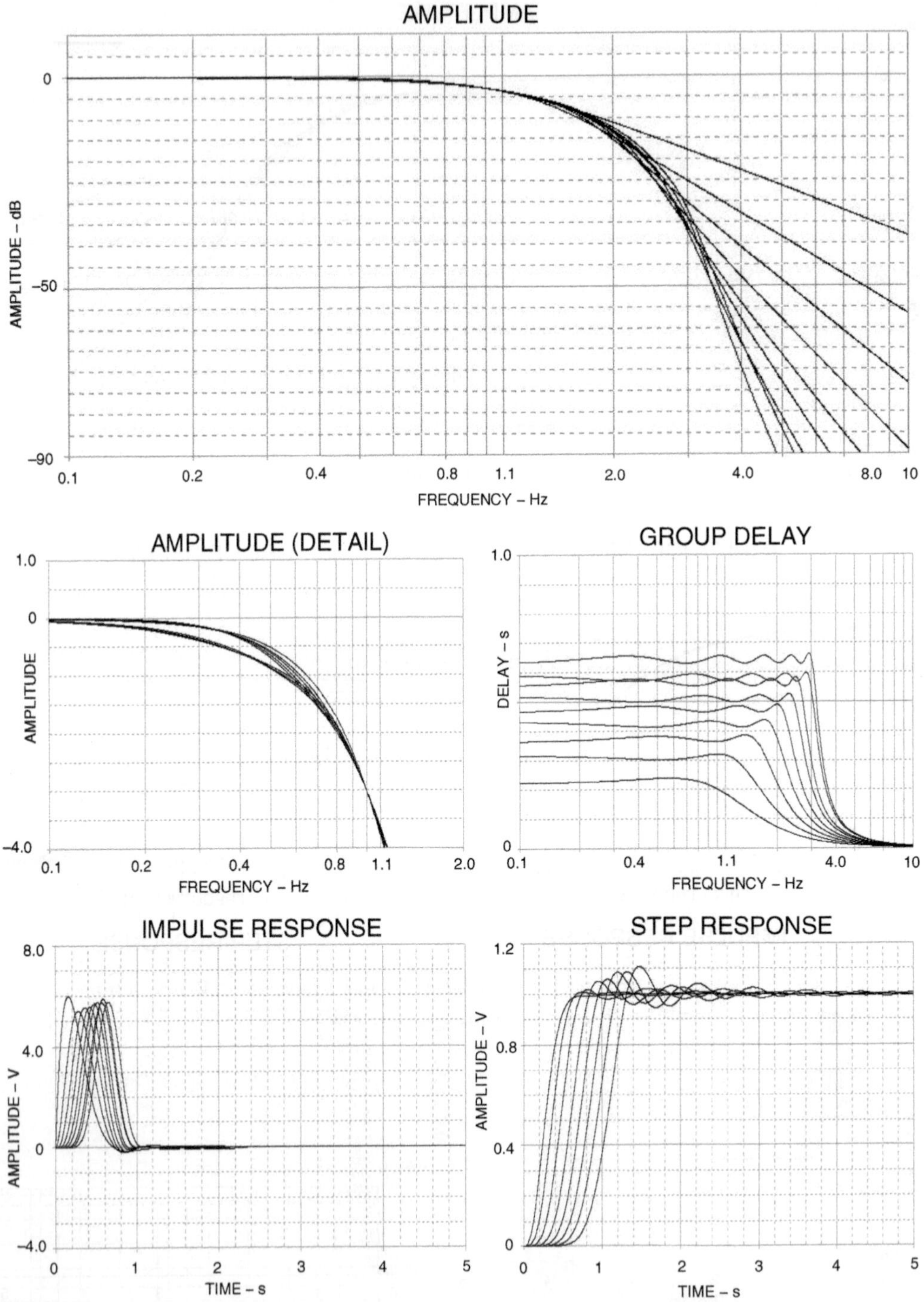

Figure 5-23: Linear phase response with equiripple error of 0.5°

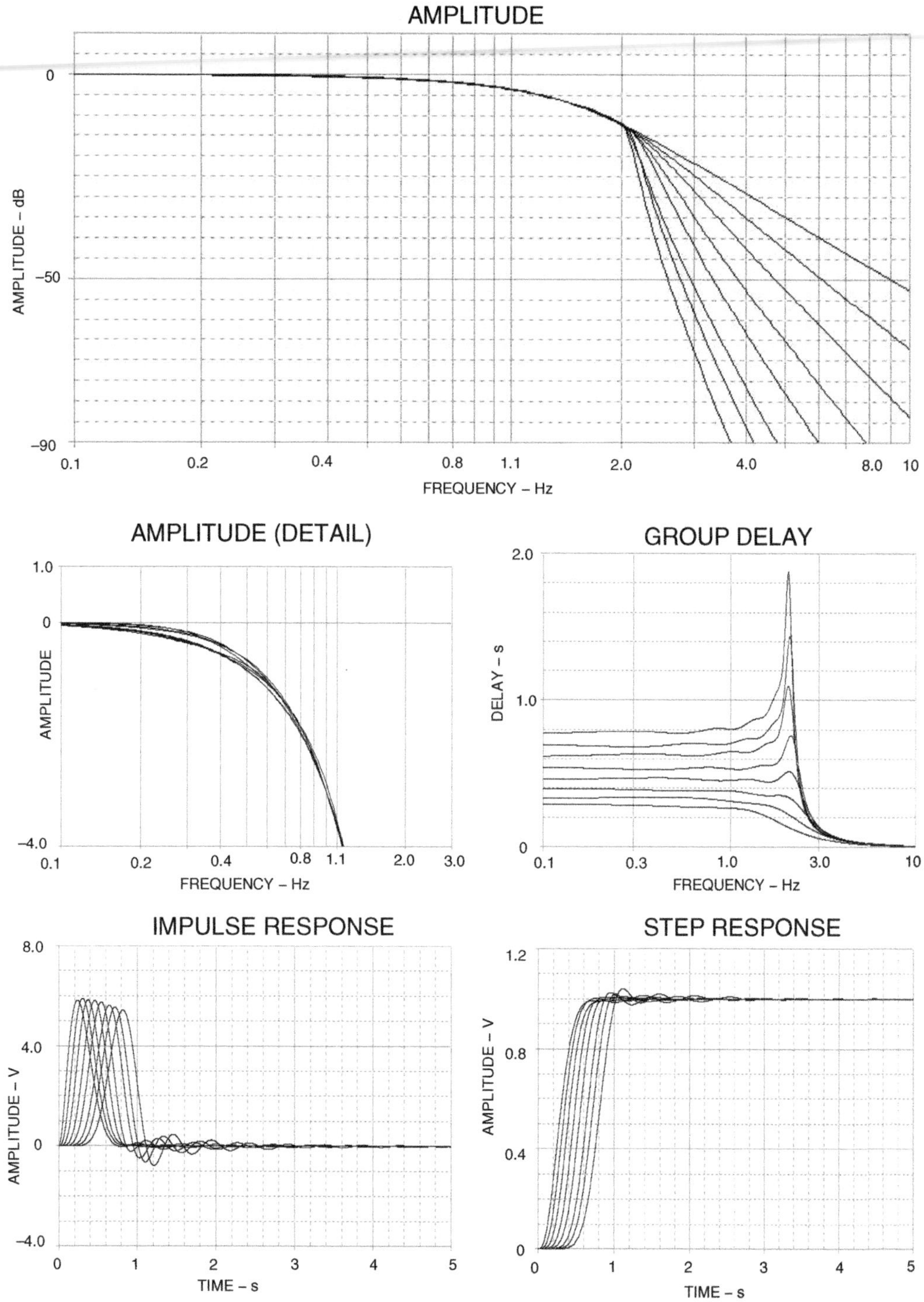

Figure 5-24: Gaussian to 12 dB response

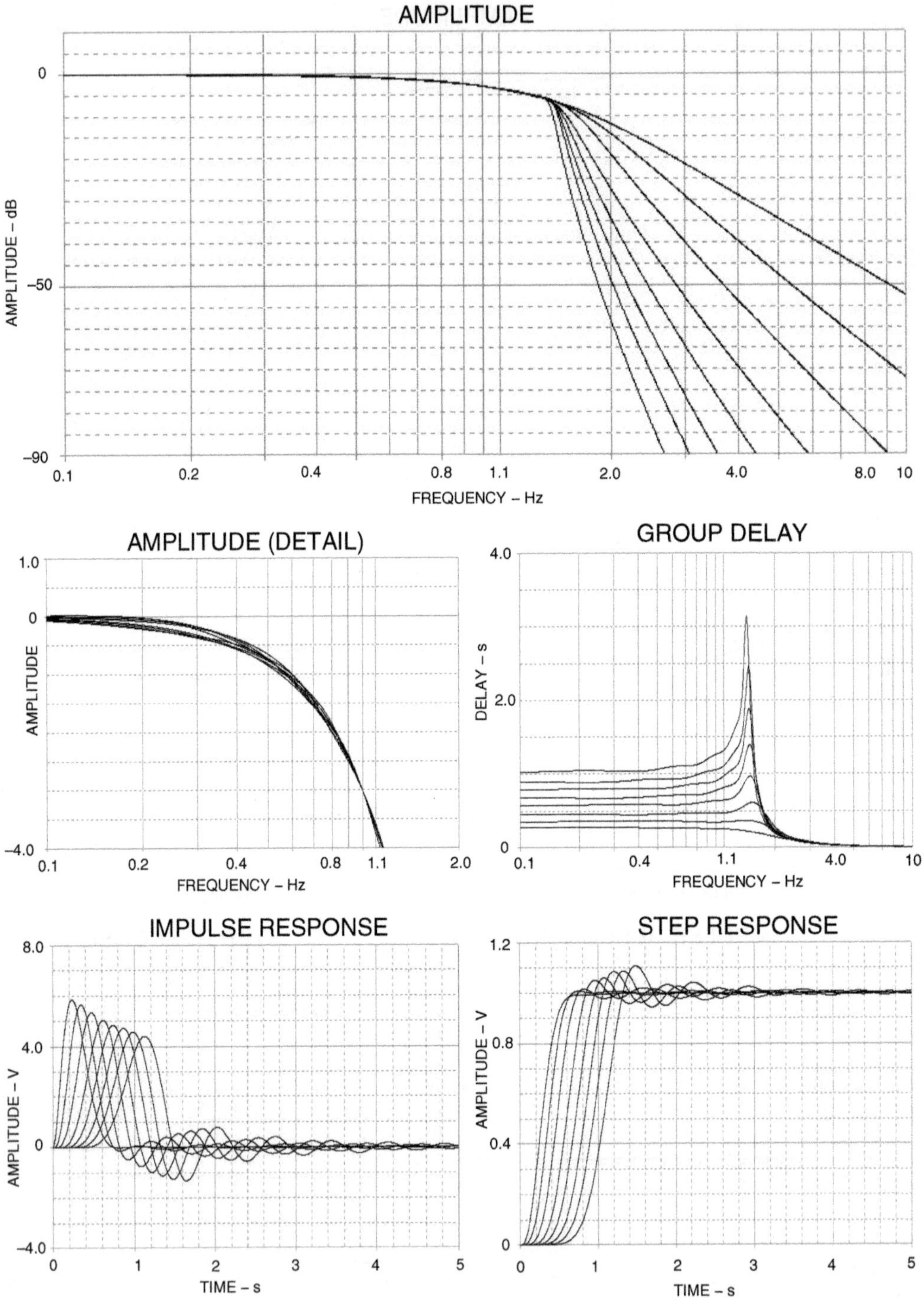

Figure 5-25: Gaussian to 6 dB response

ORDER	SECTION	REAL PART	IMAGINARY PART	F_0	α	Q	-3 dB FREQUENCY	PEAKING FREQUENCY	PEAKING LEVEL
2	1	0.7071	0.7071	1.0000	1.4142	0.7071	1.0000		
3	1	0.5000	0.8660	1.0000	1.0000	1.0000		0.7071	1.2493
	2	1.0000		1.0000			1.0000		
4	1	0.9239	0.3827	1.0000	1.8478	0.5412	0.7195		
	2	0.3827	0.9239	1.0000	0.7654	1.3065		0.8409	3.0102
5	1	0.8090	0.5878	1.0000	1.6180	0.6180	0.8588		
	2	0.3090	0.9511	1.0000	0.6180	1.6182		0.8995	4.6163
	3	1.0000		1.0000			1.0000		
6	1	0.9659	0.2588	1.0000	1.9319	0.5176	0.6758		
	2	0.7071	0.7071	1.0000	1.4142	0.7071	1.0000		
	3	0.2588	0.9659	1.0000	0.5176	1.9319		0.9306	6.0210
7	1	0.9010	0.4339	1.0000	1.8019	0.5550	0.7449		
	2	0.6235	0.7818	1.0000	1.2470	0.8019		0.4717	0.2204
	3	0.2225	0.9749	1.0000	0.4450	2.2471		0.9492	7.2530
	4	1.0000		1.0000			1.0000		
8	1	0.9808	0.1951	1.0000	1.9616	0.5098	0.6615		
	2	0.8315	0.5556	1.0000	1.6629	0.6013	0.8295		
	3	0.5556	0.8315	1.0000	1.1112	0.9000		0.6186	0.8876
	4	0.1951	0.9808	1.0000	0.3902	2.5628		0.9612	8.3429
9	1	0.9397	0.3420	1.0000	1.8794	0.5321	0.7026		
	2	0.7660	0.6428	1.0000	1.5320	0.6527	0.9172		
	3	0.5000	0.8660	1.0000	1.0000	1.0000		0.7071	1.2493
	4	0.1737	0.9848	1.0000	0.3474	2.8785		0.9694	9.3165
	5	1.0000		1.0000			1.0000		
10	1	0.9877	0.1564	1.0000	1.9754	0.5062	0.6549		
	2	0.8910	0.4540	1.0000	1.7820	0.5612	0.7564		
	3	0.7071	0.7071	1.0000	1.4142	0.7071	1.0000		
	4	0.4540	0.8910	1.0000	0.9080	1.1013		0.7667	1.8407
	5	0.1564	0.9877	1.0000	0.3128	3.1970		0.9752	10.2023

Figure 5-26: Butterworth design table

ORDER	SECTION	REAL PART	IMAGINARY PART	F_0	α	Q	-3 dB FREQUENCY	PEAKING FREQUENCY	PEAKING LEVEL
2	1	0.6743	0.7075	0.9774	1.3798	0.7247		0.2142	0.0100
3	1	0.4233	0.8663	0.9642	0.8780	1.1389		0.7558	2.0595
	2	0.8467		0.8467			0.8467		
4	1	0.6762	0.3828	0.7770	1.7405	0.5746	0.6069		
	2	0.2801	0.9241	0.9656	0.5801	1.7237		0.8806	5.1110
5	1	0.5120	0.5879	0.7796	1.3135	0.7613		0.2889	0.0827
	2	0.1956	0.9512	0.9711	0.4028	2.4824		0.9309	8.0772
	3	0.6328		0.6328			0.6328		
6	1	0.5335	0.2588	0.5930	1.7995	0.5557	0.4425		
	2	0.3906	0.7072	0.8079	0.9670	1.0342		0.5895	1.4482
	3	0.1430	0.9660	0.9765	0.2929	3.4144		0.9554	10.7605
7	1	0.4393	0.4339	0.6175	1.4229	0.7028	0.6136		
	2	0.3040	0.7819	0.8389	0.7247	1.3798		0.7204	3.4077
	3	0.1085	0.9750	0.9810	0.2212	4.5208		0.9689	13.1578
	4	0.4876		0.4876			0.4876		
8	1	0.4268	0.1951	0.4693	1.8190	0.5498	0.3451		
	2	0.3168	0.5556	0.6396	0.9907	1.0094		0.4564	1.3041
	3	0.2418	0.8315	0.8659	0.5585	1.7906		0.7956	5.4126
	4	0.0849	0.9808	0.9845	0.1725	5.7978		0.9771	15.2977
9	1	0.3686	0.3420	0.5028	1.4661	0.6821	0.4844		
	2	0.3005	0.6428	0.7096	0.8470	1.1807		0.5682	2.3008
	3	0.1961	0.8661	0.8880	0.4417	2.2642		0.8436	7.3155
	4	0.0681	0.9848	0.9872	0.1380	7.2478		0.9824	17.2249
	5	0.3923		0.3923			0.3923		
10	1	0.3522	0.1564	0.3854	1.8279	0.5471	0.2814		
	2	0.3178	0.454	0.5542	1.1469	0.8719		0.3242	0.5412
	3	0.2522	0.7071	0.7507	0.6719	1.4884		0.6606	3.9742
	4	0.1619	0.891	0.9056	0.3576	2.7968		0.8762	9.0742
	5	0.0558	0.9877	0.9893	0.1128	8.8645		0.9861	18.9669

Figure 5-27: 0.01dB Chebyshev design table

ORDER	SECTION	REAL PART	IMAGINARY PART	F_0	α	Q	-3 dB FREQUENCY	PEAKING FREQUENCY	PEAKING LEVEL
2	1	0.6104	0.7106	0.9368	1.3032	0.7673		0.3638	0.0999
3	1	0.3490	0.8684	0.9359	0.7458	1.3408		0.7952	3.1978
	2	0.6970		0.6970			0.6970		
4	1	0.2177	0.9254	0.9507	0.4580	2.1834		0.8994	7.0167
	2	0.5257	0.3833	0.6506	1.6160	0.6188	0.5596		
5	1	0.3842	0.5884	0.7027	1.0935	0.9145		0.4457	0.7662
	2	0.1468	0.9521	0.9634	0.3048	3.2812		0.9407	10.4226
	3	0.4749		0.4749			0.4749		
6	1	0.3916	0.2590	0.4695	1.6682	0.5995	0.3879		
	2	0.2867	0.7077	0.7636	0.7509	1.3316		0.6470	3.1478
	3	0.1049	0.9667	0.9724	0.2158	4.6348		0.9610	13.3714
7	1	0.3178	0.4341	0.5380	1.1814	0.8464		0.2957	0.4157
	2	0.2200	0.7823	0.8126	0.5414	1.8469		0.7507	5.6595
	3	0.0785	0.9755	0.9787	0.1604	6.2335		0.9723	15.9226
	4	0.3528		0.3528			0.3528		
8	1	0.3058	0.1952	0.3628	1.6858	0.5932	0.2956		
	2	0.2529	0.5558	0.6106	0.8283	1.2073		0.4949	2.4532
	3	0.1732	0.8319	0.8497	0.4077	2.4531		0.8137	7.9784
	4	0.0608	0.9812	0.9831	0.1237	8.0819		0.9793	18.1669
9	1	0.2622	0.3421	0.4310	1.2166	0.8219		0.2197	0.3037
	2	0.2137	0.6430	0.6776	0.6308	1.5854		0.6064	4.4576
	3	0.1395	0.8663	0.8775	0.3180	3.1450		0.8550	10.0636
	4	0.0485	0.9852	0.9864	0.0982	10.1795		0.9840	20.1650
	5	0.2790		0.2790			0.2790		
10	1	0.2493	0.1564	0.2943	1.6942	0.5902	0.2382		
	2	0.2249	0.4541	0.5067	0.8876	1.1266		0.3945	1.9880
	3	0.1785	0.7073	0.7295	0.4894	2.0434		0.6844	6.4750
	4	0.1146	0.8913	0.8986	0.2551	3.9208		0.8839	11.9386
	5	0.0395	0.9880	0.9888	0.0799	12.5163		0.9872	21.9565

Figure 5-28: 0.1 dB Chebyshev design table

ORDER	SECTION	REAL PART	IMAGINARY PART	F_0	α	Q	-3 dB FREQUENCY	PEAKING FREQUENCY	PEAKING LEVEL
2	1	0.5621	0.7154	0.9098	1.2356	0.8093		0.4425	0.2502
3	1	0.3062	0.8712	0.9234	0.6632	1.5079		0.8156	4.0734
	2	0.6124		0.6124			0.6124		
4	1	0.4501	0.3840	0.5916	1.5215	0.6572	0.5470		
	2	0.1865	0.9272	0.9458	0.3944	2.5356		0.9082	8.2538
5	1	0.3247	0.5892	0.6727	0.9653	1.0359		0.4917	1.4585
	2	0.1240	0.9533	0.9613	0.2580	3.8763		0.9452	11.8413
	3	0.4013		0.4013			0.4013		
6	1	0.3284	0.2593	0.4184	1.5697	0.6371	0.3730		
	2	0.2404	0.7083	0.7480	0.6428	1.5557		0.6663	4.3121
	3	0.0880	0.9675	0.9715	0.1811	5.5205		0.9635	14.8753
7	1	0.2652	0.4344	0.5090	1.0421	0.9596		0.3441	1.0173
	2	0.1835	0.7828	0.8040	0.4565	2.1908		0.7610	7.0443
	3	0.0655	0.9761	0.9783	0.1339	7.4679		0.9739	17.4835
	4	0.2944		0.2944			0.2944		
8	1	0.2543	0.1953	0.3206	1.5862	0.6304	0.2822		
	2	0.2156	0.5561	0.5964	0.7230	1.3832		0.5126	3.4258
	3	0.1441	0.8323	0.8447	0.3412	2.9309		0.8197	9.4683
	4	0.0506	0.9817	0.9830	0.1029	9.7173		0.9804	19.7624
9	1	0.2176	0.3423	0.4056	1.0730	0.9320		0.2642	0.8624
	2	0.1774	0.6433	0.6673	0.5317	1.8808		0.6184	5.8052
	3	0.1158	0.8667	0.8744	0.2649	3.7755		0.8589	11.6163
	4	0.0402	0.9856	0.9864	0.0815	12.2659		0.9848	21.7812
	5	0.2315		0.2315			0.2315		
10	1	0.2065	0.1565	0.2591	1.5940	0.6274	0.2267		
	2	0.1863	0.4543	0.4910	0.7588	1.3178		0.4143	3.0721
	3	0.1478	0.7075	0.7228	0.4090	2.4451		0.6919	7.9515
	4	0.0949	0.8915	0.8965	0.2117	4.7236		0.8864	13.5344
	5	0.0327	0.9883	0.9888	0.0661	15.1199		0.9878	23.5957

Figure 5-29: 0.25 dB Chebyshev design table

ORDER	SECTION	REAL PART	IMAGINARY PART	F_0	α	Q	-3 dB FREQUENCY	PEAKING FREQUENCY	PEAKING LEVEL
2	1	0.5129	0.7225	1.2314	1.1577	0.8638		0.7072	0.5002
3	1	0.2683	0.8753	1.0688	0.5861	1.7061		0.9727	5.0301
	2	0.5366		0.6265			0.6265		
4	1	0.3872	0.3850	0.5969	1.4182	0.7051	0.5951		
	2	0.1605	0.9297	1.0313	0.3402	2.9391		1.0010	9.4918
5	1	0.2767	0.5902	0.6905	0.8490	1.1779		0.5522	2.2849
	2	0.1057	0.9550	1.0178	0.2200	4.5451		1.0054	13.2037
	3	0.3420		0.3623			0.3623		
6	1	0.2784	0.2596	0.3963	1.4627	0.6836	0.3827		
	2	0.2037	0.7091	0.7680	0.5522	1.8109		0.7071	5.5025
	3	0.0746	0.9687	1.0114	0.1536	6.5119		1.0055	16.2998
7	1	0.2241	0.4349	0.5040	0.9161	1.0916		0.3839	1.7838
	2	0.1550	0.7836	0.8228	0.3881	2.5767		0.7912	8.3880
	3	0.0553	0.9771	1.0081	0.1130	8.8487		1.0049	18.9515
	4	0.2487		0.2562			0.2562		
8	1	0.2144	0.1955	0.2968	1.4779	0.6767	0.2835		
	2	0.1817	0.5565	0.5989	0.6208	1.6109		0.5381	4.5815
	3	0.1214	0.8328	0.8610	0.2885	3.4662		0.8429	10.8885
	4	0.0426	0.9824	1.0060	0.0867	11.5305		1.0041	21.2452
9	1	0.1831	0.3425	0.3954	0.9429	1.0605		0.2947	1.6023
	2	0.1493	0.6436	0.6727	0.4520	2.2126		0.6374	7.1258
	3	0.0974	0.8671	0.8884	0.2233	4.4779		0.8773	13.0759
	4	0.0338	0.9861	1.0046	0.0686	14.5829		1.0034	23.2820
	5	0.1949		0.1984			0.1984		
10	1	0.1736	0.1566	0.2338	1.4851	0.6734	0.2221		
	2	0.1566	0.4545	0.4807	0.6515	1.5349		0.4267	4.2087
	3	0.1243	0.7078	0.7186	0.3459	2.8907		0.6968	9.3520
	4	0.0798	0.8919	0.8955	0.1782	5.6107		0.8883	15.0149
	5	0.0275	0.9887	0.9891	0.0556	17.9833		0.9883	25.1008

Figure 5-30: 0.5 dB Chebyshev design table

ORDER	SECTION	REAL PART	IMAGINARY PART	F_0	α	Q	-3 dB FREQUENCY	PEAKING FREQUENCY	PEAKING LEVEL
2	1	0.4508	0.7351	0.8623	1.0456	0.9564		0.5806	0.9995
3	1	0.2257	0.8822	0.9106	0.4957	2.0173		0.8528	6.3708
	2	0.4513		0.4513			0.4513		
4	1	0.3199	0.3868	0.5019	1.2746	0.7845		0.2174	0.1557
	2	0.1325	0.9339	0.9433	0.2809	3.5594		0.9245	11.1142
5	1	0.2265	0.5918	0.6337	0.7149	1.3988		0.5467	3.5089
	2	0.0865	0.9575	0.9614	0.1800	5.5559		0.9536	14.9305
	3	0.2800		0.2800			0.2800		
6	1	0.2268	0.2601	0.3451	1.3144	0.7608		0.1273	0.0813
	2	0.1550	0.7106	0.7273	0.4262	2.3462		0.6935	7.6090
	3	0.0608	0.9707	0.9726	0.1249	8.0036		0.9688	18.0827
7	1	0.1819	0.4354	0.4719	0.7710	1.2971		0.3956	2.9579
	2	0.1259	0.7846	0.7946	0.3169	3.1558		0.7744	10.0927
	3	0.0449	0.9785	0.9795	0.0918	10.8982		0.9775	20.7563
	4	0.2019		0.2019			0.2019		
8	1	0.1737	0.1956	0.2616	1.3280	0.7530		0.0899	0.0611
	2	0.1473	0.5571	0.5762	0.5112	1.9560		0.5373	6.1210
	3	0.0984	0.8337	0.8395	0.2344	4.2657		0.8279	12.6599
	4	0.0346	0.9836	0.9842	0.0702	14.2391		0.9830	23.0750
9	1	0.1482	0.3427	0.3734	0.7938	1.2597		0.3090	2.7498
	2	0.1208	0.6442	0.6554	0.3686	2.7129		0.6328	8.8187
	3	0.0788	0.8679	0.8715	0.1809	5.5268		0.8643	14.8852
	4	0.0274	0.9869	0.9873	0.0555	18.0226		0.9865	25.1197
	5	0.1577		0.1577			0.1577		
10	1	0.1403	0.1567	0.2103	1.3341	0.7496		0.0698	0.0530
	2	0.1266	0.4548	0.4721	0.5363	1.8645		0.4368	5.7354
	3	0.1005	0.7084	0.7155	0.2809	3.5597		0.7012	11.1147
	4	0.0645	0.8926	0.8949	0.1441	6.9374		0.8903	16.8466
	5	0.0222	0.9895	0.9897	0.0449	22.2916		0.9893	26.9650

Figure 5-31: 1 dB Chebyshev design table

ORDER	SECTION	REAL PART	IMAGINARY PART	F_0	α	Q	-3 dB FREQUENCY	PEAKING FREQUENCY	PEAKING LEVEL
2	1	1.1050	0.6368	1.2754	1.7328	0.5771	1.0020		
3	1	1.0509	1.0025	1.4524	1.4471	0.6910	1.4185		
	2	1.3270		1.3270			1.3270		
4	1	1.3596	0.4071	1.4192	1.9160	0.5219	0.9705		
	2	0.9877	1.2476	1.5912	1.2414	0.8055		0.7622	0.2349
5	1	1.3851	0.7201	1.5611	1.7745	0.5635	1.1876		
	2	0.9606	1.4756	1.7607	1.0911	0.9165		1.1201	0.7768
	3	1.5069		1.5069			1.5069		
6	1	1.5735	0.3213	1.6060	1.9596	0.5103	1.0638		
	2	1.3836	0.9727	1.6913	1.6361	0.6112	1.4323		
	3	0.9318	1.6640	1.9071	0.9772	1.0234		1.3786	1.3851
7	1	1.6130	0.5896	1.7174	1.8784	0.5324	1.2074		
	2	1.3797	1.1923	1.8235	1.5132	0.6608	1.6964		
	3	0.9104	1.8375	2.0507	0.8879	1.1262		1.5961	1.9860
	4	1.6853		1.6853			1.6853		
8	1	1.7627	0.2737	1.7838	1.9763	0.5060	1.1675		
	2	0.8955	2.0044	2.1953	0.8158	1.2258		1.7932	2.5585
	3	1.3780	1.3926	1.9591	1.4067	0.7109		0.2011	0.0005
	4	1.6419	0.8256	1.8378	1.7868	0.5597	1.3849		
9	1	1.8081	0.5126	1.8794	1.9242	0.5197	1.2774		
	2	1.6532	1.0319	1.9488	1.6966	0.5894	1.5747		
	3	1.3683	1.5685	2.0815	1.3148	0.7606		0.7668	0.0807
	4	0.8788	2.1509	2.3235	0.7564	1.3220		1.9632	3.0949
	5	1.8575		1.8575			1.8575		
10	1	1.9335	0.2451	1.9490	1.9841	0.5040	1.2685		
	2	1.8467	0.7335	1.9870	1.8587	0.5380	1.4177		
	3	1.6661	1.2246	2.0678	1.6115	0.6205	1.7848		
	4	1.3648	1.7395	2.2110	1.2346	0.8100		1.0785	0.2531
	5	0.8686	2.2994	2.4580	0.7067	1.4150		2.1291	3.5944

Figure 5-32: Bessel design table

ORDER	SECTION	REAL PART	IMAGINARY PART	F_0	α	Q	-3 dB FREQUENCY	PEAKING FREQUENCY	PEAKING LEVEL
2	1	1.0087	0.6680	1.2098	1.6675	0.5997	0.9999		
3	1	0.8541	1.0725	1.3710	1.2459	0.8026		0.6487	0.2232
	2	1.0459		1.0459			1.0459		
4	1	0.9648	0.4748	1.0753	1.7945	0.5573	0.8056		
	2	0.7448	1.4008	1.5865	0.9389	1.0650		1.1864	1.6286
5	1	0.8915	0.8733	1.2480	1.4287	0.6999	1.2351		
	2	0.6731	1.7085	1.8363	0.7331	1.3641		1.5703	3.3234
	3	0.9430		0.9430			0.9430		
6	1	0.8904	0.4111	0.9807	1.8158	0.5507	0.7229		
	2	0.8233	1.2179	1.4701	1.1201	0.8928		0.8975	0.6495
	3	0.6152	1.9810	2.0743	0.5932	1.6859		1.8831	4.9365
7	1	0.8425	0.7791	1.1475	1.4684	0.6810	1.1036		
	2	0.7708	1.5351	1.7177	0.8975	1.1143		1.3276	1.9162
	3	0.5727	2.2456	2.3175	0.4942	2.0233		2.1713	6.3948
	4	0.8615		0.8615			0.8615		
8	1	0.8195	0.3711	0.8996	1.8219	0.5489	0.6600		
	2	0.7930	1.1054	1.3604	1.1658	0.8578		0.7701	0.4705
	3	0.7213	1.8134	1.9516	0.7392	1.3528		1.6638	3.2627
	4	0.5341	2.4761	2.5330	0.4217	2.3713		2.4178	7.6973
9	1	0.7853	0.7125	1.0604	1.4812	0.6751	1.0102		
	2	0.7555	1.4127	1.6020	0.9432	1.0602		1.1937	1.6005
	3	0.6849	2.0854	2.1950	0.6241	1.6024		1.9697	4.5404
	4	0.5060	2.7133	2.7601	0.3667	2.7274		2.6657	8.8633
	5	0.7983		0.7983			0.7983		
10	1	0.7592	0.3413	0.8324	1.8241	0.5482	0.6096		
	2	0.7467	1.0195	1.2637	1.1818	0.8462		0.6941	0.4145
	3	0.7159	1.6836	1.8295	0.7826	1.2778		1.5238	2.8507
	4	0.6475	2.3198	2.4085	0.5377	1.8598		2.2276	5.7152
	5	0.4777	2.9128	2.9517	0.3237	3.0895		2.8734	9.9130

Figure 5-33: Linear phase with equiripple error of 0.05° design table

ORDER	SECTION	REAL PART	IMAGINARY PART	F_0	α	Q	-3 dB FREQUENCY	PEAKING FREQUENCY	PEAKING LEVEL
2	1	0.8590	0.6981	1.1069	1.5521	0.6443	1.0000		
3	1	0.6969	1.1318	1.3292	1.0486	0.9536		0.8918	0.9836
	2	0.8257		0.8257			0.8257		
4	1	0.7448	0.5133	0.9045	1.6468	0.6072	0.7597		
	2	0.6037	1.4983	1.6154	0.7475	1.3379		1.3713	3.1817
5	1	0.6775	0.9401	1.1588	1.1693	0.8552		0.6518	0.4579
	2	0.5412	1.8256	1.9041	0.5684	1.7592		1.7435	5.2720
	3	0.7056		0.7056			0.7056		
6	1	0.6519	0.4374	0.7850	1.6608	0.6021	0.6522		
	2	0.6167	1.2963	1.4355	0.8592	1.1639		1.1402	2.2042
	3	0.4893	2.0982	2.1545	0.4542	2.2016		2.0404	7.0848
7	1	0.6190	0.8338	1.0385	1.1922	0.8388		0.5586	0.3798
	2	0.5816	1.6455	1.7453	0.6665	1.5004		1.5393	4.0353
	3	0.4598	2.3994	2.4431	0.3764	2.6567		2.3549	8.6433
	4	0.6283		0.6283			0.6283		
8	1	0.5791	0.3857	0.6958	1.6646	0.6007	0.5764		
	2	0.5665	1.1505	1.2824	0.8835	1.1319		1.0014	2.0187
	3	0.5303	1.8914	1.9643	0.5399	1.8521		1.8155	5.6819
	4	0.4148	2.5780	2.6112	0.3177	3.1475		2.5444	10.0703
9	1	0.5688	0.7595	0.9489	1.1989	0.8341		0.5033	0.3581
	2	0.5545	1.5089	1.6076	0.6899	1.4496		1.4033	3.7748
	3	0.5179	2.2329	2.2922	0.4519	2.2130		2.1720	7.1270
	4	0.4080	2.9028	2.9313	0.2784	3.5923		2.8740	11.1925
	5	0.5728		0.5728			0.5728		
10	1	0.5249	0.3487	0.6302	1.6659	0.6003	0.5215		
	2	0.5193	1.0429	1.1650	0.8915	1.1217		0.9044	1.9598
	3	0.5051	1.7264	1.7988	0.5616	1.7806		1.6509	5.3681
	4	0.4711	2.3850	2.4311	0.3876	2.5802		2.3380	8.3994
	5	0.3708	2.9940	3.0169	0.2458	4.0681		2.9709	12.2539

Figure 5-34: Linear phase with equiripple error of 0.5° design table

ORDER	SECTION	REAL PART	IMAGINARY PART	F_0	α	Q	-3 dB FREQUENCY	PEAKING FREQUENCY	PEAKING LEVEL
3	1	0.9360	1.2168	1.5352	1.2194	0.8201		0.7775	0.2956
	2	0.9360		0.9360			0.9360		
4	1	0.9278	1.6995	1.9363	0.9583	1.0435		1.4239	1.5025
	2	0.9192	0.5560	1.0743	1.7113	0.5844	0.8582		
5	1	0.8075	0.9973	1.2832	1.2585	0.7946		0.5853	0.1921
	2	0.7153	0.2053	0.7442	1.9224	0.5202	0.5065		
	3	0.8131		0.8131			0.8131		
6	1	0.7019	0.4322	0.8243	1.7030	0.5872	0.6627		
	2	0.6667	1.2931	1.4549	0.9165	1.0911		1.1080	1.7809
	3	0.4479	2.1363	2.1827	0.4104	2.4366		2.0888	7.9227
7	1	0.6155	0.7703	0.9860	1.2485	0.8010		0.4632	0.2168
	2	0.5486	1.5154	1.6116	0.6808	1.4689		1.4126	3.8745
	3	0.2905	2.1486	2.1681	0.2680	3.7318		2.1289	11.5169
	4	0.6291		0.6291			0.6291		
8	1	0.5441	0.3358	0.6394	1.7020	0.5876	0.5145		
	2	0.5175	0.9962	1.1226	0.9220	1.0846		0.8512	1.7432
	3	0.4328	1.6100	1.6672	0.5192	1.9260		1.5507	5.9962
	4	0.1978	2.0703	2.0797	0.1902	5.2571		2.0608	14.4545
9	1	0.4961	0.6192	0.7934	1.2505	0.7997		0.3705	0.2116
	2	0.4568	1.2145	1.2976	0.7041	1.4203		1.1253	3.6221
	3	0.3592	1.7429	1.7795	0.4037	2.4771		1.7055	8.0594
	4	0.1489	2.1003	2.1056	0.1414	7.0704		2.0950	17.0107
	5	0.5065		0.5065			0.5065		
10	1	0.4535	0.2794	0.5327	1.7028	0.5873	0.4283		
	2	0.4352	0.8289	0.9362	0.9297	1.0756		0.7055	1.6904
	3	0.3886	1.3448	1.3998	0.5552	1.8011		1.2874	5.4591
	4	0.2908	1.7837	1.8072	0.3218	3.1074		1.7598	9.9618
	5	0.1136	2.0599	2.0630	0.1101	9.0802		2.0568	19.1751

Figure 5-35: Gaussian to 12 dB design table

ORDER	SECTION	REAL PART	IMAGINARY PART	F_0	α	Q	-3 dB FREQUENCY	PEAKING FREQUENCY	PEAKING LEVEL
3	1	0.9622	1.2214	1.5549	1.2377	0.8080		0.7523	0.2448
	2	0.9776	0.5029	1.0994	1.7785	0.5623	0.8338		
4	1	0.7940	0.5029	0.9399	1.6896	0.5919	0.7636		
	2	0.6304	1.5407	1.6647	0.7574	1.3203		1.4058	3.0859
5	1	0.6190	0.8254	1.0317	1.1999	0.8334		0.5460	0.3548
	2	0.3559	1.5688	1.6087	0.4425	2.2600		1.5279	7.3001
	3	0.6650		0.6650			0.6650		
6	1	0.5433	0.3431	0.6426	1.6910	0.5914	0.5215		
	2	0.4672	0.9991	1.1029	0.8472	1.1804		0.8831	2.2992
	3	0.2204	1.5067	1.5227	0.2895	3.4545		1.4905	10.8596
7	1	0.4580	0.5932	0.7494	1.2223	0.8182		0.3770	0.2874
	2	0.3649	1.1286	1.1861	0.6153	1.6253		1.0680	4.6503
	3	0.1522	1.4938	1.5015	0.2027	4.9328		1.4860	13.9067
	4	0.4828		0.4828			0.4828		
8	1	0.4222	0.2640	0.4979	1.6958	0.5897	0.4026		
	2	0.3833	0.7716	0.8616	0.8898	1.1239		0.6697	1.9722
	3	0.2678	1.2066	1.2360	0.4333	2.3076		1.1765	7.4721
	4	0.1122	1.4798	1.4840	0.1512	6.6134		1.4755	16.4334
9	1	0.3700	0.4704	0.5985	1.2365	0.8088		0.2905	0.2480
	2	0.3230	0.9068	0.9626	0.6711	1.4901		0.8473	3.9831
	3	0.2309	1.2634	1.2843	0.3596	2.7811		1.2421	9.0271
	4	0.0860	1.4740	1.4765	0.1165	8.5804		1.4715	18.6849
	5	0.3842		0.3842			0.3842		
10	1	0.3384	0.2101	0.3983	1.6991	0.5885	0.3212		
	2	0.3164	0.6180	0.6943	0.9114	1.0972		0.5309	1.8164
	3	0.2677	0.9852	1.0209	0.5244	1.9068		0.9481	5.9157
	4	0.1849	1.2745	1.2878	0.2871	3.4825		1.2610	10.9284
	5	0.0671	1.4389	1.4405	0.0931	10.7401		1.4373	20.6296

Figure 5-36: Gaussian to 6 dB design table

Frequency Transformations

Until now, only filters using the low pass configuration have been examined. In this section, transforming the low-pass prototype into the other configurations: high pass, band pass, bandreject (notch) and all pass will be discussed.

Low Pass to High Pass

The low-pass prototype is converted to high-pass filter by scaling by 1/s in the transfer function. In practice, this amounts to capacitors becoming inductors with a value 1/C, and inductors becoming capacitors with a value of 1/L for passive designs. For active designs, resistors become capacitors with a value of 1/R, and capacitors become resistors with a value of 1/C. This applies only to frequency setting resistor, not those only used to set gain.

Another way to look at the transformation is to investigate the transformation in the s plane. The complex pole pairs of the low-pass prototype are made up of a real part, α, and an imaginary part, β. The normalized high-pass poles are then given by:

$$\alpha_{HP} = \frac{\alpha}{\alpha^2 + \beta^2} \qquad \text{Eq. 5-43}$$

and:

$$\beta_{HP} = \frac{\beta}{\alpha^2 + \beta^2} \qquad \text{Eq. 5-44}$$

A simple pole, α_0, is transformed to:

$$\alpha_{\omega,HP} = \frac{1}{\alpha_0} \qquad \text{Eq. 5-45}$$

Low-pass zeros, $\omega_{z,lp}$, are transformed by:

$$\omega_{Z,HP} = \frac{1}{\omega_{Z,LP}} \qquad \text{Eq. 5-46}$$

In addition, a number of zeros equal to the number of poles are added at the origin. After the normalized low pass prototype poles and zeros are converted to high pass, they are then denormalized in the same way as the low pass, that is, by frequency and impedance.

As an example a 3-pole 1dB Chebyshev low-pass filter will be converted to a high-pass filter.

From the design tables of the last section:

$$\alpha_{LP1} = 0.2257$$
$$\beta_{LP1} = 0.8822$$
$$\alpha_{LP2} = 0.4513$$

This will transform to:

$$\alpha_{HP1} = 0.2722$$
$$\beta_{HP1} = 1.0639$$
$$\alpha_{HP2} = 2.2158$$

Which then becomes:

$$F_{01} = 1.0982$$
$$\alpha = 0.4958$$
$$Q = 2.0173$$

$$F_{02} = 2.2158$$

A worked out example of this transformation will appear in a later section.

A high-pass filter can be considered to be a low-pass filter turned on its side. Instead of a flat response at dc, there is a rising response of n × (20 dB/decade), due to the zeros at the origin, where n is the number of poles. At the corner frequency a response of n × (–20 dB/decade) due to the poles is added to the above rising response. This results in a flat response beyond the corner frequency.

Low Pass to Band Pass

Transformation to the band pass response is a little more complicated. Band-pass filters can be classified as either wideband or narrowband, depending on the separation of the poles. If the corner frequencies of the band pass are widely separated (by more than two octaves), the filter is wideband and is made up of separate low pass and high pass sections, which will be cascaded. The assumption made is that with the widely separated poles, interaction between them is minimal. This condition does not hold in the case of a narrowband band-pass filter, where the separation is less than two octaves. We will be covering the narrow-band case in this discussion.

As in the high pass transformation, start with the complex pole pairs of the low-pass prototype, α and β. The pole pairs are known to be complex conjugates. This implies symmetry around dc (0 Hz.). The process of transformation to the band pass case is one of mirroring the response around dc of the low-pass prototype to the same response around the new center frequency F_0.

This clearly implies that the number of poles and zeros is doubled when the band pass transformation is done. As in the low pass case, the poles and zeros below the real axis are ignored. So an nth order low-pass prototype transforms into an nth order band pass, even though the filter order will be 2 n. An nth order band-pass filter will consist of n sections, versus n/2 sections for the low-pass prototype. It may be convenient to think of the response as n poles up and n poles down.

The value of Q_{BP} is determined by:

$$Q_{BP} = \frac{F_0}{BW} \qquad \text{Eq. 5-47}$$

where *BW* is the bandwidth at some level, typically –3 dB.

A transformation algorithm was defined by Geffe (Reference 16) for converting low-pass poles into equivalent band-pass poles.

Given the pole locations of the low-pass prototype:

$$-\alpha \pm j\beta \quad \text{Eq. 5-48}$$

and the values of F_0 and Q_{BP}, the following calculations will result in two sets of values for Q and frequencies, F_H and F_L, which define a pair of band-pass filter sections.

$$C = \alpha^2 + \beta^2 \quad \text{Eq. 5-49}$$

$$D = \frac{2\alpha}{Q_{BP}} \quad \text{Eq. 5-50}$$

$$E = \frac{C}{Q_{BP}{}^2} + 4 \quad \text{Eq. 5-51}$$

$$G = \sqrt{E^2 - 4D^2} \quad \text{Eq. 5-52}$$

$$Q = \sqrt{\frac{E+G}{2D^2}} \quad \text{Eq. 5-53}$$

Observe that the Q of each section will be the same.

The pole frequencies are determined by:

$$M = \frac{\alpha Q}{Q_{BP}} \quad \text{Eq. 5-54}$$

$$W = M + \sqrt{M2 - 1} \quad \text{Eq. 5-55}$$

$$F_{BP1} = \frac{F_0}{W} \quad \text{Eq. 5-56}$$

$$F_{BP2} = W F_0 \quad \text{Eq. 5-57}$$

Each pole pair transformation will also result in two zeros that will be located at the origin.

A normalized low-pass real pole with a magnitude of α_0 is transformed into a band pass section where:

$$Q = \frac{Q_{BP}}{\alpha_0} \quad \text{Eq. 5-58}$$

and the frequency is F_0.

Each single pole transformation will also result in a zero at the origin.

Elliptical function low-pass prototypes contain zeros as well as poles. In transforming the filter the zeros must be transformed as well. Given the low-pass zeros at $\pm j\omega_Z$, the band pass zeros are obtained as follows:

$$M = \frac{\alpha Q}{Q_{BP}} \qquad \text{Eq. 5-59}$$

$$W = M + \sqrt{M2 - 1} \qquad \text{Eq. 5-60}$$

$$F_{BP1} = \frac{F_0}{W} \qquad \text{Eq. 5-61}$$

$$F_{BP2} = W F_0 \qquad \text{Eq. 5-62}$$

Since the gain of a band-pass filter peaks at F_{BP} instead of F_0, an adjustment in the amplitude function is required to normalize the response of the aggregate filter. The gain of the individual filter section is given by:

$$A_R = A_0 \sqrt{1 + Q^2 \left(\frac{F_0}{F_{BP}} - \frac{F_{BP}}{F_0} \right)^2} \qquad \text{Eq. 5-63}$$

Where:

A_0 = gain a filter center frequency
A_R = filter section gain at resonance
F_0 = filter center frequency
F_{BP} = filter section resonant frequency.

Again using a 3 pole 1dB Chebychev as an example:

$$\alpha_{LP1} = 0.2257$$
$$\beta_{LP1} = 0.8822$$
$$\alpha_{LP2} = 0.4513$$

A 3 dB bandwidth of 0.5 Hz. with a center frequency of 1 Hz. is arbitrarily assigned. Then:

$$Q_{BP} = 2$$

Going through the calculations for the pole pair the intermediate results are:

$C = 0.829217$ $\quad D = 0.2257$
$E = 4.207034$ $\quad G = 4.098611$
$M = 1.01894$ $\quad W = 1.214489$

and:

$F_{BP1} = 0.823391$ $\quad F_{BP2} = 1.214489$

$Q_{BP1} = Q_{BP2} = 9.029157$

And for the single pole:

$F_{BP3} = 1$ $\quad Q_{BP3} = 4.431642$

Again a full example will be worked out in a later section.

Low Pass to Bandreject (Notch)

As in the band pass case, a bandreject filter can be either wideband or narrowband, depending on whether or not the poles are separated by two octaves or more. To avoid confusion, the following convention will be adopted. If the filter is wideband, it will be referred to as a bandreject filter. A narrowband filter will be referred to as a notch filter.

One way to build a notch filter is to construct it as a band-pass filter whose output is subtracted from the input (1 – BP). Another way is with cascaded low pass and high pass sections, especially for the bandreject (wideband) case. In this case, the sections are in parallel, and the output is the difference.

Just as the band pass case is a direct transformation of the low-pass prototype, where dc is transformed to F_0, the notch filter can be first transformed to the high-pass case, and then dc, which is now a zero, is transformed to F_0.

A more general approach would be to convert the poles directly. A notch transformation results in two pairs of complex poles and a pair of second order imaginary zeros from each low-pass pole pair.

First, the value of Q_{BR} is determined by:

$$Q_{BR} = \frac{F_0}{BW} \qquad \text{Eq. 5-64}$$

where *BW* is the bandwidth at –3 dB.

Given the pole locations of the low-pass prototype

$$-\alpha \pm j\beta \qquad \text{Eq. 5-65}$$

and the values of F_0 and Q_{BR}, the following calculations will result in two sets of values for Q and frequencies, F_H and F_L, which define a pair of notch filter sections.

$$C = \alpha^2 + \beta^2 \qquad \text{Eq. 5-66}$$

$$D = \frac{\alpha}{Q_{BR} C} \qquad \text{Eq. 5-67}$$

$$E = \frac{\beta}{Q_{BR} C} \qquad \text{Eq. 5-68}$$

$$F = E^2 - 4D^2 + 4 \qquad \text{Eq. 5-69}$$

$$G = \sqrt{\frac{F}{2} + \sqrt{\frac{F2}{4} + D^2 E^2}} \qquad \text{Eq. 5-70}$$

$$H = \frac{D\,E}{G} \qquad \text{Eq. 5-71}$$

$$K = \frac{1}{2}\sqrt{(D+H)^2 + (E+G)^2} \qquad \text{Eq. 5-72}$$

$$Q = \frac{K}{D+H} \qquad \text{Eq. 5-73}$$

The pole frequencies are given by:

$$F_{BR1} = \frac{F_0}{K} \qquad \text{Eq. 5-74}$$

$$F_{BR2} = K\,F_0 \qquad \text{Eq. 5-75}$$

$$F_Z = F_0 \qquad \text{Eq. 5-76}$$

$$F_0 = \sqrt{F_{BR1} * F_{BR2}} \qquad \text{Eq. 5-77}$$

where F_0 is the notch frequency and the geometric mean of F_{BR1} and F_{BR2}. A simple real pole, α_0, transforms to a single section having a Q given by:

$$Q = Q_{BR}\,\alpha_0 \qquad \text{Eq. 5-78}$$

with a frequency $F_{BR} = F_0$. There will also be transmission zero at F_0.

In some instances, such as the elimination of the power line frequency (hum) from low level sensor measurements, a notch filter for a specific frequency may be designed.

Assuming that an attenuation of A dB is required over a bandwidth of B, the required Q is determined by:

$$Q = \frac{\omega_0}{B\sqrt{10^{.1A} - 1}} \qquad \text{Eq. 5-79}$$

A 3-pole 1 dB Chebychev is again used as an example:

$$\alpha_{LP1} = 0.2257$$
$$\beta_{LP1} = 0.8822$$
$$\alpha_{LP2} = 0.4513$$

A 3 dB bandwidth of 0.1 Hz with a center frequency of 1 Hz is arbitrarily assigned. Then:

$$Q_{BR} = 10$$

Going through the calculations for the pole pair yields the intermediate results:

$$C = 0.829217 \qquad D = 0.027218$$
$$E = 0.106389 \qquad F = 4.079171$$
$$G = 2.019696 \qquad H = 0.001434$$
$$K = 1.063139$$

and:

$$F_{BR1} = 0.94061 \qquad F_{BR2} = 1.063139$$
$$Q_{BR1} = Q_{BR2} = 37.10499$$

And for the single pole:

$$F_{BP3} = 1 \qquad Q_{BP3} = 4.431642$$

Once again a full example will be worked out in a later section.

Low Pass to All Pass

The transformation from low pass to all pass involves adding a zero in the right-hand side of the s plane corresponding to each pole in the left-hand side.

In general, however, the all-pass filter is usually not designed in this manner. The main purpose of the all-pass filter is to equalize the delay of another filter. Many modulation schemes in communications use some form or another of quadrature modulation, which processes both the amplitude and phase of the signal.

All-pass filters add delay to flatten the delay curve without changing the amplitude. In most cases a closed form of the equalizer is not available. Instead the amplitude filter is designed and the delay calculated or measured. Then graphical means or computer programs are used to figure out the required sections of equalization.

Each section of the equalizer gives twice the delay of the low-pass prototype due to the interaction of the zeros. A rough estimate of the required number of sections is given by:

$$n = 2\,\Delta_{BW}\,\Delta_T + 1$$

where Δ_{BW} is the bandwidth of interest in hertz and Δ_T is the delay distortion over Δ_{BW} in seconds.

Filter Realizations

Now that it has been decided *what* to build, it must be decided *how* to build it. That means it is necessary to decide which of the filter topologies to use. Filter design is a two-step process where it is determined what is to be built (the filter transfer function) and then how to build it (the topology used for the circuit).

In general, filters are built from one-pole sections for real poles, and two-pole sections for pole pairs. While a filter can be built from three-pole, or higher order sections, the interaction between the sections increases, and therefore, component sensitivities go up.

It is better to use buffers to isolate the various sections. In addition, it is assumed that all filter sections are driven from a low impedance source. Any source impedance can be modeled as being in series with the filter input.

In all of the design equation figures the following convention will be used:

H = circuit gain in the passband or at resonance

F_0 = cutoff or resonant frequency in Hertz

ω_0 = cutoff or resonant frequency in radians/sec

Q = circuit "quality factor." Indicates circuit peaking

α = $1/Q$ = damping ratio

It is unfortunate that the symbol α is used for damping ratio. It is not the same as the α that is used to denote pole locations ($\alpha \pm j\beta$). The same issue occurs for Q. It is used for the *circuit* quality factor and also the *component* quality factor, which are not the same thing.

The circuit Q is the amount of peaking in the circuit. This is a function of the angle of the pole to the origin in the s plane. The component Q is the amount of loss in what should be lossless reactances. These losses are the parasitics of the components; dissipation factor, leakage resistance, ESR (equivalent series resistance), and so forth, in capacitors and series resistance and parasitic capacitances in inductors.

Single-Pole RC

The simplest filter building block is the passive RC section. The single pole can be either low pass or high pass. Odd order filters will have a single pole section.

The basic form of the low pass RC section is shown in Figure 5-37(A). It is assumed that the load impedance is high (> ×10), so that there is no loading of the circuit. The load will be in parallel with the shunt arm of the filter. If this is not the case, the section will have to be buffered with an op amp. A low-pass filter can be transformed to a high-pass filter by exchanging the resistor and the capacitor. The basic form of the high-pass filter is shown in Figure 5-37(B). Again it is assumed that load impedance is high.

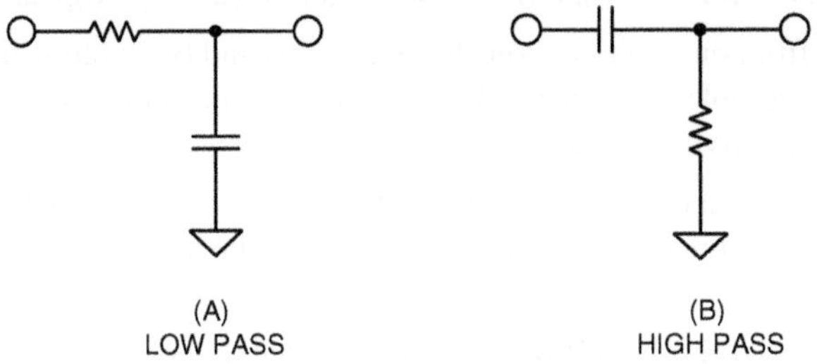

Figure 5-37: Single pole sections

The pole can also be incorporated into an amplifier circuit. Figure 5-38(A) shows an amplifier circuit with a capacitor in the feedback loop. This forms a low-pass filter since as frequency is increased, the effective feedback impedance decreases, which causes the gain to decrease.

Figure 5-38(B) shows a capacitor in series with the input resistor. This causes the signal to be blocked at dc. As the frequency is increased from dc, the impedance of the capacitor decreases and the gain of the circuit increases. This is a high-pass filter.

The design equations for single pole filters appear in Figure 5-66.

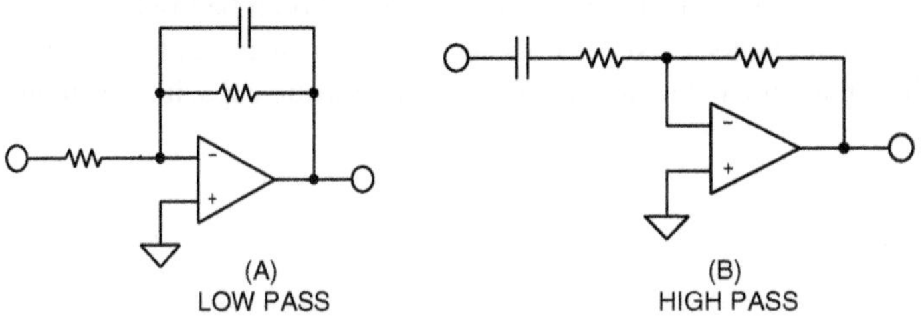

Figure 5-38: Single pole active filter blocks

Passive LC Section

While not strictly a function that uses op amps, passive filters form the basis of several active filters topologies and are included here for completeness.

As in active filters, passive filters are built up of individual subsections. Figure 5-39 shows low-pass filter sections. The full section is the basic two-pole section. Odd order filters use one half section which is a single pole section. The m-derived sections, shown in Figure 5-40, are used in designs requiring transmission zeros as well as poles.

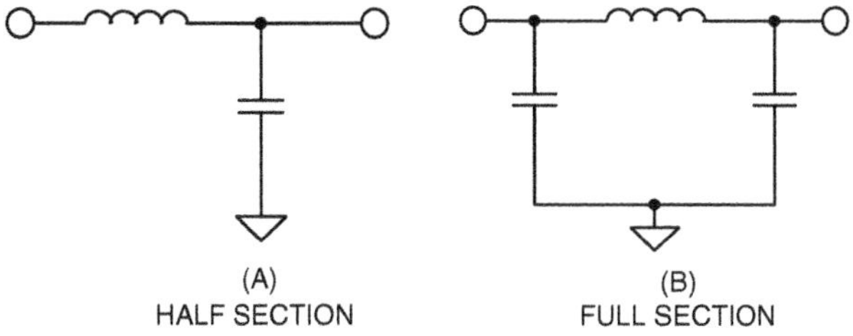

Figure 5-39: Passive filter blocks (low pass)

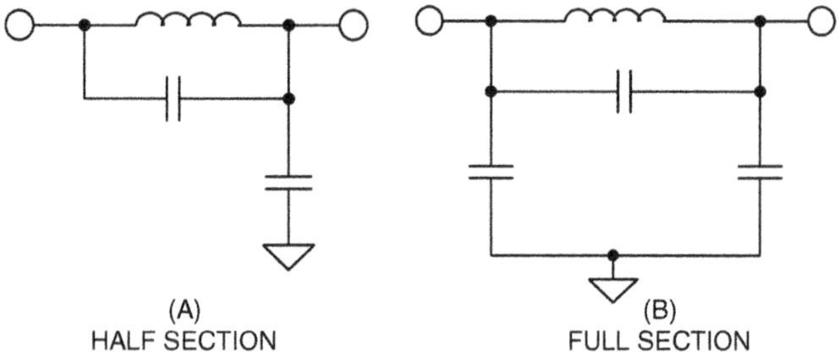

Figure 5-40: Passive filter blocks (low pass m-derived)

A low-pass filter can be transformed into a high-pass (see Figures 5-41 and 5-42) by simply replacing capacitors with inductors with reciprocal values and vise versa so:

$$L_{HP} = \frac{1}{C_{LP}} \qquad \text{Eq. 5-80}$$

and

$$C_{HP} = \frac{1}{L_{LP}} \qquad \text{Eq. 5-81}$$

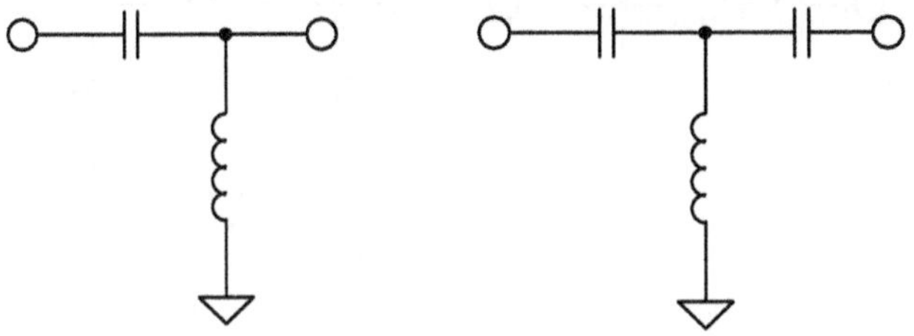

Figure 5-41: Passive filter blocks (high pass)

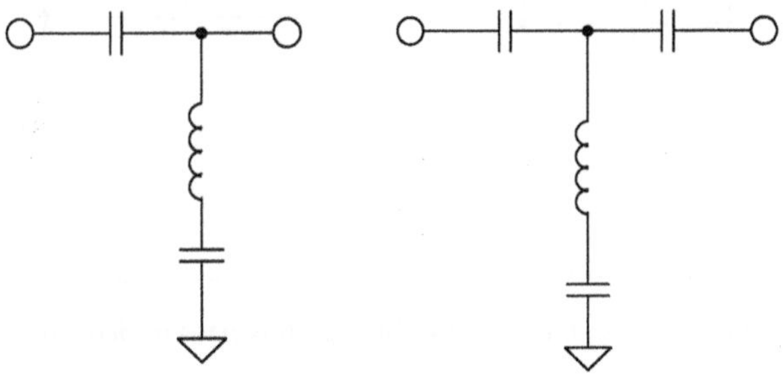

Figure 5-42: Passive filter blocks (high pass m-derived)

Transmission zeros are also reciprocated in the transformation so:

$$\omega_{Z,HP} = \frac{1}{\omega_{Z,LP}} \qquad \text{Eq. 5-82}$$

The low-pass prototype is transformed to band pass and bandreject filters as well by using the table in Figure 5-43.

For a passive filter to operate, the source and load impedances must be specified. One issue with designing passive filters is that in multipole filters each section is the load for the preceding sections and also the source impedance for subsequent sections, so component interaction is a major concern. Because of this, designers typically make use of tables, such as in Williams's book (Reference 2).

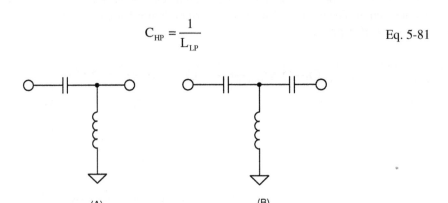

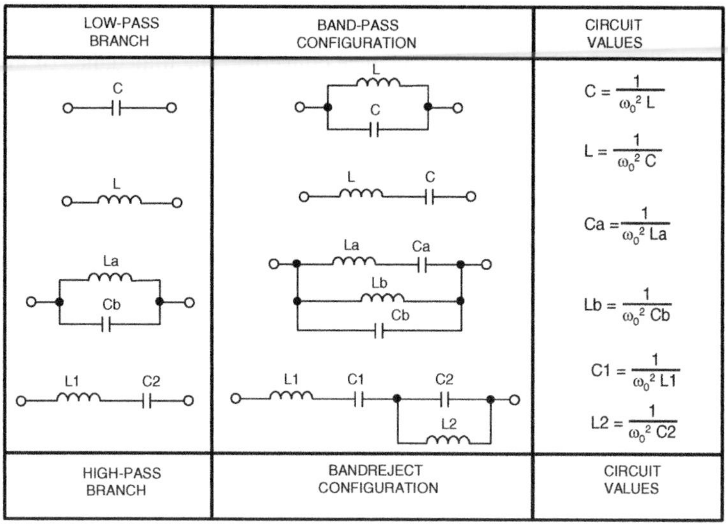

Figure 5-43: Low pass → band pass and high pass → bandreject transformation

Integrator

Any time a frequency-dependent impedance is put in a feedback network the inverse frequency response is obtained. For example, if a capacitor, which has a frequency dependent impedance that decreases with increasing frequency, is put in the feedback network of an op amp, an integrator is formed, as in Figure 5-44.

The integrator has high gain (i.e., the open-loop gain of the op amp) at dc. An integrator can also be thought of as a low-pass filter with a cutoff frequency of 0 Hz.

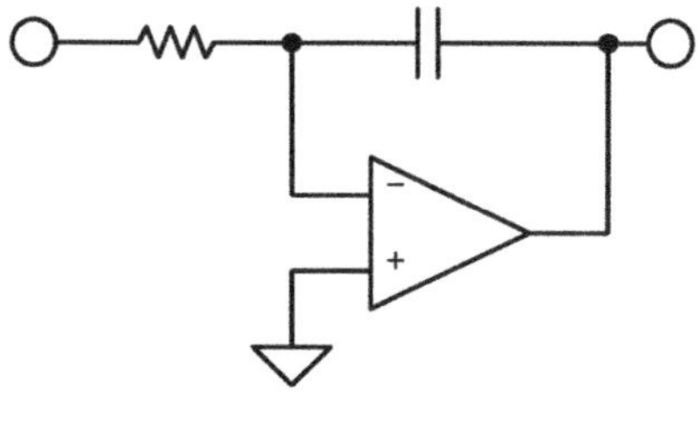

Figure 5-44: Integrator

General Impedance Converter

Figure 5-45 is the block diagram of a general impedance converter. The impedance of this circuit is:

$$Z = \frac{Z1\ Z3\ Z5}{Z2\ Z4} \qquad \text{Eq. 5-83}$$

By substituting one or two capacitors into appropriate locations (the other locations being resistors), several impedances can be synthesized (see Reference 25).

One limitation of this configuration is that the lower end of the structure must be grounded.

Active Inductor

Substituting a capacitor for Z4 and resistors for Z1, Z2, Z3, and Z5 in the GIC results in an impedance given by:

$$Z_{11} = \frac{sC\ R1\ R3\ R5}{R2} \qquad \text{Eq. 5-84}$$

By inspection it can be shown that this is an inductor with a value of:

$$L = \frac{C\ R1\ R3\ R5}{R2} \qquad \text{Eq. 5-85}$$

This is just one way to simulate an inductor as shown in Figure 5-46.

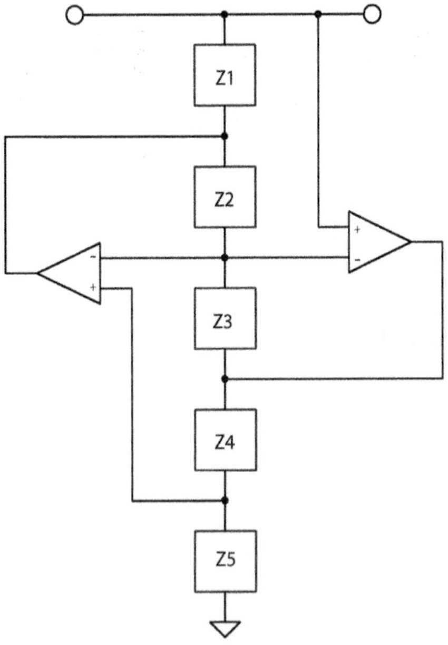

Figure 5-45: General impedance converter

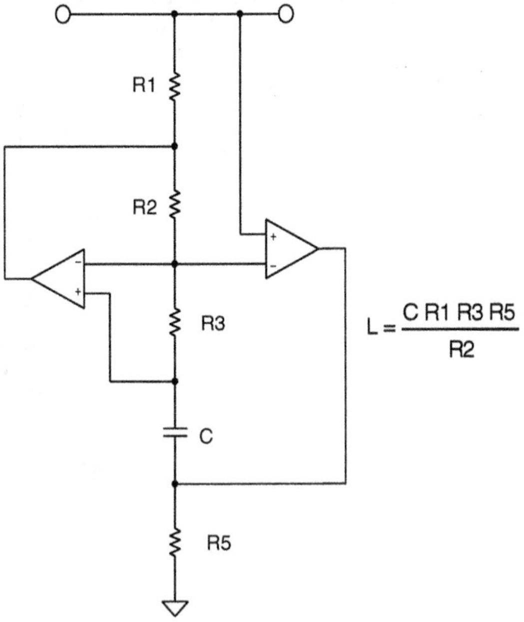

Figure 5-46: Active inductor

Frequency-Dependent Negative Resistor (FDNR)

By substituting capacitors for two of the Z1, Z3, or Z5 elements, a structure known as a frequency-dependant negative resistance (FDNR) is generated. The impedance of this structure is

$$Z_{11} = \frac{sC^2\ R2\ R4}{R5} \qquad \text{Eq. 5-86}$$

This impedance, which is called a D element, has the value:

$$D = C^2\ R4 \qquad \text{Eq. 5-87}$$

assuming

$$C1 = C2 \text{ and } R2 = R5 \qquad \text{Eq. 5-88}$$

The three possible versions of the FDNR are shown in Figure 5-47.

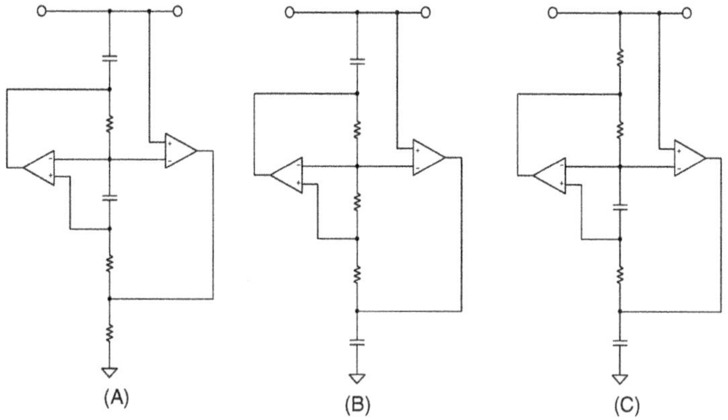

Figure 5-47: Frequency-dependent negative resistor blocks

There is theoretically no difference in these three blocks, and so they should be interchangeable. In practice, though, there may be some differences. Circuit (a) is sometimes preferred because it is the only block to provide a return path for the amplifier bias currents.

For the FDNR filter (see Reference 24), the passive realization of the filter is used as the basis of the design. As in the passive filter, the FDNR filter must then be denormalized for frequency and impedance. This is typically done before the conversion by 1/s. First take the denormalized passive prototype filter and transform the elements by 1/s. This means that inductors, whose impedance is equal to sL, transform into a resistor with an impedance of L. A resistor of value R becomes a capacitor with an impedance of R/s; and a capacitor of impedance 1/sC transforms into a frequency dependent resistor, D, with an impedance of $1/_s{}^2C$. The transformations involved with the FDNR configuration and the GIC implementation of the D element are shown in Figure 5-48. We can apply this transformation to low-pass, high-pass, band-pass, or notch filters, remembering that the FDNR block must be restricted to shunt arms.

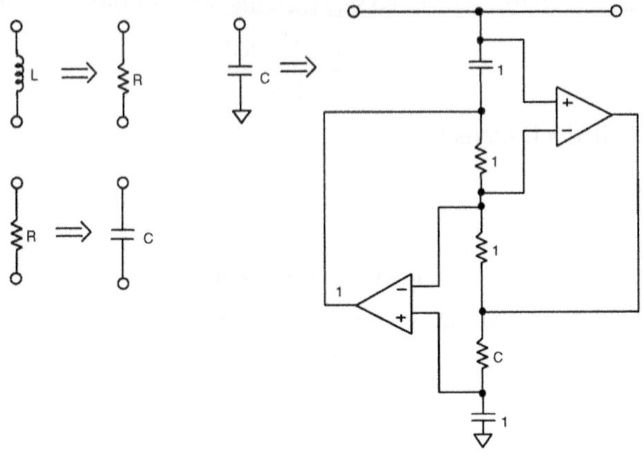

Figure 5-48: 1/s transformation

A worked out example of the FDNR filter is included in the next section.

A perceived advantage of the FDNR filter in some circles is that there are no op amps in the direct signal path, which can add noise and/or distortion, however small, to the signal. It is also relatively insensitive to component variation. These advantages of the FDNR come at the expense of an increase in the number of components required.

Sallen-Key

The Sallen-Key configuration, also known as a voltage control voltage source (VCVS), was first introduced in 1955 by R. P. Sallen and E. L. Key of MIT's Lincoln Labs (see Reference 14). It is one of the most widely used filter topologies and is shown in Figure 5-49. One reason for this popularity is that this configuration shows the least dependence of filter performance on the performance of the op amp. This is

Figure 5-49: Sallen-Key low-pass filter

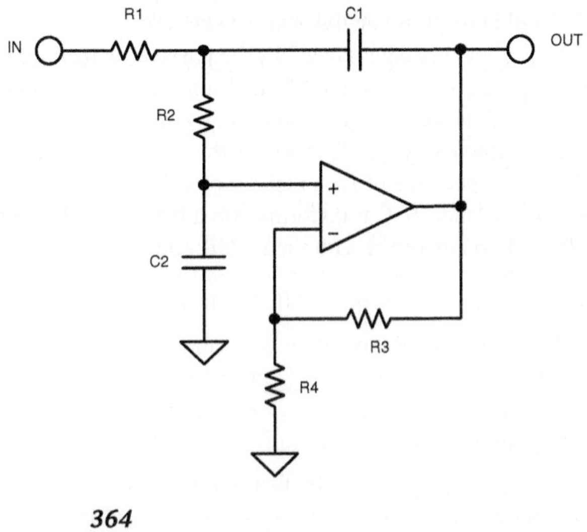

due to the fact that the op amp is configured as an amplifier, as opposed to an integrator, which minimizes the gain-bandwidth requirements of the op amp. This infers that for a given op amp, it is possible to design a higher frequency filter than with other topologies since the op amp gain bandwidth product will not limit the performance of the filter as it would if it were configured as an integrator. The signal phase through the filter is maintained (noninverting configuration).

Another advantage of this configuration is that the ratio of the largest resistor value to the smallest resistor value and the ratio of the largest capacitor value to the smallest capacitor value (component spread) are low, which is good for manufacturability. The frequency and Q terms are somewhat independent, but they are very sensitive to the gain parameter. The Sallen-Key is very Q-sensitive to element values, especially for high Q sections. The design equations for the Sallen-Key low pass are shown in Figure 5-67.

There is a special case of the Sallen-Key low-pass filter. If the gain is set to 2, the capacitor values, as well as the resistor values, will be the same.

While the Sallen-Key filter is widely used, a serious drawback is that the filter is not easily tuned, due to interaction of the component values on F_0 and Q.

To transform the low pass into the high pass we simply exchange the capacitors and the resistors in the frequency determining network (i.e., *not* the op amp gain resistors). This is shown in Figure 5-50. The comments regarding sensitivity of the filter given above for the low pass case apply to the high pass case as well. The design equations for the Sallen-Key high pass are shown in Figure 5-68.

The band-pass case of the Sallen–Key filter has a limitation (see Figure 5-51). The value of Q will determine the gain of the filter, i.e., it can not be set independently, as in the low pass or high pass cases. The design equations for the Sallen-Key band-pass are shown in Figure 5-69.

A Sallen-Key notch filter may also be constructed, but it has a large number of undesirable characteristics. The resonant frequency, or the notch frequency, cannot be adjusted easily due to component interaction. As in the band pass case, the section gain is fixed by the other design parameters, and there is a wide spread in component values, especially capacitors. Because of this, and the availability of easier to use circuits, it is not covered here.

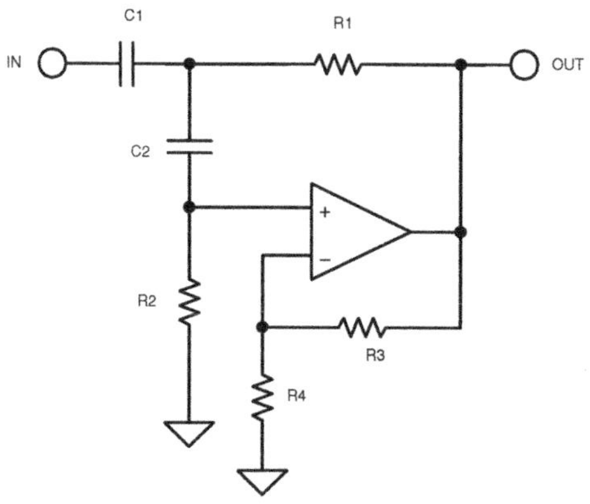

Figure 5-50: Sallen-Key high-pass filter

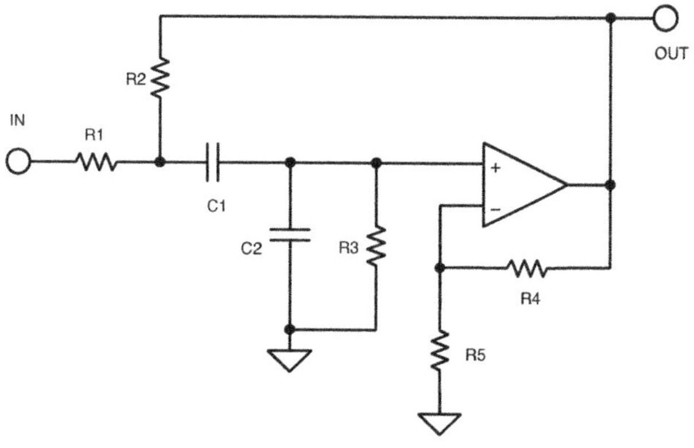

Figure 5-51: Sallen-Key band-pass filter

Multiple Feedback

The multiple feedback filter uses an op amp as an integrator as shown in Figure 5-52. Therefore, the dependence of the transfer function on the op amp parameters is greater than in the Sallen-Key realization. It is hard to generate high Q, high frequency sections due to the limitations of the open-loop gain of the op amp. A rule of thumb is that the open-loop gain of the op amp should be at least 20 dB (×10) above the amplitude response at the resonant (or cutoff) frequency, including the peaking caused by the Q of the filter. The peaking due to Q will cause an amplitude, A_0:

$$A_0 = H\,Q \qquad \text{Eq. 5-89}$$

where H is the gain of the circuit. The multiple feedback filter will invert the phase of the signal. This is equivalent to adding the resulting 180° phase shift to the phase shift of the filter itself.

The maximum to minimum component value ratios is higher in the multiple feedback case than in the Sallen-Key realization. The design equations for the multiple feedback low pass are given in Figure 5-70.

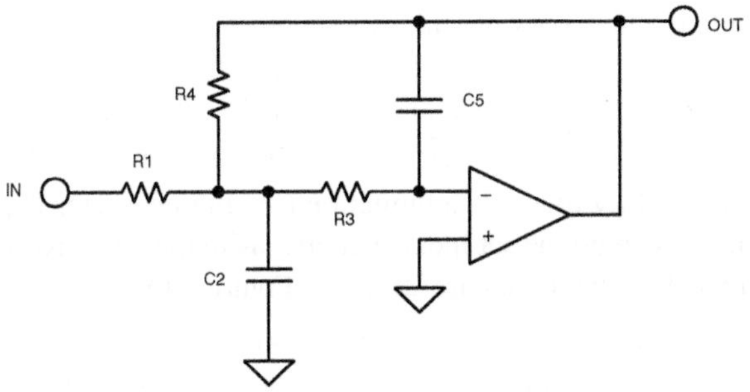

Figure 5-52: Multiple feedback low pass

Comments made about the multiple feedback low pass case apply to the high-pass case as well (see Figure 5-53). Note that resistors and capacitors are again swapped to convert the low pass case to the high pass case. The design equations for the multiple feedback high pass are given in Figure 5-71.

The design equations for the multiple feedback band pass case (see Figure 5-54) are given in Figure 5-72.

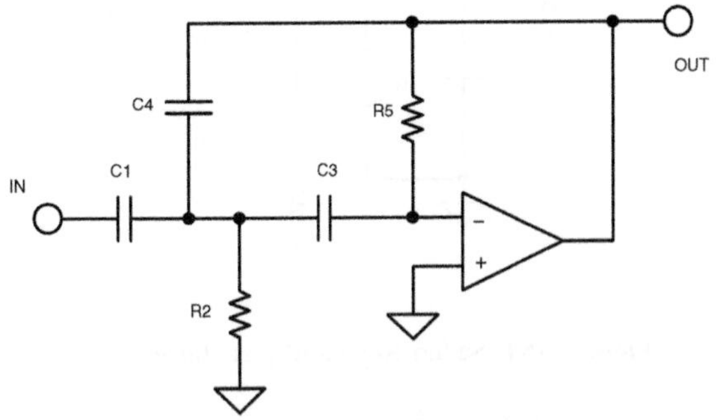

Figure 5-53: Multiple feedback high pass

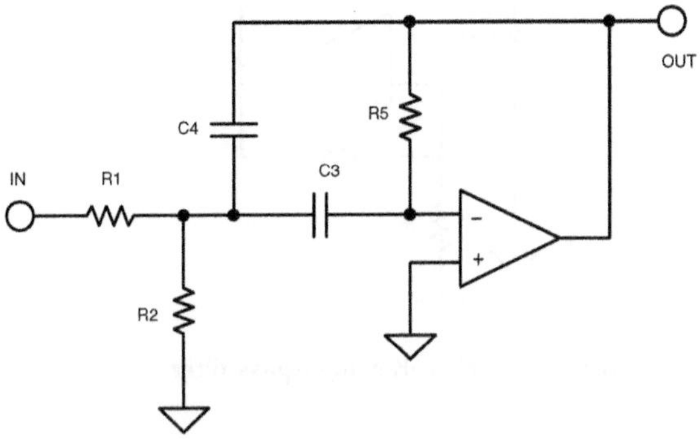

Figure 5-54: Multiple feedback band pass

This circuit is widely used in low Q (< 20) applications. It allows some tuning of the resonant frequency, F_0, by making R2 variable. Q can be adjusted (with R5) as well, but this will also change F_0.

Tuning of F_0 can be accomplished by monitoring the output of the filter with the horizontal channel of an oscilloscope, with the input to the filter connected to the vertical channel. The display will be a Lissajous pattern. This pattern will be an ellipse that will collapse to a straight line at resonance, since the phase shift will be 180°. Output could also be adjusted for maximum output, which will also occur at resonance, but this is usually not as precise, especially at lower values of Q where there is a less pronounced peak.

State Variable

The state-variable realization (see Reference 11) is shown in Figure 5-55, along with the design equations in Figure 5-73. This configuration offers the most precise implementation, at the expense of many more circuit elements. All three major parameters (gain, Q and ω_0) can be adjusted independently, and low pass, high pass, and band pass outputs are available simultaneously. Note that the low pass and high pass outputs are inverted in phase while the band pass output maintains the phase. The gain of each of the outputs of the filter is also independently variable. With an added amplifier section summing the low pass and high pass sections the notch function can also be synthesized. By changing the ratio of the summed sections, low pass notch, standard notch and high pass notch functions can be realized. A standard notch may also be realized by subtracting the band pass output from the input with the added op amp section. An all-pass filter may also be built with the four amplifier configuration by subtracting the band pass output from the input. In this instance, the band pass gain must equal 2.

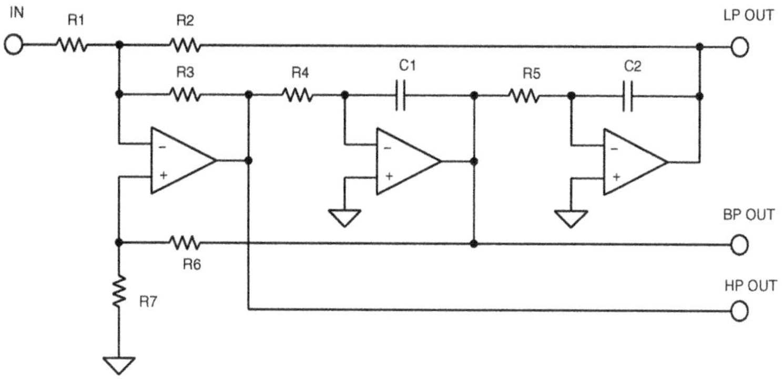

Figure 5-55: State-variable filter

Since all parameters of the state-variable filter can be adjusted independently, component spread can be minimized. Also, variations due to temperature and component tolerances are minimized. The op amps used in the integrator sections will have the same limitations on op amp gain-bandwidth as described in the multiple feedback section.

Tuning the resonant frequency of a state-variable filter is accomplished by varying R4 and R5. While it is not necessary to tune both, if varying over a wide range it is generally preferable. Holding R1 constant, tuning R2 sets the low pass gain and tuning R3 sets the high-pass gain. Band pass gain and Q are set by the ratio of R6 and R7.

Since the parameters of a state variable filter are independent and tunable, it is easy to add electronic control of frequency, Q and ω_0. This adjustment is accomplished by using an analog multiplier, multiplying DACs (MDACs) or digital pots, as shown in one of the examples in a later section. For the integrator sections adding the analog multiplier or MDAC effectively increases the time constant by dividing the voltage driving the resistor, which, in turn, provides the charging current for the integrator capacitor. This in effect raises the resistance and, in turn, the time constant. The Q and gain can be varied by changing the ratio of the various feedback paths. A digital pot will accomplish the same feat in a more direct manner, by directly changing the resistance value. The resultant tunable filter offers a great deal of utility in measurement and control circuitry. A worked out example is given in Section 8 of this chapter.

Biquadratic (Biquad)

A close cousin of the state variable filter is the biquad as shown in Figure 5-56. The name of this circuit was first used by J. Tow in 1968 (Reference 11) and later by L. C. Thomas in 1971 (see Reference 12). The name derives from the fact that the transfer function is a quadratic function in both the numerator and the denominator. Hence the transfer function is a biquadratic function. This circuit is a slight rearrangement of the state variable circuit. One significant difference is that there is not a separate high pass output. The band pass output inverts the phase. There are two low pass outputs, one in phase and one out of phase. With the addition of a fourth amplifier section, high pass, notch (low-pass, standard and high-pass) and all-pass filters can be realized. The design equations for the biquad are given in Figure 5-74.

Referring to Figure 5-74, the all-pass case of the biquad, R8 = R9/2 and R7 = R9. This is required to make the terms in the transfer function line up correctly. For the high pass output, the input, band pass and second low pass outputs are summed. In this case the constraints are that R1 = R2 = R3 and R7 = R8 = R9.

Like the state variable, the biquad filter is tunable. Adjusting R3 will adjust the Q. Adjusting R4 will set the resonant frequency. Adjusting R1 will set the gain. Frequency would generally be adjusted first followed by Q and then gain. Setting the parameters in this manner minimizes the effects of component value interaction.

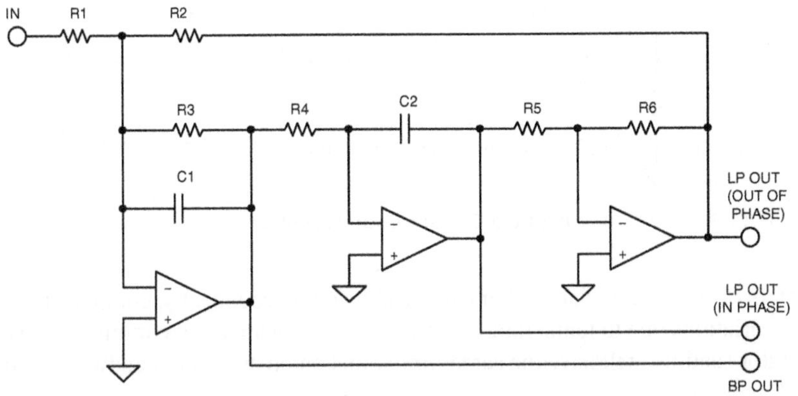

Figure 5-56: Biquad filter

Dual Amplifier Band Pass (DAPB)

The Dual Amplifier band-pass filter structure is useful in designs requiring high Qs and high frequencies. Its component sensitivity is small, and the element spread is low. A useful feature of this circuit is that the Q and resonant frequency can be adjusted more or less independently.

Referring to Figure 5-57, the resonant frequency can be adjusted by R2. R1 can then be adjusted for Q. In this topology it is useful to use dual op amps. The match of the two op amps will lower the sensitivity of Q to the amplifier parameters.

It should be noted that the DABP has a gain of 2 at resonance. If lower gain is required, resistor R1 may be split to form a voltage divider. This is reflected in the addendum to the design equations of the DABP, Figure 5-75.

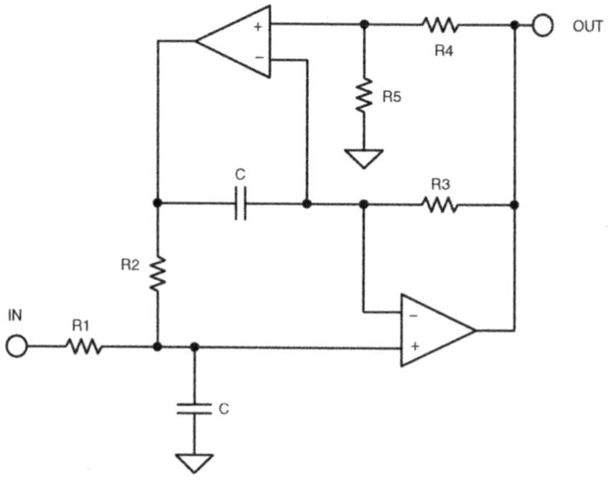

Figure 5-57: Dual amplifier band-pass filter

Twin-T Notch

The Twin T is widely used as a general-purpose notch circuit as shown in Figure 5-58. The passive implementation of the twin T (i.e., with no feedback) has a major shortcoming of having a Q that is fixed at 0.25. This issue can be rectified with the application of positive feedback to the reference node. The amount of the signal feedback, set by the R4/R5 ratio, will determine the value of Q of the circuit, which, in turn, determines the notch depth. For maximum notch depth, the resistors R4 and R5 and the associated op amp can be eliminated. In this case, the junction of C3 and R3 will be directly connected to the output.

Tuning is not easily accomplished. Using standard 1% components a 60 dB notch is as good as can be expected, with 40 dB–50 dB being more typical.

The design equations for the Twin T are given in Figure 5-76.

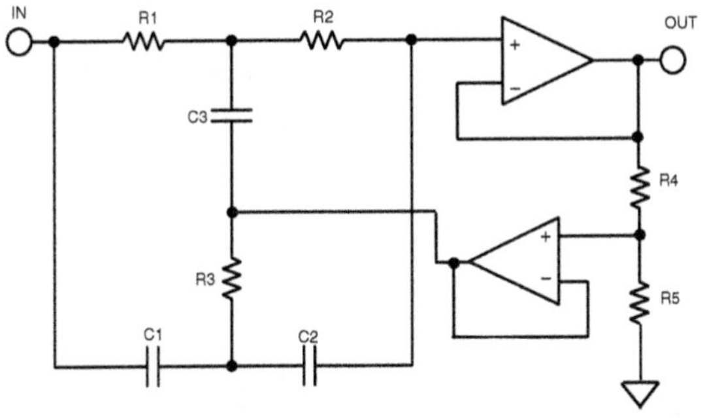

Figure 5-58: Twin-T notch filter

Bainter Notch

A simple notch filter is the Bainter circuit (see Reference 21). It is composed of simple circuit blocks with two feedback loops as shown in Figure 5-59. Also, the component sensitivity is very low.

This circuit has several interesting properties. The Q of the notch is not based on component matching as it is in every other implementation, but is instead only dependant on the gain of the amplifiers. Therefore, the

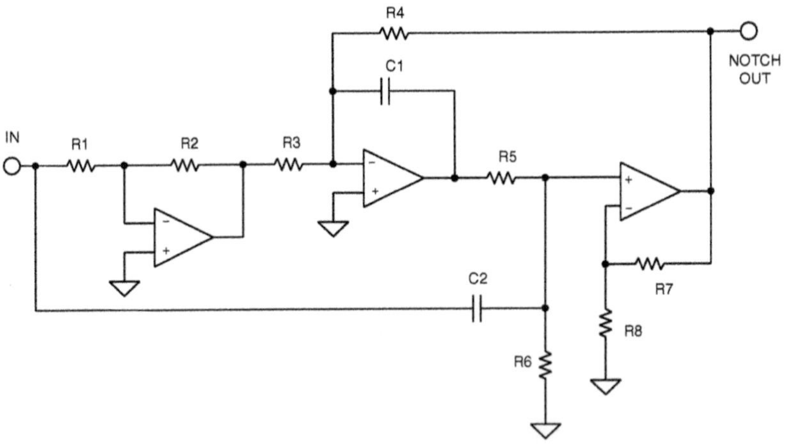

Figure 5-59: Bainter notch filter

notch depth will not drift with temperature, aging, and other environmental factors. The notch frequency may shift, but not the depth.

Amplifier open loop gain of 10^4 will yield a Q_z of > 200. It is capable of orthogonal tuning with minimal interaction. R6 tunes Q and R1 tunes ω_Z. Varying R3 sets the ratio of ω_0/ω_Z produces low pass notch (R4 > R3), notch (R4 = R3) or high pass notch (R4 < R3).

The design equations of the Bainter circuit are given in Figure 5-77.

Boctor Notch

The Boctor circuits (see References 22, 23), while moderately complicated, uses only one op amp. Due to the number of components, there is a great deal of latitude in component selection. These circuits also offer low sensitivity and the ability to tune the various parameters more or less independently.

There are two forms, a low pass notch (Figure 5-60) and a high-pass notch (Figure 5-61). For the low pass case, the preferred order of adjustment is to tune ω_0 with R4, then Q_0 with R2, next Q_z with R3 and finally ω_z with R1.

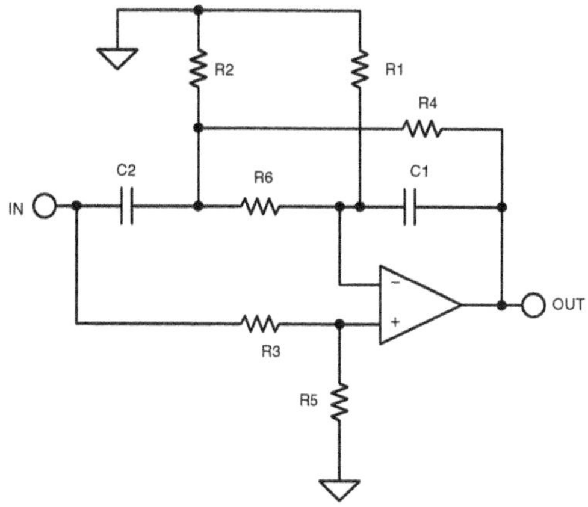

Figure 5-60: Boctor low-pass notch filter

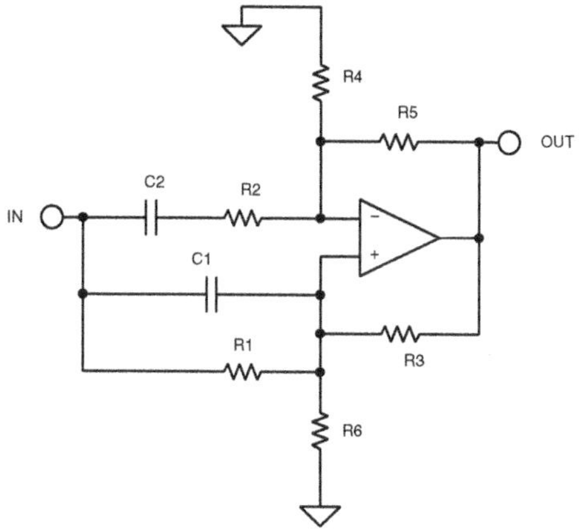

Figure 5-61: Boctor high-pass filter

In order for the components to be realizable we must define a variable, k1, such that:

$$\frac{\omega_0^2}{\omega_Z^2} < k1 < 1 \qquad \text{Eq. 5-90}$$

The design equations are given in Figure 5-78 for the low pass case and in Figure 5-79 for the high pass case.

In the high-pass case circuit gain is required and it applies only when

$$Q < \frac{1}{1 - \frac{\omega_Z^2}{\omega_0^2}} \qquad \text{Eq. 5-91}$$

but a high pass notch can be realized with one amplifier and only two capacitors, which can be the same value. The pole and zero frequencies are completely independent of the amplifier gain. The resistors can be trimmed so that even 5% capacitors can be used.

"1 – Band Pass" Notch

As mentioned in the state variable and biquad sections, a notch filter can be built as 1 – BP. The band pass section can be any of the all-pole band pass realizations discussed above, or any others. Keep in mind whether the band pass section is inverting as shown in Figure 5-62 (such as the multiple feedback circuit) or noninverting as shown in Figure 5-63 (such as the Sallen-Key), since we want to subtract, not add, the band pass output from the input.

It should be noted that the gain of the band pass amplifier must be taken into account in determining the resistor values. Unity gain band pass would yield equal values.

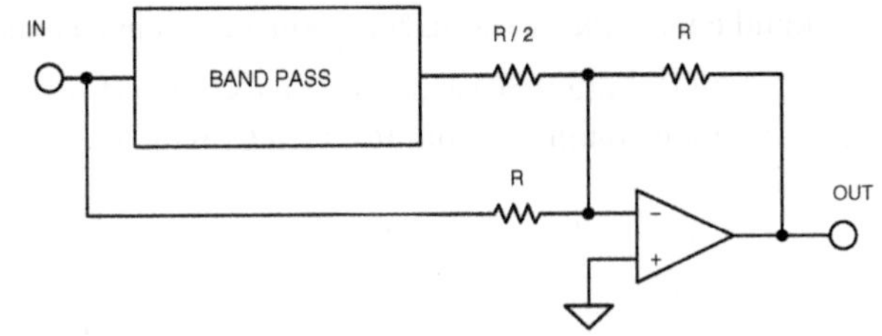

Figure 5-62: 1 – BP filter for inverting band pass configurations

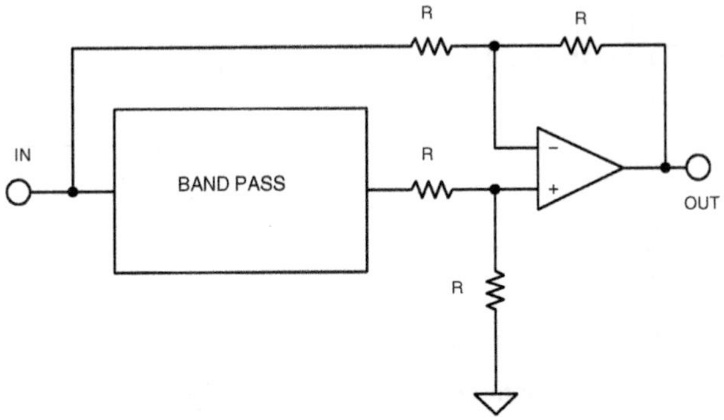

Figure 5-63: 1 – BP filter for noninverting band pass configurations

First Order All-Pass

The general form of a first order all-pass filter is shown in Figure 5-64. If the function is a simple RC high-pass (Figure 5-64A), the circuit will have a have a phase shift that goes from –180° at 0 Hz. and 0° at high frequency. It will be –90° at $\omega = 1/RC$. The resistor may be made variable to allow adjustment of the delay at a particular frequency.

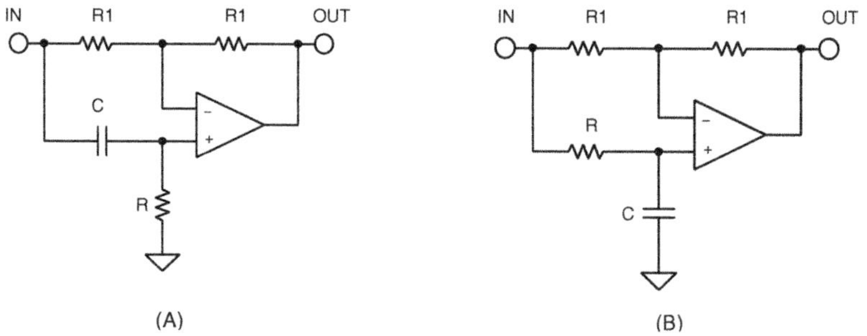

Figure 5-64: First order all-pass filters

If the function is changed to a low-pass function (Figure 5-64B), the filter is still a first order all-pass and the delay equations still hold, but the signal is inverted, changing from 0° at dc to –180° at high frequency.

Second Order All-Pass

A second order all-pass circuit shown in Figure 5-65 was first described by Delyiannis (see Reference 17). The main attraction of this circuit is that it requires only one op amp. Remember also that an all-pass filter can also be realized as 1 – 2BP.

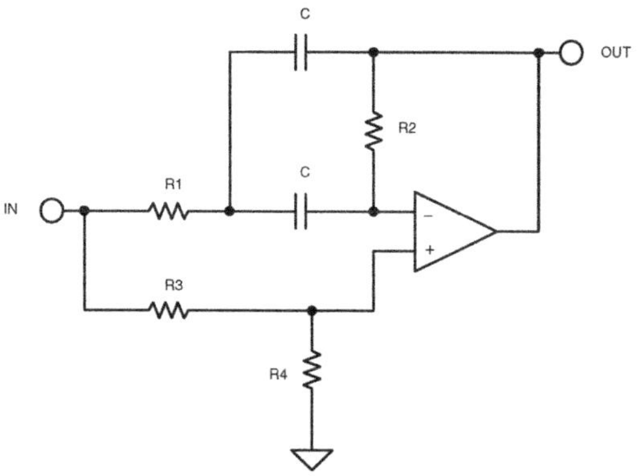

Figure 5-65: Second order all-pass filter

Any of the all-pole realizations discussed above may be used to build the filter, but be aware of whether or not the BP inverts the phase. Be aware also that the gain of the BP section must be 2. To this end, the DABP structure is particularly useful, since its gain is fixed at 2.

Figures 5-66 through 5-81 following summarize design equations for various active filter realizations. In all cases, H, ω_o, Q, and α are given, being taken from the design tables.

SINGLE POLE

LOW PASS

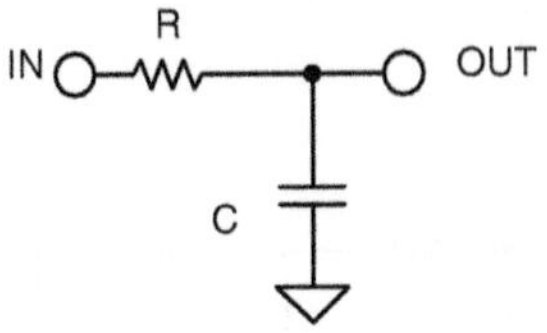

$$\frac{V_O}{V_{IN}} = \frac{1}{sC\,R+1}$$

$$F_o = \frac{1}{2\pi R\,C}$$

HIGH PASS

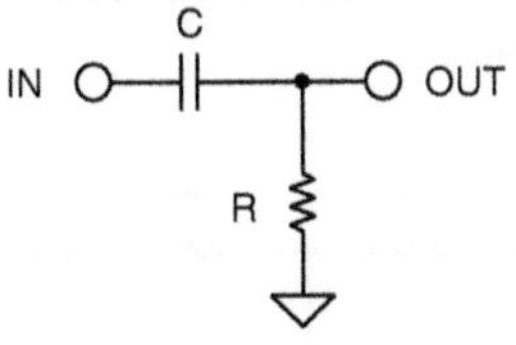

$$\frac{V_O}{V_{IN}} = \frac{sC\,R}{sC\,R+1}$$

$$F_o = \frac{1}{2\pi\,R\,C}$$

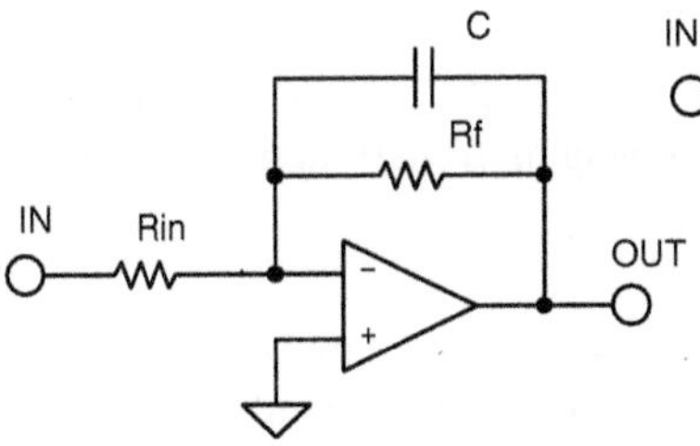

$$\frac{V_O}{V_{IN}} = -\frac{Rf}{Rin}\frac{1}{sC\,R2+1}$$

$$H_o = -\frac{Rf}{Rin}$$

$$F_o = \frac{1}{2\pi Rf C}$$

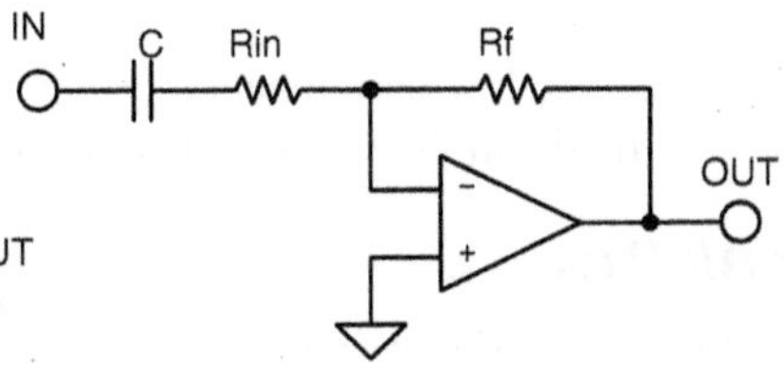

$$\frac{V_O}{V_{IN}} = -\frac{Rf}{Rin}\frac{sC\,R1}{sC\,R1+1}$$

$$H_o = -\frac{Rf}{Rin}$$

$$F_o = \frac{1}{2\pi\ Rin\ C}$$

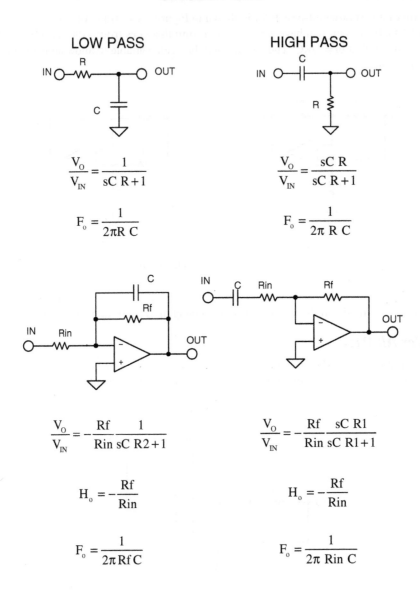

Figure 5-66: Single-pole filter design equations

SALLEN-KEY LOW PASS

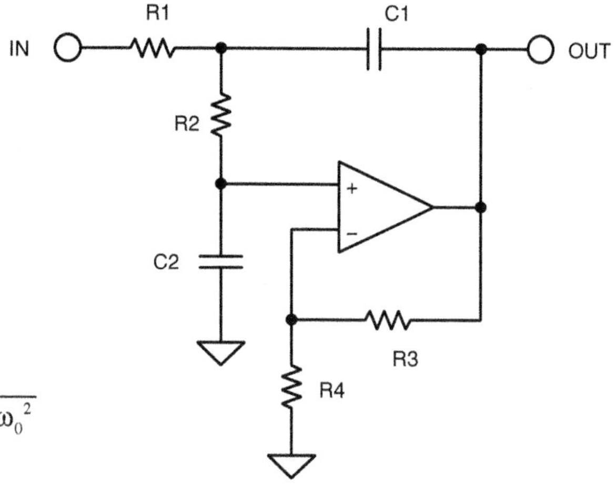

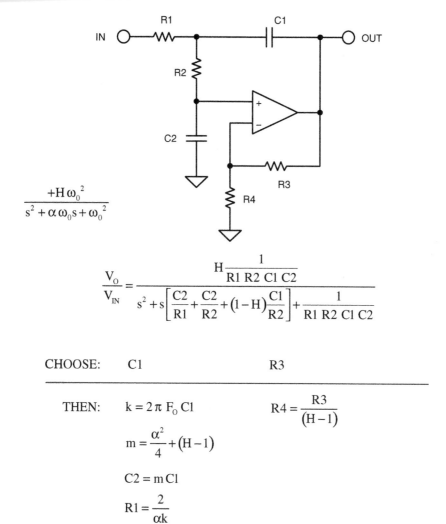

$$\frac{+H\,\omega_0^{\,2}}{s^2 + \alpha\,\omega_0 s + \omega_0^{\,2}}$$

$$\frac{V_O}{V_{IN}} = \frac{H\dfrac{1}{R1\,R2\,C1\,C2}}{s^2 + s\left[\dfrac{C2}{R1} + \dfrac{C2}{R2} + (1-H)\dfrac{C1}{R2}\right] + \dfrac{1}{R1\,R2\,C1\,C2}}$$

CHOOSE:	C1	R3
THEN:	$k = 2\pi\,F_0\,C1$	$R4 = \dfrac{R3}{(H-1)}$
	$m = \dfrac{\alpha^2}{4} + (H-1)$	
	$C2 = m\,C1$	
	$R1 = \dfrac{2}{\alpha k}$	
	$R2 = \dfrac{\alpha}{2mk}$	

Figure 5-67: Sallen-Key low pass design equations

SALLEN-KEY HIGH PASS

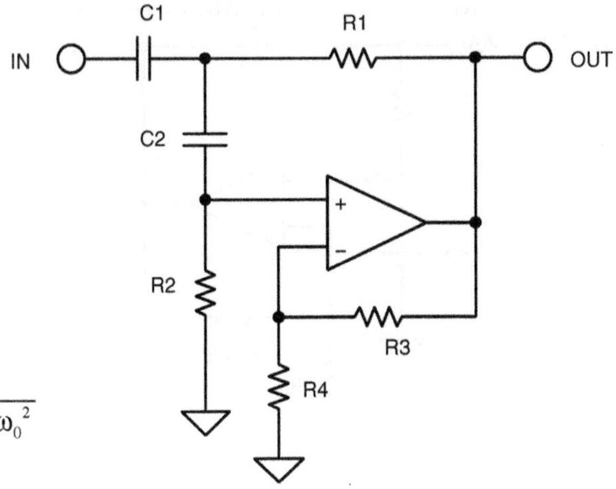

$$\frac{+H s^2}{s^2 + \alpha\, \omega_0\, s + \omega_0^{\ 2}}$$

$$\frac{V_O}{V_{IN}} = \frac{H\, s^2}{s^2 + s\left[\dfrac{\dfrac{C2}{R2} + \dfrac{C1}{R2} + (1-H)\dfrac{C2}{R1}}{C1\, C2}\right] + \dfrac{1}{R1\, R2\, C1\, C2}}$$

CHOOSE:	C1	R3
THEN:	$k = 2\, \pi\, F_0\, C1$	$R4 = \dfrac{R3}{(H-1)}$

$$C2 = C1$$

$$R1 = \frac{\alpha + \sqrt{\alpha^2 + (H-1)}}{4k}$$

$$R2 = \frac{4}{\alpha + \sqrt{\alpha^2 + (H-1)}} * \frac{1}{k}$$

Figure 5-68: Sallen-Key high pass design equations

SALLEN-KEY BAND PASS

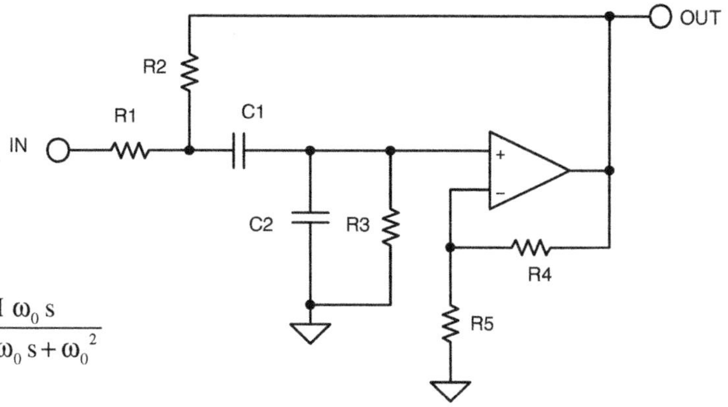

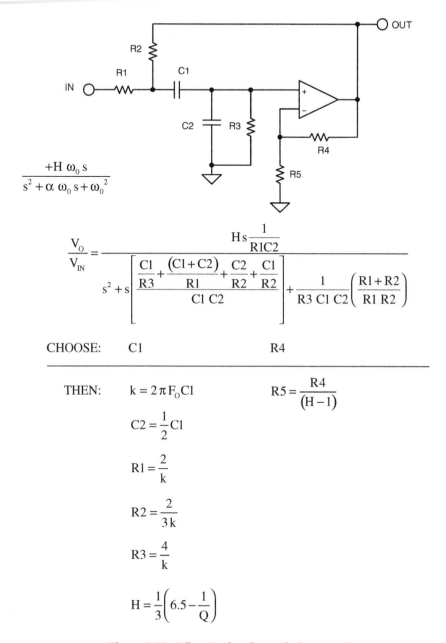

$$\frac{+H\,\omega_0\,s}{s^2+\alpha\,\omega_0\,s+\omega_0^{\,2}}$$

$$\frac{V_O}{V_{IN}}=\frac{Hs\frac{1}{R1C2}}{s^2+s\left[\frac{\frac{C1}{R3}+\frac{(C1+C2)}{R1}+\frac{C2}{R2}+\frac{C1}{R2}}{C1\,C2}\right]+\frac{1}{R3\,C1\,C2}\left(\frac{R1+R2}{R1\,R2}\right)}$$

CHOOSE: C1 R4

THEN:

$k=2\pi F_O C1$

$R5=\frac{R4}{(H-1)}$

$C2=\frac{1}{2}C1$

$R1=\frac{2}{k}$

$R2=\frac{2}{3k}$

$R3=\frac{4}{k}$

$H=\frac{1}{3}\left(6.5-\frac{1}{Q}\right)$

Figure 5-69: Sallen-Key band pass design equations

MULTIPLE FEEDBACK LOW PASS

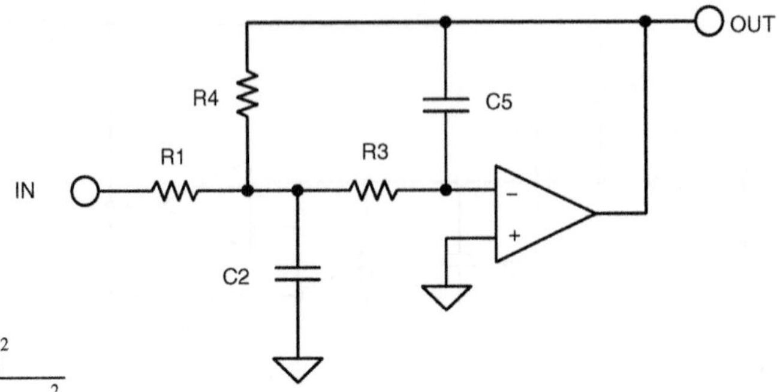

$$\frac{-H\omega_0^{\;2}}{s^2+\alpha\omega_0 s+\omega_0^{\;2}}$$

$$\frac{V_O}{V_{IN}}=\frac{-H\dfrac{1}{R3\ R4\ C2\ C5}}{s^2+s\dfrac{1}{C2}\left(\dfrac{1}{R1}+\dfrac{1}{R3}+\dfrac{1}{R4}\right)+\dfrac{1}{R3\ R4\ C2\ C5}}$$

CHOOSE: C5

THEN: $k=2\pi F_O C5$

$$C2=\frac{4}{\alpha 2}(H+1)C5$$

$$R1=\frac{\alpha}{2\ H\ k}$$

$$R3=\frac{\alpha}{2(H+1)k}$$

$$R4=\frac{\alpha}{2k}$$

Figure 5-70: Multiple feedback low pass design equations

MULTIPLE FEEDBACK HIGH PASS

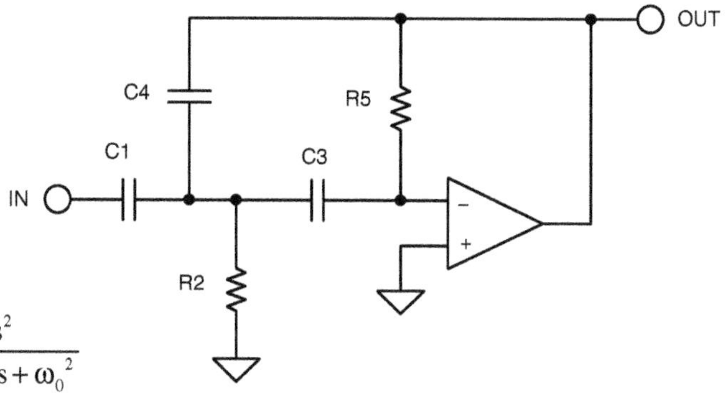

$$\frac{-H s^2}{s^2 + \alpha \omega_0 s + \omega_0^2}$$

$$\frac{V_O}{V_{IN}} = \frac{-s^2 \dfrac{C1}{C4}}{s^2 + s\dfrac{(C1 + C3 + C4)}{C3\ C4\ R5} + \dfrac{1}{R2\ R5\ C3\ C4}}$$

CHOOSE: C1

THEN: $k = 2\pi F_O C1$

$C3 = C1$

$$C4 = \frac{C1}{H}$$

$$R2 = \frac{\alpha}{k\left(2 + \dfrac{1}{H}\right)}$$

$$R = \frac{H\left(2 + \dfrac{1}{H}\right)}{\alpha k}$$

Figure 5-71: Multiple feedback high pass design equations

MULTIPLE FEEDBACK BAND PASS

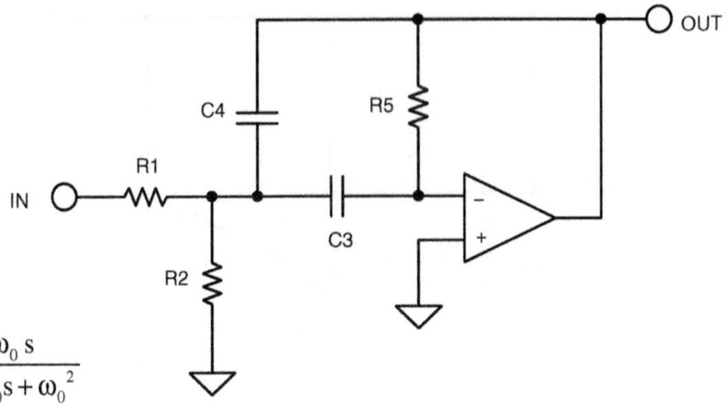

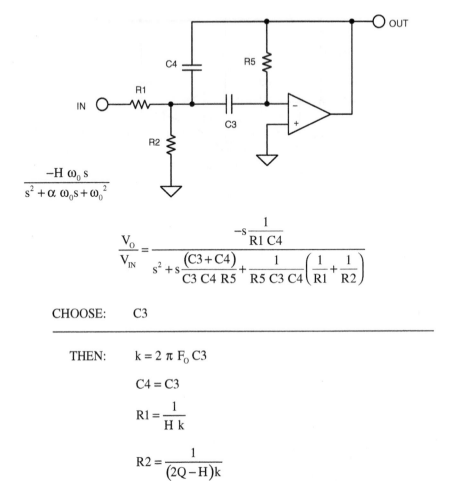

$$\frac{-H\,\omega_0\,s}{s^2 + \alpha\,\omega_0 s + \omega_0^2}$$

$$\frac{V_O}{V_{IN}} = \frac{-s\dfrac{1}{R1\,C4}}{s^2 + s\dfrac{(C3 + C4)}{C3\,C4\,R5} + \dfrac{1}{R5\,C3\,C4}\left(\dfrac{1}{R1} + \dfrac{1}{R2}\right)}$$

CHOOSE: C3

THEN: $k = 2\,\pi\,F_0\,C3$

$C4 = C3$

$$R1 = \frac{1}{H\,k}$$

$$R2 = \frac{1}{(2Q - H)k}$$

$$R5 = \frac{2Q}{k}$$

Figure 5-72: Multiple feedback band-pass design equations

STATE VARIABLE (A)

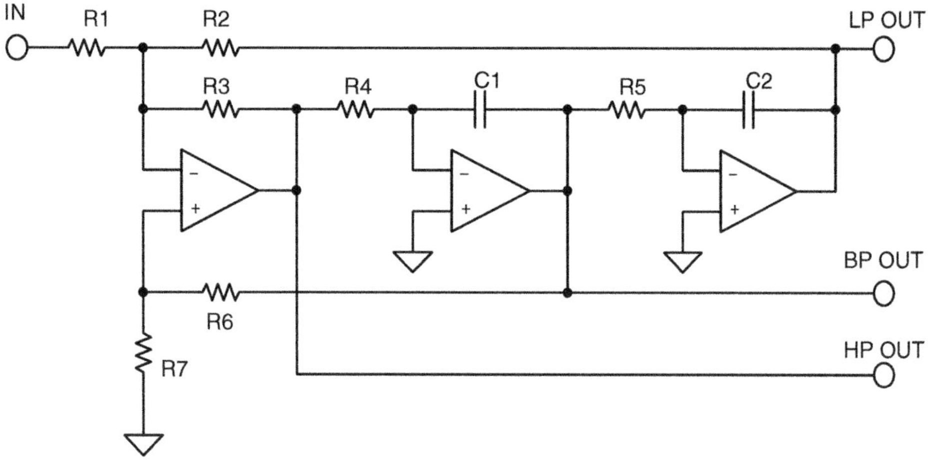

$$A_{LP(s=0)} = -\frac{R2}{R1}$$

$$A_{HP(s=\infty)} = -\frac{R3}{R1}$$

$$\omega_0 = \sqrt{\frac{R3}{R2\ R4\ R5\ C1\ C2}}$$

$$\text{LET } R4 = R5 = R, C1 = C2 = C$$

$$A_{BP(s=\omega_0)} = \frac{\frac{R6+R7}{R7}}{R1\left(\frac{1}{R1}+\frac{1}{R2}+\frac{1}{R3}\right)}$$

CHOOSE R1:

$$R2 = A_{LP} R1$$

$$R3 = A_{HP} R1$$

CHOOSE C:

$$R = \frac{2\pi F_O}{C}\sqrt{\frac{A_{HP}}{A_{LP}}}$$

CHOOSE R7:

$$R6 = R7\sqrt{R2\,R3}\ Q\left(\frac{1}{\frac{1}{R1}+\frac{1}{R2}+\frac{1}{R3}}\right)$$

Figure 5-73A: State-variable design equations

STATE VARIABLE (B)

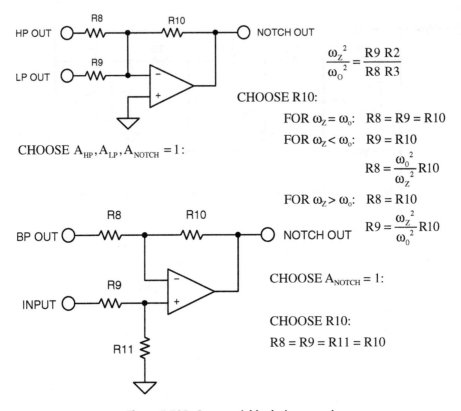

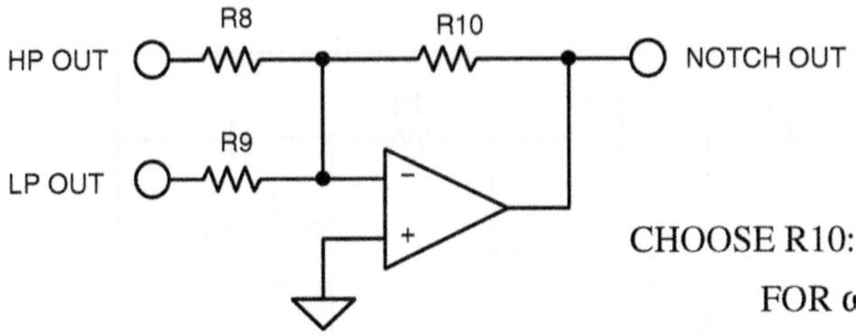

$$\frac{\omega_Z^{\,2}}{\omega_O^{\,2}} = \frac{R9\,R2}{R8\,R3}$$

CHOOSE R10:

FOR $\omega_Z = \omega_o$: $R8 = R9 = R10$

FOR $\omega_Z < \omega_o$: $R9 = R10$

$$R8 = \frac{\omega_0^{\,2}}{\omega_Z^{\,2}} R10$$

FOR $\omega_Z > \omega_o$: $R8 = R10$

$$R9 = \frac{\omega_Z^{\,2}}{\omega_0^{\,2}} R10$$

CHOOSE $A_{HP}, A_{LP}, A_{NOTCH} = 1$:

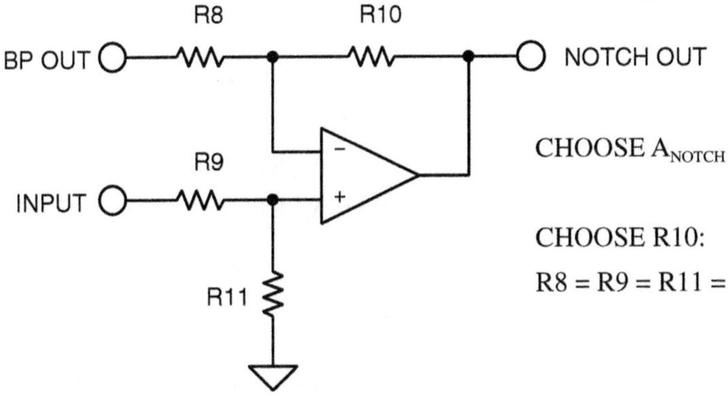

CHOOSE $A_{NOTCH} = 1$:

CHOOSE R10:

$R8 = R9 = R11 = R10$

Figure 5-73B: State-variable design equations

STATE VARIABLE (C)

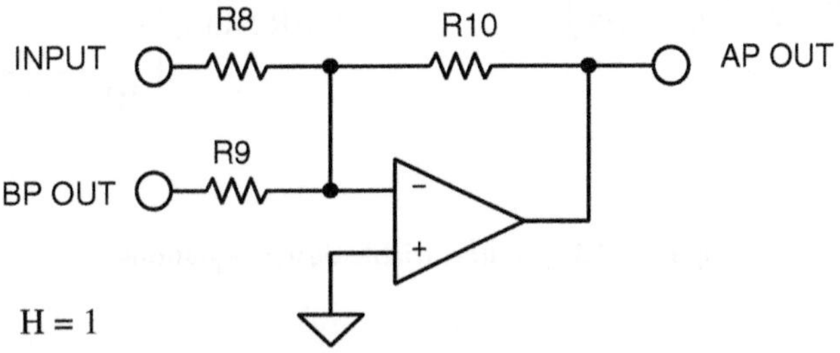

$H = 1$

$R8 = R10$

$R9 = R8/2$

Figure 7-73C: State-variable design equations

BIQUADRATIC (A)

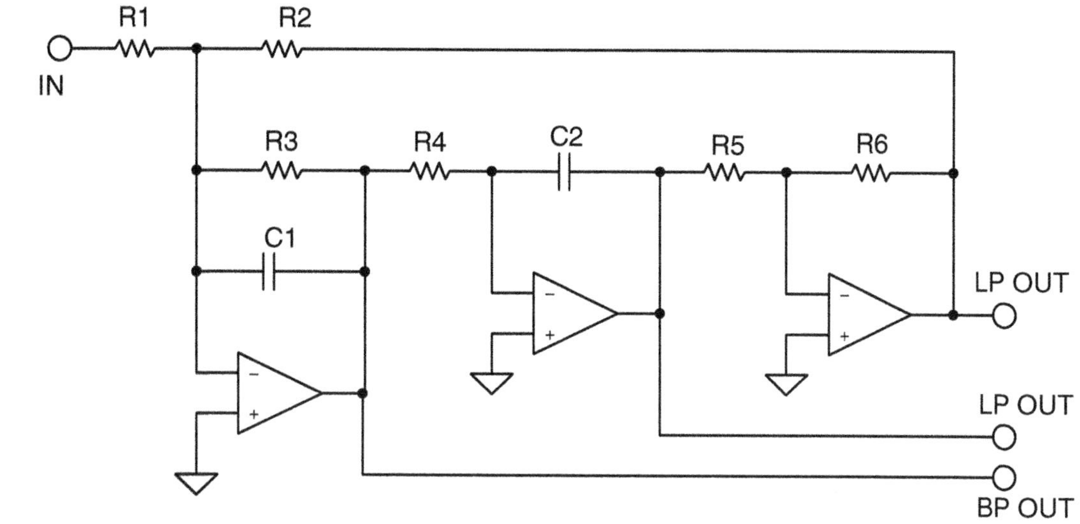

CHOOSE C, R2, R5

$$K = 2\pi f_0 C$$

$$C1 = C2 = C$$

$$R1 = \frac{R2}{H}$$

$$R3 = \frac{1}{k\alpha}$$

$$R4 = \frac{1}{k^2 R2}$$

CHOOSE C, R5, R7

$$K = 2\pi f_0 C$$

$$C1 = C2 = C$$

$$R1 = R2 = R3 = \frac{1}{k\alpha}$$

$$R4 = \frac{1}{k^2 R2}$$

$$R6 = R5$$

HIGH PASS

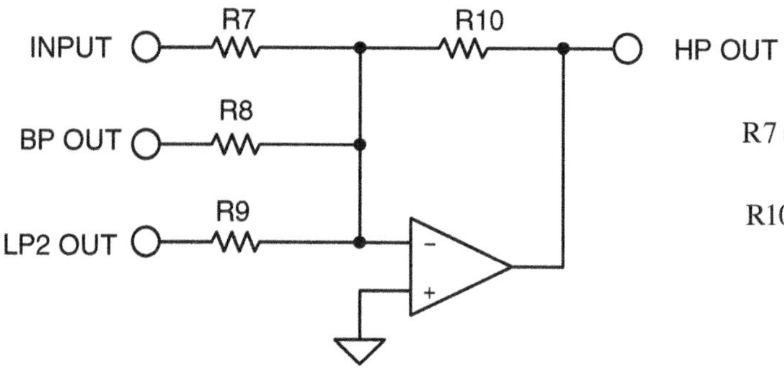

$$R7 = R8 = R9 = R$$

$$R10 = \frac{R}{H}$$

Figure 5-74A: Biquad design equations

BIQUADRATIC (B)

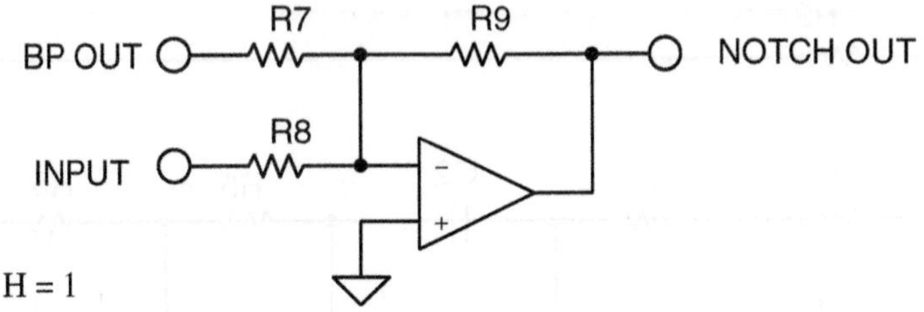

H = 1

R7 = R8 = R9

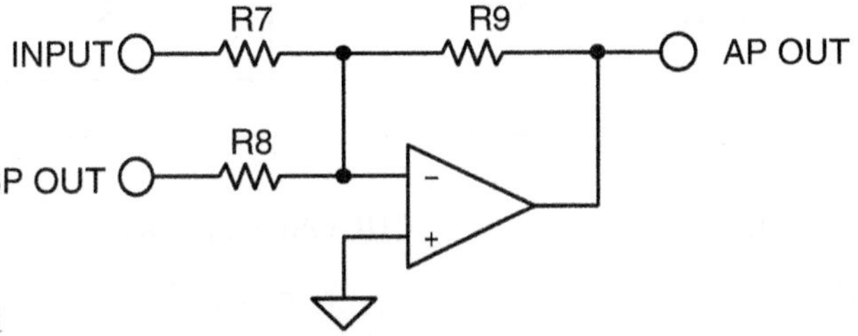

H = 1

R7 = R9

R8 = R7/2

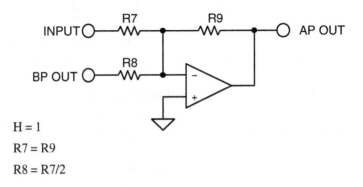

Figure 5-74B: Biquad design equations

DUAL AMPLIFIER BANDPASS

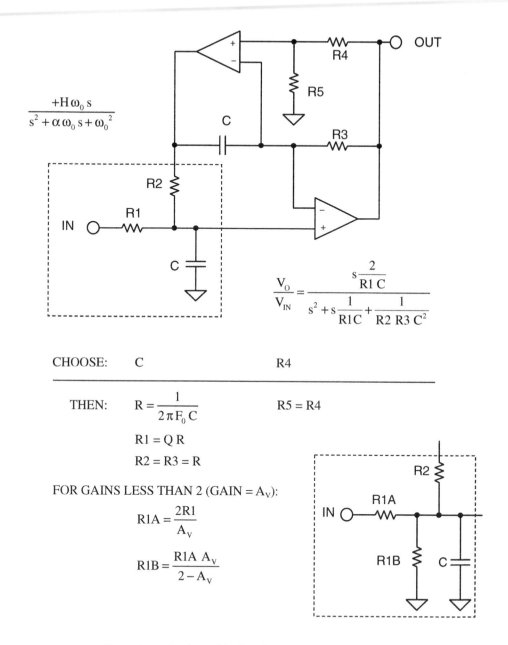

$$\frac{+H\omega_0 s}{s^2 + \alpha\omega_0 s + \omega_0^2}$$

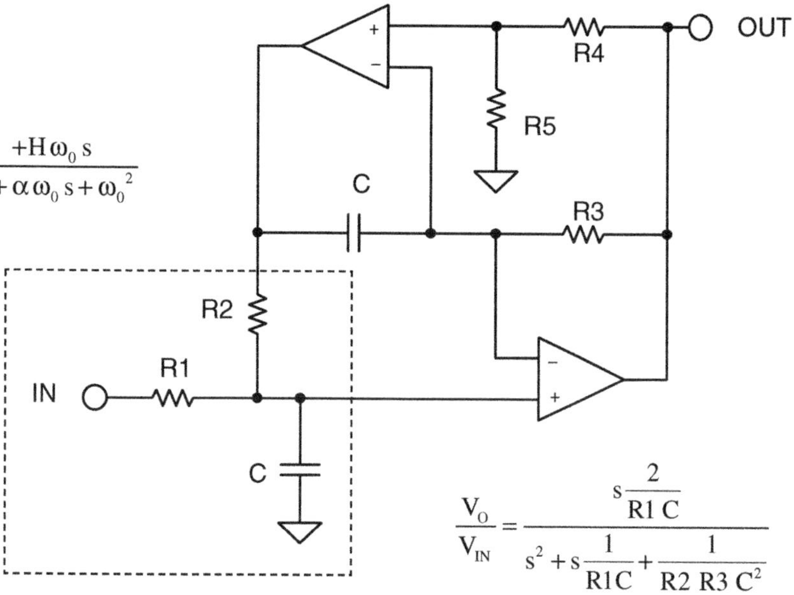

$$\frac{V_O}{V_{IN}} = \frac{s\dfrac{2}{R1\,C}}{s^2 + s\dfrac{1}{R1C} + \dfrac{1}{R2\,R3\,C^2}}$$

CHOOSE:	C	R4
THEN:	$R = \dfrac{1}{2\pi F_0 C}$	R5 = R4

$R1 = Q\,R$

$R2 = R3 = R$

FOR GAINS LESS THAN 2 (GAIN = A_V):

$$R1A = \frac{2R1}{A_V}$$

$$R1B = \frac{R1A\;A_V}{2 - A_V}$$

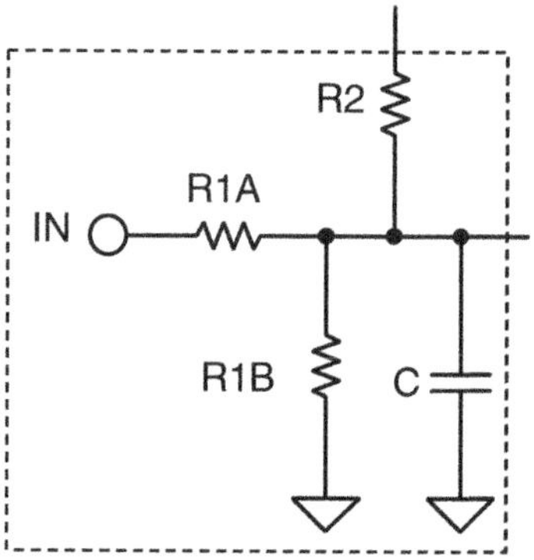

Figure 5-75: Dual amplifier band pass design equations

TWIN-T NOTCH

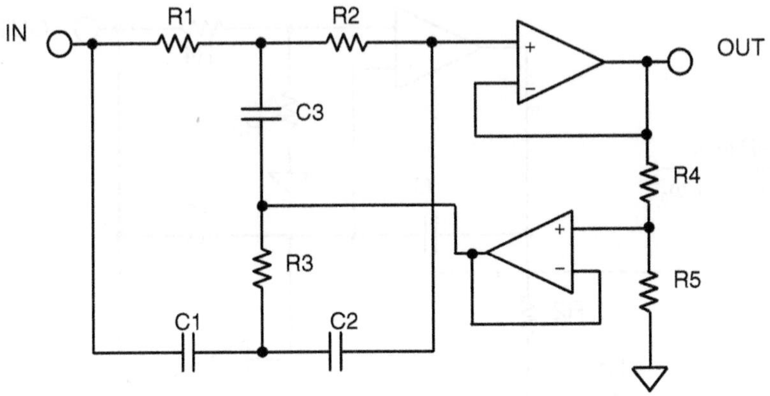

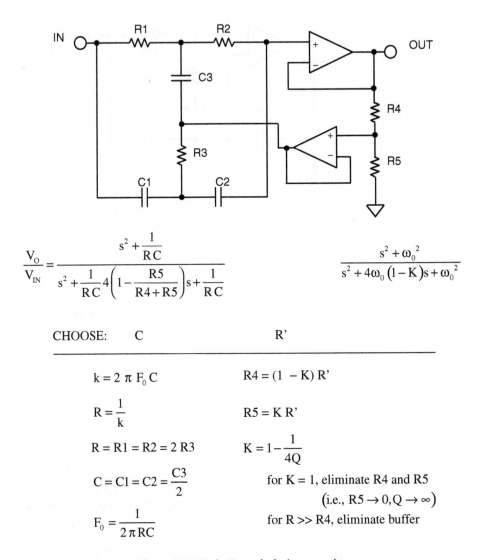

$$\frac{V_O}{V_{IN}} = \frac{s^2 + \frac{1}{RC}}{s^2 + \frac{1}{RC}4\left(1 - \frac{R5}{R4+R5}\right)s + \frac{1}{RC}} \qquad \frac{s^2 + \omega_0^2}{s^2 + 4\omega_0(1-K)s + \omega_0^2}$$

CHOOSE: C R'

$k = 2\pi F_0 C$

$R = \frac{1}{k}$

$R = R1 = R2 = 2\,R3$

$C = C1 = C2 = \frac{C3}{2}$

$F_0 = \frac{1}{2\pi RC}$

$R4 = (1 - K)\,R'$

$R5 = K\,R'$

$K = 1 - \frac{1}{4Q}$

for $K = 1$, eliminate R4 and R5 $\left(\text{i.e., } R5 \to 0, Q \to \infty\right)$

for $R >> R4$, eliminate buffer

Figure 5-76: Twin-T notch design equations

BAINTER NOTCH

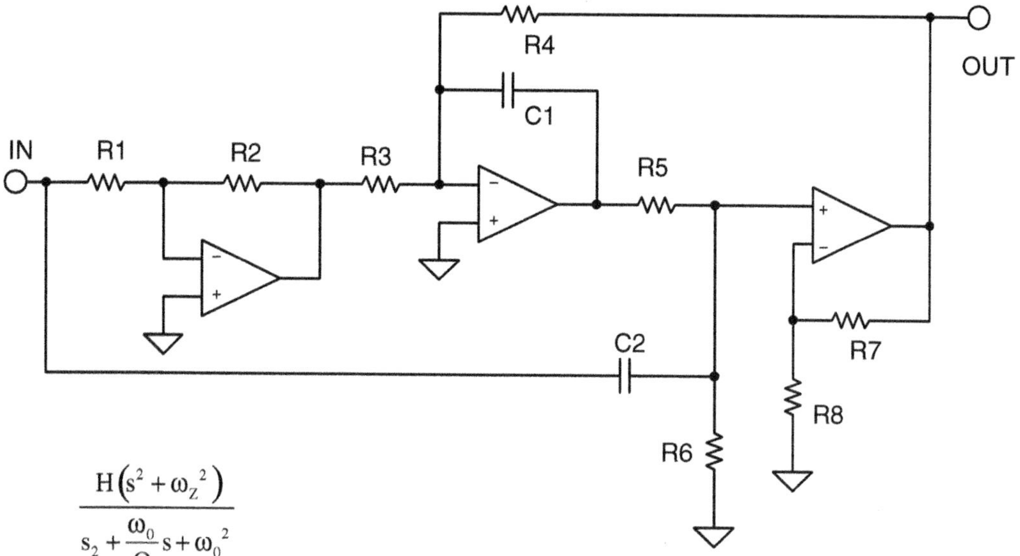

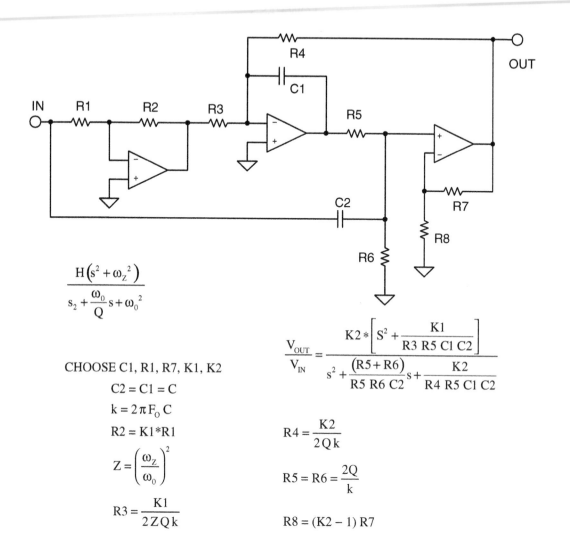

$$\frac{H\left(s^2+\omega_Z^{\ 2}\right)}{s_2+\frac{\omega_0}{Q}s+\omega_0^{\ 2}}$$

$$\frac{V_{OUT}}{V_{IN}}=\frac{K2*\left[S^2+\frac{K1}{R3\ R5\ C1\ C2}\right]}{s^2+\frac{(R5+R6)}{R5\ R6\ C2}s+\frac{K2}{R4\ R5\ C1\ C2}}$$

CHOOSE C1, R1, R7, K1, K2

$$C2 = C1 = C$$

$$k = 2\pi F_O C$$

$$R2 = K1*R1$$

$$Z=\left(\frac{\omega_Z}{\omega_0}\right)^2$$

$$R3=\frac{K1}{2ZQk}$$

$$R4=\frac{K2}{2Qk}$$

$$R5 = R6 = \frac{2Q}{k}$$

$$R8 = (K2 - 1)\ R7$$

Figure 5-77: Bainter notch

BOCTOR NOTCH

LOW PASS

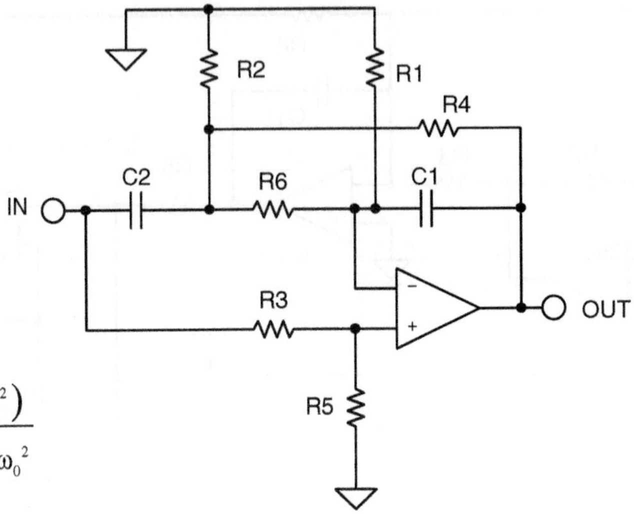

$$\frac{H\left(s^2+\omega_Z{}^2\right)}{s^2+\frac{\omega_0}{Q}s+\omega_0{}^2}$$

$$\frac{V_{OUT}}{V_{IN}}=H\frac{s^2+\frac{R1+(R2\,\|\,R4)+R6}{R1(R2\,\|\,R4)R6\;C1\;C2}}{s^2+\left(\frac{1}{R6\;C2}+\frac{1}{(R2\,\|\,R4)C2}\right)s+\frac{1}{R4\;R6\;C1\;C2}}$$

GIVEN ω_0, ω_Z, Q_0

CHOOSE R6 R5 C1

$$R4=\frac{1}{\omega_0 C1\;2Q_0}$$

$$R2=\frac{R4\;R6}{R4+R6}$$

$$R1=\frac{1}{2}\left(\frac{R6\;\omega_Z{}^2}{R4\;\omega_0{}^2}-1\right)$$

$$R3=\left(\frac{R6}{R1}+2\frac{C1}{C2}\right)R5$$

$$C2=4Q_0{}^2\frac{R4}{R6}C1$$

Figure 5-78: Boctor notch, low pass

BOCTOR NOTCH

HIGH PASS (A)

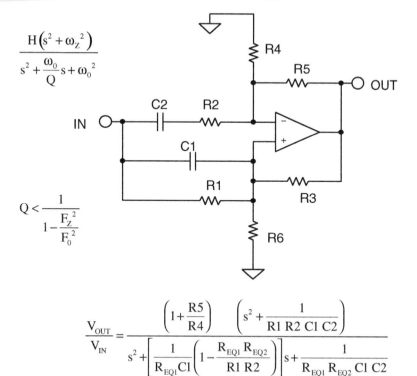

$$\frac{H\left(s^2+\omega_Z{}^2\right)}{s^2+\frac{\omega_0}{Q}s+\omega_0{}^2}$$

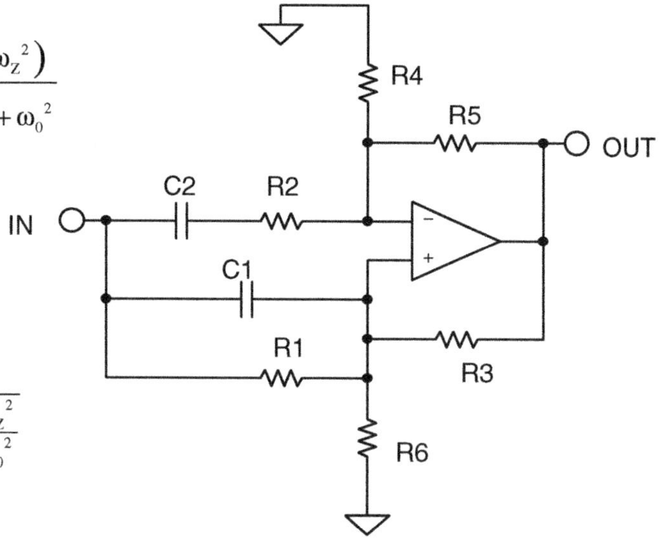

$$Q<\frac{1}{1-\frac{F_Z{}^2}{F_0{}^2}}$$

$$\frac{V_{OUT}}{V_{IN}}=\frac{\left(1+\frac{R5}{R4}\right)\left(s^2+\frac{1}{R1\,R2\,C1\,C2}\right)}{s^2+\left[\frac{1}{R_{EQ1}C1}\left(1-\frac{R_{EQ1}\,R_{EQ2}}{R1\,R2}\right)\right]s+\frac{1}{R_{EQ1}\,R_{EQ2}\,C1\,C2}}$$

WHERE: $R_{EQ1}=R1\,\|\,R3\,\|\,R6$

$R_{EQ2}=R2+\left(R4\,\|\,R5\right)$

GIVEN: F_Z F_0 H or F_Z Q H

$$Q=\frac{1}{\sqrt{2\left(\frac{F_Z{}^2}{F_0{}^2}-1\right)}}$$

$$F_0=F_Z\sqrt{\frac{1}{1-\frac{1}{2Q^2}}}$$

$$Y=\frac{1}{Q\left(1-\frac{F_Z{}^2}{F_0{}^2}\right)}$$

Figure 5-79A: Boctor high pass design equations

BOCTOR NOTCH
HIGH PASS (B)

GIVEN: C, R2, R3

$$C1 = C2 = C$$

$$R_{EQ1} = \frac{1}{C\,Y\,2\pi\,F_0}$$

$$R_{EQ2} = Y^2 R_{EQ1}$$

$$R4 = R_{EQ2} - R2\left(\frac{H}{H-1}\right)$$

$$R5 = (H-1)\,R4$$

$$R1 = \frac{1}{(2\pi F_0)^2\,R2\,C^2}$$

$$R6 = REQ1$$

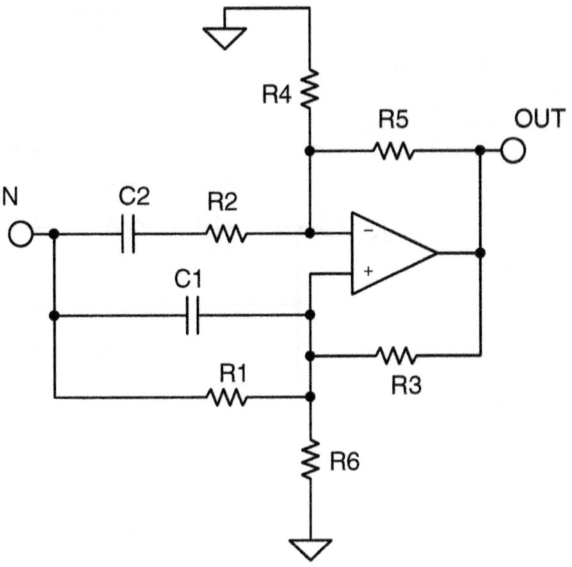

Figure 5-79B: Boctor high-pass design equations (continued)

FIRST ORDER ALL PASS

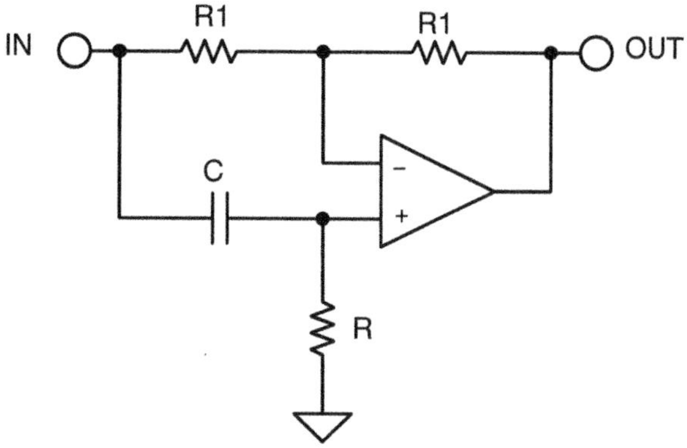

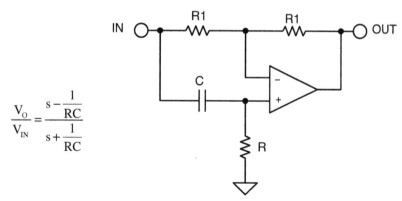

$$\frac{V_O}{V_{IN}} = \frac{s - \frac{1}{RC}}{s + \frac{1}{RC}}$$

$$\text{PHASE SHIFT } (\phi) = -2\ \text{TAN}^{-1}\left(\frac{RC}{2\pi F}\right)$$

$$\text{GROUP DELAY} = \frac{2RC}{(2\pi F R C)^2 + 1}$$

GIVEN A PHASE SHIFT OF ϕ AT A FREQUENCY = F

$$RC = 2\pi F\ \text{TAN}\left(-\frac{\phi}{2}\right)$$

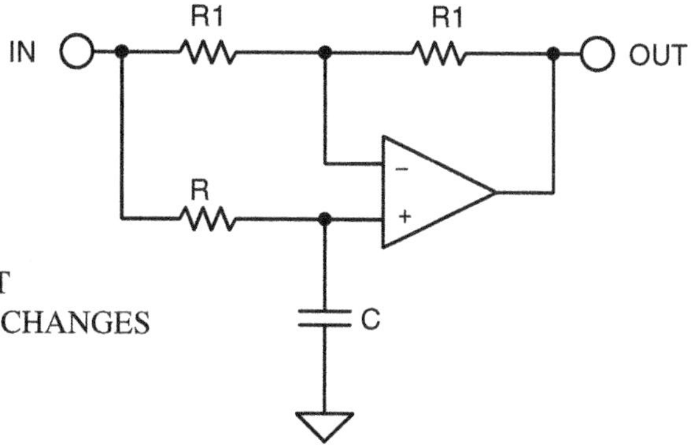

DESIGN AS ABOVE EXCEPT
THE SIGN OF THE PHASES CHANGES

Figure 5-80: First order all pass design equations

SECOND ORDER ALL PASS

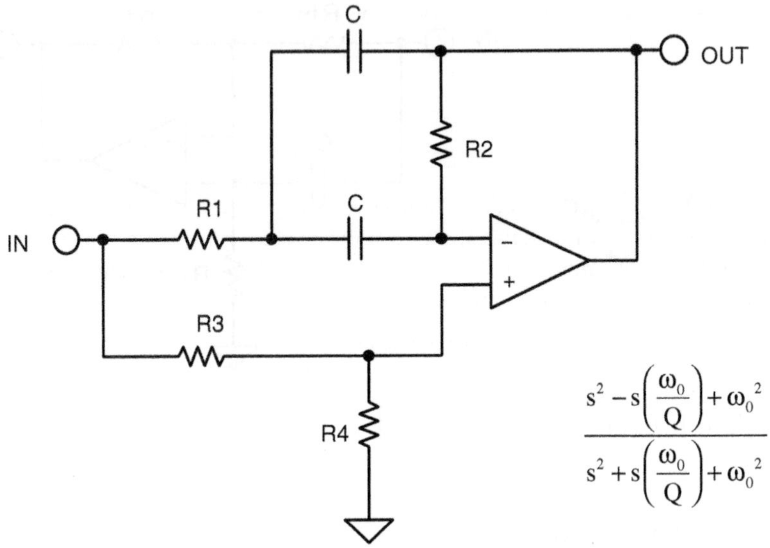

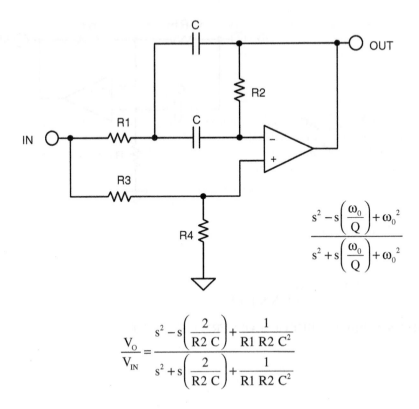

$$\frac{s^2 - s\left(\frac{\omega_0}{Q}\right) + \omega_0^{\ 2}}{s^2 + s\left(\frac{\omega_0}{Q}\right) + \omega_0^{\ 2}}$$

$$\frac{V_O}{V_{IN}} = \frac{s^2 - s\left(\frac{2}{R2\ C}\right) + \frac{1}{R1\ R2\ C^2}}{s^2 + s\left(\frac{2}{R2\ C}\right) + \frac{1}{R1\ R2\ C^2}}$$

CHOOSE: C

$$k = 2\ \pi\ F_0\ C$$

$$R2 = \frac{2Q}{k}$$

$$R1 = \frac{1}{2kQ}$$

$$R3 = R1$$

$$R4 = \frac{Q}{2}$$

Figure 5-81: Second order all pass

Practical Problems in Filter Implementation

In the previous sections filters were dealt with as mathematical functions. The filter designs were assumed to have been implemented with "perfect" components. When the filter is built with real-world components, design trade-offs must typically be made.

In building a filter with an order greater the two, multiple second and/or first order sections are used. The frequencies and Qs of these sections must align precisely or the overall response of the filter will be affected. For example, the antialiasing filter design example in the next section is a 5th order Butterworth filter, made up of a second order section with a frequency (F_0) = 1 and a Q = 1.618, a second order section with a frequency (F_0) = 1 and a Q = 0.618, and a first order section with a frequency (F_0) = 1 (for a filter normalized to 1 rad/sec). If the Q or frequency response of any of the sections is off slightly, the overall response will deviate from the desired response. It may be close, but it won't be exact. As is typically the case with engineering, a decision must be made as to which trade-offs should be made. For instance, do we really need a particular response exactly? Is there a problem if there is a little more ripple in the pass band? Or if the cutoff frequency is at a slightly different frequency? These are the types of questions that face a designer, and will vary from design to design.

Passive Components (Resistors, Capacitors, Inductors)

Passive components are the first problem. When designing filters, the calculated values of components will most likely not be available commercially. Resistors, capacitors, and inductors come in standard values. While custom values can be ordered, the practical tolerance will probably still be ±1% at best. An alternative is to build the required value out of a series and/or parallel combination of standard values. This increases the cost and size of the filter. Not only is the cost of components increased, so are the manufacturing costs, both for loading and tuning the filter. Furthermore, success will be still limited by the number of parts that are used, their tolerance, and their tracking, both over temperature and time.

A more practical way is to use a circuit analysis program to determine the response using standard values. The program can also evaluate the effects of component drift over temperature. The values of the sensitive components are adjusted using parallel combinations where needed, until the response is within the desired limits. Many of the higher end filter CAD programs include this feature.

The resonant frequency and Q of a filter are typically determined by the component values. Obviously, if the component value is drifting, the frequency and the Q of the filter will drift. This, in turn, will cause the frequency response to vary. This is especially true in higher order filters.

Higher order implies higher Q sections. Higher Q sections means that component values are more critical, since the Q is typically set by the ratio of two or more components, typically capacitors.

In addition to the initial tolerance of the components, the effects of temperature/time drift must be evaluated. The temperature coefficients of the various components may be different in both magnitude and sign. Capacitors, especially, are difficult in that not only do they drift, but the temperature coefficient (TC) is also a function of temperature, as shown in Figure 5-82. This represents the temperature coefficient of a (relatively) poor film capacitor, which might be typical for a Polyester or Polycarbonate type. *Linear TC* in film capacitors can be found in the polystyrene, polypropylene, and Teflon dielectrics. In these types TC is on the order of 100 ppm/°C–200 ppm/°C, and if necessary, this can be compensated with a complementary TC elsewhere in the circuit.

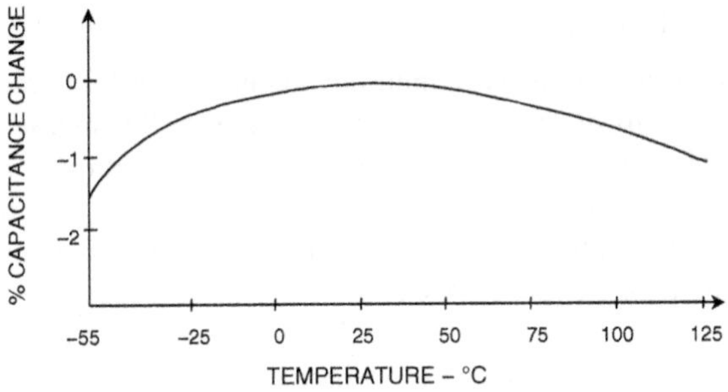

Figure 5-82: A poor film capacitor temperature coefficient

The lowest TC dielectrics are NPO (or COG) ceramic (±30 ppm/°C), and polystyrene (–120 ppm/°C). Some capacitors, mainly the plastic film types, such as polystyrene and polypropylene, also have a limited temperature range.

While there is infinite choice of the values of the passive components for building filters, in practice there are physical limits. Capacitor values below 10 pF and above 10 μF are not practical. Electrolytic capacitors should be avoided. Electrolytic capacitors are typically very leaky. A further potential problem is if they are operated without a polarizing voltage, they become nonlinear when the ac voltage reverse biases them. Even with a dc polarizing voltage, the ac signal can reduce the instantaneous voltage to 0 V or below. Large values of film capacitors are physically very large.

Resistor values of less than 100 Ω should be avoided, as should values over 1 MΩ. Very low resistance values (under 100 Ω) can require a great deal of drive current and dissipate a great deal of power. Both of these should be avoided. And low values and very large values of resistors may not be as readily available. Very large values tend to be more prone to parasitics since smaller capacitances will couple more easily into larger impedance levels. Noise also increases with the square root of the resistor value. Larger value resistors also will cause larger offsets due to the effects of the amplifier bias currents.

Parasitic capacitances due to circuit layout and other sources affect the performance of the circuit. They can form between two traces on a PC board (on the same side or opposite side of the board), between leads of adjacent components, and just about everything else you can (and in most cases can't) think of. These capacitances are usually small, so their effect is greater at high impedance nodes. Thus, they can be controlled most of the time by keeping the impedance of the circuits down. Remember that the effects of stray capacitance are frequency-dependent, being worse at high frequencies because the impedance drops with increasing frequency.

Parasitics are not just associated with outside sources. They are also present in the components themselves.

A capacitor is more than just a capacitor in most instances. A real capacitor has inductance (from the leads and other sources) and resistance as shown in Figure 5-83. This resistance shows up in the specifications

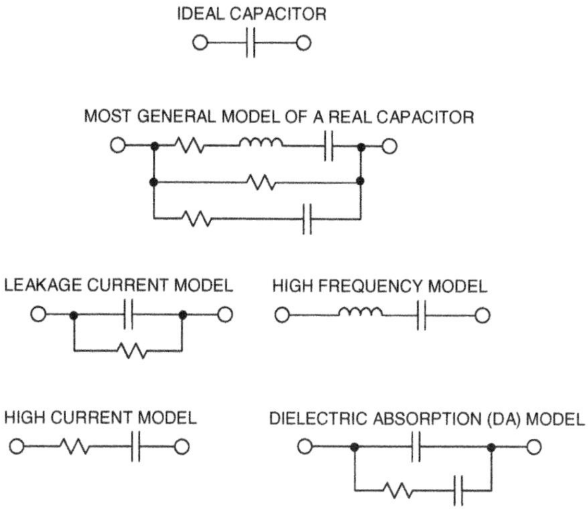

Figure 5-83: Capacitor equivalent circuit

as leakage and poor power factor. Obviously, users would like capacitors with very low leakage and good power factor (see Figure 5-84).

CAPACITOR COMPARISON CHART

TYPE	TYPICAL DA	ADVANTAGES	DISADVANTAGES
Polystyrene	0.001% to 0.02%	Inexpensive Low DA Good Stability (~120ppm/°C)	Damaged by Temperature > 85°C Large High Inductance Vendors Limited
Polypropylene	0.001% to 0.02%	Inexpensive Low DA Stable (~200ppm/°C) Wide Range of Values	Damaged by Temperature > 105°C Large High Inductance
Teflon	0.003% to 0.02%	Low DA Available Good Stability Operational above 125 °C Wide Range of Values	Expensive Large High Inductance
Polycarbonate	0.1%	Good Stability Low Cost Wide Temperature Range Wide Range of Values	Large DA Limits to 8-Bit Applications High Inductance
Polyester	0.3% to 0.5%	Moderate Stability Low Cost Wide Temperature Range Low Inductance (Stacked Film)	Large DA Limits to 8-Bit Applications High Inductance (Conventional)
NP0 Ceramic	<0.1%	Small Case Size Inexpensive, Many Vendors Good Stability (30ppm/°C) 1% Values Available Low Inductance (chip)	DA Generally Low (May not be Specified) Low Maximum Values (≤ 10nF)
Monolithic Ceramic (High K)	>0.2%	Low Inductance (chip) Wide Range of Values	Poor Stability Poor DA High Voltage Coefficient
Mica	>0.003%	Low Loss at HF Low Inductance Good Stability 1% Values Available	Quite Large Low Maximum Values (≤ 10nF) Expensive
Aluminum Electrolytic	Very High	Large Values High Currents High Voltages Small Size	High Leakage Usually Polarized Poor Stability, Accuracy Inductive
Tantalum Electrolytic	Very High	Small Size Large Values Medium Inductance	High Leakage Usually Polarized Expensive Poor Stability, Accuracy

Figure 5-84: Capacitor comparison chart

In general, it is best to use plastic film (preferably Teflon or polystyrene) or mica capacitors and metal film resistors, both of moderate to low values in our filters.

One way to reduce component parasitics is to use surface-mounted devices. Not having leads means that the lead inductance is reduced. Also, being physically smaller allows more optimal placement. A disadvantage is that not all types of capacitors are available in surface mount. Ceramic capacitors are popular surface-mount types, and of these, the NPO family has the best characteristics for filtering. Ceramic capacitors may also be prone to microphonics. Microphonics occurs when the capacitor turns into a motion sensor, similar to a strain gage, and turns vibration into an electrical signal, which is a form of noise.

Resistors also have parasitic inductances due to leads and parasitic capacitance. The various qualities of resistors are compared in Figure 5-85.

RESISTOR COMPARISON CHART

	TYPE	**ADVANTAGES**	**DISADVANTAGES**
DISCRETE	Carbon Composition	Lowest Cost High Power/Small Case Size Wide Range of Values	Poor Tolerance (5%) Poor Temperature Coefficient (1500 ppm/°C)
	Wirewound	Excellent Tolerance (0.01%) Excellent TC (1ppm/°C) High Power	Reactance is a Problem Large Case Size Most Expensive
	Metal Film	Good Tolerance (0.1%) Good TC (<1 to 100ppm/°C) Moderate Cost Wide Range of Values Low Voltage Coefficient	Must be Stabilized with Burn-In Low Power
	Bulk Metal or Metal Foil	Excellent Tolerance (to 0.005%) Excellent TC (to <1ppm/°C) Low Reactance Low Voltage Coefficient	Low Power Very Expensive
	High Megohm	Very High Values (10^8 to $10_{14}\Omega$) Only Choice for Some Circuits	High Voltage Coefficient (200ppm/V) Fragile Glass Case (Needs Special Handling) Expensive
NETWORKS	Thick Film	Low Cost High Power Laser-Trimmable Readily Available	Fair Matching (0.1%) Poor TC (>100ppm/°C) Poor Tracking TC (10ppm/°C)
	Thin Film	Good Matching (<0.01%) Good TC (<100ppm/°C) Good Tracking TC (2ppm/°C) Moderate Cost Laser-Trimmable Low Capacitance Suitable for Hybrid IC Substrate	Often Large Geometry Limited Values and Configurations

Figure 5-85: Resistor comparison chart

Limitations of Active Elements (Op Amps) in Filters

The active element of the filter will also have a pronounced effect on the response. In developing the various topologies (Multiple Feedback, Sallen-Key, State-Variable, and so forth), the active element was always modeled as a "perfect" operational amplifier. That is, it has:

1) infinite gain

2) infinite input impedance

3) zero output impedance

none of which vary with frequency. While amplifiers have improved a great deal over the years, this model has not yet been realized.

The most important limitation of the amplifier has to do with its gain variation with frequency. All amplifiers are band limited. This is due mainly to the physical limitations of the devices with which the amplifier is constructed. Negative feedback theory tells us that the response of an amplifier must be first order (–6 dB per octave) when the gain falls to unity in order to be stable. To accomplish this, a real pole is usually introduced in the amplifier so the gain rolls off to <1 by the time the phase shift reaches 180° (plus some phase margin, hopefully). This roll-off is equivalent to that of a single-pole filter. So in simplistic terms, the transfer function of the amplifier is added to the transfer function of the filter to give a composite function. How much the frequency-dependent nature of the op amp affects the filter is dependent on which topology is used as well as the ratio of the filter frequency to the amplifier bandwidth.

The Sallen-Key configuration, for instance, is the least dependent on the frequency response of the amplifier. All that is required is for the amplifier response to be flat to just past the frequency where the attenuation of the filter is below the minimum attenuation required. This is because the amplifier is used as a gain block. Beyond cutoff, the attenuation of the filter is reduced by the roll-off of the gain of the op amp. This is because the output of the amplifier is phase-shifted, which results in incomplete nulling when fed back to the input. There is also an issue with the output impedance of the amplifier rising with frequency as the open loop gain rolls off. This causes the filter to lose attenuation.

The state-variable configuration uses the op amps in two modes, as amplifiers and as integrators. As amplifiers, the constraint on frequency response is basically the same as for the Sallen-Key, which is flat out to the minimum attenuation frequency. As an integrator, however, more is required. A good rule of thumb is that the open-loop gain of the amplifier must be greater than 10 times the closed-loop gain (including peaking from the Q of the circuit). This should be taken as the absolute minimum requirement. What this means is that there must be 20 dB loop gain, minimum. Therefore, an op amp with 10 MHz unity gain bandwidth is the minimum required to make a 1 MHz integrator. What happens is that the effective Q of the circuit increases as loop gain decreases. This phenomenon is called *Q enhancement*. The mechanism for Q enhancement is similar to that of slew rate limitation. Without sufficient loop gain, the op amp virtual ground is no longer at ground. In other words, the op amp is no longer behaving as an op amp. Because of this, the integrator no longer behaves like an integrator.

The multiple feedback configuration also places heavy constraints on the active element. Q enhancement is a problem in this topology as well. As the loop gain falls, the Q of the circuit increases, and the parameters of the filter change. The same rule of thumb as used for the integrator also applies to the multiple feedback topology (loop gain should be at least 20 dB). The filter gain must also be factored into this equation.

In the FDNR realization, the requirements for the op amps are not as clear. To make the circuit work, we assume that the op amps will be able to force the input terminals to be the same voltage. This implies that the loop gain be a minimum of 20 dB at the resonant frequency.

Also it is generally considered to be advantageous to have the two op amps in each leg matched. This is easily accomplished using dual op amps. It is also a good idea to have low bias current devices for the op amps so, all other things being equal, FET input op amps should be used.

In addition to the frequency-dependent limitations of the op amp, other of its parameters may be important to the filter designer.

One is input impedance. We assume in the "perfect" model that the input impedance is infinite. This is required so that the input of the op amp does not load the network around it. This means that we probably want to use FET amplifiers with high impedance circuits.

There is also a small frequency-dependent term to the input impedance, since the effective impedance is the real input impedance multiplied by the loop gain. This is not usually a major source of error, since the network impedance of a high frequency filter should be low.

Distortion resulting from Input Capacitance Modulation

Another subtle effect can be noticed with FET input amps. The input capacitance of a FET changes with the applied voltage. When the amplifier is used in the inverting configuration, such as with the multiple feedback configuration, the applied voltage is held to 0 V. Therefore there is no capacitance modulation. However, when the amplifier is used in the noninverting configuration, such as in the Sallen-Key circuit, this form of distortion can exist.

There are two ways to address this issue. The first is to keep the equivalent impedance low. The second is to balance the impedance seen by the inputs. This is accomplished by adding a network into the feedback leg of the amplifier which is equal to the equivalent input impedance. Note that this will only work for a unity gain application.

As an example, which is taken from the OP176 data sheet, a 1 kHz high-pass Sallen-Key filter is shown (Figure 5-86). Figure 5-87 shows the distortion for the uncompensated version (curve A1) as well as with the compensation (curve A2). Also shown is the same circuit with the impedances scaled up by a factor of 10 (B1 uncompensated, B2 compensated). Note that the compensation improves the distortion, but not as much as having low impedance to start with.

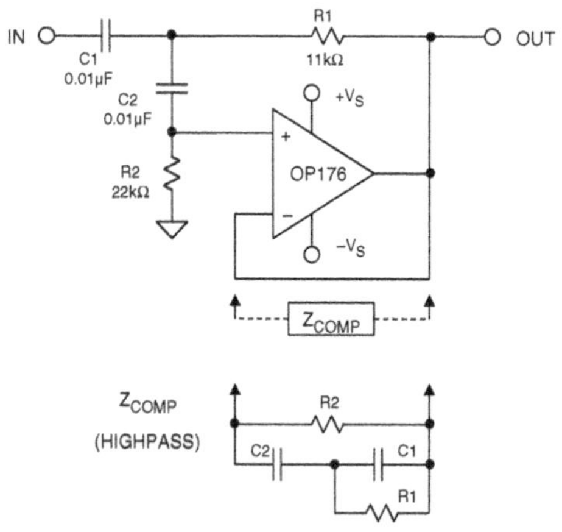

Figure 5-86: Compensation for input capacitance voltage modulation

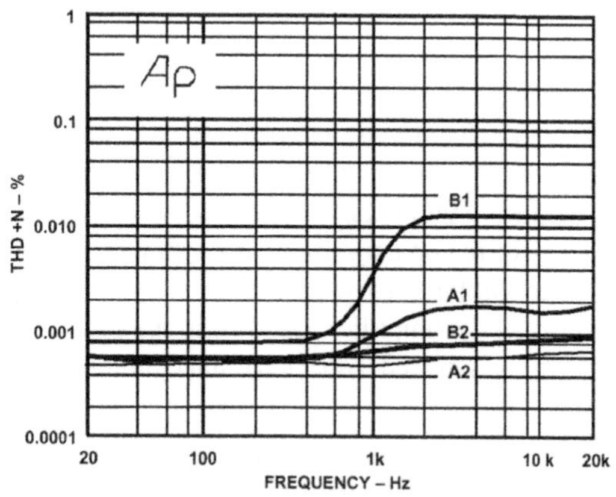

Figure 5-87: Distortion due to input capacitance modulation

Similarly, the op amp output impedance affects the response of the filter. The output impedance of the amplifier is divided by the loop gain, therefore the output impedance will rise with increasing frequency. This may have an effect with high frequency filters if the output impedance of the stage driving the filter becomes a significant portion of the network impedance.

The fall of loop gain with frequency can also affect the distortion of the op amp, since there is less loop gain available for correction. In the multiple feedback configuration the feedback loop is also frequency-dependent, which may further reduce the feedback correction, resulting in increased distortion. This effect is counteracted somewhat by the reduction of distortion components in the filter network (assuming a low-pass or band-pass filter).

All of the discussion so far is based on using classical voltage feedback op amps. Current feedback, or transimpedance, op amps offer improved high frequency response, but are unusable in any topologies discussed except the Sallen-Key. The problem is that capacitance in the feedback loop of a current feedback amplifier usually causes it to become unstable. Also, most current feedback amplifiers will drive only a small capacitive load. Therefore, it is difficult to build classical integrators using current feedback amplifiers. Some current feedback op amps have an external pin that may be used to configure them as a very good integrator, but this configuration does not lend itself to classical active filter designs.

Current feedback integrators tend to be noninverting, which is not acceptable in the state variable configuration. Also, the bandwidth of a current feedback amplifier is set by its feedback resistor, which would make the Multiple Feedback topology difficult to implement. Another limitation of the current feedback amplifier in the Multiple Feedback configuration is the low input impedance of the inverting terminal. This would result in loading of the filter network. Sallen-Key filters are possible with current feedback amplifiers, since the amplifier is used as a noninverting gain block. New topologies that capitalize on the current feedback amplifiers' superior high frequency performance, and compensate for its limitations, will have to be developed.

The last thing to be aware of is exceeding the dynamic range of the amplifier. Qs over 0.707 will cause peaking in the response of the filter (see Figures 5-5 through 5-7). For high Qs, this could cause overload of the input or output stages of the amplifier with a large input. Note that relatively small values of Q can cause significant peaking. The Q times the gain of the circuit must stay under the loop gain (plus some margin; again, 20 dB is a good starting point). This holds for multiple amplifier topologies as well. Be aware of internal node levels, as well as input and output levels. As an amplifier overloads, its effective Q decreases, so the transfer function will appear to change even if the output appears undistorted. This shows up as the transfer function changing with increasing input level.

We have been dealing mostly with low-pass filters in these discussions, but the same principles are valid for high pass, band pass, and bandreject as well. In general, things like Q enhancement and limited gain/bandwidth will not affect high-pass filters, since the resonant frequency will hopefully be low in relation to the cutoff frequency of the op amp. Remember, though, that the high-pass filter will have a low pass section, by default, at the cutoff frequency of the amplifier. Band-pass and bandreject (notch) filters will be affected, especially since both tend to have high values of Q.

The general effect of the op amp's frequency response on the filter Q is shown in Figure 5-88.

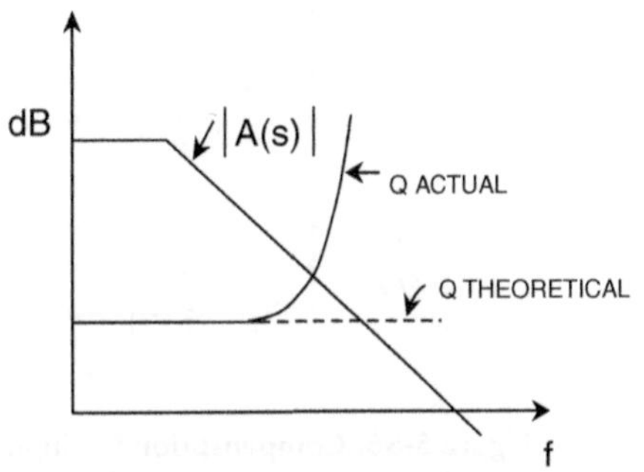

Figure 5-88: Q enhancement

As an example of the Q enhancement phenomenon, consider the Spice simulation of a 10 kHz band pass Multiple Feedback filter with Q = 10 and gain = 1, using a good high frequency amplifier (the AD847) as the active device. The circuit diagram is shown in Figure 5-89. The open loop gain of the AD847 is greater than 70 dB at 10 kHz as shown in Figure 5-91(A). This is well over the 20 dB minimum, so the filter works as designed as shown in Figure 5-90.

Figure 5-89: 1 kHz multiple feedback band-pass filter

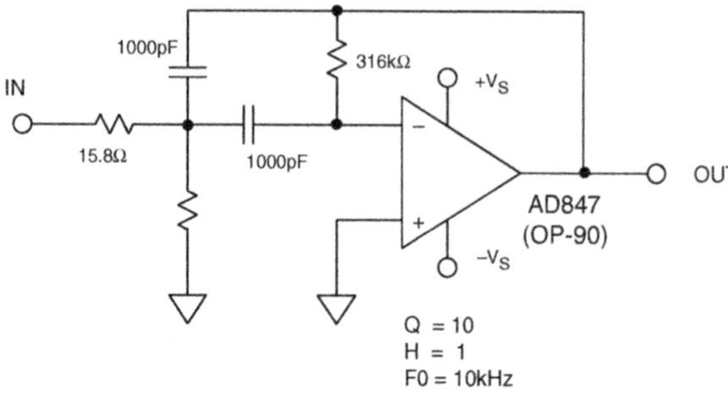

Figure 5-90: Effects of "Q enhancement"

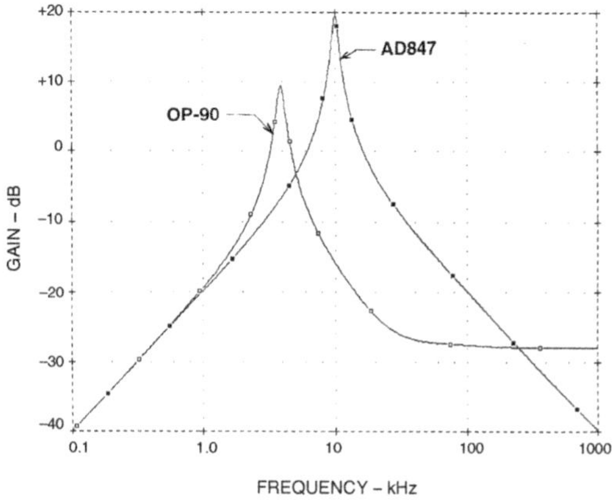

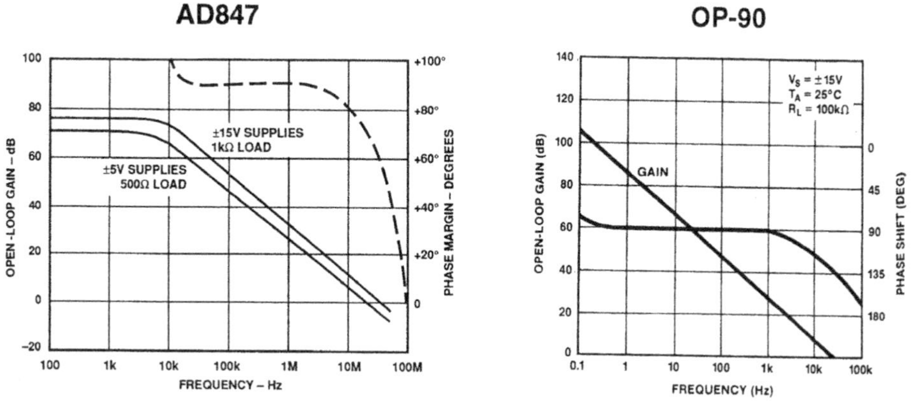

Figure 5-91: AD847 and OP-90 Bode plots

We now replace the AD847 with an OP-90. The OP-90 is a dc precision amplifier and so has a limited bandwidth. In fact, its open loop gain is less than 10 dB at 10 kHz (see Figure 5-91(B)). This is not to imply that the AD847 is in all cases better than the OP-90. It is a case of misapplying the OP-90.

From the output for the OP-90, also shown in Figure 5-90, it can be seen that the magnitude of the output has been reduced, and the center frequency has shifted downward.

Design Examples

Several examples will now be worked out to demonstrate the concepts previously discussed

Antialias Filter

As an example, passive and active antialiasing filters will now be designed based upon a common set of specifications. The active filter will be designed in four ways: Sallen-Key, Multiple Feedback, State Variable, and Frequency-Dependent Negative Resistance (FDNR).

The specifications for the filter are given as follows:

1) The cutoff frequency will be 8 kHz.
2) The stopband attenuation will be 72 dB. This corresponds to a 12-bit system.
3) Nyquist frequency of 50 kSPS.
4) The Butterworth filter response is chosen in order to give the best compromise between attenuation and phase response.

Consulting the Butterworth response curves (Figure 5-14, reproduced in Figure 5-92), it can be seen that for a frequency ratio of 6.25 (50 kSPS/8 kSPS), a filter order of 5 is required.

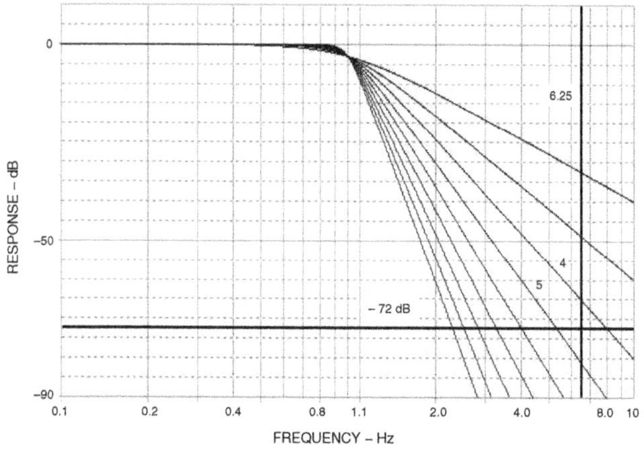

Figure 5-92: Determining filter order

Now consulting the Butterworth design table (Figure 5-25), the normalized poles of a 5th order Butterworth filter are:

STAGE	F_0	α
1	1.000	1.618
2	1.000	0.618
3	1.000	------

The last stage is a real (single) pole, thus the lack of an alpha value. It should be noted that this is not necessarily the order of implementation in hardware. In general, you would typically put the real pole last and put the second order sections in order of decreasing alpha (increasing Q) as we have done here. This will avoid peaking due to high Q sections possibly overloading internal nodes. Another feature of putting the single pole at the end is to bandlimit the noise of the op amps. This is especially true if the single pole is implemented as a passive filter.

For the passive design, we will choose the zero input impedance configuration. While "classic" passive filters are typically double terminated, that is with termination on both source and load ends, the concern here is with voltage transfer not power transfer so the source termination will not be used. From the design table (see Reference 2, p. 313), we find the normalized values for the filter (see Figure 5-93).

Figure 5-93: Normalized passive filter implementation

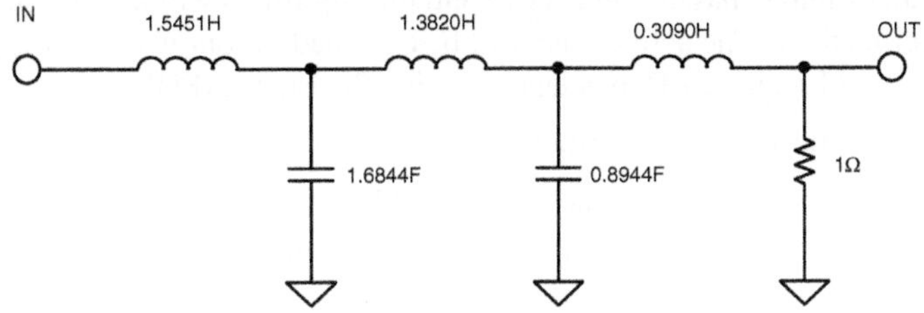

These values are normalized for a 1 rad/s filter with a 1 Ω termination. To scale the filter we divide all reactive elements by the desired cutoff frequency, 8 kHz (= 50265 rad/sec, = $2\pi\ 8\times10^3$). This is commonly referred to as the frequency scale factor (FSF). We also need to scale the impedance.

For this example, an arbitrary value of 1000 Ω is chosen. To scale the impedance, we multiply all resistor and inductor values and divide all capacitor values by this magnitude, which is commonly referred to as the impedance scaling factor (Z).

After scaling, the circuit looks like Figure 5-94.

Figure 5-94: Passive filter implementation

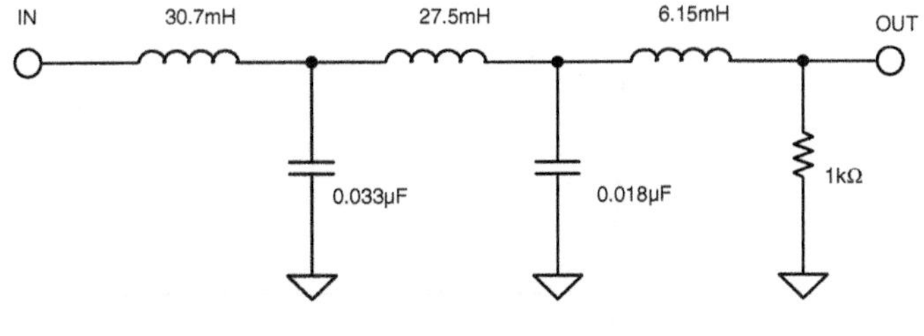

For the Sallen-Key active filter, use the design equations shown in Figure 5-49. The values for C1 in each section are arbitrarily chosen to give reasonable resistor values. The implementation is shown in Figure 5-95.

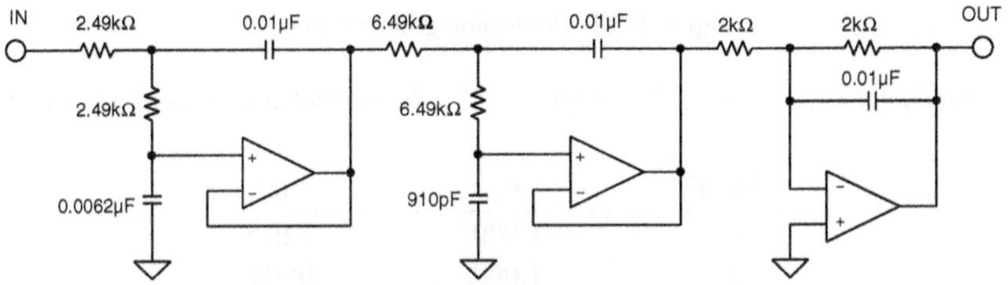

Figure 5-95: Sallen-Key implementation

The exact values have been rounded to the nearest standard value. For most active realization to work correctly, it is required to have a zero-impedance driver, and a return path for dc due to the bias current of the op amp. Both of these criteria are approximately met when you use an op amp to drive the filter.

In the above example the single pole has been built as an active circuit. It would have been just as correct to configure it as a passive RC filter. The advantage to the active section is lower output impedance, which may be an advantage in some applications, notably driving an ADC input that uses a switched capacitor structure.

This type of input is common on sigma-delta ADCs as well as many other CMOS type of converters. It also eliminates the loading effects of the input impedance of the following stage on the passive section.

Figure 5-96 shows a multiple feedback realization of the filter. It was designed using the equations in Figure 5-52. In this case, the last section is a passive RC circuit.

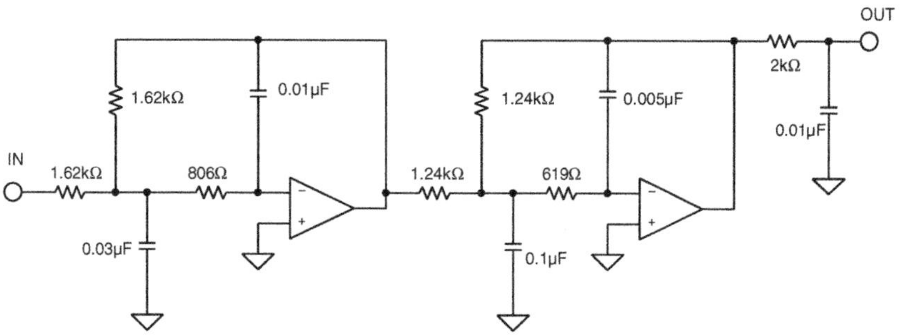

Figure 5-96: Multiple feedback implementation

An optional buffer could be added after the passive section, if desired. This would give many of the advantages outlined above, except for bandlimiting the noise of the output amp. By using one of the above two filter realizations, we have both an inverting and a noninverting design.

The state-variable filter, shown in Figure 5-97, was designed with the equations in Figure 5-55. Again, we have rounded the resistor values to the nearest standard 1% value.

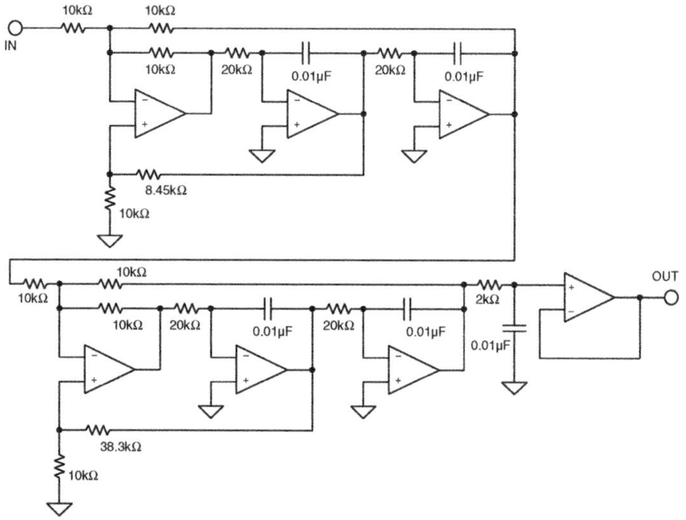

Figure 5-97: State variable implementation

Obviously this filter implementation has many more parts than either the Sallen-Key or the multiple feedback. The rational for using this circuit is that stability is improved and the individual parameters are independently adjustable.

The frequency dependent negative resistance (FDNR) realization of this filter is shown in Figure 5-98.

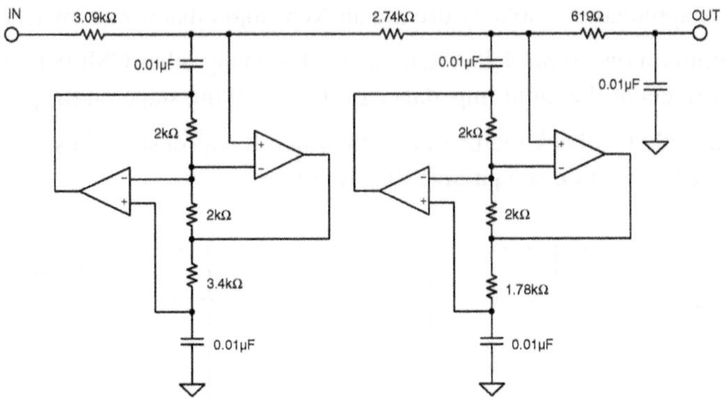

Figure 5-98: FDNR Implementation

In the conversion process from passive to FDNR, the D element is normalized for a capacitance of 1 F. The filter is then scaled to a more reasonable value (0.01 µF in this case).

In all of the above implementations standard values were used instead of the calculated values. Any variation from the ideal values will cause a shift in the filter response characteristic, but often the effects are minimal. The computer can be used to evaluate these variations on the overall performance and determine if they are acceptable.

To examine the effect of using standard values, take the Sallen-Key implementation. Figure 5-99 shows the response of each of the three sections of the filter. While the Sallen-Key was the filter used, the results from any of the other implementations will give similar results.

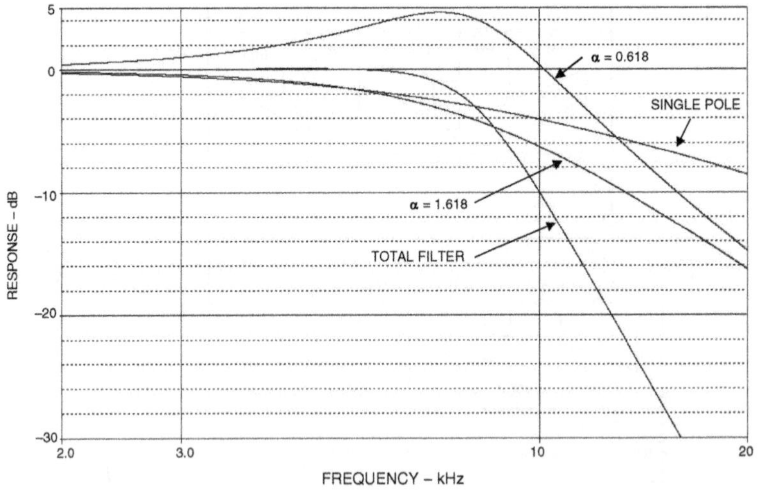

Figure 5-99: Individual section response

Figure 5-100 then shows the effect of using standard values instead of calculated values. Notice that the general shape of the filter remains the same, just slightly shifted in frequency. This investigation was done only for the standard value of the resistors. To understand the total effect of component tolerance the same type of calculations would have to be done for the tolerance of all the components and also for their temperature and aging effects.

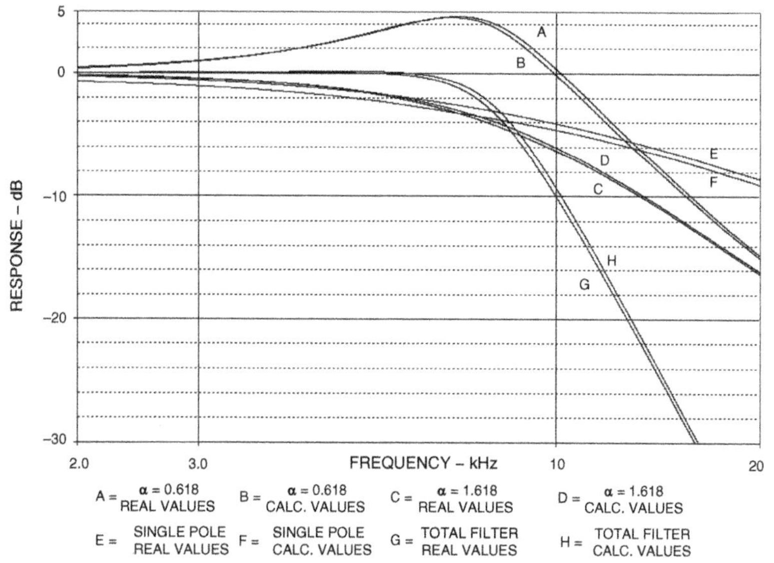

Figure 5-100: Effect of using standard value resistors

In active filter applications using op amps, the dc accuracy of the amplifier is often critical to optimal filter performance. The amplifier's offset voltage will be passed by the low-pass filter and may be amplified to produce excessive output offset. For low frequency applications requiring large value resistors, bias currents flowing through these resistors will also generate an output offset voltage.

In addition, at higher frequencies, an op amp's dynamics must be carefully considered. Here, slew rate, bandwidth, and open-loop gain play a major role in op amp selection. The slew rate must be fast as well as symmetrical to minimize distortion.

Transformations

In the next example the transformation process will be investigated.

As mentioned earlier, filter theory is based on a low pass prototype, which is then manipulated into the other forms. In these examples the prototype that will be used is a 1 kHz, 3-pole, 0.5 dB Chebyshev filter. A Chebyshev was chosen because it would show more clearly if the responses were not correct, a Butterworth would probably be too forgiving in this instance. A 3-pole filter was chosen so that a pole pair and a single pole would be transformed.

The pole locations for the LP prototype were taken from Figure 5-30. They are:

STAGE	α	β	F_O	α
1	0.2683	0.8753	1.0688	0.5861
2	0.5366		0.6265	

The first stage is the pole pair and the second stage is the single pole. Note the unfortunate convention of using α for 2 entirely separate parameters. The α and β on the left are the pole locations in the s-plane. These are the values that are used in the transformation algorithms. The α on the right is 1/Q, which is what the design equations for the physical filters want to see.

The Sallen-Key topology will be used to build the filter. The design equations in Figure 5-67 (pole pair) and Figure 5-66 (single pole) were then used to design the filter. The schematic is shown in Figure 5-101.

Figure 5-101: Low pass prototype

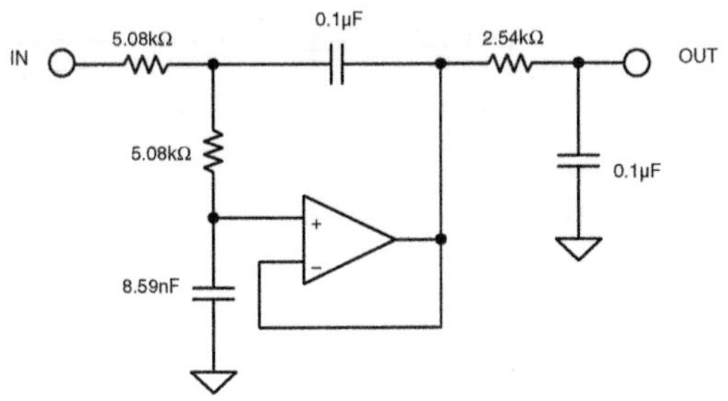

Using the equation string described in Section 5, the filter is now transformed into a high-pass filter. The results of the transformation are:

STAGE	α	β	F_O	α
1	0.3201	1.0443	0.9356	0.5861
2	1.8636		1.596	

A word of caution is warranted here. Since the convention of describing a Chebyshev filter is to quote the end of the error band instead of the 3 dB frequency, the F_0 must be divided (for high pass) by the ratio of ripple band to 3 dB bandwidth (Table 1, Section 4).

The Sallen-Key topology will again be used to build the filter. The design equations in Figure 5-68 (pole pair) and Figure 5-66 (single pole) where then used to design the filter. The schematic is shown in Figure 5-102.

Figure 5-102: High pass transformation

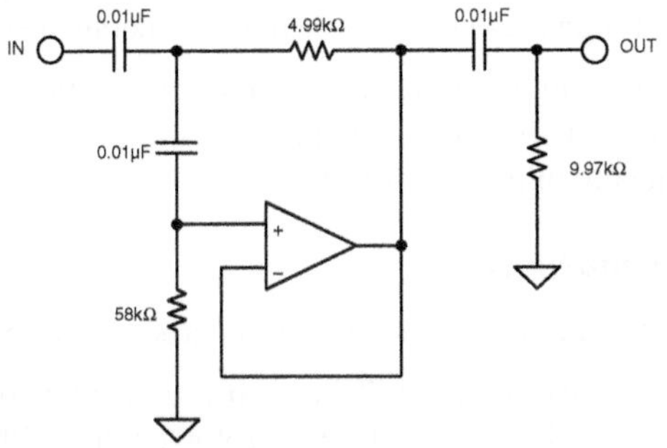

Figure 5-103 shows the response of the low pass prototype and the high pass transformation. Note that they are symmetric around the cutoff frequency of 1 kHz. Also note that the errorband is at 1 kHz, not the –3 dB point, which is characteristic of Chebyshev filters.

The low pass prototype is now converted to a band-pass filter. The equation string outlined in Section 5-5 is used for the transformation. Each pole of the prototype filter will transform into a pole pair. Therefore the 3-pole prototype, when transformed, will have six poles (3-pole pairs). In addition, there will be six zeros at the origin.

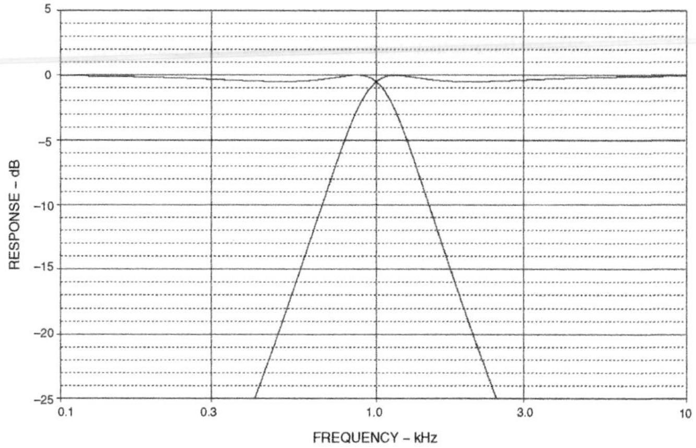

Figure 5-103: Low pass and high pass response

Part of the transformation process is to specify the 3 dB bandwidth of the resultant filter. In this case this bandwidth will be set to 500 Hz. The results of the transformation yield:

STAGE	F_0	Q	A_0
1	804.5	7.63	3.49
2	1243	7.63	3.49
3	1000	3.73	1

The reason for the gain requirement for the first two stages is that their center frequencies will be attenuated relative to the center frequency of the total filter. Since the resultant Q_s are moderate (less than 20) the Multiple Feedback topology will be chosen. Figure 5-72 was then used to design the filter sections.

Figure 5-104 is the schematic of the filter and Figure 5-105 shows the filter response.

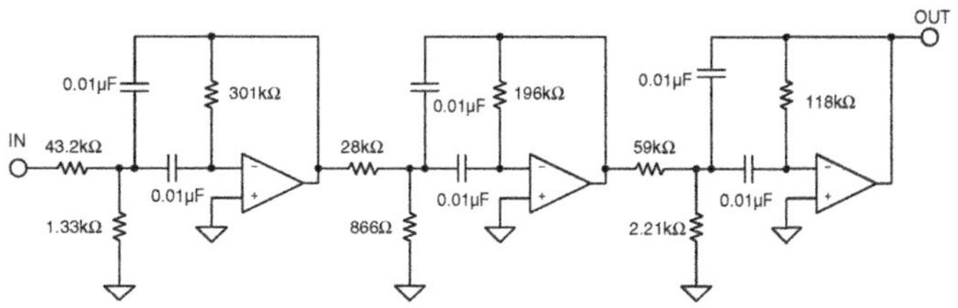

Figure 5-104: Band pass transformation

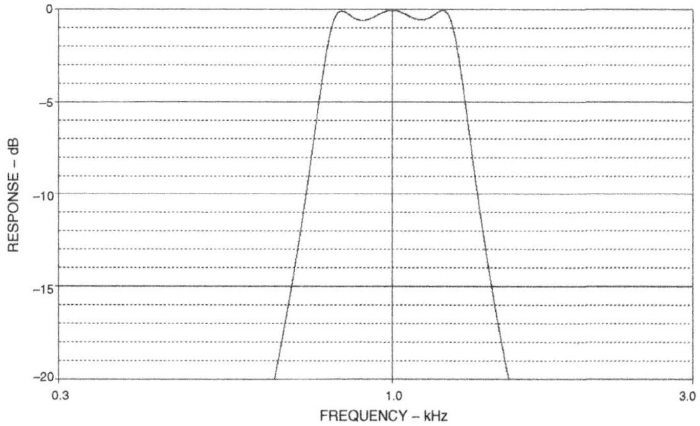

Figure 5-105: Band pass filter response

Note that again there is symmetry around the center frequency. Also the 500 Hz bandwidth is not 250 Hz either side of the center frequency (arithmetic symmetry). Instead the symmetry is geometric, which means that for any two frequencies (F_1 and F_2) of equal amplitude are related by:

$$F_0 = \sqrt{F_1 * F_2}$$

Lastly the prototype will be transformed into a bandreject filter. For this the equation string in Section 5-5 is used. Again, each pole of the prototype filter will transform into a pole pair. Therefore, the 3-pole prototype, when transformed, will have six poles (3-pole pairs).

As in the band pass case, part of the transformation process is to specify the 3 dB bandwidth of the resultant filter. Again in this case this bandwidth will be set to 500 Hz. The results of the transformation yield:

STAGE	F_0	Q	F_{0Z}
1	763.7	6.54	1000
2	1309	6.54	1000
3	1000	1.07	1000

Note that there are three cases of notch filters required. There is a standard notch ($F_0 = F_Z$, section 3), a low pass notch ($F_0 < F_Z$, section 1) and a high pass notch ($F_0 > F_Z$, section 2). Since there is a requirement for all three types of notches, the Bainter Notch is used to build the filter. The filter is designed using Figure 5-77. The gain factors K1 and K2 are arbitrarily set to 1. Figure 5-106 is the schematic of the filter.

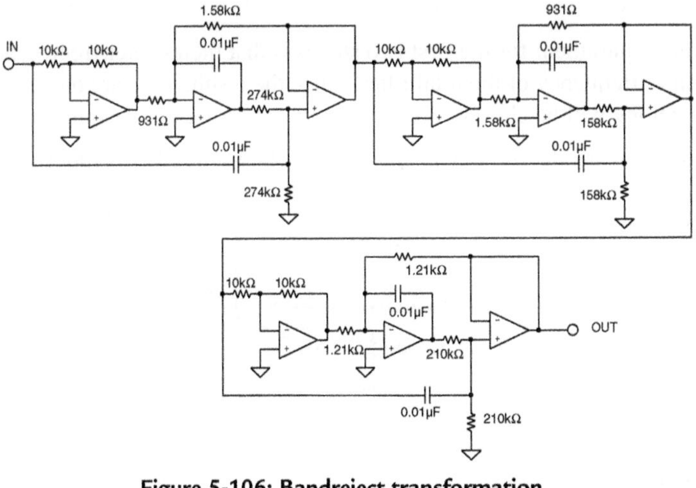

Figure 5-106: Bandreject transformation

The response of the filter is shown in Figure 5-107 and in detail in Figure 5-108. Again, note the symmetry around the center frequency. Again the frequencies have geometric symmetry.

CD Reconstruction Filter

This design was done for a magazine article describing a high quality outboard D/A converter for use with digital audio sources (see Reference 26).

A reconstruction filter is required on the output of a D/A converter because, despite the name, the output of a D/A converter is not really an analog voltage but, instead, a series of steps. The converter will put out a discrete voltage, which it will then hold until the next sample is asserted. The filter's job is to remove the

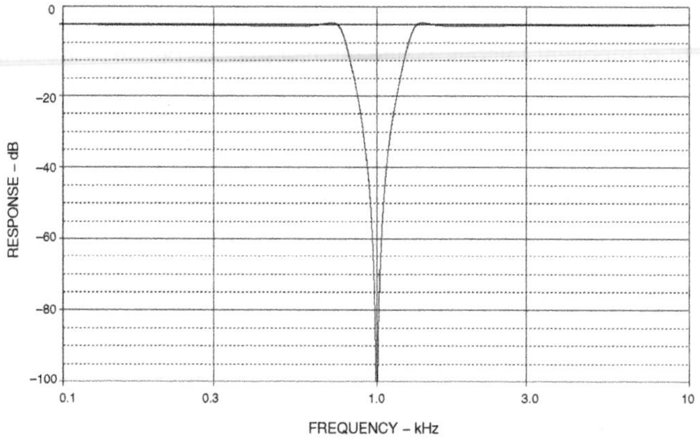

Figure 5-107: Bandreject response

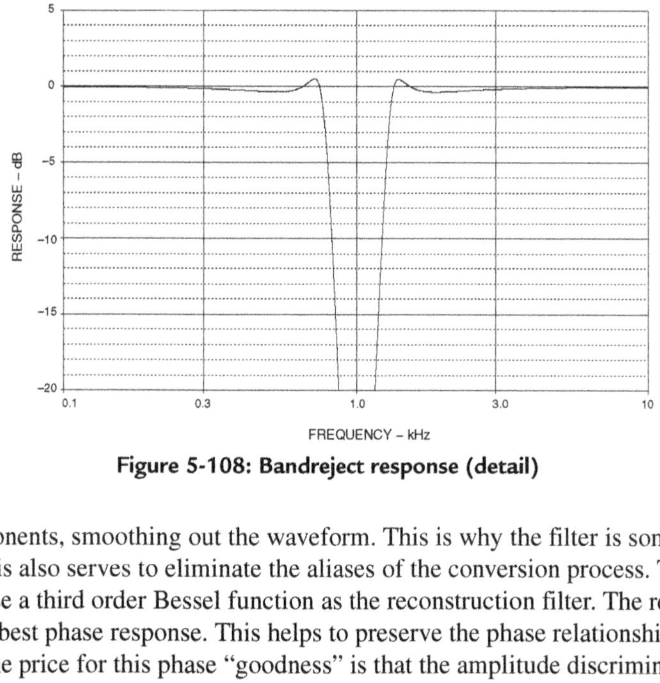

Figure 5-108: Bandreject response (detail)

high frequency components, smoothing out the waveform. This is why the filter is sometimes referred to as a smoothing filter. This also serves to eliminate the aliases of the conversion process. The "standard" in the audio industry is to use a third order Bessel function as the reconstruction filter. The reason to use a Bessel filter is that it has the best phase response. This helps to preserve the phase relationship of the individual tones in the music. The price for this phase "goodness" is that the amplitude discrimination is not as good as some other filter types. If we assume that we are using 8× oversampling of the 48 kSPS data stream in the D/A converter the aliases will appear at 364 kHz (8 × 48 k – 20 k). The digital filter that is used in the interpolation process will eliminate the frequencies between 20 kHz and 364 kHz. If we assume that the bandedge is 30 kHz, we have a frequency ratio of approximately 12 (364 ÷30). We use 30 kHz as the band edge, rather than 20 kHz to minimize the roll-off due to the filter in the pass band. In fact, the complete design for this filter includes a shelving filter to compensate for the pass band roll-off. Extrapolating from Figure 5-20, a third order Bessel will only provide on the order of 55 dB attenuation at 12 × Fo. This is only about 9-bit accuracy.

By designing the filter as 7th order, and by designing it as a linear phase with equiripple error of 0.05°, the stopband attenuation can be increased to about 120 dB at 12 × Fo. This is close to the 20-bit system.

The filter will be designed as an FDNR type. This is an arbitrary decision. Reasons to choose this topology are its low sensitivities-to-component tolerances and the fact that the op amps are in the shunt arms rather than in the direct signal path.

The first step is to find the passive prototype. To do this, use the charts in Williams's book. Then get the circuit shown in Figure 5-109A. Next perform a translation in the s-plane. This gives the circuit shown in Figure 5-109B. This filter is scaled for a frequency of 1 Hz and an impedance level of 1 Ω. The D structure of the converted filter is replaced by a GIC structure that can be physically realized. The filter is then denormalized by frequency (30 kHz) and impedance (arbitrarily chosen to be 1 kΩ). This gives a frequency-scaling factor (FS) of 1.884×10^5 (= $2\pi (3 \times 10^4)$). Next, arbitrarily choose a value of 1 nF for the capacitor. This gives an impedance-scaling factor (Z) of 5305 (= (C_{OLD}/C_{NEW})/ FSF).

Then multiply the resistor values by Z. This results in the resistors that had the normalized value of 1 Ω will now have a value of 5.305 kΩ. For the sake of simplicity adopt the standard value of 5.36 kΩ. Working backwards, this will cause the cutoff frequency to change to 29.693 kHz. This slight shift of the cutoff frequency will be acceptable.

The frequency scaling factor is then recalculated with the new center frequency and this value is used to denormalize the rest of the resistors. The design flow is illustrated in Figure 5-109. The final schematic is shown it Figure 5-109D.

Figure 5-109A: CD Reconstruction filter—passive prototype

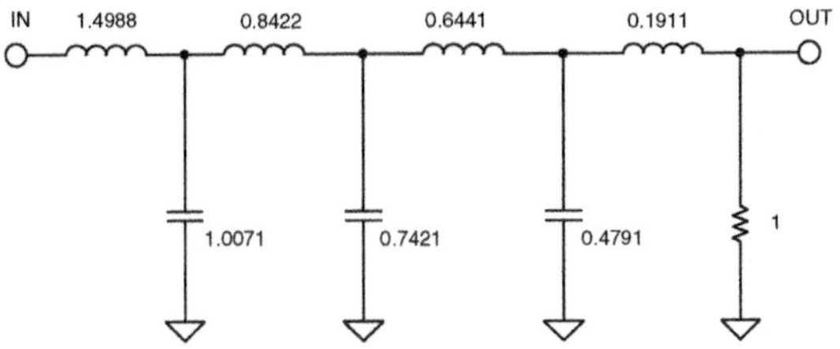

Figure 5-109B: CD Reconstruction filter—transformation in s-plane

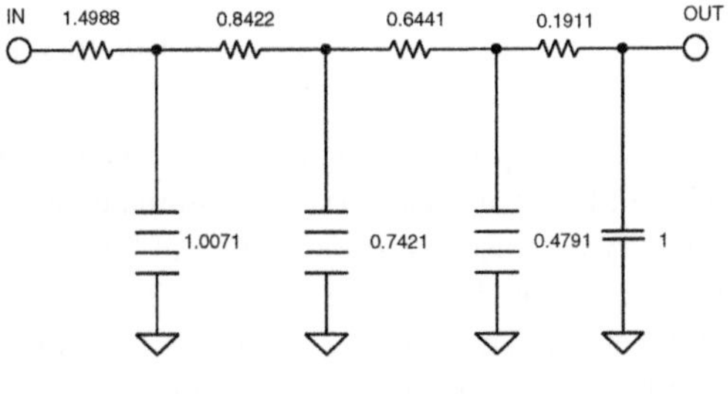

Figure 5-109C: CD Reconstruction filter—normalized FDNR

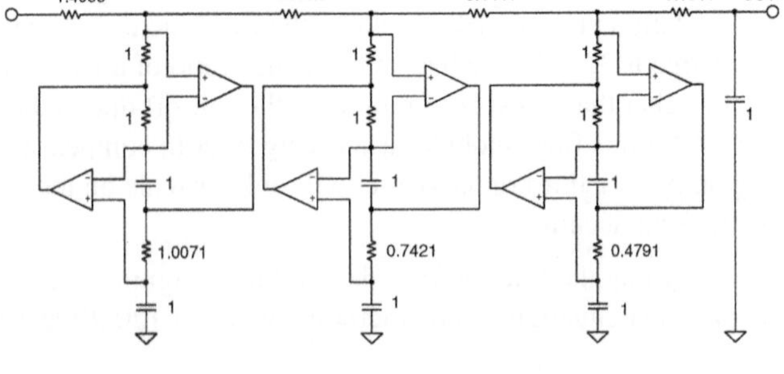

Figure 5-109D: CD Reconstruction filter—final filter

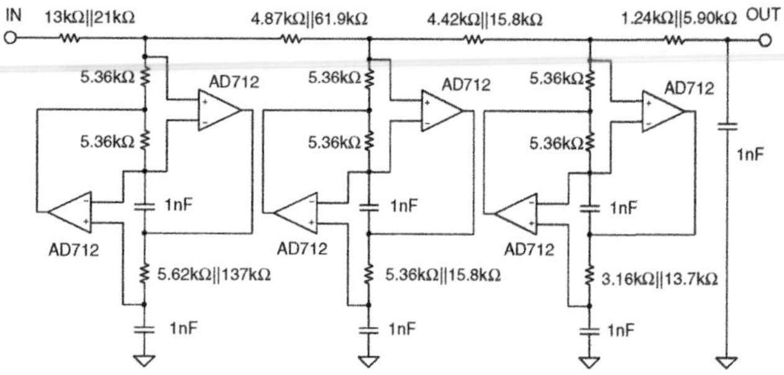

The performance of the filter is shown in Figure 5-110(A–D).

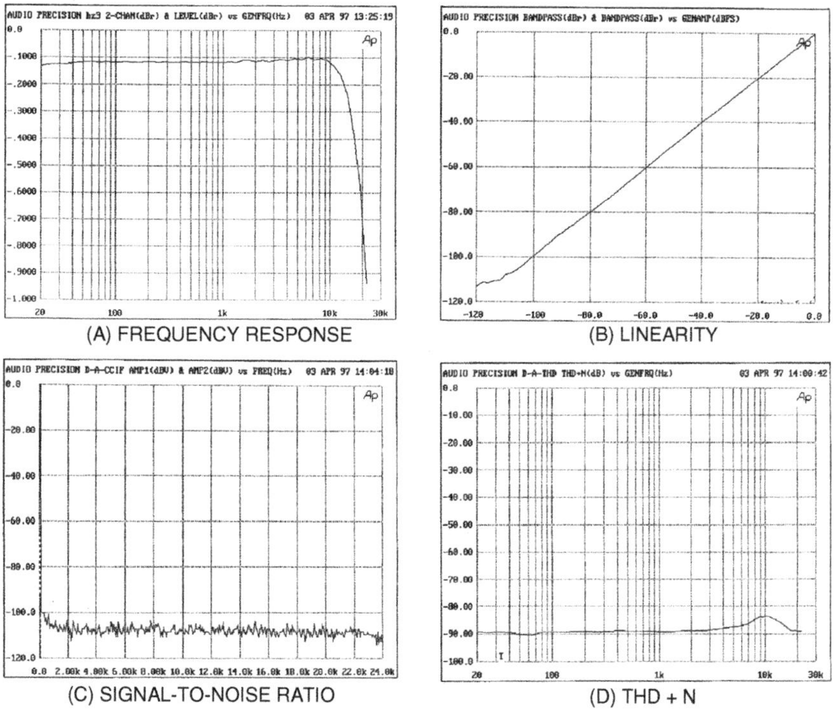

Figure 5-110: CD filter performance

Digitally Programmable State Variable Filter

One of the attractive features of the state variable filter is that the parameters (gain, cutoff frequency, and "Q") can be individually adjusted. This attribute can be exploited to allow control of these parameters.

To start, the filter is slightly reconfigured. The resistor divider that determines Q (R6 & R7 of Figure 5-54) is changed to an inverting configuration. The new filter schematic is shown in Figure 5-111. The resistors R1, R2, R3, and R4 (of Figure 5-111) are then replaced by CMOS multiplying DACs. Note that R5 is implemented as the feedback resistor implemented in the DAC. The schematic of this circuit is shown in Figure 5-112.

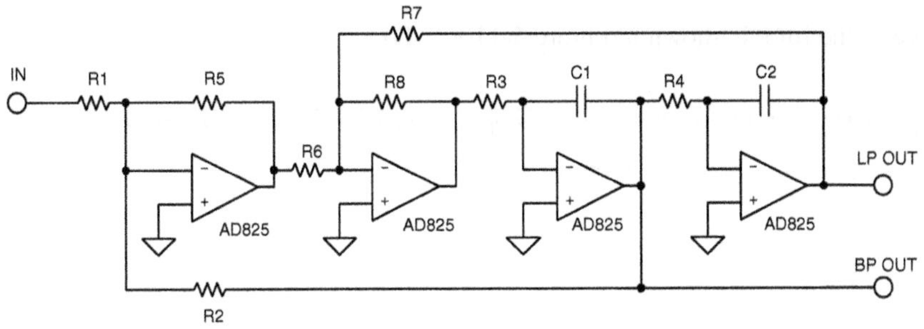

Figure 5-111: Redrawn state-variable filter

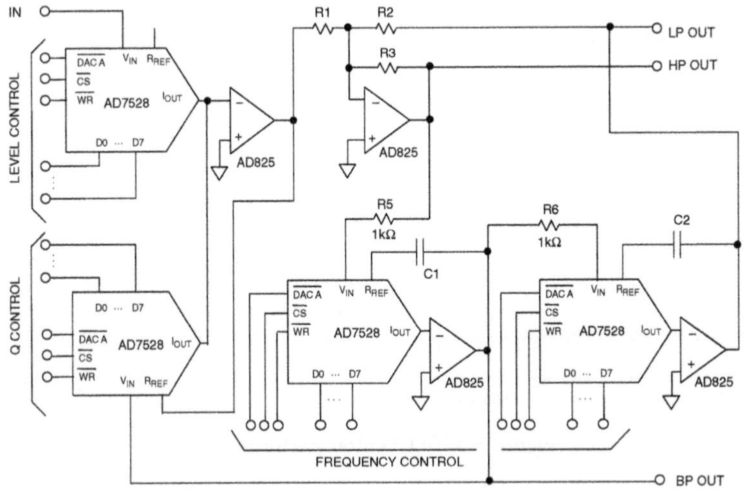

Figure 5-112: Digitally controlled state variable filter

The AD7528 is an 8-bit dual MDAC. The AD825 is a high speed FET input op amp. Using these components the frequency range can be varied from around 550 Hz to around 150 kHz (Figure 5-113). The Q can be varied from approximately 0.5 to over 12.5 (Figure 5-114). The gain of the circuit can be varied from 0 dB to –48 dB (Figure 5-115).

Figure 5-113: Frequency response versus DAC control word

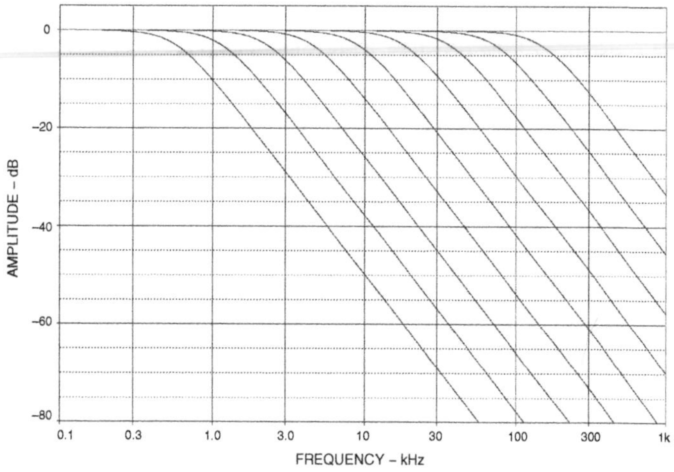

Figure 5-114: Q Variation versus DAC control word

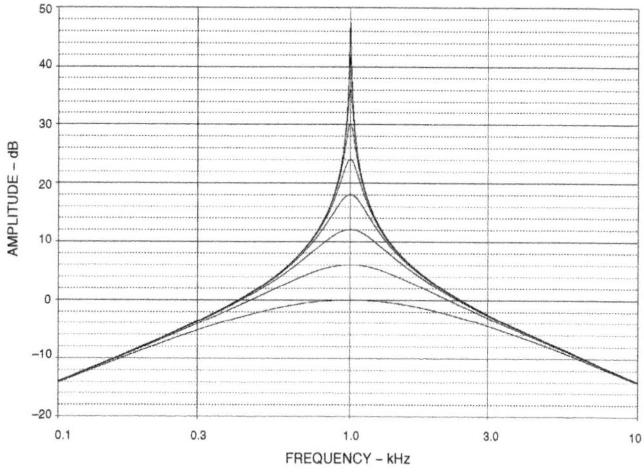

Figure 5-115: Gain variation versus DAC control word

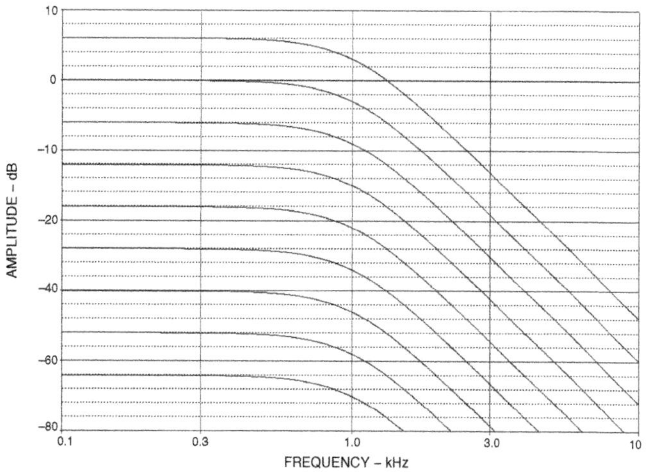

The operation of the DACs in controlling the parameters can best be thought of as the DACs changing the effective resistance of the resistors. This relationship is

$$\text{DAC EQUIVALENT RESISTANCE } \frac{256 * \text{DAC RESISTANCE}}{\text{DAC CODE (DECIMAL)}}$$

This, in effect, varies the resistance from 11 kΩ to 2.8 MΩ for the AD7528.

One limitation of this design is that the frequency is dependent on the ladder resistance of the DAC. This particular parameter is not controlled. DACs are trimmed so that the ratios of the resistors, not their absolute values, are controlled. In the case of the AD7528, the typical value is 11 kΩ. It is specified as 8 kΩ min. and 15 kΩ max. A simple modification of the circuit can eliminate this issue. The cost is two more op amps (Figure 5-116). In this case, the effective resistor value is set by the fixed resistors rather than the DAC's resistance. Since there are two integrators the extra inversions caused by the added op amps cancel.

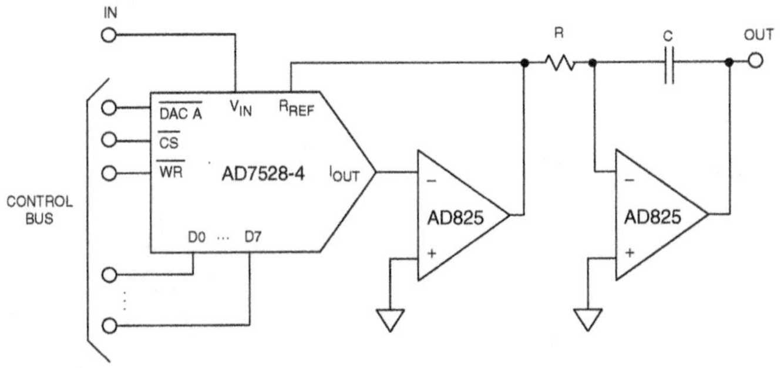

Figure 5-116: Improved digitally variable integrator

As a side note, the multiplying DACs could be replaced by analog multipliers. In this case the control would obviously be an analog rather than a digital signal. We also could just as easily have used a digital pot in place of the MDACs. The difference is that instead of increasing the effective resistance, the value of the pot would be the maximum.

60 Hz Notch Filter

A very common problem in instrumentation is that of interference of the telemetry that is to be measured. One of the primary sources of this interference is the power line. This is particularly true of high impedance circuits. Another path for this noise is ground loops. One possible solution is to use a notch filter to remove the 60 Hz. component. Since this is a single frequency interference, the Twin-T circuit will be used.

Since the maximum attenuation is desired and the minimum notch width is desired, the maximum Q of the circuit is desired. This means the maximum amount of positive feedback is used (R5 open and R4 shorted). Due to the high impedance of the network, a FET input op amp is used.

The filter is designed using Figure 5-78. The schematic is shown in Figure 5-117 and the response in Figure 5-118.

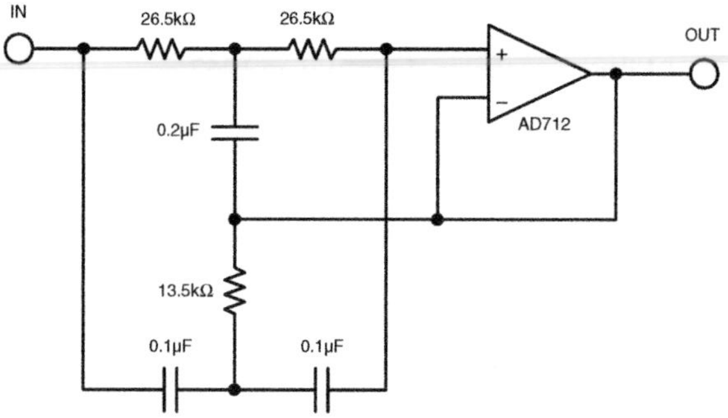

Figure 5-117: 60 Hz Twin-T notch filter

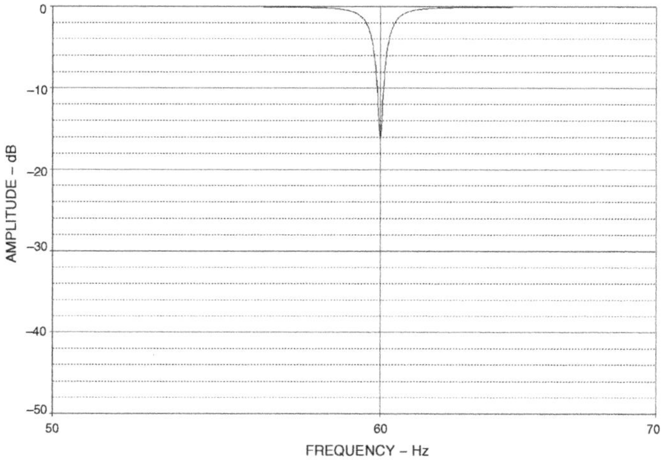

Figure 5-118: 60 Hz notch response

References: Analog Filters

1. A. I. Zverev, **Handbook of Filter Synthesis**, John Wiley, 1967.
2. A. B. Williams, **Electronic Filter Design Handbook**, McGraw-Hill, 1981, ISBN: 0-07-070430-9.
3. M. E. Van Valkenburg, **Analog Filter Design**, Holt, Rinehart & Winston, 1982.
4. M. E. Van Valkenburg, **Introduction to Modern Network Synthesis**, John Wiley and Sons, 1960.
5. Zverev and H. J. Blinchikoff, **Filtering in the Time and Frequency Domain**, John Wiley and Sons, 1976.
6. S. Franco, **Design with Operational Amplifiers and Analog Integrated Circuits**, McGraw-Hill 1988, ISBN: 0-07-021799-8.
7. W. Cauer, **Synthesis of Linear Communications Networks**, McGraw-Hill, New York, 1958.
8. Budak, **Passive and Active Network Analysis and Synthesis**, Houghton Mifflin Company, Boston, 1974.
9. L. P. Huelsman and P. E. Allen, **Introduction to the Theory and Design of Active Filters**, McGraw Hill, 1980, ISBN: 0-07-030854-3.
10. R. W. Daniels, **Approximation Methods for Electronic Filter Design**, McGraw-Hill, New York, 1974.
11. J. Tow, "Active RC Filters – a State-space Realization," **Proc. IEEE**, 1968, vol.56, pp. 1137–1139.
12. L.C. Thomas, "The Biquad: Part I – Some Practical Design Considerations," **IEEE Trans. Circuits and Systems**, Vol. CAS-18, 1971, pp. 350–357.
13. L.C. Thomas, "The Biquad: Part I – A Multipurpose Active Filtering System," **IEEE Trans. Circuits and Systems**, Vol. CAS-18, 1971, pp. 358–361.
14. R. P. Sallen and E. L. Key, "A Practical Method of Designing RC Active Filters," **IRE Trans. Circuit Theory**, Vol. CT-2, 1955, pp. 74–85.
15. P. R. Geffe, "How to Build High-Quality Filters out of Low-Quality Parts," **Electronics**, Nov. 1976, pp. 111–113.
16. P. R. Geffe, "Designers Guide to Active Band-pass Filters," **EDN**, Apr. 5 1974, pp. 46–52.
17. T. Delyiannis, "High-Q Factor Circuit with Reduced Sensitivity," **Electronics Letters**, 4, Dec. 1968, p. 577.
18. J. J. Friend, "A Single Operational-Amplifier Biquadratic Filter Section," **IEEE ISCT Digest Technical Papers**, 1970 p. 189.
19. L. Storch, "Synthesis of Constant-Time-Delay Ladder Networks Using Bessel Polynomials," **Proceedings of IRE**, Vol. 42, 1954, pp. 1666–1675.
20. K. W. Henderson and W. H. Kautz, "Transient Response of Conventional Filters," **IRE Trans. Circuit Theory**, Vol. CT-5, 1958, pp. 333–347.
21. J. R. Bainter, "Active Filter Has Stable Notch and Response Can be Regulated," **Electronics**, Oct. 2 1975, pp.115-117.
22. S. A. Boctor, "Single Amplifier Functionally Tunable Low-Pass Notch Filter," **IEEE Trans. Circuits and Systems**, Vol. CAS-22, 1975, pp. 875–881.
23. S. A. Boctor, "A Novel Second-Order Canonical RC-Active Realization of High-Pass-Notch Filter," **Proc. 1974 IEEE Int. Symp. Circuits and Systems**, pp. 640–644.

24. L. T. Burton, "Network Transfer Function Using the Concept of Frequency Dependant Negative Resistance," **IEEE Trans. Circuit Theory**, Vol. CT-16, 1969, pp. 406–408.

25. L. T. Burton and D. Trefleaven, "Active Filter Design Using General Impedance Converters," **EDN**, Feb. 1973, pp.68–75.

26. H. Zumbahlen, "A New Outboard DAC, Part 2," **Audio Electronics**, Jan. 1997, pp. 26–32, 42.

27. M. Williamsen, "Notch-Filter Design," **Audio Electronics**, Jan. 2000, pp. 10–17.

28. W. Jung, "Bootstrapped IC Substrate Lowers Distortion in JFET Op Amps," **Analog Devices AN232**.

29. H. Zumbahlen, "Passive and Active Filtering," **Analog Devices AN281**.

30. P. Toomey & W. Hunt, "AD7528 Dual 8-Bit CMOS DAC," **Analog Devices AN318**.

31. W. Slattery, "8th Order Programmable Low-pass Analog Filter Using 12-Bit DACs," **Analog Devices AN209**.

32. **CMOS DAC Application Guide**, Analog Devices.

CHAPTER 6

Signal Amplifiers

CHAPTER 6

Signal Amplifiers

Walt Jung, Walt Kester

SECTION 6-1

Audio Amplifiers

Walt Jung

Audio Preamplifiers

Audio signal preamplifiers (preamps) represent the low-level end of the dynamic range of practical audio circuits using modern IC devices. In general, amplifying stages with input signal levels of 10 mV or less fall into the preamp category. This section discusses some basic types of audio preamps, which are:

Microphone—including preamps for dynamic, electret, and phantom-powered microphones, using transformer input circuits, operating from dual and single supplies.

Phonograph—including preamps for moving magnet and moving coil phono cartridges in various topologies, with detailed response analysis and discussion.

In general, when working signals drop to a level of ≈1 mV, the input noise generated by the first system amplifying stage becomes critical for wide dynamic range and good signal-to-noise ratio. For example, if internally generated noise of an input stage is 1 μV and the input signal voltage 1 mV, the best signal-to-noise ratio possible is just 60 dB.

In a given application, both the input voltage level and impedance of a source are usually fixed. Thus, for best signal-to-noise ratio, the input noise generated by the first amplifying stage must be minimized when operated from the intended source. This factor has definite implications to the preamp designer, as a "low noise" circuit for low impedances is quite different from one with low noise operating from a high impedance.

Successfully minimizing the input noise of an amplifier requires a full understanding of all the various factors that contribute to total noise. This includes the amplifier itself as well as the external circuit in which it is used; in fact, *the total circuit environment must be considered* both to minimize noise and maximize dynamic range and signal fidelity.

A further design complication is the fact that not only is a basic gain or signal scaling function to be accomplished, but *signal frequency response* may also need to be altered in a predictable manner. Microphone preamps are an example of wideband, flat frequency response, low noise amplifiers. In contrast to this, phonograph preamp circuits not only scale the signal, they also impart a specific frequency response characteristic to it. A major part of the design for the RIAA phono preamps of this section is a systematic analysis process, which can be used to predictably select components for optimum performance in frequency response terms. This leads to very precise functioning, and excellent correlation between a computer-based design and measured lab operation.

Microphone Preamplifiers

The microphone preamplifier (mic preamp) is a basic low level audio amplification requirement. Mic preamps can assume a variety of forms, considering the wide range of possible signal levels, the microphone types, and their impedances. These factors influence the optimum circuit for a specific application. Discussed in this section are mic preamps that work with both high and low impedance microphones, both with and without phantom power, and with transformer input stages.

Single-Ended, Single-Supply High-Impedance Mic Preamp

A very simple form of mic preamp is shown in Figure 6-1. This is a noninverting stage with a single-ended input, most useful with high-impedance microphones such as dynamic and piezoelectric types. As shown, it has adjustable gain of 20 dB–40 dB via R_{GAIN}, and is useful with audio sources with 600 Ω or greater source impedances.

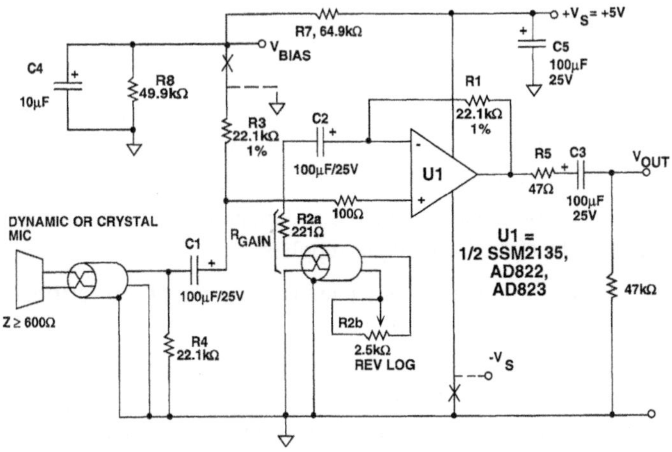

Figure 6-1: A single-ended, single-supply mic preamp

The U1 op amp can greatly affect the overall performance, not only in general amplification terms but also in suitability for single supply operation (as shown here). In terms of noise performance, the U1 device should have a low input noise with ≥500 Ω sources, with the external circuit values adjusted so that the source impedance (microphone) dominates the overall source resistance.

For very low noise on 5 V supplies, very few devices are suitable. Among these the dual SSM2135 or the OP213, and AD822/AD823 stand out, and are recommended as first choices. For very low power, minimal quiescient current parts like the AD8541 can be considered. Many other low noise devices can also work well in this circuit for total supply voltages of 10 V or more, for example the OP275, and OP270/OP470 types. The circuit is also easily adapted for dual supply use, as noted below.

In this circuit, gain-determining resistors R1||R2 (where R2a + R2b = R_{GAIN}) are scaled such that their total resistance is less than the expected source impedance, that is 1 kΩ or less. This minimizes the contribution of the gain resistors to input noise, at high gain. As noted, gain of the circuit is adjusted in the feedback path via resistor R_{GAIN}. In a system sense, control of a microphone or other low level channel signal level is preferably done *after* it has undergone some gain, as the case here. R_{GAIN} can of course be a fixed value.

Because of the single supply operation, input/output coupling is via polar capacitors, namely C1, C2, and C3. C4 is a noise filter, and C5 a bypass. For lowest noise in the circuit, the amplifier biasing must also be *noiseless*; that is, free from noise added directly or indirectly by the biasing (see Reference 1). Resistors with dc across them should have low excess noise (film types), or be ac-bypassed. Thus R1, R2, R3, R4, R7, and R8 are preferably metal films, with R7–R8 bypassed. A 2.2 V bias provided from R7–R8 biases the output of U1 to near midsupply. If higher supply voltage is used, R7–R8 can be adjusted for maximum output with a particular amplifier. For example, with low bias current, rail-rail output op amps, R7 and R8 should be high, equal values (≥100 kΩ).

While the OP213 or SSM2135 for U1 is optimum when operating from lower impedance sources, FET input types such as the AD82x families (or a select CMOS part) is preferable for high impedance sources, such as crystal or ceramic mics. To adapt the circuit for this, R3 and R4 should be 1MΩ or more, and C1 a 0.1 μF film capacitor.

Bandwidth using the OP213 or SSM2135 is about 30 kHz at maximum gain, or about 20 kHz for similar conditions with the AD822 (or AD820). Distortion and noise performance will reflect the U1 device and source impedance. With a shorted input, an SSM2135 measures output noise of about 110 μV rms at a gain of 100, with a 1 kHz THD + N of 0.022% at 1 V rms into a 2 kΩ load. The AD820 measures about 200 μV rms with 0.05% THD + N for similar conditions. For both, the figures improve at lower gains.

The circuit of Figure 6-1 is a good one if modest performance and simplicity are required, but requires attention to details. The input cable to the microphone must be shielded, and no longer than required. Similar comments apply to a cable for R_{GAIN} (if remote).

To adapt this circuit for dual supply use, R3 is returned to ground as noted, plus the bias network of R7, R8, and C4 is eliminated. U1 is operated on symmetric supplies (±5 V, ±15 V, and so forth), with the $-V_S$ rail bypassed similar to $+V_S$. Coupling caps C1, C2, and C3 are retained, but must be polarized to matched the amplifier used (or nonpolar types). Although microphones with output impedances of less than 600 Ω can be used with this circuit, the noise performance will not be optimum. Also, many of these typically require a balanced input interface. Subsequent circuits show methods of optimizing noise with low impedance, balanced output microphones, as suited for professional applications.

Electret Mic Preamp Interface

A popular mic type for speech recording and other noncritical applications is the electret type. This is a permanently polarized condenser mic, typically with a built-in common-source FET amplifier. The amplified output signal is taken from the same single ended lead which supplies the microphone with dc power, typically from a 3-10 V dc source.

Figure 6-2 illustrates a basic interface circuit that is useful in powering and scaling the output signal of an electret mic for further use. In this case the scaled output signal from this interface is fed into the LEFT and RIGHT inputs of a 5 V supply powered CODEC for digitization and processing. Dc phantom power is fed to the mic capsules by the R_A-C_A-R_B decoupling network from the 5 V supply, and the ac output signal is tapped off by C_{IN}-R2, and fed to U1. The R_B resistors will vary with different mics and supply voltages, and the values shown are typical. For a quiet mic supply voltage, a filtered/scaled V_R can be generated by the optional U2 connection shown.

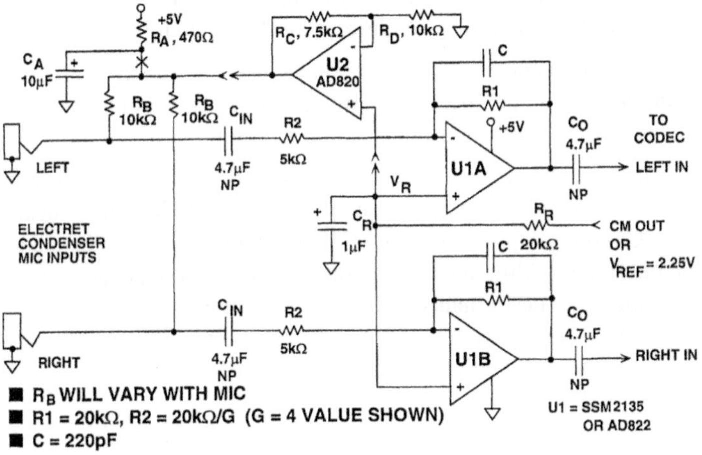

Figure 6-2: An electret mic interface for 5 V powered CODECs

The U1 dual scaling amplifier is an SSM2135 or AD822, and is used to normalize the mic signal to either a 1 V rms line level or 100 mV rms mic level typically required by CODEC inputs, and also to low-pass filter it prior to digitization. With a wide variety of electret mics and operating parameters, some signal level scaling is often required.

The scaling gain is simply R1/R2, and R2 is selected to provide a gain "G," to yield 0.1 V rms at the mic inputs of the CODEC, with the rated output from the mic. The U1 stages are inverting, so G can be greater or less than unity, i.e., other than 4 as is shown here, to normalize any practical input signal to an optimum CODEC level. The amplifier's low-pass corner frequency is set by the time constant R1-C, which results in a –3 dB point of 36 kHz. Bias for the U1 stages is provided from the CODEC, via the reference or CMOUT pins, typically a 2.25 V–2.5 V reference voltage. The low frequency time constants C_{IN}-R_B/R2 and C_O-20 kΩ are wideband to minimize LF phase shift. These (nonpolar) capacitors can be reduced to 1 µF or less, for narrowband uses.

Transformer-Coupled Low-Impedance Microphone Preamps

For any op amp, the best noise performance is attained when the characteristic *noise resistance* of the amplifier, R_n, is equal to the source resistance, R_s. Examples of microphone preamps that make use of this factor are discussed in this section. They utilize an input matching transformer to more closely optimize an amplifier to a source impedance which is unequal to the amplifier R_n. A basic circuit operating on this principle is shown in Figure 6-3. In order to select an optimum transformer turns ratio for a given source resistance (R_s), calculate the characteristic R_n of the op amp in use.

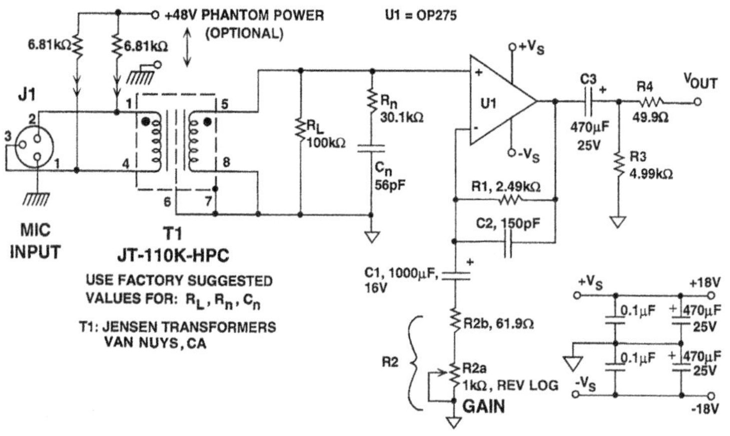

Figure 6-3: Transformer input mic preamplifier with 28 dB to 50 dB gain

R_n must first be calculated from the op amp's e_n and i_n data as:

$$R_n = \frac{e_n}{i_n} \qquad \text{Eq. 6-1}$$

where e_n is in $V/\sqrt{Hz}$ and i_n is in $A/\sqrt{Hz}$. A turns ratio for T1 may be calculated as:

$$\frac{N_s}{N_p} = \sqrt{\frac{R_n}{R_s}} \qquad \text{Eq. 6-2}$$

where N_s/N_p is the transformer secondary/primary turns ratio. For the OP275 op amp, the values of e_n and i_n are $7\ nV/\sqrt{Hz}$ and $1.5\ pA/\sqrt{Hz}$, respectively; thus,

$$R_n = \frac{e_n}{i_n} = \frac{7\times10^{-9}}{1.5\times10^{-12}} = 4.7k\Omega$$

Since both e_n and i_n vary with frequency, R_n will also vary with frequency. Therefore, a value calculated for R_n from the data sheet (such as above) is most accurate at the specified frequency. If the amplifier is to be optimized for a specific frequency, the e_n and i_n values should be for that frequency. However, audio amplifiers are wideband circuits, so latitude is due here. When available, a minimum noise-figure plot for the amplifier will allow graphical determination of the optimum source resistance for noise.

For this case, an optimum transformer turns ratio can be calculated to provide the optimum R_n to the op amp, working from a given R_s. For example, if R_s is 150 Ω, an optimum turns ratio for an OP275 (or other amplifier) with an R_n of 4.7 kΩ will be:

$$\frac{N_s}{N_p} = \sqrt{\frac{R_n}{R_s}} = \sqrt{\frac{4.7\times10^3}{1.5\times10^2}} \approx 5.6$$

Other examples matching these criteria would include OP27 family types.

Transformers are catalogued in fairly narrow and specific impedance ranges, so a unit with a rated secondary impedance in the range of 5 kΩ to 10 kΩ will be useful (the amplifier minimum noise impedance is reasonably broad). A suitable unit for this purpose is the Jensen JT-110K-HPC. Note that T1 must be adequately shielded and otherwise suitable for operation in low level environments. The use of the matching transformer allows the circuit to achieve an equivalent input noise (referred to the transformer input) that is only a few decibels above the theoretical limit, or very close to the thermal noise of the source resistance. For example, the thermal noise of a 150 Ω resistor in a 20 kHz noise bandwidth at room temperature is 219 nV. A real circuit has a higher input referred noise, due to the transformer plus op amp noise.

An additional advantage of the transformer lies in the *effective voltage gain* that it provides, due to the step up turns ratio. For a given circuit total numeric gain, G_{total}, this reduces the gain required from the op amp U1, $G_{(U1)}$, to:

$$G_{(U1)} = \frac{G_{total}}{N_s / N_p} \qquad \text{Eq. 6-3}$$

Thus, in the composite circuit of Figure 6-3 gain G_{total} is the product of the transformer step up, N_s/N_p, and (R1 + R2)/R1, which is $G_{(U1)}$. This has advantages of allowing more amplifier loop gain, thus greater bandwidth and accuracy, lower distortion, and so forth.

The transformer input example mic preamp stage of Figure 6-3 uses the JT-110K-HPC transformer for T1 with a primary/secondary ratio of about 1/8 (150 Ω/10 kΩ). The op amp section has a variable gain of about 3.3-41 times, which, in combination with the 17.8 dB transformer gain, yields a composite gain of 28 dB to 50 dB (26 to 300 times). Transient response of the transformer plus U1 amplifier is excellent. U1 here is one-half an OP275, operating on ±18 V power. Supplies should be well regulated and decoupled close to U1, particularly with low impedance loads. *Care should be used to operate U1 below maximum voltage rating*. The OP275 is rated for maximum supplies of ±22 V.

For best results, passive components should be high quality, such as 1% metal film resistors, a reverse log taper film pot for R2a, and low ESR capacitors for C1 and C3. Microphone phantom powering (see References 2 and 3) can be used, simply by adding the ±0.1% matched 6.81 kΩ resistors and a 48 V dc source, as shown. Close matching of the dc feed resistors is recommended by the transformer manufacturer whenever phantom power is used, to optimize CMR and to minimize the transformer's primary dc current flow (see Reference 4). Note that use of phantom powering has little or no effect on the preamp, since the transformer decouples the CM dc variations at the primary. CMR in an input transformer such as the JT-110K-HPC is typically 85 dB or more at 1 kHz, and substantially better at lower frequencies.

THD + N performance versus frequency of this OP275 mic preamp is shown in the family of curves in Figure 6-4. The test conditions are 35 dB gain, and successive input sweeps resulting in outputs of 0.5, 1, 2, and 5 V rms into 600 Ω. For these distortion tests as well as most of those following throughout these sections, THD + N frequency sweeps at various levels are used for sensitivity to slewing related distortions (see References 5–7), and output loaded tests are used for sensitivity to load related nonlinearities.

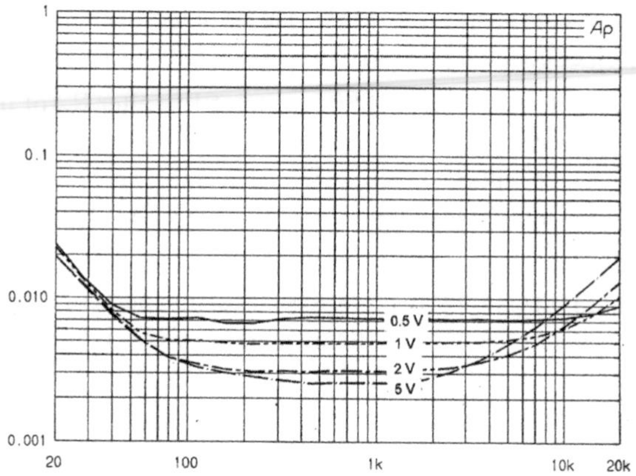

Figure 6-4: Transformer-coupled mic preamplifier THD + N (%) versus frequency (Hz) for 35 dB gain, outputs of 0.5, 1, 2, and 5 V rms into 600 Ω

For the OP275 data shown in Figure 6-4, there are three interest regions, a sub-100 Hz region where distortion is largely transformer-related, a 100 Hz–3 kHz region where distortion is lowest, and a greater-than-3 kHz region where it again rises. For most of the spectrum THD + N is ≤0.01% for medium outputs, and slightly higher at high frequencies.

The –3 dB bandwidth of this circuit is about 100 kHz, and is dominated by the JT-110K-HPC transformer and its termination network, assuming a 150 Ω source impedance. Conversely, for higher or lower source impedances, the bandwidth will lower or rise in proportion, so application of this circuit should take this into account. For example, capacitor microphone capsules with emitter follower outputs appear as a ≈15 Ω source.

Very Low Noise Transformer Coupled Mic Preamp

A high performance low noise mic preamp is shown in Figure 6-5, using a lower ratio transformer, the Jensen JT-16A. This transformer has a lower nominal step up ratio of about 2/1, and is optimized for use with

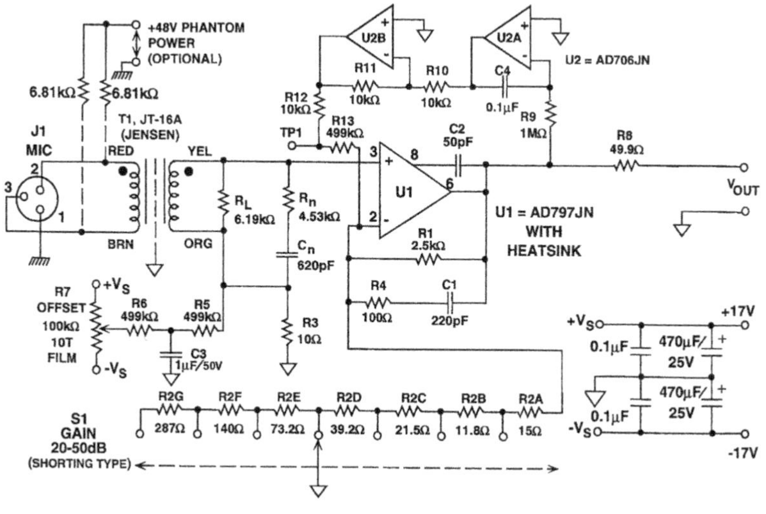

Figure 6-5: Low noise transformer input 20 dB to 50 dB gain mic preamp

lower noise resistance amplifiers such as the AD797. As can be noted from the figure, the general topology is similar to the previous transformer coupled preamp, but some details allow premium levels of performance.

This preamp has a selectable gain feature, using GAIN switch S1 to alter R2 of the feedback network. This varies U1's gain (and thus overall gain) over a range of 20 dB–50dB, making the preamp suitable for a wide range of uses. With the R2 step values shown, gain is selectable in 5 dB increments. This ranges from 50 dB with R2 (total) = 15 Ω, down to 20 dB with R2 (total) = 588.5 Ω. The transformer provides a fixed gain of about 5.6 dB.

Inasmuch as the AD797 has high precision as well as low distortion audio characteristics, this circuit can be dc-coupled quite effectively. This has the worthwhile advantage of eliminating large electrolytic coupling caps in the gain network and in the output coupling between U1 and V_{OUT}. This is accomplished as follows:

The initial device offset of the AD797 is 80 μV(max), a factor that allows a relatively simple trim by OFFSET trimmer R7 to null offset. R7 has a range of ±150 μV at the AD797 input, with noise well decoupled by C3. With the preamp warmed up well, and working at a midrange gain setting of 35 dB, the offset can be trimmed out. This is best done with the servo temporarily defeated, by grounding test point TP1. Under this condition the V_{OUT} dc level is then trimmed to <1 mV, via R7. This nulls out the residual offset of the AD797, and also ensures that the gain-range network sees minimal dc, which minimizes "pops" with gain changes. The offset shift thereafter with gain is only a few mV, and is of little concern, since the servo circuit of U2A and U2B holds the longer-term dc offset to 100 μV or less, with little gain interaction. Note that for the gain-change scheme to work properly, S1 *must* be a shorting (make-before-break) type.

THD + N performance versus frequency of this mic preamp is shown in Figure 6-6, for conditions of 35 dB gain, and successive input sweeps resulting in outputs of 0.5, 1, 2, and 5 V rms into 600 Ω. From these data it is essentially clear that the only distortion in the circuit is due to the transformer, which is small and occurs only at the low frequencies. Above 100 Hz, the apparent distortion is noise limited, to the highest frequencies.

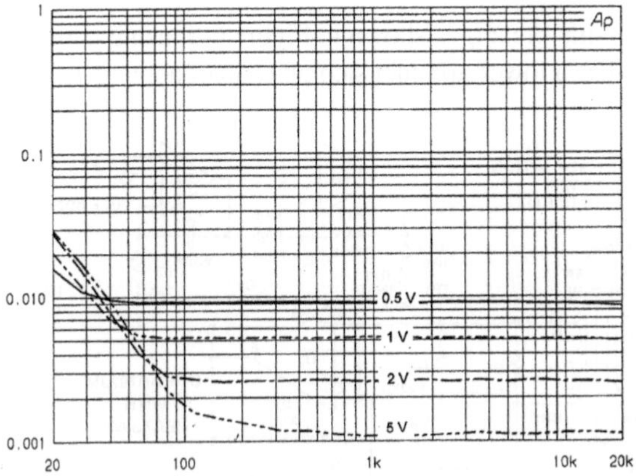

Figure 6-6: Low noise transformer input mic preamp THD + N (%) versus frequency (Hz) for 35 dB gain, outputs of 0.5, 1, 2, and 5 V rms into 600 Ω

The –3 dB bandwidth of this circuit is just under 150 kHz, and while this is essentially dominated by the JT-16A transformer and termination, bandwidth does reduce slightly at the highest gain (50 dB). Like the previous transformer coupled circuit, this circuit also assumes a 150 Ω source impedance, and similar application caveats apply.

The basic circuit as shown is single-ended with V_{OUT} taken from R8. However, a transformer can be simply added, as an option for driving balanced lines. When this is done, a nickel core type is suggested, for lowest distortion. One type suitable would be a Jensen JT-11-DM (or similar). It is coupled to the U1 output via a 10 Ω resistor.

Just as shown the circuit is suited for local, higher impedance loads of 1 kΩ and more. For very high levels of output drive or to drive long lines, a dedicated high current output driver should be used with U1, as generally described in the "Line Drivers" section. This can be most simply implemented by making U1 a composite amplifier, using a AD797 input section plus a follower-type output stage. A good choice for this would be a BUF04 IC, connected between Pin 6 of the AD797 and the remaining circuitry. The buffer will isolate the U1 stage, allowing it to operate with highest linearity with difficult loads. Note also that ±17 V supplies won't be necessary with the AD797 unless extreme voltage swings are required. More conventional (±15 V) supplies will minimize the U1 heating.

References: Microphone Preamplifiers

1. C. D. Motchenbacher, F. C. Fitchen, **Low-Noise Electronic Design**, Wiley, New York, 1973, ISBN 0-471-61950-7.
2. G. Bore, "Powering Condenser Microphones," **db**, June 1970.
3. "ANSI Standard 268-15 (Revision 1987, amendments 1989, 1990, 1991)." American National Standards Institute, 11 W. 42nd St., New York, NY, 10036.
4. Steve Hogan, "Standard Mic Input Application," Jensen Transformers Application note JT99-0003, November, 1992.
5. W. Jung, M. Stephens, C. Todd, "Slewing Induced Distortion & Its Effect on Audio Amplifier Performance—With Correlated Measurement/Listening Results," **presented at 57th AES convention**, May 1977, AES preprint # 1252.
6. W. Jung, M. Stephens, C. Todd, "An Overview of SID and TIM," Parts 1–3, **Audio**, June, July, August, 1979.
7. Walter G. Jung, **Audio IC Op Amp Applications, 3rd Ed.**, Howard W. Sams & Co., 1987, ISBN 0-672-22452-6.
8. W. Jung, A. Garcia, "A Low Noise Microphone Preamp with a Phantom Power Option," **Analog Devices AN242**, November 1992.
9. Walt Jung, "Audio Preamplifiers, Line Drivers, and Line Receivers," within Chapter 8 of Walt Kester, **System Application Guide**, Analog Devices, Inc., 1993, ISBN 0-916550-13-3, pp. 8-1 to 8-100.
10. Walt Jung, "Microphone Preamplifiers for Audio," **Analog Dialogue**, Vol. 28, No. 2, 1994, pp. 12–18.

RIAA Phono Preamplifiers

An example of an audio range preamplifier application requiring equalized frequency response is the RIAA phono preamp. While LP record sales have faded with the establishment of new digital media, for completeness equipment is still designed to include phono playback stages. RIAA preamp stages, as amplifiers with predictable, nonflat frequency response, have more general application connotations. The design techniques within this section are specific to RIAA as an example, but they are also applicable to other frequency dependent amplitude designs in general. The techniques are also useful as a study tool, considering the various approaches advanced to optimize the function of high performance gain with predictable equalization (EQ). These last two points make these discussions useful in a much broader sense.

Some RIAA Basics

The RIAA equalization curve (see Reference 1) is shown in Figure 6-7, expressed as it is relative to dc. This curve indicates maximum gain below 50 Hz (f1), with two high frequency inflection points. Above f1, the gain rolls off at 6 dB/octave until a first high frequency breakpoint is reached at 500 Hz (f2). Gain then remains relatively constant until a second high frequency breakpoint is reached at about 2.1 kHz (f3), where it again rolls off at 6 dB/octave through the remainder of the audio region and above.

Figure 6-7: Ideal RIAA de-emphasis (time constants of 3180 μs, 318 μs, 75 μs)

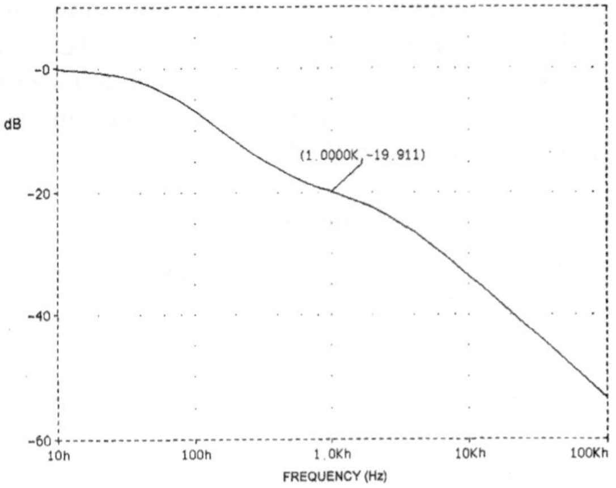

Use of a low frequency roll-off (f0, not shown) is at the option of the designer. Frequency response can be extended towards dc, or, alternately, rolled off at a low frequency below 50 Hz. When applied, this roll-off is popularly called a "rumble" filter, as it reduces turntable/record related low frequency disturbances, lessening low frequency driver overload. This roll-off may or may not coincide with a fourth time constant (below).

However, gain at the frequencies f1, f2, and f3 describes the basic RIAA curve. In the standard, this is described in terms of three corresponding time constants, T1, T2, and T3, defined as 3180 μs, 318 μs, and 75 μs, respectively (Reference 1, again). The T1–T3 are here described as they correspond to *ascending frequency*, the reverse of the terminology in Reference 1 (however, the time constants themselves are identical). In some literature one may occasionally find the frequencies corresponding to T1, T2, and T3 referenced. These exact frequencies can be found simply by the basic relationship of:

$$f = 1/(2 \bullet \pi \bullet T) \qquad \text{Eq. 6-4}$$

So, for the three time constants specified, the frequencies are:

$$f1 = 1/(2 \bullet \pi \bullet T1) = 1/(2 \bullet \pi \bullet 3180E-6) = 50\,Hz$$

$$f2 = 1/(2 \bullet \pi \bullet T2) = 1/(2 \bullet \pi \bullet 318E-6) = 500\,Hz$$

$$f3 = 1/(2 \bullet \pi \bullet T3) = 1/(2 \bullet \pi \bullet 75E-6) = 2122\,Hz$$

An IEC amendment to the basic RIAA response adds a fourth time constant of 7950 μs, corresponding to an f0 of 20 Hz when used (see Reference 2). Use of this roll-off has never been standardized in the US, and isn't treated in detail here.

The characteristic gain in dB for an RIAA preamp is generally specified relative to a 1kHz reference frequency. For convenience in evaluating the RIAA curve numerically, Figure 6-8 is a complete 10 Hz–100 kHz relative decibel table for the three basic RIAA time constants. From these data several key points

FREQ	VDB(6)[1]	VDB(5)[2]
1.000E+01	1.974E+01	–1.684E-01
1.259E+01	1.965E+01	–2.639E-01
1.585E+01	1.950E+01	–4.109E-01
1.995E+01	1.928E+01	–6.341E-01
2.512E+01	1.895E+01	–9.654E-01
3.162E+01	1.847E+01	–1.443E+00
3.981E+01	1.781E+01	–2.103E+00
5.012E+01	1.694E+01	–2.975E+00
6.310E+01	1.584E+01	–4.067E+00
7.943E+01	1.455E+01	–5.362E+00
1.000E+02	1.309E+01	–6.823E+00
1.259E+02	1.151E+01	–8.398E+00
1.585E+02	9.877E+00	–1.003E+01
1.995E+02	8.236E+00	–1.167E+01
2.512E+02	6.645E+00	–1.327E+01
3.162E+02	5.155E+00	–1.476E+01
3.981E+02	3.810E+00	–1.610E+01
5.012E+02	2.636E+00	–1.727E+01
6.310E+02	1.636E+00	–1.828E+01
7.943E+02	7.763E–01	–1.913E+01
1.000E+03	**8.338E–07**	**–1.991E+01**
1.259E+03	–7.682E–01	–2.068E+01
1.585E+03	–1.606E+00	–2.152E+01
1.995E+03	–2.578E+00	–2.249E+01
2.512E+03	–3.726E+00	–2.364E+01
3.162E+03	–5.062E+00	–2.497E+01
3.981E+03	–6.572E+00	–2.648E+01
5.012E+03	–8.227E+00	–2.814E+01
6.310E+03	–9.992E+00	–2.990E+01
7.943E+03	–1.184E+01	–3.175E+01
1.000E+04	–1.373E+01	–3.365E+01
1.259E+04	–1.567E+01	–3.558E+01
1.585E+04	–1.763E+01	–3.754E+01
1.995E+04	–1.960E+01	–3.951E+01
2.512E+04	–2.158E+01	–4.149E+01
3.162E+04	–2.357E+01	–4.348E+01
3.981E+04	–2.557E+01	–4.548E+01
5.012E+04	–2.756E+01	–4.747E+01
6.310E+04	–2.956E+01	–4.947E+01
7.943E+04	–3.156E+01	–5.147E+01
1.000E+05	–3.356E+01	–5.347E+01

Notes: (1) Denotes 1 kHz 0 dB reference
(2) Denotes dc 0 dB reference

Figure 6-8: Idealized RIAA frequency response referred to 1 kHz and to dc

can be observed: If the 1 kHz gain is taken as the zero dB reference, frequencies below or above show higher or lower dB levels, respectively (Note 1, column 2). With a dc 0 dB reference, it can be noted that the 1 kHz gain is 19.91 dB below the dc gain (Note 2, column 3).

Expressed in terms of a gain ratio, this means that in an ideal RIAA preamp the 1 kHz gain is always 0.101 times the dc gain. The constant 0.101 is unique to all RIAA preamp designs following the above curve, therefore it can be designated as "K_{RIAA}", or:

$$K_{RIAA} = 0.101 \qquad \text{Eq. 6-5}$$

This constant logically shows up in the various gain expressions of the RIAA preamp designs following. In all examples discussed here (and virtually all RIAA preamps in general), the shape of the standard RIAA curve is fixed, so specifying gain for a given frequency (1 kHz) also defines the gain for all other frequencies.

It can also be noted from the RIAA curve of Figure 6-7 that the gain characteristic continues to fall at higher frequencies. This implies that an amplifier with unity-gain stability for 100% feedback is ultimately required, which can indeed be true, when a standard feedback configuration is used. Many circuit approaches can be used to accomplish RIAA phono-playback equalization; however, all must satisfy the general frequency response characteristic of Figure 6-7.

Equalization Networks for RIAA Equalizers

Two equalization networks well suited in practice to RIAA phono reproduction are illustrated in Figure 6-9a and 6-9b, networks N1 and N2. Both networks with values as listed can yield with high accuracy the three standard RIAA time constants of 3180 μs, 318 μs, and 75 μs as outlined by network theory (see References 3-6). For convenience, both theoretical values for the ideal individual time constants are shown at the left, as well as closest fit standard "no trim" values to the right. Designers can, of course, parallel and/or series RC values as may be deemed appropriate, adhering to network theory.

There are of course an infinite set of possible RC combinations from which to choose network values, but practicality should rule any final selection. A theoretical starting point for a network value selection can begin with *any* component, but in practice the much smaller range of available capacitors suggests their

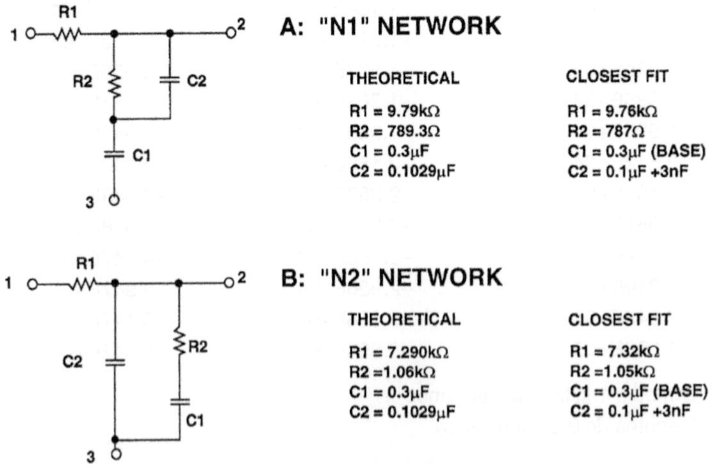

Figure 6-9: Two RIAA EQ networks (T1 = 3180 μs, T2 = 318 μs, T3 = 75 μs)

selection first, then resistors, since they have a much broader span of (stock) values. Note that precision film resistors can in fact be obtained (on special order) in virtually *any* value, up to several megohms. The values listed here are those taken as standard from the E96 series.

Very high standards of EQ accuracy are possible, to tolerances of noticeably better than ±0.1 dB (see for example data from Reference 8, also quoted in 6). In the design process, there are several distinct general aspects of EQ component selection which can impact the ultimate accuracy. These are worth placing in perspective before starting a design.

The *selection tolerance* of the component defines how far an ideal (zero manufacturing tolerance) component deviates from the theoretical value. A good design will seek to minimize this error by using either carefully selected standard values, or series and/or shunt combinations, so as to achieve selection tolerance of less than 1%, preferably zero.

The *manufacturing tolerance* of the component defines how far an otherwise ideal component deviates from its stated catalog value, such as ±1%, ±2%, and so forth. This can obviously be controlled by tighter specifications, but usually at some premium, particularly with capacitors of ±1% or less. Note that a "hidden" premium here can be long delivery times for certain values. Care should be taken to use standard stock values with capacitors—even to the extent that multiple standard values may be preferable (three times 0.01 µF for 0.03 µF, as an example).

Topology-related parasitics must also be given attention, as they can also potentially wreck accuracy. Amplifier gain-bandwidth is one possible source of parasitic EQ error. However, a more likely error source is the parasitic zero associated with active feedback equalizers. If left uncompensated below 100 kHz, this alone can be a serious error.

In any event, for high equalization accuracy to be "real," once a basic solid topology is selected, the designer must provide for the qualification of components used, by precise measurement and screening, or tight purchase tolerances. An alternative is iterative trimming against a reference standard such as that of Reference 9, but this isn't suited for production. An example is the data of Reference 8, derived with the network of Reference 9. If used, the utility of such a trim technique lies in the reduction of the equipment accuracy burden. While the comparator used needs to have high *resolution*, the accuracy is transferred to the network comparison standard used.

It should be understood that an appropriately selected high quality network will allow excellent accuracy, for example either N1 or N2 with the "closest fit" (single component) values of exact value yield a broadband error of about ±0.15 dB. Accuracy about three times better than this is achieved with the use of N1 and the composite C2, as noted. The composite C2 is strongly suggested, as without it there is a selection error of about 3%.

It is also strongly recommended that only the highest quality components be employed for use in these networks, for obvious reasons. Regardless of the quality of the remainder of the circuit, it is surely true that the equalization accuracy and fidelity can be no better than the quality of those components used to define the transfer function. Thus only the best available components are used in the N1 (or N2) RC network, selected as follows:

Capacitors—should have close initial tolerance (1%–2%), a low dissipation factor and low dielectric absorption, be noninductive in construction, and have stably terminated low-loss leads. These criteria in general are best met by capacitors of the Teflon, polypropylene and polystyrene film families, with 1%–2% polypropylene types being preferred as the most practical. Types to definitely avoid are the "high K" ceramics. In contrast, "low K" ceramic types, such as "NPO" or "COG" dielectrics, have excellent dissipation factors. See the passive component discussions of Chapter 7 on capacitors, as well as the component-specific references at the end of this section.

Resistors—should also be close tolerance (≤1%), have low nonlinearity (low voltage coefficient), be temperature stable, with solid stable terminations and low-loss noninductive leads. Types that best meet these criteria are the bulk metal foil types and selected thick films, or selected military grade RN55 or RN60 style metal film resistor types. See the passive component discussions of Chapter 7 on resistors, as well as the component-specific references at the end of this section.

It should be noted also that the specific component values suggested might not be totally optimum from a low impedance, low noise standpoint. But, practicalities will likely deter using appreciably lower ones. For example, one could reduce the input resistance of either network down to say 1 kΩ, and thus lower the input referred noise contribution of the network. But, this in turn would necessitate greater drive capability from the amplifier stage, and raise the C values up to 1 μF–3 μF, where they are large, expensive, and most difficult to obtain. This may be justified for some uses, where performance is the guiding criterion rather than cost effectiveness, or the amplifiers used are sufficiently low in noise to justify such a step. Regardless of the absolute level of impedance used, in any case the components should be adequately shielded against noise pickup, with the outside foils of C1 or C2 connected either to common or a low impedance point.

These very same N1/N2 networks can suffice for both active and passive type equalization. Active (feedback) equalizers use the network simply by returning the input resistor R1 to common, that is jumpering points 1–3, and employing the network as a two-terminal impedance between points 1+3, and 2. Passive equalizers use the same network in a three-terminal mode, placed between two wideband gain blocks.

RIAA Equalizer Topologies

Many different circuit topologies can be used to realize an RIAA equalizer. Dependent upon the output level of the phono cartridge to be used, the 1 kHz gain of the preamp can range from 30 dB to more than 50 dB.

Magnetic phono cartridges in popular use consist of two basic types: moving magnet (MM) and moving coil (MC). The moving magnet types, which are the most familiar, are suitable for the first two circuits described. The moving coil cartridge types are higher performance devices; they are less commonplace but still highly popular.

Functionally, both types of magnetic cartridges perform similarly, and both must be equalized for flat response in accordance with the RIAA characteristic. A big difference in application, however, is the fact that moving magnet types have typical sensitivities of about 1 mV of output for each cm/s of recorded velocity. In moving coil types, sensitivity on the order of 0.1 mV is more common (for a similar velocity). In application then, a moving coil RIAA preamp must have more gain than one for moving magnets. Typically, 1 kHz gains are 40 dB–50 dB for moving coils, but only 30 dB–40 dB for moving magnets. Noise performance of a moving coil preamp can become a critical performance factor however, because of low-output voltage and low impedance involved—typically this is in the range of just 3 Ω–40 Ω. The following circuit examples illustrate techniques that are useful to these requirements.

Actively Equalized RIAA Preamp Topologies

The most familiar RIAA topology is shown in general form in Figure 6-10, and is called an active feedback equalizer, as the network N used to accomplish the EQ is part of an active feedback path (see References 10, 11). In these and all of the following discussions it is assumed that the input from the pickup is appropriately terminated by R_t–C_t, which are selected for flat *cartridge* frequency response driving U1. The following discussions deal with the *amplification* frequency response, given this ideal input signal.

Assuming an adequately high gain amplifier for U1, the gain/frequency characteristics of this circuit are determined largely by the network. The gain of the stage is set by the values of the network N and R_3, and

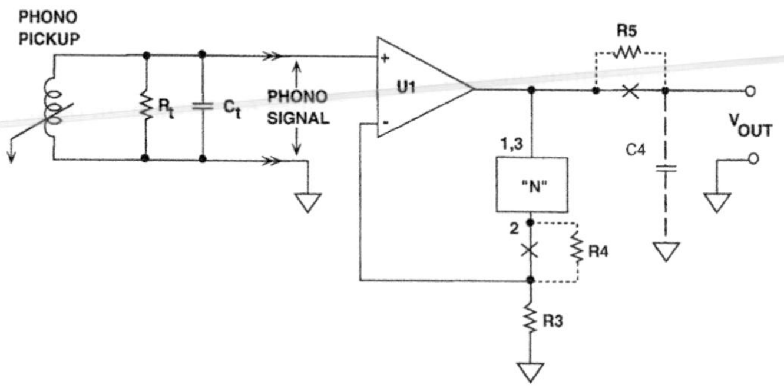

Figure 6-10: Active feedback RIAA equalizer

the U1 output is a low impedance, V_{OUT}. The 1 kHz gain of this stage is defined by the RIAA curve and resistors R_1 and R_3, and is:

$$G = 0.101 \bullet \left[1 + (R_1/R_3)\right] \qquad \text{Eq. 6-6}$$

where 0.101 is the constant K_{RIAA}. R_1 is within N; R_4 and R_5 are discussed momentarily.

As noted previously, an ideal RIAA response continues to fall with increasing frequency, and can in fact be less than unity at some high frequency (Figure 6-7, again). But, the basic U1 topology of Figure 6-10 can't achieve this, as the minimum gain seen at the output of U1 approaches unity at some (high) parasitic zero frequency, where the network equivalent series capacitive impedance of C_1 and C_2 is equal to R_3. At this zero frequency, the response from U1 simply levels off and ceases to track the RIAA curve.

However, in terms of practical consequence the error created by this zero may or may not be of significance, dependent upon where the zero falls (as determined by gain). If well above audibility (i.e., ≥ 100 kHz), it will introduce a small equalization error at the upper end of the audio range. For example, if it falls at 100 kHz, the 20 kHz error is only about 0.3 dB. Fortunately, this error is easily compensated by a simple low-pass filter after the amplifier, R_5–C_4. The filter time constant is set to match the zero T_4, which is:

$$T4 = R_3 \bullet C_{EQUIV} \qquad \text{Eq. 6-7}$$

where R_3 is the value required for a specific gain in the design.

C_{EQUIV} is the series equivalent capacitance of network capacitors C_1 and C_2, or:

$$C_{EQUIV} = (C_1 \bullet C_2)/(C_1 + C_2) \qquad \text{Eq. 6-8}$$

Here the C_{EQUIV} is 7.6 nF and R_3 200 Ω, so T4 = 1.5µs The product of R5 and C4 are set equal to T4, so picking a R5 value solves for C4 as:

$$C4 = T4/R5 \qquad \text{Eq. 6-9}$$

The 1.5 µs R_5-C_4 time constant is realized with R_5 = 499 Ω and C4 = 3 nF. This design step increases the output impedance, making it more load susceptible. This should be weighed against the added parts and loading. In general, R_5 should be low, i.e., ≤1 kΩ.

In some designs, a resistor R_4 (dotted in Figure 6-10) may be used with N (for example, for purposes of amplifier stability at a gain higher than unity). With R_4, T4 is calculated as:

$$T4 = (R_3 + R_4) \bullet C_{EQUIV} \qquad \text{Eq. 6-10}$$

The R_5–C_4 product is again chosen to be equal to this T4 (more on this below).

The next two schematics illustrate variations of the most popular approach to achieving a simple RIAA phono preamp, using active feedback, as just described. Figure 6-11 is a high performance, dc-coupled version using precision 1% metal film resistors and 1% or 2% capacitors of polystyrene or polypropylene type. Amplifier U1 provides the gain, and equalization components R1-R2-C1-C2 form the RIAA network, providing accurate realization with standard component values. N1 is the network, with 1 and 3 common.

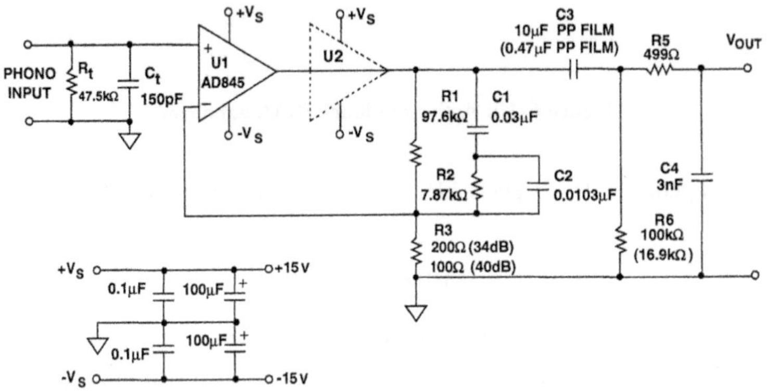

Figure 6-11: A dc-coupled active feedback RIAA moving magnet preamp

As mentioned, input RC components R_t-C_t terminate the moving magnet cartridge with recommended values (shown as typical). In terms of desired amplifier parameters for optimum performance, they are considerably demanding. For lowest noise from a cartridge's inductive source, the amplifier should have an input voltage noise density of $5\ nV/\sqrt{Hz}$ or less (favoring a bipolar), and an input current noise density of $1\ pA/\sqrt{Hz}$ or less (favoring a FET). In either case, the 1/F noise corner should be as low as possible.

For bipolar-input amplifiers, dc input-bias current can be a potential problem when direct coupling to the cartridge, so in this circuit only a very low input bias current type is suggested. If a bipolar input amplifier is used for U1, it should have an input current of <<100 nA for minimum dc offset problems (assuming a typical phono cartridge of ≈1 kΩ resistance). Examples are the OP27, OP270 families. FET-input amplifiers generally have negligible bias currents but also tend typically to have higher voltage noise. FET-input types useful for U1 are the AD845 and OP42, even though their voltage noise is not as low as the best of the bipolar devices mentioned. On the plus side, they both have a high output current and slew rate, for low distortion driving the feedback network load (approximately the R3 value at high frequencies). Of the two, the OP42 has lower noise, the AD845 higher output current and slew rate.

For high gain accuracy at high stage gains, the amplifier should have a high gain-bandwidth product; preferably >5 MHz at audio frequencies. Because of the 100% feedback through the network at high frequencies, the U1 amplifier must be unity-gain-stable. To minimize noise from sources other than the amplifier, gain resistor R_3 is set to a relatively low value, which generates a low voltage noise in relation to the amplifier.

RIAA accuracy is quite good using the stock equalizer values. A PSpice simulation run is shown in Figure 6-12 for the suggested gain of 34 dB. In this expanded scale plot over the 20 kHz–20 kHz range, the error relative to the 1 kHz gain is less than ±0.1 dB.

As can be noted from Figure 6-12, the relative amplitude is expanded, to easily show response errors. A perfect response would be a straight line at 0 dB, meaning that the circuit under test had exactly the same gain as an ideal RIAA amplifier of the same 1 kHz gain. This high sensitivity in the simulation is done via the use of a feature in PSpice allowing the direct entry of Laplace statements (see Reference 10). With this evaluation tool, the ideal transfer function of an RIAA equalizer can be readily generated. The key parameters are the three time constants described above, and the ideal dc gain.

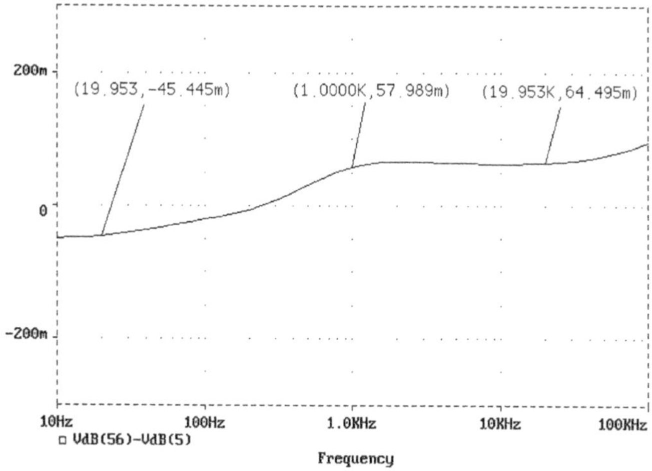

Figure 6-12: Relative error (B) versus frequency for dc-coupled active feedback RIAA moving magnet preamp, gain of 34 dB

The syntax to enable this mode of comparison is contained in the listing of Figure 6-13, which is the PSpice CIR file for the circuit of Figure 6-11. The Laplace details are all contained within the dotted box, and need only the editing of one value, "ENORM," for gain normalization from one circuit to another (see boldface). In this case ENORM is set to 490.7, to match the ideal R1 and R3 values of Figure 6-13. When the analysis is run, a difference display of the circuit-under-test and the ideal outputs (i.e., VdB(56)–VdB(5)) shows the relative response (Figure 6-12). Vertical axis scaling is easily adjusted for sensitivity, and is ±300 mB as displayed in Figure 6-12.

```
RIAA34LP: 34 dB gain RIAA preamp with AD845
*
.OPT ACCT LIST NODE OPTS NOPAGE LIBRARY
.AC DEC 10 10 100KHZ
.LIB D:\PS\ADLIB\AD_RELL.LIB
.PRINT AC VDB(5) VDB(56)
.PROBE
VIN 1 0 AC 1E-3
VCC   52  0  +15V
VEE   53  0  -15V
* ---------- V(5) = idealized RIAA frequency response ----------------
*
* Uses Laplace feature of PSpice Analog Behavioral option
* for frequency response reference.
* ENORM = ideal U1 DC gain = 1+(R1/R3) Use ideal values for R1, R3
* T1 - T3 are time constants desired (in µs).
* Input = node 1, Laplace Output = node 5
.PARAM ENORM  = {490.7}
.PARAM T1 = {3180}       ; Reference RIAA constants, do not alter!
.PARAM T2 = {318}        ; Reference RIAA constants, do not alter!
.PARAM T3 = {75}         ; Reference RIAA constants, do not alter!
*
ERIAA 5 0 LAPLACE {ENORM*V(1)}={(1+(T2*1E-6)*S)/((1+(T1*1E-6)*S)*(1+(T3*1E-6)*S))}
RDUMMY5 5 0 1E9
*
* -------------------------------------------------------------------
*
*     (+) (-) V+  V-   OUT
XU3   1  21  52  53   55 AD845
*  Active values          Theoretical values
R1 55  21  97.6K           ; 97.9k
R2 21  8   7.87K           ; 7.8931563k
C1 55  8   30NF            ; 30nF
C2 21  8   10.3NF          ; 10.2881nF
R3 21  0   200             ; 199.9148
C3 55  100 10E-6
R6 100 0   100K
R5 100 56  499
C4 56  0   3.0000E-9
.END
```

Figure 6-13: An example PSpice circuit file that uses the Laplace feature for ideal RIAA response comparison

The 1 kHz gain of this circuit can be calculated from Eq. 6-6. For the values shown, the gain is just under 50 times (≈34 dB). Higher gains are possible by decreasing R_3, but gains >40 dB may show increasing equalization errors, dependent upon amplifier bandwidth. For example, R_3 can be 100 Ω for a gain of about 100 times (≈40 dB). Note that if R_3 is changed to 100 Ω, C_4 should also be changed to 1.5 nF, to satisfy Eq. 6-9.

Dependent upon the amplifier in use, this circuit is capable of very low distortion over its entire range, generally below 0.01% at levels up to 7 V rms, assuming ±15 V supplies. Higher output with ±17 V supplies is possible, but will require a heat sink for the AD845. U2 is an optional unity-gain buffer useful with some op amps, particularly at higher gains or with a low-Z network. But this isn't likely to be necessary with U1 an AD845.

For extended low-frequency response, C_3 and R_6 are the large values, with C_3 preferably a polypropylene film type. If applied, the alternate values form a simple 6 dB per octave rumble filter with a 20 Hz corner. As can be noted from the figure's simplicity, C_3 is the only dc blocking capacitor in the circuit. Since the dc circuit gain is on the order of 54 dB, the amplifier used must be a low offset-voltage device, with an offset voltage that is insensitive to the source. Since these preamps are high gain, low level circuits (≥50 dB of gain at 50 Hz/60 Hz), supply voltages should be well regulated and noise-free, and reasonable care should be taken with the shielding and conductor routing in their layout.

Alternately, an inexpensive ac-coupled form of this circuit can be built with higher bias current, low noise bipolar op amps, for example the OP275, I_B = 350 nA(max), which would tend to make direct coupling to a cartridge difficult. This form of the circuit is shown in Figure 6-14, and can be used with many unity gain stable bipolar op amps.

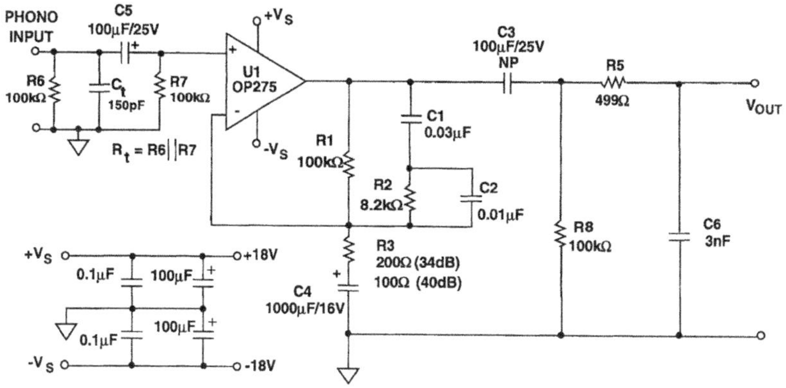

Figure 6-14: An ac-coupled active feedback moving magnet RIAA preamp

Here input ac coupling to U1 is added with C_5, and the cartridge termination resistance R_t is made up of the R_6-R_7 parallel equivalent. R_3 of the feedback network is ac-grounded via C_4, a large value electrolytic. These measures reduce the dc offset at the output of U1 to a few mV. Nearest 5% values are also used for the network components, making it easily reproducible and inexpensive. C_3 is a nonpolar electrolytic type, and the R_3-C_4 time constant as shown provides a corner frequency of ≤1 Hz at the 34 dB gain.

Frequency response of this version (not shown) isn't quite as good as that of Figure 6-11, but is still within ±0.2 dB over 20 Hz–20 kHz (neglecting the effects of the low frequency roll-off). If a tighter frequency response is desired, the N1 network values can be adjusted. With a higher rated maximum supply voltage for the OP275, the power supplies of this version can be ±21 V if desired, for outputs up to 10 V rms.

There is another, very useful variation on the actively equalized RIAA topology. This is one that operates at appreciably higher gain and with lower noise, making it suitable for operation with higher output moving coil (MC) cartridges. In this design example, shown in Figure 6-15, the basic circuit is used is quite similar to that of Figure 6-11. The lower R_t and C_t values shown are typical for moving coil cartridges. They are of course chosen per the manufacturer's recommendations (in particular the resistance).

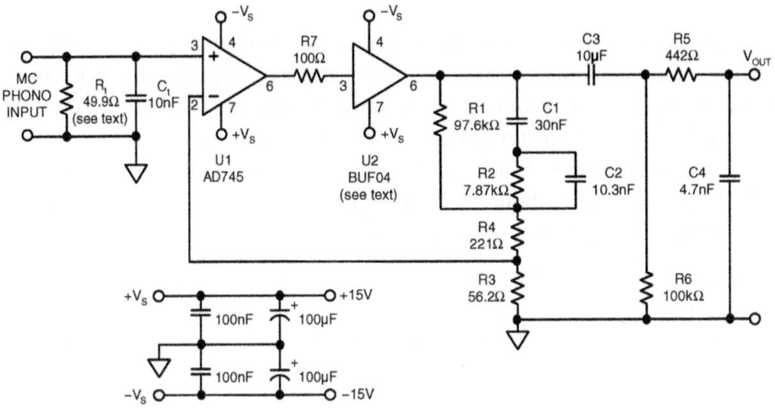

Figure 6-15: A low noise dc-coupled active feedback RIAA moving coil preamp with 45 dB of gain

To make it suitable for a high-output MC cartridge, a very low-noise FET op amp is used for U1, the AD745. The AD745 is stable at a minimum gain of five times, as opposed to the unity-gain stable op amps of the prior examples. This factor requires a modification to gain resistors R1-R3. This is the inclusion of an extra resistor, R4. With the ratio shown, R3 and R4 form a 5/1 voltage divider for the voltage seen at the bottom of network N (the R1-R2-C2 node). This satisfies U1's gain-of-five stability requirement.

In this gain setup, R3 is still used for the gain adjustment, and R1-R2-C1-C2 still form the basic N1 RIAA network. With R4 used, Eq. 6-9 is used to calculate the T4 time constant. With C4 chosen as a standard value, R5 is then calculated. With these N1 network values and a 45 dB 1 kHz gain, R3 is 56.2 Ω, which is still suitable as a low noise value operating with either an AD745 or an OP37 used for U1.

Some subtle points of circuit operation are worth noting. The dc gain of this circuit is close to 1800, which can result in saturation of U1 if offset isn't sufficiently low. Fortunately, the AD745 has a maximum offset of 1.5 mV over temperature, making the output referred offset always less than 3 V. While this may limit the maximum output swing some due to asymmetrical clipping, 5 V rms or more of swing should be available operating from ±15 V supplies. Coupling capacitor C3 decouples the dc output offset at U2, so any negative consequences of dc-coupling the U1 gain path are minimal.

For minimal loading of U1 and maximum linearity at high gains, the unity-gain buffer amplifier U2 is used, a BUF04. The BUF04 is internally configured for unity-gain operation, and needs no additional components. Note that this buffer is optional, and is not absolutely required. Other buffer amplifiers are discussed later in this chapter.

This Figure 6-15 circuit was analyzed with PSpice using the Laplace comparison technique earlier described, and the results are displayed in Figure 6-16. As was true previously, the vertical scaling of this

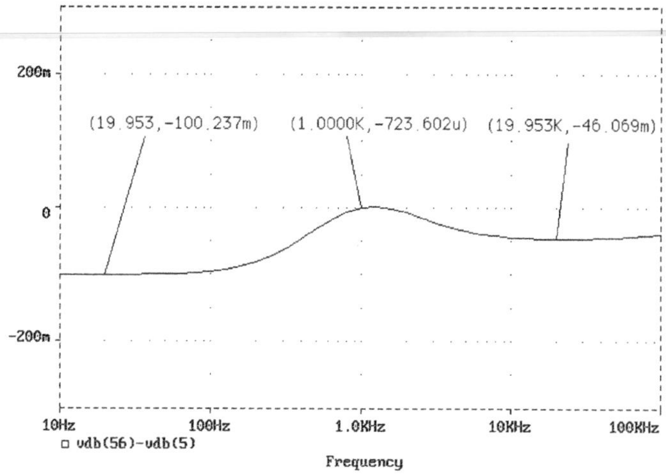

Figure 6-16: Relative error (B) versus frequency for dc-coupled active feedback RIAA moving coil preamp, gain of 45 dB (simulation)

display is very sensitive; ±300 mB (or ±0.3 dB). Thus, placed in context, gain errors relative to 1 kHz over 20 kHz–20 kHz are extremely small, ≈0.1 dB. Lab measurements of the circuit were also consistent with the simulation. Of course in terms of audible effects, errors of ±0.1 dB or less aren't likely to be apparent.

Distortion/noise measurements of the circuit are essentially dominated by noise (as opposed to actual distortion) measuring ~0.01% THD + N or less, over output levels ranging from 0.5 to 5 V rms, from 20 Hz–20 kHz. Of course, as with any high gain circuit, layout and lead dress into the circuit are extremely critical to noise, and must be arranged for minimum susceptibility. Supply voltages must be low in noise, and well regulated.

This exercise has illustrated both the basic design process of the active RIAA equalizer, as well as a convenient SPICE analysis method to optimize the design for best frequency response. It is not suggested that the exact network values shown of the examples are the only ones suitable. To the contrary, great many sets of values can be used with success comparable to that shown above.

This final active equalizer circuit example is the best of the bunch, and has a virtue of being easily adapted for other operating conditions; i.e., higher gain, other networks, and so forth. For example, note that even lower noise MC operation is possible, by using the $\leq 1\ nV/\sqrt{Hz}$ AD797 for U1, and scaling the N1 RC components further downward. This will have the desirable effect of making R3 lower than 50 Ω, which minimizes the R1–R4 network's noise. Note that gains of 50 dB or more are also possible, suitable for very low output moving coil cartridges (given suitable attention to worst-case U1 offsets).

Passively Equalized RIAA Preamp Topologies

Another RIAA design approach is the so-called *passively equalized* preamp (see Reference 11). This topology consists of two high quality, wideband gain blocks, separated by a three terminal passive network, N (N can be either network N1 or N2). The gain blocks are assumed very wide in bandwidth, so in essence the preamp's entire frequency response is defined by the passive network, thus the name passively equalized.

A circuit topology useful for such RIAA phono applications is shown in Figure 6-17. This circuit consists of two high-quality wide bandwidth gain blocks, U1 and U2, as discussed above. Selection of these amplifiers and their operating conditions optimizes the preamp for gain, noise, and overload characteristics. The circuit can be set up for either MM or MC operation by simple value changes and op amp selection.

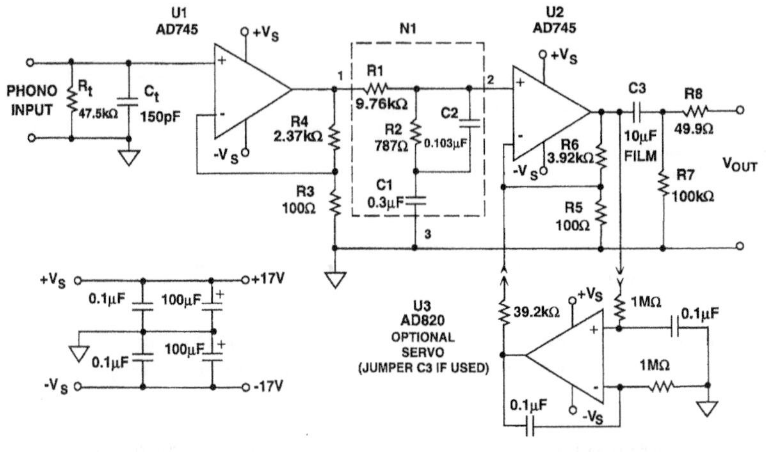

Figure 6-17: A passively equalized RIAA preamp with 40 dB gain

The gain stages are set up for the required total gain, via R_4-R_3 and R_6-R_5. In general, the total 1 kHz gain of this circuit G is:

$$G = 0.101 \bullet \left[1 + (R_4/R_3)\right] \bullet \left[1 + (R_6/R_5)\right] \qquad \text{Eq. 6-11}$$

The op amp gain blocks could be made identical for purposes of simplicity but are not necessarily so for the following reasons. A preamplifier topology such as this must be carefully optimized for signal-handling capability, both from an overload standpoint and from a low-noise viewpoint. Stage U1 is chosen for a gain sufficiently high that the input-referred noise will be predominantly due to this stage and the cartridge, but not so high that it will readily clip at high level high frequency inputs. Amplifiers with a ≈10 V rms output capability allow U1 to accept ≈400 mV rms at high frequencies using ±18 V supplies, while still operating with useful gain (about 25 times).

The gain of the two blocks are set by R_4-R_3 and R_6-R_5, as defined by Eq. 6-11. The gain values shown yield a 1 kHz gain that is the product of the U1-U2 stage gains (24.7 times 40.2), times that of the interstage network N (0.101). This yields an overall 40 dB 1 kHz gain. Other gains are realized most simply by changes to R_5 or R_3.

As previously noted, a passively equalized preamplifier such as this must be carefully optimized both from an overload standpoint and from a low noise viewpoint. Stage U1 is chosen for a gain sufficiently high that the input-referred noise will be predominantly due to this stage (and the cartridge, when connected), but not so high that it will readily clip at high level high frequency inputs. To aid this objective, maximum supply voltage and a high output capability amplifier should be used for U1.

Note that U1 operates at relatively high gain, but it needn't be unity gain stable. Decompensated low noise op amps such as the OP37 and the FET input AD745 will provide best signal/noise ratio here. For other FET-input types, the AD845, as well as the OP17 family types, will also yield good performance, but with higher noise levels.

In general, the preceding factors dictate that gain distribution between U1 and U2 be LOW/HIGH from an overload standpoint, but HIGH/LOW from a noise standpoint. Practically, these conflicting requirements can be mitigated by choosing the highest allowable supply voltage for U1, as well as a low noise device. Because of nearly 40 dB loss in the network N at 20 kHz, the output overload of the circuit will be noted at high frequencies first. With the gain distribution shown, the circuit allows a 3 V rms undistorted output to 20 kHz with ±15 V supplies, or more with higher supply voltages.

The equalization network N following U1 should use the lowest impedance values practical from the standpoint of low noise, as the noise output at Pin 2 of the network is equivalent to the input referred noise of A2. The network of Figure 6-17 uses the "N1" RC values of R_1-R_2-C_1-C_2 of Figure 6-9a. As noted, scaling can be applied to either network of Figure 6-9 for component selection, as long as the same ratios are maintained.

Noise in amplifier U2 is less critical than U1 at low frequencies, but is still not negligible. A low voltage noise device is very valuable to the U1 and U2 positions, as is a relatively low input current noise. If extremely low noise performance is sought, such as for a moving coil preamp, the N1 values can be reduced further, and R3 be lowered for lower noise and additional gain. For example, a 45 dB gain preamp could be realized by just dropping R3 to 56.2 Ω, and using an OP37 for U1.

As mentioned before, a low bias current device is appropriate to U1 using bipolar amplifiers. With a 100nA or less bias current device, direct coupling to a moving magnet phono cartridge is practical. For example, the 80 nA (maximum) bias current of the OP37 will induce only an additional 80 µV–160 µV input voltage offset at U1 for a typical 1 kΩ–2 kΩ cartridge resistance. For lower dc resistance MC cartridges, this will be much less of course. Similarly, the bias current induced offset voltage of U2, from the 10 kΩ dc resistance of R_1 will also be low relative to the amplified offset of U1. As a result, the worst-case overall output dc offset using two AD745s can be held to under 2 V for a 40 dB gain, allowing a single C_3 coupling capacitor for dc blocking purposes.

Frequency response of this passively equalized preamp tends to be better than that of the active versions, because of less interaction with the amplifier(s) as compared to the active preamps. It can approach the inherent accuracy of the network components in the audio range, with potentially greater errors at higher frequencies.

Figure 6-18 illustrates this point, in a simulation of the Figure 6-17 circuit using the OP37 models. The midband error is on the order of ±0.02 dB with the N1 network composite values. For practical purposes

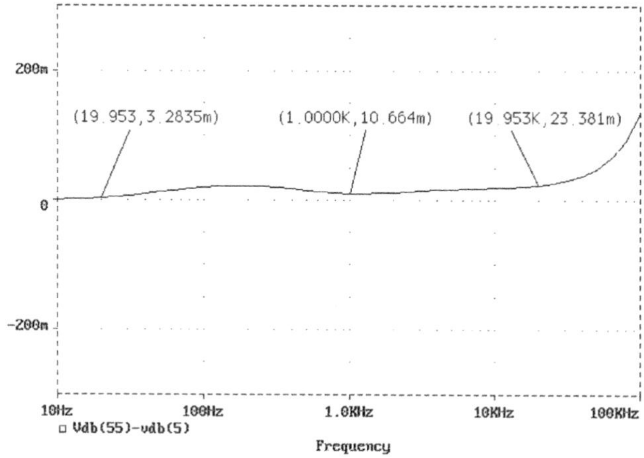

Figure 6-18: Relative error (B) versus frequency for passively equalized RIAA preamp, gain of 40 dB (simulation)

then, the frequency response errors of this circuit will be governed by the tolerances of the network components used within it.

This circuit also can be optionally adapted to servo control of the output offset. This is accomplished by deleting coupling capacitor C_3, substituting a jumper in its place, and using the noninverting servo integrator U3 around stage U2. This is shown as an option within Figure 6-17. A general-purpose noninverting servo can be used for U3, along with a low-offset op amp, such as the AD820, or the OP97.

References: RIAA Phono Preamplifiers

1. RIAA, "Standard Recording and Reproducing Characteristic, Bulletin E1," November 6, 1978, RIAA, 1 E. 57th St, NY, NY, 10022.
2. IEC, "Publication 98 (1964), Amendment #4," September 1976.
3. F. Bradley, R. McCoy, "Driftless DC Amplifier," **Electronics**, April 1952.
4. G. Korn, T. Korn, **Electronic Analog and Hybrid Computers, 2nd Ed.**, McGraw-Hill, 1972, ISBN 0-07-035363-8.
5. D. Stout, M. Kaufman, **Handbook of Operational Amplifier Circuit Design**, McGraw-Hill, 1976, ISBN 0-07-061797-X.
6. S. Lipshitz, "On RIAA Equalization Networks," **JAES**, Vol. 27 #6, June 1979, pp. 458–481.
7. P. Baxendall, "Comments on 'On RIAA Equalization Networks'," **JAES**, Jan/Feb 1981. See also: S. Lipshitz, "Author's Response."
8. Walt Jung, "The PAT-5/WJ-1 Equalization Errors" (Letters), **The Audio Amateur**, issue 3/78, pp. 49–53.
9. S. Lipshitz, W. Jung, "A High Accuracy Inverse RIAA Network," **The Audio Amateur**, issue 1/80, pp. 22-24.
10. Walt Jung, "SPICE Technique Compares Frequency Responses," **EDN**, November 25, 1993, pp. 188 and 190.
11. Walt Jung, "Topology Considerations for RIAA Phono Preamplifiers," **67th AES Convention**, November 1980, preprint #1719.
12. G. Erdi, et al, "Op Amps Tackle Noise and for Once Noise Loses," **Electronic Design**, December 20, 1980, pp. 65–71.
13. Walt Jung, "Audio Preamplifiers, Line Drivers, and Line Receivers," within Chapter 8 of Walt Kester, **System Application Guide**, Analog Devices, Inc., 1993, ISBN 0-916550-13-3, pp. 8-1 to 8-100.
14. W. Jung, R. Marsh, "Picking Capacitors," Parts 1&2, **Audio**, February & March 1980.
15. B. Duncan, "With a Strange Device," multiple part series on capacitors, **Hi-Fi News And Record Review**, April-November 1985.
16. B. Duncan, M. Colloms, "Pièce de Résistance, Parts 1–3," **Hi-Fi News And Record Review**, March, April 1987 (Duncan); June 1987 (Colloms).

Audio Line Level Stages

Audio line level stages represent an intermediate level in dynamic range for practical audio circuits using modern IC devices. Line level amplifying stages generally work with single-ended or balanced input/output signal levels of 1 V–10 V, and at medium levels of power. This section discusses some basic types of audio line stages which are:

Line receivers—including line receiver stages that accept single-ended or balanced line level signals with maximum noise immunity, and provide scaled outputs for further processing.

Line amplifiers—including amplifiers that scale a received signal in single-ended form and feature low distortion designs.

Line drivers—including single-ended and balanced drivers capable of driving appreciable output levels in terms of voltage, swing, current levels, and/or difficult loads, such as capacitive lines.

Some general concepts of line driving and buffer amplifier design have been covered previously, with emphasis on video applications (see References 1–3). Some of the material in this section continues and expands on those themes with audio line-receiver and line-driver discussions in a wide variety of applications. Video applications for line driving and receiving is discussed in detail within Section 6-2.

Audio transmission systems, unlike their video counterparts, do *not* use terminated transmission lines as a rule, so long transmission lines usually appear capacitive. Therefore, the concepts of capacitive load isolation are also important to audio drivers. In general, when building practical audio circuits of any type, "housekeeping" rules of layout and bypassing are strongly recommended, particularly so for audio buffer and line driver circuits. They are discussed in further detail in that section.

The function of sending/receiving audio signals between various system components has traditionally involved trade-offs of one form or another. Fully differential or *balanced* transmission systems are best at rejecting low frequency and RF noise, so they are used for highest performance, and are discussed in some detail following.

A typical audio system block diagram using differential or balanced transmission is shown in Figure 6-19. In concept, a balanced transmission system like this could use several input/output coupling schemes within the driver and receiver. Some major points distinguishing coupling method details are discussed briefly, before addressing actual circuits.

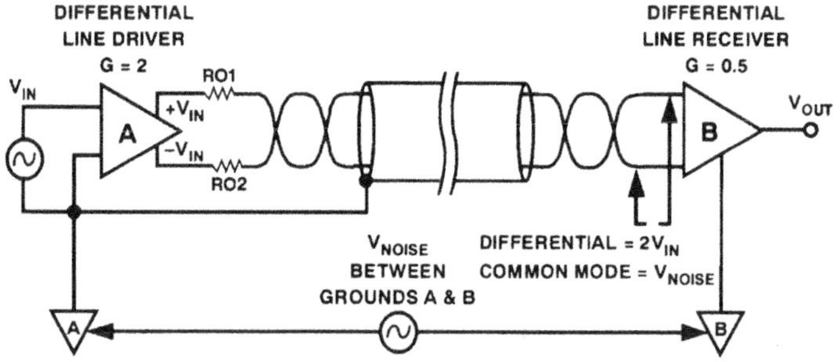

Figure 6-19: An audio balanced transmission system

Transformers (see References 4–7) have been a traditional audio line coupling element. They can be used at either input or output stage. They also have well-known problems with noise pickup, frequency response, distortion, and operating level. While these problems are soluble to some degree, answers are usually costly. Nevertheless, transformers are unexcelled in notable areas.

The single most outstanding virtue of transformer operation lies in the ability to isolate while transmitting the signal, which shows up in two regards. The first of these is that transformers, which transmit an audio signal between the end terminals of an isolated winding, thus *galvanically isolate* driver and receiver stages. This is accomplished at common-mode voltage (ground difference) levels up to the breakdown potential of the windings, allowing signals to be transmitted across very high ground potential differences (tens or hundreds of volts). This feature is very difficult to achieve with solid-state circuitry. Secondly, suitably designed transformers can have very high common-mode rejection (CMR) over the audio range, ≥100 dB in some cases, a factor basically intrinsic to their nature. Some transformer-isolated stages are described in this section.

In practice the general system of Figure 6-19 can use either transformer-based or active stage coupling to the line, at either end. The goal for either approach is to reproduce a final signal V_{OUT} equal to V_{IN}, while rejecting noise between grounds A and B by a factor of 80 dB–100 dB. Typically, a unity gain (overall) design uses a balanced line drive of $\pm V_{IN}$, followed by a receiver gain "G" of ½, which maximizes the receiver CM range.

A point worth noting that the ± voltage drive to the line need *not* be exactly balanced to reap the benefits of balanced transmission. In fact the drive can be asymmetrical to some degree, and the signal will still be received at V_{OUT} with correct amplitude, and with good noise rejection. What *does* need to be provided is two well-balanced line-driving impedances, R_{O1} and R_{O2}. Also, in conjunction with these balanced drive impedances, the associated (+) and (–) receiver input impedances should also be equal. The technical reasons for this will be apparent shortly.

Audio Line Receivers

A brief review of the topologies and application points of audio line receivers helps in understanding their evolution, and more importantly, how their audio performance differs with topological changes. Figure 6-20 is a diagram of a classic 4-resistor differential amplifier. This general circuit is also known as the most simple instrumentation amplifier form, even though its performance as an in amp has severe limitations. Within audio applications, this and related circuits are called "line receivers" for the sake of brevity. Various in amp type topologies have developmental histories dating from the late sixties up to and including today's modern in amp ICs (see References 8–12 and Chapter 2).

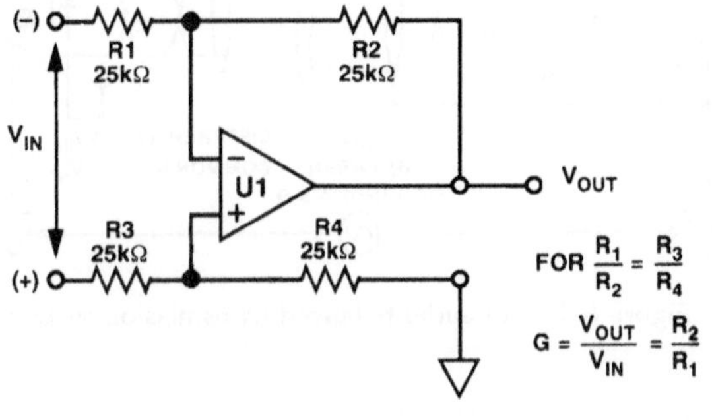

Figure 6-20: A simple line receiver using a 4-resistor differential amplifier

In today's professional audio world, signals by and large get transmitted in balanced mode (Figure 6-19). This fact is simply due to the much greater noise immunity of this method, vis-à-vis the more simple, but highly noise-susceptible single ended method.

Yet, even within the professional audio world there is no real unanimity on signal driver and receiver circuits for use within balanced circuits—they take on many forms and have differing performance, and a wide variety of circuits find use. Before getting into actual receiver circuitry, it is helpful to take a brief look at some problems impacting circuit performance in terms of common mode (CM) noise susceptibility. This will illustrate how careful hardware choices lowers system cost/size, and maintains excellent performance. Conversely, simple receivers can also be used, for modest performance.

Source-Load Interactions within Balanced Systems

Some recent attention has focused on the general problem of noise susceptibility in audio system interfacing (see References 13 and 14). The discussions below are concerned with how balanced system drivers and receivers interact fundamentally to produce undesired side effects of noise susceptibility. Suggestions for practical solutions are then offered.

In most simple form, a balanced audio transmission system consists of a differential output driver, an interconnecting cable, and a differential input receiver, such as shown in Figure 6-19 (again). The driver produces nominally equal and out-of-phase output signals, with some characteristic (and matched) source impedances, R_{OUT1} and R_{OUT2}. As will be seen, from a noise susceptibility standpoint, it is highly desirable that these two impedances be well balanced, i.e., matched. The driver is connected to the input end of a balanced transmission line, typically a shielded twisted pair. At the opposite end of this line, a differential input receiver receives the balanced signal, and (ideally) rejects the CM noise voltage V_{NOISE}.

The design of both the driver and the receiver has great influence upon how well the overall scheme works in transmitting a noise-free audio signal from driver to receiver. References 14 and 15 discuss different driver and receiver types, active and passive. These papers bring out the inherent degradation in noise sensitivity active receivers can trigger, if they do not have input characteristics which are an appropriate complement to the system driving impedances.

From a noise introduction point of view, the balanced transmission system under discussion can be analyzed as a bridge circuit, as shown in Figure 6-21. Here the two source resistances R_{OUT1} and R_{OUT2} correspond to the output resistances of the differential driver voltage sources. Similarly, input resistances R_{IN1}

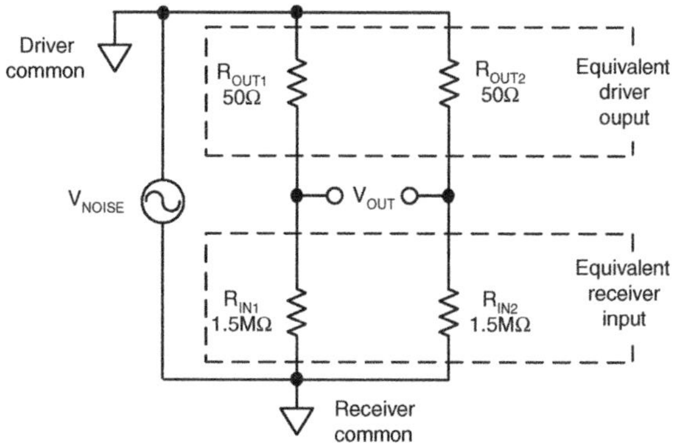

Figure 6-21: A conceptual driver/receiver diagram of a balanced line audio system with key impedances and CM noise

and R_{IN2} correspond to the input resistances of the differential receiver. V_{OUT} represents the output of this bridge, which is due solely to the bridge mismatching and CM noise V_{NOISE}.

Such a bridge as Figure 6-21, when *maximized* for output sensitivity, will produce a differential output V_{OUT} which is highest as a function of element unbalance when all four resistances are of the same order.

The following general expression illustrates the intrinsic common-mode rejection (CMR) of this bridge as a function of the resistance values and their deviation, K_R:

$$CMR(dB) = 20 \bullet \log_{10}\left[\left(1 + \left(R_{IN}/R_{OUT}\right)\right)/K_R\right] \quad \text{Eq. 6-12}$$

Some sample calculations with this relationship show that *CMR* is a *minimum* for a given change in K_R (the total resistor deviation expressed in fractional form) when $R_{IN} \approx R_{OUT}$. A CMR minimum is simply another way of saying that the bridge is most sensitive to the excitation voltage V_{NOISE} when $R_{IN} \approx R_{OUT}$. To place this in perspective, a conventional instrumentation bridge operates thusly, with all four arms nominally equal. This yields the highest sensitivity to the applied voltage (see Chapter 4).

On the other hand, CMR is maximum and bridge sensitivity is *minimized,* when the upper and lower arm resistances differ widely. This improves substantially as R_{IN} becomes >> R_{OUT}. Or, within an audio system, as the driver R_{OUT} is by design made much less than the receiver R_{IN}. With the example values above, there is a 1/30,000 ratio between the R_{OUT}/R_{IN} upper/lower elements. This factor makes relatively high percentage changes in either the upper (or the lower) arm resistances a somewhat harmless phenomenon. In other words, small R_{OUT} or R_{IN} changes will then have little CMR effect upon the output.

For example, taking the Figure 6-21 values and assuming a 10% change in R_{OUT}, will produce an output which is about 110 dB down from the noise voltage V_{NOISE}. By contrast, if all the bridge values were to be equal, the same 10% deviation would produce an output only 26 dB down. Note that there are two control point towards this. One can *lower* R_{OUT}, for a given R_{IN}, and increase CMR. Or, one can achieve the same effect by *increasing* R_{IN}, for a given R_{OUT}. This makes the point that *a high ratio between R_{IN} and R_{OUT}* aids in maintaining high CMR, as is shown by Figure 6-22.

In a real transmission system, there will be inevitable noise potentials developed between the respective driver and receiver chassis common points, since they are located separately and are powered with different power sources. The resulting noise voltage can be predicted with the aid of Figure 6-22. As a minimum, a good system should maintain an R_{IN}/R_{OUT} of at least 1000, with 10 k or more a goal. Under such conditions, with a bridge unbalance of 10%, this will still allow a theoretical CMR of 80 dB to 100 dB (see center column, with cited examples shown in boldface).

R_{IN}/R_{OUT}	CMR(dB) for K_R = 0.1	CMR(dB) for K_R = 0.01
10	40.8	60.8
100	60.1	80.1
400	**72.1**	92.1
1k	**80**	100
10k	**100**	120
30k	**109.5**	129.5
100k	120	140

Figure 6-22: High R_{IN}/R_{OUT} minimizes sensitivity to CM noise, bridge imbalance

As noted, dependent upon the bridge impedance-related sensitivity, some fraction of the CM V_{NOISE} appears as V_{OUT}. The basic process of the conversion of the CM noise voltage into a differential voltage is called *mode conversion*. It is important to understand that mode conversion can only be prevented, not fixed after the fact. Once the noise voltage appears as a differential signal, no receiver can distinguish it from a valid signal.

Finally, it is important to realize that what has been discussed thus far addresses the most basic portion of this system and the impact on CMR. The line receiver circuitry itself obviously also has a big influence, as it determines R_{IN}. This is discussed next.

The Simple Line Receiver

The simple line receiver circuit of Figure 6-23 below uses four matched resistors and an op amp for gain. Such bridge-based difference amplifiers are *critically* dependent upon the resistor ratio matching for good performance, an enormously important point. The amplifier can also be critical, but most practical limitations of this topology arises from *un*-balancing of the R_1-R_4 bridge (for either/both ac and dc).

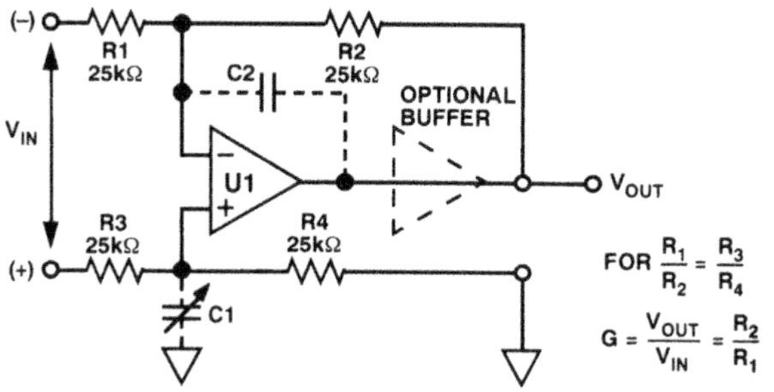

Figure 6-23: A simple line receiver with optional HF trim and buffered output

The circuit appears somewhat trivial, as the minimum ingredients are four matched film resistors and a good audio op amp. While this works functionally, how well it works in rejecting noise is another thing. C1 and C2 are optional, and can be used to trim HF CMR. Also optional is the use of an in-the-loop unity-gain buffer (more below).

A main purpose of this circuit and all line receivers is to reject CM noise, as discussed above. But even with a high quality op amp for U1, noise rejection is only as good as the resistor matching. More precisely, the resistor *ratios* R_1/R_2 and R_3/R_4 must match extremely well to reject noise (the absolute values are relatively unimportant).

With care, picking four 1% resistors from a same-vendor, same-batch lot is a step that can yield ratio matching of, say, 0.1% total error, and will achieve a common-mode rejection (CMR) of 66 dB. Four 1% tolerance resistors with just one off by 1% will yield about 46 dB CMR. In general, the worst-case CMR of a circuit of this type is:

$$\mathrm{CMR\,(dB)} = 20 \bullet \log_{10}\left[\left(1+\left(R_2/R_1\right)\right)/4K_R\right] \qquad \text{Eq. 6-13}$$

where here "K_R" is the *individual resistor tolerance* in fractional form. This form of the expression is most useful for cases using four discrete resistors (see Reference 8). More likely, a single component network with a *net matching tolerance* of K_R would be used for this function, in which case the expression then becomes:

$$\mathrm{CMR\,(dB)} = 20 \bullet \log_{10}\left[\left(1+\left(R_2/R_1\right)\right)/K_R\right] \qquad \text{Eq. 6-14}$$

In either case, this assumes a significantly higher *amplifier* CMR, such as ≥100 dB).

Eq. 6-13 shows that the worst-case CMR due to tolerance build-up for four unselected 1% resistors to be much worse, 34 dB in fact. Clearly for high and stable noise rejection, circuits such as these need four single-substrate resistors, made/trimmed simultaneously. Networks using thick-film and thin-film technology are available from companies such as Caddock and Vishay-Ohmtek, in ratio matches of 0.01% or better.

Some simulations may bring this point of critical matching home more clearly. Figure 6-24 shows the effect of various dc-matching of a differential amplifier such as Figure 6-23, with a single resistor mismatch in R_1 of 0.1% (top trace), 0.01% (middle trace) and perfect matching (bottom). Also, this display of CM error versus frequency indicates a reactive imbalance for the perfect dc-balanced (bottom) case. This is due to an intentional capacitive mismatch in the circuit, representing stray capacitance imbalance.

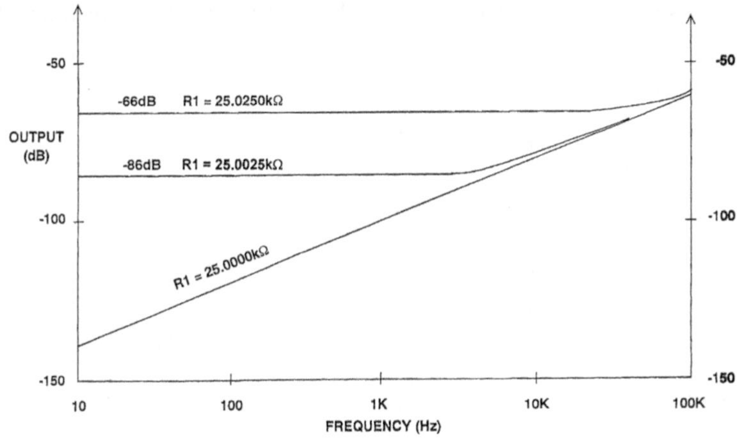

Figure 6-24: Simple line receiver CM rejection versus frequency for various R1 trims (simulation)

The effect of ac-matching on this differential amplifier using matched resistances, with C_2 matched to C_1 (10pF) results in essentially flat CM error versus frequency (not shown). In contrast, a 10% capacitive mismatch results in a CM degradation as low as 10 kHz. Clearly then, for wideband audio uses, the bridge ratio needs to be maintained for ac as well as dc, to achieve flat CMR versus frequency. Capacitances from the R_2/R_1 and R_4/R_3 nodes need to be balanced. In practice, this is best achieved with very low and/or balanced parasitic capacitances at C_1-C_2.

It is worthy of note that this circuit also has a highly desirable side property; that is, it *divides down* the input CM voltage. Thus it is inherently protected against overvoltage. In general, practical line receivers *require* some sort of input protection, for safe use in harsh environments, and to allow CM voltages to exceed the supply rails. The receiver gain-set resistors double in performing this function here. The working CM input range of Figure 6-23 is $(1+(R_3/R_4)) \times V_{CM(U1)}$, and differential input resistance is $R_1 + R_3$. Circuit gain isn't easily changed, because of the matched R_1–R_4 ratios.

Implementing the Simple Line Receiver Function

To offer a reasonably high impedance to the line, simple receivers such as these typically use input resistances of 25 kΩ or more. When working from 50 Ω sources, this allows a basic R_{IN}/R_{OUT} ratio of 500 (see previous discussion and Figure 6-22, again). Given a well-matched resistor network and low or balanced parasitic capacitances, suggested amplifiers for U1 are the AD711, AD744 (singles), and the AD712, AD746, OP249, OP275 (duals). With 10 kΩ–25 kΩ resistances, extremely low voltage noise in the amplifier isn't critical. High slew rate (SR) and output drive allows clean high frequency, high amplitude levels, and 600 Ω load capability. If low impedance loads need to be driven, output current of a standard op amp can be boosted with an in-the-loop unity gain buffer, connected between U1 and the load/feedback point. Devices such as the BUF04 or a (follower-connected) AD811 can serve well here (discussed in the later sections).

From a dc and ac trim/balance perspective, the Figure 6-23 topology is most effective with resistors and amplifier made simultaneously in a single monolithic IC. The ADI SSM2141 and SSM2143 are such ICs, characterized as low distortion, high CMR audio line receivers with net gains of unity (SSM2141), and 0.5 (SSM2143). The SSM2141 has resistors as shown in Figure 6-23, while the SSM2143 uses 12 kΩ/6 kΩ resistors.

In applying circuits of the Figure 6-23 type (or other topologies which resistively load the source), a designer must bear in mind that all *external* resistances added to the four resistances can potentially degrade CMR, unless kept to proportional value increases. To place this in perspective, a 2.5 Ω or 0.01% mismatch can easily occur with wiring, and if not balanced out, this mismatch will degrade the CMR of otherwise perfectly matched 25 kΩ resistors to 86 dB. These circuits are therefore best fed from balanced, low impedance drive sources, preferably 25 Ω or less.

Other Issues with the Simple Line Receiver

An application point that becomes relevant for large high performance systems with multiple balanced pair lines is the issue of receiver *load balance*. Ideally, an audio line receiver should exhibit equal ac loading at the two inputs. With the simple line receiver of Figure 6-23 (and all similar circuits), this goal isn't met—i.e., the basic circuit does *not* present balanced loading to the two input lines. It is important to note at this point that this is not a function of the devices used to implement the circuit, it is more a function of the architecture itself.

When Figure 6-23 is driven from complementary sources V_{IN} and $-V_{IN}$, the simple line receiver exhibits a property of unbalanced input currents in the R_1 and R_3 legs, due generally to feedback action. For the like values of Figure 6-23, the current in R_1 is three times that in R_3. Thus the inputs load the two input lines differently, as noted.

In large systems with multiple balanced transmission line pairs, the current imbalance in the input lines is potentially serious, as associated fields will not cancel as they do for completely balanced loading. Thus there is potential for crosstalk impairment in such systems using the simple line receiver topology.

On the other hand, while not optimum from a large system and/or line balance viewpoint, the simple line receiver is nevertheless useful in more modest situations. With resistors R_1–R_3 relatively high (20 k or more), it is adequate for small-scale or confined systems where I/O lines are relatively short, few in number, they are not cabled, and the source impedances are low. In such uses, devices like the SSM2141 and SSM2143 can serve well as efficient, single IC line receiver solutions.

Balanced Line Receivers

For highest performance uses, it is a key point that audio line receivers exhibit equal loading to the source at both inputs—i.e., they should be truly balanced. At least two topologies meet this criteria and are thus suited for professional use in balanced systems.

Balanced Feedback Differential Line Receiver

David Birt of the BBC has analyzed the simple line receiver topology and presented a modified and balanced form, as shown here in Figure 6-25 (see Reference 7). Here U1 uses an 4-resistor network identical to that of Figure 6-23, while a second feedback path from unity gain inverter U2 drives the previously grounded R_4 reference terminal. This has two basic effects overall; the input currents in the R_1–R_3 input legs become equal in magnitude, or balanced, and the gain of the stage is halved.

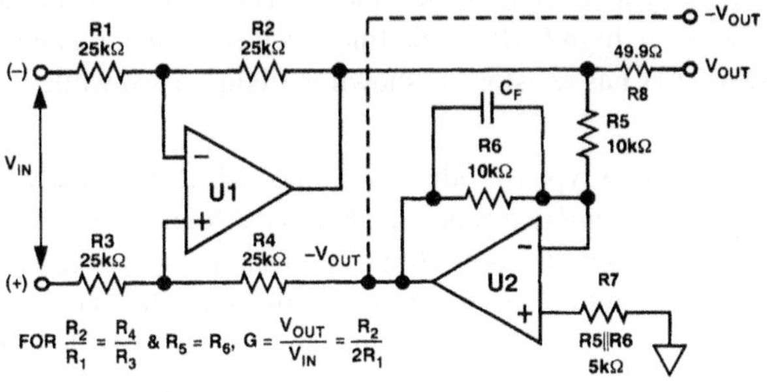

Figure 6-25: Balanced line receiver using push-pull feedback

Compared to Figure 6-23, and for like resistor ratios, the Figure 6-25 gain from V_{IN} to V_{OUT} is ½, or a gain of –6 dB (0.5) as shown. However it also offers an optional complementary output from U2, $-V_{OUT}$. Like Figure 6-23, the gain of this circuit is not easily changed, as it also involves precise resistor ratios.

Because of the two feedback paths, this circuit holds the inputs of U1 at a null for differential input signals. However CM signals are still seen by U1, and the CM range of the circuit is $(1+(R_3/R_4)) \times V_{CM(U1)}$. Differential input resistance is $R_1 + R_3$. As can be noted from the figure, the circuit can be broken into a simple line receiver (left), plus an inverter (right). Existing line receivers like Figure 6-23 can be converted to a balanced topology by adding inverter U2. A performance example of this is discussed below.

Alternate Balanced Line Receivers

Other instrumentation amplifier types can achieve the goal of fully balanced input loading, but may not be desirable for other reasons. For example, there are standard in amp circuits not shown here which use either two or three amplifiers and have properties of high input impedance, due to the use of noninverting inputs (see References 8–11). The drawbacks of these topologies as audio line receivers lie in limited gain and CM range. Also, importantly, they require four resistors beyond those for gain, just for input overload protection. Since these resistors also influence gain and CMR, they must also be precision ratio matched types. As a net result, workable audio line receivers using these in amps aren't really highly practical (eight or more matched resistors, plus two or three op amps).

An "All Inverting" Balanced Line Receiver

Figure 6-26 is an elegantly attractive topology that seems well-suited to audio line receiver use. Using all amplifiers as inverting stages, this circuit can be configured for very high CM voltage range and high input resistance. With the resistor ratios matched as shown, the CMR of this circuit can be better than the others for a given resistor match, since both amplifiers see no CM voltage. The CM range of this circuit is set as $(R_1/R_2) \times V_{OUT(MAX)U1}$. The differential input resistance is $R_1 + R_3$.

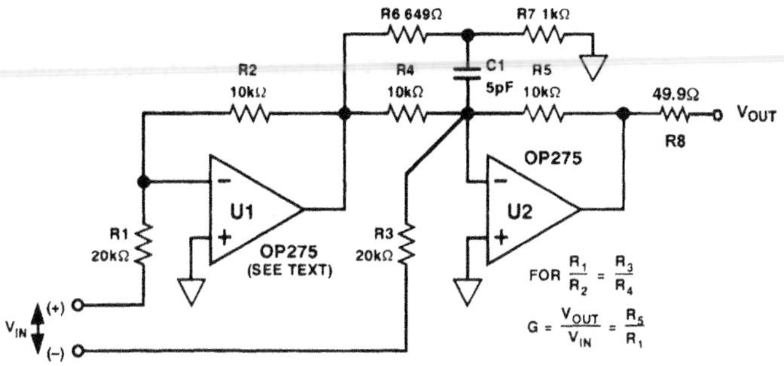

Figure 6-26: "All Inverting" balanced line receiver

The circuit has the unusual and desirable property of single resistor gain adjustment via R_5, *without* any CMR interaction. Gain can also range from below to above unity, making it flexible in that regard. As shown, it is driven with a balanced signal, but note that it can also be driven with single-ended sources at either the (+) or (–) terminal, with no gain interaction from the opposite input port, due to the use of inverting amplifiers. Multiple inputs can be summed, with additional ratio matched input resistor pairs (not shown). In this example gain is set at 0.5, consistent with general line receiver system requirements.

This circuit is also well known in basic form (see References 8–11). Note, however, that in this configuration, optional phase lead compensation is used to enhance high frequency CMR. A small capacitor shunting R_4 with a value chosen to compensate for the gain-bandwidth of U1 compensates for the lag through U1, and maximizes phase matching of the ± CM signals at U2. However, for op amps with gain bandwidths above a few MHz and practical resistor values, this can result in difficult-to-control small capacitor values.

The R_6-R_7-C_1 tee network reduces the effective value of C_1, by dividing the applied voltage. A nominal division ratio can be approximated by this expression:

$$K_C = 1/\left(2\pi\ BW_{(U1)}R_4C_1\right) \qquad \text{Eq. 6-15}$$

where K_C is the division ratio of R_6-R_7. For this example, with BW(U1) about 5 MHz (the closed loop bandwidth of U1), K_C is about 0.6, making C_C effectively about 3 pF. Circuit parasitics, loading effects and part variations make this inexact; however, once nominal compensation for a given layout and devices is achieved, a 30 dB CMR improvement in 10 kHz CMR is possible (vis-à-vis no phase compensation). This trim network isn't necessary to the circuit's basic function, but is nevertheless useful for audio applications.

Performance of Balanced Line Receivers

The balanced line receiver configurations of Figure 6-25 and Figure 6-26 were tested for CM performance, with common conditions of G = 0.5, Vs = ±18 V, and a 10 V rms input sweep, 20 Hz–50 kHz, filter bandwidth of 80 kHz. The Figure 6-25 topology was implemented with an SSM2141 for U1, with U2 an OP275 inverter, with C_F = 68 pF. Figure 6-26 was implemented with an OP275 and a resistor network matched to 0.005%, with the C_1 network trimmed as shown. These results are shown in Figure 6-27.

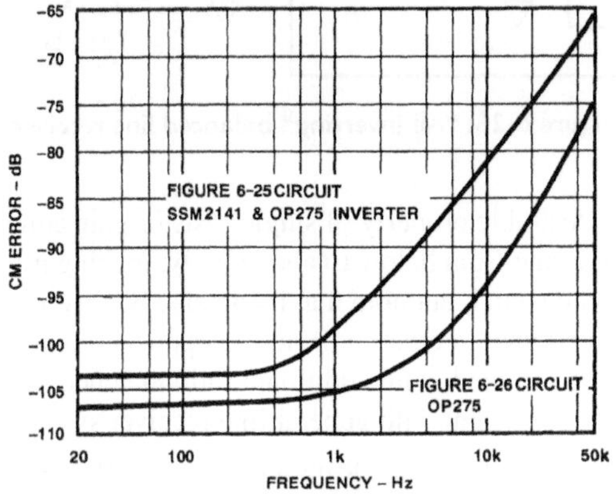

Figure 6-27: Balanced line receivers CM error versus frequency

Both circuits show excellent results, with ≤1 kHz CM errors of –100 dB or lower. The Figure 6-26 topology offers better results at the higher frequencies, perhaps due to the trimming technique used (not applicable to the Figure 6-25 circuit). In the worst case, the CM errors are no poorer than –80 dB at 10 kHz, still very good for an untrimmed circuit.

THD + N data was taken on both circuits and, while dominated by the noise floor at many levels, there are some differences worth noting between the two. Figure 6-28 shows THD + N performance of the Figure 6-25 SSM2141/OP275 circuit for loading conditions of 100 kΩ, successive input sweeps of 1, 2, 5 and 10 V rms, and ±18 V supplies. The lower level sweeps are noise dominated, while the 5 and 10 V sweeps show some distortion rise at high frequencies. Distortion of this circuit also rises with loading of 600 Ω (not shown).

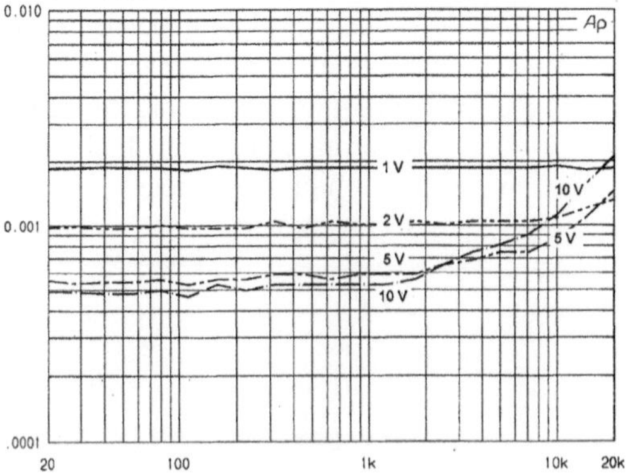

Figure 6-28: Balanced line receiver of Figure 6-25, THD + N versus frequency

The performance of the Figure 6-26 circuit for similar input drive and power supply conditions is shown in Figure 6-29, and for conditions of 600 Ω loading. These data indicate less loading and frequency dependence, due primarily to the OP275's higher slew rate, and greater available output drive into 600 Ω loads.

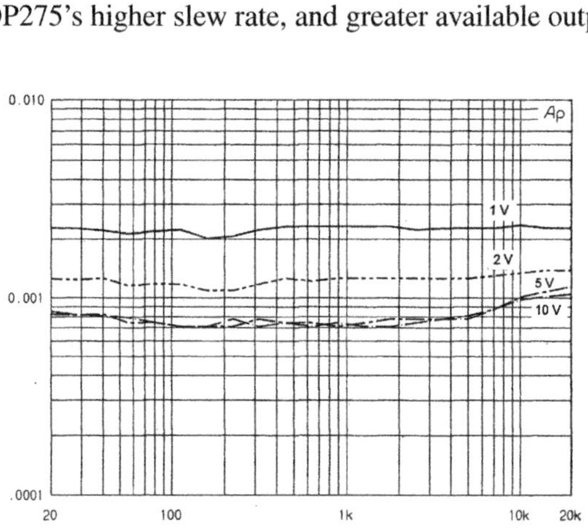

Figure 6-29: Balanced line receiver of Figure 6-26, THD + N versus frequency

In the all-inverting circuit of Figure 6-26, THD + N performance is much more limited by noise than actual distortion, over frequency. The circuit is a very flexible one, and can be set up for a variety of other op amps and input resistances, as well as the already mentioned single-resistor gain change operation.

The line receivers covered above offer good performance. However, with input resistances on the order of 25 kΩ, they can still be subject to CM errors due to driver impedance mismatches. This is not an issue that can be dealt with cleanly, as the designer of a line receiver circuit doesn't necessarily have any pre-knowledge of the worst-case driver impedance characteristics. So, to guarantee high CMR performance even in the instance of substantial driver impedance and/or mismatching, there are two possible solutions. One is to make the line receiver circuit input impedance as high as practical, which then allows good CM performance with impedance mismatches on the order of 10% (Figure 6-22). Alternately, a line receiver can utilize a line transformer, which offers high CM rejection and galvanic isolation. Both approaches are discussed next.

A Buffered Input Balanced Line Receiver

The circuit of Figure 6-30 represents an example of a classic three-op-amp instrumentation amplifier (in amp) topology, dressed up and optimized for use as an audio line receiver (see Reference 16). The use of FET input stage buffers in amplifiers U1A and U1B allows megohm-level bias resistance to be used for R_{IN1} and R_{IN2}, which greatly desensitizes this receiver against loading of the source and CM errors. Optional resistor R_{IN3} terminates the line differentially. Protection resistors R_{P1} and R_{P2} allow over-voltages at the two inputs, by limiting amplifier fault currents to safe levels. The input stage can use either dual or single amplifiers, with performance options described below.

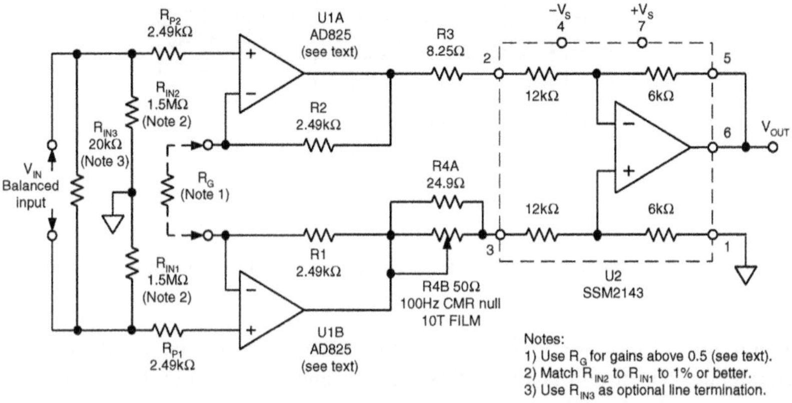

Figure 6-30: A buffered input balanced line receiver

Within the circuit, the *differential* gain of stage U1A, U1B (or G_1) is set by R_1-R_2-R_G, as:

$$G1 = 1 + \left[(2R_1)/R_G\right] \qquad \text{Eq. 6-16}$$

where $R_1 = R_2$, and R_G is used for high gains. Without R_G, the first stage gain is unity.

While the differential gain of U1 is as noted, the CM gain is unity, since the connection simply passes CM signals to the output. Thus both differential and CM forms of signal are presented to the inputs of stage U2. Note, however, because differential and CM signals are scaled differently by U1, there can be a net potential gain in CMR. Practically, it means that this overall configuration can achieve useful CMR figures higher than the intrinsic CMR of U2, whatever that figure may be.

The U2 stage, a pre-trimmed 4 resistor in amp, suppresses the CM component from U1A/U1B, while amplifying differential signals by a factor of 1/2. For an overall net gain G higher than 0.5 from this line receiver, the value of R_G is:

$$R_G = R_1/(G - 0.5) \qquad \text{Eq. 6-17}$$

For net overall gains of 1, 2 and 4 times, the required gain resistance R_G works out to be 4.99 kΩ, 1.69 kΩ, and 715 Ω, respectively (using closest standard values).

Seasoned analog designers may wonder what's so new about this circuit, as it has been around for more than 30 years in solid-state form (Reference 9). While true, some refinements here lend it worthwhile audio utility. First, as mentioned, FET input op amps for the U1 stages allow very low bias current, and load the inputs infinitesimally. Source loading will be essentially determined by the 1% resistances used for R_{IN1} and R_{IN2}.

While FET amplifiers are most useful here, a serious selection caveat is in order. The types used for U1A and U1B must *not* be general-purpose types prone to sign-reversal, which could possibly come about with combined large signal and CM inputs (see Chapter 7 overvoltage discussions). All of the types tested for Figure 6-30 have FET input stages, with CM ranges of at least ±10 V on ±15 V supplies. Note that any op amp can mis-behave if severely overdriven at the input (i.e., beyond the rails). Of the three types tested, the AD825 is the most robust for overload, while the AD845 is less robust, but offers best CM performance. Wideband operation is a virtue, allowing better high frequency performance before degradation sets in. Finally, an FET input structure is less susceptible to RF rectification problems, which can be critically important in an audio line receiver used within an RF environment (see Chapter 7 RFI discussions).

Selection of the U2 device also has a great bearing on CMR. Although there are a number of unity gain 4 resistor in-amps available for the U2 function, the choice here is for less than unity gain (in this case 0.5). To extract the highest possible CMR performance, the U2 network balance is externally trimmed by the R_3-R_4 resistances. R_{4B} can be either the film trimmer noted, or a selected fixed resistor. The values shown allow a trim range of more than ±0.05%, sufficient to trim any SSM2143 part to a null. In the performance data following, the SSM2143 used for U2 reflected such a trim, with a basic low frequency CMR of ~110 dB for the stage.

Buffered Input Balanced Line Receiver Performance

To demonstrate these concepts, a number of measurements were made on the Figure 6-30 circuit, using a number of single and dual IC op amps for the U1A and B positions, and an SSM2143 for U2. Although this basic 3 amplifier in amp structure can in principle offer potential gains in CM performance over the intrinsic CMR of U2, this phenomenon is less pronounced at relatively low overall gains as true here (i.e., gains of 1, 2, or 4 times). And, it is also dependent upon the specific U1 and U2 performance. Thus the CMR of both the U1 and U2 stage devices can effect the measured CM performance.

The test setup used employs an Audio Precision System 1 in a modified crosstalk test mode, where channel A drives the test circuit, which in turn has its output monitored by channel B. This allows a swept narrow band tracking analysis, over a dynamic range of 130 dB or more at low frequencies, and a frequency range of 20 Hz to 200 kHz. In the results following, the CM error curves displayed are referenced to a 0 dB calibrated output level from U2 of 1 V rms, with the circuit set for a gain of 0.5 (i.e., R_G open). The drive to the circuit was 2 V rms, and power supplies were ±13 V.

Figure 6-31 shows CM error results for paired (2) single samples of the AD825, the AD845, and the dual AD823. All devices have CM errors at or below a –90 dB level below 3 kHz, with low frequency errors of the AD845s well below –100 dB. The CM corner for all devices is enough to achieve –80 dB or better 20 kHz CM error.

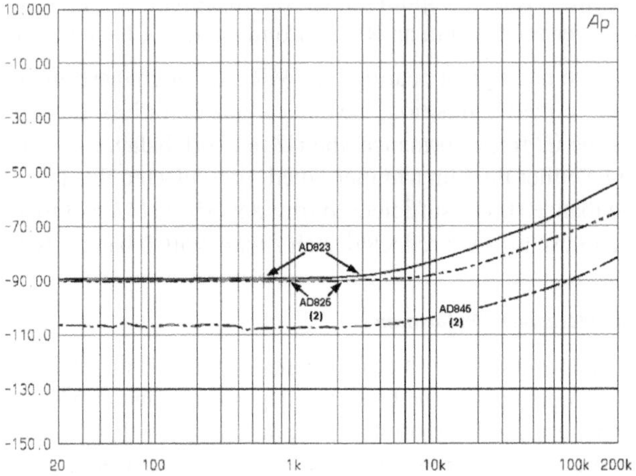

Figure 6-31: CM error (dB) versus frequency (Hz), for various U1A and U1B devices within the circuit of Figure 6-30, gain = 0.5.

While these results show generally what the various devices can do in this circuit, the data should not be taken as an absolute indication of future results. Sample-sample variations of a few dB will exist, and this should be taken into account. One point worth noting, however, is that devices with intrinsically high CMR input stages perform best. An example of this is the AD845, which uses cascode inputs for high CMR, and shows a CMR of 100 dB or more to above 20 kHz. It does however have less input dynamic range than the AD823 and AD825, because of the bias headroom required for the cascode stage.

A criticism directed towards active line receiver circuits such as the SSM2141 and SSM2143 has been their high sensitivity to source resistance mismatches. However, as the discussions above have shown, the problem comes from the relative source/load impedances, and the degree by which these are mismatched. One can specify a very well-controlled, high CMR receiver, such as, for example, the SSM2141 or SSM2143, and have the as-used CMR degrade simply due to uncontrolled source impedance(s).

In the relatively uncontrolled environment of real-world audio system interfacing, source resistance mismatches of a few ohms can be typical. This mismatch level is sufficient to ruin the CMR performance of a simple line receiver, if the receiver uses resistances on the order of 20 k, and is fed from a source resistance of 50 Ω or more. This can be readily illustrated by a sample calculation using the bridge circuit CMR relationship of Eq. 6-12, and plugging in R_{OUT} resistances of 50 Ω and 55 Ω (a 10% mismatch), using an R_{IN} resistance of 20 kΩ. This degree of mismatch for 20 kΩ loading conditions destroys CMR, as it degrades to 72 dB with a 50 Ω source mismatched 10% (see Figure 6-22).

The buffered topology of Figure 6-30 directly addresses this issue, as shown in the dual matched/mismatched source resistance CM plots of Figure 6-32. In these tests, the Figure 6-30 circuit is again exercised with the AD825 and AD845 op amp paired samples at an overall gain of 0.5, and R_{IN3} = 20 kΩ. The circuit is fed from a source resistance of 50 Ω, and is operated under both matched and mismatched source resis-

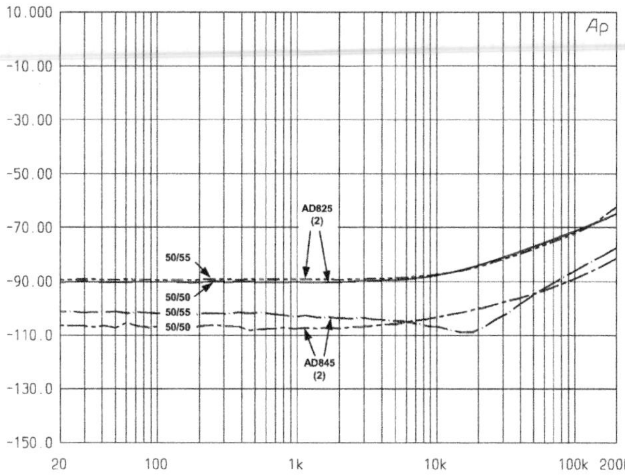

Figure 6-32: CM error (dB) versus frequency (Hz), for AD825 and AD845 pairs, nominally 50 Ω source impedances matched/mismatched 10%

tance conditions. This allows the degradation with mismatching to be clearly shown as separation between the 50/50 (matched) and 50/55 (mismatched) paired curves for each device type.

In the AD825 pair curves, the CM degradation is less than 1 dB, even for the test condition of the relatively high 10% source resistance mismatch. The AD845 device has better overall CMR than the AD825, which shows up as low frequency errors of better than –100 dB with the matched impedances. While mis-matching degrades CMR by a few dB, it is still –100 dB or below as high as 20 kHz. The AD845 can be used for U1A and U1B in the Figure 6-30 circuit for highest wideband CMR (do, however, be careful to note the device dynamic range limits). Considering its superior overload behavior, the AD825 is the best choice, and is thus recommended.

These tests make clear that higher R_{IN} aids in desensitizing system CMR degradation against source mismatching. The designer has the option of using even higher input resistance for R_{IN1} and R_{IN2}, to further reduce the source sensitivity.

Transformer-Input Line Receiver

The classic solution to the CM isolation of audio signals is the *line input* transformer (see References 4,5). This device, usually a 1:1 ratio unit, offers galvanic isolation and very high CM voltage breakdown ratings. It is a preferred (or only) solution where true galvanic isolation is a necessity. Transformers are also useful for high and consistent low to middle audio frequency CM performance, both from unit-unit, and also when high immunity to varying differential source resistance is sought. These features do come at some cost however. Quality transformers are pricey, at about 10–20 times a single IC's cost. They also occupy a relatively large package size vis-à-vis a solid-state equivalent.

All of the various factors above are the designer's ultimate decision points, dependent upon the exact system requirements. When optimized for high performance, it is not likely that either a completely solid-state or a completely transformer-based line receiver solution will be considered either simple, or low in cost. Performance comes with a price.

Interestingly, when a near-ultimate in low frequency CM rejection is required, or completely tweak-free operation is sought, a hybrid line receiver solution may be a good choice. An example of a line transformer buffered by a simple line receiver is shown in Figure 6-33. This circuit combines good features of the simple line receiver and the transformer, and offers outstanding performance with attractive simplicity.

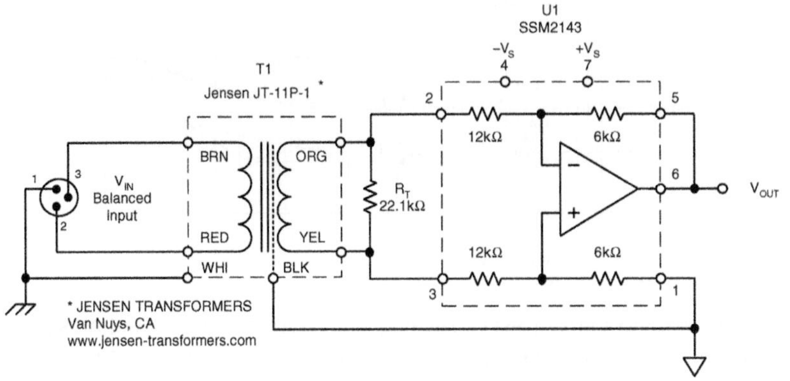

Figure 6-33: A transformer-input line receiver circuit

The circuit has only three components; T1, the secondary termination resistor, and U1. It can be noted that this combination can operate the transformer in either a balanced mode (just as shown) or in an alternate single-ended mode (with U1-2 grounded). The choice of which mode is used does make a big CMR performance difference, as will be apparent. The transformer secondary is terminated in a net 10 kΩ resistance, which here is comprised of R_T and the 12 kΩ input resistances of U1. If other forms of termination are used, or an alternate U1 part, R_T should be adjusted accordingly (see Reference 17). The net gain of the circuit is product of the loaded transformer primary/secondary voltage ratio, and the voltage gain of U1 (0.5). Inasmuch as T1 shows a voltage loss for such loaded conditions, the overall gain of the circuit is approximately 0.35 (or –9 dB). If more gain is desired, an alternate SSM2143 operational mode is possible. The 12 kΩ/6 kΩ input output resistors can be reversed, which will then allow it to operate at a gain of 2.

The CMR test results for the circuit of Figure 6-33 are shown in Figure 6.34. Conditions for this test setup are similar to those of the buffered balanced line receiver test, described above. They are referenced to a 1 V operating level at the output of U1.

Several important points should be apparent from these results. In the single-ended mode (upper curve), the basic data sheet performance of the JT-11P-1 transformer can be seen. This includes an approximate 60 Hz CM error of –107 dB. While this simply buffered operating mode does offers excellent low frequency CMR, it can also be noted that this also degrades with rising frequency. At 20 kHz the single-ended mode error is just slightly more than –50 dB—good, but not superlative. Any of the ICs tested for the Figure 6-30 circuit better this performance, by about 30 dB or more.

On the other hand, by simply letting the SSM2143 operate in a balanced mode, the lower curve CMR performance results. This is clearly a major improvement vis-à-vis the single-ended case, with the low frequency CM error reduced to –130 dB or better, approaching the noise floor of the instrumentation. The errors aren't quite as low at the higher frequencies, with a 20 kHz CMR of ~60 dB.

A point worth noting is the transformer-based line receiver cannot really compete with the buffered balanced receiver of Figure 6-30 for high frequency CMR. This is because the transformer interwinding

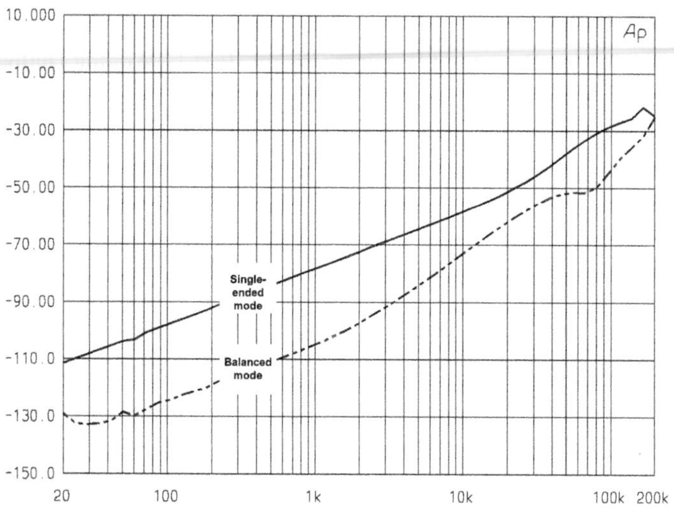

Figure 6-34: CMR errors for Figure 6-33 transformer-input line receiver circuit, operating in single-ended and balanced modes

capacitance acts as a high-pass filter for CM noise. This has the effect of passing CM noise to the output at the higher frequencies. The degree of severity for this phenomenon will of course vary with the design of the specific transformer.

Nevertheless, it is readily apparent from the performance data that the preferred method of transformer operation is to operate it in a balanced secondary mode (i.e., as shown in Figure 6-33), which does mitigate the CMR loss with frequency somewhat.

A Summary of Line Receivers

Both active and passive solutions for minimizing balanced transmission system CM noise have been explored, each with their own set of characteristics. Readers should take these data for the general trends they convey, not any absolute performance levels.

To summarize, the buffered balanced line receiver of Figure 6-30 offers excellent broadband CMR with low differential source resistance sensitivity, and can be user-customized in a variety of ways, including gain, CM input impedance, and so forth. The solid-state line receiver approach here has virtues of the best high frequency CMR in absolute terms, as well as the better CMR versus frequency flatness. While the example circuit shown works well, optimization for a production role may need some enhancement for worst-case minimum CMR. This can be done via careful trim or selection of the U2 circuit, and/or selection of an optimum pair of singles for U1A and U1B. Input dynamic range of the circuit can be optimized by selection of the U1A-U1B pair types, and the supply voltages used. All types tested can be operated at supplies of up to ±18 V (maximum), or as low as ±13 V. The typically used ±15 V supplies will provide both excellent performance and high input dynamic range.

The transformer-input approach to a line receiver offers good to superlative low frequency CMR, combined with "no tweak" operation. When the transformer used is combined with an in amp for balanced mode secondary buffering, as in Figure 6-33, further CMR reduction is possible over a range of low to middle frequencies. On the downside, there are negatives of cost, size, and eventual CMR degradation with frequency.

References: Audio Line Receivers

1. Walt Jung, "Op Amps in Line-Driver and Receiver Circuits, Part 1," **Analog Dialogue**, 26-2, 1992.
2. W. Jung, A. Garcia, "Op Amps in Line-Driver and Receiver Circuits, Part 2," **Analog Dialogue**, 27-1, 1993.
3. Walt Jung, "Applications for Amplifiers in Audio," Ch. 5 within Walt Kester, Editor, **1992 Amplifier Applications Guide**, Analog Devices, Inc., Norwood, MA, 1992, ISBN 0-916550-10-9.
4. Deane Jensen, "Transformer Application Notes (various)," Jensen Transformers, 7135 Hayvenhurst Avenue, Van Nuys, CA, 91406, (213) 876-0059.
5. Bruce Hofer, "Transformers in Audio Design," **Sound & Video Contractor**, March 15, 1986.
6. Henry Ott, Noise Reduction Techniques in Electronic Systems, 2d Ed., Wiley, 1988.
7. David Birt, "Electronically Balanced Analogue-Line Interfaces," **Proceedings of Institute of Acoustics Conference**, Windermere, U.K., Nov. 1990.
8. Robert Demrow, "Narrowing the Margin of Error," **Electronics**, April 15, 1968.
9. Robert Demrow, "Evolution from Operational Amplifier to Data Amplifier," **Analog Devices Application Note**, September, 1968.
10. Walter Borlase, "Application/Analysis of the AD520 Monolithic Data Amplifier," **Analog Devices Application Note**, 1972.
11. Jeff Riskin, "A User's Guide to IC Instrumentation Amplifiers, **Analog Devices AN244**, January, 1978.
12. S. Wurcer, L. Counts, "A Programmable Instrumentation Amplifier for 12-Bit Resolution Systems," **IEEE Journal of Solid-State Circuits**, Vol. SC-17 #6, Dec. 1982.
13. N. Muncy, "Noise Susceptibility in Analog and Digital Signal Processing Systems," **JAES**, Vol 43, No 6, June, 1995.
14. B. Whitlock, "Balanced Lines in Audio—Fact, Fiction, and Transformers," **JAES**, Vol 43, No 6, June, 1995.
15. B. Whitlock, "A New Balanced Audio Input Circuit For Maximum Common-Mode Rejection In Real-World Environments," **presented at 101st AES Convention**, November 1996, preprint #4372.
16. Walt Jung, "Practical Circuits for Quiet Audio Transmissions," **Electronic Design Analog Applications Issue**, November 17, 1997, pp. 45–49.
17. "JT-11P-1 Line Input Transformer Data Sheet," Jensen Transformers, 7135 Hayvenhurst Avenue, Van Nuys, CA, 91406, 213-876-0059.

Audio Buffers and Line Drivers

Audio line drivers and buffer amplifiers can take on a wide variety of forms. These include both single-ended and differential output drivers, as well as transformer isolated drivers. Within these general formats there are also many different performance options, and many of these are covered in this section. Note that a later section of this chapter also discusses buffers for a general context, as for video and instrumentation applications.

Many op amps useful as video drivers and/or buffers do well for audio drivers/buffers, because of the high current output stages necessary for good linearity over video bandwidths (see References 1–4). Some examples of video IC amplifiers that are audio-useful are the AD810, AD811 and AD812, AD815, AD817 and AD826, AD818, AD829, AD845, and AD847. Other types notable for either high or unusually linear output drives or other performance features useful towards audio are the AD797 and the OP275.

Some High Current Buffer Basics

As a preliminary to detailed application discussions, some basic circuit principles germane to high current buffers and drivers should be treated first. With output currents up to 100 mA or more, "housekeeping" details of bypassing, grounding and wiring also become important, and must be considered to achieve high performance. These are briefly discussed here in the context of high current audio buffers, using the unity gain buffer circuit of Figure 6-35, as a point of departure.

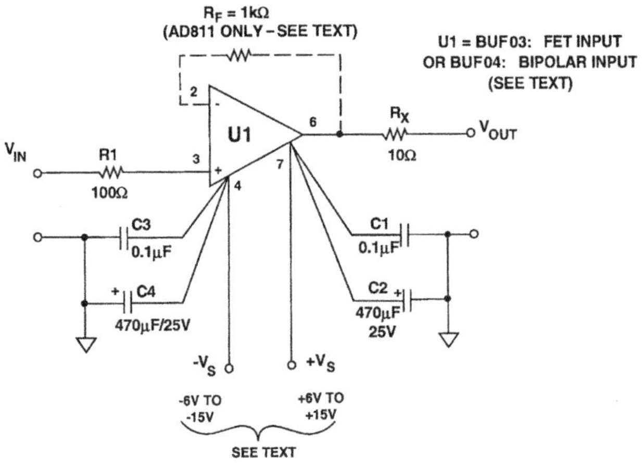

Figure 6-35: A unity-gain, standalone buffer circuit

First, despite which IC is used for U1, close attention should be given to making buffer stages free from parasitic effects, at both input, output, and supplies. Physical construction of buffer-drivers and other high current stages should be in accordance with high speed rules. A heavy copper ground plane is preferred, and circuit layout should be compact, with low capacitance high-Z nodes. Signal and ground runs should be laid out with signal coupling and load current flow in mind (see References 5–7 and Chapter 7).

In addition, the power supplies should be well bypassed close to the high current supply pins. In Figure 6-35 this is indicated by the Kelvin connections of C_1–C_4 to the U1 ±Vs pins. This should be used as standard practice for all high current stages, and is intended as a given for all the driver applications of this section.

As a minimum, local low inductance/low ESR RF bypass caps should used within 0.25" of the device supply pins, shown as C_1 and C_3. These are preferably 0.1 µF stacked polyester film, or other low inductance capacitor type, preferably films. In addition, for high peak current loads, the high frequency bypasses are paralleled by local, short lead/large value, low ESR electrolytics such as C_2 and C_4, in a range of 470 μF/25 V and up. Note that capacitor ESR reduces in inverse proportion to electrical size and voltage rating, so larger size and/or voltage units help. These capacitors carry transient output currents, and should be aluminum electrolytic types rated for high frequency use, that is switching supply types. Such types tend to have a broad range of lowest high frequency minimum impedance and are thus less likely to cause power line resonance than are tantalum units.

Dc power management and dissipation can also be important with buffer ICs. For example, the BUF03 and the AD811 ICs can dissipate fairly large power levels even with light loading, for supplies above ±12 V. This is because the quiescent current of these devices is 15 mA–18 mA, relatively independent of operating voltage.

As a conservative general rule of reliability, any IC with a power dissipation above 300 mW should not be used without a heat sink. For buffer or driver circuits using this power or more, use the lowest thermal resistance package possible, and add the appropriate heatsink (Thermalloy 2227 for the BUF03 or other TO-99 ICs, or Aavid #5801 for the BUF04, AD811 or other high dissipation 8-pin DIP ICs).

Output resistor R_X in this circuit should be 10 Ω or more, to isolate the buffer from capacitive loading (more on this elsewhere in this chapter). For an extra safety margin against possible destabilization due to capacitive loads, make this resistor as high as feasible from a voltage loss point of view.

The input resistor R_1 is a "bullet-proof" safety item, and can serve two purposes. One is as a parasitic suppression device, which may be required for stability with some amplifiers (not absolutely essential for those here). A subtler feature of this resistance comes about when the buffer is operated within a feedback loop, and is driven from an op amp output. Internally, many buffer ICs have clamping diodes from input to output, and under overload conditions, these diodes act to clamp overdrive. With the inclusion of R_1, this prevents excess current drive into the buffer IC under this clamping condition.

Because of this stage's very high bandwidth, low phase shift, and low output impedance, fast buffers such as this can be used both "stand alone" just as shown, or as a more conventional "in loop" buffer as well, to minimize loading of a weaker, slower amplifier. The improvement raises the linear output up to ±100 mA with the AD811 or the BUF04, while maximizing linearity, preserving gain, and lowering distortion.

Buffer THD + N Performance

Operating in a pure standalone mode, THD + N tests on several unity gain buffers are shown in Figure 6-36. These tests were for common conditions of 10 V rms output into a 600 Ω load, operating from ±18 V power supplies.

The BUF03, an open-loop design, shows a distortion for these conditions of about 0.15%. The BUF04, a closed-loop current feedback design buffer, shows a very low distortion of about 0.004%. The AD811 is also a current feedback amplifier, but it is externally configured as a unity gain follower, with R_F = 1 kΩ. Note that *all current feedback ICs will require such a resistor*, but the value required may vary part to part. The AD811 shows an intermediate distortion level, under 0.01%.

As a choice among these types, both the BUF04 and AD811 are capable of more than ±100 mA of output, with input currents on the order of 1 μA–2 μA. The BUF03 has a lower output current (±70 mA), but the advantage of a much lower input current (~200 pA).

Dual Amplifier Buffers

In addition to standard operation of the various single op amps as unity-gain buffers, certain high output current *dual* op amp ICs also work exceedingly well as buffers. Using a dual IC to buffer a signal has the advantage of doubling the output drive while using basically the same package size, an obvious benefit.

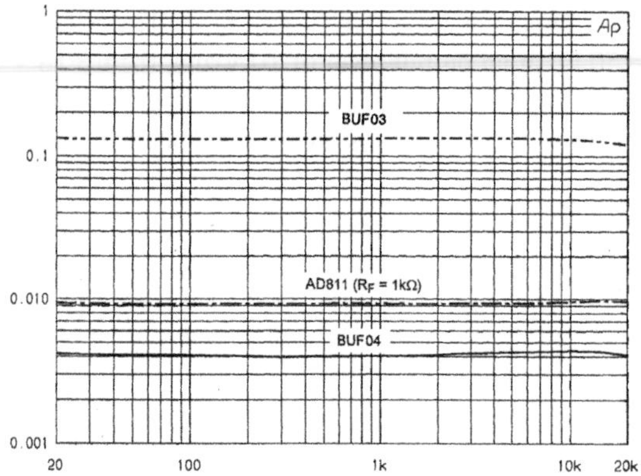

Figure 6-36: THD + N (%) versus frequency (Hz) for various buffer ICs, for VOUT = VIN = 10 V rms, RLOAD = 600 Ω, VS ±18 V

Two design steps allow this to be successfully implemented. The first is the selection of a basically linear single device that also is available as a dual. The second is to devise a method of combining the outputs of the two op amps in a linear fashion, without any side effects.

Among the suitable candidates for this task are the dual version of the AD817, the AD826, as well as the AD811 dual, the AD812. Figure 6-37 shows a hookup that is useful towards increasing buffer output current to more than 100 mA.

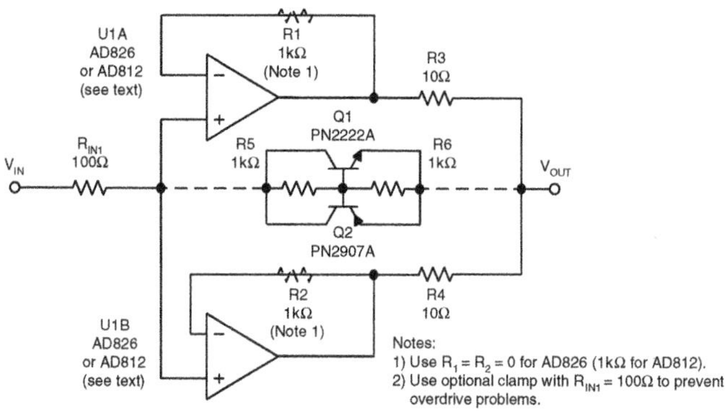

Figure 6-37: Dual op amp buffer circuit raises output current to more than 100 mA with low distortion

Ignoring Q1-Q2 for the moment, the circuit can be seen as a pair of unity-gain followers paralleled at the output, through small value resistors R_3 and R_4. These resistors provide balanced drive from the U1A and B sections, linearly combining the signals. With the use of the voltage feedback AD826 op amp, the circuit is quite simple, since R_1 and R_2 reduce to zero. If the AD812 current feedback device is used, these two resistors should be 1 kΩ (shown dotted). The Q1-Q2 bidirectional clamp circuit is optional, and when used can provide protection against input overdrive, and/or adjustable current limiting via R_5.

This circuit offers excellent performance, as shown in Figure 6-38. These THD + N plots show performance with loads of both 600 Ω and 150 Ω, at an output level of 10 Vrms, from ±18 V supplies (without clamping active). The AD826 offers the lowest distortion, due to the voltage feedback architecture, with less than 0.01% THD + N, even when driving 150 Ω, which is an approximate 100 mA combined peak output.

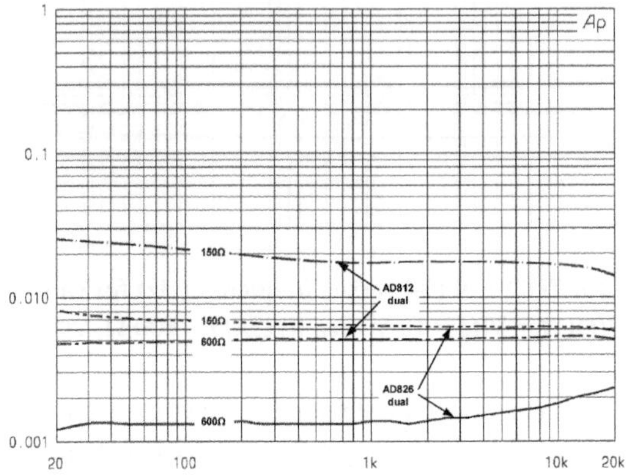

Figure 6-38: THD + N (%) versus frequency (Hz) for AD826 and AD812 dual buffer ICs, for $V_{OUT} = V_{IN}$ = 10 V rms, R_{LOAD} = 150 Ω/600 Ω, V_S ±18 V

Capacitive Loading Issues

Audio driver output stages are typically operated as voltage sources feeding high impedance loads. When connected via long transmission lines between stages, the result is that the driver sees an unterminated line, which can appear highly capacitive. Audio driver stability with capacitive loading can be a difficult design issue, but for good reason—it isn't always an easy thing to achieve. If easy, it may be at the expense of performance or circuit complexity. Fortunately, some standard techniques exist for stabilizing op amp drivers with capacitive loads, and these can be implemented in a reasonably direct fashion. These are covered in detail within the next section of this chapter. The discussions immediately following emphasize driver linearity.

Op Amp Device/Topology Related Distortions

Single-ended audio drivers can be built using a linear, noninverting gain stage as a starting point. Indeed such a circuit, given appropriate op amp choice and gain scaling, can well serve as a basic audio driver. Topologically, a noninverting gain stage is preferable, since it loads the signal source less, and, it also adds no sign inversion. However, this configuration is subject to certain distortions, which should be understood in order to extract the best performance in an application. Distortion performance for a number of audio op amps in such a line driver circuit are now discussed, in this context.

The circuit of Figure 6-39 is a test configuration that loads the U.U.T. op amp with 500 Ω and 1 nF. This is a reasonably stressful test load, which can differentiate the distortion of various devices with outputs of 7 V rms or more. A gain of 2 is used, which subjects the U.U.T. to a relatively high input CM voltage, thus this configuration is sensitive to CM distortion in the amplifier. For the following tests of this section (except as noted to the contrary), V_S = ±18 V, and the analyzer bandwidth is 10 Hz–80 kHz.

Given amplifiers with sufficient load drive and output stage linearity in this circuit, there can still be nonlinear effects due to the CM voltage. This distortion is due to the nonlinear C-V characteristic seen at the

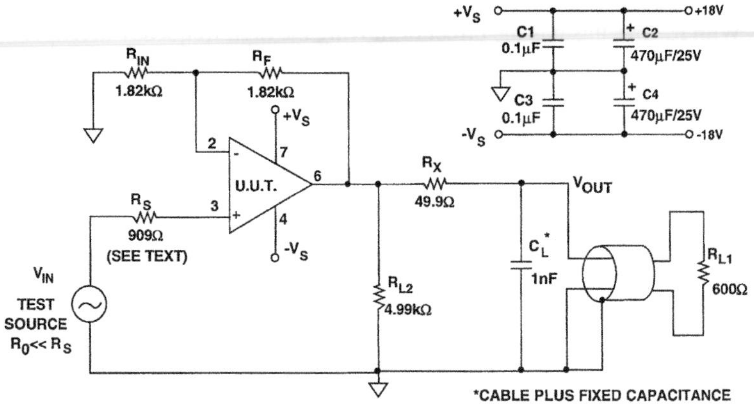

Figure 6-39: Test circuit for audio line driver amplifiers

two amplifier inputs, and can be minimized by matching the two impedances seen at the respective (+) and (–) inputs (see Reference 8). When this is done, the differential component of the error is minimized, and the distortion seen in V_{OUT} falls to a minimum.

This general point is illustrated by Figure 6-40, a family of plots for an OP275 op amp within the circuit of Figure 6-39. The OP275 use junction FET devices in the input stage, which have appreciable (nonlinear) capacitance to the substrate. The test is done with various values of source resistance R_S. As noted, distortion is lowest when R_S is equal to the parallel equivalent of R_F and R_{IN} or, in this case, about 910 Ω. For either higher or lower values of R_S, distortion rises. Appreciably higher source impedance (10 kΩ) can cause the distortion to rise lower in frequency, making performance much worse overall.

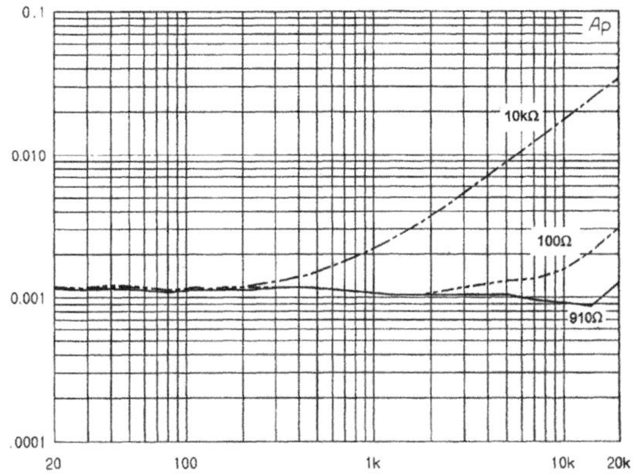

Figure 6-40: Follower mode R_S sensitivity of OP275 bipolar/JFET input op amp THD + N (%) versus frequency (Hz), V_{OUT} = 7 V rms, R_L = 500 Ω, V_S = ±18 V

It is therefore suggested that, whenever possible, amplifiers operated as voltage followers should have their source impedances balanced for lowest distortion. Note that the OP275 device is just one example, and its sensitivity to CM distortion effects is not at all unique in this regard.

While the balancing of the two source impedances is most helpful, *lowering the absolute value* can also minimize this distortion. With R_S low, this has the effect of moving the high frequency breakpoint of the distortion rise upwards in the spectrum, were it is less likely to be harmful. The best overall control of this distortion mechanism with an amplifier subject to it is the use of the lowest practical, balanced source impedances.

It is important to understand that virtually all IC op amps, *particularly those using JFET inputs*, as well as discrete JFET and bipolar transistors are subject to nonlinear C-V effects, to some degree. In the tests of other amplifiers within the Figure 6-39 circuit, R_S was maintained at 910 Ω, to minimize the effects of this distortion mechanism.

With high output, high slew rate linear amplifiers, the distortion generated for these test conditions can parallel that of the test equipment residual, as shown in Figure 6-41. Here the AD817, AD818 and AD845 amplifiers show THD + N which is essentially equal to the residual for these conditions, and appreciably below 0.001%.

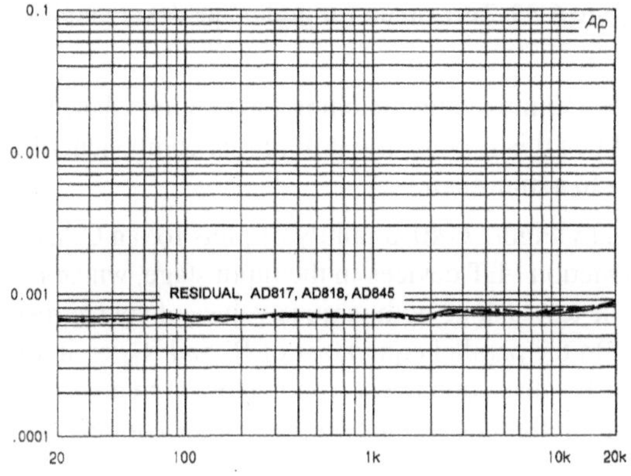

Figure 6-41: A driver group, THD + N (%) versus frequency (Hz), for V_{OUT} = 7 V rms, R_S = 909 Ω, R_L = 500 Ω, V_S = ±18 V

Amplifier types expressly designed for audio use also do well for these THD + N tests, as shown in Figure 6-42. The industry standard 5534 is near or just above the residual level, while the OP275 plot falls just above the 0.001% level and the 5532 is slightly higher.

These tests reflect performance of a variety of single amplifiers, as exercised for matched-source test conditions, with medium output loading of 600 Ω. Varying test conditions may change the absolute levels of performance. So also may different samples, or in the case of industry standard parts, alternate vendors.

The data of Figures 6-41 and 6-42 reflect older (but available) op amp devices capable of very high performance in these tests. Several more recent devices also do well for this driver test. Figure 6-43 shows performance of newer FET input op amps, the AD825, the AD8610 and the AD8065. The AD825 was tested under conditions identical to those of Figure 6-41 and 6-42. The AD8610 and AD8065 were tested under similar conditions, but with ±13 V power supplies, reflecting a lower maximum supply rating.

Note that the latter two amplifiers still can accommodate more than a 7 V rms output swing, even with the reduced supplies. Under these conditions, the AD8065 distortion is near the test set residual and the AD8610 slightly higher at high frequencies. The AD825 has somewhat higher distortion, but this is almost frequency-independent.

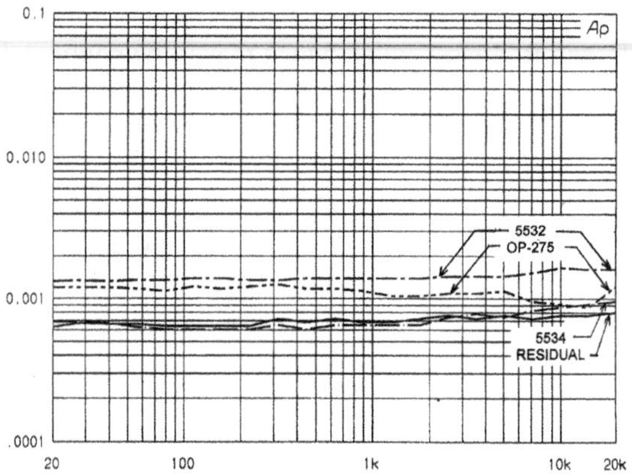

Figure 6-42: B driver group, THD + N (%) versus frequency (Hz), for V_{OUT} = 7 V rms, R_S = 909 Ω, R_L = 500Ω, V_S = ±18 V

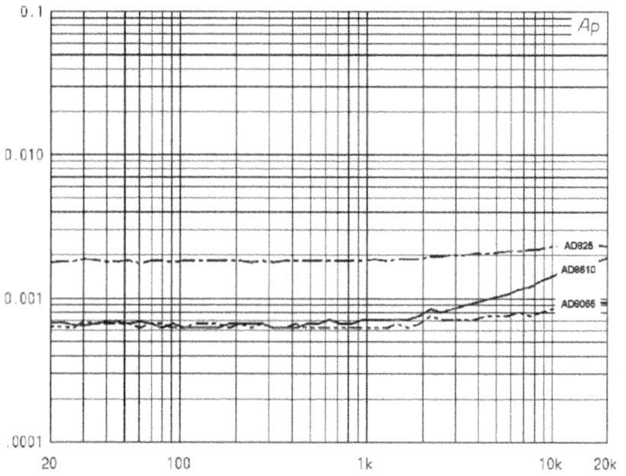

Figure 6-43: C driver group, THD + N (%) versus frequency (Hz), for V_{OUT} = 7 V rms, R_S = 909 Ω, R_L = 500 Ω, V_S = ±13 V or ±18 V

Single-Ended Line Drivers

This section discusses a variety of line driver circuit examples that drive single-ended lines, optimized for different operating environments, supply voltages, and performance.

Consumer Equipment Line Driver

One common driver application is a line output stage for consumer preamps, CD, and DVD players, and so forth. This is typically an economical audio stage with a nominal gain of 5 to 10 times, operating from supplies of ±10 V to ±18 V, usually with a rated output of 2 V rms–3 V rms, and a capability of driving loads of 10 kΩ or more.

For simplicity of biasing and minimum output dc offset, ac-coupling is used, and the circuit is typically fed from a volume control. For stereo operation, a dual channel device is typically sought for this type application, one which is also optimized for audio uses.

Such a stage is shown in Figure 6-44, and uses an OP275 dual op amp as the gain element. In this circuit input and feedback resistors R_1 and R_3 are set equal, which makes the nominal dc bias at U1's output close to zero. The U1 bias current flowing in these resistors also serves to polarize coupling capacitors C_1 and C_2 positively, as noted.

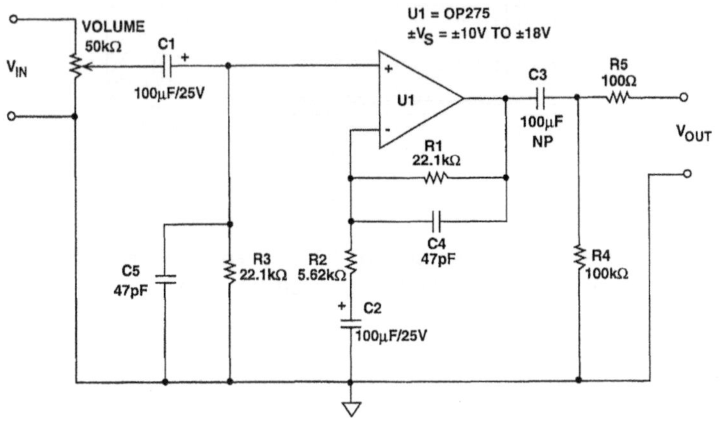

Figure 6-44: Consumer equipment line driver stage

This bias is due to the sign of the OP275's PNP input stage bias currents, so reverse C_1-C_2 if an NPN input amplifier is used.

R_2 sets the gain of the stage in conjunction with R_1. The stage gain is nominally five times for the values shown. C_2 sets the low frequency roll-off along with R_2, which in this case is ≈0.3 Hz. Although this frequency is quite low, it does allow some range of gain increase if desired, simply by lowering R_2. Output capacitor C_3 must be nonpolarized, since the worst-case dc at the output of U1 is ≤10 mV (and can be bipolar). Typically it will be about one-quarter this, so if a few mV can be tolerated, C_3 can be eliminated.

THD + N performance of the stage (not shown) was measured for outputs of 1 V rms–3 Vrms into a 10 kΩ/600 pF load, using ±18 V supplies with an R_S of 1 kΩ. At lower output levels performance is noise limited, measuring less than 0.002%. At the 3 V output level a slight increase in high frequency distortion is noted. Although this application is an example where the amplifier ± source impedances cannot be matched (due to the variations of the volume control), nevertheless the performance is still quite good.

Noise is the limiting factor for lower level signals so, if lower noise is desired, R_2 can be reduced. The ultimate practical limit to noise is the volume control's finite output impedance. This causes higher noise at positions of high output resistance, interacting with the noise current from U1. For example if the effective R_S from the volume control is 10 kΩ, a 1.2 $pA/\sqrt{Hz}$ noise current from U1 will produce an input referred 12 $nV/\sqrt{Hz}$ noise voltage, from this source alone. The Figure 6-44 driver is a flexible one, and operates at supplies as low as ±10 V with outputs up to 3 V rms, with slight distortion increases. With ±5 V supplies up to 2 V rms is available, with higher distortion (but still ≤0.01%).

Paralleled Output Line Driver

Often a modest increase in output may be needed for a driver, but circumstances may not warrant the use of additional buffer devices. Figure 6-45 shows how a second section of a dual op amp can be used to provide additional load drive.

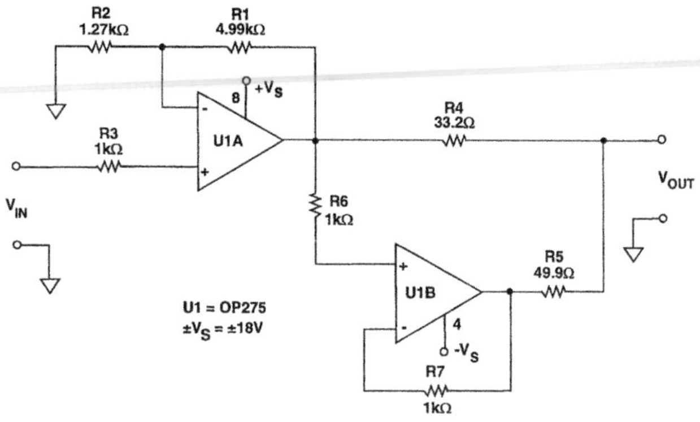

Figure 6-45: Paralleled output dual op amp line driver

In this circuit, using an OP275 dual op amp, the U1A section is a gain-of-five voltage amplifier, while the U1B section is a voltage follower, used simply to provide additional current to the load. Current sharing is determined by output summer resistors R_4 and R_5, and the parallel stage drives 600 Ω loads with less distortion than a single OP275 section.

THD + N performance data is shown in Figure 6-46, with the driver operating from ±18 V supplies, and for output levels of 1, 2, 5, and 9 V rms into 600 Ω.

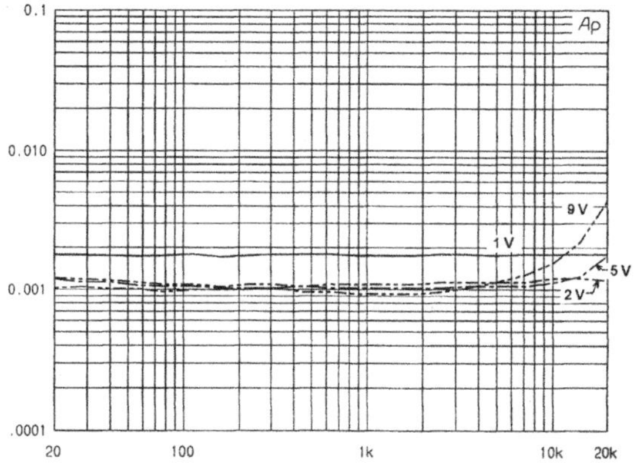

Figure 6-46: Paralleled output dual op amp line driver, THD + N (%) versus frequency (Hz), for V_{OUT} = 1, 2, 5, 9 V rms, R_L = 500 Ω, V_S = ±18 V

This general scheme can be used with any unity gain stable dual op amp, and also can be adapted for various gain levels, via R_1-R_2. For different devices and/or gains, the ratio of R_4 and R_5 may need adjustment, for lowest distortion into the load.

A Wide Dynamic Range Ultralow Distortion Driver

Single-ended line drivers are simple conceptually, but when pushed to performance limits in dynamic range and distortion, they challenge device choice. The AD797 answers this challenge with its input noise of $\leq 1\ nV/\sqrt{Hz}$ and a distortion canceling output stage. These features allow low and high extremes of dynamic range to be pushed simultaneously.

The AD797 uses a single voltage gain stage, comprised of a folded cascode input combined with a boot strapped current mirror load, allowing the high incremental impedance necessary for a 146 dB gain. This buffered single-stage topology is a departure from past devices using multiple stages, with performance benefits in terms of bandwidth, phase margin, settling time, and input noise (see Reference 9).

For standard uses, the AD797 is employed like any 5-pin op amp, such as shown in Figure 6-47 (neglecting the capacitors for the moment). From the A/B part of the table, relatively low values for resistors R_1-R_2 are recommended for lowest noise. Selecting these resistors should be done with care, since values ≥100 Ω will degrade noise performance. Suggested values for gains of G = 10 – 1000 are noted. The AD797 can drive loads of up to 50 mA, and is rated for distortion driving loads of 600 Ω.

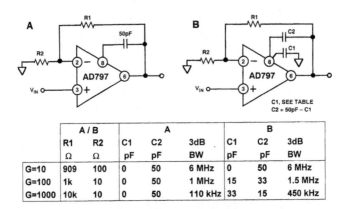

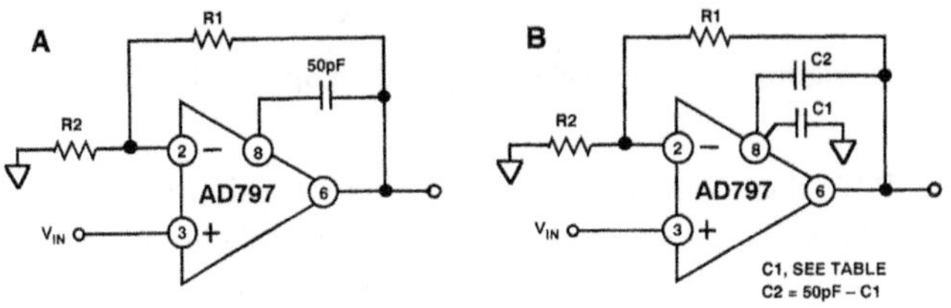

	A / B		A			B		
	R1 Ω	R2 Ω	C1 pF	C2 pF	3dB BW	C1 pF	C2 pF	3dB BW
G=10	909	100	0	50	6 MHz	0	50	6 MHz
G=100	1k	10	0	50	1 MHz	15	33	1.5 MHz
G=1000	10k	10	0	50	110 kHz	33	15	450 kHz

Figure 6-47: Recommended AD797 connections for distortion cancellation and/or bandwidth enhancement

For amplifier applications requiring more top grade performance, optional capacitors C_1 and C_2 can be used. In the Figure 6-47A configuration, superior performance is realized due to distortion cancellation with the use of a single extra capacitor, enabled simply by adding the 50 pF unit as shown. This provides compensation for output stage distortion without effecting the forward gain path, effective over the range of gains noted.

An additional option with the AD797 is the use of *controlled decompensation*, available with the Figure 6-47B option and the use of capacitors C_1, C_2. At gains of 100 or more, adding C_1 as in column B of the table enhances amplifier open loop bandwidth, allowing very high gain-bandwidths to be achieved—150 MHz at G = 100, and 450 MHz at G = 1000. For high gain operation this extra gain-bandwidth can be very effective.

A family of distortion curves for various AD797 gain configurations driving 600 Ω is shown in Figure 6-48. As noted, at low frequencies the data is limited by noise, while at high frequencies distortion is measurable, but still extremely low. The distortion for a gain of 10 times at 20 kHz for example, is on the order of ≈0.0001%, implying a dynamic range of about 120 dB re 3 V rms, or even more for higher level signals.

An additional point worth making at this point is one regarding the AD797's special distortion cancellation ability. Referring to Figure 6-47A, it should be noted that the 50 pF capacitor is connected between Pins 8 and 6 of the AD797. Pin 6 is of course the output of the device for standard hookups. However, in special situations, even greater output current may be required, and a unity gain buffer amplifier can be added between Pin 6 and the load. For example, one of the buffers of Figure 6-36 or 6-37 could be used to extend output current to ≥100 mA.

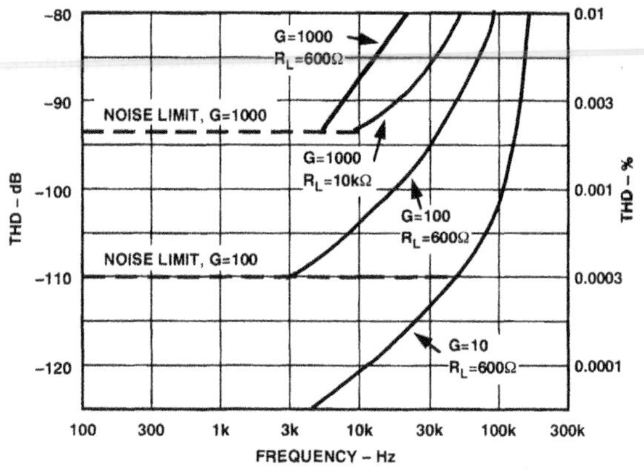

Figure 6-48: THD versus frequency at 3 V rms output for AD797 distortion cancellation and/or bandwidth enhancement circuit of Figure 6-47

The special point worth noting for this situation is that the 50 pF distortion cancellation capacitor should then be connected between the AD797 Pin 8 *and the output of the buffer* (not the AD797 Pin 6). This step allows the distortion correction to be applied not just to the AD797 internal circuits, but also extends it to include the buffer.

A case in point were this would come to useful purpose lies with the AD797 mic preamp, discussed in some detail earlier in the chapter (Figure 6-5). As mentioned therein, a BUF04 would work well as just such a buffered output option for the AD797 preamp. It would be connected as described above, with the 50 pF distortion cancellation capacitor. Details of this are left as a reader exercise (but should even so be obvious).

Current Boosted Buffered Line Drivers

When load drive capability suitable for less than 600 Ω in impedance is required, it is most likely outside the output current and/or linearity rating of even the best op amps. For such cases, a current-boosted (buffered) driver stage can be used, allowing loads down to as low as 150 Ω (or less) to be driven. Another example would a driver for long audio lines, i.e., lines more than several hundred feet in length.

Figure 6-49 is a high quality current-boosted driver example, using an AD845 at U1 as a gain stage and voltage driver, in concert with a unity voltage gain current booster stage, U2. The overall voltage gain is five times as shown, but this is easily modifiable via alternate values for R_1 and R_2. In any case, for lowest CM distortion effects, input resistor R_3 should be set equal to $R_1||R_2$ (this assumes a low impedance source for V_{IN}).

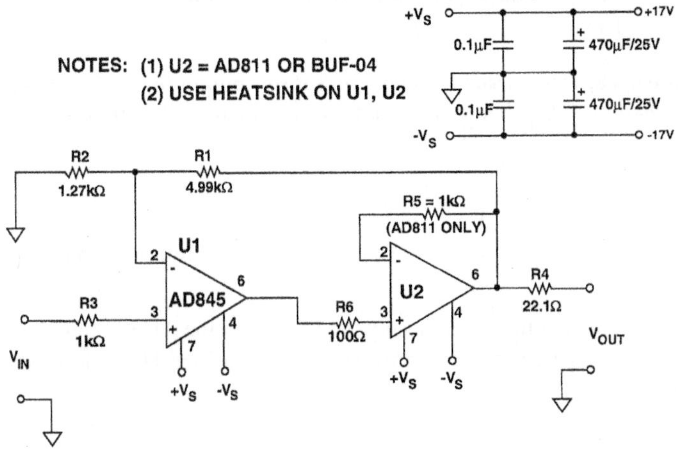

Figure 6-49: Current-boosted line driver

The amplifier used for U2 can be either the AD811 op amp, or the BUF04 buffer for simplicity. If the AD811 or similar CFB op amp is used (AD812, and so forth), it needs to be configured as a follower, with R_5 connected as shown. Since the BUF04 is internally connected as a follower, it doesn't need the R_5 external feedback resistor.

Because of the high internal dissipation of the AD845 and AD811, these devices *must* be used with a heat sink on supplies of ±17 V. However, such high supplies are only justified for extreme outputs. Supplies of ±12 V also work, and will eliminate need for a heat sink (with lower maximum outputs). In any case, power supplies should be well bypassed.

A special note is applicable here—Always *observe maximum device breakdown voltage ratings within applications. Production versions of this circuit should use supplies of ±17 V or less, for 36 V(max) rated parts. Similarly, 24 V(max) rated parts should use supplies of ±12 V or less. In all cases, use only enough supply voltage to achieve low distortion at the maximum required output swing.*

For loads of 150Ω, the output series isolation resistor R_4 is lowered to 22.1Ω to minimize power loss, and to allow levels of 7 V rms or more. The THD + N data for this circuit is shown in Figure 6-51, using an AD811 as the U2 buffer. The test conditions are input sweeps resulting in 1, 2, 4 and 8 V rms output, using ±18 V power supplies.

For the AD811 operating as U2, the Figure 6-50 data below shows THD + N dominated by noise and residual distortion at nearly all levels and frequencies driving 150Ω, up to 8 V rms. At this level, a slight distortion rise is noted above 10 kHz, yet it is still ≈0.001%. With the BUF04 as U2 (not shown) THD + N is comparable at lower output levels, but does show a distortion rise with 8 V rms output at high frequencies (yet still below 0.01%).

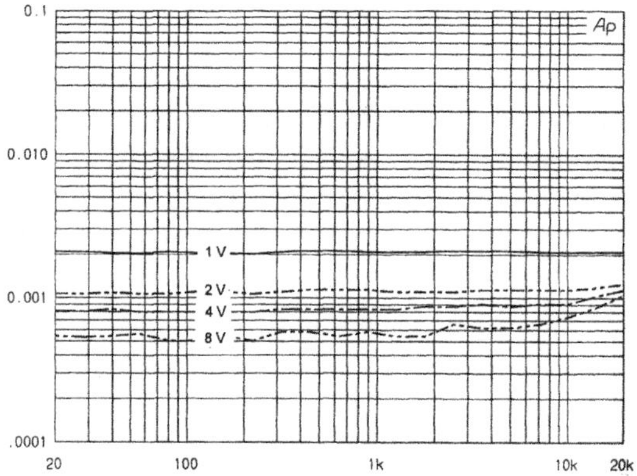

Figure 6-50: Current boosted driver of Figure 6-49 using AD811 as U2, THD + N (%) versus frequency (Hz), for V_{OUT} = 1, 2, 4, 8 V rms, R_L = 150 Ω, V_S = ±18 V

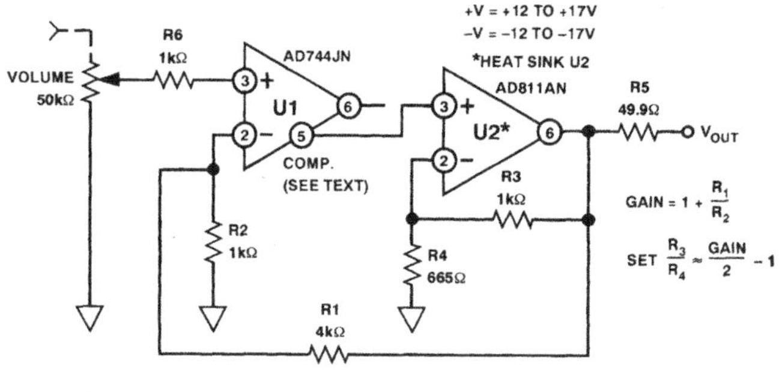

Figure 6-51: Composite current-boosted line driver one

There is a power/performance tradeoff involved with the choice between the two mentioned U2 devices which should be understood. The BUF04 has a standby dissipation of about 200 mW on ±15 V, while the AD811 is more than double this dissipation, at 500 mW. Therefore, while the AD811 does yield the lower distortion, it also should be operated more conservatively from a power standpoint. As noted above, only the minimum (±) supply voltage required to sustain a given output should be used with the circuit in general, and particularly with the AD811 employed at U2.

As for U1 in this circuit, other amplifiers can be used, but only with due caution against poor performance. Quite simply, it is difficult to improve upon the AD845's performance in this application. Three possible candidates would include the "Group C" op amps of Figure 6-43, operating on suitable power supplies;

±17 V or less for the AD825, and ±12 V or less for the AD8610 and AD8065. Of these, the AD8065 would seem to hold the greatest promise, having shown the lowest wideband distortion in the Figure 6-43 tests.

As considered within the buffered driver circuit of Figure 6-50, however, the AD8065 will be operated in an even more linear fashion; that is, it is operating essentially unloaded at the output. This is a key factor towards highest performance, as it moves the burden of linear load drive to the buffer stage. Further variations of this circuit technique will be reprised later, within other driver applications to be discussed.

Composite Current-Boosted Drivers

Another useful current-boosted circuit technique combines the positive aspects of two different amplifiers into a single composite amp structure, producing a very high performance line driver (see References 10–13 for several variations of this basic circuit). With an FET input IC used as the input stage, dc offset change from source resistance variations of a typical volume control of ≈50 kΩ is nil, allowing total direct coupling. As previously noted, with a high current, wideband booster output stage, line impedances down to 150 Ω can be driven with excellent linearity.

This type of composite amplifier allows good features of two dissimilar ICs to be exploited; each optimized for the respective input and output tasks. Figure 6-51 shows a low distortion composite amplifier using two op amp ICs with such performance.

A factor here aiding performance is that the U1 AD744 stage operates unloaded, and also that the AD744's compensation pin (5) drives U2. This step (unique to the AD744) removes any possibilities of U1 class AB output stage distortion. Another key point is that the overall gain bandwidth and SR of U1 are boosted by a factor equal to the voltage gain of U2, an AD811 op amp, which itself operates at a voltage gain. These factors enhance this circuit by providing both high and linear load current capability, providing a composite equivalent of an FET input power op amp.

This design operates at an *overall* voltage gain set by R_1 and R_2 (just as a conventional noninverting amplifier) which in this case is five times. Since the circuit also uses a local loop around stage U2, the R_3/R_4 ratio setting the U2 stage gain should be selected as noted. This complements the overall gain set by R_1 and R_2, and optimizes loop stability.

Also note that the U2's feedback resistor R_3 has a preferred minimum value for stability purposes (again, as is unique to CFB amplifier types). Here with the AD811, a 1 kΩ value suffices, so this value is fixed. R_4 is then chosen for the required U2 stage gain. Further design details are contained in the original references (see References 2 and 10).

The composite amplifier performance for a typical audio load of 600 Ω, THD + N at output levels of 1, 2, 4 and 8 V rms is shown in Figure 6-52, while operating from supplies of ±18 V for this test. The apparent distortion is noise or residual limited at almost all levels, rising just slightly at the higher frequencies.

Lower impedance loads can also be driven with this circuit, down to 150 Ω. Note that for operating supply voltages of more than ±12 V, a clip on heat sink is recommended for U2, as previously discussed for the AD811. Practical versions of this circuit can readily use supplies of ±12 V, and still operate very well.

The circuit of Figure 6-51 is a very flexible one, and can also be adapted in a variety of ways. Although the original version shown uses the AD744 compensation pin (5) to drive the output stage U2 device, conventional internally compensated op amps can also be used for U1, and still realize the many features of the architecture.

The ability to adapt the topology to differing devices in single and dual op amp formats allows such dual FET devices as the AD823 to be usefully employed in a stereo realization. Similarly, dual CFB op amps such as the AD812 can be used in the U2 output stage. Thus a complete stereo version of the circuit can be efficiently built, based on only two IC packages.

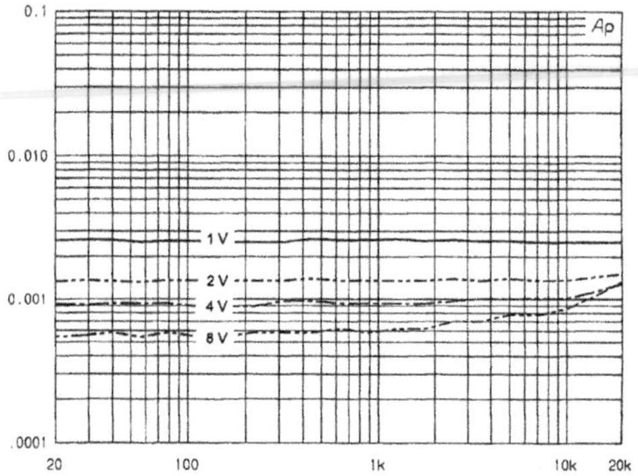

Figure 6-52: Composite current-boosted driver one of Figure 6-51, THD + N (%) versus frequency (Hz), for V_{OUT} = 1, 2, 4, 8 V rms, R_L = 600 Ω, V_S = ±18 V

This topology's flexibility also opens up a diversity of other applications beyond the basic line driver. For example, using a power-packaged dual CFB op amp such as the AD815 for U2, allows very low impedance loads such as headphones to be driven, down to as low as 10 Ω (see Reference 12).

The composite current-boosted line driver two, shown in Figure 6-53, summarizes a number of the above-mentioned options, and adds some other features as well.

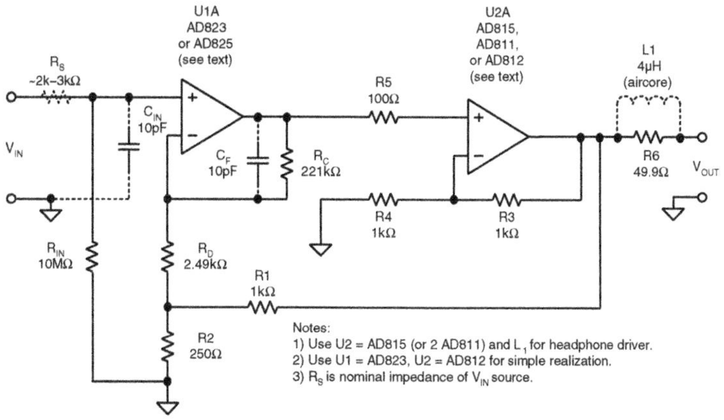

Figure 6-53: Composite current-boosted line driver two

Similarities within this circuit to the predecessor are resistances R_1–R_4, which perform similar functions to the previous version. Overall gain is again calculated via R_1-R_2, while output stage gain is set via R_3, R_4, and so forth.

Note that an additional pair of resistances, R_C and R_D, form a local feedback path around stage U1. This addition allows the effective open-loop bandwidth of U1 as it operates within the overall loop to be increased. For the values shown, using an AD823 for U1, the open-loop bandwidth is about 100 kHz. This means that the open-loop bandwidth of the entire circuit is greater than the audio bandwidth, which means phase errors within the pass band will be minimized.

An optional small capacitance (C_F, 10 pF–20 pF) can be useful for stabilizing the U1 stage, particularly if it employs a wide bandwidth device such as the AD825. When C_F is used, a like capacitor C_{IN} can also be used, to preserve high frequency impedance matching.

The primary input impedance balancing of the circuit is accomplished via resistance R_D, which has a dual role. External resistance R_S is the nominal output resistance of a volume control (typical for a 50 kΩ audio taper control at listening level). R_D is chosen to match R_S, and R_C will be approximately 100 times the R_D value when using the AD823.

The necessity of inductor output L1 depends upon whether the circuit is to be used with low impedance loads. For headphones, the L1 choke is necessary to prevent excessive voltage loss from a simple R_6 connection. R_6 is used in either a headphone or line driver case. If configured as a headphone driver, the circuit should use several square inches of PCB area around U2, to heat sink the AD815 device (see device data sheet). The AD811 and AD812 can also be used to drive higher impedance phones, such as 100 Ω or more.

Because of the vast number of options with this circuit, no performance is presented here. However, some insight into headphone driver performance is contained in Reference 12.

Differential Line Drivers

Unlike differential line receivers, a standard circuit topology for differential line *drivers* isn't nearly so clear-cut. A variety of different circuit types for driving audio lines in a balanced mode are discussed in this section, with their contrasts in performance and complexity. The virtues of balanced audio line operation are many. The largest and most obvious advantage is the inherent rejection of inevitable system ground noises, between the driver and receiver equipment locations.

There are also more subtle advantages to balanced line operation. Differential drivers tend to inject less noise onto the power supply rails. Related to this, they also produce inherently less noise onto the ground system, since by definition the return path for a differential signal is not ground. This can be a significant advantage when high currents need to be driven into a long audio line, as it can reduce multiple channel crosstalk. The circuits that follow illustrate a variety of methods for differential line driving.

"Inverter-Follower" Differential Line Driver

A straightforward approach to developing a differential drive signal of 2 V_{IN} is to amplify in complementary fashion a single-ended input V_{IN}, with equal gain inverter and follower op amp stages. With op amp gains of ±1, this develops outputs $-V_{IN}$ and V_{IN} with respect to common, or $V_{OUT} = 2\ V_{IN}$ differentially. This "inverter/follower" driver is easily accomplished with a dual op amp such as the OP275, plus an 8 × 20k film resistor network (or discrete), as shown in Figure 6-54. Here U1A provides the gain of –1 channel, while U1B operates at a gain of +1. The differential output signal across the balanced output line is 2 V_{IN}, and the differential output impedance is equal to $R_A + R_B$, or 100 Ω. The output resistors $R_A + R_B$ should be well matched, for reasons discussed earlier.

Use of like-value gain resistors around the U1 sections makes the respective channel noise gains match, and also makes their purchase easy. In addition, this forces the source impedances seen by the op amp ± inputs to be matched. Capacitors C_1-C_2 provide a ultrasonic roll-off, and enhance stability into capacitive lines. Overall, this circuit is high in performance for its cost and simplicity. Note that if a resistor network is used for R_1– R_8, the entire circuit can be built with only eight components.

THD + N performance of the Figure 6-54 circuit operating on ±18 V supplies is shown in Figure 6-55, for a series of successive sweeps resulting in output levels of 1, 2, 5 and 10 V rms across 600 Ω. The distortion in most instances is about 0.001%, and somewhat higher at a 1 V output level (noise limited at this level). Maximum output level is about 12 V rms into 600 Ω before clipping (not shown).

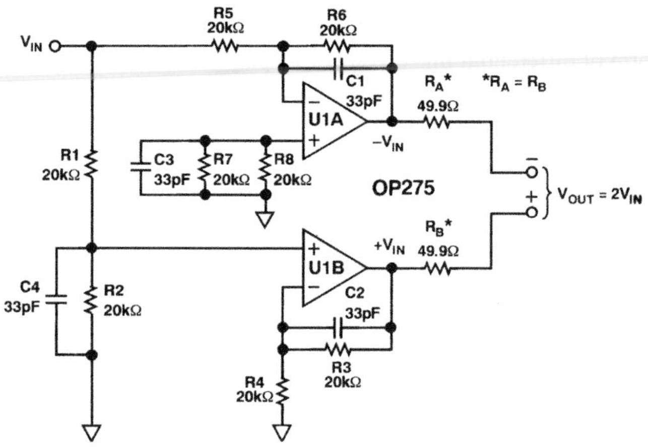

Figure 6-54: An "inverter-follower" differential line driver

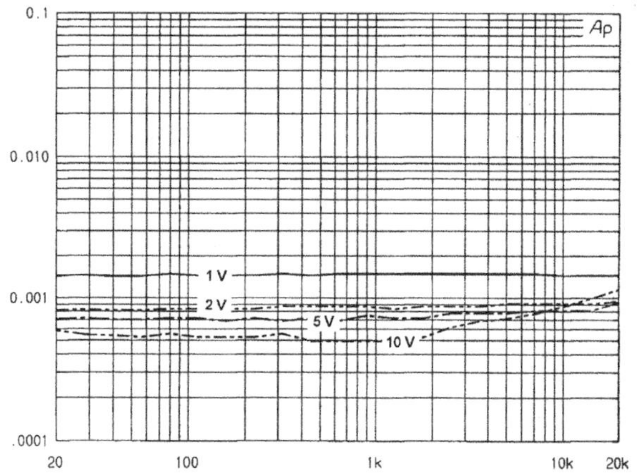

Figure 6-55: Inverter-follower driver of Figure 6-54, THD + N (%) versus frequency (Hz), for V_{OUT} = 1, 2, 5, 10 V rms, R_L = 600 Ω, V_S = ±18 V

In system terms, this type of differential line driver can potentially run into application problems, and should be used with some caveats in mind. In reality, this driver circuit uses two mirror-imaged, single-ended drivers, and they produce voltage output signals with respect to the source (V_{IN}) common point.

At the load end of a cable being driven, if the receiver used has a high impedance differential input (such as discussed in the line receiver section) there is no real problem in application for this driver circuit. However, it should be noted that one side of the differential output from Figure 6-54 *cannot be grounded without side effect.* This is because the source drive V_{OUT} is *not* truly floating, as would be in the case of a transformer winding.

In this sense, the circuit is *pseudo differential,* and it shouldn't be used indiscriminately. Nevertheless, within small and defined systems, is still has the obvious advantage of simplicity and, as noted, it can achieve high performance. Note also that with the matched source resistances R_A and R_B of 49.9 Ω as shown, nothing will be damaged even if one output is shorted—other than a loss of half the signal. Finally, if balanced high impedance differential loading is used at the receiver, there will be no side effects.

Cross-Coupled Differential Line Driver

A more sophisticated form of differential line driver uses a pair of *cross-coupled* op amps with both positive and negative feedback paths. The general form of this type of circuit is a cross-coupled Howland circuit, after the classic resistor bridge based current pump. The cross-coupled form was described by Pontis in a solid-state transformer emulator for high performance instrumentation (see Reference 13).

Application-wise, this configuration provides maximum flexibility, allowing a differential output signal V_{OUT} to be maintained constant and independent of the load common connections. This means that either side can be shorted to common without loss of signal level, i.e., as can be done with a transformer.

Figure 6-56 shows the SSM2142 balanced line driver IC in an application. The SSM2142 consists of two Howland circuits A2 and A3, cross-coupled as noted, plus an input buffer (A1). The trimmed multiple resistor array and trio of op amps shown is packaged in an 8-pin miniDIP IC with the pinout noted.

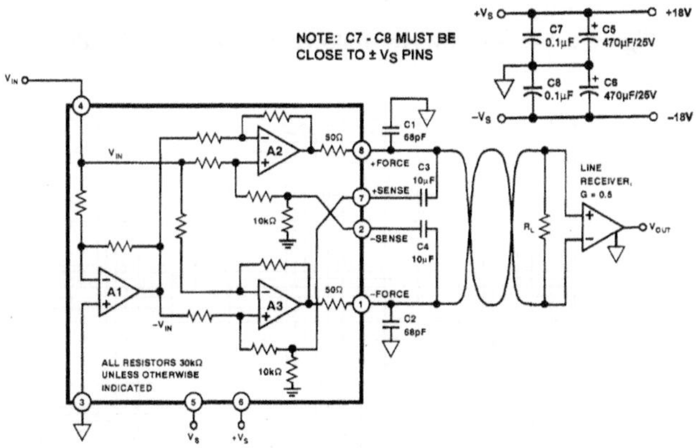

Figure 6-56: SSM2142 cross-coupled differential line driver used within balanced driver/receiver system

The SSM2142 line driver is designed for a single-ended to differential gain of 2 working into a 600 Ω load. In the simplest use, it is strapped with the respective output FORCE/SENSE pins tied together (7-8, 1-2). Small film capacitors C_1-C_2 preload the IC for stability against varying cable lengths. To decouple line dc offsets, the optional capacitors C_3-C_4 are used as shown, and should be nonpolar types, preferably films.

An additional "housekeeping" caveat with the SSM2142 involves the high frequency power supply bypassing. The 0.1 µF low inductance bypass caps C_7 and C_8 must be within 0.25" of power supply Pins 5 and 6, as noted in the figure. If this bypassing is compromised by long lead lengths, excessive THD will be evident.

In a system application, the SSM2142 is used with a complementary gain of 0.5 receiver, either an SSM2143, or one of the other line receivers discussed previously. The complete hookup of Figure 6-56 comprises an entire single-ended-to-differential and back to single-ended transmission system, with noise isolation and a net end-to-end unity gain.

Figure 6-57 shows the THD + N performance of the SSM2142 driver portion of Figure 6-56, for sweeps yielding output levels of 1, 2, 5 and 10 V rms across 600 Ω. While performance is noise limited for the 1 V output curve, distortion drops to ≤0.001% and near residual for most higher levels, rising only with higher frequencies and the 10 V output curve.

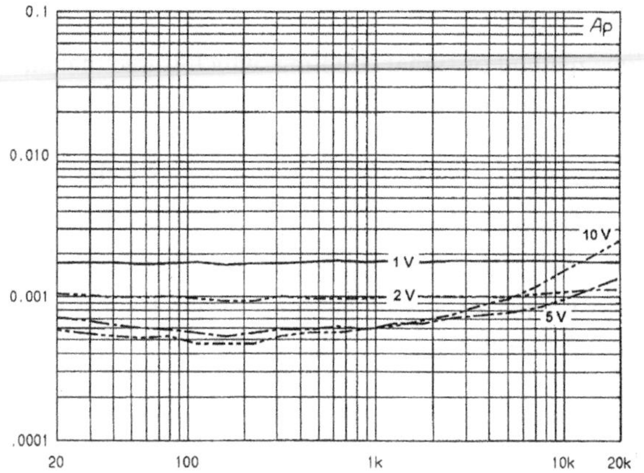

Figure 6-57: SSM2142 driver portion of Figure 6-56, THD + N (%) versus frequency (Hz), for V_{OUT} = 1, 2, 5, 10 V rms, R_L = 600 Ω, V_S = ±18 V

These two differential drivers are suited for 600 Ω or higher loads, and, within those constraints, perform well.

As should be obvious, these drivers do *not* offer galvanic isolation, which means that in all applications there must be a dc current path between the grounds of the driver and the final receiver. In practice however this isn't necessarily a problem.

The following circuits illustrate differential drivers that do offer galvanic isolation, and can therefore be used with ground potential differences up to several hundred volts (or the actual voltage breakdown rating of the transformer in use).

Transformer-Coupled Line Drivers

Transformers provide a unique method of signal coupling, which is one that allows completely isolated common potentials, i.e., *galvanic isolation*. As noted previously in the line receiver section, transformers are not without their technical and practical limitations, but their singular ability to galvanically isolate grounds maintains a place for them in difficult application areas (see References 15 and 16).

Basic Transformer-Coupled Line Driver

The circuit of Figure 6-58 uses some previously described concepts to form a basic low dc offset, high linearity driver using a high quality nickel core output transformer. U1 and U2 form a high current driver, similar to the Figure 6-49 current-boosted driver.

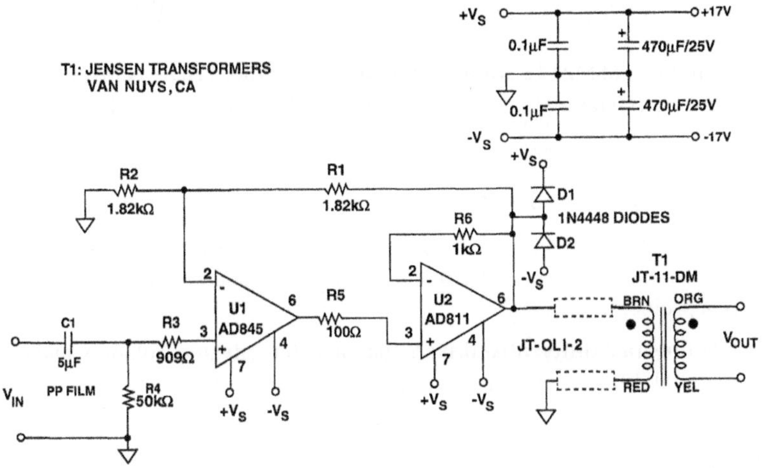

Figure 6-58: A basic transformer-coupled line driver

In this circuit U1 is a low offset voltage FET input op amp, for the purpose of holding the dc offset seen at the primary of T1 to a minimum (±12.5 mV maximum as shown, typically less). Dc current flowing into the primary winding of a transformer should be minimized, for lowest distortion. C_1, a high quality film capacitor, decouples any dc offset present on V_{IN}, for similar reasons.

The U1-U2 device combination is capable of ±100 mA or more of output, which aids greatly in the ability of this circuit to drive low impedances. The buffering of U2 is recommended for long lines, or for the absolute lowest distortion. Although T1 is shown with a 1:1 coupling ratio, other winding configurations are possible with transformer variations, that is step-up or step-down, allowing either 600 Ω or 150 Ω loads.

As can be noted, the T1 primary isn't driven directly, but is isolated by two series isolation devices, Jensen JT-OLI-2s. Each of these is an LR shunt combination of about 39 Ω and 3.7 µH. The net impedance offers a very low DCR, and an increasing impedance above 1.5 MHz for load isolation (see device data sheet and Reference 17). The use of two isolators as shown offers best output CMR rejection for the transformer, but one will also work (with less CMR performance), as will a single 10 Ω resistor.

THD + N performance for this driver-transformer combination is shown in Figure 6-59, for supplies of ±18 V and successive input sweeps, resulting in outputs of 1, 2, 4, and 8 V rms into 600 Ω. These data were taken with a single series resistance of 10 Ω driving T1 (which could be conservative compared to operation with two isolators).

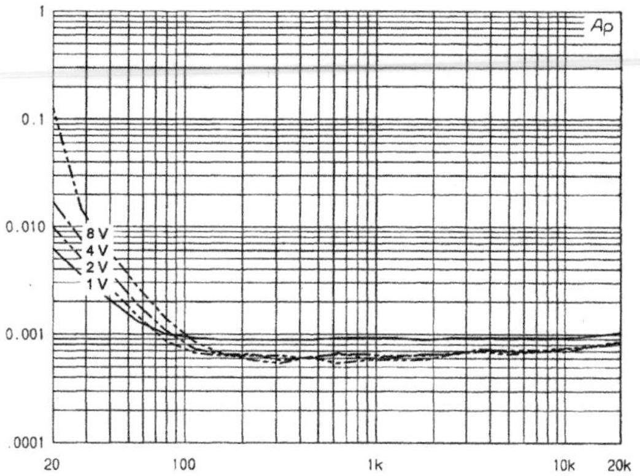

Figure 6-59: Transformer driver of Figure 6-58, THD + N (%) versus frequency (Hz), for V_{OUT} = 1, 2, 4, 8 V rms, R_L = 600 Ω, V_S = ±18 V

As with the 2x and 5x basic drivers previously described, these data are essentially distortion free above 100 Hz. At lower frequencies there is seen a level dependent, inverse-frequency dependent distortion. The measured distortion reaches a maximum at 20 Hz with output levels of 8 V rms (≈20 dBm), while at lower levels it is substantially less.

To one degree or another, this distortion phenomenon is basic to audio transformers. It is lessened (but not totally eliminated) in the higher quality transformer types, such as the nickel-core unit used in the Figure 6-58 circuit.

In practice, there are some factors that tend to mitigate the seriousness of the low frequency distortion seen in the performance data of Figure 6-59. First, rarely will maximum audio levels ever be seen at 20 Hz. Thus suitably derated operation of T1 will strongly reduce the incidence of this distortion.

However, if the lowest distortion possible independent of level is desired, then some additional effort will need to be expended on making the transformer driver more sophisticated. This can take the form of actively applying feedback around the transformer, to lower its nonlinearity to negligible levels. This is design approach is discussed with the next driver circuits.

Feedback Transformer-Coupled Line Drivers

While nonpremium core transformers are more economical than the nickel core types, as a trade-off they do have much higher distortion. To further complicate the design issue, the distortion characteristics of most transformers varies with level and frequency in complex ways, rising more rapidly at higher levels and lower frequencies. This behavior is even less forgiving than that of the nickel core types, and somewhat complicates the application of audio transformers. While a nickel core transformer has distortion characteristics sufficiently low as to allow their use without distortion correction (Figure 6-58) the same simply isn't true for other core materials.

A family of distortion curves for another transformer type illustrates this behavior, shown in Figure 6-60. This series of plots is for a Lundahl LL1517 silicon iron C core unit, with successive output levels of 0.5, 1, 2 and 5 V rms into a 600 Ω load. Individual device samples will vary, but the general pattern is typical of many audio transformers.

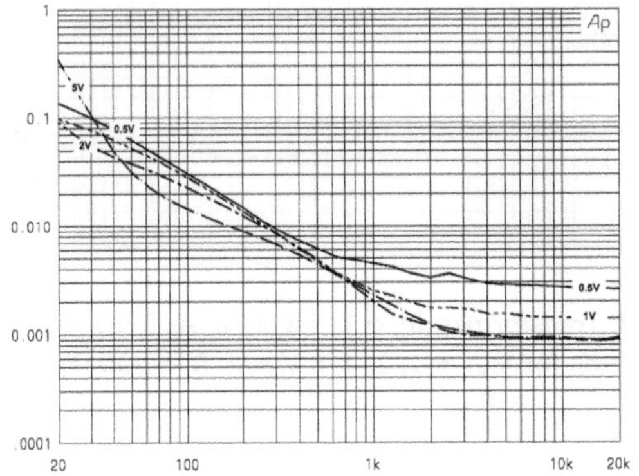

Figure 6-60: Lundahl LL1517 transformer and driver (without feedback), THD + N (%) versus frequency (Hz), for V_{OUT} = 0.5, 1, 2, 5 V rms, R_L = 600 Ω

Werner Baudisch (see Reference 18) developed a very effective driver technique for minimization of transformer distortion. The technique involves the use of a drive amplifier, connected to the transformer primary in a direct manner. The amplifier uses conventional negative feedback for gain stabilization. In addition, a primary sensing resistance develops a voltage sample proportional to primary current, and the voltage thus derived is also fed back to the amplifier. This second feedback path is positive feedback, so the arrangement is also known as a *mixed feedback* driver (see Reference 19).

This very useful technique of the mixed feedback driver can be used to advantage to integrate a line driver with the transformer primary within a feedback loop, which cancels the bulk of the objectionable distortion. In practice, with careful driver adjustment it is possible to reduce the distortion of the transformer plus driver almost to *that of the driver stage, operating without the transformer.* The beauty of the principle is that the inherent floating transformer operation is not lost, and is still effectively applied in a highly linear mode. Due to the action of the mixed feedback, the transformer primary resistance is effectively cancelled, thus appreciably lowering the net secondary output impedance.

The circuit of Figure 6-61 is a basic, single-ended mixed feedback driver using either a Lundahl LL1582 or LL2811 transformer as T1, and an AD845 or an OP275 as the amplifier. These transformers have two 1:1 primaries, as well as two 1:1 secondaries. As used, both primaries are connected in series, and the T1 net voltage transfer is unity.

To enable correct mixed feedback operation, two key ratios within the circuit must be set to match. One ratio is between the net T1 primary resistance, $R_{PRIMARY}$ and sample resistor R_4, and the other is R_2 and R_1. This relationship is:

$$R_{PRIMARY}/R_4 = R_2/R_1 \qquad \text{Eq. 6-18}$$

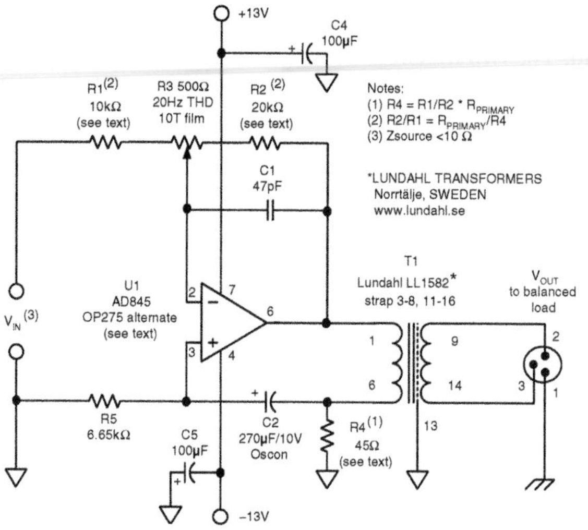

Figure 6-61: A basic single-ended mixed feedback transformer driver

It is important to note that $R_{PRIMARY}$ is the total effective dc resistance of T1. As used here, two series 45 Ω primaries are used, so $R_{PRIMARY}$ is 90 Ω. Gain of the driver circuit is established as in a standard inverter, or the R_2-R_1 ratio. For a gain of 2 ×, R_2 is then simply 2 times R_1, i.e., 20 kΩ and 10 kΩ. R_4 may then be selected as:

$$R_4 = R_1/R_2 * R_{PRIMARY} \qquad \text{Eq. 6-19}$$

With the R_1/R_2 ratio of 0.5, this makes R_4 simply one-half $R_{PRIMARY}$, or in this case 45 Ω.

Note the value of R_1 is critical, thus the V_{IN} source impedance must be low (<10 Ω). This and other subtleties are effective performance keys. One is the sensitivity of the ratio match described by Eq. 6-18. Only when trimmed optimally will the lowest frequency THD be minimum. Thus a multiturn film trimmer R_3 is used to trim out the various tolerances and the winding resistance of T1. Further, the positive feedback path is ac-coupled via C_2. This provision prevents dc latchup, should positive feedback override the negative. However, a simple time constant of, say, 8 ms (corresponding to 20 Hz) is *not sufficient* for lowest low frequency THD. To counteract this, the C_2-R_5 time constant is set quite long (≈1.8 seconds), which enables lowest possible 20 Hz THD. With the suggested AD845 for U1, distortion is lowest, as it is also with an Oscon capacitor for C_2. A larger value ordinary aluminum electrolytic can also be used for C_2, with a penalty of somewhat high distortion. Alternately, an OP275 can also be used for U1 (see following).

With the AD845 FET input op amp used for U1, the maximum dc at the T1 primary is essentially the amplifier Vos times the stage's 3× noise gain, or ≤7.5 mV. Since the AD845 can also dissipate ≈250 mW, the lowest possible supplies help keep the offset change with temperature as low as possible.

Lab THD + N measurements of Figure 6-61 were made using an LL2811, a transformer like the LL1582, but without a Faraday shield. The two transformers are very similar, but the LL1582 is recommended for single-ended drive circuits. The performance of this feedback driver is shown in Figure 6-62, for successive output levels of 0.5, 1, 2, and 5 V rms into a 600 Ω load. Comparison of these data with Figure 6-60 bears out the utility of the distortion reduction; it is decreased by orders of magnitude. More importantly, the level dependence with decreasing frequency is essentially eliminated.

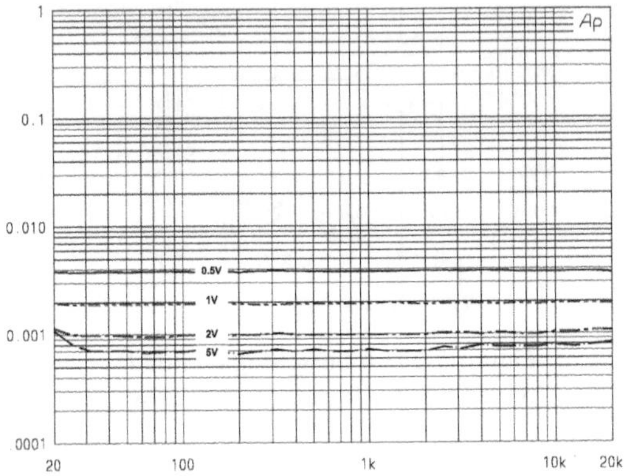

Figure 6-62: Figure 6-61 driver with Lundahl LL2811 transformer and AD845, THD + N (%) versus frequency (Hz), for V_{OUT} = 0.5, 1, 2, 5 V rms, R_L = 600 Ω

These data do in fact represent almost an ideal THD + N pattern; the distortion level is flat with frequency, and it decreases with increasing output level. An extremely slight increase in THD + N can just be discerned at 20 Hz in the 5 V curve. The alternate OP275 for U1 also works well, but does have slightly higher distortion (not shown).

Although directly comparable data is not presented for it, it is worth noting that the LL1517 transformer can also be used with the Figure 6-61 circuit, with R_4 = 9.2 Ω, and the two primaries connected in series. However, some additional data on a circuit quite similar to Figure 6-61 does reveal a potential limitation for this type of driver.

Figure 6-63 shows a set of high level THD + N curves for a mixed feedback driver using an AD8610 op amp for U1, and the LL1517 transformer. Three THD + N sweeps are made, with the lowest THD + N curve representing the best possible null. The other two curves show increased THD + N at low frequencies, for conditions of R_2/R_1 ratio mismatches of 1% and 5%, respectively. This demonstrates how critical a proper null is towards achieving the lowest possible distortion at the low end of the audio band.

It is possible to tweak the ratio via R_3 for an excellent 20 Hz high level null at room temperature, and this is recommended to get the most from one of these circuits. It must also be remembered that the TC of the T1 copper windings is about 0.39%/°C. Therefore, only a 10°C ambient temperature change would be sufficient to degrade the best null by nearly 5%. The resulting performance would then roughly represent the upper curve of Figure 6-63—still quite good, but just not quite as good as possible in absolute terms.

For the best and most consistent performance, wide temperature range applications of this type of circuit should therefore employ some means of temperature compensation for the copper winding(s) of T1. One means of achieving this would be to employ a thermally sensitive device to track the copper TC of T1. The

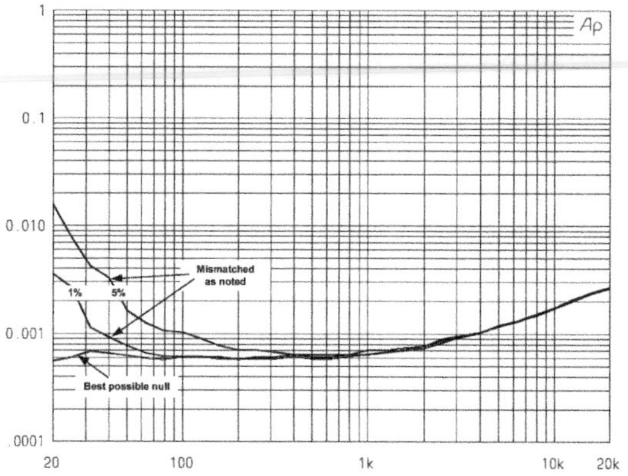

Figure 6-63: Lundahl LL1517 transformer with mixed feedback AD8610 driver, THD + N (%) versus frequency (Hz) for various null accuracies

net goal should be to hold the $R_{PRIMARY}/R_4$ ratio constant over temperature. It should also be noted that for this approach to work, it is assumed that the R_2/R_1 ratio is temperature independent. This is possible with the use of close tolerance, low TC metal film resistors, i.e., 50 ppm/°C or better (or the use of a low tracking TC network).

It should also be noted that the *output balance* of an audio transformer is a very important factor when designing audio line drivers. Poor transformer balance can lead to mode conversion of CM signals on the output line (see Reference 20). The result is that a spurious differential mode signal can be created due to poor balance. A transformer can attain good balance (i.e., 60 dB or better) by the use of sophisticated winding techniques, or the use of a Faraday shield, as is true in the case of the LL1582 and the LL1517.

Transformer drivers can of course also be operated in a balanced drive fashion. This has the advantage of doubling the available drive voltage for given supply voltages, plus lowering the distortion produced. Mixed feedback principles can be extended to a balanced arrangement, which lowers the distortion in the same manner as for the single-ended circuit just described. An example circuit is shown in Figure 6-64.

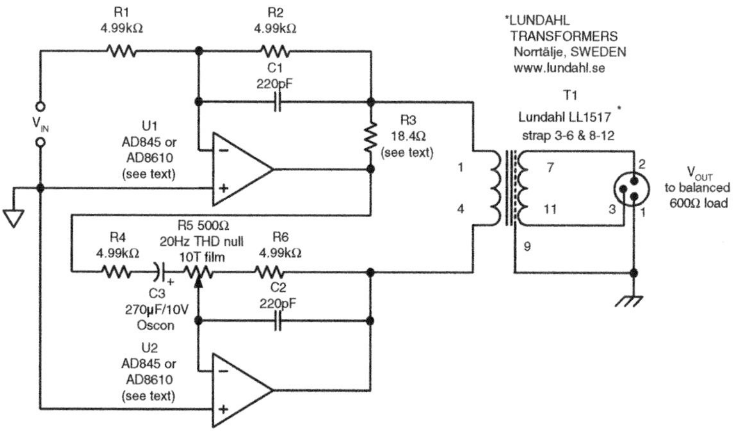

Figure 6-64: A balanced transformer driver circuit that applies mixed feedback principles of distortion minimization

In Figure 6-64, a U1-U2 low distortion op amp pair drive an LL1517 transformer. U1 is an inverting gain circuit as defined by gain resistors R_1-R_2, which drives the top of T1. Placed in series with the T1 primary, R_3 acts as a current sampling resistor, and develops a correction voltage to drive the second inverter, U2, through R_4.

This scheme is adapted from the mixed feedback balanced driver circuit of Arne Offenberg (see Reference 21). There are, however, two main differences in this version. One is operation of the U1 stage as an inverter, which eliminates any CM distortion effects in U1, and the second being the ability to easily set the overall driver gain via R_1-R_2. Within this circuit, it should be noted that the resistances R_1-R_2 do *not* affect the distortion null (as they do in the simpler circuit of Figure 6-61).

The distortion null in this form of the circuit occurs when the ratios $R_3/R_{PRIMARY}$ and R_4/R_6 match. For simplicity, the second inverter gain is set to unity, so R_3 is selected as:

$$R_3 = R_4/R_6 * R_{PRIMARY} \qquad \text{Eq. 6-20}$$

For the R_4–R_6 values shown, R_3 then is simply equal to $R_{PRIMARY}$, or 18.4 Ω when used with the LL1517 transformer with series connected primaries.

As with any of these driver circuits, the exact op amp selection has a great bearing on final performance. Within a circuit using two amplifiers, dual devices are obviously attractive. The distortion testing below discusses amplifier options.

THD + N performance data for the balanced transformer driver of Figure 6-64 is shown in Figure 6-65, using an LL1517 transformer with successive output levels of 0.5, 1, 2, and 5 V rms into a 600 Ω load. For these tests, the supply voltages were ±13 V, and the U1-U2 op amp test devices were pairs of either the AD8610 or the AD845 (note—two AD8610 singles are comparable to a single AD8620 dual).

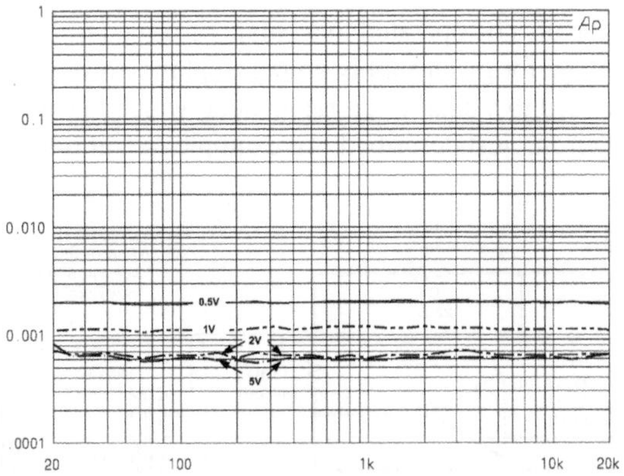

Figure 6-65: Figure 6-64 balanced driver with Lundahl LL1517 transformer and two AD8610s, THD + N (%) versus frequency (Hz), for V_{OUT} = 0.5, 1, 2, 5 V rms, R_L = 600 Ω

An interesting thing about these plots is the fact that the THD + N is both low and essentially unchanged with frequency, which is again near ideal. Low frequency nulling of the distortion is almost as critical in this circuit as in the previous, and a slight upturn in THD + N can be seen at 20 Hz, for the highest level (5 V).

For the AD8610 devices shown by these data, the wideband THD + N was slightly lower than a comparable test with the AD845 pair (the latter not shown). Both amplifier sets show essentially flat THD + N versus frequency characteristics.

Because of the balanced drive nature of this circuit, the realization offers lower distortion than the simpler single-ended version of Figure 6-61, plus a buffering of the distortion null sensitivity against the input source impedance and gain adjustment. It can thus be considered a more robust method of transformer distortion minimization. For these reasons, the balanced form of driver is recommended for professional or other high performance requirements. Note, however, that similar caveats do apply with regard to stabilizing the distortion null against temperature.

This driver can be used with the LL1517 and many other transformers with, of course, an appropriate choice of R_3. Note that the performance data above reflects use of the op amps operating unbuffered. For very low impedance loads and/or long lines, a pair of the previously described unity gain buffers should be considered, and both the U1 and U2 stages operated with output buffering. This will allow the retention of THD + N performance as is shown in Figure 6-65, but in the face of more difficult loads.

References: Audio Line Drivers

1. Walt Jung, "Op Amps in Line-Driver and Receiver Circuits, Part 1," **Analog Dialogue**, 26-2, 1992.
2. W. Jung, A. Garcia, "Op Amps in Line-Driver and Receiver Circuits, Part 2," **Analog Dialogue**, 27-1, 1993.
3. Walt Jung, "Walt's Tools & Tips: 'Op Amp Audio: Buffers Part I'" **Electronic Design**, September 1, 1998, pp. 165, 166.
4. Walt Jung, "Walt's Tools & Tips: 'Op Amp Audio: Buffers Part II'" **Electronic Design**, October 1, 1998, pp. 103, 104.
5. Paul Brokaw, "An IC Amplifer User's Guide to Decoupling, Grounding, And Making Things Go Right For a Change," **Analog Devices AN202**.
6. P. Brokaw, J. Barrow, "Grounding for Low and High Frequency Circuits," **Analog Dialogue**, 23-3 (1989).
7. Alan Rich, "Shielding and Guarding," **Analog Dialogue**, 17-1 (1983).
8. Scott Wurcer, "Input Impedance Compensation," discussion within **AD743 data sheet**, Analog Devices.
9. Scott Wurcer, "An Operational Amplifier Architecture With a Single Gain Stage and Distortion Cancellation," presented at **92nd Audio Engineering Society Convention**, March 1992, preprint #3231.
10. W. Jung, S. Wurcer, 'A High Performance Audio Composite Line Stage' within "Applications for Amplifiers in Audio," Ch. 5 in W. Kester, Editor, **1992 Amplifier Applications Guide**, Analog Devices, Inc., Norwood, MA, 1992, ISBN 0-916550-10-9.
11. Walt Jung, 'High Performance Audio Stages Using Transimpedance Amplifiers,' within Gary Galo, "POOGE-5: Rite of Passage for the DAC960," **The Audio Amateur**, issue 2, 1992.
12. Walt Jung, "Composite Line Driver with Low Distortion" **Electronic Design Analog Special Issue**, June 24, 1996, p. 78.
13. Walt Jung, "Walt's Tools & Tips: 'Op Amp Audio: Minimizing Input errors'" **Electronic Design**, December 14, 1998, pp. 80–82.
14. George Pontis, "Floating a Source Output," **HP Journal**, August 1980.
15. "Transformer Application Notes (various)," Jensen Transformers, 7135 Hayvenhurst Avenue, Van Nuys, CA, 91406, (213) 876-0059.
16. Bruce Hofer, "Transformers in Audio Design," **Sound & Video Contractor**, March 15, 1986.
17. Deane Jensen, "Some Tips on Stabilizing Operational Amplifiers," Jensen Transformers AN-001, 7135 Hayvenhurst Avenue, Van Nuys, CA, 91406, 213-876-0059.
18. Werner Baudisch, "Schaltungsanordnung mit Verstärker mit Ausgangsübartrager," **German patent DE2901567**, issued July 24, 1980.
19. Per Lundahl, "Mixed Feedback Drive Circuits For Audio Output Transformers," Lundahl Transformers, Norrtälje, Sweden, www.lundahl.se.
20. Per Lundahl, "Winding Arrangements of Output Transformers," Lundahl Transformers, Norrtälje, Sweden, www.lundahl.se.
21. Arne Offenberg, "Mixed Feedback Balanced Driver Circuit," LL2811 Audio Output Transformer data sheet, Lundahl Transformers, Norrtälje, Sweden, www.lundahl.se.

Buffer Amplifiers and Driving Capacitive Loads

Walt Jung, Walt Kester

Buffer Amplifiers

In the early days of high speed circuits, simple emitter followers were often used as high speed buffers. The term *buffer* was generally accepted to mean a unity-gain, open-loop amplifier. With the availability of matching PNP transistors, a simple emitter follower can be improved, as shown in Figure 6-66A. This complementary circuit offers first-order cancellation of dc offset voltage, and can achieve bandwidths greater than 100 MHz. Typical offset voltages without trimming are usually less than 50 mV, even with unmatched discrete transistors. The HOS-100 hybrid amplifier from Analog Devices represented an early implementation of this circuit. This device was a popular building block in early high speed ADCs, DACs, sample-and-holds, and multiplexers.

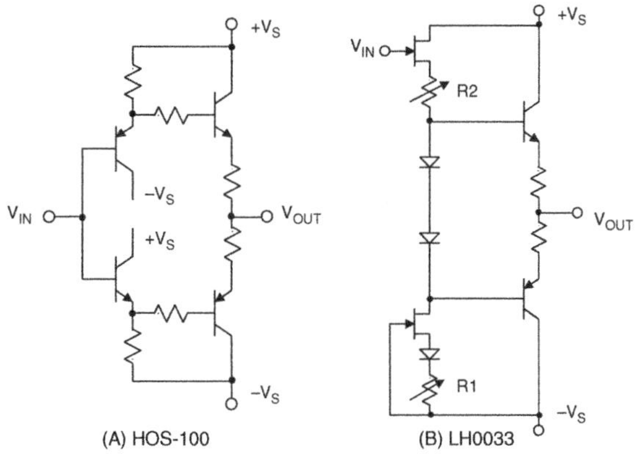

Figure 6-66: Early open-loop hybrid buffer amplifiers: (A) HOS-100 bipolar, (B) LH0033 FET input

If high input impedance is required, a dual FET can be used as an input stage ahead of a complementary emitter follower, as shown in Figure 6-66B. This form of the buffer circuit was implemented by both National Semiconductor Corporation as the LH0033, and by Analog Devices as the ADLH0033.

In the realizations of these hybrid devices, thick film resistors were laser trimmed to minimize input offset voltage. For example, in the Figure 6-66(B) circuit, R1 is first trimmed to set the bias current in the dual matched FET pair, which is from the 2N5911 series of parts. R2 is then trimmed to minimize the buffer input-to-output offset voltage.

Circuits such as these achieved bandwidths of about 100 MHz at fairly respectable levels of harmonic distortion, typically better than –60 dBc. However, they suffered from dc and ac nonlinearities when driving loads less than 500 Ω.

One of the first totally monolithic implementations of these functions was the Precision Monolithics, Inc. BUF03 shown in Figure 6-67 (see Reference 1). PMI is now a division of Analog Devices. This open-loop IC buffer achieved a bandwidth of about 50 MHz for a 2 V peak-to-peak signal.

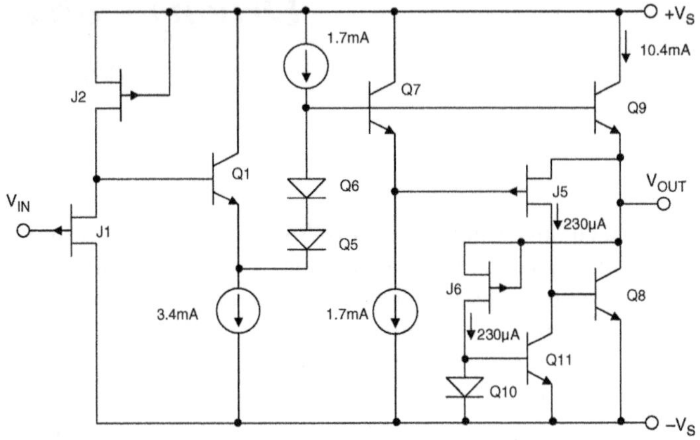

Figure 6-67: BUF03 monolithic open-loop buffer 1979 vintage

The BUF03 circuit is interesting because it demonstrates techniques that eliminated the requirement for the slow, bandwidth-limited vertical PNP transistors associated with most IC processes available at the time of the design (approximately 1979).

Within the circuit, input transistor J1 is a FET source follower that is biased by an identical FET J2, thereby making the gate-to-source voltage of J1 nominally zero. The output of J1 is applied to emitter follower Q1, and diodes Q5 and Q6 compensate for the combined base-emitter drops of Q1 and Q7/Q9.

The current through Q7 is held at a constant 1.7 mA, therefore its V_{BE} is constant. Transistors Q7 and Q9 are scaled such that the current in Q9 is six times that of Q7 for equal V_{BE} drops. If the load current changes, and Q9 is required to source more or less current, its V_{BE} attempts to increase or decrease. This change is applied between the gate and source of J5, which then reduces/increases current in the base of Q8 to maintain the current in Q9 at six times that of Q7. The localized feedback works for load currents up to ±10 mA. The current in Q9 is therefore held constant at 10.4 mA (independent of load) because its V_{BE} drop does not change with either output voltage or load current.

With a 1 kΩ load, and an output voltage of +10 V, transistor Q8 must sink 0.2 mA, and Q9 supplies 10 mA to the load, 0.2 mA to J5, and 0.2 mA to J6. For an output of –10 V, Q8 must sink 20.2 mA so that the net current delivered to the load is –10 mA. In addition to achieving a bandwidth of approximately 50 MHz (2 V peak-to-peak output), on-chip zener-zap trimming was used to achieve a dc offset of typically less than 6 mV.

One of the problems with all the open-loop buffers discussed thus far is that although high bandwidths can be achieved, the devices discussed don't take advantage of negative feedback. Distortion and dc performance suffer considerably when open-loop buffers are loaded with typical video impedance levels of 50 Ω, 75 Ω, or 100 Ω. The solution is to use a properly compensated wide bandwidth op amp in a unity-gain follower configuration. In the early days of monolithic op amps, process limitations prevented this, so the open-loop approach provided a popular interim solution.

Today, however, practically all unity-gain-stable voltage or current feedback op amps can be used in a simple follower configuration. Usually, however, the general-purpose op amps are compensated to operate over a wide range of gains and feedback conditions. Therefore, bandwidth suffers somewhat at low gains, especially in the unity-gain noninverting mode, and additional external compensation is usually required.

A practical solution is to compensate the op amp for the desired closed-loop gain, while including the gain setting resistors on-chip, as shown in Figure 6-68. Note that this form of op amp, internally configured as a buffer, may typically have no feedback pin. Also, putting the resistors and compensation on-chip also serves to reduce parasitics.

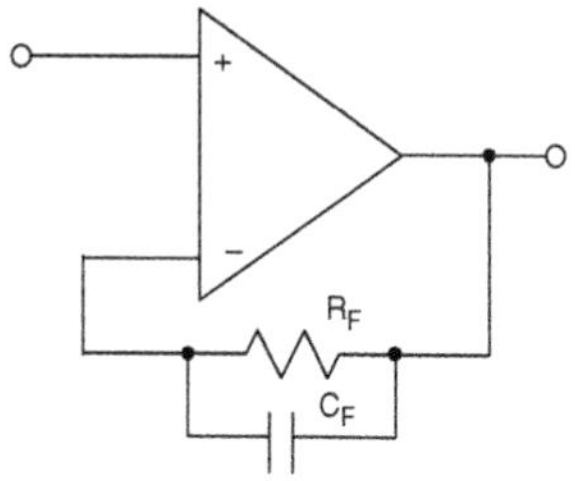

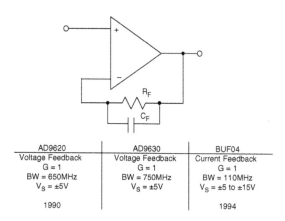

AD9620	AD9630	BUF04
Voltage Feedback	Voltage Feedback	Current Feedback
G = 1	G = 1	G = 1
BW = 650MHz	BW = 750MHz	BW = 110MHz
V_S = ±5V	V_S = ±5V	V_S = ±5 to ±15V
1990		1994

Figure 6-68: Early closed-loop unity-gain monolithic buffers

A number of op amps are optimized in this manner. Roy Gosser's AD9620 (see Reference 2) was probably the earliest monolithic implementation. The AD9620 was a 1990 product release, and achieved a bandwidth of 600 MHz using ±5 V supplies. It was optimized for unity gain, and used the voltage feedback architecture. A newer design based on similar techniques is the AD9630, which achieves a 750 MHz bandwidth.

The BUF04 unity gain buffer (see Reference 3) was released in 1994 and achieves a bandwidth of 120 MHz. This device was optimized for large signals and operates on supplies from ±5 V to ±15 V. Because of the wide supply range, the BUF04 is useful not only as a standalone unit-gain buffer, but also within a feedback loop with a standard op amp, to boost output (see discussions within "Audio Amplifiers" portion of this chapter).

Closed-loop buffers with a gain of two find wide applications as transmission line drivers, as shown below in Figure 6-69. The internally configured fixed gain of the amplifier compensates for the loss incurred by the source and load termination. Impedances of 50 Ω, 75 Ω, and 100 Ω are popular cable impedances. The AD8074/AD8075 500 MHz triple buffers are optimized for gains of 1 and 2, respectively. The dual AD8079A/AD8079B 260 MHz buffer is optimized for gains of 2 and 2.2, respectively.

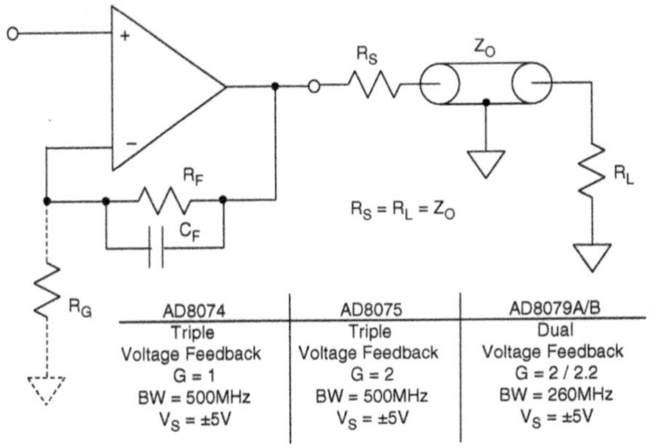

Figure 6-69: Fixed-gain video transmission line drivers

The buffer amplifiers discussed above are all dedicated to either unity or some higher fixed-gain setting. As wide bandwidth fixed-gain blocks they are simply applied, without the need for additional gain configuration components. They will, of course (as with any high-speed amplifier), require supply bypassing components as well as appropriate layout.

Buffers can also be implemented with almost any unity-gain-stable voltage or current feedback op amp. Examples that come to mind for voltage feedback devices are the single AD817 (or the dual counterpart AD826), or for current feedback devices the AD811, AD8001, AD8015, along with their dual-device cousins (as is applicable). In addition, there are op amps with feature rail-rail outputs as well as operation at low supply voltages—the AD8031/AD8032 and AD8041/AD8042 are examples.

In implementing a high speed unity-gain buffer with a voltage feedback op amp, there will typically be no resistor required in the feedback loop, which considerably simplifies the circuit. Note that this isn't a 100% hard-and-fast rule, however, so always check the device data sheet to be sure. A unity-gain buffer with a current feedback op amp will *always* require a feedback resistor, typically in the range of 500 Ω–1000 Ω. So, be sure to use a value appropriate not only to the basic part, but also the specific power supplies in use.

Driving Capacitive Loads

From either a system or signal fidelity point of view, transmission line coupling between stages is best, and is described in some detail in the next section. However, complete transmission line system design may not always be possible or practical. In addition, various other parasitic issues need careful consideration in high performance designs. One such problem parasitic is amplifier *load capacitance*, which potentially comes into play for all wide bandwidth situations that do not use transmission line signal coupling.

A general design rule for wideband linear drivers is that capacitive loading (cap loading) effects should *always* be considered. This is because PC board capacitance can build up quickly, especially for wide and long signal runs over ground planes insulated by a thin, higher K dielectric. For example, a 0.025" PC trace using a G-10 dielectric of 0.03" over a ground plane will run about 22 pF/foot (see References 4 and 5). Even relatively small load capacitance (i.e., <100 pF) can be troublesome, since while not causing outright oscillation, it can still stretch amplifier settling time to greater than desirable levels for a given accuracy.

The effects of cap loading on high speed amplifier outputs are not simply detrimental, they are actually an anathema to high quality signals. However, before-the-fact designer knowledge still allows high circuit performance by employing various tricks of the trade to combat the capacitive loading. If it is not driven via a transmission line, remote signal circuitry should be very carefully checked for capacitive loading, and characterized as well as possible. Drivers that face poorly defined load capacitance should be bulletproofed accordingly with an appropriate design technique from the options list below.

Short of a true matched transmission line system, a number of ways exist to drive a load that is capacitive in nature, while still maintaining amplifier stability.

Custom capacitive load (cap load) compensation includes two possible options, namely a) overcompensation, and b) an intentionally forced high loop noise gain allowing crossover in a stable region. Both of these steps can be effective in special situations, as they reduce the amplifier's effective closed-loop bandwidth, so as to restore stability in the presence of cap loading.

Overcompensation of the amplifier, when possible, reduces amplifier bandwidth so that the additional load capacitance no longer represents a danger to phase margin. As a practical matter, however, amplifier compensation nodes to allow this are available on very few of the newer high speed amplifiers.

Nevertheless, there are still useful examples, and one is the AD829, compensated by a single capacitor to ac-common, at Pin 5. A more recent and analogous part is the AD8021, which is similarly compensated. In general, almost any amplifier using external compensation can always be over compensated to reduce bandwidth. This will restore stability against cap loads, by lowering the amplifier's unity gain frequency.

Forcing a high noise gain is shown in Figure 6-70, where the left side capacitively loaded amplifier with a noise gain of unity is unstable, due to a $1/\beta$ – open-loop roll-off intersection on the Bode diagram in a –12 dB/octave rolloff region. For such a case, introducing higher noise gain can restore often stability, so that the critical intersection occurs in a stable –6 dB/octave region, as depicted at the right diagram and Bode plot.

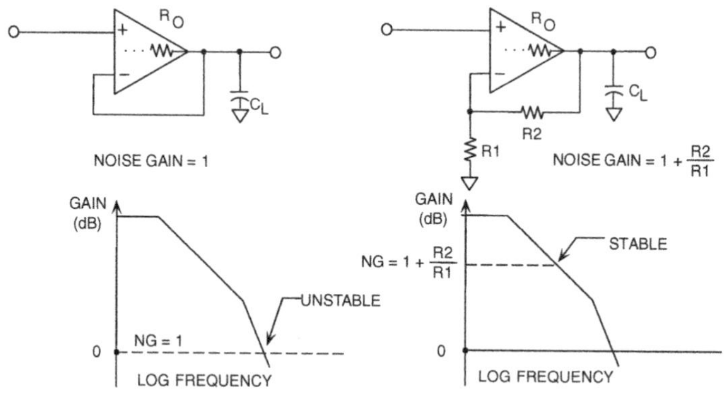

Figure 6-70: Effect of capacitive loading on op amp stability

To enable a higher noise gain (which does not necessarily need to be the same as the stage's *signal gain*), use is made of resistive or RC pads at the amplifier input, as in Figure 6-71. This trick is broader in scope than overcompensation, and has the advantage of not requiring access to any internal amplifier nodes. This generally allows use with any amplifier setup, even voltage followers (left) or inverters (right). An extra resistor, R_D, is added which works against R_F to force the noise gain of the stage to a level appreciably higher than the signal gain (unity for both cases here).

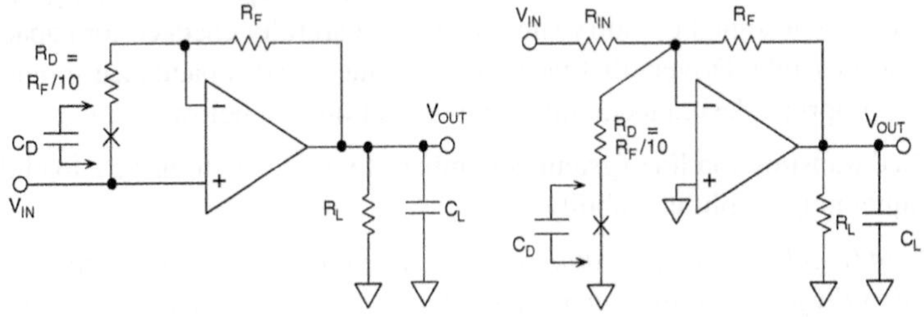

Figure 6-71: Raising noise gain (dc or ac) for follower (A) or inverter (B) stability

Assuming C_L is a value that produces a parasitic pole slightly above or near the amplifier's natural crossover, this loading combination would lead to oscillation due to the excessive phase lag. However, with R_D connected, the forced higher amplifier noise gain produces a new $1/\beta$ and open-loop roll-off intersection, purposely set about a decade lower in frequency. This is low enough that the extra phase lag from C_L near the amplifier's natural unity-gain crossover is no longer a problem, and stability is restored.

A drawback to this trick is that both the dc offset and input noise of the amplifier are raised by the value of the noise gain, when R_D is dc-connected. But, when C_D is used in series with R_D, the offset voltage of the amplifier is not raised, and the gained-up ac noise components are confined to a frequency region above $1/(2\pi \cdot R_D \cdot C_D)$. A further caution is that this technique can be somewhat tricky when separating these operating dc and ac regions, and should be applied carefully with regard to settling time (see Reference 6). Note that these simplified examples are generic, and in practice the absolute component values should be matched to a specific amplifier.

"Passive" cap load compensation, shown in Figure 6-72, is the most simple (and most popular) isolation technique available. It uses a simple "out-of-the-loop" series resistor R_X to isolate the cap load, and can be used with any amplifier, current or voltage feedback, FET or bipolar input.

Figure 6-72: Open-loop series resistance isolates capacitive load

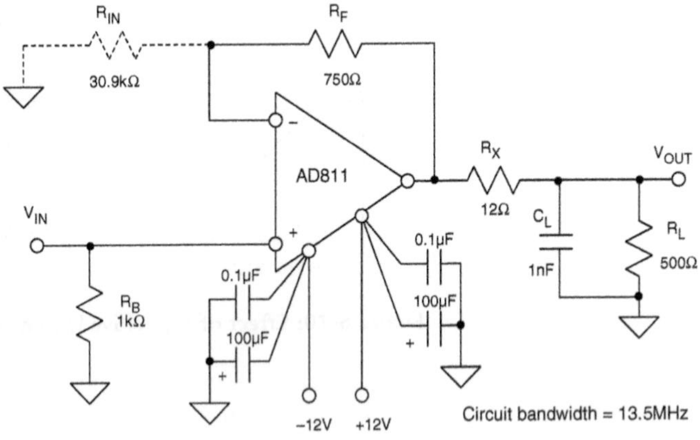

As noted, since this technique applies to just about any amplifier, it is a major reason why it is so useful. It is shown here with a current feedback amplifier suitable for high current line driving, the AD811, and it consists of just the simple (passive) series isolation resistor, R_X. This resistor's minimum value for stability will vary from device to device, so the amplifier data sheet should be consulted for other ICs. Generally, information will be provided as to the amount of load capacitance tolerated, and a suggested minimum resistor value for stability purposes.

Drawbacks of this approach are the loss of bandwidth as R_X works against C_L, the loss of voltage swing, a possible lower slew rate limit due to I_{MAX} and C_L, and a gain error due to the R_X-R_L division. The gain error can be optionally compensated with R_{IN}, which is ratioed to R_F as R_L is to R_X. In this example, a ±100 mA output from the op amp into C_L can slew V_{OUT} at a rate of 100 V/µs, far below the intrinsic AD811 slew rate of 2500 V/µs. Although the drawbacks are serious, this form of cap load compensation is nevertheless useful because of its simplicity. If the amplifier isn't otherwise protected, then an R_X resistor of 50 Ω–100 Ω should be used with virtually any amplifier facing capacitive loading. Although a noninverting amplifier is shown, the technique applies equally to inverters.

With very high speed amplifiers, or in applications where lowest settling time is critical, even small values of load capacitance can be disruptive to frequency response, but are nevertheless sometimes inescapable. One case in point is an amplifier used for driving ADC inputs. Since high speed ADC inputs quite often look capacitive in nature, this presents an oil/water type of problem. In such cases the amplifier *must* be stable driving the capacitance, but it must also preserve its best bandwidth and settling time characteristics. To address this cap load case, R_S and C_L data for a specified settling time is appropriate.

Some applications, in particular those that require driving the relatively high impedance of an ADC, do not have a convenient back termination resistor to dampen the effects of capacitive loading. At high frequencies, an amplifier's output impedance is rising with frequency and acts like an inductance which, in combination with C_L, causes peaking or, even worse, oscillation. When the bandwidth of an amplifier is an appreciable percentage of device F_t, the situation is complicated by the fact that the loading effects are reflected back into its internal stages. In spite of this, the basic behavior of most very wide bandwidth amplifiers such as the AD8001 is very similar.

In general, a small damping resistor (R_S) placed in series with C_L will help restore the desired response (see Figure 6-73). The best choice for this resistor's value will depend upon the criterion used in determining the desired response. Traditionally, simple stability or an acceptable amount of peaking has been used, but a

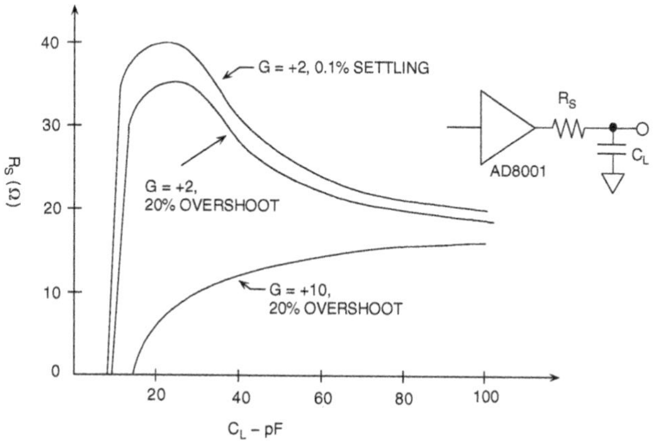

Figure 6-73: AD8001 R_S required for various C_L values

more strict measure such as 0.1% (or even 0.01%) settling will yield different values. For a given amplifier, a family of R_S-C_L curves exists, such as those of Figure 6-73. These data will aid in selecting R_S for a given application.

The basic shape of this curve can be easily explained. When C_L is very small, no resistor is necessary. When C_L increases to some threshold value an R_S becomes necessary. Since the frequency at which the damping is required is related to the $R_S \bullet C_L$ time constant, the R_S needed will initially increase rapidly from zero, and then will decrease as C_L is further increased. A relatively strict requirement, such as for 0.1%, settling will generally require a larger R_S for a given C_L, giving a curve falling higher (in terms of R_S) than that for a less stringent requirement, such as 20% overshoot. For the common gain configuration of +2, these two curves are plotted in the figure for 0.1% settling (upper-most curve) and 20% overshoot (middle curve). It is also worth mentioning that higher closed-loop gains dramatically lessen the problem, and will require less R_S for the same performance. The third (lowermost) curve illustrates this, demonstrating a closed-loop gain of 10 R_S requirement for 20% overshoot for the AD8001 amplifier. This can be related to the earlier discussion associated with Figure 6-70.

The recommended values for R_S will optimize response, but it is important to note that generally C_L will degrade the maximum bandwidth and settling time performance which is achievable. In the limit, a large $R_S \bullet C_L$ time constant will dominate the response. In any given application, the value for R_S should be taken as a starting point in an optimization process which accounts for board parasitics and other secondary effects.

Active or "in-the-loop" cap load compensation can also be used as shown in Figure 6-74, and this scheme modifies the passive configuration, providing feedback correction for the dc and low frequency gain error associated with R_X. In contrast to the passive form, active compensation can only be used with voltage feedback amplifiers, because current feedback amplifiers do not allow the integrating connection of C_F.

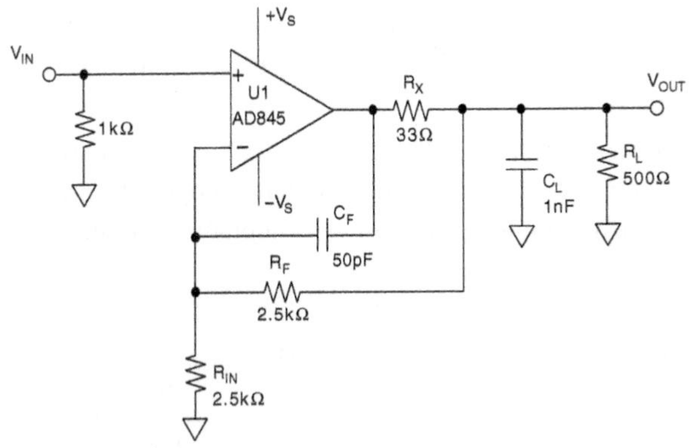

Figure 6-74: Active "in-the-loop" capacitive load compensation corrects for dc and LF gain errors

This circuit returns the dc feedback from the output side of isolation resistor R_X, thus correcting for errors. Ac feedback is returned via C_F, which bypasses R_X/R_F at high frequencies. With an appropriate value of C_F (which varies with C_L for fixed resistances) this stage can be adjusted for a well damped transient response (see References 6 and 7). There is still a bandwidth reduction, a headroom loss, and also (usually) a slew rate reduction, but the dc errors can be very low. A drawback is the need to tune C_F to C_L, as even if this is done well initially, any change to C_L will alter the response away from flat. The circuit as shown is useful for voltage feedback amplifiers only, because capacitor C_F provides integration around U1. It also can be implemented in inverting fashion, by driving the bottom end of R_{IN}, while grounding the op amp (+) input.

Internal cap load compensation involves the use of an amplifier that has topological provisions for the effects of external cap loading. To the user, this is the most transparent of the various techniques, as it works for any feedback situation, for any value of load capacitance. Drawbacks are that it produces higher distortion than does an otherwise similar amplifier without the network, and the compensation against cap loading is somewhat signal level dependent.

The internal cap load compensated amplifier sounds at first like the best of all possible worlds, since the user need do nothing at all to set it up. Figure 6-75 is a simplified diagram of an AD817 amplifier with internal cap load compensation. The cap load compensation is the C_F-resistor network, which is highlighted by the dotted area within the unity gain output stage of the amplifier. It is important to note at this point that *this RC network only makes its presence felt for certain load conditions.*

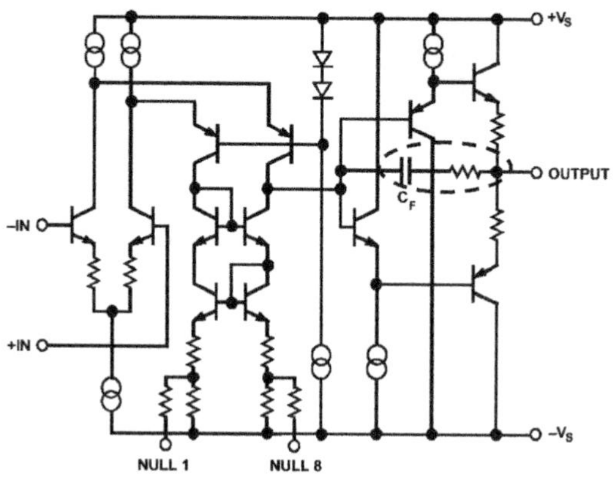

Figure 6-75: AD817 simplified schematic illustrates internal compensation for driving capacitive loads

Under normal (noncapacitive or light resistive) loading, there is limited input/output voltage error across the output stage, so the C_F network then sees a relatively small voltage drop, and has little or no effect on the AD817's high impedance compensation node. However when a capacitor (or other heavy) load is present, the high currents in the output stage produce a voltage difference across the C_F network, which effectively adds capacitance to the compensation node. With this relatively heavy loading, a net larger compensation capacitance results, and reduces the amplifier speed in a manner that is adaptive to the external capacitance, C_L. *As a point of reference, note that it requires 6.3 mA peak current to support a 2 Vp-p swing across a 100 pF load at 10 MHz.*

Since this mechanism is resident in the amplifier output stage and it dynamically affects the overall compensation characteristics, it acts independent of the specific external feedback hookup, as well as the external capacitor's size. In other words, it can be transparent to the user in the sense that no specific design conditions need be set to make it work (other than selecting the right IC). Some amplifiers using internal cap load compensation are the AD817, the AD847, and their dual equivalents, AD826 and AD827.

There are, however, some caveats also associated with this internal compensation scheme. As with the passive compensation techniques, bandwidth decreases as the device slows down to prevent oscillation with higher load currents. Also, this adaptive compensation network has its greatest effect when enough output current flows to produce significant voltage drop across the C_F network. Conversely, at small signal levels, the effect of the network on speed is less, so greater ringing may actually be possible for some circuits, with lower-level outputs.

The dynamic nature of this internal cap load compensation is illustrated in Figure 6-76, which shows an AD817 unity gain inverter being exercised at both high (left) and low (right) output levels, with common conditions of $V_S = \pm 15$ V, $R_L = 1$ kΩ, $C_L = 1$ nF, and using 1 kΩ input/feedback resistors. In both photos the input signal is on the top trace, the output signal is on the bottom trace, and the time scale is fixed.

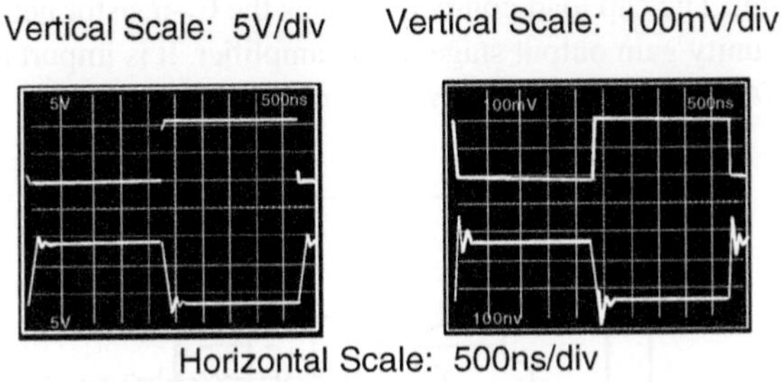

Figure 6-76: Response of internal cap load compensated amplifier varies with signal level

In the 10 V p-p output left photo, the output has slowed down appreciably to accommodate the capacitive load, but settling is still relatively clean, with a small percentage of overshoot. For this high level case, the bandwidth reduction due to C_L is most effective.

In the right photo, the 200 mV p-p output shows greater overshoot and ringing for the lower level signal. The point is that the performance of the cap load compensated amplifier is signal-dependent, but is always stable with any cap load.

Finally, because the circuit is based on a nonlinear principle, the internal network affects distortion performance and load drive ability, and these factors influence amplifier performance in video applications. Though the network's presence does not by any means make devices like the AD817 or AD847 unusable for video, it does not permit the very lowest levels of distortion and differential gain and phase that are achievable with amplifiers without this network, but otherwise comparable.

While the individual techniques for countering cap loading outlined above have various specific trade-offs as noted, all of the techniques have a common drawback of reducing speed (both bandwidth and slew rate). If these parameters cannot be sacrificed, a matched transmission line system is the solution, and is discussed in more detail, in the "Video Amplifiers" portion of this chapter.

As for choosing among the cap load compensation schemes, it would seem on the surface that amplifiers using the internal form offer the best possible solution to the problem—just pick the right amplifier and simply forget about it. And indeed, that would seem the "panacea" for all cap load situations—if you use the "right" amplifier you never need to think about cap loading again. Could there be more to it?

Yes. The "gotcha" of internal cap load compensation is subtle, and lies in the fact that the dynamic adaptive nature of the compensation mechanism actually can produce higher levels of distortion, vis-à-vis an otherwise similar amplifier, *without* the C_F-resistor network. Like the old saying about no free lunches, to attain top-notch levels of high frequency ac performance, give the issue of whether to use an internally compensated cap load amplifier more serious thought than simply picking a trendy device. For example, the AD818, which is a gain-of-two stable video op amp, offers excellent performance in terms of video gain and phase measurements. It is simply a gain-of-two stable AD817 op amp, but without the internal cap load compensation network. For similar video stage driver applications, the AD817 will not perform as well as the more suitable AD818.

On the other hand, if you have no requirements for the lowest levels of distortion, then such an amplifier as the AD817 could be a very good choice. Such amplifiers are certainly easier to use, and are relatively forgiving about output loading issues.

References: Buffers and Driving Cap Loads

1. George Erdi, "A 300 V/µs Monolithic Voltage Follower," **IEEE Journal of Solid-State Circuits**, Vol. SC-14, No. 6, December, 1979, pp. 1059–1065.
2. Royal A. Gosser, "Wideband Transconductance Generator," **US Patent 5,150,074**, Filed May 3, 1991, issued September 22, 1992.
3. Derek. F. Bowers, "A 6.8mA Closed-Loop Monolithic Buffer with 120MHz Bandwidth, 4000 V/µs Slew Rate, and ±12 V Signal Compatibility," **1994 Bipolar/BiCMOS Circuits and Technology Meeting 1.3**, pp. 23–26.
4. Walt Kester, "Maintaining Transmission Line Impedances on the PC Board," within Chapter 11 of Walt Kester, Editor, **System Application Guide**, Analog Devices, Inc., 1993, ISBN 0-916550-13-3.
5. William R. Blood, Jr., **MECL System Design Handbook**, (HB205, Rev.1), Motorola Semiconductor Products, Inc., 1988.
6. Joe Buxton, "Careful Design Tames High-Speed Op Amps," **Electronic Design**, April 11, 1991.
7. Walt Jung, "Op Amps in Line-Driver and Receiver Circuits, Part 1," **Analog Dialogue**, Vol. 26-2, 1992.
8. Dave Whitney, Walt Jung, "Applying a High-Performance Video Operational Amplifier," **Analog Dialogue**, 26-1, 1992.

SECTION 6-3

Video Amplifiers

Walt Kester

Video Signals and Specifications

Before discussing some video applications for op amps, it is a good idea to review some basics regarding video signals and specifications. The standard video format is the specification of how the video signal looks from an electrical point of view. Light strikes the surface of an image sensing device within the camera, producing a voltage level corresponding to the amount of light hitting a particular spatial region of the surface. This information is then placed into the standard format and sequenced out of the camera. Along with the actual light and color information, synchronization pulses are added to the signal to allow the receiving device—a television monitor, for instance—to identify where the sequence is in the frame data.

A standard video format image is read out on a line-by-line basis from left to right, top to bottom. A technique called *interlacing* refers to the reading of all even-numbered lines, top to bottom, followed by all odd lines as shown in Figure 6-77.

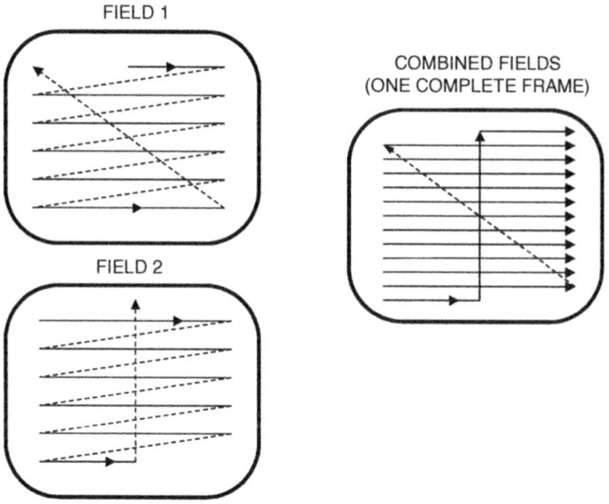

Figure 6-77: Standard broadcast television interlace format

The television picture *frame* is thus divided into even and odd *fields*. Interlacing is used to produce an apparent update of the entire frame in half the time that a full update actually occurs. This results in a television image with less apparent flicker. Typical broadcast television frame update rates are 30 Hz and 25 Hz, depending upon the line frequency. It should be noted that interlacing is not always required in graphics display systems where the refresh rate is usually greater (typically 60 Hz).

The original black and white, or *monochrome*, television specification in the USA is the EIA RS-170 specification that prescribes all timing and voltage level requirements for standard commercial broadcast video signals. The standard American specification for color signals, NTSC, modifies RS-170 to work with color signals by adding color information to the signal which otherwise contains only brightness information.

A video signal comprises a series of analog television lines. Each line is separated from the next by a synchronization pulse called the *horizontal sync*. The fields of the picture are separated by a longer synchronization pulse, called the *vertical sync*. In the case of a monitor receiving the signal, its electron beam scans the face of the display tube with the brightness of the beam controlled by the amplitude of the video signal. A single line of an NTSC color video signal is shown in Figure 6-78.

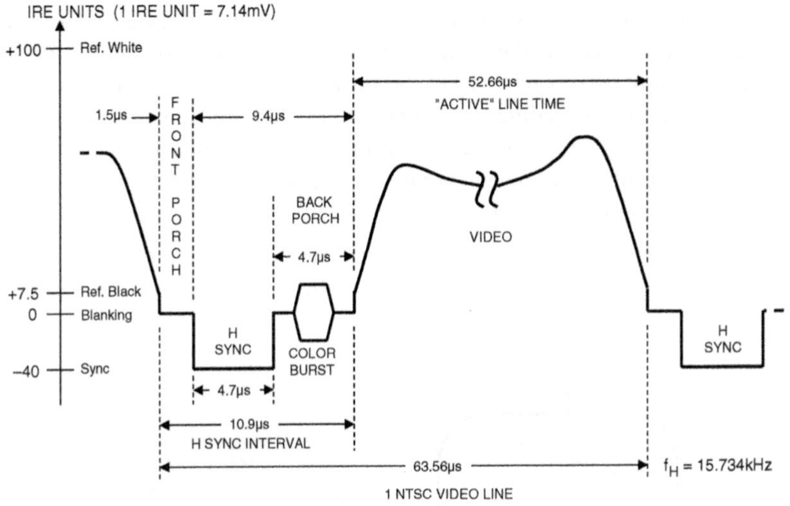

Figure 6-78: NTSC composite color video line

Whenever a horizontal sync pulse is detected, the beam is reset to the left side of the screen and moved down to the next line position. A vertical sync pulse, indicated by a horizontal sync pulse of longer duration, resets the beam to the top left point of the screen to a line centered between the first two lines of the previous scan. This allows the current field to be displayed between the previous one.

A simplified block diagram of the NTSC color processing system is shown in Figure 6-79. The three color signals (RGB: red, green, and blue) from the color camera are combined in a *matrix* unit to produce what is called the *luminance* signal (Y) and two color difference signals (I and Q). These *components* are further combined to produce what is called the *composite* color signal.

In the NTSC system (used in the U.S. and Japan), the color subcarrier frequency is 3.58 MHz. The PAL system (used in the U.K. and Germany) and SECAM system (used in France) use a 4.43 MHz color subcarrier.

In terms of their key frequency differences, a comparison between the NTSC system and the PAL system are given in Figure 6-80.

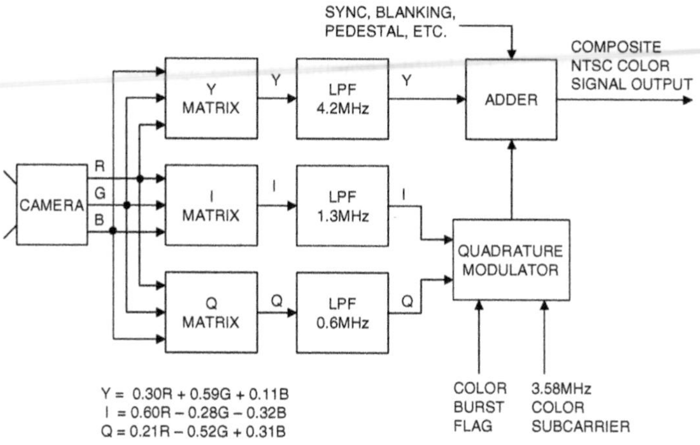

Figure 6-79: Generating the composite NTSC color signal

	NTSC	PAL
Horizontal Lines	525	625
Color Subcarrier Frequency	3.58MHz	4.43MHz
Frame Frequency	30Hz	25Hz
Field Frequency	60Hz	50Hz
Horizontal Sync Frequency	15.734kHz	15.625kHz

Figure 6-80: NTSC and PAL signal characteristics

Differential Gain And Phase Specifications

The color (or *chrominance*) information in the composite video signal is contained in the amplitude and phase of the subcarrier. The *intensity* or *saturation* of the color is determined by the amplitude of the subcarrier signal, and the precise color displayed (i.e., red, green, blue, and combinations) is determined by the phase of the subcarrier signal with respect to the phase of the color burst. The chrominance signal modulates the luminance signal which determines the relative blackness or whiteness of the color. To preserve color fidelity, it is important that the amplitude and phase of a constant-amplitude and phase color subcarrier remain constant across the range of black to white.

Any variation of the *amplitude* of the color subcarrier from black to white levels is called *differential gain* (expressed in %), and any variation in phase with respect to the color subcarrier is called *differential phase* (expressed in degrees). Degradations of up several percent of differential gain and several of degrees differential phase are acceptable for home viewing purposes, but individual components in the video signal path (amplifiers, switches, and so forth) must meet much tighter specifications. This is because the signal must pass through many circuits from the camera to the home. As a result, individual professional video systems have stringent requirements for differential gain and phase, usually limiting changes to less than 0.1% and 0.1°.

These system specifications mandate even more stringent standards for individual components, with the differential gain and differential phase requirements for op amps approaching 0.01% and 0.01°.

Video Formats in Graphics Display Systems

Several system architectures may be used to build a graphics display system. The most general approach is illustrated in Figure 6-81. It consists of a host microprocessor, a graphics controller, three color memory banks or frame buffer, one for each of the primary colors red, green, and blue (only one for monochrome systems). The microprocessor provides the image information to the graphics controller. This information typically includes position and color information. The graphics controller is responsible for interpreting this information and adding the required output signals such as sync, blanking, and memory management signals.

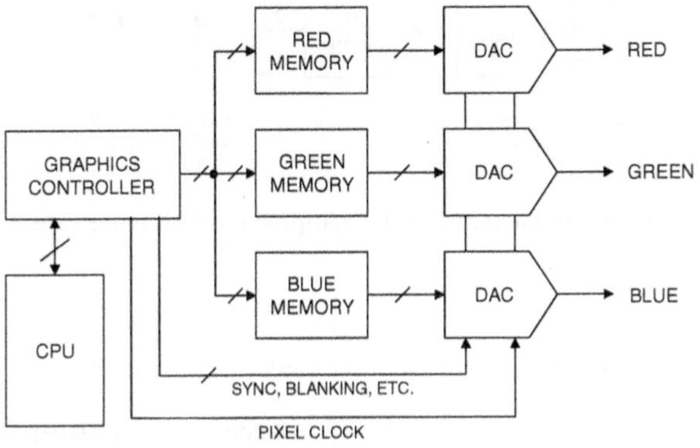

Figure 6-81: Simplified graphics control system for generating RGB signals

Unlike broadcast video, the horizontal and vertical resolution, as well as the refresh rate in a graphics display system, can vary widely depending upon the desired performance. The resolution in such a system is defined in terms of *pixels*: the number of horizontal lines (expressed as pixels) and the number of pixels in each line. For instance, a 640 × 480 monitor has 480 horizontal lines, and each horizontal line is divided into 640 pixels. So a single frame would contain 307,200 pixels. In a color system, each pixel requires RGB intensity data. This data is generally stored as 8- or 10-bit words in the memory.

The memory holds the intensity information for each pixel. The DACs use the words in the memory and information from the memory controller to write the pixel information to the monitor. Special video DACs called "RAMDACs" greatly simplify the storage of the pixel data by using color lookup tables. These DACs also have inputs to facilitate the generation of the sync and blanking signals.

Figure 6-82 shows some typical resolutions and pixel rates for common display systems, assuming a 60 Hz, noninterlaced refresh rate. Standard computer graphics monitors, like television monitors, use a display technique known as *raster scan*. This technique writes information to the screen line by line, left to right, top to bottom, as has been previously discussed. The monitor must receive a great deal of information to display a complete picture. Not only must the intensity information for each pixel be present in the signal, but information must be provided to determine when a new line needs to start (HSYNC) and when a new picture frame should start (VSYNC). The computer industry has generally standardized on formats defined in EIA video standard RS-343A. Unlike broadcast video, the refresh rate can also vary and interlacing may or may not be utilized. The pixel clock frequency gives a good idea of the settling time and bandwidth requirements for any analog component, such as the DAC, which is placed in the path of the RGB signals. The pixel clock frequency can be estimated by finding the product of the horizontal resolution times the vertical resolution times the refresh rate. An additional 30%, called the retrace factor, should be added to allow for overhead.

RESOLUTION	PIXEL RATE
640 × 480	25MHz
800 × 600	38MHz
1024 × 768	65MHz
1280 × 1024	105MHz
1500 × 1500	180MHz
2048 × 2048	330MHz

Pixel Rate ≈ Vertical Resolution × Horizontal Resolution × Refresh Rate × 1.3

Figure 6-82: Typical graphics resolution and pixel rates for 60 Hz noninterlaced refresh rate

Bandwidth Considerations in Video Applications

The bandwidth of an op amp used in a video application must be sufficient so that the video signal is not attenuated or significantly shifted in phase. This generally implies that the bandwidth of the op amp be much greater than that of the maximum video frequency. It is not uncommon to require that amplifiers in the signal path in video equipment such as switchers or special effects generators have 0.1 dB bandwidths of 50 MHz or greater. High definition television requires even higher 0.1 dB bandwidth. Circuit parasitics as well as the load impedance can significantly affect the 0.1 dB bandwidth at high frequencies. This implies careful attention to layout, decoupling, and grounding as well as the use of transmission line techniques at the op amp output. It is common to use source and load terminations with high quality 75 Ω coaxial cable so that the load presented to the op amp output appears as a 150 Ω resistive load. Maintaining accurate control of 0.1 dB bandwidth is almost impossible with reactive loads.

Achieving the highest 0.1 dB bandwidth flatness is therefore important in many video applications. Voltage feedback op amps can be optimized for maximum 0.1dB bandwidth provided the closed-loop gain and load conditions are known. In video applications, closed-loop gains of +1 and +2 are the most common with a 100 Ω or 150 Ω output load, representing the impedance of 50 Ω or 75 Ω source and load terminated cables.

As an example, the AD8074 (G = +1) and AD8075 (G = +2) are triple video buffers optimized for driving source and load terminated 75 Ω cables. These devices use a voltage feedback architecture and have on-chip gain-setting resistors. Figure 6-83 shows the frequency response of the AD8075 buffer on two vertical scales: 1 dB/division and 0.1 dB/division. The plots labeled "GAIN" show a 3 dB bandwidth of 350 MHz (□), and the plots labeled "FLATNESS" show a 0.1 dB bandwidth of 70 MHz (○). Note that the small-signal (200 mV p-p) and large-signal (2 V p-p) bandwidths are approximately equal.

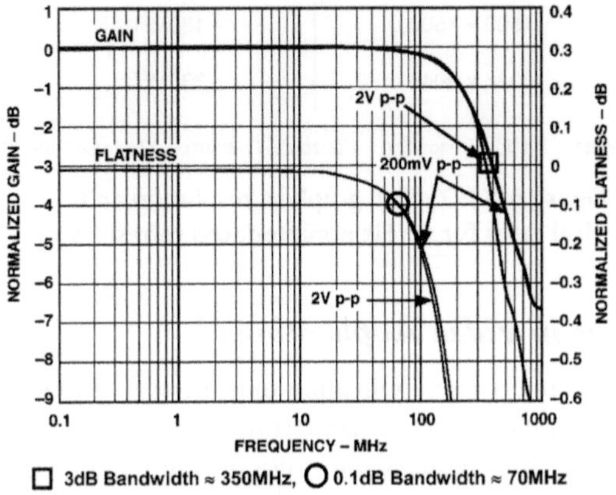

Figure 6-83: AD8075 triple video buffer gain and gain flatness, G = +2, R_L = 150 Ω

Voltage feedback op amps are optimized for bandwidth flatness by adjusting both the compensation capacitor, which sets the dominant pole, and the external feedback network. However, because of the critical relationship between the feedback resistor and the bandwidth of a current feedback op amp, optimum bandwidth flatness is highly dependent on the feedback resistor value, the resistor parasitics, as well as the op amp package and PCB parasitics. Figure 6-84 shows the bandwidth flatness (0.1 dB/division) plotted versus the feedback resistance for the AD8001 in a noninverting gain of 2. The 100 Ω load resistor represents a source and load terminated 50 Ω cable. These plots were made using the AD8001 evaluation board with surface mount resistors.

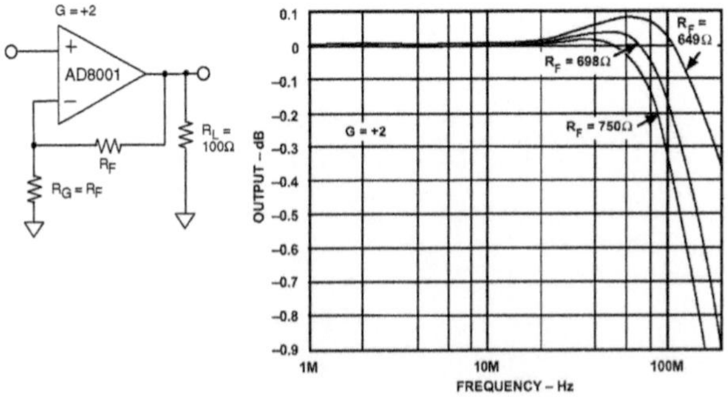

Figure 6-84: AD8001 current feedback op amp bandwidth flatness versus feedback resistor value

It is recommended that once the optimum resistor values have been determined, 1% tolerance values should be used. In addition, resistors of different construction have different associated parasitic capacitance and inductance. Surface-mount resistors are an optimum choice, thus leaded components aren't recommended for high frequency use.

Slightly different resistor values may be required to achieve optimum performance of the AD8001 in the DIP versus the SOIC packages (see Figure 6-85). The SOIC package exhibits slightly lower parasitic capacitance and inductance than the DIP. The data shows the optimum feedback (R_G) and feedforward (R_F) resistors for highest 0.1 dB bandwidth for the AD8001 in the DIP and the SOIC packages. As you might suspect, the SOIC package can be optimized for higher 0.1 dB bandwidth because of lower parasitics.

AD8001AN
(DIP)

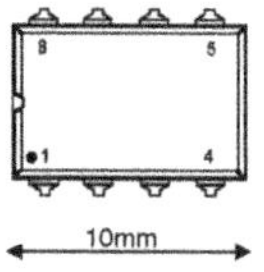

GAIN	−1	+1	+1
R_F	649Ω	1050Ω	750Ω
R_G	649Ω	-	750Ω
0.1dB Flatness	105MHz	70MHz	105MHz

AD8001AR
(SOIC)

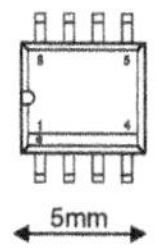

GAIN	−1	+1	+1
R_F	604Ω	953Ω	681Ω
R_G	604Ω	-	681Ω
0.1dB Flatness	130MHz	100MHz	120MHz

Figure 6-85: Optimum values of R_F and R_G for AD8001 DIP and SOIC packages for maximum 0.1 dB bandwidth

As has been discussed, the current feedback op amp is relatively insensitive to capacitance on the inverting input when it is used in the inverting mode (as in an I/V application). This is because the low inverting input impedance is in parallel with the external capacitance and tends to minimize its effect. In the noninverting mode, however, even a few picofarads of stray inverting input capacitance may cause peaking and instability. Figure 6-86 shows the effects of adding summing junction capacitance to the inverting input of the

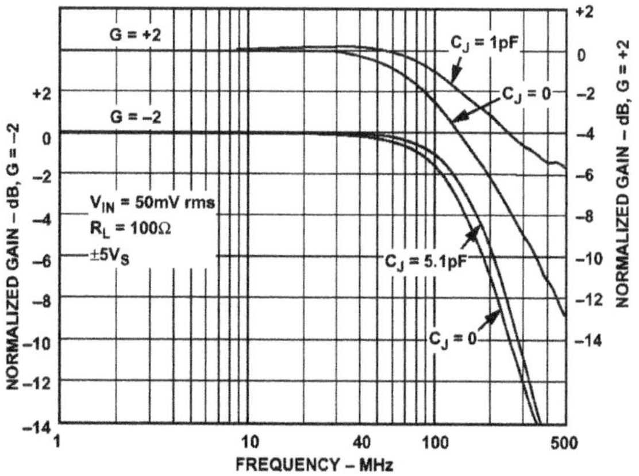

Figure 6-86: AD8004 current feedback op amp sensitivity to inverting input capacitance for G = +2, G = −2

AD8004 (SOIC package) for G = +2. Note that only 1 pF of added inverting input capacitance (C_J) causes a significant increase in bandwidth and an increase in peaking. For G = –2, however, 5 pF of additional inverting input capacitance causes only a small increase in bandwidth and no significant increase in peaking.

It should be noted that high-speed voltage feedback op amps are sensitive to stray inverting input capacitance when used in either the inverting or noninverting mode, because both positive and negative inputs are high impedance.

Video Signal Transmission

High quality video signals are best transmitted over terminated coaxial cable having a controlled characteristic impedance. The characteristic impedance is given by the equation $Zo = \sqrt{(L/C)}$ where L is the distributed inductance per foot, and C is the distributed capacitance per foot. Popular values are 50 Ω, 75 Ω, and 93 Ω or 100 Ω.

If a length of coaxial cable is properly terminated, it presents a *resistive* load to the driver. If left unterminated, however, it may present a predominately capacitive load to the driver depending on the output frequency. If the length of an unterminated cable is much less than the wavelength of the output frequency of the driver, the load appears approximately as a lumped capacitance. For instance, at the audio frequency of 20 kHz (wavelength ≈ 50,000 feet, or 9.5 miles), a 5-foot length of unterminated 50 Ω coaxial cable would appear as a lumped capacitance of approximately 150 pF (the distributed capacitance of coaxial cable is about 30 pF/ft).

At 100 MHz (wavelength ≈ 10 feet), however, the unterminated coax must be treated as a transmission line in order to calculate the standing wave pattern and the voltage at the unterminated cable output. Figure 6-87 summarizes transmission line behavior for different frequencies.

- All interconnections are really transmission lines which have a characteristic impedance (even if not controlled).
- The characteristic impedance is equal to $\sqrt{(L/C)}$, where L and C are the distributed inductance and capacitance.
- Correctly terminated transmission lines have impedances equal to their characteristic impedance.
- Unterminated transmission lines behave approximately as lumped capacitance if the wavelength of the output frequency is much greater than the length of the cable.
  - Example: At 20kHz (wavelength = 9.5 miles), 5 feet of unterminated 50Ωcable (30pF/ft) appears like a 150pF load
  - Example: At 100MHz, (wavelength = 10 feet), 5 feet of 50Ω must be properly terminated to prevent reflections and standing waves.

Figure 6-87: Driving cables

Because of skin effect and wire resistance, coaxial cable exhibits a loss that is a function of frequency. This varies considerably between cable types. For instance at 100 MHz the attenuation RG188A/U is 8 dB/100 ft, RG58/U is 5.5 dB/100 ft, and RG59/U 3.6 dB/100 ft (see Reference 4). Skin effect also affects the pulse response of long coaxial cables. The response to a fast pulse will rise sharply for the first 50% of the output swing, then taper off during the remaining portion of the edge. Calculations show that the 10% to 90% waveform risetime is 30 times greater than the 0% to 50% rise time when the cable is skin effect limited (Reference 4).

Transmission Line Driver Lab

It is useful to examine the fidelity of a pulse signal, for conditions of proper/improper transmission line source/load terminations. Some lab experiments were set up to do this.

To illustrate the behavior of a high speed op amp driving a coaxial cable, consider the circuit of Figure 6-88. Here the AD8001 drives 5 feet of 50 Ω coaxial cable, which is load-end terminated in the characteristic impedance of 50 Ω. No termination is used at the amplifier (driving) end. The pulse response is also shown in the figure.

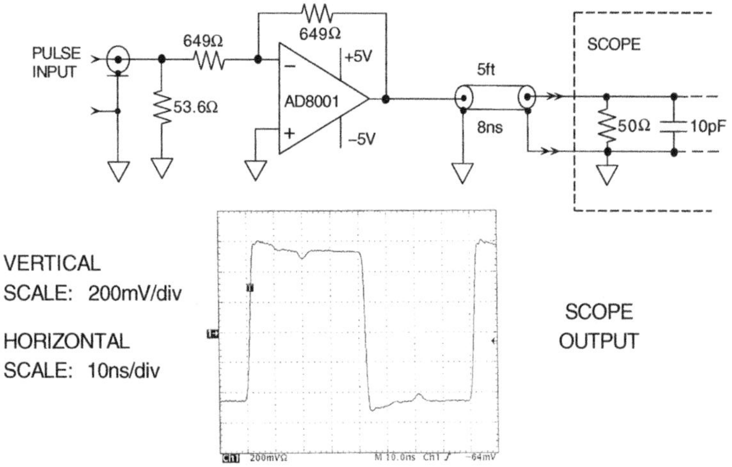

Figure 6-88: Pulse response of AD8001 driving 5 feet of load-only terminated 50 Ω coaxial cable

The output of the cable was measured by connecting it directly to the 50 Ω input of a 500 MHz Tektronix 644A digitizing oscilloscope. The 50 Ω resistor termination is actually the input of the scope. However, this 50 Ω load is not a perfect line termination, it is lower at high frequencies (due to the scope shunt input capacitance of about 10 pF).

As a consequence some of the positive going pulse edge is reflected out of phase to the source. When this reflection reaches the op amp, it sees the op amp closed-loop output impedance, which, at 100 MHz, is approximately 100 Ω (higher than line impedance).

Upon arriving at the op amp output, the negative-going reflection from the load is then rereflected back towards the load, without undergoing another phase reversal. This then accounts for the negative going "blip" seen on the upper plateau of the waveform, which occurs approximately 16 ns after the leading edge. This time difference is equal to the round-trip delay of the cable (2 • 5ft • 1.6 ns/ft = 16 ns). An additional point worth noting is that, in the frequency domain (which is not shown by these tests) the cable mismatch will also cause a loss of bandwidth flatness at the load.

Figure 6-89 shows a second case, the results of driving the same coaxial cable, but now used with both a 50 Ω source-end as well as the 50 Ω load-end termination at the scope. It should be noted that this case is the preferred way to drive a transmission line, because a portion of the reflection from the load impedance mismatch is absorbed by the amplifier's source termination resistor of 50 Ω. A disadvantage is that there is a gain loss of 6 dB, because of the 2/1 voltage division which occurs between the equal value source and load terminations, i.e., 50 Ω/50 Ω.

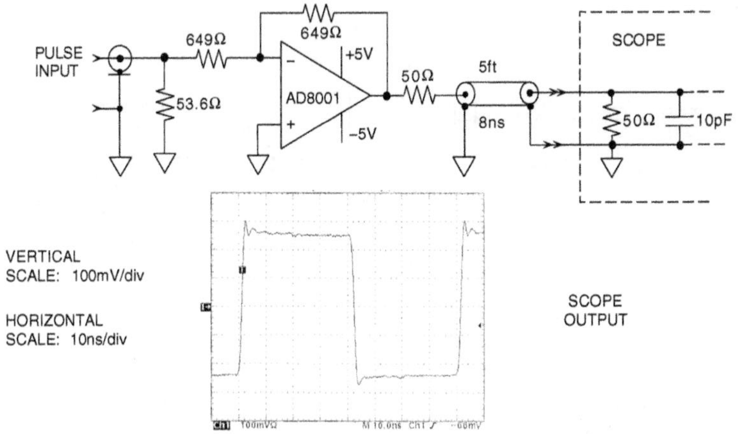

Figure 6-89: Pulse response of AD8001 driving 5 feet of source and load terminated 50 Ω coaxial cable

However, a major positive attribute of this configuration, with the line impedance matched source and load terminations in conjunction with a low-loss cable, is that *the best bandwidth flatness is ensured*, especially at lower operating frequencies. In addition to this, the amplifier is operated with a near optimum load condition, i.e., into a resistive load. The load in this case is 50 Ω plus 50 Ω, or 100 Ω. In general, it will be twice the impedance of the transmission line in use, i.e., 150 Ω for a 75 Ω line, and so forth.

In practice, the gain loss associated with the 2/1 source/load impedance is easily made up, simply by operating the line driver stage at a gain of 2x. Typically, video driver stages are noninverting to preserve the waveform sign, and operate at a fixed and precise gain of 2 times. Thus they will inherently provide a net signal transfer gain of unity, as measured from input to the final end-of-line load termination (this neglects any associated transmission line losses, and assumes precise resistor ratios for the gain resistors). Another very practical point is that the same driver can be used for a wide variety of transmission lines, simply by changing the value of the source termination resistor.

Source-end (only) terminations can also be used as shown in Figure 6-90, where the op amp is now source-terminated by the 50 Ω resistor which drives the cable. At the load end, the scope is set for 1 MΩ input impedance, which represents an approximate open circuit. The initial leading edge of the pulse at the op amp output sees a 100 Ω load (the 50 Ω source resistor in series with the 50 Ω coax impedance. When the pulse reaches the load, a large portion is reflected in phase, because of the high load impedance, resulting in a full-amplitude pulse at the load. When the reflection reaches the source-end of the cable, it sees the 50 Ω source resistance in series with the op amp closed-loop output impedance (approximately 100 Ω at the frequency represented by the 2 ns rise time pulse edge). The rereflected portion remains in phase, and then appears at the scope input as the positive going "blip," approximately 16 ns after the leading edge.

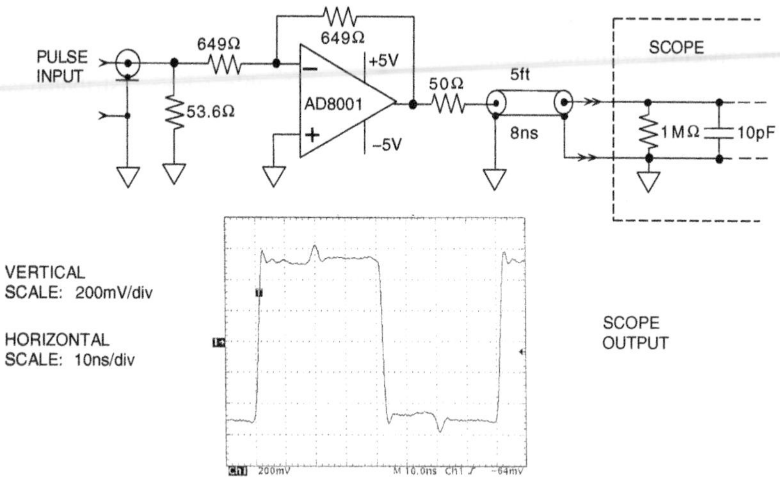

Figure 6-90: Pulse response of AD8001 driving 5 feet of source-only terminated 50 Ω coaxial cable

From these experiments, one can easily see that the preferred method for minimum reflections (and therefore maximum bandwidth flatness) is to use both source and load terminations and try to minimize any reactance associated with the load. The experiments represent a worst-case condition, where the frequencies contained in the fast edges are greater than 100 MHz. (using the rule-of-thumb that bandwidth = 0.35/rise time).

At less demanding video frequencies, either load-only, or source-only terminations may give acceptable results, but the op amp data sheet should always be consulted to determine the op amp's closed-loop output impedance at the maximum frequency of interest; i.e., is it less than the line impedance? A major disadvantage of the source-only termination is that it requires a truly high impedance load (high resistance and minimal parasitic capacitance) for minimum absorption of energy. It also places a burden on the driving amplifier, to maintain the low output impedance at high frequencies.

Now, for a truly worst-case, let us replace the 5 feet of coaxial cable with an uncontrolled-impedance cable (one that is largely capacitive with little inductance). Also, let's use a capacitance of 150 pF to simulate the cable (corresponding to the total capacitance of 5 feet of coaxial cable, whose distributed capacitance is about 30 pF/foot). Figure 6-91 shows the output of the AD8001 op amp, driving a lumped 160 pF capacitance (including the scope input capacitance of 10 pF).

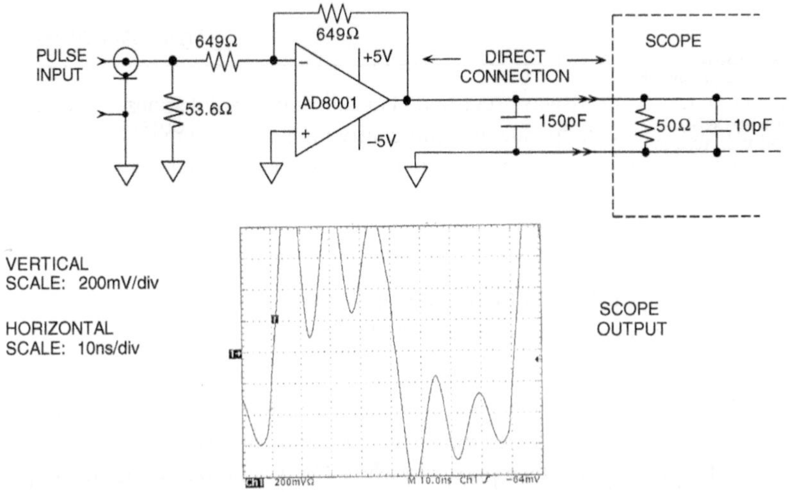

Figure 6-91: Pulse response of AD8001 driving 160 pF||50 Ω load

Overshoot and severe ringing on the pulse waveform is noted, due to the capacitive loading. This example illustrates the need to use good quality controlled-impedance coaxial cable in the transmission of high frequency signals, even over short distances. Failure to adhere to controlled-impedance lines for signal distribution can result in severe loss of pulse waveform fidelity, and loss of bandwidth flatness in the frequency domain.

To summarize, transmission line driver circuits should use proper line terminations for best response. The ideal method of line termination is matching line-impedance-value resistances at both source and load end (Figure 6-89). The associated 6dB gain loss is easily made up in the amplifier. Next best is a source-only termination (Figure 6-90), with due care towards maintaining a high impedance at the load end, and a low drive impedance amplifier. This type of termination provides near-full amplitude level at the load end, making the gain of the driver less critical. Load-only termination can also be used (Figure 6-88), but may be more critical of load end parasitic effects and the amplifier performance. It also provides near-full amplitude level at the load end.

Direct drive of uncontrolled load impedances, especially lumped capacitive lines, should be avoided wherever signal fidelity is important (Figure 6-91).

Video Line Drivers

The AD8047 and AD8048 voltage feedback op amps have been optimized to offer outstanding performance as video line drivers. They utilize the "quad core" g_m stage as previously described for high slew rate and low distortion (see Chapter 1). The AD8048 (optimized for G = +2) has a differential gain of 0.01% and a differential phase of 0.02°, making it well suited for HDTV applications.

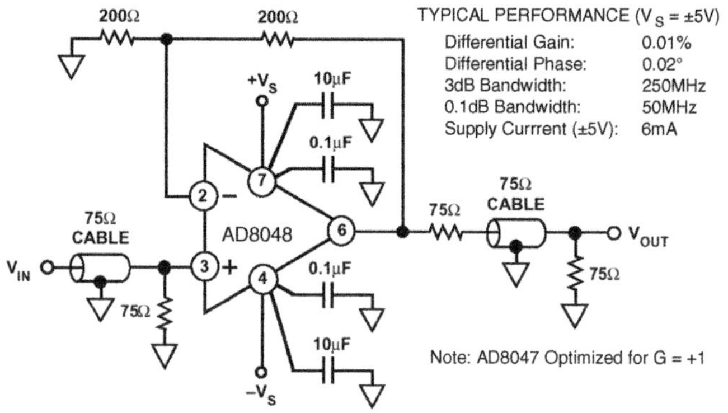

Figure 6-92: High performance video line driver using the AD8048

In the configuration using the AD8048 shown in Figure 6-92, the 0.1 dB bandwidth is 50 MHz for ±5 V supplies, slew rate is 1000 V/µs, and 0.1% settling time is 13 ns. The total quiescent current is 6 mA (±5 V), and quiescent power dissipation 60mW. Performance of this circuit will be optimum with the gain-of-two stable AD8048 op amp, as its parameters have been optimized for this gain. Alternately, if a gain-of-one stable op amp is desired, the AD8047 can be used.

Note that a very wide variety of both voltage feedback and current feedback devices can be used similarly as a gain-of-two line driver (although the required feedback resistances may vary by device). Examples would be the AD818/AD828, AD8055/AD8056, AD8057/AD8058, and AD8061/AD8062/AD8063 families of voltage feedback op amps, and the AD811/AD812/AD813, AD8001/AD8002, AD8012 families as a partial list. There are differences among all of these devices for applicable supply ranges, single-supply compatibility, and so forth, so consult device data sheets.

It is often desirable to drive more than one coaxial cable, which represents a dc load of 150 Ω to a driver. The typical maximum video signal level is 1 V into 75 Ω, which represents 2 V at the output of the driver, and a current of 13.3 mA. Thus a 50 mA output current video op amp such as the AD8047 or AD8048 would theoretically be capable of driving three source and load-terminated 75 Ω loads. But, there are other important subtle considerations for this application. Differential gain and phase may be degraded for high output currents. Also, the op amp closed-loop output impedance affects crosstalk between the driven output channels. So it is often better to select a video driver fully specified for the required fan-out and load, especially if the fan-out is greater than two.

Video Distribution Amplifier

The AD8010 op amp is optimized for driving multiple video loads in parallel. Video performance of 0.02% differential gain and 0.03% differential phase is maintained, while driving eight 75 Ω source and load-terminated video lines. The AD8010 uses the current feedback architecture and has a 0.1 dB bandwidth of 60 MHz with eight video loads. Typical supply current (neglecting load current) is 15.5 mA on ±5 V supplies. A typical connection diagram is shown in Figure 6-93. The AD8010 is offered in three packages: an 8-lead DIP (θ_{JA} = 90°C/W), 16-lead wide body SOIC (θ_{JA} = 73°C/W), and a low thermal resistance, 8-lead SOIC (θ_{JA} = 122°C/W).

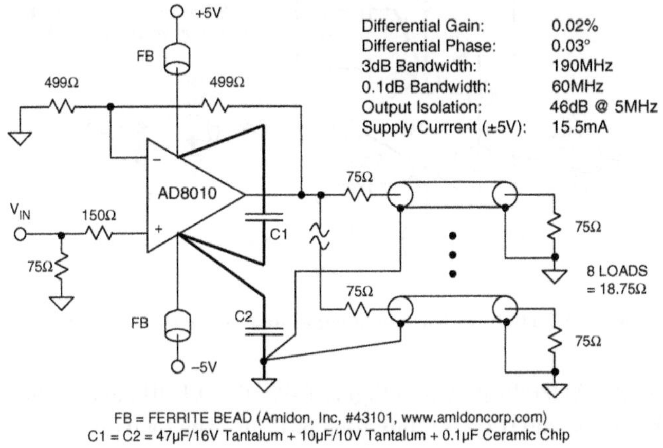

Figure 6-93: The AD8010 video distribution amplifier

The power supply decoupling scheme used for the AD8010 requires special attention. The conventional technique of bypassing each power supply pin individually to ground can have an adverse effect on the differential phase error of the circuit. This is because there is an internal compensation capacitor in the AD8010 that is referenced to the negative supply. The recommended technique shown in Figure 6-93 is to connect three parallel bypass capacitors from the positive supply to the negative supply, and then to bypass the negative supply to ground with a similar set, as shown. For high frequency decoupling, 0.1 µF ceramic surface-mount capacitors are recommended. The high currents that can flow through the power supply pins require additional large tantalum electrolytic decoupling capacitors. As shown, a 47 µF/16 V tantalum in parallel with a 10 µF/10 V tantalum capacitor is desirable. The grounded side of the C2 capacitors bypassing the negative supply should be brought to a single-point output return ground. In addition to the bypass capacitors described above, ferrite beads such as those noted should be placed in series with both positive and negative supplies for further decoupling.

Another important consideration for driving multiple cables is high frequency isolation between the outputs. Due largely to its low output impedance, the AD8010 achieves better than 46 dB output-to-output isolation at 5 MHz, while driving 75 Ω source and load-terminated cables.

Differential Line Drivers/Receivers

There are a number of applications for differential signal drivers and receivers. Among these are analog-digital-converter (ADC) input buffers, where differential operation can provide lower levels of second-order distortion for certain converters. Other uses include high frequency bridge excitation, and drivers for balanced transmission twisted pair lines such as in ADSL and HDSL.

The transmission of high quality signals across noisy interfaces (either between individual PC boards or between racks) has always been a challenge to designers. Differential techniques using high common-mode rejection ratio (CMRR) instrumentation amplifiers largely solves the problem at low frequencies. Examples of this have already been discussed, under the "Audio Amplifiers" portion of this chapter.

At audio frequencies, transformers, or products such as the SSM2142 balanced line driver and SSM2141/SSM2143 line receivers offer outstanding CMRRs and the ability to transmit low level signals in the presence of large amounts of noise, as noted. At high frequencies, small bifilar-wound toroid transformers are effective.

In contrast to this, the problem of signal transmission at video frequencies is a more complex one. Transformers suitable for video coupling aren't very effective, because the baseband video signal has low frequency components down to a few tens of Hz, and an upper bandwidth limit that can be in the tens (hundreds) of MHz. This make a workable video transformer an item extremely difficult to make.

Another point is that video signals are generally processed in single-ended form, and therefore don't adapt easily to balanced transmission line techniques. Related to this, shielded twin-conductor coaxial cable with good bandwidth is usually somewhat bulky and expensive, and has not found great acceptance.

As a result of these factors, designing high bandwidth, low distortion differential video drivers and receivers with high CMRR at high frequencies is an extremely difficult task.

Nevertheless, even in the face of all of the above problems, there are various differential techniques available right now that offer distinct advantages over single-ended methods. Some of these techniques make use of discrete components, while others utilize state-of-the-art video differential amplifiers.

Approaches To Video Differential Driving/Receiving

Two solutions to differential transmission and reception are shown in Figure 6-94. One is the ideal case (top), where a balanced differential driver drives a balanced twin-conductor coaxial cable, which then drives a terminated differential line receiver. However, as discussed, this circuit is difficult to implement fully at video frequencies.

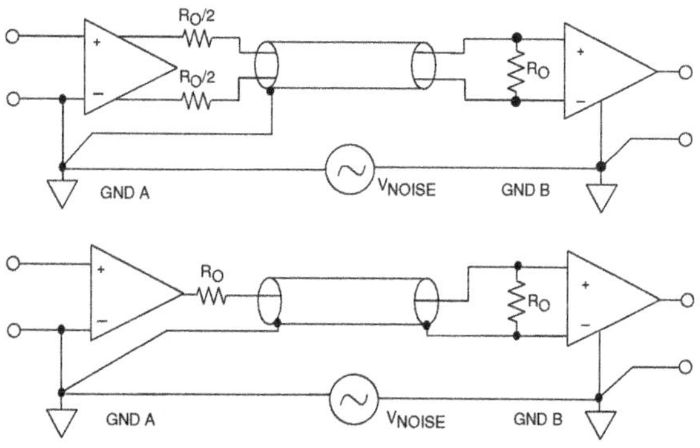

Figure 6-94: Two approaches to differential line driving and receiving

A second, most often used approach uses a single-ended driver driving a source-terminated coaxial cable (bottom), with the cable shield grounded at the transmitter. At the receiver, the coaxial cable is terminated in its characteristic impedance, but the shield is left floating in order to prevent a ground loop between the two systems. Common-mode ground noise is rejected by the CMRR of the differential line receiver.

Inverter-Follower Differential Driver

The circuit of Figure 6-95 is a useful differential driver for high speed 10-12-bit ADCs, differential video lines, and other balanced loads at 1–4 V rms output levels.

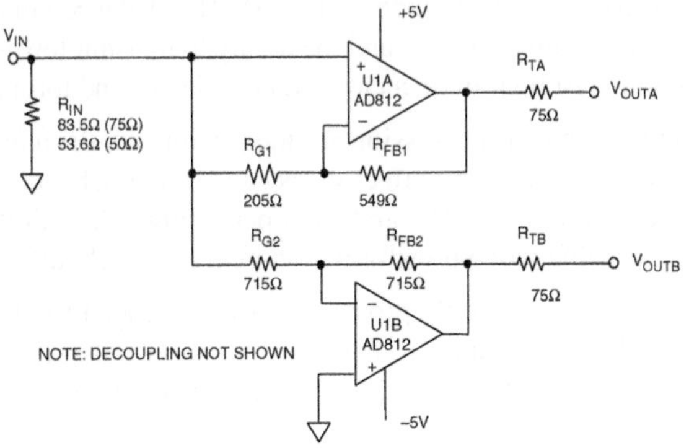

Figure 6-95: Differential driver using an inverter and a follower

It is shown operating from ±5 V supplies, but it can also be adapted to supplies in the range of ±5 to ±15 V. When operated directly from ±5 V as here, it minimizes potential for destructive ADC overdrive when higher supply voltage buffers drive a ±5 V powered ADC, in addition to also minimizing driver power.

In many of these differential drivers the performance criteria is often high. In addition to low output distortion, the two signals should maintain gain and phase flatness. In this topology, two sections of an AD812 dual current feedback amplifier are used for the channel A and B buffers. This provides inherently better open-loop bandwidth matching than using two singles, where bandwidth varies between different manufacturing lots. The two buffers here operate with precise gains of ±1, as defined by their respective feedback and input resistances. Channel B buffer U1B is conventional, and uses a matched pair of 715 Ω resistors—an optimum value for the AD812 on ±5 V supplies.

In channel A, noninverting buffer U1A has an inherent signal gain of 1, by virtue of the bootstrapped feedback network R_{FB1} and R_{G1} (see Reference 5). It also has a higher noise gain, for phase matching. Normally a current feedback amplifier operating as a simple unity gain follower would use one (optimum) resistor R_{FB1}, and no gain resistor at all. Here, with input resistor R_{G1} added, a U1A noise gain like that of U1B results. Due to the bootstrap connection of R_{FB1}-R_{G1}, the signal gain is maintained at unity. Given the matched open-loop bandwidths of U1A and U1B, similar noise gains in the A-B channels provide closely matched output bandwidths between the driver sides, a distinction that greatly impacts overall matching performance.

In setting up a design for the driver, the effects of resistor gain errors should be considered for R_{G2}-R_{FB2}. Here a worst-case 2% mismatch will result in less than 0.2 dB gain error between channels A and B. This error can be improved simply by specifying tighter resistor ratio matching, avoiding trimming.

If desired, phase match can be trimmed via R_{G1}, so that the phase of channel A matches that of B. This can be done by using a pair of closely matched (0.1% or better) resistors to sum the A and B channels, as R_{G1} is adjusted for the best null conditions at the sum node. The A-B gain and phase matching is quite effective in this driver; the test results of the circuit as shown 0.04 dB and 0.1° between the A and B output signals at 10 MHz, when operated into dual 150 Ω loads. The 3 dB bandwidth of the driver is about 60 MHz.

Net input impedance of the circuit is set to a standard line termination value such as 75 Ω (or 50 Ω), by choosing R_{IN} so that the desired value results when paralleled with $R_{G2.}$ In this example, an R_{IN} value of 83.5Ω provides a standard input impedance of 75 Ω when paralleled with 715 Ω. For the circuit just as shown, dual voltage feedback amplifier types with sufficiently high speed and low distortion can also be used. This allows greater freedom with regard to resistor values using such devices as the AD826 and AD828.

Gain of the circuit can be changed if desired, but this isn't totally straightforward. An easy step to satisfy diverse gain requirements is to simply use a triple amplifier such as the AD813 or AD8013, with the third channel as a variable gain input buffer. Note that if an amplifier is used with specifications substantially different than the AD812, some adjustment of resistor values may be necessary.

Cross-Coupled Differential Driver

Another differential driver approach uses cross-coupled feedback to get very high CMR and complementary outputs at the same time. In Figure 6-96, AD8002 dual current feedback amplifier sections are used as cross-coupled inverters, the outputs are forced equal and opposite, assuring zero output common-mode voltage (see Reference 6).

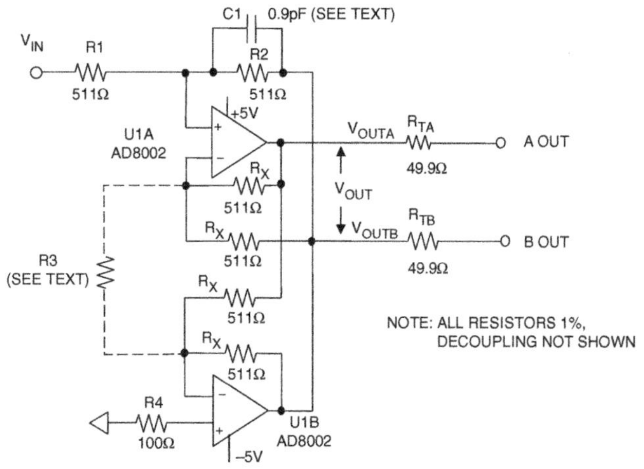

Figure 6-96: Cross-coupled differential driver provides balanced outputs and 250 MHz bandwidth

The gain cell that results, U1A and U1B plus cross-coupling resistances R_X, is fundamentally a differential input/output topology, but it behaves as a voltage feedback amplifier with regard to the feedback port at the U1A (+) node. The V_{IN} to V_{OUT} gain is:

$$G = \frac{V_{OUT}}{V_{IN}} = \frac{2R2}{R1} \qquad \text{Eq. 6-21}$$

where V_{OUT} is the differential output, equal to $V_{OUTA}–V_{OUTB}$.

The relationship of Eq. 6-21 may not be obvious, so it can be derived as follows:

Using the conventional inverting op amp gain equation, the input voltage V_{IN} develops an output voltage V_{OUTB} given by:

$$V_{OUTB} = -V_{IN} \frac{R2}{R1} \qquad \text{Eq. 6-22}$$

Also, $V_{OUTA} = -V_{OUTB}$, because V_{OUTA} is inverted by U1B.

However, $V_{OUT} = V_{OUTA} - V_{OUTB} = -2\ V_{OUTB}$.

Therefore,

$$V_{OUT} = -2\left(-V_{IN}\frac{R2}{R1}\right) = 2V_{IN}\frac{R2}{R1} \qquad \text{Eq. 6-23}$$

and

$$\frac{V_{OUT}}{V_{IN}} = \frac{2R2}{R1}. \qquad \text{Eq. 6-24}$$

This circuit has some unique benefits. First, the differential voltage gain is set by a single resistor ratio, so there is no necessity for side-side resistor matching with gain changes, as is the case for conventional differential amplifiers (see line receivers, below). Second, because the (overall) circuit emulates a voltage feedback amplifier, these gain resistances are not as restrictive as in the case of a conventional current feedback amplifier. Thus, they are not highly critical as to value as long as the equivalent resistance seen by U1A is reasonably low (≤1 kΩ in this case).

A third and important advantage is that cell bandwidth can be optimized to a desired gain by a single optional resistor, R3, as follows. If, for instance, a gain of 20 is desired (R2/R1=10), the bandwidth would otherwise be reduced by roughly this amount, since without R3, the cell operates with a constant gain-bandwidth product (voltage feedback mode). With R3 present however, advantage can be taken of the AD8002 current feedback amplifier characteristics. Additional internal gain is added by the connection of R3, which, given the appropriate value, effectively raises gain-bandwidth to a level so as to restore the bandwidth which would otherwise be lost by the higher closed loop gain.

In the circuit as shown, no R3 is necessary at the low working gain of 2, since the 511 Ω R_X resistors are already optimized for maximum bandwidth. Note that these four matched R_X resistances are somewhat critical, and will change in absolute value with the use of another current feedback amplifier. At higher gain closed loop gains, R3 can be chosen to optimize the working transconductance in the input stages of U1A and U1B, as follows:

$$R3 \cong \frac{R_X}{(R2/R1)-1} \qquad \text{Eq. 6-25}$$

As in any high speed inverting feedback amplifier, a small high-Q chip type feedback capacitance, C1, may be needed to optimize flatness of frequency response. In this example, a 0.9 pF value was found optimum for minimizing peaking. In general, provision should be made on the PC layout for an NPO chip capacitor in the range of 0.5 pF–2 pF. This capacitor is then value selected at board characterization for optimum frequency response.

Performance for the circuit of Figure 6-96 was examined with a dual trace, 1 MHz–500 MHz swept frequency response plot, as is shown in Figure 6-97. The test output levels were 0 dBm into matched 50 Ω loads, through back termination resistances R_{TA} and R_{TB}, as measured at V_{OUTA} and V_{OUTB}.

In this plot the vertical scale is 2 dB/div, and it shows the 3 dB bandwidth of the driver measuring about 250 MHz, with peaking about 0.1 dB. The four R_X resistors, along with R_{TA} and R_{TB} control low frequency amplitude matching, which was within 0.1 dB in the lab tests, using 511 Ω 1% resistor types. For tightest amplitude matching, these resistor ratios can be more closely controlled.

Due to the very high gain bandwidths involved with the AD8002, the construction of this circuit should only be undertaken by following RF rules. This includes the use of a heavy ground plane, and the use of

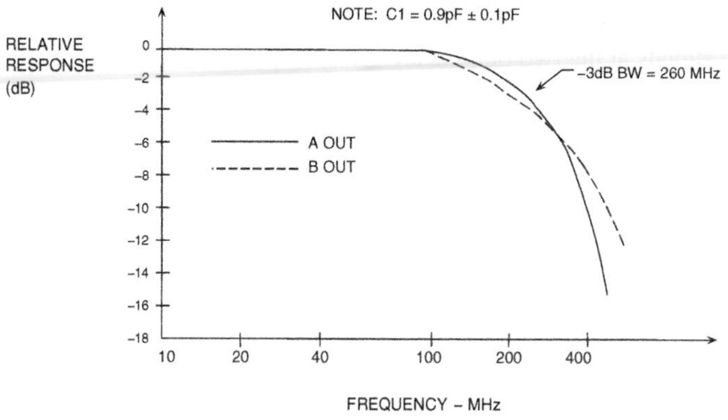

Figure 6-97: Frequency response of AD8002 cross-coupled driver

chip bypass capacitors of zero lead length at the ±5 V supply pins. For lowest parasitic effect and low inductance, chip style resistors are also recommended for this circuit (see Reference 7). The optimization of C1 has already been noted, above. While a chip style NP0 is good in general for C1, a small film trimmer may also be useful, as it will allow optimizing peaking on an individual circuit basis.

Although this circuit example illustrates a wideband video driver, it should be noted that lower bandwidth applications could also find this push-pull topology useful. An audio frequency application could, for example, use an AD812 for U1A and U1B, or pair of AD811s. Operating on ±15 V, these will allow a high level of balanced, linear output.

Fully Integrated Differential Drivers

A block diagram of the new AD813X family of fully differential amplifiers optimized for differential driving is shown in Figure 6-98. Figure 6-98A shows the details of the internal circuit, and Figure 6-98B shows the equivalent circuit. The gain is set by the external R_F and R_G resistors, and the common-mode voltage is

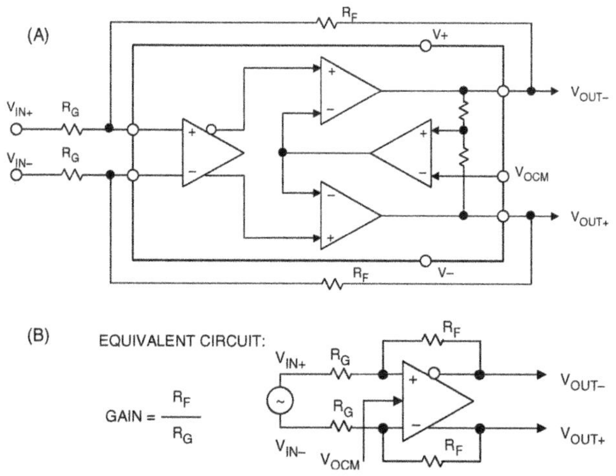

Figure 6-98: AD8138 differential driver amplifier functional schematic (A) and equivalent circuit (B)

set by the voltage on the V_{OCM} pin. The internal common-mode feedback forces the V_{OUT+} and V_{OUT-} outputs to be balanced, i.e., the signals at the two outputs are always equal in amplitude but 180° out of phase per the equation:

$$V_{OCM} = (V_{OUT+} + V_{OUT-})/2 \qquad \text{Eq. 6-26}$$

The circuit can be used with either a differential or a single-ended input, and the voltage gain is equal to the ratio of R_F to R_G.

The AD8138 has a 3 dB small-signal bandwidth of 320 MHz (G = +1) and is designed to give low harmonic distortion as an ADC driver (see Chapter 3). The circuit provides excellent output gain and phase matching, and the balanced structure suppresses even-order harmonics.

It should be noted that the AD8131 differential driver is a sister device to the AD8138 in terms of the function illustrated in Figure 6-98A, and includes internal gain-set resistors.

A 4-Resistor Differential Line Receiver

Figure 6-99 shows a low cost, medium performance line receiver using a high speed op amp that is rated for video use. It is actually a standard 4-resistor difference amplifier optimized for high speed, with a differential to single-ended gain of R2/R1. Using low value, dc-accurate, ac-trimmed resistances for R1–R4, and a high speed, high CMR op amp, provides the good performance.

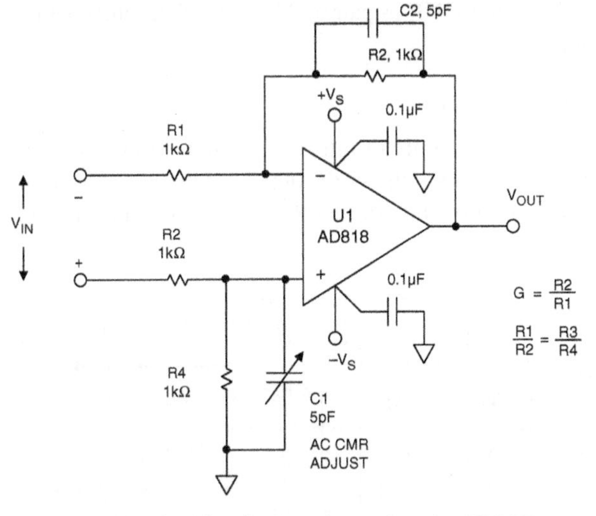

Figure 6-99: Simple video line receiver using the AD818 op amp

Practically speaking, however, at low frequencies resistor matching can be more critical to overall CMR than the rated CMR of the op amp. For example, the worst-case CMR (in dB) of this circuit due to resistor mismatch is:

$$\text{CMR} = 20\log_{10}\left(\frac{1+\dfrac{R2}{R1}}{4Kr}\right) \qquad \text{Eq. 6-27}$$

In this expression "Kr" is a single resistor tolerance in fractional form (1% = 0.01, and so forth), and it is assumed the amplifier has significantly higher CMR (100 dB). Using discrete 1% metal films for R1/R2 and R3/R4 yields a worst-case CMR of 34 dB, 0.1% types 54 dB, and so forth. Of course four random 1% resistors will on the average yield a CMR better than 34 dB, but not dramatically so. A single substrate dual matched pair thin film network is preferred, for reasons of best noise rejection and simplicity. One type suitable is the Vishay VTF series part 1005, (see Reference 7) which has a ratio match of 0.1%, and will provide a worst-case low frequency CMR of 66 dB.

This circuit has an interesting and desirable side property. Because of the resistors it divides down the input voltage, and the amplifier is protected against overvoltage. This allows CM voltages to exceed ±5 V supply rails in some cases without hazard. For operation with ±15 V supplies, inputs should not exceed the supply rails.

At frequencies above 1 MHz, the bridge balance is dominated by ac effects, and a C1-C2 capacitive balance trim should be used for best performance. The C1 adjustment is intended to allow this, providing for the cancellation of stray layout capacitance(s) by electrically matching the net C1-C2 values.

Within a given PC layout with low and stable parasitic capacitance, C1 is best adjusted once in 0.5 pF increments, for best high frequency CMR. Using designated PC pads, production values would then use the trimmed value. Good ac matching is essential to achieving good high frequency CMR. C1-C2 should be physically similar types, such as NPO ceramic chip capacitors.

While the circuit as shown has unity gain, it can be gain-scaled in discrete steps, as long as the noted resistor ratios are maintained. In practice, this means using taps on a multi-ratio network for gain change, so as to raise both R2 and R4, in identical proportions. There is no other simple way to change gain in this receiver circuit.

Alternately, a scheme for continuous gain control without interaction with CMR is to simply follow this receiver with a scaling amplifier/driver, with adjustable gain. The use of an AD828 amplifier (AD818 dual) allows this, with the addition of only two resistors.

Video gain/phase performance of this stage is dependent upon the device used for U1 and the operating supply voltages. Suitable voltage feedback amplifiers work best at supplies of ±10 to ±15 V, which maximizes op amp bandwidth. And, while many high speed amplifiers function in this circuit, those expressly designed with very low distortion video operation perform best.

The circuit just as shown can be used with supplies of ±5 V to ±15 V, but lowest NTSC video distortion occurs for supplies of ±10 V or more using the AD818, where differential gain/differential phase errors are less than 0.01%/0.05°. With the AD818 operating at ±5 V supplies, the distortion rises somewhat, but the lowest power drain of 70 mW occurs. If low distortion and lowest power operation on ±5 V is important, the use of an AD8055 (or AD8056 section) should be considered for the U1 function; they will dissipate 50 mW. A drawback to this circuit is that it does load a 75 Ω video line to some extent, and so should be used with this loading taken into account. On the plus side, it has wide dynamic range for both signal and CM voltages, plus the inherent overvoltage protection.

Active Feedback Differential Line Receiver

The AD8129/AD8130 differential line receivers, along with their predecessor the AD830, utilize a novel amplifier topology called *active feedback* (see Reference 8). A simplified block diagram of these devices is shown in Figure 6-100.

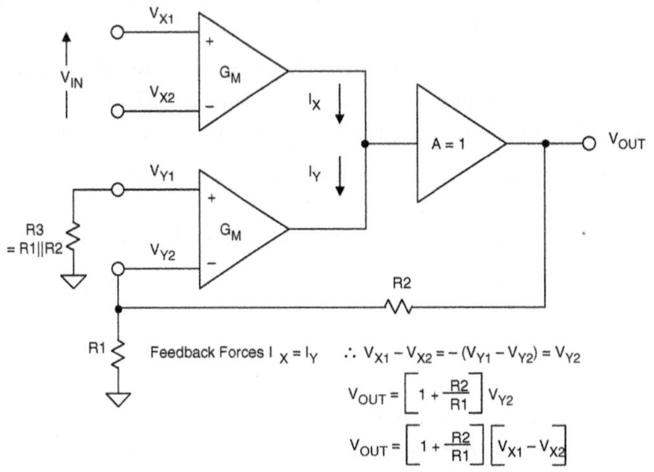

Figure 6-100: The AD830/AD8129/AD8130 active feedback amplifier topology

The AD830 and the AD8129/AD8130 have two sets of fully differential inputs, available at V_{X1}–V_{X2} and V_{Y1}–V_{Y2}, respectively. Internally, the outputs of the two GM stages are summed and drive a buffer output stage.

In this device the overall feedback loop forces the internal currents I_X and I_Y to be equal. This condition forces the differential voltages V_{X1}–V_{X2} and V_{Y1}–V_{Y2} to be equal and opposite in polarity. Feedback is taken from the output back to one input differential pair, while the other pair is driven directly by an input differential input signal.

An important point of this architecture is that high CM rejection is provided by the two differential input pairs, so *CMR is not dependent on resistor bridges* and their associated matching problems. The inherently wideband balanced circuit and the quasi-floating operation of the driven input provide the high CMR, which is typically 100 dB at dc.

The general expression for the stage's gain "G" is like a non-inverting op amp, or:

$$G = \frac{V_{OUT}}{V_{IN}} = 1 + \frac{R2}{R1} \qquad \text{Eq. 6-28}$$

As should be noted, this expression is identical to the gain of a noninverting op amp stage, with R2 and R1 in analogous positions.

The AD8129 is a low noise high gain (G = 10 or greater) version of this family, intended for applications with very long cables where signal attenuation is significant. The related AD8130 device is stable at a gain of one. It is used for those applications where lower gains are required, such as a gain-of-two, for driving source and load terminated cables.

The AD8129 and AD8130 have a wide power supply range, from single +5 V to ±12 V, allowing wide common-mode and differential-mode voltage ranges. The wide common-mode range enables the driver/receiver pair to operate without isolation transformers in many systems where the ground potential difference between driver and receiver locations is several volts. Both devices include a logic-controlled power-down function.

Both devices have high, balanced input impedances, and achieve 70 dB CMR @ 10MHz, providing excellent rejection of high frequency common-mode signals. Figure 6-101 shows AD8130 CMR for various supplies. As can be noted, it can be as high as 95 dB at 1 MHz, an impressive figure considering that no trimming is required.

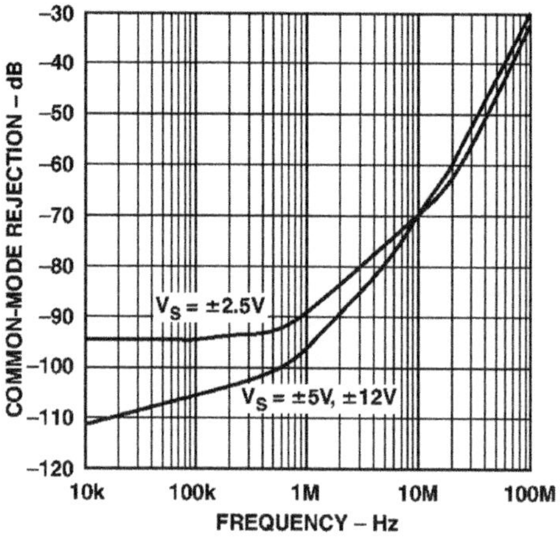

Figure 6-101: AD8130 common-mode rejection versus frequency for ±2.5 V, ±5 V, and ±12 V supplies

The typical 3 dB bandwidth for the AD8129 is 200 MHz, while the 0.1 dB bandwidth is 30 MHz in the SOIC package, and 50 MHz in the µSOIC package. The conditions for these specifications are for V_S = ±5 V and G = 10.

The typical 3 dB bandwidth for the AD8130 is 270 MHz, and the 0.1 dB bandwidth is 45 MHz, in either package. The conditions for these specifications are for V_S = ±5 V and G = 1. Typical differential gain and phase specifications for the AD8130 for G = 2, V_S = ±5 V, and R_L = 150 Ω are 0.13% and 0.15°, respectively.

A Cable-Tap or Loopthrough Amplifier

Figure 6-102 shows an example of a video *cable-tap* amplifier (or *loopthrough*) connection where the input signal is tapped from a coax line and applied to one input stage of the AD8130, with the output signal tied to the second input stage. The net gain is unity. Functionally, the input and local grounds are isolated by the CMR of the AD8130, which is typically 70 dB at 10 MHz. Note that in order to provide a dc path for the input bias currents of the upper stage, there must be a common path between the source and local grounds (shown as Z_{CM}). This impedance is not critical, but must be low enough that 60 Hz noise and other voltage components remain within the AD8130's CM range.

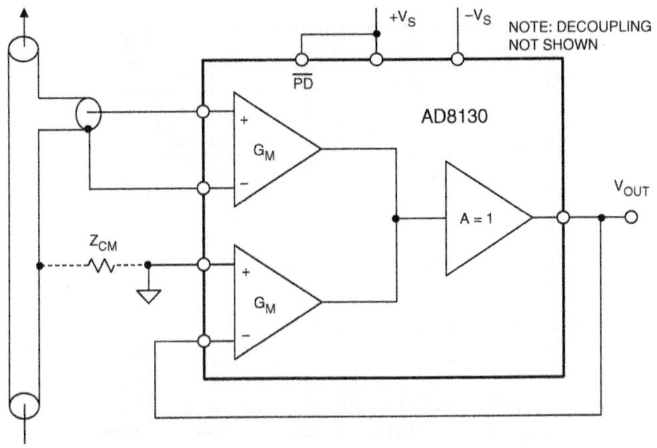

Figure 6-102: Video "cable-tap" amplifier using the AD8130

The circuit is efficient with the simplicity as shown, and requires no gain set resistors, and so forth to implement. Normal bypass capacitors and supply decoupling must of course be used, as in any high speed circuit. Other than the necessary dc path for the two inputs, it has little affect on the video cable it is monitoring, due to the high impedance AD8130 inputs. The circuit just as shown operates on supplies of ±5 V to ±12 V, but a ±15 V version can also be implemented by using the AD830 (without the AD8130's power-down function).

This circuit can also act as a video repeater, by connecting equal value feedforward and feedback resistors to implement a gain-of-two, for driving a source and load-terminated video cable (i.e., R2 and R1, as in Figure 6-100).

Further application examples of this family of active feedback amplifiers are contained in the "Amplifier Ideas" section of this chapter plus, of course the device data sheets.

High Speed Clamping Amplifiers

There are many situations where it is desirable to *clamp* the output of an op amp, to prevent overdriving following circuitry. Specially designed high speed, fast recovery clamping amplifiers offer an attractive alternative to designing external clamping/protection circuits. The AD8036/AD8037 low distortion, wide bandwidth clamp amplifiers represent a significant breakthrough in this technology. These devices allow the designer to specify a high (V_H) and low (V_L) clamp voltage. The output of the device clamps when the input exceeds either of these two levels. The AD8036/AD8037 offer superior clamping performance compared to competing devices that use output-clamping. Recovery time from overdrive is less than 5 ns, and small signal bandwidth is 240 MHz (AD8036) and 270 MHz (AD8037).

The key to the AD8036 and AD8037's fast, accurate clamp and amplifier performance is their proprietary input clamp architecture. This new design reduces clamp errors by more than 10 times over previous output clamp based circuits, as well as substantially increasing the bandwidth, precision, and versatility of the clamp inputs.

Figure 6-103 is an idealized block diagram of the AD8036 clamp amplifier, connected as a unity gain voltage follower. The primary signal path comprises A1 (a 1200 V/μs, 240 MHz high voltage gain, differential-to-single-ended amplifier) and A2 (a G = + 1 high current gain output buffer). The AD8037 differs from the AD8036 only in that A1 is optimized for closed-loop gains of two or greater.

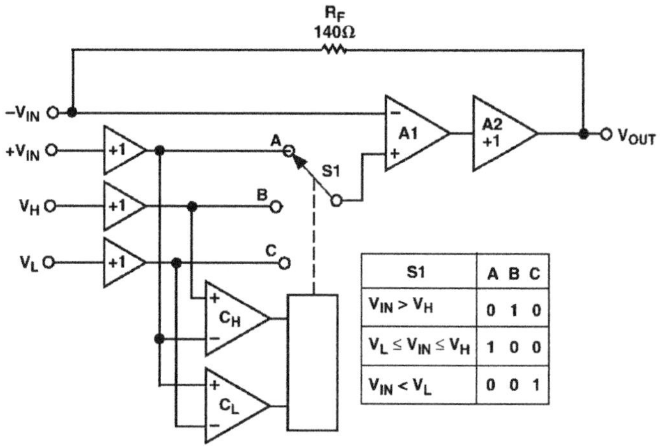

Figure 6-103: AD8036/AD8037 clamp amplifier equivalent circuit

The input clamp section is comprised of comparators C_H and C_L, which drive switch S1 through a decoder. The unity-gain buffers before the $+V_{IN}$, V_H, and V_L inputs isolate the input pins from the comparators and S1, without reducing bandwidth or precision. The two comparators have about the same bandwidth as A1 (240 MHz), so they can keep up with signals within the useful bandwidth of the AD8036. To illustrate the operation of the input clamp circuit, consider the case where V_H is referenced to 1 V, V_L is open, and the AD8036 is set for a gain of +1 by connecting its output back to its inverting input through the recommended 140 Ω feedback resistor. Note that the main signal path always operates closed loop, since the clamping circuit only affects A1's noninverting input.

If a 0 V to +2 V voltage ramp is applied to the AD8036's $+V_{IN}$ for the connection, V_{OUT} should track $+V_{IN}$ perfectly up to +1 V, then limit at exactly 1 V as $+V_{IN}$ continues to +2 V. In practice, the AD8036 comes close to this ideal behavior. As the $+V_{IN}$ input voltage ramps from zero to 1 V, the output of the high limit comparator C_H starts in the off state, as does the output of C_L. When $+V_{IN}$ just exceeds V_H (practically, by about 18 mV), C_H changes state, switching S1 from "A" to "B" reference level. Since the + input of A1 is now connected to V_H, further increases in $+V_{IN}$ have no effect on the AD8036's output. The AD8036 is now operating as a unity-gain buffer for the V_H input, as any variation in V_H, for $V_H > 1$ V, will be faithfully produced at V_{OUT}.

AD8036 operation for negative inputs and negative V_L clamp levels is similar, with comparator C_L controlling S1. Since the comparators see the voltage on the $+V_{IN}$ pin as their common reference level, the voltage

V_H and V_L are defined as "High" or "Low" with respect to $+V_{IN}$. For example, if V_{IN} is zero volts, V_H is open, and V_L is +1 V, comparator C_L will switch S1 to "C," and the AD8036 will buffer the V_L voltage and ignore $+V_{IN}$.

The AD8036/AD8037 performance closely matches the described ideal. The comparator's threshold extends from 60 mV inside the clamp window defined by the voltages on V_L and V_H to 60 mV beyond the window's edge. Switch S1 is implemented with current steering, so that A1's + input makes a continuous transition from, say, V_{IN} to V_H as the input voltage traverses the comparator's input threshold from 0.9 V to 1.0 V for V_H = 1.0 V.

The practical effect of the nonideal operation softens the transition from amplification to clamping modes, without compromising the absolute clamp limit set by the input clamping circuit. Figure 6-104 is a graph of V_{OUT} versus V_{IN} for the AD8036 and a typical *output* clamp amplifier. Both amplifiers are set for G = +1 and V_H = +1 V. The worst-case error between V_{OUT} (ideally clamped) and V_{OUT} (actual) is typically 18 mV times the amplifier closed-loop gain. This occurs when V_{IN} equals V_H (or V_L). As V_{IN} goes above and/or below this limit, V_{OUT} will stay within 5 mV of the ideal value.

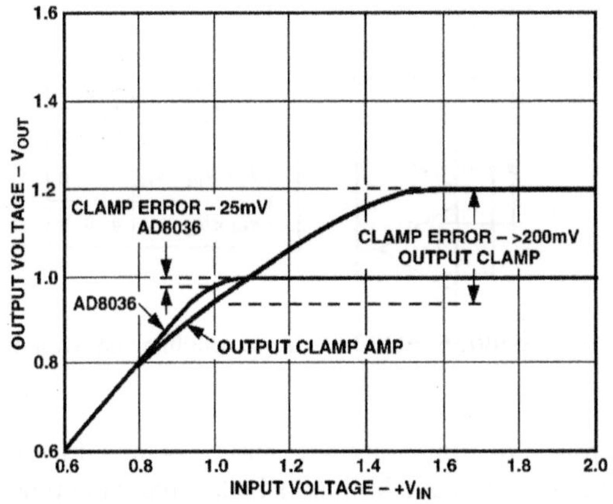

Figure 6-104: Comparison between input and output clamping

In contrast, the output clamp amplifier's transfer curve typically will show some compression starting at an input of 0.8 V, and can have an output voltage as far as 200 mV over the clamp limit. In addition, since the output clamp causes the amplifier to operate open-loop in the clamp mode, the amplifier's output impedance will increase, potentially causing additional errors, and the recovery time is significantly longer.

Flash Converter with Clamp Amp Input Protection

Figure 6-105 shows the AD9002 8-bit, 125 MSPS flash converter driven by the AD8037 (240 MHz bandwidth) clamping amplifier. The clamp voltages on the AD8037 are set to +0.55 V and –0.55 V, referenced to the ±0.5 V input signal, with the twin 806 Ω/100 Ω external resistive dividers. The AD8037 also supplies a gain of two, and an offset of –1 V (using the AD780 voltage reference), to match the 0 V to –2 V input range of the AD9002 flash converter. The output signal is clamped at +0.1 V and –2.1 V.

This multifunction clamping circuit therefore performs several important functions as well as preventing damage to the flash converter (which would otherwise occur should the input exceed 0.5 V, thereby forward

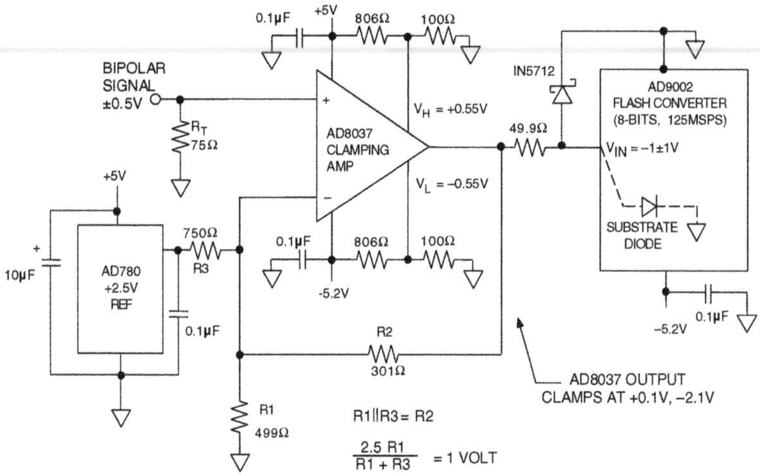

Figure 6-105: AD9002 8-bit, 125MSPS flash converter driven by AD8037 clamp amplifier

biasing the substrate diode). The 1N5712 Schottky diode is a safety-valve device, adding further protection for the flash converter during power-up.

Multiple criteria must be met in designing the feedback network around the AD8037. These are a specified gain and a fixed offset which will enable the output swing of the clamped amplifier to match the target input range of the converter.

The feedback resistor, R2 = 301 Ω, is selected for optimum bandwidth per the data sheet recommendation. For a gain of two, the parallel combination of R1 and R3 must also equal R2:

$$\frac{R1 \cdot R3}{R1 + R3} = R2 = 301\ \Omega \qquad \text{Eq. 6-29}$$

(nearest 1% standard resistor value).

In addition, the Thevenin equivalent output voltage from the AD780 +2.5 V reference and the R3/R1 divider must be +1 V, to provide the required –1 V offset at the output of the AD8037. This will cause the output swing of the AD8037 to be biased at –1 V when V_{IN} is zero, and to range from 0 to –2 V as V_{IN} ranges from –0.5 V to 0.5 V.

$$\frac{2.5 \cdot R1}{R1 + R3} = 1V \qquad \text{Eq. 6-30}$$

Solving these equations yields resistance values of R1 = 499 Ω, R3 = 750 Ω, using the nearest 1% standard values.

Other input and output voltages ranges can also be accommodated, by appropriate changes in the external resistors.

Further fast clamping op amp application examples are given in Reference 9, and the "Amplifier Ideas" section of this chapter (plus, of course, the device data sheets).

High Speed Video Multiplexing with Op Amps Utilizing Disable Function

A common video circuit function is the multiplexer, a stage which selects one of "n" video inputs and transmits a buffered version of the selected signal to the output. A number of video op amps (AD810, AD813, AD8013, AD8074/AD8075) have a *disable* mode which, when activated by applying the appropriate control level to a pin on the package, disables the op amp output stage and drops the power to a lower value.

In the case of the AD8013 (triple current-feedback op amp), asserting any one of the disable pins about 1.6 V from the negative supply will put the corresponding amplifier into a disabled, powered-down state. In this condition, the amplifier's quiescent current drops to about 0.3 mA, its output becomes a high impedance, and there is a high level of isolation from the input to the output. In the case of the gain-of-two line driver, for example, the impedance at the output node will be about equal to the sum of the feedback and feedforward resistors (1.6 kΩ) in parallel with about 12 pF capacitance. Input-to-output isolation is about 66 dB at 5 MHz.

Leaving the disable pin disconnected (floating) will leave the corresponding amplifier operational (i.e., enabled). The input impedance of the disable pin is about 40 kΩ in parallel with 5 pF. When driven to 0 V, with the negative supply at –5 V, about 100 μA flows into the disable pin.

When the disable pins are driven by CMOS logic, on a single +5 V supply, the disable and enable times are about 50 ns. When operated on dual supplies, level shifting will be required from standard logic outputs to the disable pins.

The AD8013's input stages include protection from the large differential voltages that may be applied when disabled. Internal clamps limit this voltage to about ±3 V. The high input-to-output isolation will be maintained for voltages below this limit.

Wiring the amplifier outputs together as shown in Figure 6-106 forms a 3:1 multiplexer with about 50 ns switching time between channels. The 0.1 dB bandwidth of the circuit is 35 MHz, and the OFF channel isolation is 60 dB at 10 MHz. The simple logic level-shifting circuit shown on the diagram does not significantly affect switching time.

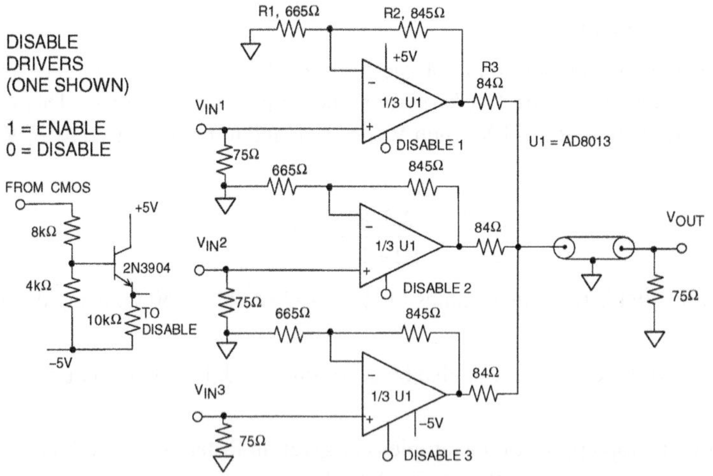

Figure 6-106: AD8013 3:1 video multiplexer switches in 50 ns

Setting up this amplifier is not entirely straightforward, and some explanation will help with subtleness. The feedback resistor R2 of 845 Ω was chosen first, to allow optimum bandwidth of the AD8013 current feedback op amp. The analogous resistors of the other channels use an identical value, for similar reasons.

Note that when any given channel is ON, it must drive both the termination resistor R_L, and the net dummy resistance, $R_X/2$, where R_X is an equivalent series resistance equal to R1 + R2 + R3. To provide a net overall gain of unity, as well as an effective source resistance of 75 Ω, the other resistor values must be as shown. In essence, the Thevenin equivalent value of $R_X/2$ and R3 should equal the desired source termination impedance of 75 Ω (which it does).

It is also desirable that the ON channels have a net gain of 2x, as seen behind the 75 Ω output impedance. The lower value of R1 vis-à-vis R2, along with the above relationship, allows these mutual criteria to be met.

Configuring two amplifiers of an AD8013 as unity gain followers with the third to set the gain results in a high performance 2:1 multiplexer, as shown in Figure 6-107.

This circuit takes advantage of the low crosstalk between the amplifiers, and achieves an OFF channel isolation of 50 dB at 10 MHz. The differential gain and phase performance of the circuit is 0.03% and 0.07°, respectively. The output stage operates at a gain of 2x, and can drive a 75 Ω source terminated line if desired.

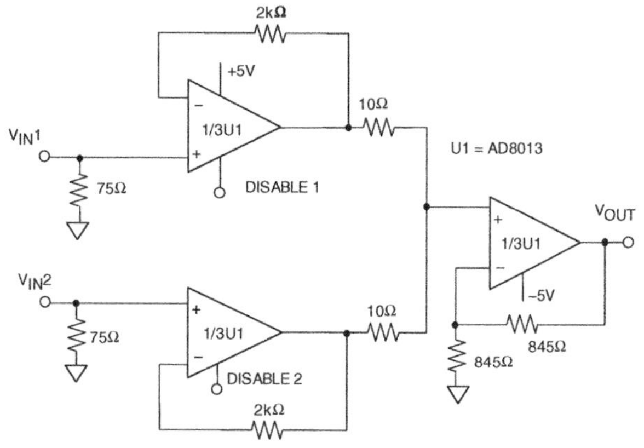

Figure 6-107: 2:1 video multiplexer based on the AD8013

Programmable Gain Amplifier using the AD813 Current Feedback Video Op Amp

Closely related to the multiplexers described above is a programmable gain video amplifier, or PGA, as shown in Figure 6-108. In the case of the AD813, the individual channels are disabled by pulling the disable pin about 2.5 V below the positive supply. This puts the corresponding amplifier in its powered down state. In this condition, the amplifier's quiescent supply current drops to about 0.5 mA, its output becomes a high impedance, and there is a high level of isolation between the input and the output.

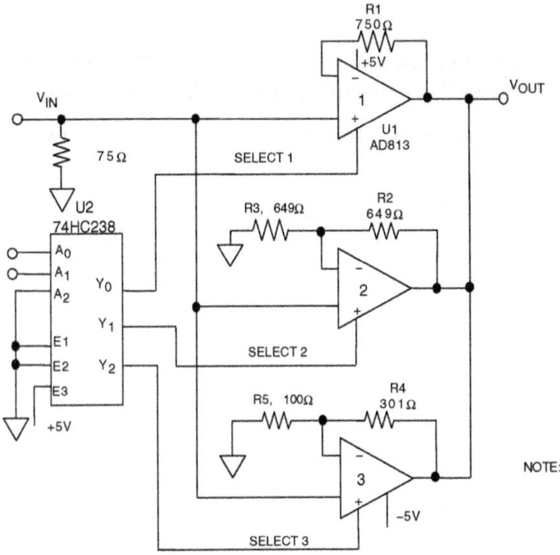

OUTPUT TABLE

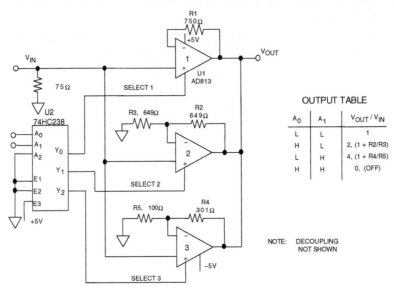

A_0	A_1	V_{OUT}/V_{IN}
L	L	1
H	L	2, (1 + R2/R3)
L	H	4, (1 + R4/R5)
H	H	0, (OFF)

NOTE: DECOUPLING NOT SHOWN

Figure 6-108: Programmable gain video amplifier using the AD813 triple current feedback amplifier

Leaving the disable pin disconnected (floating) will leave the amplifier operational, in the enabled state. When grounded, about 50 μA flows out of a disable pin when operating on ±5 V supplies. The switching threshold is such that the disable pins can be driven directly from +5 V CMOS logic as shown, with no level shifting (as in the previous example).

With a two-line digital control input, this circuit can be set up to provide three different gain settings. This makes it a useful circuit in various systems that can employ signal normalization or gain ranging prior to A/D conversion, such as CCD systems, ultrasound, and so forth. The gains can be binary-related as here, or they can be arbitrary. An extremely useful feature of the AD813 current feedback amplifier is the fact that the bandwidth does not reduce as gain is increased. Instead, it stays relatively constant as gain is raised. Thus, more useful bandwidth is available at the higher programmed gains than would be true for a fixed gain-bandwidth product VFB amplifier type.

In the circuit, channel 1 of the AD813 is a unity gain channel, channel 2 has a gain of 2, and channel 3 a gain of 4, while the fourth control state is OFF. As is indicated by the table, these gains can easily be varied by adjustment of the R2/R3 or R4/R5 ratios. For the gain range and values shown, the PGA will be able to maintain a 3 dB bandwidth of about 50 MHz or more for loading as shown (a high impedance load of 1 kΩ or more is assumed). Fine tuning the bandwidth for a given gain setting can be accomplished by lowering the resistor values at the higher gains, as shown in the circuit, where for G = 1, R1 = 750 Ω, for G = 2, R2 = 649 Ω, and for G = 4, R4 = 301 Ω.

Integrated Video Multiplexers and Crosspoint Switches

Traditional CMOS switches and multiplexers suffer from several disadvantages at video frequencies. Their switching time (typically 100 ns or so) is not fast enough for today's applications, and they require external buffering in order to drive typical video loads. In addition, the small variation of the CMOS switch "on" resistance with signal level (*R_{ON} modulation*) introduces unwanted distortion in differential gain and phase. Multiplexers based on complementary bipolar technology offer a better solution at video frequencies.

Functional block diagrams of the AD8170/AD8174/AD8180/AD8182 bipolar video multiplexer are shown in Figure 6-109. The AD8183/AD8185 video multiplexer is shown in Figure 6-110. These devices offer a high degree of flexibility and are ideally suited to video applications, with excellent differential gain and phase specifications. Switching time for all devices in the family is 10 ns to 0.1%.

The AD8170/AD8174 series of muxes include an on-chip current feedback op amp output buffer whose gain can be set externally. Off channel isolation and crosstalk are typically greater than 80 dB for the entire family.

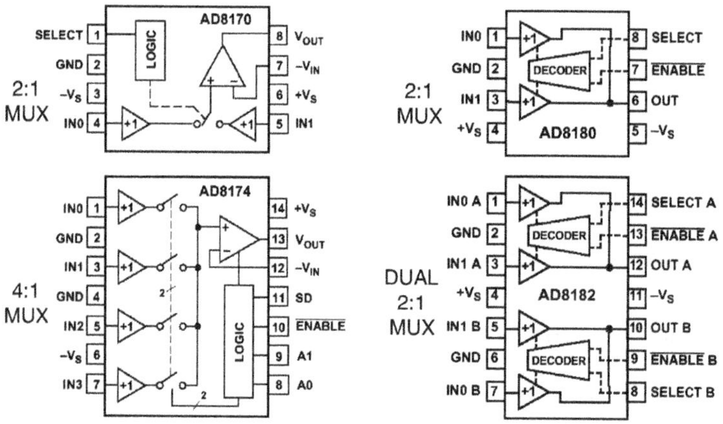

Figure 6-109: AD8170/AD8174/AD8180/AD8182 bipolar video multiplexers

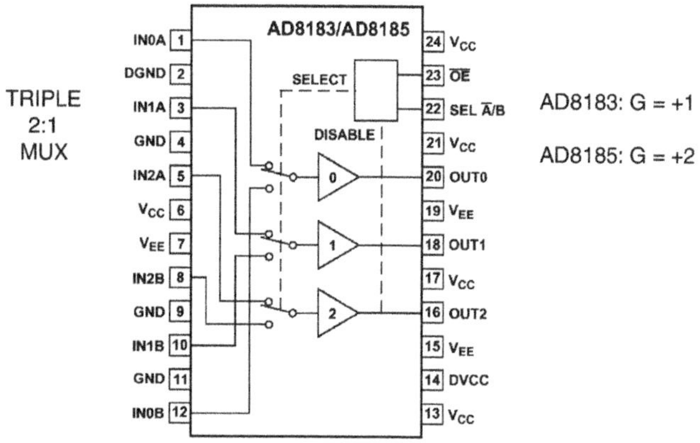

Figure 6-110: AD8183/AD8185 triple 2:1 video multiplexers

Dual RGB Source Video Multiplexer

Figure 6-111 shows an application circuit for three AD8170 2:1 muxes, where a single RGB monitor is switched between two RGB computer video sources.

In this setup, the overall effect is that of a three-pole, double-throw switch. The three video sources constitute the three poles, and either the upper or lower of the video sources constitute the two switch states.

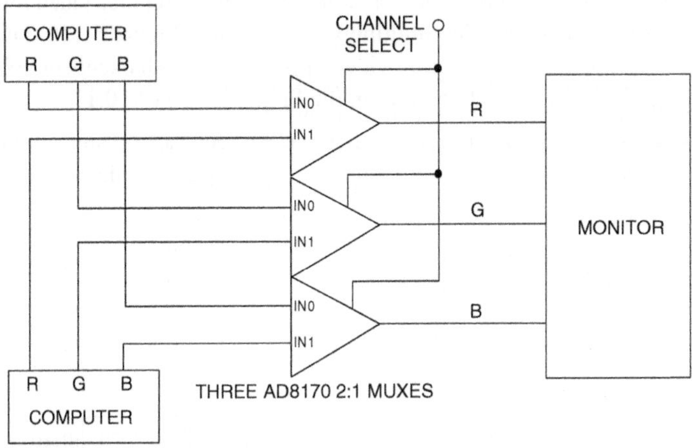

Figure 6-111: Dual source RGB multiplexer using three 2:1 muxes

Digitizing RGB Signals with One ADC

The AD8174 4:1 mux is used in Figure 6-112, to allow a single high speed ADC to digitize the RGB outputs of a scanner.

The RGB video signals from the scanner are fed in sequence to the ADC, and digitized in sequence, making efficient use of the scanner data with one ADC.

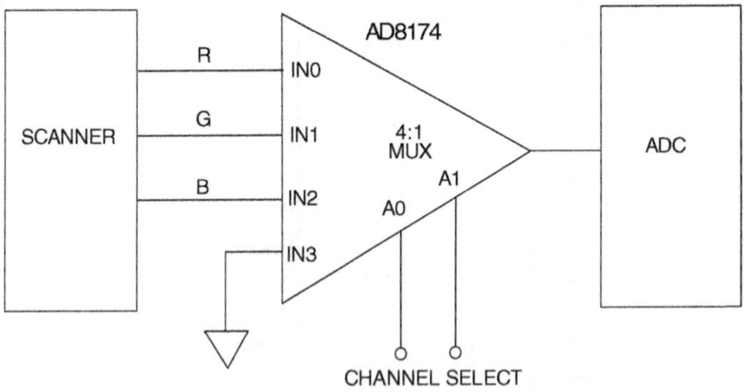

Figure 6-112: Digitizing RGB signals with one ADC and a 4:1 mux

Figure 6-113 shows two AD8174 4:1 muxes functionally expanded into an 8:1 multiplexer. The A0 and A1 inputs are conventional, with complemented Enable inputs.

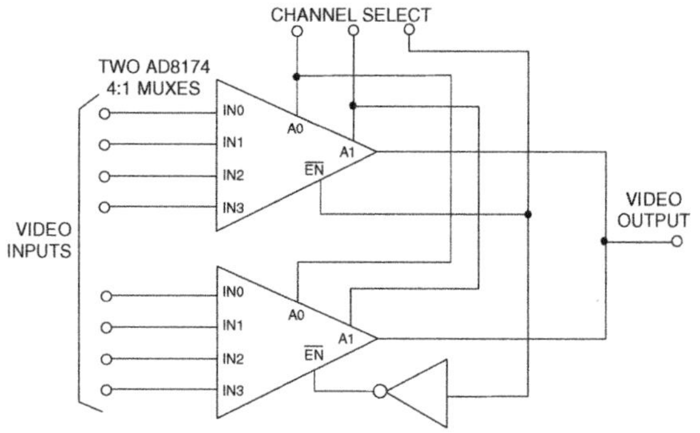

Figure 6-113: Expanding two 4:1 muxes into an 8:1 mux

The AD8116 extends the mux concepts to a fully integrated, 16×16 buffered video crosspoint switch matrix (Figure 6-114). The 3 dB bandwidth is greater than 200 MHz, and the 0.1 dB gain flatness extends to 60 MHz. Channel switching time is less than 30 ns to 0.1%. Channel-to-channel crosstalk is –70 dB measured at 5 MHz. Differential gain and phase is 0.01% and 0.01° for a 150 Ω load. Total power dissipation is 900 mW on ±5 V.

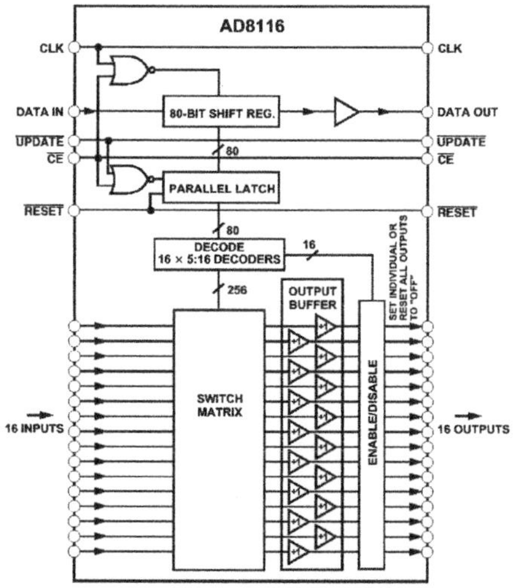

Figure 6-114: AD8116 16×16 200 MHz buffered video crosspoint switch

The AD8116 includes output buffers that can be put into a high impedance state for paralleling crosspoint stages so that the off channels do not load the output bus. The channel switching is performed via a serial digital control that can accommodate "daisy chaining" of several devices. The AD8116 package is a

128-pin 14 mm × 14 mm LQFP. Other members of the crosspoint switch family include the AD8110/AD8111 260 MHz 16 × 8 buffered crosspoint switch, the AD8113 audio/video 60 MHz 16 × 16 crosspoint switch, and the AD8114/AD8115 low cost 225 MHz 16 × 16 crosspoint switch.

Single Supply Video Applications

Optimum video performance in terms of differential gain and phase, bandwidth flatness, and so forth, is generally achieved using dual supplies of ±5 V or ±12 V. In many applications, however, stringent broadcast standards are not required, and single-supply operation may be desirable from a cost and power standpoint. This section illustrates a few op amp single-supply applications. All of the op amps are fully specified for both ±5 V and +5 V (and +3 V where the design supports it). Both rail-to-rail and nonrail-to-rail applications are shown (details of rail-to-rail op amp topologies are discussed in Chapter 1).

Single-Supply RGB Buffer

Op amps such as the AD8041/AD8042 and AD8044 can provide buffering of RGB signals that include ground, while operating from a single +3 V or +5 V supply. The signals that drive an RGB monitor are usually supplied by current output DACs that operate from a single +5 V supply. Examples are triple video DACs such as the ADV7120/ADV7121/ADV7122 from Analog Devices.

During the horizontal blanking interval, the current output of the DACs goes to zero, and the RGB signals are pulled to ground by the termination resistors. If more than one RGB monitor is desired, it cannot simply be connected in parallel because this would be a mis-termination. Therefore, buffering must be provided before connecting a second monitor.

RGB signals include ground as part of their dynamic output range. Previously a dual supply op amp had been required for this buffering, with sometimes this being the only component requiring a negative supply. This makes it quite inconvenient to incorporate a multiple monitor feature. Figure 6-115 shows a diagram of one channel of a single supply op amp gain-of-two buffer, for driving a second RGB monitor. No current is required when the amplifier output is at ground. The termination resistor at the monitor helps pull the output down at low voltage levels.

Note that the input and output are at ground during the horizontal blanking interval. The RGB signals are specified to output a maximum of 700 mV peak. The peak output of the AD8041 is 1.4 V, with the termination resistors providing a divide-by-two. All three channels (RGB) signals can be buffered in a like manner with duplication of this circuit. Another possibility is to use three sections of the (similar) quad AD8044 op amp.

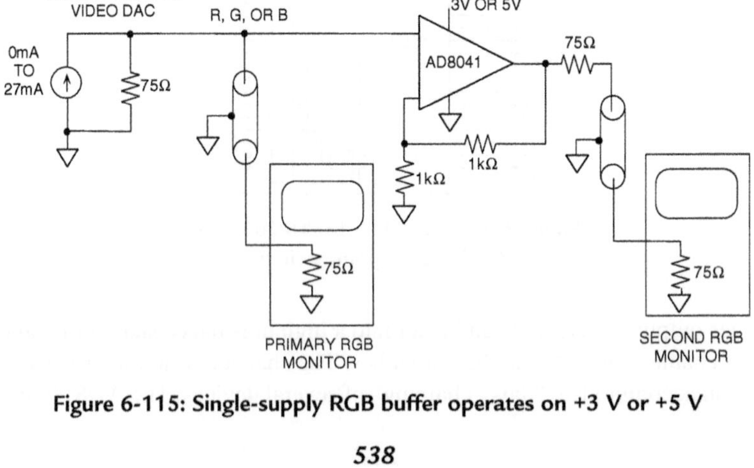

Figure 6-115: Single-supply RGB buffer operates on +3 V or +5 V

Single-Supply Sync Stripper

Some RGB monitors use only three cables total and carry the synchronizing signals and the Green (G) signal on the same cable (Green-with-sync). The sync signals are pulses that go in the negative direction from the blanking level of the G signal.

In some applications, for example prior to digitizing component video signals with ADCs, it is desirable to remove or strip the sync portion from the G signal. Figure 6-116 is a circuit using the AD8041 running on a single 5 V supply to perform this function. The signal at V_{IN} is the Green-with-sync signal from an ADV7120, a single supply triple video DAC.

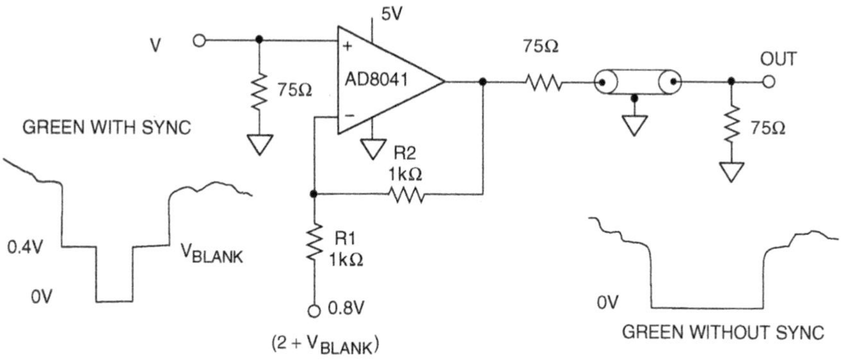

Figure 6-116: Single-supply video sync stripper

Because of the fact that the DAC used is single supply, the lowest level of the sync tip is at ground or slightly above. The AD8041 is set for a gain of two to compensate for the divide-by-two of the output terminations.

In this setup, the op amp used must have a CM capability that includes zero (as is true for the AD8041 family). For voltages *above* one-half the 0.8 V reference level applied to R1, the op amp operates as a linear amplifier, going positive from ground level at the output. For inputs *below* the reference level, the op amp saturates, with the output going to ground as used here. The result is that the negative sync tips are removed.

The reference voltage for R1 is twice the dc blanking level of the G signal; normally, this is 2 × 0.4 V = 0.8 V. Alternately, if the blanking level is at ground and the sync tip is negative (as in some dual supply systems), then R1 is tied to ground. The resulting VOUT will have the sync removed, and the blanking level at ground, as noted.

A Low Distortion, Single-Supply Video Line Driver with Zero-Volt Output

When operated with a single supply, the AD8031 80 MHz rail-to-rail voltage feedback op amp has optimum distortion performance when the signal has a common mode level of $V_S/2$, and when there is also about 500 mV of headroom to each rail. If this rule is violated, distortion performance suffers. But, if low distortion is required for signals close to ground, a level-shifting emitter follower can be used at the op amp output.

Figure 6-117 shows an AD8031 op amp, configured as a single supply gain-of-two line driver. With the output driving a back terminated 50 Ω line, the overall gain is unity from V_{IN} to V_{OUT}. In addition to minimizing reflections, the 50 Ω back termination resistor protects the transistor from damage if the cable is short circuited.

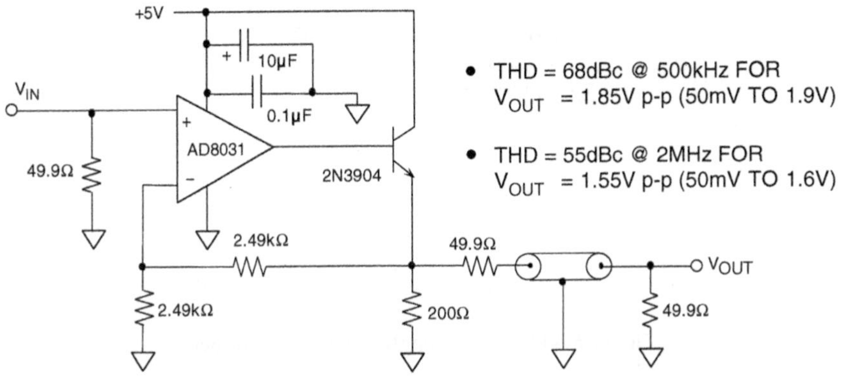

Figure 6-117: Low distortion zero-volt output single-supply line driver using the AD8031

The 2N3904 emitter follower inside the feedback loop ensures that the output voltage from the AD8031 always stays about 700 mV (or more) above ground, which minimizes distortion. Excellent distortion is obtained using this circuit, even when the output signal swings to within 50 mV of ground.

The circuit was tested at 500 kHz and 2 MHz using a single 5 V supply. For the 500 kHz signal, THD was 68 dBc with a peak-to-peak swing at a V_{OUT} of 1.85 V (50 mV to +1.9 V). This corresponds to a signal at the emitter follower output of 3.7 V p-p (100 mV to 3.8 V). Data was taken with an output signal of 2 MHz, and a THD of 55 dBc was measured with a V_{OUT} of 1.55 V p-p (50 mV to 1.6 V).

This circuit can also be used to drive the analog input of a single supply high speed ADC whose input voltage range is ground-referenced. In this case, the emitter of the external transistor is connected to the ADC input, and the termination resistor is deleted. In this case, peak positive voltage swings of approximately 3.8 V are possible before significant distortion begins to occur.

Headroom Considerations in ac-Coupled Single-Supply Video Circuits

The ac coupling of arbitrary waveforms can actually introduce problems that don't exist at all in dc-coupled or dc restored systems. These problems have to do with the waveform duty cycle, and are particularly acute with signals that approach the rails, as they can in ac-coupled, low supply voltage systems.

In Figure 6-118(A), an example of a 50% duty cycle square wave of about 2 V p-p level is shown, with the signal swing biased symmetrically between the upper and lower clip points of a 5 V supply amplifier.

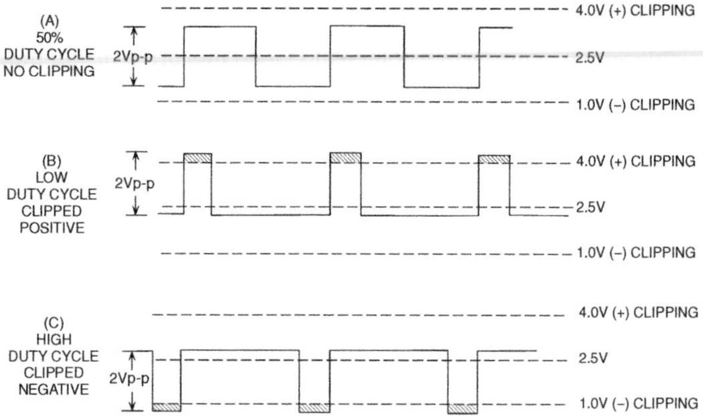

Figure 6-118: Waveform duty cycle taxes headroom in ac-coupled single-supply op amps

Assume that the amplifier has a complementary emitter follower output and can only swing to the limited dc levels as marked, about 1 V from either rail. In cases (B) and (C), the duty cycle of the input waveform is adjusted to both low and high duty cycle extremes *while maintaining the same peak-to-peak input level.* At the amplifier output, the waveform is seen to clip either negative or positive, in (B) and (C), respectively.

Since standard video waveforms *do* vary in duty cycle as the scene changes, the point is made that low distortion operation on ac-coupled single supply stages must take the duty cycle headroom degradation effect into account. If a stage has a 3 V p-p output swing available before clipping, and it must cleanly reproduce an *arbitrary* waveform, then the maximum allowable amplitude is less than one-half this 3 V p-p swing, that is <1.5 V p-p.

An example of violating this criteria are the 2 V p-p waveforms of Figure 6-118(B) and (C), which clip for both the low and high duty cycles. Note that the criteria set down above is based on avoiding hard clipping, while subtle distortion increases may in fact take place at lower levels. This suggests an even more conservative criterion for lowest distortion operation, such as in composite NTSC video amplifiers.

Single-Supply AC-Coupled Composite Video Line Driver

Figure 6-119 shows a single supply gain-of-two composite video line driver using the AD8041. Since the sync tips of a composite video signal extend below ground, the input must be ac-coupled and shifted positively to prevent clipping during negative excursions. The input is terminated in 75 Ω and ac-coupled via the 47 μF to a voltage divider that provides the dc bias point to the input. Setting the optimal common-mode bias voltage requires some understanding of the nature of composite video signals and the video performance of the AD8041.

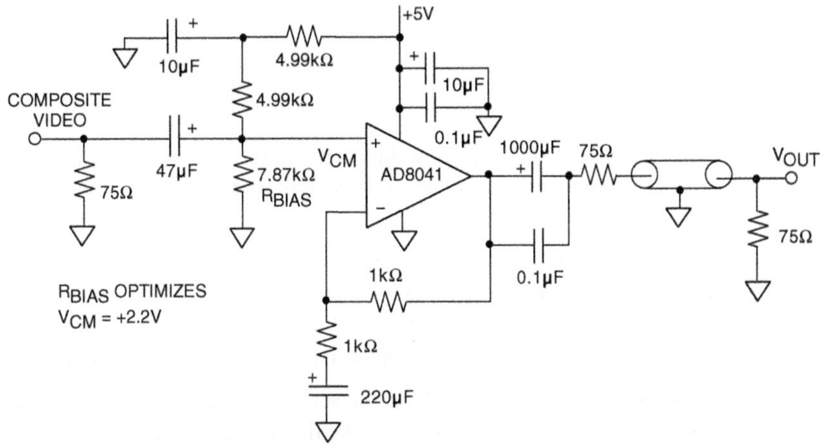

Figure 6-119: Single-supply ac-coupled composite video line driver has ΔG = 0.06% and Δϕ = 0.06°

As discussed above, signals of bounded peak-to-peak amplitude that vary in duty cycle require larger dynamic swing capability than their peak-to-peak amplitude after ac coupling. As a worst case, the dynamic signal swing required will approach twice the peak-to-peak value. The two bounding cases are for a duty cycle that is mostly low, but occasionally goes high at a fraction of a percent duty cycle, and vice versa.

Composite video is not quite this demanding. One bounding extreme is for a signal that is mostly black for an entire frame, but occasionally has a white (full intensity), minimum width spike at least once per frame.

The other extreme is for a video signal that is full white everywhere. The blanking intervals and sync tips of such a signal will have negative going excursions in compliance with composite video specifications. The combination of horizontal and vertical blanking intervals limit such a signal to being at its highest level (white) for about 75% of the time.

As a result of the duty cycle variations between the extremes presented above, a 1 V p-p composite video signal that is multiplied by a gain-of-two requires about 3.2 V p-p of dynamic voltage swing at the output for the op amp, to pass a composite video signal of arbitrary duty cycle without distortion.

The AD8041 device family not only has ample signal swing capability to handle the dynamic range required, but also has excellent differential gain and phase when buffering these signals in an ac-coupled configuration.

To test this, the differential gain and phase were measured for the AD8041 while the supplies were varied. As the lower supply is raised to approach the video signal, the first effect is that the sync tips become compressed before the differential gain and phase are adversely affected. Thus, there must be adequate swing in the negative direction to pass the sync tips without compression.

As the upper supply is lowered to approach the video, the differential gain and phase were not significantly adversely affected until the difference between the peak video output and the supply reached 0.6 V. Thus, the highest video level should be kept at least 0.6 V below the positive supply rail.

Taking the above into account, it was found that the optimal point to bias the noninverting input was at +2.2 V dc. Operating at this point, the worst case differential gain was 0.06% and the differential phase 0.06°.

The ac coupling capacitors used in the circuit may at first glance appear quite large. There is a reason for this. Note that a composite video signal has a lower frequency band edge of 30 Hz. The resistances at the various ac coupling points—especially at the output—are quite small. In order to minimize phase shifts and baseline tilt, the large value capacitors shown are required for best waveform reproduction.

For video system performance that is not to be of the highest quality, the value of these capacitors can be reduced by a factor of up to five with only a slight observable change in the picture quality.

Single-Supply AC-Coupled Single-Ended-to-Differential Driver

The circuit shown in Figure 6-120 provides a flexible solution to differential line driving in a single-supply application and utilizes the dual AD8042. The basic operation of the cross-coupled configuration has been described earlier in this section. The input, V_{IN}, is a single-ended signal that is capacitively coupled into the feedforward resistor, R1. The noninverting inputs of each half of the AD8042 are biased at 2.5 V.

The gain from single-ended input to the differential output is equal to 2R2/R1, as noted in the figure. If desired, this gain can be varied by simply changing one resistor (either R1 or R2). The input capacitor may need increase, for the processing of low frequency information with low phase shift.

It should also be noted that there is no output coupling capacitor, as none is required for differentially connected loads. The output terminals will be biased at approximately 2.5 V.

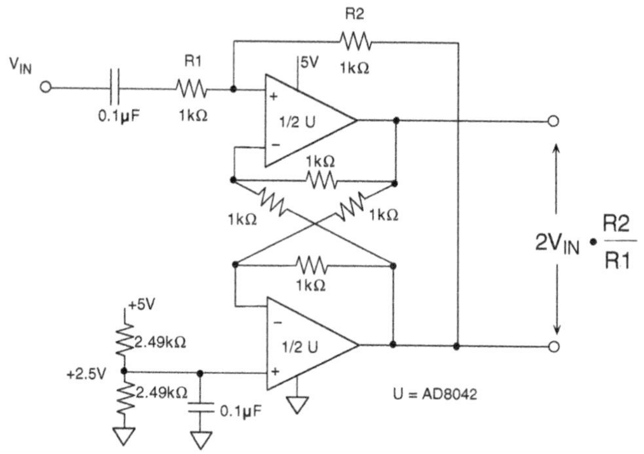

Figure 6-120: Single-supply ac-coupled differential driver

References: Video Amplifiers

1. W. A. Kester, "PCM Signal Codecs for Video Applications," **SMPTE Jour**nal, No. 88, November 1979, pp. 770–778.
2. "IEEE Standard for Performance Measurements of A/D and D/A Converters for PCM Television Circuits," **IEEE Standard 746-1984**.
3. Walt Kester, "Maintaining Transmission Line Impedances on the PC Board," within Chapter 11 of Walt Kester, Editor, **System Application Guide**, Analog Devices, Inc., 1993, ISBN 0-916550-13-3.
4. William R. Blood, Jr., **MECL System Design Handbook** (HB205, Rev.1), Motorola Semiconductor Products, Inc., 1988.
5. Dave Whitney, Walt Jung, "Applying a High-Performance Video Operational Amplifier," **Analog Dialogue**, 26-1, 1992.
6. Walt Jung, Scott Wurcer, "Design Video Circuits Using High-Speed Op-Amp Systems," **Electronic Design Analog Applications Issue**, November 7, 1994.
7. Vishay chip resistors and type VTF networks, www.vishay.com.
8. Walt Kester, "Video Line Receiver Applications Using the AD830 Active Feedback Amplifier Topology," within Chapter 11 of Walt Kester, Editor, **System Application Guide**, Analog Devices, Inc., 1993, ISBN 0-916550-13-3.
9. Peter Checkovich, "Understanding and Using High-Speed Clamping Amplifiers," **Analog Dialogue**, Vol. 29-1, 1995.
10. "Chapters 1, 2, and 4," within Walt Kester, Editor, **Practical Analog Design Techniques**, Analog Devices, Inc., Norwood, MA, 1995, ISBN 0-916550-16-8.

SECTION 6-4

Communication Amplifiers

Walt Kester

Components used in the signal path in communications systems must have wide dynamic range at high frequencies. Dynamic range is primarily limited by distortion and noise introduced by the active elements in amplifiers, mixers, and so forth. In the past, amplifiers for communications applications consisted primarily of "gain blocks" with appropriate specifications. Typically such amplifiers are specified for gain, bandwidth, distortion, and so forth, as a system is designed, and purchased as a self-contained package. This package itself is actually a communications amplifier subsystem.

Today, however, op amps with bandwidths of hundreds of megahertz, low noise, high dynamic range and flexible supply voltages also make popular building blocks in communications systems. They are easily configured for a given gain, and can deliver good performance.

Communications-Specific Specifications

As a necessity, this means that high frequency op amps must be fully specified not only in terms of traditional op amp ac specifications (bandwidth, slew rate, settling time), but also in terms of *communications-specific specifications*. These latter specifications would include performance for harmonic distortion, spurious free dynamic range (SFDR), intermodulation distortion, intercept points (IP2, IP3), noise, and noise figure (NF). Figure 6-121 illustrates these specifications.

- Noise
  - Noise referred to input (RTI)
  - Noise referred to output (RTO)
- Distortion
  - Second and third order intercept points (IP2, IP3)
  - Spurious free dynamic range (SFDR)
  - Harmonic distortion
    - Single-tone
    - Multi-tone
    - Out-of-band
  - Multitone Power Ratio (MTPR)
  - Noise Factor (NF), Noise Figure (NF)

Figure 6-121: Dynamic range specifications in communications systems

This portion of the chapter will examine these specifications, and how they apply to the amplifiers used in wireless and wired communications systems. In addition, several application specific amplifiers such as variable gain amplifiers (VGAs), CATV drivers, and xDSL drivers will also be discussed.

Distortion Specifications

When a spectrally pure sinewave passes through an amplifier (or other active device), various harmonic distortion products are produced, depending upon the nature and the severity of the nonlinearity. However, simply measuring harmonic distortion produced by single tone sinewaves of various frequencies does not give all the information required to evaluate the amplifier's potential performance in a communications application. In most communications systems there are a number of channels which are "stacked" in frequency. It is often required that an amplifier be rated in terms of the intermodulation distortion (IMD) produced with two or more specified tones applied.

Intermodulation distortion products are of special interest in the IF and RF area, and a major concern in the design of radio receivers. Rather than simply examining the harmonic distortion or total harmonic distortion (THD) produced by a single tone sinewave input, it is often useful to look at the distortion products produced by two tones.

As shown in Figure 6-122, two tones will produce second and third order intermodulation products. The example shows the second and third order products produced by applying two frequencies, f_1 and f_2, to a nonlinear device. The second order products located at f_2+f_1 and $f_2–f_1$ are located far away from the two tones, and may be removed by filtering. The third order products located at $2f_1+f_2$ and $2f_2+f_1$ may likewise be filtered. The third order products located at $2f_1–f_2$ and $2f_2–f_1$, however, are close to the original tones, and filtering them is difficult.

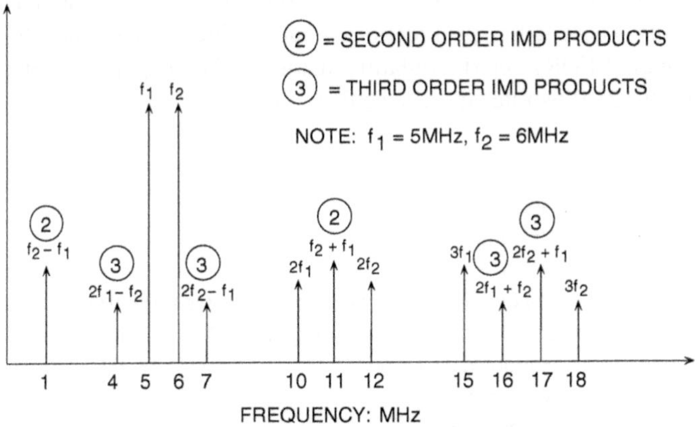

Figure 6-122: Second and third order intermodulation distortion products

Third order IMD products are especially troublesome in multichannel communications systems where the channel separation is constant across the frequency band. Third order IMD products can mask out small signals in the presence of larger ones.

Third order IMD is often specified in terms of the *third order intercept* point, as is shown by Figure 6-123. Two spectrally pure tones are applied to the system. The output signal power in a single tone (in dBm) as well as the relative amplitude of the third order products (referenced to a single tone) is plotted as a function of input signal power. The fundamental is shown by the *slope = 1* curve in the diagram. If the system nonlinearity is approximated by a power series expansion, it can be shown that second order IMD amplitudes increase 2 dB for every 1 dB of signal increase, as represented by *slope = 2* curve in the diagram.

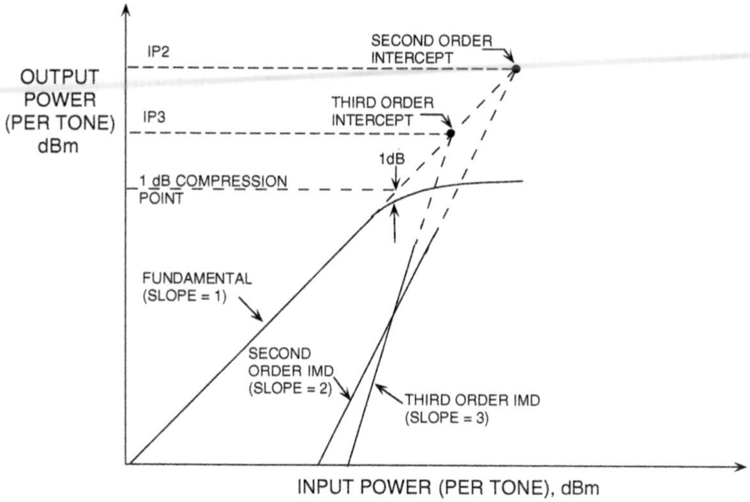

Figure 6-123: Intercept points and 1 dB compression point

Similarly, the third order IMD amplitudes increase 3 dB for every 1 dB of signal increase, as indicated by the *slope = 3* plotted line. With a low level two-tone input signal, and two data points, one can draw the second and third order IMD lines as they are shown in Figure 6-123 (using the principle that a point and a slope define a straight line).

Once the input reaches a certain level, however, the output signal begins to soft-limit, or compress. A parameter of interest here is the *1 dB compression point*. This is the point at which the output signal is compressed 1 dB from an ideal input/output transfer function. This is shown in Figure 6-123 within the region where the ideal slope = 1 line becomes dotted, and the actual response exhibits compression (solid).

Nevertheless, both the second and third order intercept lines may be extended, to intersect the (dotted) extension of the ideal output signal line. These intersections are called the *second* and *third order intercept points*, respectively, or IP2 and IP3. These power level values are usually referenced to the output power of the device delivered to a matched load (usually, but not necessarily 50 Ω) expressed in dBm.

It should be noted that IP2, IP3, and the 1 dB compression point are all a function of frequency, and as one would expect, the distortion is worse at higher frequencies.

For a given frequency, knowing the third order intercept point allows calculation of the approximate level of the third order IMD products as a function of output signal level. Figure 6-124 shows the third order intercept value as a function of frequency for a typical wideband low-distortion amplifier.

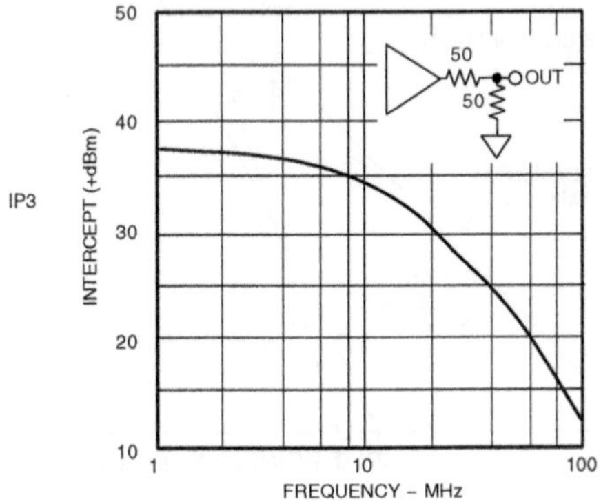

Figure 6-124: Third order intercept point (IP3) versus frequency for a low distortion amplifier

Assume the op amp output signal is 5 MHz and 2 V peak-to-peak into a 100 Ω load (50 Ω source and load termination). The voltage into the 50 Ω load is therefore 1 V peak-to-peak, corresponding to +4dBm. From Figure 6-124, the value of the third order intercept at 5 MHz is 36 dBm. The difference between +36 dBm and +4 dBm is 32 dB. This value is then multiplied by 2 to yield 64 dB (the value of the third-order intermodulation products referenced to the power in a single tone). Therefore, the intermodulation products should be –64 dBc (dB below carrier frequency), or at an output power level of –60 dBm.

Figure 6-125 shows the graphical analysis for this example. A similar analysis can be performed for the second-order intermodulation products, using data for IP2.

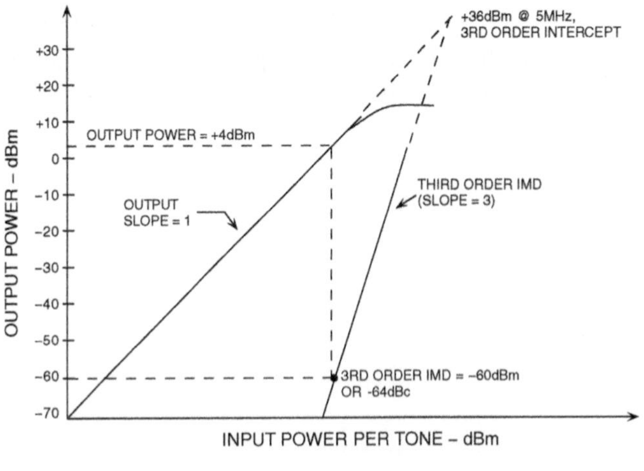

Figure 6-125: Using IP3 to calculate the third order IMD product amplitude

Another popular specification in communications systems is *spurious free dynamic range*, or SFDR. Figure 6-126 shows two variations of this specification. Single-tone SFDR (left) is the ratio of the signal (or carrier) to the *worst* spur in the bandwidth of interest. This spur may or may not be harmonically related to the signal. SFDR can be referenced to the signal or carrier level (dBc), or to full scale (dBFS).

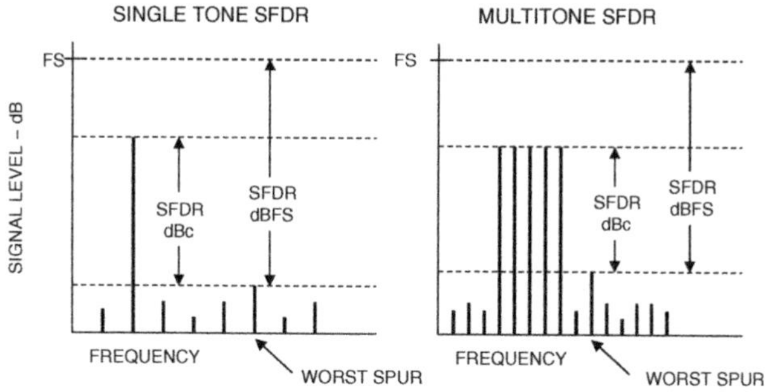

Figure 6-126: Spurious free dynamic range (SFDR) in communications systems

Because most amplifiers are soft limiters, the dBc unit is more often used. However, in systems that have a hard-limiter that precisely defines full scale (such as with ADCs), both dBc and dBFS may be used. It is important to understand that they both describe the worst spur amplitude. SFDR can also be specified for two tones or multitones (right), thereby simulating complex signals that contain multiple carriers and channels.

Multitone power ratio is another way of describing distortion in a multichannel communication system. Figure 6-127 shows the frequency partitioning in an xDSL system. The QAM signals in the upstream data path are represented by a number of equal amplitude tones, separated equally in frequency. One channel is completely eliminated from the input signal (shown as an empty bin), but intermodulation distortion caused by the system nonlinearity will cause a small signal to appear in that bin.

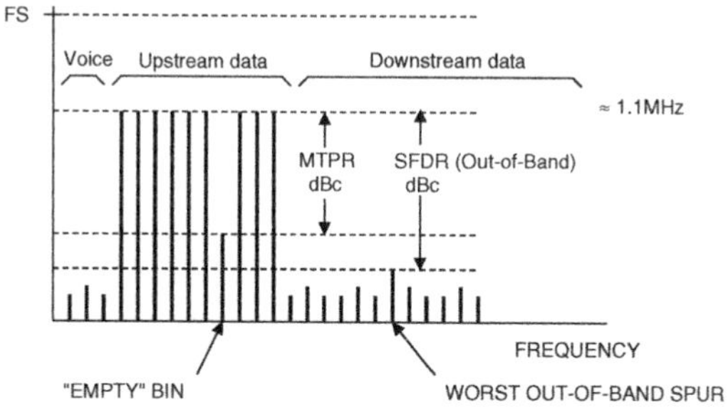

Figure 6-127: Multitone power ratio (MTPR) and out-of-band SFDR in xDSL applications

The ratio of the tone amplitude to the amplitude of the unwanted signal in the empty bin is defined as the multitone power ratio, or MTPR. It is equally important that the amplitude of the intermodulation products caused by the multitone signal (simulating multiple channels) not interfere with signals in either the voice band or the downstream data band. The amplitude of the worst spur produced in these bands to the amplitude of the multitone signal is therefore defined as the *out-of-band SFDR*.

Noise Specifications

Op amp noise is generally specified in terms of input current and voltage noise, as previously discussed in Chapter 1 of this book. In communications systems, however, noise is often specified in terms of *noise figure* (NF)—see Figure 6-128. This can lead to confusion, especially when op amps are used as gain blocks and the noise figure of the op amp is not specified for the specific circuit conditions. In order to understand how to apply noise figure to op amps, we will first review the basic theory behind noise figure.

- NF is usually specified for matched input/output conditions, but this is not always a system requirement
- Noise Figure is a popular figure of merit in RF applications: LNAs, Mixers, etc.
- Difficulties arise when applying NF to op amps. NF is dependent on
  - Impedance levels
  - Feedback network
  - Closed loop gain
- Other difficulties arise due to different definitions of NF as found in various textbooks
- We will start with the basics and work up to the op amp issues

Figure 6-128: Noise figure in communications applications

The first concept is that of *available power* from a source. The available power of a source is the maximum power that can be drawn from the source. Figure 6-129 shows a resistor of value R as the noise source. The thermal noise of this source is $\sqrt{(4kTBR)}$. The maximum noise that can be transferred to an ideal noiseless load occurs when the load resistance is also equal to R.

The *available power*, P_a, of a source is the maximum power that can be drawn from the source.
This occurs when the load resistance is equal to the source resistance.

$$P_a = \frac{v_n^2}{4R} = kTB$$

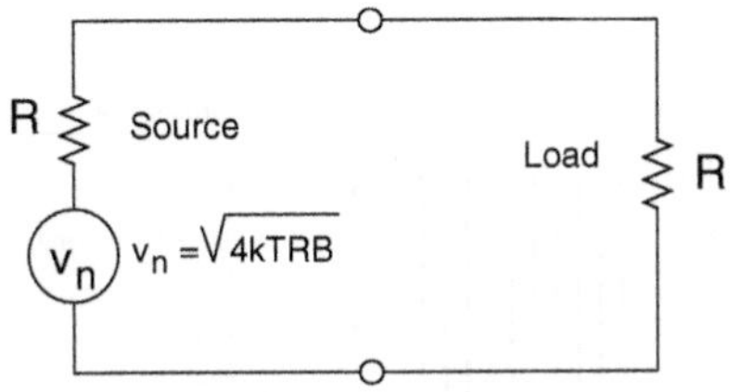

k = 1.38 × 10^{-23} Joules/K (Boltzman's Constant)
T = Temperature (assume 300K, room temperature)
B = Noise bandwidth (Hz)

$$P_{a\,(dBm)} = -174dBm + 10 \log B$$

Figure 6-129: Available noise power from a source

Under these conditions, the maximum available noise power from the source reduces to kTB, where k is Boltzmann's constant, T is the absolute temperature, and B is the noise bandwidth. Note that this power is independent of the value of the source resistance, R.

The next important concept is that of *available power gain* of a two-port network, as shown in Figure 6-130. The two-port network is driven from a signal source having an impedance. The equations show the available signal power from the source and the available signal power from the output of the network. The available power gain is simply the ratio of the available output power to the available power from the source.

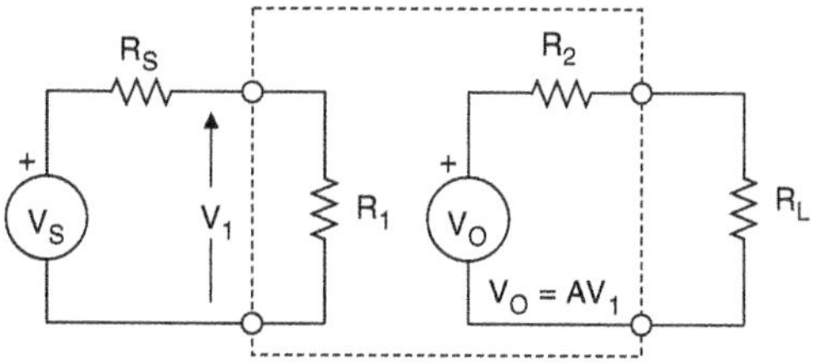

Available signal power from source $= P_{as} = \dfrac{V_S^2}{4R_S}$

Available signal power at output $= P_{ao} = \dfrac{V_O^2}{4R_2}$

Available power gain $= G_a = \dfrac{P_{ao}}{P_{as}} = \dfrac{V_O^2 R_S}{V_S^2 R_2}$

Figure 6-130: Available power gain of a two-port network

The gain and the noise of a two-port network can now be defined in terms of the available power gain, G, and the noise factor, F, as shown in Figure 6-131. The noise factor, F, is defined as the ratio of the total available output noise power to the available output noise power due to the source only. For a resistive source, the available noise power from the source is simply kTB, and the output noise power due to the source only is G · kTB.

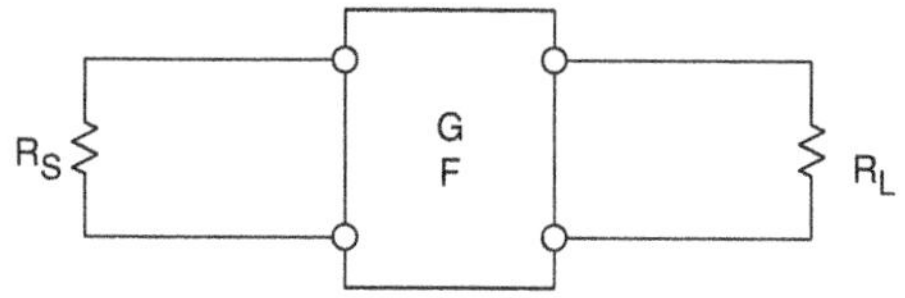

G = Available Power Gain of Network

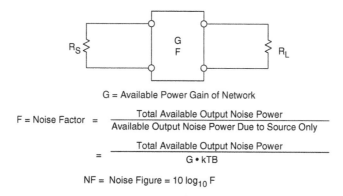

$$F = \text{Noise Factor} = \frac{\text{Total Available Output Noise Power}}{\text{Available Output Noise Power Due to Source Only}}$$

$$= \frac{\text{Total Available Output Noise Power}}{G \cdot kTB}$$

$$NF = \text{Noise Figure} = 10\log_{10} F$$

Figure 6-131: Definition of noise factor and noise figure for a two-port noisy network.

Note that the noise factor, F, is expressed as a ratio, and the noise figure, NF, is simply the ratio F expressed in dB. An ideal noiseless two-port network therefore has noise factor F = 1, and a noise figure NF = 0 dB.

These same definitions can be used to calculate the NF of an op amp circuit; however, it is much easier to work in terms of the square of voltage noise spectral density and current noise spectral density, rather than power or power spectral density (see Figure 6-132). Also, unmatched conditions are easier to deal with using this approach. *The noise factor F for an op amp is simply the ratio of the square of the total output noise spectral density to the square of the output noise spectral density due to the source only.* The noise figure NF = 10 • logF.

- With op amps, it is easier to work with voltage and current noise spectral density, rather than power or power spectral density.
- Unmatched conditions are more easily dealt with using voltage noise spectral density analysis.
- Voltage noise spectral densities add using root-sum-squares (RSS).
- A 1000Ωresistor has a voltage noise spectral density of 4nV/√Hz @ 25°C (300K). (This is good to remember!)
- The basic definition of Noise Fact or and Noise Figure in terms of voltage noise spectral density becomes:

$$\text{Noise Factor} = F = \frac{(\text{Total Output Voltage Noise Spectral Density})^2}{(\text{Output Voltage Noise Spectral Density Due to Source Only})^2}$$

$$\text{Noise Figure} = NF = 10 \log_{10} F$$

Figure 6-132: Noise figure for op amps

In RF or IF gain blocks, the input impedance is defined. However, when using an op amp in the noninverting mode as a gain block, the input impedance is high (relative to transmission line impedances), and there are several options regarding the input termination which affect the noise figure. These options have been generalized to cover any two-port network with optional input terminations in Figure 6-133.

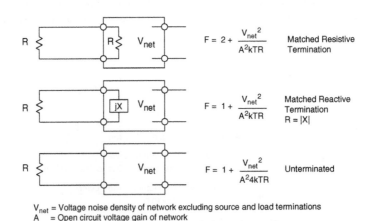

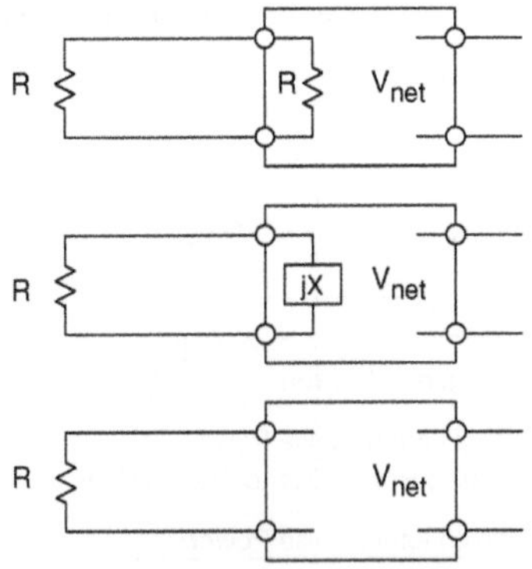

$F = 2 + \frac{V_{net}^2}{A^2 kTR}$ Matched Resistive Termination

$F = 1 + \frac{V_{net}^2}{A^2 kTR}$ Matched Reactive Termination R = |X|

$F = 1 + \frac{V_{net}^2}{A^2 4kTR}$ Unterminated

V_{net} = Voltage noise density of network excluding source and load terminations
A = Open circuit voltage gain of network

Figure 6-133: Noise factor for resistive, reactive, and unterminated conditions

Assume that the open-circuit voltage gain of the network is A and that the total output noise spectral density (excluding that due to the source resistance and the input termination) is equal to V_{net}.

The top diagram of Figure 6-133 shows the traditional matched case where the input is resistively terminated to match the source impedance. In this case, the input termination resistor not only attenuates the voltage noise of the source by a factor of 2, but also contributes noise due to its own thermal resistance.

The middle diagram of this figure shows the case of a reactive matched termination. Reactive terminations are often used where the bandwidth is limited but centered on a high frequency carrier. In this case, the source voltage noise is attenuated by a factor of 2, but the reactive termination adds no additional noise of its own to the total output noise.

The bottom diagram in Figure 6-133 shows the case of an unmatched, unterminated input. In this case, the voltage noise of the source is not attenuated, and there is obviously no additional noise due to the input termination because there is no input termination. Although this situation is not likely in a system using RF/IF gain blocks that generally require impedance matching at all interfaces, it is a possibility when using an op amp as the gain block, since the noninverting configuration input impedance is relatively high.

If we assume that the noise of the network, V_{net}, is very small relative to the source noise, then it is obvious that the input termination resistor adds 3 dB to the overall noise figure as well as reduces the overall voltage gain by a factor of 2. This is compared to the lowest noise case where there is no input termination. In fact, the lowest possible noise figure for a noiseless network with only an input resistive matched termination is 3 dB. Lower noise figures can only be obtained by using matched reactive terminations.

On the other hand, if the noise of the network, V_{net}, is very large with respect to the source noise, adding the resistive termination increases the overall noise figure by 6 dB compared to the unmatched unterminated case.

Summarizing, it is interesting to note that using large source resistances will decrease the noise figure but increase overall circuit noise. This illustrates the important fact that *noise figures can be compared only if they are specified at the same impedance level.* In Figure 6-134 these effects of amplifier input terminations on overall circuit noise and noise figure are summarized.

- For a low noise network, adding the matching input termination resistor makes the noise figure 3dB worse. The voltage gain is also reduced by a factor of 2.
- For a high noise network, adding the matching termination resistor makes the noise figure 6dB worse.
- Reactive matched terminations are often used at fixed IF/RF frequencies in LNAs, mixers, etc.
- Using large source and termination resistors decreases noise figure but increases overall circuit noise.
- Noise figures should only be compared at the same impedance level.

Figure 6-134: Effects of input termination on noise figure

Op amp noise has two components: low frequency noise whose spectral density is inversely proportional to the square root of the frequency, and white noise at medium and high frequencies. The low frequency noise is known as 1/f noise (the noise *power* obeys a 1/f law—the noise voltage or noise current is proportional to $1/\sqrt{f}$). The frequency at which the 1/f noise spectral density equals the white noise is known as the "1/f Corner Frequency." This is an important figure of merit for op amps, with low values indicating better performance. Values of the 1/f corner frequency vary from a few Hz for the most modern low noise low frequency amplifiers, to several hundreds, or even thousands of Hz for some high-speed op amps.

In most applications of high speed op amps, the total output RMS noise is generally of interest. Because of the high bandwidths, the chief contributor to the output RMS noise is the white noise, and that of the 1/f noise is negligible.

In order to better understand the effects of noise in high speed op amps, we use the classical noise model shown in Figure 6-135. This diagram identifies all possible white noise sources, including the external noise in the source and the feedback resistors.

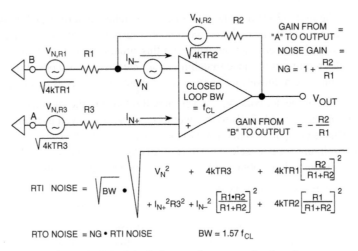

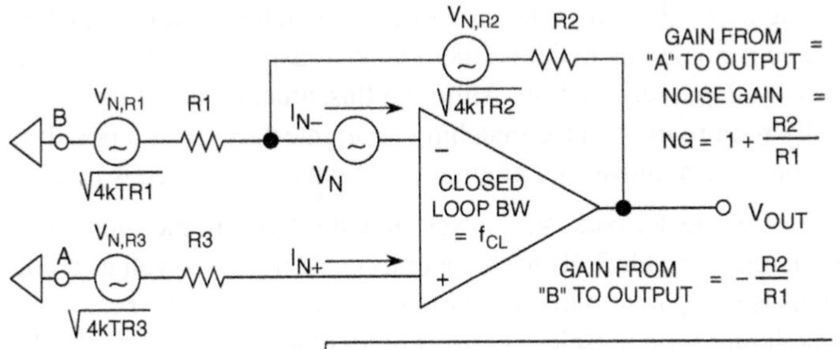

$$\text{RTI NOISE} = \sqrt{BW}\bullet\sqrt{V_N^2 + 4kTR3 + 4kTR1\left[\frac{R2}{R1+R2}\right]^2 + I_{N+}^2R3^2 + I_{N-}^2\left[\frac{R1\bullet R2}{R1+R2}\right]^2 + 4kTR2\left[\frac{R1}{R1+R2}\right]^2}$$

$$\text{RTO NOISE} = NG\bullet\text{RTI NOISE} \qquad BW = 1.57\,f_{CL}$$

Figure 6-135: Calculating total op amp circuit noise

The equation in the figure allows for calculation of the total output RMS noise over the closed-loop bandwidth of the amplifier. This formula works quite well when the frequency response of the op amp is relatively flat. If there are more than a few dB of high frequency peaking, however, the actual noise will be greater than the predicted because the contribution over the last octave before the 3 dB cutoff frequency will dominate.

In most applications, the op amp feedback network is designed so that the bandwidth is relatively flat, and the formula provides a good estimate. Note that BW in the equation is the equivalent *noise bandwidth,* which, for a single-pole system, is obtained by multiplying the closed-loop 3 dB bandwidth by 1.57.

Figure 6-136 is a table that indicates how the individual noise contributors of Figure 6-135 are referred to the output. After calculating the individual noise spectral densities in this table, they can be squared, added, and then the square root of the sum of the squares yields the RSS value of the output noise spectral density, since all the sources are uncorrelated. This value is multiplied by the square root of the noise bandwidth (noise bandwidth = closed-loop 3 dB bandwidth multiplied by a correction factor of 1.57) to obtain the final value for the output RMS noise.

Typical high speed op amps have bandwidths greater than 150 MHz or so, and bipolar input stages have input voltage noises ranging from about 2 to $20\ \text{nV}/\sqrt{\text{Hz}}$. To put voltage noise in perspective, let's look at the Johnson noise spectral density of a resistor:

$$v_n = \sqrt{4kTR \times BW}$$

where *k* is Boltzmann's constant, *T* is the absolute temperature, *R* is the resistor value, and *BW* is the equivalent noise bandwidth of interest. (The equivalent noise bandwidth of a single-pole system is 1.57 times the 3 dB frequency). Using the formula, a 100 Ω resistor has a noise density of $1.3\ \text{nV}/\sqrt{\text{Hz}}$, and a 1000 Ω resistor about $4\ \text{nV}/\sqrt{\text{Hz}}$ (values are at room temperature: 27°C, or 300K).

NOISE SOURCE EXPRESSED AS A VOLTAGE	MULTIPLY BY THIS FACTOR TO REFER TO THE OP AMP OUTPUT
Johnson Noise in R3: $\sqrt{(4kTR3)}$	Noise Gain = 1 + R2/R1
Noninverting Input Current Noise Flowing in R3: $I_{n+}R3$	Noise Gain = 1 + R2/R1
Input Voltage Noise: V_n	Noise Gain = 1 + R2/R1
Johnson Noise in R1: $\sqrt{(4kTR1)}$	–R2/R1 (Gain from input of R1, "B," to Output)
Johnson Noise in R2: $\sqrt{(4kTR2)}$	1
Inverting Input Current Noise Flowing in R2: $I_{n-}R2$	1

Figure 6-136: Referring all noise sources to the output

The base-emitter in a bipolar transistor has an equivalent noise voltage source that is due to the "shot noise" of the collector current flowing in the transistor's (noiseless) incremental emitter resistance, r_e. The current noise is proportional to the square root of the collector current, Ic. The emitter resistance, on the other hand, is inversely proportional to the collector current, so *the shot-noise voltage is inversely proportional to the square root of the collector current.*

Voltage noise in FET input op amps tends to be larger than for bipolar ones, but current noise is extremely low (generally only a few tens of $\text{fA}/\sqrt{\text{Hz}}$) because of the low input bias currents. However, FET inputs are not generally required for op amp applications requiring bandwidths greater than 100 MHz.

Op amps also have input current noise on each input. For high speed FET input op amps, the gate currents are so low that input current noise is almost always negligible (measured in $\text{fA}/\sqrt{\text{Hz}}$).

For a voltage feedback (VFB) op amp, the inverting and noninverting input current noise are typically equal, and almost always uncorrelated. Typical values for wideband VFB op amps range from $0.5\ \text{pA}/\sqrt{\text{Hz}}$ to $5\ \text{pA}/\sqrt{\text{Hz}}$. The input current noise of a bipolar input stage is increased when input bias-current cancellation generators are added, because their current noise is not correlated, and therefore adds (in an RSS manner) to the intrinsic current noise of the bipolar stage.

The input voltage noise in current feedback (CFB) op amps tends to be lower than for VFB op amps having the same approximate bandwidth. This is because the input stage in a CFB op amp is usually operated at a higher current, thereby reducing the emitter resistance and hence the voltage noise. Typical values for CFB op amps range from about 1 to $5\ \text{nV}/\sqrt{\text{Hz}}$.

The input current noise of CFB op amps tends to be larger than for VFB op amps because of the generally higher bias current levels. The inverting and noninverting current noise of a CFB is usually different because of the unique input architecture, and are specified separately. In most cases, the inverting input current noise is the larger of the two. Typical input current noise for CFB op amps ranges from 5 to $40\ \text{pA}/\sqrt{\text{Hz}}$.

The general principle of noise calculation is that uncorrelated noise sources add in a root-sum-squares manner, which means that if a noise source has a contribution to the output noise of a system which is less than

20% of the amplitude of the noise from other noise source in the system, then its contribution to the total system noise will be less than 2% of the total, and that noise source can almost invariably be ignored—in many cases, noise sources smaller than 33% of the largest can be ignored. This can simplify the calculations using the formula, assuming the correct decisions are made regarding the sources to be included and those to be neglected.

The sources which dominate the output noise are highly dependent on the closed-loop gain of the op amp. Notice that for high values of closed loop gain, the op amp voltage noise will tend be the chief contributor to the output noise. At low gains, the effects of the input current noise must also be considered, and may dominate, especially in the case of a CFB op amp.

Feedforward/feedback resistors in high speed op amp circuits may range from less than 100 Ω to more than 1 kΩ, so it is difficult to generalize about their contribution to the total output noise without knowing the specific values and the closed loop gain. The best way to make the calculations is to write a simple computer program that performs the calculations automatically and includes all noise sources (see Reference 1 for one example). In most high speed applications, the source impedance noise can be neglected for source impedances of 100 Ω or less.

Figure 6-137 shows an example calculation of total output noise for the AD8011 (300 MHz, 1 mA) CFB op amp. All six possible sources are included in the calculation. The appropriate multiplying factors which reflect the sources to the output are also shown on the diagram. For G = 2, the close-loop bandwidth of the AD8011 is 180 MHz. The correction factor of 1.57 in the final calculation converts this single-pole bandwidth into the circuit's equivalent noise bandwidth.

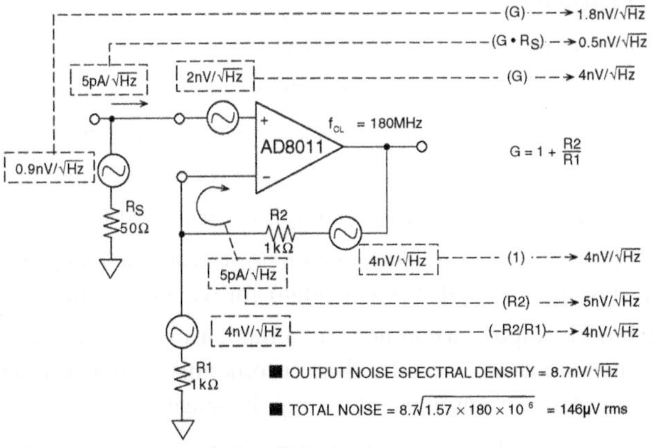

Figure 6-137: AD8011 output noise analysis

Now that the total output noise has been calculated, the issue of noise figure can be addressed. Figure 6-138 shows two cases for the AD8011 circuit: the top diagram corresponds to an unterminated input condition, and the bottom diagram corresponds to a terminated input condition.

For the unterminated case (top), the total output noise from the previous diagram (i.e., Figure 6-137) was $8.7\ \text{nV}/\sqrt{\text{Hz}}$. Note this includes the noise of the 50 Ω source. The output noise due only to the source is simply the noise gain multiplied by the noise of the source, or $G\sqrt{(4kTR_S)} = 1.8\ \text{nV}/\sqrt{\text{Hz}}$. The noise figure is simply NF = 20 log(8.7/1.8) = 13.7 dB. For the terminated case (bottom), the total output noise is still approximately $8.7\ \text{nV}/\sqrt{\text{Hz}}$. Note that the input noise current (I_{n+}) now actually flows through 25 Ω

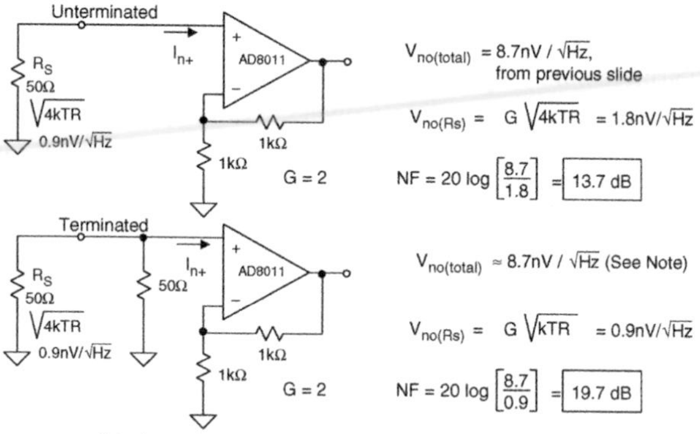

Note: Input noise current (I_{n+}) flows through 50Ω (unterminated case)
or 25Ω (terminated case), but the overall effect of this is negligible.

Figure 6-138: AD8011 noise figure for unterminated and terminated input conditions

rather than 50 Ω for the unterminated case, but the overall effect of this difference on the total output noise calculation is negligible.

The noise of the source, however, is now $\sqrt{(kTR)}$ due to the 50 Ω divider network, and reflected output, it becomes $G\sqrt{(kTR)} = 0.9\ nV/\sqrt{Hz}$. The noise figure is calculated as NF = 20 log(8.7/0.9) = 19.7 dB. Notice that the terminated case yields a noise figure that is approximately 6 dB worse than the unterminated case.

Finally, it should be noted that noise figure is actually a function of frequency. Figure 6-139 shows the noise figure of the AD8350 measured with a spot noise meter, as a function of frequency over the bandwidth 10 MHz to 1 GHz. The top curve is the noise figure, and the bottom curve is the closed-loop gain flatness.

In most cases, the approximations such as those used in the example of Figures 6-135 through 6-137 will give sufficient accuracy, provided the closed-loop bandwidth is relatively flat. However, using the actual spot noise figure may be desirable in high frequency narrowband applications involving specific carrier frequencies.

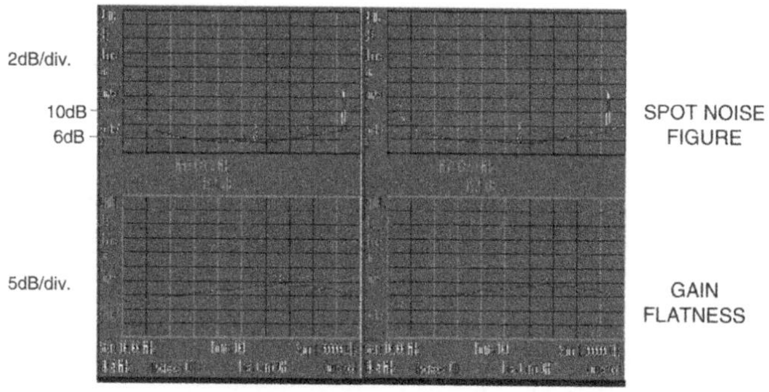

Spot Noise Figure shows variations over frequency: 10MHz to 1GHz
Measurement bandwidth set to 4MHz for above data
Integration required to get average noise power

Figure 6-139: AD8350 spot noise figure and gain versus frequency from 10 MHz to 1 GHz

Variable Gain Amplifiers (VGAs) in Automatic Gain Control (AGC)

Wideband, low distortion variable gain amplifiers find wide applications in communications systems. One example is automatic gain control (AGC) in radio receivers. Typically, the received energy exhibits a large dynamic range due to the variability of the propagation path, requiring dynamic range compression in the receiver. In this case, the wanted information is in the modulation envelope (whatever the modulation mode), not in the absolute magnitude of the carrier. For example, a 1 MHz carrier modulated at 1 kHz to a 30% modulation depth would convey the same information, whether the received carrier level is at 0 dBm or –120 dBm. Some type of automatic gain control (AGC) in the receiver is generally utilized to restore the carrier amplitude to some normalized reference level, in the presence of large input fluctuations. AGC circuits are dynamic-range compressors which respond to some signal metric (often mean amplitude) acquired over an interval corresponding to many periods of the carrier.

Consequently, they require time to adjust to variations in received signal level. The time required to respond to a sudden increase in signal level can be reduced by using peak detection methods, but with some loss of robustness, since transient noise peaks can now activate the AGC detection circuits. Nonlinear filtering and the concept of "delayed AGC" can be useful in optimizing an AGC system. Many trade-offs are found in practice; Figure 6-140 shows a basic AGC system.

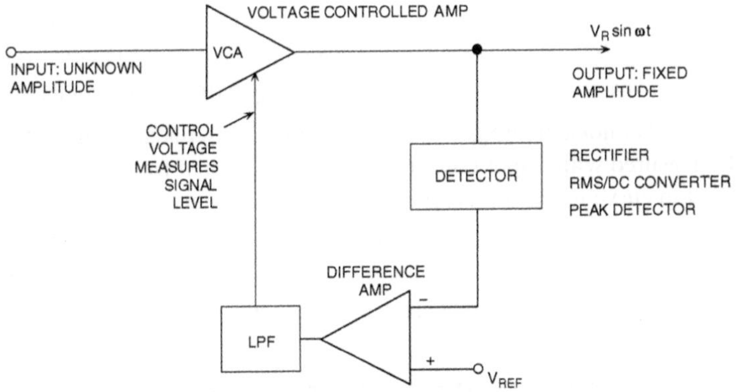

Figure 6-140: A typical automatic gain control (AGC) system

It is interesting to note that an AGC loop actually has *two* outputs. The more obvious output is of course the amplitude-stabilized signal. The less obvious output is the control voltage to the VCA. In reality, this voltage is a measure of the average amplitude of the input signal. If the system is precisely scaled, the control voltage may be used as a measure of the input signal, which is sometimes also known as a *received signal strength indicator* (RSSI).

This latter point, given a suitably precise VCA gain control law, allows implementation of a receiving system that is calibrated for incoming signal level.

Voltage Controlled Amplifiers (VCAs)

An analog multiplier can be used as a variable-gain amplifier, as shown in Figure 6-141 below. The control voltage is applied to one input, and the signal to the other. In this configuration, the gain is directly proportional to the control voltage.

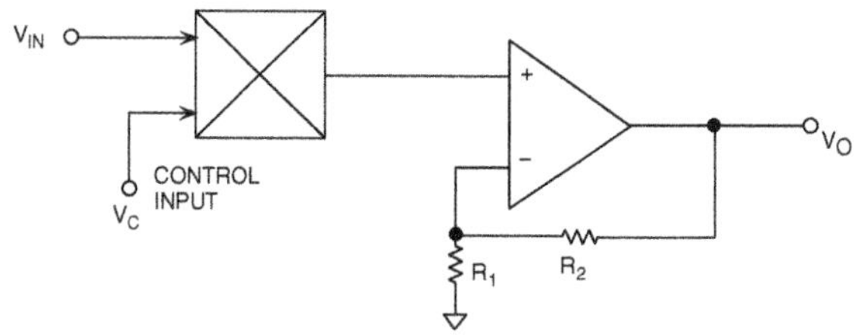

$$V_O = \frac{V_{IN}}{K} \times \left(1 + \frac{R_2}{R_1}\right) V_C$$

Figure 6-141: Using a multiplier as a voltage-controlled amplifier (VCA)

Most VCAs made with analog multipliers have gain that is *linear in volts* with respect to the control voltage, and they tend to be noisy. There is a demand, however, for a VCA that combines a wide gain range with constant bandwidth and phase, low noise with large signal-handling capabilities, and low distortion with low power consumption, while providing accurate, stable, *linear-in-dB* gain. The AD600, AD602, AD603, AD604, AD605, and AD8367 achieve these demanding and conflicting objectives with a unique and elegant solution—the X-AMP® (for *exponential amplifier* see Reference 2).

The concept is simple: a fixed-gain amplifier follows a passive, broadband attenuator, with special means to alter voltage-controlled attenuation, as in Figure 6-142.

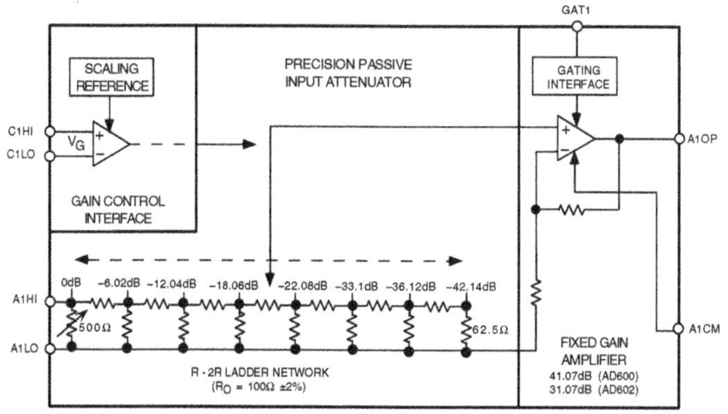

Figure 6-142: Single channel of the dual 30 MHz AD600/AD602 X-AMP

The AD600/AD602 amplifier stage is optimized for low input noise, and negative feedback is used to accurately define its moderately high gain (30 dB–40 dB) and minimize distortion. Since amplifier gain is fixed, so also are its ac and transient response characteristics, including distortion and group delay. As its gain is high, the input is never driven beyond a few millivolts, always operating within a small signal response range.

The attenuator network is a 7-section (8-tap) R-2R ladder. The ratio between adjacent taps is exactly 2, or 6.02 dB, providing the basis for precise linear-in-dB behavior, while overall attenuation is 42.14 dB. As will be shown, the amplifier's input can be connected to any one of these taps, or even *interpolated* between them, with only a small deviation error of about ±0.2 dB. Overall gain can be varied from the fixed (maximum) gain, to a value 42.14 dB less. In the AD600, the fixed gain is 41.07 dB (voltage gain of 113); using this choice, the full gain range is –1.07 dB to +41.07 dB. The gain is related to control voltage by the relationship $G_{dB} = 32\ V_G + 20$ where V_G is in volts. For the AD602, the fixed gain is 31.07 dB (voltage gain of 35.8), and the gain is given by $G_{dB} = 32\ V_G + 10$.

The gain at $V_G = 0$ is laser trimmed to an absolute accuracy of ±0.2 dB. The gain scaling is determined by an on-chip bandgap reference (shared by both channels), laser trimmed for high accuracy and low temperature coefficient. Figure 6-143 shows the gain versus the differential control voltage for both the AD600 and the AD602. Deviation from an ideal control law is only a fraction of a dB over a large part of the dynamic range.

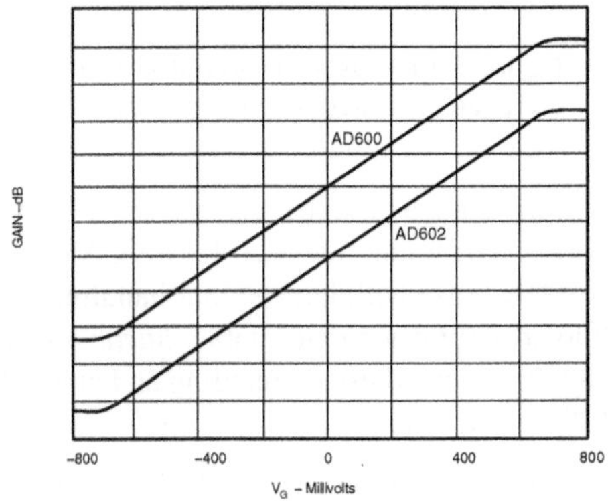

Figure 6-143: Gain of the AD600/AD602 as a function of control voltage

In order to understand the operation of the X-AMP, consider the simplified diagram shown in Figure 6-144. Note that each of the eight taps is connected to an input of one of eight bipolar differential pairs, used as current-controlled transconductance (g_m) stages; the other input of all these g_m stages is connected to the amplifier's gain-determining feedback network, R_{F1}/R_{F2}. When emitter bias current I_E is directed to one of the eight transistor pairs (not shown here), it becomes the complete amplifier input stage.

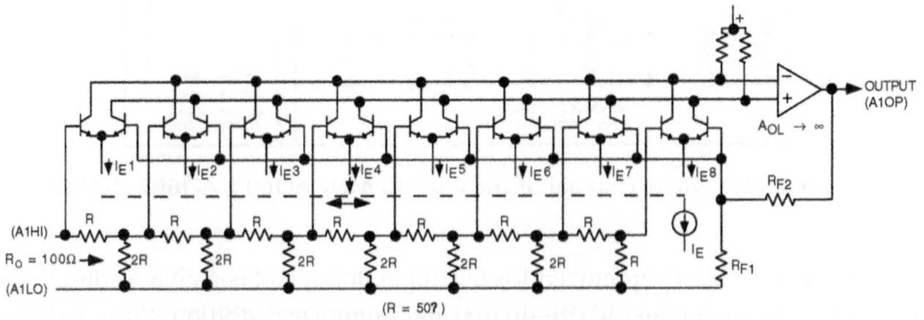

Figure 6-144: Continuous interpolation between taps using current-controlled g_m stages in the X-AMP

When I_E is connected to the left-most pair, the signal input is connected directly to the amplifier, giving maximum gain. The distortion is very low, even at high frequencies, due to the careful open-loop design, aided by the negative feedback. If I_E were to be abruptly switched to the second pair, the overall gain would drop by exactly 6.02 dB, and the distortion would remain low, because only one g_m stage remains active.

In reality, the bias current is *gradually* transferred from the first pair to the second. When I_E is equally divided between two g_m stages, both are active, and the situation arises where we have an op amp with two input stages fighting for control of the loop, one getting the full signal, and the other getting a signal exactly half as large.

Analysis shows that the effective gain is reduced, not by 3 dB, as one might first expect, but rather by 20 log1.5, or 3.52 dB. This error, when divided equally over the whole range, would amount to a gain ripple of ±0.25 dB; however, the interpolation circuit actually generates a Gaussian distribution of bias currents, and a significant fraction of I_E always flows in adjacent stages. This smoothes the gain function and actually lowers the ripple. As I_E moves further to the right, the overall gain progressively drops.

The key features of the X-AMP product family are summarized in Figure 6-145. Note the other members of the family beyond the described AD600/AD602.

	BANDWIDTH	DISTORTION	NOISE	INPUT Z	SUPPLY
AD600/602	35MHz	–60dBc @ 10MHz	1.4nV/√Hz	100Ω	±5V
AD603	90MHz	–60dBc @ 10MHz	1.3nV/√Hz	100Ω	±5V
AD604	40MHz	–43dBc @ 10MHz	0.8nV/√Hz	300kΩ	±5V
AD605	40MHz	–51dBc @ 10MHz	1.8nV/√Hz	200Ω	+5V
AD8367	500MHz	IP3 = +31.5dBm @140MHz	NF = 7.8dB @140MHz	200Ω	+2.7 to +5V

Figure 6-145: X-AMP family key specifications

The total input-referred noise of the AD600/AD602 X-AMP is ($1.4\ \text{nV}/\sqrt{\text{Hz}}$ at 25°C); only slightly more than the thermal noise of a 100 Ω resistor ($1.29\ \text{nV}/\sqrt{\text{Hz}}$ at 25°C). The input-referred noise is constant regardless of the attenuator setting, therefore the output noise is always constant and independent of gain.

For the AD600, the amplifier gain is 113 and the output noise spectral density is therefore $1.4\ \text{nV}/\sqrt{\text{Hz}} \times 113$, or $158\ \text{nV}/\sqrt{\text{Hz}}$. Referred to its maximum output of 2 V rms, the signal-to-noise ratio would be 82 dB in a 1 MHz bandwidth. The corresponding signal-to-noise ratio of the AD602 is 10 dB greater, or 92 dB.

Digitally Controlled Variable Gain Amplifiers for CATV Upstream Data Line Drivers

Cable modems offer much higher data rates than standard dial-up connections and have become very popular. In addition to receiving data (*downstream*), the cable modem also transmits data (*upstream*). This requires a low distortion digitally controlled variable gain amplifier capable of driving the 75 Ω coaxial cable at a nominal level of 1 V rms (+11.2 dBm, or 60 dBmV). The AD8323 is a member of a family of CATV upstream line drivers suitable for this application. The AD8323 gain is controlled by an 8-bit serial word that determines the desired gain over a 53.5 dB range, resulting in gain changes of 0.7526 dB/LSB. The AD8323 block diagram is shown Figure 6-146.

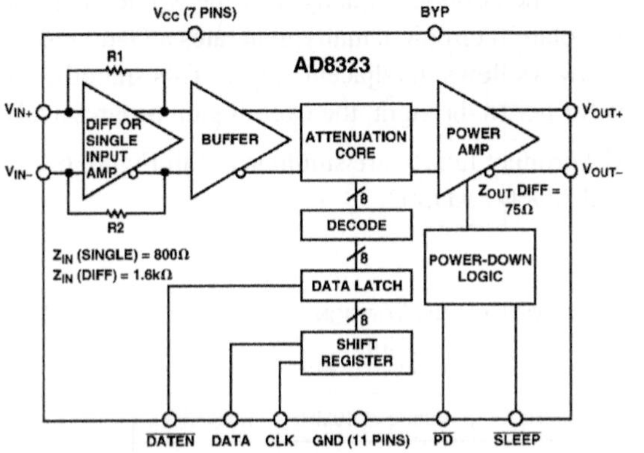

Figure 6-146: AD8323 CATV digitally controlled variable gain amplifier

The AD8323 has a variable attenuator core where the attenuation is digitally controlled from 0 dB to –53.5 dB. The input buffer has a gain of approximately + 27.5 dB; therefore, the resulting overall gain range is from –26 dB to +27.5 dB. The AD8323 is composed of four analog functions in the power-up mode. The input amplifier (preamp) can be used either single-ended or differentially. The preamp stage drives a vernier stage that provides the fine tune gain adjustment. The 0.7526 dB/LSB resolution is implemented in this stage and provides a total of approximately 5.25 dB of attenuation. After the vernier stage, a DAC provides the bulk of the AD8323's attenuation (8 bits, or 48 dB).

The signals in the preamp and vernier gain blocks are differential to improve the PSRR and linearity. A differential current is fed from the DAC to the output stage, which amplifies these currents to the appropriate levels necessary to drive the 75 Ω load.

A key performance and cost advantage of the AD8323 results from the ability to maintain a constant dynamic output impedance of 75 Ω during power-up/power-down conditions. The output stage uses negative feedback to implement a differential 75 Ω dynamic output impedance. This eliminates the need for an external 75 Ω termination, resulting in twice the effective output voltage when compared to a standard op amp.

These features allow the AD8323 to operate on a single 5 V supply and still deliver the required output power. Distortion performance of –56 dBc is achieved with an output level up to 1 V rms (+11.2 dBm, or 60 dBmV) at a 21 MHz bandwidth.

The key specifications for the AD8323 are shown in Figure 6-147.

- Supports cable modem DOCSIS (Data Over Cable Service) standard for upstream path transmission
- Gain/attenuation programmable in 0.7526dB steps over a 53.5dB range:
  –26dB to +27.5dB
- 3-wire SPI digital interface
- Bandwidth: > 100MHz (all gains)
- Low distortion @ 1V RMS (+11.2dBm, +60dBmV) output into 75Ω
  - –56dBc SFDR @ 21MHz
  - –55dBc SFDR @ 42MHz
- Single 5V supply (133mA)
- Power-down mode (35mA), sleep mode (4mA)
- 75Ω dynamic output impedance in power-up or power-down modes

Figure 6-147: AD8323 CATV line driver key specifications

xDSL Upstream Data Line Drivers

Various versions of DSL are now used to provide fast internet connections. The upstream data path requires the transmission of +13 dBm discrete multitone (DMT) signals occupying a bandwidth between approximately 144 kHz and 500 kHz. The DMT signal can have a crest factor as high as 5.3, requiring the line driver to provide peak power of +27.5 dBm, which translates into 7.5 V peak voltage on the 100 Ω telephone line.

DMT modulation appears in the frequency domain as power contained in several individual frequency subbands, sometimes referred to as tones or bins, each of which is uniformly separated in frequency. A quadrature amplitude modulated (QAM) signal occurs at the center of each subband or tone. Difficulties will exist when decoding these subbands if a signal from one subband is corrupted by the signal from other subbands, regardless of whether the corruption comes from an adjacent subband or harmonics of other subbands.

Conventional methods of expressing the output signal integrity of line drivers, such as single-tone harmonic distortion or THD, two-tone intermodulation distortion (IMD), and third order intercept (IP3), become significantly less meaningful when amplifiers are required to process DMT and other heavily modulated waveforms.

A typical ADSL upstream DMT signal can contain as many as 27 carriers (subbands or tones) of QAM signals (as shown in Figure 6-148). Multitone power ratio (MTTR) is the relative difference between the measured power in a typical subband (at one tone or carrier) versus the power at another subband specifically

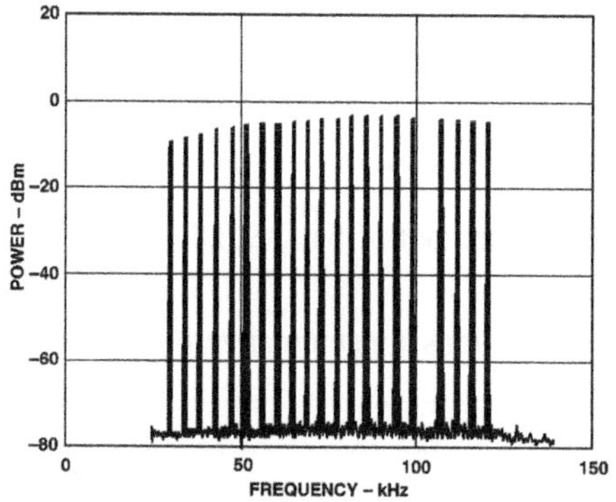

Figure 6-148: Discrete multitone (DMT) signal in the frequency domain

selected to contain no QAM data. In other words, a selected subband (or tone) remains open and void of intentional power (without a QAM signal), yielding an empty frequency bin. MTPR, sometimes referred to as the "empty bin test," is typically expressed in dBc and is a key specification for all types of DSL systems.

Another important specification for an xDSL line driver is *out-of-band* SFDR. Spurs produced by distortion of the DMT upstream data can fall in the downstream frequency regions and distort voiceband and downstream data.

Figure 6-149 shows an xDSL line driver application circuit based on the AD8018 line driver (one member of a family of Analog Devices' DSL line drivers). The peak DMT signal can be 7.5 V on the 100 Ω telephone line.

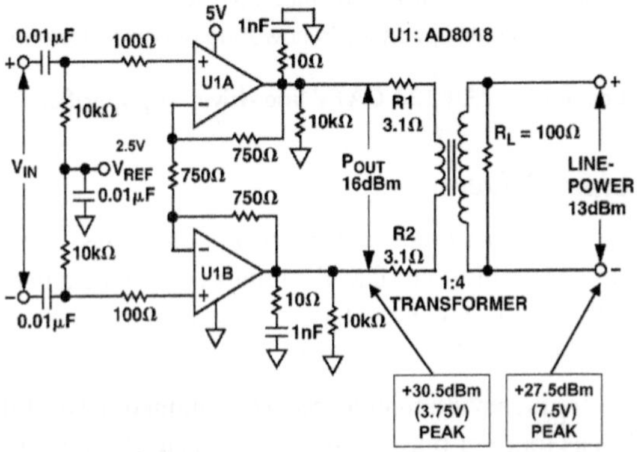

Figure 6-149: AD8018 xDSL upstream data line driver application

Assuming maximum low distortion output swing available from the AD8018 line driver on a single 5 V supply is 4 V, taking into account the power lost due to the two 3.1 Ω back-termination resistors, a transformer with a 1:4 or greater step-up is needed.

The AD8018 is therefore coupled to the phone line through a step-up transformer with a 1:4 turns ratio. R1 and R2 are back-termination or line-matching resistors, each 3.1 Ω. The total differential load presented to the AD8018 output is 12.5 Ω, including the termination resistors. Even under these conditions, the AD8018 provides low distortion signals to within 0.5 V or the power rails.

The transformer circuit presents a complex impedance to the AD8018 output, and therefore for stability, a series R-C network should be connected between each amplifier's output and ground. The recommended values are 10 Ω for the resistor and 1 nF for the capacitor to create a low impedance path to ground at frequencies above 16 MHz. The 10 kΩ output resistors connected to ground are added to improve common-mode stability.

For the AD8018 circuit of Figure 6-149, the out-of-band SFDR versus upstream line power is shown in Figure 6-150 for various supply voltages.

Some key AD8018 features and specifications are summarized in Figure 6-151.

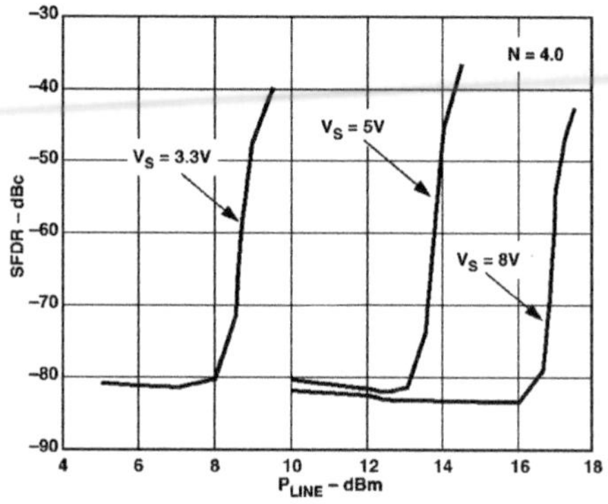

Figure 6-150: Out-of-band SFDR versus upstream line power, 144 kHz to 500 kHz

- Dual current feedback amplifiers
- Bandwidth: 130MHz (–3dB)
- Slew rate: 300V/µs
- Rail-to-rail output stage (swings within 0.5V of rails for R_L = 5Ω)
  - +16dBm into 12.5Ωload
  - +30.5dBm peak power (3.75V) with +5V supply
- MTPR: –70dBc (25kHz to 138kHz)
- Maintains –82dBc out-of-band SFDR, 144kHz to 500kHz, for output power = +16dBm, R_L = 12.5Ω
- Input voltage noise: 4.5nV/√Hz @100kHz
- Low supply current: 9mA/amplifier (full power mode)
- Standby mode (4.5mA/amplifier)
- Shutdown mode (0.3mA/amplifier)

Figure 6-151: AD8018 xDSL line driver key specifications

References: Communications Amplifiers

1. Bob Clarke, "Find Op Amp Noise with Spreadsheet," **Electronic Design**, December 13, 1990, or Analog Devices AN253.
2. Barrie Gilbert, "A Low Noise Wideband Variable-Gain Amplifier Using an Interpolated Ladder Attenuator," **IEEE ISSCC Technical Digest**, 1991, pp. 280, 281, 330.

Amplifier Ideas

Walt Jung, Walt Kester

This section of the chapter features miscellaneous op amp applications, within a format of *amplifier ideas*. They range broadly across the spectrum, illustrating many innovative op amp uses that don't otherwise fit categories. Some of the concepts have been inspired by publication elsewhere. In such cases, an appropriate original reference is given.

High Efficiency Line Driver

Conventional video line drivers use a series or *back-termination* resistor, selected to match the transmission line characteristic impedance. Although simple, this scheme is inherently inefficient, as both load and series termination resistors drop the same voltage. This isn't usually a problem with 1 V p-p video signals operating on high voltage supplies, such as ±12 V or ±15 V. However, with lower voltage supplies, particularly 5 V or less, driver headroom is definitely an issue. For such conditions, a conventional driver may simply not be able to accommodate a signal of twice V_{OUT} without distortion.

Figure 6-152 illustrates a solution to this driver efficiency problem. In this line driver (adapted from a circuit by Victor Koren, see Reference 1), a Howland type of feedback configuration is used. This allows the series termination resistor R5 to be appreciably smaller, thus dropping less voltage and improving stage efficiency. Both positive and negative loop feedback paths are used around the op amp, R3 and R4, plus R1 and R2. An AD817 is chosen for its video characteristics, and line driving capability. The circuit also works with many other op amps, provided they have sufficient output drive.

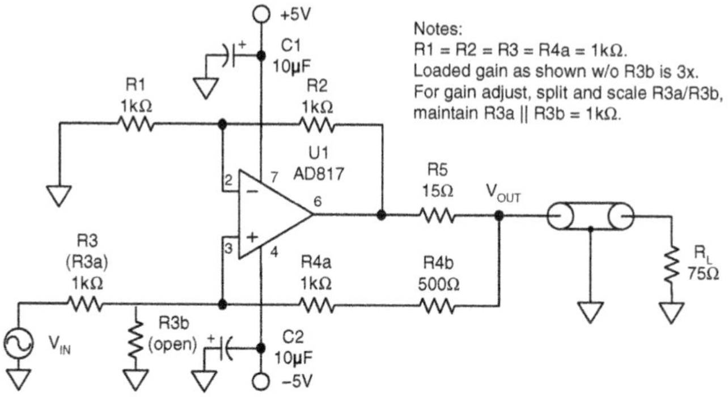

Figure 6-152: A high efficiency video line driver

In this example, a 75 Ω line is being driven, and R5 is set to 15 Ω. With the scaling chosen, this produces 1/5 the voltage drop of a more conventional 75 Ω resistor. For every volt of V_{OUT}, the amplifier needs only to produce 20% more, i.e., 1.2 V per V of V_{OUT}. This allows the design to operate easily on 5 V or even lower supplies, and still provide undistorted 1 V p-p video signals at V_{OUT}. The ± feedback paths produce the proper synthesized source impedance when the R1–R5 resistors are properly selected.

Given the desired output impedance R_O, R5 is related by a scaling factor, so that $R5 < R_O$. A direct design approach is to simply set R5 at some fraction of R_O, which then leads to a R1 through R4 resistor set that will provide the proper R_O. In this example design, R5 is set at 1/5 R_O as noted earlier, or 15 Ω.

As per the notes of the figure, a major simplifying design step is to make four of the feedback resistors equal, namely R1 through R3 and R4a. It also helps further to make these a common, readily available value. This should be a value moderately higher than the target load impedance. In this case, a 1 kΩ base value is chosen.

This defines the R4 value (R4a + R4b) as:

$$R4 = (R5 \times R1 + R_O \times R2)/(R_O - R5) \qquad \text{Eq. 6-31}$$

R4b is then simply R4-R4a. The design is further simplified with all of the noted resistors part of a single common array, *including* R4b (which is made from two parallel 1 kΩ resistors in this case). Note that R4b won't necessarily be so easily achieved in other design examples. Nevertheless, it is desirable for as many of the R1 through R4 resistors as possible be part of a common array, matched to 1% or better.

Gain of the stage just as shown is about 3× with the output loaded. If gain must be adjusted, there is a specific procedure to be followed. This is a necessary condition for proper stage function (for any gain), and is needed to maintain the synthesized R_O. For example, if a unity (1×) overall gain is desired, R3 can be changed to two resistors, i.e., R3a = 3 kΩ, R3b = 1.5 kΩ. Note that this reduces the drive to the op amp, but it also maintains the same 1 kΩ Thevenin impedance for R3 (where R3 is the resistance looking back to the input from R4a). Similarly, equal value 2 kΩ resistors could be used, which provides a net loaded stage gain of 1.5×. Of course, for arbitrary gains, a common array may not be possible, and ordinary 1% metal film types can also be used.

Also related to the above, the driving source at V_{IN} must be a very low impedance with respect to 1 kΩ, again to maintain the synthesized impedance relations. This is best achieved by use of an R3 driving source direct from an op amp output. Alternately, if the V_{IN} driving impedance is both fixed and known, it can be subtracted from R3.

A more general caveat (which applies to all Howland circuits) is that the design environment must maintain this source driving impedance *for all conditions*, as the circuit itself is *not* open source stable. For example, if R3 is opened, the positive feedback can override the negative feedback via R1 and R2, and the circuit could latch up.

Finally, although this design illustrates a driver oriented to a video standard of 75 Ω with 1 V p-p signals, there is no reason why the same design principles cannot be applied to other impedances and/or signal levels.

A Simple Wide Bandwidth Noise Generator

While most electronic designs seek noise minimization, there are occasions where a known quantity and/or quality of spectrally flat (white) noise is desirable. One such example is a dither source for enhanced dynamic resolution A/D conversion. For such applications, it is useful to be able to predict the output of a noise generator. It turns out that a carefully chosen decompensated op amp set up to amplify its own input noise is very useful as a wideband noise generator (see Reference 2).

Figure 6-153 illustrates this technique, which simply employs the op amp U1 as a fixed gain stage, amplifying its input noise by the stage factor G, where G = 1 + (R1/R2). This process is made easier by some simplifying assumptions, described next.

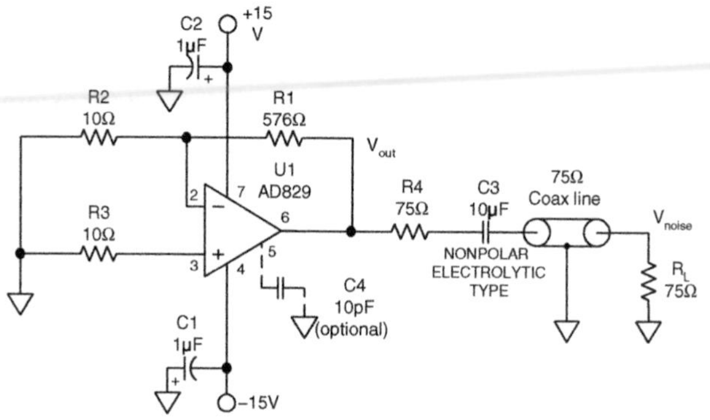

Figure 6-153: A simple wideband noise generator

By purposely selecting R2 and R3 values of 10 Ω or less, their Johnson noise contribution is forced to be less than the voltage noise of the amplifier. Similarly, the amplifier's current noise components in R2-R3, when converted to voltage noise, are also negligible. Thus the dominant circuit noise is reduced to the input voltage noise of U1.

To scale the amplifier noise to a given level of V_{noise} across R_L, select a stage gain which produces a noise density at V_{OUT} which is 2G times the typical U1 noise of $1.7\,nV/\sqrt{Hz}$. This will produce a V_{OUT} twice V_{noise}. For example, for a V_{noise} of $50\,nV/\sqrt{Hz}$, using a fixed R2 value of 10 Ω, the required R_1 is:

$$R1 = 10 \times \left(\left(\left(2 \times V_{NOISE}\right)/1.7\right) - 1\right) \qquad \text{Eq. 6-32}$$

Where V_{noise} is in $nV/\sqrt{Hz}$, and the 1.7 is the U1 voltage noise ($nV/\sqrt{Hz}$). This computes to 576 Ω (nearest standard value) for a wideband $50\,nV/\sqrt{Hz}$. Alternately, an audio range noise source of $1000\,nV/\sqrt{Hz}$ with several hundred kHz of bandwidth is achieved with R1 = 11.8 kΩ, and C3 = 100 µF. By choosing a bipolar-input, voltage feedback U1 device, with a single effective gain stage, a major performance point is achieved. Such an amplifier has a flat, frequency-independent input voltage noise response (i.e., a white noise characteristic). Many of the ADI high speed amplifiers use this topology within a folded cascode architecture.

In contrast to this, multiple stage, pole-zero compensated amplifiers such as the OP27 (and other similar architectures) can have peaks in the output noise response. This is due to the frequency compensation method used, and the associated gain distributions in the signal path. When picking U1, look for a noise characterization plot that shows flat input-referred voltage noise over several decades.

For the AD829 device used, input voltage noise is flat from below 100 Hz to more than 10 MHz, as is noted in Figure 6-154. Within the actual circuit, the upper bandwidth limit will be gain/compensation dependent, which can be controlled as described next.

The output is coupled through a nonpolar capacitor, C3, which removes any amplified dc offset at V_{out}. The C3 value should be large enough to pass the lowest noise frequencies of interest. As shown the response of this network is –3 dB at about 100 Hz, but C3 can be

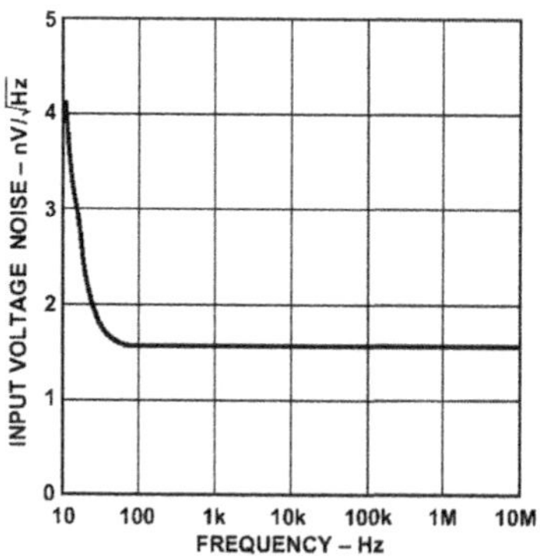

Figure 6-154: AD829 input voltage noise spectral density

changed for other low frequency limits. Source termination resistance R4 allows standard 75 Ω cables to be driven, providing distribution to a remote 75 Ω load, R_L.

In general, this noise generator's utility is greatest with a decompensated (or externally compensated) op amp, to take advantage of the maximum bandwidth possible. For the AD829, bandwidth is highest with lowest Pin 5 capacitance (i.e., no PCB Pin 5 pad, or Pin 5 cutoff). Conversely, C4 can be used to reduce bandwidth, if desired. With minimum Pin 5 capacitance, the AD829 gain bandwidth can be above 500 MHz, allowing extended response. In any case, the stage's effective –3 dB bandwidth varies inversely proportional to stage gain G. Some noise variations can be expected from IC sample-sample, so an R1 trim method can be used to set a output calibration level. Alternately, if ultrawide bandwidth noise isn't required, another op amp to consider is the AD817.

A final note—bipolar input amplifiers such as the AD829 typically use PTAT biasing for the input's differential stage tail current. Since equivalent input noise varies as the square root of this tail current, this can make noise output vary somewhat with temperature. The net effect causes noise to change less than 1 dB for a 50°C temperature change.

Single-Supply Half- and Full-Wave Rectifier

There are a number of ways to construct half- and full-wave rectifiers using combinations of op amps and diodes, but the circuit shown in Figure 6-155 requires only a dual op amp, two resistors, and operates on a single supply (see Reference 3).

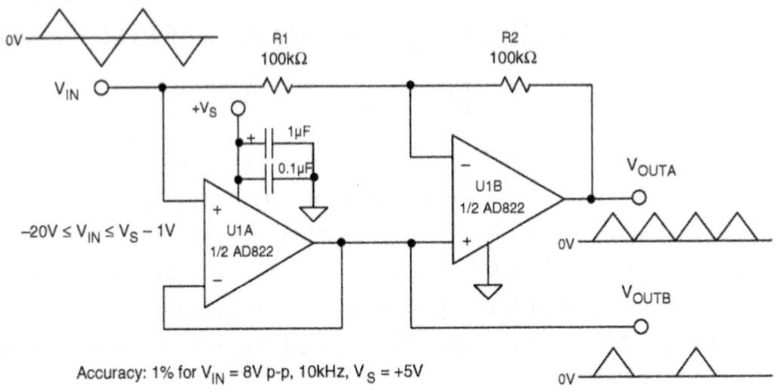

Figure 6-155: Single-supply half- and full-wave rectifier uses no diodes

The circuit will work with any single-supply op amp whose inputs can withstand being pulled below ground. The AD820 (single) or AD822 (dual) op amps have N-channel JFET inputs, which allow the input voltage to go to 20 V below the negative supply.

The output stage of these op amps is a complementary bipolar common emitter rail-to-rail stage with an output resistance of approximately 40 Ω when sourcing current and 20 Ω when sinking current. As a result of this stage, the outputs can go within a few millivolts of the supply rails under light loading.

When the input signal is above ground, unity-gain follower U1A and the loop of the amplifier U1B bootstrap R1. This bootstrapping forces the inputs of U1B to be equal. Thus, no current flows in R1 or R2, and the output V_{OUTA} tracks the input. Conversely, when the input is negative, the output of U1A is forced to zero (saturated). The noninverting input of U1B sees the ground-level output of U1A, and during this phase operates as a unity-gain inverter, rectifying the negative portion of the input V_{IN}.

The net output at V_{OUTA} is therefore a full-wave rectified version of V_{IN}. In addition, a half-wave rectified version is obtained at the output of U1A (V_{OUTB}) if desired.

The circuit operates with a single power supply of 3 V to 20 V. The circuit will maintain an accuracy of better than 1% over a 10 kHz bandwidth for inputs of 8 V p-p on a +5 V supply. The input should not go more than 20 V below the negative supply, or closer than 1 V to the positive supply. Inputs of ±18 V can be rectified using a single +20 V supply.

Paralleled Amplifiers Drive Loads Quietly

Paralleling op amps is a method to increase load drive while keeping output impedance low, and also to reduce noise voltage. Figure 6-156A shows a classic stacked-amplifier circuit. This configuration halves the input voltage noise of a single op amp, and quadruples load drive. However, it does have several weaknesses.

First, it is necessary to *individually* set the correct gain for each amplifier. Second, series resistors must be added to each output, to ensure equal load current distribution among the op amps. Third, the input range can become limited at high gains because of the inherent offset of any of the amplifiers.

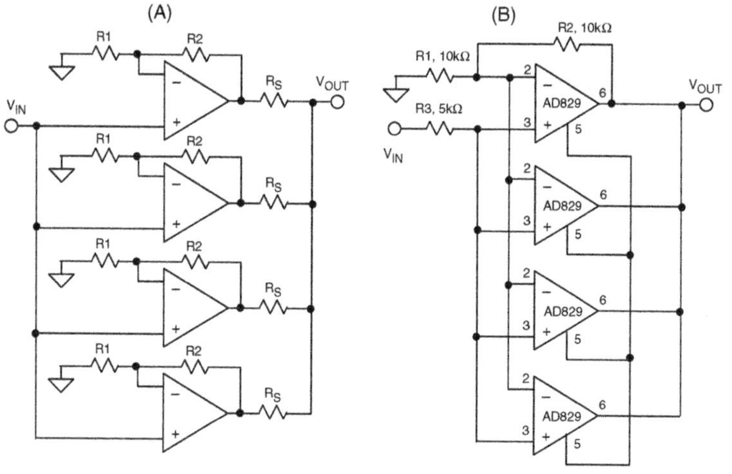

Figure 6-156: Paralleled amplifiers drive loads quietly

The circuit shown in Figure 6-156B also has half the noise voltage of an individual amplifier, and it also quadruples the load drive. But in so doing, it reduces the component count from twelve resistors to three (see Reference 4). In addition, the circuit has a gain-bandwidth product of about 750 MHz. Although the topology of Figure 6-156B is generally applicable to all externally compensated amplifiers (i.e., those with a pinned-out high impedance node before the output driver stage—such as the AD844 and AD846), the AD829 op amp is particularly well suited to video and other broadband applications.

Note that in the B circuit, external R_S load sharing resistors aren't required, because the only voltage difference between the individual outputs is due to the slight offset mismatch between the AD829 complementary emitter follower output driver stages, and internal 15 Ω emitter resistors ensure equal output current distribution. The closed-loop gain has no effect on this small offset voltage.

The end result of the four paralleled output stages in Figure 6-156B is a composite amplifier with both greater load drive and lower noise, but using only the conventional feedback components. The circuit in B increases the drive current by a factor of four, similar to A, but with a vast difference in the parts count.

In order to understand how the circuit reduces noise, let the RTI voltage noise of the individual op amps be V_{N1}, V_{N2}, V_{N3}, and V_{N4}, and let the total noise voltage be V_N.

Because all the inputs are connected in parallel, as well as the high impedance nodes, then:

$$(V_N - V_{N1})gm + (V_N - V_{N2})gm + (V_N - V_{N3})gm + (V_N - V_{N4})gm = 0 \qquad \text{Eq. 6-33}$$

$$V_N = \tfrac{1}{4}(V_{NI} + V_{N2} + V_{N3} + V_{N4}). \qquad \text{Eq. 6-34}$$

But because the voltage noise of the amplifiers is uncorrelated, and the noise spectral density for each amplifier is the same:

$$V_N = \tfrac{1}{4}\sqrt{\left[4(V_{N1})^2\right]} \qquad \text{Eq. 6-35}$$

$$V_N = V_{N1}/2 \qquad \text{Eq. 6-36}$$

This result also implies that all uncorrelated parameters such as input offset voltage, input offset voltage drift, CMRR, PSRR, and so forth, will also approach their true mean values, thus reducing effects arising from the variability of the devices.

The AD829 is flexible and can operate on supply voltages from ±5 V to ±15 V. It's uncompensated gain-bandwidth product is 750 MHz. Nominal output current for rated performance is 20 mA, so in the circuit shown, 80 mA is available to drive the load.

The input voltage noise of a single AD829 is $1.7\ \text{nV}/\sqrt{\text{Hz}}$, so the parallel circuit has an input voltage noise of approximately $0.85\ \text{nV}/\sqrt{\text{Hz}}$. In order to take advantage of this low voltage noise, however, the circuit must be driven from a relatively low source impedance, because the input current noise of a single AD829 is $1.5\ \text{pA}/\sqrt{\text{Hz}}$. In the parallel circuit, the input current noise is therefore $3\ \text{pA}/\sqrt{\text{Hz}}$.

Notice that in the AD829 circuit, R3 = R1||R2 for bias current cancellation. This works because the input bias currents of the AD829 are not internally compensated: they are approximately equal, and of the same sign. If amplifiers with internal bias current compensation or current feedback op amps are used, the input bias currents may not be equal or of the same sign, and R3 should be made equal to zero.

Power-Down Sequencing Circuit for Multiple Supply Applications

The operating time of battery operated portable equipment can be extended by using power-down techniques. Many new components offer a power-down feature to implement this function. However, there may be times when this feature is not offered and other means must be devised. The solution may also require proper power supply sequencing in multiple supply systems.

Figure 6-157 shows a single-supply op amp powered from 15 V driving an single-supply ADC powered from 5 V. In many cases, the same 5 V can supply the op amp and the ADC, and in that case, there is no sequencing problem. However, in some cases better system performance is obtained by driving the op amp with a higher supply voltage.

In the circuit, MOSFETs Q1 and Q2 switch the 5 V and the 15 V to the devices in the proper sequence. On power-up, the voltage to the ADC must be supplied first; and on power-down, the voltage to the ADC must be removed last. This is to ensure that the V_{IN} input to the ADC is never more than 0.3 V above the V_{DD} positive supply or more than 0.3 V below the negative supply, thereby preventing damage and possible latch-up.

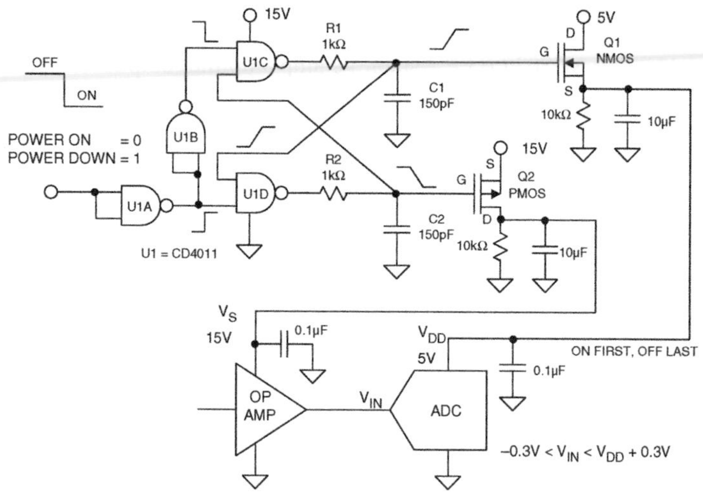

Figure 6-157: Power-down sequencing circuit for multiple supply applications

The MOSFETs, Q1 and Q2 switch the 5 V and the 15 V to the ADC and the op amp, respectively, in a sequence controlled by two cross-coupled CD4011 CMOS NAND gates (U1C and U1D). The gates are powered from the 15 V supply so that sufficient gate drive voltage is available to turn Q1 and Q2 on and off.

To initiate the power-on mode, a logic 0 is applied to the input of U1A, forcing its output high. This forces the output of U1B low, which causes U1C's output to go high. The R1-C1 time constant delays the application of the 15 V to the gate of Q1 which ultimately turns Q1 on with 5 V at its source. The delayed output of U1C is also applied to an input of U1D which forces its output low. U1D's output is delayed by the R2-C2 time constant, and ultimately forces the gate of Q2 to zero, which applies 15 V to the drain of Q2, and to the op amp V_S supply.

To initiate the power-down mode, a logic 1 is applied to the input of U1A which forces its output to zero, and the output of U1D is forced high, ultimately turning off Q2, the 15 V supply. The delayed output of U1D is applied to an input of U1C, the output of U1C goes low, and ultimately the gate of Q1 is forced to zero, turning off the 5 V to the V_{DD} input of the ADC.

It is important to note that when system power is applied to the overall circuit, the 5 V should come up either before, or simultaneously with, the 15 V. Similarly, when system power is removed, the 15 V should be removed first or simultaneously with the 5 V.

This circuit is based on a modification of the one described in Reference 5, where the desired sequencing is the reverse of the one described here. The reverse sequence (15 V turned on before 5 V, and 5 V turned off before 15 V) can be easily achieved by replacing the CD4011 NAND gates with CD4001 NOR gates, reversing the "sense" of the power-on/power-down control input, and swapping the gate drive signals to Q1 and Q2.

Programmable Pulse Generator Using the AD8037 Clamping Amplifier

The AD8036 (G ≥ 1 stable) and AD8037 (G ≥ 2 stable) clamp amplifier outputs can be set accurately to well controlled flat levels determined by the clamping voltages. This, along with wide bandwidth and high slew rate suits them well for numerous applications.

A basic description of the AD8036/AD8037 operation can be found in the Video Applications section of this chapter. Figure 6-158 is a diagram of a programmable level pulse generator (see Reference 6).

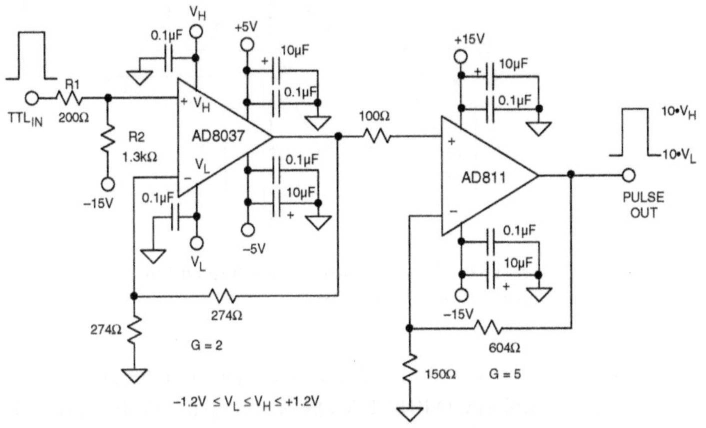

Figure 6-158: Programmable pulse generator using AD8037 clamping amplifier

The circuit accepts a TTL timing signal for its input and generates pulses at the output up to 24 V p-p with 2500 V/µs slew rate. The output levels can be programmed to anywhere in the range between –12 V to +12 V.

The circuit uses an AD8037 operating at a gain of two with an AD811 to boost the output to the ±12 V range. The AD811 was chosen for its ability to operate with ±15 V supplies and its high slew rate. R1 and R2 level shift the TTL input signal level approximately 2 V negative, making it symmetrical above and below ground. This ensures that both the high and low logic levels will be clamped by the AD8037. For well controlled signal levels in the output pulse, the high and low output levels result from the clamping action of the AD8037 and aren't controlled by either the high/low logic levels passing through a linear amplifier. For good output rise/fall times, logic with high edge speed should be used.

The high logic levels are clamped at two times the voltage at V_H, while the low logic levels are clamped at two times the voltage at V_L. The output of the AD8037 is amplified by the AD811 operating at a gain of 5. The overall gain of 10 will cause the high output level to be 10 times the voltage at V_H, and the low output level 10 times the voltage at V_L. For this gain, the clamping levels for a ±12 V output pulse are V_H = +1.2 V and V_L = –1.2 V.

Full-Wave Rectifier Using the AD8037 Clamping Amplifier

The clamping inputs can be used as additional inputs to the AD8036/AD8037. As such, they have an input bandwidth comparable to the amplifier inputs and lend themselves to some unique functions when they are driven dynamically.

Figure 6-159 is a schematic for a full-wave rectifier, also called an absolute value generator (Reference 6). It works well up to 20 MHz and can operate at significantly higher frequencies with some performance degradation. The distortion performance is significantly better than diode-based full-wave rectifiers, especially at high frequencies.

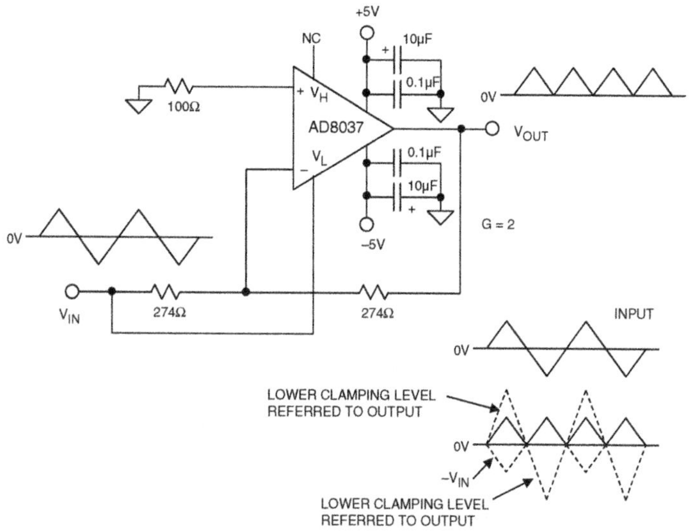

Figure 6-159: Full-wave rectifier using the AD8037 clamping amplifier

The AD8037 is configured as an inverting amplifier with a gain of unity. The V_{IN} input drives the inverting amplifier and also directly drives V_L, the lower level clamping input. The high level clamping input, V_H, is left floating and plays no role in the circuit.

When the input is negative, the amplifier acts as a unity-gain inverter and outputs a positive signal at the same amplitude as the input, with opposite polarity. V_L is driven negative by V_{IN}, so it performs no clamping action, because the positive output signal is always higher than the negative level driving V_L.

When the input is positive, the output result is the sum of two separate effects. First, the inverting amplifier multiplies the V_{IN} input by –1, because of the unity-gain inverting configuration. This effectively produces an offset at the output, but with a dynamic level that is equal to –1 times the input. Second, although the positive input is grounded (through 100 Ω), the output is clamped at two times the voltage applied to V_L (a positive, dynamic voltage in this case). The factor of two is because the 2x amplifier noise gain.

The sum of these two actions results in an output that is equal to unity times the input signal for positive input signals, as shown in Figure 6-159. Thus, for either positive or negative input signals, the output is unity times the absolute value of the input signal. The circuit can be easily configured to produce the negative absolute value of the input by applying the input to V_H rather than V_L.

The circuit can get to within about 40 mV of ground during the time when the input crosses zero. This voltage is fixed over a wide frequency range, and is a result of the switching between the conventional op amp input and the clamp input. However, because there are no diodes to rapidly switch from forward to reverse bias, the performance far exceeds diode-based full-wave rectifiers. Signals up to 20 MHz can be rectified with minimal distortion.

If desired, the 40 mV offset can be removed by adding an offset to the circuit, with little additional complexity. A 27.4 kΩ input resistor to the inverting input will have a gain of 0.01, while changing the gain of the circuit by only 1%. A plus or minus 4 V dc level (depending on the polarity of the rectifier) fed into this resistor will then compensate for the offset.

Full-wave rectifiers are useful in many applications including AM signal detection, high frequency ac voltmeters, and various arithmetic operations.

AD8037 Clamping Amplifier Amplitude Modulator

The AD8037 can also be configured as an amplitude modulator as shown in Figure 6-160 (Reference 6). The positive input of the AD8037 is driven with a square wave of sufficient amplitude to produce clamping action at both the high and low levels set by V_H and V_L. This is the higher frequency carrier signal.

The modulation signal is applied to both the input of a unity gain inverting amplifier and to V_L, the lower clamping input. V_H is biased at 0.5 V for the example to be discussed but can assume other values.

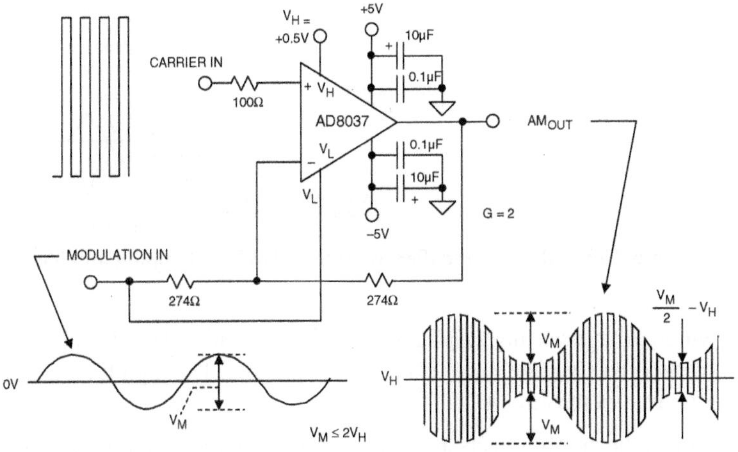

Figure 6-160: AD8037 clamping amplifier amplitude modulator

To understand the circuit operation, it is helpful to first consider a simpler circuit. If both V_H and V_L are dc-biased at +0.5 V and −0.5 V, respectively, and the carrier and modulation inputs driven as above, the output would be a 2 V p-p square wave at the carrier frequency riding on a waveform at the modulating frequency.

The inverting input (modulation signal) is creating a varying offset to the 2 V p-p square wave at the output. Both the high and low levels clamp at twice the input levels on the clamps because the noise gain of the circuit is two.

When V_L is driven by the modulation signal instead of being held at dc level, a more complicated situation results. The resulting waveform is composed of an upper envelope and a lower envelope with the carrier square wave in between. The upper and lower envelopes are 180° out of phase as in a typical AM waveform.

The upper envelope is produced by the upper clamp level being offset by the waveform applied to the inverting input. This offset is the opposite polarity of the input waveform because of the inverting configuration.

The lower envelope is produced by the sum of two effects. First, it is offset by the waveform applied to the inverting input as in the case of the simpler circuit above. The polarity of this offset is in the same direction as the upper envelope.

Second, the output is driven in the opposite direction of the offset at twice the offset voltage by the modulation signal being applied to V_L. This results from the noise gain being equal to two, and since there is no inversion in this connection, it is opposite in polarity from the offset.

The result at the output for the lower envelope is the sum of these two effects, which produces the lower envelope of an AM waveform. The depth of modulation can be modified by changing the amplitude of the modulation signal. This changes the amplitude of the upper and lower envelope waveforms.

The modulation depth can also be changed by changing the dc bias applied to V_H. In this case, the amplitudes of the upper and lower envelope waveforms stay constant, but the spacing between them changes. This alters the ratio of the envelope amplitude to the amplitude of the overall waveform.

For $V_H = +0.5$ V, 100% modulation occurs when the peak-to-peak amplitude of the modulation input $V_M = 1$ V. The AM output is always offset by V_H for a bipolar modulation input. In general, for a peak-to-peak modulation amplitude of V_M, the two output modulated envelopes are separated by an amount equal to $V_M/2 - V_H$.

Sync Inserter Using the AD8037 Clamping Amplifier

Video signals typically combine an active video region with both horizontal and vertical blanking intervals during their respective retrace times. A sync signal is required during the blanking intervals in some systems. In RGB systems, the sync is usually inserted on the Green signal. In composite video systems, it is inserted during blanking on the single-channel composite signal, or on the luminance (or Y) signal in an S-video system. Further details on video signals can be found in the Video Applications section of this chapter.

The AD8037 input clamping amplifier can be used to make a video sync inserter that does not require accuracy in the amplitude or shape of the sync pulse (see Reference 7). The circuit shown in Figure 6-161 uses the AD8037 to create the proper amplitude sync insertion, and the dc levels of the sync pulse do not affect the active video level. The circuit is also noninverting with a gain-of-two, which allows for driving a back-terminated cable with no loss of amplitude.

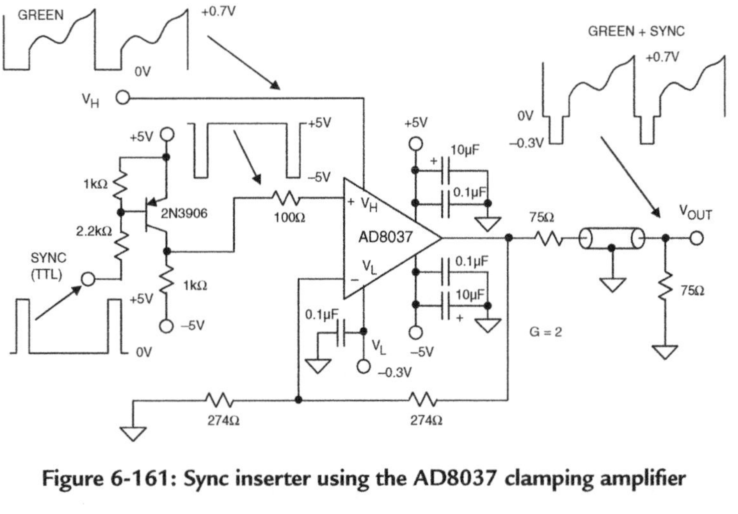

Figure 6-161: Sync inserter using the AD8037 clamping amplifier

The Green video signal is applied to the V_H input of the AD8037. This signal has a blanking level of 0 V and an approximate full-scale value of 0.7 V. The TTL-level sync pulse is applied to the base of the 2N3906 transistor. The signal at the collector of the 2N3906 are inverted sync pulses with an amplitude of 10 V p-p which are applied to the noninverting input of the AD8037.

The amplitude of the signal into the noninverting input of the AD8037 is 5 V during the active video portion, and since this is greater than the maximum positive Green signal excursion of 0.7 V on the V_H input, the Green signal is passed through to the output of the AD8037 with a gain of +2.

During the blanking interval, the sync signal into the noninverting input of the AD8037 goes to –5 V, and the output of the AD8037 is clamped to a value which is two times the dc level on the V_L input. Nominally, the sync should be –0.3 V referenced to a 0 V blanking level, and this level is applied to V_L from a voltage reference, or a simple divider.

The high and low levels of the sync pulse generated by the 2N3906 can be relatively loosely defined. The value of the high-level sync input to the noninverting input of the AD8037 must be higher than the active video signal; the value of the low level sync input must be lower than the dc voltage on the V_L input. The rising and falling edges of the sync pulse input determine the timing of the inserted sync, but the dc level at the V_L input of the AD8037 will always determine the sync amplitude.

The 2N3906 PNP transistor serves as a level translator and simply provides an appropriate drive signal from a TTL source of positive-going sync. The sync input to the noninverting input of the AD8037 neither influences the dc level of the output video nor determines the amplitude of the inserted sync.

AD8037 Clamped Amplifier As Piecewise Linear Amplifier

Piecewise linear amplifiers are often implemented using diodes in the feedback loop of an op amp. When the diodes become forward biased, they switch in resistors that alter the closed-loop gain of the amplifier.

This approach has three disadvantages. First, the diode's forward voltage drop (even with Schottky diodes) reduces accuracy and speed during the switching region. Second, diode stray capacitance can limit bandwidth. Third, the 2 mV/°C drift of the diode's forward bias voltage introduces errors in the transfer function. The circuit shown in Figure 6-162 avoids these problems by using the fast and accurate clamping function of the AD8037 to set the breakpoints (see Reference 8).

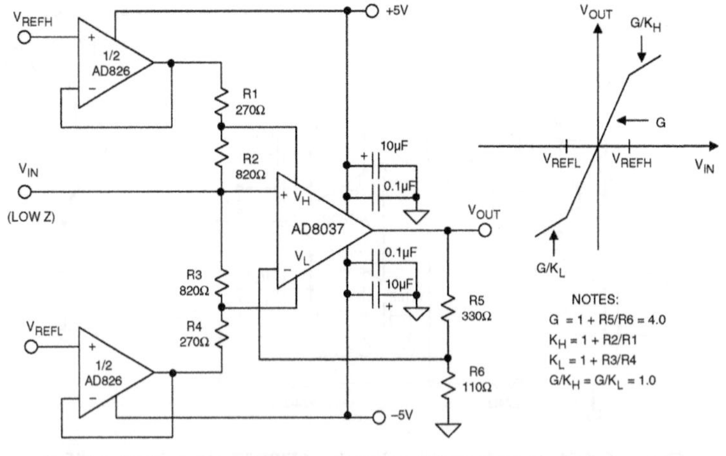

Figure 6-162: Piecewise linear amplifier uses AD8037 clamping amplifier rather than diodes

If the V_{IN} signal applied to the noninverting input of the AD8037 lies between the clamp voltages (set by V_{REFH} and V_{REFL}), the AD8037 works as a standard op amp with a gain = G = 1 + R5/R6. If the input signal is greater than the upper clamp voltage, V_H, the amplifier disconnects the input signal, and V_H becomes the noninverting signal input. Likewise, if the signal at the noninverting input of the AD8037 is below the lower clamp voltage, V_L, the amplifier also disconnects the signal input, and V_L becomes the noninverting signal input.

Figure 6-162 also graphically illustrates the operation of the circuit. When V_{IN} is between V_{REFH} and V_{REFL}, the circuit is a standard noninverting op amp with a gain G = 1 + R5/R6. When V_{IN} is greater than V_{REFH}, V_H becomes the noninverting input to the amplifier.

The transfer function from V_{IN} to V_{OUT} comprises two parts under this condition. From V_{IN} to V_H, the signal is attenuated by a factor $K_H = 1 + R2/R1$. From V_H to V_{OUT}, the gain remains G = 1 + R5/R6. This leads to an overall gain of G/K_H in this region. The circuit behaves similarly when V_{IN} is below V_{REFL}. The gain in this condition is G/K_L, where $K_L = 1 + R3/R4$.

Careful layout ensures adherence to the desired nonlinear transfer function over a 5 MHz bandwidth. The stability of the breakpoints is determined by the tracking of the resistor temperature coefficients, the 10 µV/°C offset voltage drift of the AD8037, and the temperature stability of the reference voltages.

The reference voltages can be generated using precision voltage references or DACs. To maintain accuracy, the reference voltages should be buffered with a fast op amp, such as the dual AD826, to provide a low source impedance throughout the input signal bandwidth.

The analog input voltage, V_{IN}, should also be driven from a low impedance source, such as an op amp, to prevent errors due to the loading effect of the R1-R2-R3-R4 network.

Using the AD830 Active Feedback Amplifier as an Integrator

The active feedback amplifier topology used in the AD830/AD8130 can be used to produce a precision voltage-to-current converter which, in turn, makes possible the creation of grounded-capacitor integrators (see Reference 9).

The design discussed here uses the AD830 to deliver a bipolar output current at high impedance (see Figure 6-163). Using R = 1 kΩ, the output current is simply equal to V_{IN}/R, or 1 mA per volt of input. The maximum output current is limited to ±30 mA by the output drive capability of the AD830.

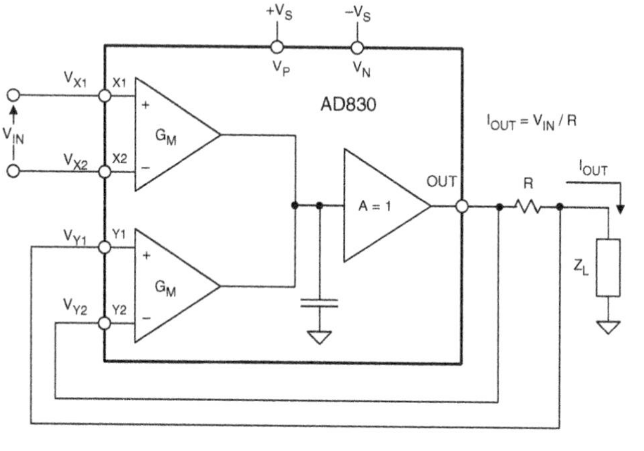

Figure 6-163: Constant current source using the AD830 active feedback amplifier

The output resistance is determined by the CMR performance of the AD830. A CMR of 60 dB yields an effective output resistance of 1000 × R. The output impedance at any frequency can be determined by consulting the CMR data provided in the data sheet. The compliance range on the output is ± (V_S – 2 V) reduced or increased by V_{IN}.

Figure 6-164A shows a standard op amp integrator circuit using the AD825, and Figure 6-164B shows the improved AD830 grounded-capacitor circuit. The dc operating point for testing purposes is determined by R1 in the AD825 op amp circuit, and by R1 and C1 in the AD830 circuit. R2 and C2 determine the integrator time constant in both circuits.

If the op amp in Figure 6-164A is assumed to be ideal, i.e., zero output impedance, and infinite input impedance, then the only difference between the two circuit topologies is the finite input resistance of the op amp based integrator as set by R2.

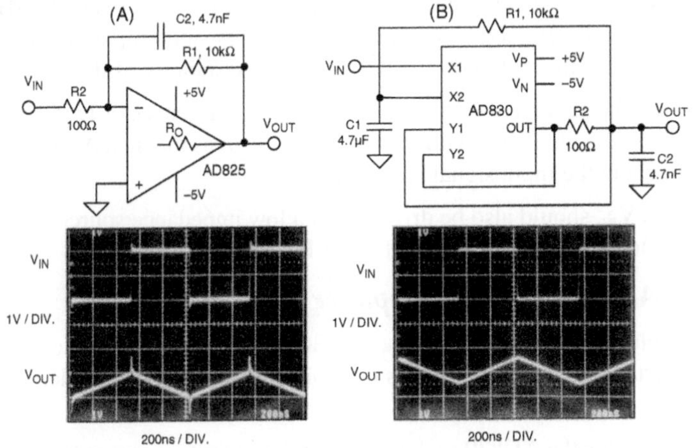

Figure 6-164: Traditional op amp integrator versus grounded capacitor integrator

However, in a real op amp, the output resistance is finite and increases with signal frequency as the open-loop gain decreases. This causes the "ground" at the output of the op amp to degrade at high frequencies. The result is a relatively large spike on the output voltage waveform whenever the input switches.

This can be explained as follows. Assume the input V_{IN} switches between $-V_A$ and $+V_A$. When the input is at $-V_A$ long enough for the op amp to settle, the current in resistor R2 is $-V_A$/R2 and the output increases due to C2 being charged by the op amp. As V_{IN} suddenly switches to $+V_A$, the voltage across C2 cannot change instantaneously, and neither can the op amp's output because it behaves itself as an integrator. This implies that the change in input voltage will be impressed upon the voltage divider formed by the R_O of the op amp and R2. This change in voltage at V_{OUT} will also be coupled by C2 to the summing node at the inverting input of the op amp.

If V_{IN} is generated by a source with finite source resistance, this voltage spike will also appear at the input. Only after the amplifier settles will the external components again define the integrator time constant and the circuit function as desired.

It can be seen by comparing the waveforms of (A) and (B) that no spike develops in the output waveform produced by the grounded capacitor integrator using the AD830. This is because the integrating capacitor is connected to a true ground. In addition, the input is completely isolated from the output. Therefore, if an aberration did occur, it would not be coupled back to the driving source.

Various active filter topologies can be realized from this fundamental integrator building block. For example, two such sections can implement a biquad. An example of a simple all-pass filter using the AD830 is described in Reference 10.

Instrumentation Amplifier with 290MHz Gain-Bandwidth

The circuit shown in Figure 6-165 combines a dual AD828 op amp with the AD830 active feedback difference amplifier to form a high frequency instrumentation amplifier (see Reference 11). The circuit's performance for ±5 V supplies for gains of 10 and 50 are shown in the figure, along with appropriate component values.

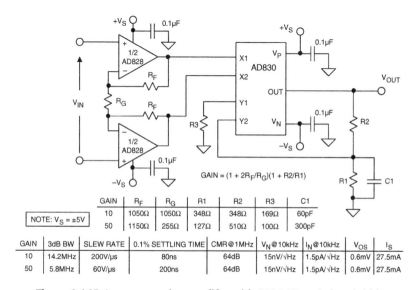

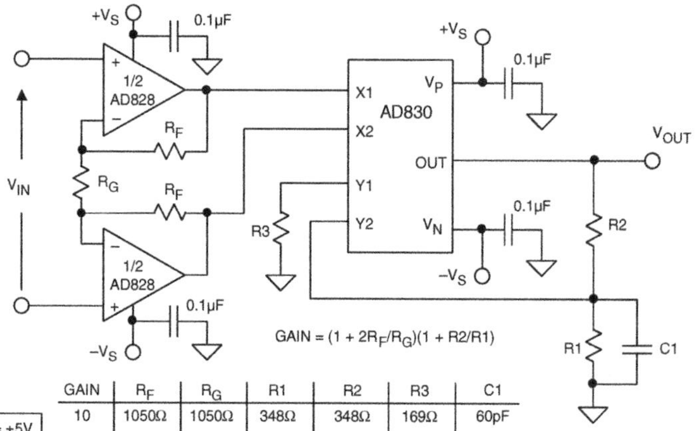

NOTE: $V_S = \pm 5V$

GAIN	R_F	R_G	R1	R2	R3	C1
10	1050Ω	1050Ω	348Ω	348Ω	169Ω	60pF
50	1150Ω	255Ω	127Ω	510Ω	100Ω	300pF

GAIN	3dB BW	SLEW RATE	0.1% SETTLING TIME	CMR@1MHz	V_N@10kHz	I_N@10kHz	V_{OS}	I_S
10	14.2MHz	200V/μs	80ns	64dB	15nV/√Hz	1.5pA/√Hz	0.6mV	27.5mA
50	5.8MHz	60V/μs	200ns	64dB	15nV/√Hz	1.5pA/√Hz	0.6mV	27.5mA

Figure 6-165: Instrumentation amplifier with 290 MHz gain bandwidth

The circuit can be configured for different gains, and will operate on supplies ranging from ±4 V to ±16.5 V.

The gain is proportioned between the AD828 stage and the AD830 stage, such that the closed-loop bandwidths of both stages are approximately equal. Under these conditions, a gain bandwidth of 290 MHz is obtained.

The input AD828 stage dominates the effective referred-to-input (RTI) input voltage noise and offset voltage. Capacitor C1 causes gain peaking in the AD830 which compensates for the AD828 input stage roll-off. The optimum value for C1 must be determined experimentally in a prototype or by a careful SPICE evaluation.

Note that R3 is made equal to the parallel combination of R1 and R2 to provide first-order input bias current cancellation at the Y1-Y2 input of the AD830.

Programmable Gain Amplifier with Arbitrary Attenuation Step Size

The R/2R ladder is a popular resistor topology often used to implement a current or voltage 6 dB step attenuator. However, if the resistors are appropriately scaled, the network can be modified to provide any desired attenuation step.

A programmable gain amplifier (PGA) can be made with a attenuating ladder network followed by CMOS multiplexer and a fixed gain amplifier, as in Figure 6-166. This circuit has several advantages (see Reference 12).

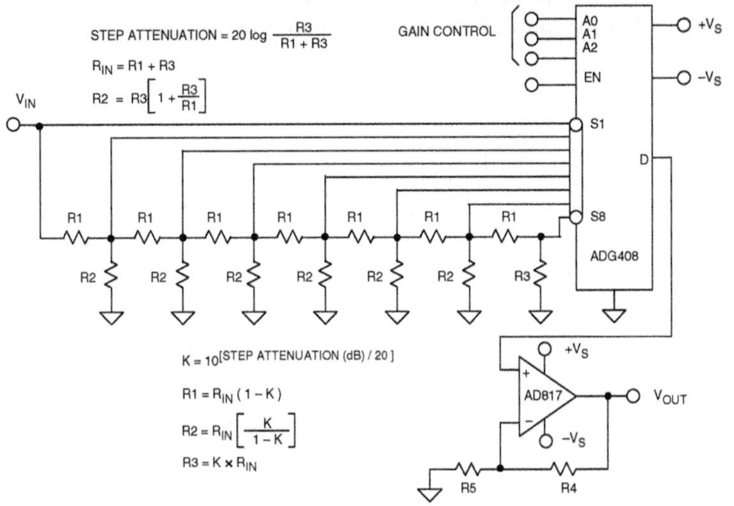

Figure 6-166: Programmable gain amplifier with arbitrary attenuation step size

First, as stated previously, the attenuation step size doesn't have to be 6 dB. Manipulating the resistor ratios, as described below, can easily change it. Second, the bandwidth of the circuit is always the same, regardless of the attenuation, due to the fact that the op amp buffer operates at a fixed gain. Third, the circuit is flexible, because practically any CMOS multiplexer and op amp can be used. The bandwidth of the circuit is determined primarily by the output op amp. Switching time between gain settings is determined by the multiplexer switching time and the op amp settling time.

The resistor ladder as shown uses three different resistor values: R1, R2, and R3. The step attenuation in dB is given by

$$\text{Step Attenuation (dB)} = 20\log\left[R3/(R1+R3)\right] \qquad \text{Eq. 6-37}$$

Also, the following relationships apply:

$$R_{IN} = R1 + R3 \qquad \text{Eq. 6-38}$$

$$R2 = R3\left[1 + R3/R1\right] \qquad \text{Eq. 6-39}$$

If R1 = R3, then R2 = 2 × R1. In this case, the R-2R network provides 6 dB step attenuation.

To determine the resistor values for a specific step attenuation and input resistance, use the formulas as follows:

$$K = 10^{[\text{Step Attenuation (dB)}/20]} \qquad \text{Eq. 6-40}$$

where the step attenuation is entered as a negative number.

Then, the following equations complete the design:

$$R1 = R_{IN}(1-K) \qquad \text{Eq. 6-41}$$

$$R2 = R_{IN} \times K/(1-K) \qquad \text{Eq. 6-42}$$

$$R3 = K \times R_{IN} \qquad \text{Eq. 6-43}$$

For example, to implement a resistor ladder with a –1.5 dB step attenuation and a 500 Ω input impedance: K = 0.8414, R1 = 79.3 Ω, R2 = 2653 Ω, and R3 = 420.7 Ω using the above equations.

The gain of the op amp is equal to 1 + R4/R5. The overall gain of the PGA is equal to the op amp gain minus the attenuation setting.

Finally, it is interesting to note that the AD60x-series of X-Amps discussed in the previous "Communications Amplifiers" section of this chapter uses the same basic approach described above. In the AD60x X-Amp series however, attenuation is continuously variable, because an interpolation circuit rather than a multiplexer is used to connect the individual taps of the network to the op amp input.

A Wideband In Amp

Some op amps with provisions for external offset trim can be used in unusually creative ways. In fact, if the two offset null inputs are considered as an additional differential signal input pair, this point becomes more clear. Although designed principally for adjustment of device V_{OS}, the null inputs can often be used for additional signals. An example is the wideband in amp of Gerstenhaber and Gianino (see Reference 13). In the circuit of Figure 6-167, the op amp used is the AD817. Designed for low distortion video circuits, it has a relatively high resistance between the input differential pair emitters, R_E, approximately 1 kΩ. It also has internal, large-value 8 kΩ resistors in series with the V_{OS} nulling terminals at Pins 1 and 8, labeled here as R_1.

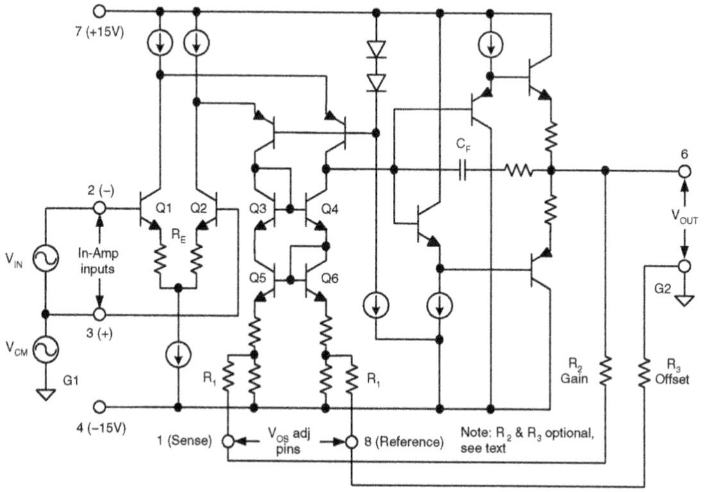

Figure 6-167: An AD817 wideband in amp configuration

Functioning here as an in amp, the AD817 is operated unconventionally. No feedback is used to inverting input Pin 2. Instead, the V_{IN} differential signal is applied between Pins 2 and 3, as noted. Typically there is also an associated CM noise, V_{CM}. Note that there must be a return path between the input ground G1 and output ground G2, to allow bias current flow (as with standard in amps). The input differential pair stage Q1 and Q2 produces an output signal current, driving quad-connected current mirror stage Q3, Q4, Q5, Q6. At the bottom of the current mirror is the balancing resistor network, functioning here as a signal current input. The connection from Pin 1 to the amplifier output closes a negative feedback loop, to the output at Pin 6. A balancing reference input is applied to Pin 8, either ground (as shown) or a variable offset voltage. Differential gain of the circuit, G, is:

$$G = V_{OUT}/V_{IN} = R_1/R_E \qquad \text{Eq. 6-44}$$

For the values noted, gain is about 8×, and bandwidth is 5 MHz. CMR is excellent, measuring more than 80 dB at 1 MHz. Optional trim resistors R_2 and R_3 are used to adjust gain via R_2, or, alternately, offset via R_3. For best CMR, the values should be the same.

Negative Resistance Buffer

There is often a requirement for driving a lower load impedance than a given op amp may be capable of meeting. This can be particularly true for precision op amps in general and, more specifically, with rail-rail output types. The latter class of op amps typically can have an output impedance on the order of several kΩ, which can limit load drive and lower open loop gain when driving low impedances. A straightforward way of addressing this problem is a unity gain buffer, which will work in almost all cases. But ordinary op amps can also be used for the buffer function. An interesting method is to use a second op amp as a negative resistance generator, to synthesize a negative resistance whose value is set equal to the load resistance. When this is carefully done the load disappears, as a parallel connection of R_L and $-R_L$ is infinite. Figure 6-168 illustrates this technique, in both basic and practical forms.

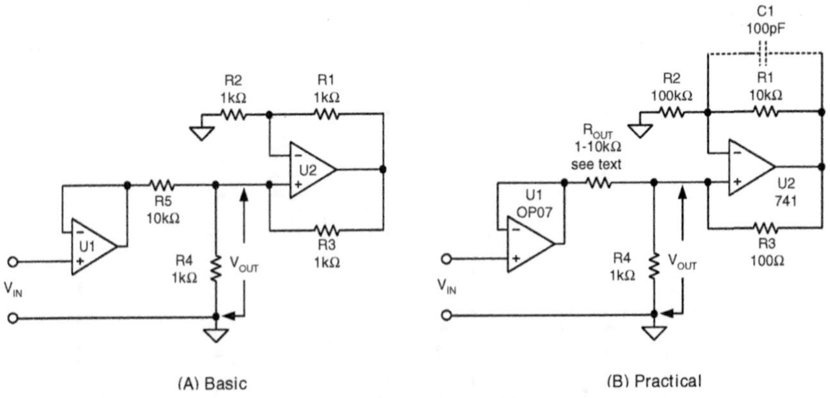

Figure 6-168: Negative resistance buffer circuits

A basic form of the circuit is shown in Figure 6-168A to illustrate the concept. Here op amp U1 is intended to drive load resistor R4, but would normally be prevented from doing so with high precision by the output resistance represented by R5. But, due to the connection of the U2 stage the voltage V_{OUT} is amplified by a factor of 1 + R1/R2. This amplified voltage is fed back to the V_{OUT} node by R3. With the R1–R4 values scaled as shown, this produces a negative resistance of –R4 at the V_{OUT} node. Thus the driving amplifier U1 does not see the real resistance R4 loading, which can be confirmed by examining the (small) current in R5. If operation is not apparent, the circuit can be analyzed by viewing it as a balanced bridge, with ratios of R1/R2 matching R3/R4.

But the Figure 6-168A circuit isn't very efficient, as twice the load voltage must be developed for operation, and double the load current flows in U2. The same principles are employed in the more practical Figure 6-168B version, with the R1–R4 values rescaled to reduce power and to gain headroom in U2, *while maintaining the same ratios*. In a real circuit there is likely no need for R5, and U1 can drive the load directly. It does so taking full advantage of the precision characteristics of the U1 type. U2 can be almost any ordinary op amp capable of the load current required.

Cross-Coupled In Amps Provide Increased CMR

A primary in amp benefit is the ability to reject CM signals in the process of amplifying a low-level differential signal. While most in amps perform well below about 100 Hz, their CM rejection degrades rapidly with frequency. The circuit in Figure 6-169 is a composite in amp with much increased CMR vis-à-vis more conventional hookups (see Reference 15). It consists of three in amps, with unity-gain connected U1 and U2 cross-coupled at their inputs. In amp U3 amplifies the difference between V_{O1} and V_{O2}, while rejecting CM signals. The in amps are AD623s, but the scheme works with other devices.

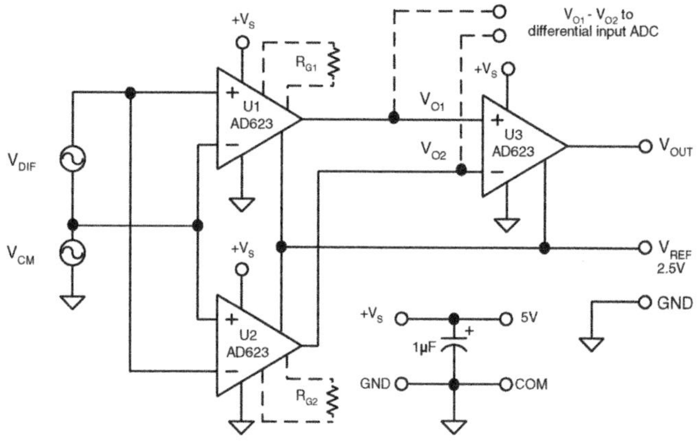

Figure 6-169: Two cross-coupled and similar in amp devices followed by a third provides much increased CMR with frequency

Because of the fact that U1 and U2 have CM responses that are correlated (by the nature of their design), their output CM errors due to V_{CM} will be similar. For U3, this CM error appears as a CM signal, and is rejected further. Meanwhile, the desired differential signal, equal to 2 V_{DIF} appears as $V_{O1} - V_{O2}$, and is amplified by U3 at unity gain. Overall gain is 2× as shown, but can be raised by a gain factor programmable by R_{G1} and R_{G2}. Note that due to the fact that the U1 and U2 CM errors correlate, their matching isn't necessary.

A big advantage of this scheme is the extended frequency range over which the composite in amp has good CMR. For example, at the gain of 2 as shown, CMR as measured at either V_{O1} or V_{O2} will be on the order of 60 dB or more at 10 kHz. At V_{OUT} however (the output of U3), the CMR is increased to about 85 dB, or more than 20 dB. The low frequency CMR corner of the composite in amp is about 6 kHz, as opposed to about 500 Hz as measured at either V_{O1} or V_{O2}. At higher gains, for example a gain of 100 (as set by R_{G1} and

R_{G2} = 2.05 kΩ), CMR increases to more than 110 dB at low frequencies, and a corner frequency of about 2 kHz is noted, while 10 kHz CMR is more than 100 dB. For these measurements $+V_S$ was 5 V, and the V_{REF} applied to all devices was 2.5 V.

Although the example shown is single-supply, it is also useful with dual-supply in amps. Another possible mode is to use V_{O1} and V_{O2} to drive a differential input ADC, which eliminates a need for in amp U3. ADC scaling can be matched via R_G, and the V_{REF} used.

References: Amplifier Ideas

1. Victor Koren, "Line Driver Economically Synthesizes Impedance," **EDN**, January 6, 1994 p. 79. See also: "Feedback and Amplification," **EDN**, May 26, 1994 pp. 106.
2. Walt Jung, "Simple Wideband Noise Generator," **Electronic Design**, October 1, 1996, p. 102.
3. Lewis Counts, Mark Murphy, JoAnn Close, "Diode-less Rectifier Takes Rail-to-Rail Input," **EDN**, October 28, 1993.
4. Moshe Gerstenhaber, Mark Murphy, "Paralleled Amplifiers Drive Loads Quietly," **EDN**, April 23, 1992, p. 171.
5. John Wynne, "Simple Circuit Adds Power Down," **Electronic Design**, January 7, 1993, p. 116.
6. "AD8036/AD8037 APPLICATIONS," within **Data Sheet for AD8036/AD8037 Low Distortion, Wide Bandwidth Voltage Feedback Clamp Amps**, www.analog.com.
7. Peter Checkovich, "Clamp Amp Serves as Sync Inserter," **Electronic Design**, October 14, 1996, pp. 132, 134.
8. Brian Harrington, "Piecewise Linear Amplifier Eschews Diodes," **EDN**, October 12, 1995, pp. 112–113.
9. Eberhard Bruner, "Turn Feedback Amp Into Integrator," **Electronic Design**, July 10, 1995, pp. 101–102.
10. Eberhard Bruner, "Simple All-Pass Filter," **Electronic Design**, June 26, 1995, pp. 106–108.
11. Paul Hendricks, "Instrumentation Amplifier Has 290MHz GBW," **EDN**, May 12, 1994, p. 86.
12. Victor Koren, "Programmable-Gain Amp Uses Arbitrary-Attenuation Step Ladder," **Electronic Design**, April 16, 2001, p. 99.
13. Moshe Gerstenhaber, Mike Gianino, "Op Amp Doubles As Instrumentation Amplifier," **EDN**, September 15, 1994, p. 164.
14. Elliott Simons, "Negative Resistor Cancels Op Amp Load," **EDN**, May 24, 2001, pp. 108.
15. Moshe Gerstenhaber, Chau Tran, "Composite Instrumentation Amp Extends CMRR Frequency Range 10×," **Electronic Design**, February 4, 2002, pp. 65, 66.

SECTION 6-6

Composite Amplifiers

Walt Jung

The term "composite op amp" can mean a variety of things. In the most general sense of the word, any additional circuitry at either the input or the output of an op amp could make the combination of what is termed a composite amplifier. This can be a valuable thing, as often such enhancements allow new performance levels to be realized from the resultant amplifier.

Some straightforward op amp performance enhancements of this type of have already been treated elsewhere in this book. For example, within the "Buffer Amplifiers" section of this chapter, as well as some of the specialized buffers in the "Audio" section of this chapter are found what could be termed composite op amps. In these examples, a standard output stage buffering design step is to utilize a unity-gain buffer, running on the same supplies as the op amp being buffered. So long as this buffer has sufficient bandwidth, this is an easy and straightforward step—insert the buffer between the op amp and the load, connect the feedback around the op amp plus buffer, and that's it.

A very useful means of increasing op amp performance can be obtained by blending the performance advantages of two ICs, or a standard op amp IC and discrete transistors. Such a combination is known as a *composite amplifier.* In special situations, a well-designed composite amp can often outperform standard op amps. The reason this is true is that the composite amplifier can be optimized for a unique and specialized performance, a combination that may not be available (or practical) in a standard op amp.

However, whenever an input (or output) circuit is added to an op amp that provides additional voltage gain, then the open-loop gain/phase characteristics of the composite op amp may need to be examined for possible stability problems. Note that this applies even when a unity-gain stable op amp is used within the composite, because the additional voltage gain raises the net open-loop gain of the combination. This will be made clearer by some circuit examples that follow.

In this section composite op amp circuits are described which fall into these categories:

- Multiple Op Amp Composite Amplifiers
- Voltage-Boosted Output Composite Amplifiers
- Gain-Boosted Input Composite Amplifiers
- A Nostalgia Composite Op Amp

These sections follow, with one or more circuit examples of each type.

Multiple Op Amp Composite Amplifiers

The simplest composite amplifier form utilizes two (or more) op amps, merged into a single equivalent composite. This is usually done for reasons of offset control, although in some instances it may be for increased gain capability, more output swing, and so forth.

Two-Op-Amp Composite Amplifier

The most flexible composite amplifier version combines two op amps in such a way that both signal inputs are still accessible to an application. A good example of this is the circuit shown in Figure 6-170 (see Reference 1).

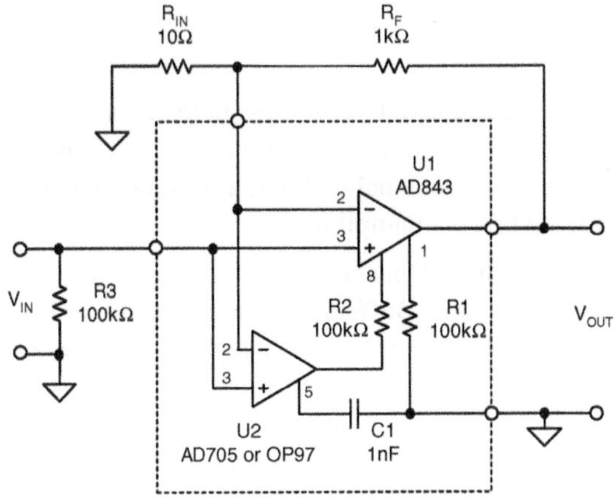

Figure 6-170: Low noise, low drift two-op-amp composite amplifier

In this circuit, U1 is a high speed FET input op amp, the AD843. While FET input devices are typically excellent for fast data acquisition applications, their offset and drift are often higher than the best bipolar op amps. By combining the fast AD843 with a low offset and low drift, super-β input device for U2, the best of both worlds is achieved. Offset and drift are reduced essentially to the maximum U2 specification levels—an offset of 60 μV, a drift of 0.6 μV/°C, and 100pA of bias current (for the OP97E). The composite op amp formed is the dotted box outline, and is applied as a 4-pin op amp.

Both the U1 and U2 op amps have inputs connected in parallel, and both amplify the signal. Device U1 drives the load and feedback loop directly. U2, however, drives the offset null input of U1, via a 100 kΩ resistor connected to Pin 8, R2. R1 provides a complementary resistance at the opposite offset input, Pin 1. C1 is used to over-compensate U2, at Pin 5. Note that the three just described pins are unique to the AD843, and either the AD705 or the OP97 devices. The U1/U2 signal inputs, output, and power supply pins are all standard, and the circuit operates on conventional ±15 V supplies.

The circuit as shown has a noninverting gain of 101×, as determined by R_F and R_{IN}. However, other applications are also possible, both inverting and noninverting in style. A detailed technical analysis of this circuit was presented in Reference 2.

Low Voltage Single-Supply to High Output Voltage Interface

There are numerous cases when an op amp designed for low (or single) supply voltage operation might need to be interfaced into a system operating on higher voltage and/or dual power supplies. An example would be the numerous low voltage chopper-stabilized op amps, which, without some means of easy interfacing, would simply not be available for use on high voltage supplies.

The circuit of Figure 6-171 shows how a low voltage, single-supply chopper-stabilized amplifier, the AD8551, can be used on a ±15 V supply system.

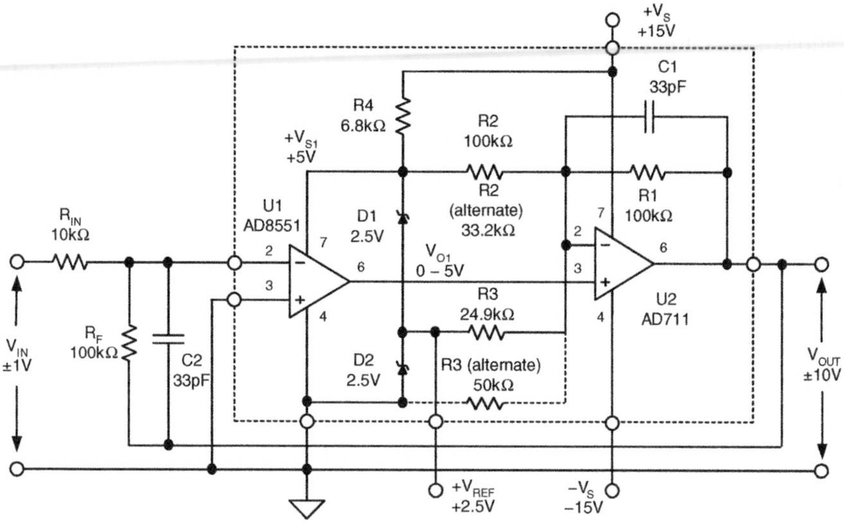

Figure 6-171: Chopper-stabilized 160 dB gain, low voltage single-supply to high output voltage composite amplifier

In this circuit, the U1 AD8551 operates as a precision input stage of the composite amplifier, working from a local 5 V supply generated from the main 15 V rail by reference diodes D1, D2. This satisfies the supply requirements for the U1 stage, with the R4 value selected to supply the required current.

To interface the 0 V to 5 V output swing of U1 to a ±10 V range, the U2 output stage operates as both a level shifter and a gain stage. A nominal gain of 6x is provided, with a dc offset providing the required level shifting. With the R1–R3 resistors and the 5 V supply used as a reference, the gain and level-shifting is accomplished. The gain of 6x translates back to a 0.833 V to 4.167 V positive output swing from U1, a range that even non rail-rail output op amps can most likely accommodate.

A virtue of this circuit is that the output of U1 is not loaded, and thus it operates at its full characteristic gain. For the AD8551, this is typically 145dB. When the additional 15.6dB of the U2 stage is added, the net open-loop gain of the composite amplifier formed is more than 160 dB. Further, this high gain will be maintained for relatively lower impedance loads, by virtue of the fact that a standard emitter-follower type output stage is used within U2. So, the dc accuracy of this composite amplifier will be very high, and will also be well-maintained over a wide range of loads due to the buffering of stage U2.

The voltage-offsetting network as shown uses two 2.5 V reference diodes, which provides a 2.5 V V_{REF} output from D2. This is usually a handy asset to have in any 5 V supply device system. Alternately, a single 5 V reference diode can be used, with the alternate values and connections for R2 and R3 substituted.

To compensate the system for the additional voltage gain of U2, two feedback capacitors are used, C1 and C2. *Note that for anything other than this exact circuit, one or both of these capacitors may require adjustment.* This is best done by applying a low-level square wave to the input such that the final output is on the order of 100 mV p-p or less, and verifying that the output step response is well damped, with minimal overshoot.

C1 ensures stability for stage U2, and C2 provides overall bandwidth control for the main feedback loop. In general, the U2 amplifier should have more bandwidth than the U1 stage. However, the relative dc accuracy of U2 is not at all critical. The AD711 is shown as one possible choice, but many other types can also be used. While not shown for simplicity, conventional supply bypassing of the composite amplifier should be used.

System-wise, this composite amplifier behaves as a ±15 V powered op amp with an input CM range equal to the specification of the U1 device in use. Overall loop feedback is provided as with any conventional feedback stage, i.e., by R_F, R_{IN} and C2. The circuit is applied by treating the parts within the dotted box as a single op amp, with the external components adjusted to suit a given application. An application caveat is that saturation of U1–U2 should be avoided, due to the longer overload recovery. This can be addressed with a 11 V–12 V back-back clamping network across the feedback impedance.

Although the example hookup shown is a gain-of-10 inverter, other inverting configurations such as integrators and also noninverting stages are possible. The caveat here is that the CM range of U1 must be observed. However this is likely no handicap; as such an amplifier is most likely to be used with very high gain and low CM input voltages. It could for example be used as a 5 V-powered bridge amplifier with a ±10 V output range.

There are of course any number of other op amp input and output devices that will work within this general setup. A more general-purpose low voltage part for U1 would be the AD8541. Offset voltage will be higher, and gain less, vis-à-vis the AD8551. For optimum dynamic range and linearity, the biasing of the U1 stage output is centered within the total U1 supply voltage V_{S1}, which can be different than 5 V if needed. This feature is provided by R1–R3. The absolute values of these resistors aren't critical, but they should be maintained as to their ratio. They can be part of a common 100 kΩ array for simplicity.

Voltage-Boosted Output Composite Amplifiers

A number of schemes are useful towards boosting the output swing of standard op amps. This can be either to achieve greater swing (i.e., closer to the rails), or, to develop swings greater than normally possible with standard ICs, i.e., ≈40 V swings. In both cases it may also be desirable to increase load drive to 100mA or more.

Voltage Boosted, Rail-Rail Output Driver

A common requirement in modern system is the rail-rail capable op amp. But all op amps aren't designed with rail-rail outputs, so this may not be possible in all instances. Of course, it makes good sense to utilize standard off-the-shelf rail-rail IC op amps, whenever they meet the application requirements. Nevertheless, it is possible to add an output stage to a standard op amp device that may itself not be rail-rail in function. By using common-emitter (or common-source) discrete transistors external to the op amp, a rail-rail capability is realized. An example designed in this fashion is Figure 6-172.

Within this circuit Q1 and Q2 are the complementary buffer transistors that provide the rail-rail output swing. The circuit works as follows: Q1 is driven by the voltage drop across R4, and diode-connected Q3. This voltage is developed from the positive rail supply terminal of U1, so the quiescent bias current of Q1 will be related to the quiescent current of U1. Similarly, Q2 is driven from R3 and Q4, via the negative rail terminal of U1. The Q1–Q3 and Q2–Q4 pairs make up current mirrors, developing a quiescent bias current that flows in Q1-Q2. The U1 quiescent current is about 400 μA, and with the resistance values shown, the Q1-Q2 bias current is about 10 mA.

The output stage added to the U1 op amp adds additional voltage gain, and a current gain boost of 25 times, essentially the ratio of R4/R9 and R3/R10. Thus for a 100 mA output from Q1-Q2, U1 only supplies 4 mA. The swing across R2 is relatively low, allowing operation on low voltage supplies of ±6 V, or up to ±15 V.

The simulation data of Figure 6-173 illustrates some salient characteristics of the composite op amp while driving a load of 85 Ω. The open loop gain of the circuit is shown by the topmost, or composite gain curve, which indicates a low frequency gain of over 130 dB, crossing unity gain at about 630 kHz. The intermediate curve is the OP97 op amp gain characteristics. The difference between this and the upper curve is the added

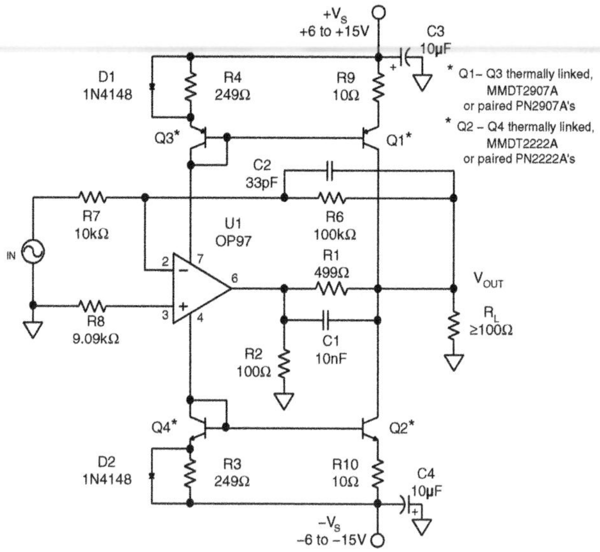

Figure 6-172: Voltage boosted rail-rail output composite op amp

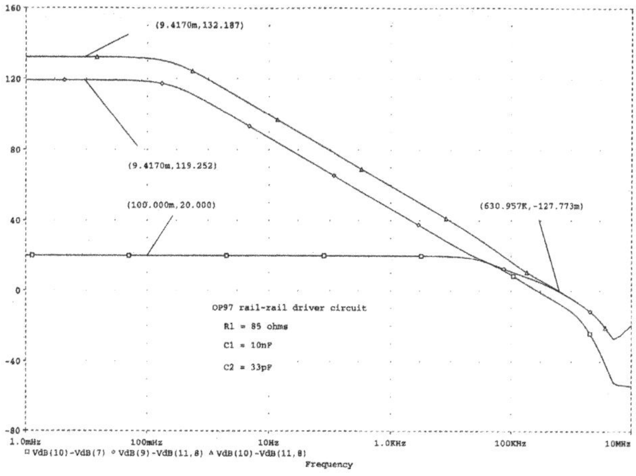

Figure 6-173: Gain (dB) versus frequency characteristics of Figure 6-172 composite op amp

gain, which is about 13 dB. The lowest curve indicates the closed-loop gain versus frequency characteristics of the composite op amp, which is 20 dB in this case, as set by R6 and R7 (as in a standard inverter).

There are a couple of critical points in setting up this circuit. Bandwidth can be controlled by C1 and C2. C1 reduces the added gain at high frequencies, which can be noticed from the composite gain curve, starting below 100 kHz. C2 reduces the closed-loop gain, starting about 50 kHz. For greater closed-loop gains, C2 may not be needed at all.

Bias control is achieved by the use of thermal coupling between the dual current mirror transistors. The easiest way to accomplish this is to use packaged dual types, either SOT-363 or SM-8 devices (see References

3 and 4). Alternately, TO92 equivalent PN2222A and PN2907A types can be used, with the two flat sides facing and clipped together.

The circuit as shown drives a 100Ω load to within 2 V of the rails, limited by the drop across R9 and R10. Current limiting is provided by a shunt silicon diode across R4 and R3, either with 1N4148 diodes, or diode-connected transistors. This limits peak output current to about ±60 mA. More output current is possible, by adding additional like devices in parallel to Q1 and Q2, with additional 10 Ω emitter resistors for each.

A point that should be noted about the booster circuit of Figure 6-172 is that *the biasing is dependent upon the quiescent current of the op amp*. Thus, this current must be stable within certain bounds, otherwise the idle current in Q1-Q2 could deviate—either too low (causing excess distortion), or too high, causing overheating. So, changing the U1 op amp isn't recommended, unless the biasing loop is re-analyzed for the new device.

Another point is that this type of circuit, which uses the power pins of the op amp for a signal path, *may not model at all in SPICE.* This is due to the fact that many op amp SPICE models do *not* model power supply currents so as to reflect output current—so be forewarned. However, the ADI OP97 model does happen to model these currents correctly, so the reader can easily replicate this circuit with the OP97 (as well as many other ADI models). Discussion of these models can be found in Chapter 7 of this book.

High Voltage Boosted Output Driver

With some subtle but key changes to the basic voltage-boosted composite amplifier of Figure 6-172, output swing can be extended even higher, more than double the standard ±10 V swing for ±15 V rails. A basic circuit that does this is shown in Figure 6-174.

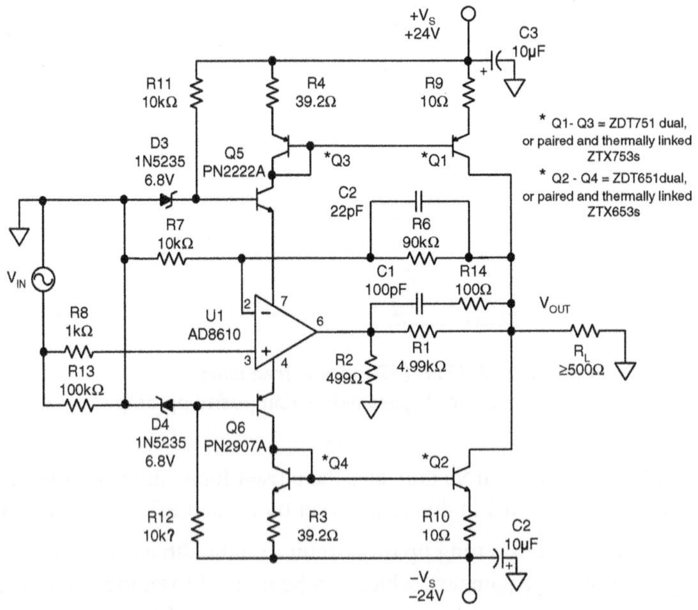

Figure 6-174: High voltage boosted rail-rail output composite op amp

This circuit can readily be recognized as being similar to the lower voltage counterpart of Figure 6-172. To achieve higher voltage capability, the U1 op amp is operated from a pair of combination level-shift/regulator transistors, Q5 and Q6. These are biased in turn from the D3 and D4 zener diodes at their bases, to ±6.8 V, respectively. The op amp rails are then simply ±6.2 V, while the main rails $+V_S$ and $-V_S$ can be virtually any potential, as will be ultimately limited by the Q1 and Q2 voltage/power ratings.

The op amp supply current passes through Q5 and Q6, driving the bases of Q1-Q2 as previously. To accommodate the higher ±24 V supplies, the standard 2222A and 2907A parts used in Figure 6-172 are replaced by higher voltage parts, the dual ZDT751 and ZDT651 (see Reference 4). Thermal matching is best maintained by using these dual types, but comparable TO92 pairs can also be used, for Q1-Q3, ZTX753s, and for Q2-Q4, ZTX653s (see References 5 and 6). In any case, a large area PCB land (i.e., 1–2 square inches) should be used to for the Q1-Q2 collectors for heat sinking purposes.

In this new circuit, an AD8610 op amp is used for U1, offering very low offset voltage, and higher speed. The quiescent current of the AD8610 is typically 2.5 mA. A 4/1 gain is used in this circuit, as established by R4/R9 and R3/R10. The idle current in Q1-Q2 is therefore about 10 mA, leading to a ≈240 mW dissipation each, on ±24 V supplies. This is low enough to not require a heat sink. However, the copper land area described above should be provided on the PCB around Q1-Q2 for heat sink purposes, tied electrically to their collectors. These measures, plus the active current limiting, help protect the output devices against shorts.

This circuit has a novel method of current limiting. As operated on ±6.2 V, the AD8610 will swing just over ±5 V. In driving R2 to this limit, ±10 mA of current will be delivered to the current mirrors, resulting in a maximum output current four times this, or ±40 mA. This is just about the dc safe-area limit of the Q1 and Q2 devices as used on ±24 V. For low impedance loads below 500 Ω, the maximum output voltage is a product of the 40mA limit and the load (for example 40 mA into 100 Ω yields 4 V peak). The maximum voltage swing into a 500 Ω load is then about ±20 V, again, as determined by the current limiting. Into higher impedance loads, the swing is proportionally greater, up until the point Q1-Q2 reach their saturation limits.

Although the circuit is quite versatile as shown, many other options are also possible. Other op amps can be used but, as noted before, the idle current should be taken into account. This is even more critical on higher voltage supplies, as it directly affects the power dissipated in Q1 and Q2.

For higher output currents from Q1-Q2, additional similar transistors can be paralleled, each with individual emitter resistors like R9 and R10. This will be practical for scaling up current by a factor of two to three times (assuming one additional package of the ZDT751 and ZDT651 types).

For ampere level current outputs, an additional current gain stage in the form of a complementary emitter follower can be added, driven from the Q1-Q2 collectors, with a 1:1 gain in the current mirror, and appropriate emitter follower biasing. With this step, the circuit will have been converted into a complete power amplifier. Details of this are left as an exercise for the reader. However, a good starting point might be the Alexander power amplifier topology (see Reference 7).

Gain-Boosted Input Composite Amplifiers

One of the most popular configurations used to enhance op amp performance is the gain-boosted input composite op amp. Here, a preamp gain stage is added ahead of a standard IC op amp, allowing greater open-loop gain, lower noise, and other performance enhancements. Another worthy improvement is the thermal isolation between the critical input stage, and the IC output stage that delivers the load current. The preamp can be a matched pair of bipolar transistors (NPN or PNP), or JFETs of either N or P types.

Prototype Bipolar Transistor Gain-Goosted Input Composite Amplifier

For illustration of the basics, a prototype example composite amplifier is the two-stage op amp of Figure 6-175. This circuit uses a matched NPN differential pair as a preamp stage ahead of U1, a standard AD711 type op amp. The preamp stage adds voltage gain to that of U1, making the overall gain higher, thus lowering gain-related errors.

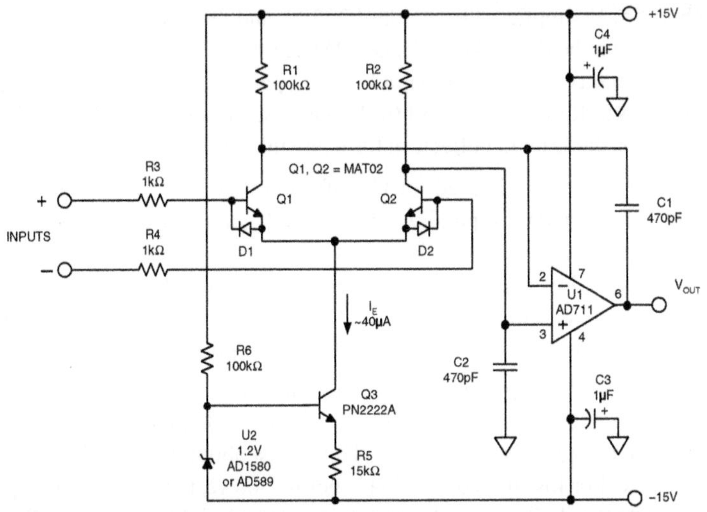

Figure 6-175: Bipolar transistor gain-boosted input composite op amp

Because of the added gain, the relative precision of the output op amp isn't very critical. It can be selected for sufficient output drive, slew rate and bandwidth. Within a given application the composite amplifier has overall feedback around both stages. Note—for this and following circuit examples, the op amp is *uncommitted* (i.e., external feedback).

In this example, a bipolar transistor differential pair, Q1-Q2 is loaded by a stable, matched load resistor pair, R1 and R2 (where R1 = R2 = R_L). The exact value of these resistors isn't overly critical, but they should match and track well. R1 and R2 are selected to drop 2 V–3 V at ½ I_E. For a 2 V drop, a suitable R_L is then:

$$R_L = \frac{4}{I_E} \qquad \text{Eq. 6-45}$$

Here, Q1 and Q2 operate at ≈20 µA each, so 100 kΩ values work for R1-R2. Note that in operation, *the second stage op amp must be capable of operating with input CM voltages of 2 V below the* $+V_S$ *rail.* This criterion is fine for many PFET input amplifiers such as the AD711, but others should be checked for CM input voltage compatibility. Note that similar preamp stages can also be built with PNP bipolars, or with JFETs, and some of these will be described later.

The dc or low frequency gain of the preamp stage, G1, can be quite high with bipolar transistors, since their g_m is high. G1 can be expressed as:

$$G1 = \frac{R_L I_E}{2V_T} \qquad \text{Eq. 6-46}$$

where V_T = KT/q (≈0.026 V at 27°C).

In this example, at 27°C, G1 is about 77 times (37.7 dB). Overall numeric gain is, of course, the product of the preamp gain and the U1 op amp gain. The minimum AD711 dc gain is 150,000, so the gain of the composite is more than 11.5 million (≥141 dB).

As the preamp stage provides additional gain, this extra gain must be phase controlled at high frequencies for unity-gain stability of the composite amplifier with applied feedback. In this circuit, compensation caps C1 and C2 provide this function with U1 connected as a differential integrator.

The unity gain frequency, Fu, can be expressed approximately as:

$$F_u \approx \frac{I_E}{4\pi C_C V_T} \qquad \text{Eq. 6-47}$$

where π is 3.14, C1 = C2 = C_C.

The performance of this composite op amp is illustrated in the gain and phase versus frequency simulation plot of Figure 6-176. The additional gain of the preamp raises the net dc gain to ≈149 dB, and the unity gain crossover frequency is shown to be ≈252 kHz, both of which generally agree with the estimated figures. The phase margin ϕ_m at 252 kHz is about 75 degrees, which is conservative. This op amp should be stable for all closed-loop gains down to unity (in fact, C1 and C2 could possibly be lowered).

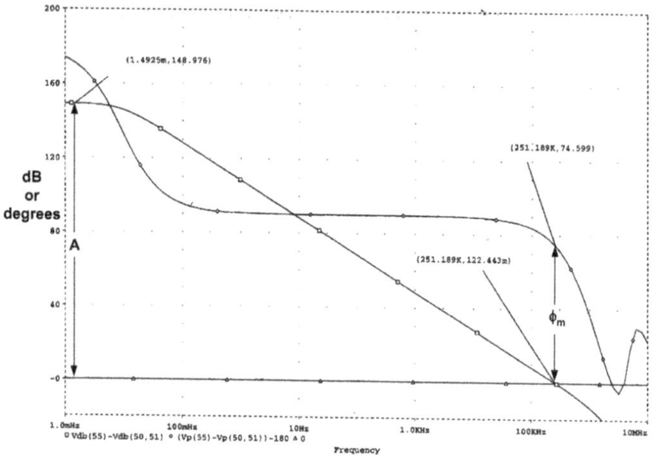

Figure 6-176: Gain/phase versus frequency for Figure 6-175 composite op amp

Slew rate of the composite op amp can be no higher than the specified SR of output stage U1. For cases where the effective SR is to be lower, it can be estimated as:

$$SR \approx \frac{I_E}{C_C} \qquad \text{Eq. 6-48}$$

Using Eqs. 6-47 and 6-48, the chosen values calculate a bandwidth of 260 kHz, and an SR of just under 0.1 V/µs. Actual bandwidths of 236 kHz–238 kHz were measured on 4 op amps for U1 (AD711, AD820, LM301A and LF356), while SR was +0.085 V/µs and –0.087 V/µs. As would be expected, the least bandwidth was measured with the lowest bandwidth U1 device, an LM301A. This demonstrates the relative insensitivity to U1 bandwidth.

With high gain input transistors, the bias current can be low. Generally, this will be:

$$I_B = \frac{I_E}{2H_{FE}} \qquad \text{Eq. 6-49}$$

Where I_B is the bias current of either Q1 or Q2, and H_{FE} is their dc gain. The MAT02 diodes protect against E-B reverse voltage, while the 1 kΩ resistors limit diode current.

Bias currents of 30 nA were measured with a MAT02 for Q1-Q2. Similar results can be obtained with high gain discrete transistors, such as 2N5210s. Offset voltage, however, is a different story. Monolithic duals such as the MAT02 will be far superior for offset voltage, with a V_{OS} specification of 50 µV. Nonmonolithic packaged duals will also function in this circuit, but with degradation of dc parameters versus a monolithic device such as the MAT02. As can be noted from the numbers quoted above, speed isn't a major asset of this amplifier. However, the dc performance is excellent, as noted, placing it in an OP177 class for gain.

The Q1-Q2 emitter current, I_E, can be established by a variety of means. The most general form is the U2, Q3, and R5 arrangement. This works for a wide range of inputs, and also offers relatively flat gain for a bipolar Q1-Q2 gain stage, since the PTAT current from Q3 compensates the temperature–related gain (Eq. 6-46). For those applications where the input of the amplifier is operating in an inverting mode, a more simple solution would be a resistor of 332 kΩ from the Q1-Q2 emitters to $-V_S$.

Of course, as a practical matter one wouldn't use the complex Figure 6-175 circuit, if an OP177 (or another standard device) could do the job more simply or inexpensively. Nevertheless, the above discussion illustrates how one can *tailor* a composite op amp's characteristics, to get exactly what is needed. The composite op amp circuit of Figure 6-175 could be used with a rail-rail output stage device for U1 (AD820), or with a very high current output stage (AD817, AD825), or any other performance niche not available from standard devices. Examples of these performance options follow.

Low Noise, Gain-Boosted Input Composite Amplifier

One of the more sound reasons for adding a preamp stage before a standard op amp is to lower the effective input noise, to a level lower than that of readily available IC devices. Figure 6-177 shows how this can be achieved within the same basic topology as described above for the Figure 6-175 prototype composite. As

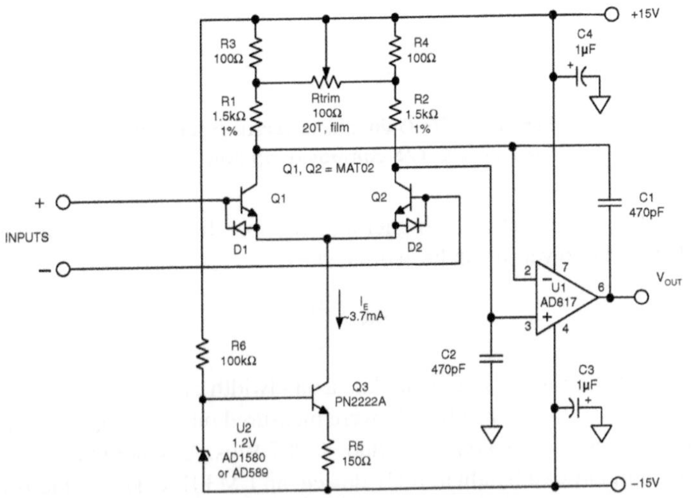

Figure 6-177: Low noise gain-boosted input composite op amp

will be noticed, this circuit is similar to the prototype, with the exception of the added offset trim network, a higher bias level for I_E, and a faster output op amp, U1. Rtrim nulls the offset for best dc accuracy in critical applications. If this isn't necessary, connect R1-R2 as in Figure 6-175.

By raising the current level of Q1-Q2 by roughly a factor of 100x compared to the prototype, the effective input noise of Q1-Q2 is dramatically lowered. At the operating current level of 3.7 mA, the MAT02 achieves an input voltage noise density $< 1\ nV/\sqrt{Hz}$ (see Reference 10). To eliminate their added noise, the series base resistors are dropped. Both bandwidth and SR are also improved in this circuit, since they are both proportional to I_E. The estimated bandwidth and SR for this circuit are 24 MHz and 7.9 V/μs, respectively. Measurements show about 21.2 MHz for bandwidth, and a SR close 7.6 V/μs. Both the actual bandwidth and SR are less than the AD817 specifications of 50 MHz and 350 V/μs. The circuit as shown is close to unity-gain stable, with 44° of phase margin at the unity gain frequency. Of course, very low noise amplifiers such as this will often be applied at some appreciably higher gain, for example 10, 100, or more. When this is the case, then C1 and C2 can be reduced, allowing greater bandwidth and SR.

An even lower noise op amp can be achieved simply by adding one or more low noise pairs parallel to Q1-Q2, and operating the combination at 6 mA of current. See References 11 and 12 for examples to achieve $0.5\ nV/\sqrt{Hz}$ or less noise.

JFET Transistor Gain-Boosted Input Composite Amplifier

The circuit of Figure 6-178 illustrates an alternative compensation method for composite op amps. This technique has the advantage of simplicity, but also the disadvantage of being conditionally stable. This technique goes back to the very earliest days of IC op amps, when discrete or monolithic matched FET pairs were used ahead of a standard IC op amp such as the 741, 709, and so forth. Further details are contained within References 13 and 14. The example here isn't offered as a practical example, inasmuch as so many superior IC FET op amps are available today. However, it does give insight into this type of compensation, which is applicable either to FET or bipolar input stages.

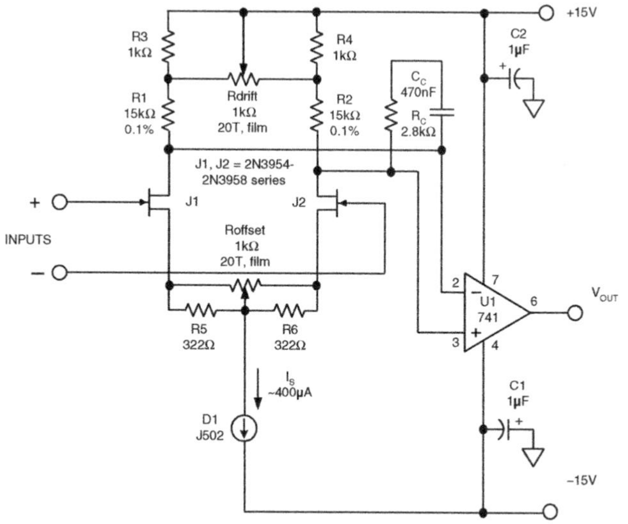

Figure 6-178: JFET transistor gain-boosted input composite amplifier

There are practical reasons why this type of FET input composite amplifier isn't used today. One is that to do it correctly involves many involved trims, another is that it requires a considerable number of parts. Dual FET devices don't come with sub-mV offsets, as do bipolars, so there is the need to trim out offset. Roffset does this, for J1-J2 V_{OS} up to 50 mV. For lowest drift, the drain currents should also be trimmed, via Rdrift.

N-channel JFET duals such as the 2N3954 and J401 series are specified for operation at a total I_S of 400 µA, or 200 µA/side. Their transconductance is much lower than a bipolar; for these conditions; it is typically ≈1400 µS. Therefore the gain of this preamp will typically be much lower than would a bipolar stage. With matched load resistors, gain is:

$$G1 = \frac{R_L \bullet g_{fs}}{2} \qquad \text{Eq. 6-50}$$

where g_{fs} is the specified JFET transconductance at $I_S/2$.

For the conditions shown, G1 works out to be 10.5 (20.4 dB). Note—if used, the Roffset network reduces gain somewhat, and Eq. 6-50 *doesn't* take this into account.

Compensation for this composite amplifier is via the RC network, R_C-C_C. This network reduces the gain of the preamp to unity above the zero frequency, which allows the aggregate open-loop response to then assume that of the U1 amplifier before the unity-gain crossover. It is chosen by setting R_C as:

$$R_C = \frac{4}{g_{fs}} \qquad \text{Eq. 6-51}$$

where g_{fs} is again the specified JFET transconductance at $I_S/2$.

In this case, R_C works out to be 2.8 kΩ. C_C is then chosen to provide a zero at some frequency that should be a very small fraction of the U1 op amp's unity gain frequency. The importance of this point will be made clearer by various open-loop response shapes.

It should be recalled that the classic open-loop response of an unconditionally stable op amp is a constant –6 dB/octave for gain, with an associated 90° phase shift. For such a device, any 1/β closed-loop response that intersects this open-loop response will be stable. For example, the 741 response (Δ), as so marked in Figure 6-179, is such a characteristic. But, as can also be noted from Figure 6-179, the gain/phase response of a composite op amp compensated as in Figure 6-178 just isn't a simple matter.

In the case of the added preamp stage and the R_C-C_C network compensation, the composite gain response (□) assumes *a multiple-slope response.* Associated with this gain response, note also that *the phase characteristic (o) varies radically with frequency*. In particular, the phase dip around 46 Hz signifies a frequency where a loop closure could be problematic, as the phase margin is only 40° at this point.

Faced with this type of open-loop gain/phase response, a designer needs to careful in crafting the closed-loop gain configuration. The first step is to decide what level of closed-loop gain is required by the application. Given that, an ideal 1/β curve can be drawn on a Bode diagram, to determine the rate-of-closure at the intersection. For optimum stability, it is desirable that this intersection occurs with a relative –6 dB/octave between the open-loop gain curve and an ideal 1/β curve. Note that if such a 1/β curve were drawn on Figure 6-179 at a 100 dB gain, it would intersect in a –12 dB/octave region. This is because C_C is 470 nF in this example, which places the phase dip at 46Hz, with a composite gain curve which, as noted, is dropping at a rate greater than 6dB/octave. So, with the proposed gain curve intersecting in this region, stability could be marginal.

On the other hand however, with the R_C-C_C "phase funnies" forced down to a low frequency that corresponds to very high closed-loop gains (i.e., ≈100 dB), the practical potential for instability is minimized.

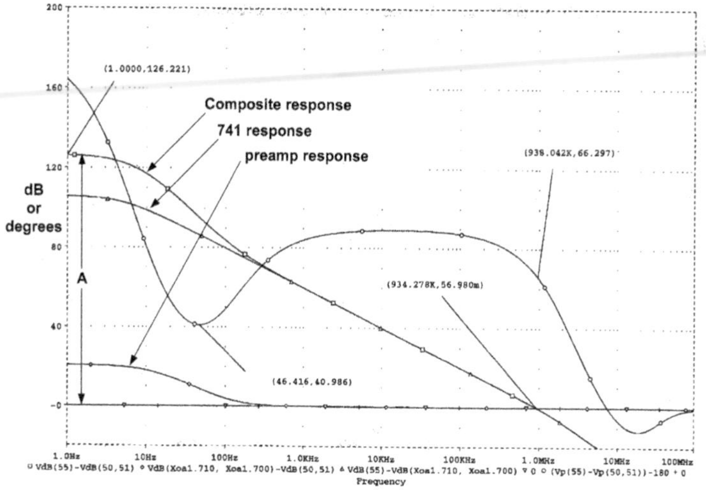

Figure 6-179: Gain/phase versus frequency for Figure 6-178 composite amplifier

Note also that it doesn't make good sense to build a 100 dB gain feedback amplifier based on limited open-loop gain such as Figure 6-179.

On the other hand, if C_C were smaller, this wouldn't necessarily be the case, because the associated phase dip would then move upward in frequency. This could wreak havoc with loop closures at more practical closed-loop gains. In contrast to this, with C_C sized as shown at 470 nF, the phase anomalies are confined to very low frequencies, yet the added dc gain of the preamp is still available. Loop closures at frequencies above ≈200 Hz (at closed-loop gains of 80 dB or less) see a high frequency composite response closely resembling the 741 (□ and Δ, respectively), and should thus be stable.

Watch out for the time domain response!

It should also be noted that there is a more subtle side effect related to the R_C-C_C method of compensation, as illustrated in Figure 6-178. Simply put, this is the fact that *the time domain response of the resulting composite amplifier will be marred, compared to that of a classic 6 dB/octave roll off* (see Reference 15).

So, wherever time domain response is critical, then the more conservative, unconditionally stable compensation method of Figure 6-175 should be used. A case in point using this method with a FET preamp is the next composite amplifier.

In summary, for the ac performance characteristics as a composite amplifier, the circuit of Figure 6-178 offers a gain raised higher that that of U1, or by about 20 dB (126 dB total) using the 741. At high frequencies, the overall gain bandwidth properties of this composite mimics the U1 amplifier, when the R_C-C_C time constant is relatively large.

Dc performance limitations

The dc input characteristics of this circuit will be those specified for the J1-J2 pair, with typical room temperature bias currents of 50 pA or less. Common-mode rejection will be limited by the J1-J2 specifications, and typically no more than about 80 dB, over a limited CM range. This could be improved by cascoding the J1-J2 pair, but again, given the availability of such FET-input IC amplifiers as the AD8610, this would be a questionable design for a precision FET amplifier. In Figure 6-178, the current source used for I_S is a simple

FET current limiter diode (see Reference 16). This offers simple, two-terminal operation, at a current level optimum for the J1-J2 pair.

Low Noise JFET Gain-Boosted Input Composite Amplifier

An alternative method of executing a JFET input gain-boosted composite is to operate the input differential pair into an output op amp stage that acts as a differential integrator (i.e., similar to the Figure 6-175 prototype, insofar as the compensation). The circuit of Figure 6-180 is such an example, one that is also optimized for low noise operation, medium speed, and higher output current.

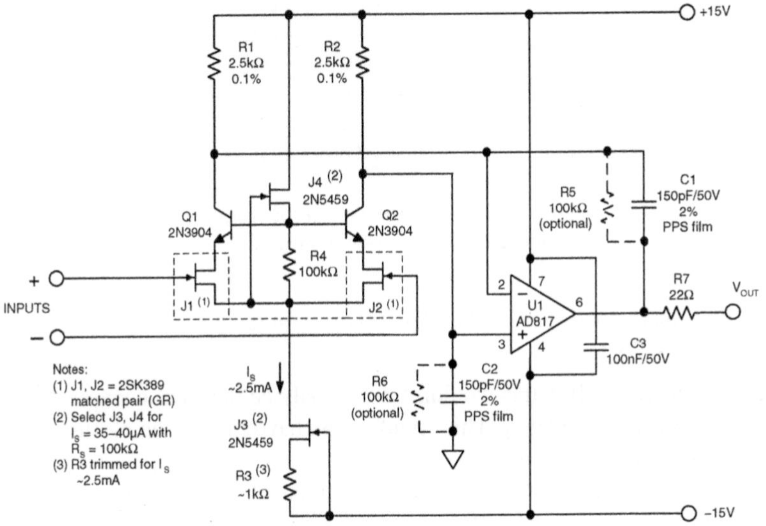

Figure 6-180: Low noise JFET gain-boosted input composite amplifier

The design uses a low noise JFET pair as the gain stage, the 2SK389 device (see Reference 17). Biased for drain currents of more than 2 mA, this device pair is capable of achieving an input voltage noise density of less than 1.5 $nV/\sqrt{Hz}$. The basic device is available in three I_{DSS} grades, GR (2.6–6.5 mA), BL (6–12 mA), and V (12–20 mA). The lowest noise will be found with use of the highest I_{DSS} parts, at the expense of course, of supply current. This design example can use any grade, by biasing J1-J2 for the GR parts (at an I_{DSS} of 2.5 mA). This still gives good noise performance for an FET-input amplifier (about 1.8 $nV/\sqrt{Hz}$), but at a still reasonable power supply drain.

A byproduct of the large geometry devices of this devices series is a relatively high capacitance. If this factor is not addressed, this large and nonlinear capacitance could cause distortion, for applications operating the circuit as a follower. To counteract this, the input stage of J1-J2 is cascoded, by the Q1-Q2 and J4 arrangement. This removes the major degradation of operation due to the J1-J2 capacitance, and it also stabilizes the dc operating points of J1-J2. From the output collectors of Q1-Q2 onward, the amplifier operates generally as the Figure 6-177 circuit previously described. An AD817 is used for U1, so as to take advantage of its wide bandwidth and high output current.

The unity-gain bandwidth of this circuit is about 15 MHz, but the open-loop gain is user selectable, by virtue of optional resistors R5 and R6. With these resistors connected, the composite amplifier open-loop bandwidth is ≈10 kHz, and open-loop gain is about 63 dB. These attributes make it well-suited for audio applications, for example. Without R5 and R6, the open-loop gain is more like that of a conventional op amp, with a gain of more than 100 dB at low frequencies.

The open-loop response for R5 and R6 open is shown in Figure 6-181. In this simulation the load resistance was 600 Ω. As can be noted, the response is clean, without phase aberrations. Phase margin at the unity-gain crossover frequency is about 63°, and the low frequency gain is about 104 dB.

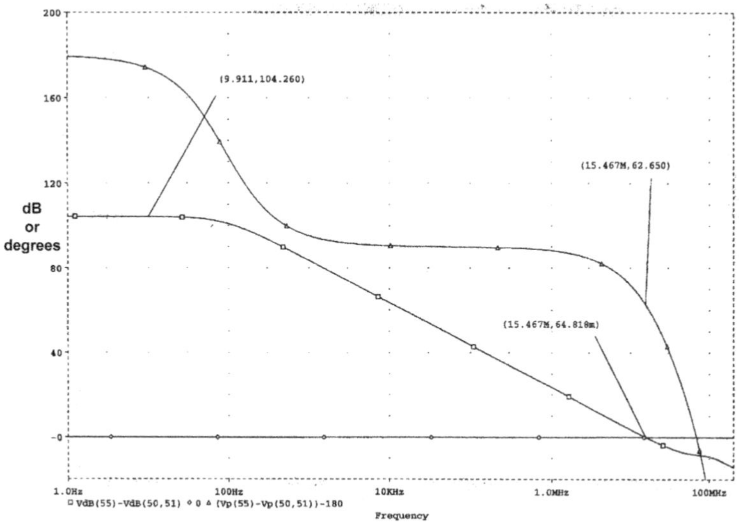

Figure 6-181: Gain/phase versus frequency for Figure 6-180 composite amplifier

Although this circuit does have some excellent ac characteristics, it should be noted that it is *not* a general-purpose op amp circuit. One reason for this is that the cascode input stage is a two-edged sword. While it reduces capacitance and improves distortion, it also limits the allowable CM input range. The positive swing headroom is limited by roughly the dc drop across R1 and R2, plus that of the cascode, or 3 V + 4.5 V. This means the most positive CM input should be less than about 5 V peak, or 3.5 V rms. This of course won't be a practical limitation for noninverting amplifiers with noise gains of 5x or more, or for low level preamps with high gains of 100x or 1000x.

As compensated in Figure 6-180, the composite op amp should be unity-gain stable. At closed-loop gains appreciably higher than about 5x, a reduction of C1-C2 can be considered, which will allow greater bandwidth and SR to be realized. Offset of the J1-J2 pair can be as high as 20 mV, so offset trim may be in order for dc-coupled applications. For lower noise, a high I_{DSS} grade for J1-J2 should be used, with I_S raised to 5 mA or more. R1-R2 will need to be lowered, and C1-C2 raised, in proportion.

"Nostalgia" Vacuum Tube Input/Output Composite Op Amp

In keeping with the theme of this book's History section, the final composite amplifier design for this section uses venerable vacuum tube devices, which formed the basis of the first ever op amps. Today however, designing a vacuum tube op amp has some advantages, vis-à-vis the early days—transistors didn't exist. Thus the nostalgia op amp shown in Figure 6-182 incorporates techniques of both today as well as yesteryear.

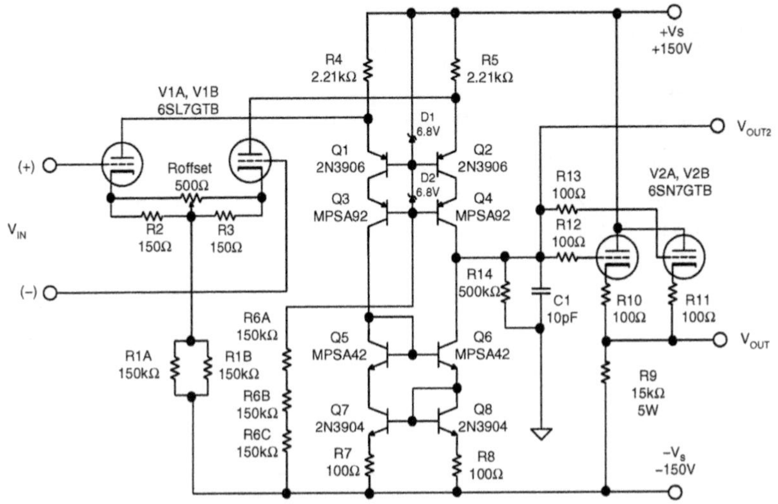

Figure 6-182: "Nostalgia" vacuum tube input/output composite op amp

Note that this particular circuit should be taken as a design exercise rather than a practical example. Yet, it was still a lot of fun to design using available SPICE models (see Reference 18). As such, it offers some insights not available to early op amp designers.

Here V1 is a dual triode input stage, using the high gain 6SL7 (or alternately, the close cousin 12AX7 miniature). It is operated here as a linear transconductance, long-tailed differential pair. Rather than using conventional plate loading, the output signal current from V1A–V1B is passed to a folded cascode stage Q1–Q4, which is loaded by a linear, high voltage current mirror, Q5–Q8. The transistors of the current mirror are also cascaded, both for higher output impedance as well as for required high voltage capability. A regulated 6.3 V heater supply for V1 is suggested, for highest stability.

Voltage gain of this one stage op amp is approximately equal to the V1A–V1B transconductance times the nodal impedance seen at V_{OUT2}. With R14 open, this impedance is very high, so gain can also be quite high (≈77 dB simulated). With R14 500 k, gain is about 53 dB. Open-loop bandwidth is established by the shunt capacitance at the high-Z node and R14, and measures about 7 MHz gain bandwidth in simulation. A 6SN7 dual cathode follower output stage for V2 allows up to 10 mA of load drive. Laboratory test results for this design are left as an exercise for the interested reader, and feedback is welcome.

References: Composite Amplifiers

1. Moshe Gerstenhaber, Mark Murphy, Scott Wurcer, "Composite Amp Has Low Noise, Drift," **Electronic Design**, January 21, 1993, pp. 62, 63.
2. Paul Brokaw, "Composite Amp Has Low Noise, Drift (Update)," **Electronic Design**, June 5, 2000.
3. Data sheet for **MMDT2907A Dual PNP Small Signal Surface Mount Transistor**, Rev B-2 and Data Sheet for **MMDT2222A Dual NPN Small Signal Surface Mount Transistor**, Rev C-2, www.diodes.com/products/.
4. Data sheet for **ZDT751 SM-8 Dual PNP Medium Power Transistors,** August 1997, and data sheet for **ZDT651 SM-8 Dual NPN Medium Power Transistors,** August 1997, www.zetex.com.
5. Data sheet for **ZTX752, ZDX753 PNP Silicon Planar Medium Power Transistors**, July 1994, www.zetex.com.
6. Data sheet for **ZTX652, ZDX653 NPN Silicon Planar Medium Power Transistors**, July 1994, www.zetex.com.
7. Mark Alexander, "The Alexander Current-Feedback Audio Amplifier," **Analog Devices AN211**. See also: Mark Alexander, "Current Feedback Audio Power Amplifier," **US Patent 5,097,223**, filed May 22, 1990, issued March 17, 1992.
8. Barry Kline, "Enhanced Op Amp Delivers 100 V p-p," **EDN**, September 5, 1985, pp. 309, 311–312.
9. Walter G. Jung, "Chapter 5, Amplifier Circuit Techniques," **IC Array Cookbook,** Hayden Book Company, 1980, ISBN 0-8104-0762-0.
10. Data sheet for **MAT02 Low Noise, Matched Dual Monolithic Transistor**, www.analog.com.
11. Andrew Jenkins, Derek Bowers, "NPN Pairs Yield Ultra-Low-Noise Op Amp," **EDN**, May 3, 1984.
12. Data sheet for **MAT03 Low Noise, Matched Dual PNP Transistor**, www.analog.com.
13. "Choosing and Using N-Channel Dual J-FETs," **Analog Dialogue,** Vol. 4, No. 2, pp. 4–9.
14. "TDN: Temperature Drift Nonlinearity—A New Dual-FET Specification," **Analog Dialogue,** Vol. 6, No. 1, pp. 13–14.
15. Robert I. Demrow, "Settling Time of Operational Amplifiers," **Analog Dialogue**, Vol. 4, No. 1. See also ADI application note AN359.
16. "The FET Constant-Current Source/Limiter," Siliconix AN103, March 10, 1997, www.vishay.com/brands/siliconix/SSFan.html.
17. Data sheet for **2SK389 Dual FET, Silicon Monolithic N-Channel Junction Type**, www.semicon.toshiba.co.jp/eng/solution/audio/pdf/e001543.pdf.
18. Duncan Munro's SPICE vacuum tube models, www.duncanamps.com.

CHAPTER 7

Hardware and Housekeeping Techniques

- Section 7-1: Passive Components
- Section 7-2: PCB Design Issues
- Section 7-3: Op Amp Power Supply Systems
- Section 7-4: Op Amp Protection
- Section 7-5: Thermal Considerations
- Section 7-6: EMI/RFI Considerations
- Section 7-7: Simulation, Breadboarding and Prototyping

CHAPTER 7

Hardware and Housekeeping Techniques

Walt Kester, James Bryant, Walt Jung, Joe Buxton, Wes Freeman

This chapter, one of the longer of those within the book, deals with topics just as important as all of those basic circuits immediately surrounding the op amp, discussed earlier. The chapter deals with various and sundry circuit/system issues that fall under the guise of system *hardware and housekeeping techniques*. In this context, the hardware and housekeeping may be all those support items surrounding an op amp, excluding the op amp itself. This includes issues of passive components, printed circuit design, power supply systems, protection of op amp devices against overvoltage and thermal effects, EMI/RFI issues, and finally, simulation, breadboarding, and prototyping. Some of these topics aren't directly involved in the actual signal path of a design, but they are every bit as important as choosing the correct device and surrounding circuit values.

SECTION 7-1

Passive Components

James Bryant, Walt Jung, Walt Kester

Introduction

When designing with op amps and other precision analog devices, it is critical that users avoid the pitfall of poor passive component choice. In fact, the wrong passive component can derail even the best op amp or data converter application. This section includes discussion of some basic traps of choosing passive components for op amp applications.

So, good money has been spent for a precision op amp or data converter, only to find that, when plugged into the board, the device doesn't meet spec. Perhaps the circuit suffers from drift, poor frequency response, and oscillations—or simply doesn't achieve expected accuracy. Well, before blaming the device, closely examine the passive components—including capacitors, resistors, potentiometers and, yes, even the printed circuit boards. In these areas, subtle effects of tolerance, temperature, parasitics, aging, and user assembly procedures can unwittingly sink a circuit. All too often these effects go unspecified (or underspecified) by passive component manufacturers.

In general, if using data converters having 12 bits or more of resolution, or op amps that cost more than a few dollars, pay very close attention to passive components. Consider the case of a 12-bit DAC, where ½ LSB corresponds to 0.012% of full scale, or only 122 ppm. A host of passive component phenomena can accumulate errors far exceeding this. But, buying the most expensive passive components won't necessarily solve the problems either. Often, a *correct* 25-cent capacitor yields a better-performing, more cost-effective design than a premium-grade part. With a few basics, understanding and analyzing passive components may prove rewarding, albeit not easy.

Capacitors

Most designers are generally familiar with the range of capacitors available. But the mechanisms by which both static and dynamic errors can occur in precision circuit designs using capacitors are sometimes easy to forget, because of the tremendous variety of types available. These include dielectrics of glass, aluminum foil, solid tantalum and tantalum foil, silver mica, ceramic, Teflon, and the film capacitors, including polyester, polycarbonate, polystyrene, and polypropylene types. In addition to the traditional leaded packages, many of these are now also offered in surface-mount styles.

Figure 7-1 is a workable model of a nonideal capacitor. The nominal capacitance, C, is shunted by a resistance R_P, which represents *insulation resistance* or leakage. A second resistance, R_S—*equivalent series resistance*, or ESR—appears in series with the capacitor and represents the resistance of the capacitor leads and plates.

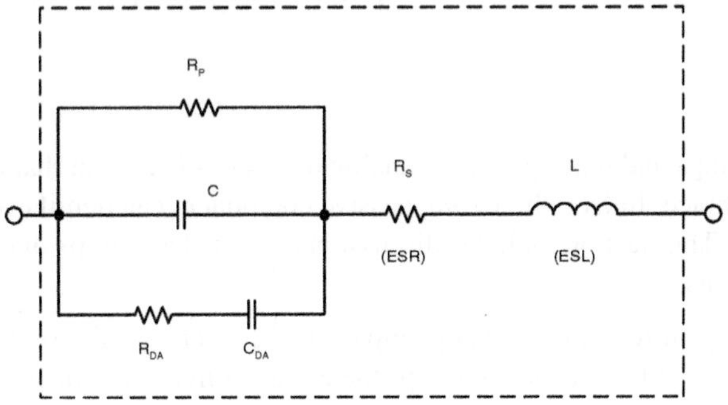

Figure 7-1: A nonideal capacitor equivalent circuit includes parasitic elements

Note that capacitor phenomena aren't that easy to separate out. The matching of phenomena and models is for convenience in explanation. Inductance, L—the *equivalent series inductance*, or ESL—models the inductance of the leads and plates. Finally, resistance R_{DA} and capacitance C_{DA} together form a simplified model of a phenomenon known as *dielectric absorption,* or DA. It can ruin fast and slow circuit dynamic performance. In a real capacitor, R_{DA} and C_{DA} extend to include multiple parallel sets. These parasitic RC elements can act to degrade timing circuits substantially, and the phenomenon is discussed further below.

Dielectric Absorption

Dielectric absorption, which is also known as "soakage" and sometimes as "dielectric hysteresis," is perhaps the least understood and potentially most damaging of various capacitor parasitic effects. Upon discharge, most capacitors are reluctant to give up all of their former charge, due to this memory consequence.

Figure 7-2 illustrates this effect. On the left of the diagram, after being charged to the source potential of V volts at time t_0, the capacitor is shorted by the switch S1 at time t_1, discharging it. At time t_2, the capacitor is then open-circuited; a residual voltage slowly builds up across its terminals and reaches a nearly constant value. This error voltage is due to DA, and is shown in the right figure, a time/voltage representation of the charge/discharge/recovery sequence. Note that the recovered voltage error is proportional to both the original charging voltage V, as well as the rated DA for the capacitor in use.

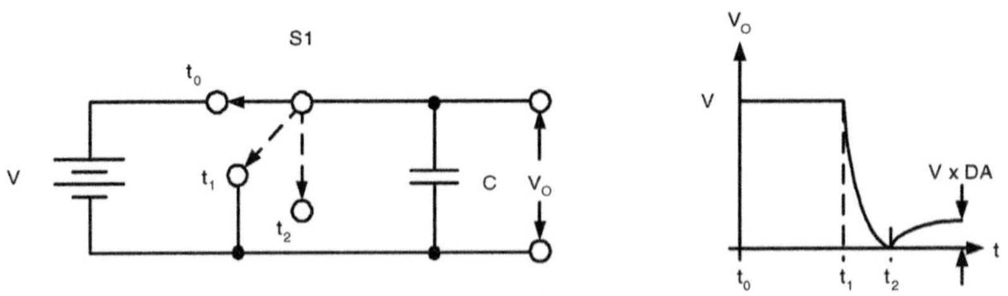

Figure 7-2: A residual open-circuit voltage after charge/discharge characterizes capacitor dielectric absorption

Standard techniques for specifying or measuring dielectric absorption are few and far between. Measured results are usually expressed as the percentage of the original charging voltage that reappears across the capacitor. Typically, the capacitor is charged for a long period, then shorted for a shorter established time. The capacitor is then allowed to recover for a specified period, and the residual voltage is then measured (see Reference 8 for details). While this explanation describes the basic phenomenon, it is important to note that real-world capacitors vary quite widely in their susceptibility to this error, with their rated DA ranging from well below to above 1%, the exact number being a function of the dielectric material used.

In practice, DA makes itself known in a variety of ways. Perhaps an integrator refuses to reset to zero, a voltage-to-frequency converter exhibits unexpected nonlinearity, or a sample-hold (SH) exhibits varying errors. This last manifestation can be particularly damaging in a data-acquisition system, where adjacent channels may be at voltages which differ by nearly full scale, as shown below.

Figure 7-3 illustrates the case of DA error in a simple SH. On the left, switches S1 and S2 represent an input multiplexer and SH switch, respectively. The multiplexer output voltage is V_X, and the sampled voltage held on C is V_Y, which is buffered by the op amp for presentation to an ADC. As can be noted by the timing diagram on the right, a DA error voltage, ϵ, appears in the hold mode, when the capacitor is effectively open circuit. This voltage is proportional to the difference of voltages V1 and V2, which, if at opposite extremes of the dynamic range, exacerbates the error. As a practical matter, the best solution for good performance in terms of DA in a SH is to use only the best capacitor.

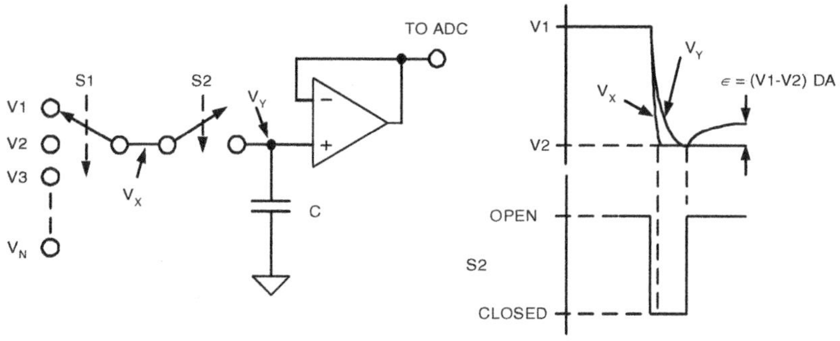

Figure 7-3: Dielectric absorption induces errors in SH applications

The DA phenomenon is a characteristic of the dielectric material itself, although inferior manufacturing processes or electrode materials can also affect it. DA is specified as a percentage of the charging voltage. It can range from a low of 0.02% for Teflon, polystyrene, and polypropylene capacitors, up to a high of 10% or more for some electrolytics. For some time frames, the DA of polystyrene can be as low as 0.002%.

Common high-K ceramics and polycarbonate capacitor types display typical DA on the order of 0.2%, it should be noted this corresponds to ½ LSB at only 8 bits. Silver mica, glass, and tantalum capacitors typically exhibit even larger DA, ranging from 1.0% to 5.0%, with those of polyester devices failing in the vicinity of 0.5%. As a rule, if the capacitor spec sheet doesn't specifically discuss DA *within your time frame and voltage range*, exercise caution. Another type with lower *specified* DA is likely a better choice.

DA can produce long tails in the transient response of fast-settling circuits, such as those found in high-pass active filters or ac amplifiers. In some devices used for such applications, Figure 7-1's R_{DA}-C_{DA} model of DA can have a time constant of milliseconds. Much longer time constants are also quite usual. In fact, several paralleled R_{DA}-C_{DA} circuit sections with a wide range of time constants can model some devices. In fast-charge, fast-discharge applications, the behavior of the DA mechanism resembles "analog memory"; the capacitor in effect tries to remember its previous voltage.

The effects of DA can be compensated for in some designs if it is simple and easily characterized, and the user is willing to do custom tweaking. In an integrator, for instance, the output signal can be fed back through a suitable compensation network, tailored to cancel the circuit equivalent of the DA by placing a negative impedance effectively in parallel. Such compensation has been shown to improve SH circuit performance by factors of 10 or more (Reference 6).

Capacitor Parasitics and Dissipation Factor

In Figure 7-1, a capacitor's leakage resistance, R_P, the effective series resistance, R_S, and effective series inductance, L, act as parasitic elements, which can degrade an external circuit's performance. The effects of these elements are often lumped together and defined as a dissipation factor, or D_F.

A capacitor's leakage is the small current that flows through the dielectric when a voltage is applied. Although modeled as a simple insulation resistance (R_P) in parallel with the capacitor, the leakage actually is nonlinear with voltage. Manufacturers often specify leakage as a megohm-microfarad product, which describes the dielectric's self-discharge time constant, in seconds. It ranges from a low of 1 s or less for high-leakage capacitors, such as electrolytic devices, to the hundreds of seconds for ceramic capacitors. Glass devices exhibit self-discharge time-constants of 1,000 or more; but the best leakage performance is shown by Teflon and the film devices (polystyrene, polypropylene), with time constants exceeding 1,000,000 megohm-microfarads. For such a device, external leakage paths—created by surface contamination of the device's case or in the associated wiring or physical assembly—can overshadow the internal dielectric-related leakage.

Effective series inductance, ESL (Figure 7-1) arises from the inductance of the capacitor leads and plates which, particularly at the higher frequencies, can turn a capacitor's normally capacitive reactance into an inductive reactance. Its magnitude strongly depends on construction details within the capacitor. Tubular wrapped-foil devices display significantly more lead inductance than molded radial-lead configurations. Multilayer ceramic and film-type devices typically exhibit the lowest series inductance, while ordinary tantalum and aluminum electrolytics typically exhibit the highest. Consequently, standard electrolytic types, if used alone, usually prove insufficient for *high speed* local bypassing applications. Note however that there also are more specialized aluminum and tantalum electrolytics available, which may be suitable for higher speed uses. These are the types generally designed for use in switch-mode power supplies, which are covered more completely in a following section.

Manufacturers of capacitors often specify effective series impedance by means of impedance-versus-frequency plots. Not surprisingly, these curves show graphically a predominantly capacitive reactance at low frequencies, with rising impedance at higher frequencies because of the effect of series inductance.

Effective series resistance, ESR (resistor R_S of Figure 7-1), is made up of the resistance of the leads and plates. As noted, many manufacturers lump the effects of ESR, ESL, and leakage into a single parameter called *dissipation factor*, or DF. Dissipation factor measures the basic inefficiency of the capacitor. Manufacturers define it as the ratio of the energy lost to energy stored per cycle by the capacitor. The ratio of ESR to total capacitive reactance—at a specified frequency—approximates the dissipation factor, which turns out to be equivalent to the reciprocal of the figure of merit, Q. Stated as an approximation, $Q \approx 1/DF$ (with DF in numeric terms). For example, a DF of 0.1% is equivalent to a fraction of 0.001; thus the inverse in terms of Q would be 1000.

Dissipation factor often varies as a function of both temperature and frequency. Capacitors with mica and glass dielectrics generally have DF values from 0.03% to 1.0%. For ordinary ceramic devices, DF ranges from a low of 0.1% to as high as 2.5% at room temperature. And electrolytics usually exceed even this level. The film capacitors are the best as a group, with DFs of less than 0.1%. Stable-dielectric ceramics, notably the NP0 (also called COG) types, have DF specs comparable to films (more below).

Tolerance, Temperature, and Other Effects

In general, precision capacitors are expensive and—even then—not necessarily easy to buy. In fact, choice of capacitance is limited both by the range of available values and by tolerances. In terms of size, the better performing capacitors in the film families tend to be limited in practical terms to 10 µF or less (for dual reasons of size and expense). In terms of low value tolerance, ±1% is possible for NP0 ceramic and some film devices, but with possibly unacceptable delivery times. Many film capacitors can be made available with tolerances of less than ±1%, but on a special order basis only.

Most capacitors are sensitive to temperature variations. DF, DA, and capacitance value are all functions of temperature. For some capacitors, these parameters vary approximately linearly with temperature, in others they vary quite nonlinearly. Although it is usually not important for SH applications, an excessively large *temperature coefficient* (TC, measured in ppm/°C) can prove harmful to the performance of precision integrators, voltage-to-frequency converters, and oscillators. NP0 ceramic capacitors, with TCs as low as 30 ppm/°C, are the best for stability, with polystyrene and polypropylene next best, with TCs in the 100–200 ppm/°C range. On the other hand, when capacitance stability is important, one should stay away from types with TCs of more than a few hundred ppm/°C, or in fact any TC that is nonlinear.

A capacitor's maximum working temperature should also be considered, in light of the expected environment. Polystyrene capacitors, for instance, melt near 85°C, compared to Teflon's ability to survive temperatures up to 200°C.

Sensitivity of capacitance and DA to applied voltage, expressed as *voltage coefficient*, can also hurt capacitor performance within a circuit application. Although capacitor manufacturers don't always clearly specify voltage coefficients, the user should always consider the possible effects of such factors. For instance, when maximum voltages are applied, some high-K ceramic devices can experience a decrease in capacitance of 50% or more. This is an inherent distortion producer, making such types unsuitable for signal path filtering, for example, and better suited for supply bypassing. Interestingly, NP0 ceramics, the stable dielectric subset from the wide range of available ceramics, do offer good performance with respect to voltage coefficient.

Similarly, the capacitance, and dissipation factor of many types vary significantly with frequency, mainly as a result of a variation in dielectric constant. In this regard, the better dielectrics are polystyrene, polypropylene, and Teflon.

Assemble Critical Components Last

The designer's worries don't end with the design process. Some common printed circuit assembly techniques can prove ruinous to even the best designs. For instance, some commonly used cleaning solvents can infiltrate certain electrolytic capacitors—those with rubber end caps are particularly susceptible. Even worse, some of the film capacitors, polystyrene in particular, actually melt when contacted by some solvents. Rough handling of the leads can damage still other capacitors, creating random or even intermittent circuit problems. Etched-foil types are particularly delicate in this regard. To avoid these difficulties it may be advisable to mount especially critical components as the last step in the board assembly process—if possible.

Table 7-1
CAPACITOR COMPARISON CHART

TYPE	TYPICAL DA	ADVANTAGES	DISADVANTAGES
Polystyrene	0.001% to 0.02%	Inexpensive Low DA Good Stability (~120ppm/°C)	Damaged by Temperature >85°C Large High Inductance Vendors Limited
Polypropylene	0.001% to 0.02%	Inexpensive Low DA Stable (~200ppm/°C) Wide Range of Values	Damaged by Temperature >105°C Large High Inductance
Teflon	0.003% to 0.02%	Low DA Available Good Stability Operational Above 125°C Wide Range of Values	Expensive Large High Inductance
Polycarbonate	0.1%	Good Stability Low Cost Wide Temperature Range Wide Range of Values	Large DA Limits to 8-Bit Applications High Inductance
Polyester	0.3% to 0.5%	Moderate Stability Low Cost Wide Temperature Range Low Inductance (Stacked Film)	Large DA Limits to 8-Bit Applications High Inductance (Conventional)
NP0 Ceramic	<0.1%	Small Case Size Inexpensive, Many Vendors Good Stability (30ppm/°C) 1% Values Available Low Inductance (Chip)	DA Generally Low (May Not be Specified) Low Maximum Values (≤10nF)
Monolithic Ceramic (High K)	>0.2%	Low Inductance (Chip) Wide Range of Values	Poor Stability Poor DA High Voltage Coefficient
Mica	>0.003%	Low Loss at HF Low Inductance Good Stability 1% Values Available	Quite Large Low Maximum Values (≤10nF) Expensive
Aluminum Electrolytic	Very High	Large Values High Currents High Voltages Small Size	High Leakage Usually Polarized Poor Stability, Accuracy Inductive
Tantalum Electrolytic	Very High	Small Size Large Values Medium Inductance	High Leakage Usually Polarized Expensive Poor Stability, Accuracy

Table 7-1 summarizes selection criteria for various capacitor types, arranged roughly in order of decreasing DA performance. In a selection process, the general information of this table should be supplemented by consultation of current vendor's catalog information (see References at end of section).

Designers should also consider the natural failure mechanisms of capacitors. Metallized film devices, for instance, often self-heal. They initially fail due to conductive bridges that develop through small perforations in the dielectric film. But, the resulting fault currents can generate sufficient heat to destroy the bridge, thus returning the capacitor to normal operation (at a slightly lower capacitance). Of course, applications in high-impedance circuits may not develop sufficient current to clear the bridge, so the designer must be wary here.

Tantalum capacitors also exhibit a degree, of self-healing but, unlike film capacitors, the phenomenon depends on the temperature at the fault location rising slowly. Therefore, tantalum capacitors self-heal best in high impedance circuits which limit the surge in current through the capacitor's defect. Use caution therefore, when specifying tantalums for high-current applications.

Electrolytic capacitor life often depends on the rate at which capacitor fluids seep through end caps. Epoxy end seals perform better than rubber seals, but an epoxy sealed capacitor can explode under severe reverse-voltage or overvoltage conditions. Finally, *all* polarized capacitors must be protected from exposure to voltages outside their specifications.

Resistors and Potentiometers

Designers have a broad range of resistor technologies to choose from, including carbon composition, carbon film, bulk metal, metal film, and both inductive and noninductive wire-wound types. As perhaps the most basic—and presumably most trouble-free—of components, resistors are often overlooked as error sources in high performance circuits.

An improperly selected resistor can subvert the accuracy of a 12-bit design by developing errors well in excess of 122 ppm (½ LSB).

Consider the simple circuit of Figure 7-4, showing a noninverting op amp where the 100× gain is set by R1 and R2. The TCs of these two resistors are a somewhat obvious source of error. Assume the op amp gain errors to be negligible, and that the resistors are perfectly matched to a 99/1 ratio at 25°C. If, as noted, the resistor TCs differ by only 25 ppm/°C, the gain of the amplifier changes by 250 ppm for a 10°C temperature change. This is about a 1 LSB error in a 12-bit system, and a major disaster in a 16-bit system. Temperature changes, however, can limit the accuracy of the Figure 7-4 amplifier in several ways. In this circuit (as well as many op amp circuits with component-ratio defined gains), the *absolute* TC of the resistors is less

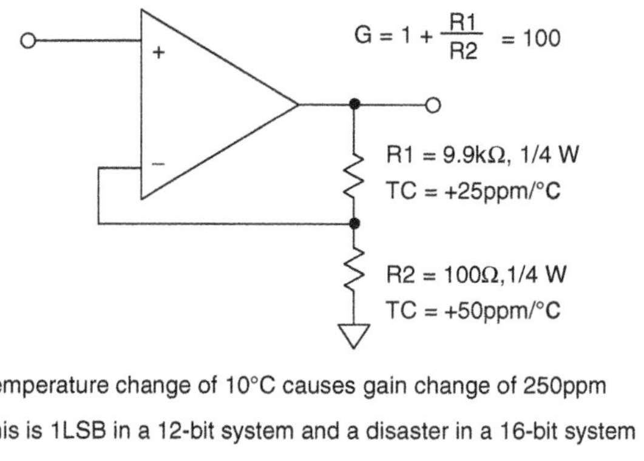

G = 1 + R1/R2 = 100
+
–
R1 = 9.9kΩ, 1/4 W
TC = +25ppm/°C
R2 = 100Ω,1/4 W
TC = +50ppm/°C
Temperature change of 10°C causes gain change of 250ppm
This is 1LSB in a 12-bit system and a disaster in a 16-bit system

Figure 7-4: Mismatched resistor TCs can induce temperature-related gain errors

important—*as long as they track one another in ratio.* But even so, some resistor types simply aren't suitable for precise work. For example, *carbon composition* units—with TCs of approximately 1,500 ppm/°C, won't work. Even if the TCs could be matched to an unlikely 1%, the resulting 15 ppm/°C differential still proves inadequate—an 8°C shift creates a 120 ppm error.

Many manufacturers offer metal film and bulk metal resistors, with absolute TCs ranging between ±1 and ±100 ppm/°C. Be aware, though; TCs can vary a great deal, particularly among discrete resistors from different batches. To avoid this problem, more expensive matched resistor pairs are offered by some manufacturers, with temperature coefficients that track one another to within 2 to 10 ppm/°C. Low priced thin-film networks have good relative performance and are widely used.

Suppose, as shown in Figure 7-5, R1 and R2 are ¼W resistors with identical 25 ppm/°C TCs. Even when the TCs are identical, there can still be significant errors. When the signal input is zero, the resistors dissipate no heat. But, if it is 100 mV, there is 9.9 V across R1, which then dissipates 9.9 mW. It will experience a temperature rise of 1.24°C (due to a 125°C/W ¼W resistor thermal resistance). This 1.24°C rise causes a resistance change of 31 ppm, and thus a corresponding gain change. But R2, with only 100mV across it, is only heated a negligible 0.0125°C. The resulting 31 ppm net gain error represents a full-scale error of ½ LSB at 14 bits, and is a disaster for a 16-bit system.

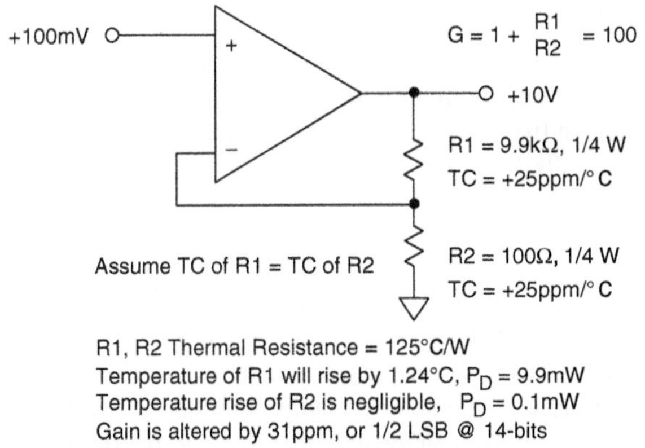

Figure 7-5: Uneven power dissipation between resistors with identical TCs can also introduce temperature-related gain errors

Even worse, the effects of this resistor self-heating also create easily calculable *nonlinearity errors*. In the Figure 7-5 example, with one-half the voltage input, the resulting self-heating error is only 15 ppm. In other words, the stage gain is not constant at ½ and full scale (nor is it so at other points), as long as uneven temperature shifts exist between the gain-determining resistors. This is by no means a worst-case example; physically smaller resistors would give worse results, due to higher associated thermal resistance.

These, and similar errors, are avoided by selecting critical resistors that are accurately matched for both value and TC, are well derated for power, and have tight thermal coupling between those resistors where matching is important. This is best achieved by using a resistor network on a single substrate—such a network may either be within an IC, or a separately packaged thin-film resistor network.

When the circuit resistances are very low (≤10 Ω), *interconnection stability* also becomes important. For example, while often overlooked as an error, the resistance TC of typical copper wire or printed circuit traces can add errors. The TC of copper is typically ~3,900 ppm/°C. Thus a precision 10 Ω, 10 ppm/°C wirewound resistor with 0. 1 Ω of copper interconnect effectively becomes a 10.1 Ω resistor with a TC of nearly 50 ppm/°C.

One final consideration applies mainly to designs that see widely varying ambient temperatures: a phenomenon known as *temperature retrace* describes the change in resistance which occurs after a specified number of cycles of exposure to low and high ambients with constant internal dissipation. Temperature retrace can exceed 10 ppm/°C, even for some of the better thin-film components.

In summary, to design resistance-based circuits for minimum temperature-related errors, consider the points noted in Figure 7-6 (along with their cost).

- Closely match resistance TCs
- Use resistors with low absolute TCs
- Use resistors with low thermal resistance (higher power ratings, larger cases)
- Tightly couple matched resistors thermally (use standard common-substrate networks)
- For large ratios consider using stepped attenuators

Figure 7-6: A number of points are important towards minimizing temperature- related errors in resistors

Resistor Parasitics

Resistors can exhibit significant levels of parasitic inductance or capacitance, especially at high frequencies. Manufacturers often specify these parasitic effects as a reactance error, in % or ppm, based on the ratio of the difference between the impedance magnitude and the dc resistance, to the resistance, at one or more frequencies.

Wirewound resistors are especially susceptible to difficulties. Although resistor manufacturers offer wirewound components in either normal or noninductively wound form, even noninductively wound resistors create headaches for designers. These resistors still appear slightly inductive (of the order of 20 μH) for values below 10 kΩ. Above 10 kΩ the same style resistors actually exhibit 5 pF of shunt capacitance.

These parasitic effects can raise havoc in dynamic circuit applications. Of particular concern are applications using wirewound resistors with values both greater than 10 kΩ. Here it isn't uncommon to see peaking, or even oscillation. These effects become more evident at low kHz frequency ranges.

Even in low-frequency circuit applications, parasitic effects in wirewound resistors can create difficulties. Exponential settling to 1 ppm may take 20 time constants or more. The parasitic effects associated with wirewound resistors can significantly increase net circuit settling time to beyond the length of the basic time constants.

Unacceptable amounts of parasitic reactance are often found even in resistors that aren't wirewound. For instance, some metal-film types have significant interlead capacitance, which shows up at high frequencies. In contrast, when considering this end-end capacitance, carbon resistors do the best at high frequencies.

Thermoelectric Effects

Another more subtle problem with resistors is the *thermocouple effect*, also sometimes referred to as *thermal EMF*. Wherever there is a junction between two different metallic conductors, a thermoelectric voltage results. The thermocouple effect is widely used to measure temperature, as described in detail within Chapter 4. However, in any low level precision op amp circuit it is also a potential source of inaccuracy, since wherever two different conductors meet, a thermocouple is formed (whether we like it or not). In fact, in many cases, it can easily produce the dominant error within an otherwise precision circuit design.

Parasitic thermocouples will cause errors when and if the various junctions forming the parasitic thermocouples are at different temperatures. With two junctions present on each side of the signal being processed within a circuit, by definition at least one thermocouple pair is formed. If the two junctions of this thermocouple pair are at different temperatures, there will be a net temperature dependent error voltage produced. Conversely, if the two junctions of a parasitic thermocouple pair are kept at an identical temperature, then the net error produced will be zero, as the voltages of the two thermocouples effectively will be canceled.

This is a critically important point, since in practice we cannot avoid connecting dissimilar metals together to build an electronic circuit. But, what we can do is carefully control temperature differentials across the circuit, so such that the undesired thermocouple errors cancel one another.

The effect of such parasitics is very hard to avoid. To understand this, consider a case of making connections *with copper wire only*. In this case, even a junction formed by different copper wire alloys can have a thermoelectric voltage that is a small fraction of 1 μV/°C. And, taking things a step further, even such apparently benign components as resistors contain parasitic thermocouples, with potentially even stronger effects.

For example, consider the resistor model shown in Figure 7-7. The two connections between the resistor material and the leads form thermocouple junctions, T1 and T2. This thermocouple EMF can be as high as 400 μV/°C for some carbon composition resistors, and as low as 0.05 μV/°C for specially constructed resistors (see Reference 15). Ordinary metal film resistors (RN-types) are typically about 20 μV/°C.

Note that these thermocouple effects are relatively unimportant for ac signals. Even for dc-only signals, they will nicely cancel one another if, as noted above, the entire resistor is at a uniform temperature. However, if there is significant power dissipation in a resistor, or if its orientation with respect to a heat source is nonsymmetrical, this can cause one of its ends to be warmer than the other, causing a net thermocouple

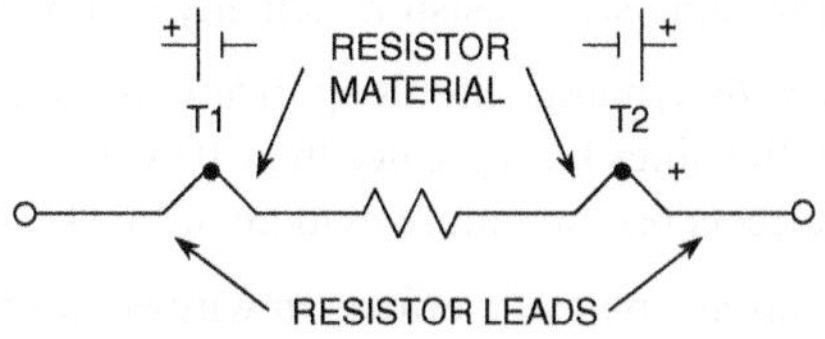

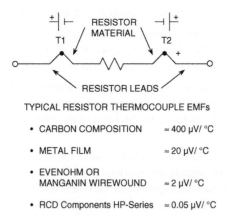

TYPICAL RESISTOR THERMOCOUPLE EMFs

- CARBON COMPOSITION ≈ 400 μV/ °C
- METAL FILM ≈ 20 μV/ °C
- EVENOHM OR MANGANIN WIREWOUND ≈ 2 μV/ °C
- RCD Components HP-Series ≈ 0.05 μV/ °C

Figure 7-7: Every resistor contains two thermocouples, formed between the leads and resistance element

error voltage. Using ordinary metal film resistors, an end-to-end temperature differential of 1°C causes a thermocouple voltage of about 20 μV. This error level is quite significant compared to the offset voltage drift of a precision op amp like the OP177, and extremely significant when compared to chopper-stabilized op amps, with their drifts of <1 μV/°C.

Figure 7-8 shows how resistor orientation can make a difference in the net thermocouple voltage. In the left diagram, standing the resistor on end in order to conserve board space will invariably cause a temperature gradient across the resistor, especially if it is dissipating any significant power. In contrast, placing the resistor flat on the PC board as shown at the right will generally eliminate the gradient. An exception might occur, if there is end-to-end resistor airflow. For such cases, orienting the resistor axis perpendicular to the airflow will minimize this source of error, since this tends to force the resistor ends to the same temperature.

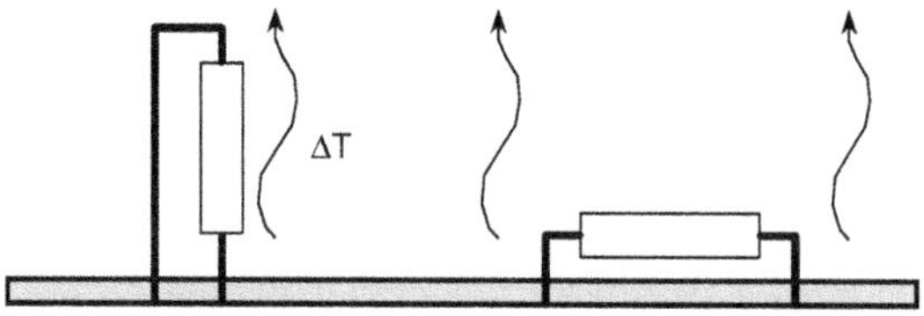

Figure 7-8: The effects of thermocouple EMFs generated by resistors can be minimized by orientation that normalizes the end temperatures

Note that this line of thinking should be extended to include orientation of resistors on a vertically mounted PC board. In such cases, natural convection air currents tend to flow upward across the board. Again, the resistor thermal axis should be perpendicular to convection, to minimize thermocouple effects. With tiny surface-mount resistors, the thermocouple effects can be less problematic, due to tighter thermal coupling between the resistor ends.

In general, designers should strive to avoid thermal gradients on or around critical circuit boards. Often this means thermally isolating components that dissipate significant amounts of power. Thermal turbulence created by large temperature gradients can also result in dynamic noise-like low frequency errors.

Voltage Sensitivity, Failure Mechanisms, and Aging

Resistors are also plagued by changes in value as a function of applied voltage. The deposited-oxide high megohm type components are especially sensitive, with voltage coefficients ranging from 1 ppm/V to more than 200 ppm/V. This is another reason to exercise caution in such precision applications as high-voltage dividers.

The normal failure mechanism of a resistor can also create circuit difficulties, if not carefully considered beforehand. For example, carbon-composition resistors fail safely, by turning into open circuits. Consequently, in some applications, these components can play a useful secondary role, as a fuse. Replacing such a resistor with a carbon-film type can possibly lead to trouble, since carbon-films can fail as short circuits. (Metal-film components usually fail as open circuits.)

All resistors tend to change slightly in value with age. Manufacturers specify long-term stability in terms of change—ppm/year. Values of 50 or 75 ppm/year are not uncommon among metal film resistors. For critical applications, metal-film devices should be burned in for at least one week at rated power. During burn-in, resistance values can shift by up to 100 or 200 ppm. Metal film resistors may need 4–5000 operational hours for full stabilization, especially if deprived of a burn-in period.

Resistor Excess Noise

Most designers have some familiarity with thermal, or Johnson, noise that occurs in resistors. But a less widely recognized secondary noise phenomenon is associated with resistors, and it is called *excess noise*. It can prove particularly troublesome in precision op amp and converter circuits, as it is evident only when current passes through a resistor.

To review briefly, thermal noise results from thermally induced random vibration of charge resistor carriers. Although the average current from the vibrations remains zero, instantaneous charge motions result in an instantaneous voltage across the terminals.

Excess noise on the other hand, occurs primarily when dc flows in a discontinuous medium—for example the conductive particles of a carbon composition resistor. The current flows unevenly through the compressed carbon granules, creating microscopic particle-to-particle "arcing." This phenomenon gives rise to a 1/f noise-power spectrum, in addition to the thermal noise spectrum. In other words, the excess spot noise voltage increases as the inverse square-root of frequency.

Excess noise often surprises the unwary designer. Resistor thermal noise and op amp input noise set the noise floor in typical op amp circuits. Only when voltages appear across input resistors and causes current to flow does the excess noise become a significant—and often dominant—factor. In general, carbon composition resistors generate the most excess noise. As the conductive medium becomes more uniform, excess noise becomes less significant. Carbon film resistors do better, with metal film, wirewound and bulk-metal-film resistors doing better yet.

Manufacturers specify excess noise in terms of a noise index—the number of microvolts of rms noise in the resistor in each decade of frequency per volt of dc drop across the resistor. The index can rise to 10 dB (3 microvolts per dc volt per decade of bandwidth) or more. Excess noise is most significant at low frequencies, while above 100 kHz thermal noise predominates.

Potentiometers

Trimming potentiometers (trimpots) can suffer from most of the phenomena that plague fixed resistors. In addition, users must also remain vigilant against some hazards unique to these components.

For instance, many trimpots aren't sealed, and can be severely damaged by board washing solvents, and even by excessive humidity. Vibration—or simply extensive use—can damage the resistive element and wiper terminations. Contact noise, TCs, parasitic effects, and limitations on adjustable range can all hamper trimpot circuit operation. Furthermore, the limited resolution of wirewound types and the hidden limits to resolution in cermet and plastic types (hysteresis, incompatible material TCs, slack) make obtaining and maintaining precise circuit settings anything but an "infinite resolution" process. Given this background, two rules are suggested for the potential trimpot user. Rule 1: Use infinite care and infinitesimal adjustment range to avoid infinite frustration when applying manual trimpots. Rule 2: *Consider the elimination of manual trimming potentiometers altogether, if possible.* A number of digitally addressable potentiometers (RDACs) are now available for direct application in similar circuit functions as classic trimpots (see Reference 17). There are also many low cost multi-channel voltage output DACs expressly designed for system voltage trimming.

Table 7-2 summarizes selection criteria for various fixed resistor types, both in discrete form and as part of networks. In a selection process, the general information of this table should be supplemented by consultation of current vendor's catalog information (see References at end of section).

Table 7-2
RESISTOR COMPARISON CHART

	TYPE	ADVANTAGES	DISADVANTAGES
DISCRETE	Carbon Composition	Lowest Cost High Power/Small Case Size Wide Range of Values	Poor Tolerance (5%) Poor Temperature Coefficient (1500 ppm/°C)
	Wirewound	Excellent Tolerance (0.01%) Excellent TC (1ppm/°C) High Power	Reactance is a Problem Large Case Size Most Expensive
	Metal Film	Good Tolerance (0.1%) Good TC (<1 to 100ppm/°C) Moderate Cost Wide Range of Values Low Voltage Coefficient	Must be Stabilized with Burn-In Low Power
	Bulk Metal or Metal Foil	Excellent Tolerance (to 0.005%) Excellent TC (to <1ppm/°C) Low Reactance Low Voltage Coefficient	Low Power Very Expensive
	High Megohm	Very High Values (10^8 to $10^{14}\Omega$) Only Choice for Some Circuits	High Voltage Coefficient (200ppm/V) Fragile Glass Case (Needs Special Handling) Expensive
NETWORKS	Thick Film	Low Cost High Power Laser-Trimmable Readily Available	Fair Matching (0.1%) Poor TC (>100ppm/°C) Poor Tracking TC (10ppm/°C)
	Thin Film	Good Matching (<0.01%) Good TC (<100ppm/°C) Good Tracking TC (2ppm/°C) Moderate Cost Laser-Trimmable Low Capacitance Suitable for Hybrid IC Substrate	Often Large Geometry Limited Values and Configurations

Inductance

Stray Inductance

All conductors are inductive, and at high frequencies, the inductance of even quite short pieces of wire or printed circuit traces may be important. The inductance of a straight wire of length L mm and circular cross-section with radius R mm in free space is given by the first equation shown in Figure 7-9.

The inductance of a strip conductor (an approximation to a PC track) of width W mm and thickness H mm in free space is also given by the second equation in Figure 7-9.

In real systems, these formulas both turn out to be approximate, but they do give some idea of the order of magnitude of inductance involved. They tell us that 1 cm of 0.5 mm o.d. wire has an inductance of 7.26 nH, and 1 cm of 0.25 mm PC track has an inductance of 9.59 nH. These figures are reasonably close to measured results.

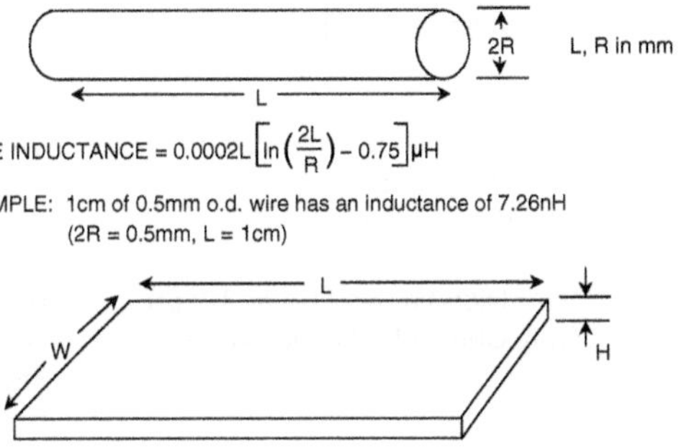

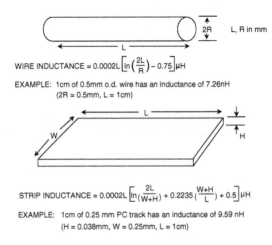

$$\text{WIRE INDUCTANCE} = 0.0002L\left[\ln\left(\frac{2L}{R}\right) - 0.75\right]\mu H$$

EXAMPLE: 1cm of 0.5mm o.d. wire has an inductance of 7.26nH
(2R = 0.5mm, L = 1cm)

$$\text{STRIP INDUCTANCE} = 0.0002L\left[\ln\left(\frac{2L}{W+H}\right) + 0.2235\left(\frac{W+H}{L}\right) + 0.5\right]\mu H$$

EXAMPLE: 1cm of 0.25 mm PC track has an inductance of 9.59 nH
(H = 0.038mm, W = 0.25mm, L = 1cm)

Figure 7-9: Wire and strip inductance calculations

At 10 MHz, an inductance of 7.26 nH has an impedance of 0.46 Ω, and so can give rise to 1% error in a 50 Ω system.

Mutual Inductance

Another consideration regarding inductance is the separation of outward and return currents. As we shall discuss in more detail later, Kirchoff's Law tells us that current flows in closed paths—there is always an outward and return path. The whole path forms a single-turn inductor.

This principle is illustrated by the contrasting signal trace routing arrangements of Figure 7-10. If the area enclosed within the turn is relatively large, as in the upper "nonideal" picture, the inductance (and hence the ac impedance) will also be large.

On the other hand, if the outward and return paths are closer together, as in the lower "improved" picture, the inductance will be much smaller.

Note that the nonideal signal routing case of Figure 7-10 has other drawbacks—the large area enclosed within the conductors produces extensive external magnetic fields, which may interact with other circuits,

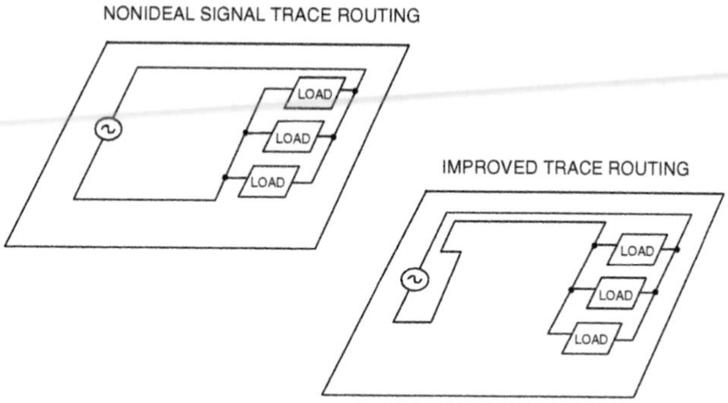

Figure 7-10: Nonideal and improved signal trace routing

causing unwanted coupling. Similarly, the large area is more vulnerable to interaction with external magnetic fields, which can induce unwanted signals in the loop.

The basic principle is illustrated in Figure 7-11, and is a common mechanism for the transfer of unwanted signals (noise) between two circuits.

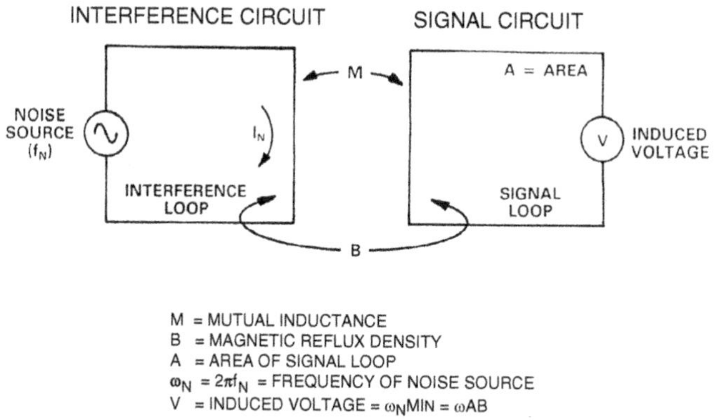

Figure 7-11: Basic principles of inductive coupling

As with most other noise sources, as soon as we define the working principle, we can see ways of reducing the effect. In this case, reducing any or all of the terms in the equations in Figure 7-11 reduces the coupling. Reducing the frequency or amplitude of the current causing the interference may be impracticable, but it is frequently possible to reduce the mutual inductance between the interfering and interfered with circuits by reducing loop areas on one or both sides and, possibly, increasing the distance between them.

A layout solution is illustrated by Figure 7-12. Here two circuits, shown as Z1 and Z2, are minimized for coupling by keeping each of the loop areas as small as is practical.

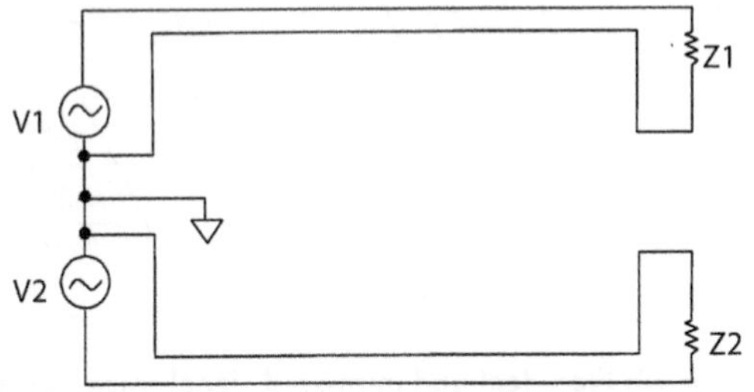

Figure 7-12: Proper signal routing and layout can reduce inductive coupling

As also illustrated in Figure 7-13, mutual inductance can be a problem in signals transmitted on cables. Mutual inductance is high in ribbon cables, especially when a single return is common to several signal circuits (top). Separate, dedicated signal and return lines for each signal circuit reduces the problem (middle). Using a cable with twisted pairs for each signal circuit as in the bottom picture is even better (but is more expensive and often unnecessary).

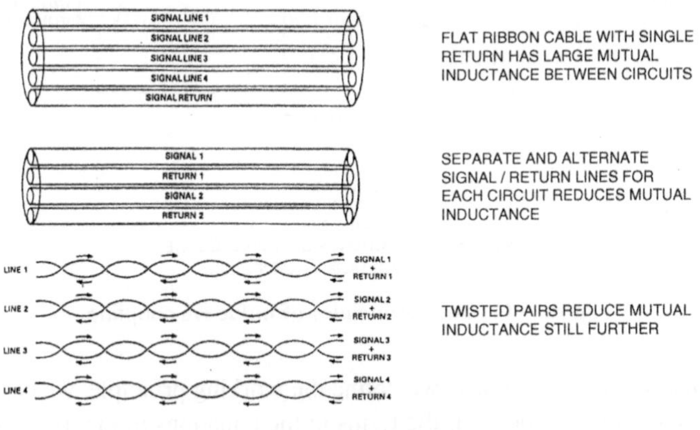

Figure 7-13: Mutual inductance and coupling within signal cabling

Shielding of magnetic fields to reduce mutual inductance is sometimes possible, but is by no means as easy as shielding an electric field with a Faraday shield (following section). HF magnetic fields are blocked by conductive material provided the skin depth in the conductor at the frequency to be screened is much less than the thickness of the conductor, and the screen has no holes (Faraday shields can tolerate small holes, magnetic screens cannot). LF and dc fields may be screened by a shield made of mu-metal sheet. Mu-metal is an alloy having very high permeability, but it is expensive, its magnetic properties are damaged by mechanical stress, and it will saturate if exposed to too high fields. Its use, therefore, should be avoided where possible.

Ringing

An inductor in series or parallel with a capacitor forms a resonant, or "tuned," circuit, whose key feature is that it shows marked change in impedance over a small range of frequency. Just how sharp the effect is depends on the relative Q of the tuned circuit. The effect is widely used to define the frequency response of narrow-band circuitry, but can also be a potential problem source.

If stray inductance and capacitance (which may or may not be stray) in a circuit should form a tuned circuit, that tuned circuit may be excited by signals in the circuit, and ring at its resonant frequency.

An example is shown in Figure 7-14, where the resonant circuit formed by an inductive power line and its decoupling capacitor may possibly be excited by fast pulse currents drawn by the powered IC.

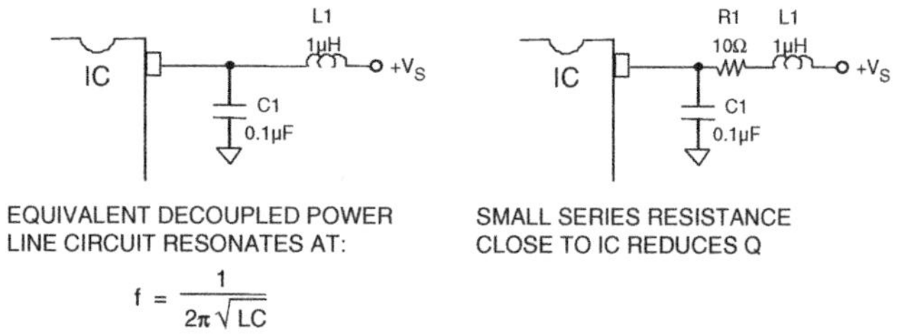

Figure 7-14: Resonant circuit formed by power line decoupling

While normal trace inductance and typical decoupling capacitances of 0.01 μF–0.1 μF will resonate well above a few MHz, an example 0.1 μF capacitor and 1 μH of inductance resonates at 500 kHz. Left unchecked, this could present a resonance problem, as shown in the left case. Should an undesired power line resonance be present, the effect may be minimized by lowering the Q of the inductance. This is most easily done by inserting a small resistance (~10 Ω) in the power line close to the IC, as shown in the right case.

Parasitic Effects in Inductors

Although inductance is one of the fundamental properties of an electronic circuit, inductors are far less common as components than are resistors and capacitors. As for precision components, they are even more rare. This is because they are harder to manufacture, less stable, and less physically robust than resistors and capacitors. It is relatively easy to manufacture stable precision inductors with inductances from nH to tens or hundreds of μH, but larger valued devices tend to be less stable, and large.

As we might expect in these circumstances, circuits are designed, where possible, to avoid the use of precision inductors. We find that stable precision inductors are relatively rarely used in precision analog circuitry, except in tuned circuits for high frequency narrow band applications. Of course, they are widely used in power filters, switching power supplies and other applications where lack of precision is unimportant (more on this in a following section).

The important features of inductors used in such applications are their current carrying and saturation characteristics, and their Q. If an inductor consists of a coil of wire with an air core, its inductance will essentially be unaffected by the current it is carrying. On the other hand, if it is wound on a core of a magnetic material (magnetic alloy or ferrite), its inductance will be nonlinear since, at high currents, the core will start to saturate. The effects of such saturation will reduce the efficiency of the circuitry employing the inductor and is liable to increase noise and harmonic generation.

As mentioned above, inductors and capacitors together form tuned circuits. Since all inductors will also have some stray capacity, all inductors will have a resonant frequency (which will normally be published on their data sheet), and should only be used as precision inductors at frequencies well below this.

Q or "Quality Factor"

The other characteristic of inductors is their Q (or "Quality Factor"), which is the ratio of the reactive impedance to the resistance, as indicated in Figure 7-15.

- $Q = 2\pi f\ L/R$
- The Q of an inductor or resonant circuit is a measure of the ratio of its reactance to its resistance.
- The resistance is the HF and NOT the DC value.
- The 3 dB bandwidth of a single tuned circuit is Fc/Q where Fc is the center frequency.

Figure 7-15: Inductor Q or quality factor

It is rarely possible to calculate the Q of an inductor from its dc resistance, since skin effect (and core losses if the inductor has a magnetic core) ensure that the Q of an inductor at high frequencies is always lower than that predicted from dc values.

Q is also a characteristic of tuned circuits (and of capacitors—but capacitors generally have such high Q values that it may be disregarded, in practice). The Q of a tuned circuit, which is generally very similar to the Q of its inductor (unless it is deliberately lowered by the use of an additional resistor), is a measure of its bandwidth around resonance. LC tuned circuits rarely have Q of much more than 100 (3 dB bandwidth of 1%), but ceramic resonators may have a Q of thousands, and quartz crystals tens of thousands.

Don't Overlook Anything

Remember, if a precision op amp or data-converter-based design does not meet specification, try not to overlook anything in trying to find the error sources. Analyze both active *and* passive components, trying to identify and challenge any assumptions or preconceived notions that may obscure to the facts. Take nothing for granted.

For example, when not tied down to prevent motion, cable conductors, moving within their surrounding dielectrics, can create significant static charge buildups that cause errors, especially when connected to high-impedance circuits. Rigid cables, or even costly low noise Teflon-insulated cables, are expensive alternative solutions.

As more and more high precision op amps become available, and system designs call for higher speed and increased accuracy, a thorough understanding of the error sources described in this section (as well those following) becomes more important.

Some additional discussions of passive components within a succeeding power supply filtering section complements this one. In addition, the very next section on PCB design issues also complements many points within this section. Similar comments apply to the section on EMI/RFI.

References: Passive Components

1. James E. Buchanan, "Dielectric Absorption—It Can Be a Real Problem In Timing Circuits," **EDN**, January 20, 1977, p. 83.
2. Lew Counts and Scott Wurcer, "Instrumentation Amplifier Nears Input Noise Floor," **Electronic Design,** June 10, 1982.
3. W. Doeling, W. Mark, T. Tadewald, and P. Reichenbacher, "Getting Rid of Hook: The Hidden PC-Board Capacitance," **Electronics,** October 12, 1978, pp. 111–117.
4. Tarlton Fleming, "Data-Acquisition System (DAS) Design Considerations," **WESCON '81 Professional Program Session Record No. 23**.
5. Walter G. Jung and Richard Marsh, "Picking Capacitors, Parts I and II," **Audio**, February and March, 1980.
6. Robert A. Pease, "Understand Capacitor Soakage to Optimize Analog Systems", **EDN**, October 13, 1982, p. 125.
7. Andy Rappaport, "Capacitors" **EDN**, October 13, 1982, p. 105.
8. Specification MIL-PRF-19978G, Capacitors, Fixed, Plastic (or Paper-Plastic) Dielectric (Hermetically Sealed in Metal, Ceramic or Glass Cases), Established and Non-Established Reliability General Specification for, May 27, 1999.
9. Specification MIL-PRF-123B, Capacitors, Fixed, Ceramic Dielectric, (Temperature Stable and General Purpose), High Reliability, General Specification for, August 6, 1990.
10. **Tantalum and Ceramic Surface Mount Capacitor Catalog**, Kemet Electronics Corporation, P.O. Box 5928, Greenville, SC, 29606, 864-963-6300.
11. A general capacitor information resource: www.faradnet.com/.
12. Southern and F-Dyne film capacitors, Southern Electronics, 215 Research Drive, Milford, CT, 06460, 203-876-7488.
13. Wesco film capacitors, Wesco Electrical Company, 201 Munson Street, Greenfield, MA, 01301, 413- 774-4358.
14. Doug Grant and Scott Wurcer, "Avoiding Passive Component Pitfalls," **The Best of Analog Dialogue**, Analog Devices, 1991, pp. 143–148.
15. RCD Components, Inc., 520 E. Industrial Park Drive, Manchester NH, 03109, 603-669-0054, www.rcd-comp.com.
16. Steve Sockolov and James Wong, "High-Accuracy Analog Needs More Than Op Amps," **Electronic Design**, October 1, 1992, p. 53.
17. Selection guide for digital potentiometers: www.analog.com/support/standard_linear/selection_guides/dig_pot.pdf>
18. Precision Resistor Co., Inc., 10601 75th St. N., Largo, FL, 33777-1427, 727-541-5771, www.precisionresistor.com
19. Ohmite Victoreen MAXI-MOX Resistors, 3601 Howard Street, Skokie, IL 60076, 847-675-2600, www.ohmite.com/victoreen/.
20. Vishay/Dale Resistors, 2300 Riverside Blvd., Norfolk, NE, 68701-2242, 402-371-0800, www.vishay.com.
21. Beyschlag Resistor Products, PO Box 1220, D-25732 Heide, Germany, www.beyschlag.com.

22. B. I. & B. Bleaney, **Electricity & Magnetism**, Oxford at the Clarendon Press, 1957, pp. 23, 24, and 52.
23. Henry W. Ott, **Noise Reduction Techniques in Electronic Systems, 2nd Edition**, John Wiley, Inc., 1988, ISBN: 0-471-85068-3.
24. G. W. A. Dummer, **Materials for Conductive and Resistive Functions**, Hayden, 1970.

PCB Design Issues

James Bryant

Printed circuit boards (PCBs) are by far the most common method of assembling modern electronic circuits. Comprised of a sandwich of insulating layer (or layers) and one or more copper conductor patterns, they can introduce various forms of errors into a circuit, particularly if the circuit is operating at either high precision or high speed. PCBs then, act as "unseen" components, wherever they are used in precision circuit designs. Since designers don't always consider the PCB electrical characteristics as additional components of their circuit, overall performance can easily end up worse than predicted. This general topic, manifested in many forms, is the focus of this section.

PCB effects that are harmful to precision circuit performance include leakage resistances, spurious voltage drops in trace foils, vias, and ground planes, the influence of stray capacitance, dielectric absorption (DA), and the related "hook." In addition, the tendency of PCBs to absorb atmospheric moisture, *hygroscopicity,* means that changes in humidity often cause the contributions of some parasitic effects to vary from day to day.

In general, PCB effects can be divided into two broad categories—those that most noticeably affect the static or dc operation of the circuit, and those that most noticeably affect dynamic or ac circuit operation.

Another very broad area of PCB design is the topic of grounding. Grounding is a problem area in itself for all analog designs, and it can be said that implementing a PCB based circuit doesn't change that fact. Fortunately, certain principles of quality grounding, namely the use of ground planes, are intrinsic to the PCB environment. This factor is one of the more significant advantages to PCB-based analog designs, and appreciable discussion of this section is focused on this issue.

Some other aspects of grounding that must be managed include the control of spurious ground and signal return voltages that can degrade performance. These voltages can be due to external signal coupling, common currents, or simply excessive IR drops in ground conductors. Proper conductor routing and sizing, as well as differential signal handling and ground isolation techniques enables control of such parasitic voltages.

One final area of grounding to be discussed is grounding appropriate for a mixed-signal, analog/digital environment. Although this isn't the specific overall focus of the book, it is certainly true that interfacing with ADCs (or DACs) is a major task category of op amps, and thus it shouldn't be overlooked. Indeed, the single issue of quality grounding can drive the entire layout philosophy of a high performance mixed signal PCB design—as it well should.

Resistance of Conductors

Every engineer is familiar with resistors—little cylinders with wire or tab ends—although perhaps fewer are aware of their idiosyncrasies, as generally covered in section 7-1. But far too few engineers consider that all the wires and PCB traces with which their systems and circuits are assembled are also resistors. In higher precision systems, even these trace resistances and simple wire interconnections can have degrading effects. Copper is *not* a superconductor—and too many engineers appear to think it is.

$$R = \frac{\rho Z}{XY}$$

ρ = RESISTIVITY

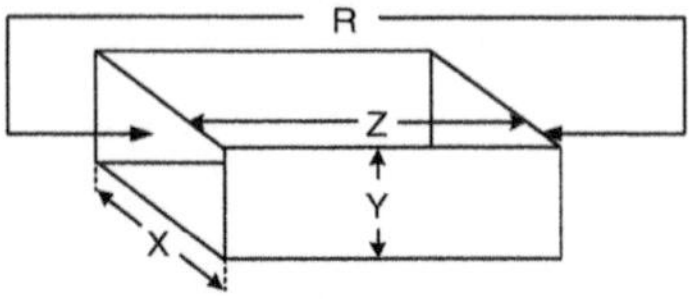

SHEET RESISTANCE CALCULATION FOR
1 OZ. COPPER CONDUCTOR:

ρ = 1.724 X 10-6 Ωcm, Y = 0.0036cm

$R = 0.48 \frac{Z}{X} \text{m}\Omega$

$\frac{Z}{X}$ = NUMBER OF SQUARES

R = SHEET RESISTANCE OF 1 SQUARE (Z = X)
= 0.48mΩ/SQUARE

Figure 7-16: Calculation of sheet resistance and linear resistance for standard copper PCB conductors

Figure 7-16 illustrates a method of calculating the sheet resistance R of a copper square, given the length Z, the width X, and the thickness Y.

At 25°C the resistivity of pure copper is 1.724E-6 ohm cm. The thickness of standard 1 ounce PCB copper foil is 0.036 mm (0.0014"). Using the relations shown, the resistance of such a standard copper element is therefore 0.48 mΩ /square. One can readily calculate the resistance of a linear trace, by effectively "stacking" a series of such squares end-end, to make up the line's length. The line length is Z and the width is X, so the line resistance R is simply a product of Z/X and the resistance of a single square, as noted in the figure. For a given copper weight and trace width, a resistance/length calculation can be made. For example, the 0.25mm (10 mil) wide traces frequently used in PCB designs equates to a resistance/length of about 19 mΩ/cm (48 mΩ /inch), which is quite large. Moreover, the temperature coefficient of resistance for copper is about 0.4%/°C around room temperature. This is a factor that shouldn't be ignored, in particular within low impedance precision circuits, where the TC can shift the net impedance over temperature.

As shown in Figure 7-17, PCB trace resistance can be a serious error when conditions aren't favorable. Consider a 16-bit ADC with a 5 kΩ input resistance, driven through 5 cm of 0.25 mm wide 1 oz. PCB track between it and its signal source. The track resistance of nearly 0.1 Ω forms a divider with the 5 kΩ load, creating an error. The resulting voltage drop is a gain error of 0.1/5 k (~0.0019%), well over 1 LSB (0.0015% for 16 bits).

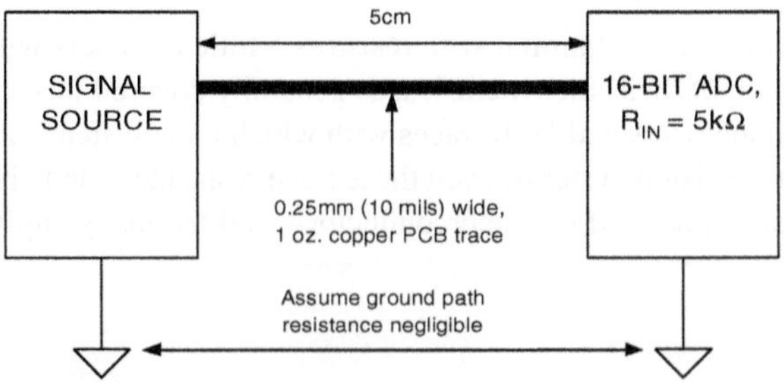

Figure 7-17: Ohm's law predicts >1 LSB of error due to drop in PCB conductor

So, when dealing with precision circuits, the point is made that even simple design items such as PCB trace resistance cannot be dealt with casually. There are various solutions that can address this issue, such as wider traces (which may take up excessive space), the use of heavier copper (which may be too expensive), or simply choosing a high impedance converter. But, the most important thing is to think it all through, avoiding any tendency to overlook items appearing innocuous on the surface.

Voltage Drop in Signal Leads—"Kelvin" Feedback

The gain error resulting from resistive voltage drop in PCB signal leads is important only with high precision and/or at high resolutions (Figure 7-17 example), or where large signal currents flow. Where load impedance is constant and resistive, adjusting overall system gain can compensate for the error. In other circumstances, it may often be removed by the use of "Kelvin" or "voltage sensing" feedback, as shown in Figure 7-18.

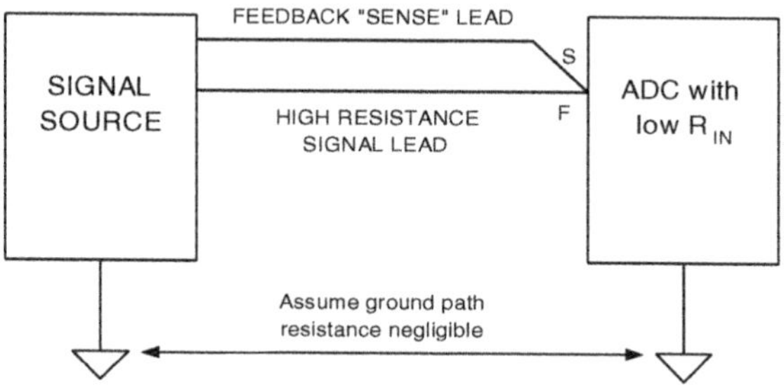

Figure 7-18: Use of a sense connection moves accuracy to the load point

In this modification to the case of Figure 7-17, a long resistive PCB trace is still used to drive the input of a high resolution ADC, with low input impedance. In this case, however, the voltage drop in the signal lead does *not* give rise to an error, as feedback is taken directly from the input pin of the ADC, and returned to the driving source. This scheme allows full accuracy to be achieved in the signal presented to the ADC, despite any voltage drop across the signal trace.

The use of separate force (F) and sense (S) connections at the load removes any errors resulting from voltage drops in the force lead but, of course, may be used only in systems where there is negative feedback. It is also impossible to use such an arrangement to drive two or more loads with equal accuracy, since feedback may only be taken from one point. Also, in this much-simplified system, errors in the common lead source/load path are ignored, the assumption being that ground path voltages are negligible. In many systems this may not necessarily be the case, and additional steps may be needed, as noted next.

Signal Return Currents

Kirchoff's Law tells us that at any point in a circuit the algebraic sum of the currents is zero. This tells us that all currents flow in circles and, particularly, that the return current must always be considered when analyzing a circuit, as is illustrated in Figure 7-19 (see References 7 and 8).

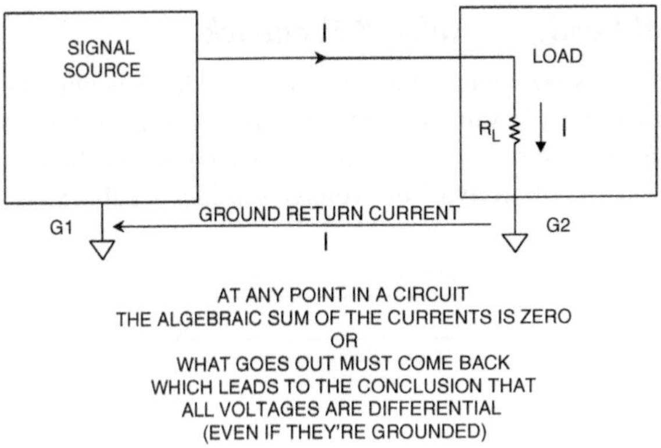
SIGNAL
SOURCE
I
LOAD
R_L
I
G1
GROUND RETURN CURRENT
I
G2
AT ANY POINT IN A CIRCUIT
THE ALGEBRAIC SUM OF THE CURRENTS IS ZERO
OR
WHAT GOES OUT MUST COME BACK
WHICH LEADS TO THE CONCLUSION THAT
ALL VOLTAGES ARE DIFFERENTIAL
(EVEN IF THEY'RE GROUNDED)

Figure 7-19: Kirchoff's law helps in analyzing voltage drops around a complete source/load coupled circuit

In dealing with grounding issues, common human tendencies provide some insight into how the correct thinking about the circuit can be helpful towards analysis. Most engineers readily consider the ground return current "I," *when they are considering a fully differential circuit.*

However, when considering the more usual circuit case, where a single-ended signal is referred to "ground," it is common to assume that all the points on the circuit diagram where ground symbols are found are at the same potential. Unfortunately, this happy circumstance just is not necessarily so.

This overly optimistic approach is illustrated in Figure 7-20, where, if it really should exist, "infinite ground conductivity" would lead to zero ground voltage difference between source ground G1 and load ground G2. Unfortunately this approach isn't a wise practice and, when dealing with high precision circuits, it can lead to disasters.

A more realistic approach to ground conductor integrity includes analysis of the impedance(s) involved, and careful attention to minimizing spurious noise voltages.

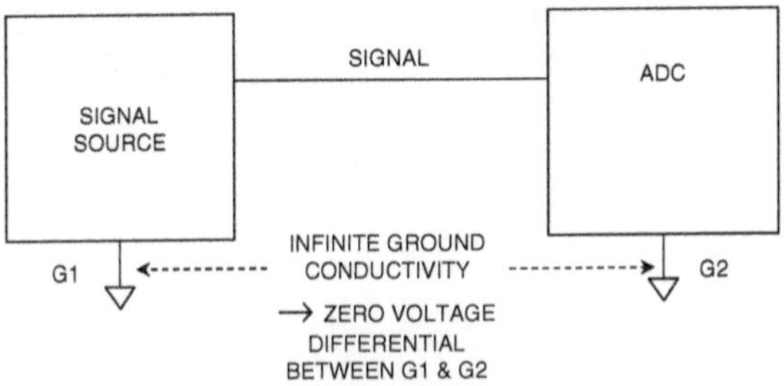
SIGNAL
ADC
SIGNAL
SOURCE
INFINITE GROUND
CONDUCTIVITY
G1
G2
→ ZERO VOLTAGE
DIFFERENTIAL
BETWEEN G1 & G2

Figure 7-20: Unlike this optimistic diagram, it is unrealistic to assume infinite conductivity between source/load grounds in a real-world system

Ground Noise and Ground Loops

A more realistic model of a ground system is shown in Figure 7-21. The signal return current flows in the complex impedance existing between ground points G1 and G2 as shown, giving rise to a voltage drop ΔV in this path. But it is important to note that additional *external* currents, such as I_{EXT}, may also flow in this same path. It is critical to understand that such currents may generate uncorrelated noise voltages between G1 and G2 (dependent upon the current magnitude and relative ground impedance).

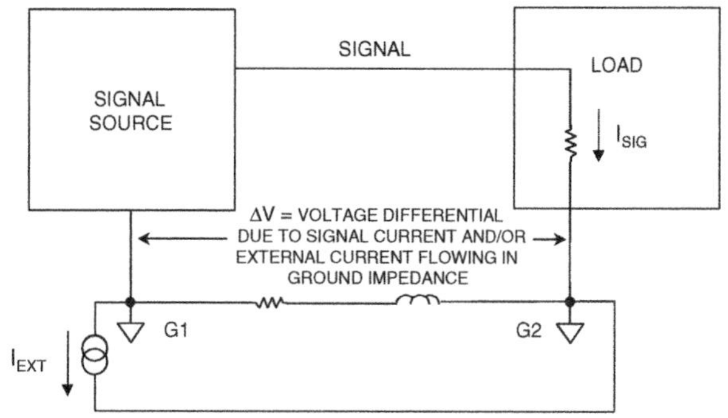

Figure 7-21: A more realistic source-to-load grounding system view includes consideration of the impedance between G1-G2, plus the effect of any nonsignal-related currents

Some portion of these undesired voltages may end up being seen at the signal's load end, and they can have the potential to corrupt the signal being transmitted.

It is evident, of course, that other currents can only flow in the ground impedance if there is a current path for them. In this case, severe problems can be caused by a high current circuit sharing an *unlooped* ground return with the signal source.

Figure 7-22 shows just such a common ground path, shared by the signal source and a high current circuit, which draws a large and varying current from its supply. This current flows in the common ground return, causing an error voltage ΔV to be developed.

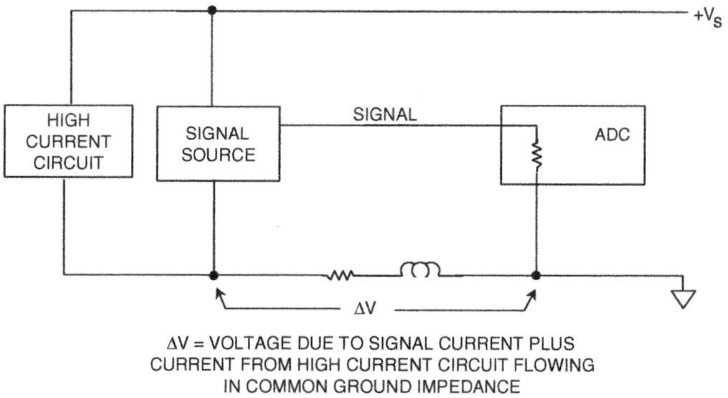

Figure 7-22: Any current flowing through a common ground impedance can cause errors

From Figure 7-23, it is also evident that if a ground network contains *loops,* or circular ground conductor patterns (with S1 closed), there is an even greater danger of it being vulnerable to EMFs induced by external magnetic fields. There is also a real danger of ground-current-related signals "escaping" from the high current areas, and causing noise in sensitive circuit regions elsewhere in the system.

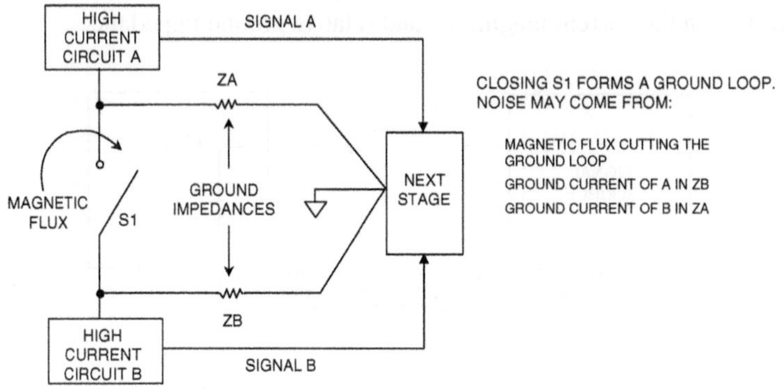

Figure 7-23: A ground loop

For these reasons ground loops are best avoided, by wiring all return paths within the circuit by separate paths back to a common point, i.e., the common ground point towards the mid-right of the diagram. This would be represented by the S1 open condition.

There are a number of possible ways of attacking the ground noise problem, apart from the presently impracticable one of using superconducting grounds. It is rare for any single method to be used to the exclusion of all others, and real systems generally contain a mixture. For descriptive purposes each approach is addressed separately.

Star Grounds

The "star" ground philosophy builds on the theory that there is one single ground point in a circuit to which all voltages are referred. This is known as the *star ground* point. It can be better understood by a visual analogy—the multiple conductors extending radially from the common schematic ground resemble a star. This can be appreciated by regarding Figure 7-23, considering many more ground returns from the common point. Note that the star point need not look like a star—it may be a point on a ground plane—but the key feature of the star ground system is that all voltages are measured with respect to a particular point in the ground network, not just to an undefined "ground" (i.e., wherever one can clip a probe). Figure 7-24 succinctly summarizes the philosophy.

- IF ALL SIGNAL VOLTAGES WITHIN A SYSTEM ARE MEASURED WITH RESPECT TO A SINGLE POINT, THAT POINT IS SAID TO BE THE SYSTEM STAR GROUND.

Figure 7-24: The star ground concept

This star grounding philosophy is reasonable theoretically, but can encounter practical difficulties. For example, if we design a star ground system, drawing out all signal paths to minimize signal interaction and the effects of high impedance signal or ground paths, we often find implementation problems. When the power supplies are added to the circuit diagram, they either add unwanted ground paths or their supply currents flowing in the existing ground paths are sufficiently so large or noisy (or both), they can corrupt the signal transmission. This particular problem can often be avoided by having separate power supplies (and thus separate ground returns) for the various circuit portions. For example, separate analog and digital supplies with separate analog and digital grounds, joined at the star point, are common in mixed signal applications.

Separate Analog and Digital Grounds

As a fact of life, digital circuitry is noisy. Saturating logic draws large, fast current spikes from its supply during switching. However, logic stages, with hundreds of millivolts (or more) of noise immunity, usually have little need for high levels of supply decoupling. On the other hand, analog circuitry is quite vulnerable to noise on both power supply rails and grounds. So, it is very sensible to separate analog and digital circuitry, to prevent digital noise from corrupting analog performance. Such separation involves separation of both ground returns *and* power rails, which is inconvenient in a mixed-signal system. Nevertheless, if a mixed-signal system is to deliver full performance capability, it is often essential to have separate analog and digital grounds, and separate power supplies. The fact that some analog circuitry will "operate" (i.e., function) from a single 5 V supply does *not* mean that it may safely be operated from the same noisy 5 V supply as the microprocessor and dynamic RAM, the electric fan, and the solenoid jackhammer. What is required is that the analog portion *operate with full performance from such a low voltage supply*, not just be functional. This distinction will by necessity require quite careful attention to both the supply rails and the ground interfacing.

Figures 7-25 and 7-26 summarize some analog and digital power supply and grounding concepts which are useful to bear in mind as systems are designed.

Figure 7-25: Some power supply and ground noise concepts appropriate for mixed-signal systems

- DIGITAL CIRCUITRY IS *NOISY*
- ANALOG CIRCUITRY IS *QUIET*
- CIRCUIT NOISE FROM DIGITAL CIRCUITRY CARRIED BY POWER AND GROUND LEADS CAN CORRUPT PRECISION ANALOG CIRCUITRY
- IT IS ADVISABLE TO SEPARATE THE POWER AND GROUND OF THE DIGITAL AND ANALOG PARTS OF A SYSTEM
- ANALOG AND DIGITAL GROUNDS MUST BE JOINED AT ONE POINT

Figure 7-26: Treatment of analog and digital grounds with data converters of mixed-signal systems

- MONOLITHIC AND HYBRID ADCS FREQUENTLY HAVE SEPARATE *AGND* AND *DGND* PINS, WHICH MUST BE JOINED TOGETHER AT THE DEVICE.
- THIS ISN'T DONE TO BE DIFFICULT, BUT BECAUSE BONDWIRE VOLTAGE DROPS ARE TOO LARGE TO ALLOW INTERNAL CONNECTION.
- THE BEST SOLUTION TO THE GROUNDING PROBLEM ARISING FROM THIS REQUIREMENT IS TO CONNECT BOTH PINS TO SYSTEM "ANALOG GROUND."
- IT IS LIKELY THAT NEITHER THE DIGITAL NOISE SO INTRODUCED IN THE SYSTEM *AGND*, NOR THE SLIGHT LOSS OF DIGITAL NOISE IMMUNITY WILL SERIOUSLY AFFECT THE SYSTEM PERFORMANCE.

Note that analog and digital ground in a system must be joined at some point, to allow signals to be referred to a common potential. This star point, or analog/digital common point, is chosen so that it does not introduce digital currents into the ground of the analog part of the system—it is often convenient to make the connection at the power supplies.

Note also that many ADCs and DACs have separate *analog ground* (AGND) and *digital ground* (DGND) pins. On the device data sheets, users are often advised to connect these pins together at the package. This seems to conflict with the advice to connect analog and digital ground at the power supplies, and, in systems with more than one converter, with the advice to join the analog and digital ground at a single point.

There is, in fact, no conflict. The labels "analog ground" and "digital ground" on these pins refer to the parts of the converter to which the pins are connected, and not to the system grounds to which they must go. For example, with an ADC, generally these two pins should be joined together and to the *analog* ground of the system. It is not possible to join the two pins within the IC package, because the analog part of the converter cannot tolerate the voltage drop resulting from the digital current flowing in the bond wire to the chip. But they can be so tied, *externally*.

Figure 7-27 illustrates this concept of ground connections for an ADC. If these pins are connected in this way, the digital noise immunity of the converter is diminished somewhat by the amount of common-mode noise between the digital and analog system grounds. However, since digital noise immunity is of the order of hundreds or thousands of millivolts, this factor is unlikely to be important.

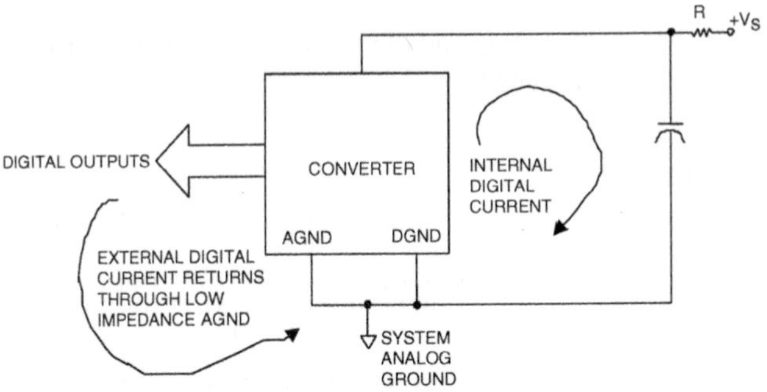

Figure 7-27: Analog (AGND) and digital ground (DGND) pins of a data converter should be returned to system analog ground

The analog noise immunity is diminished only by the external digital currents of the converter itself flowing in the analog ground. These currents should be kept quite small, and this can be minimized by ensuring that the converter outputs don't see heavy loads. A good solution towards this is to use a low input current buffer at the ADC output, such as a CMOS buffer-register IC.

If the logic supply to the converter is isolated with a small resistance and decoupled to analog ground with a local 0.1 μF capacitor, all the fast-edge digital currents of the converter will return to ground through the capacitor, and will not appear in the external ground circuit. If the analog ground impedance is maintained low, as it should be for adequate analog performance, additional noise due to the external digital ground current should rarely present a problem.

Ground Planes

Related to the star ground system discussed earlier is the use of a *ground plane*. To implement a ground plane, one side of a double-sided PCB (or one layer of a multilayer one) is made of continuous copper and used as ground. The theory behind this is that the large amount of metal will have as low a resistance as is possible. It will, because of the large flattened conductor pattern, also have as low an inductance as possible. It then offers the best possible conduction, in terms of minimizing spurious ground difference voltages across the conducting plane.

Note that ground plane concept can also be extended to include *voltage planes*. A voltage plane offers advantages similar to a ground plane, i.e., a very low impedance conductor, but is dedicated to one (or more) of the system supply voltages. Thus a system can have more than one voltage plane, as well as a ground plane.

It has been sometimes argued that ground planes shouldn't be used, as they are liable to introduce manufacture and assembly problems. Such an argument may have had limited validity some years ago when PCB adhesives were less well developed, wave-soldering less reliable, and solder resist techniques less well understood, but not today.

A summary of key points related to the construction and operation of ground planes is contained in Figure 7-28.

- ONE ENTIRE PCB SIDE (OR LAYER) IS A CONTINUOUS GROUNDED CONDUCTOR.
- THIS GIVES MINIMUM GROUND RESISTANCE AND INDUCTANCE, BUT ISN'T ALWAYS SUFFICIENT TO SOLVE ALL GROUNDING PROBLEMS.
- BREAKS IN GROUND PLANES CAN IMPROVE OR DEGRADE CIRCUIT PERFORMANCE — THERE IS NO GENERAL RULE.
- YEARS AGO GROUND PLANES WERE DIFFICULT TO FABRICATE. TODAY THEY AREN'T.
- MULTI-LAYER, GROUND AND VOLTAGE PLANE PCB DESIGNS ARE STANDARD

Figure 7-28: Characteristics of ground planes

While ground planes solve many ground impedance problems, it should still be understood they aren't a panacea. Even a continuous sheet of copper foil has residual resistance and inductance, and in some circumstances, these can be enough to prevent proper circuit function. Designers should be wary of injecting very high currents in a ground plane, as they can produce voltage drops that interfere with sensitive circuitry.

Skin Effect

At high frequencies, also consider *skin effect*, where inductive effects cause currents to flow only in the outer surface of conductors. Note that this is in contrast to the earlier discussions of this section on dc resistance of conductors.

The skin effect has the consequence of increasing the resistance of a conductor at high frequencies. Note also that this effect is separate from the increase in impedance due to the effects of the self-inductance of conductors as frequency is increased.

Skin effect is quite a complex phenomenon, and detailed calculations are beyond the scope of this discussion. However, a good approximation for copper is that the skin depth in centimeters is $6.61/\sqrt{f}$, (f in Hz).

A summary of the skin effect within a typical PCB conductor foil is shown in Figure 7-29. Note that this copper conductor cross-sectional view assumes looking into the *side* of the conducting trace.

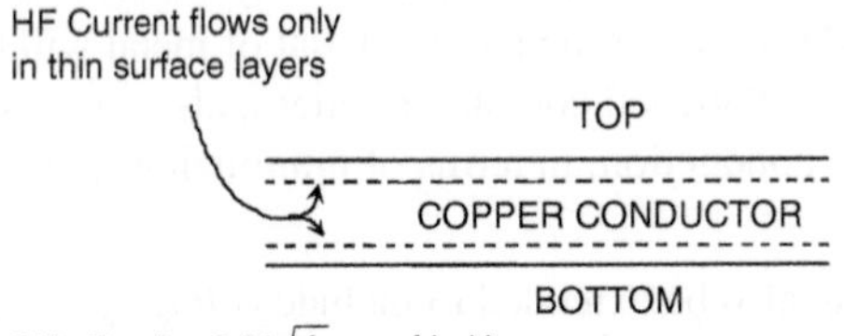

- Skin Depth: $6.61\sqrt{f}$ cm, f in Hz
- Skin Resistance: $2.6 \times 10^{-7}\sqrt{f}$ Ω per square, f in Hz
- Since skin currents flow in both sides of a PC track, the value of skin resistance in PCBs must take account of this

Figure 7-29: Skin depth in a PC conductor

Assuming that skin effects become important when the skin depth is less than 50% of the thickness of the conductor, this tells us that for a typical PC foil, we must be concerned about skin effects at frequencies above approximately 12 MHz.

Where skin effect is important, the resistance for copper is $2.6\times10^{-7}\ \sqrt{f}\ \Omega$ per square, (f in Hz). This formula is invalid if the skin thickness is greater than the conductor thickness (i.e., at dc or LF).

Figure 7-30 illustrates a case of a PCB conductor with current flow, as separated from the ground plane underneath.

In this diagram, note the (dotted) regions of HF current flow, as reduced by the skin effect. When calculating skin effect in PCBs, it is important to remember that current generally flows in both sides of the PC foil (this is not necessarily the case in microstrip lines, see below), so the resistance per square of PC foil may be half the above value.

Transmission Lines

We earlier considered the benefits of outward and return signal paths being close together so that inductance is minimized. As shown previously in Figure 7-30, when an HF signal flows in a PC track running over a ground plane, the arrangement functions as a *microstrip* transmission line, and the majority of the return current flows in the ground plane underneath the line.

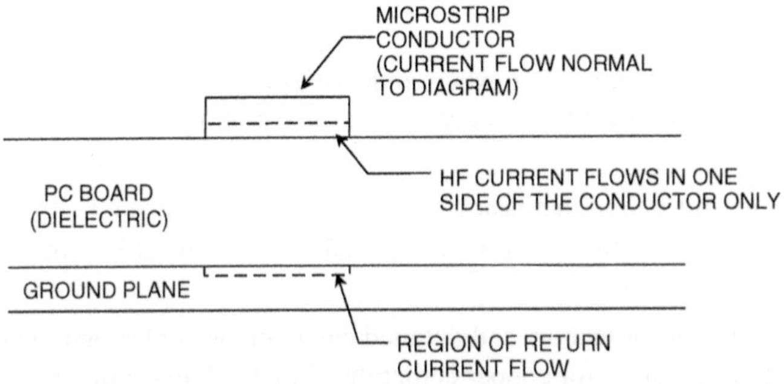

Figure 7-30: Skin effect with PC conductor and ground plane

Figure 7-31 shows the general parameters for a microstrip transmission line, given the conductor width, w, dielectric thickness, h, and the dielectric constant, E_r.

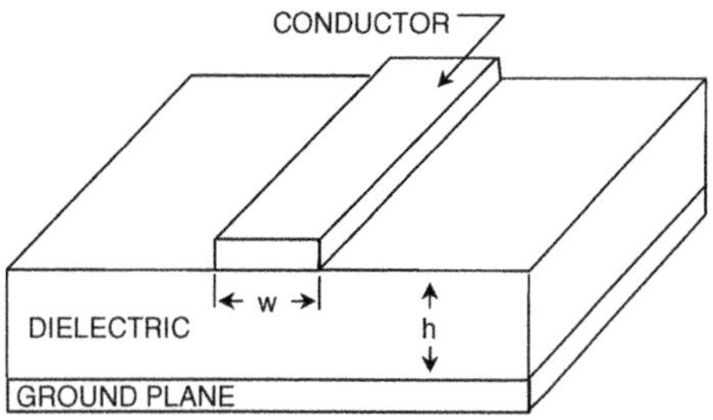

Figure 7-31: A PCB microstrip transmission line is an example of a controlled impedance conductor pair

The characteristic impedance of such a microstrip line will depend upon the width of the track and the thickness and dielectric constant of the PCB material. Designs of microstrip lines are covered in more detail within section six of this chapter.

For most dc and lower frequency applications, the characteristic impedance of PCB traces will be relatively unimportant. Even at frequencies where a track over a ground plane behaves as a transmission line, it is not necessary to worry about its characteristic impedance or proper termination if the free space wavelengths of the frequencies of interest are greater than ten times the length of the line.

However, at VHF and higher frequencies it is possible to use PCB tracks as microstrip lines within properly terminated transmission systems. Typically the microstrip will be designed to match standard coaxial cable impedances, such as 50 Ω, 75 Ω or 100 Ω, simplifying interfacing.

Note that if losses in such systems are to be minimized, the PCB material must be chosen for low high frequency losses. This usually means the use of Teflon or some other comparably low-loss PCB material. Often, though, the losses in short lines on cheap glass-fiber board are small enough to be quite acceptable.

Be Careful with Ground Plane Breaks

Wherever there is a break in the ground plane beneath a conductor, the ground plane return current must by necessity flow *around* the break. As a result, both the inductance and the vulnerability of the circuit to external fields are increased. This situation is diagrammed in Figure 7-32, where conductors A and B must cross one another.

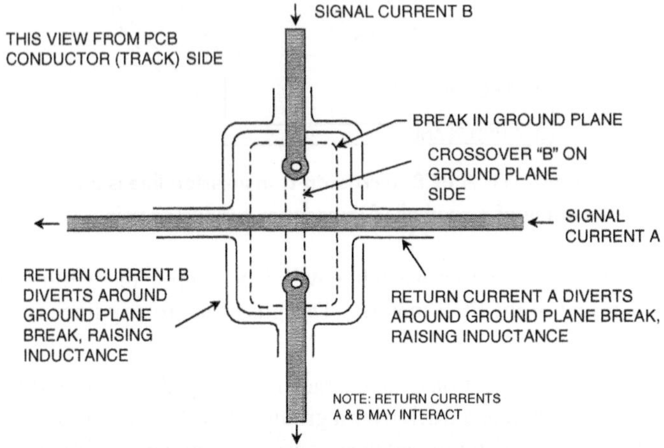

Figure 7-32: A ground plane break raises circuit inductance and increases vulnerability to external fields

Where such a break is made to allow a crossover of two perpendicular conductors, it would be far better if the second signal were carried across both the first and the ground plane by means of a piece of wire. The ground plane then acts as a shield between the two signal conductors, and the two ground return currents, flowing in opposite sides of the ground plane as a result of skin effects, do not interact.

With a multilayer board, both the crossover and the continuous ground plane can be accommodated without the need for a wire link. Multilayer PCBs are expensive and harder to trouble-shoot than more simple double-sided boards, but do offer even better shielding and signal routing. The principles involved remain unchanged but the range of layout options is increased.

The use of double-sided or multilayer PCBs with at least one continuous ground plane is undoubtedly one of the most successful design approaches for high performance mixed signal circuitry. Often the impedance of such a ground plane is sufficiently low to permit the use of a single ground plane for both analog and digital parts of the system. However, whether or not this is possible does depend upon the resolution and bandwidth required, and the amount of digital noise present in the system.

Ground Isolation Techniques

While the use of ground planes does lower impedance and helps greatly in lowering ground noise, there may still be situations where a prohibitive level of noise exists. In such cases, the use of ground error minimization and isolation techniques can be helpful.

Another illustration of a common-ground impedance coupling problem is shown in Figure 7-33. In this circuit, a precision gain-of-100 preamp amplifies a low level signal V_{IN}, using an AD8551 chopper-stabilized amplifier for best dc accuracy. At the load end, the signal V_{OUT} is measured with respect to G2, the local

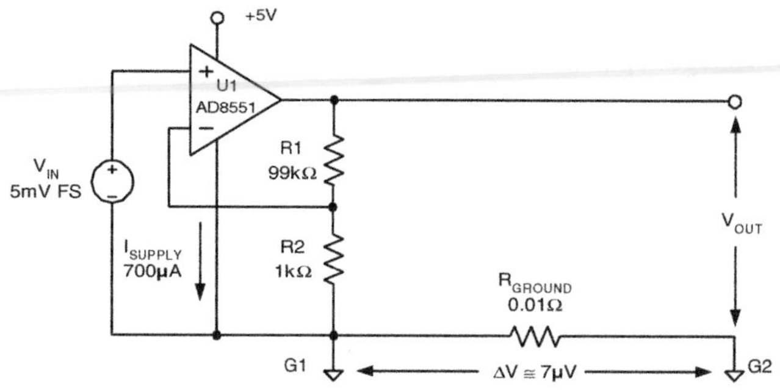

Figure 7-33: Unless care is taken, even small common ground currents can degrade precision amplifier accuracy

ground. Because of the small 700 µA I_{SUPPLY} of the AD8551 flowing between G1 and G2, there is a 7 µV ground error—about seven times the typical input offset expected from the op amp.

This error can be avoided simply by routing the negative supply pin current of the op amp back to star ground G2 as opposed to ground G1, by using a separate trace. This step eliminates the G1-G2 path power supply current, and so minimizes the ground leg voltage error. Note that little error will be developed in the "hot" V_{OUT} lead, as long as the current drain at the load end is small.

In some cases, there may be simply unavoidable ground voltage differences between a source signal and the load point where it is to be measured. Within the context of this "same-board" discussion, this might require rejecting ground error voltages of several tens-of-mV. Or, should the source signal originate from an "off-board" source, the magnitude of the common-mode voltages to be rejected can easily rise into a several volt range (or even tens-of-volts).

Fortunately, full signal transmission accuracy can still be accomplished in the face of such high noise voltages, by employing a principle discussed earlier. This is the use of a differential-input, *ground isolation* amplifier. The ground isolation amplifier minimizes the effect of ground error voltages between stages by processing the signal in differential fashion, thereby rejecting CM voltages by a substantial margin (typically 60 dB or more).

Two ground isolation amplifier solutions are shown in Figure 7-34. This diagram can alternately employ either the AD629 to handle CM voltages up to ±270 V, or the AMP03, which is suitable for CM voltages up to ±20 V.

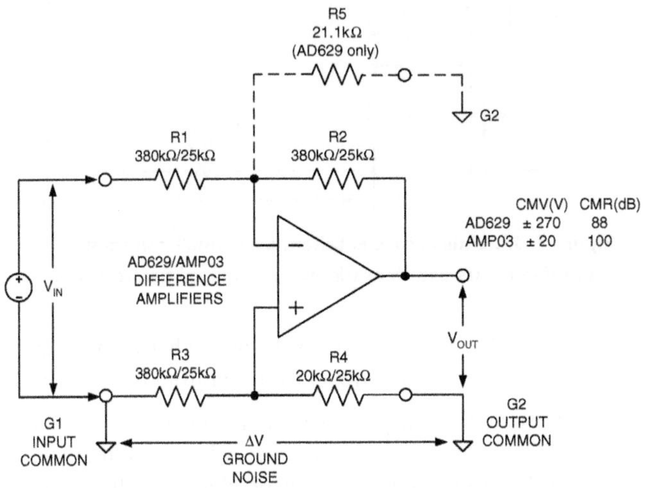

Figure 7-34: A differential input ground isolating amplifier allows high transmission accuracy by rejecting ground noise voltage between source (G1) and measurement (G2) grounds

In the circuit, input voltage V_{IN} is referred to G1, but must be measured with respect to G2. With the use of a high CMR unity-gain difference amplifier, the noise voltage ΔV existing between these two grounds is easily rejected. The AD629 offers a typical CMR of 88 dB, while the AMP03 typically achieves 100 dB. In the AD629, the high CMV rating is done by a combination of high CM attenuation, followed by differential gain, realizing a net differential gain of unity. The AD629 uses the first listed value resistors noted in the figure for R1–R5. The AMP03 operates as a precision four-resistor differential amplifier, using the 25 kΩ value R1–R4 resistors noted. Both devices are complete, one package solutions to the ground-isolation amplifier.

This scheme allows relative freedom from tightly controlling ground drop voltages, or running additional and/or larger PCB traces to minimize such error voltages. Note that it can be implemented with either the fixed gain difference amplifiers shown, or with a standard in amp IC, configured for unity gain. The AD623, for example, also allows single-supply use. In any case, signal polarity is also controllable by simple reversal of the difference amplifier inputs.

In general terms, transmitting a signal from one point on a PCB to another for measurement or further processing can be optimized by two key interrelated techniques. These are the use of high-impedance, differential signal-handling techniques. The high impedance loading of an in amp minimizes voltage drops, and differential sensing of the remote voltage minimizes sensitivity to ground noise.

When the further signal processing is A/D conversion, these transmission criteria can be implemented *without* adding a differential ground isolation amplifier stage. Simply select an ADC that operates differentially. The high input impedance of the ADC minimizes load sensitivity to the PCB wiring resistance. In addition, the differential input feature allows the output of the source to be sensed directly at the source output terminals (even if single-ended). The CMR of the ADC then eliminates sensitivity to noise voltages between the ADC and source grounds.

An illustration of this concept using an ADC with high impedance differential inputs is shown in Figure 7-35. Note that the general concept can be extended to virtually any signal source, driving any load. All loads, even single-ended ones, become differential-input by adding an appropriate differential input stage.

The differential input can be provided by either a fully developed high-Z in amp, or in many cases it can be a simple subtractor stage op amp, such as Figure 7-34.

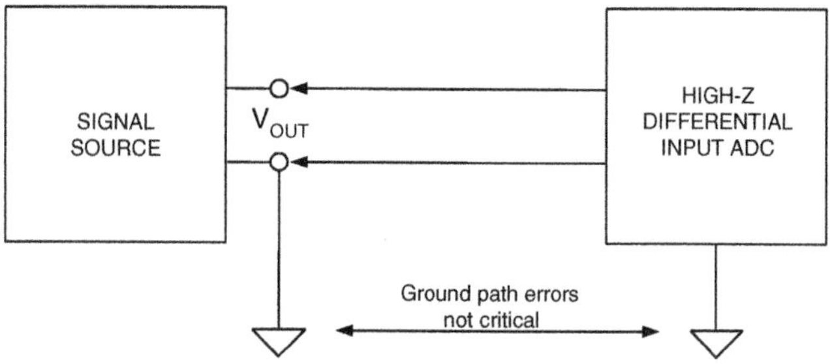

Figure 7-35: A high-impedance differential input ADC also allows high transmission accuracy between source and load

Static PCB Effects

Leakage resistance is the dominant static circuit board effect. Contamination of the PCB surface by flux residues, deposited salts, and other debris can create leakage paths between circuit nodes. Even on well-cleaned boards, it is not unusual to find 10 nA or more of leakage to nearby nodes from 15 V supply rails. Nanoamperes of leakage current into the wrong nodes often cause volts of error at a circuit's output; for example, 10 nA into a 10 megohm resistance causes 0.1 V of error. Unfortunately, the standard op amp pinout places the $-V_S$ supply pin next to the + input, which is often hoped to be at high impedance. To help identify nodes sensitive to the effects of leakage currents ask the simple question: If a spurious current of a few nanoamperes or more were injected into this node, would it matter?

If the circuit is already built, moisture sensitivity can be localized to a suspect node with a classic test. While observing circuit operation, blow on potential trouble spots through a simple soda straw. The straw focuses the breath's moisture which, with the board's salt content in susceptible portions of the design, disrupts circuit operation upon contact.

There are several means of eliminating simple surface leakage problems. Thorough washing of circuit boards to remove residues helps considerably. A simple procedure includes vigorously brushing the boards with isopropyl alcohol, followed by thorough washing with deionized water and an 85°C bakeout for a few hours. Be careful when selecting board-washing solvents, though. When cleaned with certain solvents, some water-soluble fluxes create salt deposits, exacerbating the leakage problem.

Unfortunately, if a circuit displays sensitivity to leakage, even the most rigorous cleaning can offer only a temporary solution. Problems soon return upon handling, or exposure to foul atmospheres, and high humidity. Some additional means must be sought to stabilize circuit behavior, such as conformal surface coating.

Fortunately, there is an answer to this, namely *guarding*, which offers a fairly reliable and permanent solution to the problem of surface leakage. Well-designed guards can eliminate leakage problems, even for circuits exposed to harsh industrial environments. Two schematics illustrate the basic guarding principle, as applied to typical inverting and noninverting op amp circuits.

Figure 7-36 illustrates an inverting mode guard application. In this case, the op amp reference input is grounded, so the guard is a grounded ring surrounding all leads to the inverting input, as noted by the dotted line.

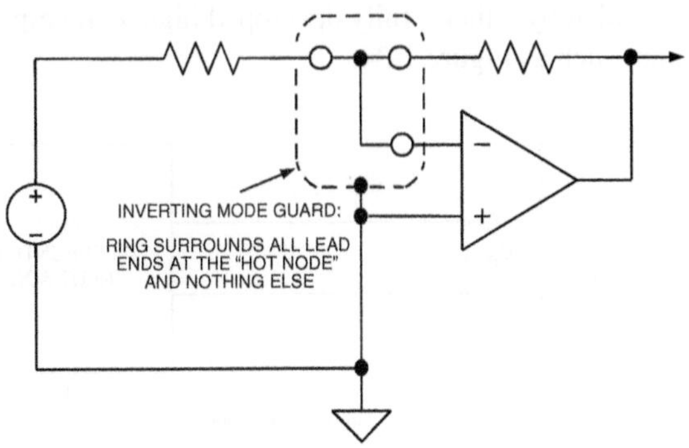
INVERTING MODE GUARD:
RING SURROUNDS ALL LEAD ENDS AT THE "HOT NODE" AND NOTHING ELSE

Figure 7-36: Inverting mode guard encloses all op amp inverting input connections within a grounded guard ring

Guarding basic principles are simple: *Completely* surround sensitive nodes with conductors that can readily sink stray currents, and maintain the guard conductors at the exact potential of the sensitive node (as otherwise the guard will serve as a leakage source rather than a leakage sink). For example, to keep leakage into a node below 1 pA (assuming 1000-megohm leakage resistance) the guard and guarded node must be within 1 mV. Generally, the low offset of a modern op amp is sufficient to meet this criterion.

There are important caveats to be noted with implementing a true high-quality guard. For traditional through-hole PCB connections, the guard pattern should appear on *both* sides of the circuit board, to be most effective. And, it should also be connected along its length by several vias. Finally, when either justified or required by the system design parameters, do make an effort to include guards in the PCB design process from the outset—there is little likelihood that a proper guard can be added as an afterthought.

Figure 7-37 illustrates the case for a noninverting guard. In this instance the op amp reference input is directly driven by the source, which complicates matters considerably. Again, the guard ring completely surrounds all of the input nodal connections. In this instance, however, the guard is driven from the low impedance feedback divider connected to the inverting input.

Usually the guard-to-divider junction will be a direct connection, but in some cases a unity gain buffer might be used at "X" to drive a cable shield, or also to maintain the lowest possible impedance at the guard ring.

In lieu of the buffer, another useful step is to use an additional, directly grounded screen ring, "Y," which surrounds the inner guard and the feedback nodes as shown. This step costs nothing except some added layout time, and will greatly help buffer leakage effects into the higher impedance inner guard ring.

Of course what hasn't been addressed to this point is just how the op amp itself is connected into these guarded islands without compromising performance. The traditional method, using a TO-99 metal can package device, was to employ double-sided PCB guard rings, with both op amp inputs terminated within the guarded ring.

The high impedance sensor discussions in Chapter 4 use the above-described method. The section immediately following illustrates how more modern IC packages can be mounted to PCB patterns, and take advantage of guarding and low-leakage operation.

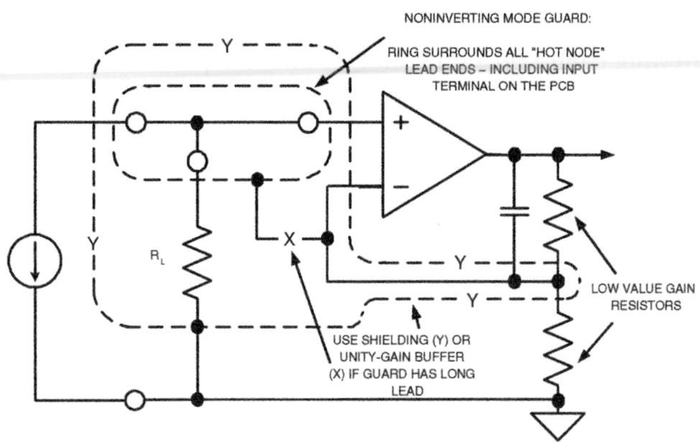

Figure 7-37: Noninverting mode guard encloses all op amp noninverting input connections within a low impedance, driven guard ring

Sample MINIDIP and SOIC Op Amp PCB Guard Layouts

Modern assembly practices have favored smaller plastic packages such as 8-pin MINIDIP and SOIC types. Some suggested partial layouts for guard circuits using these packages is shown in Figures 7-38 and 7-39. While guard traces may also be possible with even more tiny op amp footprints, such as SOT23 and so forth, the required trace separations become even more confining, challenging the layout designer as well as the manufacturing processes.

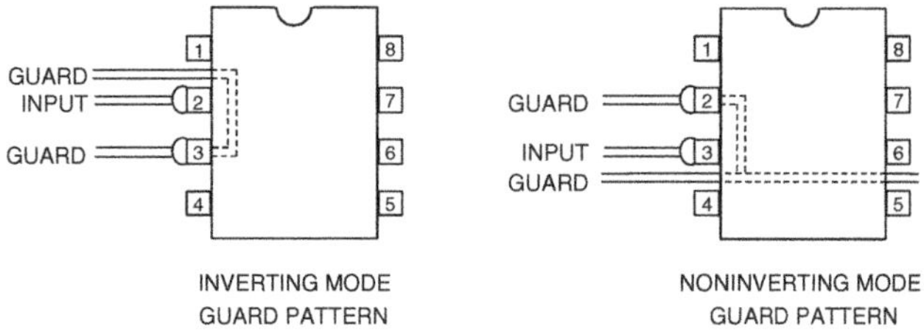

Figure 7-38: PCB guard patterns for inverting and noninverting mode op amps using 8-pin MINIDIP (N) package

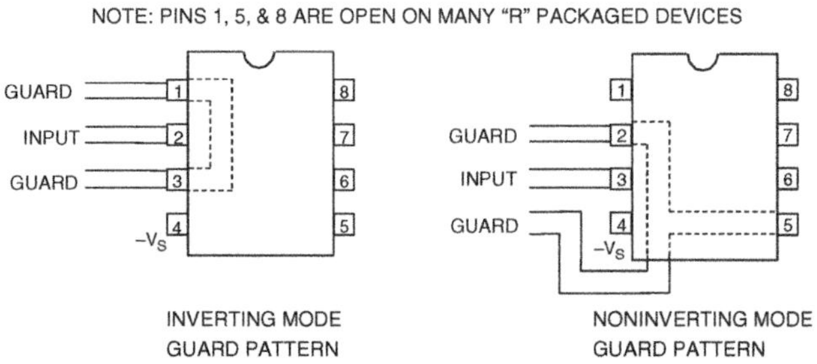

Figure 7-39: PCB guard patterns for inverting and noninverting mode op amps using 8-pin SOIC (R) package

For the ADI "N" style MINIDIP package, Figure 7-38 illustrates how guarding can be accomplished for inverting (left) and noninverting (right) operating modes. This setup would also be applicable to other op amp devices where relatively high voltages occur at Pin 1 or 4. Using a standard 8-pin DIP outline, it can be noted that this package's 0.1" pin spacing allows a PC trace (here, the guard trace) to pass between adjacent pins. This is the key to implementing effective DIP package guarding, as it can adequately prevent a leakage path from the $-V_S$ supply at Pin 4, or from similar high potentials at Pin 1.

For the left-side inverting mode, note that the Pin 3 connected and grounded guard traces surround the op amp inverting input (Pin 2), and run parallel to the input trace. This guard would be continued out to and around the source and feedback connections of Figure 7-36 (or other similar circuit), including an input pad in the case of a cable. In the right-side noninverting mode, the guard voltage is the feedback divider voltage to Pin 2. This corresponds to the inverting input node of the amplifier, from Figure 7-37.

Note that in both cases of Figure 7-38, the guard physical connections shown are only partial—an actual layout would include all sensitive nodes within the circuit. In both the inverting and the noninverting modes using the MINIDIP or other through-hole style package, the PCB guard traces should be located on both sides of the board, with top and bottom traces connected with several vias.

Things become slightly more complicated when using guarding techniques with the SOIC surface-mount ("R") package, as the 0.05" pin spacing doesn't easily allow routing of PCB traces between the pins. But, there is still an effective guarding answer, at least for the inverting case. Figure 7-39 shows guards for the ADI "R" style SOIC package.

Note that for many single op amp devices in this SOIC "R" package, Pins 1, 5, and 8 are "no connect" pins. For such instances, this means that these locations can be employed in the layout to route guard traces. In the case of the inverting mode (left), the guarding is still completely effective, with the dummy Pin 1 and Pin 3 serving as the grounded guard trace. This is a fully effective guard without compromise. Also, with SOIC op amps, much of the circuitry around the device will not use through-hole components. So, the guard ring may only be necessary on the op amp PCB side.

In the case of the follower stage (right), the guard trace must be routed around the negative supply at Pin 4, and thus Pin 4 to Pin 3 leakage isn't fully guarded. For this reason, a precision high impedance follower stage using an SOIC package op amp isn't generally recommended, as guarding isn't possible for dual supply connected devices.

However, an exception to this caveat does apply to the use of a *single-supply* op amp as a noninverting stage. For example, if the AD8551 is used, Pin 4 becomes ground, and some degree of intrinsic guarding is then established by default.

Dynamic PCB Effects

Although static PCB effects can come and go with changes in humidity or board contamination, problems that most noticeably affect the dynamic performance of a circuit usually remain relatively constant. Short of a new design, washing or any other simple fixes can't fix them. As such, they can permanently and adversely affect a design's specifications and performance. The problems of stray capacitance, linked to lead and component placement, are reasonably well known to most circuit designers. Since lead placement can be permanently dealt with by correct layout, any remaining difficulty is solved by training assembly personnel to orient components or bend leads optimally.

Dielectric absorption (DA), on the other hand, represents a more troublesome and still poorly understood circuit-board phenomenon. Like DA in discrete capacitors, DA in a printed-circuit board can be modeled by a series resistor and capacitor connecting two closely spaced nodes. Its effect is inverse with spacing and linear with length.

As shown in Figure 7-40, the RC model for this effective capacitance ranges from 0.1 pF to 2.0 pF, with the resistance ranging from 50 MΩ to 500 MΩ. Values of 0.5 pF and 100 MΩ are most common. Consequently, circuit-board DA interacts most strongly with high-impedance circuits.

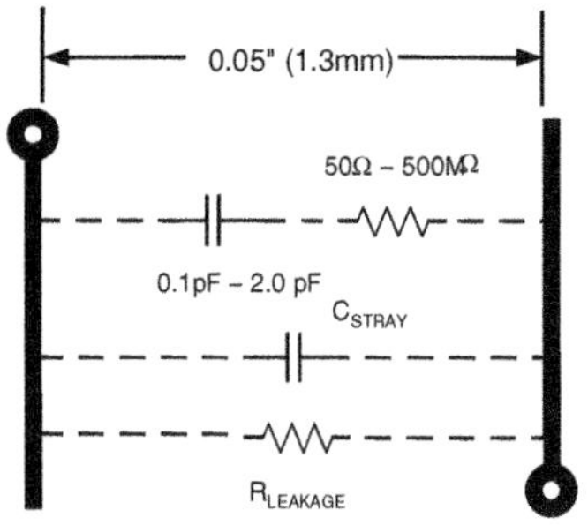

Figure 7-40: DA plagues dynamic response of PCB-based circuits

PCB DA most noticeably influences dynamic circuit response, for example, settling time. Unlike circuit leakage, the effects aren't usually linked to humidity or other environmental conditions but, rather, are a function of the board's dielectric properties. The chemistry involved in producing plated-through holes seems to exacerbate the problem. If circuits don't meet expected transient response specs, consider PCB DA as a possible cause.

Fortunately, there are solutions. As in the case of capacitor DA, external components can be used to compensate for the effect. More importantly, surface guards that totally isolate sensitive nodes from parasitic coupling often eliminate the problem (note that these guards should be duplicated on both sides of the board, in cases of through-hole components). As previously noted, low loss PCB dielectrics are also available.

PCB "hook," similar if not identical to DA, is characterized by variation in effective circuit-board capacitance with frequency (see Reference 1). In general, it affects high impedance circuit transient response where board capacitance is an appreciable portion of the total in the circuit. Circuits operating at frequencies below 10 kHz are the most susceptible. As in circuit board DA, the board's chemical makeup very much influences its effects.

Stray Capacitance

When two conductors aren't short-circuited together, or totally screened from each other by a conducting (Faraday) screen, there is a capacitance between them. So, on any PCB, there will be a large number of capacitors associated with any circuit (which may or may not be considered in models of the circuit). Where high frequency performance matters (and even dc and VLF circuits may use devices with high Ft and therefore be vulnerable to HF instability), it is very important to consider the effects of this stray capacitance.

Any basic textbook will provide formulas for the capacitance of parallel wires and other geometric configurations (see References 9 and 10). The example to be considered in this discussion is the parallel plate capacitor, often formed by conductors on opposite sides of a PCB. The basic diagram describing this capacitance is shown in Figure 7-41.

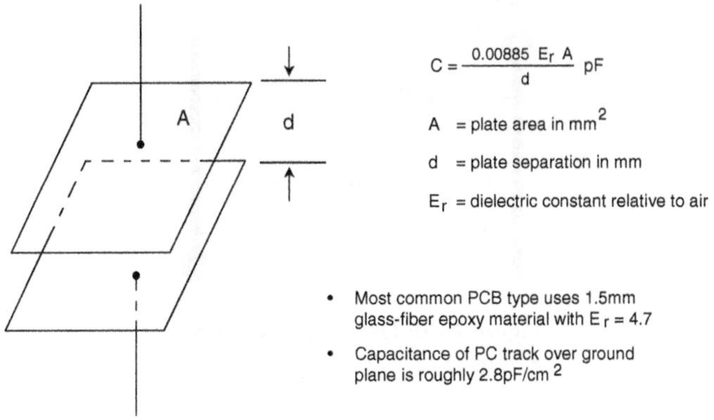

A
d
C = 0.00885 Er A / d pF
A = plate area in mm2
d = plate separation in mm
Er = dielectric constant relative to air
• Most common PCB type uses 1.5mm glass-fiber epoxy material with Er = 4.7
• Capacitance of PC track over ground plane is roughly 2.8pF/cm 2

Figure 7-41: Capacitance of two parallel plates

From this formula, it can be calculated that for general-purpose PCB material ($E_r = 4.7$, d = 1.5mm), the capacitance between conductors on opposite sides of the board is just under $3pF/cm^2$. In general, such capacitance will be parasitic, and circuits must be designed so that it does not affect their performance.

While it is possible to use PCB capacitance in place of small discrete capacitors, the dielectric properties of common PCB substrate materials cause such capacitors to behave poorly. They have a rather high temperature coefficient and poor Q at high frequencies, which makes them unsuitable for many applications. Boards made with lower-loss dielectrics such as Teflon are expensive exceptions to this rule.

Capacitive Noise and Faraday Shields

There is a capacitance between any two conductors separated by a dielectric (air or vacuum are dielectrics). If there is a change of voltage on one, there will be a movement of charge on the other. A basic model for this is shown in Figure 7-42.

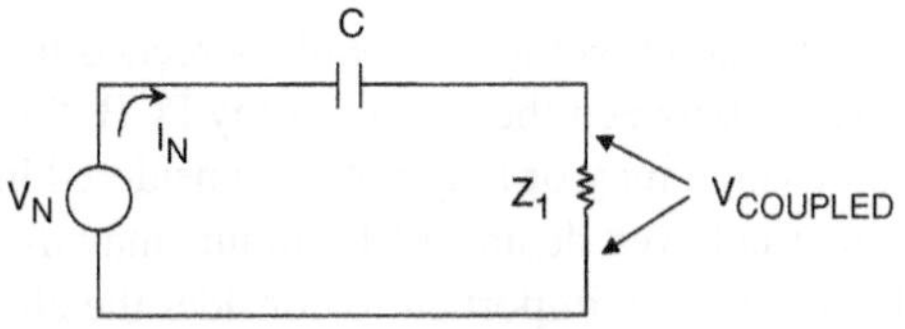

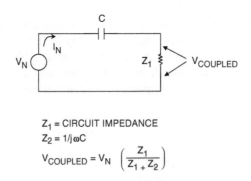

Z_1 = CIRCUIT IMPEDANCE

$Z_2 = 1/j\omega C$

$$V_{COUPLED} = V_N \left(\frac{Z_1}{Z_1 + Z_2}\right)$$

Figure 7-42: Capacitive coupling equivalent circuit model

It is evident that the noise voltage, $V_{COUPLED}$, appearing across Z_1, may be reduced by several means, all of which reduce noise current in Z_1. They are reduction of the signal voltage V_N, reduction of the frequency involved, reduction of the capacitance, or reduction of Z_1 itself. Unfortunately, however, often none of these circuit parameters can be freely changed, and an alternate method is needed to minimize the interference. The best solution towards reducing the noise coupling effect of C is to insert a grounded conductor, also known as a *Faraday shield*, between the noise source and the affected circuit. This has the desirable effect of reducing Z_1 noise current, thus reducing $V_{COUPLED}$.

A Faraday shield model is shown by Figure 7-43. In the left picture, the function of the shield is noted by how it effectively divides the coupling capacitance, C. In the right picture the net effect on the coupled voltage across Z_1 is shown. Although the noise current I_N still flows in the shield, most of it is now diverted away from Z_1. As a result, the coupled noise voltage $V_{COUPLED}$ across Z_1 is reduced.

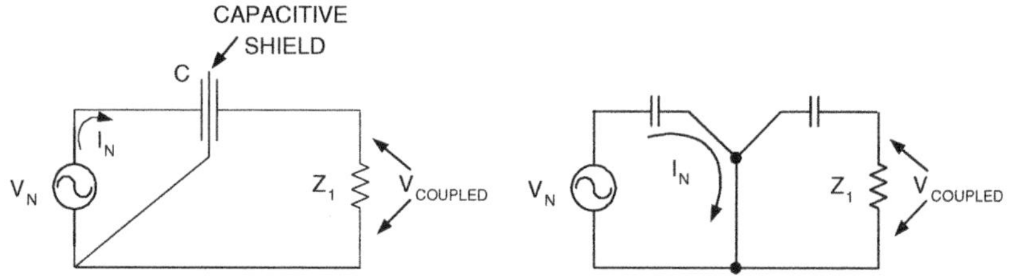

Figure 7-43: An operational model of a Faraday shield

A Faraday shield is easily implemented and almost always successful. Thus capacitively coupled noise is rarely an intractable problem. However, to be fully effective, a Faraday shield must completely block the electric field between the noise source and the shielded circuit. It must also be connected so that the displacement current returns to its source, without flowing in any part of the circuit where it can introduce conducted noise.

The Floating Shield Problem

It is quite important to note here that *a conductor that is intended to function as a Faraday shield must never be left floating, as this almost always increases capacity and exacerbates the noise problem.*

An example of this "floating shield" problem is seen in side-brazed ceramic IC packages. These DIP packages have a small square conducting Kovar lid soldered onto a metallized rim on the ceramic package top. Package manufacturers offer only two options: the metallized rim may be connected to one of the corner pins of the package, or it may be left unconnected.

Most logic circuits have a ground pin at one of the package corners, and therefore the lid is grounded. Alas, many analog circuits don't have a ground pin at a package corner, and the lid is left floating—acting as an antenna for noise. Such circuits turn out to be far more vulnerable to electric field noise than the same chip in a plastic DIP package, where the chip is completely unshielded.

Whenever practical, it is good practice for the user to ground the lid of any side-brazed ceramic IC where the lid is not grounded by the manufacturer, thus implementing an *effective* Faraday shield. This can be done with a wire soldered to the lid (this will not damage the device, as the chip is thermally and electrically isolated from the lid). If soldering to the lid is unacceptable, a grounded phosphor-bronze clip may be used to make the ground connection, or conductive paint from the lid to the ground pin.

A safety note is appropriate at this point. Never attempt to ground such a lid without first verifying that it is unconnected. Occasionally device types are found with the lid connected to a power supply rather than to ground.

A case where a Faraday shield is impracticable is between IC chip bondwires, which has important consequences. The stray capacitance between chip bondwires and associated leadframes is typically ≈ 0.2 pF, with observed values generally between 0.05 pF and 0.6 pF.)

Buffering ADCs against Logic Noise

If a high resolution data converter (ADC or DAC) is connected to a high speed data bus that carries logic noise with a 2 V/ns–5 V/ns edge rate, this noise is easily connected to the converter analog port via stray capacitance across the device. Whenever the data bus is active, intolerable amounts of noise are capacitively coupled into the analog port, thus seriously degrading performance.

This particular effect is illustrated by the diagram of Figure 7-44, where multiple package capacitors couple noisy edge signals from the data bus into the analog input of an ADC.

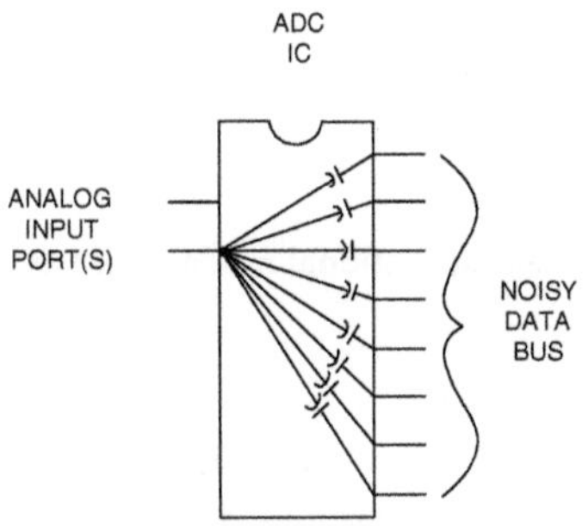

Figure 7-44: A high speed ADC IC sitting on a fast data bus couples digital noise into the analog port, thus limiting performance

Present technology offers no cure for this problem, within the affected IC device itself. The problem also limits performance possible from other broadband monolithic mixed-signal ICs with single-chip analog and digital circuits. Fortunately, this coupled noise problem can be avoided by *not* connecting the data bus directly to the converter.

Instead, *use a CMOS latched buffer as a converter-to-bus interface*, as shown by Figure 7-45. Now the CMOS buffer IC acts as a Faraday shield and dramatically reduces noise coupling from the digital bus. This solution costs money, occupies board area, reduces reliability (very slightly), consumes power, and complicates the design—but it does improve the signal-to-noise ratio of the converter. The designer must decide whether it is worthwhile for individual cases, but in general it is highly recommended.

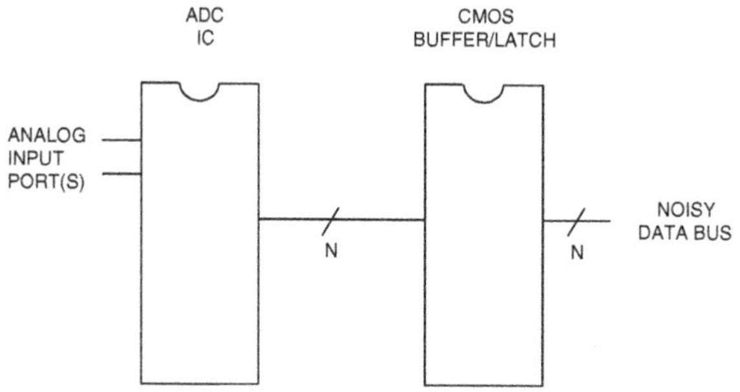

- THE OUTPUT BUFFER/LATCH ACTS AS A FARADAY SHIELD BETWEEN "N" LINES OF A FAST, NOISY DATA BUS AND A HIGH PERFORMANCE ADC
- THIS MEASURE ADDS COST, BOARD AREA, POWER CONSUMPTION, RELIABILITY REDUCTION, DESIGN COMPLEXITY, AND MOST IMPORTANTLY, IMPROVED PERFORMANCE

Figure 7-45: A high speed ADC IC using a CMOS buffer/latch at the output shows enhanced immunity of digital data bus noise

References: PCB Design Issues

1. W. Doeling, W. Mark, T. Tadewald, and P. Reichenbacher, "Getting Rid of Hook: The Hidden PC-Board Capacitance," **Electronics,** October 12, 1978, pp. 111–117.
2. Alan Rich, "Shielding and Guarding," **Analog Dialogue,** Vol. 17, No. 1, 1983, p. 8.
3. Ralph Morrison, **Grounding and Shielding Techniques in Instrumentation**, 3rd Edition, John Wiley, Inc., 1986, ISBN: 0-471-83805-5.
4. Henry W. Ott, **Noise Reduction Techniques in Electronic Systems, 2nd Edition**, John Wiley, Inc., 1988, ISBN: 0-471-85068-3.
5. Paul Brokaw, "An IC Amplifier User's Guide to Decoupling, Grounding and Making Things Go Right for a Change," **Analog Devices AN202**.
6. Paul Brokaw, "Analog Signal-Handling for High Speed and Accuracy," **Analog Devices AN342**.
7. Paul Brokaw and Jeff Barrow, "Grounding for Low- and High-Frequency Circuits," **Analog Devices AN345.**
8. Jeff Barrow, "Avoiding Ground Problems in High Speed Circuits," **RF Design**, July 1989.
9. B. I. & B. Bleaney, **Electricity & Magnetism**, Oxford at the Clarendon Press, 1957, pp. 23, 24, and 52.
10. G. W. A. Dummer, H. Nordenberg, **Fixed and Variable Capacitors**, McGraw-Hill, 1960, pp. 11–13.

Op Amp Power Supply Systems

Walt Jung, Walt Kester

Op amp circuits have traditionally been powered from well-regulated, low noise linear power supplies. This type of power system is typically characterized by medium-to-low power conversion efficiency. Such linear regulators usually excel in terms of self-generated and radiated noise components. If the designer's life were truly simple, it might continue with such familiar designs offering good performance and minimal side effects.

But, the designer's life is hardly so simple. Modern systems may allow using linear regulators, but multiple output levels and/or polarities are often required. There may also be some additional requirements set for efficiency, which may dictate the use of dc-dc conversion techniques, and, unfortunately, their higher associated noise output.

This section addresses power supply design issues for op amp systems, taking into account the regulator types most likely to be used. The primary dc power sources are assumed to be either rectified and smoothed ac sources (i.e., mains derived), a battery stack, or a switching regulator output. The latter example could be fed from either a battery or a mains-derived dc source.

As noted in Figure 7-46, linear mode regulation is generally recommended as an optimum starting point in all instances (first bullet). Nevertheless, in some cases, a degree of hybridization between fully linear and switching mode regulation may be required (second bullet). This could be either for efficiency or other diverse reasons.

- High performance analog power systems use linear regulators, with primary power derived from:
  - AC line power
  - Battery power systems
  - DC- DC power conversion systems
- Switching regulators should be avoided if at all possible, but if not…
  - Apply noise control techniques
  - Use quality layout and grounding
  - Be aware of EMI

Figure 7-46: Regulation priorities for op amp power supply systems

Whenever switching-type regulators are involved in powering precision analog circuits, noise control is very likely to be a design issue. Therefore some focus of this section is on minimizing noise when using switching regulators.

Linear IC Regulation

Linear IC voltage regulators have long been standard power system building blocks. After an initial introduction in 5 V logic voltage regulator form, they have since expanded into other standard voltage levels spanning from 3 V to 24 V, handling output currents from as low as 100 mA (or less) to as high as 5 A (or more). For several good reasons, linear style IC voltage regulators have been valuable system components since the early days. As mentioned above, a basic reason is the relatively low noise characteristic vis-à-vis the switching type of regulator. Others are a low parts count and overall simplicity compared to discrete solutions. But, because of their power losses, these linear regulators have also been known for being relatively inefficient. Early generation devices (of which many are still available) required 2 V or more of unregulated input above the regulated output voltage, making them lossy in power terms.

More recently however, linear IC regulators have been developed with more liberal (i.e., lower) limits on minimum input-output voltage. This voltage, known more commonly as *dropout* voltage, has led to what is termed the *Low DropOut* regulator, or more simply, the LDO. Dropout voltage (V_{MIN}) is defined simply as that minimum input-output differential where the regulator undergoes a 2% reduction in output voltage. For example, if a nominal 5.0 V LDO output drops to 4.9 V (–2%) under conditions of an input-output differential of 0.5 V, by this definition the LDO's dropout voltage is 0.5 V.

Dropout voltage is extremely critical to a linear regulator's power efficiency. The lower the voltage allowable across a regulator while still maintaining a regulated output, the less power the regulator dissipates as a result. A low regulator dropout voltage is the key to this, as it takes a lower dropout to maintain regulation as the input voltage lowers. In performance terms, the bottom line for LDOs is simply that more useful power is delivered to the load and less heat is generated in the regulator. LDOs are key elements of power systems providing stable voltages from batteries, such as portable computers, cellular phones, and so forth. This is because they maintain a regulated output down to lower points on the battery's discharge curve. Or, within classic mains-powered raw dc supplies, LDOs allow lower transformer secondary voltages, reducing system shutdowns under brownout conditions, as well as allowing cooler operation.

Some Linear Voltage Regulator Basics

A brief review of three terminal linear IC regulator fundamentals is necessary before understanding the LDO variety. Most (but not all) of the general three-terminal regulator types available today are *positive leg, series style* regulators. This simply means that they control the regulated voltage output by means of a pass element in series with the positive unregulated input. And, although they are fewer in number, there are also *negative leg* series style regulators, which operate in a fashion complementary to the positive units.

A basic hookup diagram of a three terminal regulator is shown in Figure 7-47. In terms of basic functionality, many standard voltage regulators operate in a series mode, three-terminal form, just as shown here. As can be noted from this figure, the three I/O terminals are V_{IN}, GND (or Common), and V_{OUT}. Note also that this regulator block, in the absence of any assigned voltage polarity, could in principle be a positive type regulator. Or, it might also be a negative style of voltage regulator—the principle is the same for both—a common terminal, as well as input and output terminals.

In operation, two power components become dissipated in the regulator, one a function of $V_{IN} - V_{OUT}$ and I_L, plus a second, which is a function of V_{IN} and I_{GROUND}. The first of these is usually dominant. Analysis of the situation will reveal that as the dropout voltage V_{MIN} is reduced, the regulator is able to deliver a higher percentage of the input power to the load, and is thus more efficient, running cooler and saving power. This is the core appeal of the modern LDO type of regulator (see Reference 1).

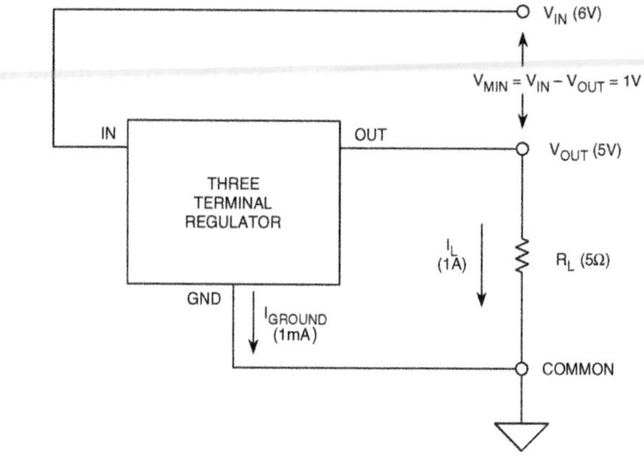

Figure 7-47: A basic three-terminal regulator hookup (either positive or negative)

A more detailed look within a typical regulator block diagram reveals a variety of elements, as is shown in Figure 7-48. Note that all regulators will contain those functional components connected via solid lines. The connections shown dotted indicate options, which might be available when more than three I/O pins are available.

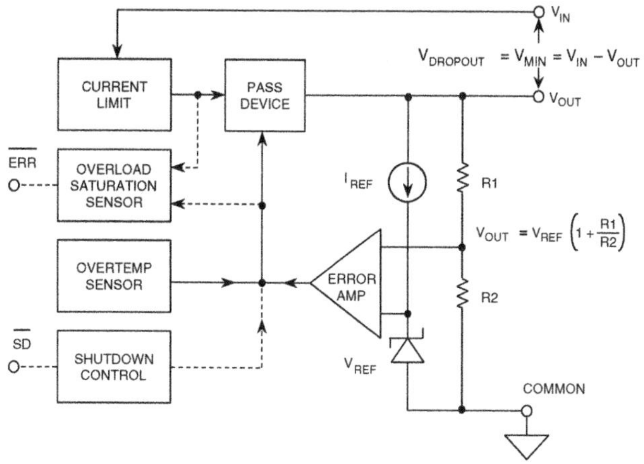

Figure 7-48: Functional diagram of a typical voltage regulator

In operation, a voltage reference block produces a stable voltage V_{REF}, which is almost always a voltage based on the bandgap voltage of silicon, typically ~1.2 V (see Reference 2). This allows output voltages of 3 V or more from supplies as low as 5 V. This voltage drives one input of an error amplifier, with the second input connected to the divider, R1-R2. The error amplifier drives the pass device, which in turn controls the output. The resulting regulated voltage is then simply:

$$V_{OUT} = V_{REF}\left(1+\frac{R1}{R2}\right) \qquad \text{Eq. 7-1}$$

Pass Devices

The pass device is a foremost regulator part, and the type chosen here has a major influence on almost all regulator performance issues. Most notable among these is dropout voltage, V_{MIN}. Analysis shows that the use of an *inverting* mode pass transistor allows the pass device to be effectively saturated, thus minimizing the associated voltage losses. Therefore, this factor makes the two most desirable pass devices for LDO use a PNP bipolar, or a PMOS transistor. These device types achieve the lowest levels of V_{IN}–V_{OUT} required for LDO operation. In contrast, NPN bipolars are poor as pass devices in terms of low dropout, particularly when they are Darlington connected.

Standard fixed-voltage IC regulator architectures illustrate this point regarding pass devices. For example, the fixed-voltage LM309 5 V regulators and family derivatives such as the 7805, 7815, et al, (and their various low and medium current alternates) are poor in terms of dropout voltage. These designs use a Darlington pass connection, not known for low dropout (~1.5 V typical), or for low quiescent current (~5 mA).

±15 V Regulator Using Adjustable Voltage ICs

Later developments in references and three-terminal regulation techniques led to the development of the *voltage-adjustable* regulator. The original IC to employ this concept was the LM317, a positive regulator. The device produces a fixed reference voltage of 1.25 V, appearing between the V_{OUT} and ADJ pins of the IC. External scaling resistors set up the desired output voltage, adjustable in the range of 1.25 V–30 V. A complementary device, the LM337, operates in similar fashion, regulating negative voltages.

An application example using standard *adjustable* three terminal regulators to implement a ±15 V linear power supply is shown in Figure 7-49. This is a circuit that might be used for powering traditional op amp supply rails. It is capable of better line regulation performance than would an otherwise similar circuit, using standard fixed-voltage regulator devices, such as for example 7815 and 7915 ICs. However, in terms of power efficiency it isn't outstanding, due to the use of the chosen ICs, which require 2 V or more of headroom for operation.

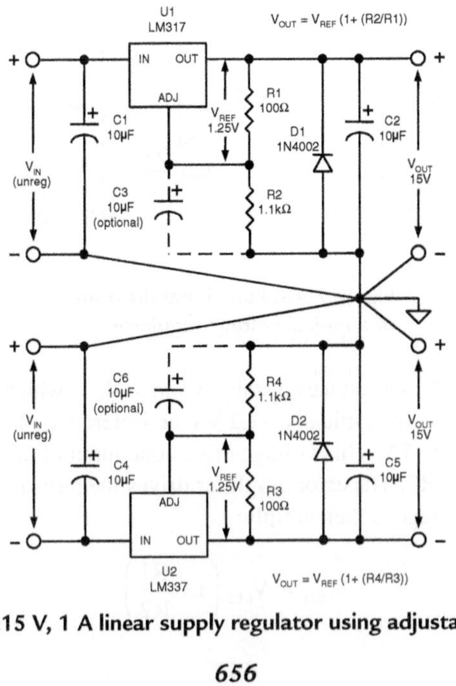

Figure 7-49: A classic ±15 V, 1 A linear supply regulator using adjustable voltage regulator ICs

In the upper portion of this circuit an LM317 adjustable regulator is used, with R2 and R1 chosen to provide a 15 V output at the upper output terminal. If desired, R2 can easily be adjusted for other output levels, according to the figure's V_{OUT} expression. Resistor R1 should be left fixed, as it sets the minimum regulator drain of 10 mA or more.

In this circuit, capacitors C1 and C2 should be tantalum types, and R1-R2 metal films. C3 is optional, but is highly recommended if the lowest level of output noise is desired. The normally reverse biased diode D1 provides a protective output clamp, for system cases where the output voltage would tend to reverse, if one supply should fail. The circuit operates from a rectified and filtered ac supply at V_{IN}, polarized as shown. The output current is determined by choosing the regulator IC for appropriate current capability.

To implement the negative supply portion, the sister device to the LM317 is used, the LM337. The bottom circuit section thus mirrors the operation of the upper, delivering a negative 15 V at the lowest output terminal. Programming of the LM337 for output voltage is similar to that of the LM317, but uses resistors R4 and R3. R4 should be used to adjust the voltage, with R3 remaining fixed. C6 is again optional, but is recommended for reasons of lowest noise.

Low Dropout Regulator Architectures

In contrast to traditional three terminal regulators with Darlington or single-NPN pass devices, low dropout regulators employ lower voltage threshold pass devices. This basic operational difference allows them to operate effectively down to a range of 100 mV–200 mV in terms of their specified V_{MIN}. In terms of use within a system, this factor can have fairly significant operational advantages.

An effective implementation of some key LDO features is contained in the Analog Devices series of anyCAP LDO regulators. Devices of this ADP330X series are so named for their relative insensitivity to the output capacitor, in terms of both its size and ESR. Available in power efficient packages such as the ADI Thermal Coastline (and other thermally-enhanced packages), they come in both stand-alone LDO and LDO controller forms (used with an external PMOS FET). They also offer a wide span of fixed output voltages from 1.8 V to 5 V, with rated current outputs up to 500 mA. User-adjustable output voltage versions are also available. A basic simplified diagram for the family is shown schematically in Figure 7-50.

One of the key differences in the ADP330X LDO series is the use of a high gain vertical PNP pass device, Q1, allowing typical dropout voltages for the series to be on the order of 1 mV/mA for currents of 200 mA or less.

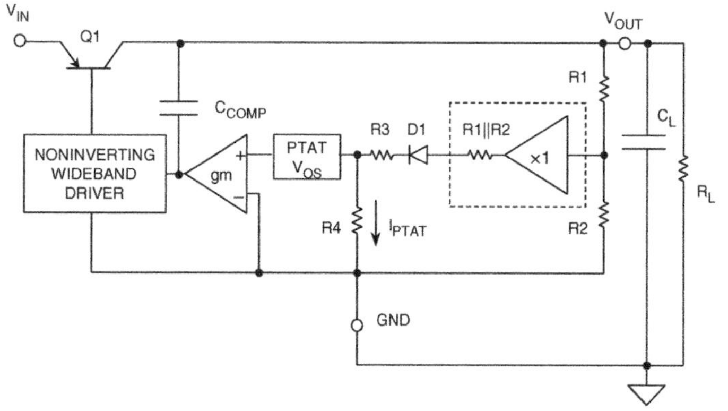

Figure 7-50: The ADP330X anyCAP LDO architecture has both dc and ac performance advantages

In circuit operation, V_{REF} is defined as a reference voltage existing at the output of a zero impedance divider of ratio R1/R2. In Figure 7-50, this is depicted symbolically by the (dotted) unity gain buffer amplifier fed by R1/R2, which has an output of V_{REF}. This reference voltage feeds into a series connection of (dotted) R1||R2, then actual components D1, R3, R4, and so forth. The regulator output voltage is:

$$V_{OUT} = V_{REF}\left(1+\frac{R1}{R2}\right) \qquad \text{Eq. 7-2}$$

In the various devices of the ADP330X series, the R1-R2 divider is adjusted to produce standard output voltages of 1.8 V, 2.5 V, 2.7 V, 3.0 V, 3.2 V, 3.3 V, and 5.0 V. The regulator behaves as if the entire error amplifier has simply an offset voltage of V_{REF} volts, as seen at the output of a conventional R1-R2 divider.

While the above described dc performance enhancements of the ADP330x series are worthwhile, more dramatic improvements come in areas of ac-related performance. Capacitive loading, and the potential instability it brings, is a major deterrent to easy LDO applications. One method of providing some measure of immunity to variation in an amplifier response pole is the use of a frequency compensation technique called *pole splitting*. In the Figure 7-50 circuit, C_{COMP} functions as the pole splitting capacitor, and provides benefits of a buffered, C_L independent single-pole response. As a result, frequency response is dominated by the regulator's internal compensation, and becomes relatively immune to the value and ESR of load capacitor C_L.

This feature makes the design tolerant of virtually any output capacitor type. C_L, the load capacitor, can be as low as 0.47 μF, and it can also be a multilayer ceramic capacitor (MLCC) type, allowing a very small physical size for the entire regulation function.

Fixed Voltage, 50/100/200/500 mA LDO Regulators

A basic regulator application diagram common to various fixed voltage devices of the ADP330X device series is shown by Figure 7-51. Operation of the various pins and internal functions is discussed next. To adapt this general diagram to a specific current and voltage requirement, select a basic device for output current from the table in the diagram. Then select the output voltage by the part number suffix, consistent with the table.

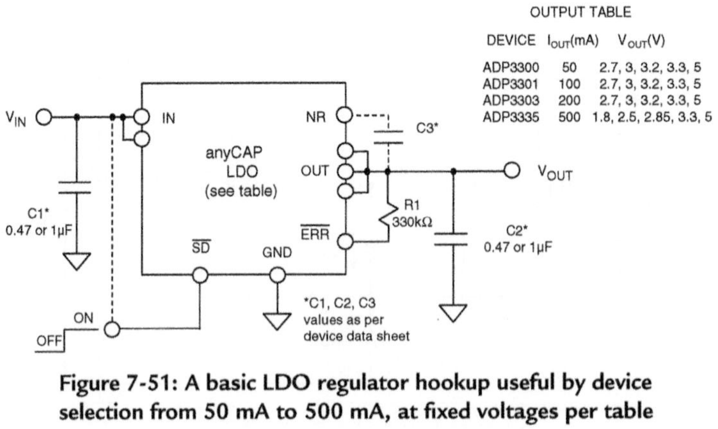

Figure 7-51: A basic LDO regulator hookup useful by device selection from 50 mA to 500 mA, at fixed voltages per table

This circuit is a general one, illustrating common points. For example, the ADP3300 is a 50 mA basic LDO regulator device, designed for those fixed output voltages as noted. An actual ADP3300 device ordered would be ADP3300ART-YY, where the "YY" is a voltage designator suffix such as 2.7, 3, 3.2, 3.3, or 5, for the respective table voltages. The "ART" portion of the part number designates the package (SOT23 6-lead). To produce 5 V from the circuit, use the ADP3300ART-5. Similar comments apply to the other devices, insofar as part numbering. For example, an ADP3301AR-5 depicts an SO-8 packaged 100 mA device, producing 5 V output.

In operation, the circuit produces rated output voltage for loads under the max current limit, for input voltages above $V_{OUT} + V_{MIN}$ (where V_{MIN} is the dropout voltage for the specific device used, at rated current). The circuit is ON when the shutdown input is in a HIGH state, either by a logic HIGH control input to the $\overline{SD}$ pin, or by simply tying this pin to V_{IN} (shown dotted). When $\overline{SD}$ is LOW or grounded, the regulator shuts down, and draws a minimum quiescent current.

The anyCAP regulator devices maintain regulation over a wide range of load, input voltage and temperature conditions. Most devices have a combined error band of ±1.4% (or less). When an overload condition is detected, the open collector $\overline{ERR}$ goes to a LOW state. R1 is a pullup resistor for the $\overline{ERR}$ output. This resistor can be eliminated if the load provides a pullup current.

C3, connected between the OUT and NR pins, can be used for an optional noise reduction (NR) feature. This is accomplished by bypassing a portion of the internal resistive divider, which reduces output noise ~10 dB. When exercised, only the recommended low leakage capacitors as specific to a particular part should be used.

The C1 input and C2 output capacitors should be selected as either 0.47 µF or 1 µF values respectively, again, as per the particular device used. For most devices of the series 0.47 µF suffices, but the ADP3335 uses the 1 µF values. Larger capacitors can also be used, and will provide better transient performance.

Heat sinking of device packages with more than five pins is enhanced, by use of multiple IN and OUT pins. All of the pins available should therefore be used in the PCB design, to minimize layout thermal resistance.

Adjustable Voltage, 200 mA LDO Regulator

In addition to the fixed output voltage LDO devices discussed above, adjustable versions are also available, to realize nonstandard voltages. The ADP3331 is one such device, and shown in Figure 7-52, configured as a 2.8 V output, 200 mA LDO application.

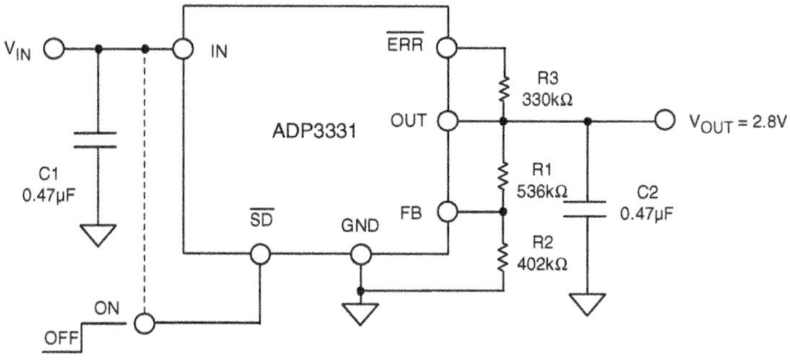

Figure 7-52: An adjustable 200 mA LDO regulator set up for a 2.8 V output

The ADP3331 is generally similar to other anyCAP LDO parts, with two notable exceptions. It has a lower quiescent current (~34 μA when lightly loaded) and most importantly, the output voltage is user-adjustable. As noted in the circuit, R1 and R2 are external precision resistors used to define the regulator operating voltage.

The output of this regulator is V_{OUT}, which is related to feedback pin FB voltage V_{FB} as:

$$V_{OUT} = V_{FB}\left(1+\frac{R1}{R2}\right) \qquad \text{Eq. 7-3}$$

where V_{FB} is 1.204 V. Resistors R1 and R2 program V_{OUT}, and their parallel equivalent should be kept close to 230 kΩ for best stability.

To select R1 and R2, first calculate their ideal values, according to the following two expressions:

$$R1 = 230\left(\frac{V_{OUT}}{V_{FB}}\right)k\Omega \qquad \text{Eq. 7-4}$$

$$R2 = \frac{230}{\left(1+\frac{V_{FB}}{V_{OUT}}\right)}k\Omega \qquad \text{Eq. 7-5}$$

In the example circuit, V_{OUT} is 2.8 V, which yields R1 = 534.9 kΩ, and R2 = 403.5 kΩ. As noted in the figure, closest standard 1% values are used, which provides an output of 2.8093 V (perfect resistors assumed). In practice, the resistor tolerances should be added to the ±1.4% tolerance of the ADP3331 for an estimation of overall error.

To complement the above-discussed anyCAP series of standalone LDO regulators, there is the LDO *regulator controller*. The regulator controller IC picks up where the standalone regulator stops for either load current or power dissipation, using an external PMOS FET pass device. As such, the current capability of the LDO can be extended to several amps. An LDO regulator controller application is shown later in this discussion. The application examples above illustrate a subset of the entire anyCAP family of LDOs. Further information on this series of standalone and regulator controller LDO devices can be found in the references at the end of the section.

Charge-Pump Voltage Converters

Another method for developing supply voltage for op amp systems employs what is known as a *charge-pump* circuit (also called switched capacitor voltage conversion). Charge-pump voltage converters accomplish energy transfer and voltage conversion using charges stored on capacitors, thus the name, charge-pump.

Using switching techniques, charge pumps convert supply voltage of one polarity to a higher or lower voltage, or to an alternate polarity (at either higher or lower voltage). This is accomplished with only an array of low resistance switches, a clock for timing, and a few external storage capacitors to hold the charges being transferred in the voltage conversion process. No inductive components are used, thus EMI generation is kept to a minimum. Although relatively high currents are switched internally, the high current switching is localized, and therefore the generated noise is not as great as in inductive type switchers. With due consideration towards component selection, charge-pump converters can be implemented with reasonable noise performance.

The two common charge-pump voltage converters are the *voltage inverter* and the *voltage doubler* circuits. In a voltage inverter, a charge pump capacitor is charged to the input voltage during the first half of the switching cycle. During the second half of the switching cycle the input voltage stored on the charge pump capacitor is inverted and applied to an output capacitor and the load. Thus the output voltage is essentially the negative of the input voltage, and the average input current is approximately equal to the output current. The switching frequency impacts the size of the external capacitors required, and higher switching frequencies allow the use of smaller capacitors. The duty cycle—defined as the ratio of charge pump charging time to the entire switching cycle time—is usually 50%, which yields optimal transfer efficiency.

A voltage doubler works similarly to the inverter. In this case the pump capacitor accomplishes a voltage doubling function. In the first phase it is charged from the input, but in the second phase of the cycle it appears in series with the output capacitor. Over time, this has the effect of doubling the magnitude of the input voltage across the output capacitor and load. Both the inverter and voltage doubler circuits provide no voltage regulation in basic form. However, techniques exist to add regulation (discussed below).

There are advantages and disadvantages to using charge-pump techniques, compared to inductor-based switching regulators. An obvious key advantage is the elimination of the inductor and the related magnetic design issues. In addition, charge-pump converters typically have relatively low noise and minimal radiated EMI. Application circuits are simple, and usually only two or three external capacitors are required. Because there are no inductors, the final PCB height can generally be made smaller than a comparable inductance-based switching regulator. Charge-pump inverters are also low in cost, compact, and capable of efficiencies greater than 90%. Obviously, current output is limited by the capacitor size and the switch capacity. Typical IC charge-pump inverters have 150 mA maximum outputs.

On their downside, charge-pump converters don't maintain high efficiency for a wide voltage range of input to output, unlike inductive switching regulators. Nevertheless, they are still often suitable for lower current loads where any efficiency disadvantages are a small portion of a larger system power budget. A summary of general charge-pump operating characteristics is shown in Figure 7-53.

An example of charge-pump applicability is the voltage inverter function. Inverters are often useful where a relatively low current negative voltage (i.e., –3 V) is required, in addition to a primary positive voltage (such as 5 V). This may occur in a single supply system, where only a few high performance parts require the negative voltage. Similarly, voltage doublers are useful in low current applications, where a voltage greater than a primary supply voltage is required.

- No Inductors
- Minimal Radiated EMI
- Simple Implementation: Two External Capacitors (Plus an Input Capacitor)
- Efficiency > 90% Achievable
- Low Cost, Compact, Low Profile (Height)
- Optimized for Doubling or Inverting Supply Voltage:
  - ADM660 or ADM8660
- Voltage Regulated Output Devices Available:
  - ADP3603/ADP3604/ADP3605/ADP3607

Figure 7-53: Some general charge-pump characteristics

Unregulated Inverter and Doubler Charge Pumps

Illustrating these principles are a pair of basic charge-pump ICs from Analog Devices, shown in Figure 7-54. The ADM660 is a popular charge-pump IC, and is shown here operating as both a voltage inverter (left) and the doubler (right). Switching frequency of this IC is selectable between 25 kHz and 120 kHz using the FC input pin. With the FC input is open as shown, the switching frequency is 25 kHz; with it connected to the V+ pin, frequency increases to 120 kHz. Generally, efficiency is greater when operating at the higher frequency. Only two external electrolytic capacitors are required for operation, C1 and C2 (ESR should be <200 mΩ). The value of these capacitors is flexible. For a 25 kHz switching frequency 10 µF tantalum types are recommended; for 120 kHz operation 2.2 µF provides comparable performance. Larger values can also be used, and will provide lower output ripple (at the expense of greater size and cost).

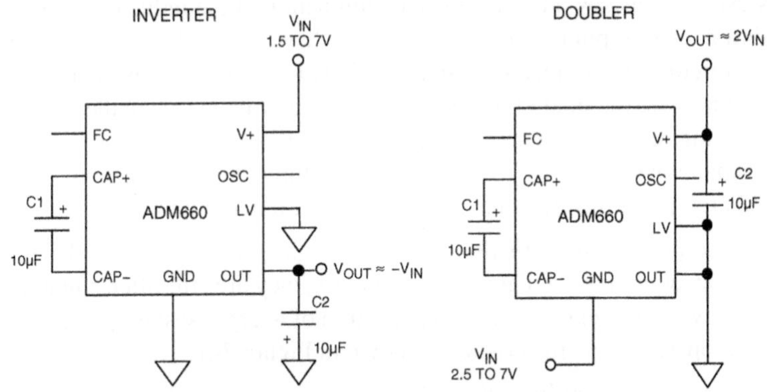

Figure 7-54: ADM660 IC functions as a supply inverter (left) or doubler (right)

These circuits accept V_{IN} inputs over the ranges noted, and deliver a nominal voltage output tracking the input voltage in magnitude, as noted in the output expressions. Although the output voltage is not regulated in these basic designs, it is still relatively low in impedance, due to the nominal 9 Ω resistance of the IC switches.

Efficiency of these circuits using the ADM660/ADM8660 can be 90% or more, for output currents up to 50 mA at a 120 kHz frequency. The ADM8660 is a device similar to the ADM660, however it is optimized for inverter operation, and includes a shutdown feature which reduces the quiescent current to 5 µA.

Regulated Output Charge-Pump Voltage Converters

Adding regulation to a simple charge-pump voltage converter function greatly enhances its usefulness for most applications. There are several techniques for adding regulation to a charge-pump converter. The most straightforward is to follow the charge-pump inverter/doubler with an LDO regulator. The LDO provides the regulated output, and can also reduce the charge-pump converter's ripple. This approach, however, adds complexity and reduces the available output voltage by the dropout voltage of the LDO (~200 mV). These factors may or may not be a disadvantage.

By far the simplest and most effective method for achieving regulation in a charge-pump voltage converter is simply to use a charge-pump design with an internal error amplifier, to control the on-resistance of one of the switches.

This method is used in the ADP3603/ADP3604/ADP3605 voltage inverters, devices offering regulated outputs for positive input voltage ranges. The output is sensed and fed back into the device via a sensing pin, V_{SENSE}. Key features of the series are good output regulation, 3% in the ADP3605, and a high switching frequency of 250 kHz, good for both high efficiency and small component size.

An example circuit for the ADP3605 IC from this series is shown in Figure 7-55. The application is a 5 V to –3 V inverter, with the output regulated ±3% for currents up to 60 mA. In normal operation, the SHUTDOWN pin is connected to ground (as shown dotted). Alternately, a logic HIGH at this pin shuts the device down to a standby current of 2 μA.

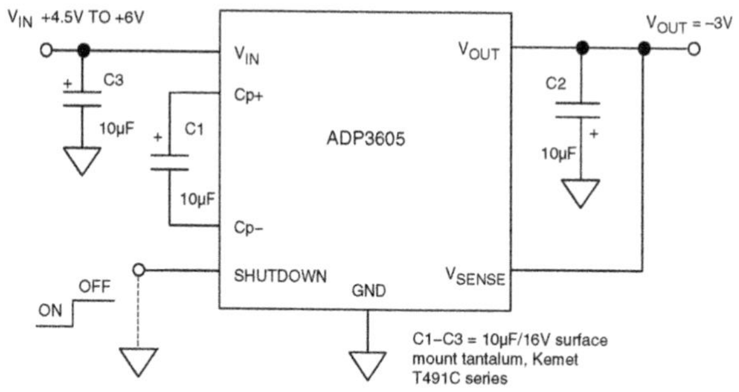

Figure 7-55: ADP3605 5 V to –3 V, 60 mA regulated supply inverter

The 10 μF capacitors for C1–C3 should have ESRs of less than 150 mΩ (4.7 μF can be used at the expense of slightly higher output ripple voltage). C1 is the most critical of the three, because of its higher current flow. The tantalum type listed is recommended for lowest output ripple.

With values as shown, typical output ripple voltage ranges up to approximately 60 mV as the output current varies over the 60 mA range. Although output is regulated for currents up to 60 mA, higher currents of up to 100 mA are also possible with further voltage deviation, and proportionally greater ripple.

These application examples illustrate a subset of the entire charge-pump IC family. Further information on these devices can be found in the end-of-section references.

Linear Post Regulator for Switching Supplies

Another powerful noise reduction option that can be utilized in conjunction with a switching type supply is the option of a *linear post regulator* stage. This is at best an LDO type of regulator, chosen for the desired clean analog voltage level and current. It is preceded by a switching stage, which might be a buck or boost type inductor-based design, or it may also be a charge-pump. The switching converter allows the overall design to be more power-efficient, and the linear post regulator provides clean regulation at the load, reducing the noise of the switcher. This type of regulator can also be termed *hybrid regulation*, since it combines both switching and linear regulation concepts.

An example circuit is shown in Figure 7-56, which features a 3.3 V/1 A low noise, analog-compatible regulator. It operates from a nominal 9 V supply, using a buck or step-down type of switching regulator, as the first stage at the left. The switcher output is set for a few hundred mV above the desired final voltage output, minimizing power in the LDO stage at the right. This feature can eliminate need for a heat sink on the LDO pass device.

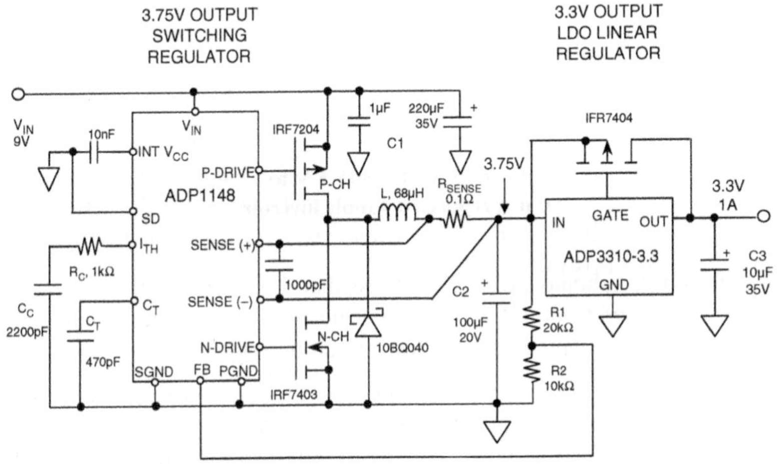

Figure 7-56: A linear post-regulator operating after a switching/ linear regulator is capable of low noise, as well as good dc efficiency

In this example the ADP1148 IC switcher is set up for a 3.75 V output by R1-R2 but, in principle, this voltage can be anything suitable to match the headroom of the companion LDO (within specification limits, of course). In addition, the principle extends to any LDO devices and other current levels, and other switching regulators. The ADP3310-3.3 is a fixed-voltage LDO controller, driving a PMOS FET pass device, with a 3.3 V output.

The linear post regulation stage provides both noise-reduction (in this case about 14 dB), as well as good dc regulation. To realize best results, good grounding practices must be followed. In tests, noise at the 3.3 V output was about 5mV p-p at the 150 kHz switcher frequency. Note that the LDO noise rejection for such relatively high frequencies is much less than at 100 Hz/120 Hz. Note also that C2's ESR will indirectly control the final noise output. The ripple figures given are for a general-purpose C2 part, and can be improved.

Power Supply Noise Reduction and Filtering

During the last decade or so, switching power supplies have become much more common in electronic systems. As a consequence, they also are being used for analog supplies. Good reasons for the general popularity include their high efficiency, low temperature rise, small size, and light weight.

In spite of these benefits, switchers *do* have drawbacks, most notably high output noise. This noise generally extends over a broad band of frequencies, resulting in both conducted and radiated noise, as well as unwanted electric and magnetic fields. Voltage output noise of switching supplies are short-duration voltage transients, or spikes. Although the fundamental switching frequency can range from 20 kHz to 1 MHz, the spikes can contain frequency components extending to 100 MHz or more. While specifying switching supplies in terms of RMS noise is common vendor practice, as a user you should also specify the *peak* (or p-p) amplitudes of the switching spikes, with the output loading of your system.

This section discusses filter techniques for rendering a switching regulator output *analog ready*, that is sufficiently quiet to power precision op amp and other analog circuitry with relatively small loss of dc terminal voltage. The filter solutions presented are generally applicable to all power supply types incorporating switching element(s) in their energy path. This includes charge-pump as well as other switching type converters and supplies. This section focuses on reducing *conducted type* switching power supply noise with external post filters, as opposed to radiated type noise.

Tools useful for combating high frequency switcher noise are shown by Figure 7-57. These differ in electrical characteristics as well as practicality towards noise reduction, and are listed roughly in an order of priorities. Of these tools, L and C are the most powerful filter elements, and are the most cost-effective, as well as small in size.

- Capacitors
- Inductors
- Ferrites
- Resistors
- Linear Post Regulation
- Proper Layout and Grounding
- Physical Separation

Figure 7-57: Tools useful in reducing power supply noise

Capacitors

Capacitors are probably the single most important filter component for reducing switching-related noise. As noted in the first section of this chapter, there are many different types of capacitors. It is also quite true that understanding of their individual characteristics is absolutely mandatory to the design of effective and practical power supply filters. There are generally three classes of capacitors useful in 10 kHz–100 MHz filters, broadly distinguished as the generic dielectric types; *electrolytic*, *film*, and *ceramic*. These discussions complement earlier ones, focusing on power-related concepts. With any dielectric, a major potential filter loss element is ESR (equivalent series resistance), the net parasitic resistance of the capacitor. ESR provides an ultimate limit to filter performance, and requires more than casual consideration, because it can vary both with frequency and temperature in some types. Another capacitor loss element is ESL (equivalent series inductance). ESL determines the frequency where the net impedance characteristic switches from capacitive to inductive. This varies from as low as 10 kHz in some electrolytics to as high as 100 MHz or more in chip ceramic types. Both ESR and ESL are minimized when a leadless package is used. All capacitor types mentioned are available in surface mount packages, preferable for high speed uses.

The *electrolytic* family provides an excellent, cost-effective low frequency filter component, because of the wide range of values, a high capacitance-to-volume ratio, and a broad range of working voltages. It includes *general-purpose aluminum electrolytic* types, available in working voltages from below 10 V up to about 500 V, and in size from one to several thousand μF (with proportional case sizes). All electrolytic capacitors are polarized, and cannot withstand more than a volt or so of reverse bias without damage.

A subset of the general electrolytic family includes *tantalum* types, generally limited to voltages of 100 V or less, with capacitance of 500 μF or less (see Reference 7). In a given size, tantalums exhibit a higher capacitance-to-volume ratios than do general purpose electrolytics, and have both a higher frequency range and lower ESR. They are generally more expensive than standard electrolytics, and must be carefully applied with respect to surge and ripple currents.

A subset of aluminum electrolytic capacitors is the *switching* type, designed for handling high pulse currents at frequencies up to several hundred kHz with low losses (see Reference 8). This capacitor type can compete with tantalums in high frequency filtering applications, with the advantage of a broader range of values.

A more specialized high performance aluminum electrolytic capacitor type uses an organic semiconductor electrolyte (see Reference 9). The *OS-CON* capacitors feature appreciably lower ESR and higher frequency range than do other electrolytic types, with an additional feature of minimal low temperature ESR degradation.

Film capacitors are available in very broad value ranges and an array of dielectrics, including polyester, polycarbonate, polypropylene, and polystyrene. Because of the low dielectric constant of these films, their volumetric efficiency is quite low, and a 10 μF/50 V polyester capacitor (for example) is actually a handful. Metalized (as opposed to foil) electrodes do help to reduce size, but even the highest dielectric constant units among film types (polyester, polycarbonate) are still larger than any electrolytic, even using the thinnest films with the lowest voltage ratings (50 V). Where film types excel is in their low dielectric losses, a factor that may not necessarily be a practical advantage for filtering switchers. For example, ESR in film capacitors can be as low as 10 mΩ or less, and the behavior of films generally is very high in terms of Q. In fact, this can cause problems of spurious resonance in filters, requiring damping components.

As typically constructed using wound layers, film capacitors can be inductive, which limits their effectiveness for high frequency filtering. Obviously, only noninductively made film caps are useful for switching regulator filters. One specific style which is noninductive is the *stacked-film* type, where the capacitor plates are cut as small overlapping linear sheet sections from a much larger wound drum of dielectric/plate material. This technique offers the low inductance attractiveness of a plate sheet style capacitor with conventional leads (see References 8 and 10). Obviously, minimal lead length should be used for best high frequency effectiveness. Very high current polycarbonate film types are also available, specifically designed for switching power supplies, with a variety of low inductance terminations to minimize ESL (see Reference 11). Dependent upon their electrical and physical size, film capacitors can be useful at frequencies to above 10 MHz. At the highest frequencies, only stacked film types should be considered. Leadless surface-mount packages are now available for film types, minimizing inductance.

Ceramic is often the capacitor material of choice above a few MHz, due to its compact size, low loss, and availability up to several μF in the high-K dielectric formulations (X7R and Z5U), at voltage ratings up to 200 V (see ceramic families of Reference 7).

Multilayer ceramic "chip caps" are very popular for bypassing and/or filtering at 10 MHz or more, simply because their very low inductance design allows near optimum RF bypassing. For smaller values, ceramic chip caps have an operating frequency range to 1 GHz. For high frequency applications, a useful selection can be ensured by selecting a value that has a self-resonant frequency *above* the highest frequency of interest.

The capacitor model and waveforms of Figure 7-58 illustrate how the various parasitic model elements become dominant, dependent upon the operating frequency. Assume an input current pulse changing from 0 to 1 A in 100 ns, as noted in the figure, and consider what voltage will be developed across the capacitor.

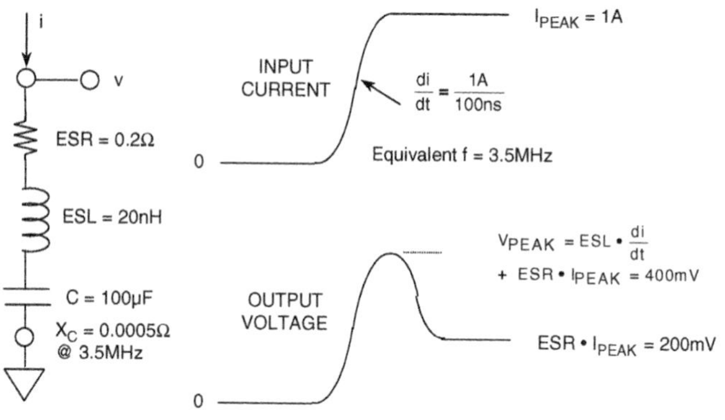

Figure 7-58: Capacitor equivalent circuit and response to input current pulse

The fast-rising edge of the current waveform shown results in an initial voltage peak across the capacitor, which is proportional to the ESL. After the initial transient, the voltage settles down to a longer duration level which is proportional to the ESR of the capacitor. Thus the ESL determines how effective a filter the capacitor is for the fastest components of the current signal, and the ESR is important for longer time frame components. Note that an overall time frame of a few microseconds (or even less) is relevant here. As things turn out, this means switching frequencies in the 100 kHz to 1 MHz range. Unfortunately, however, this happens to be the region where most electrolytic types begin to perform poorly.

All electrolytics will display impedance curves similar in general shape to that of Figure 7-59. In a practical capacitor, at frequencies below about 10 kHz the net impedance seen at the terminals is almost purely capacitive (C region). At intermediate frequencies, the net impedance is determined by ESR, for example

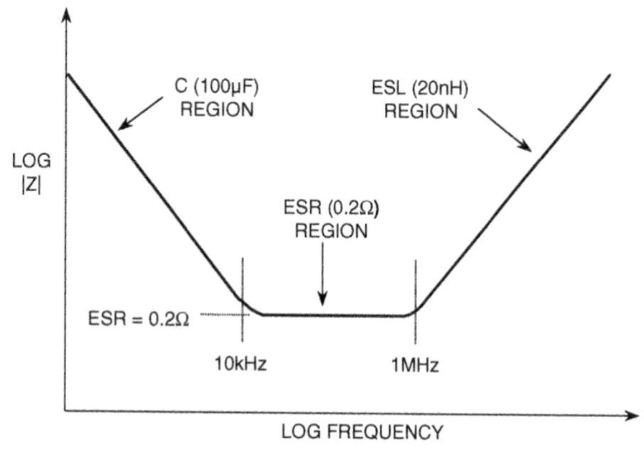

Figure 7-59: Electrolytic capacitor impedance versus frequency

about 0.1 Ω to 0.5 Ω at ~125 kHz, for several types (ESR region). Above about several hundred kHz to 1 MHz these capacitor types become inductive, with net impedance rising (ESL region).

The minimum impedance within the 10 kHz – 1 MHz range will vary with the magnitude of the capacitor's ESR. This is the primary reason why ESR is the most critical item in determining a given capacitor's effectiveness as a switching supply filter element. Higher up in frequency, the inductive region will vary with ESL (which in turn is also strongly affected by package style). It should go without saying that a wideband impedance plot for a capacitor being considered for a filter application will go a long way towards predicting its potential value, as well as for comparing one type against another.

It should be understood that all real-world capacitors have some finite ESR. While it is usually desirable for filter capacitors to possess low ESR, this isn't always so. In some cases, the ESR may actually be helpful in reducing resonance peaks in filters, by supplying "free" damping. For example, in most electrolytic types, a nominally flat broad series resonance region can be noted in an impedance versus frequency plot. This occurs where $|Z|$ falls to a minimum level, nominally equal to the capacitor's ESR at that frequency. This low Q resonance can generally be noted to cover a relatively wide frequency range of several octaves. Contrasted to the high Q sharp resonances of film and ceramic caps, electrolytic's low Q behavior can be useful in controlling resonant peaks.

Ferrites

A second important filter element is the inductor, available in various forms. The use of *ferrite* core materials is prevalent in inductors most practical for power supply filtering. Regarding inductors, ferrites, which are nonconductive ceramics manufactured from the oxides of nickel, zinc, manganese, and so forth., are extremely useful in power supply filters (see Reference 12). Ferrites can act as either inductors or resistors, dependent upon their construction and the frequency range. At low frequencies (<100 kHz), inductive ferrites are useful in low-pass LC filters. At higher frequencies, ferrites become resistive, which can be an important characteristic in high-frequency filters. Again, exact behavior is a function of the specifics. Ferrite impedance depends on material, operating frequency range, dc bias current, number of turns, size, shape, and temperature. Figure 7-60 summarizes a number of ferrite characteristics.

- Ferrites Good for Frequencies Above 25kHz
- Many Sizes / Shapes Available Including Leaded "Resistor Style"
- Ferrite Impedance at High Frequencies Primarily Resistive — Ideal for HF Filtering
- Low DC Loss: Resistance of Wire Passing Through Ferrite is Very Low
- High Saturation Current Versions Available
- Choice Depends Upon:
  - Source and Frequency of Interference
  - Impedance Required at Interference Frequency
  - Environmental: Temperature, AC and DC Field Strength, Size and Space Available
- Always Test the Design

Figure 7-60: A summary of ferrite characteristics

Several ferrite manufacturers offer a wide selection of ferrite materials from which to choose, as well as a variety of packaging styles for the finished network (see References 13 and 14). A simple form is the *bead* of ferrite material, a cylinder of the ferrite which is simply slipped over the power supply lead to the decoupled stage. Alternately, the *leaded ferrite bead* is the same bead, premounted on a length of wire and used as a component (see Reference 14). More complex beads offer multiple holes through the cylinder for increased decoupling, plus other variations. Surface-mount beads are also available. PSpice models of Fair-Rite ferrites are available, allowing ferrite impedance estimations (see Reference 15). The models match measured rather than theoretical impedances.

Selecting the proper ferrite is not straightforward. A ferrite's impedance is dependent upon a number of interdependent variables, and is difficult to quantify analytically. However, knowing the following system characteristics will make selection easier. First, determine the frequency range of the noise to be filtered. Second, the expected temperature range of the filter should be known, as ferrite impedance varies with temperature. Third, the dc current flowing through the ferrite must be known, to ensure that the ferrite does not saturate. Although models and other analytical tools may prove useful, the general guidelines given above, coupled with actual filter experimentation connected under system load conditions, should lead to a proper ferrite selection.

Card Entry Filter

Using proper component selection, low and high frequency band filters can be designed to smooth a noisy switching supply output to produce an *analog ready* supply. It is most practical to do this over two (and sometimes more) stages, each stage optimized for a range of frequencies.

A basic stage can be used to carry the entire load current, and filter noise by 60 dB or more up to a 1 MHz–10 MHz range. Figure 7-61 illustrates this type of filter, which is used as a *card entry filter,* providing broadband filtering for all power entering a PC card.

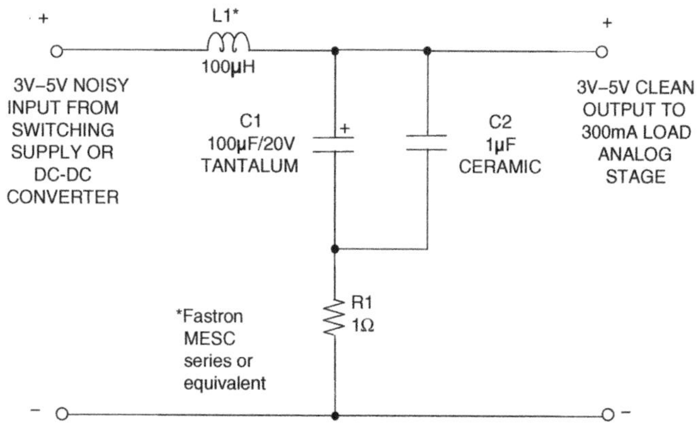

Figure 7-61: A card-entry filter is useful for low-medium frequency power line noise filtering in analog systems

In this filter, L1 and C1 perform the primary filtering, which provides a corner frequency of about 1.6 kHz. With the corner thus placed well below typical switching frequencies, the circuit can have good attenuation up to 1 MHz, where the typical attenuation is on the order of 60 dB. At higher frequencies parasitics limit performance, and a second filter stage will be more useful.

The ultimate level of performance available from this filter will be related to the components used within it. L1 should be derated for the operating current, thus for 300 mA loads it is a 1 A type. The specified L1 choke has a typical DCR of 0.65 Ω, for low drop across the filter (see Reference 16). C1 can be either a tantalum or an aluminum electrolytic, with moderately low ESR. For current levels lower than 300 mA, L1 can be proportionally downsized, saving space. The resistor R1 provides damping for the LC filter, to prevent possible ringing. R1 can be reduced or even possibly eliminated, if the ESR of C1 provides a comparable impedance.

While the example shown is a single-supply configuration, obviously the same filter concepts apply for dual supplies.

Rail Bypass/Distribution Filter

A complement to the card-entry filter is the rail-bypass filter scheme of Figure 7-62. When operating from relatively clean power supplies, the heavy noise filtering of the card entry filter may not be necessary. However, some sort of low frequency bypassing with appreciable energy storage is almost always good, and this is especially true if high currents are being delivered by the stages under power.

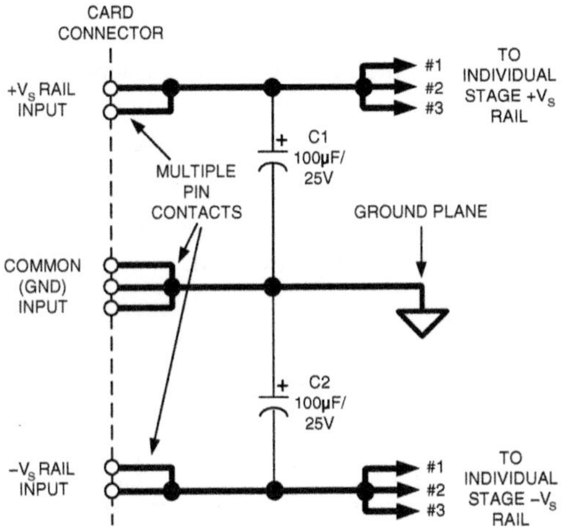

Figure 7-62: Dual-supply low frequency rail bypass/distribution filter

In such cases, some lumped low frequency bypassing is appropriate on the card. Although these energy storage filters need not be immediately adjacent to the ICs they serve, they should be within a few inches. This type of bypassing scheme should be considered a minimum for powering any analog circuit. The exact capacitor values aren't critical, and can vary appreciably. The most important thing is to avoid leaving them out.

The circuit shown uses C1 and C2 as these bypasses in a dual-rail system. Note that multiple card contacts are recommended for the I/O pins, especially ground connection. From the capacitors outward, supply rail traces are distributed to each stage as shown, in "star" distribution fashion. Note: while this is the optimum method to minimize inter-stage crosstalk, in practice some degree of "daisy chaining" is often difficult to avoid. A prudent designer should therefore carefully consider common supply currents effects in designing these PCB distribution paths.

Wider than normal traces are recommended for these supply rails, especially those carrying appreciable current. If the current levels are in the ampere region, then star-type supply distribution with ultrawide traces should be considered mandatory. In extreme cases, a dedicated power plane can be used. The impedance of the ground return path is minimized by the use of a ground plane.

Local High Frequency Bypass/Decoupling

At each individual analog stage, further local, high-frequency-only filtering is used. With this technique, used in conjunction with either the card-entry filter or the low frequency bypassing network, such smaller and simpler local filter stages provide optimum high frequency decoupling. *These stages are provided directly at the power pins, of* all *individual analog stages.*

Figure 7-63 shows this technique, in both correct (left) as well as incorrect example implementations (right). In the left example, a typical 0.1 μF chip ceramic capacitor goes directly to the opposite PCB side ground plane, by virtue of the via, and on to the IC's GND pin by a second via. In contrast, the less desirable setup at the right adds additional PCB trace inductance in the ground path of the decoupling cap, reducing effectiveness.

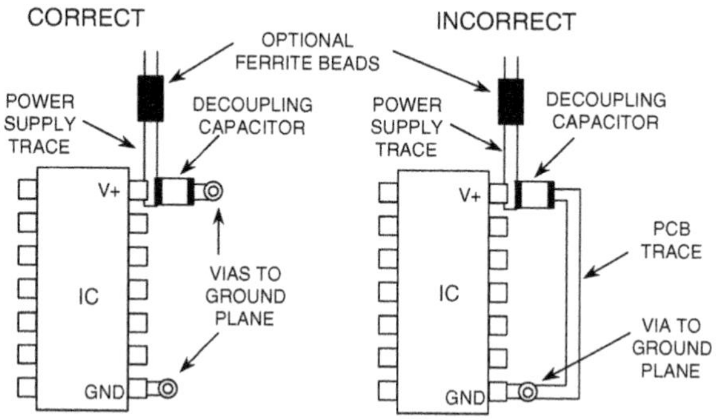

Figure 7-63: Localized high frequency supply filter(s) provides optimum filtering and decoupling via short low-inductance path (ground plane)

The general technique is shown here as suitable for a single-rail power supply, but the concept obviously extends to dual rail systems. Note: if the decoupled IC in question is an op amp, the GND pin shown is the $-V_S$ pin. For dual supply op amp uses, there is no op amp GND pin per se, so the dual decoupling networks should go directly to the ground plane when used, or other local ground.

All high frequency (i.e., ≥10 MHz) ICs should use a bypassing scheme similar to Figure 7-63 for best performance. Trying to operate op amps and other high performance ICs without local bypassing is almost always folly. It *may* be possible in a few circumstances, *if* the circuitry is strictly micropower in nature, and the gain-bandwidth in the kHz range. To put things into an overall perspective however, note that a pair of 0.1 μF ceramic bypass caps cost less than 25 cents. Hardly a worthy saving compared to the potential grief and lost time of troubleshooting a system without bypassing.

In contrast, the ferrite beads aren't 100% necessary, but they will add extra HF noise isolation and decoupling, which is often desirable. Possible caveats here would be to verify that the beads never saturate, when the op amps are handling high currents.

Note that with some ferrites, even before full saturation occurs, some beads can be nonlinear, so if a power stage is required to operate with a low distortion output, this should also be lab checked.

Figure 7-64 summarizes the previous points of this section regarding power supply conditioning techniques for op amp circuitry.

- Use Proper Layout and Grounding Techniques
- At HF Local Decoupling at IC Power Pins is Mandatory
- At HF Ground Planes are Mandatory
- External LC Filters Very Effective in Reducing Ripple
- Low ESR/ESL Capacitors Give Best Results
- Parallel Caps Lower ESR/ESL and Increase C
- Linear Post Regulation Effective for Noise Reduction and Best Regulation
- Completely Analytical Approach Difficult
  - Prototyping Required for Optimum Results

Once Design is Final, Don't Switch Vendors or Substitute Parts
  - Without First Verifying Performance within the Circuit

Figure 7-64: A summary of power supply conditioning techniques for high performance op amp circuitry

References: Op Amp Power Supply Systems

1. Walt Jung, "References and Low Dropout Linear Regulators," Section 2 within Walt Kester, Ed., **Practical Design Techniques for Power and Thermal Management**, Analog Devices, Inc., 1998, ISBN 0-916550-19-2.
2. Paul Brokaw, "A Simple Three-Terminal IC Bandgap Voltage Reference," **IEEE Journal of Solid-State Circuits**, Vol. SC-9, December, 1974.
3. Frank Goodenough, "Vertical-PNP-Based Monolithic LDO Regulator Sports Advanced Features," **Electronic Design**, May 13, 1996.
4. Frank Goodenough, "Low Dropout Regulators Get Application Specific," **Electronic Design**, May 13, 1996.
5. Walt Kester, Brian Erisman, Gurgit Thandi, "Switched Capacitor Voltage Converters," Section 4 within Walt Kester, Editor, **Practical Design Techniques for Power and Thermal Management**, Analog Devices, Inc., 1998, ISBN 0-916550-19-2.
6. Walt Jung, Walt Kester, Bill Chestnut, "Power Supply Noise Reduction and Filtering," portion of Section 8 within Walt Kester, Editor, **Practical Design Techniques for Power and Thermal Management**, Analog Devices, Inc., 1998, ISBN 0-916550-19-2.
7. **Tantalum Electrolytic and Ceramic Capacitor Families**, Kemet Electronics, Box 5928, Greenville, SC, 29606, 803-963-6300.
8. Type HFQ Aluminum Electrolytic Capacitor and Type V Stacked Polyester Film Capacitor, Panasonic, 2 Panasonic Way, Secaucus, NJ, 07094, 201-348-7000.
9. **OS-CON Aluminum Electrolytic Capacitor Technical Book**, Sanyo, 3333 Sanyo Road, Forrest City, AK, 72335, 501-633-6634.
10. Ian Clelland, "Metallized Polyester Film Capacitor Fills High Frequency Switcher Needs," **PCIM**, June, 1992.
11. Type 5MC Metallized Polycarbonate Capacitor, Electronic Concepts, Inc., Box 1278, Eatontown, NJ, 07724, 908-542-7880.
12. Henry W. Ott, **Noise Reduction Techniques in Electronic Systems, 2nd Edition**, John Wiley, Inc., 1988, ISBN: 0-471-85068-3.
13. **Fair-Rite Linear Ferrites Catalog**, Fair-Rite Products, Box J, Wallkill, NY, 12886, 914-895-2055.
14. Type EXCEL leaded ferrite bead EMI filter, and Type EXC L leadless ferrite bead, Panasonic, 2 Panasonic Way, Secaucus, NJ, 07094, 201-348-7000.
15. Steve Hageman, "Use Ferrite Bead Models to Analyze EMI Suppression," **The Design Center Source,** MicroSim Newsletter**,** January, 1995.
16. "MESC series RFI suppression chokes," FASTRON GmbH, Zum Kaiserblick 25, 83620 Feldkirchen-Westerham, Germany, www.fastron.de.

SECTION 7-4

Op Amp Protection

Walt Jung, Walt Kester, James Bryant,
Joe Buxton, Wes Freeman

Frequently, op amps and other analog ICs require protection against destructive potentials at their input and output terminals. One basic reason behind this is that these ICs are by nature relatively fragile components. Although designed to be as robust as possible *for normal signals*, there are nevertheless certain application and/or handling conditions where they can see voltage transients beyond their ratings. This situation can occur for either of two instances. The first of these is *in-circuit,* that is, operating within an application circuit. The second instance is *out-of-circuit,* which might be at anytime after receipt from a supplier, but prior to final assembly and mounting of the IC. In either case, under over-voltage conditions, it is a basic fact of life that unless the designer limits the fault currents at the input (or possibly output) of the IC, it can be damaged or destroyed.

So, obviously, the designer should fully understand all of the fault mechanisms internal to those ICs that may require protection. This then allows design of networks that can protect the in-circuit IC throughout its lifetime, without undue compromise of speed, precision, and so forth. Or, for the out-of-circuit IC, it can help define proper protective handling procedures until it reaches its final destination. This section of the chapter examines a variety of protection schemes to ensure adequate protection for op amps and other analog ICs for in-circuit applications, as well as for out-of-circuit environments.

In-Circuit Overvoltage Protection

There are many common cases that stress op amps and other analog ICs at the input, while operating within an application, i.e., in-circuit. Since these ICs must often interface with the outside world, this may entail handling voltages exceeding their absolute maximum ratings. For example, sensors are often placed in environments where a fault condition can expose the circuit to a dangerously high voltage. With the sensor connected to a signal processing amplifier, the input then sees excessive voltages during a fault.

General Input Common-Mode Limitations

Whenever an op amp input common-mode (CM) voltage goes outside its supply range, the op amp can be damaged, even if the supplies are turned off. Accordingly, the absolute maximum input ratings of almost all op amps limits the greatest applied voltage to a level equal to the positive and negative supply voltage, plus about 0.3 V beyond these voltages (i.e., $+V_S + 0.3$ V, or $-V_S - 0.3$ V). While some exceptions to this general rule might exist it is important to note this: *Most IC op amps require input protection when over-voltage of more than 0.3 V beyond the rails occurs.*

A safe operating rule is to always keep the applied op amp CM voltage between the rail limits. Here, "safe" implies prevention of outright IC destruction. As will be seen later, there are also intermediate "danger-zone" CM conditions between the rails with certain op amps, which can invoke dangerous (but not necessarily destructive) behavior.

Speaking generally, it is important to note that almost *any* op amp input will break down, given sufficient overvoltage to the positive or negative rail. Under breakdown conditions high and uncontrolled current can flow, so the danger is obvious. The exact breakdown voltage is entirely dependent on the individual op amp input stage. It may be a 0.6 V diode drop, or a process-related breakdown of 50 V or more. In many cases, overvoltage stress can result in currents over 100 mA, which destroys a part almost instantly.

Therefore, unless otherwise stated on the data sheet, op amp input fault current should be limited to ≤5 mA to avoid damage. This is a conservative rule of thumb, based on metal trace widths in a typical op amp input. Higher levels of current can cause *metal migration*, a cumulative effect, which, if sustained, eventually leads to an open trace. Should a migration situation be present, failure may only appear after a long time due to multiple overvoltages, a very difficult failure to identify. So, even though an amplifier may appear to withstand overvoltage currents well above 5 mA for a short time period, it is important to limit the current to 5 mA (or preferably less) for long term reliability.

Figure 7-65 illustrates an external, general-purpose op amp CM protection circuit. The basis of this scheme is the use of Schottky diodes D1 and D2, plus an external current limiting resistor, R_{LIMIT}. With appropriate selection of these parts, input protection for a great many op amps can be ensured. Note that an op amp may also have *internal* protection diodes to the supplies (as shown) which conduct at about 0.6 V forward drop above or below the respective rails. In this case however, the external Schottky diodes effectively parallel any internal diodes, so the internal units never reach their threshold. Diverting fault currents externally eliminates potential stress, protecting the op amp.

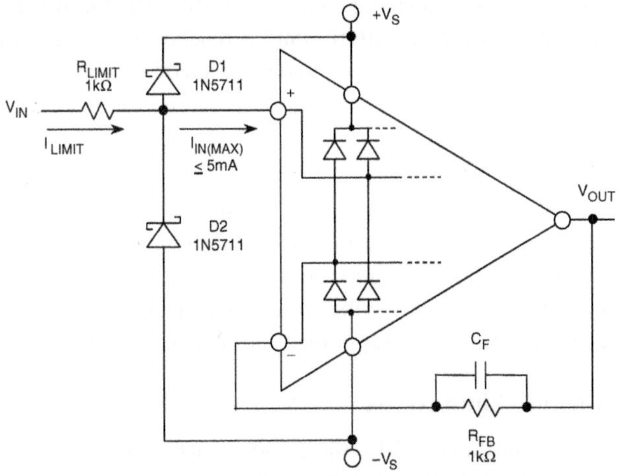

Figure 7-65: A general-purpose op amp CM overvoltage protection network using Schottky clamp diodes with current limit resistance

The external diodes also allow other degrees of freedom, some not so obvious. For example, if fault current is allowed to flow in the op amp, R_{LIMIT} must then be chosen so that the maximum current is no more than 5 mA for the worst case V_{IN}. This criterion can result in rather large R_{LIMIT} values, and the associated increase in noise and offset voltage may not be acceptable. For instance, to protect against a V_{IN} of 100 V with the 5 mA criterion, R_{LIMIT} must be ≥20 kΩ. However with external Schottky clamping diodes, this allows R_{LIMIT} to be governed by the maximum allowable D1-D2 current, which can be larger than 5 mA. However, care must be used here, for at very high currents the Schottky diode drop may exceed 0.6 V, possibly activating internal op amp diodes.

It is very useful to keep the R_{LIMIT} value as low as possible, to minimize offset and noise errors. R_{LIMIT}, in series with the op amp input, produces a bias-current-proportional voltage drop. Left uncorrected, this voltage appears as an increase in the circuit's offset voltage. Thus for op amps where the bias currents are moderate and approximately equal (most bipolar types), compensation resistor R_{FB} balances the dc effect, and minimizes this error. For low bias current op amps (Ib ≤10 nA, or FET types) it is likely R_{FB} won't be necessary. To minimize noise associated with R_{FB}, bypass it with a capacitor, C_F.

Clamping Diode Leakage

For obvious reasons, it is critical that diodes used for protective clamping at an op amp input have a leakage sufficiently low to not interfere with the bias level of the application.

Figure 7-66 illustrates how some well-known diodes differ in terms of leakage current, as a function of the reverse bias voltage, Vbias.

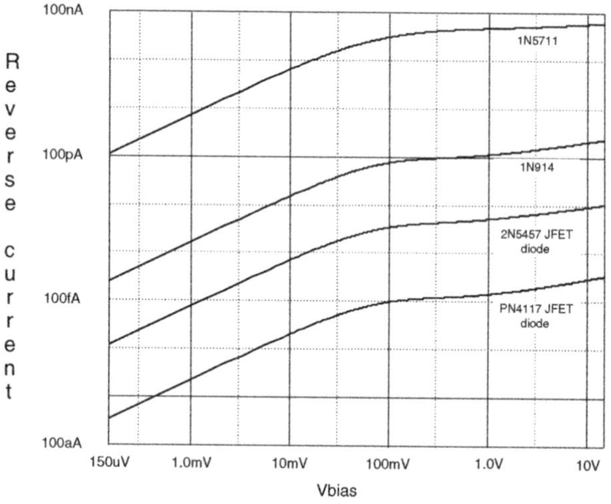

Figure 7-66: Reverse bias current characteristics for diodes useful in protective clamping networks (PSpice simulation)

In this chart, a 25°C simulation using PSpice diode models, it is easy to see that not only is the diode type critical, so is the reverse bias. The 1N5711 Schottky type for example, has a leakage of nearly 100 nA at a reverse bias of 15 V, as it would typically be used with a ±15 V powered op amp. With this level of leakage, such diodes will only be useful with op amps with bias currents of several μA. For protection of appreciably lower bias current op amps (particularly most FET input devices) much lower leakage is necessary.

As the data of Figure 7-66 shows, not only does selecting a better diode help control leakage current, but operating it at a low bias voltage condition reduces leakage substantially. For example, while an ordinary 1N914 or 1N4148 diode may have 200 pA of leakage at 15 V, this is reduced to slightly more than a pA with bias controlled to 1 mV. But there is a caveat here. When used in a high impedance clamp circuit, glass diodes such as the 1N914/1N4148 families should either be shielded from incident light, or use opaque packages. This is necessary to minimize parasitic photocurrent from the surrounding light, which effectively appears as diode leakage current.

Specialty diodes with much lower leakage are also available, such as diode-connected FET devices characterized as protection diodes (see DPAD series of Reference 2). Within the data of Figure 7-66, the 2N5457 is a general purpose JFET, and the 2N4117/PN4117 family consists of parts designed for low current levels. Other low leakage and specialty diodes are described in References 3 and 4.

A Flexible Voltage Follower Protection Circuit

Of course, it isn't a simple matter to effectively apply protective clamping to op amp inputs, while reducing diode bias level to a sub-mV level.

The circuit of Figure 7-67 shows low leakage input clamping and other means used with a follower connected FET op amp, with protection at input and output, for both power on or off conditions.

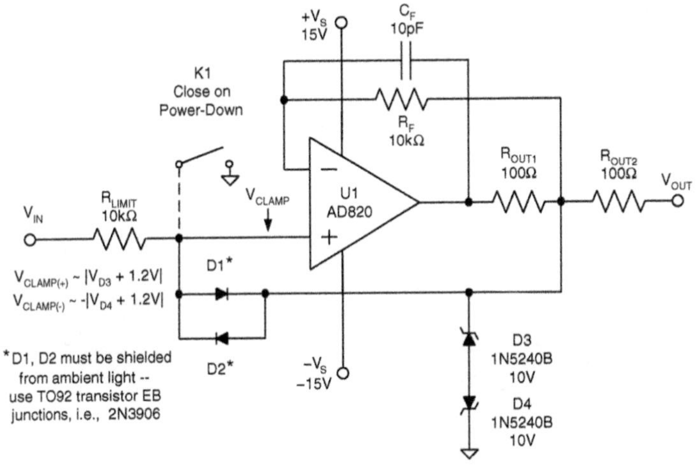

Figure 7-67: Bootstrapping the D1-D2 protection network reduces diode leakage to negligible levels, and is voltage-programmable for clamp level.

Disregarding the various diodes momentarily, this circuit is an output-current-limited voltage follower. With the addition of diodes D1-D2 and D3-D4, it has both a voltage-limited output, and an overvoltage protected input. Operating below the voltage threshold of output series-connected zener diodes D3-D4, the circuit behaves as a precision voltage follower. Under normal follower operation, that is at input/output voltages $< | V_Z + 0.6 |$ volts (where V_Z is the breakdown voltage of D3 or D4), diodes D1-D2 see only the combined offset and CM voltage errors of U1 as bias voltage. This reduces the D1-D2 leakage to very low levels, consistent with the pA level bias current of a FET input op amp. Note that D1-D2 *must* be prevented from photoconduction, and one direct means of this is to use opaque package diodes, such as the 2N3906 EB junctions discussed by Pease (see References 3 and 4). If 1N914s are used they must be light shielded. In either case, bootstrapping greatly reduces the effective D1-D2 leakage.

For input/output voltage levels greater than $V_Z + 0.6$ V, zener diodes D3-D4 break down. This action clamps both the V_{OUT} output node and the V_{CLAMP} node via D1-D2. The input of the op amp is clamped to either polarity of the two input levels of V_{CLAMP}, as indicated within the figure. Under clamp conditions, input voltage V_{IN} can rise to levels beyond the supply rails of U1 without harm, with excess current limited by R_{LIMIT}. If sustained high-level (~100 V) inputs will be applied, R_{LIMIT} should be rated as a 1–2W (or fusible) type.

This circuit has very good dc characteristics, due to the fact that the clamping network is bootstrapped. This produces very low input/output errors below the V_{CLAMP} threshold (consistent with the op amp specifications, of course). Note that this bootstrapping has ac benefits as well, as it reduces the D1-D2 capacitance seen by the source. While the ~100 pF capacitance of D3-D4 might cause a loading problem with some op amps, this is mitigated by the isolating effect of R_{OUT1}, plus the feedback compensation of C_F. Both R_{OUT1} and R_{OUT2} protect the op amp output.

The input voltage clamping level is also programmable, and is set by the choice of zener voltage V_Z. This voltage plus 1.2 V should be greater than the maximum input, but below the rail voltage, as summarized in the figure. The example uses 10 V ±5% zener diodes, so input clamping typically will occur at ±11.2 V, allowing ±10 V swings.

An important caveat to the above is that it applies for *power-on* conditions. With *power-off*, D1–D4 still clamp to the noted levels, but this now produces a condition whereby the U1 input and output voltage can exceed the rails.

Note that this could be dangerous, for a given U1 device. If so, an optional and simple means towards providing a lower, safe clamping level for power off conditions is to use a relay at the V_{CLAMP} node. The contacts are open with power applied, and closed with power absent. With attention paid to an overall PCB layout, this can preserve a pA level bias current of FET op amps used for U1.

CM Over-Voltage Protection Using CMOS Channel Protectors

A much simpler alternative for overvoltage protection is the CMOS *channel protector*. A channel protector is a device in series with the signal path; for example, preceding an op amp input. It provides overvoltage protection by dynamically altering its resistance under fault conditions. Functionally, it has the distinct advantage of affording protection for sensitive components from voltage transients, whether the power supplies are present or not. Representative devices are the ADG465/ADG466/ADG467, which are channel protectors with single, triple, and octal channel options. Because this form of protection works whether supplies are present or not, the devices are ideal for use in applications where input overvoltages are common, or where correct power sequencing can't always be guaranteed. One such example is within hot-insertion rack systems.

An application of a channel protector for overvoltage protection of a precision buffer circuit is shown in Figure 7-68. A single channel device, the ADG465 at U2, is used here at the input of the U1 precision op amp buffer, an OP777.

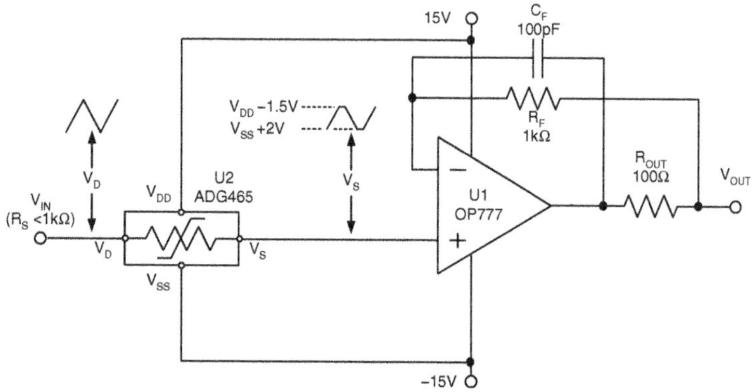

Figure 7-68: Using an ADG465 channel protector IC with a precision buffer offers great simplicity of protection and fail-safe operation during power off.

In operation, a channel protector behaves just like a series resistor of 60 Ω to 80 Ω in normal operation (i.e., nonfault conditions). Consisting of a series connection of multiple P and N MOSFETs, the protector dynamically adjusts channel resistance according to the voltage seen at the V_D terminal. Normal conduction occurs with V_D more than a threshold level above or below the rails, i.e., $(V_{SS} + 2\text{ V}) < V_D < (V_{DD} - 1.5\text{ V})$. For fault conditions the analog input voltage exceeds this range, causing one of the series MOSFETs to switch off, thus raising the channel resistance to a high level. This clamps the V_S output at one extreme range, either $V_{SS} + 2$ V or $V_{DD} - 1.5$ V, as shown in Figure 7-68.

A major channel protector advantage is the fact that both circuit and signal source protection are provided, in the event of overvoltage or power loss. Although shown here operating from op amp ±15 V supplies, these channel protectors can handle total supplies of up to 40 V. They also can withstand overvoltage inputs from $V_{SS} - 20$ V to $V_{DD} + 20$ V with power on (or ±35 V in the circuit shown). With power off ($V_{DD} = V_{SS} = 0$ V), maximum input voltage is ±35 V. Maximum room temperature channel leakage is 1 nA, making them suitable for op amps and in-amps with bias currents of several nA and up.

Related to the ADG46X series of channel protectors are several *fault-protected multiplexers*, for example the ADG508F/ADG509F, and the ADG438F/ADG439F families. Both the channel protectors and the fault-protected multiplexers are low power devices, and even under fault conditions, their supply current is limited to sub microampere levels. A further advantage of the fault-protected multiplexer devices is that they retain proper channel isolation, even for input conditions of one channel seeing an overvoltage; that is, the remaining channels still function.

CM Overvoltage Protection Using High CM Voltage In Amp

The ultimate simplicity for analog channel overvoltage protection is achieved with resistive input attenuation ahead of a precision op amp. This combination equates to a high voltage capable in amp, such as the AD629, which is able to linearly process differential signals riding upon CM voltages of up to ±270 V. Further, and most important to overvoltage protection considerations, the on-chip resistors afford protection for either common mode or differential voltages of up to ±500 V. All of this is achieved by virtue of a precision laser-trimmed thin-film resistor array and op amp, as shown in Figure 7-69.

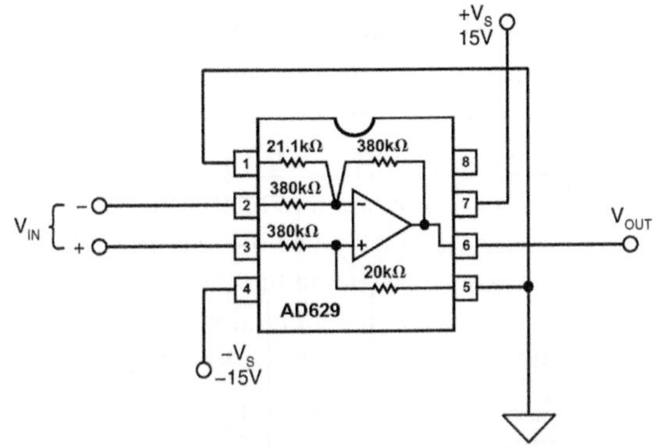

Figure 7-69: The AD629 high voltage in amp IC offers ± 500 V input overvoltage protection, one-component simplicity, and fail-safe power-off operation

Examination of this topology shows that the resistive network around the AD629's precision op amp acts to divide down the applied CM voltage at V_{IN} by a factor of 20/1. The AD629 simultaneously processes the input differential mode signal V_{IN} to a single-ended output referred to a local ground, at a gain factor of unity. Gain errors are no more than ±0.03% or 0.05%, while offset voltage is no more than 0.5 mV or 1 mV (grade dependent). The AD629 operates over a supply range of ±2.5 V to ±18 V.

These factors combine to make the AD629 a simple, one-component choice for the protection of off-card analog inputs that can potentially see dangerous transient voltages. Due to the relatively high resistor values used, protection of the device is also inherent with no power applied, since the input resistors safely limit fault currents. In addition, it offers those operating advantages inherent to an in amp: high CMR (86 dB minimum at 500 Hz), excellent overall dc precision, and the flexibility of simple polarity changes. On the flip side of performance issues, several factors make the AD629's output noise and drift relatively high, if compared to a lower gain in amp configuration such as the AMP03. These are the Johnson noise of the high value resistors, and the high noise gain of the topology (21×). These factors raise the op amp noise and drift along with the resistor noise by a factor higher than typical. Of course, whether or not this is an issue relevant to an individual application will require evaluation on a case-by-case basis.

Inverting Mode Op Amp Protection Schemes

There are some special cases of overvoltage protection requirements that don't fit into the more general CM protection schemes above. Figure 7-70 is one such example, a low bias current FET input op amp I/V converter.

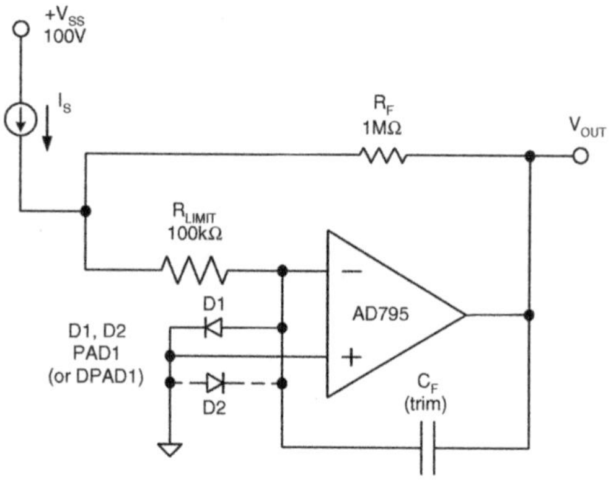

Figure 7-70: A low bias current FET input op amp I/V converter with overvoltage protection network RLIMIT and D1

In this circuit the AD795 1 pA bias current op amp is used as a precision inverter. Some current-source nature signals can originate from a high voltage potential, such as the 100 V V_{SS} level shown. As such, they have the potential of developing fault voltage levels beyond the op amp rails, producing fault current into the op amp well above safe levels. To prevent this, protection resistor R_{LIMIT} is used inside the feedback loop as shown, along with voltage clamp D1 (D2).

For normal signal condition (i.e., $I_S \leq 10$ uA) the op amp's inverting node is very close to ground, with just a tiny voltage drop across R_{LIMIT}. Normal I/V conversion takes place, with gain set by R_F. For protection, D1 is a special low leakage diode, clamping any excess voltage at the (–) node to ~0.6 V, thus protecting the op amp. The value of R_{LIMIT} is chosen to allow a 1 mA max current under fault conditions. Bootstrapping the

D1 (and/or D2) clamp diodes as shown minimizes the normal operating voltage across the inverting node, keeping the diode leakage low (see Figure 7-66). Note that for a positive source voltage as shown, only positive clamping is needed, so just one diode suffices.

Only the lowest leakage diodes (≤1 pA) such as the PAD1 (or the DPAD1 dual) should be used in this circuit. As previously noted, any clamping diode used here should either be shielded from light (or use opaque packaging), to minimize photocurrent from ambient light. Even so, the diode(s) will increase the net input current and shunt capacitance, and feedback compensation C_F will likely be necessary to control response peaking. C_F should be a very low leakage type. Also, with the use of very low input bias current devices such as the AD795, it isn't possible to use the same level of internal protection circuitry as with other ADI op amps. This factor makes the AD795 more sensitive to handling, so ESD precautions should be taken.

Amplifier Output Voltage Phase-Reversal

As alluded to above, there are "gray-area" op amp groups that have anomalous CM voltage zones, falling between the supply rails. As such, protection for these devices cannot be guaranteed by simply ensuring that the inputs stay between the rails—they must additionally stay *entirely* within their rated CM range, for consistent behavior.

Peculiar to some op amps, this misbehavior phenomenon is called *output voltage phase-reversal.* It is seen when one or both of the op amp inputs exceed their allowable input CM voltage range. Note that the inputs may still be well within the extremes of rail voltage, but simply below one specified CM limit. Typically, this is towards the negative range. Phase-reversal is most often associated with JFET and/or BiFET amplifiers, but some bipolar single-supply amplifiers are also susceptible to it.

The Figure 7-71 waveforms illustrate this general phenomenon, with an overdriven voltage follower input on the left, and the resulting output phase-reversal at the right.

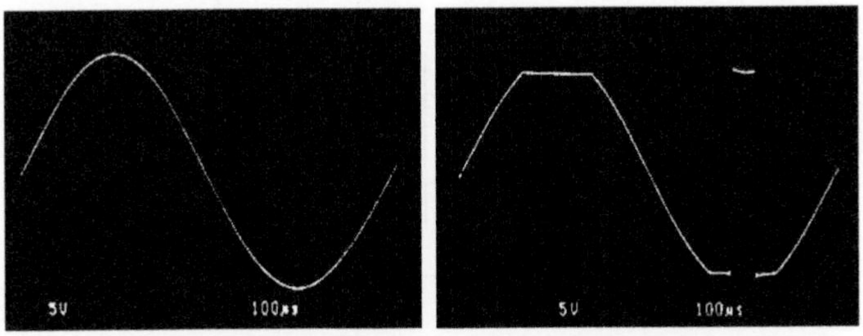

Figure 7-71: An illustration of input overdriving waveform (left) and the resulting output phase-reversal (right), using a JFET input op amp

While the specific details of the internal mechanism may vary with individual op amps, it suffices to say that the output phase-reversal occurs when a critical section of the amplifier front end saturates, causing the input-output sign relationship to temporarily reverse. Under this condition, when the CM range is exceeded, the negative-going input waveform in Figure 7-71 (left) does not continue going more negative in the output waveform, Figure 7-71 (right). Instead, the input-output relationship *phase-reverses*, with the output suddenly going positive, i.e., the spike. It is important to note that this is *not* a latching form of phase-reversal, as the output will once again continue to properly track the input, when the input returns to the CM range. In Figure 7-71, this can be seen in the continuance of the output sine wave, after the positive-going phase-reversal spike settles.

In most applications, this output voltage phase-reversal does no harm to the op amp, nor to the circuit where it is used. Indeed, since it is triggered when the CM limit is exceeded, noninverting stages with appreciable signal gain never see it, since their applied CM voltage is too small.

Note that with inverting applications the output phase-reversal problem is nonexistent, as the CM range isn't exercised. So, although a number of (mostly older) op amps suffer from phase-reversal, it still is rarely a serious problem in system design.

Nevertheless, when and if a phase-reversal susceptible amplifier used in a servo loop application sees excess CM voltage, the effect can be disastrous—it goes **Bang!** So, the best advice is to be forewarned.

An Output Phase-Reversal Do-it-Yourself Test

Since output phase-reversal may not always be fully described on a data sheet, it is quite useful to test for it. This is easily done in the lab, by driving a questionable op amp as a unity-gain follower, from a source impedance (R_{LIMIT}) of ~1kΩ. It is helpful to make this a variable, 1 kΩ –100 kΩ range resistance.

With a low resistance setting (1 kΩ), while bringing the driving signal level slowly up towards the rail limits, observe the amplifier output. If a phase-reversal mechanism is present when the CM limit of the op amp is exceeded, the output will suddenly reverse (see Figure 7-71, right). If there is no phase-reversal present in an amplifier, the output waveform will simply clip at the limits of its swing. It may prove helpful to have a well-behaved op amp available for this test, to serve as a performance reference. One such device is the AD8610.

Note that, in general, some care should be used with this test. Without a series current-limit resistor, if the generator impedance is too low (or level too high), it could possibly damage an internal junction of the op amp under test. So, obviously, caution is best for such cases.

Once a suitable R_{LIMIT} resistance value is found, well-behaved op amps will simply show a smooth, bipolar range, clipped output waveform when overdriven. This clipping will appear more like the *upper (positive swing)* portion of the waveform within Figure 7-71, right.

Fixes For Output Phase–Reversal

An op amp manufacturer might not always give the R_{LIMIT} resistance value appropriate to prevent output phase-reversal. But, the value can be determined empirically with the driving method mentioned above. Most often, the R_{LIMIT} resistor value providing protection against phase-reversal will also safely limit fault current through any input CM clamping diodes. If in doubt, a nominal value of 1 kΩ is a good starting point for testing.

Typically, FET input op amps will need only the current limiting series resistor for protection, but bipolar input devices are best protected with this same limiting resistor, *along with a Schottky diode* (i.e., R_{LIMIT} and D2, of Figure 7-65).

For a more detailed description of the output voltage phase-reversal effect, see References 7 and 8. Figure 7-72 summarizes a number of the key points relating to output voltage phase-reversal.

- Nonlatching Inversion of Transfer Function, Triggered by Exceeding Common-Mode Limit
- Sometimes Occurs in FET and Bipolar (Single-Supply) Op Amps
- Doesn't Harm Amplifier... but *Disastrous* for Servo Systems
- Not Usually Specified on Data Sheet, so Amplifier Must be Checked
- Easily Prevented:
  - All op amps: Limit applied CM voltage by clamping or other means
  - BiFETs: Add series input resistance, R_{LIMIT}
  - Bipolars: R_{LIMIT} and Schottky clamp diode to rail

Figure 7-72: A summary of key points regarding output phase-reversal in FET and bipolar input op amps

Alternately, any of the several previously mentioned CM clamping schemes can be used to prevent output phase-reversal, by setting the clamp voltage to be less than the amplifier CM range limit where phase-reversal occurs. For example, Figure 7-67 would operate to prevent phase-reversal in FET amplifiers susceptible to it, if the negative clamp limit is set so that $V_{CLAMP(-)}$ never exceeds the typical negative CM range of –11 V on a –15 V rail.

For validation of this or any of the previous overvoltage protection schemes, the circuit should be verified on a number of op amps, over a range of conditions as suitable to the final application environment.

Input Differential Protection

The discussions thus far have been on overvoltage common-mode conditions, which is typically associated with forward biasing of PN junctions inherent in the structure of the input stage. There is another equally important aspect of protection against overvoltage, which is that due to excess *differential* voltages. Excessive differential voltage, when applied to certain op amps, can lead to degradation of their operating characteristics.

This degradation is brought about by *reverse junction breakdown,* a second case of undesirable input stage conduction, occurring under conditions of *differential* over-voltage. However, in the case of reverse breakdown of a PN junction, the problem can be more subtle in nature. It is illustrated by the partial op amp input stage in Figure 7-73.

This circuit, applicable to a low noise op amp such as the OP27, is also typical of many others using low noise bipolar transistors for differential pair Q1-Q2. In the absence of any protection, it can be shown that voltages above about 7 V between the two inputs will cause a reverse junction breakdown of either Q2 or Q1 (dependent upon relative polarity). Note that in cases of E-B breakdown, even small reverse currents can cause degradation in both transistor gain and noise (see Reference 6). After E-B breakdown occurs, op amp parameters such as the bias currents and noise may well be out of specification. This is usually permanent, and it can occur gradually and quite subtly, particularly if triggered by transients. For these reasons,

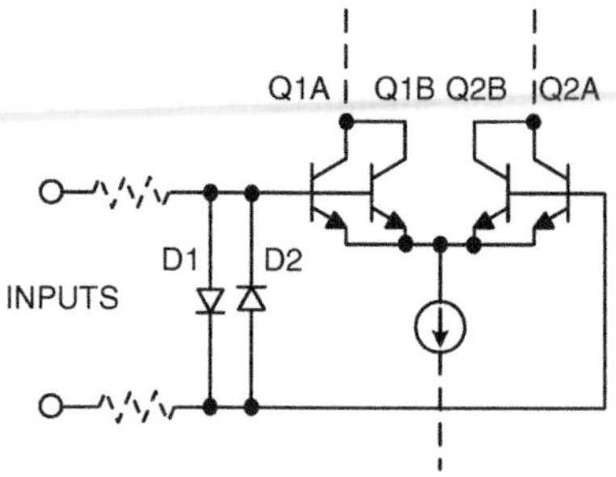

Figure 7-73: An op amp input stage with D1-D2 input differential overvoltage protection network

virtually all low noise op amps, whether NPN or PNP based, utilize protection diodes such as D1-D2 across the inputs. These diodes conduct for applied voltages greater than ±0.6 V, protecting the transistors.

The dotted series resistors function as current limiters (protection for the protection diodes) but aren't used in all cases. For example, the AD797 doesn't have the resistors, simply because they would degrade the part's specified noise of $1\ nV/\sqrt{Hz}$. Note: when the resistors are absent internally, some means of external current limiting must be provided, when and if differential overvoltage conditions do occur. Obviously, this is a trade-off situation, so the confidence of full protection must be weighed against the noise degradation. Note that an application circuit itself may provide sufficient resistance in the op amp inputs, such that additional resistance isn't needed.

In applying a low noise bipolar input stage op amp, first check the chosen part's data sheet for internal protection. When necessary, protection diodes D1-D2, if not internal to the op amp, should be added to guarantee prevention of Q1-Q2 E-B breakdown. If differential transients of more than 5 V can be seen by the op amp in the application, the diodes are in order. Ordinary low capacitance diodes will suffice, such as the 1N4148 family. Add current limiting resistors as necessary, to limit diode current to safe levels.

Other IC device junctions, such as base-collector and JFET gate-source junctions don't exhibit the same degradation in performance upon breakdown, and for these the input current should be limited to 5mA, unless the data sheet specifies a different value.

Protecting In Amps Against Overvoltage

From a protection standpoint, instrumentation amplifiers (in amps) are similar in many ways to op amps. Like op amps, their absolute maximum ratings must be observed for both common and differential mode input voltages.

A much simplified schematic of the AD620 in amp is shown in Figure 7-74, showing the input differential transistors and their associated protection parts.

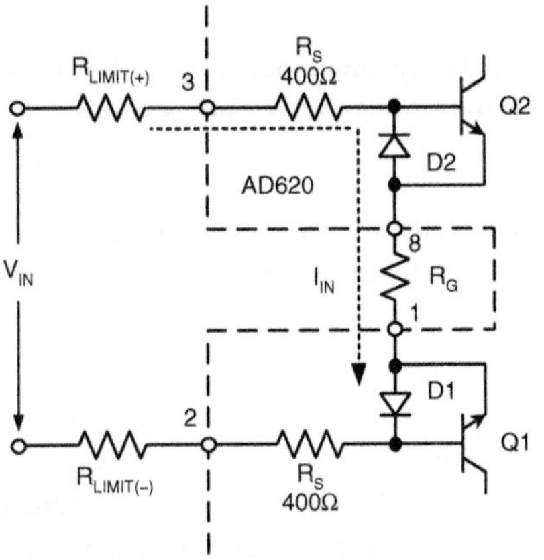

Figure 7-74: The AD620 in amp input internally uses D1-D2 and series resistors R_S for protection (additional protection can be added externally)

An important point, unique to the AD620 device, is the fact that the 400 Ω internal R_S protection resistors are *thin-film types*. Therefore these resistors don't show symptoms of diode-like conduction to the IC substrate (as would be the case were they diffused resistors). Practically, this means that the input ends of these resistors (Pins 3 and 2) can go above or below the supplies. Differential fault currents will be limited by the combination of twice the internal R_S plus the external gain resistance, R_G. Excess applied CM voltages will show current limited by R_S.

In more detail, it can be noted that input transistors Q1 and Q2 have protection diodes D1 and D2 across their base-emitter junctions, to prevent reverse breakdown. For differential voltages, analysis of shows that a fault current, I_{IN}, flows through the external R_{LIMIT} resistors (if present), the internal R_S resistors, the gain-setting resistor R_G, and two diode drops (Q2, D1). For the AD620 topology, R_G varies inversely with gain, and a worst-case (lowest resistance) occurs with the maximum gain of 1000, when R_G is 49.9 Ω. Therefore the lowest total internal path series resistance is about 850 Ω.

For the AD620, any combination of CM and differential input voltages should be limited to levels that limit the input fault current to 20 mA, maximum. A purely differential voltage of 17 V would result in this current level, for the lowest resistance case. For CM voltages that may go beyond either rail, an internal diode not shown in Figure 7-74 conducts, effectively clamping the driven input to either $+V_S$ or $-V_S$ at the R_S inner end. For this overvoltage CM condition, the 400 Ω value of R_S and the excess voltage beyond the rail determines the current level. If, for example, V_{IN} is 23 V with $+V_S$ at 15 V, 8 V appears across R_S, and the 20 mA current rating is reached. Higher fault voltages can be dealt with by adding R_{LIMIT} resistance, to maintain fault current at 20 mA or less.

A more generalized external voltage protection circuit for an in amp like the AD620 is shown in Figure 7-75.

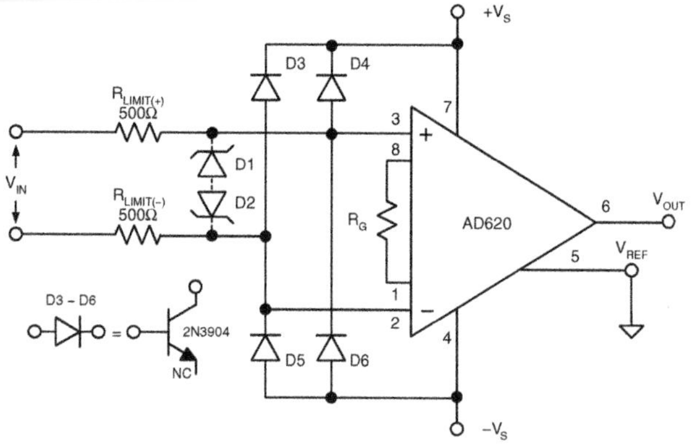

Figure 7-75: A generalized diode protection circuit for the AD620 and other in amps uses D3–D6 for CM clamping and series resistors R_{LIMIT} for protection

In this circuit, low-leakage diodes D3–D6 are used as CM clamps. Since the in amp bias current may be only 1 nA or so (for the AD620), a low leakage diode type is mandatory. As can be noted from the topology, diode bootstrapping isn't possible with this configuration.

It should be noted that not only must the diodes have basically low leakage, they must also maintain low leakage at the highest expected temperature. This suggests either FET type diodes (see Figure 7-66), or the transistor C-B types shown. The R_{LIMIT} resistors are chosen to limit the maximum diode current under fault conditions. If additional *differential* protection is used, either back-back zener or Transzorb clamps can be used, shown as D1-D2. If this is done, leakage of these diodes should be carefully considered.

The protection scheme of Figure 7-75, while effective using appropriate parts, has the downside of being busy for components. A much more simple in amp protection using fault protected devices is shown in Figure 7-76. Although shown with an AD620, this circuit is useful with many other dual-supply in amps with bias currents of 1 nA or more. It can use two-thirds of a triple ADG466 channel protector for the in amp differential inputs, or a pair of ADG465 single channel parts as shown.

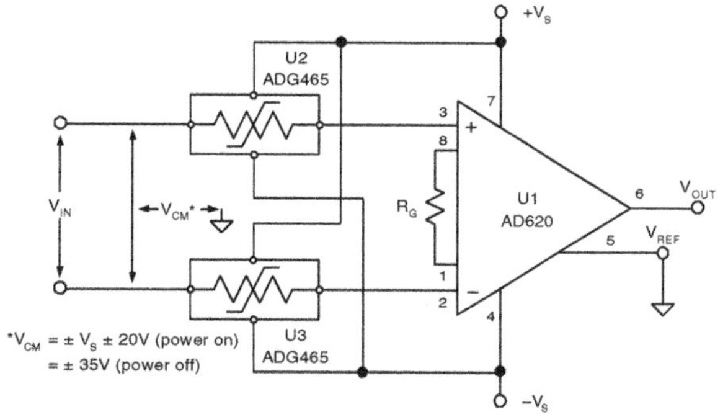

Figure 7-76: A channel protector device (or fault-protected multiplexer) provides protection for dual-supply in amps with a minimum of extra parts

Because the nature of a channel protection device is to turn off as V_{IN} approaches either rail, the scheme of Figure 7-76 doesn't function with rail sensing single-supply in amps. If near-rail operation and protection is required in an in amp application, an alternative method is necessary. Many single-supply in-amps are topologically similar to the two-amplifier in amp circuit which is shown within the dotted box of Figure 7-77.

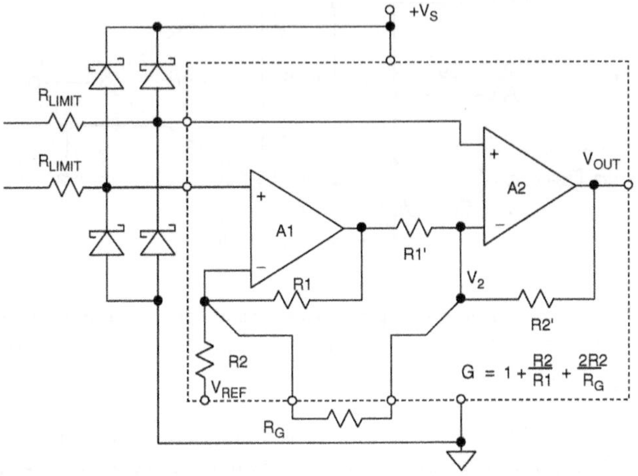

Figure 7-77: Single-supply in amps may or may not require external protection in the form of resistors and clamp diodes—if so, they can be added as shown

In terms of the necessity for externally added protection components, a given in amp may or may not require them. Each case needs to be considered individually. For example, some in amps have clamp diodes as shown, but *internal to the device*. The AD623 is such a part, but it lacks the series resistors, which can be added externally when and if necessary. Note that this approach allows the R_{LIMIT} value to be optimized for protection, with negligible impact on noise for those applications not needing the protection.

Also, some in amp devices have both internal protection resistors *and* clamping diodes, an example here is the AD627. In this device, the internal protection is adequate for transients up to 40 V beyond the supplies (a 20 mA fault current in the internal resistors). For overvoltage levels higher than this, external R_{LIMIT} resistors can be added.

The use of the Schottky diodes as shown at the two inputs is an option for in amp protection. If no clamping is specifically provided internally, then they are applicable. Their use is generally similar to the op amp protection case of Figure 7-65, with comparable caveats as far as leakage. Note that in many cases, due to internal protection networks of modern in amps, these diodes just won't be necessary. But again, there aren't hard rules on this, so always check the data sheet before finalizing an application.

To summarize, Figure 7-78 reviews the major points of the in-circuit overvoltage issues discussed in this section.

If these varied overvoltage precautions for op amps and in amps seem complex, they are. Whenever op amp (or in amp) inputs (and outputs) go outside equipment boundaries, dangerous or destructive things can happen to them. Obviously, for highest reliability these potentially hazardous situations should be anticipated.

Fortunately, most applications are contained entirely within the equipment, and usually see inputs and outputs to/from other ICs on the same power system. Therefore clamping and protection schemes typically aren't necessary for these cases.

- INPUT VOLTAGES MUST NOT EXCEED ABSOLUTE MAXIMUM RATINGS (Usually Specified with Respect to Supply Voltages)
- Requires $V_{IN(CM)}$ Stay Within a Range Extending to ≤0.3V Beyond Rails ($-V_S-0.3V \geq V_{IN} \leq +V_S+0.3V$)
- IC Input Stage Fault Currents *Must* Be Limited (≤ 5mA Unless Otherwise Specified)
- Avoid Reverse-Bias Breakdown in Input Stage Junctions!
- Differential and Common-Mode Ratings Often Differ
- No Two Amplifiers are Exactly the Same
- Watch Out for Output Phase-Reversal in JFET and SS Bipolar Op Amps
- Some ICs Contain *Internal* Input Protection
  - Diode Voltage Clamps, Current Limiting Resistors (or both)
  - Absolute Maximum Ratings Must Still Be Observed

Figure 7-78: A summary of in-circuit overvoltage points

Out-of-Circuit Overvoltage Protection

Linear ICs such as op amps and in amps must also be protected prior to the time they are mounted to a printed circuit board, that is an *out-of-circuit* state. In such a condition, ICs are completely at the mercy of their environment as to what stressful voltage surges they may see. Most often the harmful voltage surges come from *electrostatic discharge,* or, as more commonly referenced, ESD. This is a single, fast, high current transfer of electrostatic charge resulting from one of two conditions. These conditions are:

1) *Direct contact transfer between two objects at different potentials (sometimes called contact discharge)*

2) *A high electrostatic field between two objects when they are in close proximity (sometimes called air discharge)*

The prime sources of static electricity are mostly insulators and are typically synthetic materials, e.g., vinyl or plastic work surfaces, insulated shoes, finished wood chairs, Scotch tape, bubble pack, soldering irons with ungrounded tips, and so forth. Voltage levels generated by these sources can be extremely high since their charge is not readily distributed over their surfaces or conducted to other objects.

The generation of static electricity caused by rubbing two substances together is called the *triboelectric* effect. Some common examples of ordinary acts producing significant ESD voltages are shown in Figure 7-79.

- Walking across a Carpet
  1000V – 1500V
- Walking across a Vinyl Floor
  150V – 250V
- Handling Material Protected by Clear Plastic Covers
  400V – 600V
- Handling Polyethylene Bags
  1000V – 2000V
- Pouring Polyurethane Foam into a Box
  1200V – 1500V
- Note: Above Assumes 60% RH. For Low RH (30%) Voltages Can Be > 10 Times

Figure 7-79: ESD voltages generated by various ordinary circumstances

ICs can be damaged by the high voltages and high peak currents generated by ESD. Precision analog circuits, often featuring very low bias currents, are more susceptible to damage than common digital circuits, because traditional input-protection structures that protect against ESD damage increase input leakage—and thus can't be used.

For the design engineer or technician, the most common manifestation of ESD damage is a catastrophic failure of the IC. However, exposure to ESD can also cause increased leakage or degrade other parameters. If a device appears to not meet a data sheet specification during evaluation, the possibility of ESD damage should be considered. Figure 7-80 outlines some relevant points on ESD induced failures.

- ESD Failure Mechanisms:
  - Dielectric or junction damage
  - Surface charge accumulation
  - Conductor fusing
- ESD Damage Can Cause:
  - Increased leakage
  - Degradation in performance
  - Functional failures of ICs
- ESD Damage is often Cumulative:
  - For example, each ESD "zap" may increase junction damage until, finally, the device fails.

Figure 7-80: Understanding ESD damage

All ESD-sensitive devices are shipped in protective packaging. ICs are usually contained in either conductive foam or antistatic shipping tubes, and the container is then sealed in a static-dissipative plastic bag. The sealed bag is marked with a distinctive sticker, such as in Figure 7-81, which outlines the appropriate handling procedures.

The presence of outside package notices such as those shown in Figure 7-81 is notice to the user that device handling procedures appropriate for ESD protection are necessary.

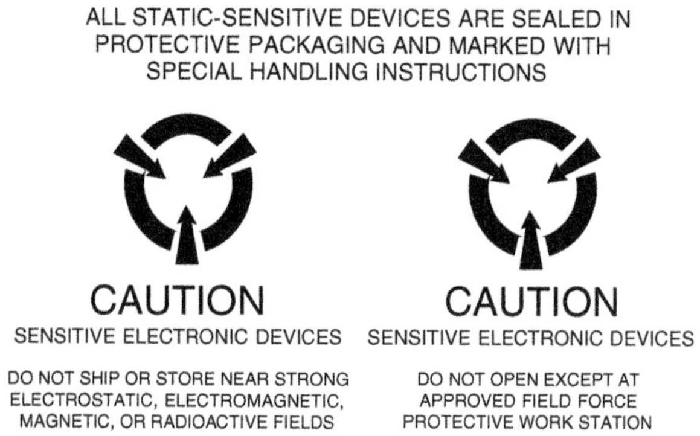

ALL STATIC-SENSITIVE DEVICES ARE SEALED IN PROTECTIVE PACKAGING AND MARKED WITH SPECIAL HANDLING INSTRUCTIONS
CAUTION
SENSITIVE ELECTRONIC DEVICES
DO NOT SHIP OR STORE NEAR STRONG ELECTROSTATIC, ELECTROMAGNETIC, MAGNETIC, OR RADIOACTIVE FIELDS
CAUTION
SENSITIVE ELECTRONIC DEVICES
DO NOT OPEN EXCEPT AT APPROVED FIELD FORCE PROTECTIVE WORK STATION

Figure 7-81: Recognizing ESD-sensitive devices by package and labeling

In addition, data sheets for ESD-sensitive ICs generally have a bold statement to that effect, as shown in Figure 7-82.

Once ESD-sensitive devices are identified, protection is relatively easy. Obviously, keeping ICs in their original protective packages as long as possible is a first step. A second step is discharging potentially damaging ESD sources before IC damage occurs. Discharging such voltages can be done quickly and safely, through a high impedance.

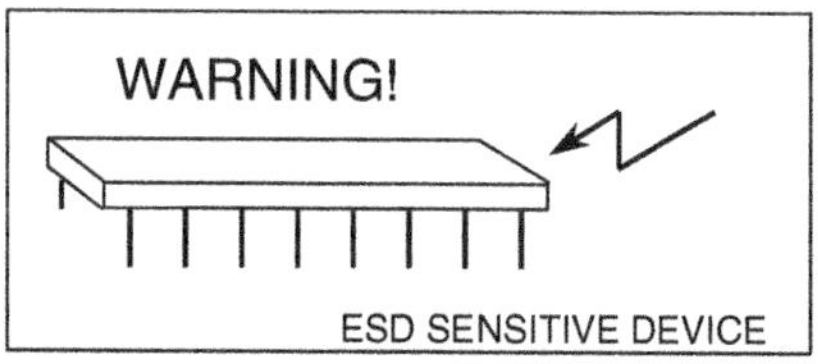

CAUTION

ESD (Electrostatic Discharge) sensitive device. Electrostatic charges as high as 4000 V readily accumulate on the human body and test equipment and can discharge without detection. Although the ADxxx features proprietary ESD protection circuitry, permanent damage may occur on devices subjected to high energy electrostatic discharges. Therefore, proper ESD precautions are recommended to avoid performance degradation or loss of functionality.

Figure 7-82: ESD data sheet statement for linear ICs

A key component required for ESD-safe IC handling is a workbench with a static-dissipative surface, shown in the workstation of Figure 7-83. The surface is connected to ground through a 1 MΩ resistor, which dissipates any static charge, while protecting the user from electrical ground fault shock hazards. If existing bench tops are nonconductive, a static-dissipative mat should be added, along with the discharge resistor.

Note that the surface of the workbench has a moderately high sheet resistance. It is neither necessary nor desirable to use a low resistance surface (such as a sheet of copper-clad PC board) for the work surface. Remember, a high peak current may flow if a charged IC is discharged through a low impedance. This is precisely what happens when a charged IC contacts a grounded copper clad board. When the same charged IC is placed on the high impedance surface of Figure 7-83, however, the peak current isn't high enough to damage the device.

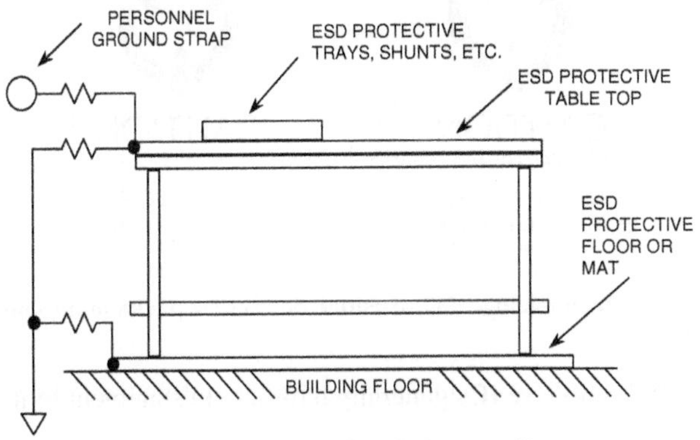

Figure 7-83: A workstation environment suitable for handling ESD-sensitive ICs

Several personnel-handling techniques are keys to minimizing ESD-related damage. At the workstation, a conductive wrist strap is recommended while handling ESD-sensitive devices. The wrist strap ensures that normal tasks, such as peeling tape from packages, won't cause IC damage. Again, a 1 MΩ resistor, from the wrist strap to ground, is required for safety. When building prototype breadboards or assembling PC boards that contain ESD-sensitive ICs, all passive components should be inserted and soldered before the ICs. This minimizes the ESD exposure of the sensitive devices. The soldering iron must, of course, have a grounded tip.

Protecting ICs from ESD requires the participation of both the IC manufacturer and the customer. IC manufacturers have a vested interest in providing the highest possible level of ESD protection for their products. IC circuit designers, process engineers, packaging specialists and others are constantly looking for new and improved circuit designs, processes, and packaging methods to withstand or shunt ESD energy.

A complete ESD protection plan, however, requires more than building ESD protection into ICs. The users of ICs must also provide their employees with the necessary knowledge of and training in ESD handling procedures, so that protection can be built in at all key points along the way, as outlined in Figure 7-84.

Special care should be taken when breadboarding and evaluating ICs. The effects of ESD damage can be cumulative, so repeated mishandling of a device can eventually cause a failure. Inserting and removing ICs from a test socket, storing devices during evaluation, and adding or removing external components on the breadboard should all be done while observing proper ESD precautions. Again, if a device fails during a prototype system development, repeated ESD stress may be the cause.

ANALOG DEVICES:

- Circuit Design and Fabrication

↓ Design and manufacture products with the highest level of ESD
↓ protection consistent with required analog and digital performance.

- Pack and Ship

↓ Pack in static dissipative material. Mark packages with ESD warning.

CUSTOMERS:

- Incoming Inspection

↓ Inspect at grounded workstation. Minimize handling.

- Inventory Control

↓ Store in original ESD-safe packaging. Minimize handling.

- Manufacturing

↓ Deliver to work area in original ESD-safe packaging. Open packages only at
↓ grounded workstation. Package subassemblies in static dissipative packaging.

- Pack and Ship

Pack in static dissipative material if required. Replacement or optional boards may require special attention.

Figure 7-84: ESD protection requires a partner relationship between ADI and the end customer with control at key points

The key word to remember with respect to ESD is *prevention*. There is no way to undo ESD damage, or to compensate for its effects.

ESD Models and Testing

Some applications have higher sensitivity to ESD than others. ICs located on a PC board surrounded by other circuits are generally much less susceptible to ESD damage than circuits that must interface with other PC boards or the outside world. These ICs are generally not specified or guaranteed to meet any particular ESD specification (with the exception of MIL-STD-883 Method 3015 classified devices). A good example of an ESD-sensitive interface is the RS-232 interface port ICs on a computer, which can easily be exposed to excess voltages. In order to guarantee ESD performance for such devices, the test methods and limits must be specified.

A host of test waveforms and specifications have been developed to evaluate the susceptibility of devices to ESD. The three most prominent of these waveforms currently in use for semiconductor or discrete devices are: The Human Body Model (HBM), the Machine Model (MM), and the Charged Device Model (CDM). Each of these models represents a fundamentally different ESD event, consequently, correlation between the test results for these models is minimal.

Since 1996, all electronic equipment sold to or within the European Community must meet Electromechanical Compatibility (EMC) levels as defined in specification IEC1000-4-x. Note that this does not apply to individual ICs, *but to the end equipment.* These standards are defined along with test methods in the various IEC1000 specifications, and are listed in Figure 7-85.

- IEC1000-4 Electromagnetic Compatibility EMC
- IEC1000-4-1 Overview of Immunity Tests
- IEC1000-4-2 Electrostatic Discharge Immunity (ESD)
- IEC1000-4-3 Radiated Radio-Frequency Electromagnetic Field Immunity
- IEC1000-4-4 Electrical Fast Transients (EFT)
- IEC1000-4-5 Lightening Surges
- IEC1000-4-6 Conducted Radio Frequency Disturbances above 9kHz
- Compliance Marking: CE

Figure 7-85: A listing of the IEC standards applicable to ESD specifications and testing procedures

IEC1000-4-2 specifies compliance testing using two coupling methods, *contact discharge* and *air-gap discharge*.

Contact discharge calls for a direct connection to the unit being tested. Air-gap discharge uses a higher test voltage, but does not make direct contact with the unit under test. With air discharge, the discharge gun is moved toward the unit under test, developing an arc across the air gap, hence the term air discharge. This method is influenced by humidity, temperature, barometric pressure, distance and rate of closure of the discharge gun. The contact-discharge method, while less realistic, is more repeatable and is gaining acceptance in preference to the air-gap method.

Although very little energy is contained within an ESD pulse, the extremely fast rise time coupled with high voltages can cause failures in unprotected ICs. Catastrophic destruction can occur immediately as a result of arcing or heating. Even if catastrophic failure does not occur immediately, the device may suffer from parametric degradation, which may result in degraded performance. The cumulative effects of continuous exposure can eventually lead to complete failure.

I-O lines are particularly vulnerable to ESD damage. Simply touching or plugging in an I-O cable can result in a static discharge that can damage or completely destroy the interface product connected to the I-O port (such as RS-232 line drivers and receivers).

Traditional ESD test methods such as MIL-STD-883B Method 3015.7 do not fully test a product's susceptibility to this type of discharge. This test was intended to test a product's susceptibility to ESD damage during handling. Each pin is tested with respect to all other pins. There are some important differences between the MIL-STD-883B Method 3015.7 test and the IEC test, noted as follows:

1) *The IEC test is much more stringent in terms of discharge energy. The peak current injected is over four times greater.*

2) *The current rise time is significantly faster in the IEC test.*

3) *The IEC test is carried out while power is applied to the device.*

It is possible that ESD discharge could induce latch-up in the device under test. This test is therefore more representative of a real-world I-O discharge where the equipment is operating normally with power applied. For maximum confidence, however, both tests should be performed on interface devices, thus ensuring maximum protection both during handling, and later, during field service.

A comparison of the test circuit values for the IEC1000-4-2 model versus the MIL-STD-883B Method 3015.7 Human Body Model is shown in Figure 7-86.

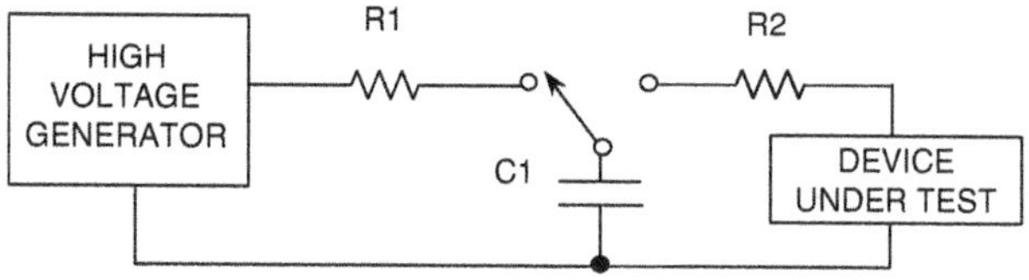

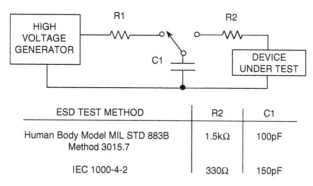

ESD TEST METHOD	R2	C1
Human Body Model MIL STD 883B Method 3015.7	1.5kΩ	100pF
IEC 1000-4-2	330Ω	150pF

NOTE: CONTACT DISCHARG E VOLTAGE SPEC FOR IEC 1000-4-2 IS ±8kV

Figure 7-86: ESD test circuits and values

The ESD waveforms for the MIL-STD-883B, METHOD 3015.7 and IEC 1000-4-2 tests are compared in Figure 7-87, left and right, respectively.

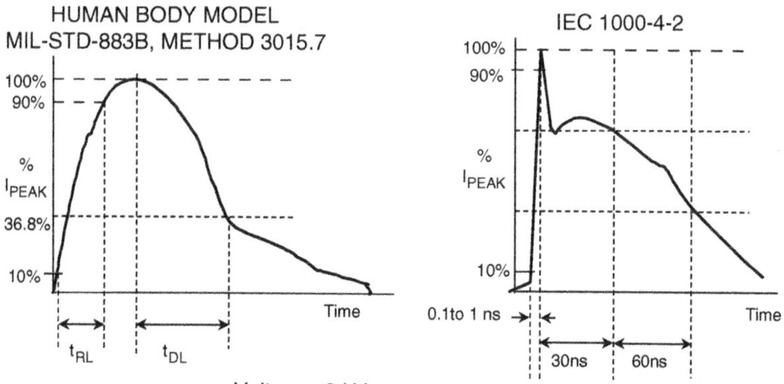

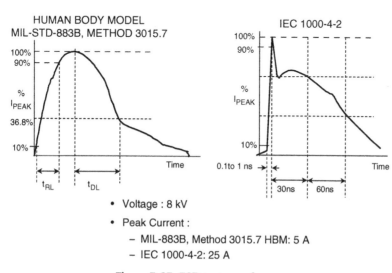

- Voltage : 8 kV
- Peak Current :
  - MIL-883B, Method 3015.7 HBM: 5 A
  - IEC 1000-4-2: 25 A

Figure 7-87: ESD test waveforms

Suitable ESD-protection design measures are relatively easy to incorporate, and most of the overvoltage protection methods already discussed in this section will help.

Additional protection can also be obtained. For RS-232 and RS-485 drivers and receivers, the ADMXXX-E series is supplied with guaranteed 15 kV (HBM) ESD specifications.

For more general uses, the addition of TransZorbs at appropriate places in a system can provide protection against ESD (see References).

Figure 7-88 summarizes the major points about ESD prevention, from both an out-of-circuit as well as an in-circuit perspective.

- Observe all Absolute Maximum Ratings on Data Sheet
- Read ADI AN-397 (See Reference 16)
- Purchase ESD-Specified Digital Interface Devices
  - ADMXXX-E Series of RS-232/RS-485 Drivers/Receivers (See Reference 18)
- Follow General Overvoltage Protection Recommendations
  - Add Series Resistance to Limit Currents
  - Add Zeners or Transient Voltage Suppressors (TVS) for Extra Protection (See Reference 19)

Figure 7-88: A summary of ESD points

References: Op Amp Protection

1. Walt Kester, Wes Freeman, Joe Buxton, "Overvoltage Protection," portion of Section 10 within Walt Kester, Editor, **Practical Design Techniques for Sensor Signal Conditioning**, Analog Devices, Inc., 1999, ISBN 0-916550-20-6.
2. Siliconix PAD/JPAD/SSTPAD series Low leakage Pico-Amp Diodes, Vishay/Siliconix, www.vishay.com/brands/siliconix/SSFsglld.html.
3. Bob Pease, "Bounding, Clamping Techniques Improve Circuit Performance," **EDN**, November 10, 1983, p. 277.
4. Bob Pease, "Understanding Diodes and Their Problems," Chapter 6 within **Troubleshooting Analog Circuits**, Butterworth-Heinemann, 1991, ISBN 0-7506-9184-0.
5. Dov Kurz, Avner Cohen, "Bootstrapping Reduces Amplifier Input Capacitance," **EDN**, March 20, 1978.
6. C.D. Motchenbacher, J.A. Connelly, Chapter 5, within **Low Noise Electronic System Design**, John Wiley, 1993.
7. Adolfo Garcia, "Operational Amplifier Output Voltage Phase Reversal," Section 11, pp. 1–10 within Walt Kester, Editor, **1992 Amplifier Applications Guide**, Analog Devices, Inc., Norwood, MA, 1992, ISBN 0-916550-10-9.
8. Adolfo Garcia, Wes Freeman, "Overvoltage Effects On Analog Integrated Circuits," Section 7 within Walt Kester, Editor, **Practical Analog Design Techniques**, Analog Devices, Inc., Norwood, MA, 1995, ISBN 0-916550-16-8.
9. Charles Kitchen, Lew Counts, **A Designer's Guide to Instrumentation Amplifiers**, Analog Devices, Inc., 2000.
10. Walt Kester, Wes Freeman, James Bryant, "Electrostatic Discharge," portion of Section 10 within Walt Kester, Editor, **Practical Design Techniques for Sensor Signal Conditioning**, Analog Devices, Inc., 1999, ISBN 0-916550-20-6.
11. MIL-STD-883 Method 3015, "Electrostatic Discharge Sensitivity Classification." Available from Standardization Document Order Desk, 700 Robbins Ave., Building #4, Section D, Philadelphia, PA, 19111-5094.
12. EIAJ ED-4701 Test Method C-111, "Electrostatic Discharges." Available from the Japan Electronics Bureau, 250 W 34th St., New York NY 10119, Attn.: Tomoko.
13. ESD Association Standard S5.2 for "Electrostatic Discharge (ESD) Sensitivity Testing -Machine Model (MM)-Component Level." Available from the ESD Association, Inc., 200 Liberty Plaza, Rome, NY 13440.
14. ESD Association Draft Standard DS5.3 for "Electrostatic Discharge (ESD) Sensitivity Testing – Charged Device Model (CDM) Component Testing." Available from the ESD Association, Inc., 200 Liberty Plaza, Rome, NY 13440.
15. **ESD Prevention Manual**, Analog Devices, Inc.
16. Niall Lyne, "Electrically Induced Damage to Standard Linear Integrated Circuits: The Most Common Causes and the Associated Fixes to Prevent Reoccurrence," **Analog Devices AN397**.
17. Mike Byrne, "How to Reliably Protect CMOS Circuits Against Power Supply Overvoltaging," **Analog Devices AN311**.
18. Data sheet for **ADM3311E RS-232 Port Transceiver**, Analog Devices, Inc., www.analog.com.
19. TransZorbs are available from General Semiconductor, Inc., 10 Melville Park Road, Melville, NY, 11747-3113, 631-847-3000.

Thermal Considerations

Walt Jung

For reliability reasons, op amp systems handling appreciable power are increasingly called upon to observe *thermal management.* All semiconductors have some specified safe upper limit for junction temperature (T_J), usually on the order of 150°C (sometimes 175°C). Like maximum power supply voltages, maximum junction temperature is a worst-case limitation which shouldn't be exceeded. In conservative designs, it won't be approached by less than an ample safety margin. Note that this is critical, since semiconductor lifetime is inversely related to operating junction temperature. Simply put, the cooler op amps are, the more they can approach their maximum life.

This limitation of power and temperature is basic, and is illustrated by a typical data sheet statement as in Figure 7-89. In this case it is for the AD8017AR, an 8-pin SOIC device.

The maximum power that can be safely dissipated by the AD8017 is limited by the associated rise in junction temperature. The maximum safe junction temperature for plastic encapsulated device is determined by the glass transition temperature of the plastic, approximately +150°C. Temporarily exceeding this limit may cause a shift in parametric performance due to a change in the stresses exerted on the die by the package. Exceeding a junction temperature of +175°C for an extended period can result in device failure.

Figure 7-89: Maximum power dissipation data sheet statement for the AD8017AR, an ADI thermally-enhanced SOIC packaged device

Tied to these statements are certain conditions of operation, such as the power dissipated by the device, and the package mounting specifics to the printed circuit board (PCB). In the case of the AD8017AR, the part is rated for 1.3 W of power at an ambient of 25°C. This assumes operation of the 8-lead SOIC package on a two-layer PCB with about 4 inches^2 (~2500 mm^2) of 2 oz. copper for heat sinking purposes. Predicting safe operation for the device under other conditions is covered next.

Thermal Basics

The symbol θ is generally used to denote *thermal resistance*. Thermal resistance is in units of °C/watt (°C/W). Unless otherwise specified, it defines the resistance heat encounters transferring from a hot IC junction to the ambient air. It might also be expressed more specifically as θ_{JA}, for *thermal resistance, junction-to-ambient*. θ_{JC} and θ_{CA} are two additional θ forms used. Following is a further explanation.

In general, a device with a thermal resistance θ equal to 100°C/W will exhibit a temperature differential of 100°C for a power dissipation of 1 W, as measured between two reference points. Note that this is a linear relationship, so 1 W of dissipation in this part will produce a 100°C differential (and so on, for other powers). For the AD8017AR example, θ is about 95°C/W, so 1.3 W of dissipation produces about a 124°C junction-to-ambient temperature differential. It is of course this rise in temperature that is used to predict the internal temperature, in order to judge the thermal reliability of a design. With the ambient at 25°C, this allows an internal junction temperature of about 150°C. In practice, most ambient temperatures are above 25°C, so less power can then be handled.

For any power dissipation P (in watts), one can calculate the effective temperature differential (ΔT) in °C as:

$$\Delta T = P \times \theta \qquad \text{Eq. 7-6}$$

where θ is the total applicable thermal resistance.

Figure 7-90 summarizes a number of basic thermal relationships.

- θ = Thermal Resistance (°C/W)
- P = Total Device Power Dissipation (W)
- T = Temperature (°C)
- ΔT = Temperature Differential = $P \times \theta$
- θ_{JA} = Junction-Ambient Thermal Resistance
- θ_{JC} = Junction-Case Thermal Resistance
- θ_{CA} = Case-Ambient Thermal Resistance
- $\theta_{JA} = \theta_{JC} + \theta_{CA}$
- $T_J = T_A + (P \times \theta_{JA})$
- Note: $T_{J(Max)}$ = 150°C (Sometimes 175°C)

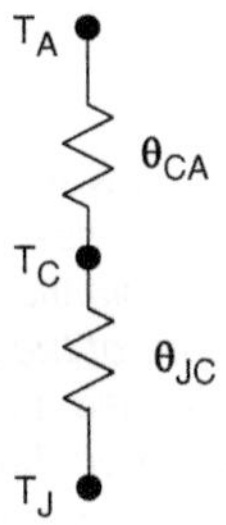

Figure 7-90: Basic thermal relationships

Note that series thermal resistances, such as the two shown in Figure 7-90, model the total thermal resistance path a device may see. Therefore the total θ for calculation purposes is the sum, i.e., $\theta_{JA} = \theta_{JC}$ and θ_{CA}. Given the ambient temperature T_A, P, and θ, then T_J can be calculated. As the relationships signify, to maintain a low T_J, either θ or the power being dissipated (or both) must be kept low. A low ΔT is the key to extending semiconductor lifetimes, as it leads to lower maximum junction temperatures.

In ICs, one temperature reference point is always the device junction, taken to mean the hottest spot inside the chip operating within a given package. The other relevant reference point will be either T_C, the case of the device, or T_A, that of the surrounding air. This leads in turn to the above-mentioned individual thermal resistances, θ_{JC} and θ_{JA}.

Taking the simplest case first, θ_{JA} is the thermal resistance of a given device measured between its *junction* and the *ambient* air. This thermal resistance is most often used with small, relatively low power ICs such as op amps, which often dissipate 1 W or less. Generally, θ_{JA} figures typical of op amps and other small devices are on the order of 90°C/W–100°C/W for a plastic 8-pin DIP package, as well as the better SOIC packages.

It should be clearly understood that these thermal resistances are *highly* package-dependent, as different materials have different degrees of thermal conductivity. As a general rule of thumb, thermal resistance of conductors is analogous to electrical resistances, that is, copper is the best, followed by aluminum, steel, and so on. Thus copper lead frame packages offer the highest performance, i.e., the lowest θ.

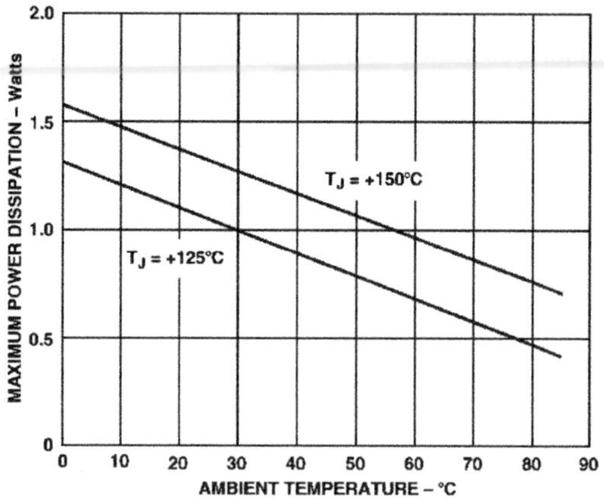

Figure 7-91: Thermal rating curves for AD8017AR op amp

Heat Sinking

By definition, a *heat sink* is an added low thermal resistance device attached to an IC to aid heat removal. A heat sink has additional thermal resistance of its own, θ_{CA}, rated in °C/W. However, most current op amp packages don't easily lend themselves to heat sink attachment (exceptions are older TO99 metal can types). Devices meant for heat sink attachment will often be noted by a θ_{JC} dramatically lower than the θ_{JA}. In this case θ will be composed of more than one component. Thermal impedances add, making a net calculation relatively simple. For example, to compute a net θ_{JA} given a relevant θ_{JC}, the thermal resistance of the heat sink, θ_{CA}, or *case* to *ambient* is added to the θ_{JC} as:

$$\theta_{JA} = \theta_{JC} + \theta_{CA} \qquad \text{Eq. 7-7}$$

and the result is the θ_{JA} for that specific circumstance.

More generally however, modern op amps *don't* use commercially available heat sinks. Instead, when significant power needs to be dissipated, such as ≥1 W, low thermal resistance copper PCB traces are used as the heat sink. In such cases, the most useful form of manufacturer data for this heat sinking are the boundary conditions of a sample PCB layout, and the resulting θ_{JA} for those conditions. This is, in fact, the type of specific information supplied for the AD8017AR, as mentioned earlier. Applying this approach, example data illustrating thermal relationships for such conditions is shown by Figure 7-91. These data apply for an AD8017AR mounted to a heat sink with an area of about 4 square inches on a 2-layer, 2-ounce copper PCB.

These curves indicate the maximum power dissipation versus temperature characteristic for the AD8017, for maximum junction temperatures of both 150°C and 125°C. Such curves are often referred to as *derating* curves, since allowable power decreases with ambient temperature.

With the AD8017AR, the proprietary ADI *Thermal Coastline* IC package is used, which allows additional power to be dissipated with no increase in the SO-8 package size. For a $T_{J(max)}$ of 150°C, the upper curve shows the allowable power in this package, which is 1.3 W at an ambient of 25°C. If a more conservative $T_{J(max)}$ of 125°C is used, the lower of the two curves applies.

A performance comparison for an 8-pin standard SOIC and the ADI Thermal Coastline version is shown in Figure 7-92. Note that the Thermal Coastline provides an allowable dissipation at 25°C of 1.3 W, whereas a standard package allows only 0.8 W. In the Thermal Coastline heat transferal is increased, accounting for the package's lower θ_{JA}.

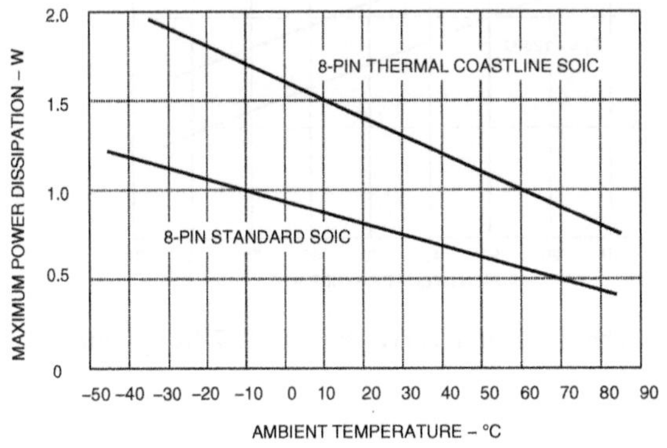

Figure 7-92: Thermal rating curves for standard (lower) and ADI Thermal Coastline (upper) 8-pin SOIC packages

Even higher power dissipation is possible, with the use of IC packages better able to transfer heat from chip to PCB. An example is the AD8016 device, available with two package options rated for 5.5 W and 3.5 W at 25°C, respectively, as shown in Figure 7-93.

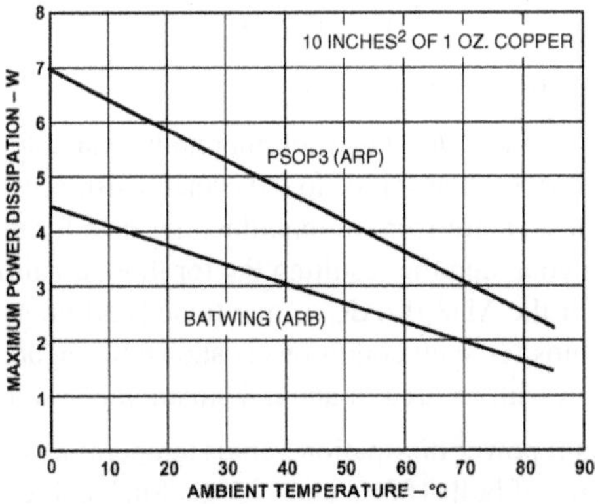

Figure 7-93: Thermal characteristic curves for the AD8016 BATWING (lower) and PSOP3 (upper) packages, for TJ(max) equal to 125°C

Taking the higher rated power option, the AD8016ARP PSOP3 package, when used with a 10-inch^2 1 oz. heat sink plane, the combination is able to handle up to 3 W of power at an ambient of 70°C, as noted by the upper curve. This corresponds to a θ_{JA} of 18°C/W, which in this case applies for a maximum junction temperature of 125°C.

The reason the PSOP3 version of the AD8016 is better able to handle power lies with the use of a large area copper slug. Internally, the IC die rest directly on this slug, with the bottom surface exposed as shown in Figure 7-94. The intent is that this surface be soldered directly to a copper plane of the PCB, thereby extending the heat sinking.

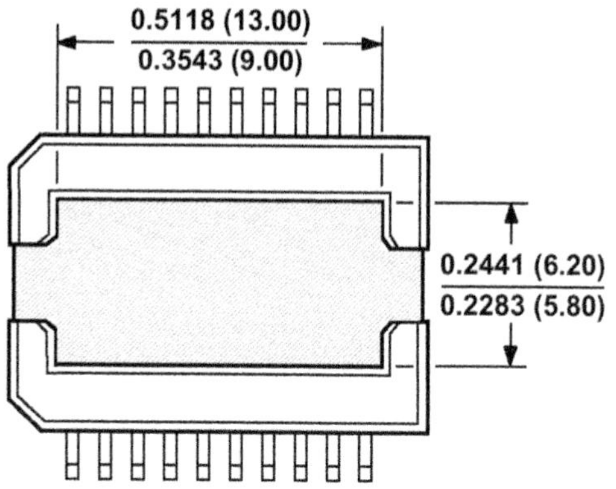

Figure 7-94: Bottom view of AD8016 20-lead PSOP3 package showing copper slug for aid in heat transfer (central gray area)

Both of AD8016 package options are characterized for both still and moving air, but the thermal information given above applies *without* the use of directed airflow. Therefore, adding additional airflow lowers thermal resistance further (see Reference 2).

For reliable, low thermal resistance designs with op amps, several design *Do's and Don'ts* are listed below. Consider all of these points, as may be practical.

1) *Do use as large an area of copper as possible for a PCB heat sink, up to the point of diminishing returns.*
2) *In conjunction with 1), do use multiple (outside) PCB layers, connected together with multiple vias.*
3) *Do use as heavy copper as is practical (2 oz. or more preferred).*
4) *Do provide sufficient natural ventilation inlets and outlets within the system, to allow heat to freely move away from hot PCB surfaces.*
5) *Do orient power-dissipating PCB planes vertically, for convection-aided airflow across heat sink areas.*
6) *Do consider the use of external* power buffer *stages, for precision op amp applications.*
7) *Do consider the use of* forced air, *for situations where several watts must be dissipated in a confined space.*
8) *Don't use solder mask planes over heat dissipating traces.*
9) *Don't use excessive supply voltages on ICs delivering power.*

For the most part, these points are obvious. However, one that could use some elaboration is number 9. Whenever an application requires only modest *voltage* swings (such as for example standard video, 2 V p-p) a wide supply voltage range can often be used. But, as the data of Figure 7-95 indicates, operation of an op amp driver on higher supply voltages produces a large IC dissipation, even though the load power is constant.

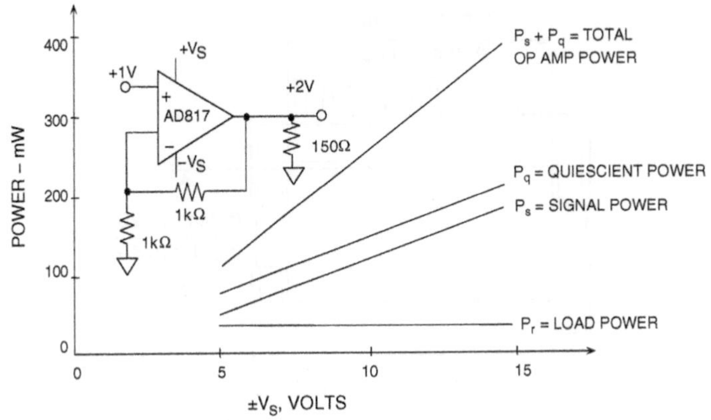

Figure 7-95: Power dissipated in video op amp driver for various supply voltages with low voltage output swing

In such cases, as long as the distortion performance of the application doesn't suffer, it can be advantageous to operate the IC on lower supplies, say ±5 V, as opposed to ±15 V. The above example data was calculated on a dc basis, which will generally tax the driver more in terms of power than a sine wave or a noise-like waveform, such as a DMT signal (see Reference 2). The general principles still hold for these ac waveforms, i.e., the op amp power dissipation is high when load current is high and the voltage low.

While there is ample opportunity for high power handling with the thermally enhanced packages described above for the AD8016 and AD8017, the increasingly popular smaller IC packages actually move in an opposite direction. Without question, it is true that today's smaller packages do noticeably sacrifice thermal performance. But, it must be understood that this is done in the interest of realizing a smaller size for the packaged op amp, and, ultimately, a much greater final PCB density for the overall system.

These points are illustrated by the thermal ratings for the AD8057 and AD8058 family of single and dual op amp devices, as is shown in Figure 7-96. The AD8057 and AD8058 op amps are available in three different packages. These are the SOT-23-5, and the 8-pin μSOIC, along with standard SOIC.

As the data shows, as the package size becomes smaller and smaller, much less power is capable of being removed. Since the lead frame is the only heatsinking possible with such tiny packages, their thermal performance is thus reduced. The θ_{JA} for the packages mentioned is 240°C/W, 200°C/W, and 160°C/W, respectively. Note this is more of a *package* than *device* limitation. Other ICs with the same packages have similar characteristics.

These discussions on the thermal application issues of op amps haven't dealt with the classic techniques of using clip-on (or bolt-on) type heat sinks. They also have not addressed the use of forced air cooling, generally considered only when tens of watts must be handled. These omissions are mainly because these approaches are seldom possible or practical with today's op amp packages.

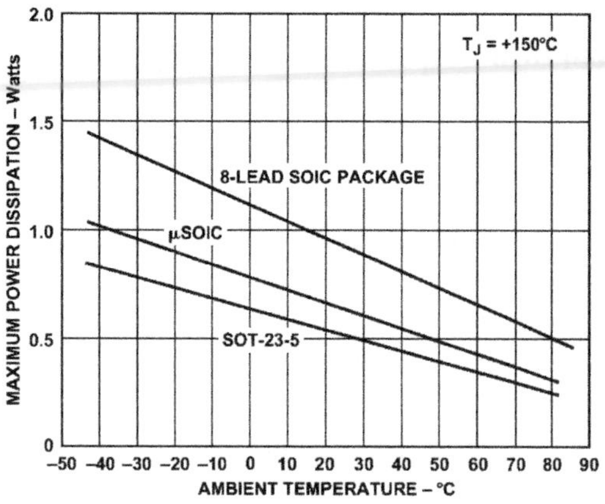

Figure 7-96: Comparative thermal performance for several AD8057/58 op amp package options

The more general discussions within References 4-7 can be consulted for this and other supplementary information.

References: Thermal Considerations

1. Data sheet for **AD8017 Dual High Output Current, High Speed Amplifier**, Analog Devices, Inc., www.analog.com.
2. Data sheet for **AD8016 Low Power, High Output Current, xDSL Line Driver**, Analog Devices, Inc., www.analog.com.
3. "Power Consideration Discussions," data sheet for **AD815 High Output Current Differential Driver**, Analog Devices, Inc., www.analog.com.
4. Walt Jung, Walt Kester, "Thermal Management," portion of Section 8 within Walt Kester, Editor, **Practical Design Techniques for Power and Thermal Management**, Analog Devices, Inc., 1998, ISBN 0-916550-19-2.
5. General Catalog, **AAVID Thermal Technologies, Inc.**, One Kool Path, Laconia, NH, 03246, 603-528-3400.
6. Seri Lee, "How to Select a Heat Sink," **Aavid Thermal Technologies**, www.aavid.com.
7. Seri Lee, "Optimum Design and Selection of Heat Sinks," **11th IEEE SEMI-THERM™ Symposium**, 1995, www.aavid.com.

EMI/RFI Considerations

James Bryant, Walt Jung, Walt Kester

Analog circuit performance is often adversely affected by high frequency signals from nearby electrical activity. And, equipment containing analog circuitry may also adversely affect systems external to it. Reference 1 (page 4) describes this complementary transmission of undesirable high frequency signals from or into local equipment as per an IEC50 definition. These corresponding aspects of broad arena of *electromagnetic compatibility*, better known as EMC, are:

- *It describes the ability of electrical and electronic systems to operate without interfering with other systems . . .*
- *It also describes the ability of such systems to operate as intended within a specified electromagnetic environment.*

Complete EMC assurance would indicate that the equipment under design should neither produce spurious signals, nor be vulnerable to out-of-band external signals (i.e., those outside its intended frequency range). It is the latter class of EMC problem to which analog equipment most often falls prey. It is the graceful handling of these spurious signals that are emphasized within this section.

The externally produced electrical activity may generate noise, and is referred to either as electromagnetic interference (EMI), or radio frequency interference (RFI). In this section, we will refer to EMI in terms of both electromagnetic and radio frequency interference. One of the more challenging tasks of the analog designer is the control of equipment against undesired operation due to EMI. It is important to note that in this context, *EMI and or RFI is almost always detrimental.* Once given entrance into the equipment, it can and will degrade its operation, quite often considerably.

This section is heavily oriented towards minimizing undesirable analog circuit operation due to the *receipt* of EMI/RFI. Misbehavior of this sort is also known as EMI or RFI *susceptibility*, indicating a tendency towards anomalous equipment behavior when exposed to EMI/RFI. There is, of course, a complementary EMC issue, namely with regard to spurious *emissions*. However, since analog circuits typically involve fewer of pulsed, high speed, high current signal edges that give rise to such spurious signals (compared to high speed logic, for example), this aspect of EMC isn't as heavily treated here. Nevertheless, the reader should bear in mind that it can be important, particularly if the analog circuitry is part of a mixed-signal environment along with high speed logic.

Since all of these various EMC design points can be critical, *the end-of-chapter references are strongly recommended for supplementary study*. Indeed, for a thorough, fully competent design with respect to EMI, RFI, and EMC, the designer will need to become intimately acquainted with one or more of these references (see References 1–6). As for the material following, it is best viewed as an introduction to this extremely broad but increasingly important topic.

EMI/RFI Mechanisms

To understand and properly control EMI and RFI, it is helpful to first segregate it into manageable portions. Thus it is useful to remember that when EMI/RFI problems do occur, they can be fundamentally broken down into a *Source*, a *Path*, and a *Receiver*. As a systems designer, you have under your direct control the receiver part of this landscape, and perhaps some portion of the path. But seldom will the designer have control over the actual source.

EMI Noise Sources

There are countless ways in which undesired noise can couple into an analog circuit to ruin its accuracy. Some of the many examples of these noise sources are listed in Figure 7-97.

- EMI/RFI noise sources can couple from anywhere
- Some common sources of externally generated noise:
  - Radio and TV Broadcasts
  - Mobile Radio Communications
  - Cellular Telephones
  - Vehicular Ignition
  - Lightning
  - Utility Power Lines
  - Electric Motors
  - Computers
  - Garage Door Openers
  - Telemetry Equipment

Figure 7-97: Some common EMI noise sources

Since little control is possible over these sources of EMI, the next best management tool to exercise over them is to recognize and understand the possible paths by which they couple into the equipment under design.

EMI Coupling Paths

The EMI coupling paths are actually very few in terms of basic number. Three very general paths are by:

1. *Interference due to conduction (common-impedance)*
2. *Interference due to capacitive or inductive coupling (near-field interference)*
3. *Electromagnetic radiation (far-field interference)*

Noise Coupling Mechanisms

EMI energy may enter wherever there is an impedance mismatch or discontinuity in a system. In general this occurs at the interface where cables carrying sensitive analog signals are connected to PC boards, and through power supply leads. Improperly connected cables or poor supply filtering schemes are often perfect conduits for interference.

Conducted noise may also be encountered when two or more currents share a common path (impedance). This common path is often a high impedance "ground" connection. If two circuits share this path, noise currents from one will produce noise voltages in the other. Steps may be taken to identify potential sources of this interference (see References 1 and 2, plus Section 2 of this chapter).

Figure 7-98 shows some of the general ways noise can enter a circuit from external sources.

- Impedance mismatches and discontinuities
- Common-mode impedance mismatches → Differential Signals
- Capacitively Coupled (Electric Field Interference)
  - dV/dt → Mutual Capacitance → Noise Current
    (Example: 1V/ns produces 1mA/pF)
- Inductively Coupled (Magnetic Field)
  - di/dt → Mutual Inductance → Noise Voltage
    (Example: 1mA/ns produces 1mV/nH)

Figure 7-98: How EMI finds paths into equipment

There is a capacitance between any two conductors separated by a dielectric (air and vacuum are dielectrics, as well as all solid or liquid insulators). If there is a change of voltage on one conductor there will be change of charge on the other, and a *displacement current* will flow in the dielectric. Where either the capacitance or the dV/dT is high, noise is easily coupled. For example, a 1 V/ns rate-of-change gives rise to displacement currents of 1 mA/pF.

If changing magnetic flux from current flowing in one circuit threads another circuit, it will induce an emf in the second circuit. Such *mutual inductance* can be a troublesome source of noise coupling from circuits with high values of dI/dT. As an example, a mutual inductance of 1 nH and a changing current of 1 A/ns will induce an emf of 1 V.

Reducing Common-Impedance Noise

Steps to be taken to eliminate or reduce noise due to the conduction path sharing of impedances, or *common-impedance noise* are outlined in Figure 7-99.

- Common-impedance noise
  - Decouple op amp power leads at LF and HF
  - Reduce common-impedance
  - Eliminate shared paths
- Techniques
  - Low impedance electrolytic (LF) and local low inductance (HF) bypasses
  - Use ground and power planes
  - Optimize system design

Figure 7-99: Some solutions to common-impedance noise

These methods should be applied in conjunction with all of the related techniques discussed earlier within Section 2 of this chapter.

Power supply rails feeding several circuits are good common-impedance examples. Real-world power sources may exhibit low output impedance, or may they not—especially over frequency. Furthermore, PCB traces used to distribute power are both inductive and resistive, and may also form a ground loop. The use of power and ground planes also reduces the power distribution impedance. These dedicated conductor layers in a PCB are continuous (ideally, that is) and, as such, offer the lowest practical resistance and inductance.

In some applications where low level signals encounter high levels of common-impedance noise it will not be possible to prevent interference and the system architecture may need to be changed. Possible changes include:

1. *Transmitting signals in differential form*
2. *Amplifying signals to higher levels for improved S/N*
3. *Converting signals into currents for transmission*
4. *Converting signals directly into digital form*

Noise Induced by Near-Field Interference

Crosstalk is the second most common form of interference. In the vicinity of the noise source, i.e., near-field, interference is not transmitted as an electromagnetic wave, and the term crosstalk may apply to either inductively or capacitively coupled signals.

Reducing Capacitance-Coupled Noise

Capacitively-coupled noise may be reduced by reducing the coupling capacity (by increasing conductor separation), but is most easily cured by shielding. A conductive and grounded shield (known as a *Faraday shield*) between the signal source and the affected node will eliminate this noise, by routing the displacement current directly to ground.

With the use of such shields, it is important to note that it is always *essential* that a Faraday shield be grounded. A floating or open-circuit shield almost invariably increases capacitively-coupled noise. For a brief review of this shielding, consult Section 2 of this chapter, and see References 2 and 3 at the end of this section.

Methods to eliminate capacitance-coupled interference are summarized in Figure 7-100.

- Reduce Level of High dV/dt Noise Sources
- Use Proper Grounding Schemes for Cable Shields
- Reduce Stray Capacitance
  - Equalize Input Lead Lengths
  - Keep Traces Short
  - Use Signal-Ground Signal-Routing Schemes
- Use Grounded Conductive Faraday Shields to Protect Against Electric Fields

Figure 7-100: Methods to reduce capacitance-coupled noise

Reducing Magnetically-Coupled Noise

Methods to eliminate interference caused by magnetic fields are summarized in Figure 7-101.

To illustrate the effect of magnetically-coupled noise, consider a circuit with a closed-loop area of A cm^2 operating in a magnetic field with an rms flux density value of B gauss. The noise voltage V_n induced in this circuit can be expressed by the following equation:

$$V_n = 2\pi f B A \cos\theta \times 10^{-8} V \qquad \text{Eq. 7-8}$$

In this equation, f represents the frequency of the magnetic field, and θ represents the angle of the magnetic field B to the circuit with loop area A. Magnetic field coupling can be reduced by reducing the circuit loop

- Careful Routing of Wiring
- Use Conductive Screens for HF Magnetic Shields
- Use High Permeability Shields for LF Magnetic Fields (Mu-Metal)
- Reduce Loop Area of Receiver
  - Twisted Pair Wiring
  - Physical Wire Placement
  - Orientation of Circuit to Interference
- Reduce Noise Sources
  - Twisted Pair Wiring
  - Driven Shields

Figure 7-101: Methods to reduce magnetically-coupled noise

area, the magnetic field intensity, or the angle of incidence. Reducing circuit loop area requires arranging the circuit conductors closer together. Twisting the conductors together reduces the loop net area. This has the effect of canceling magnetic field pickup, because the sum of positive and negative incremental loop areas is ideally equal to zero. Reducing the magnetic field directly may be difficult. However, since magnetic field intensity is inversely proportional to the cube of the distance from the source, physically moving the affected circuit away from the magnetic field has a very great effect in reducing the induced noise voltage. Finally, if the circuit is placed perpendicular to the magnetic field, pickup is minimized. If the circuit's conductors are in parallel to the magnetic field the induced noise is maximized because the angle of incidence is zero.

There are also techniques that can be used to reduce the amount of magnetic-field interference, *at its source*. In the previous paragraph, the conductors of the receiver circuit were twisted together, to cancel the induced magnetic field along the wires. The same principle can be used on the source wiring. If the source of the magnetic field is large currents flowing through nearby conductors, these wires can be twisted together to reduce the net magnetic field.

Shields and cans are not nearly as effective against magnetic fields as against electric fields, but can be useful on occasion. At low frequencies magnetic shields using high permeability material such a Mu-metal can provide modest attenuation of magnetic fields. At high frequencies simple conductive shields are quite effective provided that the thickness of the shield is greater than the skin depth of the conductor used (at the frequency involved). Note—copper skin depth is $6.6/\sqrt{f}$ cm, with f in Hz.

Passive Components: Arsenal Against EMI

Passive components, such as resistors, capacitors, and inductors, are powerful tools for reducing externally induced interference when used properly.

Simple RC networks make efficient and inexpensive one-pole, low-pass filters. Incoming noise is converted to heat and dissipated in the resistor. But note that a fixed resistor does produce thermal noise of its own. Also, when used in the input circuit of an op amp or in amp, such resistor(s) can generate input-bias-current induced offset voltage. While matching the two resistors will minimize the dc offset, the noise will remain. Figure 7-102 summarizes some popular low-pass filters for minimizing EMI.

LP Filter Type	ADVANTAGE	DISADVANTAGE
RC Section	Simple Inexpensive	Resistor Thermal Noise $I_B \times R$ Drop → Offset Single-Pole Cutoff
LC Section (Bifilar)	Very Low Noise at LF Very Low IR Drop Inexpensive Two-Pole Cutoff	Medium Complexity Nonlinear Core Effects Possible
π Section (C-L-C)	Very Low Noise at LF Very Low IR Drop Pre-packaged Filters Multiple-Pole Cutoff	Most Complex Nonlinear Core Effects Possible Expensive

Figure 7-102: Using passive components within filters to combat EMI

In applications where signal and return conductors aren't well coupled magnetically, a common-mode (CM) choke can be used to increase their mutual inductance. Note that these comments apply mostly to in amps, which naturally receive a balanced input signal (whereas op amps are inherently unbalanced inputs—unless one constructs an in amp with them). A CM choke can be simply constructed by winding several turns of the differential signal conductors together through a high permeability (> 2000) ferrite bead. The magnetic properties of the ferrite allow differential-mode currents to pass unimpeded while suppressing CM currents.

Capacitors can also be used before and after the choke, to provide additional CM and differential-mode filtering, respectively. Such a CM choke is cheap and produces very low thermal noise and bias current-induced offsets, due to the wire's low DCR. However, there is a field around the core. A metallic shield surrounding the core may be necessary to prevent coupling with other circuits. Also, note that high current levels should be avoided in the core as they may saturate the ferrite.

The third method for passive filtering takes the form of packaged π-networks (C-L-C). These packaged filters are completely self-contained and include feedthrough capacitors at the input and the output as well as a shield to prevent the inductor's magnetic field from radiating noise. These more expensive networks offer high levels of attenuation and wide operating frequency ranges, but the filters must be selected so that for the operating current levels involved the ferrite doesn't saturate.

Reducing System Susceptibility to EMI

The general examples discussed above and the techniques illustrated earlier in this section outline the procedures that can be used to reduce or eliminate EMI/RFI. Considered on a *system* basis, a summary of possible measures is given in Figure 7-103.

- Always Assume that Interference Exists
- Use Conducting Enclosures Against Electric and HF
- Magnetic Fields
- Use Mu-Metal Enclosures against LF Magnetic Fields
- Implement Cable Shields Effectively
- Use Feedthrough Capacitors and Packaged PI Filters

Figure 7-103: Reducing system EMI/RFI susceptibility

Other examples of filtering techniques useful against EMI are illustrated later in this section, under "Reducing RFI rectification within op amp and in amp circuits."

The section immediately below further details shielding principles.

A Review of Shielding Concepts

The concepts of shielding effectiveness presented next are background material. Interested readers should consult References 4–9 cited at the end of the section for more detailed information.

Applying the concepts of shielding effectively requires an understanding of the source of the interference, the environment surrounding the source, and the distance between the source and point of observation (the receiver). If the circuit is operating close to the source (in the *near*, or induction-field), the field characteristics are determined by the source. If the circuit is remotely located (in the *far*, or radiation-field), the field characteristics are determined by the transmission medium.

A circuit operates in a near-field if its distance from the source of the interference is less than the wavelength (λ) of the interference divided by 2 π, or $\lambda/2\ \pi$. If the distance between the circuit and the source of the interference is larger than this quantity, then the circuit operates in the far field. For instance, the interference caused by a 1 ns pulse edge has an upper bandwidth of approximately 350 MHz. The wavelength of a 350 MHz signal is approximately 32 inches (the speed of light is approximately 12"/ns). Dividing the wavelength by 2 π yields a distance of approximately 5 inches, the boundary between near- and far-field. If a circuit is within 5 inches of a 350 MHz interference source, then the circuit operates in the near-field of the interference. If the distance is greater than 5 inches, the circuit operates in the far-field of the interference.

Regardless of the type of interference, there is a characteristic impedance associated with it. The characteristic, or wave impedance of a field is determined by the ratio of its electric (or E-) field to its magnetic (or H-) field. In the far field, the ratio of the electric field to the magnetic field is the characteristic (wave impedance) of free space, given by $Z_o = 377\ \Omega$. In the near field, the wave-impedance is determined by the nature of the interference and its distance from the source. If the interference source is high current and low voltage (for example, a loop antenna or a power line transformer), the field is predominately magnetic and exhibits a wave impedance less than 377 Ω. If the source is low current and high voltage (for example, a rod antenna or a high speed digital switching circuit), the field is predominately electric and exhibits a wave impedance greater than 377 Ω.

Conductive enclosures can be used to shield sensitive circuits from the effects of these external fields. These materials present an *impedance mismatch* to the incident interference, because the impedance of the shield is lower than the wave impedance of the incident field. The effectiveness of the conductive shield depends on two things: First is the loss due to the *reflection* of the incident wave off the shielding material. Second is the loss due to the *absorption* of the transmitted wave *within* the shielding material. The amount of reflection loss depends upon the type of interference and its wave impedance. The amount of absorption loss, however, is independent of the type of interference. It is the same for near- and far-field radiation, as well as for electric or magnetic fields.

Reflection loss at the interface between two media depends on the difference in the characteristic impedances of the two media. For electric fields, reflection loss depends on the frequency of the interference and the shielding material. This loss can be expressed in dB, and is given by:

$$R_e\,(\mathrm{dB}) = 322 + 10\log_{10}\left[\frac{\sigma_r}{\mu_r f^3 r^2}\right] \qquad \text{Eq. 7-9}$$

where σ_r = relative conductivity of the shielding material, in Siemens per meter;

μ_r = relative permeability of the shielding material, in Henries per meter;

f = frequency of the interference, and

r = distance from source of the interference, in meters

For magnetic fields, the loss depends also on the shielding material and the frequency of the interference. Reflection loss for magnetic fields is given by:

$$R_m\,(\mathrm{dB}) = 14.6 + 10\log_{10}\left[\frac{fr^2\sigma_r}{\mu_r}\right] \qquad \text{Eq. 7-10}$$

and, for plane waves ($r > \lambda/2\pi$), the reflection loss is given by:

$$R_{pw}\,(\mathrm{dB}) = 168 + 10\log_{10}\left[\frac{\sigma_r}{\mu_r f}\right] \qquad \text{Eq. 7-11}$$

Absorption is the second loss mechanism in shielding materials. Wave attenuation due to absorption is given by:

$$A\,(\mathrm{dB}) = 3.34\ t\sqrt{\sigma_r \mu_r f} \qquad \text{Eq. 7-12}$$

where t = thickness of the shield material, in inches. This expression is valid for plane waves, electric and magnetic fields. Since the intensity of a transmitted field decreases exponentially relative to the thickness of the shielding material, the absorption loss in a shield one skin-depth (δ) thick is 9 dB. Since absorption loss is proportional to thickness and inversely proportional to skin depth, increasing the thickness of the shielding material improves shielding effectiveness at high frequencies.

Reflection loss for plane waves in the far field decreases with increasing frequency because the shield impedance, Z_s, increases with frequency. Absorption loss, on the other hand, increases with frequency because skin depth decreases. For electric fields and plane waves, the primary shielding mechanism is reflection loss, and at high frequencies, the mechanism is absorption loss.

Thus for high frequency interference signals, lightweight, easily worked high conductivity materials such as copper or aluminum can provide adequate shielding. At low frequencies however, both reflection and absorption loss to magnetic fields is low. It is thus very difficult to shield circuits from low frequency magnetic fields. In these applications, high permeability materials that exhibit low reluctance provide the best protection. These low reluctance materials provide a magnetic shunt path that diverts the magnetic field away from the protected circuit.

To summarize the characteristics of metallic materials commonly used for shielded purposes: Use high conductivity metals for HF interference, and high permeability metals for LF interference.

A properly shielded enclosure is very effective at preventing external interference from disrupting its contents as well as confining any internally-generated interference. However, in the real world, openings in the shield are often required to accommodate adjustment knobs, switches, connectors, or to provide ventilation. Unfortunately, these openings may compromise shielding effectiveness by providing paths for high-frequency interference to enter the instrument.

The longest dimension (not the total area) of an opening is used to evaluate the ability of external fields to enter the enclosure, because the openings behave as slot antennas. Eq. 7-13 can be used to calculate the shielding effectiveness, or the susceptibility to EMI leakage or penetration, of an opening in an enclosure:

$$\text{Shielding Effectiveness (dB)} = 20\log_{10}\left(\frac{\lambda}{2\cdot L}\right) \qquad \text{Eq. 7-13}$$

where λ = wavelength of the interference and

L = maximum dimension of the opening

Maximum radiation of EMI through an opening occurs when the longest dimension of the opening is equal to one half-wavelength of the interference frequency (0 dB shielding effectiveness). A rule of thumb is to keep the longest dimension less than 1/20 wavelength of the interference signal, as this provides 20 dB shielding effectiveness.

Furthermore, a few small openings on each side of an enclosure is preferred over many openings on one side. This is because the openings on different sides radiate energy in different directions and, as a result, shielding effectiveness is not compromised. If openings and seams cannot be avoided, then conductive gaskets, screens, and paints alone or in combination should be used judiciously to limit the longest dimension of any opening to less than 1/20 wavelength. Any cables, wires, connectors, indicators, or control shafts penetrating the enclosure should have circumferential metallic shields physically bonded to the enclosure at the point of entry. In those applications where unshielded cables/wires are used, filters are recommended at the shield entry point.

General Points on Cables and Shields

Although covered in detail elsewhere, it is worth noting that the improper use of cables and their shields can be a significant contributor to both radiated and conducted interference. Rather than developing an entire treatise on these issues, the interested reader should consult References 2, 3, 5, and 6 for background.

As shown in Figure 7-104, proper cable/enclosure shielding confines sensitive circuitry and signals *entirely within the shield*, with no compromise to shielding effectiveness.

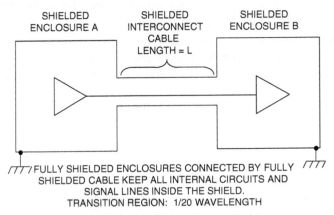

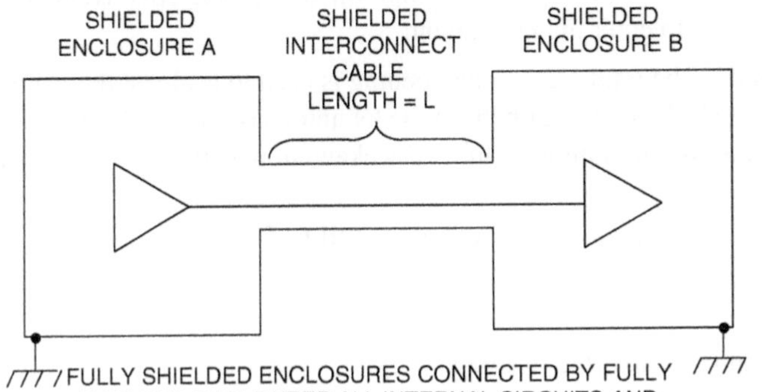
SHIELDED
ENCLOSURE A
SHIELDED
INTERCONNECT
CABLE
LENGTH = L
SHIELDED
ENCLOSURE B
FULLY SHIELDED ENCLOSURES CONNECTED BY FULLY
SHIELDED CABLE KEEP ALL INTERNAL CIRCUITS AND
SIGNAL LINES INSIDE THE SHIELD.
TRANSITION REGION: 1/20 WAVELENGTH

Figure 7-104: Shielded interconnect cables are either electrically long or short, depending upon the operating frequency

As can be noted by this diagram, the enclosures and the shield must be properly grounded, otherwise they can act as an antenna, thereby making the radiated and conducted interference problem worse (rather than better).

Depending on the type of interference (pickup/radiated, low/high frequency), proper cable shielding is implemented differently and is very dependent on the length of the cable. The first step is to determine whether the length of the cable is *electrically short* or *electrically long* at the frequency of concern. A cable is considered electrically short if the length of the cable is less than 1/20 wavelength of the highest frequency of the interference. Otherwise, it is considered to be electrically long.

For example, at 50 Hz/60 Hz, an electrically short cable is any cable length less than 150 miles, where the primary coupling mechanism for these low frequency electric fields is capacitive. As such, for any cable length less than 150 miles, the amplitude of the interference will be the same over the entire length of the cable.

In applications where the length of the cable is electrically long, or protection against high frequency interference is required, the preferred method is to connect the cable shield to low impedance points, *at both ends*. As will be seen shortly, this can be a direct connection at the driving end, and a capacitive connection at the receiver. If left ungrounded, unterminated transmission lines effects can cause reflections and standing waves along the cable. At frequencies of 10 MHz and above, circumferential (360°) shield bonds and metal connectors are required to main low impedance connections to ground.

In summary, for protection against low frequency (<1 MHz), electric-field interference, grounding the shield at one end is acceptable. For high frequency interference (>1 MHz), the preferred method is grounding the shield at both ends, using 360° circumferential bonds between the shield and the connector, and maintaining metal-to-metal continuity between the connectors and the enclosure.

In practice, however, there is a caveat involved with directly grounding the shield at both ends. When this is done, it creates a low frequency ground loop, shown in Figure 7-105.

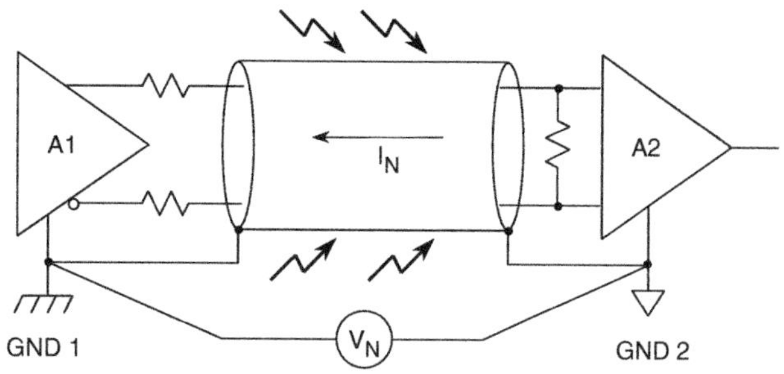

- V_N Causes Current in Shield (Usually 50Hz/60Hz)
- Differential Error Voltage is Produced at Input of A2 unless:
  - A1 Output is Perfectly Balanced and
  - A2 Input is Perfectly Balanced and
  - Cable is Perfectly Balanced

Figure 7-105: Ground loops in shielded twisted pair cable can cause errors

Whenever two systems A1 and A2 are remote from each other, there is usually a difference in the ground potentials at each system, i.e., V_N. The frequency of this potential difference is generally the line frequency (50 Hz or 60 Hz) and multiples thereof. But, if the shield is directly grounded at both ends as shown, noise current I_N flows in the shield. In a perfectly balanced system, the common-mode rejection of the system is infinite, and this current flow produces no differential error at the receiver A2. However, perfect balance is never achieved in the driver, its impedance, the cable, or the receiver, so a certain portion of the shield current will appear as a differential noise signal, at the input of A2. The following illustrates correct shield grounding for various examples.

As noted above, cable shields are subject to both low and high frequency interference. Good design practice requires that the shield be grounded at both ends if the cable is electrically long to the interference frequency, as is usually the case with RF interference.

Figure 7-106 shows a remote passive RTD sensor connected to a bridge and conditioning circuit by a shielded cable. The proper grounding method is shown in the upper part of the figure, where the shield is grounded at the receiving end.

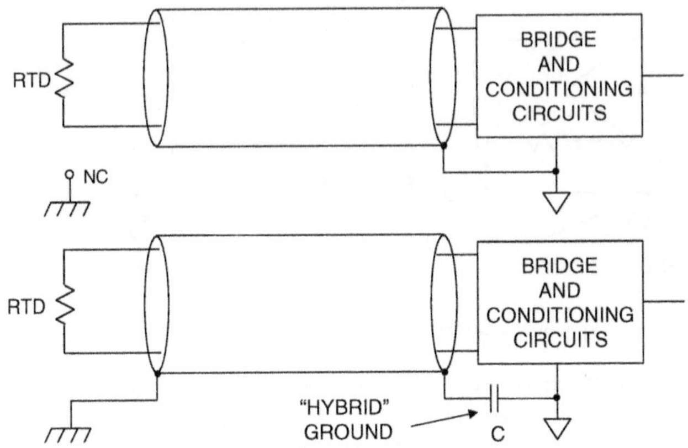

Figure 7-106: Hybrid grounding of shielded cable with passive sensor

Safety considerations may require that the remote end of the shield also be grounded. If this is the case, the receiving end can be grounded with a low inductance ceramic capacitor (0.01 µF to 0.1 µF), still providing high frequency grounding. The capacitor acts as a ground to RF signals on the shield but blocks low frequency line current to flow in the shield. This technique is often referred to as a *hybrid ground.*

A case of an active remote sensor and/or other electronics is shown Figure 7-107. In both situations, a hybrid ground is also appropriate, either for the balanced (upper) or the single-ended (lower) driver case. In both instances the capacitor "C" breaks the low frequency ground loop, providing effective RF grounding of the shielded cable at the A2 receiving end at the right side of the diagram.

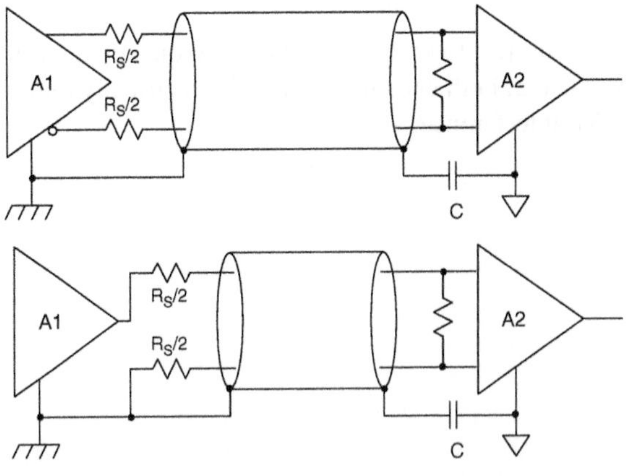

Figure 7-107: Impedance-balanced drive of balanced shielded cable aids noise-immunity with either balanced or single-ended source signals

There are also more subtle points that should be made with regard to the source termination resistances used, R_S. In both the balanced as well as the single-ended drive cases, the driving signal seen on the balanced line originates from a net impedance of R_S, which is split between the two twisted pair legs as twice $R_S/2$. In the upper case of a fully differential drive, this is straightforward, with an $R_S/2$ valued resistor connected in series with the complementary outputs from A1.

In the bottom case of the single-ended driver, note that there are still two $R_S/2$ resistors used, one in series with both legs. Here the grounded dummy return leg resistor provides an impedance-balanced ground connection drive to the differential line, aiding in overall system noise immunity. Note that this implementation is only useful for those applications with a balanced receiver at A2, as shown.

Coaxial cables are different from shielded twisted pair cables in that the signal return current path is through the shield. For this reason, the ideal situation is to ground the shield at the driving end and allow the shield to float at the differential receiver (A2) as shown in the upper portion of Figure 7-108. For this technique to work, however, the receiver must be a differential type with good high frequency CM rejection.

However, the receiver may be a single-ended type, such as typical of a standard single op amp type circuit. This is true for the bottom example of Figure 7-108, so there is no choice but to ground the coaxial cable shield at both ends for this case.

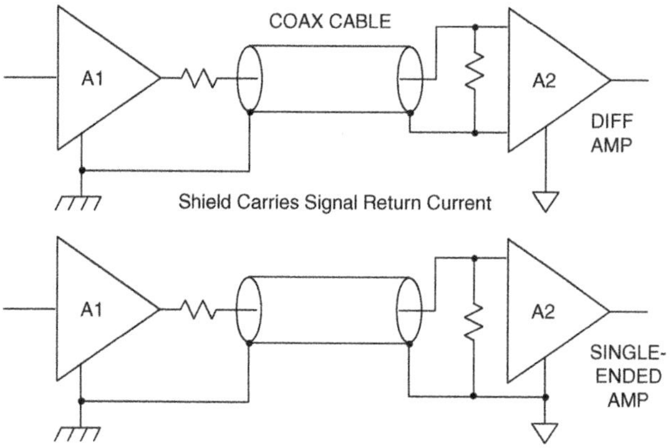

Figure 7-108: Coaxial cables can use either balanced or single-ended receivers

Input-Stage RFI Rectification Sensitivity

A well-known but poorly understood phenomenon in analog integrated circuits is *RFI rectification*, specifically as it occurs in op amps and in amps. While amplifying very small signals these devices can rectify large-amplitude, out-of-band HF signals, i.e., RFI. As a result, dc errors appear at the output in addition to the desired signal. The undesired HF signals can enter sensitive analog circuits by various means. Conductors leading into and out of the circuit provide a path for interference coupling into a circuit. These conductors pick up noise through capacitive, inductive, or radiation coupling, as discussed earlier. The spurious signals appear at the amplifier inputs, along with the desired signal. The spurious signals can be several tens of mV in amplitude, however, which causes problems. Simply stated, it cannot be assumed that a sensitive, low bandwidth dc amplifier will always reject out-of-band spurious signals. While this would be the case for a simple linear low-pass filter, op amp and in amp devices actually rectify high level HF signals, leading to nonlinearities and anomalous offsets. Methods of analysis for, as well as the prevention of, RFI rectification are discussed in this section.

Background: Op amp and In Amp RFI Rectification Sensitivity Tests

Just about all in amp and op amp input stages use emitter-coupled BJT or source-coupled FET differential pairs of some type. Depending on the device operating current, the interfering frequency and its relative amplitude, these differential pairs can behave as high frequency detectors. As will be shown, the detection process produces spectral components at the harmonics of the interference, as well at dc. It is the detected dc component of the interference that shifts amplifier bias levels, leading to inaccuracies.

The effect of RFI rectification within op amps and in amps can be evaluated with relatively simple test circuits, as described for the *RFI Rectification Test Configuration* (see page 1-38 of Reference 10). In these tests, an op amp or in amp is configured for a gain of –100 (op amp), or 100 (in amp), with dc output measured after a 100 Hz low-pass filter, preventing interference from other signals. A 100 MHz, 20 mV p-p signal is the test stimulus, chosen to be well above test device frequency limits. In operation, the test evaluates dc output shift observed under stimulus presence. While an ideal dc shift for this measurement would be zero, the actual dc shift of a given part indicates the relative RFI rectification sensitivity. Devices using both BJT and FET technologies can be tested by this method, as can devices operating at either low or high supply current levels.

In the original op amp test device set of Reference 10, some FET-input devices (OP80, OP42, OP249 and AD845) exhibited no observable shift in their output voltages, while several others showed shifts of less than 10 μV referred to the input. Of the BJT-input op amps, the amount of shift decreased with increasing device supply current. Only two devices showed no observable output voltage shift (AD797 and AD827), while others showed shifts of less than 10 μV referred to the input (OP200 and OP297). For other op amps, it is to be expected that similar patterns would be shown under such testing.

From these tests, some generalizations on RFI rectification can be made. First, device susceptibility appears to be inversely proportional to supply current; that is, devices biased at low quiescent supply currents exhibit greatest output voltage shift. Second, ICs with FET-input stages appeared to be less susceptible to rectification than those with BJTs. Note that these points are independent of whether the device is an op amp or an in amp. In practice this means that the lower power op amps *or* in amps will tend to be more susceptible to RFI rectification effects. And, FET-input op amps (or in amps) will tend to be *less* susceptible to RFI, especially those operating at higher currents.

Based on these data and from the fundamental differences between BJTs and FETs, we can summarize what we know. Bipolar transistor action is controlled by a forward-biased p-n junction (the base-emitter junction) whose I-V characteristic is exponential and quite nonlinear. FET behavior, on the other hand, is controlled by voltages applied to a reverse-biased p-n junction diode (the gate-source junction). The I-V characteristic of FETs is a square-law, and thus it is inherently more linear than that of BJTs.

For the case of the lower supply current devices, transistors in the circuit are biased well below their peak f_T collector currents. Although the ICs may be constructed on processes whose device f_Ts can reach hundreds of MHz, charge transit times increase, when transistors are operated at low current levels. The impedance levels used also make RFI rectification in these devices worse. In low power op amps, impedances are on the order of hundreds to thousands of kΩs, whereas in moderate supply-current designs impedances might be no more than just a few kΩ. Combined, these factors tend to degrade a low-power device's RFI rectification sensitivity.

Figure 7-109 summarizes these general observations on RFI rectification sensitivity, and is applicable to both op amps and in-amps.

- BJT input devices *rectify* readily
  - Forward-biased B-E junction
  - Exponential I-V Transfer Characteristic
- FET input devices less sensitive to rectifying
  - Reversed-biased p-n junction
  - Square-law I-V Transfer Characteristic
- Low I_{supply} devices versus High I_{supply} devices
  - Low I_{supply} ⇒ Higher rectification sensitivity
  - High I_{supply} ⇒ Lower rectification sensitivity

Figure 7-109: Some general observations on op amp and in amp input stage RFI rectification sensitivity

An Analytical Approach: BJT RFI Rectification

While lab experiments can demonstrate that BJT-input devices exhibit greater RFI rectification sensitivity than comparable devices with FET inputs, a more analytical approach can also be taken to explain this phenomenon.

RF circuit designers have long known that p-n junction diodes are efficient rectifiers because of their nonlinear I-V characteristics. A spectral analysis of a BJT transistor current output for a HF sinewave input reveals that, as the device is biased closer to its "knee," nonlinearity increases. This, in turn, makes its use as a detector more efficient. This is especially true in low power op amps, where input transistors are biased at very low collector currents.

A rectification analysis for the collector current of a BJT has been presented in Reference 10, and will not be repeated here except for the important conclusions. These results reveal that the original quadratic second-order term can be simplified into a frequency-dependent term, $\Delta i_c(ac)$, at twice the input frequency and a dc term, $\Delta i_c(dc)$. The latter component can be expressed as noted in Eq. 7-14, the final form for the rectified dc term:

$$\Delta i_C(DC) = \left(\frac{V_X}{V_T}\right)^2 \bullet \frac{I_C}{4} \qquad \text{Eq. 7-14}$$

This expression shows that the dc component of the second-order term is directly proportional to the *square* of the HF noise amplitude V_X, and, also, to I_C, the quiescent collector current of the transistor. To illustrate this point on rectification, note that the change in dc collector current of a bipolar transistor operating at an I_C of 1 mA with a spurious 10 mV_{peak} high frequency signal impinging upon it will be about 38 uA.

Reducing the amount of rectified collector current is a matter of reducing the quiescent current, or the magnitude of the interference. Since the op amp and in amp input stages seldom provide adjustable quiescent collector currents, reducing the level of interfering noise V_X is by far the best (and almost always the only) solution. For example, reducing the amplitude of the interference by a factor of 2, down to 5 mV_{peak} produces a net 4 to 1 reduction in the rectified collector current. Obviously, this illustrates the importance of keeping spurious HF signals away from RFI sensitive amplifier inputs.

An Analytical Approach: FET RFI Rectification

A rectification analysis for the drain current of a JFET has also been presented in Reference 10, and isn't repeated here. A similar approach was used for the rectification analysis of a FET's drain current as a function of a small voltage V_X, applied to its gate. The results of evaluating the second-order rectified term for the FET's drain current are summarized in Eq. 7-15. Like the BJT, an FET's second-order term has an ac and a dc component. The simplified expression for the dc term of the rectified drain current is given here, where the rectified dc drain current is directly proportional to the square of the amplitude of V_X, the spurious signal. However, Eq. 7-15 also reveals a very important difference between the *degree* of the rectification produced by FETs relative to BJTs.

$$\Delta i_D(DC) = \left(\frac{V_X}{V_P}\right)^2 \bullet \frac{I_{DSS}}{2} \qquad \text{Eq. 7-15}$$

Whereas in a BJT the change in collector current has a direct relationship to its quiescent collector current level, the change in a JFET's drain current is proportional to its drain current at zero gate-source voltage, I_{DSS}, and inversely proportional to the square of its channel pinch-off voltage, V_P—parameters that are geometry and process dependent. Typically, JFETs used in the input stages of in amps and op amps are biased with their quiescent current of ~$0.5 \bullet I_{DSS}$. Therefore, the change in a JFET's drain current is independent of its quiescent drain current; hence, independent of the operating point.

A quantitative comparison of second-order rectified dc terms between BJTs and FETs is illustrated in Figure 7-110. In this example, a bipolar transistor with a unit emitter area of 576 μm² is compared to a unit-area JFET designed for an I_{DSS} of 20 μA and a pinch-off voltage of 2 V. Each device is biased at 10 μA and operated at $T_A = 25°C$.

- BJT:
  - Emitter area = 576 μm²
  - I_C = 10μA
  - V_T = 25.68mV @ 25°C
  - $$\Delta i_C = \left(\frac{V_X}{V_T}\right)^2 \bullet \frac{I_C}{4} = \frac{V_X^2}{264}$$
- JFET:
  - I_{DSS} = 20μA (Z/L=1)
  - V_P = 2V
  - I_D = 10μA
  - $$\Delta i_D = \left(\frac{V_X}{V_P}\right)^2 \bullet \frac{I_{DSS}}{2} = \frac{V_X^2}{400 \times 10^3}$$
- Conclusion: BJTs ~1500 more sensitive than JFETs

Figure 7-110: Relative sensitivity comparison – BJT versus JFET

The important result is that, under identical quiescent current levels, the change in collector current in bipolar transistors is about 1500 times greater than the change in a JFET's drain current. This explains why FET-input amplifiers behave with less sensitivity to large amplitude HF stimulus. As a result, they offer more RFI rectification immunity.

What all this boils down to is this: Since a user has virtually no access to the amplifier's internal circuitry, the prevention of IC circuit performance degradation due to RFI is left essentially to those means which are external to the ICs.

As the analysis above shows, regardless of the amplifier type, *RFI rectification is directly proportional to the square of the interfering signal's amplitude*. Therefore, to minimize RFI rectification in precision amplifiers, the level of interference must be reduced or eliminated, *prior to the stage*. The most direct way to reduce or eliminate the unwanted noise is by proper filtering.

This topic is covered in the section immediately following.

Reducing RFI Rectification within Op Amp and In Amp Circuits

EMI and RFI can seriously affect the dc performance of high accuracy analog circuits. Because of their relatively low bandwidth, precision op amps and in amps simply won't accurately amplify RF signals in the MHz range. However, if these out-of-band signals are allowed to couple into a precision amplifier through either its input, output, or power supply pins, they can be internally rectified by various amplifier junctions, ultimately causing an undesirable dc offset at the output. The previous theoretical discussion of this phenomenon has shown its basic mechanisms. The logical next step is to show how proper filtering can minimize or eliminate these errors.

Elsewhere in this chapter we have discussed how proper supply decoupling minimizes RFI on IC power pins. Further discussion is required with respect to the amplifier inputs and outputs, *at the device level*. It is assumed at this point that system level EMI/RFI approaches have already been implemented, such as an RFI-tight enclosure, properly grounded shields, power rail filtering, and so forth. The steps following can be considered as circuit-level EMI/RFI prevention.

Op Amp Inputs

The best way to prevent input stage rectification is to use a low-pass filter located close to the op amp input as shown in Figure 7-111. In the case of the inverting op amp at the left, filter capacitor C is placed between equal-value resistors R1-R2. This results in a simple corner frequency expression, as shown in the figure. At very low frequencies or dc, the closed loop gain of the circuit is –R3/(R1+R2). Note that C cannot be connected directly to the inverting input of the op amp, since that would cause instability. The filter bandwidth can be chosen at least 100 times the signal bandwidth to minimize signal loss.

For the noninverting case on the right, capacitor C can be connected directly to the op amp input as shown, and an input resistor with a value "R" yields the same corner frequency as the inverting case. In both cases low inductance chip-style capacitors should be used, such as NPO ceramics. The capacitor should in any case be free of losses or voltage coefficient problems, which limits it to either the NPO mentioned, or a film type.

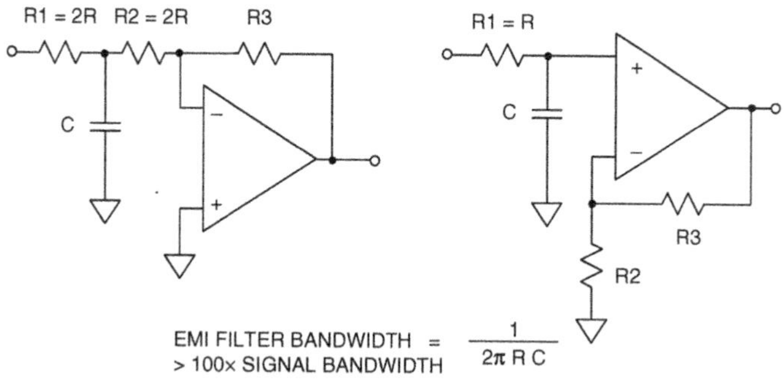

Figure 7-111: Simple EMI/RFI noise filters for op amp circuits

It should be noted that a ferrite bead can be used instead of R1, however ferrite bead impedance is not well controlled and is generally no greater than 100 Ω at 10 MHz to 100 MHz. This requires a large value capacitor to attenuate lower frequencies.

In Amp Inputs

Precision in amps are particularly sensitive to dc offset errors due to the presence of CM EMI/RFI. This is very much like the problem in op amps. And, as is true with op amps, the sensitivity to EMI/RFI is more acute with the lower power in amp devices.

A general-purpose approach to proper filtering for device level application of in amps is shown in Figure 7-112. In this circuit the in amp could, in practice, be any one of a number of devices. The relatively complex balanced RC filter preceding the in amp performs all of the high frequency filtering. The in amp would be programmed for the gain required in the application, via its gain-set resistance (not shown).

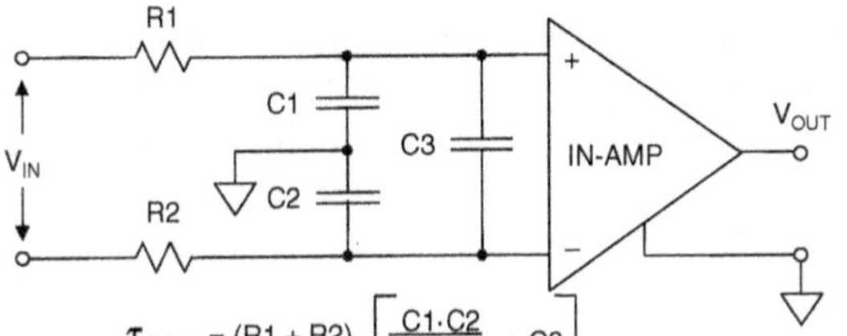

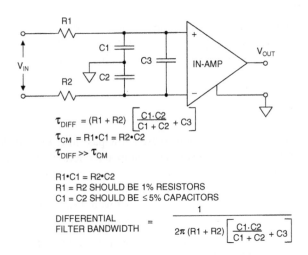

$$\tau_{DIFF} = (R1 + R2)\left[\frac{C1 \cdot C2}{C1 + C2} + C3\right]$$

$$\tau_{CM} = R1 \cdot C1 = R2 \cdot C2$$

$$\tau_{DIFF} >> \tau_{CM}$$

R1•C1 = R2•C2
R1 = R2 SHOULD BE 1% RESISTORS
C1 = C2 SHOULD BE ≤5% CAPACITORS

$$\text{DIFFERENTIAL FILTER BANDWIDTH} = \frac{1}{2\pi (R1 + R2)\left[\frac{C1 \cdot C2}{C1 + C2} + C3\right]}$$

Figure 7-112: A general-purpose common-mode/ differential-mode RC EMI/RFI filter for in amps

Within the filter, note that fully balanced filtering is provided for both CM (R1-C1 and R2-C2) as well as differential mode (DM) signals (R1+R2, and C3 || the series connection of C1-C2). If R1-R2 and C1-C2 aren't well matched, some of the input common-mode signal at V_{IN} will be converted to a differential mode signal at the in amp inputs. For this reason, C1 and C2 should be matched to within at least 5% of each other. Also, to aid this matching, R1 and R2 should be 1% metal film resistors. It is assumed that the source resistances seen at the V_{IN} terminals are low with respect to R1-R2, and matched. In this type of filter, C3 should be chosen much larger than C1 or C2 (C3 ≥ C1, C2), in order to suppress spurious differential signals due to CM⇒DM conversion resulting from mismatch of the R1-C1 and R2-C2 time constants.

The overall filter bandwidth should be at least 100 times the input signal bandwidth. Physically, the filter components should be symmetrically mounted on a PC board with a large area ground plane and placed close to the in amp inputs for optimum performance.

Figure 7-113 shows a family of these filters, as suited to a range of different in amps. The RC components should be tailored to the different in amp devices, as per the table. These filter components are selected for a reasonable balance of low EMI/RFI sensitivity and a low increase in noise (vis-à-vis that of the related in amp, without the filter).

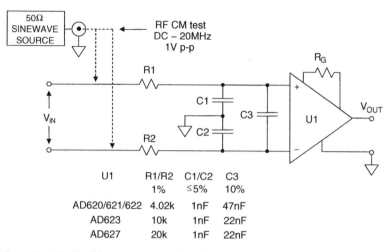

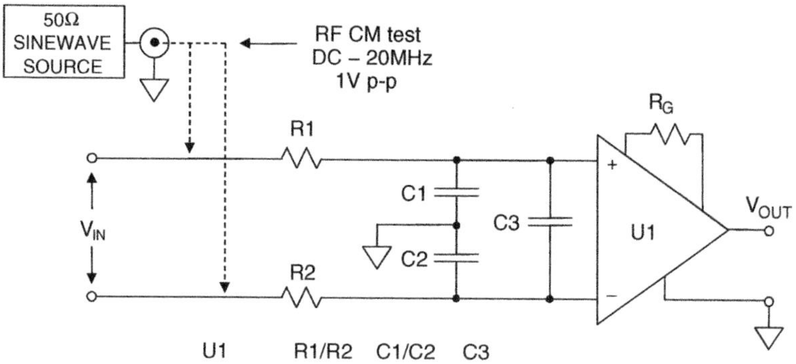

U1	R1/R2 1%	C1/C2 ≤5%	C3 10%
AD620/621/622	4.02k	1nF	47nF
AD623	10k	1nF	22nF
AD627	20k	1nF	22nF

Figure 7-113: Flexible common-mode and differential-mode RC EMI/RFI filters are useful with the AD620 series, the AD623, AD627, and other in amps

To test the EMI/RFI sensitivity of the configuration, a 1 V p-p CM signal can be applied to the input resistors, as noted. With a typically used in amp such as the AD620 working at a gain of 1000, the maximum RTI input offset voltage shift observed was 1.5 μV over the 20 MHz range. In the AD620 filter example, the differential bandwidth is about 400 Hz.

Common-mode chokes offer a simple, one-component EMI/RFI protection alternative to the passive RC filters, as shown in Figure 7-114.

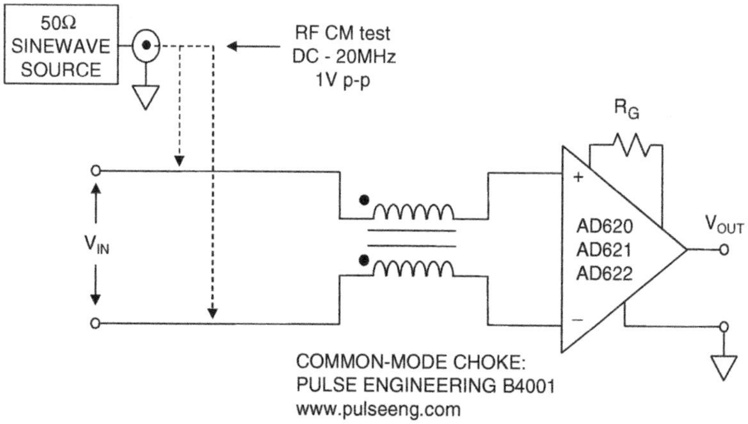

Figure 7-114: For simplicity as well as lowest noise EMI/RFI filter operation, a common-mode choke is useful with the AD620 series in amp devices

In addition to being a low component count approach, choke-based filters offer low noise, by dispensing with the resistances. Selecting the proper common-mode choke is critical, however. The choke used in the circuit of Figure 7-114 is a Pulse Engineering B4001. The maximum RTI offset shift measured from dc to 20 MHz at G = 1000 was 4.5 μV. Either an off-the-shelf choke such as the B4001 can be used for this filter, or, alternately one can be constructed. Since balance of the windings is important, bifilar wire is suggested. The core material must of course operate over the expected frequency band. Note that, unlike the Figure 7-113 family of RC filters, a choke-only filter offers no differential filtration. Differential mode filtering can be optionally added, with a second stage following the choke, by adding the R1-C3-R2 connections of Figure 7-112.

For further information on in amp EMI/RFI filtering, see References 10, and 12 – 15.

Amplifier Outputs and EMI/RFI

In addition to filtering the input and power pins, amplifier *outputs* also need to be protected from EMI/RFI, especially if they must drive long lengths of cable, which act as antennas. RF signals received on an output line can couple back into the amplifier input where it is rectified, and appears again on the output as an offset shift.

A resistor and/or ferrite bead, or both, in series with the output is the simplest and least expensive output filter, as shown in Figure 7-115 (upper circuit).

Adding a resistor-capacitor-resistor "T" circuit as shown in Figure 7-115 (lower circuit) improves this filter with just slightly more complexity. The output resistor and capacitor divert most of the high frequency energy away from the amplifier, making this configuration useful even with low power active devices. Of course, the time constant of the filter parts must be chosen carefully, to minimize any degradation of the desired output signal. In this case the RC components are chosen for an approximate 3 MHz signal bandwidth, suitable for instrumentation or other low bandwidth stages.

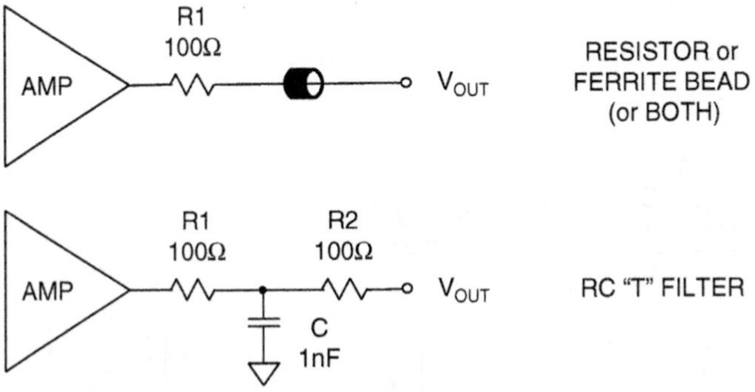

Figure 7-115: Op amp and in amp outputs should be protected against EMI/RFI, particularly if they drive long cables.

Printed Circuit Board Design for EMI/RFI Protection

This section summarizes general points on EMI/RFI with respect to the printed circuit board (PCB) layout. It complements earlier chapter discussions on general PCB design techniques. When a PCB design has not been optimized in terms of EMI/RFI, system performance can be compromised. This is true not only for signal-path performance, but also for the system's susceptibility to EMI, plus the degree of EMI radiated by the system. Failure to implement sound PCB layout techniques will very likely lead to system/instrument EMC failures.

To summarize earlier points of this section, a real-world PCB layout may allow multiple paths through which high-frequency noise can couple/radiate into and/or out of the circuit.

This is especially true for digital circuitry, operating at high *edge rates*. It is the rapid changes of logic state ($1 \Rightarrow 0$ or $0 \Rightarrow 1$), i.e., the edge rate that contains the HF energy which can easily radiate as EMI. While similar points are applicable to precision high-speed analog or mixed analog/digital circuits, logic devices are by far the worst potential EMI offenders. Identifying critical circuits and paths helps in designing the PCB for both low emissions and susceptibility to radiated and conducted external and internal noise sources.

Carefully Choose Logic Devices

Logic-family speaking, a key point in minimizing system noise problems is to *choose devices no faster than actually required by the application.* Many designers assume that faster is always better—fast logic is better than slow, high bandwidth amplifiers better than low bandwidth, and fast DACs and ADCs are better, even if the speed isn't required by the system. Unfortunately, faster is *not* better, and actually may be worse for EMI concerns.

Many fast DACs and ADCs have digital inputs and outputs with edge rates in the 1 ns/V region. Because of this wide bandwidth, the sampling clock and the digital inputs can respond to any form of high frequency noise, even glitches as narrow as 1 ns to 3 ns. These high speed data converters and amplifiers are thus easy prey for the high frequency noise of microprocessors, digital signal processors, motors, switching regulators, hand-held radios, electric jackhammers, and so forth. With some of these high speed devices, a small amount of input/output filtering may be required to desensitize the circuit from its EMI/RFI environment. A ferrite bead just before the local decoupling capacitor is very effective in filtering high frequency noise on supply lines. Of course, with circuits requiring bipolar supplies, this technique should be applied to both positive and negative supply lines.

To help reduce emissions generated by extremely fast moving digital signals at DAC inputs or ADC outputs, a small resistor or ferrite bead may be required at each digital input/output.

Design PCBs Thoughtfully

Once the system's critical paths and circuits have been identified, the next step in implementing sound PCB layout is to partition the printed circuit board according to circuit function. This involves the appropriate use of power, ground, and signal planes. Good PCB layouts also isolate critical analog paths from sources of high interference (I/O lines and connectors, for example). High frequency circuits (analog and digital) should be separated from low frequency ones. Furthermore, automatic signal routing CAD layout software should be used with extreme caution. Critical signal paths should be routed by hand, to avoid undesired coupling and/or emissions.

Properly designed multilayer PCBs can reduce EMI emissions and increase immunity to RF fields, by a factor of 10 or more, compared to double-sided boards. A multilayer board allows a complete layer to be used for the ground plane, whereas the ground plane side of a double-sided board is often disrupted with signal crossovers, and so forth. If the system has separate analog and digital ground and power planes, the analog ground plane should be underneath the analog power plane, and similarly, the digital ground plane should be underneath the digital power plane. There should be no overlap between analog and digital ground planes, nor analog and digital power planes.

Designing Controlled Impedances Traces on PCBs

A variety of trace geometries are possible with controlled impedance designs, and they may be either integral to or allied to the PCB pattern. In the discussions below, the basic patterns follow those of the IPC, as described in standard 2141 (see Reference 16).

Note that following figures use the term "ground plane." It should be understood that this plane is in fact a large area, low impedance *reference* plane. In practice it may actually be either a ground plane or a power plane, both of which are assumed to be at zero ac potential.

The first of these is the simple wire-over-a-plane form of transmission line, also called a *wire microstrip.* A cross-sectional view is shown in Figure 7-116. This type of transmission line might be a signal wire used within a breadboard, for example. It is composed simply of a discrete insulated wire spaced a fixed distance over a ground plane. The dielectric would be either the insulation wall of the wire, or a combination of this insulation and air.

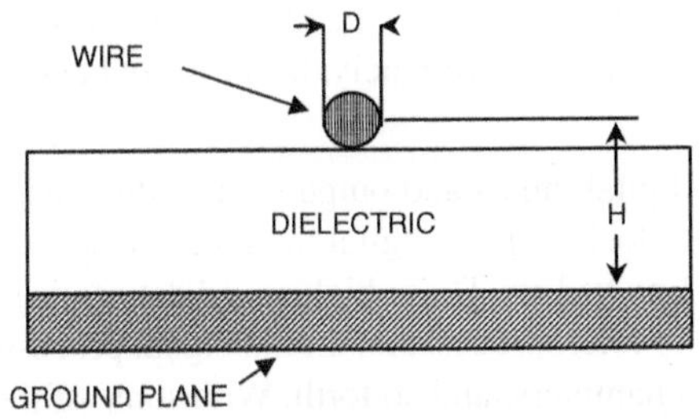

Figure 7-116: A wire microstrip transmission line with defined impedance is formed by an insulated wire spaced from a ground plane

The impedance of this line in ohms can be estimated with Eq. 7-16. Here D is the conductor diameter, H the wire spacing above the plane, and ε_r the dielectric constant.

$$Z_O(\Omega) = \frac{60}{\sqrt{\varepsilon_r}} \ln\left[\frac{4H}{D}\right] \qquad \text{Eq. 7-16}$$

For patterns integral to the PCB, there are a variety of geometric models from which to choose, single-ended and differential. These are covered in some detail within IPC standard 2141 (see Reference 16), but information on two popular examples is shown here.

Before beginning any PCB-based transmission line design, it should be understood that there are abundant equations, all claiming to cover such designs. In this context, "Which of these is accurate?" is an extremely pertinent question. The unfortunate answer is, *none is perfectly so.* All of the existing equations are approximations, and thus accurate to varying degrees, depending upon specifics. The best known and most widely quoted equations are those of Reference 16, but even these come with application caveats.

Reference 17 has evaluated the Reference 16 equations for various geometric patterns against test PCB samples, finding that predicted accuracy varies according to target impedance. Reference 18 also evaluates the Reference 16 equations, offering an alternative and even more complex set (see Reference 19). The equations quoted below are from Reference 16, and offered here as a starting point for a design, subject to further analysis, testing, and design verification. The bottom line is, study carefully and take PCB trace impedance equations with a proper dose of salt.

Microstrip PCB transmission lines

For a simple two-sided PCB design where one side is a ground plane, a signal trace on the other side can be designed for controlled impedance. This geometry is known as a *surface microstrip,* or more simply, *microstrip.*

A cross-sectional view of a two-layer PCB illustrates this microstrip geometry as shown in Figure 7-117.

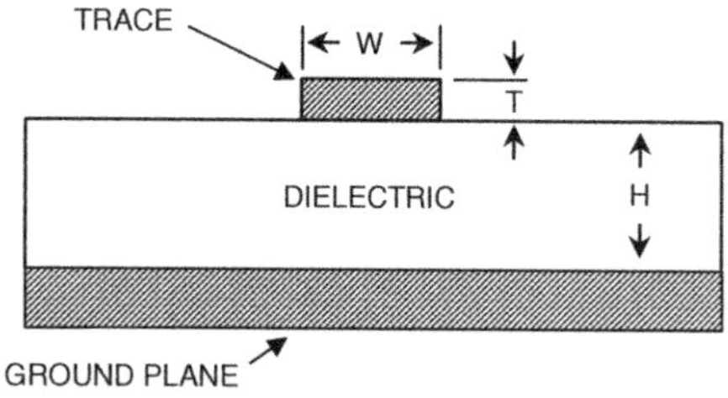

Figure 7-117: A microstrip transmission line with defined impedance is formed by a PCB trace of appropriate geometry, spaced from a ground plane

For a given PCB laminate and copper weight, note that all parameters will be predetermined except for W, the width of the signal trace. Eq. 7-17 can then be used to design a PCB trace to match the impedance required by the circuit. For the signal trace of width W and thickness T, separated by distance H from a ground (or power) plane by a PCB dielectric with dielectric constant ε_r, the characteristic impedance is:

$$Z_O(\Omega) = \frac{87}{\sqrt{\varepsilon_r + 1.41}} \ln\left[\frac{5.98H}{(0.8W + T)}\right] \qquad \text{Eq. 7-17}$$

Note that in these expressions, measurements are in common dimensions (mils).

These transmission lines will have not only a characteristic impedance, but also capacitance. This can be calculated in terms of pF/in as shown in Eq. 7-18.

$$C_O(pF/in) = \frac{0.67(\varepsilon_r + 1.41)}{\ln[5.98H/(0.8W + T)]} \qquad \text{Eq. 7-18}$$

As an example including these calculations, a 2-layer board might use 20 mil wide (W), 1 ounce (T = 1.4) copper traces separated by 10 mil (H) FR-4 (ε_r = 4.0) dielectric material. The resulting impedance for this microstrip would be about 50 Ω. For other standard impedances, for example the 75 Ω video standard, adjust "W" to about 8.3 mils.

Some Microstrip Rules of Thumb

This example touches an interesting and quite handy point. Reference 17 discusses a useful rule of thumb pertaining to microstrip PCB impedance. For a case of dielectric constant of 4.0 (FR-4), it turns out that when W/H is 2/1, the resulting impedance will be close to 50 Ω (as in the first example, with W = 20 mils).

Careful readers will note that Eq. 7-17 predicts Z_o to be about 46 Ω, generally consistent with accuracy quoted in Reference 17 (>5%). The IPC microstrip equation is most accurate between 50 Ω and 100 Ω, but is substantially less so for lower (or higher) impedances. Reference 20 gives tabular results of various PCB industry impedance calculator tools.

The propagation delay of the microstrip line can also be calculated, as per Eq. 7-19. This is the one-way transit time for a microstrip signal trace. Interestingly, for a given geometry model, *the delay constant in ns/ft is a function only of the dielectric constant, and not the trace dimensions* (see Reference 21). Note that this is quite a convenient situation. It means that, with a given PCB laminate (and given ε_r), the propagation delay constant is fixed for various impedance lines.

$$t_{pd}\left(\text{ns/ft}\right) = 1.017\sqrt{0.475\varepsilon_r + 0.67} \qquad \text{Eq. 7-19}$$

This delay constant can also be expressed in terms of ps/in, a form which will be more practical for smaller PCBs. This is:

$$t_{pd}\left(\text{ps/in}\right) = 85\sqrt{0.475\varepsilon_r + 0.67} \qquad \text{Eq. 7-20}$$

Thus for an example PCB dielectric constant of 4.0, it can be noted that a microstrip's delay constant is about 1.63 ns/ft, or 136 ps/in. These two additional rules of thumb can be useful in designing the timing of signals across PCB trace runs.

Symmetric Stripline PCB Transmission Lines

A method of PCB design preferred from many viewpoints is a multilayer PCB. This arrangement *embeds* the signal trace between a power and a ground plane, as shown in the cross-sectional view of Figure 7-118. The low impedance ac ground planes and the embedded signal trace form a *symmetric stripline* transmission line.

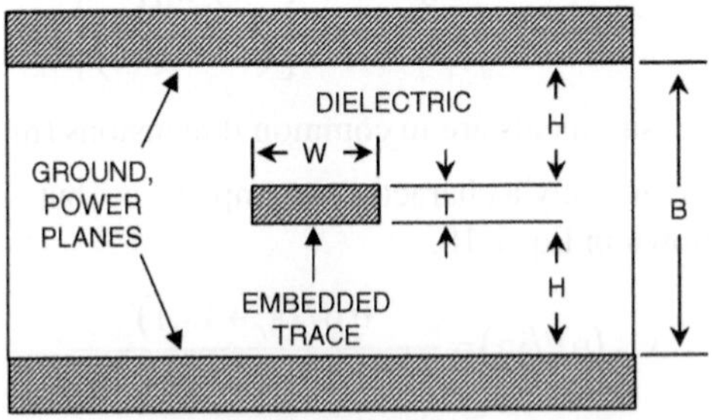

Figure 7-118: A symmetric stripline transmission line with defined impedance is formed by a PCB trace of appropriate geometry embedded between equally spaced ground and/or power planes

As can be noted from the figure, the return current path for a high frequency signal trace is located directly above and below the signal trace on the ground/power planes. The high frequency signal is thus contained entirely inside the PCB, minimizing emissions, and providing natural shielding against incoming spurious signals.

The characteristic impedance of this arrangement is again dependent upon geometry and the ε_r of the PCB dielectric. An expression for Z_O of the stripline transmission line is:

$$Z_O(\Omega) = \frac{60}{\sqrt{\varepsilon_r}} \ln\left[\frac{1.9(B)}{(0.8W+T)}\right] \qquad \text{Eq. 7-21}$$

Here, all dimensions are again in mils, and B is the spacing between the two planes. In this symmetric geometry, note that B is also equal to 2H + T. Reference 17 indicates that the accuracy of this Reference 16 equation is typically on the order of 6%.

Another handy rule of thumb for the symmetric stripline in an $\varepsilon_r = 4.0$ case is to make B a multiple of W, in the range of 2 to 2.2. This will result in an stripline impedance of about 50 Ω. Of course this rule is based on a further approximation, by neglecting T. Nevertheless, it is still useful for ballpark estimates.

The symmetric stripline also has a characteristic capacitance, which can be calculated in terms of pF/in as shown in Eq. 7-22.

$$C_O(\text{pF/in}) = \frac{1.41(\varepsilon_r)}{\ln[3.81H/(0.8W+T)]} \qquad \text{Eq. 7-22}$$

The propagation delay of the symmetric stripline is shown in eq. 7-23.

$$t_{pd}(\text{ns/ft}) = 1.017\sqrt{\varepsilon_r} \qquad \text{Eq. 7-23}$$

or, in terms of ps:

$$t_{pd}(\text{ps/in}) = 85\sqrt{\varepsilon_r} \qquad \text{Eq. 7-24}$$

For a PCB dielectric constant of 4.0, it can be noted that the symmetric stripline's delay constant is almost exactly 2 ns/ft, or 170 ps/in.

Some Pros and Cons of Embedding Traces

The above discussions allow the design of PCB traces of defined impedance, either on a surface layer or embedded between layers. There are, of course, many other considerations beyond these impedance issues.

Embedded signals do have one major and obvious disadvantage—the debugging of the hidden circuit traces is difficult to impossible. Some of the pros and cons of embedded signal traces are summarized in Figure 7-119.

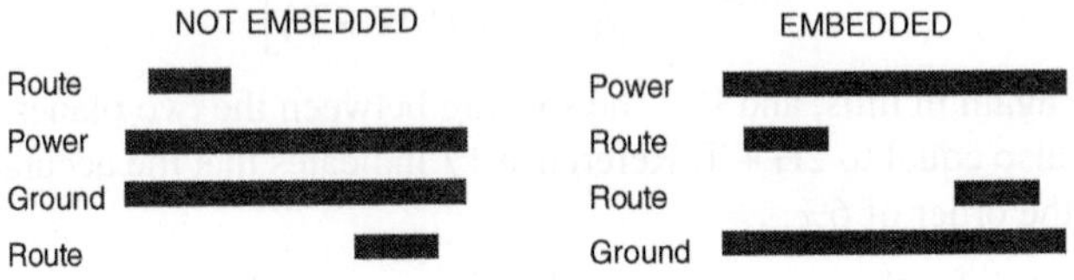

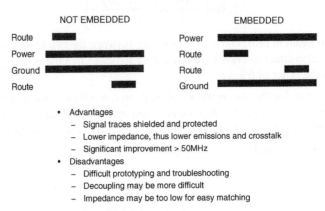

- Advantages
  - Signal traces shielded and protected
  - Lower impedance, thus lower emissions and crosstalk
  - Significant improvement > 50MHz
- Disadvantages
  - Difficult prototyping and troubleshooting
  - Decoupling may be more difficult
  - Impedance may be too low for easy matching

Figure 7-119: The pros and cons of not embedding versus the embedding of signal traces in multilayer PCB designs

Multilayer PCBs can be designed *without* the use of embedded traces, as shown in the left cross-sectional example. This embedded case could be considered as a doubled two-layer PCB design (i.e., four copper layers overall). The routed traces at the top form a microstrip with the power plane, while the traces at the bottom form a microstrip with the ground plane. In this example, the signal traces of both outer layers are readily accessible for measurement and troubleshooting purposes. But, the arrangement does nothing to take advantage of the shielding properties of the planes.

This nonembedded arrangement will have greater emissions and susceptibility to external signals, vis-à-vis the embedded case at the right, which uses the embedding, and does take full advantage of the planes. As in many other engineering efforts, the decision of embedded versus not-embedded for the PCB design becomes a trade-off, in this case one of reduced emissions versus ease of testing.

Transmission Line Termination Rule of Thumb

Much has been written about terminating PCB traces in their characteristic impedance, to avoid signal reflections. A good rule of thumb to determine when this is necessary is as follows: *Terminate the transmission line in its characteristic impedance when the one-way propagation delay of the PCB track is equal to or greater than one-half the applied signal rise/fall time (whichever edge is faster)*. For example, a 2-inch microstrip line over an $E_r = 4.0$ dielectric would have a delay of ~270 ps. Using the above rule strictly, termination would be appropriate whenever the signal rise time is < ~500 ps. A more conservative rule is to use a 2-inch (PCB track length)/nanosecond (rise/fall time) rule. If the signal trace exceeds this trace-length/speed criterion, termination should be used.

For example, PCB tracks for high-speed logic with rise/fall time of 5 ns should be terminated in their characteristic impedance if the track length is equal to or greater than 10 inches (where measured length *includes* meanders).

In the analog domain, it is important to note that this same 2-inch/nanosecond rule of thumb should also be used with op amps and other circuits, to determine the need for transmission line techniques. For instance, if an amplifier must output a maximum frequency of f_{max}, then the equivalent risetime t_r is related to this f_{max}. This limiting risetime, t_r, can be calculated as:

$$t_r = 0.35/f_{max} \qquad \text{Eq. 7-25}$$

The maximum PCB track length is then calculated by multiplying t_r by 2-inch/nanosecond. For example, a maximum frequency of 100 MHz corresponds to a risetime of 3.5 ns, so a 7-inch or more track carrying this signal should be treated as a transmission line.

The best ways to keep sensitive analog circuits from being affected by fast logic are to physically separate the two by the PCB layout, and to use no faster logic family than is dictated by system requirements. In some cases, this may require the use of several logic families in a system. An alternative is to use series resistance or ferrite beads to slow down the logic transitions where highest speed isn't required.

A general method of doing this is to use a series R at a logic driver output, and a shunt C at a CMOS gate input. The series resistance and the net input capacitance of the gate form a lowpass filter. Typical CMOS input capacitance is 10 pF. Locate the series resistor close to the driving gate, adding an additional small capacitance, as needed. The resistor minimizes transient switching currents, and may also eliminate the necessity for transmission line techniques. The value of the resistor should be chosen such that the rise and fall times at the receiving gate are fast enough to meet system requirement, but no faster. Also, make sure that the resistor is not so large that the logic levels at the receiver are out of specification because of the source and sink current which must flow through the resistor. Use of CMOS logic will simplify this, since the input currents are so low.

References: EMI/RFI Considerations

1. Tim Williams, **EMC for Product Designers, 2nd Ed.**, Newnes, Oxford, 1996, ISBN: 0-7506-2466-3.
2. Henry Ott, **Noise Reduction Techniques In Electronic Systems, 2nd Ed.**, John Wiley & Sons, New York, 1988, ISBN 0-471-85068-3.
3. Mark Montrose, **EMC and the Printed Circuit Board**, IEEE Press, 1999, ISBN 0-7803-4703-X.
4. Ralph Morrison, **Grounding And Shielding Techniques in Instrumentation, 3rd Ed.**, John Wiley & Sons, New York, 1986, ISBN 0-471-83805-5.
5. Daryl Gerke and William Kimmel, "Designer's Guide to Electromagnetic Compatibility," **EDN**, January 20, 1994.
6. **Designing for EMC** (Workshop Notes), Kimmel Gerke Associates, Ltd., 1994.
7. Daryl Gerke and William Kimmel, "EMI and Circuit Components," **EDN**, September 1, 2000.
8. Alan Rich, "Understanding Interference-Type Noise," **Analog Dialogue**, Vol. 16, No. 3, 1982, pp. 16–19 (also available as application note AN346).
9. Alan Rich, "Shielding and Guarding," **Analog Dialogue**, Vol. 17, No. 1, 1983, pp. 8–13 (also available as application note AN347).
10. James Wong, Joe Buxton, Adolfo Garcia, James Bryant, "Filtering and Protection Against EMI/RFI" and "Input Stage RFI Rectification Sensitivity," Chapter 1, pp. 21–55 of **Systems Application Guide**, 1993, Analog Devices, Inc., Norwood, MA, ISBN 0-916550-13-3.
11. Adolfo Garcia, "EMI/RFI Considerations," Chapter 7, pp. 69–88 of **High Speed Design Techniques**, 1996, Analog Devices, Inc., Norwood, MA, 1993, ISBN 0-916550-17-6.
12. Walt Kester, Walt Jung, Chuck Kitchen, "Preventing RFI Rectification," Chapter 10, pp. 10.39–10.43 of **Practical Design Techniques for Sensor Signal Conditioning**, Analog Devices, Inc., Norwood, MA, 1999, ISBN 0-916550-20-6.
13. Charles Kitchen, Lew Counts, **A Designer's Guide to Instrumentation Amplifiers**, Analog Devices, Inc., 2000.
14. **B4001 and B4003 Common-Mode Chokes**, Pulse Engineering, Inc., 12220 World Trade Drive, San Diego, CA, 92128, 619-674-8100, www.pulseeng.com.
15. **Understanding Common-Mode Noise**, Pulse Engineering, Inc., 12220 World Trade Drive, San Diego, CA, 92128, 619-674-8100, www.pulseeng.com.
16. Standard IPC-2141, "Controlled Impedance Circuit Boards and High Speed Logic Design," 1996, Institute for Interconnection and Packaging Electronic Circuits, 2215 Sanders Road, Northbrook, IL, 60062-6135, 847-509-9700, www.ipc.org.
17. Eric Bogatin, "Verifying the Accuracy of 2D Field Solvers for Characteristic Impedance Calculation," Ansoft Seminar, October 11, 2000, www.bogatinenterprises.com.
18. Andrew Burkhardt, Christopher Gregg, Alan Staniforth, "Calculation of PCB Track Impedance," Technical Paper S-19-5, presented at the IPC Printed Circuits Expo '99 Conference, March 14–18, 1999.
19. Brian C. Wadell, **Transmission Line Design Handbook,** Artech House, Norwood, MA, 1991, ISBN: 0-89006-436-9.
20. Eric Bogatin, "No Myths Allowed," "Impedance Calculations," "A Chip Center Column," November 1, 1999, www.chipcenter.com/signalintegrity.

21. William R. Blood, Jr., **MECL System Design Handbook** (HB205/D, Rev. 1A May 1988), ON Semiconductor, August, 2000.

22. Paul Brokaw, "An IC Amplifier User Guide To Decoupling, Grounding, And Making Things Go Right For A Change," **Analog Devices AN202**.

Some useful EMC and signal integrity related URLs:

Eric Bogatin website, www.bogatinenterprises.com

Chip Center's "Signal Integrity" page, www.chipcenter.com/signalintegrity

Kimmel Gerke Associates website, www.emiguru.com

Henry Ott website, www.hottconsultants.com

IEEE EMC website, www.ewh.ieee.org/soc/emcs

Mark Montrose website, www.montrosecompliance.com/index.html

Tim Williams website, www.elmac.co.uk

Simulation, Breadboarding and Prototyping

Joe Buxton, Walt Kester, James Bryant, Walt Jung

In this final section of the chapter, the practical aspects of assembling hardware for op amp and other analog functions are brought to play. Various experimental techniques are useful towards verifying the integrity of a design. These include electronic *analog circuit simulation* programs, to be used with (but not to the exclusion of) the allied lab processes of *breadboarding* and *prototyping*.

Analog Circuit Simulation

In the past decade, circuit simulation has taken on an increasingly important role within analog circuit design. The most popular simulation tool for this is SPICE, which is available in multiple forms for various computer platforms (see References 1 and 2). However, to achieve meaningful simulation results, designers need accurate models of many system components. The most critical of these are realistic models for ICs, the active devices that drive modern designs. In the early 1990s, Analog Devices developed an advanced op amp SPICE model, which is, in fact, still in use today (see References 3 and 4). Within this innovative open amplifier architecture, gain and phase response can be fully modeled, enabling designers to accurately predict ac, dc, and transient performance behavior. This modeling methodology has also been extended to include other devices such as in amps, voltage references, and analog multipliers.

Figure 7-120 lists some major SPICE simulation objectives. The popularity of SPICE simulation has led to many op amp macromodel releases, which (ideally) software-mimic amplifier electrical performance. With numerous models available, several confusions are possible. There may be uncertainty as to what is/isn't modeled, plus a basic question of *model accuracy*. All of these points are important, in order to place confidence in simulation results. So, *verification* of a model is important, checking it by comparison to the actual device performance conditions, before trusting it for serious designs.

- Understand Realistic Simulation Goals
- Evaluate Available Models Accordingly
- Know the Capabilities for Each Competing Op Amp Model
- Following Simulation, *Breadboarding is Always Desirable and Necessary*

Figure 7-120: Used wisely, simulation is a powerful design tool

Of course, a successful first design step using an accurate op amp model by itself doesn't necessarily guarantee totally valid simulations. A simulation based on incomplete information has limited value. All parts of a target circuit should be modeled, including the surrounding passive components, various parasitic effects, and temperature changes. Then, the circuit needs to be verified in the lab by breadboarding and prototyping. A breadboard circuit is a quickly executed mockup of a circuit design using a semi-permanent lab platform, i.e., one that is in less than final physical form. It is intended to show real performance, but without the total physical environment. A good breadboard can often reveal behavior not predicted by SPICE, either because of an incomplete model, external circuit parasitics, or numerous other reasons. However, by using SPICE along with intelligent breadboarding techniques, a circuit can be efficiently and quickly designed with reasonably good assurance of working properly on a prototype version, or even a final PCB. The following prototype phase is just one step removed from a final PCB, and may in fact be an actual test PCB, with nearly all design components incorporated, and with close to full performance.

The breadboard/prototype design steps are closely allied to simulation, usually following it in the overall design process. These are more fully discussed in subsequent sections.

Macromodel versus Micromodel

The distinction between *macromodel* and *micromodel* is often unclear. A micromodel uses the actual *transistor level* and other SPICE models of an IC device, with all active and passive parts fully characterized according to the manufacturing process. In differentiating this type of model from a macromodel, some authors use the term *device level model* to describe the resulting overall op amp model (see Reference 5). Typically, a micromodel is used in the actual design process of an IC.

A macromodel takes another route in emulating op amp performance. Taking into consideration final device performance, it uses ideal native SPICE elements to model observed behavior—as many as necessary. In developing a macromodel, a real device is measured in terms of lab and data sheet performance, and the macromodel is adjusted to match this behavior. Some aspects of performance may be sacrificed in doing this. Figure 7-121 compares the major pros and cons between macromodels and micromodels.

	METHODOLOGY	ADVANTAGES	DISADVANTAGES
MACROMODEL	Ideal Elements Model Device Behavior	Fast Simulation Time, Easily Modified	May Not Model All Characteristics
MICROMODEL	Fully Characterized Transistor Level Circuit	Most Complete Model	Slow Simulation, Convergence Difficulty, Nonavailability

Figure 7-121: Differentiating the Macromodel and Micromodel

There are advantages and disadvantages to both approaches. A micromodel can give a complete and accurate model of op amp circuit behavior under almost all conditions. But, because of a large number of transistors and diodes with nonlinear junctions, simulation time is very long. Of course, manufacturers are also reluctant to release such models, since they contain proprietary information. And, even though all transistors may be included, this isn't a guarantee of total accuracy, as the transistor models themselves don't cover all operational regions precisely. Furthermore, with a high node count, SPICE can have convergence difficulties, causing a failed simulation. This point would make a micromodel virtually useless for multiple amplifier active filters, for example.

On the other hand, a carefully developed macromodel can produce both accurate results and simulation time savings. In more advanced macromodels such as the ADSpice model described, transient and ac device performance can be closely replicated. Op amp nonlinear behavior can also be included, such as output voltage and current swing limits.

However, because these macromodels are still simplifications of real devices, all nonlinearities aren't modeled. For example, not all ADSpice models include common-mode input voltage range, or noise (while more recent ones do). Typically, in model development parameters are optimized as may be critical to the intended application; for example, ac and transient response. Including every possible characteristic could lead to cumbersome macromodels that may even have convergence problems. Thus, ADSpice macromodels include those op amp behavior characteristics critical to intended performance for normal operating conditions, but not necessarily all nonlinear behavior.

The ADSpice Op Amp Macromodels

The basic ADSpice model was developed as an op amp macromodeling advance, and as an improved design tool for more accurate application circuit simulations. Since being introduced in 1990, it has become a standard op amp macromodel topology, as evidenced by industry adoption of the frequency shaping concepts (see References 6 and 7).

Prior to about 1990, a dominant op amp model architecture was the Boyle model (see Reference 8). This macromodel, developed in the early 70s, cannot accurately model higher speed amplifiers. The primary reason for this is that it has limited frequency shaping ability—only two poles and no zeroes. In contrast, the ADSpice model topology has a flexible and open architecture, allowing virtually unlimited pole and zero frequency shaping stages to be cascaded. This key difference provides much more accurate ac and transient response, vis-à-vis the more simplistic Boyle model topology.

An ADSpice model is comprised of three main portions, described as follows. The first of these is a combined input and gain stage, which will include transistor models as appropriate to the device being modeled (NPN or PNP bipolar, JFET, MOSFET, and so forth). Next are the synthetic pole and zero stages, which are comprised of ideal SPICE native elements. There may be only a few of these or there may be many, dependent on the complexity of the op amp's frequency response. Finally, there is an output stage, which couples the first two sections to the outside world.

Before describing these sections in detail, it is important to realize that many variations upon what is shown do in fact exist. This is due to not just differences from one op amp model to another, but also to evolutionary topology developments in op amp hardware, which in turn has led to corresponding modeling changes. For example, modern op amps often include either rail-rail output or input stages, or both. Consequently more recent developments in the ADSpice models have addressed these issues, along with corresponding model developments.

Furthermore, although the Boyle model and the original ADSpice models were designed to support *voltage feedback* op amp topologies, subsequent additions have added *current feedback* amplifier topologies. In fact, Reference 9 describes an ADSpice current feedback macromodel which appeared just shortly after the voltage feedback model of Reference 3. These current feedback macromodels are discussed in more detail later on.

Input and Gain/Pole Stages

A basic ADSpice voltage feedback op amp macromodel input stage is shown in Figure 7-122. As noted, it uses what are (typically) the only transistors in the entire model, in this example the Q1-Q2 NPN pair, to the left on the diagram. These are needed to properly model an op amp's differential input stage characteristics. A basic tenet of this model topology is that this stage is designed for unity gain, by the proper choice of Q1-Q2 operating current and gain-setting resistors R3-R4 and R5-R6.

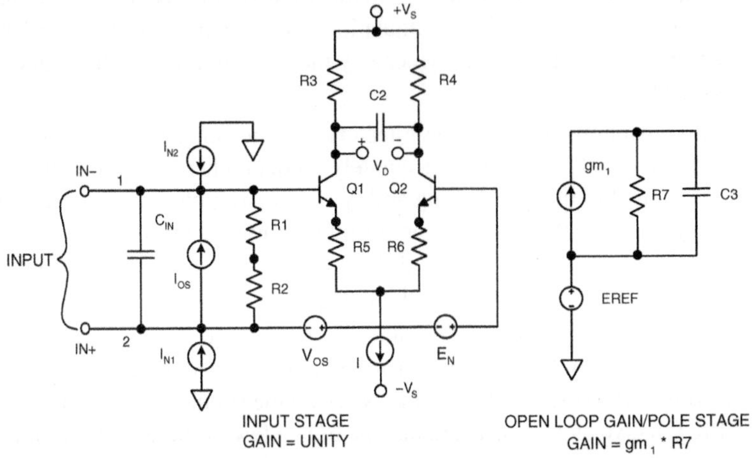

Figure 7-122: Input and gain/pole stages of ADSpice macromodel

Although this example uses NPN transistors, the input stage is easily modified to use PNP bipolars, JFET, or MOSFET devices. The rest of the input stage uses simple SPICE elements such as resistors, capacitors, and controlled sources.

The open-loop gain versus frequency characteristics of the modeled op amp is provided by the gain stage, to the right in the diagram. Here controlled source gm_1 senses the differential collector voltage V_D from the input stage, converting this voltage to a proportional current. The gm_1 output current flows in load resistor R_7, producing a single ended voltage referenced to an internal voltage, EREF. Typically, this voltage is derived as a supply voltage midpoint, and is used throughout the model.

By simply making the gm_1–R_7 product equal to the specified gain of the op amp, this stage produces the entire open-loop gain of the macromodel. This design factor means that all other model stages operate at unity gain, a feature leading to significant flexibility in adding and deleting subsequent stages. This approach allows the quick synthesis of the complex ac characteristics typical of high performance, high speed op amps. In addition, this stage also provides the dominant pole of the amplifier's ac response. The open-loop pole frequency is set by selection of capacitor C3, as noted in the diagram.

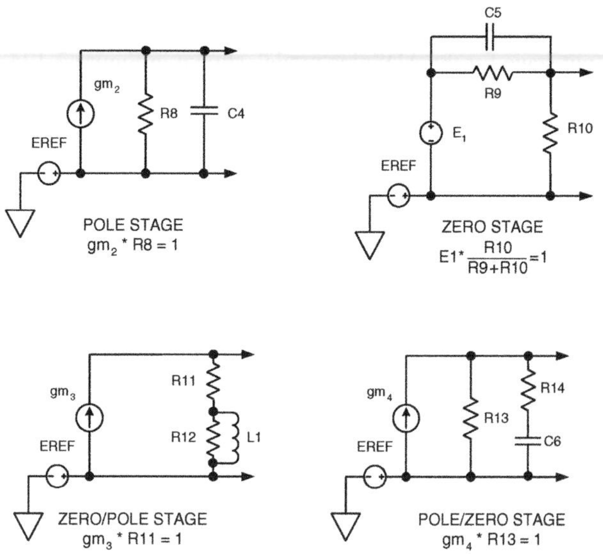

Figure 7-123: The frequency-shaping stages possible within the ADSpice model

Frequency-Shaping Stages

Following the gain stage of the macromodel is a variable but unlimited number of pole and/or zero stages which, in combination, provide frequency response shaping. Typical topologies for these stages are as shown in the Figure 7-123 diagram. The stages can be either a single pole or a single zero, or combined pole/zero or zero/pole stages. All such stages have a dc transfer gain of unity, and a given amplifier type can have all or just a few of these stages, as may be require to synthesize its response.

The pole or zero frequency is set by the combination of the resistor(s) and capacitor, or resistor(s) and inductor, as may be the case. Because an infinite number of values are possible in SPICE, choice of RC values is somewhat arbitrary, and a wide range work. Early ADSpice models used relatively high values, while later ones employ lower values to reduce noise (described in more detail later). In all instances, it is assumed that each stage provides zero loading to the driving stage. The stages shown reflect no particular op amp, but example principles can be found within the OP27 model (see Reference 10).

Because all of these frequency-shaping stages are dc-coupled and have unity gain, any number of them can be added or deleted, with no affect on the model's low frequency response. Most importantly, the high frequency gain and phase response can be precisely tailored to match a real amplifier's response. The benefits of this frequency-shaping flexibility are especially apparent in performance comparisons of the ADSpice model closed loop pulse response and stability analysis, versus that of a more simplistic model. This point is demonstrated by a later example.

Macromodel Output Stages

A general form of the output stage for the ADSpice model, shown in Figure 7-124, models a number of important op amp characteristics. The Thevenin equivalent resistance of R_{O1} and R_{O2} mimics the op amp's dc open-loop output impedance, while inductor L_O models the rise in impedance at high frequencies. A unity gain characteristic for the stage is set by the g_7–R_{O1} and g_8–R_{O2} products.

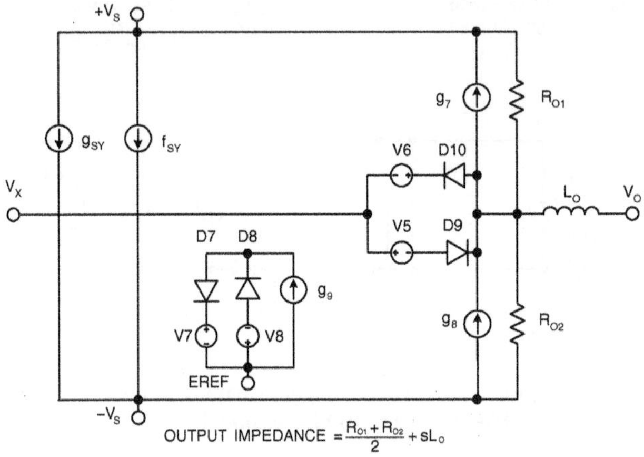

Figure 7-124: General-purpose macromodel output stage

Additionally, output load current is correctly reflected in the supply currents. This feature is a significant improvement over the Boyle model, because the power consumption of the loaded circuit can be analyzed accurately. Furthermore, circuits using the op amp supply currents as part of the signal path can also be correctly simulated. The output stage shown is not intended to reflect any particular op amp, but close similarity is found within the AD817 model (see Reference 11).

With the recent advent of numerous rail-rail output stage op amps, a number of customized model topologies have been developed. This expands the ADSpice library to include rail-rail model behavior, matching op amp architectures using P and N MOSFET devices, as well as bipolar devices. Characteristically, a rail-rail output stage includes several key differentiating performance points. First and foremost is the ability to swing the op amp output to within a few mV of both supplies. A second point is the fact that such an output stage has a voltage gain greater than one, and a third is the relatively high output impedance (high as contrasted to traditional emitter follower outputs).

Examples of several modeling approaches to rail-rail output stages are found in the ADI SPICE macromodel library. Reference 12 employs CMOS devices to realize a rail-rail output, while Reference 13 uses bipolar devices to the same end. The macromodels of References 14 and 15 use synthesis techniques to model rail-rail outputs. References 16–18 utilize combinations of selected discrete device models and synthesis techniques, to realize rail-rail output operation for both op amp and in amp devices.

In addition to rail-rail output operation, many modern op amps also feature rail-rail *input stages*. Such stages essentially duplicate, for example, an NPN-based differential stage with a complementary PNP stage, both stages operating in parallel. This allows the op amp to provide a CM range that includes both supply rails. This performance feature can also be accomplished within CMOS op amps, using both a P and N type MOS differential pairs. Model examples reflecting rail-rail input stages include References 13, 14, and 17.

Model Transient Response

The performance advantage of the multiple pole/zero stages is readily demonstrated in a transient pulse response test, as in Figure 7-125. This figure compares an actual OP249 op amp, the ADSpice model, and the Boyle model. It reveals the improved execution resulting from the unlimited number of poles and zeros in this model.

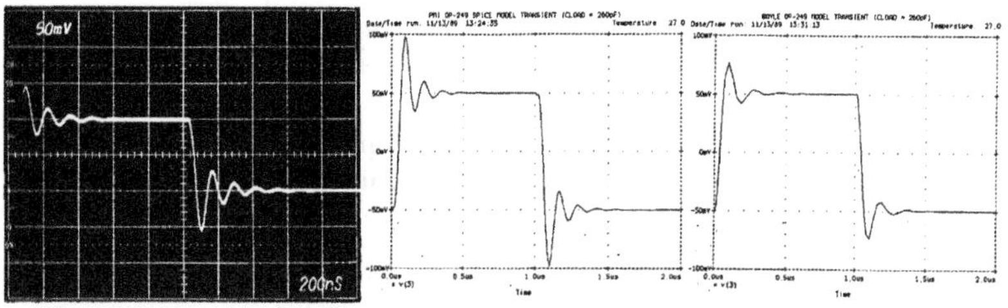

VERTICAL SCALE: 50mV / div.
HORIZONTAL SCALE: 200ns / div.

LOAD = 260pF

Figure 7-125: A pulse response comparison of an OP249 follower (left) model favors the ADSpice model in terms of fidelity (center), but not the Boyle (right)

The difference is easily apparent from this transient analysis plot for a unity gain follower circuit. An OP249 amplifier was used, with the output connected to the inverting input, and a 260 pF capacitive load.

This results in ringing, as seen in the op amp response (left). Note that the ADSpice model accurately predicts the amount of overshoot and frequency of the damped ringing (center). In contrast, the Boyle model (right) predicts about half the overshoot and significantly less ringing.

The Noise Model

An important enhancement to the ADSpice model is the ability to realistically model noise performance of an op amp. The capability to model a circuit's noise in SPICE can be appreciated by anyone who has tried to analyze noise by hand. A complete analysis is an involved and tedious task that involves adding all the individual noise contributions from all active devices and all resistors, and referring them to the input.

To aid this task, the ADSpice model was enhanced to include noise generators that accurately mimic the broadband and 1/f noise of an actual op amp. Conceptually, this involves first making an existing model noiseless, and then adding discrete noise generators, so as to emulate the target device. As noted earlier, all ADI models aren't necessarily designed for this noise-accurate performance. Selected device models are designed for noise however, when their typical uses include low noise applications.

The first step is an exercise in scaling down the model internal impedances. For example, by reducing the resistances in the pole/zero stages from a base resistance of 1E6 Ω to 1 Ω, total noise is reduced dramatically, as figure 7-126 illustrates.

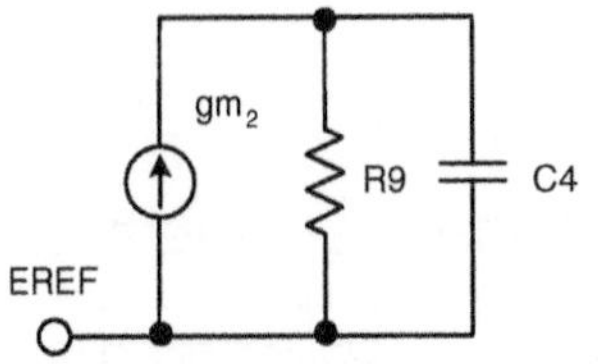

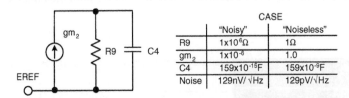

	CASE	
	"Noisy"	"Noiseless"
R9	$1\times10^{6}\Omega$	1Ω
gm_2	1×10^{-6}	1.0
C4	159×10^{-15}F	159×10^{-9}F
Noise	$129nV/\sqrt{Hz}$	$129pV/\sqrt{Hz}$

Figure 7-126: Towards achieving low noise operation, a first design step is the reduction of pole/zero cell impedances to low values

For the "Noisy" column of the table, the noise from the pole stage shown with a large R9 resistor value is 129 $nV/\sqrt{Hz}$. But when this resistor is scaled down by a factor of 10^6, to 1 Ω, as in the "Noiseless" column, stage noise is 129 $pV/\sqrt{Hz}$. Note also that transconductance and capacitance values are also scaled by the same factor, maintaining the same gain and pole frequency. To make the model's input stage noiseless, it is operated at a high current and with reduced load resistances, making noise contributions negligible. Extending these techniques to the entire model renders it essentially noiseless.

Once global noise reduction is achieved, independent noise sources are added, one for voltage noise and two for current noise. The basic noise source topology used is like Figure 7-127, and it can be set up to produce both voltage and current noise outputs.

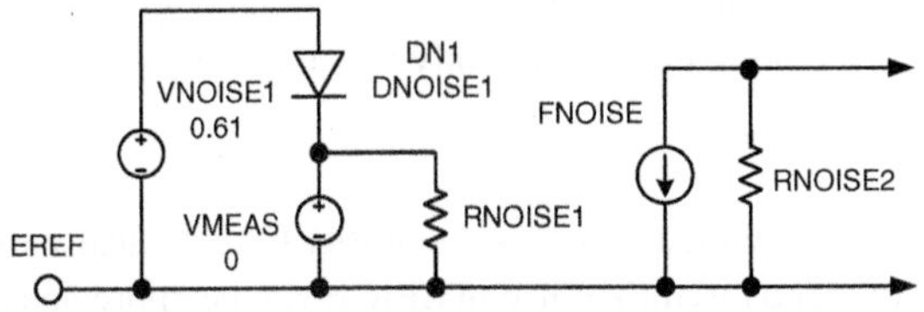

Figure 7-127: A basic SPICE noise generator is formed with diodes, resistors, and controlled sources

Note that, within SPICE, semiconductor models can generate 1/f (flicker) noise. The noise generators use diodes such as DN1 to produce this portion of the noise, modeling the 1/f noise of the op amp. By properly specifying diode model parameters and bias voltage VNOISE1, the 1/f noise is tailored to match the op amp. The noise current from DN1 passes through a zero voltage source. Here VMEAS is being used as a measurement device, combining the 1/f noise from DN1 and the broadband noise from RNOISE1.

RNOISE1 is selected for a value providing an appropriate broadband noise. The combined noise current in VMEAS is monitored by FNOISE, and appears as a voltage across RNOISE2. This voltage is then injected in series with one amplifier input via a controlled voltage source, such as E_N of Figure 7-122. Either FNOISE or a controlled voltage source coefficient can be used for overall noise voltage scaling.

Current noise generation is similar to the above, except that the RNOISE2 voltage producing resistor isn't used, and two current-controlled sources drive the amplifier inputs. With all noise generators symmetrical about ground, dc errors aren't introduced.

Current Feedback Amplifier Models

As noted previously, a new model topology was developed for current feedback amplifiers, to accommodate their unique input stage structure (see Reference 9). The model uses a topology as shown in Figure 7-128 for the input and gain stages. The remaining model portions (not shown) contain multiple pole/zero stages and the output stage, and are essentially the same as voltage feedback amplifiers, described above.

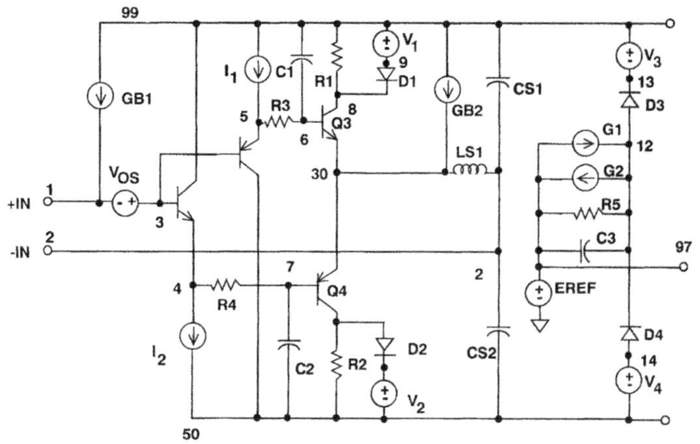

Figure 7-128: Input and gain stages of current feedback op amp macromodel

The four bipolar transistor input stage resembles actual current feedback amplifiers, with a high impedance noninverting input (+IN) and a low impedance inverting input (–IN). In current feedback amplifiers, the maximum slew rate is very high, because dynamic slew current isn't limited to a differential pair tail current (as in voltage feedback op amps). In current feedback op amp designs, much larger amounts of error current can flow in the inverting input, as developed by the feedback network. Internally, this current flows in either Q3 or Q4, and charges compensation capacitor C3 via current mirrors.

The current mirrors of the ADSpice model are actually voltage controlled current sources in the gain stage, G1 and G2. They sense voltage drops across input stage resistors R1 and R2, and translating this into a C3 charging current. By making the value of G1 and G2 equal to the R1-R2 reciprocal, the slew currents will be identical. By clamping the R1-R2 voltage drops via D1–V1 and D2–V2, the maximum current is limited, which thus sets the highest slew rate. Open loop gain or transresistance of the model is set by R5, and the open loop pole frequency by C3–R5 (as described previously, Figure 7-122). The output from across R5–C3 (node 12) drives the model's succeeding frequency-shaping stages, and EREF is again an internal reference voltage.

One of the unique properties of current feedback amplifiers is that bandwidth is a function of the feedback resistor and the internal compensation capacitor, C3. The lower the feedback resistor, the greater the bandwidth, until a practical lower limit is reached, i.e., the value at which the part oscillates. As the model includes a low impedance inverting input, it accurately mimics real part behavior as R_F is altered. Figure 7-129 compares the ADSpice model to the actual device for an AD811 video amplifier. As shown, the model accurately predicts the gain roll-off at the much lower frequency for the 1 kΩ feedback resistor as opposed to the 500 Ω resistor.

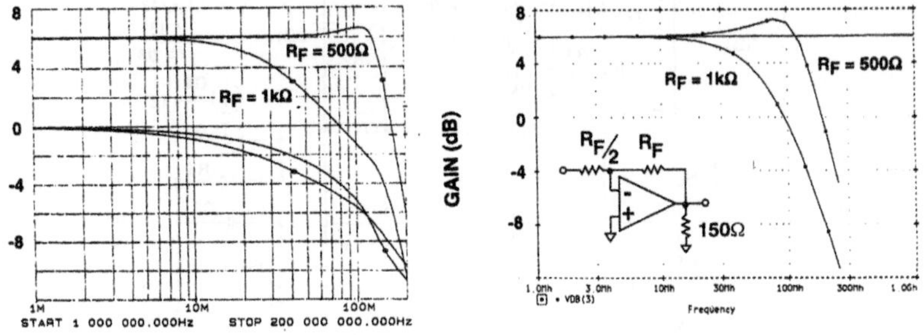

Figure 7-129: Comparison of a real AD811 current feedback op amp (left) with macromodel (right) shows similar characteristics as feedback resistance is varied

The current feedback amplifier input and gain stage is an enhancement to the ADSpice model that increases flexibility in modeling different op amp devices, and provides a net increase in design cycle speed.

Simulation Must Not Replace Breadboarding

No matter how accurate the models, or how much confidence you have with simulations, *SPICE analysis alone should never totally replace breadboarding*. As part of a layout and the actual devices existing within a real world PCB assembly, there are second and third order effects that can easily become relevant to performance. By and large, SPICE will never "know" of such things, unless explicitly entered into the SPICE netlist. However, this may be either difficult or outright impossible. The user may not be aware of some things before a PCB is built and tested within the final system—spurious signal coupling, the effects of crosstalk, the inevitable parasitic capacitance, inductance, and resistance—on and on goes the list. It is virtually impossible to include all of these effects in a simulation. Without actually building a PCB, and operating it under the intended conditions, the user will not have any data whatsoever on the magnitudes involved.

Furthermore, remember the fact that no macromodel includes all op amp characteristics. For example, exceeding the input voltage range can cause nonlinear behavior in an op amp, which is not necessarily included in its model. Because of such effects that a simulation might not predict, it is necessary to breadboard the circuit.

Even with models as comprehensive as those of the ADSpice library, external effects can easily cause a circuit to work improperly. As noted, PCB parasitics can significantly alter the frequency performance in high speed designs. Such parasitics are easily overlooked in a SPICE simulation, but a breadboard will reveal the problems.

Ultimately, simulation and breadboarding should be used together to maximize the design efficiency. Figure 7-130 summarizes these pro and con points of analog simulations.

- Understand What's Real (Hardware), and What Isn't (SPICE)
- Use Breadboarding/Prototyping as Final Design Verification
- Be Aware of Non-Modeled Op Amp Characteristics
- Pay Attention to PCB Parasitics Impacting Circuit Behavior

Figure 7-130: Some analog simulation caveats

Obviously, the designer needs to be wary of what SPICE can/cannot do, and the necessity of closely allying simulations with breadboarding and prototyping.

Simulation is a Tool to be Used "Wisely"

It must be remembered that while simulation is an extremely powerful tool, it must be used wisely to realize its full benefits. This includes knowing models well, understanding PCB and other parasitic effects, and anticipating the results. For example, consider a simple differential amplifier comprised of an op amp and four equal resistors, to be analyzed for common-mode rejection ratio (CMRR) performance. At low frequencies, CMRR will be dominated by resistor mismatch, while at higher frequencies it is dominated by op amp CMRR performance. However, a SPICE simulation will only show this if the external resistors are realistically mismatched, and the op amp model used also properly treats not only dc CMRR, but also CMRR reduction at higher frequencies. If these critically important points are overlooked in the analysis, then an optimistic result will shows excellent CMRR performance over the entire circuit bandwidth. Unfortunately, this is simply wrong. Alternatively, substituting into the netlist resistors mismatched by their specified tolerances as well as an ADSpice model (which *does* have CMRR frequency effects modeled) the end results will be quite different. CMRR performance at low frequencies will be limited by resistor mismatch errors, and will degrade at higher frequencies, as would a real op amp device with CMRR versus frequency effects.

Know the Models

Using various dc and ac tests, any op amp macromodel can be checked for accuracy and functional completeness. Specialized test simulations can also be devised for other op amp parameters important for a particular analysis. All this is critically important, as knowing a model's capabilities ahead of time can help prevent many headaches later.

Understand PCB Parasitics

Even if the model passes all preliminary tests, caution still should be exercised. As noted, PCB parasitics can have significant impact on a circuit's performance. This is especially true for high speed circuits. A few picofarads of capacitance on the input node can make the difference between a stable circuit and one that oscillates. Thus, these effects need to be carefully considered when simulating the circuit to achieve meaningful results.

To illustrate the impact of PCB parasitics, the simple voltage follower circuit of Figure 7-131 (left) was built twice. The first time this was on a carefully laid out PCB, and the second time on a component plug-in type of prototype board. An AD847 op amp is used because of its 50 MHz bandwidth, which makes the parasitic effects much more critical (smaller C values will have a greater effect on results).

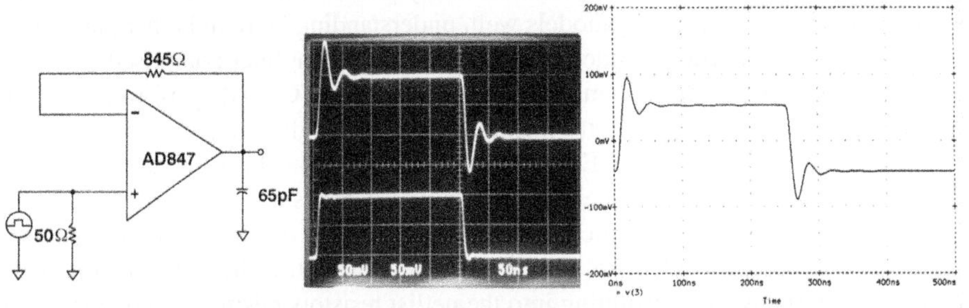

Figure 7-131: With care and low parasitic effects in the PCB layout, results of lab testing (center) and simulation (right) can converge

As the results above indicate, this circuit executed on a properly laid out PCB has a clean response with minor overshoot and ringing (center picture). The SPICE model results also closely agree with the real part, showing a corresponding simulation (right picture).

On the other hand, the same circuit built on the plug-in prototype board shows distinctly different results. In general, it shows much worse performance, due to the relatively high nodal capacitances around the op amp inputs, which degrade the square wave response to severe ringing, much less than full capability of the part.

This is shown in Figure 7-132 in the center and right pictures, respectively. The voltage follower circuit on the left shows the additional capacitances as inherent to the prototype board. With this test circuit and corre-

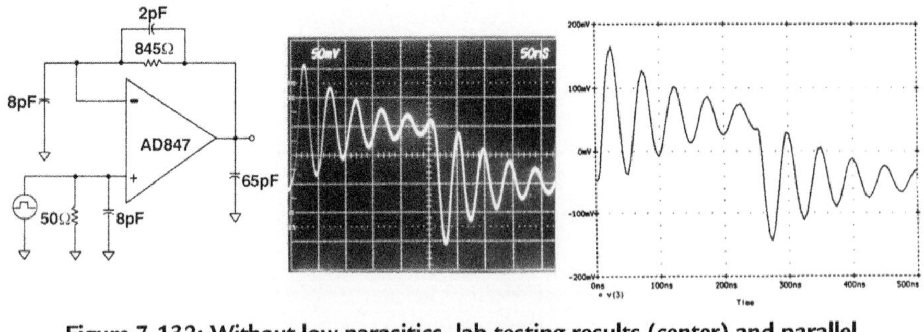

Figure 7-132: Without low parasitics, lab testing results (center) and parallel simulation (right) still show convergence—with a poorly damped response

sponding analysis, there was (initially) no agreement between the poor lab test, and the parallel SPICE test. However, when the relevant PCB parasitic capacitances are included in the SPICE file, then the simulation results do agree with the real circuit, as noted in the right picture.

This example illustrates several key points. One, PCB parasitics can easily make a high speed circuit behave much differently from a simplistic SPICE analysis. Secondly, when the SPICE netlist is adjusted to more reasonably reflect the parasitic elements of a PCB, the simulation results do compare with the actual lab test. Finally, a point that should be obvious, a clean PCB layout with minimal parasitics is critically important to high speed designs. To put this in a broader perspective, op amps of today are capable of operating to 1 GHz or more.

Another interesting point is that the simulation can be used as a rough measure of the PCB layout design. If the simulation, without any parasitics, agrees with the PCB, there is a reasonable assurance that PCB is laid out well.

Parasitic PCB elements are not the only area that may cause differences between the simulation and the breadboard. A circuit may exhibit nonlinear behavior during power-on that will cause a device to lock up. Or, a device may oscillate due to insufficient power supply decoupling or lead inductance. SPICE circuits need *no* bypassing, *but real world ones always do*. It is, practically speaking, impossible to anticipate all normal or abnormal operating conditions to which an amplifier might be subjected.

Thus, it is always important that circuits be breadboarded and thoroughly checked in the lab. Careful forethought in these stages of design helps minimize any unknown problems from showing up when the final PCBs are manufactured.

Simulation Speeds the Design Cycle

Simulation is very effective in the initial design phase, to try out different ideas and circuit configurations. When a circuit topology has been decided upon and tested in SPICE, a breadboard can be built. If the simulation was done carefully, the breadboard has good likelihood of working correctly without significant modifications.

When the simulation and the actual results correlate, the circuit can easily be altered in SPICE to perform many different types of analysis. For example, it is much easier to try to optimize the circuit while working within SPICE, as opposed to repeatedly modifying a breadboard. Quick substitutions of the op amps and components can be made in SPICE and the results immediately viewed.

Worst-case and sensitivity analyses are also done in SPICE much easier than on paper, and with multiple SPICE runs, the sensitivity to a certain parameter can be determined. Consider for example an analysis of a multistage active filter, for all possible combinations of component values. This is a nightmare, if not impossible, either by hand or in the lab, but valid results for response extremes can be obtained relatively easily via a SPICE Monte Carlo option, providing greater design confidence.

Some general SPICE-related points are useful towards an overall healthy perspective on this, as shown in Figure 7-133.

- Quickly Check Circuit Ideas
- Eases Circuit Optimization
- Allows Component Alteration for Worst Case and Sensitivity Analyses
- Allows Quick Comparison of Different Op Amps

Figure 7-133: Some useful points on using SPICE simulations

While simulation cannot reasonably be allowed to replace breadboarding, the two can and should be used together, to increase the efficiency of a design cycle.

SPICE Support

A variety of industry vendors offer SPICE analysis packages for various computer platforms, including the PC. The first of these and among the most popular is PSpice®, a commercial program that now includes allied packages for both schematic capture and PCB layout (see Reference 19). In addition, many vendors also offer low or no cost limited capability student versions of their SPICE programs.

Model Support

The ADSpice model library is available in several different forms. Included within it are models of several IC device types, in addition to the op amps discussed above. These are for in amps, analog multipliers, voltage references, analog switches, analog multiplexers, matched transistors, and buffers. Individual op amp models are available as listings on many data sheets. Electronic ASCII text files of the model library are found from either the ADI website (see References), the Analog Devices Literature Center via 1-800-ANALOGD (1-800-262-5643), or on the ADI support CD.

Acknowledgments:

There have been numerous model authors of SPICE macromodels for the ADI library. These include Derek Bowers, Eberhard Brunner, Joe Buxton, Vic Chang, Bob Day, Wes Freeman, Adolfo Garcia, Antonio Germano, John Hayes, John McDonald, Troy Murphy, Al Neves, Steve Reine, Bill Tolley, Tim Watkins, and James Wong.

Breadboard and Prototyping Techniques

A basic principle of a breadboard or a prototype structure is that it is a *temporary* one, designed to test the performance of an electronic circuit or system. By definition it must therefore be easy to modify, particularly so for a breadboard.

There are many commercial prototyping systems, but unfortunately for the analog designer, almost all of them are designed for prototyping *digital* systems. In such environments, noise immunities are hundreds of millivolts or more. Prototyping methods commonly used include noncopper-clad Matrix board, Vectorboard, wire-wrap, and plug-in breadboard systems. Quite simply, these all are unsuitable for high performance or high frequency analog prototyping, because of their excessively high parasitic resistance, inductance, and capacitance levels. Even the use of standard IC sockets is inadvisable in many prototyping applications (more on this below).

Figure 7-134 summarizes a number of key points on selecting a useful analog breadboard and/or prototyping system, which follows.

- Always Use a Ground Plane for Precision or High Frequency Circuits
- Minimize Parasitic Resistance, Capacitance, and Inductance
- If Sockets Are Required, Use "Pin Sockets" ("Cage Jacks")
- Pay Equal Attention to Signal Routing, Component Placement, Grounding, and Decoupling in Both the Prototype and the Final Design
- Popular Prototyping Techniques:
  - Freehand "Deadbug" Using Point-to-Point Wiring
  - "Solder-mount"
  - Milled PC Board from CAD Layout
  - Multilayer Boards: Double-Sided with Additional Point-to-Point Wiring

Figure 7-134: A summary of analog prototyping system key points

One of the more important considerations in selecting a prototyping method is the requirement for a large-area ground plane. This is required for high frequency circuits as well as low speed precision circuits, especially when prototyping circuits involving ADCs or DACs. The differentiation between *high speed* and *high precision* mixed-signal circuits is difficult to make. For example, 16+-bit ADCs (and DACs) may operate on high speed clocks (>10 MHz) with rise and fall times of less than a few nanoseconds, while the effective throughput rate of the converters may be less than 100 kSPS. Successful prototyping of these circuits requires that equal (and thorough) attention be given to good high speed and high precision circuit techniques.

Deadbug Prototyping

A simple technique for analog prototyping uses a solid copper-clad board as a ground plane (see References 20 and 21). In this method, the ground pins of the ICs are soldered directly to the plane, and the other components are wired together above it. This allows HF decoupling paths to be very short indeed. All lead lengths should be as short as possible, and signal routing should separate high level and low level signals. Connection wires should be located close to the surface of the board to minimize the possibility of stray inductive coupling. In most cases, 18-gauge or larger insulated wire should be used. Parallel runs should not be "bundled" because of possible coupling. Ideally the layout (at least the relative placement of the components on the board) should be similar to the layout to be used on the final PCB. This approach is often referred to as *deadbug prototyping*, because the ICs are often mounted upside down with their leads up in the air (with the exception of the ground pins, which are bent over and soldered directly to the ground plane). The upside-down ICs look like deceased insects, hence the name.

Figure 7-135 shows a hand-wired "deadbug" analog breadboard. This circuit uses two high speed op amps and, in fact, gives excellent performance in spite of its lack of esthetic appeal. The IC op amps are mounted upside down on the copper board with the leads bent over. The signals are connected with short point-to-point wiring. The characteristic impedance of a wire over a ground plane is about 120 Ω, although this may vary as much as ±40%, depending on the distance from the plane. The decoupling capacitors are connected directly from the op amp power pins to the copper-clad ground plane. When working at frequencies of several hundred MHz, it is a good idea to use only one side of the board for ground. Many people drill holes in the board and connect the sides together by soldering short pieces of wire. If care isn't taken, this may result in unexpected ground loops between the two sides of the board, especially at RF frequencies.

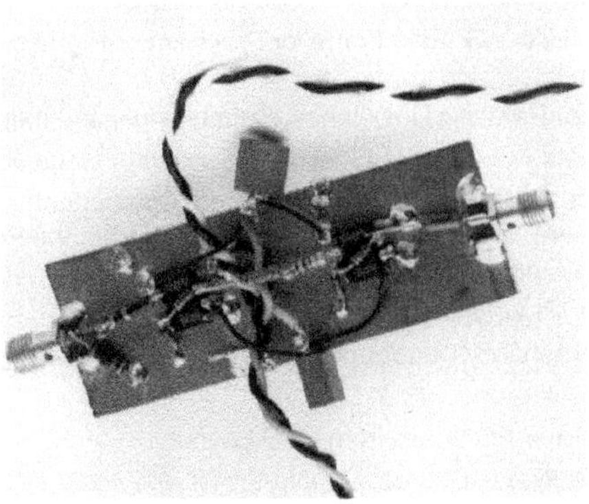

Figure 7-135: A "deadbug" analog breadboard

Pieces of copper-clad board may be soldered at right angles to the main ground plane to provide screening, or circuitry may be constructed on both sides of the board (with through-hole connections) with the board itself providing screening. For this, the board will need corner standoffs to protect underside components from being crushed.

When the components of a breadboard of this type are wired point-to-point in the air (a type of construction strongly advocated by Bob Pease (see Reference 21) and sometimes known as "bird's nest" construction) there is always the risk of the circuitry being crushed and resulting short-circuits. Also, if the circuitry rises high above the ground plane, the screening effect of the ground plane is diminished, and interaction between different parts of the circuit is more likely. Nevertheless, the technique is very practical and widely used because the circuit may easily be modified (this of course assumes the person doing the modifications is adept with soldering techniques).

Another prototype breadboard variation is shown in Figure 7-136. Here the single-sided copper-clad board has pre-drilled holes on 0.1" centers (see Reference 22). Power buses are used at the top and bottom of the board. The decoupling capacitors are used on the power pins of each IC. Because of the loss of copper area due to the predrilled holes, this technique does not provide as low a ground impedance as a completely covered copper-clad board of Figure 7-135, so be forewarned.

Figure 7-136: A "deadbug" prototype using 0.1" predrilled single-sided, copper-clad printed board material

In a variation of this technique, the ICs and other components are mounted on the noncopper-clad side of the board. The holes are used as vias, and the point-to-point wiring is done on the copper-clad side of the board. Note that the copper surrounding each hole used for a via must be drilled out, to prevent shorting. This approach requires that all IC pins be on 0.1" centers. For low frequency circuits, low profile sockets can be used, and the socket pins then will allow easy point-to-point wiring.

Solder-Mount Prototyping

There is a commercial breadboarding system that has most of the advantages of the above techniques (robust ground, screening, ease of circuit alteration, low capacitance, and low inductance) and several additional advantages: it is rigid, components are close to the ground plane and, where necessary, node capacitances and line impedances can be easily calculated. This system is made by Wainwright Instruments and is available in Europe as "Mini-Mount" and in the USA (where the trademark "Mini-Mount" is the property of another company) as "Solder-Mount" (see References 23 and 24).

Solder-Mount consists of small pieces of PCB with etched patterns on one side and contact adhesive on the other. These pieces are stuck to the ground plane, and components are soldered to them. They are available in a wide variety of patterns, including ready-made pads for IC packages of all sizes from 8-pin SOICs to 64-pin DILs, strips with solder pads at intervals (which intervals range from 0.040" to 0.25", the range includes strips with 0.1" pad spacing which may be used to mount DIL devices), strips with conductors of the correct width to form microstrip transmission lines (50 Ω, 60 Ω, 75 Ω or 100 Ω) when mounted on the ground plane, and a variety of pads for mounting various other components. Self-adhesive tinned copper strips and rectangles (LO-PADS) are also available as tie-points for connections. They have a relatively high capacitance to ground and therefore serve as low inductance decoupling capacitors. They come in sheet form and may be cut with a knife or scissors.

The main advantage of Solder-Mount construction over "bird's nest" or "deadbug" is that the resulting circuit is far more rigid and, if desired, may be made far smaller (the latest Solder-Mounts are for surface-mount devices and allow the construction of breadboards scarcely larger than the final PCB, although it is generally more convenient if the prototype is somewhat larger). Solder-Mount is sufficiently durable that it may be used for small quantity production as well as prototyping.

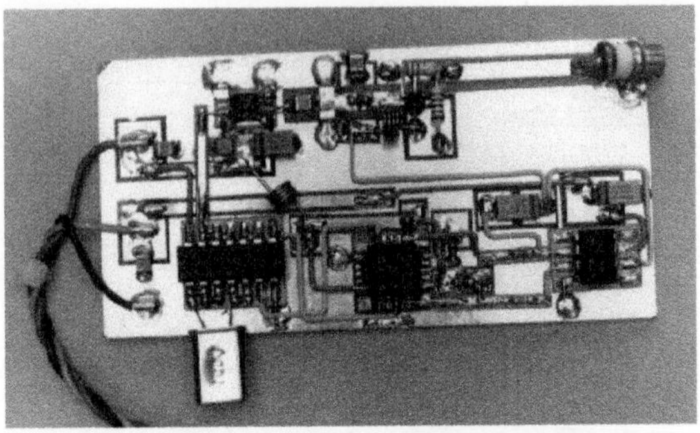

Figure 7-137: A "Solder-Mount" constructed prototype board

Figure 7-137 shows an example of a 2.5 GHz phase-locked-loop prototype, built with Solder-Mount techniques. While this is a high speed circuit, the method is equally suitable for the construction of high resolution low frequency analog circuitry.

A particularly convenient feature of Solder-Mount at VHF is the relative ease with which transmission lines can be formed. As noted earlier, if a conductor runs over a ground plane, it forms a microstrip transmission line. The Solder-Mount components include strips that form microstrip lines when mounted on a ground plane (they are available with impedances of 50 Ω, 60 Ω, 75 Ω, and 100 Ω). These strips may be used as transmission lines for impedance matching or, alternately, more simply as power buses. Note that glass fiber/epoxy PCB is somewhat lossy at VHF/ UHF, but losses will probably be tolerable if microstrip runs are short.

Milled PCB Prototyping

Both "deadbug" and "Solder-Mount" prototypes become tedious for complex analog circuits, and larger circuits are better prototyped using more formal layout techniques.

There is a prototyping approach that is but one step removed from conventional PCB construction, described as follows. This is to actually lay out a double-sided board, using conventional CAD techniques. PC-based software layout packages offer ease of layout as well as schematic capture to verify connections (see References 25 and 26). Although most layout software has some degree of autorouting capability, this feature is best left to digital designs. The analog traces and component placements should be done by hand, following the rules discussed elsewhere in this chapter. After the board layout is complete, the software verifies the connections per the schematic diagram net list.

Many designers find that they can make use of CAD techniques to lay out simple boards. The result is a pattern-generation tape (or Gerber file) which would normally be sent to a PCB manufacturing facility where the final board is made.

Rather than use a PCB manufacturer, however, automatic drilling and milling machines that accept the PG tape directly are available (see References 27 and 28). An example of such a prototype circuit board is shown in Figure 7-138 (top view).

These systems produce either single or double-sided circuit boards directly, by drilling all holes and using a milling technique to remove conductive copper, thus creating the required insulation paths and, finally, the finished prototype circuit board. The result can be a board functionally quite similar to a final manufactured double-sided PCB.

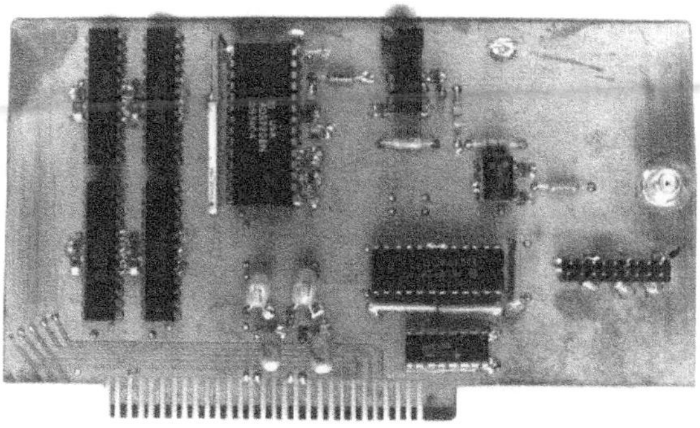

Figure 7-138: A milled circuit construction prototype board (top view)

However, it should be noted that a chief caveat of this method is that there is no "plated-through" hole capability. Because of this, any conductive "vias" required between the two layers of the board must be manually wired and soldered on both sides.

Minimum trace widths of 25 mils (1 mil = 0.001") and 12 mil spacing between traces are standard, although smaller trace widths can be achieved with care. The minimum spacing between lines is dictated by the size of the milling bit used, typically 10 to 12 mils.

A bottom-side view of this same milled prototype circuit board is shown in Figure 7-139. The accessible nature of the copper pattern allows access to the traces for modifications.

Perhaps the greatest single advantage of the milled circuit type of prototype circuit board is that it approaches the format of the final PCB design most closely. By its very nature, however, it is basically limited to only single or double-sided boards.

Figure 7-139: A milled circuit construction prototype board (bottom view)

Beware of Sockets

IC sockets can degrade the performance of high speed or high precision analog ICs. Although they make prototyping easier, even *low profile* sockets often introduce enough parasitic capacitance and inductance to degrade the performance of a high speed circuit. If sockets must be used, a socket made of individual *pin sockets* (sometimes called *cage jacks*) mounted in the ground plane board may be acceptable, as in Figure 7-140.

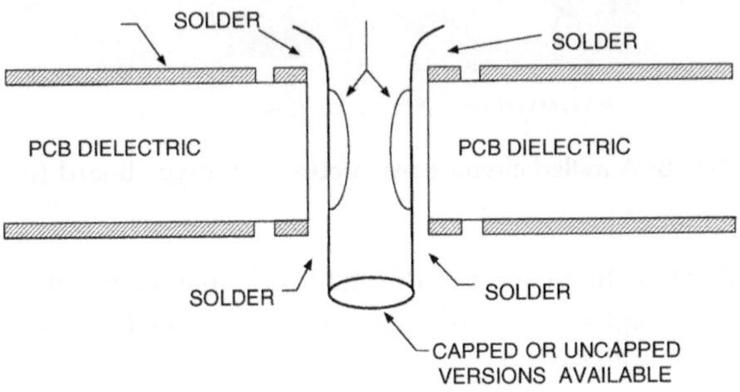

Figure 7-140: When necessary, use pin sockets for minimal parasitic effects

To use this technique, clear the copper (on both sides of the board) for about 0.5 mm around each ungrounded pin socket, Then solder the grounded socket pins to ground, on both sides of the board.

Both capped and uncapped versions of these pin sockets are available (AMP part numbers 5-330808-3, and 5-330808-6, respectively). The pin sockets protrude through the board far enough to allow point-to-point wiring interconnections.

Because of the spring-loaded gold-plated contacts within the pin socket, there is good electrical and mechanical connection to the IC pins. Multiple insertions, however, may degrade the performance of the pin socket, so this factor should be kept in mind.

Note also that the uncapped versions allow the IC pins to extend out the bottom of the socket. This feature leads to an additional useful function. Once a prototype using the pin sockets is working and no further changes are to be made the IC pins can be soldered directly to the bottom of the socket. This establishes a rugged, permanent connection.

Some Additional Prototyping Points

The prototyping techniques discussed so far have been limited to single or double-sided PCBs. Multilayer PCBs do not easily lend themselves to standard prototyping techniques. If multilayer board prototyping is required, one side of a double-sided board can be used for ground and the other side for power and signals. Point-to-point wiring can be used for additional runs which would normally be placed on the additional layers provided by a multilayer board. However, it is difficult to control the impedance of the point-to-point wiring runs, and the high frequency performance of a circuit prototyped in this manner may differ significantly from the final multilayer board.

Other difficulties in prototyping may occur with op amps or other linear devices having bandwidths greater than a few hundred megahertz. Small variations in parasitic capacitance (<1 pF) between the prototype and the final board can cause subtle differences in bandwidth and settling time.

Sometimes, prototyping is done with DIP packages, when the final production package is an SOIC. *This is not recommended.* At high frequencies, small package-related parasitic differences can account for different performance, between prototype and final PCB. To minimize this effect, always prototype with the final packages.

Evaluation Boards

Most manufacturers of analog ICs provide *evaluation boards,* usually at a nominal cost. These boards allow customers to evaluate ICs without constructing their own prototypes. Regardless of the product, the manufacturer has taken proper precautions regarding grounding, layout, and decoupling to ensure optimum device performance. Where applicable, the evaluation PCB artwork is usually made available free of charge, should a customer wish to copy the layout directly or make modifications to suit an application.

General-Purpose Op Amp Evaluation Boards

Evaluation boards can either be dedicated to a particular IC, or they can be general-purpose. With op amps the most universal linear IC, it is logical that evaluation boards be developed for them, to aid easy applications. However, it is also important that a good quality evaluation board avoid the parasitic effects discussed above. An example is the general-purpose dual amplifier evaluation board of in Figure 7-141 (see Reference 29).

Figure 7-141: A general-purpose op amp evaluation board allows fast, easy configuration of low frequency op amp circuits

This board uses pin sockets for any standard dual op amp pinout device, and a flexible set of component jumper locations allows it to be setup for inverting or noninverting amplifiers. Various gains can be configured by choice of the component values, in either ac- or dc-coupled configurations.

The card design provides signal coupling via BNC connectors at input and output. It also uses external lab power supplies, which are wired to the lug terminals at the top. The card does, however, contain local supply voltage decoupling and bypassing components.

These general-purpose boards are intended for medium to high precision uses at frequencies below 10 MHz, with moderate op amp input currents. For higher operating speeds, a dedicated, device-specific evaluation board is likely to be a better choice.

Dedicated Op Amp Evaluation Boards

In high speed/high precision ICs, special attention must be given to power supply decoupling. For example, fast slewing signals into relatively low impedance loads produce high speed transient currents at the power supply pins of an op amp. The transient currents produce corresponding voltages across any parasitic impedance that may exist in the power supply traces. These voltages, in turn, may couple to the amplifier output, because of the op amp's finite power supply rejection at high frequencies.

The AD8001 high speed current-feedback amplifier is a case in point, and a dedicated evaluation board is available for it. A bottom side view of this SOIC board is shown in Figure 7-142. A triple decoupling scheme was chosen, to ensure a low impedance ground path at all transient frequencies. Highest frequency transients are shunted to ground by dual 1000 pF/0.01 μF ceramic chip capacitors, located as close to the power supply pins as possible to minimize series inductance and resistance. With these surface-mount components, there is minimum stray inductance and resistance in the ground plane path. Lower frequency transient currents are shunted by the larger 10 μF tantalum capacitors.

Figure 7-142: A high speed op amp such as the AD8001 requires a dedicated evaluation board with suitable ground planes and decoupling (bottom view)

The input and output signal traces of this board are 50 Ω microstrip transmission lines, as can be noted towards the right and left. Gain-set resistors are chip-style film resistors, which have low parasitic inductance. These can be seen in the center of the photo, mounted at a slight diagonal.

Note also that there is considerable continuous ground plane area on both sides of the PCB. Plated-through holes connect the top and bottom side ground planes at several points, in order to maintain lowest possible impedance and best high frequency ground continuity.

Input and output connections to the card are provided via the SMA connectors as shown, which terminate the input/output signal transmission lines. The board's power connection from external lab supplies is made via solder terminals, which are seen at the ends of the broad supply line traces.

Some of these points are more easily seen in a topside view of the same card, which is shown in Figure 7-143. This AD8001 evaluation board is a noninverting signal gain stage, optimized for lowest parasitic capacitance. The cutaway area around the SOIC outline of the AD8001 provides lowest stray capacitance, as can be noted in this view.

In this view is also seen the virtually continuous ground plane and the multiple vias, connecting the top/bottom planes.

Figure 7-143: The AD8001 evaluation board uses a large area ground plane as well as minimal parasitic capacitance (top view)

Summary

In summary, good analog designers utilize as many tools as possible to ensure that the final system design performs correctly. The first step is the intelligent use of IC op amp and other macromodels, where available, to simulate the circuit. The second step is the construction of a prototype board to further verify the design, and to validate the simulation. The final PCB layout should then be based on the prototype layout as much as possible, with careful attention to parasitic effects.

References: Simulation, Breadboarding and Prototyping

1. L. W. Nagel, "SPICE2: A Computer Program to Simulate Semiconductor Circuits," May 1975, UCB/ERL M75/520, Univ. of California, Berkeley, CA, 94720.
2. A.Vladimirescu, K.Zhang, A.R.Newton, D.O.Pederson, "SPICE Version 2G User's Guide," August 1981, Department of Electrical Engineering and Computer Sciences, Univ. of California, Berkeley, CA, 94720.
3. Mark Alexander, Derek Bowers, "SPICE-Compatible Op Amp Macromodels," **EDN**, February 15, 1990 and March 1, 1990 (available as Analog Devices, Inc. AN138).
4. Joe Buxton, "Analog Circuit Simulation," Chapter 13 of **Amplifier Applications Guide**, 1992, Analog Devices, Inc., Norwood, MA, ISBN 0-916550-10-9.
5. Andrei Vladimerescu, **The SPICE Book**, John Wiley & Sons, New York, 1994, ISBN 0-471-60926-9.
6. "Development of an Extensive SPICE Macromodel for "Current-Feedback Amplifiers," National Semiconductor AN-840, July 1992.
7. David Hindi, "A SPICE-Compatible Macromodel for CMOS Operational Amplifiers," National Semiconductor AN-856, September 1992.
8. Boyle, et al, "Macromodelling of Integrated Circuit Operational Amplifiers," **IEEE Journal of Solid-State Circuits**, Vol. SC-9, no.6, December 1974.
9. Derek Bowers, Mark Alexander, Joe Buxton, "A Comprehensive Simulation Macromodel for 'Current Feedback' Operational Amplifiers," **IEE Proceedings**, Vol. 137, Pt. G, # 2, April 1990.
10. Joe Buxton, "OP27 Op Amp Macromodel, Rev B," Analog Devices, Inc., SPICE model library, December 1990, www.analog.com.
11. Antonio Germano, "AD817 Op Amp Macromodel, Rev A," Analog Devices, Inc., SPICE model library, November 1992, www.analog.com.
12. Antonio Germano, "OP295 Op Amp Macromodel, Rev B," Analog Devices, Inc., SPICE model library, February 1995, www.analog.com.
13. Antonio Germano, "OP284 Op Amp Macromodel, Rev B," Analog Devices, Inc., SPICE model library, November 1995, www.analog.com.
14. Steve Reine, "AD8031A Op Amp Macromodel, Rev C," Analog Devices, Inc., SPICE model library, August 1996, www.analog.com.
15. Steve Reine, "AD823AN Op Amp Macromodel, Rev C," Analog Devices, Inc., SPICE model library, April 1997, www.analog.com.
16. John Hayes, "AD8051 Op Amp Macromodel, Rev 0," Analog Devices, Inc., SPICE model library, September 1998, www.analog.com.
17. Troy Murphy, "AD8552 Op Amp Macromodel, Rev 1.0," Analog Devices, Inc., SPICE model library, July, 1999, www.analog.com.
18. John Hayes, Tim Watkins, "AD623 In Amp Macromodel, Rev B," Analog Devices, Inc., SPICE model library, September 2000, www.analog.com.
19. PSpice Simulation software, 1-888-671-9500, www.pspice.com.
20. Jim Williams, "High Speed Amplifier Techniques," Linear Technology AN-47, August, 1991.

21. Robert A. Pease, **Troubleshooting Analog Circuits**, Butterworth-Heinemann, 1991, ISBN 0-7506-9184-0.

22. Vector Electronic Company, 12460 Gladstone Ave., Sylmar, CA 91342, Tel. 818-365-9661.

23. Wainwright Instruments Inc., 69 Madison Ave., Telford, PA, 18969-1829, 215-723-4333.

24. Wainwright Instruments GmbH, Widdersberger Strasse 14, DW-8138 Andechs-Frieding, Germany. +49-8152-3162.

25. PADS Software, Advanced CAM Technologies, Inc., 16450 Los Gatos Blvd., Suite 110, Los Gatos, CA 95032, www.ecam.com/.

26. ACCEL Technologies, Inc., 17140 Bernardo Center Drive, Suite 100, San Diego, CA 92128, www.acceltech.com/.

27. LPKF Laser & Electronics, 28220 SW Boberg Rd., Wilsonville, OR 97020, 800-345-LPKF or 503-454-4200, www.lpkfcadcam.com.

28. T-Tech, Inc., 5591-B New Peachtree Road, Atlanta, GA, 34341, 800-370-1530 or 770-455-0676, www.T-Tech.com.

29. Adolfo Garcia, "Evaluation Boards for Single, Dual and Quad Operational Amplifiers," **Analog Devices AN398**, January 1996.

CHAPTER 8

Op Amp History

CHAPTER 8

Op Amp History

Walt Jung

The theme of this chapter is to provide the reader with a more comprehensive *historical background of the operational amplifier* (op amp for short—see below). This story begins back in the vacuum tube era and continues until today (2004). While most of today's op amp users are probably somewhat familiar with integrated circuit (IC) op amp history, considerably fewer are familiar with the non-IC solid-state op amp. Even more likely, very few are familiar with the origins of the op amp in vacuum tube form, even if they are old enough to have used some of those devices in the '50s or '60s. This introduction addresses these issues with a narrative of not only how op amps originated and evolved, but also what key factors gave rise to the op amp's origin in the first place.[1]

A developmental background of the op amp begins early in the twentieth century, starting with certain fundamental beginnings. Of these, there were two key inventions very early in the century. The first was not an amplifier, but a two-element vacuum-tube-based rectifier, the "Fleming diode," by J. A. Fleming, patented in 1904 (see Reference 1). This was an evolutionary step beyond Edison's filament-based lamp, by virtue of the addition of a *plate* electrode, which (when positively biased) captured electrons emitted from the filament (*cathode*). Since this device passed current in one direction only, it performed a rectification function. This patent was the culmination of Fleming's earlier work in the late years of the nineteenth century.

A second development (and one more germane to amplification) was the invention of the three-element triode vacuum tube by Lee De Forest, the "Audion," in 1906. This was the first active device capable of signal amplification (see Reference 2). De Forest added a control grid electrode, between the diode filament and plate, and an amplifier device was born. While these first tubes of the twentieth century had their drawbacks, the world of modern electronics was being born, and more key developments were soon to follow.

For op amps, the invention of the feedback amplifier principle at Bell Telephone Laboratories (Bell Labs) during the late 1920s and early '30s was truly an enabling development. This landmark invention led directly to the first phase of vacuum tube op amps, a general-purpose form of feedback amplifier using vacuum tubes, beginning in the very early 1940s and continuing through the World War II years.

After World War II there was a transition period as vacuum tube op amps were improved and refined, at least in circuit terms. But these amplifiers were fundamentally large, bulky, power-hungry devices. So, after a decade or more, vacuum tube op amps began to be replaced by miniaturized solid-state op amps in the 1950s and 1960s.

A final major transitional phase of op amp history began with the development of the first IC op amp, in the mid 1960s. Once IC technology became widely established, things moved quickly through the latter of the twentieth century years, with milestone after milestone of progress being made in device performance.

---

[1] Note—this chapter of the book is not necessarily required for the use of op amps, and can be optionally skipped. Nevertheless, it should offer interesting background reading, as it provides a greater appreciation of current devices once their beginnings are more fully understood.

A Definition for the Fledgling Op Amp

Although it may seem inappropriate at this point in the book to define what an op amp was in those early days, it is necessary to do so, albeit briefly, because what is commonly known today as an op amp is different in some regards from the very first op amps. The introductory section of Chapter 1, where the discussion is more closely oriented around today's op amp definition, supplements the meaning below.

The very first op amps were not even called such, nor were they even called "operational amplifiers." The naming of the device came after the war years, in 1947.

For this historical discussion, it may be more clear to call one of these first op amps a *general-purpose, dc-coupled, high gain, inverting feedback amplifier*. This of course is a loose definition, but it nevertheless fits what transpired.

- *General-purpose* may be interpreted to mean that such an amplifier (or multiple amplifiers) operates on bipolar power supplies, with input and output signal ranges centered around 0V (ground).
- *Dc-coupled* response implies that the signals handled include steady-state or dc potentials, as well as ac signals.
- *High gain* implies a magnitude of dc gain in excess of 1000 × (60 dB) or more, as may be sufficient to make system errors low when driving a rated load impedance.
- *Inverting mode operation* means that this feedback amplifier had, in effect, one signal input node, with the signal return being understood as ground or common. Multiple signals were summed at this input through resistors, along with the feedback signal, via another resistor. *Note that this single-ended operating mode is a major distinction from today's differential input op amps.* Operation of these first feedback amplifiers in only a single-ended mode was, in fact, destined to continue for many years before differential input operation became more widespread.
- *A feedback amplifier* of this type could be used in a variety of ways, dependent upon the nature of the feedback element used with it. This capability of satisfying a variety of applications was later to give rise to the name.

So, given this background, op amp history can now be explored.

Introduction

Setting the Stage for the Op Amp

Op amps are high gain amplifiers, and are used almost invariably with overall loop feedback. The principle of the feedback amplifier has to rank as one of the more notable developments of the twentieth century—right up there with the automobile or airplane for breadth of utility and general value to engineering. Most importantly, such feedback systems, although originally conceived as a solution to a communications problem, operate today in more diverse situations. This is a clear tribute to the concept's fundamental value.

Today the application of negative feedback is so common that it is often taken for granted. But this wasn't always the case. Working as a young Western Electric Company engineer on telephone channel amplifiers, Harold S. Black first developed feedback amplifier principles. Note that this was far from a brief inspirational effort, or narrow in scope. In fact, it took some nine years after the broadly written 1928 patent application, until the 1937 issuance (see Reference 3). Additionally, Black outlined the concepts in a **Bell System Technical Journal** article, and, much later, in a 50th anniversary piece where he described the overall timeline of these efforts (see References 4 and 5).

Like circumstances surrounding other key inventions, there were others working on negative feedback amplifier applications. One example would be Paul Voigt's mid-1920s work (see References 6 and 7).[1] The prolific British inventor Alan Blumlein did 1930s feedback amplifier work, using it to control amplifier output impedance (see Reference 8).[2] Finally, a research group at N. V. Philips in the Netherlands is said to have been exploring feedback amplifiers within roughly the same time frame as Black (late 20s to early 30s). In 1937 B. D. H. Tellegen published a paper on feedback amplifiers, with attributions to K. Posthumus and Black (see References 9 and 10).[3] In Tellegen's paper are the same equations as those within Black's (substituting A for Black's μ).

It isn't the purpose here to challenge Black's work, rather to note that sometimes overlapping but independent parallel developments occur, even for major inventions. Other examples of this will be seen shortly, in the development of differential amplifier techniques. In the long run, a broad-based, widely accepted body of work tends to be seen as the more significant effort. In the case of Black's feedback amplifier, there is no doubt that it is a most significant effort. It is also both broad-based and widely accepted.

There are also many earlier *positive* feedback uses; a summary is found in Reference 11.

---

1 Some suggest Paul Voigt as the true feedback amplifier inventor, not Black (see Ref. 6, 7). Examination of Voigt's UK patent 231,972 fails to show a feedback amplifier theory comparable to Black's detailed exposition of Ref. 3 and 4. In fact, there are no equations presented to describe Voigt's system behavior.

2 Blumlein's UK patent 425,553 is focused on controlling amplifier output impedance through voltage and/or current feedback, not addressing in detail the broader ramifications of feedback.

3 Examination of UK patent 323,823 fails to find reference to K. Posthumus, apparently a practice with N. V. Philips UK patents of that period. The patent does show a rudimentary feedback amplifier, but unfortunately the overall clarity is marred by various revisions and corrections, to both text and figures.

Black's Feedback Amplifier

The basis of Black's feedback amplifier lies in the application of a portion of the output back to the input, so as to reduce the overall gain. When properly applied, this provides the resultant amplifier with characteristics of enhanced gain stability, greater bandwidth, lower distortion, and usefully modified stage input and output impedance(s).

A block diagram of Black's basic feedback amplifier system is shown in Figure 8-1 below. Note that Black's "μ" for a forward gain symbol is today typically replaced by "A." As so used, the feedback network β defines the overall transfer expression of the amplifier. Thus a few passive components, typically just resistors or sometimes reactive networks, set the gain and frequency response characteristics of this system.

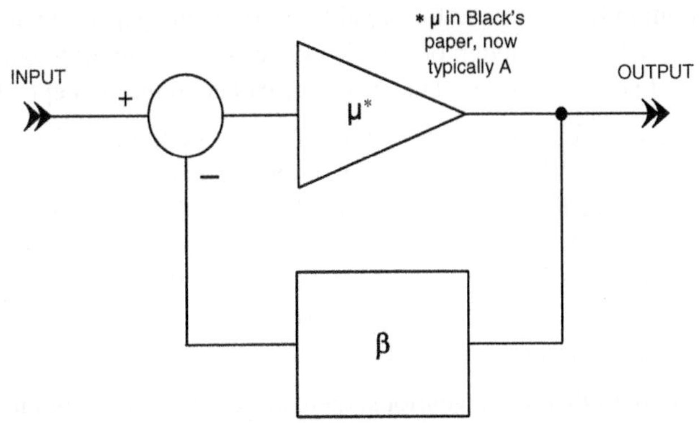

Figure 8-1: A block diagram of Black's feedback amplifier, comprised of a forward gain "μ" and a feedback path of "β"

At the time Black's work was initiated, the problem he faced was how to make practical a series signal connection of hundreds (if not thousands) of telephone system repeater amplifiers using directly heated triode tubes. The magnitude of this problem becomes obvious when it is considered that each amplifier alone couldn't be held more stable to less than 1 dB of gain variation, and even under the best of conditions, the stage distortion was unacceptable.

Black's feedback amplifier invention led not just to better repeater amplifiers for Western Electric, but to countless millions of other widely varying applications. Almost every op amp application ever conceived uses feedback. So, given the fact that modern op amp types number in the dozens (in individual models, many thousands), it isn't hard to appreciate the importance feedback principles take on for today's designs.

A significant reason that Black's feedback concept took root and prospered wasn't simply because it was a useful and sound idea. That it was, but it was also *different*, and many experienced engineers fought the idea of "throwing away gain." However, Black did have help in selling the radically new concept, help that was available to few other inventors. By this help what is meant is that he had the full backing of the Bell Telephone System, and all that this implied towards forging and promoting a new technical concept. An interesting narrative of the feedback amplifier's development and the interplay of Black and his coworkers can be found in David Mindell's paper, "Opening Black's Box: Rethinking Feedback's Myth of Origin" (see Reference 12).

The 1930s and 1940s at Bell Labs could very well be regarded as golden years. They produced not just Black's feedback amplifier, but also other key technical developments that expanded and supported the amplifier. This support came from some of the period's finest engineers—not just the finest Bell Labs engineers, but the world's finest.

To quote Black's own words on the Bell Labs support activity related to his landmark invention, *"Within a few years, Harry Nyquist would publish his generalized rule for avoiding instability in a feedback amplifier, and Hendrick W. Bode would spearhead the development of systematic techniques of design whereby one could get the most out of a specified situation and still satisfy Nyquist's criterion."* (see Reference 5).

The feedback amplifier papers and patents of Harry Nyquist and Hendrick Bode (see References 13 and 14), taken along with the body of Black's original work, form solid foundations for modern feedback amplifier design. Bode later published a classic feedback amplifier textbook (see Reference 15). Later on, he also gave a talk summarizing his views on the feedback amplifier's development (see Reference 16).

In addition to his famous stability criteria, Nyquist also supplied circuit-level hardware concepts, such as a patent on direct-coupled amplifier interstage coupling (see Reference 17). This idea was later to become a standard coupling method for vacuum tube op amps.

Outside Bell Labs, other engineers were also working on feedback amplifier applications of their own, affirming the concept in diverse practical applications. Frederick Terman was among the first to publicize the concept for ac feedback amplifiers, in a 1938 article (see Reference 18).

For single-ended signal path dc amplifiers, there were numerous landmark papers during the World War II period. Stewart Miller's 1941 article offered techniques for high and stable gain with response to dc (see Reference 19). This article introduced what later became a standard gain stabilization concept, called "cathode compensation," where a second dual triode section is used for desensitization of heater voltage variations. Ginzton's 1944 amplifier article employed Miller's cathode compensation, as well as Nyquist's level-shifting method (see Reference 20). The level shifter is attributed to Brubaker (who apparently duplicated Nyquist's earlier work). Artzt's 1945 article surveys various dc amplifier techniques, with emphasis on stability (see Reference 21).

After World War II, the MIT Radiation Laboratory textbook series documented many valuable electronic techniques, including a volume dedicated to vacuum tube amplifiers. The classic Valley-Wallman volume number 18 is not only generally devoted to amplifiers, it includes a chapter on dc amplifiers (see Reference 22). While this book doesn't discuss op amps by name, it does include dc feedback circuitry examples. Op amps did exist, and had even been named as of 1947, just prior to the book's publication.

References: Introduction

(Note: Appended annotations indicate relevance to op amp history.)

Early Vacuum Tube and Feedback Amplifier Developments

1. J. A. Fleming, "Instrument for Converting Alternating Electric Currents into Continuous Currents," **US Patent 803,684**, filed April 19, 1905, issued Nov. 7, 1905. See also UK Patent 24,850, filed Nov. 16, 1904. *(The 'Fleming diode' or the first vacuum tube rectifier.)*
2. Lee de Forest, "Device for Amplifying Feeble Electrical Currents," **US Patent 841,387**, filed October 25, 1906, issued January 15, 1907. *(The triode vacuum tube, or 'Audion,' the first amplifying device.)*
3. H. S. Black, "Wave Translation System," **US Patent 2,102,671**, filed August 8, 1928, issued December 21, 1937. *(The basis of feedback amplifier systems.)*
4. H. S. Black, "Stabilized Feedback Amplifiers," **Bell System Technical Journal**, Vol. 13, No. 1, January 1934, pp. 1–18. *(A practical summary of feedback amplifier systems.)*
5. Harold S. Black, "Inventing the Negative Feedback Amplifier," **IEEE Spectrum**, December, 1977. *(Inventor's 50th anniversary story on the invention of the feedback amplifier.)*
6. Geoffrey Horn, "Voigt, not Black," **Stereophile**, Letters, April 1998, pp. 18, 21.
7. Paul G. A. H. Voigt, "Improvements in or Relating to Thermionic Amplifying Circuits for Telephony," **UK Patent 231,972**, filed January 29, 1924, issued April 16, 1925. *(A motional feedback loudspeaker/amplifier system.)*
8. A. D. Blumlein, "Improvements in and relating to Thermionic Valve Amplifiers," **UK Patent 425,553**, filed Sept. 8, 1933, issued March 18, 1935. *(Use of feedback to control amplifier output impedance.)*
9. B. D. H. Tellegen, "Inverse Feedback," **Philips Technical Review**, Vol. 2, No. 10, October, 1937. *(Another feedback amplifier development, generally paralleling Black's.)*
10. "Improvements in or Relating to Arrangements for Amplifying Electrical Oscillations," **UK Patent 323,823**, filed October 18, 1928, issued January 16, 1930 (original filing), filed July 18, 1929, final approval January 1938 and February 1939 (amended filing). *(A simple one stage feedback amplifier system.)*
11. D. G. Tucker, "The History of Positive Feedback," **Radio and Electronic Engineer**, Vol. 42, No. 2, February 1972, pp. 69–80.
12. David A. Mindell, "Opening Black's Box: Rethinking Feedback's Myth of Origin," **Technology and Culture**, Vol. 41, July, 2000, pp. 405–434. *(A perspective discussion of the inter-related events and cultures surrounding the feedback amplifier's invention.)*
13. Harry Nyquist, "Regeneration Theory," **Bell System Technical Journal**, Vol. 11, No. 3, July, 1932, pp. 126–147. See also: "Regenerative Amplifier," **US Patent 1,915,440**, filed May 1, 1930, issued June 27, 1933 *(The prediction of feedback amplifier stability by means of circular gain-phase plots.)*
14. Hendrick Bode, "Relations Between Attenuation and Phase In Feedback Amplifier Design," **Bell System Technical Journal**, Vol. 19, No. 3, July, 1940. See also: "Amplifier," **US Patent 2,123,178**, filed June 22, 1937, issued July 12, 1938. *(The prediction of feedback amplifier stability by means of semi-log gain-phase plots.)*
15. Hendrick Bode, **Network Analysis and Feedback Amplifier Design**, Van Nostrand, 1945. *(Bode's classic text on network analysis, as it relates to the design of feedback amplifiers.)*

16. Hendrick Bode, "Feedback—the History of an Idea," **Proceedings of the Symposium on Active Networks and Feedback Systems**, Polytechnic Press, 1960. Reprinted within **Selected Papers on Mathematical Trends in Control Theory**, Dover Books, 1964. *(A perspective historical summary of the author's thoughts on the development of the feedback amplifier.)*

17. Harry Nyquist, "Distortionless Amplifying System," **US Patent 1,751,527**, filed November 24, 1926, issued March 25, 1930. *(A means of direct-coupling multiple amplifier stages via resistance networks for interstage coupling.)*

18. F. E. Terman, "Feedback Amplifier Design," **Electronics**, January 1937, pp. 12–15, 50. *(Some practical ac-coupled topologies for implementing feedback amplifiers.)*

19. Stewart E. Miller, "Sensitive DC Amplifier with AC Operation," **Electronics**, November, 1941, pp. 27–31, 106–109. *(Design example of a stable, high-gain direct-coupled amplifier including 'cathode-compensation' against variations in filament voltage, use of glow tube inter-stage coupling, and a stable line-operated DC supply.)*

20. Edward L. Ginzton, "DC Amplifier Design Techniques," **Electronics**, March 1944, pp. 98–102. *(Various design means for improving direct-coupled amplifiers.)*

21. Maurice Artzt, "Survey of DC Amplifiers," **Electronics**, August, 1945, pp. 112–118. *(Survey of direct-coupled amplifier designs, both single-ended and differential, with emphasis on high stability.)*

22. George E. Valley, Jr., Henry Wallman, **Vacuum Tube Amplifiers**, MIT Radiation Labs Series No. 18, McGraw-Hill, 1948. *(A classic WWII Radiation Lab development team textbook. Chapter 11, by John W. Gray, deals with direct-coupled amplifiers.)*

Vacuum Tube Op Amps

Development of Differential Amplifier Techniques

While amplifiers using both feedback and nonfeedback topologies were being refined in the late 1930s and throughout the 1940s, there were some very interesting developments within the realm of differential amplifiers.

Today's op amps utilize differential topologies to a high degree, but the reader should understand that this wasn't universally so back in the days of vacuum tube amplifiers. In fact, vacuum tube op amp topologies that fully utilized differential techniques never really became well established before the breed began dying off. Nevertheless, it is still a useful thing to examine some key differential amplifier publications up through about 1950, at which point it represented a maturing of the art. The fully differential, defined gain precision dc amplifier was, of course, the forerunner of what we know today as the *instrumentation amplifier* (see Chapter 2 of this book).

The earliest vacuum tube differential amplifiers were reported well back in the 1930s, and evolved steadily over the next 15–20 years. Many of these authors addressed the problems of low level instrumentation amplifier circuitry used in obtaining signals from living tissue, thus the apparatus involved was often called a "biological amplifier."

One of the early authors in this field was B. H. C. Matthews, writing on a special differential input amplifier in 1934 (see Reference 1). Matthews' amplifier did indeed have differential inputs, but since the common cathodes were tied directly to the power supply common, it wasn't optimized towards minimizing response to common-mode (CM) inputs. Note that in those days CM signals were often referred to as *push-push* signals, to denote signals in-phase at both inputs.

Alan Blumlein's UK patent 482,470 of 1936 went a step further in this regard, by biasing the common-cathodes of a differential pair through a common resistance to ground (see dual triodes within Figure 2 of Reference 2). Blumlein's patent was concerned with wideband signals, not biological ones, using ac-coupling. Nevertheless, it was a distinct improvement over the Matthews amplifier, since it provided bias conditions more amenable to CM signal rejection.

In 1937 Franklin Offner discussed a variety of differential amplifiers, and among them is found one similar to Blumlein's configuration (see Figure 3 of Reference 3). Like those of Blumlein, Offner's circuits also used ac-coupling. A useful technique that appears in this paper is the use of *common-mode feedback* to increase CM rejection (Figure 4 of Reference 3). To enable this, a CM sample from a downstream stage is feedback to an earlier stage. This feedback decreases the CM gain, and thus improves CM rejection.

Otto Schmitt discussed a common cathode, cathodes-to-common dual pentode circuit in 1937 (see Reference 4). This circuit, while novel in the operation of the pentode screens, didn't minimize response to CM input signals (similar to the Matthews circuit, above).

In 1938 J. F. Toennies discussed what might be the first form of what has subsequently come to be known as the *long-tailed pair* (see Reference 5). In this form of differential input amplifier, the push-pull input signals are applied to the dual grids of the stage, and the common cathodes are returned to a high negative voltage, through a high value common resistance. Toennies' fundamental circuit (Figure 1 of Reference 5) used dual triodes with a plate supply of 135 V, and a cathode bias supply of –90 V.

The action of the large value cathode resistance biased to a high negative voltage acts to optimize the differential coupling of the stage, while at the same time minimizing the CM response, as noted in Figure 8-2 below. This may intuitively be appreciated by considering the effect of the large cathode resistance to a high negative voltage $-V_S$, as in B, versus the simple cathode-coupled pair as in A. In A, the cathode resistance R_K is returned to ground, the same point common to the grids (the return for the $+V_S$ supply).

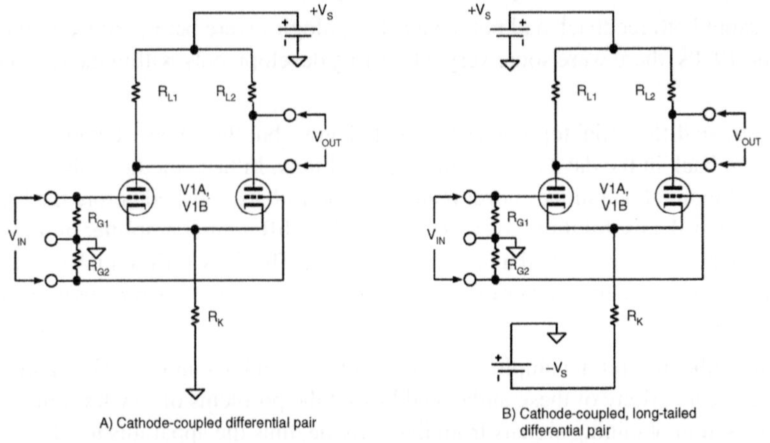

Figure 8-2: A comparison of simple differential pair biasing in A (left) and long-tailed biasing as in B (right)

The constant current action of the long-tailed biasing shown in 8-2B tends to minimize response to CM inputs (while not impairing differential response). Later on, some more advanced designs were even to go as far as using a pentode tube for the "long tail" common-cathode bias, capitalizing on a pentode's high incremental resistance.

In 1938 Otto Schmitt also discussed a long-tailed pair form of amplifier (see Reference 6). The context of his discussion was not so much aimed towards optimizing CM rejection, but rather using such a stage as a phase inverter. With one input grid of such a stage grounded, and the opposite grid driven in single-ended fashion, out-of-phase signals result with equal plate load resistors. Schmitt was a prolific inventor, and was to return later on (below).

Lionel Jofeh, within UK patent 529,044 in 1939, offered a complete catalog of eight forms of cathode-coupled amplifiers (see Reference 7).

Harold Goldberg presented a complete multistage, direct-coupled differential amplifier in 1940 (see Reference 8). Using power pentodes within a unique low voltage differential input stage, Goldberg reported an equivalent input noise of 2 μV for the circuit. This work parallels some of the earlier work mentioned above, apparently developed independently.

In 1941, Otto Schmitt published another work on the differential amplifier topic, going into some detail of analysis (see Reference 9). In this work he clearly outlines the advantages of the long-tailed pair, in terms of the stage's phase-inversion properties. He also covers the case of a degenerated long-tailed pair, where a common cathode-cathode resistance is used for gain adjustment, and the individual cathodes are biased to a negative voltage with resistors of values twice that of a single cathode-coupled stage.

Walther Richter wrote on cathode follower and differential circuits in 1943 (see Reference 10). While primarily focused on single-ended cathode followers, this article also does an analysis of the long-tailed pair.

Harold Goldberg wrote again on his multistage differential amplifier, in 1944 (see Reference 11). The 1944 version still used batteries for most of the power, but did add a pentode to supply the bias current of the first stage long-tailed pair.

Writing in 1944, G. Robert Mezger offered a differential amplifier design with a new method of interstage level-shift coupling (see Reference 12). Previous designs had used either a resistive level-shift like Nyquist, or the more recent glow-tube technique of Miller. Mezger's design used a 12J5 triode as the bottom level-shift element, which acts as constant current source. Working against a fixed resistance at the top, this allows a wideband level shift. Good overall stability was reported in a design that used both differential and CM feedback. Regulation was used for plate and critical heater circuits.

Franklin Offner wrote a letter to the editor in 1945, expressing dissatisfaction with other differential amplifier authors (see Reference 13). In this work he comments on the work of Toennies (Reference 5, again), "...merely an application of in-phase degeneration by the use of a large cathode resistor,..." Offner also overlooked Blumlein's patent.

D. H. Parnum published a two-part survey of differential amplifier techniques in 1945 (see Reference 14). This work analyzed some previously published designs, and presented two differential-throughout amplifier examples, both dc- and ac-coupled.

In a comprehensive study of differential amplifier designs from 1947, Denis L. Johnston presented a three-part article on design techniques, with a finished design example (see Reference 15). This article is notable not only for the wealth of detailed information, but it also contains a bibliography of 61 references to related works.

The input stage of Johnston's design example amplifier used an input long-tailed pair based on the 6CS7 dual triode, with the cathode current supplied by a 6J7G pentode (see Figure 10a of Reference 15). The second stage was also a long-tailed pair, directly coupled to the first stage, with CM feedback. Multiple stages of supply regulation are used.

D. H. Parnum also published another work on differential amplifiers, in 1950 (see Reference 16). In this paper he presented a critique of the input stage design of the Johnston design (Reference 15), pointing out necessary conditions for optimizing CM rejection for multiple stage amplifiers.

The P. O. Bishop and E. J. Harris design paper of 1950 is similar in overall scope to the Johnston work noted above (see Reference 17). It reviews the work of many other designers in the biological area, and presents a sophisticated example design. In this circuit (Figure 3 of Reference 17) a 954 pentode pair is used for input cathode followers, driving a 6J6 dual triode long-tailed pair. Both the input stages as well as the next two stages used 12SH7 pentodes for the tail current sources. Highly stabilized power supplies are used for the plate supplies, with critical heaters also stabilized.

In 1950 Richard McFee published some modifications useful to improve the CM rejection of a single dual triode stage (see Reference 18).

One of the better overview papers for this body of work appeared in 1950, authored by Harry Grundfest (see Reference 19). This paper also gives greater insight into how the biological amplifiers were being used at this time, and offers many references to other differential amplifier work.

It is notable that Grundfest credits Offner (Reference 3, again) with the invention of the long-tailed pair. However, it can be argued that it isn't apparent from Offner's schematics that a true long-tailed pair is actually being used (there being no negative supply for the cathode resistance). The type of biasing that Offner (and Blumlein) use is a simple resistor from the common cathodes to circuit common, which would typically have just a few volts of bias across it, and, more importantly, would have a value roughly comparable in magnitude to the cathode impedance.

Unfortunately, Grundfest also overlooks Blumlein (Reference 2), who preceded Offner with a similar circuit. This similarity is apparent if one compares Blumlein's Figure 2 against Offner's Figure 3, in terms of how the biasing is established.

One of the deepest technical discussions on the topic of DC differential amplifiers can be found in C. M.Verhagen's paper of 1953 (see Reference 20). Verhagen goes into the electron physics of the vacuum tube itself, as well as the detailed circuitry around it, as to how they both effect stability of operation. This paper includes detailed mathematical expressions and critiques of prior work. Many other topical papers are referenced, including some of those above.

The above discussion is meant as a prefacing overview of dc differential amplifiers, as this technology may impact op amp designs. It isn't totally comprehensive, so there are likely other useful papers on the topic. Nevertheless, this discussion should serve to orient readers on many of the general design practices for stable dc differential amplifiers.

Op Amp and Analog Computing Developments

Some of the differential amplifier work described above did find its way into op amps. But, there was also much other significant amplifier work being done, at Bell Labs and elsewhere in the US, as well around the world. The narrative of op amp development now focuses on the thread of *analog computing*, which was the first op amp application.

In the late 1930s George A. Philbrick, at Foxboro Corporation, was developing analog process control simulation circuits with vacuum tubes and passive parts. Philbrick developed many interesting circuits, and some were op amp forebears (see Reference 21).

In fact, within this article, he describes a single tube circuit that performs some op amp functions (Figure 3A). This directly coupled circuit develops an operating relationship between input and output voltages, producing a voltage output proportional to the ratio of two impedances. While this circuit (using floating batteries for power) can't be termed a general-purpose op amp circuit, it nevertheless demonstrates some of the working principles. In Reference 22, Per Holst further describes this early Philbrick work. Within a decade Philbrick was to start his own company supplying vacuum tube op amps and other components used within analog simulation schemes (see below).

The very first vacuum tube amplifiers fitting the introductory section op amp definition came about early in the 1940s wartime period. The overall context was the use of this amplifier as a building block within the Bell Labs-designed M9 gun director system used by WWII Allied Forces. These op amp circuits were general-purpose, using bipolar supply voltages for power, handling bipolar input/output signals with respect to a common voltage (ground). As true to the definition, the overall transfer function was defined by the externally connected input and feedback impedances (more on this follows).

These early amplifiers were part of a specialized analog computer system that was designed to calculate proper gun aiming for fire upon enemy targets. The work on this project started in 1940, and was pioneered by Clarence A. Lovell, David Parkinson, and many other engineers of the Bell Labs staff. Their efforts have been chronicled in great detail by Higgins, et al, as well by James S. Small (see References 23 and 24).

This Bell Labs design project resulted in a prototype gun director system that was called the T10, first tested in December of 1941. While the T10 was the first sample gun director, in later production the gun director was known as the Western Electric M9 (see Reference 25). Further documentation of this work is found in US Patents 2,404,081 and 2,404,387 (see References 26 and 27), plus a related paper by Lovell (see Reference 28). The patents illustrate many common feedback amplifier examples in varied tasks.

However, in terms of an overall technical view of the M9, perhaps the most definitive discussion can be found within "Artillery Director," US Patent 2,493,183, by William Boghosian, Sidney Darlington, and Henry Och of Bell Labs (see Reference 29). This key document breaks down the design of the analog computation scheme into the numerous subsystems involved. Op amps can be found throughout the patent figures, performing functions of buffers, summers, differentiators, inverters, and so forth.

Karl Swartzel's Op Amp

In terms of op amp details, the Boghosian et al patent references another crucial patent document. Many other Bell Labs M9-related patents, underscoring its seminal nature, also referenced this latter work. The patent in question here is US Patent 2,401,779, "Summing Amplifier" by Karl D. Swartzel Jr. of Bell Labs (see Reference 30), and a design that could well be the genesis of op amps. Ironically, Swartzel's work was never given due publicity by Bell Labs.[1] Filed May 1, 1941, it languished within the system during the war, finally being issued in 1946. Of course, the same could be said about many wartime patents—in fact many other Bell Labs patents met similar fates.

A schematic diagram for Swartzel's op amp is shown in Figure 8-3, which includes a table of values taken from the patent text. Although the context of the patent is an application as a summing amplifier, it is also obvious that this is a general-purpose, high-gain amplifier, externally configured for a variety of tasks by the use of suitable feedback components—the crux of the matter regarding the op amp function.

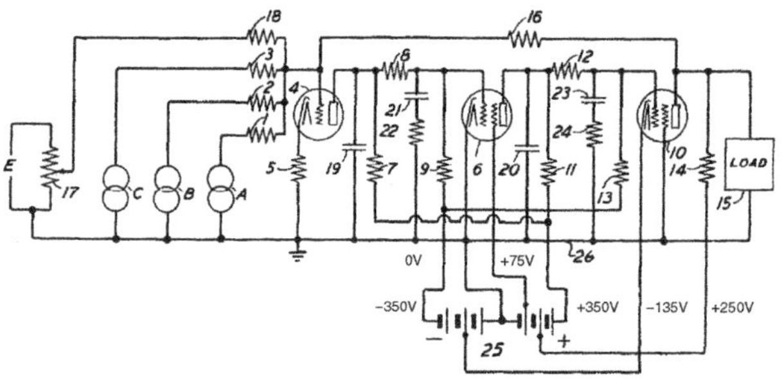

A) Schematic diagram

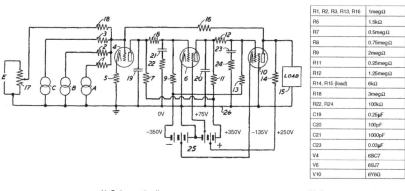

R1, R2, R3, R13, R16	1megΩ
R5	1.5kΩ
R7	0.5megΩ
R8	0.75megΩ
R9	2megΩ
R11	0.25megΩ
R12	1.25megΩ
R14, R15 (load)	6kΩ
R18	3megΩ
R22, R24	100kΩ
C19	0.25µF
C20	100pF
C21	1000pF
C23	0.03µF
V4	6SC7
V6	6SJ7
V10	6Y6G

B) Component values

Figure 8-3: Schematic diagram and component values for "Summing Amplifier" (US Patent 2,401,779, assigned to Bell Telephone Laboratories, Inc.)

[1] Bell Labs documented virtually everything on the M9 in its **Bell Laboratories Record**, as can be noted by the section references. But no op amp schematics were included in this long string of articles.

In the schematic of this "summing amplifier" there can be noted a number of key points. Three directly coupled tubes provided a high overall gain, with a net sign inversion with respect to the grid of input tube 4. Positive and negative power supplies are provided by the tapped battery, 25. The amplifier output swing at the load 15 is bipolar with respect to the common terminal, 26. In this instance of use, resistor R16 applies the feedback, and three signals are being summed via input resistors R1, R2, R3, with the input common to terminal 26. The input via resistor R18 was used for offset control.

The amplifier gain quoted in the patent was 60,000 (95 dB), and as noted, the circuit could drive loads of 6 kΩ, which is quite an achievement. It operated from supplies of ±350 V, with intermediate voltages as noted. The inter-stage level shift networks are the form described by Nyquist, and there are several RC networks used for stabilization purposes.

Over time, changes changes were made to this basic design, which will be described later in this section. The Swartzel op amp was truly a seminal work, as it allowed the creation of a complete, highly sophisticated analog computer system critical to WWII defense. It also spawned numerous other amplifier designs derived from its basic topology.

The M9 From A Bell Labs View

Because of a general embargo on the publication of defense-related technical information during WWII, a great deal of work came to light quite some time after the original development. Many M9 project details on its various components fell into this category, which included op amp diagrams.

Nonetheless, Bell Labs did begin documenting some of the M9 work, even before the war ended. The **Bell Laboratories Record** of December 1943, published a tribute to Harold Black, for his work on the feedback amplifier (see Reference 31). In the same and subsequent issues, there were published two stories on a public demonstration of the M9 system, as well as its development (see Reference 32).

Several key developments in electrical components were also documented in the **Bell Laboratories Record,** on capacitors, resistor networks, resistors, and precision potentiometers (see References 33–36).

There was also fitting recognition of M9 designers Lovell, Parkinson and Kuhn within the **Bell Laboratories Record**, on the occasion of their Medal for Merit award in April of 1947 (see Reference 37). The importance of this work in the view of Bell Labs is underscored by some of the distinguished names associated with the project. Contributors beyond those mentioned above also included Hendrick Bode, Claude Shannon, and other notables of the Bell Labs staff. Finally, there was a then-unrecognized importance: It established the utility of the (yet-to-be-named) op amp concept—op amps were born!

The M9 From A World View

Viewed historically, the work of Lovell, Parkinson and other Bell Labs designers assumes broad significance, since it was war-needs driven, and the outcome literally affected millions of lives. The work provided an analog control computer for a gun director system instrumental to the war effort, achieving high hit rates against incoming targets—up to 90% by some accounts. The work of the M9 system teamed with the SCR584 radar system was highly successful at its mission—indeed fortunate for world freedom.

Robert Buderi wrote a detailed narrative of WWII radar developments, and his book contains an interesting account of the M9's role (see Reference 38). A broad, single-source computing, control, and historical perspective is found in David Mindell's thesis, "**Datum for its Own Annihilation: "Feedback, Control and Computing, 1916–1945**, (see Reference 39). In Chapter 8, Mindell also has an account of the M9/SCR584 success.

Naming the Op Amp

Further wartime op amp development work was carried out in the labs of Columbia University of New York, and was documented in 1947 by the program's research head, Professor John Ragazzini (see Reference 40). This often-cited key paper is perhaps best known for coining the term *operational amplifier,* which, of course, we now shorten to the more simple *op amp*. Quoting from this paper on the naming:

> "*As an amplifier so connected can perform the mathematical operations of arithmetic and calculus on the voltages applied to its input, it is hereafter termed an 'operational amplifier'.*"

The Ragazzini paper outlines a variety of ways that op amps can be used, along with their defining mathematical relationships. This paper also references the Bell Labs work on what became the M9 gun director, specifically mentioning the op amp circuits used.

The work that gave rise to the above paper was a late WWII NDRC Division 7 contract with Columbia University.[1] At that time, Loebe Julie was a bright young research engineer in the Columbia University Labs. Julie did work on these early op amps, which were aimed at improvements to the M9 gun director system, stated contractually as "Fire Control Electronics."

Reportedly working against the wishes of Ragazzini, Julie was engaged to do this work at the behest of analog computer engineer George A. Philbrick, part of the Division 7 team (see References 41 and 42). Julie completed a two-tube op amp design, using a pair of 6SL7 dual triodes in a full differential-in/differential-out arrangement (see Reference 41, again). For whatever the reason, his lab boss Ragazzini gave Julie's amplifier work but a minor acknowledgment at the very end of the paper.

The op amp schematic shown in the Ragazzini paper (Figure 1 of Reference 40) doesn't match the schematic attributed to Julie (Reference 41). Ragazzini doesn't cite any specifications for this circuit, so the origins and intent aren't clear, unless it was intended as a modest performance example. It doesn't seem as if it could be an M9 system candidate, for a couple of reasons.

For example, briefly analyzing the Figure 1 Ragazzini op amp, it seems doubtful that this particular design was really intended to operate in the same environment as the original M9 op amp (Reference 30, or Figure 8-3). Swartzel's three-stage circuit used a triode and two pentodes, with one of the latter a power output stage. So, Ragazzini's circuit wouldn't appear to match the gain characteristics of Swartzel's design, as it used three cascaded triodes. It also wouldn't be capable of the same output drive, by virtue of its use of a 6SL7 output stage, loaded with 300 kΩ.

Evolution of the Vacuum Tube Op Amp

Nevertheless, Julie's op amp design was notable in some regards. It had a better input stage—due to the use of a long-tailed 6SL7 dual triode pair, with balanced loads. This feature would inherently improve drift over previous single-ended triodes or pentodes.

A truly key feature that Julie's circuit held over previous single-ended input designs was the basic fact that it offered *two signal inputs* (inverting and noninverting) as opposed to the single inverting input (Figure 8-3). The active use of both op amp inputs allows much greater signal interface freedom. In fact, this feature is today a hallmark of what can be called a functionally complete op amp—nearly 60 years later. The differential input stage not only improved the drift performance, but it made the op amp immeasurably

[1] Mindell (reference 39) lists in his Table 6-1, a contract No.76 for "Fire Control Electronics," with Ragazzini as investigator, running from November 15, 1943 to September 30, 1945, at a cost of $85,000. The Division 7 supervisor for this Columbia University project is listed as SHC, for Samuel H. Caldwell.

easier to apply. Ironically, however, some time passed before the application of op amps caught up with the availability of that second input.

Much other work was also done on the improvement of direct-coupled amplifiers during the war years and shortly afterwards. Stewart Miller, Edward Ginzton, and Maurice Artzt wrote papers on the improvement of direct-coupled amplifiers, addressing such concerns as input stage drift stabilization against heater voltage variations, interstage coupling and level-shifting schemes, and control of supply impedance interactions (see References 43–45). Some additional examples of improved dc amplifiers can be found in the Valley-Wallman book (see Reference 46).

Before the 1940s ended, companies were already beginning to capitalize on op amp and analog computing technology. Seymour Frost wrote about an analog computer developed at Reeves Instrument Corporation, called REAC (see Reference 47). This computer used as its nucleus an op amp circuit similar to the Swartzel M9 design. In the Reeves circuit the first stage was changed to a 6SL7 dual triode, used in a Miller-compensated low drift setup (Reference 43).

Chopper Stabilization of the Vacuum Tube Op Amp

Even with the use of balanced dual triode input stages, drift was still a continuing problem of early vacuum tube op amps. Many users sought means to hold the input-referred offset to a sub mV level, as opposed to the tens to hundreds of mV typically encountered. The drift had two components, warm-up related, and random or longer term, both of which necessitated frequent rezeroing of amplifiers. This problem was at least partially solved in 1949, with Edwin A. Goldberg's invention of the *chopper-stabilized* op amp (see Reference 48).

The chopper-stabilized op amp employs a second, high gain, ac-coupled amplifier. It is arranged as a side-path to the main amplifier. The chopper channel is arranged with the input signal path AC-coupled to the inverting input of the main DC-coupled amplifier, and a 60 Hz or 400 Hz switch periodically commutating to ground. The switching action chops the small dc input signal to ac, which is greatly amplified (1000 or more). The ac output of the chopper path is synchronously rectified, filtered, and applied to the main amplifier second input. In the resulting composite amplifier, main amplifier drift is reduced by a factor roughly equal to the chopper gain.

With chopper stabilization, op amps could have offset voltages stable to a few μV, and long term drift sufficiently low that manual zeroing wasn't required. Another key benefit was that the dc and low frequency gain was also boosted, by an amount equal to the additional gain factor provided by the chopper channel. By this means, the dc open-loop gain of a chopper amplifier could easily exceed 100,000 times (100 dB). Goldberg's amplifier of Ref. 48 for example, had a dc gain of 150,000,000, or 163 dB.

As a consequence of the above, the dc gain-related precision of a chopper amplifier is much higher than that of a conventional op amp, due to the additional open-loop gain. This basic point, combined with the "zero offset, zero drift" operating feature, made chopper-stabilized vacuum tube op amps a standard choice for precision analog work. This point is one that, generally speaking, is also essentially true even today, with IC chopper op amps readily available. *Note—although today's chopper amplifiers operate by a different method, the net effect is still big improvements in dc offset, drift, and gain.*

There were, however, some serious downsides to these early chopper amplifiers. The basic chopper architecture described above essentially "uses up" the noninverting second op amp input of a dual triode pair, to apply the dc offset correction signal. Thus all of the early chopper op amps operated in an *inverting-only* mode. In time, improved chopper architectures were developed to overcome this limitation, and the very high dc precision was made available for all modes of use.

A second limitation was the fact that the first chopping devices used were mechanical switches (vibrators). As such, they were failure-prone, often before the tubes used alongside. In time all solid-state chopping devices were to be developed, but this didn't impact vacuum tube chopper amps.

Frank Bradley and Rawley McCoy of the Reeves Instrument Corporation discussed yet another variation on the M9 op amp design in 1952 (see Reference 49). In the Bradley-McCoy circuit, a circuit similar to the M9 topology (but with a dual triode front end) was augmented by the addition of a chopper side path. The resulting amplifier had a DC gain of 30,000,00 (150 dB), and very low drift and offset voltage.

By the mid to late fifties many companies began offering solutions to analog computing using chopper-stabilized op amps. However, this wasn't true right after the chopper amplifier became available in the early 1950s—it came about later on.

Shortly after 1950 Granino and Theresa Korn published the first of their textbooks on analog computing, **Electronic Analog Computers** (see Reference 50). This book, along with the second edition in 1956, became the early op amp user's standard reference work. The op amp example fifth chapter of the first edition shows a few chopper-stabilized examples, and Goldberg's work is mentioned. By the time the second edition came out in 1956, chopper amplifiers dominated the examples.

Among the circuits presented in the Korn and Korn **Electronic Analog Computers**, first edition was a later version of the Bell Labs-designed op amp for the M9. A distinct evolutionary path can be noted in this schematic, shown in Figure 8-4.

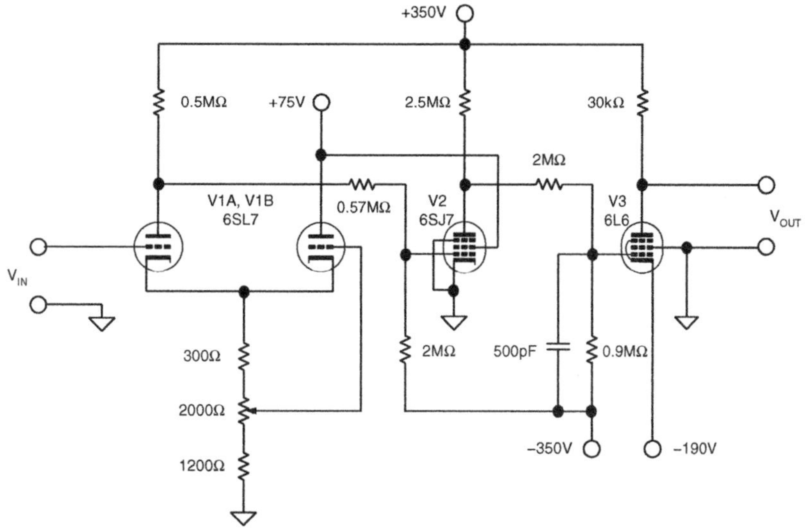

Figure 8-4: Schematic diagram of late M9 system op amp designed at Bell Telephone Laboratories

This version changes the input stage from a single pentode to a dual triode, with Miller compensation added for improved stability (adjustable by the 2 kΩ potentiometer). A close comparison finds that the Frost amplifier (Reference 47) is topologically almost identical with this M9 op amp version. And, vis-à-vis the original Swartzel design of Figure 8-3, further subtle changes to be noted are different ac compensation networks.

Use of the Noninverting Op Amp Input

One aspect of things that did not change was the use of the op amp signal input. All of the examples in the Korn and Korn book (Reference 50) use the op amp with the single-input, parallel feedback mode. In fact, although some of the op amp circuit examples shown in the book have balanced, dual triode inputs, engineering practice out in the world was still in the inverting-only mode. A glance at a topology such as Figure 8-4 reveals the difficulty with applying CM inputs—the amplifier simply was not designed to handle such signals. This was to change, but not very rapidly.

There are, of course, sound technical reasons why op amps didn't get much use in a noninverting mode. Probably the biggest single reason would be the fact that it was much more difficult to make an op amp work over a high CM range (such as ±100 V) which was then used with many circuits. This would require a major redesign of the front end and most likely would also have eliminated the use of chopper stabilization.

Despite that, one early reference to the use of the op amp in a noninverting signal manner was by Omar Patterson, in a patent filed in 1951 (see Reference 51). Although Patterson's patent is a broad array of analog computing circuits, it does utilize a common op amp structure, which is detailed as his Figure 1.

In this design Patterson uses a fully balanced dual triode front end, with the long-tailed pair's cathode current being established by a triode tube. With the balanced plate loading and regulated cathode current, the topology would have good CM response, and be capable of handling a fairly wide range of CM voltages. This op amp was reported to have a gain as high as 10,000 (80 dB), so it was capable of reasonable accuracy.

Patterson goes on in the patent to outline a voltage follower gain stage using this op amp (his Figure 10). In the extreme case of 100% feedback, the feedback stage's gain would be unity, with high input impedance. This is quoted for use of the circuit as an improved cathode follower. Quoting directly, "The advantages of this circuit lie in its extremely low output impedance, and its high degree of independence of tube characteristics."

George Philbrick and GAP/R

After WWII, George Philbrick also continued with op amp development work. Shortly thereafter he formed a company bearing his name, George A. Philbrick Researches, Inc., in 1946 (GAP/R). In many regards, Philbrick's work was instrumental in the development of op amp technology. His company was to see growth over the span of the vacuum tube technology days and well into the solid-state era.

Not too long after forming GAP/R, Philbrick introduced the world's first commercially available op amp, known as the K2-W. This modular 8-pin octal plug-in op amp was developed in 1952, and appeared in January 1953 (see Reference 52). A photo and schematic of this $20 op amp are shown in Figure 8-5.

The K2-W used two 12AX7 dual triodes, with one of the two tubes operated as a long-tailed pair input stage, which offered fully differential operation at the input. With the K2-W operating on ±300 V supplies, the input stage's 220 kΩ tail resistor was returned to the –300 V supply, fulfilling the long-tailed pair biasing requirement.

Half of the remaining 12AX7 dual triode was operated as a second gain stage, which in turn drove the remaining section as a cathode follower output, through a level shifter part 8355037 (typically thyrite devices). Overall gain of the K2-W was enhanced by positive feedback through the 150 kΩ resistor, connected back to the cathode of the second stage. Operating from the ±300 V power supplies @ 4.5 mA, the K2-W was able to achieve a ±50 V rated signal range at both input and output. DC gain was typically 15,000, and the entire circuit was packaged in a convenient, plug-in octal tube-based package.

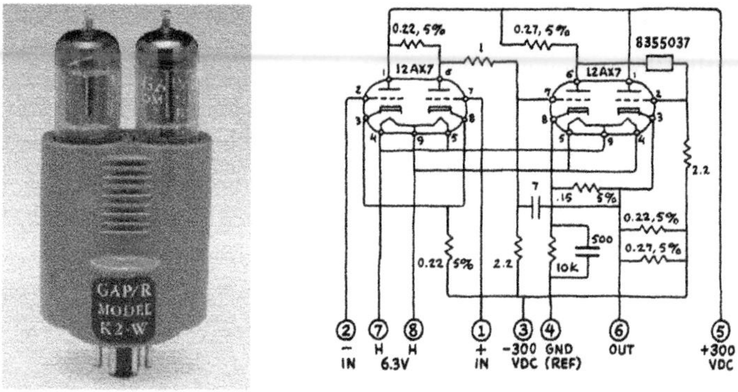

Figure 8-5: The GAP/R K2-W op amp, photo and schematic diagram (courtesy of GAP/R alumnus Dan Sheingold; schematic values in megohms and pF.)

Some vacuum tube op amp manufacturers provided chopper-stabilizer "add-on" units. In the case of GAP/R, this was the GAP/R K2-P. This unit, when used with the K2-W, formed a combination low drift, high gain op amp (see Reference 53).

Early on GAP/R set an excellent standard for application information, publishing a popular 1956 manual for the K2-W and related amplifiers which went through at least 10 printings by 1963 (Reference 53). GAP/R not only made available applications literature for their devices; they published a periodical devoted to analog computation, the **Lightning Empiricist**. It contained technical articles and new product information.

GAP/R also published what is now a classic set of reprints, the "Palimpsest" (see Reference 54). Some researchers see George A. Philbrick as a veritable op amp founding father. For example, Roedel, in his "An Introduction to Analog Computors" (Reference 54, again), gives Philbrick and Lovell credit for being the first op amp users.

It is also undoubtedly true that the GAP/R organization produced some of the best documentation and application support for op amps, both vacuum tube and solid state.

Because of the longevity of so many op amp principles, much of the wisdom imparted in the GAP/R app notes is still as valid today as it was in the 1950s. Although it did not appear until several years later, the best example of this is GAP/R's classic 1965 op amp book edited by Dan Sheingold, **Applications Manual for Computing Amplifiers...** (see Reference 55).

Armed with this book (and perhaps a copy of Korn and Korn's **Electronic Analog Computers**), the op amp user of the late 1960s was well prepared to face op amp circuit hardships. This was not just with analog computation tasks, but also the growing list of diverse applications into which op amps were finding new homes.

The Twilight Years of Vacuum Tube Op Amps

In the late 1950s and early 1960s, vacuum tube op amps had more or less reached their peak of technical sophistication, at least in terms of the circuitry within them. Packaging and size issues of course made a big impact on the overall appeal for the system designer, and work was done in these areas.

An interesting design using three 9-pin miniature dual triodes is shown in Figure 8-6. This compact design was done by Bela (Bel) Losmandy of Op Amp Labs (see Reference 56), then working with Micro-Gee Products, Inc. in 1956 (see Reference 57).

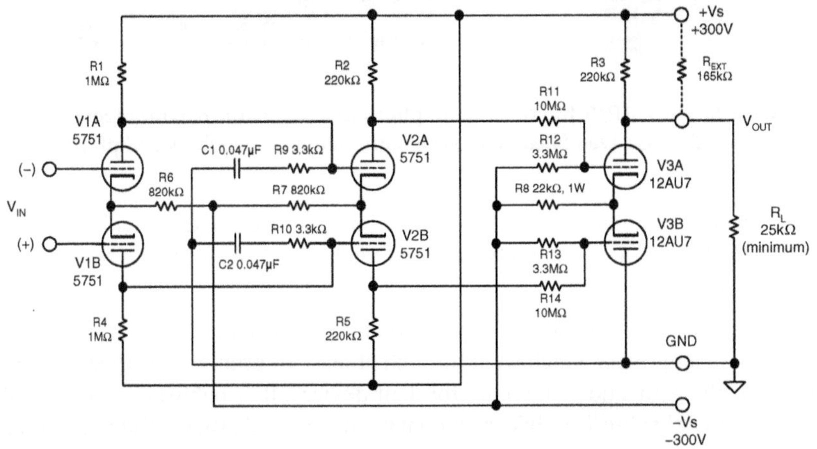

Figure 8-6: A fully differential op amp design by Bela Losmandy, for Micro-Gee Products, Inc.

There are several interesting aspects of this circuit. First, it is entirely differential, right up to the 12AU7 output stage. This allows it to handle CM inputs with lower errors, and improves the drift characteristics. As can also be noted, there is only one (dual) level shift circuit, following the V1–V2 directly coupled differential stages. This minimizes gain loss, and improves the overall performance. The entire amplifier operates on supplies of ±300 V @ 8 mA, and has a gain of more than 10,000 operating into a 25 kΩ load.

In the late 1950's and 1960's, two more publications appeared chronicling op amp developments. One was a long overview paper by Konigsberg, which appeared in 1959 (see Reference 59), the other was the analog and digital oriented computing handbook by Harry Husky and Granino Korn, **Computer Handbook**, in 1962 (see Reference 60). While this book was perhaps one of the last hurrahs for the vacuum tube op amp, it does contain a wealth of detailed design information on them.

By the time the 1960s rolled around, the solid-state era was already in progress. Vacuum tube op amps were on the wane, and smaller, low power, solid-state devices would soon take over op amp applications.

References: Vacuum Tube Op Amps

(Note—appended annotations indicate relevance to op amp history.)

Development of Differential Amplifier Techniques

1. B. H. C. Matthews, "A Special Purpose Amplifier," Proceedings of the Physiological Society, Vol. 81, March 1934, pp. 28, 29. *(A dual-triode differential amplifier, with dual triode common-cathodes biased to ground through a high resistance, ac-coupled at the output.)*
2. A. D. Blumlein, "Improvements in or Relating to Thermionic Valve Amplifying Circuit Arrangements," **UK Patent 482,740**, filed July 4, 1936, issued April 4, 1938. *(A basic common-cathode/resistance-biased-to-common, dual-triode differential amplifier, ac-coupled at input and output.)*
3. Franklin Offner, "Push-Pull Resistance Coupled Amplifiers," **Review of Scientific Instruments**, Vol. 8, January, 1937, pp. 20, 21. *(AC-coupled, cascade differential amplifier circuits, including the use of the dual-triode common-cathodes biased to ground through a high resistance.)*
4. O. H. Schmitt, "A Simple Differential Amplifier," **Review of Scientific Instruments**, Vol. 8, No. 4, April, 1937, pp. 126, 127. *(A basic common-cathode/cathode-to-common, dual-pentode differential amplifier, with plate-screen coupling, single-ended dc-coupling at the output.)*
5. J. F. Toennies, "Differential Amplifier," **Review of Scientific Instruments**, Vol. 9, March, 1938, pp. 95–97. *(A differential amplifier circuit that uses a triode and pentode in a common-cathode configuration, biased to a negative voltage through a high resistance.)*
6. O. H. Schmitt, "Cathode Phase Inversion," **Journal of Scientific Instruments**, Vol. 15, No. 3, 1938, pp. 100, 101. *(A circuit is presented as a single-input phase inverter, but is also a true dc-coupled, differential-input/differential-output amplifier topology. The dual triodes have their common cathodes biased to a negative voltage through a high resistance.)*
7. Lionel Jofeh, "Improvements in Electric Circuits Comprising Electronic Discharge Devices," **UK Patent 529,044**, filed May 9, 1939, issued Nov. 13, 1940. *(A catalog of the various forms of dual triode, dual pentode differential amplifier circuits, with variations in methods of cathode coupling.)*
8. Harold Goldberg, "A High-Gain DC Amplifier for Bio-Electric Recording," **Transactions AIEE**, Vol. 59, January 1940, pp. 60–64. *(A completely differential, multistage, direct-coupled, high gain amplifier for biomedical work. Illustrates the use of common-cathode bias resistance for reducing sensitivity to in-phase inputs.)*
9. O. H. Schmitt, "Cathode Phase Inversion," **Review of Scientific Instruments**, Vol. 12, No. 11, November 1941, p. 548. *(An expansion of previous work of the same name; includes a true dc-coupled, differential-input/differential-output amplifier topology, with the dual-triode common-cathodes biased to a negative voltage through a high resistance.)*
10. Walther Richter, "Cathode Follower Circuits," **Electronics**, November 1943, pp. 112–117, 312. *(Analysis of various cathode follower circuits, including the long-tailed differential pair.)*
11. Harold Goldberg, "Bioelectric-Research Apparatus," **Proceedings of IRE**, Vol. 32, June 1944, pp. 330–336. *(Another differential, multi-stage, direct-coupled, high gain amplifier for biomedical work. Illustrates the use of pentode bias resistance for long-tailed pair; uses battery operated front end.)*
12. G. Robert Mezger, "A Stable Direct-Coupled Amplifier," **Electronics**, July, 1944, pp. 106–110, 352, 353. *(A stable, high gain, differential-input/differential-output, direct-coupled amplifier. Uses novel active device interstage coupling technique.)*

13. Franklin Offner, "Balanced Amplifiers," **Proceedings of IRE**, Vol. 33, March 1945, p. 202. *(A criticism of other differential amplifier circuit papers.)*

14. D. H. Parnum, "Biological Amplifiers, Parts 1 and 2," **Wireless World**, Nov. 1945, pp. 337–340, and Dec. 1945, pp. 373-376. *(A survey of various differential amplifier techniques, with a multistage ac-coupled design example.)*

15. Denis L. Johnston, "Electro-Encephalograph Amplifier, Parts 1–3," **Wireless Engineer**, August 1947, pp. 231–242, Sept. 1947, pp. 271–277, and October 1947, pp. 292–297. *(A comprehensive overview of various biological differential amplifier techniques with many example circuits, both dc- and ac-coupled. Includes 61 references.)*

16. D. H. Parnum, "Transmission Factor of Differential Amplifiers," **Wireless Engineer**, April 1950, pp. 125–129. *(An analysis of the common-mode and differential signal response of differential amplifiers.)*

17. P. O. Bishop, E. J. Harris, "A DC Amplifier for Biological Application," **Review of Scientific Instruments**, Vol. 21, No. 4, April 1950, pp. 366–377. *(A very sophisticated differential amplifier system, employing common-mode feedback, active differential pair current sources, input shield bootstrapping, power supply stabilization, and so forth. Many references cited.)*

18. Richard McFee, "Improving Differential Amplifier Rejection Ratio," **Review of Scientific Instruments**, Vol. 21, No. 8, August 1950, pp. 770–771. *(A feedback modification to a standard dual triode differential stage to improve common-mode rejection.)*

19. Harry Grundfest, "Biological Requirements for the Design of Amplifiers," **Proceedings of the IRE**, Vol. 38, September 1950, pp. 1018–1028. *(An overview of amplifier requirements for biological measurements; includes commentary on the history of differential amplifier development.)*

20. C. M. Verhagen, "A Survey of the Limits in DC Amplification," **Proceedings of the IRE,** Vol. 41, May 1953, pp. 615–630. *(A technical discussion of both tube and circuit parameters which impact DC stability.)*

Op Amp and Analog Computing Developments

21. George A. Philbrick, "Designing Industrial Controllers by Analog," **Electronics**, June 1948. *(Figure 3b example of an early analog computation circuit.)*

22. Per A. Holst, "George A. Philbrick and Polyphemus—The First Electronic Training Simulator," **IEEE Annals of the History of Computing**, Vol. 4, No. 2, April 1982, pp. 143–156. *(George A. Philbrick's analog computing circuitry at Foxboro, late 1930s.)*

23. H. C. Higgins, et al, "Defense Research at Bell Laboratories: Electrical Computers for Fire Control," **IEEE Annals of the History of Computing**, Vol. 4, No. 3, July, 1982, pp. 218–236. See also: M. D. Fagen, Ed. **A History of Engineering and Science in the Bell System**, Vol. 2, "National Service in War and Peace, 1925–1975," Bell Telephone Laboratories, 1978, ISBN 0-0932764-001-2. *(A summary of various Bell Labs fire control analog computer developments, including op amps and feedback components. Note—the IEEE article is reprinted from the broader, more detailed Bell Telephone Laboratories volume.)*

24. James S. Small, "General-Purpose Electronic Analog Computing: 1945–1965," **IEEE Annals of the History of Computing**, Vol. 15, No. 2, 1993, pp. 8–18. *(An overview of analog computing, includes discussion of early operational amplifiers.)*

25. David A. Mindell, "Automation's Finest Hour: Bell Labs and Automatic Control in World War II," **IEEE Control Systems**, December 1995, pp. 72–80. *(Narrative of the T10 computer system and the M9 gun director developments at Bell Labs.)*

26. C. A. Lovell, et al, "Artillery Predictor," **US Patent 2,404,081**, filed May 1, 1941, issued September 24, 1946. *(The mathematics of analog computer system using op amps for functions of repeating, inverting, and summing amplifiers, plus differentiation.)*

27. C. A. Lovell, et al, "Electrical Computing System," **US Patent 2,404,387**, filed May 1, 1941, issued July 23, 1946. *(An analog computer system using op amps for control.)*

28. C. A. Lovell, "Continuous Electrical Computation," **Bell Laboratories Record**, 25, March, 1947, pp. 114–118. *(An overview of various fire-control analog computational circuits of the T10 and M9 systems, many illustrating uses of op amps.)*

29. W. H. Boghosian, et al, "Artillery Director," **US Patent 2,493,183**, filed May 21, 1942, issued Jan. 3, 1950. *(An artillery fire control system using op amps for control.)*

30. K. D. Swartzel, Jr., "Summing Amplifier," **US Patent 2,401,779**, filed May 1, 1941, issued July 11, 1946. *(The first operational amplifier, used as a summing amplifier.)*

31. Recognition of Harold Black, "Historic Firsts: The Negative Feedback Amplifier," **Bell Laboratories Record**, Vol. 22, No. 4, December, 1943, p. 173. *(A Bell Laboratories tribute to the Black negative feedback amplifier invention.)*

32. "Electrical Gun Director Demonstrated," **Bell Laboratories Record**, Vol. 22, No. 4, December 1943, pp. 157–167. See also: "Development of the Electrical Director," **Bell Laboratories Record**, Vol. 22, No. 5, January 1944, pp. 225–230. *(Bell Laboratories narratives of the M9 gun director system demonstration and development.)*

33. J. R. Weeks, "Polystyrene Capacitors," **Bell Laboratories Record**, 24, March, 1946, pp. 111–115. *(Development of a new capacitor film dielectric for electronic analog computer networks.)*

34. E. C. Hageman, "Precision Resistance Networks for Computer Circuits," **Bell Laboratories Record**, 24, December, 1946, pp. 445–449. *(Development of precision resistor networks for electronic analog computers.)*

35. C. Pfister, "Precision Carbon Resistor," **Bell Laboratories Record**, 26, October, 1946, pp. 401–406. *(Development of deposited carbon resistors for electronic analog computers.)*

36. D. G. Blattner, "Precision Potentiometers for Analog Computers," **Bell Laboratories Record**, 32, May, 1954, pp. 171–177. *(Development of precision wire wound potentiometers for use in electronic analog computers.)*

37. Recognition of M9 Designers C. A. Lovell, D. B. Parkinson, and J. J. Kuhn, "Medals for Merit," **Bell Laboratories Record**, Vol. 27, May, 1947, p. 208. *(The Medal for Merit awarded to the M9 designers on April 8, 1947 is the nation's highest civilian award.)*

38. Robert Buderi, **The Invention That Changed the World**, Simon and Schuster, 1996, ISBN: 0-684-81021-2. *(A marvelous account of radar development during WWII, centered largely on the MIT Radiation Lab team—includes a narrative on the integration of the M9 fire control system with the SCR-584 radar, and the system's operational success.)*

39. David A. Mindell, **"Datum for its Own Annihilation:" Feedback, Control and Computing 1916–1945**, PhD thesis, MIT, May 2, 1996. *(Historical survey of computing control systems. Chapter 8 of this work is 'Radar and System Integration,' which covers the Bell Labs work on the T10 and M9 gun director projects.)*

Naming the Op Amp

40. John R. Ragazzini, Robert H. Randall and Frederick A. Russell, "Analysis of Problems in Dynamics by Electronic Circuits," **Proceedings of the IRE**, vol. 35, May 1947, pp. 444–452. *(An overview of operational amplifier uses, and first formal definition of the term.)*

41. George Rostky, "Unsung Hero Pioneered Op Amp," **EE Times**, March 24, 1997, or, www.eetonline.com/anniversary/designclassics/opamp.html. *(A narrative on the op amp work of Loebe Julie at Columbia University during WWII [and footnoted in the Ragazzini paper]. A schematic diagram of the Julie op amp is included.)*

42. Bob Pease, "What's All This Julie Stuff, Anyhow?," **Electronic Design**, May 3, 1999, or, www.elecdesign.com/Articles/ArticleID/6071/6071.html *(Another narrative on the op amp work of Loebe Julie at Columbia University during WWII [and footnoted in the Ragazzini paper].)*

Evolution of the Vacuum Tube Op Amp

43. Stewart E. Miller, "Sensitive DC Amplifier with AC Operation," **Electronics**, November, 1941, pp. 27– 31, 106–109. *(Design example of a stable, high-gain direct-coupled amplifier including 'cathode-compensation' against variations in filament voltage, use of glow tube inter-stage coupling, and a stable line-operated dc supply.)*

44. Edward L. Ginzton, "DC Amplifier Design Techniques," **Electronics**, March 1944, pp. 98–102. *(Various design means for improving direct-coupled amplifiers.)*

45. Maurice Artzt, "Survey of DC Amplifiers," **Electronics**, August, 1945, pp. 112–118. *(Survey of direct-coupled amplifier designs, both single-ended and differential, with emphasis on high stability.)*

46. George E. Valley, Jr., Henry Wallman, **Vacuum Tube Amplifiers**, MIT Radiation Labs Series No. 18, McGraw-Hill, 1948. *(A classic WWII Radiation Lab development team textbook. Chapter 11, by John W. Gray, deals with direct-coupled amplifiers.)*

47. Seymour Frost, "Compact Analog Computer," **Electronics**, July, 1948, pp. 116–120, 122. *(A description of the Reeves Electronic Analog Computer [REAC], which used as the computing amplifier a circuit similar to the M9 op amp. Uses a Miller-compensated triode input stage.)*

48. E. A. Goldberg, "Stabilization of Wide-Band Direct-Current Amplifiers for Zero and Gain," **RCA Review**, June 1950, pp. 296–300. See also: "Stabilized Direct Current Amplifier," **US Patent 2,684,999**, filed April 28, 1949, issued July 27, 1954. *(A system for lowering vacuum tube op amp offset voltage, drift, and gain errors, by means of an ac-coupled parallel-path, with synchronous rectification, and dc signal reinsertion—in short, the* chopper-stabilized *op amp.)*

49. Frank R. Bradley, Rawley McCoy, "Driftless DC Amplifier," **Electronics**, April 1952, pp. 144–148. *(A description of the Reeves A-105 Analog Computer, which used as the computing amplifier a circuit partially similar to the M9 op amp. Adds chopper stabilization for low drift.)*

50. Granino Korn and Theresa Korn, **Electronic Analog Computers,** McGraw-Hill, 1952. *(A classic early work on the uses and methodology of analog computing. In Chapter 5, an op amp circuit attributed to Bell Labs and the M9 project is described, along with many other examples, and design details.)*

51. Omar L. Patterson, "Computing Circuits," **US Patent 2,855,145**, filed July 30, 1951, issued Oct. 7, 1958. *(A catalog of analog computation circuits based on a fully differential input op amp circuit, including use of a voltage follower configuration.)*

52. **Data Sheet For Model K2-W Operational Amplifier**, George A. Philbrick Researches, Inc., Boston, MA, January 1953. See also "40 Years Ago," **Electronic Design**, December 16, 1995, p. 8. *(The George A. Philbrick Research dual triode K2-W, the first commercial vacuum tube op amp.)*

53. Henry Paynter, Ed., **Applications Manual for PHILBRICK OCTAL PLUG-IN Computing Amplifiers**, George A. Philbrick Researches, Inc., Boston, MA, 1956. *(The first op amp application manual.)*

54. Jerry Roedel, "An Introduction to Analog Computors," within **Palimpsest on the Electronic Analog Art**, George A. Philbrick Researches, Inc., Boston, MA,1955. *(Roedel's paper is one of many analog computing papers, [including Ragazzini et al, above].)*

55. Dan Sheingold, Ed., **Applications Manual for Operational Amplifiers for Modeling, Measuring, Manipulating, and Much Else**, George A. Philbrick Researches, Inc., Boston, MA, 1965. *(The classic GAP/R op amp application manual, now a collector's item.)*

56. **Opamp Labs Inc.,** 1033 N Sycamore Avenue, Los Angeles, CA, 90038 (323) 934-3566, www.opamplabs.com/bel.htm.

57. Bela Losmandy, **Three Stage Micro-Gee Products, Inc.Op Amp**, 1956.

58. Granino Korn and Theresa Korn, **Electronic Analog Computers, Second Edition** McGraw-Hill, 1956. *(Second edition of the classic work on the uses and methodology of analog computing. Chapter 5 features an increased number of op amp circuits, many chopper-stabilized.)*

59. R. L. Konigsberg, "Operational Amplifiers," **Advances in Electronics and Electron Physics**, Vol. 11, 1959, pp. 225–285. *(An overview of vacuum tube and [some] solid-state op amps, with 75 references.)*

60. Harry Husky, Granino Korn, **Computer Handbook**, McGraw-Hill, 1962. *(An encyclopedic reference work on analog computing. Includes detailed op amp design section by Edward Billinghurst, plus a comprehensive list of op amp circuit examples, with specifications.)*

Solid-State Modular and Hybrid Op Amps

Vacuum tube op amps continued to flourish for some time into the 1950s and 1960s, but their competition was eventually to arrive from solid-state developments. These took the form of several key innovations, all of which required a presence before effective solid-state op amp designs could be established. This discussion treats *modular and hybrid* solid-state op amps, which preceded and overlapped solid-state IC op amps.

There were three of these key developments, the invention of *the transistor*, the invention of *the integrated circuit* (IC), and the invention of *the planar IC process*. A detailed history of solid-state inventions and related process developments can be found in articles celebrating the transistor's 50th anniversary, within the Autumn 1997 **Bell Labs Technical Journal** (see References 1 and 2).

Birth of the Transistor

John Bardeen, Walter Brattain, and William Shockley of Bell Labs, working with *germanium* semiconductor materials, first discovered the transistor effect in December of 1947. Of course, this first step was a demonstration of gain via a new principle, using a semiconducting material, as opposed to a vacuum tube. But more remained to be done before commercial transistors were to appear. For their achievement, the trio received the 1956 Nobel Prize in physics. In addition to vacuum tubes, circuit designers now had a lower voltage, lower power, miniature amplifying device (see References 3 and 4).

Over the course of the next 10 years or more, various means were explored to improve the germanium transistors. The best germanium transistors were still relatively limited in terms of leakage currents, general stability, maximum junction temperature, and frequency response. Some of these problems were never to be solved. While many performance improvements were made, it was recognized early that *silicon* as a semiconductor material had greater potential, so this occupied many researchers.

In May of 1954, Gordon Teal of Texas Instruments developed a grown-junction *silicon* transistor. These transistors could operate to 150°C, far higher than germanium. They also had lower leakage, and were generally superior amplifying devices. Additional processing refinements were to improve upon the early silicon transistors, and eventually lead a path to the invention of the first integrated circuits in the late fifties.

Birth of the IC

In 1958, Jack Kilby of Texas Instruments invented the *integrated circuit*, now known universally as the IC (see Reference 5). For this effort he was ultimately to become a co-recipient of the 2000 Nobel Prize in physics.

Kilby's work, however, important as it was, could arguably be said to be nonexclusive in terms of first authorship of the integrated circuit. In early 1959, Robert Noyce, an engineer at Fairchild Semiconductor, also developed an IC concept (see Reference 6).

The nucleus of Noyce's concept was actually closer to the concept of today's ICs, as it used interconnecting metal trace layers between transistors and resistors. Kilby's IC, by contrast, used bond wires.

As might be expected from such differences between two key inventions, so closely timed in their origination, there was no instant consensus on the true "IC inventor." Subsequent patent fights between the two inventor's companies persisted into the 1960s. Today, both men are recognized as IC inventors.

The Planar Process

In general parallel with the Noyce's early IC developments, Jean Hoerni (also of Fairchild Semiconductor) had been working on means to protect and stabilize silicon diode and transistor characteristics. Until that time, the junctions of all *mesa* process devices were essentially left exposed. This was a serious limitation of the mesa process.

The mesa process is so named because the areas surrounding the central base-emitter regions are etched away, thus leaving this area exposed on a plateau, or mesa. In practice, this factor makes a semiconductor so constructed susceptible to contaminants, and as a result, inherently less stable. This was the fatal flaw that Hoerni's invention addressed.

Hoerni's solution to the problem was to re-arrange the transistor geometry into a flat, or *planar* surface, thus giving the new process its name (see References 7 and 8). However, the important distinction in terms of device protection is that within the planar process the otherwise exposed regions are left covered with silicon dioxide. This feature reduced the device sensitivity to contaminants; making a much better, more stable, transistor or IC.

With the arrangement of the device terminals on a planar surface, Hoerni's invention was also directly amenable to the flat metal conducting traces that were intrinsic to Noyce's IC invention. Furthermore, the planar process required no additional process steps in its implementation, so it made the higher performance economical as well. As time has now shown, the development of the planar process was another key semiconductor invention. It is now widely used in production of transistors and ICs.

At a time in the early 1960s shortly after the invention of the planar process, the three key developments had been made. They were the (silicon) transistor itself, the IC, and the planar process. The stage was now set for important solid-state developments in op amps. This was to take place in three stages. First, there would be *discrete transistor* and *modular* op amp versions, second there would be *hybrid op amps*, which could be produced in a couple of ways. One hybrid method utilized discrete transistors in chip form(s), interconnected to form an op amp; another was a specially matched transistor pair combined with an IC op amp for improved performance, and thirdly, the op amp finally became a complete, integral, dedicated IC—the *IC op amp*. This latter developmental stage is covered more fully within the next section of this chapter.

Of course, within these developmental stages there were considerable improvements made to device performance. And, as with the vacuum tube/solid-state periods, each stage overlapped the previous and/or the next one to a great extent.

Solid-State Modular and Hybrid Op Amp Designs

There were new as well as old companies involved in early solid-state op amps. GAP/R was already well established as a vacuum tube op amp supplier, so solid-state op amps for them were a new form of the same basic product. With GAP/R and others in the 1960s, the Boston area was to become the first center of the solid-state op amp world. Elsewhere, other companies were formed to meet market demand for the more compact transistor op amps. Burr-Brown Research Corporation in Arizona fell into this category. Formed by Robert Page Burr and Thomas Brown in 1956, Burr-Brown was an early modular op amp supplier, and supported their products with an applications book (see Reference 9). The Burr-Brown product line grew steadily over the years, emerging into a major supplier of precision amplifiers and other instrumentation ICs. Texas Instruments bought Burr-Brown in 2000, merging the product lines of the two companies.

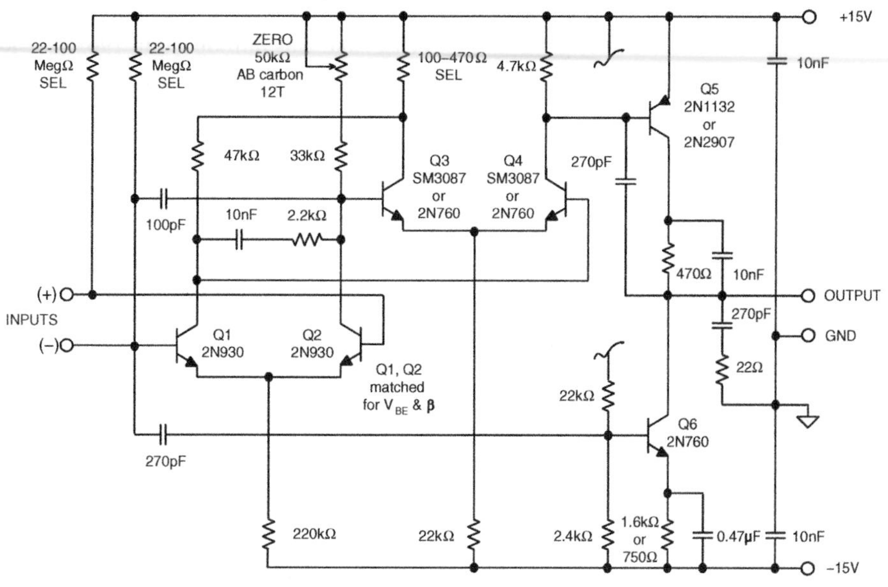

Figure 8-7: The GAP/R model P65 solid-state op amp

On the other hand, in 1960 GAP/R was a transitioning company, and while they maintained the vacuum tube op amp line for some time, they stayed away from solid-state op amps until quality silicon transistors could be found. GAP/R began to introduce solid-state op amps in the early 1960s. George Philbrick was simply unwilling to produce germanium transistor solid-state op amps, and he also had specific ideas about the optimum amplifier topology that could be used—more on this follows.

The new solid-state op amps were to transition power supply and signal range standards from ±300 V/±100 V down to ±15 V/±10 V, a standard that still exists today. And of course, new packaging for the op amps was to emerge, in several forms.

The GAP/R P65, shown above in Figure 8-7, was a general-purpose device. It was designed by Alan Pearlman, with later revisions by Bob Malter, and was produced from 1961 through 1971. The first stage Q1-Q2 used a pair of matched 2N930s, with a tail current of 66 μA, and had hand-selected bias compensation (the SEL resistors).

The second P65 stage of Q3-Q4 ran at substantially more current, and featured a gain-boosting positive feedback loop via the SEL and 47 kΩ resistors. The common-emitter output stage was PNP Q5, loaded by NPN current source Q6. The two-stage NPN differential pair cascade used in the P65 design was to become a basic part of other GAP/R op amps, such as the P45 (described below).

Small value feedforward capacitors sped up the ac response, and an output RC snubber provided stability, along with phase compensation across the Q1-Q2 collectors. The transistor types shown represented the original P65, but later on the P65A used better transistors (such as the 2N2907), and thus could deliver more output drive.

Another GAP/R solid-state op amp was the P45, shown in Figure 8-8 as a photo of the card-mounted op amp, and the schematic. The P45 was designed by Bob Pease, and was introduced in 1963 (see Reference 10). The edge connector card package shown was used with the P45 and P65, as well as many other GAP/R solid-state amplifiers.

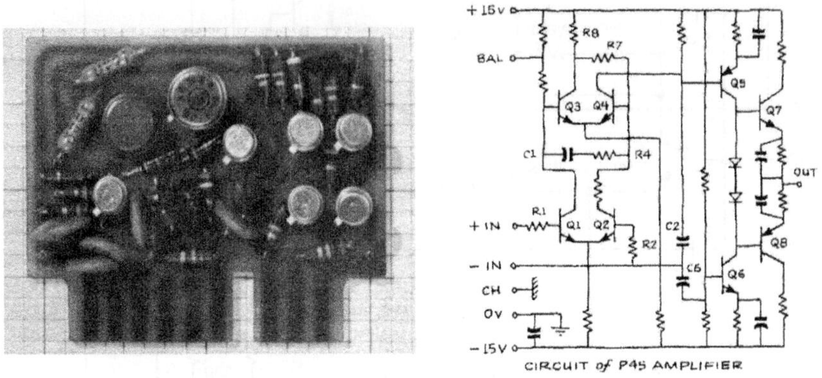

Figure 8-8: The GAP/R model P45 solid-state op amp

The P45 design was aimed at fast, inverting mode applications. With a class AB output stage, the P45A could deliver ±10 V at ±20 mA to the load. Gain was rated a minimum of 50,000 at 25°C into a load of 500 Ω. One of the more outstanding specifications of the P45 was its gain-bandwidth product of 100 MHz. In 1966, a P45A cost $118 in quantities of 1–4 (see Reference 11). Both the P65 and P45 ran on ±15 V, the new power standard, and were intended for input/output signal ranges of ±10 V.

As mentioned, the cascaded NPN differential pair topology used in the P45 and P65 designs was to become a basic part of other GAP/R op amps. A feature of the design was the controlled positive feedback path, from the Q3 collector back to the Q4 base. Offset was controlled by a potentiometer connected between the BAL pin and 15 V in the P45, with a similar arrangement used in the P65 (Figure 8-7).

In the P45, the first two gain stages are followed by PNP common-emitter stage Q5, which provides a great deal of the voltage gain. Emitter followers Q7 and Q8 buffer the high impedance node at Q5's collector, providing a low impedance source to the load.

In 1962 Alan Pearlman and partner Roger R. (Tim) Noble formed their own Boston-area company, Nexus Research Laboratory, Inc. Nexus competed with both GAP/R and Burr-Brown in the growing solid-state op amp field (and ultimately with a third local company). The Nexus mission was to deliver solid-state op amps to customers for printed circuit board mounting, thus the Nexus designs used a rectangular, potted module package. They were so popular that they influenced GAP/R to follow suit with modular designs of their own.

In 1962, George Philbrick himself did the layout of another P65 derivative, the PP65, which was one of the first GAP/R modules. Shown in Figure 8-9, this square outline, 0.2" centered, 7-pin footprint was to become more or less a modular op amp standard. It used five pins for output/power/offset on one side, with the two input pins on the opposite side.

Figure 8-9: The GAP/R model PP65 potted module solid-state op amp

It might be easy for some to dismiss the importance of a package design within a chart of op amp progress. Nevertheless, the modular package format opened up new opportunities and, for the first time, allowed the op amp to be treated *as a component*. This opening of application opportunities enhanced op amp growth significantly.[1]

Varactor Bridge Op Amps

George Philbrick championed a novel op amp type that became a GAP/R profit maker—the *varactor bridge* amplifier. In this circuit, voltage variable capacitors (varactors) are used in an input stage that processes the op amp error voltage as a phase-sensitive ac carrier. By careful bridge component arrangement, the op amp input terminals are forced to see only tiny dc leakage currents, i.e., as small as 1 pA (or in some cases, much less).

As a result, a varactor bridge op amp achieved the lowest input current of any op amp available in the solid-state period. Lower than common tubes, in fact. In addition, since there was no input dc path to common, the allowable input CM voltage of a varactor bridge op amp could go very high—to levels as high as ±200 V.

[1] This pattern was to be repeated again and again with op amps, and continues even today, with miniature SOIC packages displacing DIP and other through-hole packages.

Figure 8-10 illustrates in block diagram form a varactor bridge op amp. There are four main components, the front end composed of the bridge circuit and a high frequency oscillator, an ac amplifier to gain-up the bridge output error voltage, a synchronous phase detector to convert the amplified ac error to a corresponding dc error, and finally an output amplifier, providing additional dc gain and load drive.

The circuit worked as follows: A small dc error voltage V_{IN} applied to the matched varactor diodes D1 and D2 causes an ac bridge imbalance, which is fed into the ac amplifier. This ac voltage will be phase-sensitive, dependent upon the dc error. The remaining parts of the loop amplify and detect the dc error.

To apply the amplifier, an external feedback loop is closed from V_{OUT} back to the inverting input terminal, just as with conventional op amps. The difference in the case of the varactor bridge op amp lies in the fact that two unusual degrees of freedom existed, in terms of both bias current and CM voltage.

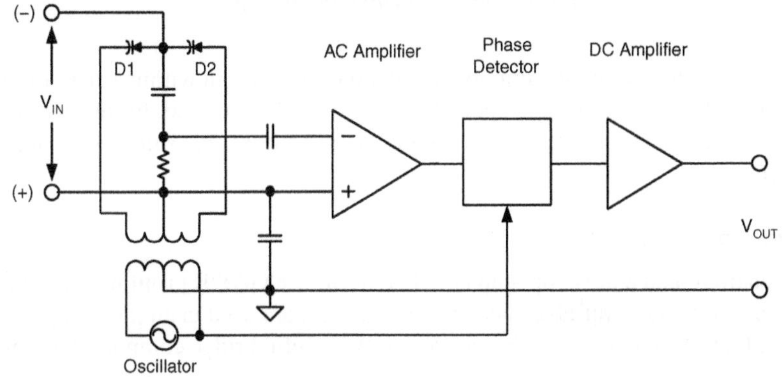

Figure 8-10: Generalized block diagram for a varactor bridge solid-state op amp

The GAP/R varactor bridge op amp model was called the P2. It was a premium part in terms of the dc specifications, but not speed. In fact, the unity-gain frequency was just 75 kHz, but the specification that people keyed on was the ±10 pA input offset current. Also, the CM range of ±200 V very likely enabled a few applications that previously might have required the use of a tube amplifier to address.

In 1966, an SP2A sold for a then astronomical price of $227 (see Reference 11). Bob Pease wrote a fascinating narrative on the P2's collaborative development at GAP/R, which was by engineers George Philbrick and Bob Malter (see Reference 12).

George Philbrick was also issued a patent on a varactor bridge amplifier, in 1968 (see Reference 13). In 1966 GAP/R and Nexus Research Laboratories were purchased by Teledyne Corporation, and the merged product line continued into hybrids and ICs.

Later on there was to be a sad note in this GAP/R history. GAP/R founder and master innovator George Philbrick passed away in late 1974, at a relatively young age of 61. A tribute to George Philbrick was offered by his partner and collaborator, Professor Henry Paynter of MIT (see Reference 14).

The Birth of ADI

The emergence of a third Boston-area op amp company took place in the mid 1960s. In January of 1965 Analog Devices Inc. (ADI) was founded by Matt Lorber and Ray Stata. Operating initially from

Cambridge, MA, op amps were the first product of the new company. Many of the early op amps were modular designs (more on this below).

Dan Sheingold has suggested that the ADI founders may have intentionally left out the word "Research" as part of the ADI name (see Reference 15). This would be to differentiate the new company from all of the (then) three competitors, Burr-Brown, GAP/R, and Nexus Research Laboratories, and thus broaden market appeal for op amps.

And, it seemed to work for the new ADI venture, with sales taking off quite soon. One of the first "products" of the new ADI was application support for op amps (also noted by Sheingold, Reference 15). In the first year, Ray Stata authored a comprehensive guide to op amps (see Reference 16). Examples of this application support were to continue through the early years and afterwards, echoing a successful business practice established by GAP/R.

The first few ADI years resulted in many new op amps, in mostly the modular package style, using both bipolar transistor and FET technologies. A complete list is much more broad than can be covered here, so just some highlights will be sampled.

Model 3xx Series Varactor Bridge Op Amps

To compete with GAP/R and their P2, ADI marketed a number of varactor bridge input op amps. The first varactor bridge op amps were the 301, 302, and 303 models, which were all similar, but differed in detail as to the input mode. They were differential (301), inverting (302), or noninverting (303). The 301 had a max input current of 2 pA, but the others got as low as 0.5 pA. The 301 sold for $198, while the 302 A and 303 A were $110.

Lewis R. Smith designed these amplifiers, as well as their successors, models 310 and 311 (see Reference 17). These latter designs were able to achieve significantly improved input currents, which were ±10 fA for the signal input of both amplifiers (just about three orders of magnitude below the GAP/R P2 series). An input current specification this low was then (and still is) a most impressive achievement. Interestingly, the 310 and 311 models were also sold for lower prices, which was $75 for the J grade.

Lewis Smith also described his varactor bridge designs in a patent (see Reference 18). It is a high tribute to the model 310 and 311 designs that they are still being produced in 2004. The devices are available through Intronics (see Reference 19).

The Many Op Amp Categories

Many of the earliest ADI modular op amps used bipolar transistors for the input stages, and they all used bipolar transistors in later stages. Matched duals of either bipolar or FET types were scarce in the early 1960s, but these were incorporated into designs soon after announcement. An early listing of ADI op amps has five categories: general purpose, low bias current, low drift, wideband, and high voltage/current (see Reference 20). This list was expanded considerably in only a year (see Reference 21).

In the general-purpose types, the models 111 and later the 118 were popular units, due to a combination of good basic specs and attractive prices. The varactor types already mentioned led performance for low bias current types, but there were also FET input types such as the early model 142 with bias currents in the tens of pA range.

In the low drift category, various chopper amplifiers such as the 210, 211, 220, and later the 232 and 233, and the 260 led in performance, There were also low drift chopperless amplifiers such as the 180 and 183, using precision bipolar transistor front ends. There was considerable support for choppers over the next few years (see References 22–24).

Model 121 Op Amp

A design done by Dick Burwen for ADI was the model 121, a fast, fully differential op amp, in 1966. This design demonstrates some useful circuit techniques in Figure 8-11.

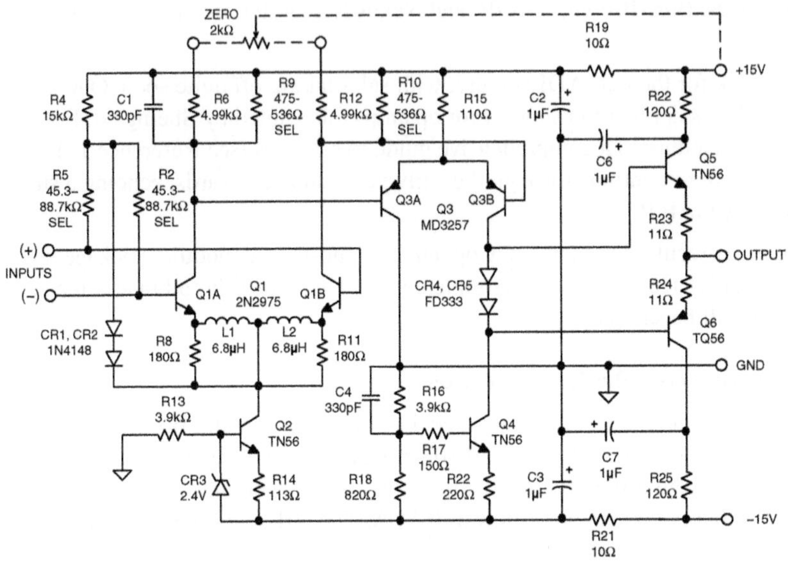

Figure 8-11: The ADI model 121 wideband DC op amp

One of the techniques is how to make a high speed, low noise input stage, which is by means of the L1–L2 chokes. At low frequencies the chokes shunt the otherwise noisy degeneration resistors, R8–R9. It also shows the use of relatively heavy bypassing and decoupling internal to the op amp (a necessary practical step, but possible within the confines of a module).

The model 121 NPN input stage runs at a high tail current, for fast slewing. But, note that the R2 and R5 resistors compensate the input bias current, which would otherwise be high. These (selected) resistors provide a temperature tracking bias from the floating diode source, CR1–CR2. This bias scheme was patented by Burwen, and was also used in other ADI op amps of the period (see Reference 25).

As can be noted, the model 121 used a current source for the 2N2975 matched NPN pair input stage, Q2, to optimize CMR. In critical locations, 1% metal film resistors are also used. The gain-of-five stable model 121 had a gain of 25,000 (or 88 dB), achieved a slew rate of 250 V/µs, and sold for $98 in 1968 (see Reference 20).

Analog Dialogue Magazine is Born

ADI's continuing thread of customer support through applications information was enhanced considerably in 1967, when the magazine **Analog Dialogue** was launched (see Reference 26). The initial charter for the magazine was stated as "A Journal for the Exchange of Operational Amplifier Technology," later on this was broadened to "A Journal for the Exchange of Analog Technology."

But, disseminate op amp info is what the early **Analog Dialogue** did, and also what it did well. The premier issue featured an op amp article by Ray Stata that is still available as an app note (see Reference 27). And, a similar comment can also be made for a subsequent Ray Stata article (see Reference 28).

A milestone in the life of the young magazine was the arrival of Dan Sheingold as editor, in 1969 (see Reference 29). Already highly experienced as a skilled op amp expert and editorial writer from vacuum tube and early solid-state years at GAP/R, Dan Sheingold brought a unique set of skills to the task of editorial guidance for **Analog Dialogue.** Dan's leadership as editor continues today. For more than 35 years his high technical communication standards have been an industry benchmark.

A Family of High Speed FET Op Amp Designs

One of the more illuminating development threads to be found within the ADI op amp portfolio is that of the high-speed FET input modular and hybrid products. This design family began with the model 45 in 1970 (see Reference 30). John Cadigan designed all of these amplifiers, and they continued evolving over the next 10 years or more.

The reasons for this product line's longevity (which extended well into the era of IC op amps) is simply that these amplifiers met difficult technical needs. These needs weren't to be solved by early IC amplifiers, and indeed were not met by ICs at all, until better processes became available. The combination of high speed (meaning here fast settling to a defined narrow error band) and excellent dc accuracy made these high speed FET amplifiers the best answer for accurately driving A/D converters and other accuracy-critical amplifier needs.

Model 45, 44 and 48 FET Op Amps

The Model 45 was out first, and was targeted for lower cost applications, with under 1μs settling time to 0.01%. The next two models of the series were the 44 and 48, as represented in a simplified schematic shown in Figure 8-12 (see References 31–34). The Model 45 schematic is similar to the 44 and 48, with some simplifications.

All of these amplifiers used FET inputs, based on a high speed matched NFET pair, Q2. For the 44 and 48, a balanced PNP current mirror loaded the input stage. The mirror output signal drove integrator stage Q6 via a Darlington buffer, Q5. The input stage ran at about 1 mA/side, and the 44/48 had slew rates of 75 V/μs and 110 V/μs, respectively.

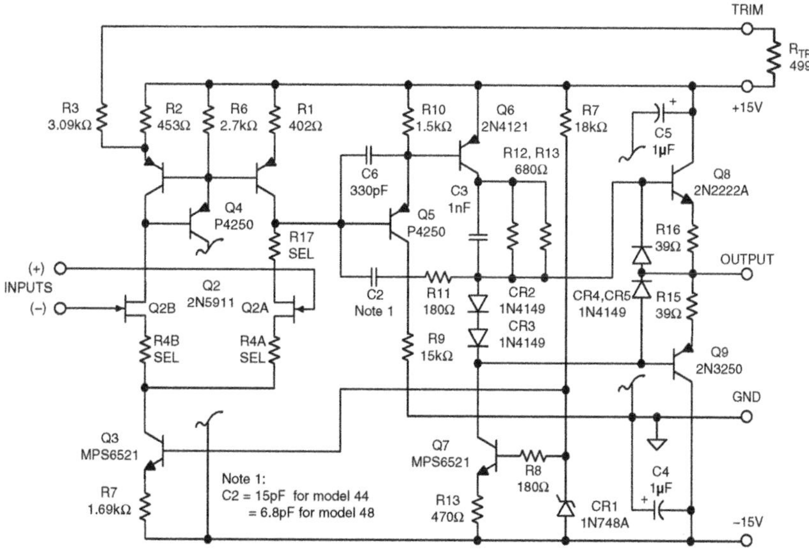

Figure 8-12: The ADI Model 44 and 48 high speed FET-input modular op amps

Like the Model 45, quick settling to a rated error band as low as 0.01% was a key feature of the 44 and 48. These amplifiers achieved 0.01% guaranteed settling of 1000 ns and 500 ns limits respectively, for up to ±10 V of output, in either inverting or noninverting modes.

In addition, these models all had class AB output stages, and were well-suited towards driving coax lines. They used a standard 1.125" square module package, with pinouts as noted. A 499 Ω trim resistor between the trim pin and $+V_S$ provided the rated dc offset without trimming, or, alternately, a 1k Ω pot was used for a more precise trim.

Development of the line continued after Models 44 and 48, and included others in the series, the Model 46, and the 47.

The Model 50 FET Op Amp

The highest speed version of the series came about with the Model 50 modular op amp, which appeared in 1973 (see References 35 and 36). Circuit details of the Model 50 will be more apparent with the discussion of the HOS-050, immediately below.

The HOS-050 FET Op Amp

Prior to being acquired by ADI in 1979, Computer Labs (Greensboro, NC) was in the business of building high speed data acquisition systems. As part of their A/D converter architectures, they routinely used fast op amps to drive the converters, employing amplifiers with characteristics like the Model 50 from ADI.

In 1977 Computer Labs developed a hybrid IC version of the ADI Model 50, calling it the HOS-050 (see Reference 37). An HOS-050 schematic is shown in Figure 8-13.

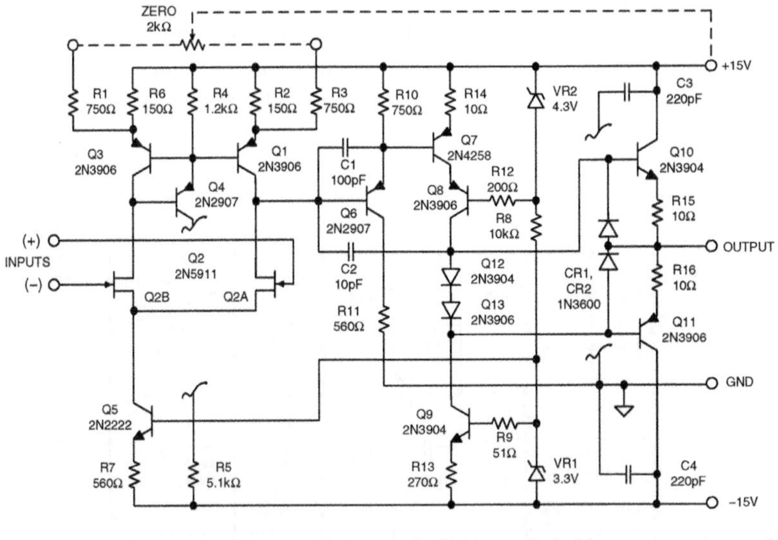

Figure 8-13: The ADI HOS-050 high speed FET-input hybrid op amp schematic

There are many similarities between this and the earlier amplifier shown in Figure 8-12, but the differences are mostly speed-related. As can be noted, the HOS-050 input stage runs at higher current, for higher slew rate and bandwidth. Likewise, the Q7 integrator stage is run at a higher level, and is also cascoded by Q8. The HOS-050 used a balanced form of dc offset trim, which, even if disconnected, allowed the op amp to function well.

The output stage used fast transistors, with a higher threshold level of current limiting. The entire circuit ran on the warm side, dissipating about 600 mW on ±15 V supplies.

The Model 50 and the HOS-050 had rated outputs of ±10 V, and ±100 mA, a 100 MHz bandwidth, and the HOS-050 settled to 0.01% in 200 ns. Both the model 50 and the HOS-050 without doubt achieved the highest levels of performance.

In addition, the HOS-050 represented perhaps one of the more impressive hybrid op amp ICs built at ADI, with its combination of excellent specs, contained within a small TO-8 package. After the acquisition of Computer Labs by ADI, there were two top quality hybrid IC production facilities available to ADI customers. One of these was in the Boston area, with the other at ADI Greensboro, the former Computer Labs site.

There were several other hybrid IC op amps manufactured by ADI in the 1970s and 1980s. Among these were the HOS-060, the ADLH0032, and the AD3554. Hybrid IC construction was the most dense form of circuit packaging available (save for a purely monolithic form of IC). Some appreciation for this high packing density can be gleaned from the photo of the HOS-050 IC op amp in Figure 8-14.

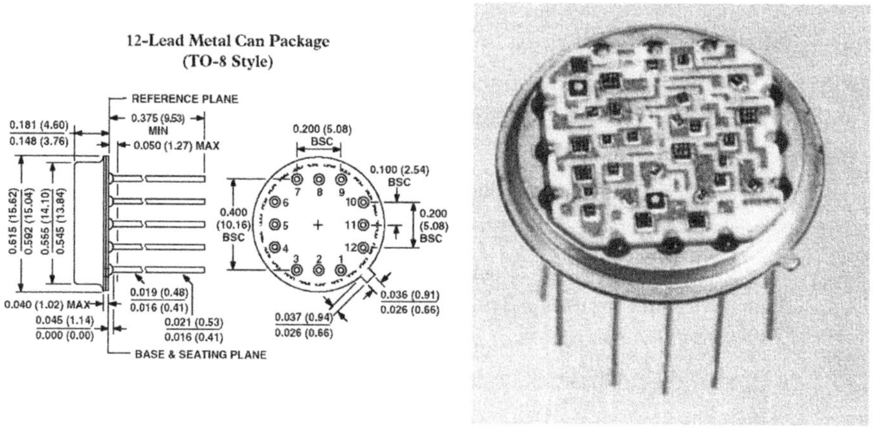

Figure 8-14: The ADI HOS-050 high speed hybrid IC op amp

In this figure, the outline for the hermetically sealed TO-8 package is shown at the left, for size reference. In the right photo of the exposed HOS-050 circuit, it can be noted that virtually 100% of the substrate area is occupied with the conductor traces and the individual circuit components, which included thin film resistors. Further, to maximize circuit area, note that the active substrate area is extended even into the four corners.

While such op amps as the Model 50 and the HOS-050 may have reached pinnacles in terms of the combined circuit performance and their complexity of packaging, this situation didn't last long. Like the fate of modular op amps, hybrid op amp lifetimes were to be relatively short. As soon as IC op amps of comparable electrical performance could be built, the market for the sophisticated but hard-to-produce hybrid ICs shrunk, leaving the hybrids to be sold only into military or other long-lifetime or specialty systems.

References: Solid-state Modular and Hybrid Op Amps

(Note: Appended annotations indicate relevance to op amp history.)

Birth of the Transistor

1. Ian M. Ross, "The Foundation of the Silicon Age," **Bell Labs Technical Journal**, Vol. 2, No. 4, Autumn 1997.
2. C. Mark Melliar-Smith et al, "Key Steps to the Integrated Circuit," **Bell Labs Technical Journal**, Vol. 2, No. 4, Autumn 1997.
3. J. Bardeen, W. H. Brattain, "The Transistor, a Semi-Conductor Triode," **Physical Review**, Vol. 74, No. 2, July 15, 1947 pp. 230–231. *(the invention of the germanium transistor.)*
4. W. Shockley, "The Theory of p-n Junctions in Semiconductors and p-n Junction Transistors," **Bell System Technical Journal**, Vol. 28, No. 4, July 1949, pp. 435–489. *(theory behind the germanium transistor.)*

Birth of the IC

5. J. S. Kilby, "Invention of the Integrated Circuit," **IRE Transactions on Electron Devices**, Vol. ED-23, No. 7, July 1976, pp. 648-654. *(Kilby's IC invention at TI.)*
6. Robert N. Noyce, "Semiconductor Device-and-Lead Structure," **US Patent 2,981,877**, filed July 30, 1959, issued April 25, 1961. *(Noyce's IC invention at Fairchild.)*

The Planar Process

7. Jean A. Hoerni, "Method of Manufacturing Semiconductor Devices," **US Patent 3,025,589**, filed May 1, 1959, issued March 20, 1962. *(the planar process—a manufacturing means of protecting and stabilizing semiconductors.)*
8. Jean Hoerni, "Planar Silicon Diodes and Transistors," **IRE Transactions on Electron Devices**, Vol. 8, March 1961, p. 168 *(technical discussion of planar processed devices).*

Solid-State Modular and Hybrid Op Amp Designs

9. Burr-Brown Applications Staff, **Handbook of Operational Amplifier Applications, First Edition,** Burr-Brown Research Corporation, 1963. *(Burr-Brown's early semiconductor oriented op amp application manual.)*
10. Robert A. Pease, "Design of a Modern High-Performance Operational Amplifier," **GAP/R Lightning Empiricist**, Vol. 11, No. 2, April 1963. *(The designer's description of the P45 op amp. Note—Q8 polarity correction via RAP to WJ email of 5/23/2001.)*
11. **Philbrick Solid-State Operational Amplifiers**, GAP/R bulletin of 3/1/1966.
12. Bob Pease, "Chapter 9, The Story of the P2," within Jim Williams, Ed., **Analog Circuit Design,** Butterworth-Heinemann**,** 1991, ISBN: 0-7506-9166-2.
13. George A. Philbrick, "Electronic Amplifier," **US Patent 3,405,366**, filed June 30, 1965, issued Oct. 8, 1968. *(A varactor bridge amplifier.)*
14. H. M. Paynter, "In Memoriam: George A. Philbrick," **ASME Journal of Systems, Measurement and Control**, June 1975, pp. 213–215.
15. Dan Sheingold, "Analog Dialectic," **Analog Dialogue**, Vol. 30, No. 3, p. 1.
16. Ray Stata, "Operational Amplifiers, Parts I and II," **Electromechanical Design**, September and November, 1965. *(The first ADI op amp application manual.)*

17. Data sheet for **Model 310, 311 Ultra Low Bias Current Varactor Bridge Operational Amplifiers**, Analog Devices, Inc., December 1969.
18. Lewis R. Smith, "Operational Amplifier with Varactor Bridge Input Circuit," **US Patent 3,530,390**, filed Dec. 28, 1966, issued Sept. 22, 1970. *(A varactor bridge amplifier.)*
19. **ADI Modular Products**, www.intronics.com/products/analogproducts/.
20. **Capsule Listing of Analog Devices Op Amps**, ADI bulletin of April 1968.
21. **Selection Handbook and Catalog Guide to Operational Amplifiers**, ADI catalog of January 1969.
22. Ray Stata, "Applications Manual 201, 202, 203 & 210 Chopper Stabilized Operational Amplifiers," ADI application note, 1967. *(Use of 200 series modular chopper amplifiers.)*
23. ADI Staff, "Circuit Advantages of Chopper Stabilized Operational Amplifiers," ADI application note, September 1970. *(Operation and use of inverting-type modular chopper stabilized op amps.)*
24. Peter Zicko, "Designing with Chopper Stabilized Operational Amplifiers," ADI application note, September 1970. *(Operation and use of noninverting-type modular chopper stabilized op amps.)*
25. Richard S. Burwen, "Input Current Compensation with Temperature for Differential Transistor Amplifier," **US Patent 3,467,908**, filed Feb. 7, 1968, issued Sept. 16, 1969. *(A method of compensating for bias current in a differential pair amplifier.)*
26. **Analog Dialogue**, Vol. 1, No. 1, April 1967. *(The premier issue of* ***Analog Dialogue****—'A Journal for the Exchange of Operational Amplifier Technology'.)*
27. Ray Stata, "Operational Integrators," **Analog Dialogue**, Vol. 1, No. 1, April, 1967. *(Reprinted as ADI AN357.)*
28. Ray Stata, "User's Guide to Applying and Measuring Operational Amplifier Specifications," **Analog Dialogue**, Vol. 1, No. 3. *(Reprinted as ADI AN356.)*
29. **Analog Dialogue**, Vol. 3, No. 1, March 1969. *(The first issue of* ***Analog Dialogue*** *under the editorial guidance of Dan Sheingold.)*
30. Data sheet for **Model 45 Fast Settling FET Operational Amplifier**, Analog Devices, Inc., May, 1970.
31. "New Modular Op Amps," **Analog Dialogue**, Vol. 5, No. 2, pp. 12 *(Model 43 and 44 op amps).*
32. "Op Amp Settles to 0.01% in 300 ns," **Analog Dialogue**, Vol. 6, No. 2, pp. 11 *(Model 48 op amp).*
33. Data sheet for **Model 48 Fast Settling Differential FET Op Amp**, Analog Devices, Inc., June 1972.
34. Data sheet for **Model 44 Fast Settling High Accuracy Op Amp**, Analog Devices, Inc., August 1972.
35. Data sheet for **Model 50 Fast Settling 100 mA Output Differential FET Op Amp**, Analog Devices, Inc., May 1973.
36. "Model 50: Wideband, Fast Settling Op Amp," **Analog Dialogue**, Vol. 7, No. 2, pp. 13. *(The Model 50* modular *op amp.)*
37. Low-Cost HOS-050C is Internally Compensated, Settles to 0.1% in 80 ns, 0.01% in 200 ns," **Analog Dialogue**, Vol. 16, No. 2, p. 24. *(The HOS-050C* hybrid *op amp, first introduced by Computer Labs in 1977.)*

IC Op Amps

Birth of the Monolithic IC Op Amp

The first generally recognized monolithic IC op amp was from Fairchild Semiconductor Corporation (FSC), the *μA702*. The μA702 was designed by a young engineer, Robert J. (Bob) Widlar. As will be seen, Bob Widlar was a man who was shortly to make an indelible mark on the IC world. But, his 1963 μA702 didn't exactly take the world by storm. It wasn't well received, due to quirky characteristics—odd supply voltages, low input/output swings, low gain, and so forth. Nevertheless, and despite these shortcomings, the μA702 established some important IC design trends. As pioneered by Bob Widlar, these concepts were to carry over to future op amps (see Reference 1). In fact, they are standard linear IC design concepts today. While the μA702 isn't covered in detail here, information on it can be found in Reference 2.

The μA709

Not long after the μA702 a major IC op amp landmark came about, specifically the introduction of another Bob Widlar op amp for Fairchild in 1965, the *μA709,* (see Reference 3). The 709[1] improved markedly on the 702; it had higher gain (45,000 or ~94 dB), greater input/output ranges (±10 V), lower input current (200 nA) and higher output current, and operated from symmetrical power supplies (±15 V). The 709 quickly became a standard, and was produced for decades. Figure 8-15 is a 709 schematic.

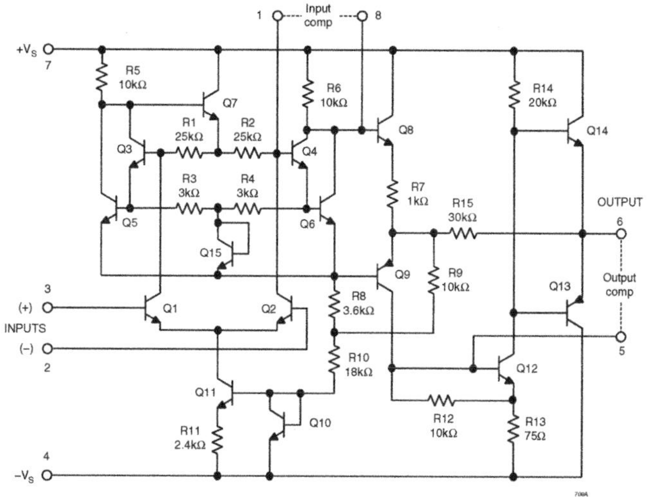

Figure 8-15: The μA709 monolithic IC op amp

[1] Although the original Fairchild designation was "μA709," the design was broadly second-sourced. The widely-used generic name became simply "709." Likewise, the μA702 is known as the "702."

So universal was the 709 that it can be regarded as an IC op amp classic. Although the individual specifications were surpassed by many subsequent designs, the 709 remains a milestone, as the first widely used monolithic IC op amp.

Many design principles from the 702 were used again in the 709, such as the use of matched transistors, for the first and second stages, and the logarithmic biased (delta-V_{BE}) current source, Q10–Q11. There were also new wrinkles added. Because the 709 used what was basically an NPN IC process, Widlar resorted to some clever tricks to create PNP functions. He used a modified NPN structure for two PNPs, the level shifter Q9 and the output PNP, Q13. The output stage operated class-B, with no Q13–Q14 bias. Local feedback around this stage via R15 minimized deadzone.

Frequency compensation for the 709 was achieved with two RC networks, between pins 1–8, and pins 6–5. The associated network values could be changed for optimum ac response, using four networks for gains of 0 dB to 60 dB.

Although the 709 was a vast improvement over the 702, it still had quirks of its own, and these gave rise to application problems. For example, without some user-added series resistance, the output stage could blow out for sustained shorts. Many saw the frequency compensation scheme as difficult, plus it took up board space. Also, the 709 could latch up whenever the input CM voltage rose high enough to saturate the input stage. And, excessive differential input voltages could blow out the input transistors. Although savvy users could work around these 709 application quirks, it sometimes took extra parts to do it. So, in one sense the above use-related issues served as a general lesson towards the necessity of bullet-proofing an IC op amp against various application stresses.

The LM101

Not content to rest on his 702 and 709 laurels, Bob Widlar moved on to another company, National Semiconductor Corporation (NSC). His next IC op amp design, the *LM101*, was introduced in 1967 (see Reference 4). This began a second IC op amp generation (the 709 is generally regarded as the first generation of IC op amps).

The LM101 family[1] used a simpler two-stage topology, one that addressed the application problems of the 709. It was also an op amp design that influenced a great many ones to follow. A simplified circuit of the 101 is shown in Figure 8-16.

The LM101 design objectives were to eliminate such 709 problems as:

- No short-circuit protection.
- Complex frequency compensation.
- Latchup with high CM inputs.
- Sensitivity to excessive differential input voltage.
- Excessive power dissipation and limited power supply range.
- Sensitivity to capacitive loads.

For the same reasons that the 709 has historical importance, so does the LM101, as it represents the next op amp technology level. In fact, the all-purpose topology used is the basis for a range of many other general-purpose devices, variants of the LM101 design.

---

[1] The LM101 family included three temperature ranges, LM101, LM201, and LM301, for military, industrial, and commercial ranges, respectively. Also known generically as 101, and so forth. Similarly, the following LM101A series devices became know as 101A, 201A, 301A, and so forth.

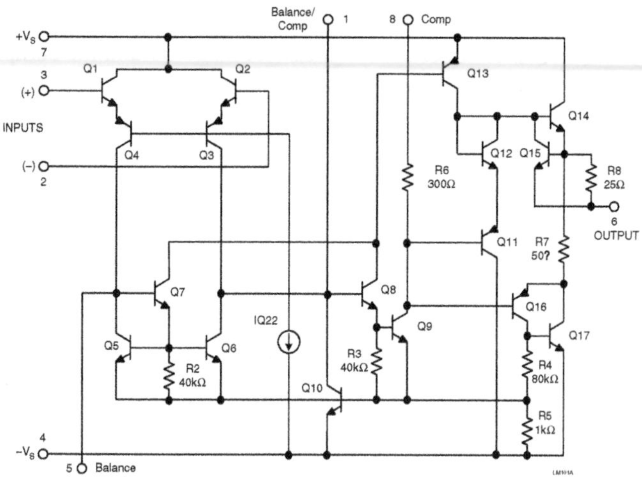

Figure 8-16: The LM101 monolithic IC op amp

The new 101 design did solve the 709's problems, and it added some further refinements. Gain was 160,000 (~104 dB), and the useful supply range increased from ±5 V to ±20 V. For easy upgrading, the 101 used the same pins as 709 for inputs, output, and power.

A major goal of the 101 design was simpler frequency compensation. To enable this, the 101 uses a *two-stage* amplifier design, as fewer stages are easier to compensate. But, to retain high voltage gain, the two stages needed the gain of the three 709 stages. In the 101, the high gain per stage is done using active loads, which increase the available gain per stage to a maximum. An example is Q13, which provides the collector load for Q9.

In the first stage of the 101, active loading is also used, Q5–Q6. Q1–Q4 and Q2–Q3 here form an equivalent PNP differential pair. Although the PNPs have low gains, they are buffered by high-gain NPNs, Q1–Q2. The net resulting input current was 120 nA.

Note that the CM input range of this stage is quite high, as Q1–Q2 can swing positive to $+V_S$. The negative CM limit is about four V_{BE} above the $-V_S$ rail. This wide CM range prevents input stage saturation and latch-up. Another feature of this composite input stage is a very high differential voltage rating, due to the PNP high base-emitter breakdowns. The input stage can safely tolerate inputs of ±30 V.

The second stage of the 101 is the common-emitter amplifier, Q9. With the above mentioned loading of Q13, this stage achieved a voltage gain of about 60 dB, and the overall gain of the op amp was typically over 100 dB. A class AB output stage is used, consisting of NPN Q14, and the equivalent PNP, Q16–Q17. These transistors were biased by Q11–Q12. Sensing resistors R7, R8, and Q15, along with an elaborate loop comprising Q16 and Q9–Q10, provided current limiting.

An important differentiation of the 101 versus the 709 was the much simpler frequency compensation. In the 101, this was accomplished by a single external 30 pF capacitor, connected between pins 1 and 8. As can be noted from the 101 internal connections, this capacitor makes the second gain stage Q9 an integrator, forcing overall gain to roll off from its maximum value of 104 dB at 10 Hz at a rate of 6 dB per octave, crossing unity gain at about 1 MHz. This compensation made a 101 device stable in any feedback configuration, down to the unity gain.

Viewed analytically, the 101 op amp topology can be seen as a two-stage voltage amplifier, formed by an input g_m stage consisting of Q1–Q6, which drives an integrator stage Q9 and the compensation capacitor, and a unity gain output buffer, Q11–Q17. This type of topology is discussed further in Chapter 1 of this book.

But, a salient point to be noted is the fact that this form of compensation takes advantage of *pole splitting* in the second stage, which results in the multiplied capacitance of the compensation capacitor to provide a stable –6 dB/octave roll-off (see Reference 5). This was a critically important point at the time, as it allowed a single small (30 pF) capacitor to provide the entire compensation. Many two-stage IC op amp architectures introduced since the original 101 use a similar signal path and compensation method.

The µA741

In less than a year's time after the 101's introduction, Fairchild introduced their answer to it, which was the *µA741* op amp. Designed by Dave Fullagar and introduced in 1968, the µA741 used a similar signal path to the LM101 (see Reference 6). A simplified schematic of the µA741 is shown in Figure 8-17.

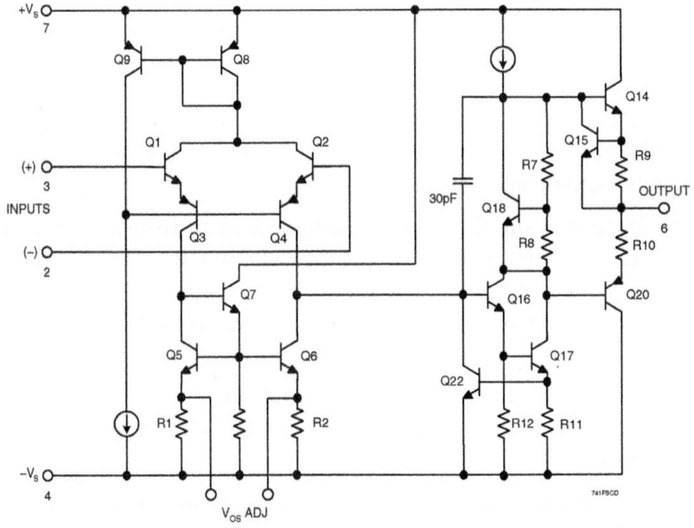

Figure 8-17: The µA741 monolithic IC op amp

Although there are obvious biasing differences, the 741 signal path is essentially equivalent to the 101, and it provides similar features in terms of short-circuit and input over voltage protection, and has a comparable bandwidth. Nevertheless, for the reason that the 741 had the 30 pF compensation capacitor *on the chip*, it became the standard.[2]

[2] George Erdi told an interesting story of the µA741's genesis while working at Fairchild and sharing an office with the designer, Dave Fullagar. It seems that shortly after the LM101 appeared, the two were discussing the reason why the required compensation cap was external. Fullagar's conclusion was that the National process in use at the time simply couldn't accommodate the internal capacitor. He said, "*Well, we can do that!*" and so, shortly afterwards, the internally compensated µA741 was born.

The moral here seems to be that ease-of-use is more valuable to users than flexibility. The 101, with the user-added capacitor, was functionally equivalent to the 741. In fact, National Semiconductor had introduced the *LH101*, a hybrid package of an LM101 chip plus a 30 pF capacitor, in early 1968. Nevertheless, the 741 became a greater standard.

The LM101A

Bob Widlar updated his basic LM101 design with the *LM101A*, which was introduced by National Semiconductor in late 1968. This was a more refined version of the 101 op amp architecture, featuring lower and more stable input bias current (see Reference 7).

At about the same time, they also introduced the *LM107*, which was an LM101A with the 30 pF compensation capacitor on the same monolithic chip. The 107 and 741 could be said to be comparable for ac specifications, but the 107 had an edge for dc parameters.

The μA748

The *μA748*, an externally compensated derivative of the μA741, was introduced by Fairchild in 1969. It was Fairchild's answer to the National LM101/LM101A series. The 748 functioned just like the 101/101A types, with an external capacitor between pins 1–8.

Multiple 741 Types, General-Purpose Single-Supply Types

With the 741 being such a popular device, it readily lent itself to dual and quad versions. Space doesn't permit discussion of all, but among the more popular were the Motorola *MC1558/1458*, a pair of 741s in an 8-pin DIP pinout. Almost since the beginning dual versions have been the more popular for IC op amps. Quad 741 types also became available, such as the Motorola *MC4741*, and the National Semiconductor *LM148*.

In 1972, Russell and Frederiksen of National Semiconductor introduced an amplifier technique suitable for operation in a single-supply environment at low voltages (see Reference 8). This amplifier, which was to become the *LM324*, became the low cost industry standard general purpose quad op amp. It was followed by a similar dual, the *LM358*. One of the key concepts used in the paper was an input stage g^m reduction method, credited to James Solomon (Reference 5).

Since this is a historical discussion of IC op amps, one would assume that all of the above general-purpose IC op amps would have long since disappeared, being 30-odd years old. But such isn't the case—many of them are still available even now, in 2004!

The AD741—A Precision 741

Neither the 741 nor the 101A were designed as true high precision amplifiers. In the years following the development of the 741 and 101A op amps, other IC manufacturers looked into refining the performance of these popular products for the precision analog marketplace. In 1971 ADI took a key step towards this, and acquired Nova Devices of Wilmington, MA. This added both technology and design capability to the ADI portfolio, which was to be immediately useful for the manufacture of linear ICs.

One of the first ADI ICs to be produced was an enhanced design 741 type op amp, the *AD741* (see Reference 9). A schematic of this circuit is shown in Figure 8-18.

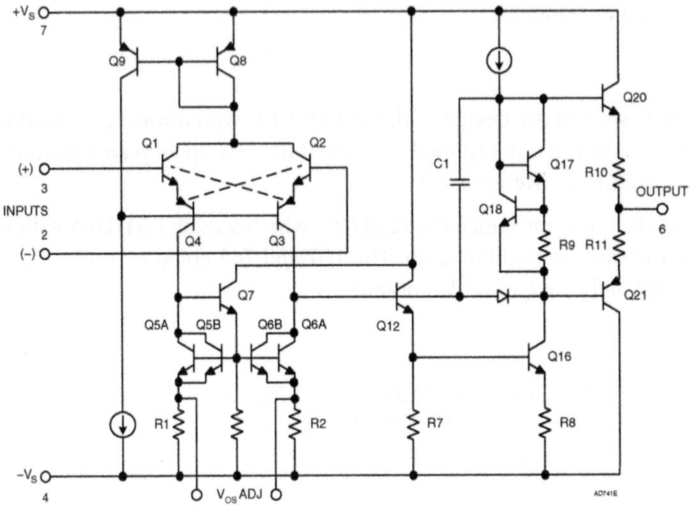

Figure 8-18: The AD741 monolithic IC op amp

Although this circuit looks deceptively like the μA741 of Figure 8-17, it should be understood that there are many subtleties affecting IC performance that don't necessarily appear on the data sheet. In this case, some key differences include a thermally balanced layout, evident from the use of a cross-quad input stage (denoted by the Q1–Q4 cross markings), and the quad operation of the current mirror load transistors, Q5–Q6. In addition, a better output stage was used, with higher efficiency transistors.

The premium version of this design was the AD741L, which achieved an offset of 500 μV (max), a drift of 5 μV/°C (max), a bias current of 50 nA (max), and a minimum gain of 50,000 (94 dB) into a 2 kΩ load. ADI also produced an improved 301A type amplifier, the AD301AL, with dc specifications similar to those of the AD741L above.

In 1973, the AD741L sold for $6.00 in 100 piece lots, while the AD741J could be purchased for just $1.25 in the same quantities.

With the establishment of general-purpose IC op amps a fact, other designers began to focus on greater precision. This was sought through the reduction of various errors; lower bias currents, lower offset voltage, higher gain, and so forth. A couple of the development paths that follow the general thread of higher IC op amp precision will now be discussed.[3]

SuperBeta IC Op Amps—LM108 to OP97

After his release of the 101 op amp, Bob Widlar began to explore the *superbeta* bipolar transistor technique.[4] A superbeta transistor is one subjected to extra diffusion steps, to raise the forward gain from a typical 200, to several thousand or more. Used in the input stage of an IC op amp, a pair of superbeta transistors can potentially reduce input currents by a factor of 10–20 times.

---

3 For coherence, this superbeta precision op amp (and other) threads will be presented in a continuous fashion. In actual time of course, each thread paralleled many other concurrent IC op amp developments.

4 To avoid confusion, the term "superbeta" should really be "super-H_{FE}"—but "superbeta" has stuck.

However, the use of superbeta transistors isn't exactly straightforward, because the super-beta process reduces the breakdown voltage to 5 V or less. This factor requires extra circuitry around the superbeta devices to buffer high voltages normal to an op amp.

The first IC to use super-beta transistors was the LM102 voltage follower of 1967, by Bob Widlar (see Reference 10), followed by the upgraded LM110 in 1970. Widlar also published a more general description of superbeta transistor operation, in Reference 11.

The 102 and 110 voltage follower ICs were somewhat specialized parts. Internally configured as unity-gain buffers, there was no user configuration needed (or possible). Nevertheless, the use of the superbeta devices at the input established their viability, at least in one context of application.

Meanwhile, in these very early years of the technology, Bob Widlar wasn't the only designer working on superbeta concepts as applied to op amps. At Motorola Semiconductor, Solomon, Davis, and Lee developed the *MC1556* op amp, reporting on it in early 1969 (see Reference 12).[5]

This two-stage op amp design used a combination super-beta NPN pair, combined with a PNP pair as the input stage. With a quoted super-beta transistor gain of 4,000, the design had a 2 nA input current. It was also known for a slew rate appreciably higher than the 741 or other devices available at the time.

In late 1969, Bob Widlar contributed another IC op amp design, the *LM108* (see Reference 13). The LM108 was the first of what turn out to be a long line of precision IC op amps with low input currents, by virtue of a super-beta input transistor front end. A simplified schematic of the LM108 is shown in Figure 8-19.

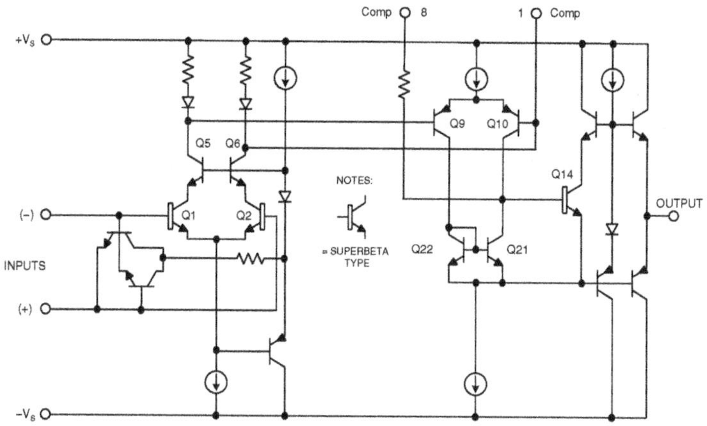

Figure 8-19: The LM108 superbeta input monolithic IC op amp

In this circuit the superbeta NPN devices are indicated by a wider base in the symbol, and the remaining transistors are high voltage types. Q1–Q2 make up the super-beta input differential pair, and are cascoded by Q5–Q6. The diode drops biasing this cascode are arranged so that Q1–Q2 see a 0 V V_{CE}. The second stage on the 108 is a PNP differential pair, Q9–Q10, with a balanced load, Q21–Q22. The output voltage is developed at the emitter of Q14, and buffered by a class AB output stage.

---

[5] Ironically, the 1556 op amp may not have gotten all the credit due, as perhaps the earliest use of super-beta devices, within a general-purpose op amp. A second irony is that the paper itself is better known (and often quoted) for the establishment of input stage g_m reduction as a means of raising slew rate.

The 108 design achieved a notably low bias current, typically under 1 nA at room temperature. Offset voltage was typically 700 μV and 2 mV (max), and gain was 300,000 (or 110 dB). It had very wide input and output ranges, typically ±14 V operating from ±15 V supplies, and it consumed 300 μA of quiescent current. Further, it could operate down to supplies of ±2 V, making it useful on 5 V rails. A point worth noting here is that the 108 differed from Widlar's previous LM101/101A designs with a load rating of 10 kΩ (whereas the 101/101A could drive 2 kΩ at rated gain). This was obviously a byproduct of the low power nature of the 108 design.

The basic LM108 design was later upgraded by National, to the *LM108A*. This was a 500 μV(max) offset voltage version of the part. An internally compensated version was also offered, the *LM112*.

Later, many other companies brought out their own competitive versions of the 108 and 112 op amps, with similar sounding names, and some with much improved performance. In the 1970s, ADI was one such company, offering the AD108 and AD108A, with specifications like the originals.

In 1969 Marv Rudin and Garth Wilson formed Precision Monolithics Incorporated (PMI), a brand new company with a charter of precision linear ICs. PMI introduced their counter to the 108A, the *OP08*, in 1976. This wasn't simply a second source to the 108A, but a revised and upgraded design by George Erdi and Larry Farnsley. Erdi was known as the father of the Fairchild μA725 (and the SSS725, at PMI). Erdi came to PMI in 1969, from Fairchild, where he had already established some key op amp design concepts (see narrative on 725 to OP07).

The new OP08 design added a thermally balanced layout, to reduce offset voltage and to increase gain. This was reflected in an offset voltage of 150 μV (max) for the best grade, a minimum gain spec of 50,000 (94 dB) into a 2 kΩ load (other specs were comparable to the 108A). At the same time the PMI *OP12* was introduced. This was a device similar to the OP08, but with internal compensation, and one which competed with the 112.

Another IC company to introduce 108/112 style designs was Linear Technology Corporation (LTC). Formed in 1981 by former National and Precision Monolithics engineers, Linear Technology introduced their own superbeta op amps, the *LT1008* and *LT1012* in 1983 (see Reference 14). Designed by a team headed by former PMI op amp designer George Erdi, the LT1008 featured trimmed offset voltage of 120 μV (max), a drift of 1.5 μV/°C (max), and a minimum gain of 120,000 (~102 dB) driving 2 kΩ. A notable feature of these amplifiers versus the earlier 108A types was the use of *input bias current cancellation*, allowing an LT1008 bias current as low as ±100 pA(max).

Precision Monolithics followed up on the OP08 and OP12 designs with the *PM1008* and *PM1012*, released in 1987. These were designed by Peter Gaussen of the Twickenham UK design center. The PM1008 had specs comparable to the LT1008, and the PM1012, to the LT1012. Also in 1987, the performance bar was raised a bit higher by the introduction of the even more tightly specified PMI *OP97*, an internally compensated super-beta input op amp functionally like the 112 or 1012. A simplified schematic of the OP97 family is shown in Figure 8-20.

The OP97 best grade (A, E) offers an offset voltage of 25 μV (max), a drift of 0.6 μV/°C (max), a bias current of ±100 pA (max), and a minimum gain of 200,000 (106 dB) driving a 2 kΩ load. It is notable that the OP97 was marketed as a "low power OP07," which of course technically speaking it isn't.[6] The OP97 uses a two-stage topology, the OP07 a three-stage. Not at all the same inside—but to many users, lower power with precision can be very important, rendering the internal differences moot.

---

[6] Former PMI and ADI op amp product line director Jerry Zis relates that while the OP97 may have been marketed with a focus on the OP07 users looking for a lower power device, this niche was nevertheless a real need. Of course, it also helps to have great specs, plus a family of dual and quad devices, which the standard OP07 never did have—but which the OP97/OP297/OP497 family eventually provided.

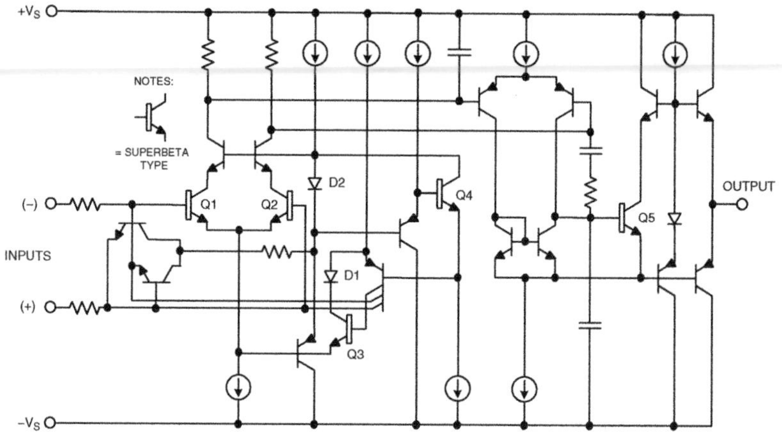

Figure 8-20: The OP97/OP297/OP497 super-beta input monolithic IC op amp

The OP97 is still available today, as are the other dual and quad family members, the OP297 (dual) and OP497 (quad) devices. The latter devices were designed by Derek Bowers, adding laser trimming (as opposed to the use of zener-zap trim on earlier family devices), and were released in 1990 and 1991.

The AD508 and AD517

ADI entered the super-beta op amp game at an early point, with their own super-beta input part, the *AD508*, an externally compensated precision device (see Reference 15). Designed by Modesto "Mitch" Maidique who came to ADI from Nova Devices, the AD508 released in 1972. It was an upgrade of his 1971 ADI precision op amp design, the *AD504*, which was a very high precision op amp in its own right (see Reference 16).

Quite unlike the 108 series of op amps topologically, the AD508 could be said to be an inherently high precision design. It featured the use of thin-film resistors, a super-beta input stage with balanced active loading, and a thermally balanced layout. The design used a two-stage double-integrator topology, with a triple buffered output, for very high load and thermal immunity. The AD508K typically achieved an open-loop gain of ~138 dB while driving 2 kΩ (much higher than any 108 or 112 topology amplifier), a bias current under 10 nA, and a low drift of 0.5 µV/°C (max) (see Reference 17).

An internally compensated version of the AD508 was introduced in 1978, the *AD517* (see Reference 18). This amplifier also used a superbeta input stage, and added the important feature of laser wafer trimming (see Reference 19). This trimming allowed offset voltage to be held as low as 25 µV (max), and drift as low as 0.5 µV/°C (max), both for the highest grade, the AD517L.

Much later on, ADI also introduced its own series of internally compensated super-beta op amps, styled along the lines of the OP97 series of devices. These were the AD705 (single), AD706 (dual) and AD704 (quad) series of op amps (see References 20–22). Designed by Reed Snyder, these op amps were introduced in 1990 and 1991.

Precision Monolithics was purchased by ADI in 1990, and the op amp product lines of the two companies were merged. Today, the product catalog of ADI includes many ADI originated (ADxxx) as well as many original PMI products (OPxxx).

Precision Bipolar IC Op Amps—μA725 to the OP07 Families

A second thread of development for precision op amps started at roughly the same time as the LM108 design, in 1969. Working then for Fairchild Semiconductor, George Erdi developed the *μA725*, the first IC op amp to be designed from the ground up with very high precision in mind.

In a rather complete technical paper on the 725 circuit and precision op amp design in general, Erdi laid down some rules that have become gospel in many terms (see Reference 23). A simplified schematic of the 725 is shown in Figure 8-21.

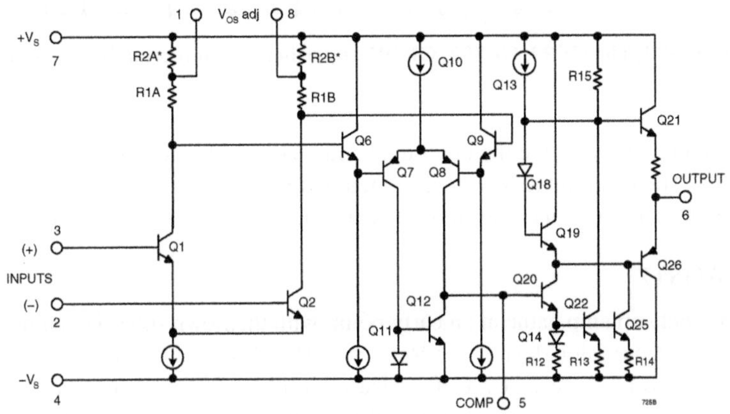

Figure 8-21: The μA725 monolithic IC op amp

The 725 is basically a three-stage design, consisting of a differential NPN input pair Q1– Q2, followed by a second differential stage, Q7–Q8, and a final single-ended output stage Q22, which is buffered by class AB emitter followers Q21 and Q26. The circuit was externally compensated by a four-component RC network at Pin 5. The three stages yielded much higher gain than previous two-stage amplifiers, but at the expense of more complex compensation.

Optional trimming of input offset voltage took place at pins 1–8, where an external 100 kΩ pot with the wiper to $+V_S$ was adjusted for lowest offset. When done in this manner, this also gave lowest drift.

Some circuit subtleties are belied by the schematic's simplicity, but yet important. Q1 and Q2 are actually a quad set (dual pairs), with the paralleled pairs straddling the chip's axis of thermal symmetry. The idea behind this was that thermal changes due to output stage dissipation would be seen as equal thermally-induced offsets by the two input stage halves, and thus be rejected. This principle, first established in the 725 design, has since become a basic precision design principle (see Reference 15, and within Reference 23, the Figure 2 chip photograph).

Another key point of 725 performance optimization concerns offset nulling for a condition of zero input offset and lowest drift, described in some detail by Erdi within References 23 and 24. The 725 had a typical offset voltage spec of 600 μV, and with the offset nulled as recommended, the resulting drift was 0.6 μV/°C. The bias current was typically 45 nA, and open-loop gain was 132 dB.

George Erdi left Fairchild in 1969, to join the newly formed PMI. At PMI, he continued with the 725 precision amplifier concept, designing the SSS725 version.[7] This op amp was identical to the original in functionality, but offered improved performance. There was also an OP06 produced at PMI later on. The OP06 was like the 725, but with the addition of differential input protection.

Not too long after the SS725 at PMI came the *OP05* op amp, in 1972 (see Reference 25). With the new OP05 design George Erdi considerably simplified application of precision op amps, making it internally compensated, adding input bias current cancellation, and differential overvoltage protection. Topologically, with these enhancements the OP05 can be said to be identical to the 725's three-stage architecture.

Precision op amp users now had a simple-to-apply device. A major system error was still left to the user to deal with: offset voltage. The OP05 used a manual trimming scheme similar to the 725 for offset adjustment, via a 20 kΩ pot. The unadjusted maximum offset for the OP05 was 500 μV, and drift was 0.6 μV/°C after null.

The OP05 was successful in its own right, but the offset voltage issue was still there. About this time, other IC companies were turning to active wafer trim schemes, such as the aforementioned ADI laser wafer trimming scheme (Reference 19). The next phase of 725 and OP05 evolution was to address active trimming of op amp offset, to deliver higher accuracy in the finished op amp device.

In 1975, Erdi reported on an offset trim technique that used 300 mA over-current pulses, to progressively short zener diodes in a string. With the zener string arranged strategically in the input stage load resistances of an op amp, this so-called "zener-zapping" could be used to trim the offset of an op amp on the wafer (see Reference 26). The first op amp to utilize this new trim technique was Erdi's *OP07*, which was introduced by PMI in 1975 (see Reference 27).

In the OP07, shown in simplified schematic form in Figure 8-22, the (not shown) zener strings are connected in parallel with segmented load resistances R2A and R2B. A simplified schematic of the scheme is shown in Reference 27, Figure 3, but in essence the series of zener diodes parallel the segmented partial load resistances, the values of which are sized to control progressively larger offsets.

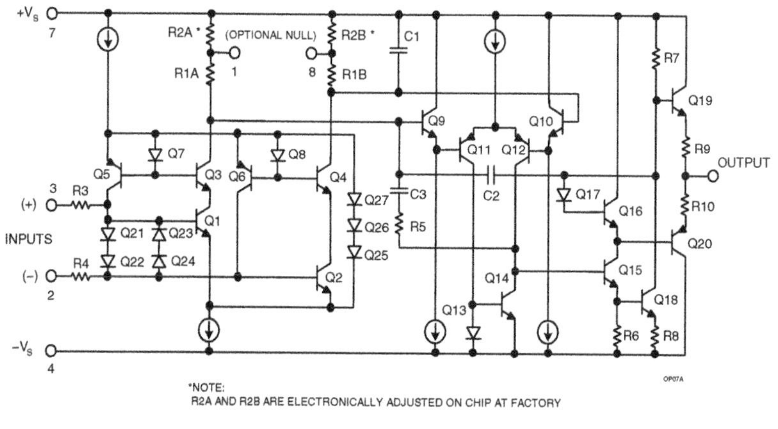

Figure 8-22: The OP07 monolithic IC op amp

[7] The "SSS" prefix was used on early PMI amplifiers, and stood for Superior Second Source. Another example was the PMI SSS741.

At trim time, a computer measures the actual op amp offset, then selects the appropriate zener to reduce it to the next level, and then zaps that zener with a high pulse of current.

This current pulse effectively shorts the zener, and so the section of load resistance in parallel. This process is iterated until the offset cannot be further reduced.

The new OP07 thus created had some impressive offset specifications. It was reported that the entire distribution of parts trimmed had offsets of 150 μV or less, and a prime grade, the OP07A was specified at 25 μV (max) for offset. Importantly, since this trim method also simultaneously reduced drift as the offset is nulled, the trimmed OP07 amplifiers had drift rates of 0.6 μV/°C (max), and typically much less than this.

The zener-zap trim technique was a valuable innovation in its own right, as it could be applied to other devices to reduce errors, and at a low additional cost to the manufacturing process. It is today one of many active trim techniques used with precision op amps (see the more detailed discussions of trimming in Chapter 1).

The OP07 went on to become the "741" of precision op amps, that is the standard device of its precision class. It was (and still is) widely second-sourced, and many spin-off devices followed it in time.

PMI went forward with the OP07 op amp evolution, and introduced the OP77, a higher open-loop gain version of the OP07 in 1988. The best grade OP77A featured a typical gain of ~142 dB, an offset of 25 μV, and a drift of 0.3 μV/°C (max). Later, an additional device was added to the roster, the OP177. This part offered similar performance to the OP77A, as the OP177F, specified over the industrial temperature range.

Prior to the 1990 acquisition of PMI by ADI, the ADI designers turned out some excellent OP07 type amplifiers in their own right. Designed by Moshe Gerstenhaber, the AD707 essentially matched the OP77 and OP177 spec-for-spec, operating over commercial and industrial ranges (see Reference 28). It was introduced in 1988. The AD708 dual was also offered in 1989, providing basically the performance of two AD707s. Moshe Gerstenhaber also designed the AD708 (see Reference 29).

The OP27 and OP37

As noted above, the OP07 lineage also included other related devices. Two such op amps, also designed by George Erdi at PMI, were the OP27 and OP37. These devices were released in 1980 (see References 30 and 31). Figure 8-23 is a simplified schematic of the OP27 and OP37 op amps.

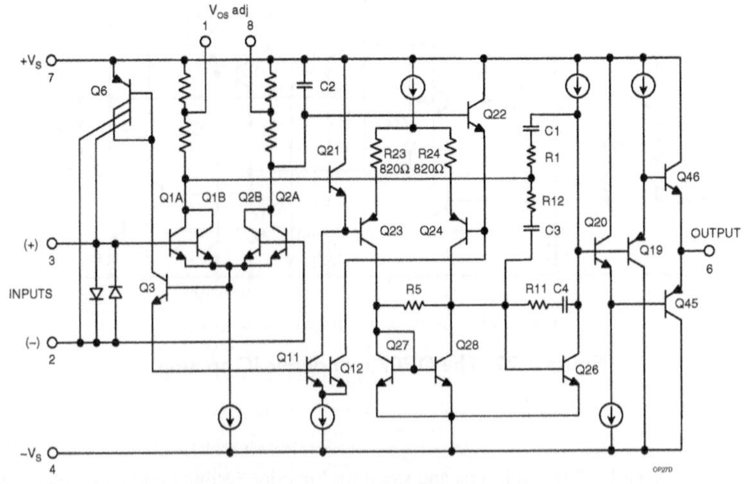

Figure 8-23: The OP27 and OP37 monolithic IC op amps

From the apparent similarity to the OP07 schematic of Figure 8-22, it might be easy to conclude that this amplifier was an adaptation of the OP07. However, the similarity ends in the fact that they are both three-stage amplifiers, and in truth the two different designs have been optimized with different end applications in mind.

In the design process of OP27/37, an examination of various noise sources was done, and the three-stage architecture is biased with the goal of both lower input noise, and higher speed (see Reference 31). Thus the stage operating currents are higher vis-à-vis the OP07, and provision for a decompensated version was also done (the OP37, stable at a gain of five). This was achieved by making the compensation cap C1 smaller on the OP37 version, while the basic OP27 is stable at unity gain. Towards the lower input noise, the current-limit protection resistors in series with the inputs were also removed.

The OP27 did achieve the goals of lower noise and greater speed, with an input noise density of 3.0 $nV/\sqrt{Hz}$ at 1 kHz, a 1/f corner of 2.7 Hz, while the slew rate was 2.8 V/μs and unity gain-bandwidth was 8 MHz. While realizing these new ac performance levels, the OP27/OP37 also retained impressive dc specifications as well. With a zener-zapped trim to the first stage, the offset was 25 μV (max), drift was 0.6 μV/°C (max), and voltage gain was typically 126 dB. The OP27 and OP37 went on to become widely second-sourced, and became standard devices for use as low noise, high dc precision amplifiers.

Single-Supply and Micropackaged OP07 Compatibles

It would be understandable for many to conclude that the high dc precision represented by the better performing versions of the OP07 and OP27 class devices would be sufficient for most applications. More recently however, the ground rules have changed.

While the high precision is still often sought, amplifier versions with single-supply capability are now in demand, as are tiny and even tinier packages. The traditional chip designs of the OP07/OP27 generation often can't work in new applications, because the circuit demands single-supply operation, and/or the package size is incompatible with the large chip size of the older products.

The small relative scale of some of these modern IC packages is shown in Figure 8-24. In the upper row, the decreasing size going from the 14-pin SOIC at the right to the SC-70 package at the left is quite clear. In the bottom portion of the figure, the SC-70 and SOT-23 packages are shown in another perspective, relative to a US one cent piece.

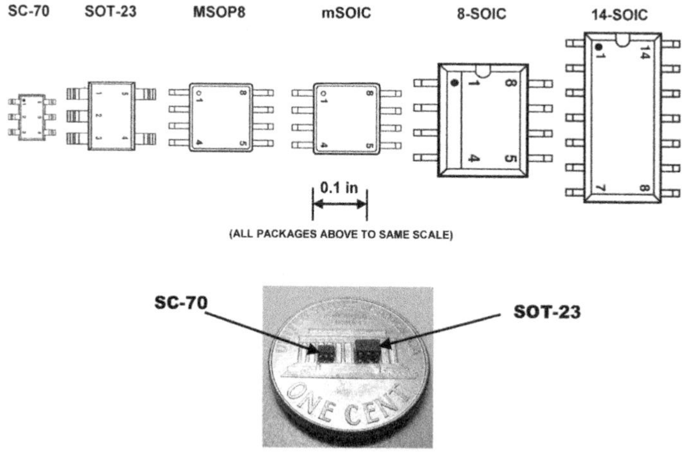

Figure 8-24: The relative scale of some modern IC op amp packages

Two very recent OP07 lineage devices from ADI address these two issues. One is the *OP777* op amp series, which includes the OP777 (single), the OP727 (dual) and OP747 (quad) devices (see Reference 32). Designed by Derek Bowers and released in 2000, these new devices feature rail-to-rail CMOS output stages, a ground sensing bipolar PNP input stage, and a 270 μA operating current. These designs operate over a supply range of 2.7 V–30 V, in MSOP, SOIC and TSSOP packages.

Even more recent is the *OP1177* series, also designed by Derek Bowers and released in 2001. This series includes the OP1177 (single), the OP2177 (dual) and OP4117 (quad) devices (see Reference 33). This design series has a slightly higher operating current than the OP777 series, at 400 μA per amplifier, and it operates from dual supplies of ±2.5 V to ±15 V. While not aimed at single-supply applications, this design does offer a wide range of small packages, with specifications applicable over a –40°C to +125°C range.

Precision JFET IC Op Amps—AD503 to the AD820/AD822/AD824 and AD823 Families

The development of FET input IC op amps was neither as rapid nor as straightforward as the growth of their bipolar IC cousins. There were numerous reasons for this, which will become apparent as this narrative progresses.

First of all, the relative scarcity of high quality FET input op amps early in the history of ICs was certainly not because no one wanted them, but rather because very few could make them. Many FET input op amps had already existed from the days of modular and hybrid types (see preceding section of this Chapter), and FET input amplifiers in general were highly sought after for fast signal processing and low current instrumentation uses. Unfortunately, the development of high performance monolithic FET IC op amps was to become a somewhat long and torturous process.

An early FET input op amp was by Douglas Sullivan and Mitch Maidique. This ADI amplifier was known as the *AD503* and *AD506*, and it was released in 1970. A schematic and photo of the chips used for this design is shown in Figure 8-25.

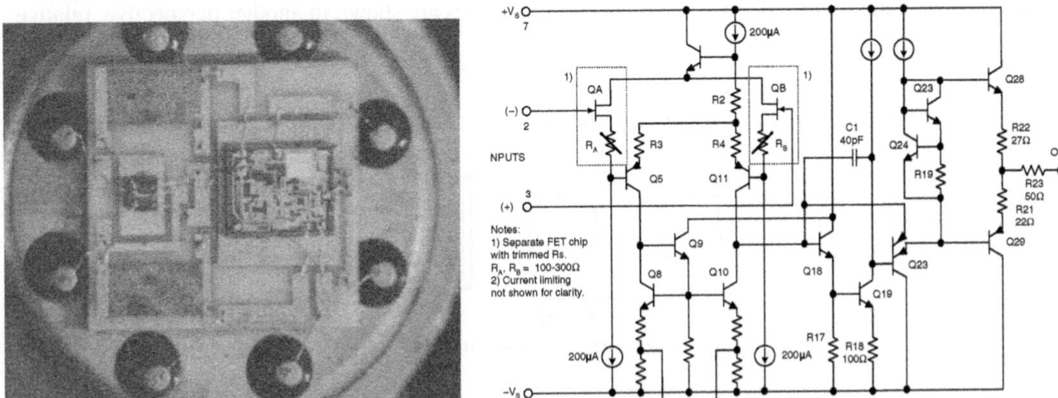

Figure 8-25: The AD503 and AD506 two-chip hybrid IC op amps

As should be evident from the schematic, this amplifier used two chips. One was a main amplifier chip, somewhat similar to a 741 after the input stage. The input stage consisted of a selected N-channel JFET pair, Q_A and Q_B. In the photo to the left, the two active chips can be noted at center right and left, respectively. Also used was a pair of trimmed resistors, R_A and R_B, shown at the left upper and lower corners of the substrate.

In the case of the AD506J and K grades, these resistors were laser trimmed for lowest offset, delivering to the user devices with maximum offset of no more than 3.5 mV and 1.5 mV, respectively. The nontrimmed AD503 was similar in function, except for higher initial offset (which could be trimmed by the user, via the offset adjust pins). Because of the bootstrapping configuration used, the design had excellent CM specifications—CMR typically was 90 dB, with constant bias current over the input range. It is worthy of note that the AD503/AD506 bias current (as well as later ADI FET input devices) was specified *after a five minute warmup period, a conservative method not used by all op amp makers.*

Operation of the AD503/AD506 family was described in a 1971 applications bulletin (see Reference 35). There were also related uncompensated amplifier types, namely the *AD513* and *AD516* (see Reference 36). Later on, an even tighter *AD506L* grade was introduced, with a 1 mV (max) offset and a 10 μV/°C (max) drift (see Reference 37).

Shortly after the time frame of the early bipolar op amps, there were also several completely monolithic FET input IC op amps, for example the Fairchild *μA740*, and the Intersil *ICL8007* (see Reference 38). The ICL8007 was perhaps the best of these early completely monolithic P-channel FET input op amps, but that isn't saying a lot. Offset voltages could be as high as tens of mV, and drifts several tens of μV/°C. Input current was low, but that was about the best that could be said of them.

The problem with all the monolithic FETs of the early seventies was simply that the FET devices themselves were poorly controlled. To make any material improvement in monolithic FET IC op amps, a fundamentally better process was needed.

In 1974, this was to happen, in the form of a paper by two National Semiconductor engineers, Rod Russell and David Culmer (see Reference 39). In this paper Russell and Culmer described a new fabrication technique for making FET devices, using *ion-implantation*. This allowed more stable P-channel JFETs to be made, along with quality NPN bipolars. The same paper also described a new series of FET input op amps, the *LF155/LF156/LF157* devices. These parts had much lower offsets and drifts than any previous all-monolithic FET op amp, 5 mV (max) for offset and a typical drift of 5 μV/°C.[8]

While the idea of ion-implantation caught on and became an industry standard method of IC fabrication, the same was not entirely true for the LF155/156/157 devices. Although they were second-sourced (and are still available), others sought a cleaner solution to a standard FET IC op amp topology. The LF155 series used an asymmetrical topology, and there was difficulty controlling the quiescent current.

At PMI, George Erdi designed an FET input op amp series to compete with the National LF155/156/157 parts, which were called the *OP15*, *OP16*, and *OP17*, respectively. They used zener-zap trimming and bias current cancellation; and the best A and E grades achieved offsets of 500 μV (max), and drifts of 5 μV/°C (max).

RCA introduced their answer for a general-purpose FET input op amp, the *CA3130*, also in 1974 (see Reference 40). Using a P-channel MOS input stage and a CMOS output stage, this device was suitable for lower voltage, single-supply uses. It was not, however, a high precision part, due mostly to the poor stability of the MOS devices used. Nevertheless, it was high on general utility, as were the *CA3140* and other spin-offs.

Texas Instruments got into the FET op amp market with their own amplifier series in 1978 (see Reference 41). These devices, in the form of singles, duals, and quads of various power ratings (and speed) did use a PFET input pair operating into a current mirror, with a conventional second stage (a la the 101 or 741, but with higher speed). This line, the *TL06x, TL07x,* and *TL08x*, became standard devices, and are still

[8] Specifications are quoted from December 2001 data sheet for LF155 and LF156 devices.

available. While the faster slew rate and symmetrical signal path of these devices helped ac applications, they weren't designed for high precision.

ADI had been working on an improved FET input monolithic IC op amp, and introduced the first of a long series of devices, the *AD542*, in 1978 (see Reference 42). This two-stage circuit design used a P-channel JFET input differential pair, followed by a second stage integrator. Careful design and laser trimming achieved a maximum offset as low as 0.5 mV in the AD542L, and a maximum drift of 10 μV/°C. While this was not as good as the best bipolar input amplifiers, it was better than any other monolithic FET had done.

Continuing along this same path were other amplifiers such as the *AD544*, a higher speed relation to the AD542, introduced in 1980 (see Reference 43). Both of these devices were designed by Lew Counts, and were aimed at fast settling data acquisition use. They were followed in 1981 by dual counterparts, the *AD642* and *AD644* (see Reference 44). All these devices had trimmed, zero TC supply and input stage currents, for overall stability and predictable slew rate. These features were retained in later precision devices.

This series of JFET input op amps reached their highest precision in 1982, with the introduction of the *AD547* (see Reference 45). This device, designed by Scott Wurcer, achieved for the first time in a monolithic FET op amp a maximum drift of 1 μV/°C, combined with a 250 μV (max) offset, for the AD547L grade of the part. The goals of such low offset and drift were met with laser trimming for both offset and drift at the wafer level. This also has become routine for all high precision ADI FET amplifiers.

The AD711/AD712/AD713 and OP249 IC Op Amps

In 1986 the *AD711/AD712* and *AD548/AD648* FET op amp families were introduced by ADI (see Reference 46). The AD711/AD712 were, respectively, single and dual parts with finely tuned specifications, designed to meet general-purpose as well as intermediate precision uses, but at a moderate cost. The AD712KN sold for $1.90 in quantities of 100, while the AD648KN sold in similar lots for $2.60.

The series featured offset voltages of 500 μV (max), a drift of 10 μV/°C (max) for the AD711K, at a quiescent current of 3 mA. The AD548K had similar offset voltage specifications, and half the drift, at a supply current of 200 μA. JoAnn Close designed the AD548/AD648 series of amplifiers, with inputs from Scott Wurcer and Lew Counts.

Scott Wurcer designed the AD711/AD712 series. The AD711 and AD712 were ultimately to be joined by a quad version, the *AD713*. This family of JFET IC op amps have been very popular since their introduction, and are still available.

Prior to the 1990 acquisition by ADI, PMI introduced their own dual JFET input IC op amp, the *OP249*. Designed by Jim Butler, this similarly specified dual op amp competed directly against the AD712.

Electrometer IC Op Amps

One area of great demand on op amp performance has traditionally been the *electrometer amplifier*, where input currents are required to be less than 1 pA. In the days of the modular op amp, such ultralow current devices as the model 310 and 311 varactor bridge amplifiers had addressed this role. (See the previous section of this chapter for a basic discussion on these amplifiers.) It should be understood that the term electrometer amplifier is here meant to imply any amplifier with ultralow bias currents. It might be a varactor bridge based design, or it might be some other type of front end allowing ultralow bias currents, for example several semiconductor types—MOSFETs, JFETs, and so forth.

The AD515 and AD545 Hybrid IC Electrometer Amplifiers

In hybrid IC form, there were a couple of early electrometer op amps from ADI. The first of these was the *AD515*, a two-chip hybrid similar in general architecture to the AD503 (discussed in conjunction with

Figure 8-25). The AD515 operated at a low power, with a quiescent current of 1.5 mA (see Reference 47). It achieved some impressively low input currents; 75 fA for the best grade AD515L, while maintaining a low offset of 1 mV(max). The AD515 was a successful product, with specifications that were not soon to be eclipsed.

Another early two-chip hybrid IC electrometer op amp was the *AD545*, introduced in 1978 (see Reference 48). This design also operated at low power like the AD515, but with a higher maximum input bias current, 1 pA for the AD545L.

Monolithic IC Electrometer Amplifiers

One of the early monolithic IC electrometer op amps, was the *OPA111*. Burr-Brown introduced this device in 1984 (see Reference 49). Designed by Steve Millaway, the OPA111 used a dielectrically-isolated process for fabrication.

The OPA111 circuit employed P-channel JFETs in the input and second stages, and a first stage cascode design for low bias current variation with input CM changes. The design addressed some of the weak points of the previous LF155/156/157 series (Reference 39). Reference 49 cited several LF15x circuit weaknesses; one was the use of current source loading for the input JFET pair, another was the means of offset trimming, and another was potential susceptibility to popcorn noise, due to the noise currents of the second stage bipolar differential pair. These points were addressed by the OPA111 design.

The OPA111 name was said to have been based on the combination of three key specs; 1 mV (max) offset, a drift of 1 μV/°C (max), and an input voltage noise of 1 μV rms in a 10 Hz–10 kHz bandwidth. This particular combination of specifications was tough to beat, and the OPA111 became a successful IC op amp.

Released in 1987, the first completely monolithic IC electrometer op amp from ADI was the *AD549*, designed by JoAnn Close and Lew Counts (see Reference 50). This op amp achieved its low bias current by virtue of the use of a new "topgate" FET, as designed by Jody Lapham and Paul Brokaw (see Reference 51), plus a sophisticated scheme of bootstrapping around the critical input P-channel JFET pair.

A schematic as adapted from the associated patent is shown in Figure 8-26 (see Reference 52). In the AD549 circuit, the input FETs are J6 and J7 with the input signals applied to their top gates at 10 and 12. The back gates BG1 and BG2 of the pair are biased at approximately the same DC level by a bootstrap loop

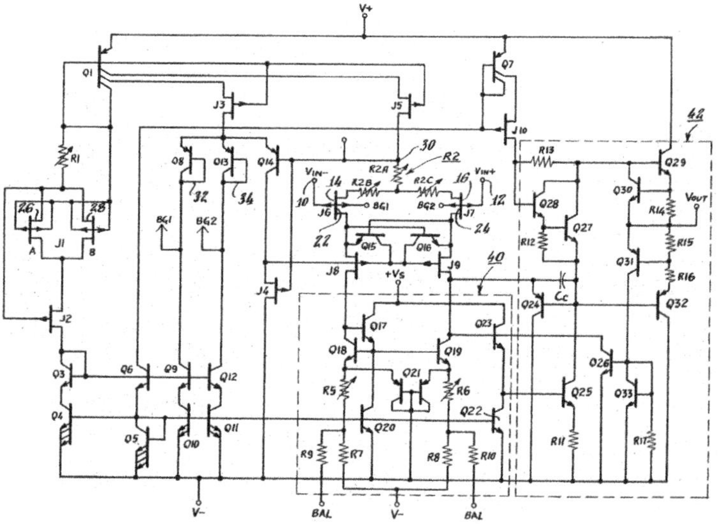

Figure 8-26: The AD549 electrometer IC op amp schematic (adapted from US Patent 4,639,683)

through Q14, and Q13– Q8. A second bootstrap loop through J4 and J8–J9 bootstraps the drains of J6–J7, thus providing for an input bias current level independent of CM voltage, over a ±10 V range.

With this circuit, built on a junction-isolated process, the AD549L was able to achieve a bias current of less than 60 fA, along with a 500 μV (max) offset and a drift of 10 μV/°C (max). It was provided in a hermetically sealed TO-99 package, with Pin 8 connected to the case for guarding within the final application circuit. The AD549L sold for $15.45 in 100 piece lots.

In 1988 ADI introduced another electrometer amplifier based on the design of the AD549, the *AD546* (see Reference 53). JoAnn Close also designed this op amp, and it was offered in a plastic package with somewhat relaxed specifications (vis-à-vis the AD549). The AD546KN had a maximum bias current of 500 fA, a maximum offset of 1 mV, and a drift of 20 μV/°C (typ). It sold for $4.50 in 100 piece lots.

The very latest electrometer amplifier in this series is the still-supplied *AD795*. It is available in an SOIC package and has bias currents of 1 pA or less (see Reference 54).

The AD743/745 Low Noise JFET IC Op Amps

Prior to about 1990, input voltage noise performance in JFET IC op amps had never been competitive with the best bipolar devices, many of which achieved noise densities of 3 $\mathrm{nV}/\sqrt{\mathrm{Hz}}$ (see earlier OP27 discussions).

In 1990, ADI introduced an answer to applications such as hydrophone amplifiers, which require simultaneously low voltage and current noise from an amplifier. The new amplifier was the *AD743* and *AD745*, designed by Scott Wurcer (see Reference 55). The design of these amplifiers attacked the voltage noise issue by the use of a quad array of very large input transistors, as described in Reference 56.[9]

The result was an input-referred noise of 2.9 $\mathrm{nV}/\sqrt{\mathrm{Hz}}$ (at 10 kHz) for the two devices, and precision dc amplifier performance specifications. The basic AD743 is a unity-gain stable part, while the faster AD745 is stable at noise gains of five or more.

The AD820/AD822/AD824 and AD823 Series JFET IC Op Amps

In the early 1990s, the first of a series of JFET op amps on the ADI CB process began to appear. This process featured comparable speed and gain NPN and PNP bipolars, designed by Jody Lapham and Brad Scharf (see References 57 and 58). It also had an N-channel FET structure, and a neat feature of this FET was that the pinch-off voltage allowed it to be used as a differential pair at the op amp front end, and the two gates could operate linearly to the negative rail. Thus with a common-emitter complementary bipolar output, a rail-to-rail output stage could be built. The combination of these two key features created a single (or dual) supply op amp with a low-current JFET input stage.

The first op amp of this type to appear was the *AD820*, a single low-power op amp, released in 1993 (see Reference 59). The AD820 was designed by JoAnn Close and Francisco dos Santos. The device architecture was very flexible, and it could be operated from single supplies as low as 3 V, or from dual supplies of up to ±18 V. The input bias current was 10 pA (max) for the AD820B, and the quiescent current was 800 μA (typical).

With the success of the AD820, a dual version was the obvious next step, and the *AD822* appeared in 1994, with specs similar to the AD820 (see Reference 60). Rounding out this family next was the *AD824*, which appeared in 1995 (see Reference 61).

---

[9] Of course, "very large" is a relative description. Nevertheless, Figure 6 of Reference 56 shows the four input stage transistors consuming about one-half of the chip area.

The AD820/AD822/AD824 were relatively low power parts, with moderate speed. In 1995 a higher speed dual using the same general topology appeared, the *AD823* (see Reference 62). Designed by Jeff Townsend, this amplifier had a 16 MHz bandwidth, and a 22 V/μs slew rate. It also operated from a wide supply range, ±1.5 V to ±18 V dual supplies, or single supplies of +3 V to 36 V.

High Speed IC Op Amps

In the earliest years of IC op amps, everyone was using essentially the same NPN bipolar process, and speed was severely limited because of the slow PNP transistors available. An early scheme to partially get around the PNP bottleneck was the *LM118/218/318*, designed by Bob Dobkin at National Semiconductor in 1971 (see Reference 63). ADI produced their own version of this op amp, the *AD518*, designed by Dave Kress. Although these amplifiers did achieve much higher slew rate and bandwidth, they did not settle fast, nor were they well-suited to driving low impedance loads.

In the early seventies, just about the only truly fast IC process was owned by Harris Semiconductor. This dielectrically isolated process produced equal speed NPN and PNPs, and the Harris HA2500 series became popular for fast settling characteristics. In 1973 ADI released the fast *AD509* op amp, a screened Harris part (see Reference 64).

Until junction isolated CB processes came on board, the dielectrically isolated parts were to dominate high speed applications. There were however, notable exceptions to this general rule. The *AD744*, designed by Scott Wurcer, was introduced in 1988 (see Reference 65). Although this op amp still used a basic NPN process, it took advantage of ion-implanted P-channel FETs for the input differential stage, and could settle quickly and cleanly, reaching a 900 ns settling time to 0.01%.

ADI introduced a high speed 36 V CB process in 1988 (see References 57 and 58, again), and with it, a host of fast IC op amps. Among these were a high speed voltage feedback group, the *AD840* series, and the *AD846* current feedback op amp, all designed by Wyn Palmer. Many other very successful op amps were to soon follow in this series, using the CB process. Notable among them were the unity-gain stable *AD847* and externally compensated *AD829*, also designed by Wyn Palmer. Later on, the *AD811* designed by Dave Whitney, was among the first high performance current feedback op amps available on the CB process, achieving very low video distortion specifications while driving 75 Ω cables (see Reference 66).

Frank Goodenough's Op Amp Reporting for Electronic Design

A notable documentation source on these high speed op amp developments was an **Electronic Design** series, by analog editor Frank Goodenough (see References 67–71). The CB process was just the beginning of ADI high speed IC op amps, and within less than a decade a further jump in performance was produced. This was the 12 V XFCB process, introduced in 1993 (see References 72 and 73). This produced such key parts as the *AD8001*, designed by Scott Wurcer (see Reference 74). The AD8001 set new performance standards, hitting a bandwidth of 800MHz on ±5 V supplies, and achieving very low video distortion.

Frank Goodenough's op amp articles continued to provide a valuable source of IC performance, as well as historical references, through the late 1990s, including other op amp categories as well (see References 75–78). He passed away in February of 1998, and was fittingly memorialized by Roger Allan of **Electronic Design** (see Reference 79).

References: IC Op Amps

(Note: Appended annotations indicate relevance to op amp history.)

Birth of the Monolithic IC Op Amp, General-Purpose IC Op Amps

1. Bob Widlar, "Design Techniques for Monolithic Operational Amplifiers," **IEEE Journal of Solid-State Circuits**, Vol. SC-4, August 1969. *(Design methods used in μA709, LM101A and LM108.)*
2. "The μA702 Wideband Amplifier," Chapter 5 within James N. Giles, Editor, **Fairchild Semiconductor Linear Integrated Circuits Handbook**, Fairchild Semiconductor, 1967, pp. 33–55. *(A detailed design and application discussion of the μA702 IC op amp.)*
3. R. J. Widlar, "A Unique Circuit Design for a High Performance Operational Amplifier Especially Suited to Monolithic Construction," **Proceedings of the NEC**, Vol. XXI, October 1965, pp.85–89. *(The μA709, the first widely used IC op amp.)*
4. Robert J. Widlar, "Monolithic Op Amp with Simplified Frequency Compensation," **EEE**, July 1967. *(The LM101 IC op amp.)*
5. James Solomon, "The Monolithic Operational Amplifier: A Tutorial Study," **IEEE Journal of Solid-State Circuits**, Vol. SC-9, No. 6, December 1974. See also National Semiconductor AN-A. *(The classic paper on IC op amp design techniques.)*
6. Dave Fullagar, "A New High Performance Monolithic Operational Amplifier," **Fairchild Semiconductor Application Brief**, May 1968. *(The μA741 IC op amp.)*
7. Robert J. Widlar, "IC Op Amp with Improved Input-Current Characteristics," **EEE**, December 1968. *(The LM101A IC op amp.)*
8. Ronald Russell, Thomas Frederiksen, "Automotive and Industrial Electronic Building Blocks," **IEEE Journal of Solid-State Circuits**, Vol. SC-7, December 1972, pp. 446–454. *(The LM324 and other single-supply ICs.)*
9. Data sheet for **AD741J, K, L, S Lowest Cost High Accuracy IC Op Amps**, Analog Devices, Inc., January 1973.

SuperBeta IC Op Amps

10. Robert J. Widlar, "The LM110 An Improved IC Voltage Follower," **National Semiconductor LB-11**, March 1970. *(The use of superbeta transistors in follower-connected IC op amps LM102 and LM110.)*
11. R. J. Widlar, "Super Gain Transistors for ICs," **IEEE Journal of Solid-State Circuits**, Vol. SC-4, August 1969 pp. 249–251. *(General principles of super-beta transistors in IC op amps.)*
12. Jim Solomon, William Davis, P. Lee, "A Self-Compensated Monolithic Operational Amplifier with Low Input Current and High Slew Rate," **ISSCC Digest of Technical Papers**, February 1969, pp.14–15. *(The use of superbeta transistors in the MC1556 IC op amp.)*
13. Bob Widlar, "IC Op Amp Beats FETs on Input Current," **EEE**, December 1969. *(The superbeta input LM108 IC op amp.)*
14. George Erdi, Jim Williams, "Precision Op Amp Serves Host of Needs," **Electronic Design**, September 1, 1983. *(The LT1008 and LT1012 superbeta input IC op amps.)*
15. Modesto Maidique, "A High Precision Super Beta Operational Amplifier," **IEEE Journal of Solid-State Circuits**, Vol. SC-7, December 1972. See also: "AD508: Monolithic Chopperless Op Amp Super β Inputs, <1 μV/°C Drift," **Analog Dialogue**, Vol. 6, No. 1, 1972. *(The superbeta input AD508 IC op amp.)*

16. Modesto Maidique, "Monolithic Operational Amplifier With 1 μV/°C Drift," **Analog Dialogue**, Vol. 5, No. 3. *(The AD504 IC op amp.)*

17. Preliminary Data sheet for **AD508J, K, L IC Chopperless Low Drift Operational Amplifier**, Analog Devices, Inc., June, 1972.

18. Doug Grant, "Low Drift Super Beta Op Amp," **Analog Dialogue**, Vol. 12, No. 1. *(The AD517 IC op amp.)*

19. Richard Wagner, "Laser-Trimming on the Wafer," **Analog Dialogue**, Vol. 9, No. 3. *(Laser wafer trimming of thin-film resistors for IC offset and gain.)*

20. "Precision Op Amp," **Analog Dialogue**, Vol. 24, No. 1. *(The AD705 super-beta precision IC op amp.)*

21. "Two Precision Dual Op Amp Families," **Analog Dialogue**, Vol. 24, No. 3. *(The AD706 and OP297 superbeta precision dual IC op amps.)*

22. "Quad Op Amp," **Analog Dialogue**, Vol. 25, No. 1. *(The AD704 superbeta precision quad IC op amp.)*

Precision Bipolar IC Op Amps

23. George Erdi, "A Low Drift, Low Noise Monolithic Operational Amplifier For Low Level Signal Processing," **Fairchild Semiconductor Application Brief APP-136**, July 1969. *(The μA725 IC op amp.)*

24. George Erdi, "Minimizing Offset Voltage Drift With Temperature In Monolithic Operational Amplifiers," **Proceedings of NEC**, 1969 pp. 121–123. *(Method of offset trim for minimum drift in precision dc amplifiers.)*

25. George Erdi, "Instrumentation Operational Amplifier With Low Noise, Drift, Bias Current," **Precision Monolithics APP Note**, 1972. *(The OP05 IC op amp.)*

26. George Erdi, "A Precision Trim Technique for Monolithic Analog Circuits," **IEEE Journal of Solid-State Circuits**, Vol. SC-10, December, 1975 pp. 412–416. *(The "zener-zap" offset trim method.)*

27. Donn Soderquist, George Erdi, "The OP-07 Ultra-Low Offset Voltage Op Amp," **Precision Monolithics AN-13**, December, 1975. *(The OP07 IC op amp.)*

28. "Precision Bipolar Op Amp Has Lowest Offset, Drift," **Analog Dialogue**, Vol. 22, No. 1. *(The AD707 precision IC op amp.)*

29. Bill Schweber, "Dual Bipolar Op Amp Provides Superior Matching of Specs," **Analog Dialogue**, Vol. 23 No. 1. *(The AD708 precision IC op amp, dual of the AD707.)*

30. George Erdi, Tom Schwartz, Scott Bernardi, and Walt Jung, "Op Amps Tackle Noise-and for Once, Noise Loses," **Electronic Design**, December 12, 1980. *(The OP27 and OP37 IC op amps.)*

31. George Erdi, "Amplifier Techniques for Combining Low Noise, Precision, and High-Speed Performance," **IEEE Journal of Solid-State Circuits**, Vol. SC-16, December, 1981 pp. 653–661. *(Design techniques of the OP27 and OP37 IC op amps.)*

32. "Precision Micropower Single-Supply Op Amps Have 100 μV max Offset," **Analog Dialogue**, Vol. 34, No. 4. *(The OP777, OP727, and OP747 precision single and dual supply, rail-rail output single/dual/quad IC op amps.)*

33. Data sheet for "**Precision Low Noise Low Input Bias Current Operational Amplifiers OP1177/OP2177**," www.analog.com.

Precision JFET IC Op Amps

34. Doug Sullivan, Modesto Maidique, "High Performance IC FET-Input Op Amp," **Analog Dialogue**, Vol. 4, No. 2, 1970. *(The AD503 and AD506 two-chip hybrid IC op amp.)*

35. Richard S. Burwen, Doug Sullivan, "AD503, AD506 IC FET Input Operational Amplifiers Technical Bulletin," **Analog Devices Application Bulletin**, August 1971. *(Design and application of the AD503 and AD506 two-chip hybrid IC op amp.)*

36. "FET-Input ICs Can Slew at 50 V/µs," **Analog Dialogue**, Vol. 5, No. 3, 1971. *(The AD513 two-chip hybrid IC op amp.)*

37. "AD506L: Economical Low-Drift FET-Input," **Analog Dialogue**, Vol. 7, No. 1, 1973. *(The AD506L two-chip hybrid IC op amp.)*

38. Dave Fullagar, "Better Understanding of FET Operation Yields Viable Monolithic JFET Op Amp," **Electronics**, November 6, 1972. *(The ICL8007 IC op amp.)*

39. Ronald W. Russell, Daniel D. Culmer, "Ion-Implanted JFET-Bipolar Monolithic Analog Circuits," **ISSCC Digest of Technical Papers**, February 1974, pp. 140–141, 243. *(Ion-implanted JFETs, and the LF155/156/157 series of FET IC op amps.)*

40. Otto Schade, Jr. "CMOS/Bipolar Linear Integrated Circuits," **ISSCC Digest of Technical Papers**, February 1974, pp. 136–137. See also: R. L. Sanquini, "Building C-MOS, Bipolar Circuits on Monolithic Chip Enhances Specs," **Electronics**, October 3, 1974. *(The CA3130 CMOS IC op amp.)*

41. Dale Pippenger, Dave May, "Put BIFETs Into Your Linear Circuits," **Electronic Design,** January 4, 1978, pp. 104–107, 109–111.

42. Lew Counts , Rich Frantz, "High-Performance, Low-Cost Bipolar-FET Op Amp," **Analog Dialogue**, Vol. 12, No. 3, 1978. *(The AD542 monolithic FET IC op amp.)*

43. "Fast FET Op Amp," **Analog Dialogue**, Vol. 14, No. 1, 1980. *(The AD544 IC op amp.)*

44. "High Performance Dual FET Op Amps," **Analog Dialogue**, Vol. 15, No. 2, 1981. *(The AD642/AD644 IC op amps.)*

45. "First Monolithic FET Op Amp with 1 µV/°C Drift," **Analog Dialogue**, Vol. 16, No. 1 1982. *(The AD547 drift and offset trimmed IC op amp.)*

46. "Highest-Performing Low-Cost BiFET Op Amps," **Analog Dialogue**, Vol. 20, No. 2, 1986, p. 22.

47. Dave Kress, "FET-Input Electrometers," **Analog Dialogue**, Vol. 10, No. 1, 1976, p.11.

48. "FET-Input AD545," **Analog Dialogue**, Vol. 12, No. 3, 1978, p.18.

49. Steve Millaway, "Monolithic Op Amp Hits Trio of Lows," **Electronic Design,** February 9, 1984. *(The OPA111 low bias current IC op amp.)*

50. JoAnn Close, Lew Counts, "Junction-Isolation Process Yields 50 fA Op Amp," **Electronic Design**, July 10, 1986, pp.99–102, 104. See also: JoAnn Close, "Monolithic Electrometer Has 60 fA Max Bias Current," **Analog Dialogue**, Vol. 21, No. 2, 1987, p. 22. *(The AD549 electrometer IC op amp.)*

51. Jerome F. Lapham, Adrian P. Brokaw, "Low-Leakage JFET," **US Patent 4,985,739**, filed April 27, 1987, issued January 15, 1991. *(Design of the 'top-gate' low-leakage JFET IC process.)*

52. Lewis Counts, JoAnn Close, "Very Low Input Current JFET Amplifier," **US Patent 4,639,683**, filed February 7, 1986, issued January 27, 1987. *(The 'top-gate' low-leakage JFET IC process and AD549 IC op amp.)*

53. "High Performance Electrometer Op Amp in Plastic 8-Pin DIP," **Analog Dialogue**, Vol. 23, No. 4, 1988, p. 12. *(The AD546 electrometer IC op amp.)*

54. "Low-Noise, Low-Drift Precision Op Amps for Instrumentation," **Analog Dialogue**, Vol. 27, No. 1, 1993. *(The AD795 low noise, low I_B and OP213 dual single-supply, precision IC op amps.)*

55. "FET-Input Op Amp Has Lowest Combined V and I Noise," **Analog Dialogue**, Vol. 25, No. 1, 1991, p. 12. *(The AD743 and AD745 low noise IC op amps.)*

56. Scott Wurcer, "A 3 nV/√Hz, DC-Precise, JFET Operational Amplifier," **Proceedings of BCTM**, 1989, pp. 116–119. *(Design principles of the AD743 and AD745 low noise IC op amps.)*

57. "Op Amps Combine Superb DC Precision and Fast Settling," **Analog Dialogue**, Vol. 22, No. 2, 1988. *(The AD846 IC op amp, the AD840 series, and the high speed CB process used.)*

58. Jerome F. Lapham, Brad W. Scharf, "Integrated Circuit with Complementary Junction-Isolated Transistors and Method of Making Same," **US Patent 4,969,823**, filed May 5, 1988, issued Nov. 13, 1990. *(Design of the ADI CB IC process.)*

59. "Single-Supply FET," **Analog Dialogue**, Vol. 27, No. 2, 1993, p. 26. *(The AD820 single-supply rail-rail output IC op amp.)*

60. "Dual FET, 3 V to ±18 V," **Analog Dialogue**, Vol. 28, No. 1, 1994, p. 24. *(The AD822 single-supply rail-rail output dual IC op amp.)*

61. "Quad JFET, Single-Supply Op Amp," **Analog Dialogue**, Vol. 29, No. 1. *(The AD824 single-supply, rail-rail output quad IC op amp.)*

62. "Dual 16MHz Rail-Rail FET," **Analog Dialogue**, Vol. 29, No. 2. *(The AD823 single-supply, rail-rail output dual IC op amp.)*

High Speed IC Op Amps

63. Bob Dobkin, "LM118 Op Amp Slews 70 V/µs," **National Semiconductor LB-17**, September 1971. *(The high slew rate LM118 IC op amp.)*

64. "AD509: Fast Op Amp 2 µs to 0.01%," **Analog Dialogue**, Vol. 7, No. 1, p. 10. *(The AD509 fast IC op amp.)*

65. Kristen Dinsmore, "Fast, Accurate BIFET Op Amp Settles to 0.01% in 900 ns Max," **Analog Dialogue**, Vol. 21, No. 2, 1988, pp. 14–15.

66. Dave Whitney and Walt Jung, "Applying a High-Performance Video Operational Amplifier," **Analog Dialogue**, Vol. 26, No. 1 1992, pp. 10–13. *(The AD811 high-speed, high output video IC op amp and applications.)*

67. Frank Goodenough, "Focus Monolithic IC Op Amps: Faster and More Precise," **Electronic Design**, February 4, 1988, pp. 127–132. *(A survey of high-speed IC op amps.)*

68. Frank Goodenough, "A Slew of New High Performance Op Amps Shatters Speed Limits," **Electronic Design**, March 3, 1988, pp. 29–32. *(High speed IC op amps and processes, including the ADI CB process.)*

69. Frank Goodenough, "New Processes, Designs Boost IC Op Amp Speeds," **Electronic Design**, April 12, 1990, pp. 45–48, 52–56. *(Survey of high speed IC op amps with emphasis on CB processes and current feedback architectures.)*

70. Frank Goodenough, "Linear ICs Attain 8 GHz NPNs, 4 GHz PNPs," **Electronic Design**, December 19, 1991, pp. 35–37, 40, 43–45. *(The high-speed 'UHF-1' DI process as used by Harris.)*

71. Frank Goodenough, "New Processes to Spawn Next-Generation Analog, Mixed-Signal, Power ICs," **Electronic Design**, January 2, 1992, pp. 59–62, 64–66, 68, 70. *(Survey of high-speed process developments.)*
72. Frank Goodenough, "Wideband IC Op Amps Reach New Bandwidth Highs," **Electronic Design**, September 2, 1993, pp. 48, 50, 52, 56, 58, 63, 64–66. *(The ADI XFCB process, and a survey of other high speed processes.)*
73. Frank Goodenough, "IC Op Amps Combine Low Cost and Performance," **Electronic Design**, September 2, 1993, pp. 39–40, 42, 46. *(Survey of high speed processes, ADI AD8001 XFCB process IC op amp and AD96XX series IC op amps.)*
74. "Fast Op Amp: High Performance, Low Power, Low Cost," **Analog Dialogue**, Vol. 28, No. 2, 1994, p. 11. *(The AD8001 high-speed IC op amp, first on the XFCB process.)*
75. Frank Goodenough, "Silicon, Analog Processes Becoming More Sophisticated," **Electronic Design**, May 2, 1994, pp. 97,98, 100–103. *(Survey of complementary and mixed signal compatible processes).*
76. Frank Goodenough, "Rail-Rail In-and-Out IC Op Amps Run Off 2.7 V," **Electronic Design**, May 16, 1994, pp. 51, 52, 54, 56, 60, 64–65. *(Survey of rail-rail, low voltage single-supply design techniques and IC op amp devices.)*
77. Frank Goodenough, "Single-Supply Op Amps Come of Age," **Electronic Design Analog Applications Issue**, November 7, 1994, pp. 52–61. *(Survey of rail-rail design techniques and IC op amp devices.)*
78. Frank Goodenough, "Op Amp, ADC Topple Old Price/Performance Marks," **Electronic Design**, December 5, 1994, pp. 75, 76, 78, 80, 81, 84, 87. *(New product highlight of high speed, low power AD8011 IC op amp and AD878 ADC.)*
79. Roger Allan, "Frank Goodenough: 1925–1998," **Electronic Design**, March 9, 1998, p. 6.

Classic Cameo
Bob Widlar—Linear IC Pioneer, Personality

Bob Widlar reviewing his LM10 op amp, circa 1977
(Photo courtesy of Bob Pease and National Semiconductor)

This history of IC op amp developments began with the work of Bob Widlar, back in the early 1960s. Starting with the first successful IC op amp, the μA709, Widlar was to author a virtually unbroken string of IC op amp successes. Only his better-known *op amp* achievements are covered here, so readers should not feel he designed only op amps. As noted earlier, many linear IC design techniques he pioneered early on became standard methods. It should also be understood that he made major contributions to other IC circuits, for example IC bandgap voltage references, and IC three-terminal voltage regulators.

Throughout his career, Widlar was known not only as an innovator, but also as a colorful personality of the first order. Some Widlar stories can be found in a remembrance offered by Bob Pease.[10] Another tribute was also offered by Jim Solomon, which includes personal views of this most fascinating designer by a number of Widlar's co-workers.[11]

Bob Widlar passed away in February of 1991, at a relatively young age of 53 years. He was running near his home in Mexico, a favorite pastime of his. It is safe to say that his work efforts (and also his play antics) will not be forgotten.

---

[10] Bob Pease. "What's All This Widlar Stuff, Anyhow?" **Electronic Design**, July 25, 1991, pp. 146, 148, 150.

[11] James E. Solomon, "A Tribute to Bob Widlar," **IEEE Journal of Solid-State Circuits**, Vol. SC-26, No. 8, August 1991, pp. 1087–1089.

INDEX

- Subject Index
- Analog Devices' Parts Index
- Standard Device Parts Index

SUBJECT INDEX

B

D

M

P

Q

R

ANALOG DEVICES' PARTS INDEX

STANDARD DEVICE PARTS INDEX

Printed and bound by CPI Group (UK) Ltd, Croydon, CR0 4YY
03/10/2024
01040334-0011